The Abel Prize

Niels Henrik Abel 1802–1829
The only contemporary portrait of Abel, painted by Johan Gørbitz in 1826

Helge Holden • Ragni Piene
Editors

The Abel Prize 2018-2022

THE ABEL PRIZE

Det Norske Videnskaps-Akademi
The Norwegian Academy of Science and Letters

Springer

Editors
Helge Holden
Department of Mathematical Sciences
Norwegian University of Science and Technology
Trondheim, Norway

Ragni Piene
Department of Mathematics Blindern
University of Oslo
Oslo, Norway

Photo of R. Langlands on front page and Curriculum Vitae by Peter Badge/Typos1 and on part title page by Heiko Junge/NTB Scanpix
Photo of K. Uhlenbeck on front page and Curriculum Vitae by Peter Badge/Typos1 and on part title page by Trygve Indrelid/Abel Prize
Photo of H. Furstenberg and G. Margulis on front page and Curriculum Vitae by Peter Badge/Typos1 and on part title page by Abel Prize
Photo of L. Lovász and A. Wigderson on front page and Curriculum Vitae by Peter Badge/Typos1 and on part title page by Tamas Szigeti (Lovász) and Abel Prize (Wigderson)
Photo of D. Sullivan on front page and Curriculum Vitae by Peter Badge/Typos1 and on part title page by Naina Helén Jåma

ISSN 2661-829X ISSN 2661-8303 (electronic)
The Abel Prize
ISBN 978-3-031-33972-1 ISBN 978-3-031-33973-8 (eBook)
https://doi.org/10.1007/978-3-031-33973-8

This work was supported by Norvegian Academy of Science and Letters

This Springer imprint is published by the registered company Springer Nature Switzerland AG
The registered company address is: Gewerbestrasse 11, 6330 Cham, Switzerland

Paper in this product is recyclable.

Au reste il me parait que si l'on veut faire des progres dans les mathematiques il faut étudier les maitres et non pas les ecoliers.

Niels Henrik Abel[†]

[†] "Finally, it appears to me that if one wants to make progress in mathematics, one should study the masters, not their students." In: "Memoires de Mathématiques par N. H. Abel", Paris, August 9, 1826, in the margin of p. 79. Original (Ms.fol. 351 A) in The National Library of Norway. Reprinted with permission.

Preface

This book constitutes the fourth volume[1] in a series on the Abel Laureates, covering the period 2018–2022.

As in previous volumes there is one part per year. Each part starts with the full citation from the Abel Committee, followed by an autobiographical piece by the laureate(s). The autobiographical pieces are enhanced by photos – old and new. Then comes a description of the scientific accomplishments of the laureate(s). The parts end with a curriculum vitae and a complete bibliography of each laureate.

In the first part, James Arthur writes on the work of Robert Langlands, while in the second part, Simon Donaldson presents the work of Karen K. Uhlenbeck. The third part contains Vitaly Bergelson, Eli Glasner, and Benjamin Weiss's article on the work of Hillel Furstenberg, as well as Alex Eskin, David Fisher, and Dmitry Kleinbock's article on the work of Gregory Margulis. In the fourth part, the work of Lázló Lovász is presented by Martin Grötschel and Jaroslav Nesetril, and that of Avi Wigderson by Boaz Barak, Yael Kalai, Ran Raz, Salil Vadhan, and Nisheeth Vishnoi. In the fifth part Edson de Faria, Sebastian von Strien, and Shmuel Weinberger write on the work of Dennis Sullivan.

The traditional award ceremonies in Oslo could not take place in 2020 and 2021 due to the COVID-19 pandemic. They were replaced by award ceremonies in the Norwegian embassies in the countries of the laureates. Lectures were given by the laureates, but there were no traditional Abel lectures. Fortunately, in 2022, we could return to the normal award ceremony in Oslo, where Dennis Sullivan received the Abel Prize from His Majesty King Harald. We were happy that also the laureates from the two previous years, Hillel Furstenberg, Gregory Margulis, Lázló Lovász, and Avi Wigderson could be present in Oslo on that occasion.

The last part is meant to give, through a collection of photos, an idea of all the activities that took place in connection with the Abel Prize during the last five years.

[1] H. Holden, R. Piene (eds.): *The Abel Prize 2003–2007. The First Five Years*, Springer, Heidelberg, 2010, *The Abel Prize 2008–2012*, Springer, Heidelberg, 2014, and *The Abel Prize 2013–2017*, Springer, Heidelberg, 2019.

The back matter contains updates regarding publications and curriculum vitae for all laureates. Finally, we list the members of the Abel Committee and the Abel Board for the period 2018–2022.

The annual interview of the Abel Laureate(s) – The Abel Prize Interviews – can be watched on the video channel of the Norwegian Academy of Science and Letters. Transcripts of the interviews have been published, and publication details can be found in the back matter.

We would like to express our gratitude to the laureates for collaborating with us on this project, especially for providing the autobiographical pieces and the photos. We would like to thank the mathematicians who agreed to write about the scientific work of the laureates, and thus are helping us in making the laureates' work known to a broader audience.

Thanks go Marius Thaule and Erlend Due Børve for their LaTeX expertise and the preparation of the bibliographies as well as copyediting the manuscripts.

The technical preparation of the manuscript was financed by the Niels Henrik Abel Board.

Trondheim, Norway
Oslo, Norway
June 6, 2023

Helge Holden
Ragni Piene

Contents

Part III Hillel Furstenberg and Grigoriy Margulis

Part IV 2021 László Lovász and Avi Wigderson

Part V 2022 Dennis P. Sullivan

Part VI Abel Activities 2018–2022

Part I
2018 Robert P. Langlands

"for his visionary program connecting representation theory to number theory"

THE
ABEL
PRIZE

Citation

The Norwegian Academy of Science and Letters has decided to award the Abel Prize for 2018 to **Robert P. Langlands**, Institute for Advanced Study, Princeton, New Jersey, USA,

> *"for his visionary program connecting representation theory to number theory."*

The Langlands program predicts the existence of a tight web of connections between automorphic forms and Galois groups. The great achievement of algebraic number theory in the first third of the 20th century was class field theory. This theory is a vast generalisation of Gauss's law of quadratic reciprocity. It provides an array of powerful tools for studying problems governed by abelian Galois groups. The non-abelian case turns out to be substantially deeper. Langlands, in a famous letter to André Weil in 1967, outlined a far-reaching program that revolutionised the understanding of this problem.

Langlands's recognition that one should relate representations of Galois groups to automorphic forms involves an unexpected and fundamental insight, now called Langlands functoriality. The key tenet of Langlands functoriality is that automorphic representations of a reductive group should be related, via L-functions, to Galois representations in a dual group.

Jacquet and Langlands were able to establish a first case of functoriality for $\mathrm{GL}(2)$, using the Selberg trace formula. Langlands's work on base change for $\mathrm{GL}(2)$ proved further cases of functoriality, which played a role in Wiles's proof of important cases of the Shimura–Taniyama–Weil conjecture.

The group $\mathrm{GL}(2)$ is the simplest example of a non-abelian reductive group. To proceed to the general case, Langland's saw the need for a stable trace formula, now established by Arthur. Together with Ngô's proof of the so-called Fundamental Lemma, conjectured by Langlands, this has led to the endoscopic classification of automorphic representations of classical groups, in terms of those of general linear groups.

Functoriality dramatically unifies a number of important results, including the modularity of elliptic curves and the proof of the Sato–Tate conjecture. It also lends weight to many outstanding conjectures, such as the Ramanujan–Peterson and Selberg conjectures, and the Hasse–Weil conjecture for zeta functions.

Functoriality for reductive groups over number fields remains out of reach, but great progress has been achieved by the work of many experts, including the Fields medallists Drinfeld, Lafforgue and Ngô, all inspired by the guiding light of the Langlands program. New facets of the theory have evolved, such as the Langlands conjectures over local fields and function fields, and the geometric Langlands program. Langlands's ideas have elevated automorphic representations to a profound role in other areas of mathematics, far beyond the wildest dreams of early pioneers such as Weyl and Harish-Chandra.

Autobiography*

Robert P. Langlands

I was born and passed the first two decades of my life in the vicinity of Vancouver, British Columbia, an area now overwhelmed by immigration from the Orient, above all, China and India. More precisely, I was born in New Westminster in 1936, spent the first few years of my childhood on the shore about seventy miles to the north, in Lang Bay close to Powell River, returned to New Westminster to begin school, then moved to White Rock, where I passed my adolescence, and then went to Vancouver, of which New Westminster is now a suburb, to the University for five years, leaving in 1958 for graduate school, never to return except for short visits. I recall first the geography of the area, then the circumstances there in the nineteenth century and the beginning of the twentieth as well as the circumstances of my family and me. It is an area that has seen changes that, although peaceful have been definitive: a semi-rural, even partly rural environment, with a population that, apart from a visible, but small, indigenous component, still had close ties to the Old Country, generally meaning Great Britain and Ireland, and to Eastern Canada, has become more urban, much more cosmopolitan, and undoubtedly much more sophisticated, and apparently, much wealthier.

Canada, like its neighbour, the United States, like most countries, was built on conquest and oppression. In the area where the two countries now meet, this conquest is often referred to as a discovery, and this discovery, in so far as overland voyages are concerned is recent: Alexander Mackenzie reached the mouth of the Bella Coola river in 1793; Lewis and Clark reached the mouth of the Columbia River in 1805; Simon Fraser reached the mouth of the Fraser river in 1808. The first and last of these rivers lie in Canada, the second reaches the ocean in the USA.

---

* Based on a manuscript published in *Langlands' Program and His Mathematical World*. Some paragraphs have been removed by the editors. Reprinted with permission.

R. Langlands
Institute for Advanced Study, 1 Einstein Drive, Princeton, NJ 08540, USA, e-mail: rpl@ias.edu

H. Holden, R. Piene (eds.), *The Abel Prize 2018-2022*, The Abel Prize, https://doi.org/10.1007/978-3-031-33973-8_1

My childhood and youth were passed in a very small compass, first New Westminster and elementary school, then White Rock and high school, then Vancouver and university, I recall the geography. Vancouver lies with its back to the northern bank of the Fraser River at its mouth; New Westminster lies just a little upstream. The mouth of the river itself is formed from alluvial islands and the region south of it, Surrey and Delta, as far as the border with the US, a matter of 15–20 miles is also alluvial. When I was child, it was almost largely farmland, although some of the farmers had employment elsewhere. Three of my uncles were longshoremen (stevedores) on the docks at New Westminster and two of these also had small farms in Surrey. As I recall the only livestock consisted of fowl. White Rock lay at the time in Surrey but on the coast, almost at the point where the border with the USA first reaches the sea, not far from the mouth of a small river, the Campbell. I had at its source a more distant relative, married to a cousin, perhaps, of my grandfather, with a genuine farm: cows to be milked, something I attempted there but never since, and fruit to be harvested. He was also a water diviner and the owner of the last horse and buggy in Langley, the municipality adjacent to Surrey.

Lang Bay, about half-way between Vancouver and Bella Coola, was isolated and on the shore. My first recollections are from there, where we lived in a rented summer house, with two neighbours, an elderly woman and her grand-daughter. My memory is largely of sea, shore, the woods, which were boggy, the neighbour's fields, and a grazing goat. Occasionally but seldom, there were visitors from the south. When I came of school age, my mother, a Catholic of Irish descent, was eager to return to New Westminster where there was a parochial school and, I suppose, to her large family, three living sisters and six brothers. I flourished in the school, appreciated the nuns, who were often young, often pretty and gentle. The traditional costume, now largely, perhaps completely, abandoned, I regarded as normal. I learned to read quickly, had no trouble with the arithmetic, and skipped a grade. I liked to read, even, under the influence of the Books of Knowledge, popular at the time and pedalled door-to-door, tried, for reasons otherwise forgotten, to learn French on my own, accompanying in the various volumes a little British family, complete with dog, on its voyage to Paris and France. Like today's more desperate voyagers, I never got beyond Calais, although I was going in the opposite direction. It was many years before I returned to the language, but then with somewhat more success. My faith was also fervent for a brief period—I even toyed at the age of seven or eight with the notion of becoming a priest, something that would have corresponded to my mother's ambitions for my recently discovered academic ability, as it would have to those of many Catholic mothers of the period, but already before leaving New Westminster my faith was failing and my desire for a greater freedom growing. In particular, I wanted to leave the parochial school for the public school.

We moved to White Rock very shortly after the war, where I spent my adolescence, arriving in 1946 and leaving for university in 1953. These were not academic years. There were very few Catholics in White Rock, indeed the children with whom I consorted never saw the inside of a church. My mother was, unfortunately attached to the Church and eager, even desperate that I remain in it. The marriage was a mixed marriage and my father, to compensate for his own sins—principally, perhaps only, gambling—would join in her efforts to bring me to church every Sunday morning, where I was a reluctant altar boy to a disagreeable priest of Irish descent. He himself stayed at home. Church meant also confession, a vicious practice, to which, for me, the only response after the age of twelve was prevarication or invention. There are, of course, many other practices that are very much worse and not far to seek. I was ashamed, as an adolescent, of my inability to resist my parents' pressure and still do not like to recall my youthful weakness. Luckily, once I left home to attend University and I was then not yet seventeen, I could abandon the Church and churches completely except for one midnight mass, the occasional funeral, and some touristic visits. Also as a concession to my wife, who was pleasing her mother, who herself appears to have been negligent about such niceties, we were married in a church, not however a Catholic church.

The Catholic Church aside, I would say that my childhood—in a society that had not yet ceased to be a frontier society, thus a society of people who had acquired independence at more of a cost to others, in this instance largely to the indigenous inhabitants, than to themselves, among people who, by and large, were indifferent to any but an extremely modest success, financial or otherwise, major success being beyond their imagination, and only a few of whom were unlucky enough to have occasion to be confronted with authority,—encouraged a natural, even if not necessarily bold, independence, a very useful characteristic for a mathematician.

I do not entirely understand my mother's relation to the Church. A good part of her large family did not take the Church nearly so seriously as she did. Her childhood had some difficult elements. Her father was, I believe, a fireman on the Canadian National Railroad, who barely survived a head-on clash with another train and did not recover from the incident, suffering in following years from epilepsy, severe psychic disorders, as well as, I understand, alcoholism. He spent a good deal of his time in an institution, although he could, apparently, return home on weekends, from Essondale to New Westminster on foot accompanied by his dog. That was a substantial trek. So my grandmother, who had been married at sixteen, was responsible for the family. She worked as a charwoman, in the houses of women who were better off. As a result, my mother, who was apparently a lively, popular young woman, on her school's basketball team, found herself wearing the cast-off clothes of her classmates. She never forgot it. My grandmother, Emily, whose maiden name was Dickson, was, so far as I know, above such feelings. She was a tremendously warm woman, beloved of her children and grandchildren. I knew her only for a few years. Dickson is not an Irish name, but my mother's ancestors seem otherwise to have been almost entirely Irish and they seem largely, perhaps entirely, to have left Ireland, usually south-east Ireland, for example, Kilkenny Co., before the famine. I believe that the South-East was not strongly affected by it. My grandfather's family name, Phelan, is distinctly Irish and has, I believe, its origins in a region farther to the west, in the town of Cork, but as I recall reading once in a history of Ireland, the tribe of the Phelans was displaced by the Norman invaders, whom it was attempting to resist, to south-eastern Ireland in the 12th Century, whence some of them came much later to Canada.

One exception to the Irish descent of my mother is an ancestor, a young German named Schildknecht who left Wittenberg or Wittenberge in Germany just before the American Revolution, in which he fought as a corporal in the South Carolina Loyalist Regiment. As compensation after Great Britain's loss, he was granted land in Ship Harbour, Nova Scotia where he settled with his wife, born in the American colonies but clearly the daughter of German emigrants. The name Schildknecht became Shellnutt and their descendants mixed with the Irish immigrants. A daughter Mary Catherine Shellnutt married an O'Bryan. My grandmother's mother was her granddaughter. They must have had a number of male children as well because the surname Shellnutt seems to be fairly common in Nova Scotia. As I observed, three of my mother's brothers were longshoreman. So was her paternal grandfather. He was killed in the famous Halifax explosion of 1917 when a French cargo ship that was carrying munitions exploded in the harbour leaving 2000 residents of the city dead and 9000 injured. He was not working at the time, rather he was, with his wife, on the way home from mass.

In these peripheral ways my mother's family was affected by the fortuities of the world's affairs. They themselves were not much concerned with these. Even the genealogical information on both sides is not traditional, but has been, by and large, the result of efforts of a later generation. My father's family were more recent immigrants. My mother's parents moved across Canada from Halifax to New Westminster, stopping for an unsuccessful attempt at farming in the province of Saskatchewan where my mother was born. My father seems to have been conceived in England and his mother, who was apparently not prepared for life in a tent in British Columbia, returned with her children to England for a couple of years not long after his birth. That he had two sisters, one his twin, the other born a very short time before him, did not make her life any easier.

She was the sixth child in a family of seven. Her paternal grandfather—I know nothing of her mother—had been a private in the British Army, who stationed for a while in Cork, Ireland met and married, either then or later, a woman called Mary. Nothing more is known about her antecedents, nor about her surname. She may have been Irish or, perhaps, the daughter of an English soldier. I cannot say. On the other hand, like my mother's mother, perhaps even more so, she seems to have been a very resourceful and courageous woman. The marriage was apparently first recorded in Hobart, Tasmania, at the time their first child was born. Her husband died in Tasmania in 1845 at the age of 36, not in the course of his military duties but of an illness. His grave and gravestone are still to be found in a famous cemetery, the Isle of the Dead, in Tasmania. His wife managed not only to find her way back to England, with at least two sons, of which my great-grandfather was the youngest and two daughters, and to enroll the sons in the Duke of York's Military School, a school near London for the orphans of soldiers. She found employment for herself in the same institution as a laundress and for her eldest daughter as well. The whole family is listed in the 1851 Census as residing in the School. The individual indications with place of birth are as follows: (i) Flowers, Elizabeth, 13, servant of Quartermaster, Enniskillin, Ireland; Flowers, Mary. F., 39, Laundress, Cork, Ireland; Richard Flowers, 10, soldier's son, Manchester, England; Robert Flowers, 10, sol-

Fig. 1: Five months old (Photo: private)

dier's son, Hobart, Van Diemen's Land, Australia. The sons were not twins, but the birth of the second was only eight months after that of the first. There was also a second sister, Mary Ann, eight years old at the time. A last son seems to have died in infancy. All in all, five children were born in about five years. Apparently the older son remained in the school and then joined the British army as a private, a rank at which he remained all his life. The younger son, Robert Flowers, my grandmother's father, asked to be released into his mother's custody, became first a draper and then an auctioneer, and with time, he became, in Newcastle upon Tyne, first a councillor and then an alderman. He seems to have been a responsible son. Although, I have no information as to his mother's fate, it is clear that his two sisters came to Newcastle upon Tyne, presumably with him, where they married and, much later, died.

My paternal grandmother was married relatively late, as she was approaching thirty, perhaps past it. My reading of the circumstances is that she expected to remain at home, in her father's house, which I believe was a substantial house in Westgate, so far as I know a well-to-do quarter in Newcastle. However at some point her father, who was a widower, decided to marry again. His new wife was considerably younger, almost thirty years, also well-to-do, and apparently took up enough space in his home that there was no longer room for my grandmother. So she herself married, a slightly younger man, my grandfather. They appear to have been members of the same Methodist congregation, but I am not certain. So far as I know, her father's marriage, however unwelcome it may have been for her, was a blessing for him in his later years.

My grandmother was the only member of her family to emigrate. From her and from my grandfather as well, not from my mother's family, I acquired the notion of the old country, a notion often invoked in their house. It was represented in their home by a bust of Kitchener, labelled Kitchener of Khartoum, invoking his famous colonial exploits in Africa. I was, myself, disabused of any notion of a special relation to the old country when I later met, as a mathematician, a number of Englishmen. I may simply have been unfortunate in my first encounters. I came to know more agreeable specimens later.

Some time after my grandparents' marriage, two or three years, the business of my paternal grandfather's father, who was a cabinet maker, seems to have collapsed, whether for general economic reasons or for illness, mental or physical, I cannot say. His whole family moved to Canada: two daughters, both of whom were married to clergymen, one apparently a missionary to the Indians of Kispiox, in northern British Columbia, where my great-grandfather is buried, and two sons, both carpenters, presumably trained in their father's shop, one of whom was later killed in an accident during the construction of the Hudson Bay building in Vancouver. My grandmother appears to have been an unhappy woman, although she was, in comparison with her husband, her children, and my mother's family, cultivated. She could play the piano and, when I began university and acquired some intellectual interests, it was from her that I borrowed the *The Imitation of Christ* by Thomas Kempis, a famous medieval work of devotional literature. I confess that I neglected to return it. So far as I know, she passed a good many hours with the Bible and other devotional matters, but by the time I became curious about the world, I had little occasion to talk to her and, she, in any case, was growing senile. Whatever cultivation she had, she had acquired, I should think, in her parents' home. I had many more maternal cousins than paternal, but there was a greater awareness of the value, at least commercial, of a university education among the paternal cousins, two had business degrees and became accountants, another had a degree in engineering from a prominent American university and worked for International Business Machines. His mother, my father's eldest sister, had been trained first as an elementary school teacher and, then, as a nurse, while her sister and her two brothers all left high-school early, presumably, at least in the case of the boys, to become apprentice carpenters. A friend of my grandfather with whom he had emigrated from Newcastle and to whom he remained close until my grandfather's death told me, at my grandfather's funeral, that he had reproached my grandfather for favouring the eldest daughter, but was told that the others would not profit so well as she from any more extensive education. What role my grandmother played in these decisions, I do not know.

I knew her as a frail, rather withdrawn woman, who had little support from her children as she aged. My father perhaps assumed more of the charge than any other, but my mother did not cooperate. New Westminster was not a large town, and my mother's brothers and sisters would have formed a large and boisterous group, not all so pious as my mother. My grandmother did not approve of the marriage and did not, I believe, disguise her feelings. My father made, however, a better marriage than he deserved. I doubt that my grandmother was aware of his more serious failings, although she expressed in the Bible she gave him on his fifteenth birthday only a feeble hope that he would read it.

I had far more cousins on my mother's side and found them more congenial, at least as an adolescent. They were easier with each other and with their mother, although my father was certainly close to his twin sister. So far as I know, apart from me, only one or two of the very youngest of my maternal cousins attended university. This did not, necessarily, prevent them from prospering. I myself, as a high-school student, had no notion whatsoever of attending university. My dream was to quit school, as one said, as soon as the law permitted, namely at the age of fifteen, and to take to the road, hitch-hiking to Toronto. Certainly, a large number of students in White Rock left at this age and found work, often seasonal, as loggers or as unskilled labourers of one kind or another. My mother, by temperament or as a consequence of her childhood experience and of my response to reading, writing, and arithmetic, will have had some ambitions for me, but more likely as a priest or a medical doctor. It was certainly noticed at the high school I attended, by observation or from the IQ tests to which we were all subject, that I had more than the usual aptitude for academic topics, but I myself was not impressed.

The few years spent in New Westminster were an occasion to meet a good many of my numerous cousins, a good proportion of whom were of about my age. New Westminster was a pleasant city in which to pass the first half of the 1940s. It was founded in 1858 as the capital of the Colony of British Columbia and remained larger than Vancouver, itself founded considerably later, until into the twentieth century. It was well and carefully planned, not large but with broad chestnut-lined streets, spacious boulevards and parks. It was a joy to be in. Thanks to the Second World War, the streets were during my early childhood almost entirely free of motorized vehicles. The port itself, on the river, did not intrude and the town, hardly more than a mile or two square was everywhere accessible to a child between the ages of six and nine. So as an introduction to urban life it could not have been gentler. I was told later by my mother that I had some trouble at first protecting myself from larger bellicose schoolmates, but I have no memory of that. The only childhood fisticuffs I remember were in White Rock. There were only two incidents. In the first, not provoked by me, I, in a burst of fury, pummelled my larger opponent and had to be pulled off him by the spectators; in the second, provoked by me, I had my nose broken.

Very few events from New Westminster are fixed in my memory: an older brother of some playmates ran away from home, and so far as I know, never returned. A classmate at the Catholic school, a girl my age, Maruka, Ukrainian I believe, whose parents were gardeners who later opened a prosperous nursery, was taken ill by one of the childhood diseases feared at the time and died within a few days. She was also a neighbour and we walked to school together. Although I was not particularly troubled by her death at the time, her image has stayed with me. I also remember her mother weeping as my own mother tried to console her, as well as a second attempt at consolation a couple of years later, just before the war's end, when a second neighbour received the notice of her son's death. My own family, uncles in particular, were not affected. Perhaps they all had children. One uncle, the youngest, served in the Air Force although he never went abroad, and a cousin, who had been born in the American state of Montana, went off to serve in the US Navy, returning to New Westminster some time later. His own father had spent some time in the USA, having run off, I believe, to join the American Navy towards the end of the First World War. My grandfather had brought him home from a similar earlier attempt, but yielded to his obstinacy. Both the father and the son married Americans.

In White Rock there was a small area of land, an Indian reserve, reserved for the members of a local tribe, of which there were only a few members remaining. The tribe, like many others, had been decimated in the nineteenth century by a disease that arrived with the colonists. The few children went to the local school but did not remain long and were pretty much ignored by the other children. This seemed to me at the time a normal state of affairs. There were also a few Métis in the town, but they were not distinguished from the rest of the population. My parent's store and our home above it were across the street from the reserve, but not from its residences. The chief came occasionally to buy lumber or other building material and would chat with my mother, who was usually at the cash register. Those boys in the town who were fond of fishing and of solitude would spend time in the reserve because the river ran through it, as did the trail to the border and the adjacent American town of Blaine.

I was nine, almost ten, years old when we arrived in White Rock, sixteen when I left it for university, returning only in the summer, and nineteen when I left it for good. It was a pleasant, but a strange, place for an adolescent. There was the ocean and the shore, although there were few boats. Except for the crab fisherman, the owners of the few boats, namely small rowboats, were usually Americans with a summer cottage. There was little exchange between the children on our side of the border and those on the other. The Americans were considered richer and were recognizable, from the front or from the back, by a slight plumpness.

Before the war the town's principal function was as a resort village in proximity to New Westminster and, to a lesser extent because of the lack of bridges and tunnels, to Vancouver. After the war, it served a different function, or so it seems to me on reflection. The summer cottages afforded inexpensive lodging. So there were a good number of families with no father, with a father who appeared only infrequently, or a feckless father. The proprietor of the local hotel or of the local dance hall were in my eyes rich. There were other families and their children as well, but those children whose families allowed them, for one reason or another, more freedom appealed to me.

My wife has a copy of the year book of the high school with photos of the classes, thus for grades 7 to 12. They suggest that there were about four hundred children in the school. Their faces and names are by and large familiar, but many of them came from the surrounding rural areas and many kept pretty much to themselves returning home directly after school, so that I knew much less about them and their circumstances than about those in the town itself, where there were also those youths who had left early to find work full or part time. Schooling was only compulsory until age 15. A hint of restiveness, a desire for independence, drew me to those who had left school or were free, for one reason or another, of parental constraints, but I was very young, not very bold, and could not entirely free myself from the interdictions imposed on me by my mother and, in her wake, my father. As a Catholic child, I believed, without any question that any sins would be observed not only by God above but by my recently deceased grandmother, whom I cherished, at his side.

It is not that I was up to the company of the children or youths whom I admired or envied. I had started school rather early and had skipped a grade, while a good number of my classmates had failed a grade, thus been kept back, and not just once but several times, so that they were substantially older than I was. Moreover they had substantially more freedom than I. They may not have been able to read or write with any ease, but I envied them, both the boys and the girls. My mother, curiously enough, because it was not shared by a good number of her brothers and sisters, had a fear of sin, in particular of books, not of course childrens' books, as a source of sin, that made life with her difficult. My father, whose Methodist/Wesleyan (in the diluted Canadian form of the United Church) background had not left him immune to sin, but had left him with a strong sense of propriety and of possible disapproval of the neighbours, provided no relief. So I had to struggle for whatever freedom I had. I did quickly take to foul language, although I could not use it at home. From the age of twelve to the age of fourteen, I could probably compete with the most imaginative or coarse of Indian taxi-drivers in Vancouver or of current American politicians, but between the ages of fourteen and sixteen my passion for this form of expression slowly dissipated. In those years I met, at one of the school dances, called mixers and introduced to encourage civilized social intercourse between the boys and the girls in the school, someone almost as timid as I but, in contrast to me, with plans for the future. This was a decisive, perhaps the decisive, event in my life.

A determining feature of those years may have been labour. The period after the war was an economically favourable period. My parents had moved to White Rock, away from New Westminster, probably at the urging of my mother, where they had founded a business—lumber and building supplies. My father provided the technical competence—he had training as a carpenter, although he never acquired his journeyman's papers—while my mother took care of the books. As would be normal at the time, perhaps today as well, my father was responsible for the collection of overdue accounts, which were frequent enough. This was, for him, not an agreeable task. For me, the fortunate aspect of the business was to provide me with an occupation to fill time that would otherwise have been idled away. Although in no

Fig. 2: Wedding 1956 (Photo: private)

way fragile, I was not a particularly strong youth, nor was I particularly athletic. I tried but I was younger than my classmates, so that I was never chosen for school athletic teams. On the other hand, in those days, kegs of nails, sacks of cement, agricultural tiles, plywood, plasterboard, and lumber of all kinds were loaded onto trucks by hand and unloaded in the same way. From the age of twelve or thirteen that was how I spent my time after school and on Saturdays. During the university years it was how I spent the summers, earning the funds to pay for the winter's food and lodging. It meant, above all, that I arrived at twenty reasonably robust, with a body that has not failed me, at least not seriously, over the next sixty years. It also meant that, without being particularly adroit physically, I could manage, although not with great skill, those household tasks associated with the building trades. However, as time went on I was ever more disinclined to undertake anything outside mathematics that demanded patience. With age, mathematics demands that quality more and more.

On reflection and I have, oddly enough, never indulged myself in reflection about these matters, after my early childhood, even during, I had little to do with my father outside these common labours. There was little disagreeable about them. Although he was occasionally impatient with my lack of dexterity, it was pleasant to work with my father. He was generous with my pay, adequate to allow me to indulge my juvenile sartorial extravagance, to go to the movies and so on, and, now and again, as a diversion, he suggested that I work as a swamper, a term used only in Canada, thus as a helper on the local light-delivery truck, which was not only more leisurely, with a good deal of time spent beside the driver watching the world go by, but entailed occasionally a trip to Vancouver or New Westminster for a load of cement, drainage tiles, or sashes and doors. This was hardly work!

The postwar years were prosperous; the business thrived, ostensibly under the hand of my father, but the determination was, I believe, my mother's, although this was not apparent to me at the time. My father had a modest taste for luxury. As the business prospered, they were soon able to construct a building to house it, with an apartment upstairs for the family. With a stone facing and large plate glass windows, it was at that time and place an imposing edifice. It now houses a restaurant. With time, my father was able to buy himself a Buick, at that time a luxury automobile, and to construct a house in the best part of town, on a cliff on the shore with a splendid view over the Strait of Juan de Fuca. Unfortunately, as my mother grew older and became ill, she was no longer able to save him from himself, and a vice—gambling—that had been present from the beginning, although hidden from the children, and a constant source of anxiety to her, took over. He, and thus they, slowly lost everything. By then, I was far away. It fell to my two sisters to do what they could, and it was considerable, to mitigate the disaster.

To return to my own development, how did it happen that rather than hitchhiking to Toronto, I went to university? It may have been the effect of my new acquaintance, but that would have been an unconscious effect, although I doubt that she would have encouraged my hitchhiking plans. She would certainly not have been willing to be a part of them. It would undoubtedly have meant that we parted ways.

There were two things. First of all, in the last year of high school, we had a new teacher, Crawford Vogler, and a new textbook, a textbook that introduced us to English literature. He was a very enthusiastic, very sympathetic teacher. I recall that he gave two or three students special assignments. I was asked to report on the novel The Ordeal of Richard Feverel by the well-known Victorian novelist George Meredith. It was all a little puzzling to me and any expectation on his part must have been disappointed. However, as I remarked above, I will have been given an IQ test and he will have been aware of any unusual talent. That will have been the source of his disappointed confidence in me. Nonetheless, he took, either then or on some other occasion, a full class period to explain to me in front of the other students that I must go to university. Going to university meant going to the University of British Columbia. There were then no other universities in the province and the possibility of going elsewhere would never have crossed my mind. I was impressed then and there by the suggestion and decided not only to take the entrance examinations but to study for them. I was successful; I even received a bursary to pay the academic fees.

Secondly, although my father had left school after nine years to become an apprentice, my wife, not at that time of course, was the child of a man with a more meagre educational background—born on Prince Edward Island, into a mixed, Franco-Irish marriage his mother died when he was two years old and he was given into the care of a Scottish family. So his initial language was Gaelic, but apparently somewhat frail and uncomfortable with the robust sons of the family, he left home at an early age, working in the logging camps of Quebec, where he learned French, as spoken by the loggers. He must have been quite agile, since his task was sometimes to clear log-jams. This was done by inserting an explosive in the jam and then running away, from one log to another, in order to be clear of the jam before the explosion freed the logs—a slip would be fatal.

What he had not learned neither with the Scots nor in the logging camps was to read and write. His chance came during the Great Depression, when various unions or political parties undertook not only to feed the unemployed but also to educate them. He had kept some books, by Frederick Engels, Karl Kautsky, August Bebel and other socialist authors, that he undertook to read at that time, most of which we still have. He liked to cite at length various passages from these books. He had a good memory but reading was always difficult for him. I think his wife, my mother-in-law, gave him further lessons after their marriage. My own father could read well enough, although writing was another matter. I do not think I ever received more than one brief note from him, in an emergency. One book in particular, I took away from my future father-in-law's library to read, *The Story of the World's Great Thinkers* by Ernest R. Trattner with biographies of many of the world's renowned thinkers, for example, Copernicus, Hutton,..., Marx, Pasteur, Freud,..., Einstein. I remember being particularly impressed by the story of Hutton and the age of the Earth. The book itself had been very popular, deservedly so, in the late thirties and early forties of the last century. So second-hand copies are still easily found.

I recalled at length various genealogical facts related to my family. I recall one or two related to my wife. They are striking. Her mother's family, like my mother's large with ten children, had emigrated not from England but from Scotland at the time of the First World War. My grandfather and his brother had returned to the Old Country as soldiers but they were sufficiently old that they never, so far as I know, saw battle. Lord Kitchener, who as I recalled was a familiar figure to me from my grandparents' dining room, was the Secretary of State for War for Great Britain at the beginning and had introduced the policy of sending brothers together to the battlefield on the principle that side-by-side they would fight better. The result was that many families lost all their sons at one stroke. My wife's grandfather's family seem to have been so affected, not her grandfather and one of his brother's who had also emigrated, who survived, although injured, as members of the Canadian army, but the four brothers who remained in Scotland and fought with the British army all perished, at least three apparently as a result of Kitchener's policy, which

I believe was finally abandoned. Part at least of her father's family had been in Canada much longer, descendants of a Basque fisherman and his Micmac wife, who are to be found in the census undertaken by the British after the conquest in 1763. My children are aware of their descent, although it is hardly apparent. Nevertheless, one of my daughters, the blonde among the four children, learned recently from her dentist that there is an aspect of her dental structure that is a sure sign of an indigenous ancestor.

Almost the first event marking the change from childhood years to university years were aptitude tests. They were followed by consultation with some member of the university's teaching staff. I was offered, given my arithmetic talent, such possibilities as accountancy although academic possibilities were also mentioned. I suppose I expressed an intention to study mathematics and physics and it was observed that in that case I might even want to take a Ph.D. degree. I listened and, my studies not having yet begun, returned to White Rock, where for some reason or other I visited the house of my future in-laws, who were in bed. I took the opportunity of asking my future father-in-law what a Ph.D. was. Somewhat surprisingly, now but not then, he knew. Sometime later, I consulted with a mathematician, Dr. Jennings, with the title for professors customary in Canadian universities, who suggested to me that as a mathematician there would be several foreign languages to be learned. I took his remark seriously, although, initially, seriously may not have entailed effectively. One year of French or some other language was a normal requirement. At the end of the first year I acquired a basic text for German grammar and reading it over the summer felt that I was adequately equipped in that direction, and in the second year moved on to a course in Russian.

In retrospect, these efforts were a little ridiculous, but I had occasion later to make more serious and more effective attempts to master these and other languages. I pity the mathematicians of today, not only the native speakers of English who have no occasion to learn the classical European languages, which offered until recently, outside mathematics and within mathematics as well, a great deal but also the European mathematicians and mathematicians from elsewhere, but especially the Europeans—the French who have all but destroyed Breton, the Germans, who have destroyed above all Yiddish but also less important languages as well, like Wendish/Sorbian—who now are assiduously rendering their own more and more difficult of access. They, with a notion that acquiring English is today still a cultural achievement, merit perhaps more contempt than pity. Some, of course, like the editors of Springer Verlag, are in it for the money. It may be that the response as to whether mathematics should be a matter of several languages or of one will ultimately be given by the Chinese.

In my first year of university, my principal preoccupation was some mastery of English. As I have implied I was in high-school negligent and had ignored the basics not of English orthography but of English grammar. I was assiduous—to the amusement of the other students, some of them adults, and the teacher, Dr. Morrison—consulting the dictionary for every unfamiliar word in every poem and learning, in particular, to avoid comma splices.

The mathematics was new to me but by today's standards elementary, largely, as I recall, trigonometry. The second year was again relatively slow, mathematics and a course of logic, a subject about which, as a mathematician, I have tried to inform myself but always unsuccessfully. The physics course, the Russian course, and the course on English literature, from the beginning to the nineteenth century, Chaucer, Fielding, and a good number of other authors, all appealed to me. We had been introduced to Shakespeare in high-school.

The third year was more interesting. For various reasons the multi-variable calculus courses were not successful, a bad and indifferent teacher undermined the efforts of a conscientious and potentially excellent teacher, Dr. Leimanis. However Marvin Marcus, now very old but still, I believe, with us, recommended Courant's classic book on differential and integral calculus, which I studied, but occasionally, as with the inverse function theorem, in too superficial a manner. Either in the second but more likely third year, I had two other courses, each, as I recall, half-year courses. Dr. Christian gave an excellent course on algebra from a well-known book of Dickson. Once again, I did not always grasp adequately the interest of various important points, for example, the theory of the cubic equation. The textbook in the other course, on linear algebra and geometry was also a widely used American text, the names of whose authors I forget, but during the summer, at the suggestion of Christian, I read Halmos's book on vector spaces, widely used at the time. I confess that I fell in love with the abstraction of his presentation, a passion good in its way for a mathematician, but it is best not to be overwhelmed by it. A book that, in some sense, was a surer sign of my fate as a specialist of automorphic forms and the Hecke theory was a book that I found on my own, a translation Modern Algebra and Matrix Theory of a once familiar German text by Schreier and Sperner where the theory of elementary divisors is treated at length.

By the fourth year, I could give myself up almost completely to mathematics. Not entirely through a fault of my own, I had abandoned any intention I may have had to become a physicist. In retrospect, I did not and do not have the right kind of imagination, but the decisive event was a course in thermodynamics in the third year. This is a difficult subject, in particular the topics of heat and entropy and in response to a homework question I wrote an extravagantly long essay, which unfortunately I did not keep. Given my age—I had just turned nineteen—and my lack of a solid pre-university education, it probably was not so bad. The teacher, an English experimentalist, chose to mock it in class. That, I think, was the turning point. Certainly, given the nature of whatever talent I had, it was for the best. I did, that year, have a physics course that I much enjoyed, the optics course, above all the experiments. I was fortunate to have a partner, Alan Goodacre, in the laboratory experiments that were an important part of the course and that in his hands always yielded the expected, thus the correct, results. There was a division of labour, in which I took the easy half, the theory, and he the difficult half, the experiments. I still have occasion to meet him occasionally in Ottawa, where he was an experimentalist with the dominion laboratories for many years.

So, in the fourth year, I focussed except for a second course in Russian on mathematics. I learned or began to learn a great deal: function theory, in particular from the third volume, which I studied on my own, of the prescribed text, a translation of a German text by Konrad Knopf, the Weierstrass theory of elliptic functions; ordinary differential equations, including something about special functions and the beginnings of spectral theory, which I supplemented later in graduate school and the years immediately following, with the book of Coddington and Levinson and with M. H. Stone's book on the spectral theory of operators in Hilbert space, both excellent preparation for the general theory of Eisenstein series, which has been a major concern of my career and which led to its major achievement, what is often referred to as the Langlands programme. Galois theory was the principal topic in another course but it went by, I am afraid, more or less unremarked. I also participated in my fourth year at university, or more likely in the following year, in a seminar on commutative algebra, based on the book *Ideal Theory* of D. G. Northcott. In any case, I managed during the following year to write, on my own initiative, a master's thesis on some idea or problem that I found in it. None of the professor's were familiar with the topic so that they were uncertain what to do. Out of the goodness of their hearts and, I suppose, because my performance was otherwise satisfactory they accepted it—even though I had had to confess at some point during the proceedings that I had found an important error in it—and let me move on to graduate school. That year 1957–58, between my four undergraduate years and my two graduate years at Yale, was one of the most demanding of my life. I had been married a year before, at a very young age, was teaching one undergraduate course, my first experience of lecturing, was taking enough courses to acquire the necessary credits for a master's degree, without which I could not move on to the important stage of a doctoral degree, and finally I was writing the master's thesis. I remember almost nothing from that year: life in a trailer with my wife; a charming girl in the freshman class I was teaching who seemed to be taken with me, an infatuation to which unmarried and otherwise uncommitted I would have been happy to respond; as well as an incident with a second professor of physics. This I remember clearly.

The occasion was the final examination of a course on mathematical methods in physics, a course for graduate students offered by this professor, an immigrant from Europe. The focus was the representation theory of finite groups, characters, orthogonality and so on. The single problem assigned was to analyze the representation of the tetrahedral group on the sum of the four tangent spaces at the vertices, each a three-dimensional space, which he thought of as provided with the natural metric. He expected the students to decompose the representation using the orthogonality relations. The best, most direct solution is of course to use a non-orthogonal basis directed away from the vertices along the edges. Then the problem is solved by inspection. He looked at all the zeros and ones in the calculation and was persuaded that I, as a thick-headed student, had inappropriately introduced the regular representation and was about to fail me, which would have meant no master's degree and no move to Yale. He seemed to be obdurate, but for unexplained reasons ultimately gave me a passing grade. Perhaps a colleague explained the solution to him.

Fig. 3: Ankara, Turkey, in 1967. (Photo: private)

All in all, the years at the University of British Columbia were very profitable. I learned something: how to write English, a beginning in three other languages, a little bit of physics that did no harm and, both in the courses and on my own, a good deal of mathematics. The campus too was a pleasure, small and forested. Photographs I have seen suggest that it is now an asphalt desert, but so are many places.

I add as well that I was by and large satisfied with my independent reading in various mathematical domains. It still seems to me that a mathematician's obligation, as much, even more, than proving theorems, an activity that sometimes demands more ingenuity than insight or foresight, is the preservation of the creations of the past, not necessarily all, but certainly those that sustain the subject's depth and intellectual pertinence. Although I had an exaggerated fondness for abstraction, one nourished by Halmos, it was sated by the book of Dixmier, *Les algèbres d'opérateurs dans l'espace hilbertien* which I believe I managed to read from beginning to end as a part of my studies for the master's degree. Nor was I able to maintain a sustained interest in logic, even the introductory text *The elements of mathematical logic* by Paul Rosenbloom defeated me.

I set off for Yale University full of hope and my wife followed in a couple of months with our first child. A second one was to be born less than a year later. At Yale I followed two or three helpful courses, one on the basics of functional analysis with Nelson Dunford, using the book *Linear Operators, Part 1*, that he wrote with Jack Schwartz. As one can see from the second volume of the book, it was in this course that I first proved something that could be called a theorem. Another course was formally given by Einar Hille and made use of his book *Functional Analysis and Semi-groups*. Hille was a consummate analyst and it was a pleasure to acquire some familiarity with various objects of classical analysis, especially the Laplace transform, from it. A third course was given by Felix Browder on partial differential equations, especially the topics of current concern to specialists. He was never well-prepared, often taking two or three runs at a proof without ever succeeding in completing it. I, however, was diligent, took my notes home and usually managed to put the collected information together to arrive at a proof, so that I gained much from his lectures.

I also read a great deal, especially from Dover paperbacks, which were available cheaply. I remember, in particular, D. V. Widder's book, *The Laplace transform*, in the Princeton University Press series, Burnside's book *The Theory of Groups of Finite Order*, and the first edition of Zygmund's book *Trigonometrical Series*, a book I read carefully. I read the second book superficially, forming the extravagant ambition of proving the Burnside conjecture on groups of odd order, proved not much later by Feit and Thompson. Zygmund's book I seemed to have read quite carefully, for it saved me from failing the oral examinations at the end of the year. I had not systematically prepared for them, thinking I could take a chance with what I knew. It turned out that I had largely forgotten what I had learned about commutative algebra in Vancouver. So things were looking bad. After algebra came analysis, and the examiner Shizuo Kakutani fortunately knew a great deal about various convexity theorems, due to one or the other of the brothers Riesz, that I too had at my fingertips—at the time, from a recent reading of Zygmund's book on trigonometrical series, but not today. So I was saved by a kind of miracle.

Sometime during the year, I solved a problem about Lie semi-groups, a topic introduced by Hille and during the summer, putting together on my own what I had learned about elliptic partial differential and about Lie semi-groups, wrote what I considered an acceptable thesis. As with my master's thesis in Vancouver, no-one on the faculty could read it, so that there was some question as to whether it could be accepted. Kakutani was opposed to this, but his colleagues decided none the less to accept it. Fortunately, I did not discover any errors and much of the material was later incorporated into the book *Elliptic Operators and Lie groups* by Derek Robinson. So it had some success.

Having prepared the doctoral thesis I was completely free for the entire second year of my stay at Yale. This was one of the happiest years of my professional life. For the first time in several years, my time was my own. There was one seminar proposed on functions of several complex variables that, as it turned out, never took place because of some discord between the organizers, and lectures by S. Gaal on the paper of Selberg on the trace formula and what Godement began to refer to as Eisenstein series, a topic begun by Hecke and Maaß, a student of Hecke. I read the articles that were to be treated in the seminar on my own. By a fortunate coincidence the material on domains of holomorphy they contained allowed me to prove some relatively simple theorems on the holomorphic continuation of the Eisenstein series, a topic in Gaal's lectures which later became central for me. I forget the titles of the articles and the names of their authors.

I would like to insert here a note of appreciation to Yale for the two years I spent there as a student. After deciding to apply for admission to a doctoral program in mathematics, I applied at Harvard, Yale, and Wisconsin. Harvard accepted me, but with no money; Yale accepted me with a bursary; and Wisconsin accepted me with a position as teaching assistant. Yale was thus the best choice. What I want to admit or to explain here is how fortunate I believe I was to be spared the trial of a sudden immersion in an atmosphere of competitive and well-trained young mathematicians who had spent their undergraduate years in major centres and to have had rather two stressless years if not to make a mathematician of myself at least to have a start at it. When I arrived at Princeton after these two years my schooling was up to the more demanding style of my contemporaries, the students, and of the young faculty.

In fact my arrival at Princeton was a matter largely of chance. I would have preferred to stay at Yale and, I believe, most of the faculty would have been happy to keep me, but Kakutani was again opposed and this time successfully. The place, Princeton, and the time were chosen by hazard. Leonard Gross, then at Yale, but who later spent his career at Cornell University, suggested to me and a friend a trip to the Institute for Advanced Study, where some of his friends from his student days in Chicago, among them Edward Nelson and Paul Cohen, were spending a year. I happened to speak briefly with Nelson about my work on Lie semi-groups. It turned out that he had investigated similar topics. He was favourably impressed and, as he was to become an assistant professor at Princeton University in the following year, suggested to his future colleagues that they appoint me as an instructor. I received the appointment with no application, no letters of recommendation, nothing, only the oral recommendation of Nelson. The lives of young mathematicians, and of their older colleagues as well, in the USA, and no doubt elsewhere as well—the USA often serves as an unfortunate model far beyond its borders—are more burdened with red tape than they once were. If I had become a mathematician in the present context, domestic and international, I would, I believe, have become quite a different one and, indeed, abandoned the undertaking, or perhaps never have begun it. Semi-groups had, however, little to do with my first years at Princeton. In part because of the work of Selberg, a good many mathematicians had returned to the study of Hecke and Robert Gunning was offering a course on his theory in Princeton which I attended. There was also a seminar on analysis, that Salomon Bochner attended and, I believe, fostered. I was asked to deliver a talk on my own work and, not having anything else to offer, discussed the somewhat accidental efforts inspired by Gaal's seminar. Bochner appreciated it, not, I would guess, so much because of the material, but because it had no relation whatsoever to my thesis, an indication of independent thought. Bochner was an analyst of broad scope, who during the early part of his career in Germany had known, I believe, Helmut Hasse, Emmy Noether and others in the Göttingen school. He encouraged me to pursue the study of these series and, thus of automorphic forms, in particular to extend the considerations from the field of rational numbers to finite extensions of this field, thus to learn about algebraic numbers. This was my first introduction to German texts, a text of Landau and that, *Vorlesungen über die Theorie der Algebraischen Zahlen*, of Hecke, and then, rather quickly as I pursued the topic of Eisenstein series, the papers of Hecke and Siegel, in which the modern theory of automorphic forms was created. Sometime during this period I also came across the monograph *The general theory of Dirichlet's series* by Hardy and M. Riesz that contains the very important theorem of Landau on Dirichlet series with positive coefficients, a theorem that was, I believe, implicit in the work of Rankin and Selberg on Ramanujan's conjecture. Rankin was a student of Hardy.

I observed in the paper *Problems in the theory of automorphic forms*, that appeared in 1970, that functoriality in my sense and the theorem of Landau taken together would yield not only Ramanujan's conjecture itself but also a very general form of it. This is a conviction I had not afterwards questioned. On writing this preface, however, I glanced again at that paper and observed that functoriality and the theorem of Landau were not in themselves sufficient; one would need in addition the principal theorem of Godement and Jacquet in their treatise *Zeta functions of simple algebras* as well as the understanding of the spectral decomposition of $L^2(\mathrm{GL}(n;F)\backslash\mathrm{GL}(n;\mathbb{A}_F))$, available in a paper of Moeglin and Waldspurger. There is no reference to these works, both of which appeared later, in my paper. I have to return to the Godement–Jacquet theorem and understand why and how it is so strong.

After Siegel the modern theory was created, hardly entirely but I would say decisively, first by Selberg, whose trace formula, although a form of the Frobenius reciprocity theorem, was at a much more difficult analytic level, and Harish-Chandra, whose work on reduction theory, although in comparison with his work on representation theory minor, transformed the theory of automorphic forms into a part of the theory not of particular reductive groups but of all reductive groups and, ultimately, an aspect of representation theory. Certainly I, as a young mathematician, moved naturally along this route. It was inevitable; as I just observed, the trace formula, which became almost immediately after its introduction central to the theory, is intrinsically representation-theoretic.

A topic that was less obvious, but in its way also inevitable, was class field theory. Not only was it in the air in Princeton, from which Artin had not long been absent, but it was apparently also a topic that Bochner felt was important, and it must have been in late summer of 1963 that he suggested that I should or declared that I must offer a course in class field theory, at the time an arcane topic, of no interest to the bulk of mathematicians. I was flabbergasted! I had hardly begun to learn algebraic number theory and the semester was about to begin. I protested that it was out of the question, but he insisted. I yielded and set about preparing the course, which placed the problem of a non-abelian class field theory, if not at the centre, certainly on the fringes of my mathematical ambition.

The academic year 1964–65 I spent at Berkeley in California with no teaching responsibilities. My initial ambition was to learn algebraic geometry. This was before Grothendieck dominated the subject and I took with me Weil's *Algebraic Geometry* and Conforto's *Abelsche Funktionen und Algebraische Geometrie*. I even organized with Phillip Griffiths a seminar on algebraic geometry from which he clearly profited much more than I. I was also infatuated during that year with Harish-Chandra's papers on spherical functions and matrix coefficients and wanted to construct that theory along the lines of the theory of hypergeometric functions with which I was taken at the time, but was unsuccessful. All in all, I was disappointed with my mathematical accomplishments during that year.

The next year, back in Princeton, was no better. I had formed two ambitions, both rather extravagant, to make some progress in non-abelian class field theory and to create a general theory of Hecke L-functions. I made no progress and by the spring of 1966 was coming to believe that mathematics was not the career for me. Indeed I had sentiments of this sort already during the year in Berkeley. The year 1963/64, a year in which I not only was learning class field theory for the course I was giving but also writing up the extremely long paper on Eisenstein series, may have caused an unrecognized exhaustion that was the source of the mathematical famine of the Berkeley year. Anyhow, as I have recounted elsewhere, a friend, Orhan Türkay, whom I had met in Princeton and who was also spending a year in Berkeley, suggested I come to Turkey, not to his university where he taught economics but to the newly created Middle East Technical University. It was not a suggestion that I initially took seriously. However, after my return to Princeton and a fruitless struggle with the two topics mentioned, I recalled his invitation and began, influenced perhaps by Agatha Christie novels, to reflect on the possibility of a romantic trip to the Middle East, today a somewhat incredible notion, especially for someone accompanied by a wife and four young children.

As a diversion, a pleasant one, during 1965–66 I offered to teach mathematics for engineers, a task that was beneath the dignity of my colleagues. I enjoyed it and the engineering students and I had, as well, the pleasure of consulting various books that appealed to me. From Maxwell's *Electricity and Magnetism* I learned methods for plotting the level lines of harmonic functions, an amusing occupation for me and for the students; Relton's book on Bessel functions was an opportunity to learn some concrete spectral theory. I do not suppose that specialized monographs were better or more readily available fifty or sixty years ago than today, but I was more likely to find the time, the energy, and the patience to read them.

Once having begun to make the arrangements for a year, or more, abroad, life became simpler and more relaxed. For lack of anything better to do, I began, perhaps in the summer or the autumn, to make some idle calculations of the constant terms of the Eisenstein series. Once calculated, it was, as I recall, almost immediately evident that they offered examples of a general notion of a Hecke L-function, what I had been searching for in vain. I did not change my plans, but did abandon my efforts to learn Turkish and to improve my knowledge of Russian, both appealing pastimes that had to be abandoned in order to investigate the implications of these calculations. By the end of the year I had a fairly clear idea of what might be done and, during a brief, unexpected conversation with André Weil began to explain what I had in mind. A more detailed hand-written explanation sent after the conversation he did not find persuasive, but he did invite me to attend a seminar in which he was explaining how he hoped to extend the Hecke theory for GL(2) to general number fields. He was having trouble with the complex case. I explained the theory in this case to him in two letters, one sent before the departure for Turkey, one after my arrival, in which, thanks to the knowledge of ordinary differential equations acquired over the years and to my own experiences in attempting to create a generalized Hecke theory, the complex case was treated. Weil was unable to read them, but turned to Jacquet for help, and this led to the joint effort with Jacquet to develop a theory of automorphic forms for GL(2) compatible with my expectations, and perhaps with Jacquet's as well.

Apart from childhood excursions across the nearby border and my years at Yale, Princeton, and Berkeley, I had never been abroad, never visited French-speaking Canada, so that Turkey was my first experience with a genuinely different country. Although it was a pleasant and instructive visit that did not interfere with my mathematical projects, I did not accomplish what I had hoped, scarcely, in spite of my intentions, learned the language, scarcely visited the country and acquired only a superficial understanding of its history. My first mistake was not to take an immediate linguistic plunge, not to understand that for an anglophone English is, in many respects, a handicap in one's encounter with the world. The language itself attracts too many people and the wrong sort; it alienates others and is, by and large, an obstacle to a genuine intimacy with a land and its people. I exaggerate but for, say, an academic to acquire English today is scarcely more difficult than learning to drive an automobile, yet there are many academics, in particular, a good number of my fellow mathematicians, who regard it as a genuine intellectual accomplishment, which they are eager to display whenever the occasion arises. They and, in particular, their excessive use of English as a vehicle of publication deprives the field of a great deal of whatever secondary intellectual pleasures it offers. Before my first visit to Turkey, I did not understand this. The situation was not so extreme in 1967 as it is today. At all events during this visit I acquired some knowledge of the language, some acquaintances among the students that, when many years later, more experienced, better prepared, I returned, served me well. A number of the former students became friends.

In the meantime, I had been more determined in my efforts, linguistic and historical, but since my major preoccupation was always mathematics, I was not able to satisfy these predilections to any satisfactory extent. I have presently one final mathematical project, but once it is accomplished, or even if I make some encouraging progress, I hope to pass the time left to me indulging them. My memory is failing; so are my energies, but I am nevertheless hopeful.

The major undertaking of my mathematical career has been the theory of automorphic forms and its many manifestations, and a secondary one has been renormalization. I think that most of my colleagues regard me as competent to discuss automorphic forms; very few will regard me as competent to discuss renormalization. I shall nevertheless attempt to do so very briefly, because I fear it is a subject with enormous potential and enormous depth although also enormously difficult that has not been met with any genuinely adequate effort on the part of mathematicians. Here, however, as with automorphic forms, the principal failing seems to be a reluctance to come to terms with the central issues, to forget that the central problem in both cases is to construct a theory, not to prove something that with a certain amount of good will passes muster as a theorem.

Automorphic forms are another matter, for the major problems have not been entirely neglected, but there is still much more that we do not understand than that we understand. At present, the subject has strong ties to arithmetic, to analytic number theory, to algebraic geometry, and to differential geometry. There are three different theories, relevant in three different contexts: the theory over an algebraic number field; the theory over a function field in one variable over a finite field; the theory over a compact Riemann surface, thus over a function field in one variable over $\mathbb{C}$. They have many similarities, but a number of important differences. There is a famous analogy of earlier forms of them, in which reductive Lie groups, apart from $\mathrm{GL}(1)$, played no role, with the Rosetta stone. Weil has written a brief and charming description of it in the essay *De la métaphysique aux mathématiques*, but the analogy is less persuasive for the modern theories. The essay is none the less well worth reading. There is an extension of the analogy in a Bourbaki lecture of Edward Frenkel, *Gauge theory and Langlands duality*, that appeared in a Bourbaki lecture in 2009. He has our three topics, which he lists as *Number theory*, *Curves over* $\mathbb{F}_q$, *Riemann surfaces* but he adds a fourth, *Quantum Physics*. I prefer to remain in this essay within the domain of pure mathematics, indeed lack of competence forces me to do so.

Fig. 4: In 1999. (Photograph by C.J. Mozzochi, courtesy of the Simons Foundation.)

It is well to recall that the first theory, which is central to the theory of algebraic numbers and diophantine equations, has a history beginning late in the eighteenth century or early in the nineteenth, with Gauss and Kummer and continuing without interruption until the middle of the last century; the second, which may ultimately form a major part of the theory of finite fields, is thus a theory that begins perhaps with Galois, but began to play a much larger role with the introduction by Hasse and Weil of an algebraic geometry over finite fields; the third is at best a part of the theory of Riemann surfaces, which itself does not, as one remarks when recalling Euler's concern with elliptic integrals, begin entirely with early nineteenth century mathematicians like Abel, Jacobi, Weierstrass, and Riemann and their study of algebraic curves or Riemann surfaces, but acquires at that time a new richness. It also entails—even without turning to the fourth, neglected topic—a relation to, perhaps speculative, physical theories, the introduction—in the definition of the geometric form of the Hecke operators—of differential geometry, curvature, and the Chern–Gauss–Bonnet theorem, which appears—in spite of its name—to be, in its present form, largely a twentieth century creation. I, myself, greatly regret as my career draws to a close not having spent enough time with the writings of the early founders of these theories.

It is certainly not necessary to understand the theories as a whole to contribute to the theory of automorphic forms. On the other hand, deliberately to restrict one's attention to a few tried methods and to a few familiar concepts, and deliberately to turn one's back on a marvelous fusion and coherence of many of the principal mathematical concepts of the past 250 years, seems to me an unpardonable form of self-mutilation, rendered more tempting by the linguistic, thus intellectual, inadequacies noted above. For many, even most, contemporary mathematicians, the mathematical past is largely inaccessible. This ignorance is unfortunate. Humans have not been on the earth for such a long time that it is appropriate to forget our history, or that of our accomplishments, mathematics among them. The indifference to the mathematical past and the distance from it will, I suppose, nevertheless become more pronounced if or when the Asian nations assume a major role in mathematics and even English ceases to be the principal, or even an adequate, medium of communication. The possibility of reducing mathematics to a trivial pursuit, a struggle for a large number of citations, for a prize, or just for a tenured position, is also there. It is difficult not to be pessimistic! In my view, or at least in a coarse simplification of my view, a great deal has been lost or destroyed in the largely successful European—at least initially, for the centre of power shifted in the last century—attempt to conquer or dominate the world and little is done to preserve what remains.

To return to the three theories. I have spent most time with the first and believe that it the most difficult, although it may now have the fewest practitioners. The two principal questions are very general: functoriality and reciprocity. Functoriality entails an answer to such questions as Ramanujan's conjecture in its general form, thus it entails establishing all the expected properties of automorphic L-functions short of the Riemann hypothesis. A first tool is the trace formula in the form established and developed extensively by James Arthur. Then it requires a development of the trace formula, in the sense of analytic number theory, say in the form with which Ali Altuğ has been struggling, whose goal would be, in my view, the formulation and the proof of the diophantine equalities, thus a comparison of the number of solutions of two different diophantine equations, necessary for the comparisons envisaged in my essay *A prologue to functoriality and reciprocity: Part 1*. This is not work that I myself have undertaken, perhaps because I have no ideas, but largely because I think of it as an undertaking that requires decades and I myself do not have decades. Once functoriality has been established or some progress with it has been accomplished, there will remain reciprocity: is the motivic galoisian group a quotient of the automorphic galoisian group? This is an even more daunting problem. There may be domains and mathematicians whose efforts suggest solutions but my ignorance is so great that I am reluctant to offer any advice. Otherwise, we have only scraps of results, scraps of a method, but they are serious scraps, prominent in, for example, the proof of the Fermat conjecture. I do not suppose we shall ever have any general information about these two groups more concrete than a statement that one is a quotient of the other. Specific information is another matter.

In the article in preparation whose title appears above, I broach the problem of an explicit description of the automorphic galoisian group for the third form of the theory of automorphic forms, thus the theory over a Riemann surface. There is no analogue of the motivic galoisian group. I confine myself in the article to unramified representations, thus to a nevertheless large and interesting quotient of the galoisian group, and define a group that I call the *AB*-group because it was introduced by Atiyah and Bott and make some effort to persuade the reader that it is indeed the unramified automorphic galoisian group for the third theory. Further reflection is, however, necessary.

Fig. 5: In 1990. (Photograph by C.J. Mozzochi, courtesy of the Simons Foundation.)

I have not yet had a chance to examine the geometric theory over a finite field, but, after hearing of the work of Vincent Lafforgue, I am tempted to ask myself whether the automorphic galoisian group over function fields over a finite field κ is not pretty much the Galois group of the function field F_κ, the field of rational functions on the curve defined over κ, although I have not yet had the leisure to study his papers.

For a period of several years, I was absorbed by an altogether different mathematical theory. Sometime in 1984/85 I had the good fortune of an enlightening conversation with the physicist Giovanni Gallavotti. As we strolled on the grounds of the Institute for Advanced Study, he described to me the mathematical problems arising from renormalization. They are very beautiful, of very general concern, and, in some sense, not sufficiently studied by mathematicians. There is, as I discovered, a great deal to learn if one wishes to understand their significance, whether in quantum field theory or fluid dynamics, or even in a more mathematical context. I was never able to come to grips with all these matters, although a colleague Yvan Saint-Aubin and I spent a great deal of time with numerical experiments. Once again I hope, after I have succeeded in explaining what I see as the expression of functoriality in the automorphic theory over (the function fields of) Riemann surfaces, to return to renormalization, not in the expectation of accomplishing anything, but rather in the desire to understand the problems in more depth than before.

Let me try to describe the relevant issues in a very simple, somewhat factitious example. I am hardly in a position to discuss quantum field theory or fluid mechanics. Renormalization is related to a change of scale. Suppose we have a cube (my personal experience is limited to dimension two) of porous material whose precise constitution is unknown, although we know the probability—referred to as a crossing probability—that water forced into the material on one small area α on the surface can make its way to another small area β, thus that there is an open channel between them. The collection of crossing probabilities $\{\pi(\alpha,\beta)\}$ is a property of the material used. An extreme case would be that the initial choice was between a solid, thus impassable, cube with probability x and an empty cube, in which all crossings are possible, with probability $1-x$. We can then think of taking eight of such cubes and placing them together to form a cube with double the original linear dimension. It will have different crossing probabilities, because in the larger cube, they are affected by the possibility of a very large number of paths, moving in and out of the eight constituent cubes. Then we change scale so that the new cube has edges of length one, thus so that the original cube has sides of length 1/2. We continue in this way. Thus we start with something very small and perhaps very simple, as nature seems to do, and arrive at something very large. What happens? We can expect that most of the time, thus for most initial probability distributions, as the size grows, complete permeability or complete impermeability become more and more likely. On the other hand, it may be that some other configurations maintain themselves or are generated in the process. There are of course an infinite number of initial distributions, even an infinite-dimensional space of initial distributions. Nevertheless there may be a tendency for all of it, or for large chunks of it, to be shrunk by an infinite number of repetitions to a point. These points and their nature are of considerable interest. What happens, for example, if we start from a point in the vicinity of one these limit points. Typically, there is a subspace of finite codimension that contracts under the operation described to the point while there is an expansion in the remaining directions, so that each orbit under repeated applications of the process described forms a kind of discrete hyperbola, except for those that start on the subspace. These dynamics are very simple.

It is, however, not even well understood how to create such systems. More to the point it is not well understood, certainly not in a mathematical sense, how to create such systems that are physically relevant nor how to demonstrate that some system, presumed to be physically relevant, has the expected mathematical properties. It would, I think, be a pleasure to reflect on these matters, to try to understand some part of what is known and something of what is not known, but with absolutely no ambitions.

The work of Robert Langlands

James G. Arthur

Foreword

A more accurate title might have been *On the Work of Robert Langlands in Representation Theory, Automorphic Forms, Number Theory and Arithmetic Geometry*. For I have left out a significant part of Langlands' work, his papers in percolation theory and in mathematical physics, published in the years 1988–2000. I have however included a brief description of his recent work on the geometric theory. It occurs near the end of §11, the last section otherwise devoted to Beyond Endoscopy.

There is more than enough to discuss in the subjects I will be considering. I hope to be able to communicate the remarkable continuity that runs throughout all of the work of Langlands, with its roots in several fundamental areas of mathematics. What is now known as the Langlands program represents a unification of some of the deepest parts of these areas.

For example, it has been suggested that the Langlands program is ultimately a theory of L-functions, with its roots in analytic number theory. With this interpretation it goes back to Euler. Others might see the Langlands program at its most striking in its discovery of the long sought reciprocity laws for nonabelian class field theory, a culmination of perhaps two centuries of study of algebraic number theory. This of course goes back to Gauss, and his law of quadratic reciprocity. Both interpretations are equally valid, and they come together in the work from the 1920s and 1930s of Emil Artin. But they are by no means the full story. The origins of the subject for Langlands, and the power he was able to bring to it right from the beginning, came from harmonic analysis and group representations, specifically the work of Harish-Chandra. He was the first mathematician to gain a deep understand-

J.G. Arthur
Department of Mathematics, University of Toronto,
40 St. George Street,
Toronto, Ontario,
Canada M5S 2E4,
e-mail: arthur@math.utoronto.ca

H. Holden, R. Piene (eds.), *The Abel Prize 2018-2022*, The Abel Prize,
https://doi.org/10.1007/978-3-031-33973-8_2

ing of the fundamental contributions of Harish-Chandra to the representation theory of semisimple/reductive Lie groups. This analytic side of the subject has continued to inform the Langlands program right up to the present time. I note, for no particular reason except perhaps for some further sense of unity, that Harish-Chandra was a graduate student of Paul Dirac at Cambridge. However, his early career as a physicist ended when he switched completely to mathematics a few years later.

My hope has been to bring the work of Langlands to a more general mathematical audience. In the attempt to emphasize the continuity of the work, I have tried to write the report as a narrative that evolves with time. This entails fewer statements of formal theorems, and more efforts to describe underlying ideas. It also includes more repetition than would a formal paper. The fundamental ideas of Langlands occur again and again, often in different guises. Seeing them appear in this way might give a broader sense of the symmetry of the subject.

For example, Section 2 on Langlands' fundamental manuscript on Eisenstein series is certainly among the more technical sides of the report. We then follow it in Section 3 with a relatively leisurely introduction to class field theory. In general, I hope that a nonspecialist reader will be encouraged by the more elementary parts, and initially at least, not feel the need to take the more difficult passages as seriously.

A reader might also find it helpful to consult Langlands' later commentary on his various papers, to be found in the different sections of his website. I have certainly gained insight from it in the preparation of this report. I should add that I know some of the papers better than others, and I apologize in advance for any misstatements in my attempts to make this deep and fundamental work as accessible as I can.

Contents

1 Group representations and harmonic analysis

As we have noted in the Foreword above, Langlands' early work came from group representations and harmonic analysis. These areas have remained at the heart of much of what is now known as the Langlands program. The analytic power in their methods has been indispensable in many of Langlands' greatest discoveries.

The groups in question are (unimodular) locally compact groups H. A unitary representation of H is a (weakly continuous) homomorphism

$$R\colon H \to \mathrm{U}(\mathscr{H})$$

from H to the group of unitary operators on a Hilbert space $\mathscr{H}$. For example, one could take H to be a product

$$H = G \times G$$

of a group G with itself, and $\mathscr{H}$ to be the Hilbert space $L^2(G)$ of square integrable complex-valued functions on G with respect to a Haar measure. One then has the regular representation

$$(R_H(y_1,y_2)\phi)(x) = \phi(y_1^{-1}xy_2), \quad \phi \in L^2(G), \quad x,y_1,y_2 \in G,$$

of $G \times G$ on $\mathscr{H}$. Another broad example is associated with a discrete subgroup Γ of $H = G$, in which one assumes that the quotient space $\Gamma \backslash H$ of right cosets has finite volume with respect to an H-invariant measure. In this case, one has the unitary representation

$$(R_\Gamma(y)\phi)(x) = \phi(xy), \quad \phi \in L^2(\Gamma \backslash H), \quad x \in \Gamma \backslash H, \quad y \in H,$$

of H by right translation on $\mathscr{H} = L^2(\Gamma \backslash H)$.

A representation π of H on a Hilbert space V is *irreducible* if V has no closed, π-invariant subspaces other than $\{0\}$ and V. Recall also that two unitary representations (π, V) and (π', V') of H are *(unitarily) equivalent* if

$$\pi'(y) = U\pi(y)U^{-1}, \quad y \in H,$$

for a unitary, linear, intertwining isomorphism U from V to V'. For any given H, one would like to classify $\Pi_{\mathrm{unit}}(H)$, the set of equivalence classes of irreducible unitary representations of H. This can be regarded as the fundamental problem in group representations.

The fundamental problem for harmonic analysis would apply to any natural unitary representation R of H, such as $R = R_H$ or $R = R_\Gamma$ as above. It is to find an *explicit* decomposition of R into irreducible representations. This presupposes a knowledge of $\Pi_{\mathrm{unit}}(H)$, or at least a subset of $\Pi_{\mathrm{unit}}(H)$ that is large enough to support the measure class of $\Pi_{\mathrm{unit}}(H)$ that governs the decomposition of R.

For example, in the special case of the additive group $H = \mathbb{R} \oplus \mathbb{R}$, the irreducible unitary representations are the one-dimensional representations

$$\pi(y_1,y_2) = \mathrm{e}^{\lambda_1 y_1}\mathrm{e}^{\lambda_2 y_2}, \quad (y_1,y_2) \in \mathbb{R} \oplus \mathbb{R},$$

where (λ_1, λ_2) ranges over the points in the imaginary space $i\mathbb{R} \oplus i\mathbb{R}$. The decomposition of R_H is just classical harmonic analysis, the *Fourier transform* $\phi \to \widehat{\phi}$ being an explicit unitary isomorphism from $L^2(\mathbb{R})$ onto $L^2(i\mathbb{R})$ such that

$$(R_H(y_1,y_2)\phi)^\wedge(\lambda) = e^{-\lambda y_1}\widehat{\phi}e^{\lambda y_2}, \quad \lambda \in i\mathbb{R}.$$

From the other broad example, consider the special case that $H = \mathbb{R}$ and $\Gamma = \mathbb{Z}$. In this case, the map that assigns Fourier coefficients to functions is an explicit unitary isomorphism from $L^2(\mathbb{Z}\setminus\mathbb{R})$ onto the Hilbert space $L^2(2\pi i\mathbb{Z})$ of functions on the subset of irreducible representations of $\mathbb{R}$ that occur in the decomposition of $\mathbb{R}$.

The mathematical area Langlands entered in 1960 was considerably more elaborate than these basic examples would suggest. In particular, the groups H were nonabelian, which meant that the irreducible unitary representations $\pi \in \Pi_{\mathrm{unit}}(H)$ were typically infinite-dimensional. One consequence of this for harmonic analysis was that the decompositions of representations R_H and R_Γ typically had both a continuous part, qualitatively like the theory of Fourier transforms, and a discrete part, like the theory of Fourier series. By 1960, the theory was already rich and sophisticated, thanks in large measure to the ongoing efforts of Harish-Chandra.

From the beginning, Harish-Chandra had limited his efforts to semisimple Lie groups, such as the special linear groups $\mathrm{SL}(n,\mathbb{R})$, the special orthogonal groups $\mathrm{SO}(p,q,\mathbb{R})$ and the symplectic groups $\mathrm{Sp}(2n,\mathbb{R})$. In contrast to abstract locally compact groups, semisimple Lie groups have a rich structure, which gives rise to an even richer structure for their representations. Harish-Chandra's goal was to establish the Plancherel formula for any such G. It includes the problem of the explicit decomposition of the regular representation $R_{G\times G}$. The problem is actually a little more precise. A solution would in fact include a natural measure, the Plancherel measure, within the measure class that gives the decomposition into irreducible representations. Harish-Chandra ultimately established the Plancherel formula around 1975, but by 1960, he was well on his way to constructing the *discrete series*[1] for G. These are irreducible representations $\pi \in \Pi_{\mathrm{unit}}(G)$ of G such that the products

$$\pi^\vee \otimes \pi \to (y_1,y_2) = {}^t\pi(y_1^{-1}) \otimes \pi(y_2)$$

give the representations that occur discretely in the decomposition of $R_{G\times G}$. His construction of the discrete series was completed around 1965, and is among Harish-Chandra's greatest achievements.

In 1960, Harish-Chandra's papers were regarded by many as being simply too difficult for anyone to read. Nonetheless, Langlands set about doing a comprehensive study of Harish-Chandra's work. Within a couple of years, he had acquired a mastery of at least part of it, including the nascent discrete series. This was demonstrated widely in three remarkable contributions to the 1965 AMS Summer Symposium in Boulder, Colorado, a broad conference designed to assess the general state of the subject.

Before I discuss Langlands' Boulder contributions, let me mention a later paper, because it bears directly on the work of Harish-Chandra. There are a couple of points to be made first. Harish-Chandra had found that, paradoxically, it was more natural to study the set $\Pi(G)$ of equivalence classes of *all* irreducible representa-

[1] Harish-Chandra no doubt appropriated this term from Valentine Bargmann, the physicist who classified the irreducible unitary representations of the group $\mathrm{SL}(2,\mathbb{R})$ in 1947.

tions of G, rather than just the unitary ones. To be sure, his Plancherel measure was to be supported on the subset $\Pi_{\text{temp}}(G)$ of tempered representations that he would introduce in 1966 [88], and these are all unitary. But much of his developing analytic power required a command of the full set $\Pi(G)$. Another curious fact is that there are interesting unitary representations $\pi \in \Pi_{\text{unit}}(G)$ that do not lie in $\Pi_{\text{temp}}(G)$. This contradicts our intuition from classical Fourier analysis, in which the irreducible representations of $\mathbb{R}$ are one-dimensional quasi-characters

$$x \to \mathrm{e}^{-\lambda x}, \quad x \in \mathbb{R}, \lambda \in \mathbb{C},$$

while both the unitary and the tempered representations coincide with the set of characters on $\mathbb{R}$, in which λ is purely imaginary.

In 1973, Langlands gave a classification of the full set of irreducible representations $\Pi(G)$ [151], modulo a knowledge of the subset $\Pi_{\text{temp}}(G)$ of tempered representations. By that time, the set $\Pi_{\text{temp}}(G)$ had been classified by Harish-Chandra up to a set of Plancherel measure 0. The remaining singular representations in $\Pi_{\text{temp}}(G)$ were classified soon afterwards by Knapp and Zuckerman [114] [115], which meant that the Langlands classification then gave all the irreducible representations. The setting in Langlands' paper was actually slightly different. He replaced Harish-Chandra's semisimple Lie groups by groups $G(\mathbb{R})$ of real points, in which G here represents a *reductive algebraic group* over $\mathbb{R}$. (A reductive algebraic group is a finite quotient of the product of a semisimple algebraic group with an algebraic torus.) The Langlands classification was soon extended to p-adic groups, in the weaker sense that it classifies all irreducible representations $\pi \in \Pi(G(\mathbb{Q}_p))$ in terms of representations that are tempered. (The tempered representations $\Pi_{\text{temp}}(G(\mathbb{Q}_p))$ of a p-adic group are another story, which we will come to in due course.) It thus pertains to all the local ingredients of general automorphic representations. For obvious reasons, the Langlands classification has become very influential. We shall return to it in Section 10, as a foundation for Langlands' later theory of endoscopy.

The three earlier contributions of Langlands to the Boulder proceedings were all striking, but one of them, on the theory of Eisenstein series [134], dominates the others in depth and importance. We leave it until the next section, and describe the other two here.

All three of Langlands' Boulder articles pertain to the quotients $\Gamma \backslash H$, which come with the associated representations R_Γ of H on $L^2(\Gamma \backslash H)$. The paper [135] concerns the space $\Gamma \backslash H$ itself. It applies to the case

$$\Gamma = G(\mathbb{Z}) \subset G(\mathbb{R}) = H$$

for a split (Chevalley) algebraic group G over $\mathbb{Q}$, such as for example the group $G = \mathrm{SL}(n)$. The problem arose from the work of Tamagawa and Weil on the volume of $\Gamma \backslash H$, with respect to a canonical measure obtained from what is known as the Tamagawa measure.

Langlands established an explicit formula for the volume of $\Gamma \backslash H$ in this case, in terms of special values of the Riemann zeta function. His short proof was an ingenious combination of an interesting contour integral motivated by his theory of Eisenstein series, a well known real variable integral formula of Gindikin and Karpelevich (which had been used by Harish-Chandra in a different context), and some standard properties of Chevalley groups. At the time there was also a famous conjecture for the volume attached to any G [251], which Weil had formulated in adelic form for simply connected groups. Langlands' paper confirmed Weil's conjecture in the special case of split groups. This was extended to quasi-split groups in 1972 by K. F. Lai [130], using the method of Langlands. Then later, following the suggestion of Jacquet and Langlands on p. 525 of [103], Kottwitz [122] extended Lai's result to arbitrary G, using a simple form [22, Corollary 23.6] of the general trace formula.[2] More precisely, Kottwitz extended the result to any G that satisfies the Hasse principle, which was known at the time for any group without factors of type E_8. The Hasse principle was later established for that last case by Chernousov [49]. Langlands' Boulder paper [135] thus became the foundation for a general proof of Weil's conjecture.

The second Boulder paper [133] to discuss here concerns the representation R_Γ. Langlands used it to introduce a version in higher rank of the Selberg trace formula for compact quotient.

Suppose that $H = G$ is a semisimple Lie group, and that Γ is a discrete subgroup with $\Gamma \backslash G$ compact. Among other things, Selberg introduced a formula that could be applied to the representation R_Γ in this case. It is an identity

$$\sum_{\{\gamma\}} \mathrm{vol}(\Gamma_\gamma \backslash G_\gamma) \int_{G_\gamma \backslash G} f(x^{-1}\gamma x)\,\mathrm{d}x = \sum_{\{\pi\}} \mathrm{mult}_\Gamma(\pi)\,\mathrm{tr}(\pi(f)), \tag{1}$$

in which $f \in C_c^\infty(G)$ is to be regarded as a general test function on G. On the left-hand side, $\{\gamma\}$ stands for a set of representations of conjugacy classes in Γ, Γ_γ is the centralizer of γ in Γ, G_γ is the centralizer of γ in G, and $\mathrm{vol}(\Gamma_\gamma \backslash G_\gamma)$ is the volume of the quotient $\Gamma_\gamma \backslash G_\gamma$ with respect to the right invariant measure defined by a fixed Haar measure on G_γ. The integral over $G_\gamma \backslash G$ is taken with respect to the quotient of a fixed Haar measure on G by the chosen measure on G_γ. On the right-hand side $\{\pi\}$ ranges over $\Pi_{\mathrm{unit}}(G)$, $\mathrm{mult}_\Gamma(\pi)$ is the multiplicity (a finite nonnegative integer) with which π occurs discretely in the irreducible decomposition of $L^2(\Gamma \backslash G)$, and $\mathrm{tr}(\pi(f))$ is the trace of the operator

$$\pi(f) = \int_G f(x)\pi(x)\,\mathrm{d}x$$

on the Hilbert space V on which π acts. The left-hand side is often called the *geometric side*, since the objects $\{\gamma\}$ have natural geometric interpretations. The right-hand side is called the *spectral side*, since the coefficients $\mathrm{mult}_\Gamma(\pi)$ concern spectral data in the decomposition of R_Γ. We note that if G is a finite group, the formula becomes the well known theorem of Frobenius reciprocity, or rather the special case that applies to the trivial one-dimensional representation of the subgroup $\Gamma \subset G$.

---

[2] For simplicity we shall often restrict references for the general trace formula to this introductory survey. The reader can then consult the original articles listed there, as needed.

In [133], Langlands derived (1) from first principles. He followed the original argument of Selberg, but the form of (1) is in some sense new. It differs from that of Selberg in that it reflects the theory of group representations, and in particular, the work of Harish-Chandra. Langlands then posed the question of using (1) to derive an explicit formula for the multiplicity $\mathrm{mult}_\Gamma(\pi)$ of π. One does not expect a closed formula for all π. Langlands was asking about the case that π lies in the subset $\Pi_2(G)$ of discrete series. Incidentally, the subscript 2 here means "square integrable", in the sense that every matrix coefficient

$$x \to (\pi(x)\phi, \psi), \quad \phi, \psi \in V, \tag{2}$$

of π is a square integrable function of $x \in G$. Harish-Chandra had earlier noted that π belongs to the discrete series if and only if it is square integrable. Armed with the simple trace formula (1), Langlands proposed letting the test function f be a matrix coefficient of π as above (with $\phi, \psi \neq 0$). The problem with this, however, was that f is not compactly supported. In particular, the integrals in (1) need not converge. Langlands added the condition that π lie in the smaller subset $\Pi_1(G)$ of *integrable* discrete series. With this assumption, the integrals do converge, and he was able to use the work of Harish-Chandra to compute the terms in (1) explicitly. He thereby obtained a simple, explicit formula for $\mathrm{mult}_\Gamma(\pi)$.

As a supplement, Langlands added a conjectural interpretation of his multiplicities $\mathrm{mult}_\Gamma(\pi)$ in terms of the cohomology of complex vector bundles. It was a generalization of the Borel–Weil formula for compact groups, suggested perhaps by a later proof of Kostant. It was very appealing at the time, especially to mathematicians with a background in complex analysis, and for a few years was sometimes known as "The Langlands Conjecture". This was of course before the sweeping conjectures that evolved into the Langlands program. The original Langlands conjecture from [133] was established a few years later by Wilfried Schmid [201].

Langlands' formula for $\mathrm{mult}_\Gamma(\gamma)$ is an explicit, finite linear combination of terms. These were attached to the values of the character of π at regular, elliptic elements, namely points γ at which the centralizer G_γ is a compact abelian subgroup of G. We shall not pause here to describe the formula precisely, or to recall how Harish-Chandra was able to construct the character of the infinite-dimensional representation π as a locally integrable class function on G. We note only that the formula obtained by Langlands, simple as it may be, is part of the foundation of the ongoing comparison of spectral data in representation theory with arithmetic spectral data in Shimura varieties. To be sure, it is the adelic version of the formula with its enrichment by Hecke operators that is relevant today, as well as a further stabilization of the formula. But it is still interesting to think that this early result of Langlands would have an implicit role in the study of Shimura varieties he began ten years later. We shall discuss these matters in Section 8.

It is also interesting that we now have a proper chain

$$\Pi_1(G) \subset \Pi_2(G) \subset \Pi_{\mathrm{temp}}(G) \subset \Pi_{\mathrm{unit}}(G) \subset \Pi(G) \tag{3}$$

of sets of (equivalence classes of) irreducible representations of G. Each of these families has its own role to play in some aspect of the theory. It is not hard to describe $\Pi_1(G)$ explicitly as a proper subset of $\Pi_2(G)$ in terms of the parametrization Harish-Chandra gave to the discrete series, a problem that arose with the publication of Langlands' paper, and that was solved shortly thereafter.

Langlands' paper on multiplicities provides a good introduction to the trace formula (1) for compact quotient. There is now a trace formula for general arithmetic quotients $\Gamma \backslash G$, such as

$$\Gamma \backslash G = \mathrm{SL}(n,\mathbb{Z}) \backslash \mathrm{SL}(n,\mathbb{R}),$$

which are typically noncompact (see [22]). This is much more difficult to establish, owing to the rather severe singularities at the boundary (which manifest themselves analytically as badly divergent integrals). However, the trace formula has been central to the subject. It represents an essential part of the work of Langlands, as a force behind proofs of fundamental theorems such as those for inner twists of GL(2) [103], base change for GL(2) [149], L-indistinguishably for $\widehat{\mathrm{SL}}(2)$ [127] and the cohomology of Shimura varieties [140], as well as foundation of broader theories, such as Endoscopy [150] and Beyond Endoscopy [155] that remain conjectural.

2 Eisenstein series

The remaining Boulder article [134] of Langlands was a concise summary (with some supplementary ideas that were later used [11] in the development of a general trace formula) of his unpublished 270-page manuscript *On the Functional Equations Satisfied by Eisenstein Series*. The manuscript was later published, with four supplementary Appendices, as the monograph [141]. It was well ahead of its time, and was described in the preface of [141] of having been "almost impenetrable". Harish-Chandra temporarily suspended his work on the Plancherel formula to study the manuscript. The result was a set of expository lecture notes [89], which were a little closer to his own perhaps more familiar style, but which did not contain the most difficult part of the manuscript, the final Chapter 7. Langlands' theory of Eisenstein series has gradually become more widely understood. There are now a number of expositions of varying length, the most comprehensive being the monograph [178] of Mœglin and Waldspurger. Our aim here is to discuss some of the background and the content of Langlands' theory.

Eisenstein series have to do with the spectral decomposition of a space $L^2(\Gamma \backslash G)$. In general, there is an orthogonal decomposition

$$L^2(\Gamma \backslash G) = \mathrm{L}^2_{\mathrm{disc}}(\Gamma \backslash G) \oplus \mathrm{L}^2_{\mathrm{cont}}(\Gamma \backslash G)$$

where $\mathrm{L}^2_{\mathrm{cont}}(\Gamma \backslash G)$ is an R_Γ-invariant subspace of $L^2(\Gamma \backslash G)$ that decomposes into a continuous direct sum (sometimes known as a direct integral), and $\mathrm{L}^2_{\mathrm{disc}}(\Gamma \backslash G)$ is a subspace that decomposes discretely. Langlands' manuscript provided an explicit description of the continuous part of R_Γ. More precisely, it gave a decomposition of the continuous spectrum $\mathrm{L}^2_{\mathrm{cont}}(\Gamma \setminus G)$ in terms of discrete spectra $\mathrm{L}^2_{\mathrm{disc}}(M \cap \Gamma \setminus M)$, for a finite set of proper subgroups M of G. Before we recall how this works, we should review the general setting of Langlands' work, and the way it is usually formulated today.

Langlands took G to be a semisimple Lie group and Γ to be a discrete subgroup such that the quotient $\Gamma \setminus G$ satisfies some general axioms. By far the most important case is that of an arithmetic quotient $\Gamma(N) \setminus G$, where G is now a *reductive* algebraic group over $\mathbb{Q}$, with a $\mathbb{Z}$-scheme structure (such as $\mathrm{SL}(n)$), and

$$\Gamma = \Gamma(N) = \{\gamma \in G(\mathbb{Z}) : \gamma \equiv 1 \pmod N\}$$

is a principal congruence subgroup of $G(\mathbb{Z})$. The right regular representation of $G(\mathbb{R})$ on $L^2(\Gamma(N) \setminus G(\mathbb{R}))$ is best studied in the modern adelic formulation, which we pause briefly to review. (The reader can also refer to Langlands' adelic reformulation of his results in Appendix II of [141].)

The real field $\mathbb{R}$ is the completion of $\mathbb{Q}$ with respect to the usual archimedean absolute value $|\cdot| = |\cdot|_\infty$. For every prime number p, there is also a p-adic absolute value $|\cdot|_p$ on $\mathbb{Q}$, defined by setting $|x|_p = p^{-r}$ if

$$x = (ab^{-1})p^r,$$

for integers a, b and r with $(a,p) = (b,p) = 1$, and $|x|_p = 0$ if $x = 0$. Like $\mathbb{R}$, its completion $\mathbb{Q}_p$ is a locally compact field in which $\mathbb{Q}$ embeds as a dense subfield, and to which $|\cdot|_p$ extends continuously. Unlike $\mathbb{R}$, $\mathbb{Q}_p$ has an open, compact subring

$$\mathbb{Z}_p = \{x_p \in \mathbb{Q}_p : |x|_p \leq 1\}.$$

The ring product of all these completions is no longer locally compact. Suppose however that

$$S \subset \{v\} = \{\infty\} \cup \{p \text{ prime}\}$$

indexes a finite subset of these valuations that contains the archimedean absolute value $|\cdot|_\infty$. The product

$$\widehat{\mathbb{Z}}^S = \prod_{p \notin S} \mathbb{Z}_p$$

is then a compact ring, while the larger product

$$\mathbb{A}_S = \left(\prod_{v \in S} \mathbb{Q}_v\right) \widehat{\mathbb{Z}}^S$$

is a locally compact ring. The topological direct limit

$$\mathbb{A} = \varinjlim_{S} \mathbb{A}_S$$

of adeles is therefore a locally compact ring. It contains the diagonal image of the field $\mathbb{Q}$ as a discrete, cocompact subring.

For the given algebraic group G over $\mathbb{Q}$, we can form the group $G(\mathbb{A})$ of points with values in the adele ring $\mathbb{A} = \mathbb{A}_{\mathbb{Q}}$ of $\mathbb{Q}$. It is a locally compact group, which contains $G(\mathbb{Q})$ as a discrete subgroup. The pair

$$\Gamma = G(\mathbb{Q}) \subset H = G(\mathbb{A})$$

is thus an example of the kind of object we have been considering, and to which we can attach a Hilbert space $L^2(G(\mathbb{Q}) \setminus G(\mathbb{A}))$. This looks very different from the concrete Hilbert space $L^2(\Gamma(N) \setminus G(\mathbb{R}))$ above. It is not.

Suppose that $K = K^\infty$ is an open compact subgroup of the nonarchimedean part

$$G(\mathbb{A}^\infty) = \left\{ x = \prod_v x_v \in G(\mathbb{A}) \, : \, x_\infty = 1 \right\}$$

of $G(\mathbb{A})$. The product $G(\mathbb{R})K$ is then an open subgroup of $G(\mathbb{A})$. Under some natural conditions on G, the set of $(G(\mathbb{Q}), G(\mathbb{R})K)$ double cosets in $G(\mathbb{A})$ is then finite. Writing

$$G(\mathbb{A}) = \coprod_{i=1}^{n} \left(G(\mathbb{Q}) \cdot x^i \cdot G(\mathbb{R})K \right),$$

for elements $x^1 = 1, x^2, \ldots, x^n$ in $G(\mathbb{A}^\infty)$, we obtain a right $G(\mathbb{R})$-invariant decomposition

$$\begin{aligned} G(\mathbb{Q}) \setminus G(\mathbb{A})/K &= \coprod_{i=1}^{n} G(\mathbb{Q}) \setminus (G(\mathbb{Q}) \cdot x^i \cdot G(\mathbb{R})K)/K \\ &\cong \coprod_{i=1}^{n} (\Gamma^i \setminus G(\mathbb{R})), \end{aligned}$$

for discrete subgroups

$$\Gamma^i = G(\mathbb{R}) \cap \left(G(\mathbb{Q}) \cdot x^i K (x^i)^{-1} \right)$$

of $G(\mathbb{R})$, and a $G(\mathbb{R})$-isomorphism of Hilbert spaces

$$L^2(G(\mathbb{Q}) \setminus G(\mathbb{A})/K) = \bigoplus_{i=1}^{n} \left(\Gamma^i \setminus G(\mathbb{R}) \right). \tag{4}$$

Each discrete group Γ^i is defined by congruence conditions, which are determined by the choice of K and x^i. Each Hilbert space on the right-hand side of (4) is thus a modest generalization of the space $L^2(\Gamma(N) \setminus G(\mathbb{R}))$ above. But we can see directly that it contains more information that just the regular representation of $G(\mathbb{R})$.

On one hand, the action of $G(\mathbb{R})$ on the spaces on either side of (4) corresponds to the action by right convolution on either space by functions in the algebra $C_c(G(\mathbb{R}))$. But there is a supplementary convolution algebra, the Hecke algebra $\mathscr{H}(G(\mathbb{A}^\infty), K)$

of compactly supported functions that are left and right invariant under translation by the group K. It also acts by right convolution on the left-hand side of (4), in a way that clearly commutes with the action of $G(\mathbb{R})$. The corresponding action of $\mathscr{H}(G(\mathbb{A}^\infty),K)$ on the right-hand side of (4) includes general analogues of the operators defined by Hecke on classical modular forms.

Hecke operators for general groups are at the heart of the theory. They contain the data that according to Langlands' later conjectures govern much of the arithmetic world. One builds them into the representation theory by setting

$$C_c^\infty(G(\mathbb{A})) = C_c(G(\mathbb{R})) \otimes C_c^\infty(G(\mathbb{A}^\infty))$$

where $C_c^\infty(G(\mathbb{A}^\infty))$ denotes the space of locally constant, complex-valued functions of compact support on $G(\mathbb{A}^\infty)$. Any function in $C_c^\infty(G(\mathbb{A}))$ is bi-invariant under translation by an open compact subgroup $K = K^\infty$ of $G(\mathbb{A}^\infty)$, and therefore acts by right convolution on the corresponding space (4). When we vary f, and hence K, we are working with the understanding that the action of the convolution algebra $C_c^\infty(G(\mathbb{A}))$ on the Hilbert space $L^2(G(\mathbb{Q}) \setminus G(\mathbb{A}))$ is equivalent to the right regular representation R_G of $G(\mathbb{A})$ on the space. This is the standard modern setting, which despite possible appearances, streamlines the joint study of the right regular representation of $G(\mathbb{R})$ and the underlying Hecke operators.

We would now like to describe Langlands' work on Eisenstein series. Following the statement in [22], we shall describe the main results in complete detail, making the rest of this section one of the more technical parts of our report. For a start, the results are formulated in terms of the basic structure of algebraic groups, which we will have to apply with only limited comments. We should first take care of the minor annoyance that $G(\mathbb{A})$ can have noncompact centre, which implies for trivial reasons that $L^2(G(\mathbb{Q}) \setminus G(\mathbb{A}))$ will have no discrete spectrum. One deals with it either by replacing $G(\mathbb{A})$ by the quotient $A_G(\mathbb{R})^0 \setminus G(\mathbb{A})$, in which A_G is the $\mathbb{Q}$-split component of the centre $Z = Z(G)$ of G, or by the subgroup

$$G(\mathbb{A})^1 = \{x \in G(\mathbb{A}) : |\chi(x)| = 1, \quad \chi \in X(G)_\mathbb{Q}\},$$

in which $X(G)_\mathbb{Q}$ is the group of characters from G to $\mathbb{G}_m = \mathrm{GL}(1)$ defined over $\mathbb{Q}$, and

$$|\chi(x)| = \prod_v |\chi(x_v)|_v$$

is the (adelic) absolute value on the group $\mathbb{A}^* = \mathrm{GL}(1,\mathbb{A})$ of ideles.

The two modifications are equivalent. It is clear that $G(\mathbb{Q})$ embeds as a discrete subgroup of either $G(\mathbb{A})^1$ or $A_G(\mathbb{R})^0 \setminus G(\mathbb{A})$. Moreover, there is a surjective homomorphism H_G from the group $G(\mathbb{A})$ onto the real vector space

$$\mathfrak{a}_G = \mathrm{Hom}(X(G)_\mathbb{Q}, \mathbb{R}),$$

defined by setting

$$\mathrm{e}^{\langle H_G(x),\chi\rangle} = |\chi(x)|, \quad x \in G(\mathbb{A}), \chi \in X(G)_\mathbb{Q},$$

whose kernel is $G(\mathbb{A})^1$, and whose restriction to the central subgroup $A_G(\mathbb{R})^0$ of $G(\mathbb{A})$ is an isomorphism onto $\mathfrak{a}_G$. It follows that

$$G(\mathbb{A}) = G(\mathbb{A})^1 \times A_G(\mathbb{R})^0.$$

In particular, there is a unitary $(G(\mathbb{A}) = G(\mathbb{A})^1 \times A_G(\mathbb{R})^0)$-equivariant isomorphism between the two Hilbert spaces

$$L^2(G(\mathbb{Q}) \setminus G(\mathbb{A})^1) \cong L^2(G(\mathbb{Q})A_G(\mathbb{R})^0 \setminus G(\mathbb{A})),$$

each of which will have nontrivial discrete spectra

$$\mathrm{L}^2_{\mathrm{disc}}(G(\mathbb{Q}) \setminus G(\mathbb{A})^1) \cong \mathrm{L}^2_{\mathrm{disc}}(G(\mathbb{Q})A_G(\mathbb{R})^0 \setminus G(\mathbb{A})),$$

(unless G is a split torus over $\mathbb{Q}$). The spaces on the left are perhaps more natural. For among other things, we can identify the irreducible unitary representations $\pi \in \Pi_{\mathrm{unit}}(G)$ of $G(\mathbb{A})^1$ with the $i\mathfrak{a}_G^*$-orbits

$$\{\pi_\lambda(x) = \pi_0(x)e^{-\lambda(H_G(x))} : x \in G(\mathbb{A}), \lambda \in i\mathfrak{a}_G^*\}$$

of irreducible unitary representations of $G(\mathbb{A})$. The base point π_0 can be any irreducible unitary representation of $G(\mathbb{A})$ whose restriction of $G(\mathbb{A})^1$ is π. A similar convention holds if π is replaced by any representation R of $G(\mathbb{A})^1$, such as for example the representation $R_{G,\mathrm{disc}}$ of $G(\mathbb{A})^1$ on $\mathrm{L}^2_{\mathrm{disc}}(G(\mathbb{Q}) \setminus G(\mathbb{A})^1)$, and λ is any point in the complex vector space $\mathfrak{a}^*_{G,\mathbb{C}} = \mathfrak{a}_G^* \otimes \mathbb{C}$, with the understanding that R_0 is a representation of $A_G(\mathbb{R})^0 \setminus G(\mathbb{A})$. These conventions are due to Harish-Chandra, and are quite natural, even if they might seem cumbersome at first.

We fix a minimal parabolic subgroup P_0 of G over $\mathbb{Q}$, together with a Levi decomposition $P_0 = M_0N_0$, where M_0 (resp. N_0) is a reductive (resp. unipotent) subgroup of P_0 over $\mathbb{Q}$. We also fix a suitable maximal compact subgroup $K_0 = \prod_v K_v$ of $G(\mathbb{A})$ [12, p. 9] with $G(\mathbb{A}) = P_0(\mathbb{A})K_0$. We shall then work in what remains of this section with the finite set of standard parabolic groups, namely the subgroups P of G which contain P_0. Any such P has a unique Levi decomposition

$$P = M_PN_P,$$

in which the Levi component M_P contains M_0. Since P contains P_0, we then have a decomposition

$$G(\mathbb{A}) = P(\mathbb{A})K_0 = N_P(\mathbb{A})M_P(\mathbb{A})K_0 = N_P(\mathbb{A})M_P(\mathbb{A})^1A_P(\mathbb{R})^0K_0,$$

where $A_P = A_{M_P}$ in the notation above (with M_P in place of G). This allows us to define a continuous mapping

$$H_P \colon G(\mathbb{A}) \to \mathfrak{a}_P, \quad \mathfrak{a}_P = \mathfrak{a}_{M_P},$$

by setting

$$H_P(nmk) = H_{M_P}(m), \quad n \in N_P(\mathbb{A}), m \in M_P(\mathbb{A}), k \in K,$$

where H_{M_P} and $\mathfrak{a}_P = \mathfrak{a}_{M_P}$ are again as above. Finally, we fix Haar measures $\mathrm{d}x$, $\mathrm{d}n$, $\mathrm{d}m$, $\mathrm{d}a$ and $\mathrm{d}k$ on the groups $G(\mathbb{A})$, $N_P(\mathbb{A})$, $M_P(\mathbb{A})^1$ (or $M_P(\mathbb{A})$, depending on the context), $A_P(\mathbb{R})^0$ and K such that for the decomposition above, we have

$$\mathrm{d}x = \mathrm{e}^{2\rho_P(H_P(a))}\,\mathrm{d}n\,\mathrm{d}m\,\mathrm{d}a\,\mathrm{d}k.$$

Here ρ_P is the familiar vector in $\mathfrak{a}_P$ such that $\mathrm{e}^{2\rho_P(H_P(\cdot))}$ is the modular function on the (nonunimodular) group $P(\mathbb{A})$. The notation is again due to Harish-Chandra, and is convenient for working with representations of $G(\mathbb{A})$ induced from $P(\mathbb{A})$.

Suppose that P is a standard parabolic subgroup of G, and that λ lies in $\mathfrak{a}^*_{P,\mathbb{C}}$. We write

$$y \to \mathscr{I}_P(\lambda, y), \quad y \in G(\mathbb{A}),$$

for the induced representation

$$\mathrm{Ind}_{P(\mathbb{A})}^{G(\mathbb{A})}(I_{N_P(\mathbb{A})} \otimes R_{M_P,\mathrm{disc},\lambda})$$

of $G(\mathbb{A})$ obtained from λ and the discrete spectrum of the reductive group M_P. This representation acts on the Hilbert space $\mathscr{H}_P$ of measurable functions

$$\phi : N_P(\mathbb{A})M_P(\mathbb{Q})A_P(\mathbb{R})^0 \backslash G(\mathbb{A}) \to \mathbb{C}$$

such that the function

$$\phi_x : m \to \phi(mx),$$

belongs to $\mathrm{L}^2_{\mathrm{disc}}(M_P(\mathbb{Q}) \backslash M_P(\mathbb{A})^1)$ for almost all $x \in G(\mathbb{A})$, and such that

$$\|\phi\|^2 = \int_K \int_{M_P(\mathbb{Q})\backslash M_P(\mathbb{A})^1} |\phi(mk)|^2\,\mathrm{d}m\,\mathrm{d}k < \infty.$$

For any $y \in G(\mathbb{A})$, $\mathscr{I}_P(\lambda, y)$ maps a function $\phi \in \mathscr{H}_P$ to the function

$$(\mathscr{I}_P(\lambda, y)\phi)(x) = \phi(xy)\mathrm{e}^{(\lambda+\rho_P)(H_P(xy))}\mathrm{e}^{-(\lambda+\rho_P)(H_P(x))}.$$

We have put the twist by λ into the operator $\mathscr{I}_P(\lambda, y)$, rather than the underlying Hilbert space $\mathscr{H}_P$, in order that $\mathscr{H}_P$ be independent of λ. The function $\mathrm{e}^{\rho_P(H_P(\cdot))}$ is included in the definition in order that the representation $\mathscr{I}_P(\lambda)$ be unitary whenever the inducing representation is unitary, which is to say, whenever λ belongs to the subset $i\mathfrak{a}^*_P$ of $\mathfrak{a}^*_{P,\mathbb{C}}$.

Suppose that

$$R_{M_P,\mathrm{disc}} \cong \bigoplus_{\pi} \pi \cong \bigoplus_{\pi} \Big(\bigotimes_{v} \pi_v \Big)$$

is the decomposition of $R_{M_P,\mathrm{disc}}$ into irreducible representations $\pi = \bigotimes_v \pi_v$ of $M_P(\mathbb{A})/A_P(\mathbb{R})^0$. The induced representation $\mathscr{I}_P(\lambda)$ then has a corresponding decomposition

$$\mathcal{I}_P(\lambda) \cong \bigoplus_{\pi} \mathcal{I}_P(\pi_\lambda) \cong \bigoplus_{\pi} \Big(\bigotimes_{v} \mathcal{I}_P(\pi_{v,\lambda}) \Big)$$

in terms of induced representations $\mathcal{I}_P(\pi_{v,\lambda})$ of local groups $G(\mathbb{Q}_v)$. This follows from the definition of induced representation, and the fact that

$$e^{\lambda(H_{M_P}(m))} = \prod_v e^{\lambda(H_{M_P}(m_v))},$$

for any point $m = \prod_v m_v$ in $M_P(\mathbb{A})$. If $\lambda \in i\mathfrak{a}_P^*$ is in general position, all of the induced representations $\mathcal{I}_P(\pi_{v,\lambda})$ are irreducible. Thus, if we understand the decomposition of the discrete spectrum of M_P into irreducible representations of the local groups $M_P(\mathbb{Q}_v)$, we understand the decomposition of the generic induced representations $\mathcal{I}_P(\lambda)$ into irreducible representations of the local groups $G(\mathbb{Q}_v)$.

The aim of the theory of Eisenstein series is to construct intertwining operators between the induced representations $\mathcal{I}_P(\lambda)$ and the continuous part of the regular representation R of $G(\mathbb{A})$. The problem includes being able to construct intertwining operators among the representations $\mathcal{I}_P(\lambda)$, as P and λ vary. The symmetries among pairs (P, λ) are given by the Weyl sets $W(\mathfrak{a}_P, \mathfrak{a}_{P'})$ of Langlands. For a given pair P and P' of standard parabolic subgroups, $W(\mathfrak{a}_P, \mathfrak{a}_{P'})$ is defined as the set of distinct linear isomorphisms from $\mathfrak{a}_P \subset \mathfrak{a}_0$ onto $\mathfrak{a}_{P'} \subset \mathfrak{a}_0$ obtained by restrictions of elements in the (restricted) Weyl group

$$W_0 = \mathrm{Norm}(G, A_0)/\mathrm{Cent}(G, A_0).$$

Suppose for example that $G = \mathrm{GL}(n)$. If P and P' correspond to the partitions $(n_1, \dots n_p)$ and $(n'_1, \dots n'_{p'})$ of n, the set $W(\mathfrak{a}_P, \mathfrak{a}_{P'})$ is empty unless $p = p'$, in which case

$$W(\mathfrak{a}_P, \mathfrak{a}_{P'}) \cong \{s \in S_p : n'_i = n_{s(i)},\ 1 \leq s \leq p\}.$$

In general, we say that P and P' are *associated* if the set $W(\mathfrak{a}_P, \mathfrak{a}_{P'})$ is nonempty. We would expect a pair of induced representations $\mathcal{I}_P(\lambda)$ and $\mathcal{I}_{P'}(\lambda')$ to be equivalent if P and P' belong to the same associated class, and $\lambda' = s\lambda$ for some element $s \in W(\mathfrak{a}_P, \mathfrak{a}_{P'})$.

The formal definitions apply to any elements $x \in G(\mathbb{A})$, $\phi \in \mathcal{H}_P$ and $\lambda \in \mathfrak{a}^*_{M,\mathbb{C}}$. The associated Eisenstein series is

$$E(x, \phi, \lambda) = \sum_{\delta \in P(\mathbb{Q}) \backslash G(\mathbb{Q})} \phi(\delta x) e^{(\lambda + \rho_P)(H_P(\delta x))}. \tag{5}$$

If s belongs to $W(\mathfrak{a}_P, \mathfrak{a}_{P'})$, the operator

$$M(s, \lambda) \colon \mathcal{H}_P \to \mathcal{H}_{P'}$$

that intertwines $\mathcal{I}_P(\lambda)$ with $\mathcal{I}_{P'}(s\lambda)$ is defined by

$$M(s, \lambda)(x) = \int \phi(w_s^{-1} n x) e^{(\lambda + \rho_P)(H_P(w_s^{-1} n x))} e^{(-s\lambda + \rho_{P'})(H_{P'}(x))}\, dn, \tag{6}$$

where the integral is taken over the quotient

$$N_{P'(\mathbb{A})} \cap w_s N_P(\mathbb{A}) w_s^{-1} \backslash N_{P'}(\mathbb{A}),$$

and w_s is any representative of s in $G(\mathbb{Q})$. A reader so inclined could motivate both definitions in terms of finite group theory. Each definition is a formal analogue of a general construction of Mackey [169] for the space of intertwining operators between two induced representations $\mathrm{Ind}_{H_1}^H(\rho_1)$ and $\mathrm{Ind}_{H_2}^H(\rho_2)$ of a finite group H.

It follows formally from the definitions that

$$E(x, \mathscr{I}_P(\lambda, y)\phi, \lambda) = E(xy, \phi, \lambda)$$

and

$$M(s, \lambda)\mathscr{I}_P(\lambda, y) = \mathscr{I}_{P'}(s\lambda, y)M(s, \lambda).$$

These are the desired intertwining properties. However, (5) and (6) are defined by sums and integrals over noncompact spaces. They do not generally converge. It is this fact that makes the theory of Eisenstein series so difficult.

Let $\mathscr{H}_P^0$ be the subspace of vectors $\phi \in \mathscr{H}_P$ that are K_0-finite, in the sense that the subset

$$\{\mathscr{I}_P(\lambda, k)\phi \,:\, k \in K_0\}$$

of $\mathscr{H}_P$ spans a finite-dimensional space, and that lie in a finite sum of irreducible subspaces of $\mathscr{H}_P$ under the action of $\mathscr{I}_P(\lambda)$ of $G(\mathbb{A})$. The two conditions do not depend on the choice of λ. Taken together, they are equivalent to the requirement that the function

$$\phi(x_\infty x^\infty), \quad x_\infty \in G(\mathbb{R}), x^\infty \in G(\mathbb{A}^\infty)$$

be locally constant in x^∞, and smooth, $K_\mathbb{R}$-finite and $\mathscr{Z}_\infty$-finite in x_∞, where $\mathscr{Z}_\infty = \mathscr{Z}_{G,\infty}$ denotes the algebra of bi-invariant differential operators on $G(\mathbb{R})$. The space $\mathscr{H}_P^0$ is dense in $\mathscr{H}_P$.

For any P, we can form the chamber

$$(\mathfrak{a}_P^*)^+ = \{\Lambda \in \mathfrak{a}_P^* \,:\, \Lambda(\alpha^\vee) > 0, \alpha \in \Delta_P\}$$

in $\mathfrak{a}_P^*$, in which Δ_P denotes the set of simple parabolic roots of (P, A_P).

Lemma (Langlands). *Suppose that $\phi \in \mathscr{H}_P^0$ and that λ lies in the open subset*

$$\{\lambda \in \mathfrak{a}_{P,\mathbb{C}}^* \,:\, Re(\lambda) \in \rho_P + (\mathfrak{a}_P^*)^+\}$$

of $\mathfrak{a}_{P,\mathbb{C}}^*$. Then the sum (5) and the integral (6) that define $E(x,\phi,\lambda)$ and $(M(s,\lambda)\phi)(x)$ both converge absolutely to analytic functions of λ.

For spectral theory, one is interested in points λ such that $\mathscr{I}_P(\lambda)$ is unitary, which is to say that λ belongs to the real subspace $i\mathfrak{a}_P^*$ of $\mathfrak{a}_{P,\mathbb{C}}^*$. This is outside the domain of absolute convergence for (5) and (6). The problem is to show that the functions $E(x,\phi,\lambda)$ and $M(s,\lambda)\phi$ have analytic continuation to this space. The following theorem summarizes Langlands' main results in Eisenstein series.

Main Theorem (Langlands).

(a) *Suppose that* $\phi \in \mathscr{H}_P^0$. *Then* $E(x,\phi,\lambda)$ *and* $M(s,\lambda)\phi$ *can be analytically continued to meromorphic functions of* $\lambda \in \mathfrak{a}_{P,\mathbb{C}}^*$ *that satisfy the functional equations*

$$E(x,M(s,\lambda)\phi,s\lambda) = E(x,\phi,\lambda) \tag{7}$$

and

$$M(ts,\lambda) = M(t,s\lambda)M(s,\lambda), \quad t \in W(\mathfrak{a}_P,\mathfrak{a}_{P'}). \tag{8}$$

If $\lambda \in i\mathfrak{a}_P^*$, *both* $E(x,\phi,\lambda)$ *and* $M(s,\lambda)$ *are analytic, and* $M(s,\lambda)$ *extends to a unitary operator from* $\mathscr{H}_P$ *to* $\mathscr{H}_{P'}$.

(b) *Given an associated class* $\mathscr{P} = \{P\}$, *define* $\widehat{\mathrm{L}}_{\mathscr{P}}$ *to be the Hilbert space of families of measurable functions*

$$F = \{F_P \colon i\mathfrak{a}_P^* \to \mathscr{H}_P, \quad P \in \mathscr{P}\}$$

that satisfy the symmetry condition

$$F_{P'}(s\lambda) = M(s,\lambda)F_P(\lambda), \quad s \in W(\mathfrak{a}_P,\mathfrak{a}_{P'}),$$

and the finiteness condition

$$\|F\|^2 = \sum_{P\in\mathscr{P}} n_P^{-1} \int_{i\mathfrak{a}_P^*} \|F_P(\lambda)\|^2 \, d\lambda < \infty,$$

where

$$n_P = \sum_{P'\in\mathscr{P}} |W(\mathfrak{a}_P,\mathfrak{a}_{P'})|$$

for any $P \in \mathscr{P}$. *Then the mapping that sends* F *to the function*

$$\sum_{P\in\mathscr{P}} n_P^{-1} \int_{i\mathfrak{a}_P^*} E(x,F_P(\lambda),\lambda)\, d\lambda, \quad x \in G(\mathbb{A}),$$

defined whenever $F_P(\lambda)$ *is a smooth, compactly supported function of* λ *with values in a finite-dimensional subspace of* $\mathscr{H}_P^0$, *extends to a unitary mapping from* $\widehat{\mathrm{L}}_{\mathscr{P}}$ *onto a closed* $G(\mathbb{A})$*-invariant subspace* $\mathrm{L}^2_{\mathscr{P}}(G(\mathbb{Q})\setminus G(\mathbb{A}))$ *of* $L^2(G(\mathbb{Q})\setminus G(\mathbb{A}))$. *Moreover, the original space* $L^2(G(\mathbb{Q})\setminus G(\mathbb{A}))$ *has an orthogonal direct sum decomposition*

$$L^2(G(\mathbb{Q})\setminus G(\mathbb{A})) = \bigoplus_{\mathscr{P}} \mathrm{L}^2_{\mathscr{P}}(G(\mathbb{Q})\setminus G(\mathbb{A})). \tag{9}$$

The theorem gives a qualitative description of the decomposition of R. It provides a finite decomposition

$$R = \bigoplus_{\mathscr{P}} R_{\mathscr{P}},$$

where $R_{\mathscr{P}}$ is the restriction of R to the invariant subspace $\mathrm{L}^2_{\mathscr{P}}(G(\mathbb{Q}) \setminus G(\mathbb{A}))$ of $L^2(G(\mathbb{Q}) \setminus G(\mathbb{A}))$. It also provides a unitary intertwining operator from $R_{\mathscr{P}}$ onto the representation $\widehat{R}_{\mathscr{P}}$ of $G(\mathbb{A})$ on $\widehat{L}_{\mathscr{P}}$ defined by

$$(\widehat{R}_P(y)F)_P(\lambda) = \mathscr{I}_P(\lambda, y)F_P(\lambda), \quad F \in \widehat{\mathrm{L}}^2_{\mathscr{P}}, P \in \mathscr{P}.$$

The theorem is thus compatible with the general intuition we retain from the theory of Fourier series and Fourier transforms.

In summary, let us say that while it might seem a little overwhelming at first, the theorem is quite comprehensive, and ultimately, remarkably simple. It is exactly what one might hope for in terms of an explicit decomposition of the continuous spectrum $\mathrm{L}^2_{\mathrm{cont}}(G(\mathbb{Q}) \setminus G(\mathbb{A}))$ into irreducible representations. On the other hand, the proof of the theorem is long and complex, to an extent that is hard to quantify in a few words. Langlands had to overcome many obstacles, the most severe being the analytic problems treated in Chapter 7 of [141]. This last chapter consists of a sophisticated residue scheme, designed to construct inaccessible constituents of discrete spectra $\mathrm{L}^2_{\mathrm{disc}}(M(\mathbb{Q}) \setminus M(\mathbb{A})^1)$ and their Eisenstein series from residues of cuspidal Eisenstein series.

We define a locally integrable function ϕ on $G(\mathbb{Q}) \setminus G(\mathbb{A})$ to be *cuspidal* if for every standard parabolic subgroup $P = MN$ distinct from G, the integral

$$\int_{N(\mathbb{Q})\setminus N(\mathbb{A})} \phi(nx)\,\mathrm{d}n, \quad x \in G(\mathbb{A}),$$

vanishes for almost all x. The main property of these functions is that they lie in the relative discrete spectrum [10], [141]. In other words, the $G(\mathbb{A})$-invariant subspace $\mathrm{L}^2_{\mathrm{cusp}}(G(\mathbb{Q}) \setminus G(\mathbb{A}))$ of cuspidal functions in $L^2(G(\mathbb{Q}) \setminus G(\mathbb{A}))$ lies in the summand $\mathrm{L}^2_G(G(\mathbb{Q}) \setminus G(\mathbb{A}))$ of the decomposition (9) attached to $\mathscr{P} = \{G\}$, which is the space of functions ϕ whose restriction to $G(\mathbb{A})^1$ lies in $\mathrm{L}^2_{\mathrm{disc}}(G(\mathbb{Q}) \setminus G(\mathbb{A})^1)$. For any P, let $\mathscr{H}^0_{P,\mathrm{cusp}}$ be the subspace of functions ϕ in $\mathscr{H}^0_P$ such that the function $\phi_x(m) = \phi(mx)$ defined above is cuspidal for almost all x. A *cuspidal Eisenstein series* is then a series (5) in which ϕ lies in the subspace $\mathscr{H}^0_{P,\mathrm{cusp}}$ of $\mathscr{H}^0_P$.

Langlands studied cuspidal Eisenstein series in the first six chapters of the volume. These objects were difficult enough, but there were some available techniques of Selberg, especially for the case of rank 1 (in which $\dim(A_P) = 1$). Langlands used these techniques, and others that he created. By the end of Chapter 6, he had established the analytic continuation and functional equations from Part (b) of the theorem, for cuspidal Eisenstein series.

Noncuspidal Eisenstein series, however, were a different matter. We now have a classification for the noncuspidal discrete spectrum for general linear groups [177], as well as a conjectural classification [18] in general, but it does not help us with their Eisenstein series. It was Langlands' indirect residue scheme that ultimately led to the required properties. His resolution was a very delicate interplay between the desired analytic continuation and the required spectral properties, all within an extended induction argument.

In general, the irreducible representations (typically induced) in the spectral decomposition of each space $L^2_{\mathscr{P}}(G(\mathbb{Q})\backslash G(\mathbb{A}))$ are called *automorphic representations* (while the associated Eisenstein series are called *automorphic forms*). Let us write $\Pi_{\mathrm{temp}}(G)$ for the set of such representations, where the subscript stands for "globally tempered", in whatever sense this term would have if there were a global Schwartz space on $G(\mathbb{Q})\backslash G(\mathbb{A})$. (It does *not* mean that the local constituents of representations are tempered.) We can also write $\Pi_2(G)$ and $\Pi_1(G)$ for the subset of automorphic representations in the decompositions of the spaces $L^2_G(G(\mathbb{Q})\backslash G(\mathbb{A}))$ and $L^2_{\mathrm{cusp}}(G(\mathbb{Q})\backslash G(\mathbb{A}))$. Finally, we should say that the formal definition of automorphic representations [35, 145] gives a wider class of irreducible representations $\Pi(G)$ of $G(\mathbb{A})$, which of course includes the subset $\Pi_{\mathrm{unit}}(G)$ of unitary automorphic representations. We obtain a proper chain

$$\Pi_1(G)\subset\Pi_2(G)\subset\Pi_{\mathrm{temp}}(G)\subset\Pi_{\mathrm{unit}}(G)\subset\Pi(G) \tag{10}$$

of sets of irreducible automorphic representations of $G(\mathbb{A})$, with obvious analogy to the chain (3) for real groups. This global notation leaves us free to write $\Pi_{\mathrm{cusp}}(G)$ for the set of *all* cuspidal automorphic representations of $G(\mathbb{A})$, thereby allowing for nonunitary central characters. We then have

$$\Pi_1(G)=\Pi_{\mathrm{cusp}}(G)\cap\Pi_2(G)=\Pi_{\mathrm{cusp}}(G)\cap\Pi_{\mathrm{temp}}(G)=\Pi_{\mathrm{cusp}}(G)\cap\Pi_{\mathrm{unit}}(G).$$

The analogy between the local and global chains (3) and (9) is a little fanciful, but suggests analogies between the two kinds of representations. The local chain (3) was actually taken for a *semisimple* Lie group G, in which the centre is finite. At this point, we would take G to be a Lie group that has been implicitly identified with the group of real points $G(\mathbb{R})$ of a *reductive* group over $\mathbb{R}$. The symbol $\Pi_2(G)$ in (3) would then stand for the *relative discrete series*, which is to say, tempered representations π of $G(\mathbb{R})$ whose restrictions to $G(\mathbb{R})^1$ are in the discrete series. The equivalent term *square integrable* would then be understood to mean square integrable modulo the centre of G. This slight generalization was in fact Harish-Chandra's original definition.

We will not try to formulate an analogue of the chain (9) for spaces $\mathscr{A}(G)$ of automorphic forms. However, this is a good opportunity to say a few words about these objects, even if we do not recall their precise definition from [35]. We are of course speaking here of the modern day generalizations of classical modular forms on the upper half plane from which the subject as a whole now takes its name. (See [31].)

Roughly speaking, an automorphic form is a function on $G(\mathbb{Q})\backslash G(\mathbb{A})$, while an (irreducible) automorphic representation is a representation π of $G(\mathbb{A})$ by right translation on a space of automorphic forms. We usually think of π as a representation (often unitary) on a Hilbert space. However, an automorphic form on $G(\mathbb{A})$ is required to satisfy finiteness conditions akin to those on the pre-Hilbert space $\mathscr{H}^0_G$ stated prior to the lemma above ($K_{\mathbb{R}}=K_\infty$-finite and $\mathscr{Z}_{\mathbb{R}}=\mathscr{Z}_\infty$-finite in the component x_∞, K^∞-finite and compactly supported in the component x^∞, of the variable $x=x_\infty x^\infty$), as well as a condition of moderate growth (slowly increasing). With these constraints, we have therefore to treat π as a representation of the "group algebra"

$$\mathscr{H} = \mathscr{H}_\infty \otimes \mathscr{H}^\infty = \mathscr{H}_\mathbb{R} \otimes \mathscr{H}^\mathbb{R}$$

of $G(\mathbb{A})$ (not to be confused with the Hilbert space $\mathscr{H}_G$ above). The factor $\mathscr{H}^\infty$ is the convolution algebra of locally constant functions of compact support on $G(\mathbb{A}^\infty)$, while $\mathscr{H}_\infty$ is the convolution algebra of distributions on the real group $G_\mathbb{R} = G_\infty$ that are supported on the maximal compact subgroup $K_\mathbb{R} = K_\infty$. They are often called Hecke algebras, and they act by right convolution on the space of automorphic forms. The algebra $\mathscr{H}_\mathbb{R} = \mathscr{H}_\infty$ is less well known, but it is quite elegant. It streamlines the archimedean actions as a $(\mathfrak{g}_\mathbb{R}, K_\mathbb{R})$-module ($\mathfrak{g}_\mathbb{R}$ is the Lie algebra of $G_\mathbb{R}$) into that of a single convolution algebra, which thus becomes parallel to the nonarchimedean action. It seems to have been first introduced by Flath [72], and was a basic part of the definitions in [35].

In practice, one generally wants to quantify the spaces of automorphic forms under consideration. Suppose that τ is the idempotent element in $\mathscr{H}_\infty$ attached to a finite set of irreducible characters on K_∞, that J is an ideal of finite codimension in $\mathscr{Z}_\infty$, that K is any open compact subgroup of $G(\mathbb{A}^\infty)$ and that N is a positive, slowly increasing function of $G_\infty = G(\mathbb{R})$. One can then write $\mathscr{A}(\tau, J, K, N)$ for the space of smooth (complex-valued) functions f on $G(\mathbb{A})$ with the following properties

(a) $f(\gamma x) = f(x), \qquad \gamma \in G(\mathbb{Q}), x \in G(\mathbb{A}),$
(b) $f * \tau = f,$
(c) $zf = 0, \qquad z \in J,$
(d) $f(xk) = f(x), \qquad x \in G(\mathbb{A}), k \in K,$
(e) $|f(x_\infty)| \leq c_f N(x_\infty), \qquad x_\infty \in G_\infty,$
for some positive constant c_f.

Then $\mathscr{A}(\tau, J, K, N)$ is a space of automorphic forms on $G(\mathbb{A})$, which is stabilized by the natural subalgebra of $\mathscr{H}$ attached to τ and K.

It is the nonarchimedean component of this subalgebra, the Hecke algebra

$$\mathscr{H}(G(\mathbb{A}^\infty), K) = \mathscr{H}(K \backslash G(\mathbb{A}^\infty)/K)$$

of compactly supported, K-biinvariant functions on $G(\mathbb{A}^\infty)$, that is particularly relevant to number theory. At the centre of its study are the unramified local factors

$$\mathscr{H}_v = \mathscr{H}(G_v, K_v) = \mathscr{H}(K_v \backslash G_v / K_v)$$

at which K_v is a *hyperspecial* maximal compact subgroup of $G_v = G(\mathbb{Q}_v)$. These *spherical* Hecke algebras are abelian. What they revealed soon after Langlands' work on Eisenstein series became a critical part of his next great discovery.

The notion of an automorphic form owes much to Harish-Chandra [89], as well as to Langlands. It is presented in [35, (1.3), (4.2)]. Borel and Jacquet followed this with the formal definition [35, (4.6)] of an automorphic representation as an irreducible constituent of a space of automorphic forms. This was supplemented with an equivalent formulation by Langlands [145, Proposition 2] as an irreducible constituent of a representation of $G(\mathbb{A})$ parabolically induced from a cuspidal automorphic representation. The passage between them was provided by Langlands' theory of Eisenstein series.

The spectral theory of Eisenstein series was initiated by Selberg [205], [206], [207]. He established versions of the Langlands Main Theorem for various noncompact quotients

$$\Gamma \backslash X_+ \cong \Gamma \backslash \mathrm{SL}(2,\mathbb{R})/\mathrm{SO}(2,\mathbb{R})$$

of the upper half plane

$$X_+ = \{z \in \mathbb{C} : \mathrm{Im}\,(z) > 0\}$$

and more generally, for the line bundles on these quotients attached to characters of the stabilizer $\mathrm{SO}(2,\mathbb{R})$ of $\sqrt{-1}$. This amounts to the theory of Eisenstein series on a noncompact quotient $\Gamma \backslash \mathrm{SL}(2,\mathbb{R})$. Selberg also included classical Hecke operators for modular forms into his constructions. This amounts in turn to the theory of Eisenstein series for the adelic quotient $\mathrm{SL}(2,\mathbb{Q}) \backslash \mathrm{SL}(2,\mathbb{A})$.

Selberg regarded Eisenstein series as a step towards a trace formula for noncompact quotient. In adelic terms, his aim was to find a concrete formula for the trace of the operator $R_{\mathrm{disc}}(f)$ obtained by restricting the right convolution operator

$$R(f) = \int_{\mathrm{SL}(2,\mathbb{A})} f(x)R(x)\,\mathrm{d}x, \quad f \in C_c^\infty(\mathrm{SL}(2,\mathbb{A})),$$

on $L^2(\mathrm{SL}(2,\mathbb{Q}) \backslash \mathrm{SL}(2,\mathbb{A}))$ to the discrete spectrum. By definition

$$R_{\mathrm{disc}}(f) = R(f) - R_{\mathrm{cont}}(f)$$

where $R_{\mathrm{cont}}(f)$ is the restriction of $R(f)$ to the continuous spectrum. The purpose of Eisenstein series was to provide an explicit construction for $R_{\mathrm{cont}}(f)$. We shall return to this topic in Section 6.

3 *L*-functions and class field theory

The next period in Langlands' work is often identified with his 1967 letter to Weil [132]. It was subsequently expanded [138] to the series of far-reaching conjectures and their consequences that became known as the Langlands program. Most immediately striking perhaps was Langlands' discovery of the long sought nonabelian class field theory. It will be a topic for the next section. In this section, we shall prepare the way. We shall review the theory of L-functions, and its relation to abelian class field theory. At the end, we shall then describe the hints Langlands found in Eisenstein series of what was lying ahead.

The last section might have seemed rather technical to a nonspecialist. We shall try to make amends in this section by moving at a more leisurely pace. In particular, we shall begin our discussion with the almost universally familiar notion of a Dirichlet series.

We recall that a *Dirichlet series* is an infinite series of the form

$$\sum_{n=1}^{\infty} a_n n^{-s}$$

for complex coefficients a_n and a complex number s. If the coefficients satisfy a bound

$$|a_n| \leq Cn^a, \quad n \in \mathbb{N},$$

for positive numbers C and a, the series converges absolutely to an analytic function of s in the right half plane $\mathrm{Re}\,(s) > a+1$. The original model is of course the Riemann zeta function

$$\zeta(s) = \sum_{n=1}^{\infty} n^{-s}.$$

It converges to an analytic function of s in the right half plane $\mathrm{Re}\,(s) > 1$. It also has an analytic continuation to a meromorphic function of $s \in \mathbb{C}$, whose only singularity is a simple pole at $s = 1$, and which satisfies a functional equation relating its values at s and $1-s$. The Riemann zeta function also has an Euler product. By the fundamental theorem of arithmetic, it can be represented as a product

$$\zeta(s) = \prod_p (1-p^{-s})^{-1} = \prod_p \left(\sum_{k=1}^{\infty} (p^k)^{-s} \right)$$

of Dirichlet series attached to prime numbers.

An *L-function* is a Dirichlet series with supplementary properties. There seems to be no universal agreement as to the definition, but let us say that an L-function is a Dirichlet series that converges in some right half plane, and that has an Euler product of the general form

$$L(s) = \prod_p \left(\sum_{k=1}^{\infty} c_{p,k} p^{-ks} \right),$$

for complex numbers $c_{p,k}$. We will not insist on analytic continuation and fundamental equation, simply because this has not been established for many of the L-functions that arise naturally, even though it is widely expected to hold.

In algebraic number theory, L-functions are used to encode arithmetic data. The coefficients $c_{p,k}$ in the Euler product turn out to provide a natural way to represent fundamental properties of the prime numbers. The essential examples are the L-functions of E. Artin, which were constructed from data that govern class field theory.

The goal of class field theory, as its name suggests, is to classify fields. More precisely, one would like to classify the Galois extensions of a given number field F. Let us review the problem in elementary terms.

We have been working in this paper with the base field $F = \mathbb{Q}$ for simplicity. Suppose then that K is a finite Galois extension of $\mathbb{Q}$, which we assume to be *monogenic*. This means that we can represent K as the splitting field of a monic, irreducible, integral polynomial

$$f(x) = x^n + a_{n-1}x^{n-1} + \cdots + a_1 x + a_0, \quad a_i \in \mathbb{Z}.$$

We then consider the factorization of $f(x)$ modulo a variable prime p. It is best to exclude the finite set of ramified primes for which $f(x)$ has repeated factors modulo p, a set S in which we also include the archimedean place ∞ to be consistent with earlier notation. For each $p \notin S$, the corresponding factorization,

$$f(x) \equiv f_1(x) \cdots f_r(x) \,(\mathrm{mod}\ p)$$

of f is then a product of distinct irreducible polynomials $f_i(x)$, whose degrees give us an (ordered) partition

$$\Pi_p = (n_1, \dots, n_r), \quad n_i = \deg(f_i),$$

of n. We thus obtain a mapping from unramified primes $p \notin S$ to partitions Π_p of n. (If $K/\mathbb{Q}$ is not monogenic, $f(x)$ need not be monic. In this case, we simply enlarge S to include all prime divisors of the leading coefficient a_n.)

The interest in this mapping is in its implication for the Galois group

$$\Gamma_{K/\mathbb{Q}} = \mathrm{Gal}(K/\mathbb{Q})$$

of K over $\mathbb{Q}$. This group is given by $f(x)$ as a conjugacy class of subgroups of the symmetric group S_n. Suppose for a moment that it equals the full symmetric group. The conjugacy classes of $\mathrm{Gal}(K/\mathbb{Q})$ are then parametrized by partitions of n. If Φ_p is the conjugacy class of Π_p in $\mathrm{Gal}(K/\mathbb{Q})$, the factorization of $f(x)$ modulo p then gives us a mapping

$$p \to \Phi_p \tag{11}$$

from primes $p \notin S$ to conjugacy classes Φ_p in $\Gamma_{K/\mathbb{Q}}$. In general, as we have said, $\mathrm{Gal}(K/\mathbb{Q})$ is determined by f only as a conjugacy class of subgroups of S_n. However, a basic construction in algebraic number theory attaches a *canonical* conjugacy class in $\mathrm{Gal}(K/\mathbb{Q})$ to the unramified prime p, the Frobenius class

$$\mathrm{Frob}_p = \Phi_p.$$

Its implication for the factorization of f is that among the conjugacy classes in $\Gamma_{K/\mathbb{Q}}$ attached to a partition Π_p, there is one that is canonical, the Frobenius class of p.

In these concrete terms, the problem for class field theory would be to characterize the prime p factorization data of $f(x)$ in independent terms. More precisely, for any Galois extension $(K/\mathbb{Q})$, any partition Π of n, and any conjugacy class c in $\mathrm{Gal}(K/\mathbb{Q})$ that maps to the conjugacy class of Π in S_n say, can one characterize the fibre

$$P^S(c, K/\mathbb{Q}) = \{p \notin S : \Phi_p = c\}$$

of c in some independent way? Of special interest is the case that c is the trivial class 1 in $\mathrm{Gal}(K/\mathbb{Q})$. In this case, the fibre

$$\mathrm{Spl}(K/\mathbb{Q}) = \mathrm{Spl}^S(K/\mathbb{Q}) = P^S(1, K/\mathbb{Q}),$$

the set of primes that *split completely* in K, is just the set of primes for which $f(x)$ breaks into linear factors modulo p. Its importance is in the fact, which can be obtained as a direct consequence of the Tchebotarev density theorem for example, that the mapping

$$K/\mathbb{Q} \to \mathrm{Spl}(K/\mathbb{Q}),$$

from finite Galois extensions of $\mathbb{Q}$ to subsets of prime numbers $\{p\}$, is *injective*. In other words, the set $\mathrm{Spl}(K/\mathbb{Q})$ of primes represents a "signature" for the extension $(K/\mathbb{Q})$, in the sense that it characterizes it completely. The problem for class field theory would then be to characterize the image of this mapping in some independent fashion. This would provide a classification of Galois extensions of $\mathbb{Q}$.

These remarks are primarily for motivation, since the subject is inevitably more subtle. Nevertheless, such considerations were behind the development of (abelian) class field theory in the early part of the twentieth century. They also led E. Artin to define the L-functions that bear his name, as a way to encode the data provided by the conjugacy classes $\{\Phi_p\}$ in $\mathrm{Gal}(K/\mathbb{Q})$. The coefficients in L-functions are of course numbers. The simplest way to attach numbers to a conjugacy class is to embed the underlying Galois group into a general linear group, and then take the coefficients of the resulting characteristic polynomials. An Artin L-function for a Galois extension $(K/\mathbb{Q})$ therefore depends on the choice of a representation

$$r\colon\ \mathrm{Gal}(K/\mathbb{Q}) \to \mathrm{GL}(n, \mathbb{C}).$$

In its simplest form, it is defined as an Euler product

$$L^S(s, r) = \prod_{p \notin S} L_p(s, r), \tag{12}$$

with local factors

$$L_p(s, r) = \det(1 - r(\Phi_p)p^{-s})^{-1},$$

built out of unramified Frobenius conjugacy classes Φ_p in $\mathrm{Gal}(K/\mathbb{Q})$. It converges for $\mathrm{Re}\,(s) > 1$ to an analytic function of s.

Artin showed that the Euler product $L^S(s, r)$ actually has analytic continuation to a meromorphic function of $s \in \mathbb{C}$, with a functional equation that relates its values at s and $1 - s$. More precisely, he defined Euler factors $L_v(s, r)$ at the finite set of places $v \in S$, including the archimedean place $v = \infty$. He then showed that the completed product

$$L(s, r) = L_S(s, r)L^S(s, r) = \prod_v L_v(s, r) \tag{13}$$

satisfies the precise functional equation

$$L(s,r) = \varepsilon(s,r)L(1-s,r^\vee), \tag{14}$$

for the contragredient representation

$$r^\vee = {}^t r(g^{-1}), \quad g \in \mathrm{Gal}(K/\mathbb{Q}),$$

of r, and a certain rather mysterious monomial

$$\varepsilon(s,r) = ab^s, \quad a \in \mathbb{C}^\times, b > 0.$$

Given the analytic continuation, Artin then made the following remarkable conjecture.

Conjecture (Artin). *Suppose that r is irreducible. Then $L(s,r)$ is an entire function of $s \in \mathbb{C}$, unless r is the trivial 1-dimensional representation, in which case $L^S(s,r) = L^\infty(s,r)$ is the Riemann zeta function.*

Artin's proof of the analytic continuation and functional equation was intimately tied to abelian class field theory. This is the study of *abelian* extensions (K/F) of a number field F, which is to say finite Galois extensions (K/F) with abelian Galois group

$$\Gamma_{K/F} = \mathrm{Gal}(K/F).$$

We shall say a few words about this fundamental subject, continuing for the moment to assume that $F = \mathbb{Q}$ in order to be as concrete as possible.

Consider the special case of Artin's construction in which the Galois group $\Gamma_{K/F} = \Gamma_{K/\mathbb{Q}}$ is abelian, and the representation r is irreducible. In other words, r is a 1-dimensional character on $\Gamma_{K/F}$. If p lies outside the finite set S of ramified places, the associated Frobenius class Φ_p in $\Gamma_{K/\mathbb{Q}}$ is simply an element in this abelian group. We can use the nonzero complex numbers

$$\{r(\Phi_p) : p \notin S\}$$

to define a character χ^S on the locally compact group

$$\mathrm{GL}(1,\mathbb{A}^S) = (\mathbb{A}^*)^S = \{x^S \in \prod_{p\notin S} \mathbb{Q}_p^* : |x_p|_p = 1 \text{ for almost all } p\} = \widetilde{\prod_{p\notin S}} \mathbb{Q}_p^*$$

by setting

$$\chi^S(x^S) = \prod_{p\notin S} \chi_p(x_p) = \prod_{p\notin S} r(\Phi_p)^{v_p(x_p)}, \tag{15}$$

for the valuation

$$v_p(x_p) = -\log_p(|x_p|_p), \quad x_p \in \mathbb{Q}_p^*.$$

Class field theory asserts that χ^S is the restriction to $(\mathbb{A}^S)^*$ of a uniquely determined automorphic representation of $\mathrm{GL}(1)$. In other words, there is a unique continuous, complex character χ on the quotient

$$C_F = F^* \backslash \mathbb{A}_F^* = \mathrm{GL}(1,F) \backslash \mathrm{GL}(1,\mathbb{A}_F), \quad F = \mathbb{Q},$$

whose restriction to the image of $(\mathbb{A}^*)^S$ equals χ^S. Moreover, with a minor adjustment in the definition, the mapping $r \to \chi$ becomes an isomorphism of abelian groups. We can think of this property as a fundamental arithmetic law of nature. We restate it formally as follows.

Let

$$\Gamma_F^{\mathrm{ab}} = \mathrm{Gal}(F^{\mathrm{ab}}/F), \quad F = \mathbb{Q},$$

be the Galois group of the maximal abelian extension F^{ab} of $F = \mathbb{Q}$ (in some fixed algebraic closure $\overline{F}$ of F). It is an inverse limit

$$\Gamma_F^{\mathrm{ab}} = \varprojlim_K \Gamma_{K/F}$$

over finite abelian extensions, and hence a compact, totally disconnected group. A continuous 1-dimensional character r in Γ_F^{ab} has cofinite kernel, and can therefore be identified with a character on the Galois group $\Gamma_{K/F}$ of a finite extension (K/F). This in turn maps to a character χ^S on the group of *S-idèles*

$$I_F^S = (\mathbb{A}_F^*)^S = \{x_p \in F_p^* : p \notin S\} \subset I_F = \mathbb{A}_F^*, \quad F = \mathbb{Q},$$

for a finite set of valuations S outside of which r is unramified. The following big theorem, in which we have taken $F = \mathbb{Q}$, is then the central assertion of class field theory.

Global Reciprocity Law. *For any r, the character χ^S on I_F^S descends to a unique character χ on the idèle class group*

$$C_F = F^* \backslash I_F = F^* \backslash \mathbb{A}_F^*, \qquad F = \mathbb{Q}.$$

The resulting mapping $r \to \chi$ becomes an isomorphism from the group of characters r on Γ_F^{ab} onto the group of characters χ of finite order on C_F. The dual (Artin) mapping

$$\theta_F : C_F \to \Gamma_F^{\mathrm{ab}}$$

is therefore a continuous surjective homomorphism, whose kernel is the connected component C_F^0 of 1 in C_F.

This assertion is a culmination of work by mathematicians over many years, from the law of quadratic reciprocity of Gauss to the full reciprocity law of Artin, which is a more precise version of the assertion. For the Artin reciprocity law also characterizes the preimage of each finite quotient $\Gamma_{K/F}$ of Γ_F^{ab} as the cokernel of the norm mapping $N_{K/F}$ from C_K to C_F. In general, the importance of the reciprocity law lies especially in the fact that it applies, as stated, if F is any number field. We need only replace prime numbers p in $\mathbb{Q}$ by prime ideals in F in the definitions above.

Needless to say, the proof of the reciprocity law is deep. The original argument had a large analytic component, based on the properties of certain L-functions. A purely algebraic proof based on the cohomology of Galois groups came later. For the field $F = \mathbb{Q}$, the proof is actually much easier than in the general case. This is reflected in the Kronecker–Weber theorem, which characterizes $\mathbb{Q}^{\mathrm{ab}}$ explicitly as the field generated by the complex numbers

$$\{e^{\frac{2\pi i}{n}} : n \in \mathbb{N}\}.$$

A similarly explicit result holds for any imaginary quadratic extension F of $\mathbb{Q}$. For in this case, the Kronecker Jugendtraum characterizes F^{ab} in terms of special values of elliptic functions attached to elliptic curves over $\mathbb{Q}$ with complex multiplication in F. We recall that Hilbert's twelfth problem was to characterize the maximal abelian extension F^{ab} of any F in terms of special values of natural analytic functions. Little progress has since been made on this problem, apart from the 1955 generalization of the Jugendtraum by Shimura and Taniyama to a totally complex extension of a totally real field.

How did Artin use the reciprocity law to prove his functional equation? We must first recall that Hecke has earlier attached an L-function $L^S(s,\chi)$ to any (quasi)character χ on the group $C_F = F^* \setminus \mathbb{A}_F^*$. We are using adelic notation here, as we did in our discussion of the reciprocity law, even though it was only later that Chevalley introduced the group of idèles $I = \mathbb{A}_F^*$. Hecke called χ a Grössencharacter, as a generalization of a Dirichlet character and of its analogue introduced by Weber for the number field F, which we can now take to be arbitrary. As a character on C_F (rather than what Hecke would have considered a generalized ideal class character), χ has an unramified part χ^S on the locally compact group

$$(\mathbb{A}_F^S)^* = \widetilde{\prod_{v \notin S}} F_v^*.$$

It takes the form

$$\chi^S(x^S) = \prod_{v \notin S} \chi_v(x_v) = \prod_{v \notin S} c_v^{v(x_v)}$$

for complex numbers $c_v \in \mathbb{C}^*$ attached to χ, and with v being the normalized valuation of F_v^*. This of course is parallel to (15), where we had $F = \mathbb{Q}$. In particular, S is a finite set of valuations of F that contains both the set V_∞ of archimedean places and the set of finite places v at which χ_v is ramified. The unramified Hecke L-function is then the infinite product

$$L^S(s,\chi) = \prod_{v \notin S} L_v(s,\chi) = \prod_{v \notin S} (1 - c_v q_v^{-s})^{-1},$$

where q_v is the order of the residue field F_v^*. After introducing this function, Hecke defined Euler factors $L_v(s,\chi)$ at the remaining places $v \in S$, and then proved that the completed product

$$L(s,\chi) = L_S(s,\chi)L^S(s,\chi) = \prod_v L_v(s,\chi)$$

satisfies the functional equation

$$L(s,\chi) = \varepsilon(s,\chi)L(1-s,\chi^\vee) \tag{16}$$

where $\chi^\vee(x) = \chi(x^{-1})$ and

$$\varepsilon(s,\chi) = ab^s, \quad a \in \mathbb{C}^*, b > 0.$$

The central tenet of (global) class field theory can be formulated as an assertion that every abelian Artin L-function $L^S(s,r)$ over F equals a Hecke L-function $L^S(s,\chi)$. This follows from the earlier definition of the mapping $r \to \chi$, the definition of the L-functions themselves, and of course, the Global Reciprocity Law. Therefore $L^S(s,r)$ inherits all the properties established by Hecke for $L^S(s,\chi)$. It has a completion that equals $L(s,\chi)$, and therefore has analytic continuation, and functional equation (16). This is a fundamental fact, that to this day has no direct proof. It was the main step in Artin's derivation of the functional equation (14) for an arbitrary (nonabelian) representation

$$r\colon \operatorname{Gal}(E/F) \to \operatorname{GL}(n,\mathbb{C}).$$

The other step was a formal decomposition

$$L^S(s,r) = \prod_i L^S(s,r_i)^{a_i},$$

for representations r_i of cyclic Galois groups $\Gamma_i \subset \operatorname{Gal}(K/F)$ and rational numbers a_i, obtained by Artin by first proving a character-theoretic decomposition

$$r = \sum_i a_i \operatorname{Ind}(\Gamma,\Gamma_i; r_i), \quad \Gamma = \operatorname{Gal}(K/F),$$

of r, and then using the compatibility of the L-functions with representation theoretic operations such as induction and direct sums. He was then able to use this to construct his completed L-function $L(s,r)$ from those of Hecke. The last step was to use the resulting formal identity of quotients

$$\frac{L(s,r)}{L(1-s,r^\vee)} = \prod_i \left(\frac{L(s,r_i)}{L(1-s,r_i^\vee)}\right)^{a_i} = \prod_i (a_i b_i^s)^{a_i}$$

to establish the identity

$$\frac{L(s,r)}{L(1-s,r^\vee)} = ab^s, \quad a \in \mathbb{C}^*, b > 0.$$

This is the functional equation (16).

We refer the reader to the article [53] of Cogdell on Artin L-functions for discussion of this and other questions. The earlier history of class field theory is clearly presented in the articles [252] and [93]. We also note that the argument sketched above becomes quite clear after the later proof of the Brauer Induction Theorem. This establishes that the rational numbers a_i may in fact be taken to be integers. We should point out, however, that this improvement gives little information about the Artin conjecture stated above. To rule out any poles of the L-function $L(s,r)$, we would need to control the zeros of the abelian L-functions $L^S(s,r_i)$ with $a_i < 0$. This of course would be a tall order.

It was with the thesis [236] of Tate that the work of Hecke was put into the adelic form that we have followed here. This made many of Hecke's arguments more transparent. For example, the functional equation (16) is seen to be a natural consequence of the Poisson summation formula on the locally compact abelian group $\mathbb{A}_F$, with respect to the discrete subgroup F. It is at this point that a new ingredient enters the theory, a nontrivial additive character ψ on $\mathbb{A}_F/F$, with decomposition $\psi = \widetilde{\prod}_v \psi_v$ into additive characters on the groups F_v. Its role is to identify the Pontryagin dual of $\mathbb{A}_F$ with $\mathbb{A}_F$ itself. The adelic formulation of [236] also leads to an important new way to see the final result. Tate showed that the global ε-factor in (16) has a canonical decomposition

$$\varepsilon(s,\chi) = \prod_{v \in S} \varepsilon_v(s,\chi_v,\psi_v) \tag{17}$$

into local ε-factors

$$\varepsilon_v(s,\chi_v,\psi_v) = \varepsilon(\chi_v,\psi_v) q_v^{-n_v(s-\frac{1}{2})}$$

that depend only in the localizations ψ_v and χ_v of ψ and χ, for nonzero complex numbers

$$\varepsilon(\chi_v,\psi_v) = \varepsilon(\tfrac{1}{2},\chi_v,\psi_v),$$

integers $n_v = n(r_v,\psi_v)$ known as conductors, and integers q_v given by the residual degree of F_v if v is nonarchimedean, and 1 if v is archimedean. The local ε-factors in the functional equation turned out to have a fundamental role in the local Langlands program.

We can now turn to the work of Langlands. The first thing to mention is the volume [81] in which Godement and Jacquet generalize Tate's thesis from GL(1) to GL(n). This was a major step forward, which was not due to Langlands. However, its possible existence was clearly part of his thinking right from the beginning. Langlands was in regular communication with Godement in the 1960s, and he mentions the extension of Tate's thesis on pp. 31–32 of his fundamental paper [138] as an essential premise for his conjectures. In fact, the special case with $n = 2$ was established in the volume [103] of Jacquet and Langlands and was announced in the original article [138]. We shall return to this briefly in Section 5, but we might as well stay with the more general case here for the motivation.

The problem in [81] was to attach an L-function $L(s,\pi)$ to every automorphic representation

$$\pi = \widetilde{\bigotimes_v} \pi_v$$

of $G(\mathbb{A}_F)$, for the group $G = \mathrm{GL}(n)$ over a number field F, and to prove that this function has analytic continuation and functional equation. The initial step will by now be quite familiar. Given π, we fix a finite set of valuations $S \supset S_\infty$ of F such that the local constituent π_v of π is unramified for any v outside of S. This means that the restriction of π_v to the maximal compact subgroup $G(\mathfrak{O}_v)$ of $G(F_v)$ contains the trivial one-dimensional representation. For any such v there is a general bijection

$$\pi_v \to c(\pi_v),$$

from unramified representations of $G(F_v)$ onto semisimple conjugacy classes in the complex group $G(\mathbb{C})$. We shall discuss this last point again, for more general groups, in the next section. Semisimple conjugacy classes in $G(\mathbb{C}) = \mathrm{GL}(n,\mathbb{C})$ can of course be identified with complex diagonal matrices, taken only up to permutation of their entries. The unramified global L-function of π is then defined as the Euler product

$$L^S(s,\pi) = \prod_{v \notin S} L_v(s,\pi),$$

where

$$L_v(s,\pi) = L(s,\pi_v) = \det(1 - c(\pi_v)q_v^{-s})^{-1}.$$

The local problem in [81] was to attach local L-functions

$$L_v(s,\pi) = L(s,\pi_v)$$

and the ε-factors

$$\varepsilon_v(s,\pi,\psi) = \varepsilon(s,\pi_v,\psi_v) = \varepsilon(\pi_v,\psi_v)q_v^{-n_v(s-\frac{1}{2})}$$

to the remaining valuations $v \in S$. This is naturally much more subtle for the non-abelian group $\mathrm{GL}(n)$, but the basic ideas resemble those for $\mathrm{GL}(1)$ in [236]. The same is true of the global problem of analytic continuation and functional equation. The essential tool was again the Poisson summation formula, this time for the additive group $\mathfrak{g}(\mathbb{A}_F) = \mathrm{M}_n(\mathbb{A}_F)$ of $(n \times n)$-adelic matrices, with respect to the discrete subgroup $\mathfrak{g}(F) = \mathrm{M}_n(F)$ of rational matrices. In the end, the authors constructed the local L-functions and ε-factors above for places $v \in S$, such that the full Euler product

$$L(s,\pi) = L_S(s,\pi)L^S(s,\pi) = \prod_v L_v(s,\pi)$$

has analytic continuation, and satisfies the functional equation

$$L(s,\pi) = \varepsilon(s,\pi)L(1-s,\pi^\vee), \tag{18}$$

for

$$\varepsilon(s,\pi) = \prod_{v\in S} \varepsilon_v(s,\pi,\psi). \tag{19}$$

This is the general analogue for $\mathrm{GL}(n)$ of the functional equation (16) established by Tate for $\mathrm{GL}(1)$. We note that the global solution in [81] was established only in the case that π is a *cuspidal* automorphic representation. However, the general case follows from this and the properties of Langlands' Eisenstein series. (See [33] and [102, §6].)

With the Artin L-functions of degree n and the automorphic L-functions of $\mathrm{GL}(n)$, our exposition has acquired a certain symmetry. Given the Global Reciprocity Law for $\mathrm{GL}(1)$, a reader could well ask whether every Artin L-function is an automorphic L-function. If so, we would have a general extension of the reciprocity law to what we could regard as nonabelian class field theory. We would also have a proof of the Artin conjecture stated above. For it is a fundamental consequence of the harmonic analysis used by Jacquet and Godement (and Tate and Hecke for $n = 1$) that the automorphic L-functions $L(s,\pi)$ are *entire*, apart from certain obvious exceptions related to unramified 1-dimensional automorphic representations. Would this then be the final word on the subject?

There are three points to consider in regards to the last question. One would be the uncomfortable prospect of having to prove such a general nonabelian reciprocity law, given the historical difficulty in establishing just the abelian theory. We could expect that nonabelian class field theory, whatever form it might take, would be difficult. It would be reassuring to think that the problem at least has some further structure. A second point concerns this last possibility. Suppose that r' is an irreducible Galois representation of degree n', and that ρ' is an irreducible n-dimensional representation of $\mathrm{GL}(n',\mathbb{C})$. The composition

$$\rho : \Gamma_F \xrightarrow{r'} \mathrm{GL}(n',\mathbb{C}) \xrightarrow{\rho'} \mathrm{GL}(n,\mathbb{C})$$

is then a Galois representation (often irreducible) of degree n. The Frobenius classes that define the L-functions satisfy the following relation

$$\rho(\Phi_v) = (\rho' \circ r')(\Phi_v), \quad v \notin S.$$

How would this structure be reflected in the corresponding automorphic representations? Finally, the work of Harish-Chandra has taught us that representations should be studied uniformly for all reductive groups. If some interesting phenomenon is discovered for one group, or one family of groups such as $\{\mathrm{GL}(n)\}$, it should be investigated for all groups. What are the implications of this for automorphic L-functions?

These considerations were undoubtedly part of the thinking of Langlands that led up to the Principle of Functoriality. However, perhaps the most decisive hints were in his theory of Eisenstein series. They came from L-functions he discovered in the global intertwining operators

$$M(w,\lambda)\colon \mathscr{H}_P \to \mathscr{H}_{P'}, \qquad w \in W(\mathfrak{a}_P, \mathfrak{a}_{P'}), \tag{20}$$

defined by (6). (We have written w here because we want to reserve s for the complex variable of an L-function.)

Suppose for a moment that G equals the group SL(2) over $\mathbb{Q}$, and that P is the Borel subgroup of upper triangular matrices, that ϕ lies in the one-dimensional space of constant functions in $\mathscr{H}_p$, and that w is the non-trivial Weyl element in $W(\mathfrak{a}_P, \mathfrak{a}_P)$. Then $M(w,\lambda)$ is a scalar multiple of ϕ given for a suitable λ by a convergent adelic integral over

$$N(\mathbb{A}) = \left\{ \begin{pmatrix} 1 & x \\ 0 & 1 \end{pmatrix} : x \in \mathbb{A} \right\}.$$

It is not hard to evaluate. Recall that $\lambda \in \mathfrak{a}^*_{P,\mathbb{C}}$ is a complex-valued linear form on $\mathfrak{a}_P$, a real 1-dimensional vector space we can in turn identify with the Cartan subalgebra

$$\left\{ \begin{pmatrix} u & 0 \\ 0 & -u \end{pmatrix} : u \in \mathbb{R} \right\}$$

of the Lie algebra of $\mathrm{SL}(2,\mathbb{R})$. The mapping

$$\lambda \to s = \lambda \begin{pmatrix} \frac{1}{2} & 0 \\ 0 & -\frac{1}{2} \end{pmatrix}$$

identifies λ with a complex number $s \in \mathbb{C}$. It can then be shown that

$$M(w,\lambda) = \frac{L(s)}{L(s+1)},$$

for the completed Riemann zeta function

$$L(s) = L_\infty(s)\zeta(s).$$

This simple, well known formula is suggestive. If the Riemann zeta function is at the heart of the global intertwining operator for SL(2), something new and interesting must surely be contained in the operators for groups of larger rank. (For further background, see [78, §2].)

In his investigation of the more general intertwining operators in [141], Langlands discovered some completely new L-functions. Suppose that G is a split, simple group, say over $\mathbb{Q}$, and that

$$P = MN \supset B$$

is a standard maximal parabolic subgroup. Suppose also that

$$\pi = \widetilde{\bigotimes_v} \pi_v = \pi_\infty \otimes \Big(\widetilde{\bigotimes_p} \pi_p \Big)$$

is a cuspidal automorphic representation of $M(\mathbb{A})$ that is unramified at every place v of $\mathbb{Q}$. The parametrization of unramified representations π_v of a local group

$M_v = M(\mathbb{Q}_v)$ is not difficult, but it was not particularly well known at the time. One can identify them either with complex-valued homomorphisms

$$\mathscr{H}(K_{M,v} \backslash M_v / K_{M,v}) \to \mathbb{C}$$

of the unramified Hecke algebra (under convolution) of functions in $C_c^\infty(M_v)$ biinvariant under a suitable maximal compact $K_{M,v}$, or with their induction parameters given by Weyl orbits of unramified characters on the Borel subgroup $B_v \cap M_v$ of M_v. Langlands investigated them in terms of Hecke algebras, and made a remarkable observation. For any v, the unramified representations π_v of M_v are parametrized by the semisimple conjugacy classes $c(\pi_v)$ in a different group, the *complex dual group* $\widehat{M}$ of M.

The dual group $\widehat{G}$ of G is a complex simple group whose Coxeter–Dynkin diagram is dual to that of G, in the sense that the directions of any arrows are reversed, and whose maximal torus

$$\widehat{T} = X^*(T) \otimes \mathbb{C}^*, \quad T \subset M \subset G,$$

is dual to that of G. The roots $\{\alpha^\vee\}$ of $(\widehat{G}, \widehat{T})$ are the co-roots of (G,T), and the corresponding fundamental dominant weights $\{\varpi^\vee\} = \{\varpi_\alpha^\vee\}$ form the dual basis of the set of co-roots $\{(\alpha^\vee)^\vee\} = \{\alpha\}$ of (G,T). Finally, to the maximal parabolic subgroup $P = MN$ of G, there corresponds a maximal parabolic subgroup $\widehat{P} = \widehat{M}\widehat{N}$ of $\widehat{G}$, whose Levi component $\widehat{M}$ represents the dual group of M. We assume for simplicity that the unipotent radical $\widehat{N}$ of $\widehat{P}$ is abelian, and we write r for the adjoint representation of $\widehat{M}$ in the Lie algebra $\widehat{\mathfrak{n}}$ of $\widehat{N}$. We shall say more about the dual group $\widehat{G}$, and its more sophisticated companion the L-group ${}^L G$, in the next section, but we will still not give the precise construction. For this, we simply refer the reader to the article [233] by Springer.

Given G, P, π and v, we thus have a semisimple conjugacy class $c(\pi_v)$, according to what Langlands discovered in [139]. In general, $\widehat{M}$ is not a general linear group, so it does not have a characteristic polynomial. However, Langlands used r to define

$$L_v(s,\pi,r) = L(s,\pi_p,r) = \det(1 - r(c(\pi_p))p^{-s})^{-1}$$

explicitly for $v = p$ nonarchimedean. He also defined

$$L_v(s,\pi,r) = L(s,\pi_\infty,r) = \prod_{\alpha^\vee} \left(\pi^{-(s+c_\alpha)/2}\Gamma\left(\tfrac{s+c_\alpha}{2}\right)\right)$$

implicitly for $v = \infty$ archimedean, for complex numbers $c_\alpha = c_\alpha(\pi_\varpi, r)$, and roots $\alpha^\vee$ of $(\widehat{G}, \widehat{T})$ that are not roots of $(\widehat{M}, \widehat{T})$. He then showed that the Euler product

$$L(s,\pi,r) = \prod_v L_v(s,\pi,r)$$

converged absolutely to an analytic function on some right half plane, which he conjectured had analytic continuation with functional equation

$$L(s,\pi,r) = L(1-s,\pi,r^\vee), \tag{21}$$

for the contragredient representation $r^\vee(x) = {}^t r(x)^{-1}$ of r.

This was suggested by Langlands' theory of Eisenstein series, specifically the operators $M(w,\lambda)$ in (20). Since P is maximal, there will be a unique nontrivial Weyl element $w \in W(\mathfrak{a}_P, \mathfrak{a}_{P'})$ (for a standard parabolic subgroup $P' \subset B$). Since π is unramified, $\mathscr{H}_P$ and $\mathscr{H}_{P'}$ both have canonical, one-dimensional subspaces of constant functions, which are preserved both under translation by w and by the operators $M(w,\lambda)$ themselves. We can therefore identify $M(w,\lambda)$ with a complex-valued scalar function, as was the case for $G = \mathrm{SL}(2)$ above. Let $\varpi_P^\vee = \varpi_\alpha^\vee$ be the fundamental dominant weight for $(\widehat{G},\widehat{T})$, attached to the simple root α of (G,T) that is nontrivial on the split component $A_P = A_{M_P}$ of P. The mapping

$$\lambda \to s = \lambda(\varpi_P^\vee)$$

then identifies λ with a complex number $s \in \mathbb{C}$. The main result of [139] is the formula

$$M(w,\lambda) = \frac{L(s,\pi,r^\vee)}{L(s+1,\pi,r^\vee)}, \quad \lambda \to s, \tag{22}$$

for the scalar-valued restriction of the operator (20). Langlands proved it for G, P and π (and, slightly modified, if $\widehat{N}$ is nonabelian) by the formula of Gindikin and Karpelevic for G_∞ [80], and analogues he derived for the p-adic groups G_p.

The formula (22) not only motivated the introduction of many new automorphic L-functions, but also raised interesting new questions on Artin L-functions. Given G and P, consider a continuous homomorphism

$$\rho: \Gamma_{\mathbb{Q}} \to \widehat{M}.$$

One could ask whether there is an automorphic representation π of $M(\mathbb{A})$, and a corresponding L-function $L(s,\pi,r^\vee)$ such that

$$L(s, r^\vee \circ \rho) = L(s,\pi,r^\vee).$$

Langlands' conjectural answer to this will be taken up in the next section.

4 Global Functoriality and its implications

The Principle of Functoriality is the centre of the Langlands program. It postulates deep relationships among automorphic representations on different groups G. These in turn tie fundamental arithmetic data from number theory to equally fundamental spectral data from harmonic analysis. The global relationships also suggest local relationships that should lead to a local classification of representations. These should in turn give rise to local L-functions and ε-factors. We shall discuss Global Functoriality in this section, and Local Functoriality in the next.

The general functoriality conjecture was introduced in the revolutionary paper [138], a preprint of which was written shortly after the letter [132] to Weil. Langlands had to anticipate a number of basic properties of automorphic representations to be able even to formulate the conjecture. The name *functoriality* itself did not appear in the paper. Nor did the term *automorphic representation*, which seems to have been first introduced by A. Borel in his Bourbaki lecture [32, (5.1)].

In the original paper [138], Langlands spoke simply of an irreducible representation π of $G(\mathbb{A})$ that "occurs in" $L^2(G(F)\setminus G(\mathbb{A}))$. This amounts to an informal definition of an automorphic representation of $G(\mathbb{A})$, with the understanding that it includes the nonunitary representations obtained by analytic continuation of their internal parameters into the complex domain. He also took for granted that any such representation could be obtained uniquely as a restricted tensor product

$$\pi = \widetilde{\bigotimes_{v}}\pi_v, \tag{23}$$

where for any v, π_v is an irreducible representation of $G(\mathbb{Q}_v)$. This brought harmonic analysis into an area that was already broad, but that had applied mainly to complex analysis, number theory and arithmetic geometry. The formal definition was given later by Borel and Jacquet [35] in terms of automorphic forms. As we noted in §2, this was accompanied by the somewhat more direct characterization by Langlands in terms of induced cuspidal representations [145]. The tensor product decomposition was established at the same time by Flath [72].

Langlands' paper [139] on Euler products, discussed very briefly at the end of the last section, was a precursor to the functoriality paper [138] we are discussing now. Both papers required some ad hoc background to elementary properties of automorphic representations that had yet to be developed. These properties are by now well understood. We shall review a few of them here from the perspective of [138], even if this entails some repetition from our last section.

We are assuming for the moment that G is a connected reductive algebraic group over the field $F = \mathbb{Q}$. A fundamental concept introduced in [138] was Langlands' notion of the *dual group* $\widehat{G}$ of G, and more generally, the associated *L-group*

$${}^{L}G = \widehat{G} \rtimes \mathrm{Gal}(K/F), \tag{24}$$

where K is a suitably large finite Galois extension of F. Once again, the names came later, from Borel [32] in the case of L-group, and Kottwitz [119] for the dual group. We recall from Section 3 that $\widehat{G}$ is a complex, connected, reductive algebraic Lie group, with maximal torus

$$\widehat{T} = X^*(T) \otimes \mathbb{C}^*$$

dual to the maximal torus T of G, in the sense that

$$X^*(\widehat{T}) = \mathrm{Hom}(X^*(T), \mathbb{Z}).$$

Langlands took care to make this construction rigid by among other things fixing Borel subgroups $B \supset T$ and $\widehat{B} \supset \widehat{T}$ of G and $\widehat{G}$ respectively. The supplementary structure is conveniently accommodated by letting G represent a *based root datum* and $\widehat{G}$ represent the canonical *dual based root datum* [233]. With this understanding, any outer automorphism $\alpha \in \mathrm{Out}(G)$ of G comes with a canonical dual outer automorphism $\widehat{\alpha} \in \mathrm{Out}(\widehat{G})$ of $\widehat{G}$. As a reductive group over $F = \mathbb{Q}$, G comes with a homomorphism

$$\mathrm{Gal}(K/F) \to \mathrm{Out}(G). \tag{25}$$

The L-group (24) is defined as the semidirect product of $\widehat{G}$ with $\mathrm{Gal}(K/F)$ under the corresponding dual homomorphism

$$\mathrm{Gal}(K/F) \to \mathrm{Out}(\widehat{G}).$$

As predicted in [138], the Langlands L-group has turned out to be exactly the right object to accommodate the parameters of automorphic representations.

We are taking for granted here some knowledge of the structure of reductive algebraic groups over number fields. The introductory article [233] quoted above is perhaps the most convenient reference. It is the first paper in the two volume proceedings from the 1977 AMS summer symposium in Corvallis, Oregon, which was the natural successor to the 1964 Boulder conference we discussed in Section 1. Its goal was primarily to bring the subsequent work of Langlands to a broader audience.

Langlands formulated the Principle of Functoriality (minus the name) in [138], in its full generality. However, it is easier to recognize its beauty and power if we first describe a special case, the unramified part of functoriality, for split groups G over $\mathbb{Q}$. This means that the image of the outer twisting homomorphism (25) is trivial. In particular, the semidirect product (24) that gives the L-group is actually a direct product. It also allows us for many purposes to take K equal to our base field $F = \mathbb{Q}$, and therefore to take ${}^L G = \widehat{G}$.

For the given split group, suppose that π is an automorphic representation of $G(\mathbb{A})$, with decomposition (23) into local constituents. The formal definition [35] of automorphic representation includes a weak continuity condition. This implies that there is a finite set S of valuations of $\mathbb{Q}$ containing the archimedean place $v = \infty$ such that for any $p \notin S$, π_p is an *unramified* representation of $G(\mathbb{Q}_p)$. Unramified in turn means that the restriction of π_p to the maximal compact subgroup $K_p = G(\mathbb{Z}_p)$ of $G(\mathbb{Q}_p)$ contains the trivial one-dimensional representation. (It is understood here that as a split group over $\mathbb{Q}$, G comes with a suitable $\mathbb{Z}$-scheme structure.) What makes automorphic representations π interesting is the fact, established now in some cases and conjectured by Langlands in others, is that for many π, there are deep and fundamental relationships among its unramified constituents $\{\pi_p : p \notin S\}$.

To better appreciate the phenomenon, we recall that there is a simple classification of the unramified representations $\{\pi_p\}$ of a split p-adic group $G(\mathbb{Q}_p)$. It takes the form of a bijective mapping

$$\pi_p \to c(\pi_p) \tag{26}$$

in which $c(\pi_p)$ ranges over the semisimple conjugacy classes in the complex dual group $\widehat{G}$ of G. For as Langlands pointed out in [138], any semisimple element $c_p \in \widehat{G}$ determines an unramified, one-dimensional quasi-character on a Borel subgroup $B(\mathbb{Q}_p)$ of $G(\mathbb{Q}_p)$. The corresponding induced representation $\tilde{\pi}_p$ of $G(\mathbb{Q}_p)$ then depends only on the conjugacy class $\tilde{c}_p$ of c_p in $\widehat{G}$. A given unramified representation π_p of $G(\mathbb{Q}_p)$ thus occurs as the unique irreducible constituent of $\tilde{\pi}_p$, for a unique semisimple conjugacy class $\tilde{c}_p = c(\pi_p)$ in $\widehat{G}$. We therefore have a mapping

$$\pi \to c^S(\pi) = \{c_p(\pi) = c(\pi_p) : p \notin S\} \tag{27}$$

from automorphic representations π of $G(\mathbb{A})$ to families of semisimple conjugacy classes in $\widehat{G}$.

The semisimple conjugacy classes $c_p(\pi)$ in $\widehat{G}$ are concrete objects. They can be described in terms of complex numbers. For example, if G equals $\mathrm{GL}(n)$, the semisimple classes in $\widehat{G} = \mathrm{GL}(n, \mathbb{C})$ are given by nonsingular complex diagonal matrices, taken up to permutation of their entries. These are in turn parametrized by their characteristic polynomials, or if one prefers, elements in the space $\mathbb{C}^{n-1} \times \mathbb{C}^*$ of coefficients defined by the characteristic polynomial. A general automorphic representation therefore gives us a family $c^S(\pi)$ of objects $c_p(\pi)$ with natural complex parameters. It is in the relations among these complex parameters, as p varies, that the most fundamental interest of automorphic representations lies.

We can now state the global Principle of Functoriality. In the most basic case under present consideration, it applies to a pair of split groups G' and G over the field $F = \mathbb{Q}$, together with a homomorphism

$$\rho' : {}^L G' \to {}^L G$$

between the corresponding dual groups $\widehat{G'} = {}^L G'$ and $\widehat{G}$ and ${}^L G$. The homomorphism defines a mapping of semisimple conjugacy classes in the two groups, which we also denote by ρ'.

Conjecture (Langlands' Principle of Global Functoriality). *Given G', G and ρ', suppose we also have an automorphic representation π' of $G'(\mathbb{A})$. Then there is an automorphic representation π of $G(\mathbb{A})$ such that*

$$c^S(\pi) = \rho'(c^S(\pi')).$$

In other words, there is a finite set S of primes outside of which π' and π are both unramified, and for which

$$c_p(\pi) = \rho'(c_p(\pi')), \quad p \notin S.$$

We can remove the finite set S from the assertion by defining an equivalence relation on the set of families c^S of semisimple conjugacy classes in $\widehat{G}$. We write $c^S \sim c_1^{S_1}$ for two such families c^S and $c_1^{S_1}$ if

$$c_p = c_{1,p},$$

for almost all $p \notin S \cup S_1$. For any π in the set

$$\Pi_{\text{aut}}(G) = \Pi(G)$$

of automorphic representations of $G(\mathbb{A})$, we then write $c(\pi)$ for the equivalence class that contains the family $c^S(\pi)$. We thus obtain a surjective mapping

$$\pi \to c(\pi), \quad \pi \in \Pi_{\text{aut}}(G),$$

from $\Pi_{\text{aut}}(G)$ onto the set

$$\mathscr{C}_{\text{aut}}(G) = \{c(\pi) : \pi \in \Pi_{\text{aut}}(G)\}.$$

If $G = \mathrm{GL}(n)$, the restriction of the mapping to the subset $\Pi_{\text{cusp}}(G)$ of cuspidal automorphic representations is injective. This is the theorem of strong multiplicity 1 for $\mathrm{GL}(n)$ [189], [108]. It gives a bijection from $\Pi_{\text{cusp}}(G)$ onto the set

$$\mathscr{C}_{\text{cusp}}(G) = \{c(\pi) : \pi \in \Pi_{\text{cusp}}(G)\},$$

and thus allows us to identify a cuspidal automorphic representation of $\mathrm{GL}(n)$ with (the equivalence class of) a family of complex characteristic polynomials. In general, however, the mapping is not injective, for a variety of reasons that range from obvious to deep. A full understanding of the fibres of the mapping remains an important unsolved problem.

Consider the category whose objects are split reductive groups G over $\mathbb{Q}$, and for which the morphisms from G' to G are the complex homomorphisms from $\widehat{G}'$ to $\widehat{G}$. To any object G, we can associate the set of (equivalence classes of) families $\mathscr{C}_{\text{aut}}(G)$. For any morphism $\rho' \colon \widehat{G}' \to \widehat{G}$, we can associate the function

$$\mathscr{C}_{\text{aut}}(\rho') \colon c' \to c = \rho'(c'), \quad c' \in \mathscr{C}_{\text{aut}(G')},$$

postulated by the Principle of Functoriality. The correspondence $\mathscr{C}_{\text{aut}}$ is then a functor from the category of groups we have just defined to the category of sets. This is essentially the origin of the term functoriality.

In his original paper [138], Langlands formulated what would become global functoriality in the setting of a general (connected) reductive algebraic group G over a number field F. We should say something about the general analogue of the assertion of functoriality above for split groups.

A general group G over F can be constructed from a canonical quasi-split group G^* over F by an inner twist of its Galois action, a standard technique in the theory of algebraic groups. The group G^* is obtained in turn from a canonical split group $G^{*,\text{spl}} = G^{\text{spl}}$ by an outer twist of its Galois action, attached to a homomorphism from a finite Galois group $\Gamma_{K/F} = \mathrm{Gal}(K/F)$ into the group $\mathrm{Out}(G^{\text{spl}})$ of outer homomorphisms of G^{spl}. It is the dual of this action on $\widehat{G^*}$ that gives the L-group

$$^{L}G^{*} = \widehat{G^{*}} \rtimes \Gamma_{K/F}, \quad \Gamma_{K/F} = \mathrm{Gal}(K/F)$$

described for $F = \mathbb{Q}$ earlier in this section. Since the dual $\widehat{G}$ of G equals the dual $\widehat{G^{*}}$ of G (as complex groups with actions of $\Gamma_{K/F}$), the L-groups $^{L}G = \widehat{G} \rtimes \Gamma_{K/F}$ and $^{L}G^{*} = \widehat{G^{*}} \rtimes \Gamma_{K/F}$ of G and G^{*} are equal. In particular, the general version of functoriality postulated by Langlands in [138] includes the case of the identity map ρ from ^{L}G to $^{L}G^{*}$, and hence a correspondence from the automorphic representations of G to those of the quasi-split group G^{*}. These days, this question is generally regarded as part of endoscopy, a separate theory proposed later by Langlands that seeks among other things to describe the precise nature of this correspondence. We are therefore free to consider functoriality in the more restricted context of quasi-split groups. (We shall discuss Langlands' theory of endoscopy in Section 10.)

Suppose then that G is a quasi-split group over a number field F, which for the moment we may again take to be $\mathbb{Q}$. Our discussion above for a split group carries over with little change. In particular, an automorphic representation π of $G(\mathbb{A})$ has a decomposition (23) such that π_p is unramified for any p outside a finite set S [72]. The classification of unramified representations for split groups extends, with one minor adjustment. In the quasi-split case, the bijection $\pi_p \to c_p$ takes the unramified representations of $G(\mathbb{Q}_p)$ onto the set of $\widehat{G}$-orbits (under conjugation) in $^{L}G = \widehat{G} \rtimes \Gamma_{K/F}$ that map to the Frobenius class Φ_p in $\Gamma_{K/F}$. This really is a generalization of the bijection from the special case of split groups. For we recall in general that $K/\mathbb{Q}$ can be any Galois extension through which the Galois action on $\widehat{G}$ factors. In particular, if G is split, we could take $^{L}G = \widehat{G} \times \Gamma_{K/F}$, a direct product of $\widehat{G}$ with the Galois group $\Gamma_{K/F} = \mathrm{Gal}(K/\mathbb{Q})$ of any convenient finite Galois extension $K/\mathbb{Q}$. In this case, $c_p(\pi)$ would just be the product of a semisimple conjugacy class in $\widehat{G}$ with an element in the Frobenius class Φ_p in $\Gamma_{K/F}$.

The assertion of Langlands' global functoriality conjecture then carries over to the quasi-split group G as stated, with one proviso. The mapping

$$\rho' : {}^{L}G' \to {}^{L}G$$

between L-groups must be an L-homomorphism, by which we mean that it commutes with the projections of $^{L}G'$ and ^{L}G onto $\Gamma_{K/F}$. With this understanding, we can also extend the category we have defined to quasi-split groups, and the functor $\mathscr{C}_{\mathrm{aut}}$ from this category to sets. Finally, everything we have described in this section remains valid if we allow the base field F to be an arbitrary number field. We took $F = \mathbb{Q}$ in order that the technical background might seem a little more concrete.

There is one further matter that requires comment. We could have referred to the Langlands conjecture above as "unramified global functoriality", since the term functoriality by itself often connotes a correspondence $\pi' \to \pi$ of automorphic representations that applies to the ramified local places $v \in S$ as well as the unramified $v \notin S$. This is certainly how Langlands introduced it in [138] (without the name). The resulting global assertion is more complicated, and presupposes Langlands' local functoriality. Langlands actually began with a question on general automorphic

L-functions, both local and global. He then presented local and global functoriality as questions motivated by a desire to understand the L-functions. We shall discuss these matters in the next section.

In this section, we have presented unramified global functoriality as the primary assertion in order to give a concrete statement, and also to follow the natural progression begun with our discussion of abelian class field theory in the last section. We must now define the corresponding unramified automorphic L-function.

Suppose again that G is a quasi-split reductive group over a number field F. An automorphic representation π of $G(\mathbb{A}_F)$ gives a family

$$c^S(\pi) = \{c_v(\pi) : v \notin S\}$$

of semisimple conjugacy classes in ${}^LG$. To attach an automorphic L-function to π, we would need a family of semisimple conjugacy classes in a general linear group $\mathrm{GL}(n,\mathbb{C})$ rather than in ${}^LG$. We therefore fix a finite-dimensional representation

$$r\colon {}^LG \to \mathrm{GL}(n,\mathbb{C})$$

in addition to the automorphic representation π of $G(\mathbb{A})$. We then define the associated unramified automorphic L-function as the Euler product

$$L^S(s,\pi,r) = \prod_{v\notin S} L_v(s,\pi,r) = \prod_{v\notin S} \det(1 - r(c_v(\pi))q_v^{-s})^{-1}, \tag{28}$$

with q_v being the degree of the residue class field of F_v. Once again, the product converges absolutely to an analytic function in some right half plane. The analogy with the Artin L-function (12) is clear. We can think of it as the same definition, but $\mathrm{Gal}(K/F)$ replaced by the group ${}^LG$, along with the extra structure provided by π.

In addition to defining automorphic L-functions and introducing the Principle of Functoriality, Langlands sketched the following four applications in his seminal paper [138].

1. Analytic continuation and functional equation. Langlands pointed out that the analytic continuation and functional equation of a general automorphic L-function would follow from functoriality and the special case that $G = \mathrm{GL}(n)$ and $r = St(n)$, the standard n-dimensional representation of $\mathrm{GL}(n)$. This special case (at least for cuspidal π) was established soon afterwards by Godement and Jacquet [81], as we saw in the last section.

2. Artin L-functions. We have noted that quasi-split groups are the natural setting for functoriality. The Galois factor $\mathrm{Gal}(K/\mathbb{Q})$ is then an essential part of the L-group ${}^LG$. In particular, the construction naturally includes the seemingly trivial case that G is the 1-element group $\{1\}$. Its L-group will then be an arbitrary finite Galois group $\mathrm{Gal}(K/\mathbb{Q})$, while r becomes simply an n-dimensional representation of $\mathrm{Gal}(K/\mathbb{Q})$. The associated automorphic L-function $L(s,\pi,r)$ (with π being of course the trivial 1-dimensional automorphic representation of G) is then just the general Artin L-function $L^S(s,r)$. The Principle of Functoriality can thus be interpreted as an identity

$$L^S(s,r) = L^S(s,\pi,St(n)) \tag{29}$$

between a general Artin L-function and a standard automorphic L-function for $\mathrm{GL}(n)$. This represents a general and entirely unexpected formulation of nonabelian class field theory. It identifies purely arithmetic objects, Artin L-functions, with objects associated with harmonic analysis, automorphic L-functions, thereby implying that the arithmetic L-functions have meromorphic continuation and functional equation, and that they are essentially entire. Functoriality would thus include a proof of the Artin conjecture stated in the last section. Abelian class field theory amounts to the special case that the dimension n of r equals 1. Its original aim was to establish that abelian L-functions are the Hecke–Tate L-functions attached to the automorphic representations of $\mathrm{GL}(1)$, and thereby have analytic continuation and functional equation.

3. Generalized Ramanujan conjecture. The generalized Ramanujan conjecture asserts that a unitary cuspidal automorphic representation $\pi = \widetilde{\bigotimes}_v \pi_v$ of $\mathrm{GL}(n)$ is *locally tempered*. This means that the character

$$f_v \to \mathrm{tr}(\pi(f_v)), \quad f_v \in C_c^\infty(\mathrm{GL}(n,F_v)),$$

of each local constituent π_v of π is tempered, in the sense that it extends to a continuous linear form on the Schwartz space $\mathscr{C}(\mathrm{GL}(n,F_v))$ of $\mathrm{GL}(n,F_v)$ defined by Harish-Chandra. We recall that the classical Ramanujan conjecture applies to the case that $n = 2$, and that π comes from the cusp form of weight 12 and level 1. It was proved by Deligne [63], who established more generally (for $n = 2$) that the conjecture holds if π is attached to any holomorphic cusp form. (The case that π comes from a Maass form remains an important open problem.) Langlands observed that functoriality, combined with expected properties of the correspondence $\pi' \to \pi$, would imply the generalized Ramanujan conjecture for $\mathrm{GL}(n)$. His representation-theoretic argument is strikingly similar to Deligne's geometric proof [63].

4. Sato–Tate conjecture. The Sato–Tate conjecture for the distribution of numbers $N_p(E)$ of solutions (mod p) of an elliptic curve E over $\mathbb{Q}$ has a general analogue for automorphic representations. Suppose for example that π is a (unitary) cuspidal automorphic representation of $\mathrm{GL}(n)$. The generalized Ramanujan conjecture of 3. above asserts that the conjugacy classes, represented by diagonal S_n-orbits

$$c_p(\pi) = \begin{pmatrix} c_{p,1}(\pi) & & 0 \\ & \ddots & \\ 0 & & c_{p,n}(\pi) \end{pmatrix} / S_n,$$

have eigenvalues of absolute value 1. The generalized Sato–Tate conjecture describes their distribution in the maximal torus $U(1)^n$ of the maximal compact subgroup $U(n)$ of the dual group $\mathrm{GL}(n,\mathbb{C})$. If π is *primitive* (a notion that requires functoriality even to define, as we will describe in Section 9), the distribution of these classes should be given by the weight function in the Weyl integration formula for the unitary group $U(n)$. Langlands sketched a rough argument for establishing such a result from general functoriality. Clozel, Harris, Shepherd-Barron and Taylor followed this argument in their proof of the original Sato–Tate conjecture, but using base change for $GL(n)$ and deformation results in place of functoriality. (See [91], [238].)

5 Local Functoriality and early results

Within a short time of his introduction [138] of functoriality, Langlands presented some striking results that in addition to the interest they held in their own right, offered evidence for the general principle. They are contained in four[3] major works: a long, unpublished manuscript [137] related to the local properties of Artin L-functions, a classification of representations of algebraic tori [153], the properties of Euler products from Eisenstein series [139] mentioned in Section 3, and the monograph [103] of Jacquet and Langlands on $\mathrm{GL}(2)$. Each of these represents a deep and original contribution to the subject, even if like functoriality itself, each one may have been ahead of its time.

We shall describe the four contributions in sequence in the latter part of the section. A later contribution, Langlands' monograph [149] on base change for $\mathrm{GL}(2)$, with its dramatic applications to Artin's conjecture, will be treated separately in Section 7. In the first part of this section, we shall discuss the local aspects of the Principle of Functoriality. We should also take the opportunity to add some further remarks about the structure of Langlands' fundamental paper [138].

It is important to remember that the Langlands conjectures were so revolutionary in 1970 that they took many years to be accepted (or even noticed) by the general mathematical public. The paper [138] was dense and difficult. It consisted of seven questions (not conjectures), three local and three global. The other question, which came first, was both local and global. It concerned the possibility of defining (completed) automorphic L-functions. It was actually presented as the central problem, with six supplementary questions related to functoriality being a strategy for attacking it. This is the opposite of our exposition here, in which (unramified) global functoriality was presented as the centre of the Langlands program. However, the logic of [138] is compelling. As we described in the last section, it represents a (nonabelian) analogue for automorphic representations of Artin's use of abelian class field theory to establish analytic continuation and functional equation for his (abelian) L-functions.

[3] We could have added Langlands' classification of representations of real groups [151] mentioned in Section 1, as the fifth work, but we are postponing its further discussion until Section 10.

Suppose that G is a reductive group over a number field F. We then have the global L-group

$$^LG = \widehat{G} \rtimes \mathrm{Gal}(K/F),$$

where (K/F) is a finite Galois extension over which G splits. For any valuation v of F, we have a local Galois extension (K_v/F_v), for which $\mathrm{Gal}(K_v/F_v)$ represents a conjugacy class of subgroups of $\mathrm{Gal}(K/F)$. We also have the local L-group

$$^LG_v = \widehat{G} \rtimes \mathrm{Gal}(K_v/F_v),$$

which comes with a conjugacy class of L-embeddings $^LG_v \hookrightarrow {}^LG$. At this point, we are free to take motivation from the special case of automorphic L-functions for $\mathrm{GL}(n)$ discussed near the end of Section 3, and the unramified automorphic L-functions $L^S(s,\pi,r)$ from Section 4. In particular we fix a finite-dimensional representation

$$r\colon {}^LG \to \mathrm{GL}(n,\mathbb{C}),$$

a nontrivial additive character

$$\psi\colon F^* \backslash \mathbb{A}_F^* \to \mathbb{C}^*,$$

and an automorphic representation π of G, objects that come with natural localizations r_v, ψ_v and π_v. Langlands' first question was then as follows.

Question 1 (Langlands [138]). *Is it possible to define local L-functions*

$$L_v(s,\pi,r) = L(s,\pi_v,r_v)$$

and local ε-factors

$$\varepsilon_v(s,\pi,r,\psi) = \varepsilon(s,\pi_v,r_v,\psi_v) = \varepsilon(\pi_v,r_v.\psi_v)q_v^{-(n_v-\frac{1}{2}s)},$$

for the valuations $v \in S$, with (G_v,r_v,π_v,ψ_v) ramified (or archimedean), such that the full Euler product

$$L(s,\pi,r) = L_S(s,\pi,r)L^S(s,\pi,r) = \prod_v L_v(s,\pi,r)$$

has analytic continuation, and satisfies the functional equation

$$L(s,\pi,r) = \varepsilon(s,\pi,r)L(1-s,\pi,r^\vee), \tag{30}$$

for

$$\varepsilon(s,\pi,r) = \prod_{v\in S} \varepsilon_v(s,\pi,r,\psi)?$$

It is remarkable to see Langlands' comments immediately following his statement of Question 1 [138, pp. 31–32]. He devotes one sentence each to: the motivation [139] he took from Eisenstein series, which we mentioned in Section 3, and which led to the Langlands–Shahidi method; his anticipation of the extension of the thesis of Tate from GL(1) to GL(n), established shortly thereafter [81]; and most striking of all, his description of his subsequent questions on functoriality as the nonabelian analogue of "the idea that led Artin to the general (abelian) reciprocity law". It is like reading fifty years of past and future history unfold in one short paragraph.

Having already stated the unramified version of global functoriality, we shall not restate all of Langlands' remaining questions. Questions 2 and 3 concern the local and global correspondence between the representations of a non-quasi-split inner twist of G and those of G, which are now regarded as part of the conjectural theory of endoscopy. It is the next two Questions 4 and 5 that introduce functoriality. To state the first of these as local functoriality, we take quasi-split groups G'_v and G_v over a localization $\mathbb{Q}_v$ of $\mathbb{Q}$, together with a local L-homomorphism

$$\rho'_v : {}^L G'_v \to {}^L G_v$$

between their L-groups.

Conjecture (Langlands' Principle of Local Functoriality). *There is a natural correspondence*

$$\pi'_v \to \pi_v, \quad \pi'_v \in \Pi(G'_v),$$

from the (equivalence classes of) irreducible representations of G'_v to those of G_v such that

$$L(s, \pi_v, r_v) = L(s, \pi'_v, \rho'_v \circ r_v)$$

and

$$\varepsilon(s, \pi_v, r_v, \psi_v) = \varepsilon(s, \pi'_v, \rho'_v \circ r_v, \psi_v),$$

for every finite-dimensional representation r_v of ${}^L G_v$ and every nontrivial additive character ψ_v on F_v.

Notice that unlike the statement of unramified Global Functoriality in Section 4, this conjecture is not rigid as stated, in that it does not characterize the correspondence. The property on the L- and ε- factors, which is based on a suitable answer to Question 1, represents a condition it must satisfy. Another condition is the global expectation that if G'_v, G_v and ρ'_v are localizations of global objects G', G and ρ for each v, and if $\pi' = \bigotimes_v \pi'_v$ is an automorphic representation of $G'(\mathbb{A}_F)$, one can then choose a representation $\pi_v \in \Pi(G_v)$ in the image of the correspondence $\pi'_v \to \pi_v$ for each v such that the product $\pi = \bigotimes_v \pi_v$ is an automorphic representation of G. Langlands actually imposes a stronger condition in Question 5, his general form of Global Functoriality. Namely, if π_v is *any* representation in the image of the correspondence $\pi'_v \to \pi_v$ for each v, is the product an automorphic representation of $G(\mathbb{A}_F)$? He soon realized that this was asking for a little too much. Langlands' later theory of endoscopy, which has now been established in a number of cases [23] and which we will discuss in Section 10, characterizes both the correspondence $\pi'_v \to \pi_v$, and which products attached to a given π' are actually automorphic representations of $G(\mathbb{A})$.

Questions 6 and 7 represent an important extension of functoriality to Weil groups. My understanding is that Langlands first took his conjectures to Weil in the form of Questions 1–5, as an automorphic analogue of Artin L-functions and the Artin reciprocity law. Weil pointed out that he had some years earlier introduced a natural generalization of Artin L-functions. Langlands was no doubt happy to find that his questions/conjectures extended seamlessly to Weil's L-functions, and that they were in fact the richer for it. Of the four major works that followed, and that we will be discussing presently, three depend intimately on the extensions of Galois groups Weil had discovered in abelian class field theory.

In the interest of making this report seem more concrete, we have sometimes avoided the most general setting. Let us now try to be a little more efficient. From this point on, we shall work over an arbitrary base field F of characteristic 0, local or global. Everything discussed earlier in the case of $\mathbb{Q}$ then carries over as stated with F in place of either $\mathbb{Q}$ or one of its completions $\mathbb{Q}_v$. We can also work over the absolute Galois group

$$\Gamma_F = \varprojlim \Gamma_{K/F}, \quad \Gamma_{K/F} = \mathrm{Gal}(K/F),$$

instead of the finite Galois group $\Gamma_{K/F}$, if we agree that earlier homomorphisms r and ρ' are to be continuous. This is of course because the kernel of r and ρ' in the totally disconnected, compact group Γ_F is a normal subgroup of Γ_K of finite index, so that the quotient

$$\Gamma_F/\Gamma_K = \Gamma_{K/F}$$

becomes the Galois group of the finite Galois extension K/F. We shall freely adopt this convention in referring to past discussion, usually without further comment.

Weil groups are variants $W_{K/F}$ and W_F of the local and global Galois groups $\Gamma_{K/F}$ and Γ_F (with K/F still being a finite Galois extension). We set C_F equal to the multiplicative group F^* if F is local, and to the idele class group $F^* \setminus \mathbb{A}_F^*$ if F is global. The relative Weil group $W_{K/F}$ is then an extension

$$1 \to C_K \to W_{K/F} \to \Gamma_{K/F} \to 1$$

in both cases. It is a locally compact group, which comes with an isomorphism $r_{K/F}\colon C_F \to W_{K/F}^{\mathrm{ab}}$ in addition to its projection onto $\Gamma_{K/F}$, while if $K = F$, $W_{K/F}$ obviously equals C_F.

In either the local or global setting, consider a field E with $F \subset E \subset K$. Then $W_{K/E}$ is a subgroup of finite index in $W_{K/F}$, and there is a bijection

$$W_{K/F}/W_{K/E} \cong \Gamma_{K/F}/\Gamma_{K/E}$$

of finite coset spaces. If K/E is Galois, the subgroup $W_{K/E}$ is then normal in $W_{K/F}$, and the quotient on the left is isomorphic to the Galois group $\Gamma_{E/F}$ that equals the quotient on the right. To obtain the Weil group $W_{E/F}$ as a quotient in this case, we need to take the commutator subgroup $W_{K/F}^c$ instead of $W_{K/F}$. We then obtain an isomorphism

$$W_{K/F}/W^c_{K/E} \cong W_{E/F}.$$

In particular, we have a continuous projection from $W_{K/F}$ onto $W_{E/F}$.

The family

$$\{W_{K/F} : K/F \text{ Galois}\}$$

is thus an inverse system. The *absolute Weil group* of F is the corresponding inverse limit

$$W_F = \varprojlim_K W_{K/F}.$$

It is a locally compact group, equipped with a continuous homomorphism

$$\phi_F \colon W_F \to \Gamma_F$$

with dense image, and an isomorphism

$$r_F \colon C_F \to W_F^{\text{ab}}.$$

If $F = \mathbb{C}$, $W_F = W_{\mathbb{C}/\mathbb{C}} = \mathbb{C}^*$. If $F = \mathbb{R}$, $W_F = W_{\mathbb{C}/\mathbb{R}}$ is the group generated by $\mathbb{C}^*$ and an element w, subject to the relations $w^2 = -1$, and $wzw^{-1} = \bar{z}$ for any $z \in \mathbb{C}^*$. If F is a local nonarchimedean field, W_F turns out to be the dense subgroup of elements Γ_F whose image in the quotient

$$\Gamma_{F,\text{un}} \cong \Gamma_F/I_F = \widehat{\mathbb{Z}}$$

of Γ_F by its inertia subgroup I_F is the dense subgroup $\mathbb{Z}$ of $\widehat{\mathbb{Z}}$. Finally, if F is global, W_F is given by a more complicated blend of these properties.

We refer the reader to the beginning [237, §1] of the article of Tate in the Corvallis proceedings. He takes the absolute Weil groups $\{W_F\}$ as the basic objects, equipped with mappings ϕ_F and r_F as above that satisfy four axioms. He then defines the relative Weil groups as quotients

$$W_{K/F} = W_F/W^c_K.$$

Tate also comments very briefly [237, (1.2)] on how class field theory is implicit in the existence of Weil groups, specifically the algebraic derivation of local and global class field theory in terms of Galois cohomology.

Observe that any continuous, complex finite-dimensional representation of Γ_F pulls back to a unique continuous representation of W_F. Representations of Weil groups are then more general than those of Galois groups. It is for this reason that mathematicians have taken to working with the absolute Weil form

$$^LG = \widehat{G} \rtimes W_F$$

of the L-group of a quasi-split group G over F, rather than the absolute Galois form $\widehat{G} \rtimes \Gamma_F$, or the original, less canonical form $\widehat{G} \rtimes \Gamma_{K/F}$. It is of course understood that the action of W_F on $\widehat{G}$ is through the pullback to Γ_F of a suitable finite quotient $\Gamma_{K/F}$.

For the assertions of functoriality, one would want to take the absolute Weil form of both L-groups ${}^LG'$ and ${}^LG$, with ρ' being an L-homomorphism from ${}^LG'$ to ${}^LG$, in the obvious sense that it commutes with the two projections onto W_F. One also has to insist that the ρ'-image in $\widehat{G}$ of any element in W_F be semisimple, since W_F is only locally compact. We shall again feel free to extend past discussions to this general context, without comment.

Langlands' final two Questions 6 and 7 apply to the Weil form of local and global functoriality, or rather the special case in which the first group G' equals the trivial group $\{1\}$. In other words, $\phi = \rho'$ is an L-homomorphism from W_F to ${}^LG$. (Langlands also replaces the complex dual group of G by a compact real form, in anticipation perhaps of his later comments on the generalized Ramanujan conjecture.) The two questions ask for a correspondence from L-homomorphisms $\phi\colon W_F \to {}^LG$ to representations of $G(F)$ if F is local and to automorphic representations of $G(\mathbb{A}_F)$ if F is global. This is where we see the advantage of the role of the Weil group. For there are many more such homomorphisms than there would be for the Galois group Γ_F. The case of a local field F is of particular importance. Indeed, the irreducible representations of $G(F)$ obtained in this way are believed to account for most (though not all) such representations. In fact, Question 6 has evolved into what is now known as the *local Langlands classification* (or the *local Langlands correspondence*). It has been established in a significant number of cases, for $G = \mathrm{GL}(n)$ in [92], [94] and [202], and for quasi-split classical groups in [23] and [181], even as it remains conjectural in general. (For the archimedean case, it is the original form of Question 6 that is relevant. It gives the Langlands classification for the real group, which we discussed in Section 1, and to which we will return later.)

We turn now to the four works mentioned in the beginning of the section. With many other things to discuss, we shall have to be brief, despite the importance of the results. We shall give a short description in each case, with further comments as needed later in the report.

1. Artin L-functions [137]. The reader might have noticed an irregularity in our claim of symmetry in Section 3 between n-dimensional Artin L-functions and the automorphic L-functions for $\mathrm{GL}(n)$ of Godement and Jacquet. The ε-factor in Artin's functional equation (14) is a nonconstructive global function, while its automorphic counterpart in (18) has a finite product decomposition (19) into purely local functions. The problem was to find a corresponding local decomposition for Artin ε-factors and of course, their generalizations for Weil groups.

There were good reasons for trying to do this. Langlands was thinking of the possibility of classifying $\Pi(G)$, the set of irreducible representations of a real or p-adic group $G(F)$. The key to this would be local functoriality, as stated earlier in this section, but with $G' = \{1\}$ and with the local Weil group W_F in place of Γ_F. In other words, $\Pi(G)$ should be closely tied to the L-homomorphisms

$$\phi\colon W_F \to {}^LG.$$

An essential condition would then be that for every representation r of ${}^LG$, the local L- and ε-functions $L(s,\pi,r)$ and $\varepsilon(s,\pi,r,\psi)$ for a representation π corresponding to ϕ, which were conjectured for any G in Question 1 and established for $G = \mathrm{GL}(n)$ and $r = \mathrm{St}(n)$ by Godement–Jacquet, would match independently defined functions $L(s, r\circ\phi)$ and $\varepsilon(s, r\circ\phi, \psi)$ for ϕ.

The local L-functions were already part of the Artin/Weil definition, but the construction of local ε-factors $\varepsilon(s, r\circ\phi, \psi)$ turned out to be very difficult. It was studied with some success by B. Dwork [70], [131] for Artin L-functions. Langlands used Dwork's results in his investigation of Weil L-functions. His goal was to characterize the local ε-factors

$$\varepsilon(s,r,\psi), \qquad r\colon W_F \to \mathrm{GL}(n,\mathbb{C}), \qquad F \text{ local},$$

as the unique family of functions that satisfies several natural conditions as F, r and ψ vary. This question gave rise in turn to four concrete, but very complex lemmas on Gaussian sums. My understanding is that Langlands obtained a complete solution, but that it was not all contained in the manuscript [137], which itself was not published. (See [137, (3.4.1)], and Langlands' later comments on [137] in his 1969 letter [136] to Deligne.)

Langlands' proof was purely local. Some time later, Deligne [62] found a striking global argument that led to a much shorter proof. The result was the desired canonical construction of local ε-factors

$$\varepsilon(s,r,\psi) = \varepsilon(r,\psi)q_F^{-n_F(s-\frac{1}{2})}, \tag{31}$$

for integers $n_F = n(r,\psi)$ and q_F, and any nontrivial additive character ψ on the multiplicative group F^* of the local field F. The global result remains the functional equation (14) for any Artin/Weil L-function

$$L(s,r), \qquad r\colon W_F \to \mathrm{GL}(n,\mathbb{C}), \qquad F \text{ global}, \tag{32}$$

but with the global ε-factor in (14) now having a product decomposition

$$\varepsilon(s,r) = \prod_{v\in S} \varepsilon(s,r_v,\psi_v)$$

into purely local ε-factors.

2. Automorphic representations of tori [153]. This paper gives a classification of representations (local and global) for *abelian* reductive algebraic groups, which is to say, for (algebraic) tori. Recall that a *torus* is an algebraic group that is isomorphic to a product $\mathrm{GL}(1)^k$ of multiplicative groups. A torus over F (our local or global field) comes also with an outer twisting over F, namely a homomorphism from a finite Galois group $\mathrm{Gal}(K/F)$ into the group of automorphisms of this product. The correspondence

$$T \to X^*(T) = \mathrm{Hom}(T, \mathrm{GL}(1))$$

gives an antiisomorphism from the category of algebraic tori over F that split over the finite Galois extension K, and the category of torsion free $\mathrm{Gal}(K/F)$-modules of finite rank. (See [233], for example.)

The paper [153] here is not nearly so complex as the last one [137]. It is still very interesting, a verification of functoriality in the simplest nontrivial case, and an elegant illustration of some of the ideas of Langlands implicit in [138]. It applies to the Weil form of functoriality in our description of Questions 6 and 7, specifically the special case that G is an algebraic torus over F, a local or global field (of characteristic 0), and that G' as before is the trivial group $\{1\}$. Functoriality then concerns the L-homomorphisms

$$\phi\colon W_F \to {}^L T = \widehat{T} \rtimes W_F.$$

The purpose of [153] is to establish functoriality in this case, and to establish further that the resulting correspondence is surjective.

To be more precise, we fix a local or global field F, and we write $\Phi(T)$ for the set of $\widehat{T}$-conjugacy classes of L-homomorphisms ϕ. As we have done earlier, we also write $\Pi(T)$ for the set of equivalence classes of irreducible representations of $T(F)$ if F is local, and of automorphic representations of $T(\mathbb{A}_F)$ if F is global. In the global case we have the localization mappings $\phi \to \phi_v$ and $\pi \to \pi_v$, from $\Phi(T)$ to $\Phi(T_v)$ and $\Pi(T)$ to $\Pi(T_v)$ respectively. Theorem 2 is the main result of [153]. It asserts that there is a canonical mapping

$$\Phi(T) \to \Pi(T), \qquad \phi \to \pi,$$

which is a bijection if F is local, and a surjection if F is global. In the global case, the mapping commutes with the associated localizations, and its fibres are the local equivalence classes in $\Phi(T)$, relative to the equivalence relation $\phi' \sim \phi$ in $\Phi(T)$ if $\phi'_v \sim \phi_v$ in $\Phi(T_v)$ for every localization F_v of F.

Langlands describes his proofs as "exercises in class field theory". Observe that a mapping ϕ in $\Phi(T)$ is completely determined by the projection of its image onto $\widehat{T}$. This leads to an isomorphism from $\Phi(T)$ (as an abelian group) onto the Galois cohomology group $H^1(W_F, \widehat{T})$ of continuous 1-cocycles from W_F to $\widehat{T}$, modulo continuous coboundaries. Such is the stuff of class field theory, in its algebraic form. It would indeed be a good exercise to study the proofs of Langlands, streamlined perhaps according to Tate–Nakayama duality [234].

3. Euler products [139]. We discussed this paper at the end of Section 3, as historical motivation for the Principle of Functoriality. It also represents concrete evidence for functoriality, specifically the properties of L-functions in Question 1.

Suppose that G, P, π and r are as in the formula (22) for the global intertwining operator (20) in terms of what was then the new L-function $L(s,\pi,r^\vee)$. We write (22) as

$$L(s,\pi,r^\vee) = L(s+1,\pi,r^\vee)M(w,\lambda),$$

where $M(w,\lambda)$ now represents a meromorphic scalar-valued function of the image $s \in \mathbb{C}$ of λ, according to the notation in Section 3 and the first assertion (a) of the

Main Theorem in Section 2. We see from this that if $L(s,\pi,r^\vee)$ is meromorphic in a right half plane $\mathrm{Re}(s) > b$, it continues to a meromorphic function of $\mathrm{Re}(s) > b-1$. Since we know that the Euler product for $L(s,\pi,r^\vee)$ converges in some right half plane, we conclude that $L(s,\pi,r^\vee)$ does have analytic continuation to a meromorphic function of s in the complex plane.

What of the proposed functional equation

$$L(s,\pi,r) = L(1-s,\pi,r^\vee)$$

conjectured in Question 1 of [138]? The analogue of the formula (22) for the intertwining operator

$$M(w^{-1},w\lambda)\colon \mathscr{H}_{P'} \to \mathscr{H}_P, \quad w^{-1} \in W(\mathfrak{a}_{P'},\mathfrak{a}_P),$$

is

$$M(w^{-1},w\lambda) = \frac{L(-s,\pi,r)}{L(-s+1,\pi,r)},$$

as one sees from the definitions [139, p. 47]. In general, the operators $M(w,\lambda)$ satisfy their own functional equation (8). In the case at hand, this becomes

$$M(w^{-1},w\lambda)M(w,\lambda) = M(w^{-1}w,\lambda) = M(1,\lambda) = 1.$$

This gives the identity

$$\frac{L(-s,\pi,r)}{L(-s+1,\pi,r)}\frac{L(s,\pi,r^\vee)}{L(s+1,\pi,r^\vee)} = 1$$

of meromorphic functions of s. On the other hand, if we substitute the conjectured functional equation for the first factor in the left, the product on the left becomes

$$\frac{L(1+s,\pi,r^\vee)}{L(s,\pi,r^\vee)}\frac{L(s,\pi,r^\vee)}{L(s+1,\pi,r^\vee)},$$

which is also equal to 1. In other words, the conjectural functional equation for $L(s,\pi,r)$ is compatible with the established functional equation for $M(w,\lambda)$, but is not implied by it. Nonetheless, this represents further evidence from Eisenstein series for Langlands' theory of automorphic L-functions and the Principle of Functoriality.

These observations were part of Langlands' article [139]. He did not assume that the dual unipotent radical $\widehat{N}$ was abelian, although M still represents a maximal Levi subgroup. For G, $P = MN$ and π as above, but without this assumption on $\widehat{N}$, the Lie algebra of $\widehat{N}$ has a decomposition

$$\widehat{\mathfrak{n}} = \bigoplus_{i=1}^{k} \widehat{\mathfrak{n}}_i,$$

for

$$\widehat{\mathfrak{n}}_i = \{U \in \widehat{\mathfrak{n}} : \mathrm{ad}(\varpi_P^\vee)U = iU\},$$

and for $1 \leq k \leq 6$ (as is well known). The generalization of (22) is the formula

$$M(w,\lambda) = \prod_{i=1}^{k} \frac{L(is,\pi,r_i^\vee)}{L(is+1,\pi,r_i^\vee)}, \qquad \lambda \to s,$$

where r_i is the adjoint representation of $\widehat{M}$ in $\widehat{\mathfrak{n}}_i$. (We have been following the notation of [78, (1.2.5.3)] which is slightly simpler than that of Langlands at the end of Section 5 of [139].) If one can show that the L-functions $L(s,\pi,r_i^\vee)$ for M have analytic continuation for $1 \leq i < k$, the argument above establishes that the same is true for $L(s,\pi,r_k^\vee)$. The extent to which this is possible is governed by the relevant Coxeter–Dynkin diagrams. At the end of the article [139], Langlands gives an extended table of such diagrams, which establish that for all but three simple groups M, there is at least one nontrivial representation r of $\widehat{M}$ for which the function $L(s,\pi,r)$ has meromorphic continuation.

Some years later, Shahidi began a sustained study that greatly expanded the theory [211], [214], [212], [213]. (See also [45], [44].) With an assumption on Whittaker models (well known for general linear groups M), he was able to treat general cuspidal automorphic representations $\pi \in \Pi_{\mathrm{cusp}}(M)$ (without the condition that they be unramified everywhere). The resulting theory, known now as the Langlands–Shahidi method, has established functional equations in addition to the meromorphic continuation for many of the L-functions $L(s,\pi,r)$ attached to Eisenstein series. A further examination of some special cases led to a remarkable application to functoriality. Partly in collaboration with H. Kim (the case $n=5$ below), Shahidi established functoriality for any $\pi' \in \Pi_{\mathrm{cusp}}(G')$, where G' equals GL(2), G equals GL(4) or GL(5), and

$$\rho' : \widehat{G'} = \mathrm{GL}(2,\mathbb{C}) \to \widehat{G} = \mathrm{GL}(n,\mathbb{C}), \quad n = 4,5,$$

is the symmetric cube or fourth power representation. This was thought to have been inaccessible, and has led to significant improvements in the bounds required by Ramanujan's conjecture for π'. (See [215], and the references therein.)

4. Automorphic representations of GL(2) **[103].** The 348-page monograph of Jacquet–Langlands was a partial exception to a remark from the beginning of this section. It really consists of two parts, the first twelve Sections 1–12 culminating in a striking application to Artin L-functions, and the last two Sections 15–16 on the comparison of representations of $G = \mathrm{GL}(2)$ with those of an inner twist, the multiplicative group G' of a quaternion algebra. The two intermediate Sections 13–14 are in some sense transitional. They extend the methods from Tate's thesis from GL(1) to GL(2) in Section 13, anticipating the later volume [81] of Godement–Jacquet for GL(n) we mentioned in our Section 3, and from GL(1) to G' in Section 14, in anticipation of the comparisons in Sections 15 and 16. The first part of the monograph was widely read by mathematicians at the time, and quickly became a basic part of their thinking. The second part was slower to be taken up, but the principal result (as opposed perhaps to its proof) did soon find important applications among number theorists.

The main theme of the first part is an extension of Hecke theory to all automorphic representations of $\mathrm{GL}(2)$. Hecke was the first to attach L-functions to what amounted to a special class of cuspidal automorphic representations π for $\mathrm{GL}(2)$. He showed that these L-functions have analytic continuation, with functional equation, to *entire* functions on $\mathbb{C}$. Hecke was of course working with classical holomorphic modular forms of weight k for, let us say $\mathrm{SL}(2,\mathbb{Z})$. It was to forms in this space that he attached his L-functions. Hecke also introduced the operators $\{T_n\}$ on the space that bear his name, and he proved that for any simultaneous eigenform of these operators, the associated L-function has an Euler product. The local components of the corresponding automorphic representation $\pi = \bigotimes_v \pi_v$ are then characterized at the p-adic places by the natural relation between the conjugacy classes $c(\pi_p)$ and the eigenvalues of T_p, and at the archimedean valuation ∞ by the requirement that π_∞ be a discrete series representation corresponding to k in the parametrization of Harish-Chandra.

Of particular relevance to [103] is the converse theorem Hecke then established. He showed that any Dirichlet series with certain properties, the main ones being an Euler product, the analytic continuation to an entire function of s, and an appropriate functional equation, gives rise to a cuspidal automorphic L-function $L(s,\pi)$. Unlike the original theorem, which was actually for modular forms of level N (that is, for congruence subgroups $\Gamma_0(N)$, or equivalently automorphic representations π with ramified components π_p for $p|N$), Hecke's converse theorem really was restricted to forms for the full modular group $\Gamma_0(1) = \mathrm{SL}(2,\mathbb{Z})$. It took thirty years for it to be extended. In 1967, Weil [250] established a converse theorem for modular forms of level N, but with more sophisticated requirements on the given Dirichlet series. In so doing, he was able to make what became known as the Shimura–Taniyama–Weil conjecture on the modularity of elliptic curves considerably more precise.

The goal of the first part, Sections 1–12 of [103], was to extend the converse theorem of Hecke and Weil for $G = \mathrm{GL}(2)$ to any number field F and to any cuspidal automorphic representation $\pi \in \Pi_{\mathrm{cusp}}(G)$. This was a bigger task than one might perhaps imagine. If F is an imaginary quadratic extension of $\mathbb{Q}$, there are no archimedean discrete series representations $\pi_\infty \in \Pi_2(G_\infty)$ of $G(F_\infty)$, and the corresponding modular forms are all Maass forms. Even if $F = \mathbb{Q}$, one wants a theory that includes all Maass forms as well as holomorphic modular forms. This leaves no choice but to extend the classical results for modular forms to a full theory for automorphic representations of the adele group $G(\mathbb{A}_F)$. In particular, one must first establish a robust theory for the irreducible representations π_v of the local components $\mathrm{GL}(F_v)$ of $G(\mathbb{A}_F)$.

Chapter I (Sections 1–8) was devoted to the local theory. For the archimedean local fields $F_v = \mathbb{R}$ or $F_v = \mathbb{C}$, the authors used basic results of Harish-Chandra to classify the irreducible representations π_v of $G(F_v)$ (Sections 5–6). In the earlier sections (1–4), they studied the irreducible representations π_v of the nonarchimedean groups $G(F_v)$ through Weil representations, Kirillov models and Whittaker models

[249], [100], [110], [79]. They constructed various examples of representations in this case, in what amounts to be a partial classification. In all cases, they constructed local L-functions $L(s,\pi_v)$ and ε-factors $\varepsilon(s,\pi_v,\psi_v)$ for each π_v.

Chapter II (Sections 9–12) is concerned with global Hecke theory. In Theorem 11.1, the authors proved that for any $\pi \in \Pi_{\text{cusp}}(G)$, the global L-function

$$L(s,\pi) = \prod_v L(s,\pi_v) \tag{33}$$

has analytic continuation to an *entire* function of s and a functional equation

$$L(s,\pi) = \varepsilon(s,\pi)L(1-s,\pi^\vee) \tag{34}$$

with

$$\varepsilon(s,\pi) = \prod_v \varepsilon(s,\pi_v,\psi_v). \tag{35}$$

In fact, they verified these properties for a larger family of L-functions $L(s,\omega\otimes\pi)$ in Corollary 11.2, where ω ranges over quasicharacters on $C_F = F^* \backslash A_F^*$, following Weil's extension of Hecke's converse theorem.

The general converse theorem is Theorem 11.3 of [103]. In common with its predecessors, it asserts that the necessary conditions are sufficient. More precisely, any representation $\pi = \bigotimes_v \pi_v$ of $G(\mathbb{A}_F)$ that satisfies the conclusions of Theorem 11.1 and Corollary 11.2, together with two further necessary conditions, is actually a cuspidal automorphic representation of $G = \mathrm{GL}(2)$. One of the extra conditions is a bound, imposed in Theorem 11.3 to ensure that the Euler product for $L(s,\pi)$ converges in a right half plane. The other is a requirement that the local constituents π_v of π all be infinite-dimensional. This is needed for the construction of a global Whittaker model for π [103, Proposition 9.2], which in turn is a foundation for the desired embedding of π into the space of cusp forms. (The last condition is known to hold for any $\pi \in \Pi_{\text{cusp}}(G)$, a property for which the reader can consult the hints in the last paragraph of Section 14.)

The culmination of what we are calling the first part of [103] (Chapters I and II) is the application in Section 12 of this converse theorem to Artin L-functions. We are speaking now of the generalizations of these objects to continuous representations of the global Weil group W_F (rather than the global Galois group Γ_F). The fundamental question is whether any irreducible two-dimensional representation

$$r\colon W_F \to \widehat{G} = \mathrm{GL}(2,\mathbb{C})$$

corresponds to a cuspidal automorphic representation $\pi \in \Pi_{\text{cusp}}(G)$, according to precepts of local and global functoriality.

The authors begin in Section 12 by appealing to [250]. This yields a global L-function $L(s,r)$ and ε-factor $\varepsilon(s,r)$ that satisfy analogues of the properties (33), (34) and (35), or rather their extensions from Corollary 11.2, but with the proviso that the global L-function need not be entire. The converse theorem actually applies to a given representation $\pi = \bigotimes \pi_v$ of $G(\mathbb{A}_F)$, so one first needs a way to construct the local components $\pi_v = \pi(r_v)$ of π from the local components r_v of r. This is exactly the form taken by the partial local classification from Chapter I. For any r_v, the authors prove that there is at most one $\pi_v = \pi(r_v)$ such that, in addition to some obvious requirements, the corresponding families of local L-functions and ε-factors are equal. (See p. 395 of [103] for a precise statement, which was extracted from the local results in Sections 4–6.) For the given two-dimensional representation r of W_F, Theorem 12.2 of [103] then asserts that if both $L(s, \omega \otimes r)$ and $L(s, \omega^{-1} \otimes r^{\vee})$ are entire functions for every quasicharacter ω on C_F, the representation $\pi_v = \pi(r_v)$ exists for every v, and $\pi = \bigotimes \pi_v$ is a cuspidal automorphic representations of $G(\mathbb{A})$.

There is a striking interpretation of this theorem. Recall the conjecture of Artin, which asserts that the L-function $L(s,r)$ attached to any irreducible representation r of dimension greater than 1 is entire. Artin's conjecture therefore implies functoriality for r when its dimension equals 2. Since this conjecture has been with us for almost a century, and is widely held to be true, we thus have strong evidence for functoriality for the case $r \to \pi$. Langlands cites Theorem 12.2, together with the partial local classification $r_v \to \pi_v$ based on L- and ε-factors, as one of the two principal contributions of [103] to the understanding of functoriality.

The other principal contribution cited by Langlands is the local and global correspondence

$$\pi' = \bigotimes_v \pi'_v \to \pi = \bigotimes_v \pi_v, \tag{36}$$

from the representations of the multiplicative group G' of a quaternion algebra M' to those of $G = \mathrm{GL}(2)$. Its apotheosis is the comparison in Section 16, but there was considerable preparation laid down earlier in the volume. In particular, the authors introduced the local correspondence $\pi'_v \to \pi_v = \pi(\pi'_v)$ at the same time and in the same spirit as their partial Galois/Weil correspondence $r_v \to \pi_v = \pi(r_v)$ described above. That is, they attached local L- and ε-factors $L(s, \omega_v \otimes \pi'_v)$ and $\varepsilon(s, \omega_v \otimes \pi'_v, \psi_v)$ to any π'_v, and then defined $\pi_v = \pi(\pi'_v)$ as the unique representation whose L- and ε-factors (which had already been constructed) match those of π'_v. (See p. 469 of [103] for the precise statement, taken from Sections 4 and 5.)

The local factors attached to G' were of necessity defined in terms of the Lie algebra of G', which is just the underlying four-dimensional quaternion algebra M' over F. These amount to the analogue for G' of the local construction from Tate's thesis. On the other hand, the local factors for G were defined, according to the requirements of the converse from Hecke theory, in terms of the Mellin transform attached to a one-dimensional vector space. The analogue for G of Tate's thesis would be based on a different transform attached to the Lie algebra of G, the four-dimensional matrix algebra M over F. It would have in many ways been more natural. In Section 13, the authors investigate the local factors for G that arise in this manner, and show in Theorem 13.1 that they do actually match the earlier local factors for G defined by Hecke theory. In Section 14, they combine the global methods of Tate's thesis with the local factors already in place for G'. Theorem 14.1 asserts that if π' is an automorphic representation of $G'(\mathbb{A}_F)$, the corresponding L-function $L(s, \pi')$ satisfies analogues of (33), (34) and (35), apart from the requirement that it be entire. But Corollary 14.3 then asserts that so long as π' is not a one-dimensional representation attached a quasicharacter χ on C_F, the global L-functions $L(s,\omega\otimes\pi')$ and $L(s,\omega^{-1}\otimes(\pi')^{\vee})$ are indeed all entire. Finally in Theorem 14.4 at the end of Section 14, the authors establish the essential property of the global correspondence (36). If we also take for granted the hints at the end of the section, as we did in the discussion of the converse theorem above, we can conclude that if π' is not the one-dimensional representation attached to a quasicharacter χ on C_F, its global image π lies in $\Pi_{\mathrm{cusp}}(G)$.

In Theorem 15.2, the authors deal with the local correspondence $\pi'_v \to \pi_v$. They prove that it maps $\Pi(G'_v)$ *injectively* onto the relative discrete series $\Pi_2(G_v)$ of G_v, a term that carries the same meaning at any local place v as that defined for $v=\mathbb{R}$ at the end of §2. Here, they use the local constructions of Tate for GL(2), and in particular, their own conclusion from Theorem 13.1 that the Tate local factors for π_v match the Hecke local factors in terms of which the local correspondence was defined. (The statement of Theorem 15.2 was actually for nonarchimedean v, but only because the authors describe the image more explicitly as the set of representations π_v that are either supercuspidal or "special", the latter being representations in the relative discrete series that are subquotients of induced representations. The result for archimedean v is simpler, and is contained in the discussion in Section 5.) The authors supplement this result with an important formula for the characters of representations, regarded as locally integrable, conjugacy invariant functions $\Theta(\pi_v,\cdot)$ on the group. In Proposition 15.5, they establish that corresponding characters satisfy the identity

$$\Theta(\pi'_v,b'_v) = -\Theta(\pi_v,b_v), \quad \pi'_v \to \pi_v, \tag{37}$$

where $b'_v \to b_v$ is the natural bijection between the regular elliptic conjugacy classes on the two groups.

This is as far as Hecke theory goes. However, the authors still had more to say. In Section 16, they used the Selberg trace formula, something entirely different, to give a remarkable characterization of the image of the global correspondence (36). Combined with their characterization of the local correspondence $\pi'_v \to \pi_v$ we have just described, the final result is the assertion that the global correspondence $\pi' \to \pi$ from G' to G is a bijection, from the set of representations $\pi' \in \Pi(G')$ not attached to a quasicharacter χ on C_F onto the set of representations $\pi \in \Pi_{\mathrm{cusp}}(G)$ with the property that for every v such that G'_v is not split, π_v lies in the relative discrete series $\Pi_2(G_v)$ of G_v.

This is what Langlands cited as the second principal contribution of [103] to the understanding of functoriality. Since it characterizes automorphic representations of G' as a natural subset of the cuspidal automorphic representations of its quasi-split inner form $G=\mathrm{GL}(2)$, it suggests that functoriality can be formulated purely in terms of of quasi-split groups, as we have done here. As we have noted, the automorphic representations of groups that are not quasi-split are now usually treated as part of a separate theory of Langlands, the theory of endoscopy. But we could still regard the global comparison between G' and G as the first result in endoscopy, established years before the theory was formally proposed. We shall postpone our description of the actual comparison to the next section, as part of a general discussion of the trace formula.

Incidentally, the complementary set of automorphic representations of G', those attached to quasicharacters χ of C_F, are just the one-dimensional automorphic representations of G'. They are naturally bijective with the set of one-dimensional automorphic representations of G, which represent the complement of $\Pi_{\text{cusp}}(G)$ in the set $\Pi_2(G)$ of representations that comprise the full automorphic (relative) discrete spectrum of G. However, the formal correspondence between these two sets is different from (36). Its analogue for general groups was introduced [13], [18] as an attempt to describe the representations in the automorphic discrete spectrum of any group that do not satisfy the general analogue of Ramanujan's conjecture. We shall return to it in later sections.

6 Trace formula and first comparison

Selberg introduced his trace formula in 1956 [205]. He might actually have discovered it earlier, but since he published very little of his work, it is difficult to say. There are actually two formulas. One is the identity for compact quotient $\Gamma \backslash H$, whose elegant proof Langlands reconstructed in [133], as we discussed in Section 1. See also [79]. The other applies to many noncompact quotients $\Gamma \backslash H$ of rank 1. The rank here means the number of degrees of freedom one has to approach infinity in $\Gamma \backslash H$. This is more difficult. It requires among other things the theory of Eisenstein series, which we discussed in Section 2, and which Selberg had introduced for this express purpose.

In particular, Selberg established an explicit trace formula for $\Gamma \backslash \mathrm{SL}(2,\mathbb{R})$ (in the sense we discussed in Section 1), where Γ equals $\mathrm{SL}(2,\mathbb{Z})$, or a congruence subgroup of $\mathrm{SL}(2,\mathbb{Z})$, or more generally, any discrete subgroup of $\mathrm{SL}(2,\mathbb{R})$ with a reasonable fundamental domain. He also established extended formulas that included the traces of the supplementary Hecke operators attached to a congruence subgroup of $\mathrm{SL}(2,\mathbb{Z})$. As we have noted, these operators are an integral part of the adelic framework that has since been adopted. Selberg used his formulas to prove striking estimates for the closed geodesics on the Riemann surface attached to $\Gamma \subset \mathrm{SL}(2,\mathbb{R})$, taken from the geometric side, and for the eigenvalues of the Laplacian on the Riemann surface, taken from the spectral side.

Langlands' interest in the trace formula was different. He saw it as an opportunity to study functoriality. Global Functoriality postulated deep reciprocity laws between automorphic representations for pairs of groups G' and G. The trace formula for any one group, especially insofar as it existed for noncompact quotient, was clearly a complex identity. But might it be possible to compare the formulas for G' and G, without having to evaluate the various terms in either case explicitly?

This brings us to the second part [103, §15–16] of the volume of Jacquet–Langlands. We put it aside in the previous section in order that it might serve us here as a simple introduction to the general comparison of trace formulas. On pp. 516–

517 of the volume, the authors stated the adelic version of Selberg's trace formula for the group GL(2). This was perhaps the first time the formula was stated in full, Selberg having limited his publications to specializations of the formula. In addition, it represents a two-fold extension of the formula, from a modular quotient of the upper half plane to a modular quotient of $\mathrm{GL}(2,\mathbb{R})$, and then to the adelic quotient $\mathrm{GL}(2,F)\setminus\mathrm{GL}(2,\mathbb{A})$ for a number field F. (In fact, the statement in [103] applies to any global field F, but as always for us, F will be of characteristic zero.) The authors also gave a clean and concise sketch of the proof. Detailed proofs of the adelic version of Selberg's formula for PSL(2) [69] and groups G of F rank 1 [9] appeared later.

Suppose for a moment that G is an arbitrary reductive group over a number field F. We could then take f to be a function in the natural space[4] of test functions

$$C_c^\infty(G(\mathbb{A}_F)) = \varinjlim_S \left(C_c^\infty(G_\infty)\otimes C_c^\infty(G_S^\infty)\otimes \mathbf{1}^S\right)$$

on $G(\mathbb{A})$. Right convolution by f in the Hilbert space $L^2(G(F)\setminus G(\mathbb{A}_F))$ obviously converges. It is easily seen to be an integral operator, with kernel

$$K(x,y) = \sum_{\gamma\in G(F)} f(x^{-1}\gamma y), \quad x,y\in G(\mathbb{A}). \tag{38}$$

If $G(F)\setminus G(\mathbb{A}_F)$ is compact, $R(f)$ is of trace class, by standard methods in functional analysis. Its trace is given by the identity (1) from Section 1, but with $G(F)$ and $G(\mathbb{A}_F)$ in place of Γ and G. Indeed, the spectral extension on the right-hand side of (1) equals the trace of $R(f)$, by definition, while the left-hand side amounts to the geometric expression for the trace derived originally by Selberg, and by Langlands in [133].

Suppose however that $G(F)\setminus G(\mathbb{A}_F)$ is not compact. Then $R(f)$ is not of trace class. The problem is with the continuous spectrum $\mathrm{L}^2_{\mathrm{cont}}(G(F)\setminus G(\mathbb{A}_F))$. The restriction $R_{\mathrm{cont}}(f)$ of $R(f)$ to this invariant subspace is no more of trace class than would be the convolution operator on $L^2(\mathbb{R})$ of a function in $C_c^\infty(\mathbb{R})$. One must subtract the contribution of $R_{\mathrm{cont}}(f)$ to the kernel $K(x,y)$ of $R(f)$ to obtain an operator that is better behaved. This is the role of Langlands' general theory of Eisenstein series. The point is that Eisenstein series provide a spectral formula for $K(x,y)$. As a formal consequence of Langlands' Main Theorem from Section 2, one obtains a spectral expansion

$$K(x,y) = \sum_P n_P^{-1}\int_{i\mathfrak{a}_P^*}\sum_{\phi\in\mathscr{B}_P} E(x,\mathscr{I}_P(\lambda,f)\phi,\lambda)\overline{E(y,\phi,\lambda)}\,\mathrm{d}\lambda \tag{39}$$

[4] Here $S\supset S_\infty$ is a finite set of valuations of F, while $C_c^\infty(G_\infty)$ is the ordinary space of test functions on the archimedean component $G_\infty = G(F_\infty)$, and $G_c^\infty(G_S^\infty)$ is the space of locally constant, complex-valued functions of compact support on the "ramified" nonarchimedean component $G_S^\infty = G(F_S^\infty)$. We write $\mathbf{1}^S$ for the characteristic function of a suitable natural compact subgroup K^S of the remaining "unramified" component $G^S = G(\mathbb{A}^S)$ of $G(\mathbb{A})$.

in which $\mathscr{B}_P$ is an orthonormal basis of the Hilbert space $\mathscr{H}_P$ on which the induced representation $\mathscr{I}_P(\lambda)$ acts, for the kernel to accompany the simpler geometric expression (38). (See for example [22, (7.6)].) The multiple integral converges absolutely, even though there are no reasonable pointwise estimates for the integrand. The argument for this, which I learned from Langlands, and is due to Selberg, is to combine the Schwartz inequality with the fact that for a positive definite function $f = h * h^*$, the diagonal value $K(x,x)$ of the kernel bounds that of the integral in (39) over any compact subset of the domain. (See for example [22].) We shall return to these matters when we discuss the general trace formula in Section 10.

For the group $G = \mathrm{GL}(2)$ in §16 of [103], there are two terms indexed by P in (39). One is given by the Borel subgroup $P = P_0 = B$ of upper triangular matrices in G. The other corresponds to $P = G$. It is the kernel of the operator $R_{\mathrm{disc}}(f)$ obtained by restricting $R(f)$ to the relative discrete spectrum.

The group $\mathrm{GL}(2)$ has a split, one-dimensional centre, the group $Z = A_G \cong \mathrm{GL}(1)$ of scalar matrices, so to have a trace at all, one must make the usual minor adjustment. However, instead of either restricting f to the subgroup $G(\mathbb{A})^1$ of $G(\mathbb{A})$ or projecting it onto a function invariant under a subgroup $A^+_{G,\infty} \cong \mathbb{R}^+$ of $A_G(F_\infty)$, according to the discussion in Section 2, the authors take f to be η^{-1}-equivariant, for a character η on the full adelic quotient $Z(F)\backslash Z(\mathbb{A}_F)$ of the centre. The kernel $K(x,y)$ is easy to adjust. For the original function f, the integral

$$\int_{Z(F)\backslash Z(\mathbb{A}_F)} K(zx,y)\eta(z)\,\mathrm{d}z$$

becomes the kernel for the η^{-1}-equivariant function

$$x \to \int f(zx)\eta(z)\,\mathrm{d}z.$$

The formulas (38) and (39) for the kernel are essentially unchanged, even if G is a general group. They can be taken as stated so long as we understand that:

(i) f now belongs to the space $C_c^\infty(G(\mathbb{A}_F),\eta^{-1})$ of η^{-1}-equivariant test functions.
(ii) $R(f)$ is the right convolution over $Z(\mathbb{A}_F)\setminus G(\mathbb{A}_F)$ of f on the space $L^2(G(F)\setminus G(\mathbb{A}_F),\eta)$ of square integrable, η-equivariant functions on $G(F)\setminus G(\mathbb{A}_F)$.
(iii) The integrals in (39) are really taken over the kernel $i\mathfrak{a}_P^{*,G}$ in $i\mathfrak{a}_P^*$ of the canonical linear projection of $i\mathfrak{a}_P^*$ into $i\mathfrak{a}_G^*$, in which the original notation holds if we treat the dependence of $\mathscr{I}_P(\lambda,f)$ on the image of λ in $i\mathfrak{a}_G^*$ as a Dirac distribution.

The two formulas (38) and (39) for $K(x,y)$ become η^{-1}-equivariant functions in x and y on $G(F)\setminus G(\mathbb{A}_F)$, making their diagonal values at $y = x$ functions on $Z(\mathbb{A}_F)G(F)\setminus G(\mathbb{A}_F)$. The trace of the restriction $R_{\mathrm{disc}}(f)$ of $R(f)$ to the η-discrete spectrum becomes the integral over this set of the expression with $P = G$ in (39).

For our group $G = \mathrm{GL}(2)$, the trace of $R_{\mathrm{disc}}(f)$ is thus the integral over $x = y$ in $Z(\mathbb{A}_F)G(F)\setminus G(\mathbb{A}_F)$ of the difference between (38) and the expression with $P = B$ in (39). Neither of these last two functions is integrable. To obtain a trace formula, one must see how their nonintegrable parts cancel, and then find an explicit formula for the integral of what remains. As we noted, Jacquet and Langlands gave a sketch of the process in §16 of [103]. The answer they obtained for the trace of $R_{\mathrm{disc}}(f)$ is given as a sum of the eight terms (i)–(viii) on p. 516–517. We shall say a few brief words about the argument, to give ourselves some general perspective.

The value of (38) at $y = x$ is the more transparent of the two expressions (and would of course be the only expression to consider in case of compact quotient). It can again be written as

$$\sum_{\{\gamma\}} \sum_{\delta \in G_\gamma(F)\backslash G(F)} f(x^{-1}\delta^{-1}\gamma\delta x),$$

where $\{\gamma\}$ is a set of representatives of $G(F)$-conjugacy classes in $A_G(F) \setminus G(F)$. If we proceed formally as if G were a group with $Z(\mathbb{A}_F)G(F) \setminus G(\mathbb{A}_F)$ compact, in which all multiple integrals are absolutely convergent, we could write the integral of this function as

$$\sum_{\{\gamma\}} \int_{Z(\mathbb{A}_F)G(F)\backslash G(\mathbb{A}_F)} \sum_{\delta \in G_\gamma(F)\backslash G(F)} f(x^{-1}\delta^{-1}\gamma\delta x)\,\mathrm{d}x, \tag{40}$$

which in turn is equal to

$$\sum_{\{\gamma\}} \int_{Z(\mathbb{A}_F)G_\gamma(F)\backslash G_\gamma(\mathbb{A}_F)} \cdot \int_{G_\gamma(\mathbb{A}_F)\backslash G(\mathbb{A}_F)} f(x^{-1}\gamma x)\,\mathrm{d}x,$$

and hence also to

$$\sum_{\{\gamma\}} \mathrm{vol}(Z(\mathbb{A}_F)G_\gamma(F) \setminus G_\gamma(\mathbb{A}_F)) \cdot \int_{G_\gamma(\mathbb{A}_F)\backslash G(\mathbb{A}_F)} f(x^{-1}\gamma x)\,\mathrm{d}x. \tag{41}$$

This would match the left-hand side of Selberg's trace formula for compact quotient, which we quoted from [133] as (1) in Section 1. But in the case $G = \mathrm{GL}(2)$ at hand, the integrals do not all converge. There are four kinds of terms in (40), two good and two bad.

If γ is a scalar matrix, it can be represented by 1. In this case, the corresponding integral in (40) gives a contribution

$$\mathrm{vol}(Z(\mathbb{A}_F)G(F) \setminus G(\mathbb{A}_F)) \cdot f(1). \tag{42}$$

to (41). This is the term (i) on p. 516 in [103]. If the characteristic polynomial of γ is irreducible over F, its eigenvalues generate a quadratic extension $E = E_\gamma$ of F, and $G_\gamma(F)$ is isomorphic to E^*. In this case the corresponding integral in (40) converges. The contribution

$$\sum_{\{\gamma\}} \mathrm{vol}(\mathbb{A}_F^* E_\gamma^* \setminus \mathbb{A}_{E_\gamma}^*) \cdot \int_{G_\gamma(\mathbb{A}_F)\backslash G(\mathbb{A}_F)} f(x^{-1}\gamma x)\,\mathrm{d}x \tag{43}$$

of these integrals to (41) gives the term (ii) on p. 516 of [103]. The term (iii) in [103] vanishes, being a sum over the empty set of inseparable quadratic extensions of the number field F. The other terms in (40) are bad. If the characteristic polynomial of γ is a product of two *distinct* linear factors over F, the corresponding integral in (40) diverges. The explanation is that in the associated summand in (41), the integral converges but the volume coefficient is infinite. The remaining terms in (40) are attached to nontrivial unipotent elements, namely the complement of the scalar matrix 1 from (42) in the classes $\{\gamma\}$, with characteristic polynomial being the square of a linear factor. The explanation here is the other way around, where the associated volume is finite, and the integral is what diverges. The two kinds of bad classes $\{\gamma\}$ ultimately contribute to the trace formula as the respective terms (iv) and (v) on p. 514 of [103].

The other half of the trace formula concerns the spectral expansion of the kernel, specifically the value at $y = x$ of the summand of $P = B$ in (39). Following Selberg's basic ideas, Jacquet and Langlands multiplied the difference between the value of $y = x$ at (38) and this spectral function by the characteristic function of a large compact subset

$$\{x : H_P(x) \leq \log c_1\}$$

of $Z(\mathbb{A}_F)G(F) \setminus G(\mathbb{A}_F)$, defined in terms of the usual fundamental domain for any large positive number c_1. This effectively led them to a cancellation of the nonintegrable parts of each function. More precisely, they showed that for each of the two truncated functions in the difference, the integral is a sum of three linear forms in f: an explicit distribution that is independent of c_1, the product of $\log c_1$ with a simpler distribution, and a distribution that appears to be quite complicated, but which approached 0 as c_1 approaches infinity. The two multiples of $\log c_1$ are easily seen to cancel. The terms that approach 0 can be ignored. This leaves only the distributions that are independent of c_1. They are equal to the sum of the terms (i)–(v) on p. 516 of [103] from the geometric kernel we have discussed above, and the terms (vi)–(viii) on p. 517 from the spectral kernel.

This completes our general remarks on the trace formula for GL(2) in [103]. It is a curious fact, observed a couple of years later, that with a minor change in the truncation process that takes into account the noncuspidal discrete spectrum (part of the term with $P = G$ in (39)), each of the two distributions that approach 0 actually vanishes if $\log c_1$ is sufficiently large. This turned out to be a general phenomenon, which carried over to the later trace formula of an arbitrary group G. In the general case, the truncation parameter $T_1 = \log c_1$ for GL(2) is replaced by a vector T in the positive chamber $\mathfrak{a}_{P_0}^+$ that is far from the walls. It gives rise to a uniform truncation operation on the diagonal values (x,x) of each of the two expansions (38) and (39) of $K(x,x)$. The integrals of the resulting two functions of x in turn come with their own expansions, whose terms are polynomials in T if T is far from the walls of $\mathfrak{a}_{P_0}^+$ (in a sense that depends only on the support of f). The general trace formula comes from this. It is the identity of distributions obtained by setting the polynomial variable T equal to a canonical point T_0 (often 0), which is determined by the choice of a suitable maximal compact subgroup $K_0 \subset G(\mathbb{A}_F)$. (See [12, §1–2] and [22, Theorem 9.1].)

There is another matter, which is a little more disturbing. Given a locally compact group H, one says that a continuous linear form F in $h \in C_c(H)$ is *invariant* if

$$F(h^u) = F(h), \qquad h^u(x) = h(u^{-1}xu),\ u \in H,$$

for every h and u. In the case of $H = \mathrm{GL}(2, \mathbb{A}_F)$, the trace $\mathrm{tr}(R_{\mathrm{disc}}(f))$ is an invariant distribution. One might therefore expect all of the terms (i)–(viii) in the formula for $\mathrm{tr}(R_{\mathrm{disc}}(f))$ also to be invariant distributions. They are not. The problem is that the truncation operation interferes with the symmetry under conjugation. In particular, it renders the geometric terms (iv) and (v) and the spectral terms (viii) noninvariant. However, there is a natural "renormalization" process, which converts these terms to invariant distributions, and which applies uniformly to any reductive group G. We refer the reader to the formula (2) of the introduction of [15] for a general statement of the final *invariant trace formula*, and to [22, §22–23] for a description of the general correction process.

These refinements are not actually needed for the comparison in §16 of [103]. Suppose again that G is a general reductive group over F, with standard parabolic subgroup $P = MN$, and that f_w is a function in $C_c^\infty(G_w)$, for a valuation w of F. It is convenient to define a supplementary function $f_{w,M}$ on $\Pi(M_w)$ (the set of irreducible representations of $M_w = M(F_w)$) by setting

$$f_{w,M}(\sigma_w) = \mathrm{tr}(\mathscr{I}_P^G(\sigma_w, f_w)), \quad \sigma_w \in \Pi(M_w),$$

the character of the representation of G_w parabolically induced from σ_w. We then say that function $f = \prod_v f_v$ in $C_c^\infty(G(\mathbb{A}_F))$ is *cuspidal* at w if $f_{w,M} = 0$ for every parabolic subgroup $P \neq G$. The term really ought to be "invariantly cuspidal" so as not to conflict with the earlier notion of a cuspidal function from Section 2. However, the context should rule out any future confusion. In any case, the relevance of the property is that if f is cuspidal at two places v_1 and v_2, the general invariant trace formula simplifies dramatically [15, Theorem 7.1]. It reduces to something close to what it would be if $G(F) \setminus G(\mathbb{A}_F)$ were compact. In the case $G = \mathrm{GL}(2)$ of present interest, the result for any such f is that the terms (iii)–(viii) all vanish, and hence that the trace of $R_{\mathrm{disc}}(f)$ equals the sum of the terms (i) and (ii). Thus, for the group GL(2) (or for any group of semisimple F-rank 1), one does not need the invariant trace formula for this simplification. It follows easily from the basic trace formula, as Jacquet and Langlands point out for GL(2) in §16.

We can now begin our brief account of the comparison theorem in §16 of [103] for the global correspondence of representations (36). We should first say something on the structure of the multiplicative group G' of a quaternion algebra. For a broader view of this, we extract what we need from the structure of general groups.

As we suggested in Section 2, there is a remarkable classification theory for general (connected) reductive groups G over local and global fields F. I take the liberty of representing it as a diagram of four steps.

$$
\begin{array}{ll}
(1) & \{\text{semisimple Lie algebras over } \mathbb{C}\} \\
& \qquad\downarrow \\
(2) & \{\text{reductive groups over } \mathbb{C}\} \\
& \qquad\downarrow \\
(3) & \{\text{split reductive groups over } F\} \\
& \qquad\downarrow \\
(4) & \{\text{quasi-split reductive groups over } F\} \\
& \qquad\downarrow \\
(5) & \{\text{general reductive groups over } F\}
\end{array}
$$

Complex semisimple Lie algebras are bijective with Coxeter–Dynkin diagrams. These in turn are finite disjoint unions of connected diagrams, of which there are four infinite families and five exceptional diagrams. This is the starting point.

The first step (1) $\rightarrow$ (2) follows from the theory of covering groups, which is easy to manage in this setting. It is founded on our explicit knowledge of the finite abelian groups $Z(G_{\mathrm{sc}}(\mathbb{C}))$, where $G_{\mathrm{sc}}(\mathbb{C})$ ranges over the simple, simply-connected complex groups attached to the connected diagrams, and the fact that any complex reductive group is a quotient of a finite direct product of groups $G_{\mathrm{sc}}(\mathbb{C})$ and a complex torus, by a finite central subgroup. The second step (2) $\rightarrow$ (3) is a bijection, obtained from a special case of the basic construction of Chevalley groups. The third step (3) $\rightarrow$ (4) is given by *outer twistings* of the Galois action on the given split group. Its main ingredient is a homomorphism (of finite image) from the Galois group $\Gamma_F = \mathrm{Gal}(\overline{F}/F)$ to the group of automorphisms of the underlying diagram. The last step (4) $\rightarrow$ (5) is based on class field theory. It is given by *inner twistings* of the Galois action on the given quasi-split group G^* over F, or in purely algebraic terms, elements in the Galois cohomology *set* $H^1(\Gamma_F, G^*_{\mathrm{ad}}(\overline{F}))$. These objects are a part of abelian class field theory, for they reduce to Galois cohomology *groups* $H^*(\Gamma_F, X)$, attached to abelian Γ_F-modules X, to which Tate–Nakayama duality applies. To view these matters in terms of Langlands dual groups, we refer the reader to [121, §1 (F local) and §2 (F global)].

This description of the general classification goes well beyond what is needed for quaternion groups, but it does represent an implicit foundation for various topics that will arise later. For the quaternion groups G', the classification is simple. In particular, one needs only the last step (4) $\rightarrow$ (5), since the underlying quasi-split group is the split group $G = \mathrm{GL}(2)$. The conclusions are as follows. If F is local, with $F \neq \mathbb{C}$, there is exactly one quaternion group up to isomorphism. (If $F = \mathbb{C}$, there are no quaternion groups.) If F is global, the isomorphism classes of quaternion groups G' over F are parametrized by *finite, even, nonempty* sets S of valuations of F (with $F_v \neq \mathbb{C}$ for each $v \in S$). For any such S, G' is characterized by the property that for *any* valuation v, G'_v equals the unique quaternion group over F_v if v lies in S, and is equal to $\mathrm{GL}(2)_v$ otherwise.

Suppose then that G' is a quaternion group over F with F global, and with centre $Z' = Z(G')$. The characteristic polynomial gives a natural bijection $b'_v \to b_v$ between the regular elliptic conjugacy classes in $G'_v = G'(F_v)$ and $G_v = G(F_v)$, as in the character formula (37). If G'_v does not split, any regular, semisimple class is elliptic. For the group G_v, however, there are also regular hyperbolic conjugacy classes a_v in G_v. We recall that according to the general definitions, a semisimple conjugacy class $\gamma = \gamma_v$ in G_v (or G'_v) is *regular* if the centralizer G_γ is a torus, and that the regular, semisimple classes in G_v form an open dense set. The class γ is elliptic if the quotient $Z(F_v) \setminus G_\gamma(F_v)$ is compact, and hyperbolic otherwise.

For the comparison of trace formulas, Jacquet and Langlands chose matching functions $f' \in C_c^\infty(G'(\mathbb{A}), \eta)$ and $f \in C_c^\infty(G(\mathbb{A}), \eta)$ in the relevant spaces of η-equivariant test functions. More precisely, given a function $f' = \prod f'_v$ for G', they chose a function $f = \prod f_v$ for G by defining the local factors f_v as follows. If G' is split at v, they simply identified f_v with f'_v under any of the isomorphisms from $G'(F_v)$ to $G(F_v)$ in the $G(F_v)$-conjugacy class of isomorphisms attached to G'_v as an inner twist. If G'_v is not split, they took f_v to be any function such that

$$\mathrm{tr}(f_v(\pi_v)) = \delta(\pi_v) \cdot \mathrm{tr}(f'_v(\pi'_v)), \tag{44}$$

where π_v is any (irreducible) tempered representation of $G(F_v)$, and $\delta(\pi_v) = 1$ if π_v is the image of π'_v under the local correspondence $\pi'_v \to \pi_v$ in (36), and $\delta(\pi_v) = 0$ otherwise. This is a spectral relation. One sees from the character formula (37) that it is equivalent to the geometric relation

$$\int_{G_\gamma(F_v)\backslash G(F_v)} f_v(x_v^{-1}\gamma x_v)\,\mathrm{d}x_v = \varepsilon(\gamma) \int_{G'_{\gamma'}(F_v)\backslash G'(F_v)} f'_v(x_v^{-1}\gamma' x_v)\,\mathrm{d}x_v \tag{45}$$

in which γ is any regular conjugacy class in $G(F_v)$, and $\varepsilon(\gamma)$ equals (-1) if $\gamma = b_v$ is the image of an (elliptic) class $\gamma' = b'_v$ in $G'(F_v)$, and equals 0 if $\gamma = a_v$ is a hyperbolic class. The function f_v is not uniquely determined by either of these relations. However, if I_v is any invariant distribution on $G(F_v)$, $I_v(f_v)$ is uniquely determined. We also note that if G'_v is not split, the function f_v is cuspidal, in the sense above.

For the group G', the quotient $Z'(\mathbb{A}_F)G'(F) \setminus G'(\mathbb{A}_F)$ is compact. The trace formula for G' is therefore the Selberg trace formula for compact quotient, discussed above and in Section 1. The trace of the operator $R(f') = R_{\mathrm{disc}}(f')$ is accordingly equal to the sum of the two terms

$$(Z'(\mathbb{A}_F)G'(F) \setminus G'(\mathbb{A}_F)) \cdot f'(1) \tag{46}$$

and

$$\sum_{\{\gamma'\}} \mathrm{vol}(\mathbf{A}_F^* E_{\gamma'}^* \setminus \mathbf{A}_{E_{\gamma'}}^*) \int_{G_{\gamma'}(\mathbb{A}_F)\backslash G'(\mathbb{A}_F)} f'((x')^{-1}\gamma' x')\,\mathrm{d}x', \tag{47}$$

given by the analogues of (42) and (43), which is to say, of the terms (i) and (ii) in Chapter 16 of [103] for G'. As for the group G, the trace of $R_{\mathrm{disc}}(f)$ equals the larger sum of terms (i)–(viii) from [103]. But since the local factor f_v is cuspidal at any place v at which G'_v does not split, and since the set S of such places is nonempty and even, f is cuspidal at (at least) two places v. Therefore, as the authors observe on p. 523, the terms (iv)–(viii) vanish for the given f. The term (iii) vanishes for any number field, so this leaves only the sum of (42) and (43), the terms (i) and (ii) for G.

The sum (47) is over regular, globally elliptic conjugacy classes γ' in $G'(F)$, which is to say, classes whose characteristic polynomial is irreducible over F. The same is true of the sum over γ in (43). As in the local case, the characteristic polynomials then give a bijection $\gamma' \to \gamma$ between the two indices of summation. Moreover, it follows from (45) (and the fact that S is even) that the summands of γ' and γ are equal. Therefore, the sums (47) and (43) are equal, as the authors note on p. 524. All that remains of the two trace formulas are the terms (42) and (46). It follows that

$$\begin{aligned}&\operatorname{tr}(R_{\mathrm{disc}}(f')) - \operatorname{tr}(R_{\mathrm{disc}}(f))\\ &= \operatorname{vol}(Z'(\mathbb{A}_F)G'(F)\setminus G'(\mathbb{A}_F))\cdot f'(1) - \operatorname{vol}(Z(\mathbb{A}_F)G(F)\setminus G(\mathbb{A}_F))\cdot f(1).\end{aligned}$$

With further simple arguments, the authors refine this formula into three identities between its basic components. For example, an approximation obtained by varying f' (and its corresponding image f) shows that each side of the last formula vanishes. Orthogonality relations for discrete series, with the corresponding Plancherel formulas for $f'(1)$ and $f(1)$, establish that $f'(1) = f(1)$. From this it follows that

$$\operatorname{vol}(Z'(\mathbb{A}_F)G'(F)\setminus G'(\mathbb{A}_F)) = \operatorname{vol}(Z(\mathbb{A}_F)G(F)\setminus G(\mathbb{A}_F)) \tag{48}$$

and that

$$\operatorname{tr}(R_{\mathrm{disc}}(f')) = \operatorname{tr}(R_{\mathrm{disc}}(f)) \tag{49}$$

(See [103, pp. 524–525].)

The volume identity (48) is of independent interest. The authors pointed out that while it was well known, the methods used to obtain it were not. They suggested that a similar comparison of the trace formula (unknown at the time) of an arbitrary group with that of a quasi-split inner form might be used to establish a similar identity in general. This could then be combined with Langlands' early formula [135] for the corresponding volume of a Chevalley group (or rather, its imputed generalization to quasi-split groups). The goal would be to give a general proof of Weil's conjecture on Tamagawa numbers. As we noted in Section 1, this is exactly what happened, with the subsequent work of Lai and Kottwitz (and the final step for E_8 by Chernousov). Weil's conjecture, incidentally, is the elegant volume formula

$$\tau(G) = \operatorname{vol}(G(F)\setminus G(\mathbb{A})) = 1,$$

for any simply connected group G over F, taken with respect to the canonical Tamagawa measure on $G(\mathbb{A})$. It is understood that our discussion has been for Haar measures on $G'(\mathbb{A})$ and $G(\mathbb{A})$ that are compatible, in the sense that they coincide under transfer by the inner twist, even though they do not have to be the Tamagawa measures.

It is of course the other identity (49) that is the main point. We have been discussing the proof sketched in [103, §16] of Theorem 16.1, the basic characterization of the global correspondence $\pi' \to \pi$. As we noted in Section 5, it is equivalent to the assertion that the mapping $\pi' \to \pi$ is a bijection between the subsets $\Pi_{\text{cusp}}(G') = \Pi(G')$ and $\Pi_{\text{cusp}}(G)$ described at the end of Section 5. Having already shown the mapping to be injective, the authors established the surjectivity assertion by combining (49) with some elementary functional analysis. (See [103, pp. 403–503].)

This at last completes our discussion of §16 from the monograph of Jacquet and Langlands. It represents the earliest comparison of adelic trace formulas. I hope that our rather extended treatment of it will serve as an introduction to the increasingly complex comparisons that followed. We note that special cases had been established earlier by Shimizu [225], [226], [227], with comparisons of Selberg's original (non-adelic) trace formulas [205], [206], [207].

7 Base change

In their introduction to [103], the authors wrote of the comparison we have just reviewed,

> *"... the theorem of §16 is important and its proof is such a beautiful illustration of the power and ultimate simplicity of the Selberg trace formula and the theory of harmonic analysis on semi-simple groups that we could not resist adding it. Although we are very dissatisfied with the methods of the first fifteen paragraphs we see no way to improve on those of §16. They are perhaps the methods with which to attack the question left unsettled in §12."*

For a short time afterwards, Langlands was apparently unhappy with the last sentence, possibly thinking that it was premature. It is ironic then that he found an supportive answer only a few years later.

It came from base change. Suppose again that $G = \mathrm{GL}(2)$ over the number field F, and that E is a cyclic extension of F of prime degree ℓ. The *restriction of scalars* functor then attaches a quasi-split reductive group $G_E^0 = \mathrm{Res}_{E/F}(G)$ over F to the pair (G,E) such that $G_E^0(F) = G(E)$. (The reason for the superscript 0 will become clear presently.) The L-group of G_E^0 can be taken to be

$$^L G_E^0 = \widehat{G}_E^0 \rtimes \Gamma_{E/F} = \underbrace{\mathrm{GL}(2,\mathbb{C}) \times \cdots \times \mathrm{GL}(2,\mathbb{C})}_{\ell} \rtimes \mathrm{Gal}(E/F),$$

where the cyclic group $\Gamma_{E/F} = \mathrm{Gal}(E/F)$ acts by permutation on the product of groups $\mathrm{GL}(2,\mathbb{C})$. Taking $^L G = \widehat{G} \times \Gamma_{E/F}$ as the L-group of G, we then have an L-homomorphism

$$\rho: g \times \sigma \to \underbrace{(g,\ldots,g)}_{\ell} \rtimes \sigma, \quad g \in \widehat{G} = \mathrm{GL}(2,\mathbb{C}), \sigma \in \Gamma_{E/F},$$

from ${}^L G = \widehat{G} \times \Gamma_{E/F}$ into ${}^L G_E^0 = \widehat{G}_E^0 \rtimes \Gamma_{E/F}$. Functoriality for ρ is what is known as *base change*, a problem that may be formulated in this way if G is any quasi-split group over F, and E/F is any finite extension.

Langlands had earlier [138] proposed base change for GL(2) as one of several natural questions chosen to illustrate the difficulty of functoriality. His unexpected solution of the problem in 1975 followed new ideas of Saito and Shintani. (The work of both Shintani and Langlands was published only later, Shintani [231] in the 1979 Corvallis proceedings, and Langlands in his 1980 monograph [149].) Motivated by the special case solved by Shintani, Langlands established a general correspondence $\pi \to \pi_E$ from the local and global representations of G to those of G_E^0. The solution represents a new comparison of trace formulas, considerably more sophisticated than that of the local and global correspondence $\pi' \to \pi$ of (36) for quaternion groups. As such, it amounts to something beyond a proof of functoriality for this case. Like the correspondence $\pi' \to \pi$, and in common with what one might hope for in any new comparison of trace formulas, the method allows also for a characterization of the *image* of the functorial correspondence.

We draw on the introduction of [149] for a brief history of the problem. Doi and Naganuma treated cases with $F = \mathbb{Q}$, E a real quadratic extension, and with the archimedean component π_∞ of π being in the discrete series [68]. Jacquet extended these results to more general F and π in [101], but as in [68], without characterizing the image of the correspondence. It was Saito [193] who introduced the twisted trace formula with respect to E/F, a new kind of trace formula that he was then able to compare with that of GL(2). He was thus able to establish base change for more general cyclic extensions E/F of prime degree, and also to characterize its image. Finally, Shintani formulated these ideas in terms of representation theory and adele groups (rather than the classical framework of holomorphic modular forms). This allowed him to deal with critical problems related to the construction of a local correspondence $\pi_v \to \pi_{v,E}$ at the ramified places of π. However, Shintani was still restricted to representations π whose local archimedean constituents π_v belonged to Harish-Chandra's discrete series, a condition that means that the local test function f_v is chosen to be cuspidal. The problem for him was in the complexity of the trace formulas that would otherwise have to be compared. We now have some idea of this difficulty, having seen the complexity of the full trace formula for GL(2), with its eight terms (i)–(viii) from [103, §16]. The restriction under which Shintani worked is essentially the condition that f be cuspidal at two places, which reduces that trace formula to the simple terms (i) and (ii), as in the comparison for quaternion groups from [103].

Langlands understood how to work with the full trace formula. By comparing the twisted trace formula for G_E^0 with the unrestricted trace formula for $G = \mathrm{GL}(2)$, he was able to establish the general base change correspondence $\pi \to \pi_E$ for any extension E/F of prime order ℓ and any π. It was only after having removed all the restrictions that had gone before that he was able to establish its remarkable applications to Artin L-functions.

Having discussed the trace formula for GL(2) and its comparisons from [103] in some detail, we shall be content with a briefer summary of the twisted trace formula. Even for GL(2), the twisted trace formula and its comparison with the ordinary trace formula are rather more technical. A reader could bypass our summary, which leads directly to the spectral comparison identity (53) (or (54), written in the notation of Langlands), and proceed to the subsequent review of the properties of the resulting base change lifting (55).

The origin of the twisted trace formula is the (algebraic) automorphism σ_E of the quasisplit group G_E^0 over F. Its action on $G_E^0(F)$ is given by the Galois automorphism σ of $G(E)$. It also acts on $G_E^0(\mathbb{A}_F)$, and on the quotient $G_E^0(F)\setminus G_E^0(\mathbb{A}_F)$, and indeed, on the adelic quotient $M_E(F)\setminus M_E(\mathbb{A}_F)$ of any σ_E-stable subgroup M_E of G_E^0 over F. Langlands introduced a σ_E-stable character ξ_E on the adelic quotient $Z_E(F)\setminus Z_E(\mathbb{A}_F)$ of the centre $Z_E = Z(G_E)$ of G_E^0. This is the setting of the trace formula for GL(2) discussed in the last section. In particular, for any ξ_E^{-1}-equivariant test function $f_E^0 \in C_c^\infty(G_E^0(\mathbb{A}_F), \xi_E^{-1})$, we have the operator $R_{\mathrm{disc}}(f_E^0)$ in the ξ_E-discrete spectrum

$$\mathrm{L}^2_{\mathrm{disc}}(\xi_E) = \mathrm{L}^2_{\mathrm{disc}}(G_E^0(F)\setminus G_E^0(\mathbb{A}_F), \xi_E)$$

whose trace is the object of the trace formula. But it is the twisted trace of $R_{\mathrm{disc}}(f_E^0)$ that is of interest here.

We form the semidirect product

$$G_E^+ = G_E^0 \rtimes \langle \sigma_E \rangle,$$

and write

$$G_E = G_E^0 \rtimes \sigma_E$$

for the connected component attached to a fixed generator σ_E of $\Gamma_{E/F}$. The action

$$(\sigma_E \phi)(y) = \phi(\sigma_E^{-1}(y)), \quad y \in G_E^0(F)\setminus G_E^0(\mathbb{A}_F),$$

of σ_E on the Hilbert space $L^2(\xi_E)$ gives a canonical extension R_E of the representation R_E^0 of $G_E^0(\mathbb{A}_F)$ to the group generated by $G_E(\mathbb{A}_F) = \sigma_E \cdot G_E^0(\mathbb{A}_F)$. In particular, for any test function $f_E \in C_c^\infty(G_E(\mathbb{A}_F), \xi_E)$, we have a unitary operator

$$R_E(f_E) = \int_{Z_E(\mathbb{A}_F)\setminus G_E(\mathbb{A}_F)} f_E(x) R_E(x)\,\mathrm{d}x$$

on $L^2(\xi_E)$. Our notation here is slightly different from that of Langlands, as has sometimes been the case in past discussion, but it makes no difference in the argument.

The *twisted trace formula* is an explicit formula for the trace of the operator $R_{\mathrm{disc}}(f_E)$. Its proof is very similar to that of the standard trace formula. For a start, $R_{\mathrm{disc}}(f_E)$ is an integral operator on $L^2(\xi_E)$, whose kernel is given by either the geometric expansion (38) or the spectral expression (39). Both expansions are actually valid as stated for any G (taken to be a connected component of a nonconnected reductive group G^+ over F), so long as we understand that x and y are variables in $G^0(\mathbb{A})$, while in (39), P ranges over standard parabolic *subsets* of G over F. (See [16, §1].) Langlands sketched the proof of the (twisted) trace formula in §10 of [149], following the derivation of the standard trace formula for GL(2) from [103]. There is a minor difference here in the standard trace formula for GL(2) used in [103]. Instead of the character η on $Z(F)\backslash Z(\mathbb{A}_F)$ in [103], Langlands here takes a character ξ on the subgroup

$$Z(F)\cap N_{E/F}(Z(\mathbb{A}_E))\backslash N_{E/F}(Z(\mathbb{A}_E))$$

of finite index, and then takes ξ_E to be the pullback of ξ to $Z_E(F)\backslash Z_E(\mathbb{A}_F)$ under the norm map $N_{E/F}$.

To compare the trace formulas, Langlands introduced a local correspondence

$$f_E = \prod_v f_{E,v} \to \prod_v f_v = f \tag{50}$$

of global test functions. For a given valuation v of F, let γ_v be a regular orbit in $G_{E,v} = G_E(F_v)$ under conjugation by the group $G^0_{E,v} = G^0_E(F_v)$. Then its ℓth power γ_v^ℓ is a subset of $G^0_{E,v}$, which Langlands intersected with the subgroup $G_v = G(F_v)$. He then proved that the mapping

$$\gamma_v \to \delta_v = G_v \cap \gamma_v^\ell$$

is a well defined injection from the regular $G^0_{E,v}$-orbits in $G_{E,v}$ to the regular conjugacy classes in G_v. (See [149, §4 and §8].) Our $G^0_{E,v}$-conjugacy in $G_{E,v}$ is the same as σ_E-conjugacy in $G^0_{E,v}$, while the power γ_v^ℓ becomes the norm N_{E_v/F_v} in the group $G^0(E_v)$. (My apologies for reversing the notation for γ and δ in [149], for the sole purpose of having the formulas (38) and (39) for the kernel extend as stated to a twisted group. It is actually quite appropriate, for it helps to unify more sophisticated notions of endoscopy.) Langlands then defined the correspondence $f_{E,v} \to f_v$ from local test functions $f_{E,v} \in C_c^\infty(G_{E,v}, \xi_{E,v})$ to functions $f_v \in C_c^\infty(G_{v,\text{reg}}, \xi_v)$ in terms of orbital integrals

$$\operatorname{Orb}(\gamma_v, f_{E,v}) = \int_{G^0_{E,\gamma_v}(F_v)\backslash G^0_E(F_v)} f_{E,v}(x_v^{-1}\gamma_v x_v)\,\mathrm{d}x_v$$

and

$$\operatorname{Orb}(\delta_v, f_v) = \int_{G_{\delta_v}(F_v)\backslash G(F_v)} f_v(x_v^{-1}\delta_v x_v)\,\mathrm{d}x_v$$

by setting

$$\operatorname{Orb}(\delta_v, f_v) = \begin{cases} \operatorname{Orb}(\gamma_v, f_{E,v}), & \text{if } \gamma_v \to \delta_v, \\ 0, & \text{if there is no such } \gamma_v. \end{cases} \tag{51}$$

This determines the orbital integrals of f_v uniquely, but of course not the function itself.

The goal of Langlands was to compare the terms in the two trace formulas, for any pair of matching global functions f_E and f in (50). There were three serious problems, and what we might call a curiosity, to be resolved along the way, none of which arose in the comparison [103, §16] for quaternion groups.

The first problem is obvious. One needs to know that for v and $f_{E,v}$, the smooth function f_v on the open subset $G_{v,\mathrm{reg}}$ of regular elements in G_v can be chosen to have an extension (a priori unique) to a test function in the space $C_c^\infty(G_v, \xi_v)$. Langlands solved the problem in §6 and §8 of [149]. He called the set of regular orbital integrals of any function $f_v \in C_c^\infty(G_v, \xi_v)$ a *Harish-Chandra family*, and the set of functions of points δ_v obtained as above (with the vanishing condition) from a function $f_{E,v}$ in $C_c^\infty(G_{E,v}, \xi_{E,v})$ a *Shintani family*. In Lemma 6.2 of [149], he proved that any Shintani family is a Harish-Chandra family, and conversely, that any Harish-Chandra family with the appropriate vanishing condition is a Shintani family. His solution required a careful comparison of the singularities of the two kinds of functions at points δ_v and γ_v near the boundary of their common domain, while keeping track of the invariant measures used to define the orbital integrals. The case that E splits at v was treated separately in §8 of [149]. It is much simpler. For in this case, there is a canonical choice for the function f_v, as a convolution of ℓ different functions in $C_c^\infty(G_v, \xi_v)$.

The second problem is what later became known as the fundamental lemma. It amounts to a more explicit version for spherical functions of the local transfer mapping we have just discussed. In its most basic form it applies to the characteristic function $\mathbf{1}_v$ of $G(\mathcal{O}_v)$ in $G(F_v)$ and the characteristic function $\mathbf{1}_{E,v}$ of $G_E(\mathcal{O}_v)$ in $G_E(F_v)$, at any unramified valuation v for E/F. The assertion is that $\mathbf{1}_v$ represents the image of $\mathbf{1}_{E,v}$ under the transfer mapping $f_{E,v} \to f_v$, or in other words, that the orbital integrals of $\mathbf{1}_{E,v}$ and $\mathbf{1}_v$ correspond as above. This can be regarded as the remaining step in the proof of the global correspondence (50). Namely, if f_E lies in $C_c^\infty(G_E(\mathbb{A}_F), \xi_E)$, the function f is itself globally smooth, in the sense that it lies in the space $C_c^\infty(G(\mathbb{A}_F), \xi)$. The fundamental lemma here was established by Langlands in §4 of [149] for general spherical functions. As a series of relations among the vertices in certain bounded subsets of the Bruhat–Tits tree for $\mathrm{SL}(2, F_v)$, it appeared at the time to be an interesting but purely combinatorial question.

The third problem concerns the more complicated "parabolic terms" in the trace formula for G (and G_E). We recall that they vanished in the comparison of [103, §16], because the test function f was cuspidal at two places. No such restriction was permitted here for what Langlands had in mind. He was able to handle the comparison of these terms by an extended analytic argument over the final three Sections 9–11 of [149]. We shall give a short description of it, if only to give a reader the chance to look up and compare the corresponding terms, (i), (ii), (iv), (v), (vi), (vii) and (viii) from [103, pp. 516–517] and (10.9), (10.8), (10.12), (10.15), (10.30), (10.28) and (10.29) from [149, §10], in the two trace formulas.

We write the difference of the two trace formulas schematically as

$$\begin{aligned}\operatorname{tr}(R_{\mathrm{disc}}(f_E)) - \operatorname{tr}(R_{\mathrm{disc}}(f)) &= [(10.9) - (\mathrm{i})] + [(10.8) - (\mathrm{ii})] + [(10.12) - (\mathrm{iv})] \\ &+ [(10.15) - (\mathrm{v})] + [(10.30) - (\mathrm{vi})] + [(10.28) - (\mathrm{vii})] + [(10.29) - (\mathrm{viii})].\end{aligned}$$

The first two square brackets contain the elliptic terms from the two trace formulas. The differences they represent are each 0. The remaining five square brackets contain the supplementary parabolic terms. Two of them, the brackets containing (vi) and (vii), still represent invariant distributions (in f_E and f). The second of these vanishes. The first one does not, but it does represent a discrete linear combination of characters (in f_E and f). We transfer it to the left-hand side of the formula, and thereby write

$$\begin{aligned}&\operatorname{tr}(R_{\mathrm{disc}}(f_E)) - \operatorname{tr}(R_{\mathrm{disc}}(f)) - [(10.30) - (\mathrm{vi})] \\ &= [(10.12) - (\mathrm{iv})] + [(10.15) - (\mathrm{v})] + [(10.29) - (\mathrm{viii})].\end{aligned} \tag{52}$$

The right-hand side now consists entirely of parabolic noninvariant distributions in f_E and f. It is where the real analytic work of Langlands began. He first wrote the sum of the noninvariant **geometric** terms, those in the brackets containing (iv) and (v), as a sum over the discrete group F^* of values of a certain function in $\mathbb{A}_F^*$. He then showed that this function (apart from some manageable error terms we shall ignore here), although not infinitely differentiable at the archimedean places, was smooth enough to apply the (multiplicative) Poisson summation formula for the discrete group F^* of $\mathbb{A}_F^*$. Langlands then studied the formula thus obtained from (52), roughly speaking, as an identity in the local spectral parameters of the matching test functions $f_E \to f$. The left-hand side is discrete in this sense, but thanks to Poisson summation, the terms on the right-hand side all become at least partly continuous. Langlands' sophisticated analytic argument allowed him to deduce at length that each side of (52) vanishes. The formula

$$\operatorname{tr}(R_{\mathrm{disc}}(f_E)) + [(\mathrm{vi}) - (10.30)] = \operatorname{tr}(R_{\mathrm{disc}}(f)) \tag{53}$$

thus followed.

The curiosity we mentioned is the anomaly $[(\mathrm{vi}) - (10.30)]$ in the last formula. It is interesting because it comes from simple but essential terms in the two trace formulas.The contribution in each case comes from the “discrete part”of the (spectral side) of the trace formula. It represents a term that comes from Eisenstein series, and is not part of the automorphic discrete spectrum. (These terms are not be confused with the one-dimensional automorphic representations that represent the noncuspidal part of the automorphic discrete spectrum.) The question is one of interpretation. What role do they play in the final formula (53)? We shall give the answer presently.

We have completed our discussion of the base change comparison of the two trace formulas. Our notation differs somewhat from [149].[5]. Langlands observes that the extra term $[(\mathrm{vi})-(10.30)]$ in (53), which vanishes unless the prime ℓ equals 2, is a twisted character in f_E. The left-hand side of (53) is therefore itself a twisted character in f_E, which Langlands wrote as

$$\operatorname{tr}(R(\phi)R(\sigma)) = \operatorname{tr}(R_{\mathrm{disc}}(f_E)) + [(\mathrm{vi})-(10.30)].$$

He also wrote

$$\operatorname{tr}(r(f)) = \operatorname{tr}(R_{\mathrm{disc}}(f))$$

for the character in f on the right-hand side of (53). The identity (53) is therefore

$$\operatorname{tr}(R(\phi)R(\sigma)) = \operatorname{tr}(r(f)), \tag{54}$$

in his notation. It was stated by Langlands as Theorem 11.1 of [149]. His proof, which includes the arguments from Chapters 9 and 10 of [149] that we have tried to summarize, was completed finally on p. 211 of Chapter 11 (at the end of the first paragraph there). The rest of Langlands' last chapter, Propositions 11.4–11.5 and Lemmas 11.6–11.7, was then given to deriving the properties of the base change lifting

$$\pi = \bigotimes_v \pi_v \to \pi_E = \bigotimes_v \pi_{E,v}, \qquad \pi \in \Pi(G), \tag{55}$$

of automorphic representations, stated in Chapter 2 of [149], and dual to the transfer $f_{E,v} \to f_v$ of functions in (50). The means for this were of course provided by the comparison identity (53) or (54).

Suppose that $\pi_v \in \Pi(G_v)$, and that $\pi_{E,v} \in \Pi(G^0_{E,v})$ is σ_E-stable, in the sense that the representation

$$(\sigma_E \pi_{E,v})(x_v) = \pi_{E,v}(\sigma_E^{-1}(x_v)), \quad x_v \in G^0_{E,v},$$

is equivalent to $\pi_{E,v}$. The theory of Whittaker models then gives a canonical intertwining operator $I_{E,v}$, with

$$(\sigma_E \pi_{E,v})(x_v) = I_{E,v} \circ \pi_{E,v}(x_v) \circ I_{E,v}^{-1}.$$

We can say that $\pi_{E,v}$ is a *local lifting* of π_v if its twisted character $\Theta(\pi_{E,v} \times I_{E,v}, \cdot)$ matches the character $\Theta(\pi_v, \cdot)$ of π_v under the norm mapping. In other words,

---

[5] Langlands wrote ϕ for a test function on $G(\mathbb{A}_E) = G^0_E(\mathbb{A}_F)$ rather than our test function f_E on $G_E(\mathbb{A}_F) = G^0_E(\mathbb{A}_F)\sigma_E$. He then took $R(\phi)$ to be the operator $R_{\mathrm{disc}}(f_E)$ on

$$\mathrm{L}^2(G^0_E(F)\backslash G^+_E(\mathbb{A}_F)) = \mathrm{L}^2_{\mathrm{disc}}(G^0_E(F)\backslash G^0_E(\mathbb{A}_F) \rtimes \Gamma_{E/F})$$

rather than the the operator $R_{\mathrm{disc}}(f_E)$ on

$$\mathrm{L}^2(G^0_E(F)\backslash G^0_E(\mathbb{A}_F)) = \mathrm{L}^2_{\mathrm{disc}}(G^0_E(F)\backslash G^0_E(\mathbb{A}_F)\sigma_E).$$

This accounts for the integer ℓ he inserted in the definition of p. 199 of [149].

$$\Theta((\pi_{E,v} \times I_{E,v}), x_v \times \sigma_E) = \Theta(\pi_v, N_v x_v), \quad N_v = N_{E_v/F_v}, \tag{56}$$

whenever $N_v x_v$ is regular. (See [149, p. 11 and Definition 6.1] as well as [27, p. 51].) The automorphic representation π_E in (55) would then be a *global lifting* of π if for each v, $\pi_{E,v}$ is a local lifting of π_v. The word "lifting" here is used interchangeably with the phrases "base change lifting" or "base change transfer", or even just "base change". The word *transfer* best describes the phenomenon in general settings.

We should remind ourselves that the base change of automorphic representations is what corresponds to the restriction of Galois (or Weil group) representations. That is, if π is the functorial image $r \to \pi$ of a two-dimensional representation r of W_F, then π_E would be the functorial image $r_E \to \pi_E$ of the restriction r_E of r to the subgroup W_E of index 2 in W_F. This follows from the functorial interpretation of base change given at the beginning of this discussion. It can also be proved directly from a comparison of global L-functions, the strong multiplicity one theorem for $\mathrm{GL}(2)$, and the local Langlands classification of representations in terms of L-functions and ε-factors.

Langlands showed that every π_v has a unique local lifting $\pi_{E,v}$, and that every π has a unique global lifting π_E. Therefore (55) is a well defined mapping of automorphic representations from $\Pi(G)$ to $\Pi(G_E^0)$, whose restriction is easily seen to map the subset $\Pi_{\mathrm{temp}}(G)$ to $\Pi_{\mathrm{temp}}(G_E^0)$. (Recall that we defined $\Pi_{\mathrm{temp}}(G)$ somewhat informally as the subset of globally tempered automorphic representations in $\Pi(G)$. It is the set of representations that occur in the spectral decomposition of $L^2(G(F)\setminus G(\mathbb{A}))$, which by the theory of Eisenstein series, is the more concrete set of irreducible representations

$$\{\mathscr{I}_P^G(\sigma) : \sigma \in \Pi_2(M), \quad P = MN\},$$

induced parabolically from unitary, automorphic representations in the (relative) automorphic discrete spectrum of M.) This represents a proof of functoriality for cyclic (prime order) base change. It is the fundamental assertion from among the various local and global properties of base change that Langlands derives from the comparison identity (54), and for which the reader can consult from the two lists in [149, §2].

The most important of the remaining properties is the characterization of the image of the mapping. This is the analogue for base change of the problem solved for quaternion groups by Jacquet and Langlands by the comparison of trace formulas in [103, §16]. To describe it, we assume that π_E belongs to the subset $\Pi_{\mathrm{temp}}(G_E^0)$ of $\Pi(G_E^0)$. With this assumption, Langlands proves that π_E is a base change lift if and only if it is σ_E-stable, in which case its preimage is a finite subset of $\Pi_{\mathrm{temp}}(G)$. Moreover, if π_E belongs to the subset

$$\Pi_1(G_E^0) = \Pi_{\mathrm{cusp},2}(G_E^0) = \Pi_{\mathrm{cusp}}(G_E^0) \cap \Pi_2(G_E^0)$$

of cuspidal unitary representations, its preimage is the set

$$\{\pi \otimes \omega_{E/F}^k : 1 \leq k \leq \ell\}$$

of order ℓ, where π lies in the associated subset

$$\Pi_1(G) = \Pi_{\text{cusp},2}(G) = \Pi_{\text{cusp}}(G) \cap \Pi_2(G)$$

of cuspidal unitary representations of G, and the $\omega_{E/F}$ is the class field character of order ℓ associated to E/F. Conversely, suppose that π is a representation in the subset $\Pi_1(G)$ of the domain $\Pi_{\text{temp}}(G)$. Then its base change image π_E lies in the subset $\Pi_1(G_E^0)$ if and only if *either* $\ell \neq 2$ *or* $\ell = 2$ but $\pi \otimes \omega_{E/F}$ is not equivalent to π. In this case, π becomes one of the ℓ representations in the fibre of π_E.

The remaining case that

$$\pi \cong \pi \otimes \omega_{E/F}, \qquad \pi \in \Pi_1(G), \ell = 2,$$

is of special interest. It is *dihedral*, in the sense that π is the image $r \to \pi$ under functoriality of an irreducible induced representation r of the Weil group attached to the quadratic extension E/F. In other words, r is induced from a character χ_E on the subgroup C_E of index 2 in $W_{E/F}$, with $\sigma_E \chi_E \neq \chi_E$. Then $L(s,r)$ equals the entire, abelian L-function $L(s,\chi_E)$. In fact it is easy to check that, conversely, any dihedral representation r satisfies all the necessary conditions of Theorem 12.2 of [103], and therefore corresponds to a cuspidal automorphic representation $\pi \in \Pi_1(G)$. The character χ_E of C_E can of course be interpreted as an automorphic representation of $\text{GL}(1)_E$. With this interpretation, the mapping $\chi_E \to \pi$ is sometimes called *automorphic induction*.

The dihedral representations π are the source of the extra term on the left side of (53). On the one hand, π contributes less than expected to the discrete spectrum in the trace formula of G, since the fibre of π_E consists of π alone (rather than a set of order 2). But on the other hand, its base change lifting

$$\pi_E = \text{Ind}_{B_E^0}^{G_E^0} \begin{pmatrix} \chi_E & 0 \\ 0 & \sigma_E \chi_E \end{pmatrix}$$

is an induced representation, and does not contribute at all to the cuspidal discrete spectrum of G_E^0 in the twisted trace formula for G_E^0. The extra term in (53) measures this discrepancy. It comes entirely from the explicit Eisenstein terms (vi) and (10.30) in the "discrete parts" of the two trace formulas. In the case of $G = \text{GL}(2)$, or even that of $\text{GL}(n)$ [27, §3.6], one can calculate the discrepancy independently of the two trace formulas. In more complex situations, however, one must undertake a full, direct computation of "discrete parts" of the relevant trace formulas. (See [23, §4].)

To summarize, base change represents a new case of functoriality, with the pair $(G_E^0, G) = (\text{Res}_{E/F}\,\text{GL}(2), \text{GL}(2))$ in the role of a general pair (G, G') from the statements in §4 and §5. But it comes with more information than would a general case. This includes the characterization of its image and the other properties we have just described, by virtue of its origin in a comparison of trace formulas. It was these supplementary properties in particular that led Langlands to the spectacular applications to Artin's conjecture and functoriality for certain two-dimensional representations ρ of W_F. They were established in §3 in [149].

There are four classes of irreducible representations

$$r\colon W_{K/F} \to \mathrm{GL}(2,\mathbb{C}),$$

for the global Weil group

$$1 \to C_K \to W_{K/F} \to \Gamma_{K/F} \to 1$$

attached to a finite Galois extension K/F of number fields. Their images are dihedral, tetrahedral, octahedral and icosahedral (in the sense of geometric symmetry described below). The ultimate goal would be to show for each r that $L(s,r)$ equals $L(s,\pi)$, for a cuspidal automorphic representation $\pi \in \Pi_1(G) = \Pi_{\mathrm{cusp},2}(G)$.

If the image of r is dihedral, we can arrange that K/F is a quadratic extension, and that $r(C_K)$ is not central in $\widehat{G} = \mathrm{GL}(2,\mathbb{C})$. It follows that r is the irreducible representation induced from a character χ_K on C_K, and is dihedral in the sense above. There is consequently an automorphic representation $\pi \in \Pi_1(G)$ with

$$L(s,\chi_K) = L(s,r) = L(s,\pi),$$

as desired.

In the remaining cases, the image $r(C_K)$ consists of scalar matrices, since it is easy to see that r would otherwise be dihedral. The composition

$$W_{K/F} \to \mathrm{GL}(2,\mathbb{C}) \to \mathrm{PGL}(2,\mathbb{C}) \xrightarrow{\sim} \mathrm{SO}(3,\mathbb{C})$$

is then a proper orthogonal representation of the Galois group $\Gamma_{K/F}$, which by contracting K if necessary, we can assume is faithful. As a finite subgroup of $\mathrm{SO}(3,\mathbb{C})$, $\Gamma_{K/F}$ becomes the group of rigid proper motions of a tetrahedron, octahedron or icosahedron, or in algebraic terms, the group A_4, S_4 or A_5. Nothing was known about the Artin conjecture in any of these cases before Langlands' base change. He was able to use base change to establish functoriality for any tetrahedral ρ. This was the first progress in Artin's conjecture in fifty years.

Langlands' argument, which is the content of Section 3 of [149], is both striking and suggestive. It is also quite compressed. We shall review it for the tetrahedral case, in which he obtains complete results. We will then say a few words about Tunnell's extension of Langlands' argument that also leads to complete results in the octahedral case. For this discussion, we shall follow the standard practice of writing $\pi = \pi(r)$ for the functorial image in $\Pi_1(G)$, if it exists, of an irreducible representation r of $W_{K/F}$. We shall also write r_E for the restriction of r to a subgroup $W_{K/E}$ of $W_{K/F}$, as we have been doing, and $\pi_E = BC_{E/F}(\pi)$ for the base change image in $\Pi_{\mathrm{temp}}(G_E^0)$ of a representation $\pi \in \Pi_{\mathrm{temp}}(G)$.

Suppose that r is tetrahedral. Since $r(C_K)$ is contained in the group of scalar matrices in $\mathrm{GL}(2,\mathbb{C})$, given that r is not dihedral, r maps the corresponding quotient $\Gamma_{K/F}$ of $W_{K/F}$ into the group $\mathrm{SO}(3,\mathbb{C}) \cong \mathrm{PGL}(2,\mathbb{C})$. Its image is equal to the tetrahedral group A_4, a group of order 12 with normal subgroup

$$V_4 = \{1, (12)(34), (13)(24), (14)(23)\} \cong (\mathbb{Z}/2\mathbb{Z}) \times (\mathbb{Z}/2\mathbb{Z})$$

of index 3. We can identify this subgroup in turn with the bijective image in $PGL(2,\mathbb{C})$ of the set

$$\left\{ \begin{pmatrix} 1 & 0 \\ 0 & 1 \end{pmatrix}, \begin{pmatrix} 1 & 0 \\ 0 & -1 \end{pmatrix}, \begin{pmatrix} 0 & 1 \\ 1 & 0 \end{pmatrix}, \begin{pmatrix} 0 & 1 \\ -1 & 0 \end{pmatrix} \right\}.$$

Let E be the Galois extension of F of degree 3 in K/F fixed by the subgroup V_4 of A_4. One sees easily that the restriction r_E of r to $W_{K/E}$ is dihedral. Indeed, the image of r_E is a semidirect product

$$\left\{ \begin{pmatrix} z & 0 \\ 0 & z\varepsilon \end{pmatrix} \right\} \rtimes \left\langle \begin{pmatrix} 0 & 1 \\ 1 & 0 \end{pmatrix} \right\rangle,$$

where $\begin{pmatrix} z & 0 \\ 0 & z \end{pmatrix}$ ranges over the nontrivial image of C_K and ε ranges over the image of the quadratic character attached to the subgroup $\left\{ \begin{pmatrix} 1 & 0 \\ 0 & \pm 1 \end{pmatrix} \right\}$ of $\Gamma_{E/F}$. It therefore has a functorial image $\pi_E = \pi(r_E)$ in $\Pi_1(G_E^0)$. It is also easy to see that $\sigma_E \pi_E$ is isomorphic to π_E, by inspection of the action of σ_E on the normal subgroup $\Gamma_{K/E}$ of $\Gamma_{K/F}$. Therefore $\pi_E = BC_{E/F}(\pi)$ is the base change lift of a cuspidal automorphic representation $\pi \in \Pi_1(G)$ of GL(2) over F. This last representation is uniquely determined only as an element in the set

$$\{\pi \otimes \omega_{E/F}^k : 1 \leq k \leq 3\}. \tag{57}$$

Now the determinant ω_r of r and the central character ω_π of π are both characters on C_F, which pull back under the norm mapping to the same character on C_E. They therefore differ by a uniquely determined power of the class field character $\omega_{E/F}$. Replacing π by its product with this power of $\omega_{E/F}$, which is to say the unique element in (57), we can assume that $\omega_r = \omega_\pi$, and hence that π is uniquely determined by r. We would expect that π equals the functorial image $\pi(r)$ of r, but perhaps surprisingly at first glance, we do not yet have enough information to prove it. In pointing this out, Langlands noted that the properties of global base change we have described above establish that if r *does* have a functorial image, it must necessarily be equal to π. (See [149, p. 25], where Langlands writes $\pi_{ps}(r) = \pi_{pseudo}(r)$ for π.)

What is missing? By strong multiplicity 1 for GL(2) [189], the representation $\pi \in \Pi_1(G)$ is uniquely determined by the family

$$\{c_v(\pi) = c(\pi_v) : v \notin S\}$$

of semisimple conjugacy classes in $\widehat{G} = GL(2,\mathbb{C})$, for a finite set $S \supset S_\infty$ of valuations of F. We can of course choose S so that r as well as π is unramified at any $v \notin S$. We then also have the family

$$\{r_v(\Phi) = r(\Phi_v) : v \notin S\},$$

where $\Phi = \Phi_v$ is the Frobenius class in $W_{K/F}$. Then π equals the functorial image $\pi(r)$ of r if and only if the two families are equal. We know that $\pi_E = \pi(r_E)$, and hence that

$$c(\pi_{E,w}) = r_E(\Phi_w)$$

for any w outside the set S_E of valuations of E over S. Moreover, if w lies above v, it follows from the definition of base change that $r_E(\Phi_w) = r(\Phi_v)^{n(w)}$ and $c(\pi_{E,w}) = c(\pi_v)^{n(w)}$, where $n(w)$ equals the degree $[E_w : F_v]$. Therefore

$$c(\pi_v)^{n(w)} = r(\Phi_v)^{n(w)}.$$

There are two possibilities for v. If v splits completely in E, $n(w) = 1$, and

$$c(\pi_v) = r(\Phi_v),$$

as required. Otherwise v is inert, in which case $n(w) = 3$, and we have only the relation

$$r(\Phi_v)^3 = c(\pi_v)^3.$$

Therefore, if $r(\Phi_v) = \begin{pmatrix} a_v & 0 \\ 0 & b_v \end{pmatrix}$, for numbers $a_v, b_v \in \mathbb{C}^*$, then $c(\pi_v)$ is conjugate to $\begin{pmatrix} \xi_1 a_v & 0 \\ 0 & \xi_2 b_v \end{pmatrix}$ with $\xi_1^3 = \xi_2^3 = 1$. But the central character ω_π of π equals the determinant of $c(\pi_v)$, which is therefore equal to the determinant ω_r of r. It follows that

$$c(\pi_v) = \begin{pmatrix} \xi a_v & 0 \\ 0 & \xi^2 b_v \end{pmatrix}, \quad v \notin S, \tag{58}$$

for a complex number $\xi = \xi_v$ with $\xi^3 = 1$.

This was as far as the purely base change argument went. It still remained to be shown that $\xi = 1$. The means to do so came from two new cases of functoriality established shortly before base change, both related to the diagram

$$\begin{array}{ccc} \mathrm{PGL}(2,\mathbb{C}) & = & \mathrm{SO}(3,\mathbb{C}) \\ \uparrow & & \downarrow \\ \mathrm{GL}(2,\mathbb{C}) & \longrightarrow & \mathrm{SL}(3,\mathbb{C}) \\ & \overset{\phi}{\searrow} & \downarrow \\ & & \mathrm{GL}(3,\mathbb{C}), \end{array}$$

for the dual groups $\widehat{G}_1 = \mathrm{PGL}(2,\mathbb{C})$, $\widehat{H}_1 = \mathrm{SL}(3,\mathbb{C})$ and $\widehat{H} = \mathrm{GL}(3,\mathbb{C})$ of $G_1 = \mathrm{SL}(2)$, $H_1 = \mathrm{PGL}(3)$ and $H = \mathrm{GL}(3)$ respectively, and for ϕ the adjoint representation of $\mathrm{GL}(2,\mathbb{C})$. The first was due to Jacquet, Piatetskii-Shapiro and Shalika [104]. It implied functoriality for the 3-dimensional Galois representation $\sigma = \phi \circ r$. In other words there is cuspidal automorphic representation $\pi^1 = \pi(\sigma)$ of $\mathrm{GL}(3)$

such that

$$c(\pi_v^1) = \sigma(\Phi_v) = \phi(r(\Phi_v))$$

for almost all v. The second was due to Gelbart and Jacquet [77]. It implied functoriality for ϕ, and the cuspidal automorphic representation $\pi = \pi_{\mathrm{ps}}(r)$ of GL(2) we described above. Namely, there is a cuspidal automorphic representation π^2 of GL(3) such that

$$c(\pi_v^2) = \phi(c(\pi_v))$$

for almost all v. Notice that these two families of conjugacy classes are obtained by composing those of r and π, the ones we are trying to see are equal, by ϕ. It was therefore expected π^1 be equivalent to π^2. But even this required something else.

Langlands established the equivalence of π^1 and π^2 by using a fundamental criterion of Jacquet, Piatetskii-Shapiro and Shalika [104], based on the Rankin–Selberg L-functions

$$L(s, \pi_1 \times \pi_2), \qquad \pi_i \in \Pi_1(\mathrm{GL}(n_i)),\, i = 1,2,$$

they had recently constructed [106]. These are the automorphic L-functions for the group $\mathrm{GL}(n_1) \times \mathrm{GL}(n_2)$ attached to the tensor product representation of degree $n_1 n_2$. As in the special case of Tate [236] (with $n_1 = n_2 = 1$) and Godement–Jacquet [81] (with $n_1 = n$ and $n_2 = 1$), the authors established the analytic continuation and functional equation, and what is relevant here, a criterion for $L(s, \pi_1 \times \pi_2)$ to have a pole. It is that $L(s, \pi_1 \times \pi_2)$ is entire unless $n_1 = n_2$ and π_2 equals the contragredient $\pi_1^\vee$ of π_1, in which case $L(s, \pi_1 \times \pi_2)$ has a simple pole at $s = 1$.

Langlands applied the criterion with $n_1 = n_2 = 3$, $\pi_1 = \pi^2$ and $\pi_2 = (\pi^1)^\vee$. The condition implies that π^2 is equivalent to π^1 if and only if

$$L(s, \pi_v^1 \times (\pi_v^1)^\vee) = L(s, \pi_v^2 \times (\pi_v^1)^\vee)$$

for almost all v, by strong multiplicity 1. That is, if and only if

$$\det(1 - |\varpi_v|^s \cdot c(\pi_v^1) \otimes {}^t c(\pi_v^1)^{-1}) = \det(1 - |\varpi_v|^s \cdot c(\pi_v^2) \otimes {}^t c(\pi_v^1)^{-1})$$

for almost all v, where ϖ_v is a uniformizing parameter for v. This would of course be implied by the desired equality of the classes $c(\pi_v^1) = \phi(r(\Phi_v))$ and $c(\pi_v^2) = \phi(r(\pi_v))$, but it is in fact something that can be checked directly. Langlands did so in [149, p. 27], appealing implicitly to the fact he had noted earlier that σ is the representation of $\Gamma_{K/F} = A_4$ induced from a character θ of order 3 on the subgroup $\Gamma_{K/F} = V_4$. This established that $\phi(c(\pi_v))$ equals $\phi(r(\Phi_v))$ for almost all v, and hence that π^1 equals π^2.

The final step was to combine this with (58). The result is that the conjugacy class

$$\phi(c(\pi_v)) = \mathrm{Ad}\begin{pmatrix} \xi a_v & 0 \\ 0 & \xi^2 b_v \end{pmatrix} = \begin{pmatrix} \xi^2 a_v^2 & 0 & 0 \\ 0 & \xi^3 a_v b_v & 0 \\ 0 & 0 & \xi^4 b_v^2 \end{pmatrix} = \begin{pmatrix} \xi^2 a_v^2 & 0 & 0 \\ 0 & a_v b_v & 0 \\ 0 & 0 & \xi b_v^2 \end{pmatrix}$$

is the same as

$$\phi(r(\Phi_v)) = \mathrm{Ad}\begin{pmatrix} a_v & 0 \\ 0 & b_v \end{pmatrix} = \begin{pmatrix} a_v^2 & 0 & 0 \\ 0 & a_v b_v & 0 \\ 0 & 0 & b_v^2 \end{pmatrix} \sim \begin{pmatrix} b_v^2 & 0 & 0 \\ 0 & a_v b_v & 0 \\ 0 & 0 & a_v^2 \end{pmatrix}.$$

As Langlands then argued on p. 28 of [149], this implies that either $\xi = 1$, from which (58) then becomes

$$c(\pi_v) = \begin{pmatrix} \xi a_v & 0 \\ 0 & \xi^2 b_v \end{pmatrix} = r(\Phi_v)$$

as required, or that

$$a^2 = \xi b^2.$$

Taking square roots, one sees that this in turn implies either that $a_v = \xi^2 b_v$, which again gives $\xi = 1$ as required, or that $a_v = -\xi^2 b_v$. But if the very last condition holds, we get

$$\phi(r(\Phi_v))^3 = \begin{pmatrix} a_v & 0 \\ 0 & b_v \end{pmatrix}^3 = \begin{pmatrix} -b_v^3 & 0 \\ 0 & b_v^3 \end{pmatrix} \neq 1.$$

This would imply that the class $\phi(r(\Phi_v))$ has order 6, which is impossible, since $r(\Phi_v)$ lies in the dihedral group A_4.

It thus follows without exception that $c(\pi_v)$ equals $r(\Phi_v)$ for almost all v. The theorem of strong multiplicity 1 then yields the following result.

Theorem (Langlands [149]). *If F is a number field and r is a two-dimensional representation of the Weil group W_F of F of tetrahedral type, there is a cuspidal automorphic representation π of $\mathrm{GL}(2, \mathbb{A}_F)$ such that $\pi = \pi(r)$. In particular, the L-function*

$$L(s, r) = L(s, \pi)$$

is entire.

The argument we have reviewed is by any account a highly sophisticated proof. We note in passing that the two supplementary cases of functoriality [104], [77] needed to complete the argument did not use the trace formula in their proof. They were both established in the spirit of [103, Theorem 12.2], with an extension to GL(3) by Piatetskii-Shapiro of the converse theorem of Hecke and Weil. However, they would probably also be consequences of two later applications of the trace formula. The first would be automorphic induction from GL(1) to GL(3) of the character θ on $\Gamma_{K/E}$ above. This is a special case of general construction founded on base change for $\mathrm{GL}(n)$ [27] that we will mention at the end of the section. The second would be a special case of the general endoscopic classification in [23]. It is the correspondence between automorphic representations of the group $\mathrm{Sp}(2) = \mathrm{SL}(2)$, with dual group $\mathrm{PGL}(2, \mathbb{C}) = \mathrm{SO}(3, \mathbb{C})$, and self-dual automorphic representations of GL(3). We will discuss the general endoscopic classification in Section 10.

Langlands' theorem for tetrahedral representations was the foundation for its extension by Tunnell to octahedral representations. Langlands himself treated some octahedral representations over $F = \mathbb{Q}$ at the end of §3 of [149], using a converse result of Deligne and Serre [67]. Shortly thereafter, Tunnell used something different, a case of nonnormal base change for GL(2) [105] to extend Langlands' result for tetrahedral representations to general octahedral representations. We shall add a few remarks on Tunnell's proof [240].

Suppose that K/F is a Galois extension of number fields, and that r is a faithful, two-dimensional representation of the Galois group $\Gamma_{K/F}$ of octahedral type. In other words, the image of r in $\mathrm{PGL}(2,\mathbb{C}) \cong \mathrm{SO}(3,\mathbb{C})$ is equal to the octahedral group S_4. There is one respect in which the octahedral case is simpler. It is that the *binary octahedral group*, the two-fold "covering group" given by its preimage in $\mathrm{Spin}(3,\mathbb{C})$ is a direct product $(S_4 \times \mathbb{Z}/2\mathbb{Z})$. The binary tetrahedral group, on the other hand, is the nonsplit extension S_4 of A_4. It was for this reason that we took r to be a tetrahedral representation of the Weil group $W_{K/F}$ earlier. In the octahedral case, we are free to let r simply be a representation of the Galois group $\Gamma_{K/F}$, as we have just done.

Let F'/F be the quadratic extension in K/F that is fixed by A_4, regarded as a normal subgroup of the Galois group $\Gamma_{K/F} \cong S_4$. Then the restriction $r' = r_{F'}$ of r to $\Gamma_{K/F'} \cong A_4$ is an (irreducible) subrepresentation of tetrahedral type. We also choose a 2-Sylow subgroup Q_8 of $\mathrm{Gal}(K/F)$ (from the three conjugate subgroups of S_4 of order 8), and take L to be the fixed field of this group in K. Finally, we let E be the composite of $L \cdot F'$. We then have the master diagram of fields (Fig. 1), with corresponding Galois groups indicated by parentheses, for which I am indebted to W. Casselman. It can perhaps serve as a mnemonic for the complex arguments we have described. (See also p. 174 of [240].)

According to Langlands' tetrahedral theorem above, there is a cuspidal automorphic representation $\pi' = \pi(r')$ of $\mathrm{GL}(2,\mathbb{A}_{F'})$ that is the functorial image of r'. There are then exactly two cuspidal automorphic representations π_1 and π_2 of $\mathrm{GL}(2,\mathbb{A}_F)$ whose base change liftings $BC_{F'/F}(\pi_i)$ are equal to π'. They are related by $\pi_2 = \pi_1 \otimes \omega_{F'/F}$. On the other hand, the quaternion group Q_8 of order 8 is nilpotent, and hence monomial. It then follows from the converse theorem [103, Theorem 12.2] that there is a cuspidal automorphic representation $\pi_L = \pi(r_L)$ of $\mathrm{GL}(2,\mathbb{A}_L)$ that is the functorial image of the restriction r_L of r to $\Gamma_{K/L}$.

Combining these ingredients with the nonnormal base change theorem of [105], Tunnell was able to show that there is a *unique* index $i = 1,2$ such that the base change lift $BC_{L/F}(\pi_i)$ of π_i equals π_L [240, Lemma, p. 174]. Following the arguments described on p. 175 of [240], he then reached the following conclusion.

Theorem (Langlands–Tunnell [240]). *If F is a number field and r is a two-dimensional representation of the Galois group Γ_F of octahedral type, there is a cuspidal automorphic representation π of* $\mathrm{GL}(2,\mathbb{A}_F)$ *such that $\pi = \pi(r)$. In particular, the L-function*

$$L(s,r) = L(s,\pi)$$

is entire.

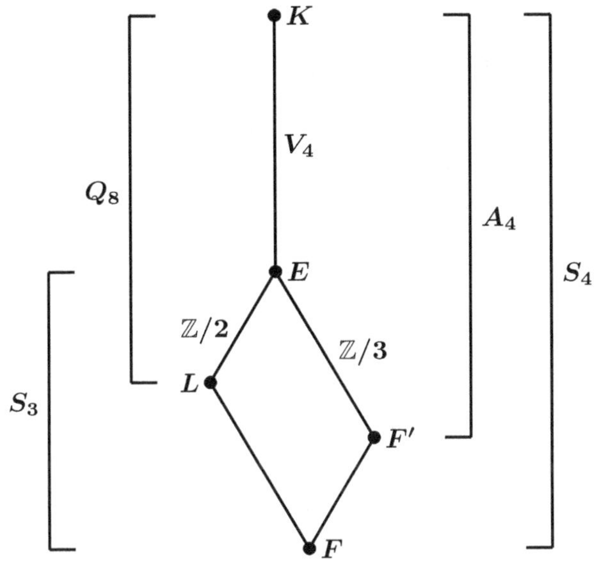

Fig. 1: Master diagram of fields

Motivated by Langlands' paper for GL(2), L. Clozel and I were later able to establish base change for the group GL(n) [27], again for a cyclic extension E/F of prime order ℓ. The general argument followed that of Langlands, but the comparison of trace formulas was more complicated. Recall that in our discussion of GL(2), there were three problems to be solved in the comparison. The first was to show that in the local transfer for functions, the image of a test function in $C_c^\infty(G_{E,v}, \xi_{E,v})$ could be chosen to lie in $C_c^\infty(G_v, \xi_v)$. Its analogue for GL(n) was established in [27, §1.3] by straightforward methods of local descent. The second problem was the explicit form of local transfer for the special case of nonarchimedean spherical functions (known later as the fundamental lemma). Its analogue for GL(n) was established in [27, §1.4]. The authors were able to rely here on a new observation [120] of Kottwitz, for the important special case of the unit functions $f_{E,v} = \mathbf{1}_{E,v}$ and $f_v = \mathbf{1}_v$. It took the form of a bijection between the summands in the two finite series that define the orbital integrals (twisted and ordinary), $\mathrm{Orb}(\gamma_v, \mathbf{1}_{E,v})$ and $\mathrm{Orb}(\delta_v, \mathbf{1}_v)$, in which the summands themselves were equal. Kottwitz was thus able to prove the required equality

$$\mathrm{Orb}(\gamma_v, \mathbf{1}_{E,v}) = \mathrm{Orb}(\delta_v, \mathbf{1}_v), \quad \gamma_v \to \delta_v$$

of the two functions without evaluating either one of them explicitly.

The relative ease with which the problems of local transfer and fundamental lemma were solved for GL(n) does not reflect how difficult they turned out to be in more general situations. However, the third problem, that of comparing the supplementary parabolic terms in the two trace formulas, does contain many of the difficulties encountered in general endoscopy [21], [179], [180]. Its solution, which took up much of Chapter 2 of [27], can be regarded as a blueprint for the situation in general. It is different from the global analytic argument used by Langlands for GL(2). The final result was a series of term by term identities (the geometric Theorem A and the spectral Theorem B, in §2.5 and §2.9 of [27]) between the constituents of the invariant trace formula for $\mathrm{GL}(n)$ (a special case of [15]) and their analogues for the twisted invariant trace formula for $\mathrm{GL}(n)_E$. Theorem B then led directly to the analogue for $\mathrm{GL}(n)$ of the spectral comparison identity (53) for GL(2).

The consequences of the base change comparison for $G = \mathrm{GL}(n)$ were derived in Chapter 3 of [27]. We shall state them for automorphic representations $\pi \in \Pi_{\mathrm{temp}}(G)$ and $\pi_E \in \Pi_{\mathrm{temp}}(G_E)$, even though they were derived in [27] only for the special case of "induced from cuspidal" automorphic representations, as opposed to automorphic representations induced from representations in the full discrete spectrum. The notion of a global lifting π_E of π is defined as for GL(2) above. Then

(i) Any π has a unique global base change lifting π_E.
(ii) A given π_E is a base change lift if and only if it is σ_E-stable, which is of course to say that the representation $\sigma_E \pi_E$ is equivalent to π_E, in which case its preimage in $\Pi_{\mathrm{temp}}(G)$ is finite.
(iii) Suppose that π_E lies in the subset $\Pi_1(G_E) = \Pi_{\mathrm{cusp},2}(G_E)$ of cuspidal representations in $\Pi_{\mathrm{temp}}(G_E)$, and that it is σ_E-stable. Then its preimage under base change is a set of order ℓ of the form

$$\{\pi \otimes \omega_{E/F}^k : 1 \leq k \leq \ell\},$$

for some $\pi \in \Pi_1(G)$.
(iv) Suppose that π_E belongs to $\Pi_1(G_E)$, but is *not* σ_E-stable. Then the induced representation

$$\Pi_E = \mathrm{Ind}_{P_E^0}^{G_E^0}(\pi_E \otimes \sigma_E \pi_E \otimes \cdots \otimes \sigma_E^{\ell-1}\pi_E),$$

which is an automorphic representation in $\Pi_{\mathrm{temp}}(\mathrm{GL}(n\ell)_E)$ by virtue of the theory of Eisenstein series, is σ_E-stable, and is a base change lifting of exactly one representation $\Pi \in \Pi_1(\mathrm{GL}(n\ell))$. The resulting map $\pi_E \to \Pi$ is an n-dimensional form of *automorphic induction*.

As an application of these results, the authors established a functorial correspondence $r \to \pi(r)$ from the irreducible n-dimensional representations r of a *nilpotent* Galois group over F to cuspidal automorphic representations $\pi = \pi(r)$ in $\Pi_1(G)$. However, since Artin's conjecture was already known for nilpotent groups, this gave no new analytic information about their L-functions.

Base change for $\mathrm{GL}(n)$ was used in the proof of R. Taylor, in partial collaboration with Clozel, M. Harris and N. Shepherd-Barron, in the proof of the Sato–Tate conjecture for many elliptic curves over any totally real field F, as we noted the end of Section 4 [238], [51], [91]. My understanding is that base change served as a substitute for functoriality in the early argument suggested by Langlands in [138].

I mention finally a recent preprint [52] of Clozel and M. S. Rajan, in which they characterize the image and fibres of solvable base change for GL(n). It will be interesting to see what applications come of it.

8 Shimura varieties

The theory of Shimura varieties has developed into a vast field, with fundamental ties to automorphic representations. I cannot do justice to the subject, or to Langlands' major contributions to it, especially in one section. I will do my best to describe some of the basic ideas and problems, with emphasis on their ties to automorphic representations. As a result, the section will perhaps be less technical, and no doubt less complete, than our earlier ones.

Shimura varieties are algebraic varieties whose complex points come from Hermitian, arithmetic, locally symmetric spaces. They are to algebraic geometry what general arithmetic locally symmetric spaces are to Riemannian geometry. The classic example is the quotient

$$\Gamma(N)\backslash X_+ = \{\gamma \in \mathrm{SL}(2,\mathbb{Z}) : \gamma \equiv 1 (\mathrm{mod}\, N)\} \backslash \{z \in \mathbb{C} : \mathrm{Im}\,(z) > 0\}$$

of the upper half plane, or more correctly, a certain disjoint union of such quotients. They can be compactified by adding a finite set of points, thereby becoming a (disjoint union of) compact Riemann surfaces. We shall give the formal definition of a Shimura variety S_K presently, noting here only that it is defined canonically over a certain number field E, known as the *reflex field* of S_K.

Langlands' interest in Shimura varieties was in their relations with automorphic forms. As was the case for number fields, Shimura varieties also come with Galois representations, and with automorphic representations to which they are supposed to be related. The Galois representations are more complicated, in the sense that they take values in a general linear group over $\mathbb{Q}_\ell$ (or rather, some finite extension L_λ of $\mathbb{Q}_\ell$), instead of over $\mathbb{C}$. (By convention, ℓ is a prime to be distinguished from p, another prime that might be engaged elsewhere.) However, the associated automorphic representations are expected to be more manageable, in the sense that their archimedean components π_∞ should typically be square integrable (as in a holomorphic modular form of weight ≥ 2), rather than an induced representation (as in a Maass form). It is for this reason that the reciprocity laws between ℓ-adic representations and automorphic forms tend to be more concrete.

Langlands' first paper on Shimura varieties (or indeed, any aspect of algebraic geometry) is the fundamental article [140] on the basic 1-dimensional Shimura varieties S_K attached to the group GL(2). The 1972 Antwerp conference at which Langlands gave the lectures was timely, coming soon after his long monograph [103] with Jacquet on the automorphic representations of GL(2). His article for these proceedings could be considered the second member of a trilogy, which also includes the monograph [149] on base change for GL(2) we have just discussed. Each of its three members is dedicated to a different but complementary side of the representation theory of GL(2).

Langlands' goal in [140] was to establish reciprocity laws between the two-dimensional λ-adic representations that arise from the étale cohomology of S_K and the automorphic representations of GL(2). This is essentially equivalent to showing that the L-functions of these λ-adic representations, defined at least at the unramified places in the same way as Artin L-functions, are equal to automorphic L-functions for GL(2). In particular, they would have analytic continuation with functional equation to entire functions of $s \in \mathbb{C}$. This in turn should then lead ultimately to explicit formulas for the Hasse–Weil zeta functions of the varieties S_K [146]. (Langlands' paper [146] is actually devoted to the compact Shimura varieties associated to various quaternion algebras, rather than the noncompact varieties S_K attached to GL(2) over $\mathbb{Q}$.) Not surprisingly, perhaps, the power to establish such things comes from the Selberg trace formula for GL(2). More precisely, the results are consequences of an intricate comparison of Selberg's formula with a completely different formula, the Lefschetz trace formula, originally for a (nonsingular, projective) algebraic variety over a finite field. Let us first say something about the general theory of Shimura varieties, after which we can return to our discussion of [140] and other papers of Langlands.

In general, a Shimura variety S_K is a quasiprojective variety with some auxiliary data, which is attached to a certain reductive group G over $\mathbb{Q}$, and which as we have noted is naturally defined over an associated number field E. The subject owes its existence to the efforts of Goro Shimura over many years. Among other things, he studied the auxiliary data to be attached to various groups G, gave a conjectural formulation of the reflex field of definition E in terms of these data, and proved the conjecture in some cases. He also studied the internal arithmetic objects of S_K (as did M. Eichler at about the same time). In some cases he was able to establish reciprocity laws between these objects and automorphic forms. His most complete results were for what are now known as Shimura curves, where

$$G = \mathrm{Res}_{F/\mathbb{Q}} G_F$$

for the multiplicative group G_F of a quaternion algebra over F (as in Sections 6 and 7 here), with the requirement that F be a totally real field that splits at exactly one archimedean place. In particular, $G = G_{\mathbb{Q}}$ could be the multiplicative group of a quaternion algebra that splits over $\mathbb{R}$. (See [230], and the references there.)

Deligne made a study of Shimura's work, on which he reported to Bourbaki in 1971 [61]. He reformulated Shimura's constructions in adelic terms. In this setting, a Shimura variety amounts to a family of complex varieties $S = \{S_K\}$ attached to a *Shimura datum* (G,X), with G being a reductive group over $\mathbb{Q}$ and X a $G(\mathbb{R})$-conjugacy class of homomorphisms

$$h \colon \mathscr{R}(\mathbb{R}) = \mathbb{C}^* \to G(\mathbb{R})$$

defined over $\mathbb{R}$, for the $\mathbb{R}$-torus

$$\mathscr{R} = \mathrm{Res}_{\mathbb{C}/\mathbb{R}}(\mathbb{G}_m), \qquad \mathbb{G}_m = \mathrm{GL}(1),$$

which satisfy the natural conditions (a), (b) and (c) on pp. 213–214 of [144]. In particular, if K_h is the centralizer of $h(\mathbb{C})$ in $G(\mathbb{R})$ for some $h \in X$, the quotient

$$G(\mathbb{R})/K_h \cong X$$

is required to have a complex Hermitian structure. The subscripts K range over open, compact, subgroups of $G(\mathbb{A}^\infty)$. For any such K, the associated variety has complex points parametrized by the adelic coset space

$$S_K(\mathbb{C}) = G(\mathbb{Q}) \backslash (X \times G(\mathbb{A}^\infty)/K) = G(\mathbb{Q}) \backslash G(\mathbb{A})/K_h K.$$

This space may or may not be compact. In any case, each of its (finitely many) connected components has a complex embedding into projective space, according to the Bailey–Borel compactification. Therefore, $S_K(\mathbb{C})$ is a complex quasiprojective manifold, and hence the set of complex points of a quasiprojective algebraic variety over $\mathbb{C}$. (See [61, §1.8], [144, §4], [176].)

The *Shimura variety* attached to the datum (G,X) is formally taken to be the inverse limit

$$S = S(G,X) = \varprojlim_K S_K.$$

It is a complex proalgebraic variety, with complex points

$$S(\mathbb{C}) = G(\mathbb{Q}) \backslash (X \times G(\mathbb{A}^\infty)) \cong G(\mathbb{Q}) \backslash G(\mathbb{A})/K_h,$$

on which the group

$$G(\mathbb{A}^\infty) = \{x \in G(\mathbb{A}) : x_\infty = x_{\mathbb{R}} = 1\}$$

acts algebraically by right translation. The quasiprojective variety attached to any $K \subset G(\mathbb{A}^\infty)$ then equals the quotient

$$S_K = S_K(G,X) = S/K.$$

It is also often called a Shimura variety, as we have already done above.

The simplest examples are given by the case that $G = T$ is a torus. For X then consists of a single point h. For an open compact subgroup U of $T(\mathbb{A}^\infty)$, the set $S_U(\mathbb{C})$ is then finite, and the corresponding Shimura variety $S_U = S_U(T,h)$ is zero-dimensional. If (G,X) is a general Shimura datum, a *special pair* for (G,X) is a pair (T,h) with $T \subset G$ and $h \in X$. Then if $U = K \cap T(\mathbb{A}^\infty)$, for an open compact subgroup $K \subset G(\mathbb{A}^\infty)$, $S_U(\mathbb{C})$ is a finite subset of $S_K(\mathbb{C})$, consisting of what are called *special points* for S_K. Similar notions apply to the proalgebraic complex varieties attached to the inverse limits over U and K.

Suppose that (G,X) is a Shimura datum. We first note that there are two simple homomorphisms

$$\mu, w\colon \mathbb{G}_m \to \mathscr{R},$$

which are basic to the general theory of Hodge structures [175, p. 214]. The first is defined over $\mathbb{C}$ by

$$\mu(z) = (z,1), \qquad z \in \mathbb{G}_m(\mathbb{C}) \cong \mathbb{C}^*.$$

The second is defined over $\mathbb{R}$ by

$$w(r) = r^{-1}, \qquad r \in \mathbb{G}_m(\mathbb{R}) \cong \mathbb{R}^*.$$

(In both cases, we want to keep in mind that

$$\mathscr{R}(\mathbb{R}) = \mathbb{C}^* \cong \{(z,\bar{z}) : z \in \mathbb{C}^*\} \subset \{(z_1,z_2) \in (C^*)^2\} \cong \mathscr{R}(\mathbb{C}),$$

and that it is in terms of these isomorphisms that the maps are defined.) Now suppose that $h \in X$. We then also have the two homomorphisms, the *cocharacter*

$$\mu_h = h \circ \mu\colon \mathbb{G}_m \to G, \quad z \to h(z,1),$$

which is defined over $\mathbb{C}$, and the *weight*

$$w_X = w_h = h \circ w\colon \mathbb{G}_m \to G, \quad r \to h(r)^{-1},$$

which is defined over $\mathbb{R}$. The cocharacter gives rise to a highest weight $\widehat{\mu}$ for $\widehat{G}$, which leads to a finite-dimensional representation $r = r_X$ of ${}^L G$. This in turn is used to form the automorphic L-functions $L(s,\pi,r)$ that should ultimately be a part of the automorphic formula for the Hasse–Weil zeta function of the Shimura varieties associated to (G,X). The weight of h depends only on the $G(\mathbb{R})$-conjugacy class X of h, since its image lies in the centre of G ([144, condition (a), p. 213]). One can therefore write w_X in place of w_h as above and call it the *weight* of (G,X). The homomorphisms μ_h and w_h are foundations for the role of Hodge structures in the moduli of Shimura varieties ([144, §4], [175, §1]).

The cocharacter μ_h of $h \in X$ also determines the reflex field $E(G,X)$ of the datum (G,X). Let C be the $G(\mathbb{C})$-conjugacy class of μ_h. Then $E(G,X) \subset \mathbb{C}$ is the field of definition of C. One can show that the intersection $C \cap G(\overline{\mathbb{Q}})$ is a $G(\overline{\mathbb{Q}})$-conjugacy class of homomorphisms $\mathbb{G}_m \to G$ over $\mathbb{Q}$, and hence that $E(G,X) \subset \overline{\mathbb{Q}}$ is also its field of definition (see [173, Proposition 4.6(c)]). In particular, $E(G,X)$ is a number field. Finally, if T is any maximal torus of G, the intersection $C \cap T(\overline{\mathbb{Q}})$ is a (finite) orbit in $T(\overline{\mathbb{Q}})$ under the Weyl group W of (G,T). It follows that $E(G,X)$ is the field of definition of this Weyl-orbit, namely the subfield of $\overline{\mathbb{Q}}$ fixed by the subgroup of elements in $\Gamma_{\mathbb{Q}} = \mathrm{Gal}(\overline{\mathbb{Q}}/\mathbb{Q})$ that stabilize the fixed finite subset $C \cap T(\overline{\mathbb{Q}})$ of $T(\overline{\mathbb{Q}})$.

We should also say something about the canonical model of $S = S(G,X)$. In general, a *model* of S over a subfield k of $\mathbb{C}$ is a scheme M over k, endowed with a right action of $G(\mathbb{A}^\infty)$ over k, and a $G(\mathbb{A}^\infty)$-equivariant isomorphism

$$S \xrightarrow{\sim} M \times_k \mathbb{C}.$$

There could be many models of S over k.

The *canonical model* is a model M over the reflex field $E(G,X)$ with special points that satisfy a certain reciprocity law based on class field theory. If (T,h) is a special pair for (G,X), the reflex field $E(T,h)$ of the associated Shimura variety is contained in the reflex field $E = E(G,X)$ of S. For any such pair, and any element $a \in G(\mathbb{A}^\infty)$, we write $[h,a]$ for the point in

$$M(\mathbb{C}) \cong S(\mathbb{C}) = G(\mathbb{Q}) \subset X \times G(\mathbb{A}^\infty)$$

attached to the product ha. We also write

$$r(T,h)\colon \mathbb{A}_E^* \to T(\mathbb{A}^\infty), \quad \mathbb{A}^\infty = \mathbb{A}_{\mathbb{Q}}^\infty,$$

for the composition of elementary maps

$$\mathbb{A}_E^* \xrightarrow{\mathrm{Res}_{E/\mathbb{Q}}(\mu_h)} T(\mathbb{A}_E) \xrightarrow{N_{E/\mathbb{Q}}} T(\mathbb{A}) \to T(\mathbb{A}^\infty),$$

obtained from the restriction of scalars functor applied to $\mu_h\colon \mathbb{G}_m \to T$, the norm map from $T(\mathbb{A}_E)$ to $T(\mathbb{A})$, and the projection $\mathbb{A} \to \mathbb{A}^\infty$ onto the finite adeles of $\mathbb{Q}$. The model M is then a *canonical model* for S if for every (T,h), and any $a \in G(\mathbb{A}^\infty)$, the following two conditions are met.

(i) The point $[h,a]$ in $M(\mathbb{C})$ is defined over the maximal abelian extension E^{ab} of $E = E(T,h)$.
(ii) The special points $[h,a]$ satisfy the reciprocity law

$$\theta_E(s)[h,a] = [h, r(s)a], \qquad s \in \mathbb{A}_E^*, r = r(T,h), \tag{59}$$

where $\theta_E(s)$ is the image of s in $\Gamma_E^{\mathrm{ab}} = \mathrm{Gal}(E^{\mathrm{ab}}/E)$ under the Artin map (from the Artin reciprocity law stated in Section 3).

The idea of a canonical model is remarkable. What makes it especially deep and interesting is the presence of the Artin map from abelian class field theory. The phenomenon was discovered by Shimura, who proved its existence in a number of cases. Deligne [61] established it in other cases, and as I understand it, the general case was established "somewhat independently" by Milne [173] and Borovoi [37] in the course of proving a conjecture of Langlands from [144]. (The later article [144] of Langlands will be our main topic for the next section.) As the name suggests, the canonical model M of a Shimura variety S is unique, up to a unique isomorphism. (See [175, Corollary 3.6].) For this reason, it is customary to identify M with S, and then simply to regard $S = S(G,X)$ and its quotients $S_K = S_K(G,X)$ as varieties over $E(G,X)$.

The basic example is the Shimura datum (G,X) in which $G = \mathrm{GL}(2)$ and X is the set of $G(\mathbb{R})$-conjugates of the $\mathbb{R}$-homomorphism

$$z = a + ib \to \begin{pmatrix} a & -b \\ b & a \end{pmatrix}, \qquad z \in \mathbb{C}^*,$$

from $\mathscr{R}(\mathbb{R}) \cong \mathbb{C}^*$ to $G(\mathbb{R})$. The corresponding Shimura varieties S_K were the objects Langlands studied in [140]. Before that, reciprocity laws between Frobenius classes and Hecke operators on modular curves were established by Eichler and Shimura [71], [228], [230]. It was an extension (to higher weight) of these results that Deligne [58] used to reduce the Ramanujan conjecture (for holomorphic modular forms) to the last of the Weil conjectures, which he later established in 1974 [63]. The congruence relations used by Eichler and Shimura do not generalize easily beyond modular curves, whereas the trace formulas extend in principle to arbitrary Shimura varieties. However, as in the cases of the Jacquet–Langlands correspondence and of base change, any attempt to exploit the trace formula presents an entirely new set of difficulties. Ihara [97] was the first to study the comparison in some cases, apparently following a suggestion of Sato. It was at this point that Langlands began his investigations.

Not surprisingly, the problems Langlands set out to solve were in the same spirit as those from [103, §16] and [149]. In particular, he wanted to extend the reciprocity laws to ramified places p. As in the Jacquet–Langlands extension of Shimizu [227] and the Shintani extension of Saito [193], this entailed reformulating the comparison in purely adelic terms. He also wanted to lay down the results for the basic noncompact varieties S_K attached to $\mathrm{GL}(2)$. We recall that the noncompactness in [103, §16] was not a problem, since the test function f for $\mathrm{GL}(2)$ was cuspidal at *two* places, forcing the extra terms on the geometric side of the trace formula to vanish. For the Langlands extension of Shintani [231] in base change, however, the noncompactness was very much a problem, since f could not be assumed to be cuspidal at any places. Its resolution required Section 9 of [149], titled "The Primitive State of our Subject Revealed", which we discussed briefly at the end of our last section. In the case here, the problem for Langlands was to extend the Lefschetz formula to the open Shimura varieties S_K for $\mathrm{GL}(2)$, and to compare the results with the Selberg formula. This was a serious task, and the main reason for the length of [140]. We do note that the difficulties for the Selberg formula were halfway between those of [103] and [149]. For in this case, the test function f is taken to be the cuspidal at *one* place.

The λ-adic representations of $\Gamma_{\mathbb{Q}} = \mathrm{Gal}(\overline{\mathbb{Q}}/\mathbb{Q})$ are on the étale cohomology groups $H^1_{\mathrm{et}}(S_K)$ of S_K. What does this have to do with automorphic representations? We have mentioned the later paper [145] of Langlands, in which he elucidated the precise relationship between automorphic representations and automorphic forms. The latter objects are well named. They are closely related to differential forms on $S_K(\mathbb{C})$, typically with values in a locally constant sheaf $\mathscr{F}$. They can therefore be used to construct de Rham cohomology groups $H^1_c(S_K(\mathbb{C}), \mathscr{F}(\mathbb{C}))$, which in turn lead to an interpretation of these groups in terms of automorphic representations.

Langlands studied the de Rham cohomology in §2 of [140]. In §3, he discussed the implications for its structure, particularly as it relates to Hecke operators. Hecke operators act on the analytic cohomology $H^1_c(S_K(\mathbb{C}),\mathscr{F}(\mathbb{C}))$ because they act on $S_K(\mathbb{C})$ as an analytic manifold. They act on the arithmetic cohomology $H^1_{\text{et}}(S_K,\mathscr{F})$ because they act on S_K as an algebraic variety over the reflex field $E(G,h)=\mathbb{Q}$. The intertwining operators between these two actions, combined with the theorem of strong multiplicity 1 for GL(2), led formally at the end of §3 to a general bijective correspondence

$$\pi \to \sigma(\pi), \tag{60}$$

between (certain) automorphic representations π of G and (certain) two-dimensional λ-adic representations σ of $\Gamma_{\mathbb{Q}}=\mathrm{Gal}(\overline{\mathbb{Q}}/\mathbb{Q})$. The reciprocity problem was then to describe the correspondence explicitly. Langlands was able to formulate this as a precise conjecture at the end of §4. The rest of the paper [140] was devoted to the proof of two (out of three) cases of the conjecture.

We are treating [140] as the foundation of Langlands' contributions to arithmetic geometry. We shall elaborate a little further on this basic paper before turning more briefly to his subsequent articles, and to some of the new ideas they represent.

We should begin by considering the de Rham cohomology for the complex variety $S_K(\mathbb{C})$ treated in [140, §2]. Langlands was of course working with the Shimura variety attached to GL(2), but since the ideas have natural and interesting generalizations, we assume for the moment that S is attached to an arbitrary Shimura datum (G,X). The space $S_K(\mathbb{C})$ is typically noncompact, so one must account for the behavior of differential forms at infinity. Langlands takes the image of the cohomology of compact support $H^*_c(\cdot)$ in the full de Rham cohomology. In general, it is better to work with L^2 (de Rham)-cohomology $H^*_{(2)}(\cdot)$, as has been the custom since the introduction of intersection cohomology, with its role in Zucker's conjecture. We write $\mathbb{A}_{\text{fin}}=\mathbb{A}^\infty$, as is convenient, and take the open compact subgroup $K\subset G(\mathbb{A}_{\text{fin}})$ to be small enough so that $S_K(\mathbb{C})$ is nonsingular.

We write $(\xi,V)=(\xi,V_\xi)$ for a fixed irreducible, finite-dimensional rational representation of G, following notation from Langlands' later article [144, §4] (rather than his notation (μ,L) from [140, §2]). Then

$$\mathscr{F}=\mathscr{F}_\xi=V_\xi\times_{G(\mathbb{Q})}(X\times G(\mathbb{A}_{\text{fin}})/K),$$

the space of $G(\mathbb{Q})$-orbits in $V\times(X\times G(\mathbb{A}_{\text{fin}})/K)$ under the action

$$\gamma\colon v\times(x,h)\to(\xi(\gamma),v)\times(\gamma x,h),\qquad \gamma\in G(\mathbb{Q}),\, h\in G(\mathbb{A}_{\text{fin}})/K,$$

is a locally constant sheaf $\mathscr{F}_\xi(\mathbb{C})$. One is interested in the L^2-cohomology

$$H^*_{(2)}(S_K(\mathbb{C}),\mathscr{F})=\bigoplus_{d=0}^{2n}H^d_{(2)}(S_K(\mathbb{C}),\mathscr{F}),\qquad n=\dim S_K(\mathbb{C}), \tag{61}$$

of $S_K(\mathbb{C})$ with coefficients in $\mathscr{F}$. We recall that this is the cohomology of the complex of $\mathscr{F}$-valued, smooth differentiable forms ω on $S_K(\mathbb{C})$ such that both ω and $d\omega$ are square integrable.

The graded, complex vector space (61) has a spectral decomposition

$$\bigoplus_{\pi} \Big(m_2(\pi) \cdot H^*(\mathfrak{g}_{\mathbb{R}}, K_{\mathbb{R}}; \pi_{\mathbb{R}} \otimes \xi) \otimes \pi_{\text{fin}}^K \Big), \tag{62}$$

where $\pi = \pi_{\mathbb{R}} \otimes \pi_{\text{fin}}$ ranges over automorphic representations of G with archimedean and nonarchimedean components $\pi_{\mathbb{R}} = \pi_\infty$ and $\pi_{\text{fin}} = \pi^\infty$ as indicated, while $m_2(\pi)$ is the multiplicity with which π occurs in the L^2-discrete spectrum (with appropriate central character determined by ξ),[6] and π_{fin}^K is the finite-dimensional space of K-invariant vectors for π_{fin}. The informal proof of this decomposition is essentially a consequence [36, VII] of the definition [36, §1.5.1] of the remaining factor, the graded, finite-dimensional, complex vector space $H^*(\mathfrak{g}_{\mathbb{R}}, K_{\mathbb{R}}; \cdot)$ of $(\mathfrak{g}_{\mathbb{R}}, K_{\mathbb{R}})$-cohomology, in which $\mathfrak{g}_{\mathbb{R}}$ is the Lie algebra of $G(\mathbb{R})$, and $K_{\mathbb{R}}$ is the stabilizer of a chosen point in X. The formal proof for L^2-cohomology is in [34].

For $(\mathfrak{g}_{\mathbb{R}}, K_{\mathbb{R}})$-cohomology, we recall that a $(\mathfrak{g}_{\mathbb{R}}, K_{\mathbb{R}})$-module is a (semisimple, locally $K_{\mathbb{R}}$-finite) complex, $K_{\mathbb{R}}$-module M, with an action of $\mathfrak{g}_{\mathbb{R}}$ that is compatible with the adjoint action of $K_{\mathbb{R}}$ on $\mathfrak{g}_{\mathbb{R}}$. As we noted in §2, this is the same thing as a module over the real Hecke algebra $\mathscr{H}_{\mathbb{R}}$. However, we retain the Lie algebra formulation here, to emphasize its relation with differential forms. In general, the $(\mathfrak{g}_{\mathbb{R}}, K_{\mathbb{R}})$-cohomology $H^*(\mathfrak{g}_{\mathbb{R}}, K_{\mathbb{R}}; M)$ of M is the $(\mathfrak{g}_{\mathbb{R}}, K_{\mathbb{R}})$-variant of the usual Lie algebra cohomology (See [36, §I.2.2].) In the case at hand, the product $\pi_{\mathbb{R}} \otimes \xi$ in (61) stands for the $(\mathfrak{g}_{\mathbb{R}}, K_{\mathbb{R}})$-module

$$M = V(\pi_{\mathbb{R}}, K_{\mathbb{R}}) \otimes V_\xi$$

where $M = V(\pi_{\mathbb{R}}, K_{\mathbb{R}})$ stands for the space of $K_{\mathbb{R}}$-finite vectors in the space $V(\pi_{\mathbb{R}})$ on which $\pi_{\mathbb{R}}$ acts. It is easily seen to vanish unless the infinitesimal character and central character of $\pi_{\mathbb{R}}$ equal those of ξ. (See [36, Theorem I.5.3].)

Consider now the case of [140], the Shimura variety attached to the group $G = \mathrm{GL}(2)$ above. We take the representation ξ of G as

$$\xi = \xi_k \otimes (\det)^m,$$

where

$$\xi_k = \mathrm{sym}^{k-1}(\mathrm{St})$$

is the $(k-1)$-symmetric power of the standard representation of G (of dimension k), and det is the 1-dimensional determinant representation, for integers $k \in \mathbb{N}$ and m, with $0 \leq m < k$. The character of ξ at a diagonal matrix then equals

[6] If ξ is nontrivial, this definition requires further comment. In general, one takes functions ϕ on $G(\mathbb{Q}) \setminus G(\mathbb{A})$ that are $\xi(z)^{-1}$-equivariant under translation by any $z \in Z(\mathbb{R})^0$. But since ξ is not generally unitary on $Z(\mathbb{R})^\circ$, one must also scale these functions by a fixed function on $G(\mathbb{Q}) \setminus G(\mathbb{A})$ whose restriction to $Z(\mathbb{R})^\circ$ equals the character $|\xi(z)|$. We shall discuss the case of GL(2) presently, following [140, p. 379].

$$\begin{aligned}\operatorname{tr}\left(\xi\begin{pmatrix}\alpha & 0\\ 0 & \beta\end{pmatrix}\right) &= (\alpha^{k-1}+\alpha^{k-2}\beta+\cdots+\alpha\beta^{k-2}+\beta^{k-1})(\alpha\beta)^m\\ &= (\alpha^{k+m-1}\beta^m+\alpha^{k+m-2}\beta^{m+1}+\cdots+\alpha^m\beta^{k+m-1})\\ &= \frac{\alpha^n\beta^m-\alpha^m\beta^n}{\alpha-\beta}, \qquad n=k+m,\end{aligned}$$

as on p. 389 of [140]. For each such ξ, the cohomology (60) is supported in degrees 0, 1 and 2, but as usual, it is the middle degree $d=1$ that is most interesting. In this case, there is precisely one irreducible representation $\pi_{\mathbb{R}}=\pi_{\mathbb{R}}(\xi)$ that contributes to the $(\mathfrak{g}_{\mathbb{R}},K_{\mathbb{R}})$-cohomology in (61). It is characterized by the properties

(i) $\pi_{\mathbb{R}}(z)=\xi(z^{-1}I), \qquad z\in Z(\mathbb{R})^\circ$,
(ii) the representation

$$x_{\mathbb{R}}\to|\xi(\det x_{\mathbb{R}})|^{\frac{1}{2}}\pi_{\mathbb{R}}(x_{\mathbb{R}}), \qquad x_{\mathbb{R}}\in G(\mathbb{R}), \tag{63}$$

is unitary.[7]

These conditions imply that the vector space

$$H^1(\pi_{\mathbb{R}},\xi)=H^1(\mathfrak{g}_{\mathbb{R}},K_{\mathbb{R}};\pi_{\mathbb{R}}\otimes\xi)$$

has dimension 2. (See [140, pp. 388–389 and Theorem 2.10].) There is thus a decomposition

$$H^1_{(2)}(S_K(\mathbb{C}),\mathscr{F})=\bigoplus_{\{\pi:\ \pi_{\mathbb{R}}=\pi_{\mathbb{R}}(\xi)\}}\left(m_2(\pi)\left(H^1(\pi_{\mathbb{R}},\xi)\otimes\pi_{\mathrm{fin}}^K\right)\right) \tag{64}$$

of the first cohomology group. The multiplicity $m_2(\pi)$ here needs to be interpreted according to the last footnote 6. For it is understood that π acts on the space of functions h on $G(\mathbb{Q})\setminus G(\mathbb{A})$ with

$$h(zx)=\xi(z)^{-1}h(x), \qquad z\in Z(\mathbb{R})^\circ, x\in G(\mathbb{A}),$$

such that the function

$$|\xi(\det x)|^{\frac{1}{2}}h(x)$$

is square integrable on $Z(\mathbb{R})^\circ G(\mathbb{Q})\setminus G(\mathbb{A})$. The representation

$$(R(y)h)(x)=|\det y|^{\frac{1}{2}}h(xy), \qquad x,y\in G(\mathbb{A}),$$

of $G(\mathbb{A})$ on this space is then unitarily equivalent to the regular representation of $G(\mathbb{A})$ on $L^2(Z(\mathbb{R})^\circ G(\mathbb{Q})\setminus G(\mathbb{A}))$. (See [140, pp. 379].) This gives the interpretation

[7] The representation in (ii) lies in the relative discrete series of $\mathrm{GL}(2,\mathbb{R})$. Even though we mentioned these objects first early in Section 1, we have so far avoided describing Harish-Chandra's general parametrization. In the case of $G=\mathrm{GL}(2,\mathbb{R})$ here, the relative discrete series with trivial central character are parametrized by the positive integers $\{k\}$. The representations $\{\pi_{\mathbb{R}}\}$ satisfying (i) and (ii) are therefore indexed by the same pairs (k,m) that parametrize $\{\xi\}$.

of the multiplicities $m_2(\pi)$. The existence of the correspondence (60) then follows as above from the global properties of λ-adic (étale) cohomology.

Langlands' conjectural reciprocity law for S_K is an explicit description of the local properties of the correspondence $\pi \to \sigma = \sigma(\pi)$. The domain consists of the automorphic representations π that give nonzero summands in (64), while the image is the corresponding set of 2-dimensional λ-adic representations σ. For any such π, Langlands sets

$$\pi'(x) = |\det x|^{-\frac{1}{2}}\pi(x), \qquad x \in G(\mathbb{A}). \tag{65}$$

His conjecture asserts roughly that the local components π'_p of π' at primes $p \neq \ell$ are images under the local Langlands correspondence[8] of Langlands parameters attached to the local components σ_p of σ. To set this up, he fixes an embedding of $\overline{\mathbb{Q}}$ into $\overline{\mathbb{Q}}_\ell$, and then takes on a supplementary conjectural assertion that for every element s in the group

$$W_{\mathbb{Q}_p} = W_{\overline{\mathbb{Q}}_p/\mathbb{Q}_p} \subset \Gamma_{\mathbb{Q}_p} = \Gamma_{\overline{\mathbb{Q}}_p/\mathbb{Q}_p},$$

the trace of $\sigma_p(s)$ lies in the subfield $\overline{\mathbb{Q}}$ of $\overline{\mathbb{Q}}_\ell$. He then uses this in §4 (pp. 403–405) to convert the $\overline{\mathbb{Q}}_\ell$-valued homomorphism

$$\sigma_p = \sigma_p(\pi)\colon \Gamma_{\mathbb{Q}_p} \to \mathrm{GL}(2, \overline{\mathbb{Q}}_\ell)$$

to a complex-valued homomorphism

$$\phi'_p = \phi'_p(\pi)\colon W_{\mathbb{Q}_p} \times \mathrm{SU}(2,\mathbb{C}) \to \mathrm{GL}(2,\mathbb{C}).$$

The conjecture stated at the end of §4 then includes the supplementary assertion, with the resulting precise statement being that for any π and p, π'_p is the image under the local Langlands correspondence of the complex-valued homomorphism ϕ'_p thus constructed.

We have largely followed the notation of Langlands from [140]. We should add a comment on the correspondence $\pi \to \pi'$ of automorphic representations in (65). It represents a transition from automorphic data to arithmetic data. Suppose for simplicity that $\xi = 1$. Consider then the archimedean parameters

$$\phi_{\mathbb{R}}, \phi'_{\mathbb{R}}\colon W_{\mathbb{R}} \to \mathrm{GL}(2,\mathbb{C})$$

---

[8] As we have noted before, the conjectural local Langlands correspondence has its origins in the Local Functoriality conjecture. For $G = \mathrm{GL}(2)$, it asserts a bijection from conjugacy classes of homomorphisms

$$\phi_v\colon L_{F_v} \to \mathrm{GL}(2,\mathbb{C}), \tag{66}$$

known now as local Langlands parameters, and irreducible representations π_v of $\mathrm{GL}(2,\mathbb{Q}_v)$, where

$$L_{\mathbb{Q}_v} = \begin{cases} W_{\mathbb{Q}_v}, & \text{if } v \text{ is archimedean,} \\ W_{\mathbb{Q}_v} \times \mathrm{SU}(2), & \text{if } v \text{ is } p\text{-adic,} \end{cases}$$

such that π_v is tempered if and only if the image of ϕ_v is bounded. (We shall discuss the general case in Section 10.)

attached to the local components $\pi_{\mathbb{R}}$ and $\pi'_{\mathbb{R}}$ of π and π', or rather, just the restriction of these parameters to the subgroup $\mathbb{C}^*$ of index 2 in $W_{\mathbb{R}}$. We can assume that the image of the restriction of each parameter lies in the group of diagonal matrices. When composed with the standard two-dimensional representation of $GL(2,\mathbb{C})$, the parameters $\phi'_{\mathbb{R}}$ gives a two-dimensional representation of the group $\mathbb{C}^* = \mathbb{S}(\mathbb{R})$, which amounts to a real Hodge structure of weight 2. It follows that

$$\phi'_{\mathbb{R}}(z) = \begin{pmatrix} z^{-1} & 0 \\ 0 & \bar{z}^{-1} \end{pmatrix}.$$

On the other hand, the preimage $\phi_{\mathbb{R}}$ of $\phi'_{\mathbb{R}}$ should be bounded, since it is supposed to be attached to a *tempered* automorphic representation π. To see that this is so, we note that for the archimedean parameters, (65) implies the identity

$$\phi'_{\mathbb{R}}(z) = \left| \det \begin{pmatrix} z & 0 \\ 0 & \bar{z} \end{pmatrix} \right|^{-\frac{1}{2}} \phi_{\mathbb{R}}(z),$$

and hence that

$$\phi_{\mathbb{R}}(z) = |z\bar{z}|^{\frac{1}{2}} \phi'_{\mathbb{R}}(z) = \begin{pmatrix} (z/\bar{z})^{-\frac{1}{2}} & 0 \\ 0 & (\bar{z}/z)^{-\frac{1}{2}} \end{pmatrix}.$$

The remaining Sections 5–7 of [140] were devoted to a new comparison of trace formulas, culminating in the proof of a significant part of the conjecture. Having spent our two last sections on the two comparisons from [103] and [149], we shall not discuss the details of the remaining comparison here, even though the Lefschetz trace formula from arithmetic geometry is quite different. Section 5 of [140] consists of some calculations on the terms in the Selberg trace formula suitable for the new comparison. Section 6 is a general description of the Selberg trace formula for GL(2), along the lines of our own discussion in the last two sections. Section 7 contains the comparison of the Selberg and Lefschetz trace formula. Langlands then uses this to prove his conjectures on the local parameters ϕ'_p in two cases. The first is that the two-dimensional representation ϕ'_p is reducible. It includes the case of good reduction, where ϕ'_p is a direct sum of two *unramified* quasicharacters on $W_{\mathbb{Q}_p}$, to which most if not all earlier work on the subject had been confined. The second is of "multiplicative reduction", where the homomorphism ϕ'_p is *special*, in the sense that it is nontrivial on the subgroup SU(2) of $L_{\mathbb{Q}_p}$. The remaining case is of "additive reduction", in which ϕ'_p is an irreducible two-dimensional representation of the subgroup $W_{\mathbb{Q}_p}$ of $L_{\mathbb{Q}_p}$, and the corresponding representation π'_p of $GL(2,\mathbb{Q}_p)$ is *supercuspidal*. Langlands did not study this case. However, it was resolved soon afterwards, with an extension of the methods of Langlands by Carayol [43]. (See [152].)

We have emphasized the Langlands Antwerp paper [140] on GL(2) over some of his later contributions to Shimura varieties for a couple of reasons. One we have already mentioned is that it joins the volumes [103] and [149] discussed in the past two sections as the third member of his GL(2)-trilogy. These works were designed to illustrate Langlands' revolutionary ideas on functoriality in the simplest of cases.

They are interrelated, and the influence they each have individually is amplified when they are taken together. Another reason is that [140] is essentially complete (given the later addition of Carayol for supercuspidal representations of GL(2)). With its focus on the noncompact variety $S_K(\mathbb{C})$ and on the reciprocity laws at places of bad reduction, it has served as a model for subsequent work on Shimura varieties, which continues to this day.

Langlands' Corvallis article [144] on motives will be our topic for the next section. Of his other articles, the most influential might be his remarkable conjectural description [142] of the set of points on a general Shimura variety modulo a prime of good reduction. It has been a foundation for a great deal of the work in in the subject since then. Other papers include the overview [143] of the question of the Hasse–Weil zeta function, and a more technical article [146] that answers the question for some simple Shimura varieties $S_K(G,X)$ for groups G related to GL(2). There is also a short article [148] on the general question of the reduction at a prime of bad reduction, the longer article [85] (with Harder and Rapoport) on the Tate conjecture for a Shimura variety attached to a Hilbert–Blumenthal surface (with G equal to $\mathrm{Res}_{F/\mathbb{Q}}\,\mathrm{GL}(2)$, for a real quadratic field F), and the important paper [164] with Rapoport that includes a motivic refinement of the conjecture from [142]. I have not studied these last three papers, interesting as they are, and will not have much further comment on them. A final paper [152] on Shimura varieties contains some later comments of Langlands, ostensibly on the problem from [143] of calculating the Hasse–Weil zeta function, but with observations on the other papers as well. It is very informative.

What is particularly far reaching in Langlands' later papers on Shimura varieties is the emergence of two fundamental phenomena that were also beginning to govern his work in the basic theory of automorphic forms [127], [150]. His discovery that they would have a parallel, central role in the theory of Shimura varieties seems to have been completely unexpected, perhaps because they were not critical in the Shimura varieties of small dimension that has been studied up until then. One is the question of the fundamental lemma, which we have already seen in the context of cyclic base change. It arises in the local geometric terms in the Selberg and Lefschetz trace formulas at the unramified place p at which one is trying to establish the reciprocity law. The other is a broader phenomenon, which includes the fundamental lemma, and is now known as *endoscopy* rather than Langlands' original term *L-indistinguishability*.[9] This affects most of the terms in the two trace formulas one is trying to compare. For the regular elliptic orbital integrals on the geometric side of the Selberg formula, it is a reflection of the fact that two elements γ_1 and γ_2 in $G(\mathbb{Q})$ over whose $G(\mathbb{A})$-conjugacy classes one would like to integrate a test function f, might be conjugate over $G(\mathbb{C})$ but not over $G(\mathbb{Q})$. The theory of endoscopy will be the topic of Section 10.

[9] "*L*-indistinguishability" is a better description of what is going on. However, the fact that it has more than twice the number of syllables than does "endoscopy", together with the increasing demands being placed on mathematicians' time, may have forced the change!

Before beginning any comparison, one must first understand the explicit form taken by the Grothendieck–Lefschetz trace formula when applied to a Shimura variety S_K. Recall that the original Lefschetz formula [167] is an identity

$$\sum_x i(\phi,x) = \sum_{k=0}^{n} (-1)^k \operatorname{tr}(H^k(\phi)) \tag{67}$$

attached to any suitable mapping ϕ, from say a compact manifold M of dimension n to itself. The (spectral) right-hand side is an alternating sum of traces of the operators

$$H^k(\phi,\mathbb{Q})\colon H^k(M,\mathbb{Q}) \to H^k(M,\mathbb{Q}), \quad 0 \leq k \leq n, \tag{68}$$

on ordinary (Betti) cohomology attached to ϕ. The (geometric) left-hand side is a sum over the fixed points x of ϕ in M of certain indices $i(\phi,x)$.

Motivated by this classical formula and the Weil conjectures [248], Grothendieck introduced his version [98] for an arithmetic variety. At its simplest, it applies to a nonsingular projective variety X over a finite field $\mathbb{F}$ of characteristic p. It is the analogue of (67) for X, with ϕ replaced by some power Φ of the Frobenius endomorphism, and $H^k(M,\mathbb{Q})$ replaced on the spectral side by $H^k(M,\mathbb{Q}_\ell)$, the ℓ-adic (étale) cohomology of X at a prime $\ell \neq p$. The geometric side becomes a sum over the finite set

$$X(\mathbb{F}'), \qquad \mathbb{F}' = (\overline{\mathbb{F}})^{\Phi}$$

of points in $X(\overline{\mathbb{F}})$ fixed by Φ.

Suppose for simplicity that the Shimura reflex field E of $S_K = S_K(G,X)$ equals $\mathbb{Q}$, and as usual, that K is small enough that $S_K(\mathbb{C})$ is nonsingular. We fix a number field L that is sufficiently large in a sense that depends in K, together with an embedding $L \subset \mathbb{C}$. The finite-dimensional complex representation ξ will then be assumed to be defined over L.

To proceed, it is necessary to have an explicit description of the set of points $S_K(\mathbb{F}_p)$ at an unramified prime p. More precisely, for any p such that

$$K = K_p K^p, \qquad K^p \subset G(\mathbb{A}^p_{\text{fin}}),$$

for an unramified maximal compact subgroup $K_p \subset G(\mathbb{Q}_p)$, one wants a suitable $\mathbb{Z}_p$-scheme structure on S_K, and a description in terms of G of the set $S_K(\overline{\mathbb{F}}_p)$ of points on S_K over the algebraic closure $\overline{\mathbb{F}}_p$, equipped with an action of the (geometric) Frobenius endomorphism Frob_p. The purpose of Langlands' paper [142] was to give a conjectural such formula under very general conditions on S_K. He arrived at it after a study of the special case of Shimura varieties of PEL (polarization, endomorphism ring, level structure) type. PEL varieties are a rather small subset of all Shimura varieties. (See [176, §9].) However, they still represent a major generalization of Shimura curves. They had been introduced by Shimura [229], but so far as I know, without any particular interest in the Lefschetz formula.

PEL varieties parametrize abelian varieties with additional structure. They include the varieties attached to GL(2) [140] we discussed above, which parametrize elliptic curves E. They also include Siegel modular varieties, the higher-dimensional generalizations attached to general symplectic groups $G = \mathrm{GSp}(2n)$.

Siegel modular varieties $S_K = S_K(G,X)$ parametrize abelian varieties A. More precisely, the set of complex points

$$S_K(\mathbb{C}) = G(\mathbb{Q}) \backslash (X \times G(\mathbb{A}_{\mathrm{fin}})/K), \quad G = \mathrm{GSp}(2n),$$

in S_K is bijective with the set of $G(\mathbb{Q})$-orbits of pairs (A,g), where A is a principally polarized abelian variety over $\mathbb{C}$ up to isogeny, and g is a K-level structure (which is to say a coset in $G(\mathbb{A}_{\mathrm{fin}})/K$). What makes the problem treated by Langlands in [142] more accessible in this case is that $S_K(G,X)$ has a canonical model, not just over the reflex field $E = \mathbb{Q}$, but also over the ring $\mathbb{Z}$, and that elevates S_K to the role of a universal modular variety over $\mathrm{Spec}(\mathbb{Z})$. As I understand it, this means that there is an isomorphism class of families

$$A_K \to S_K$$

of principally polarized abelian varieties over $\mathrm{Spec}(\mathbb{Z})$, equipped with a K-level structure, which is *universal* in the sense that the set of all such families $A'_K \to S'_K$ over any $\mathbb{Z}$-scheme S'_K is in bijection with the $\mathrm{Spec}(\mathbb{Z})$-morphisms $\phi\colon S'_K \to S_K$ under the pullback mapping

$$A'_K = \phi^* A_K.$$

I will not try to define these various terms here. But the reader can refer to [142] for the discussion of a related case, motivated by Kronecker's Jugendtraum and Hilbert's twelfth problem.

The point is that in representing a functor, S_K allows one to identify the set of points

$$S_K(\overline{\mathbb{F}}_p) = \{\overline{\phi}_p\colon \mathrm{Spec}(\overline{F}_p) \to S_K\}$$

with families $\overline{A}_p = \overline{\phi}_p^* A_K$ over $\mathrm{Spec}(\overline{F}_p)$. Equipped with the action of the Frobenius endomorphism, this set amounts in turn to a classification of isogeny classes of n-dimensional abelian varieties over $\overline{\mathbb{F}}_p$ with K^p-level structure. Such objects have been well understood for some time according to Honda–Tate theory [235], [96], and can be described explicitly. These modular properties extend to the locally constant, λ-adic sheaf $\mathscr{F}_K = \mathscr{F}_{K,\xi}$ on S_K attached to any finite-dimensional representation ξ of G over $\mathbb{Q}$. They are also compatible with the Hecke correspondence f^p on S_K defined by right translation on $S_K(\mathbb{C})$ by any element g^p in $G(\mathbb{A}^p_{\mathrm{fin}})$. (See [124, p. 375].) The classification of abelian varieties over $\overline{\mathbb{F}}_p$ then leads to a description of the set of fixed points of the composition

$$\Phi_p \circ f^p, \qquad \Phi_p = \Phi_{p,j} = (\mathrm{Frob}_p)^j,\ j \in \mathbb{N}, \tag{69}$$

acting as a correspondence on the set $S_K(\overline{\mathbb{F}}_p)$. This result can be regarded as an explicit description of of the main (elliptic) part of the geometric side of the Grothendieck–Lefschetz trace formula, for the case at hand.

These are the ideas that led Langlands in [142] to conjecture a general such formula for any Shimura variety. It was a bold step, which was refined and made clearer in his later paper [164] with Rapoport. In fact, the conjectured formula was subject to further evolution, leading up to what might be its final version in §3 in the paper [123] of Kottwitz. My understanding is that this formula is still far from proved in general. The case of general PEL-Shimura varieties has itself turned out to be a challenging problem. Progress was made by Milne [172] and Zink [256], while the special case of Siegel modular varieties discussed here was established in [123] and [174]. The proof for general PEL varieties was completed in [124], [174] and [191]. (See [50] and the introduction of [124].) As an aside, I have found it difficult at times to sort out what has been established from the literature, owing no doubt to my own imperfect grasp of the technical complexities of the subject. (See the introduction to [171], which is itself quite complex.)

With a formula (either proven or conjectured) for the geometric side of the arithmetic (Grothendieck–Lefschetz) trace formula, it would then be possible to study its comparison with the geometric side of the automorphic (Arthur–Selberg) trace formula. In particular, one could consider the problems of endoscopy that had been emphasized by Langlands in his papers following [140].

Kottwitz took up the fundamental lemma in [118]. In this paper, he was able to reduce it to a more familiar problem, the identity for twisted spherical functions required for cyclic base change. This was the problem solved by Langlands in the special case of GL(2). In a subsequent paper [120], Kottwitz reduced the problem further. For the special case of the unit function in the Hecke algebra of spherical functions, he reduced the twisted fundamental lemma to its original version for orbital integrals. The special case of GL(n) of this last reduction was, incidentally, an essential ingredient of the proof of base change for GL(n) in [27].

In each paper, Kottwitz was able to treat the problem as a natural combinatorial identity. In fact, in each case he observed that two finite series were equal simply because there was a term by term matching of their summands. However, the original fundamental lemma has turned out to be much more subtle. Even its statement draws upon the deeper notions from endoscopy, such as stability, endoscopic groups and transfer factors, that will be part of our discussion in Section 10. In particular, it is not just a combinatorial problem. Its ultimate proof by Ngô Bao Châu, drawing on the work of Waldspurger, came considerably later [186]. It required among other things, his remarkable observation that the Hitchin fibration over the field of meromorphic functions on a compact Riemann surface matches the geometric side of the trace formula (in characteristic p) over the global field of functions on a smooth projective curve in characteristic p.

The fundamental lemma is an important ingredient for our understanding of the automorphic properties of the problem. However, the full comparison requires something more, what can be called the stabilization of the Lefschetz trace formula. As such, it becomes one of a number of such constructions, beginning with the stabilization [21] of the basic automorphic trace formula (for a quasisplit group G^*), its analogue [21] for an inner form G of G^*, the stabilization of the general twisted trace formula [179], [180], the expected stabilization of a metaplectic trace formula, possible stabilizations of various relative trace formulas, and who know what else. In all such constructions, the basic ingredient is the stable trace formula for a quasisplit group G^*, which is defined by an inductive process from the basic automorphic (Arthur–Selberg) trace formula for G^*. In the other cases, the comparison is not just with this stable trace formula for G^*, but rather a linear combination of stable trace formulas for a collection of quasisplit groups G', known as endoscopic groups (attached to the problem at hand).

In [123, §4, §7], Kottwitz stabilized the elliptic part of the Lefschetz trace formula, which is to say, his proposed formula from §3 of [123]. In so doing, he appealed to his earlier results in [121] and [119], as well as Langlands' original monograph [150] on stabilization. We shall describe his construction in a way that can be compared with our general discussion of endoscopy in Section 10.

To simplify the notation, we shall allow G to represent the given Shimura datum $S_K(G,X)$, as well as the reductive group over $\mathbb{Q}$ that is its essential ingredient. This is similar to a convention from the theory of endoscopy, in which G' is often used to represent a full endoscopic datum $(G',s',\mathcal{G}',\xi')$ attached to any G under consideration, as well as the quasisplit group G' that is its main component. While we are at it, let us also write f here in place of the triplet (ξ,Φ_p,f^p) in keeping track of the dependence of the Lefschetz trace formula on these quantities. We cannot quite think of f as a test function on $G(\mathbb{A})$, which we would have if we were dealing with the automorphic trace formula. However, for any endoscopic datum G' attached to G, there is a transfer mapping

$$f \to f' = f'_\infty \cdot f'_p \cdot (f^p)'$$

from triplets f to test functions f' on $G'(\mathbb{A})$, defined by Kottwitz in [123, §7]. We shall write $\Lambda^p_{\mathrm{ell}}(f)$ for the conjectural elliptic part of the Lefschetz trace formula to exhibit its dependence on p as well as f. It is by definition given by the formula [123, (3.1)].

The stabilization in [123] is then an expansion

$$\Lambda^p_{\mathrm{ell}}(f) = \sum_{G'} \iota(G,G')\widehat{S}'_{\mathrm{ell}}(f') \tag{70}$$

of this linear form in terms of corresponding *stable*, linear forms[10] in various *stable* (automorphic) trace formulas.[10] The groups G' on the right-hand side represent equivalence classes of elliptic endoscopic data[10] for G, according to the convention above. For any such G', S'_{ell} is the elliptic part of the geometric side of the stable trace formula

$$S'_{\mathrm{geom}}(\cdot) = S'_{\mathrm{spec}}(\cdot)$$

[10] We are asking a reader unfamiliar with these other terms to wait until (or look ahead to) Section 10, where they will be discussed in somewhat greater detail.

for G'. The corresponding test function f' for G' is the transfer of f mentioned above, defined by the general transfer factors Langlands and Shelstad. It is determined only up to the values it takes when paired with a stable distribution on $G'(\mathbb{A})$. With this proviso in mind, we have written $\widehat{S}'_{\text{ell}}(f')$ for the *uniquely determined pairing* of the stable distribution $S'_{\text{ell}}(\cdot)$ with the function f'. Finally, the coefficients $\iota(G,G')$ are constructed in an elementary manner from G to G'. (See [123, Theorem 7.2].)

We should perhaps pause to remind ourselves of the ultimate goal. It is founded on the two interpretations of cohomology and the fundamental data they support. On the one hand, we have the L^2-de Rham cohomology $H^*_{(2)}(S_K(\mathbb{C}),\mathscr{F})$, and its decomposition (62) in terms of automorphic representations. On the other hand, we have the intersection cohomology $IH^*(\overline{S}_K(\mathbb{C}),\mathscr{F})$ of the Baily–Borel compactification $\overline{S}_K(\mathbb{C})$, equipped with the correspondences defined by Hecke operators. Zucker's conjecture [257], established around 1988 [168], [198], asserts that these two complex graded vector spaces are isomorphic. We are assuming that the sheaf $\mathscr{F}$, assigned as it is to the representation ξ of G, is defined over the chosen number field L. Like ordinary Betti cohomology, one can vary the coefficient field of $IH^*(\overline{S}_K(\mathbb{C}),\mathscr{F})$, taking it here to be L. If λ is any finite place of L not lying over p, the tensor product

$$IH^*_\lambda = IH^*(\overline{S}_K(\mathbb{C}),\mathscr{F}_\lambda) = IH^*(\overline{S}_K(\mathbb{C}),\mathscr{F})\otimes_L L_\lambda$$

represents a change of coefficients from L to the λ-adic field L_λ. What makes it all work is that the last space is isomorphic to the λ-adic (étale) cohomology of the variety $\overline{S}_K$ at its reduction modulo p. In particular, this λ-adic vector space comes with a representation of $\mathrm{Gal}(\overline{\mathbb{Q}}/\mathbb{Q})\times\mathscr{H}_K$, where $\mathscr{H}_K$ is the Hecke algebra of compactly supported L-valued functions in $K\backslash G(\mathbb{A}_{\text{fin}})/K$. In this setting, the spectral side of the Lefschetz trace formula is the Euler number

$$\Lambda^p_{\text{spec}}(f) = \sum_{k=0}^{m}(-1)^k\,\mathrm{tr}(IH^k_\lambda(\Phi_p\times f^p)), \tag{71}$$

for the operator on IH^k_λ attached to the composition (69). (This discussion follows the beginnings of §1 of [123].)

The actual Lefschetz formula is an identity

$$\Lambda^p_{\text{geom}}(f) = \Lambda^p_{\text{spec}}(f). \tag{72}$$

The geometric side $\Lambda^p_{\text{geom}}(f)$ includes the elliptic part $\Lambda^p_{\text{ell}}(f)$ studied by Kottwitz [123]. In general, however, there are complementary terms attached to fixed points "at infinity", or more precisely, in strata of $\overline{S}_K(\mathbb{C})$ attached to proper parabolic subgroups P of G. (The largest stratum $S_K(\mathbb{C})\subset\overline{S}_K(\mathbb{C})$ is attached to the group G itself.) These are more difficult. However, they do not occur if $S_K(\mathbb{C})$ is already compact, and are not significant in many noncompact Shimura varieties. For example, if $G=\mathrm{Res}_{F/\mathbb{Q}}(H)$, for a split reductive group H over a totally real number field

F with more than one archimedean valuation, the nonelliptic terms on the geometric side of the automorphic trace formula vanish [15, Theorem 7.1(b)], a property that should be reflected in the corresponding terms in $\Lambda^p_{\mathrm{geom}}(f)$. However, in the basic case $G = \mathrm{GSp}(2n)$ of Siegel modular varieties we have discussed, there are complementary terms in $\Lambda^p_{\mathrm{geom}}(f)$. Their analysis is part of the work of S. Morel.

At any rate, Kottwitz' stabilization (70) of $\Lambda^p_{\mathrm{ell}}(f)$ leads us to expect a similar stabilization

$$\Lambda^p_{\mathrm{geom}}(f) = \sum_{G'} \iota(G,G')\widehat{S}'_{\mathrm{geom}}(f') \tag{73}$$

of the full geometric side of the Lefschetz trace formula. We should point out that the linear forms on each side of (73) are defined, since the left-hand side equals that of (72), while the right-hand side is given by the stable trace formula of each of the groups G'. It is just that we cannot say that the two sides are equal unless $\Lambda^p_{\mathrm{ell}}(f)$ equals $\Lambda^p_{\mathrm{geom}}(f)$. (Even in this last case, the inequality still rests on Kottwitz' basic conjectural formula [123, (3.1)].) Similar comments apply to the stabilization

$$\Lambda^p_{\mathrm{spec}}(f) = \sum_{G'} \iota(G,G')\widehat{S}'_{\mathrm{spec}}(f') \tag{74}$$

of the spectral side, which would follow formally from (73), when combined with the identities $\Lambda^p_{\mathrm{spec}}(f) = \Lambda^p_{\mathrm{geom}}(f)$ (the Lefschetz formula (72)) and

$$\widehat{S}'_{\mathrm{spec}}(f') = \widehat{S}'_{\mathrm{geom}}(f')$$

(the stable formula for G'). In particular, (74) would follow from Part I of [123] (Sections 1–7) if $\Lambda^p_{\mathrm{ell}}(f)$ equals $\Lambda^p_{\mathrm{geom}}(f)$ *and* the conjectured formula [123, (3.1)] is valid.

Part II of [123] (Section 8–10) is what Kottwitz calls the destabilization of (74). The ultimate goal of the theory would be to deduce reciprocity laws for S_K, of the kind established by Langlands for $G = \mathrm{GL}(2)$ in [140], from the identity (74). But there are not shortcuts. The only way we can expect to prove (74) is to derive it from the geometric stabilization (73), and this would rest on the imposing task of proving the explicit fixed point formula [123, (3.1)] for S_K, and its extension to the compactification for S_K. After this, there would be a new set of problems on the interpretation of the right-hand side of (74). These concern a set of conjectures on the classification of the automorphic discrete spectrum that I described in [18], and applied to the cohomology of Shimura varieties in §9 of that paper.

Kottwitz assumed these conjectures in [123]. He then compared the explicit expression they yield for the right-hand side of (74) with the right-hand side of (71). The former concerns the characters of automorphic representations, the latter the virtual, λ-adic characters from intersection cohomology. By manipulating the terms in (74), he obtained a concrete formula [123, (10.5)] for the alternating sum of traces in (71) in terms of automorphic characters. This in turn suggested an explicit decomposition of the virtual, λ-adic representation of $\mathrm{Gal}(\overline{\mathbb{Q}}/\mathbb{Q}) \times \mathscr{H}_L$ on the étale cohomology of $\overline{S}_K$ in terms of the automorphic discrete spectrum of G. It is the direct sum displayed in the last paragraph of §10 of [123], but unfortunately not labeled.

This last decomposition from [123] could be regarded as the ultimate goal. It represents something beyond the character formula given by (10.5) of [123], even with all the conjectures that were taken for granted. For one thing, the integer j in (10.5) has for technical reasons to be taken to be sufficiently large. With the formula established, one could presumably extend it to all j by extrapolating the behavior of each side for j in the restricted range. This would give the reciprocity law for the unramified places p. The remaining finite set of places p with $\overline{S}_K$ ramified are more serious. They would require a solution for general G of the problems for GL(2) solved by Langlands and Carayol. Their solution would presumably give reciprocity laws for $\overline{S}_K = \overline{S}_K(G,X)$ at all places. In particular, it should ultimately express the associated motivic L-function

$$L(s, IH^k(\overline{S}_K, \mathscr{F})), \qquad 0 \leq k \leq 2d = 2\dim(S_K),$$

of weight k, and the Hasse–Weil zeta function $\zeta(s, \overline{S}_K(G,X))$ of $\overline{S}_K$, explicitly in terms of automorphic L-functions.

In the final Part III of [123], Kottwitz established the conjectural fixed point formula [123, (3.1)] for the Siegel modular varieties $S_K = S_K(G,X)$ attached to the groups $G = \mathrm{GSp}(2n)$. This represents a grounding of sorts for the conjectured identity whose general version had itself been the foundation for Parts I and II of the article, and for our discussion above. Siegel modular varieties have been studied more recently by Morel, as we have noted. She established explicit formulas for their complementary terms in $\Lambda^p_{\mathrm{geom}}(f)$ [182], those given by boundary components in $\overline{S}_K$ attached to proper parabolic subgroups P of G. The stabilization (73) of $\Lambda^p_{\mathrm{geom}}(f)$ was then given an explicit form in her paper [183]. This was based on a formula [17] for the associated nonelliptic geometric terms in the automorphic trace formula, in terms of the characters of Harish-Chandra's discrete series, and the stabilization of this formula treated in [82]. The only thing now standing in the way of a complete solution for Siegel modular varieties, or at least the immediate precursor [123, (10.5)] of a complete solution, is the "destabilization" of the expansion (73) of $\Lambda^p_{\mathrm{spec}}(f)$. This requires the local and global conjectures in [18] for the group $\mathrm{GSp}(2n)$. The conjectures have now been established for the group $\mathrm{Sp}(2n)$ [23]. It is obviously important to try to extend them to $\mathrm{GSp}(2n)$, even though there are indications that new difficulties arise.

Siegel modular varieties attached to the groups $G = \mathrm{GSp}(2n)$ represent a fundamental class of examples. We can think of their role in the general theory of Shimura varieties as something akin to that of the groups $\mathrm{GL}(N)$ in the general theory of automorphic forms. The groups $\mathrm{GL}(N)$ themselves are disqualified from this larger role since they are only attached to Shimura varieties when $N = 2$ (in which case $\mathrm{GL}(2) = \mathrm{GSp}(2n)$).

I have written more on Shimura varieties than I had originally intended. I wanted simply to give some sense of the foundations laid down by Langlands in this very rich field, following Shimura himself [230] and Deligne [61]. As we have seen, these include the comprehensive beginnings with GL(2), the conjectural formula for the points of a Shimura variety modulo p, and the recognition of the central role played by the fundamental lemma and endoscopy. Nonetheless, we shall return to the subject in the next section. In the second half of the section, we shall discuss the conclusions of Kottwitz from another point of view, which includes a conjectural derivation in terms of motives. This is turn will serve as a springboard for the introduction of further questions and problems.

9 Motives and Reciprocity

One of Grothendieck's great insights was the idea of a motive. In general terms, motives are supposed to have two simultaneous roles. On the one hand, they are to be regarded as fundamental building blocks of algebraic varieties. On the other, they represent also a universal cohomology functor for algebraic varieties, of which Betti cohomology, (algebraic) de Rham cohomology, ℓ-adic étale cohomology and crystalline cohomology become concrete realizations. Their existence was predicated by Grothendieck on a number of conjectures [112], [113] which are still largely unproved today. Perhaps for this reason, they were not widely discussed in formal mathematical circles before 1977. However, the unwritten understanding, if it existed, came to an end with the articles of Tate [237] and Langlands [144] in the 1979 Corvallis proceedings.

We are speaking of pure motives, which are the semisimple objects in the larger category of mixed motives. They would be the objects obtained from the category $(\mathrm{SProj})_F$ of smooth, projective algebraic varieties over a field F, as opposed to all varieties. Our object is to describe their role in [144], the article in which Langlands proposed what is now known as the *Reciprocity Conjecture*. Roughly speaking, Reciprocity is the analogue for general (smooth, projective) varieties of the Shimura–Taniyama–Weil conjecture for elliptic curves over $\mathbb{Q}$. Together with Functoriality, it is often regarded as one of the two fundamental pillars of the Langlands program.

Consider a pair of fields (F,Q) of characteristic 0, equipped with a complex embedding $F \subset \mathbb{C}$ for F. The category $\mathrm{Mot}_{F,Q}$ of (pure) motives over F with coefficients in Q (or just Q-motives over F) should be a semisimple, Q-linear category (which is to say, an abelian category in which short exact sequences split, and in which the abelian groups $\mathrm{Hom}(M,N)$ have been enriched to Q-vector spaces), equipped with a functor

$$(\mathrm{SProj})_F \to \mathrm{Mot}_{F,Q} \tag{75}$$

from the category of smooth projective varieties over F. It would in fact be a tensor category, equipped with a tensor product structure $\otimes$ over Q, satisfying several natural axioms. With the complex embedding $F \hookrightarrow \mathbb{C}$ of F, ordinary Betti cohomology of complex varieties with Q-coefficients gives a functor

$$H_B = H_B^*\colon (\mathrm{SProj})_F \to (\mathrm{Vect})_Q$$

with values in the category $(\mathrm{Vect})_Q$ of graded Q-vector spaces. This should factor through $\mathrm{Mot}_{F,Q}$ to a $\otimes$-functor

$$H_B = H_B^*\colon \mathrm{Mot}_{F,Q} \to (\mathrm{Vect})_Q. \tag{76}$$

It is interesting to note that the two arrows in the composition

$$(\mathrm{SProj})_F \to \mathrm{Mot}_{F,Q} \to (\mathrm{Vect})_Q \tag{77}$$

of (75) and (76), paired with the corresponding fields F and Q, illustrate the dual roles played by motives. The first arrow describes motives as building blocks of smooth projective varieties (given the grading of $\mathrm{Mot}_{F,Q}$ by weights implicit in the second one). The second arrow for its part suggests a role for $\mathrm{Mot}_{F,Q}$ as a cohomology theory for smooth projective varieties. This would be universal, in the sense that there should be similar "realizations" for all the various arithmetic cohomology theories for $(\mathrm{SProj})_F$. (See [65].) In the case of Shimura varieties, F equals the reflex field E.

Suppose now that F and Q are number fields. In fact, we might as well assume that $Q = \mathbb{Q}$, since Betti cohomology comes with a restriction of scalars functor. The functor

$$H_B\colon \mathrm{Mot}_{F,\mathbb{Q}} \to (\mathrm{Vect})_{\mathbb{Q}}$$

in (76) is called a *fibre functor* for the tensor category $\mathrm{Mot}_{F,\mathbb{Q}}$ over $\mathbb{Q}$. It gives $\mathrm{Mot}_{F,\mathbb{Q}}$ the structure of a natural (neutral) *Tannakian category*. A fundamental observation of Grothendieck was that a Tannakian category is equivalent to the category of (finite-dimensional) representations over the ground field ($\mathbb{Q}$ in this case) of an affine proalgebraic group. There would then be an (anti)equivalence from $\mathrm{Mot}_{F,\mathbb{Q}}$ to the category of representations of a proalgebraic group

$$\mathcal{G}_F = \mathcal{G}_{F,\mathbb{Q}} = \mathrm{Aut}^{\otimes}(H_B)$$

over $\mathbb{Q}$. This group would be an extension

$$\mathcal{G}_{F,\mathbb{Q}} \mapsto \Gamma_F,$$

of the absolute Galois group of F by a connected, reductive, proalgebraic group over $\mathbb{Q}$, whose finite-dimensional representations over $\mathbb{Q}$ parametrize (up to equivalence) the objects M in $\mathrm{Mot}_F = \mathrm{Mot}_{F,\mathbb{Q}}$. (See [210].)

With the Shimura–Taniyama–Weil conjecture in mind, Langlands was interested in the relations between motives and automorphic representations. Treating general automorphic representations and motives as objects over the complex numbers, he considered the complexification

$$\mathcal{G}_F = \mathcal{G}_{F,\mathbb{C}} = \mathcal{G}_{F,\mathbb{Q}} \times_{\mathbb{Q}} \mathrm{Spec}(\mathbb{C})$$

of the *motivic Galois group* above. (We are identifying the $\mathbb{Q}$-group $\mathcal{G}_F = \mathcal{G}_{F,\mathbb{Q}}$ above with its associated group $\mathcal{G}_F = \mathcal{G}_{F,\mathbb{Q}} \cong \mathcal{G}_F(\mathbb{C})$ of complex points here.) It is an extension

$$\mathscr{G}_F \to \Gamma_F$$

of Γ_F by a *complex*, reductive proalgebraic group whose finite-dimensional complex representations parametrize *complex* F-motives. Langlands then suggested that the best (and perhaps only) way to express the relations would be through a parallel *automorphic Galois group*.

It was an audacious proposal. As far as I know, nothing of the sort had ever been imagined before. Langlands had studied the L-homomorphisms

$$\phi : W_F \to {}^L G \tag{78}$$

of the global Weil group W_F ten years earlier as a means to parametrize some automorphic representations of G. But by 1977, it was well understood that these objects could not account for most automorphic representations (except in the case of a torus $G = T$ [153]). There seemed to be a general feeling that there would be nothing more in this particular direction to say.

Langlands formulated the construction of an automorphic Galois group, again as an extension

$$G_{\Pi(F)} \to \Gamma_F$$

of Γ_F by a connected, complex reductive group. Its representations of degree n would parametrize the set of automorphic representations $\Pi(F)$ of $\mathrm{GL}(n)$ over F that he called *isobaric*. These are the representations of $\mathrm{GL}(n, \mathbb{A}_F)$ denoted symbolically by

$$\pi = \pi_1 \boxplus \cdots \boxplus \pi_r, \qquad \pi_i \in \Pi_{\mathrm{cusp}}(\mathrm{GL}(n_i)), \tag{79}$$

on p. 207 of [144]. The ranks n_i correspond to a partition $(n_1, \ldots, n_r)$ of n, and π stands for a canonical irreducible constituent of the associated induced representation from the standard parabolic subgroup of $\mathrm{GL}(n, \mathbb{A}_F)$ attached to the partition.

Langlands' idea was to try to attach a Tannakian category to the representations (79). For a start, it was necessary to have a classification of automorphic representations of $\mathrm{GL}(n)$ in terms of the isobaric representations (79), which he formulated at the bottom of p. 207 of [144]. The classification was established soon afterwards by Jacquet and Shalika [107], using their theory with Piatetskii-Shapiro of Rankin–Selberg L-functions. This gave $\Pi(F)$ the structure of an abelian category. However, to obtain a tensor category from $\Pi(F)$ would require something much stronger, functoriality attached to the tensor product representations

$$\mathrm{GL}(n_i, \mathbb{C}) \times \mathrm{GL}(n_j, \mathbb{C}) \to \mathrm{GL}(n_i n_j, \mathbb{C})$$

of dual groups. This is one of the deepest cases of functoriality, and is still far from being resolved. The final ingredient would be a suitable fibre functor

$$\Pi(F) \to (\mathrm{Vect})_{\mathbb{C}}.$$

It is hard to imagine any construction of this last ingredient, but one could hope that it would be a part of the eventual proof of the cases of functoriality above. At any rate, this would essentially make the tensor category $\Pi(F)$ into a Tannakian category, of which the complex automorphic Galois group $G_{\Pi(F)}$ would then be a consequence.

The group $G_{\Pi(F)}$ would be a replacement for the Weil group, in its role (78) for parameters of automorphic representations. As we have noted, the n-dimensional representations of $G_{\Pi(F)}$ would parametrize all isobaric representations of $G = \mathrm{GL}(n)_F$. This would include all of the globally tempered representations $\Pi_{\mathrm{temp}}(G)$, which we recall are the irreducible representations of $G(\mathbb{A}_F)$ that occur in the decomposition of $L^2(G(F) \setminus G(\mathbb{A}_F))$. More generally, if G is any connected, quasisplit group over F, the algebraic L-homomorphisms

$$\phi : G_{\Pi(F)} \to {}^L G$$

over Γ_F would parametrize disjoint global packets of automorphic representations of G (L-packets) whose union contains $\Pi_{\mathrm{temp}}(G)$. The representation theory of a general reductive group G will thus be more complicated than that of $\mathrm{GL}(n)$, and will probably be best understood, through the theory of endoscopy, in terms of the theory for its quasisplit inner twist.

Langlands assumed the existence of $G_{\Pi(F)}$, and turned to the problem of relating motives to automorphic representations. His proposed solution of the problem, after the first bold step of postulating the existence of $G_{\Pi(F)}$, was elegant and simple. It was to conjecture the existence of a surjective canonical mapping

$$G_{\Pi(F)} \to \mathcal{G}_F \tag{80}$$

of complex, proalgebraic groups over Γ_F. Among other things, the mapping would be compatible with local data attached to each of the two groups, which we will discuss presently in a slightly different guise. The surjectivity of the mapping (80) was not stated explicitly in [141], but it was clearly a part of Langlands' thinking, in the Tate conjecture [152] and his implicit aim of formulating a general analogue of the Shimura–Taniyama–Weil conjecture. A complex F-motive M, being identified with a finite-dimensional complex representation of $\mathcal{G}_F$ pulls back to a complex finite-dimensional representation r_M of $G_{\Pi(F)}$. The local data in $G_{\Pi(F)}$ and $\mathcal{G}_F$ would then yield an identity

$$L(s, r_M) = L(s, M) \tag{81}$$

of L-functions. In particular, the motivic L-functions on the right would inherit the analytic continuation and functional equation from the standard automorphic L-functions on the left-hand side. It is the conjectured mapping (80) that is known today as Langlands' *Reciprocity Conjecture*.

A number of cases of the Reciprocity Conjecture have already been established, if we are prepared to state them directly as relations between motives and automorphic representations (that is, without the universal groups in (80)). For example, any complex n-dimensional representation of Γ_F is a motive, known for obvious reasons as an *Artin motive* [185, §2]. Reciprocity in this case amounts to functoriality for Artin L-functions. As we discussed in Section 7, it was established by Langlands for complex two-dimensional Galois representations of dihedral, tetrahedral and octahedral representations, with Galois groups isomorphic respectively to D_{2n}, A_4 and S_4 (the latter with Tunnell).

Shimura varieties will be another source of examples. For the Shimura varieties attached to GL(2), there are many complex two-dimensional motives, which should correspond to many cuspidal automorphic representations of GL(2). The local computations required to state Reciprocity in this case were the content of the conjecture at the end of §4 of [140]. As we discussed in Section 8, this was established later in the same article and the subsequent article of Carayol. The same phenomena occur for general Shimura varieties, although they are more subtle. For a general Shimura datum (G,X), the co-character μ_h defined in Section 8 is dual to a minuscule weight $\widehat{\mu}$ of $\widehat{G}$, which in turn gives rise to the finite-dimensional representation $r = r_X$ of the L-group ${}^LG$. This assigns motives to the various constituents of the L^2-cohomology (62) of $S_K(\mathbb{C})$ obtained from $(\mathfrak{g}_{\mathbb{R}}, K_{\mathbb{R}})$-cohomology. On the other hand, automorphic representations of G could be attached to the global parameters ϕ in the generalization of (78) if we had the group $G_{\Pi(F)}$. The reciprocity correspondence between the two kinds of objects would then be within reach if one could establish a local reciprocity law at any place v of $\mathbb{Q}$. This would be the general analogue for (G,v) of the conjecture [140, §4] for $(G,v) = (\mathrm{GL}(2),p)$. (See [144, Lemma, p. 240] and [18, Proposition 9.1] for the archimedean case $v=\infty$, topics we will take up later in this section, and the discussion in [123, §8–10] reviewed in the last section for the case $v=p$.)

The most famous example of Reciprocity is of course the Shimura–Taniyama–Weil (STW) conjecture. It applies to the motive of weight 1 (corresponding to the first cohomology group H^1) of an arbitrary elliptic curve E over $\mathbb{Q}$. The problem was to show that it corresponds to a cusp form f of weight 2 over $\mathbb{Q}$ such that

$$L(s,\pi) = L(s,E),$$

where π is the automorphic representation of GL(2) attached to f in the L-function on the left, and E is identified with its motive of weight 1 on the right. In other words, the local data of the two objects match, in the sense of [140, §4], discussed in the last section and above. This includes the requirement that the conductors of f and E, independently defined nonnegative integers, also match. We recall that this last condition was the quantitative improvement [250] Weil added to the conjecture in 1967. The problem was of course much deeper than the similar question above for a two-dimensional motive in a Shimura variety S_K attached to GL(2). For in the latter case, there was all the geometric structure of S_K that drove the Lefschetz trace formula, and its comparison with the Selberg trace formula.

The STW conjecture was established for semistable elliptic curves E by Andrew Wiles in 1995 [253], in partial collaboration with Richard Taylor [239]. *Semistable* means that for each prime p, E either has good reduction or multiplicative reduction. For example, it is included in the two conditions from [140, §4] under which

Langlands established the local reciprocity law for S_K in §7 of that paper. The authors were content to work with this restriction on E, since according to the results of Ribet [192], it was sufficient to establish Fermat's last theorem. Their proof of the STW conjecture relied on techniques [170] that were quite different from what we have been discussing in this article, as well as new methods developed expressly for the purpose. However there was one fundamental theorem from Langlands' work that was critical. It was the Langlands–Tunnell theorem, the reciprocity law for two-dimensional representations of Galois groups isomorphic to S_4. With the exceptional isomorphism

$$S_4 \cong \mathrm{PGL}_2(\mathbb{F}_3),$$

the theorem was the starting point for Wiles' extended study of deformation rings and the congruence properties for modular forms. (See [184], [83], [56] for general introductions to Wiles' proof.)

Six years after the two papers of Wiles and Taylor, C. Breuil, B. Conrad, F. Diamond and R. Taylor published a proof of the STW conjecture [40] for general E. They built on the work of various authors to remove the ramification constraints step by step, the most difficult being various calculations associated with the prime $p = 3$. The proof of the general STW conjecture has led to the proofs of unsolvability of other Fermat-like diophantine equations. Taken on its own, it represents a milestone in number theory, the proof of a longstanding fundamental case of what is now the general Reciprocity Conjecture. (See [55].)

We return now to Langlands' proposed universal automorphic Galois group $G_{\Pi(F)}$. Shortly after the publication of [144], Kottwitz pointed out that the Langlands group would be simpler if it were taken to be in the category of locally compact groups, rather than complex proalgebraic groups [119, §12]. In this formulation, the universal group would be an extension L_F of the absolute Weil group W_F by a connected compact group. It would thus take its place in a sequence

$$L_F \to W_F \to \Gamma_F$$

of three locally compact groups, all having ties to the arithmetic of the global field F.

This represented a less severe conceptual change from what had been in place before 1979. The earlier set of global Langlands parameters, as L-homomorphisms of W_F into the complex group ${}^LG$, would simply be enriched to the larger set of L-homomorphisms of L_F into ${}^LG$. Local Langlands parameters would remain the same, namely L-homomorphisms from the locally compact group

$$L_{F_v} = \begin{cases} W_{F_v}, & \text{if } v \text{ is archimedean,} \\ W_{F_v} \times \mathrm{SU}(2), & \text{if } v \text{ is nonarchimedean,} \end{cases} \tag{82}$$

into the complex group ${}^LG_v \subset {}^LG$. (Langlands had introduced the product $W_{F_v} \times \mathrm{SL}(2,\mathbb{C})$ in [144, §4] as an equivalent version of the Weil–Deligne group [237, §4]. Kottwitz chose $\mathrm{SU}(2)$ in place of $\mathrm{SL}(2,\mathbb{C})$ in order that the bounded local parameters (with respect to their images in $\widehat{G}$) would continue to be ones that correspond to locally *tempered* representations of $G(F_v)$.) The global locally compact group L_F should then be equipped with a commutative diagram

$$\begin{array}{ccccc} L_{F_v} & \longmapsto & W_{F_v} & \longrightarrow & \Gamma_{F_v} \\ \downarrow & & \downarrow & & \downarrow \\ L_F & \longmapsto & W_F & \longrightarrow & \Gamma_F \end{array} \tag{83}$$

of continuous homomorphisms for each valuation v of F. As usual, the vertical embedding on the left in (83) would be determined only up to conjugacy, and would extend the embeddings of local Weil and Galois groups.

Kottwitz proposed a set of axioms for L_F, but he seems to have been thinking of a group directly related to Langlands' construction of $G_{\Pi(F)}$. Roughly speaking, $G_{\Pi(F)}$ would be regarded as the "algebraic hull" of L_F, a proalgebraic group over F whose algebraic representations were in bijection with the continuous representations of L_F. In other words, if the proposed Tannakian category exists, thereby giving rise to a group $G_{\Pi(F)}$, the existence of a group L_F should follow formally. The algebraic hulls of the local groups L_{F_v} introduced in [144], could then revert back to the original groups (83). Finally Langlands' Reciprocity Conjecture (80) would become the existence of a continuous L-homomorphism

$$L_F \to \mathcal{G}_F. \tag{84}$$

Motivated by Langlands' paper, and the supplementary remarks by Kottwitz, I introduced a constructive version [20] of the group L_F in 2002. It is more concrete, and it leads to a correspondingly concrete description of the motivic Galois group $\mathcal{G}_F$. It also has the advantage of not requiring a Tannakian category for its existence. The construction is still conjectural. It relies on the Principle of Functoriality, as formulated in its unramified form in Section 4. Moreover, some further conditions related to functoriality would be needed for the resulting group to have the desired properties. These were discussed somewhat tentatively in §5 of [20]. They will probably be resolved one way or another by Beyond Endoscopy, the strategy proposed by Langlands around 2000 for attacking functoriality.

In Langlands' conjectural definition of $G_{\Pi(F)}$, the basic building blocks are cuspidal automorphic representations of general linear groups. This follows the implicit definition of the motivic Galois group in terms of irreducible (complex) motives. But there are automorphic representations that could be regarded as more fundamental. These would be the cuspidal, tempered automorphic representations of quasisplit groups G that are not functorial images from any smaller group. Let us be more precise.

We first recall the set

$$\mathcal{C}(G) = \mathcal{C}_{\text{aut}}(G) = \{c(\pi) : \pi \in \Pi(G)\}$$

introduced in Section 4. It consists of equivalence classes of families

$$c^S(\pi) = \{c_v(\pi) = c(\pi_v) : \pi \in \Pi(G), v \notin S\}$$

of semisimple conjugacy classes in ${}^LG$. It is best that we then consider the subset $\mathscr{C}_{\mathrm{bdd}}(G)$ of classes $c \in \mathscr{C}(G)$ that are *bounded*, in the sense that for almost all v, the projection of c_v onto $\widehat{G}$ meets a maximal compact subgroup of $\widehat{G}$. These would be the classes $c = c(\pi)$ attached to the globally tempered representations $\pi \in \Pi_{\mathrm{temp}}(G)$ (which we recall means that they should occur in the spectral decomposition of $L^2(G(F) \setminus G(\mathbb{A}_F))$, but with the further condition that they be locally tempered, in the sense that they satisfy the bounds from the general analogue of Ramanujan's conjecture. We are assuming functoriality. This implies that the unramified L-functions

$$L^S(s,c,r) = \prod_{v \notin S} L_v(s,c_v,r), \qquad c \in \mathscr{C}(G), \tag{85}$$

attached to representations r of ${}^LG$ have analytic continuation and functional equation. The L-functions attached to classes $c \in \mathscr{C}_{\mathrm{bdd}}(G)$ are the ones for which the Euler product on the right converges absolutely for $\mathrm{Re}\,(s) > 1$.

Suppose that G is simple and *simply connected*, as well as quasisplit. In this case, we say that a class $c \in \mathscr{C}_{\mathrm{bdd}}(G)$ is *primitive* if for any r,

$$\mathrm{ord}_{s=1}\, L^S(s,c,r) = [r\colon 1_{{}^LG}].$$

This amounts to asking that c not be a proper functorial image $\rho'(c')$, for some $c' \in \mathscr{C}_{\mathrm{bdd}}(G')$, and some L-homomorphism

$$\rho' \colon {}^LG' \to {}^LG \tag{86}$$

whose image in ${}^LG$ is proper. We write $\mathscr{C}_{\mathrm{prim}}(G)$ for the set of primitive classes in $\mathscr{C}_{\mathrm{bdd}}(G)$, for the simply connected group G. It is these objects, or if one prefers, corresponding automorphic representations $\pi \in \Pi_{\mathrm{prim}}(G)$, that represent the fundamental building blocks of L_F. They are the smallest family in the embedded sequence

$$\mathscr{C}_{\mathrm{prim}}(G) \subset \mathscr{C}_{\mathrm{cusp}}(G) \cap \mathscr{C}_{\mathrm{bdd}}(G) \subset \mathscr{C}_{\mathrm{bdd}}(G)$$

where $\mathscr{C}_{\mathrm{cusp}}(G)$ is the subset of classes $c = c(\pi)$ in $\mathscr{C}(G)$ for which π is cuspidal.

The main ingredient in the construction of L_F is an indexing set $\mathscr{C}_F$. It consists of isomorphism classes of pairs

$$(G,c), \qquad c \in \mathscr{C}_{\mathrm{prim}}(G),$$

with G simple and simply connected, in which (G,c) is isomorphic to (G_1,c_1) if there is an isomorphism of groups $G \to G_1$ over F, and a dual isomorphism ${}^LG_1 \to {}^LG$ that takes c_1 to c. Suppose that c belongs to $\mathscr{C}_F$ (in the sense that it represents an isomorphism class of pairs (G,c)). Since the group G is simply connected, the complex dual group $\widehat{G}$ is of adjoint type. We write K_c for a compact real form of its simply connected cover $\widehat{G}_{\mathrm{sc}}$. The Weil group W_F acquires an action on K_c from the semidirect product

$$^LG = \widehat{G} \rtimes W_F.$$

For any such c, there is then a natural extension

$$1 \to K_c \to L_c \to W_F \to 1 \tag{87}$$

of W_F by K_c, which does not generally split. There are two separate constructions of this extension, for which a reader can consult [20, §4]. The second of these includes a description of localizations

$$\begin{array}{ccc} L_{F_v} & \longmapsto & W_{F_v} \\ \downarrow & & \downarrow \\ L_c & \longmapsto & W_F \end{array} \tag{88}$$

for the group L_c, which is based on the local Langlands correspondence.

The extensions (87) and localizations (88) attached to elements $c \in \mathscr{C}_F$ are what is needed to construct the locally compact Langlands group L_F. It is defined as the fibre product

$$L_F = \prod_{c \in \mathscr{C}_F} (L_c \to W_F) \tag{89}$$

over W_F. As such, it is an extension

$$1 \to K_F \to L_F \to W_F \to 1 \tag{90}$$

of W_F by the compact simply connected group

$$K_F = \prod_{c \in \mathscr{C}_F} K_c,$$

and is hence locally compact. The required localizations (83) follow from their analogues (88) for each L_c.

We have appealed to functoriality in the definition of the sets $\mathscr{C}_{\text{prim}}(G)$, and therefore in the indexing set $\mathscr{C}_F$ used to define L_F. The expectation is that L_F will lead to a classification of automorphic representations. The best outcome would be that for any quasisplit G, the set $\Pi_{\text{bdd}}(G)$ of locally tempered representations that occur in the spectral decomposition of $L^2(G(F) \setminus G(\mathbb{A}_F))$ is a disjoint union of global L-packets, parametrized by $\widehat{G}$-conjugacy classes of L-homomorphisms

$$\phi : L_F \to {}^LG$$

whose image in $\widehat{G}$ is bounded. This reflects what might be expected for the subset of such representations attached to parameters ϕ for the Weil group W_F. As we have said, the matter would likely be resolved with a proof of functoriality by the methods of Beyond Endoscopy. In particular, if the proposed classification above needs only minor adjustments, these ought to become clear from Beyond Endoscopy.

We should emphasize that L_F represents a "thickening" of the Weil group W_F. The two groups should satisfy similar qualitative properties, such as those outlined for the Weil group by Tate in §1 of [237]. For example, if E is a finite extension of F, L_E would be the preimage of $\Gamma_E \subset \Gamma_F$ in L_F under the composition

$$L_F \to W_F \to \Gamma_F.$$

In particular, this could serve as a definition of L_F in terms of the group $L_{\mathbb{Q}}$.

In §6 of [20], there is also a tentative construction of the complex motivic Galois group $\mathscr{G}_F$ (which could conceivably serve at some point as an actual definition). It is modeled on the construction of L_F and the version (84) of Langlands' Reciprocity homomorphism. For a quasisplit group G over F, we could define a complex *G-motive* to be an L-homomorphism from $\mathscr{G}_F$ to ${}^LG$ (with the Galois form ${}^LG = \widehat{G} \rtimes \Gamma_F$ of the L-group, since $\mathscr{G}_F$ is to be regarded as a complex proalgebraic group over Γ_F). In the case that G is simple and simply connected, we would also speak of *primitive* G-motives. They would correspond to elements $c \in \mathscr{C}_{\text{prim}}(G)$ that are *algebraic* (or *motivic*). This means that if $c = c(\pi)$ for an automorphic representation π, and if

$$\phi_v \colon W_{F_v} \to {}^LG_v$$

is a Langlands parameter for π_v at an archimedean place v, the composition of ϕ_v with any finite-dimensional representation r of ${}^LG$ whose kernel contains a subgroup of finite index in W_F is of *Hodge type*. In other words, the restriction of $r \circ \phi_v$ to the subgroup $\mathbb{C}^*$ of W_{F_v} is a direct sum of (quasi)characters of the form

$$z \to z^{-p}\bar{z}^{-q}, \qquad z \in \mathbb{C}^*,\ p,q \in \mathbb{Z}. \tag{91}$$

Our restriction on r would in fact imply that each of these is of weight 0, in the sense that $p+q=0$, and hence a character of the form

$$z \to (z/\bar{z})^q.$$

The construction of $\mathscr{G}_F$ amounts to a fibre product analogous to (89), but with two changes.

(i) The indexing set $\mathscr{C}_F$ in (89) is replaced by the subset $\mathscr{C}_{F,\text{alg}}$ of algebraic indices $c \in \mathscr{C}_F$.

(ii) The diagram (87) of locally compact groups is replaced by a diagram

$$1 \to \mathscr{D}_c \to \mathscr{G}_c \to \mathscr{T}_F \to 1$$

of complex proalgebraic groups, in which $\mathscr{D}_c$ equals the simply connected complex dual $\widehat{G}_{\text{sc}}$ of the group G attached to c, and $\mathscr{T}_F$ is the Taniyama group over Γ_F introduced by Langlands in Section 5 of [144].

The automorphic Galois group would then be given as a fibre product

$$\mathscr{G}_F = \prod_{c \in \mathscr{C}_{F,\mathrm{alg}}} (\mathscr{G}_c \to \mathscr{T}_F)$$

over $\mathscr{T}_F$. The construction would thus express $\mathscr{G}_F$ explicitly in terms of the families $\mathscr{C}_{F,\mathrm{alg}}$. Otherwise said, it builds $\mathscr{G}_F$ out of primitive G-motives rather than irreducible $\mathrm{GL}(n)$-motives. With this proposed construction of $\mathscr{G}_F$, the homomorphism (84) is then defined concretely on the last page 481 of [20]. It is to be regarded as an L-homomorphism of groups over Γ_F.

Given $\mathscr{G}_F$, and any G over F, we write $\Phi_{\mathrm{alg}}(G)$ for the set of G-motives. This is a set (possibly empty) of $\widehat{G}$-conjugacy classes of proalgebraic L-homomorphisms

$$\Phi\colon \mathscr{G}_F \to {}^L G,$$

with respect to the projections of $\mathscr{G}_F$ and ${}^L G$ onto Γ_F. We can then identify these mappings with their restrictions to L_F under Langlands' proposed reciprocity mapping $\phi\colon L_F \to \mathscr{G}_F$, since the Reciprocity Conjecture should imply that the two sets are indeed in bijection. In other words, we can also regard a parameter $\Phi \in \Phi_{\mathrm{alg}}(G)$ as $\widehat{G}$-conjugacy class of L-homomorphisms

$$\Phi\colon L_F \to {}^L G$$

(with respect again to the two projections onto Γ_F) that is of Hodge type. Namely, if (r,V) is any finite-dimensional representation of ${}^L G$, and v is any archimedean place of F, the restriction of Φ_v to the subgroup $\mathbb{C}^*$ of W_{F_v} is a direct sum of characters of the form (91). For any such v, we write $\Phi_{\mathrm{alg}}(G_v)$ for the obvious local analogue of this set.

The algebraic Langlands parameters $\Phi \in \Phi_{\mathrm{alg}}(G)$ generally have nonzero weights, which means that they are nontempered. However, they can be projected naturally onto tempered parameters. To see this, we first introduce the weight homomorphism

$$w\colon C_F \to L_F$$

for L_F. It is the mapping from $C_F = F^* \backslash \mathbb{A}_F^*$ into the preimage

$$L_F^0 = K_F W_F^0$$

in L_F of the identity component W_F^0 of W_F, defined by mapping the norm $\|c\|$ for any $c \in C_F$ to the multiplicative subgroup $\mathbb{R}^+$ of W_F^0 described in [237, (1.4.4)]. We also have the *Tate homomorphism*

$$t\colon L_F \to C_F,$$

which can be expressed in terms of the composition of the projection $L_F \to W_F$ and $W_F \to C_F$. One sees that the image of w lies in the centre of L_F, and that the composition $(t \circ w)$ maps any $c \in C_F$ to $\|c\|^{-2}$. For any algebraic parameter $\Phi \in$

$\Phi_{\rm alg}(G)$, the modified Langlands parameter

$$\phi(x) = \Phi(x \cdot (w \circ t)(x)^{-\frac{1}{2}}), \qquad x \in L_F, \tag{92}$$

then lies in $\Phi_{\rm temp}(G)$. It has the property that if v is an archimedean valuation of F and r_v is a finite-dimensional representation of ${}^L G_v$, and if $(r_v \circ \Phi_v)(z)$ is a sum of algebraic characters (91), then $(r_v \circ \phi_v)(z)$ is the corresponding sum of continuous characters

$$(z/|z|^{\frac{1}{2}})^{-p}(\bar{z}/|\bar{z}|^{\frac{1}{2}})^{-q} = (z/\bar{z})^{-\frac{1}{2}(p-q)}. \tag{93}$$

We can write $\Phi_{\rm temp,alg}(G)$ for the set of parameters $\phi \in \Phi_{\rm temp}(G)$ obtained in this way, and

$$\Phi_{2,\rm alg}(G) = \Phi_2(G) \cap \Phi_{\rm temp,alg}(G)$$

for the subset of such parameters in $\Phi_2(G)$. We note that the correspondence $\phi_v \longleftrightarrow \Phi_v$ is a generalization of Langlands' dual correspondences $\phi_v \longleftrightarrow \phi'_v$ and $\pi_v \longleftrightarrow \pi'_v$ for $\mathrm{GL}(2)$ described in the last section.

We observe in passing that there is a parallel structure for the motivic Galois group $\mathscr{G}_F$. It was pointed out by Serre [210], who noted that $\mathscr{G}_F$ comes with a weight homomorphism

$$w\colon \mathbb{G}_m \to \mathscr{G}_F,$$

whose image lies in the centre of $\mathscr{G}_F$. It has the property that for a representation r of $\mathscr{G}_F$, the *weights* of the corresponding motive are given by the decomposition of $r \circ w$ into characters of $\mathbb{G}_m$. The motivic Galois group also comes with the Tate motive

$$t\colon \mathscr{G}_F \to \mathbb{G}_m$$

of weight (-2), which is to say that $t(w(x)) = x^{-2}$. These objects for $\mathscr{G}_F$ would follow immediately from their analogues for L_F and the basic properties of the Taniyama group $\mathscr{T}_F$.

Langlands' statement of the Reciprocity Conjecture actually came at the beginning of the article [144]. According to his introduction in §1, the original intent was to study two specific problems in the theory of Shimura varieties. These were taken up in the final two Sections 6 and 7 of the article. The earlier sections arose from his afterthoughts on the problems, but were for reasons of exposition presented in the reverse order. Section 2 contains the proposed mapping (80), on which our discussion to this point has been based.

Section 3 of [144] contains a brief discussion of another question, automorphic representations Langlands called anomalous. Unlike what happens for $\mathrm{GL}(n)$, these can include cuspidal automorphic representations that are not locally tempered. The first examples had been introduced for the group $\mathrm{PSp}(4)$ by Kurokawa shortly before the Corvallis conference. A second family of examples for $\mathrm{PSp}(4)$, discovered by Howe and Piatetskii-Shapiro, was discussed informally at the conference. These both turned out to be special cases of the representations for general groups introduced in [18], which we discussed in the last section as part of the conjectural stabilization (73) of spectral side of the Lefschetz trace formula. The relevant conjectures have now been established for quasisplit classical groups [23], [181].

In §4 of [144], Langlands returned to his original topic, the theory of Shimura varieties. This section again concerns motives, and as such, provides a bridge between the specific problems in the later sections of [144] and the formulation of the general Reciprocity Conjecture in Section 2. However, it applies to the motives as they are thought to appear on the geometric side of the Lefschetz trace formula, rather than the motives on the spectral side that would govern Reciprocity. This dual role would serve as a grand generalization of what happens for elliptic curves, which on the one hand represent moduli on the geometric side of the Lefschetz trace formula for GL(2), and on the other, spectral objects classified by the (now resolved) STW-conjecture. An equally fundamental analogy is the dual way of representing extensions of a number field F that is at the core of the automorphic trace formula for $\mathrm{GL}(n)$. On the one hand, they come from the irreducible polynomials of degree n over F in the elliptic part of the geometric side, as governed by the base of the Steinberg–Hitchin fibration that we will recall in the final section, and on the other, the irreducible complex representations of $\mathrm{Gal}(\overline{F}/F)$ of degree n that are conjectured to be basic components of the discrete part of the spectral side.

The conjectural formula in Section 3 of Kottwitz' paper [123], as it evolved from [142] and [164], is what allows us to understand the spectral properties of Shimura varieties S_K. However, its proof was restricted to cases in which $S_K(\mathbb{C})$ could be realized as a moduli space for geometric objects related to abelian varieties. Many, if not most, Shimura varieties are not PEL varieties, the basic moduli spaces of this sort. Abelian varieties are of course motives. The idea of §4 of [144], which Langlands learned from Deligne, was to treat an arbitrary Shimura variety $S_K(G,X)$ (apart from those that behave badly on the cocentre, that is with $G \neq G_0$ in the notation of [144, p. 217]), as a moduli space of *motives*. Deligne regarded Shimura varieties as parameter spaces for certain Hodge structures. His construction was predicated on a conjecture that these objects in turn parametrize uniquely determined motives. We shall review it as presented in [144] to get a further sense of the remarkable internal structure of a general Shimura variety.

As we have done with motives, we speak here of pure Hodge structures. Following the beginning of §4 of [144], we recall that a *real Hodge structure* V is a finite-dimensional real vector space $V_{\mathbb{R}}$ whose complexification has a decomposition

$$V_{\mathbb{C}} = \bigoplus_{p,q \in \mathbb{Z}} V^{p,q},$$

for complex subspaces $V^{p,q}$ such that $V^{q,p} = \overline{V^{p,q}}$. It is equivalent to a finite-dimensional representation σ of the group $\mathscr{R} = \mathrm{Res}_{\mathbb{C}/\mathbb{R}}(\mathbb{G}_m)$ over $\mathbb{R}$, with

$$V^{p,q} = \{v \in V_{\mathbb{C}} = V_{\mathbb{R}} \otimes \mathbb{C} : \sigma(z_1,z_2)v = z_1^{-p} z_2^{-q} v\}.$$

It is also essentially the same as an archimedean parameter

$$\phi_{\mathbb{R}} : W_{\mathbb{R}} \to \mathrm{GL}(n,\mathbb{C})$$

of Hodge type, of the kind that classifies representations of $\mathrm{GL}(n,\mathbb{R})$ of motivic significance, as we can recall from the quasicharacters (91) and the related parameters $\phi'_{\mathbb{R}}$ for GL(2) in Section 8. With either interpretation, the set of real Hodge structures is a Tannakian category, with associated group $\mathscr{R}$. A real Hodge structure V is of *weight n* if $V^{p,q}=0$ unless $p+q=n$.

Recall also that a *rational Hodge structure* V is a finite-dimensional vector space $V_{\mathbb{Q}}$ over $\mathbb{Q}$ with a direct sum decomposition

$$V_{\mathbb{Q}}=\bigoplus_{n}V_{\mathbb{Q}}^{n},$$

together with real Hodge structures of weight n on the real vector spaces $V_{\mathbb{R}}^{n}=V_{\mathbb{Q}}^{n}\otimes_{\mathbb{Q}}\mathbb{R}$. The basic example is the Tate rational Hodge structure $\mathbb{Q}(1)$ of weight -2, in which $\mathbb{Q}(1)_{\mathbb{Q}}=2\pi i\mathbb{Q}\subset\mathbb{C}$ and $\mathbb{Q}(1)_{\mathbb{C}}=\mathbb{Q}(1)^{-1,-1}$. Rational Hodge structures V contain much more information than real Hodge structures. In particular, if real Hodge structures give representations of general linear groups $\mathrm{GL}(n,\mathbb{R})$ in terms of archimedean Langlands parameters, supplementary $\mathbb{Q}$-structures should in many cases lead to motives, and therefore the enriched structure of automorphic representations of the groups $\mathrm{GL}(n)$. A necessary condition for this, however, is that V be *polarizable*, in the sense that it can be endowed with a bilinear form of the sort described on p. 215 of [144]. The situation is a generalization of the theory of abelian varieties, whereby a complex torus represents an abelian variety if and only if it has a Riemann bilinear form. The category

$$\mathrm{Hod}=\mathrm{Hod}_{\mathbb{C},\mathbb{Q}}=\{V=(V_{\mathbb{C}},V_{\mathbb{Q}})\}$$

of polarizable rational Hodge structures is Tannakian, with fibre functor $V\to V_{\mathbb{Q}}$, and a corresponding Hodge group $\mathscr{H}=\mathscr{H}_{\mathbb{C}}=\mathscr{H}_{\mathbb{C},\mathbb{Q}}$ over $\mathbb{Q}$.

Recall finally that $\mathrm{Mot}=\mathrm{Mot}_{\mathbb{C},\mathbb{Q}}$ is the Tannakian category of $\mathbb{Q}$-motives over $\mathbb{C}$, with motivic Galois group $\mathscr{G}=\mathscr{G}_{\mathbb{C}}=\mathscr{G}_{\mathbb{C},\mathbb{Q}}$. Every object in this category comes with a polarizable rational Hodge structure, according to the properties of $\mathbb{Q}$-Betti cohomology of nonsingular complex projective varieties. There is consequently a $\otimes$-functor

$$h_{\mathrm{BH}}\colon \mathrm{Mot}\to\mathrm{Hod}$$

of categories, and a corresponding group homomorphism

$$h_{\mathrm{BH}}^{*}\colon \mathscr{H}\to\mathscr{G}$$

over $\mathbb{Q}$. The Hodge conjecture implies that the functor is fully faithful, which means that no data is lost from a motive (over $\mathbb{C}$) in passing to its Hodge structure.

The construction described in §4 of [144] applies to any Shimura variety $S_K=S_K(G,X)$ with $G=G_0$. It also depends on a rational representation (ξ,V) of G. The idea is to associate to every point $x=(h,g)$ in the space

$$\mathscr{X}_K=X\times(G(\mathbb{A}_{\mathrm{fin}})/K)$$

a pair of objects $(V^x, \phi^x_{\text{fin}})$ as follows. The first component V^x is the rational Hodge structure on the vector space $V^x_{\mathbb{Q}} = V_{\mathbb{Q}} = V$ given by the representation $\xi \circ h$ of $\mathscr{R}(\mathbb{R}) = \mathbb{C}^*$ on $V_{\mathbb{C}}$. The second component ϕ^x_{fin} is the isomorphism $V^x_{\mathbb{A}_{\text{fin}}} \to V_{\mathbb{A}_{\text{fin}}}$ given by $v \to \xi(g)^{-1}v$, with the understanding that it be defined only up to composition by an element of $\xi(K)$. Each such pair is also implicitly fitted with an underlying family of polarizations $\mathscr{P}^x$ of V^x, described briefly on p. 216 of [144], with the property that if $x' = \gamma x = (\gamma h, \gamma g)$ for some $\gamma \in G(\mathbb{Q})$, there is a natural map $\gamma \colon \mathscr{P}^x \to \mathscr{P}^{x'}$.

With this machinery in place, Langlands varies the representation

$$\xi = (\xi, V) = (\xi, V(\xi)).$$

The construction then attaches to any x a functor

$$\eta^x \colon (\xi, V(\xi)) \to V^x(\xi)_{\mathbb{Q}}, \qquad V(\xi) = V,$$

from the category $\text{Rep}(G)$ of rational representations of G to the category Hod of polarizable rational Hodge structures. It is a $\otimes$-functor (commuting with tensor products), with a matching

$$\omega_{\text{Hod}} \circ \eta^x = \omega_{\text{Rep}(G)}$$

of underlying fibre functors. But Hod is equivalent to the category $\text{Rep}(\mathscr{H})$. The properties of Tannakian categories then provide a homomorphism $\phi^x \colon \mathscr{H} \to G$ over $\mathbb{Q}$ for which η^x can be identified with the functor

$$\eta^x \colon (\xi, V(\xi)) \to (\xi \circ \phi^x, V(\xi)).$$

A comparison of ϕ^x with the given mapping ϕ^x_{fin} then leads in [144] back to the original element g in the pair $x = (h, g)$.

The conclusion reached on p. 214 on [144] is that $\mathscr{X}_K$ parametrizes pairs (ϕ, g), where ϕ is a homomorphism from $\mathscr{H}$ to G over $\mathbb{Q}$, and g is an element in $G(\mathbb{A}_{\text{fin}})$ taken only up to right multiplication by an element in K. Moreover, ϕ is subject to the constraint that its composition $\phi \circ h$ with the canonical homomorphism h from $\mathscr{R}$ to $\mathscr{H}$ lies in the set X of homomorphisms from $\mathscr{R}$ to G in the original Shimura datum (G, X).

Langlands notes finally that this construction of Deligne comes with the hope/expectation that any homomorphism $\phi' \colon \mathscr{H} \to G$ over $\mathbb{Q}$ such that $\phi' \circ h$ lies in X is a composition $\phi' = \phi \circ h^*_{\text{BH}}$, for a (uniquely determined) homomorphism $\phi \colon \mathscr{G} \to G$ over $\mathbb{Q}$, which is to say, a G-motive over $\mathbb{C}$ with coefficients in $\mathbb{Q}$. The complex variety $S_K(\mathbb{C})$ would then parametrize equivalence classes of pairs

$$\{(\phi, g) : \phi \colon \mathscr{G} \to G,\ \phi \circ h \in X,\ g \in G(\mathbb{A}_{\text{fin}})/K\},$$

where (ϕ', g') is equivalent to (ϕ, g) if, for some $\gamma \in G(\mathbb{Q})$,

$$(\phi', g') = (\text{ad}(\gamma)\phi, \gamma g).$$

This completes our discussion of §4 of [144]. Langlands observed that it did not yet yield a "moduli problem in the usual sense". He also asserted that "nonetheless, there is a good deal to be learned" from the discussion we have just sketched. Indeed there is. I have yet to learn much of the subsequent history of the problem. Among other things, I am puzzled by what seems to be a paucity of later references to a construction that seems so compelling (brief as our distilled review here is). I have not yet studied the later paper [164] of Langlands and Rapoport, or the fundamental volume [66] on absolute Hodge cycles, or later papers of Milne such as [173] and [175]. I presume that this construction has had to be reformulated in terms of gerbes in order to accommodate relations among motives in different characteristic. The reduction of a moduli space modulo p was of course essential to the proof of any case of the conjectural formula [142], [164], [123] for the terms on the geometric side of the Lefschetz trace formula. A recent paper of [111] of Kisin establishes the formula for Shimura varieties of abelian type. (As I understand it [176, §9], a motive is of *abelian* type if it lies in the category generated by abelian varieties; a Shimura variety $S_K = S_K(G,X)$ with rational weight w_X is of *abelian type* if it is a moduli space in the sense of the construction above of abelian motives.)

It was in §5 of [144] that Langlands introduced the Taniyama group $\mathscr{T}_F$. We recall that it is the replacement of the Weil group W_F in the diagrams (89) and (90) for the construction we have proposed for the motivic Galois group $\mathscr{G}_F$. It is an extension

$$1 \to \mathscr{S}_F \to \mathscr{T}_F \to \Gamma_F \to 1$$

of the Galois group $\Gamma_F = \mathrm{Gal}(\overline{F}/F)$ (with F embedded in $\mathbb{C}$) by the Serre group. The Serre group $\mathscr{S}$ is in turn a complex proalgebraic torus, with a continuous action of $\Gamma_{\mathbb{Q}} = \mathrm{Gal}(\overline{\mathbb{Q}}/\mathbb{Q})$. (We are writing $\mathscr{S}_F$ for the same group, but with its Galois action restricted from $\Gamma_{\mathbb{Q}}$ to the subgroup Γ_F.) It was actually defined by Langlands in §4, following Serre's construction in [208]. The Serre group should be the commutator quotient of both $\mathscr{H}$ and $\mathscr{G}$, and thereby fit into a diagram

$$\mathscr{H} \mapsto \mathscr{G} \to \mathscr{S}$$

of complex, connected groups. The (polarizable, rational) Hodge structures and (complex) motives defined by representations of $\mathscr{S}$ are then said to be of *CM-type*.

Langlands defined $\mathscr{T}_F$ by an explicit 2-cycle from Γ_F with values in $\mathscr{S}_F$. He then used it in §6 to formulate the conjectural solution to a moduli problem he had posed at the end of §4. The problem was just one of a number of questions that would need to be resolved in order to be able to treat a general Shimura variety $S_K = S_K(G,X)$ as a moduli space, as in the special case of a PEL-variety. It concerned how the proposed parametrization of $S_K(\mathbb{C})$ by pairs (ϕ,g) changes under an automorphism τ of $\mathbb{C}$. Langlands observed in general terms that (ϕ,g) would be replaced by a pair (ϕ',g') attached to another Shimura datum $(G',X') = (G^{\tau,\phi},X^{\tau,\phi})$. The problem was to describe the group $G^{\tau,\phi}$ explicitly. We will not review Langlands' conjectural resolution, obtained as part of his construction of the Taniyama group, but we recall that the conjecture was proved soon afterwards, independently by Borovoi [37] and Milne [173]. The Taniyama group itself has remained an important part of the theory. Besides its fundamental role in the motivic Galois group $\mathscr{G}_F$, and in the volume [66] on absolute Hodge cycles and elsewhere, it has also been used in a rather different context. The paper [6] introduced an extension of the Taniyama group, which in turn has become a part of a very interesting generalization [75] of the theory of motives.

The last Section 7 of [144] contains remarks of Langlands on the cohomology of general Shimura varieties $S_K = S_K(G,X)$, with a view towards the Hasse–Weil zeta function of S_K. These were taken up, at least implicitly, in the later paper [123] of Kottwitz discussed in the last section. I will not review them here, and in particular, their bearing on the inner twist $G^{\tau,\phi}$ of G introduced by Langlands in §4, and studied further in §5 and §6 of [144].

However, I would like to draw attention to the lemma of Langlands on p. 240 of [144]. We shall use it as the starting point for a discussion related to the final formula obtained by Kottwitz in §10 of [123], which we reviewed (but did not state) at the end of the last section. We shall describe a reciprocity identity for any Shimura variety, based on the general conjectures for motives. We shall then discuss some further motivic questions suggested by the identity.

Langlands considered an archimedean parameter in the set

$$\{\phi \in \Phi(G_{\mathbb{R}}) : S_\phi^0 \subset Z(\widehat{G}_{\mathbb{R}})\}$$

such that the graded vector space

$$H^*(\phi,\xi) = \bigoplus_{\pi\in\Pi_\phi} H^*(\mathfrak{g}_{\mathbb{R}}, K_{\mathbb{R}}; \pi\otimes\xi)$$

in (62) is nonzero. He then attached two representations of the real Weil group $W_{\mathbb{R}}$ to ϕ with two different roles in mind, one motivic and one automorphic. The lemma asserts that the two representations are equivalent.

The "motivic" representation for ϕ comes from a real Hodge structure on the space $H^*(\phi,\xi)$, introduced by Langlands at the lower half of p. 239 of [144]. We note that it is to be regarded as a spectral object, in contrast to the Hodge structures from §4 of [144] we have just reviewed. As a representation of $\mathbb{C}^*$ on $H^*(\phi,\xi)$, it is defined on a subgroup of index 2 in $W_{\mathbb{R}}$. To extend it, he first considered the archimedean Weil group $W_{\mathbb{C}/\overline{E}} = W_E$, where E is the Shimura field over which S_K is defined, and $\overline{E}$ is its completion defined by the embedding $E \subset \mathbb{C}$ with which it is equipped. If $\overline{E} = \mathbb{R}$, Langlands observed that the action of $\mathbb{C}^*$ on $H^*(\phi,\xi)$ extends naturally to the Weil group $W_{\mathbb{R}} = W_{\mathbb{C}/\mathbb{R}}$. With this in hand, his representation of $W_{\mathbb{R}}$ can then be taken in general to be

$$H^*(\phi,\xi)^+ = \mathrm{Ind}(W_{\mathbb{R}}, W_{\overline{E}}, H^*(\phi,\xi)),$$

the representation on a space $H^*(\phi,\xi)^+$ obtained by inducing the subrepresentation of $W_{\overline{E}}$ on $H^*(\phi,\xi)$.

Langlands' other representation of $W_{\mathbb{R}}$, the one whose role would be automorphic, is given by the parameter

$$\phi_{\mathbb{R}} : W_{\mathbb{R}} \to {}^L G_{\mathbb{R}}$$

itself. We recall from the last section that $S_K = S_K(G,X)$ comes with a cocharacter $\mu = \mu_h$ for G, and a corresponding character $\widehat{\mu}$ that serves as the highest weight for an irreducible representation $(r, V(r))$ of $\widehat{G}$. It follows from the definition of the reflex field E of S_K that the Galois group Γ_E stabilizes the $\widehat{G}$-orbit of the minuscule weight $\widehat{\mu}$, and hence that r extends to a representation r_E of the L-group ${}^L G_E = \widehat{G} \rtimes \Gamma_E$ of E such that Γ_E acts trivially on the weight space of $\widehat{\mu}$. Langlands then introduced the representation

$$r^+ = \operatorname{Ind}({}^L G, {}^L G_E, r_E) \tag{94}$$

of the L-group ${}^L G = \widehat{G} \rtimes \Gamma_{\mathbb{Q}}$ obtained by induction from the representation r_E of ${}^L G_E$ to ${}^L G$. His second representation of $W_{\mathbb{R}}$ can then be defined as

$$r^+ \circ \Phi,$$

where

$$\Phi(w) = \phi(w)|w|^{-\frac{d}{2}}, \qquad d = \dim S_K,\ w \in W_{\mathbb{R}}.$$

It is not hard to see from the definitions that Φ is algebraic, in the sense that it is the local component of a parameter in the global set $\Phi_{\mathrm{alg}}(G)$, or equivalently, that its constituents are of the form (91). The lemma on p. 240 of [144] can be taken as the assertion that the two representations of $W_{\mathbb{R}}$ are equivalent.

There are several deeper phenomena suggested by this lemma, simple as it may be. With one representation of $W_{\mathbb{R}}$ acting on an archimedean cohomology group, and the other given explicitly in terms of an archimedean Langlands parameter, it is suggestive of the local archimedean component of the Global Reciprocity correspondence between motives and automorphic representations. We would be dealing with a specific motive here. It is represented by the de Rham cohomology (62) of $S_K(\mathbb{C})$, and can therefore simply be regarded as the motive of the Shimura variety S_K over E. We are in fact speaking of what is known as a *realization* of the motive, specifically the Hodge realization. We can think of the lemma as a property of the local archimedean part of the mapping $\Phi \to \phi$ from $\Phi_{\mathrm{alg}}(G)$ to $\Phi_{\mathrm{temp}}(G)$ in (92).

There is something else in Langlands' lemma. It suggests a broader global perspective, one that goes beyond the reflex field E and the complex embedding $E \subset \mathbb{C}$. The local archimedean reflection of this phenomenon is given by the extensions $H^*(\phi)^+$ and $V(r)^+$ of the complex vector spaces $H^*(\phi)$ and $V(r)$ on which Langlands' two representations act. The global implication is that we would need to consider an extension of the motive of S_K, with components attached to Galois conjugates $E' \subset \mathbb{C}$ of $E \subset \mathbb{C}$. What would these motives be? Are they attached to Shimura varieties $S'_{K'}$? The conjecture of Langlands of §6 of [144], established not long afterwards [37], [173] asserted that they are.

Before I try to expand on these global implications, I should first describe the local generalization [18, Proposition 9.1] that was motivated by Langlands' lemma. This in turn relies on the local conjectures of [18, §8], which together with their global counterparts were a part of our discussion of Kottwitz' conjectural spectral (de)stabilization of the Lefschetz trace formula from the last section.

The local conjectures apply to any completion F_v of a number field F. They concern enriched parameters

$$\psi_v : L_{F_v} \times \mathrm{SL}(2,\mathbb{C}) \to {}^L G_v, \qquad \psi_v \in \Psi(G_v),$$

taken as usual up to conjugacy by $\widehat{G}$, but with the property that the image of L_{F_v} projects onto a bounded subset of $\widehat{G}$. In other words, the restriction ϕ_v of ψ_v to L_{F_v} is a tempered Langlands parameter. Given ψ_v, one forms the centralizer

$$S_{\psi_v} = \mathrm{Cent}(\psi(L_{F_v} \times \mathrm{SL}(2,\mathbb{C})), \widehat{G})$$

in $\widehat{G}$ of its image, and the finite group

$$\mathbf{S}_{\psi_v} = S_{\psi_v} / S^0_{\psi_v} Z(\widehat{G})^{\Gamma_v},$$

often abelian, of connected components in S_{ψ_v} modulo the Galois invariants in the centre of $\widehat{G}$. For each ψ_v, the conjectures assert the existence of a finite set Π_{ψ_v} of representations of $G(F_v)$. This set would parametrize (in a noncanonical way) the irreducible characters

$$s_v \to \langle s_v, \pi_v \rangle, \qquad s_v \in \mathbf{S}_{\psi_v},\ \pi_v \in \Pi_{\psi_v},$$

on the group $\mathbf{S}_{\psi_v}$. However, in contrast to the (bounded) Langlands parameters ϕ_v, the representations $\pi_v \in \Pi_{\psi_v}$ need not be either tempered or irreducible. On the other hand they are conjectured to be unitary, and to be finite sums of irreducible representations. The local parameters ψ_v and packets Π_{ψ_v}, together with their global counterparts ψ and Π_ψ, are really a part of the theory of endoscopy. We have mentioned this term regularly in earlier sections, but we have not yet said what it is. We shall do so in the next section.

For the proposition from [18], we take v to be the real valuation of $\mathbb{Q}$, and start with a parameter $\psi_{\mathbb{R}}$ in the set

$$\Psi_2(G_{\mathbb{R}}) = \{\psi_{\mathbb{R}} \in \Psi(G_{\mathbb{R}}) : S^0_{\psi_{\mathbb{R}}} \subset Z(\widehat{G}_{\mathbb{R}})\}.$$

The groups $S_{\psi_{\mathbb{R}}}$ and $\mathbf{S}_{\psi_{\mathbb{R}}}$ are then abelian. However, the situation here is slightly different from that of Langlands' lemma, given the requirement that the $\psi_{\mathbb{R}}$-image of $L_{\mathbb{R}} = W_{\mathbb{R}}$ be bounded in $\widehat{G}$. We are regarding the irreducible representation ξ of G as algebraic, which means that its restriction to $G_{\mathbb{R}}$ generally has nonunitary central character. This in turn forces the cohomology $H^*(\mathfrak{g}_{\mathbb{R}}, K_{\mathbb{R}}, \pi_{\mathbb{R}} \otimes \xi)$ to vanish for representations $\pi_{\mathbb{R}}$ in the packet $\Pi_{\psi_{\mathbb{R}}}$. To rectify the problem, we write

$$\xi_{\mathbb{R}}(x_{\mathbb{R}}) = \xi(x_{\mathbb{R}})|\det(x_{\mathbb{R}})|^{\alpha_\xi}, \qquad x_{\mathbb{R}} \in G_{\mathbb{R}} = G(\mathbb{R}),$$

where $\alpha_\xi \in \mathbb{R}^+$ is the number that makes the central character of $\xi_{\mathbb{R}}$ unitary. (This is related to the earlier footnote 6 and the ensuing discussion for $\mathrm{GL}(2)$. It is also implicit in the discussion on p. 61 of [20].) With this understanding, we define $\Psi_2(G_{\mathbb{R}}, \xi_{\mathbb{R}})$ to be the subset of archimedean parameters in $\Psi_2(G_{\mathbb{R}})$ such that the graded vector space

$$H^*(\psi_{\mathbb{R}}, \xi_{\mathbb{R}}) = \bigoplus_{\pi_{\mathbb{R}} \in \Pi_{\psi_{\mathbb{R}}}} H^*(\mathfrak{g}_{\mathbb{R}}, K_{\mathbb{R}}; \pi_{\mathbb{R}} \otimes \xi_{\mathbb{R}})$$

is nonzero. The representations $\pi_{\mathbb{R}} \in \Pi_{\psi_{\mathbb{R}}}$ are interesting examples of the unitary representations with cohomology classified by Vogan and Zuckerman [244].

Consider then a parameter $\psi_{\mathbb{R}}$ in the set $\Psi_2(G_{\mathbb{R}}, \xi_{\mathbb{R}})$. The representations $\pi_{\mathbb{R}} \in \Pi_\psi$ then give rise to a real Hodge structure. This relies on the analysis of such parameters by Adams and Johnson [2], [18, §5]. It in turn gives a representation of $\mathbb{C}^*$ on $H^*(\psi_{\mathbb{R}}, \xi_{\mathbb{R}})$, but with components of the form (93) rather than (91), for our having replaced ξ by $\xi_{\mathbb{R}}$. The group $S_{\psi_{\mathbb{R}}}$ also acts on $H^*(\psi_{\mathbb{R}}, \xi_{\mathbb{R}})$, and commutes with the action of $\mathbb{C}^*$. The extra ingredient in the parameter is the group $\mathrm{SL}(2,\mathbb{C})$. It also acts on the space $H^*(\psi_{\mathbb{R}}, \xi_{\mathbb{R}})$, where it governs the grading in the manner familiar from the hard Lefschetz theorem. Since it commutes with the action of the product of $S_{\psi_{\mathbb{R}}} \times \mathbb{C}^*$, we obtain a representation of the product $S_{\psi_{\mathbb{R}}} \times \mathbb{C}^* \times \mathrm{SL}(2,\mathbb{C})$ on the space $H^*(\psi_{\mathbb{R}}, \xi_{\mathbb{R}})$. Following Langlands, we can then construct an induced representation

$$\rho^+_{\psi_{\mathbb{R}}}(s,(w,u)), \qquad (s,w,u) \in S_{\psi_{\mathbb{R}}} \times (W_{\mathbb{R}} \times \mathrm{SL}(2,\mathbb{C})),$$

of the product $S_{\psi_{\mathbb{R}}} \times (W_{\phi_{\mathbb{R}}} \times \mathrm{SL}(2,\mathbb{C}))$ on a graded vector space $H^*(\psi_{\mathbb{R}}, \xi_{\mathbb{R}})^+$ that contains $H^*(\psi_{\mathbb{R}}, \xi_{\mathbb{R}})$ and that is the analogue of the space $H^*(\phi, \xi)^+$ introduced above. This is the "motivic" representation for $\psi_{\mathbb{R}}$.

The "automorphic" representation for $\psi_{\mathbb{R}}$ is constructed as above, but in terms of the new parameter $\psi_{\mathbb{R}}$. It equals

$$\sigma^+_{\psi_{\mathbb{R}}}(s,(w,u)) = r^+_{\mathbb{R}}(\psi_{\mathbb{R}}(w,u)s), \quad (s,w,u) \in S_{\psi_{\mathbb{R}}} \times (W_{\mathbb{R}} \times \mathrm{SL}(2,\mathbb{C})), \tag{95}$$

where $r^+_{\mathbb{R}}$ is the representation of ${}^L G_{\mathbb{R}}$ attached as above to the Shimura datum (G,X). Proposition 9.1 of [18] amounts to the assertion

$$\rho^+_{\psi_{\mathbb{R}}} \cong \sigma^+_{\psi_{\mathbb{R}}} \tag{96}$$

that the two representations of $S_{\psi_{\mathbb{R}}} \times (W_{\mathbb{R}} \times \mathrm{SL}(2,\mathbb{C}))$ are equivalent. It reduces to the lemma of Langlands in the case that the parameter $\psi_{\mathbb{R}} = \phi_{\mathbb{R}}$ lies in the subset $\Phi_2(G_{\mathbb{R}})$ of $\Psi_2(G_{\mathbb{R}})$, which is to say that its restriction to $\mathrm{SL}(2,\mathbb{C})$ is trivial. (The proposition was actually formulated and proved for the smaller representations with $\mathbb{C}^*$ in place of $W_{\mathbb{R}}$, but its extension follows from the various definitions.)

We can now turn to the global implications of these local results. We shall introduce two global representations, one motivic and one automorphic, which for the moment serve simply to help us focus our thoughts. They depend on the global conjectures from [18, §8], about which we can first say a few words.

If G is any reductive group over a number field F, we write $\Psi(G)$ for the set of L-homomorphisms

$$\psi\colon L_F \times \mathrm{SL}(2,\mathbb{C}) \to {}^L G$$

such that L_F has bounded image in $\widehat{G}$, taken up to $\widehat{G}$-conjugacy. The domain here now includes the hypothetical global Langlands group L_F, in which the local groups L_{F_v} are imbedded. Any ψ would therefore give local parameters $\psi_v \in \Psi(G_v)$, local packets Π_{ψ_v} of representations, and a global packet Π_ψ of representations

$$\pi = \widetilde{\bigotimes_v} \pi_v, \qquad \pi_v \in \Pi_{\psi_v},$$

of $G(\mathbb{A})$, in which π_v is required to be unramified in a certain sense for almost all v. The natural mappings $s_v \to s$ from the local groups $\mathbf{S}_{\psi_v}$ to the corresponding global group $\mathbf{S}_\psi = S_\psi / S_\psi^0 Z(\widehat{G})^\Gamma$ will then attach a global pairing

$$s \to \langle s, \pi \rangle = \prod_v \langle s_v, \pi_v \rangle, \qquad s \in \mathbf{S}_\psi,\ \pi \in \Pi_\psi,$$

on $\mathbf{S}_\psi$ to any representation in the global packet. The factors on the right will equal 1 for almost all v, while the product of the noncanonical local pairings $\langle s_v, \pi_v \rangle$ will become canonical. The result would be a canonical finite-dimensional character $\langle \cdot, \pi \rangle$ on $\mathbf{S}_\psi$ for every representation π in the global packet Π_ψ. Finally, suppose that ψ lies in the subset $\Psi_2(G)$ of global parameters such that the connected centralizer S_ψ^0 is contained in $Z(\widehat{G})$. In general, there is a natural one-dimensional sign character ε_ψ on $\mathbf{S}_\psi$, constructed in a simple way from symplectic root numbers attached to ψ. The main conjecture is then that the automorphic discrete spectrum of G (taken modulo the centre) is a direct sum over $\psi \in \Psi_2(G)$ of representations $\pi \in \Pi_\psi$, taken with multiplicities equal to the multiplicities

$$m_2(\pi) = |\mathbf{S}_\psi|^{-1} \sum_{s \in \mathbf{S}_\psi} \varepsilon_\psi(s) \langle s, \pi \rangle, \qquad \pi \in \Pi_\psi, \tag{97}$$

of ε_ψ in $\langle \cdot, \pi \rangle$. (See [18, §8].)

We return now to the Shimura varieties $S_K = S_K(G,X)$, with F equal to either $\mathbb{Q}$ or the reflex field E. Our main global representation will be the one that is "automorphic". It is built in a natural way from the archimedean representation (95) introduced above and its nonarchimedean complement from the corresponding expression (62) for the L^2-cohomology of $S_K(\mathbb{C})$. It is the representation

$$\bigoplus_{\psi \in \Psi_2(G,\xi)} \bigoplus_{\pi \in \Pi_\psi} (\sigma_\psi^+ \otimes \pi_{\mathrm{fin}}^K)_{\varepsilon_\psi} \tag{98}$$

of $(L_{\mathbb{Q}} \times \mathrm{SL}(2,\mathbb{C})) \times \mathscr{H}_K(G)$, whose terms we describe as follows. The representation itself is a direct sum over global parameters

$$\psi\colon L_{\mathbb{Q}} \times \mathrm{SL}(2,\mathbb{C}) \to {}^L G$$

in the subset $\Psi_2(G,\xi)$ of parameters in $\Psi_2(G)$ that restrict to archimedean parameters in the subset $\Psi_2(G_{\mathbb{R}},\xi_{\mathbb{R}})$ of $\Psi_2(G_{\mathbb{R}})$, and representations π of $G(\mathbb{A})$ in the corresponding global packet Π_ψ. For any ψ and π, $(\sigma_\psi^+ \otimes \pi_{\mathrm{fin}}^K)_{\varepsilon_\psi}$ is then the representation of $(L_{\mathbb{Q}} \times \mathrm{SL}(2,\mathbb{C})) \times \mathscr{H}_K$ given by the multiplicity of the sign character ε_ψ on $\mathbf{S}_\psi$ in the representation

$$(\sigma_\psi^+ \otimes \pi_{\mathrm{fin}}^K)((w,u),s,f) = r^+(\psi(w,u)s) \otimes \langle s, \pi_{\mathrm{fin}}^K \rangle (\pi_{\mathrm{fin}}^K(f))$$

of $(L_{\mathbb{Q}} \times \mathrm{SL}(2,\mathbb{C})) \times \mathbf{S}_\psi \times \mathscr{H}_K$. In other words,

$$(\sigma_\psi^+ \otimes \pi_{\mathrm{fin}}^K)((w,u),f) = |\mathbf{S}_\psi|^{-1} \sum_{s\in\mathbf{S}_\psi} \varepsilon_\psi(s)\big(r^+(\psi(w,u)s)\big) \otimes \langle s, \pi_{\mathrm{fin}}^K \rangle \big(\pi_{\mathrm{fin}}^K(f)\big). \quad (99)$$

The representation (98) is the centre of our discussion. The essential point is that it should be equivalent to the natural representation of $(L_{\mathbb{Q}} \times \mathrm{SL}(2,\mathbb{C})) \times \mathscr{H}_K$ on our expression

$$H^*_{(2)}(S_K(\mathbb{C}),\mathscr{F})^+ = \bigoplus_{\pi} m_2(\pi)\big(H^*(\mathfrak{g}_{\mathbb{R}},K_{\mathbb{R}},\pi_{\mathbb{R}} \times \xi_{\mathbb{R}})^+ \otimes \pi_{\mathrm{fin}}^K\big) \quad (100)$$

for the extended L^2-cohomology. Indeed (100) is defined in the same way as (98), but with the archimedean representation $\sigma_{\psi_{\mathbb{R}}}^+$ of (95) replaced by the representation $\rho_{\psi_{\mathbb{R}}}^+$ on $(\mathfrak{g}_{\mathbb{R}},K_{\mathbb{R}})$-cohomology. The equivalence of the two representations (98) and (100) would then follow from the equivalence (96) of $\rho_{\psi_{\mathbb{R}}}^+$ and $\sigma_{\psi_{\mathbb{R}}}^+$, the form (97) for the multiplicity $m_2(\pi)$ in terms of the parameter ψ and the sign character, and the various definitions.

The analogue of (98) for the actual L^2-cohomology $H^*_{(2)}(S_K(\mathbb{C}),\mathscr{F})$ is a similar representation, but with the group L_E in place of $L_{\mathbb{Q}}$. It amounts to the formula stated as [18, (9.3)] (with the local representation $\rho_{\psi_{\mathbb{R}}}$ in place of $\sigma_{\psi_{\mathbb{R}}}$), where it follows from the definition of $\rho_{\psi_{\mathbb{R}}}$ and the actual assertion of Proposition 9.1 of [18]. We can think of (98) as the representation of $L_{\mathbb{Q}}$ induced from this representation of L_E. This relation is in keeping with the bijections

$$L_{\mathbb{Q}}/L_E \cong W_{\mathbb{Q}}/W_E \cong \Gamma_{\mathbb{Q}}/\Gamma_E \cong \mathrm{Hom}(E,\mathbb{C})$$

that we expect of the Langlands group, and that are stated for the Weil and Galois groups on the first page of Tate's article [237].

Remarks. 1. The representation (98) is an interesting expression on several counts. It provides some insight into automorphic representations (acting on spaces of automorphic forms) rather than the characters in terms of which they were classified in [23]. By displaying the global parameters and their corresponding families of $\mathscr{H}_K$-modules $\{\pi_{\text{fin}}^K\}$ together as tensor products, it bears a philosophical resemblance to the theta correspondence. The automorphic modules π^K are lacking the archimedean components $\pi_{\mathbb{R}}$, but these are hidden in the representations σ_ψ^+, or rather the $(\mathfrak{g}_{\mathbb{R}}, K_{\mathbb{R}})$-cohomology within the equivalent representations ρ_ψ^+.

2. The finite group $\mathbf{S}_\psi$ is abelian, since it is a subgroup of the archimedean group $\mathbf{S}_{\psi_{\mathbb{R}}}$, which in the case of the algebraic parameters ψ is itself abelian [2]. However, the other groups $\mathbf{S}_{\psi_v}$ need not be abelian, so $\langle s, \pi_{\text{fin}}^K\rangle$ can be a higher-dimensional character, which means that $\sigma_\psi^+ \otimes \pi_{\text{fin}}^K$ is not strictly a representation of the component group $\{s\} = \mathscr{S}_\psi$. It perhaps ought to be replaced in the notation by a higher-dimensional $\mathbf{S}_\psi$-module. But the formula (98) makes sense as stated, and in any case, the groups $\mathbf{S}_{\psi_v}$ are typically abelian, making $\langle s, \pi_{\text{fin}}^K\rangle$ a one-dimensional character.

3. The sign character ε_ψ is an interesting arithmetic object in its own right. That it occurs in the basic automorphic expression (98) for Shimura varieties is perhaps surprising.

4. It is the automorphic expression (98) that is closely related to the formula of Kottwitz displayed in the last paragraph of [123, §10], and that was our object of discussion at the end of the last section. His derivation of the formula by the comparison of trace formulas, even though it rests on the conjectural fixed point formula [123, (3.1)] and the conjectures of [18, §8], of course brings us closer to an actual proof of the formula than our derivation of (98) by Langlands' Functoriality and Reciprocity. Our goal for (98) has been conceptual.

5. The isomorphism of $H_{(2)}^*(S_K(\mathbb{C}), \mathscr{F})^+$ with the representation (98) could be regarded as a global counterpart of a conjectural formula of Kottwitz [190] for the representations of local groups in the cohomology of local Shimura varieties.

The automorphic representation (98) can be regarded as the primary object of this discussion, from which the others follow, and to which subsequent questions ultimately return. The representation of the product $(L_{\mathbb{Q}} \times \mathrm{SL}(2,\mathbb{C})) \times \mathscr{H}_K$ on the extended L^2-cohomology (100), while being equivalent to (98), is really secondary. For as we have noted, it is essentially a realization of a motive. The true motivic companion of (98) would be a more direct analogue. We define it formally as the algebraic representation

$$\bigoplus_{\Psi} \bigoplus_{\Pi \in \Pi_\Psi} (\Sigma_\Psi^+ \otimes \Pi_{\text{fin}}^K)_{\varepsilon_\Psi} \tag{101}$$

of $(\mathscr{G}_{\mathbb{Q}} \times \mathrm{SL}(2,\mathbb{C})) \times \mathscr{H}_K$ whose pullback to $(L_{\mathbb{Q}} \times \mathrm{SL}(2,\mathbb{C})) \times \mathscr{H}_K$ under Reciprocity equals (98). In particular, Ψ is the preimage of the parameter $\psi \in \Psi_2(G,\xi)$ from (98), under the analogue of the mapping (92). The automorphic representations in (101), which we have written as $\Pi \in \Pi_\Psi$ in place of $\pi \in \Pi_\psi$ are an interesting part of the construction. They are representations of $\mathscr{H}_K$, as an algebra of cycles in the extended cohomology space that commutes with the motivic representation of $\mathscr{G}_{\mathbb{Q}} \times \mathrm{SL}(2,\mathbb{C})$, and which in turn acts as a kind of diagonalization of this algebra.

The motivic representation (101) does not display much of its internal structure. This is because we have been treating it as a representation of the complex group

$\mathcal{G}_{\mathbb{Q}} = \mathcal{G}_{\mathbb{Q}}(\mathbb{C})$, the group of complex points of the underlying group $\mathcal{G}_{\mathbb{Q},\mathbb{Q}}$ over $\mathbb{Q}$. (We adopted the overlapping notation earlier to emphasize the parallel conjectural structure of the universal groups L_F and $\mathcal{G}_F$. Applied to $\mathcal{G}_F$ alone, the ambiguity is harmless; we can regard $\mathcal{G}_F$ either as the group of complex points of a scheme over $\mathbb{Q}$, or more traditionally, as a complex group with underlying structure as a group defined over $\mathbb{Q}$. In the case here of $F = \mathbb{Q}$, we shall often write $\mathcal{G} = \mathcal{G}_{\mathbb{Q}}$ simply for the reductive group over $\mathbb{Q}$.) It is this $\mathbb{Q}$-structure that governs the arithmetic properties of (100).

Among other things, the $\mathbb{Q}$-structure is needed to complete the Hodge realization of the Shimura variety S_K on the extended cohomology space $H^*_{(2)}(S_K(\mathbb{C}), \mathcal{F})^+$. We saw in the construction of the motivic representation (100) (using (96) and Langlands' earlier lemma) how to define the real Hodge structure on the space. To extend it to a rational Hodge structure, we would need to use the fact that the representation (100) of $\mathcal{G}_{\mathbb{Q}} \times \mathrm{SL}(2,\mathbb{C})$ as a fibre functor can be defined over $\mathbb{Q}$. The $\mathbb{Q}$-Hodge structure is required in turn to be polarizable [144, p. 215]. We would want to be able to attach an explicit polarization to the motivic parameters. This should bear a simple relation to the Lefschetz structure given by the $\mathrm{SL}(2,\mathbb{C})$-component of the parameters.

The most widely studied realization functor for motives is defined by their étale cohomology and its corresponding compatible families of ℓ-adic Galois representations. Known as the $\mathbb{A}_{\mathrm{fin}}$-realization [65], it is of obvious arithmetic importance. For the Shimura variety S_K, it was the main topic of our last section. As we recall, the Galois representations act on the ℓ-adic (étale) version of the intersection cohomology $IH^*(\overline{S}_K, \mathcal{F}_\lambda)^+$ for the Baily–Borel compactification of $\overline{S}_K$. Like the Hodge realization, it depends very much on the $\mathbb{Q}$-structure of the group $\mathcal{G}_{\mathbb{Q}}$. As a matter of fact, we cannot really speak of the $\mathbb{A}_{\mathrm{fin}}$-realization of S_K, and the individual ℓ-adic representations in particular, without this $\mathbb{Q}$-structure. For it is considerably more subtle than the $\mathbb{Q}$-structure on the Shimura group G. Without it, one has to work with compatible families of λ-adic representations, where λ ranges over the non-archimedean completions of a finite extension L of $\mathbb{Q}$ that depends on the group K. This was the point of view of Langlands in [140] and Kottwitz in [123].

We display the representation spaces we have discussed in the diagram

$$\begin{array}{ccc} H^*_{(2)}(S_K, \mathcal{F})^+ & \longleftarrow & \bigoplus_{\psi}\bigoplus_{\pi} (\sigma^+_\psi \otimes \pi^K_{\mathrm{fin}})_{\varepsilon_\psi} \qquad (98) \\ \big\updownarrow \wr & & \big\downarrow \\ IH^*(\overline{S}_K, \mathcal{F})^+ & \longleftarrow & \bigoplus_{\Psi}\bigoplus_{\Pi} (\Sigma^+_\Psi \otimes \Pi^K_{\mathrm{fin}})_{\varepsilon_\Psi} \qquad (101) \end{array}$$

The diagram is somewhat impressionistic, but it is helpful for us in keeping track of what is highly conjectural, and what is better understood and more concrete. The vertical arrow on the right is in the former category. Indeed, its domain in the upper right-hand corner is given in terms of the hypothetical automorphic Galois group L_F, whose existence is closely related to Langlands' Principle of Functoriality. The ultimate proof of this would be the goal of Beyond Endoscopy, the recent long term program of Langlands we will discuss in §11. The codomain in the lower right-hand corner depends on the hypothetical motivic Galois group $\mathscr{G}_F$, while the arrow itself is given by Langlands' Reciprocity Conjecture. There is no concrete program for the proof of this, but it will surely demand everything we can prove about Shimura varieties. The vertical arrow on the left is the isomorphism of Zucker's conjecture, which we recall has been known for twenty years. The lower horizontal arrow is given by the $\mathbb{A}_{\mathrm{fin}}$-realization of S_K^+, while its composition with the isomorphism on the left is the Hodge realization of S_K^+. The upper horizontal arrow is built out of Langlands' lemma and the proposition from [18] with which we began this discussion. It seems clear from these remarks that the automorphic representation in the upper right-hand corner can indeed be regarded as the foundation of the other spaces and arrows in the diagram.

We shall conclude this discussion with a list of five problems. These represent refinements of conjectures that would enhance our understanding, as opposed to ideas that might be applied to their eventual proofs. Some of them are accessible, requiring perhaps only a little careful thought. In fact, some of these may in fact already be known. But taken together, they present a broader picture that can only serve to help us.

Problems: 1. Realizations of S_K. We are thinking again of the Hodge and $\mathbb{A}_{\mathrm{fin}}$-realizations of the (extended) Shimura motive (101). The problem would include a more explicit description of the fibre functor, as a $\mathbb{Q}$-refinement of the complex representation (101) of $\mathscr{G}_{\mathbb{Q}} \times \mathrm{SL}(2,\mathbb{C})$ on either $H^*_{(2)}(S_K,\mathscr{F})^+$ or $IH^*(\overline{S}_K,\mathscr{F})^+$. This can be considered as a special case of the corresponding problem for the full motivic Galois group $\mathscr{G}_{\mathbb{Q}}$, which we will state as Problem 3 below. However, there are some supplementary (and simpler) questions we could ask about the case of a Shimura variety S_K here.

For example, (101) is the direct sum over a finite set of parameters $\Psi \in \Psi_{\mathrm{alg}}(G)$ (the analogue of $\Phi_{\mathrm{alg}}(G)$ for $\Phi(G)$) of complex representations of $\mathbb{G}_{\mathbb{Q}} \times \mathrm{SL}(2,\mathbb{C})$. What is the partition of this set that gives the decomposition of the sum, as a representation over $\mathbb{Q}$? Can one answer this question without a full understanding of the $\mathbb{Q}$-structure on $\mathscr{G}_{\mathbb{Q}}$? What supplementary information might be required on the complex Hecke algebra

$$\mathscr{H}_K = \mathscr{H}(K \setminus G(\mathbb{A}_{\mathrm{fin}})/K)$$

as a rational algebra of correspondences of cycles on S_K, and on its representations on the relevant space of automorphic forms $\mathscr{A}_\xi(G(\mathbb{Q}) \setminus G(\mathbb{A})/K)$. Furthermore, for each Ψ, the representation Σ_Ψ^+ in (101) is defined as in (98) in terms of an L-homomorphism of $\mathscr{G}_{\mathbb{Q}} \times \mathrm{SL}(2,\mathbb{C})$ into the complex L-group ${}^LG$. Do we need to impose a $\mathbb{Q}$-structure on ${}^LG$ to be able to ask these questions on the $\mathbb{Q}$-fibre functor? I have not thought about the problem, even to the point of being confident in posing the questions.

2. Hasse–Weil zeta function of S_K. This is of course a famous longstanding problem. It was posed in this context by Langlands [146, 159, 152], but I am not quite sure of its present status. The question here is of a conjectural formula, since the expressions (98) and (101) on which it could be based are hypothetical. The same is true of the similar formula at the end of [123, §10], even though its conjectural foundations are much less severe. The problem appears to be quite accessible, amounting no doubt to a careful collection of the relevant terms in either of the formulas above, but it would be well worth any time taken to fully understand it. For the special case of Picard modular surfaces, a family of Shimura surfaces attached to various forms of the unitary group in three variables, the answer is known, and has been fully proved. We refer the reader to the volume on the subject edited by Langlands and Ramakrishnan, and their summary [163] from the volume of its main result.

We note that there are really two zeta functions. One is the zeta function of S_K as a variety over E. The other is attached to the disconnected variety over $\mathbb{Q}$ represented by the extended cohomology spaces on the left-hand side of the diagram above. It would be a product of zeta functions taken over its components. Each factor would be the zeta function of a separate Shimura variety, obtained from S_K by the inner twist proposed by Langlands in Sections 4–6 of [144].

I will try to return to the representations in (98) and (101) elsewhere, with supplementary details and possible extensions. The goal would be to give a precise (conjectural) formula for these zeta functions.

3. $\mathbb{Q}$-structure on $\mathcal{G}_F$. We have given an explicit conjectural description of the complex motivic Galois group $\mathcal{G}_F$ over a number field F. It is a fibre product

$$\prod_c (\mathcal{G}_c \to \mathcal{T}_F),$$

over a set $\mathcal{C}_{F,\mathrm{alg}}$ of equivalence classes of pairs (G,c), of extensions

$$\mathcal{G}_c \to \mathcal{T}_F \to \Gamma_F$$

of complex simply connected groups $\mathcal{G}_c$. The problem would be to give an explicit conjectural description of the $\mathbb{Q}$-structure on $\mathcal{G}_F$ attached to the complex embedding $F \subset \mathbb{C}$ and the corresponding Betti fibre functor. (We should not forget that the two fields F and $\mathbb{Q}$ here are the two fields F and Q in (77), and have quite different sources.) Langlands' Taniyama group $\mathcal{T}_F$ is the set of complex points of a proalgebraic group that is already defined over $\mathbb{Q}$. The problem is to extend this $\mathbb{Q}$-structure to the fibre product of groups $\mathcal{G}_c$. Might the solution be given in terms of the equivalence classes of concrete families $c = \{c_v\}$ of semisimple conjugacy classes in ${}^LG$ that make up the indexing set $\mathcal{C}_{F,\mathrm{alg}}$?

There would be two steps. We have been writing $\mathcal{G}_F$ for the group over $\mathbb{Q}$ whose structure we seek. Let us write $\mathcal{G}_F^{\mathrm{spl}}$ for the same group of complex points, but with the structure of a (disconnected) split group over $\mathbb{Q}$. Intermediate between $\mathcal{G}_F$ and $\mathcal{G}_F^{\mathrm{spl}}$ would be a quasisplit group $\mathcal{G}_F^* = \mathcal{G}_F^{\mathrm{qs}}$. The first step would be to describe $\mathcal{G}_F^*$

explicitly by an outer twist of the Galois action on $\mathscr{G}_F^{\mathrm{spl}}$. The second step would be to describe $\mathscr{G}_F$ as an inner twist of $\mathscr{G}_F^*$.

How would we approach the first step? The co-ordinates of the conjugacy classes $c = \{c_v\}$ ought to be algebraic numbers. This is known in many cases, where it can be established from the "finite form" of the trace formula, with the test function f being cuspidal at an Archimedean place. It would follow in general from functoriality, which we have already taken as a prerequisite for this section. The Galois group $\Gamma_{\mathbb{Q}}$ would then act by permutation of the families c, and hence on the indices $\mathscr{C}_{F,\mathrm{alg}}$ in the fibre product that defines $\mathscr{G}_F$. It is tempting to think of using this to construct a quasisplit outer form of $\mathscr{G}_F^{\mathrm{spl}}$ over $\mathbb{Q}$.

Our concern here is the Betti realization. It would be attached to the fibre functor that assigns to a motive defined over F its Betti cohomology with $\mathbb{Q}$-coefficients. This should give a quasisplit outer form over F of the original group G. The problem is that it has also to depend on the embedding of F into $\mathbb{C}$ in order to become a group over $\mathbb{Q}$. This leads to the second step, the description of the inner twist $\mathscr{G}_F$ of $\mathscr{G}_F^*$. The problem is to describe the associated nonabelian cohomology class in $H^1(\mathbb{Q}, G_{\mathrm{ad}}^*) \cong H^1(\mathbb{Q}, \widehat{G})$ explicitly.

I regret not having had the time to think about the question (as well as many other things!), because I suspect that the answer is both simple and interesting. To exploit it, we would note that motivic Galois groups should behave like Weil groups, in the sense that as a complex group, $\mathscr{G}_F$ would be the preimage of the subgroup $\Gamma_F \subset \Gamma_Q$ in the projection $\mathscr{G}_Q \to \Gamma_Q$ for any field $Q \subset F$, and hence that

$$\mathscr{G}_Q / \mathscr{G}_F \cong \Gamma_Q / \Gamma_F \cong \mathrm{Hom}_Q(F, \mathbb{C}).$$

Taking Q to be the rational field $\mathbb{Q}$, we could then write $\mathscr{G}_{\mathbb{Q}}^*$ as a disjoint union of groups $\mathscr{G}_F^*$, taken over the embeddings that parameterize the different Betti fibre functors. They define an inner twist of $\mathscr{G}_{\mathbb{Q}}^*$ that depends only on F. The inner form $\mathscr{G}_{\mathbb{Q}}$ of $\mathscr{G}_{\mathbb{Q}}^*$ that we seek would then presumably be the direct limit over increasing fields F of the inner twists defined in this way for each F. Note that as a reductive proalgebraic group with $\mathbb{Q}$-structure (over the group $\Gamma_{\mathbb{Q}}$), $\mathscr{G}_{\mathbb{Q}}$ is completely canonical. To finish the second step, we could simply take $\mathscr{G}_F$ to be the preimage of $\Gamma_F \subset \Gamma_{\mathbb{Q}}$ in $\mathscr{G}_{\mathbb{Q}} \to \Gamma_{\mathbb{Q}}$ attached to an embedding $F \subset \mathbb{C}$.

What we have just described is related to the conjecture of Langlands stated in §6 of [144] (and proved in [37, 173]). The conjecture applies to a Shimura variety over the reflex field $E \subset \mathbb{C}$. It attaches different Shimura varieties to different complex embeddings of E, each obtained from the original one by an *explicit* inner twist. What we called the extended cohomology $H_{(2)}^*(S_K, \mathscr{F})^+$ (with locally constant sheaf $\mathscr{F}$) for the resulting motive over $\mathbb{Q}$ then becomes a disjoint union of the motives of the different Shimura varieties, or rather the Hodge realizations of these motives. It would be very interesting to compare this conjecture, and its solution, with the second step above. I hope to return to some of these questions in a future paper.

4. Realizations for $\mathscr{G}_F$. The Hodge and $\mathbb{A}_{\text{fin}}$ realizations of a Shimura variety are a fundamental part of its theory. A full solution of Problem 3 would give us a different way to view the realizations of any motive, each based on some further structure on the group $\mathscr{G}_F$.

The $\mathbb{A}_{\text{fin}}$-realization of a motive is a compatible family of ℓ-adic representations

$$\bigotimes_{\ell \neq p} r_\ell, \qquad \ell \notin S,$$

of $\Gamma_{\mathbb{Q}}$. The prime p represents a $\mathbb{Q}$-rational conjugacy class c_p, which embed diagonally in the ℓ-adic vector spaces. This formulation presupposes that as a representation of $\mathscr{G}_F$, the motive is defined over $\mathbb{Q}$. But without the $\mathbb{Q}$-structure in hand, one has to work implicitly with the groups $\mathscr{G}_F^{\text{spl}}$. As we have noted, this is what necessitates taking λ-adic representations for the completions of a finite extension L of $\mathbb{Q}$. But we are now supposing that we have the $\mathbb{Q}$-structure on $\mathscr{G}_F$ attached to a Betti fibre functor. The $\mathbb{A}_{\text{fin}}$-realization then becomes more fundamental. For it would amount to a compatible family of ℓ-adic homomorphisms from $\Gamma_{\mathbb{Q}}$ to $\mathscr{G}_F$ over $\mathbb{Q}$.

Similar comments would also apply to the Hodge realization. I have not thought precisely about how best to express them, but it would clearly be interesting to formulate the Hodge realization as further structure on the group $\mathscr{G}_F$. It is closely related to the *period realization*, which we will discuss in a moment. We should add that for any index (G,c) in $\mathscr{C}_{F,\text{alg}}$, the ramified local complements

$$\{\Phi_v \in \Phi(G_v) : v \in S\}$$

of the family $c = \{c_v : v \notin S\}$ ought to be uniquely determined by c itself. In particular, c would give us the archimedean parameter Φ_∞ on which the Hodge structure depends. This presumably follows from the theorem of strong multiplicity one for GL_N, and the fact that c is primitive.

There are other realizations for motives. One would like to understand them all in terms of $\mathscr{G}_F$. We shall say a few more words about one of them, the *period realization* (which I believe is the same as what is often called the *De Rham–Betti realization*). It is yet another extraordinary side of Grothendieck's vision for motives. It suggests a systematic approach to classical transcendental number theory. Even more remarkable is that it represents an extension of algebraic number theory to many of the classical transcendental numbers that have been with us as definite integrals or infinite series since the advent of calculus. The role of the classical Galois group Γ_F is then played by its extension given by the totally disconnected group $\mathscr{G}_F(\mathbb{Q})$. The full theory has also to include mixed motives, which we will discuss very briefly as Problem 5, but the ideas are perhaps easier to sketch in terms of pure motives. (See [116], [7].)

The basic idea comes from the familiar De Rham theorem, which asserts that for a manifold X, the pairing

$$H^k_{\text{DR}}(X,\mathbb{C}) \times H_k(X,\mathbb{C}) \to \mathbb{C}, \qquad (\phi,c) \to \int_c \phi,$$

between De Rham cohomology and complex Betti homology is nonsingular, and therefore gives an isomorphism from $H^k_{\mathrm{DR}}(X,\mathbb{C})$ to complex Betti cohomology $H^k_{\mathrm{B}}(X,\mathbb{C}) = H_k(X,\mathbb{C})^*$. Suppose now that X is a nonsingular, projective algebraic variety over $\mathbb{Q}$. The Betti cohomology of $X(\mathbb{C})$ can of course have $\mathbb{Q}$-coefficients, and so becomes a graded vector space $H_{\mathrm{B}}(X)$ over $\mathbb{Q}$. A deep theorem of Grothendieck asserts that the same is true of De Rham cohomology. Namely, there is a rational graded vector space $H_{\mathrm{DR}}(X)$ over $\mathbb{Q}$ whose complexification equals $H_{\mathrm{DR}}(X,\mathbb{C})$, together with a canonical isomorphism

$$\varpi_X : H_{\mathrm{DR}}(X)\otimes\mathbb{C} \xrightarrow{\sim} H_{\mathrm{B}}(X)\otimes\mathbb{C}.$$

The isomorphism is that of the original De Rham theorem. It assigns a complex number

$$\langle \varpi_X(\phi), c\rangle = \int_c \phi$$

to every rational differential form ϕ and every rational singular cycle c of a given degree on X. These numbers are called *periods* of X.

The main point is that this construction extends to the Tannakian category $\mathrm{Mot}_{\mathbb{Q}}$ of motives over $\mathbb{Q}$. The *period realization* of $\mathrm{Mot}_{\mathbb{Q}}$ is the $\otimes$-functor

$$M \to (H_{\mathrm{DR}}(M), H_{\mathrm{B}}(M), \varpi_M)$$

from $\mathrm{Mot}_{\mathbb{Q}}$ to the Tannakian category of triples

$$(V, W, \varpi), \qquad V, W \in \mathrm{Vect}_{\mathbb{Q}},$$

where ϖ is an isomorphism between the complex vector spaces $V_{\mathbb{C}}$ and $W_{\mathbb{C}}$. *Grothendieck's period conjecture* represents an analogue for the period realization of the Hodge conjecture for the Hodge realization, or the Tate conjecture for the $\mathbb{A}_{\mathrm{fin}}$-realization, fundamental foundations we have not been able to discuss. It implies that the period realization is fully faithful. The actual conjecture stated in [7, 4.1.1] applies to the period torsor

$$P_{\mathrm{mot}}(M) = \mathrm{Isom}^{\otimes}(H_{\mathrm{DR}|\langle M\rangle}, H_{\mathrm{B}|\langle M\rangle}),$$

where $\langle M\rangle$ is the Tannakian subcategory of $\mathrm{Mot}_{\mathbb{Q}}$ generated by a motive M, and $H_{\bullet|\langle M\rangle}$ stands for the restriction of the realization $H_{\bullet}$ to $\langle M\rangle$. The period torsor is the (noncanonical) subvariety over $\mathbb{Q}$ given by a finite-dimensional affine general linear group over $\mathbb{Q}$. It is a torsor under the motivic Galois group of M, the finite-dimensional quotient $\mathcal{G}(M) = \mathcal{G}_{\mathbb{Q}}(M)$ of $\mathcal{G} = \mathcal{G}_{\mathbb{Q}}$ attached to the subcategory $\langle M\rangle$ of $\mathrm{Mot}_{\mathbb{Q}}$. Grothendieck's conjecture asserts that the canonical complex point

$$\varpi_M \in P_{\mathrm{mot}}(M,\mathbb{C})$$

is the generic point in $P_{\mathrm{mot}}(M)$. It amounts to the assertion that the smallest algebraic subvariety of $P_{\mathrm{mot}}(\mathbb{C})$ defined over $\mathbb{Q}$ and containing ϖ_M is P_{mot} itself. It is in this form that the conjecture suggests applications to transcendental number theory (see [7, §4]).

Taking the natural extension to M of the definition for $M = X$, we define the *periods* of M to be the entries in the matrix of ϖ_M with respect to bases of the $\mathbb{Q}$-vector spaces $H_{\mathrm{DR}}(M)$ and $H_{\mathrm{B}}(M)$. (To be canonical, we allow the bases to vary, or equivalently, we take the periods to be the $\mathbb{Q}$-vector space $\mathscr{P}(M)$ generated by the periods with respect to any fixed pair of bases.) Grothendieck's period conjecture implies that any polynomial relations with rational coefficients among the periods of M are among the relations that define the variety $P_{\mathrm{mot}}(M)$. It follows easily that the algebra over $\mathbb{Q}$ generated by the periods coincides with the algebra $\mathbb{Q}[P_{\mathrm{mot}}(M)]$ over $\mathbb{Q}$. This implies in turn that $\mathbb{Q}[P_{\mathrm{mot}}(M)]$ coincides with the $\mathbb{Q}$-algebra $\mathscr{P}(\langle M\rangle)$ obtained by taking the periods of all motives in the category $\langle M\rangle$. On the other hand, the group $\mathscr{G}(M,\mathbb{Q})$ of rational points in $\mathscr{G}(M)$ acts simply transitively on the rational points $P_{\mathrm{mot}}(M,\mathbb{Q})$ in $P_{\mathrm{mot}}(M)$, and hence on the $\mathbb{Q}$-algebra

$$\mathscr{P}(\langle M\rangle) = \mathbb{Q}[P_{\mathrm{mot}}(M)].$$

Taking limits over M, we see finally that the group

$$\mathscr{G}(\mathbb{Q}) = \varprojlim_{M} \left(\mathscr{G}(M,\mathbb{Q})\right)$$

of $\mathbb{Q}$-rational points in the motivic Galois group $\mathscr{G}$ acts canonically on the $\mathbb{Q}$-algebra

$$\mathscr{P} = \varinjlim_{M} \left(\mathscr{P}(\langle M\rangle)\right)$$

of motivic periods over $\mathbb{Q}$. This is clearly a generalization of Galois theory for $\overline{\mathbb{Q}}/\mathbb{Q}$, with $\mathscr{G}(\mathbb{Q})$ being an extension of the Galois group $\Gamma_{\mathbb{Q}}$ and $\mathscr{P}$ a $\mathbb{Q}$-algebra that contains the algebraic closure $\overline{\mathbb{Q}}$ of $\mathbb{Q}$ in $\mathbb{C}$. (See [7, §5.1] for further analogies with classical Galois theory.)

I have followed the short introduction [7] in saying a few words on the Galois theory of periods. I have not stated a specific problem. Let us simply ask the same question about the period realization that we posed above for the $\mathbb{A}_{\mathrm{fin}}$ and Hodge realizations. Namely, can we formulate the theory above strictly in terms of supplementary internal structure on the motivic Galois group $\mathscr{G}$ over $\mathbb{Q}$? Even if this makes sense, it would not seem to have any immediate application. But in adding to the underlying structure of $\mathscr{G}$, it would clearly give us a broader understanding.

5. Mixed motives. Our discussion to this point has applied only to pure motives. Grothendieck's original vision was for a broader theory of mixed motives. (See [209, p. 345].) They would be attached to varieties over F that need not be either projective or nonsingular. (The case of open Shimura varieties is actually an anomaly, since L^2-cohomology and intersection cohomology take it back into the domain of pure motives.) The theory of mixed motives was subsequently developed by Deligne, initially through his extensive theory of mixed Hodge structures [59], [60], [64], and more recently, through other means such as those in [66] and [65]. It remains a major area of activity, encompassing many deep and fundamental concepts.

One of Grothendieck's basic tenets was the existence of a broader group, the mixed motivic Galois group. Over the number field F, it would be a semidirect product

$$\mathscr{G}_F^+ = \mathscr{N}_F \rtimes \mathscr{G}_F$$

of the (pure) motivic Galois group $\mathscr{G}_F$ with a proalgebraic unipotent radical $\mathscr{N}_F$. Its existence was again predicated on the theory of Tannakian categories. In particular, with a suitable fibre functor, $\mathscr{G}_F^+$ would again become a proalgebraic group over $\mathbb{Q}$. However, Grothendieck's axioms for mixed motives are deep and difficult. They generalize his standard conjectures for pure motives, which are still far from proved. My impression is that much current work in the area is to find other means to characterize mixed motives and the group $\mathscr{G}_F^+$, which are more concrete and perhaps less difficult to establish.

The problem we pose here would be to find a concrete conjectural description of $\mathscr{G}_F^+$, comparable to what we considered for pure motives in Problem 3. This would be harder than the other problems, and might seem unrealistic to some. But if we were to go ahead, there would be two possible ways to proceed. One would be to try to extend Langlands' Reciprocity Conjecture. This is the approach of Harder [84], who has studied automorphic analogues of mixed motives in terms of the (nonunitary) values of Eisenstein series. The other would be to combine a solution of Problem 3 for the group $\mathscr{G}_F$ with a description of the unipotent radical $\mathscr{N}_F$ in elementary terms. There is an explicit solution of this problem for the category of mixed Tate motives, which yields the simplest interesting mixed motivic Galois group, and is attached to the (pure) Tate motive $\mathbb{Q}(1)$ [144, p. 214]. The solution was remarkably simple, if also quite difficult to prove [65], [41]. It is perhaps a good omen.

Everything we have discussed for pure motives should extend to the theory of mixed motives. In particular, the conjectural category of mixed motives over $\mathbb{Q}$, say, would have a period resolution that adds greatly to the set of periods, a list that would then include algebraic numbers, the periods of elliptic curves over $\mathbb{Q}$ (these sets both being pure motives), the number π, values of the logarithm at rational numbers $q \notin \{-1,0,1\}$, special values of the gamma function, special values of the hypergeometric function, and perhaps most striking of all, the unknown values

$$\{\zeta(2n+1) : n \in \mathbb{N}\}$$

of the Riemann zeta function that have been a preoccupation of mathematicians since the time of Euler. (See [7, §5.2–5.7] and also [116] for more examples.)

There is one number that is conspicuously absent from the list. The exponential base e is in fact not a period. But it is an *exponential period*, a larger (countable) class of transcendental numbers attached to what are known as *exponential motives* [75].

Coda: Particle physics. There is a third conjectural universal group, in addition to the automorphic and (mixed) motivic Galois groups. This was proposed by P. Cartier, who called it the *cosmic Galois group*. It would be a quotient $\mathscr{C}^+ = \mathscr{C}^+_{\mathbb{Q}}$ of the mixed motivic Galois group $\mathscr{G}^+ = \mathscr{G}^+_{\mathbb{Q}}$. The corresponding group $\mathscr{C}^+_{\mathbb{Q}}(\mathbb{Q})$ of rational points would act like a Galois group on the $\mathbb{Q}$-vector space of periods of Feynman integrals, sums of which form the amplitudes attached to Feynman graphs [42], [54]. It is apparently unknown what this quotient should be, even as $\mathscr{C}^+$ might well turn out to be the full (mixed) motivic Galois group $\mathscr{G}^+$. This group suggests a fundamental relationship between the arithmetic Langlands program and basic particle physics, of the kind perhaps that is sometimes dreamt of. (See [188, p. 503], for example.) I am hardly a disinterested observer, and my knowledge of physics is fragmentary at the very best, but I would argue as follows.

Feynman integrals have long been a foundation for the theory of fundamental particles. In principle, they ought to give the quantum probabilities for the output data, measured from collision experiments with given input data. However, the calculations have traditionally been purely numerical, and of great difficulty. The infinite sums that go into a Feynman amplitude were originally thought to provide a convergent series. However, according to my very limited understanding, they were shown by Dyson around 1950 not to converge, but rather to give only an asymptotic formula, except in the idealized case of free particles, with input Lagrangian having only kinetic (quadratic) terms. As an approximation of this asymptotic formula, the first few terms of the infinite series, taken at points close to the origin, still give astonishing good results in the case of QED (quantum electrodynamics). However, they fail in more complex experiments. It is a fundamental problem in physics to discover a more sophisticated theory for describing quantum amplitudes in general, but which would still reduce to Feynman amplitudes in simple situations.[11]

It was shortly before the year 2000 that the physicist D. Kreimer discovered the number $\zeta(3)$ among the more complex calculations of QED. He was later joined by A. Connes, and as I understand it, they soon found that many other such calculations also gave periods of mixed motives. Moreover, the Galois action of $\mathscr{G}^+(\mathbb{Q})$ on periods, or rather its restriction to the unipotent radical $\mathscr{N}(\mathbb{Q})$, seemed to be closely related to the conceptually difficult (at least for mathematicians) physical process of renormalization.

The mixed motivic Galois group $\mathscr{G}^+$ is at the heart of much of modern arithmetic geometry. However, it has generally been regarded as inaccessible. Langlands' Reciprocity Conjecture makes it much more concrete. Combined with suitable conjectural solutions for Problems 3, 4 and 5, it would impose a rich internal automorphic structure on both $\mathscr{G}^+$ and its associated Galois group $\mathscr{G}^+(\mathbb{Q})$ for periods. Put simply, Functoriality and Reciprocity would give us the automorphic Galois group L_F, together with its close ties to the motivic Galois group $\mathscr{G}_F$. They are the centre of the Langlands program. On the other hand, Feynman diagrams have long been central to theoretical particle physics. It is expected that there will be something more

---

[11] I thank Marco Gualtieri for illuminating conversations. Any misinterpretations are entirely my doing.

fundamental that could eventually take their place. Whatever this might turn out to be, it is also reasonable to believe that the Langlands program would be a part of it.

This completes the second of our two sections on arithmetic geometry. Some of it is clearly speculative. However, I hope that the mathematical side of it at least will hold in principle, and that any inaccuracies will require only minor adjustments. In general, I will be happy if my attempts to describe some of the broader ideas behind Langlands' work and their subsequent development are some compensation for any misstatements that might also be present.

10 The theory of endoscopy

There were a number of natural questions arising from his ideas that Langlands thought deeply about in the decade of the 1970s. For example, the conjectural correspondence

$$\pi' = \bigotimes_v \pi'_v \to \bigotimes_v \pi_v = \pi$$

of functoriality (Questions 4 and 5 of [138]) was just that, a correspondence. Could it be reformulated as a well defined mapping? Compared to the explicit results for GL(2) in [103], the representation theory of the group SL(2) has more structure. What was the explanation? Also, with his more recently gained experience in the λ-adic representations of Shimura varieties, Langlands found some unexpected anomalies in the associated Hasse–Weil zeta functions [143]. Again, what was the explanation? And finally, in Harish-Chandra's classification of the discrete series representations for a real group $G(\mathbb{R})$, a monumental achievement that was ahead of its time, there were some unusual aspects of his formulas for their characters. Could they be related to Local Functoriality? Langlands confronted this last problem in the work that led to his classification [151], and in his later work [165] and [166] with Shelstad.

The questions all turned out to be related. The underlying phenomena eventually became part of Langlands' conjectural theory of endoscopy. We have mentioned endoscopy a number of times already, most notably in Kottwitz' conjectural stabilization of the Lefschetz trace formula in Section 8. In this section, we shall try to give a more systematic description of the theory, and of some of the progress that has come in its development.

Given his success with the trace formula for GL(2) (as described in the three applications from our Sections 6, 7 and 8), Langlands would of course have considered how these methods might be applied to other groups, and to more general cases of functoriality. There was no clear strategy on how to proceed. But he appears to have acquired a strong sense that the trace formula would ultimately lead to a solution, informed no doubt by his general theory of Eisenstein series, and perhaps also by a skepticism as to whether other possible approaches would have the power to treat the general case.

One might try to think about comparing trace formulas for two groups G' and G related by the L-homomorphism $\rho'\colon {}^LG' \to {}^LG$ of functoriality. The immediate question would be to relate the basic elliptic terms on the geometric side of each trace formula. The conjugacy classes $\gamma \in \Gamma_{\mathrm{ell,reg}}(G)$ that index these terms for G have coordinates defined by the algebra of G-invariant polynomials on G. One could think of using ρ' to relate these coordinates for G and G'. However, a serious problem arises immediately. The coordinates parametrize only geometric conjugacy classes, while for most groups G other than $\mathrm{GL}(n)$, there can be distinct (elliptic, regular) conjugacy classes in $G(F)$ over a ground field $F \subset \mathbb{C}$ that represent the same conjugacy class in $G(\mathbb{C})$. The theory of endoscopy begins with this problem. It brings to bear on it some sophisticated new techniques that originate with (abelian) class field theory.

Consider the example of $\mathrm{SL}(2)$, with $F = \mathbb{R}$. The regular elliptic elements

$$\left\{\gamma = \begin{pmatrix} \cos(\theta) & -\sin(\theta) \\ \sin(\theta) & \cos(\theta) \end{pmatrix},\ \delta = \begin{pmatrix} \cos(\theta) & \sin(\theta) \\ -\sin(\theta) & \cos(\theta) \end{pmatrix}\right\}, \quad \theta \in (0,\pi),$$

are not conjugate in $\mathrm{SL}(2,\mathbb{R})$. However, the matrix $g = \begin{pmatrix} i & 0 \\ 0 & -i \end{pmatrix}$ in $\mathrm{SL}(2,\mathbb{C})$ has the property that

$$g\gamma g^{-1} = \begin{pmatrix} i & 0 \\ 0 & i^{-1} \end{pmatrix} \begin{pmatrix} \cos(\theta) & -\sin(\theta) \\ \sin(\theta) & \cos(\theta) \end{pmatrix} \begin{pmatrix} i^{-1} & 0 \\ 0 & i \end{pmatrix} = \begin{pmatrix} \cos(\theta) & \sin(\theta) \\ -\sin(\theta) & \cos(\theta) \end{pmatrix} = \delta,$$

so the two elements are conjugate in $\mathrm{SL}(2,\mathbb{C})$. The regular elliptic conjugacy classes $\gamma \in \Gamma_{\mathrm{reg,ell}}(\mathrm{SL}(2))$ for $\mathrm{SL}(2,\mathbb{R})$ thus occur naturally in pairs (γ,δ) that map to the same conjugacy class in $\mathrm{SL}(2,\mathbb{C})$. The dual spectral property concerns the representations $\pi \in \Pi_2(\mathrm{SL}(2))$ in the discrete series for $\mathrm{SL}(2,\mathbb{R})$. They too occur naturally in pairs (π_n^+,π_n^-), which are parametrized by the positive integers n. Harish-Chandra's theory of infinite-dimensional characters, which we will discuss presently, shows that the two phenomena are indeed dual in a precise sense. The characters of any pair (π_n^+,π_n^-), as locally integrable functions on regular conjugacy classes, differ only on the pairs (γ,δ), and for these, only in a simple manner.

The group $G = \mathrm{SL}(2)$ is quite special. For in this case, the dual properties have formulations in terms of the real group $G'(\mathbb{R})$ with $G' = \mathrm{GL}(2)$, as well as for the complex group $G(\mathbb{C})$. Since the element $g = \begin{pmatrix} i & 0 \\ 0 & -i \end{pmatrix}$ in $\mathrm{SL}(2,\mathbb{C})$ is the product of the central element $\begin{pmatrix} i & 0 \\ 0 & i \end{pmatrix}$ in $\mathrm{GL}(2,\mathbb{C})$ with the matrix $g_1 = \begin{pmatrix} 1 & 0 \\ 0 & -1 \end{pmatrix}$ in $\mathrm{GL}(2,\mathbb{R})$, $g_1\gamma g_1^{-1}$ equals δ, and the elements γ and δ are also conjugate in $\mathrm{GL}(2,\mathbb{R})$. With this interpretation, the dual spectral property can be regarded as a very special case of Local Functoriality. It applies to $G' = \mathrm{GL}(2)$ and $G = \mathrm{SL}(2)$, with the homomorphism

$$\rho'\colon \widehat{G'} = \mathrm{GL}(2,\mathbb{C}) \to \widehat{G} = \mathrm{PGL}(2,\mathbb{C})$$

being the natural projection. In the local classification for $\mathrm{GL}(2,\mathbb{R})$, the representations $\pi'_n \in \Pi_2(G')$ in the relative discrete series for $\mathrm{GL}(2,\mathbb{R})$ (with respect to a fixed central character) are parametrized by irreducible 2-dimensional representations ϕ' of $W_{\mathbb{R}}$. These in turn are bijective with positive integers $\{n\}$. The corresponding pairs of representations $\{\pi_n^{\pm}\} \subset \Pi_2(G)$ for $\mathrm{SL}(2,\mathbb{R})$ are attached to composite homomorphisms

$$\phi = \rho' \circ \phi' \colon W_{\mathbb{R}} \to \mathrm{PGL}(2,\mathbb{C}), \qquad \phi' \in \Phi_2(G').$$

They consist simply of the irreducible constituents of the restriction of π'_n to the subgroup $\mathrm{SL}(2,\mathbb{R})$ of $\mathrm{GL}(2,\mathbb{R})$. One can obviously think of the pair $\{\pi_n^+, \pi_n^-\}$ attached in this way to ϕ as a torsor under the group

$$\mathbf{S}_\phi = \mathrm{Cent}(\phi(W_{\mathbb{R}}), \widehat{G}) = \mathbb{Z}/2\mathbb{Z}.$$

The sets $\pi_\phi = \{\pi_n^+, \pi_n^-\}$ and $\pi_{\phi'} = \{\pi'_n\}$ are called local L-packets for G and G'.

The description of $\mathbf{S}_\phi$ as a torsor is an obvious tautology in the case of $G = \mathrm{SL}(2)$, but its generalization to arbitrary groups became part of the local Langlands classification. There were still interesting questions for $\mathrm{SL}(2)$, and particularly, for groups related to $\mathrm{SL}(2)$ and its inner twists over local and global fields F. The paper [127] of Langlands with Labesse contains a comprehensive study of them.

To see how Langlands' ideas progress in general, suppose that G is a (connected) reductive group over a local or global field of characteristic 0. Quasisplit groups again play a special role in the theory, which for questions of transfer entail an underlying, fixed inner twist

$$\psi \colon G \to G^*,$$

of G to a quasisplit group G^* over F. One also has to work with classes $\gamma \in \Gamma_{\mathrm{reg}}(G)$ that are *strongly* regular, in the sense that the centralizer G_γ in G of (any representative of) γ is a maximal torus T. (Regular elements satisfy the weaker property that the identity component G_γ^0 is a maximal torus.) We may as well simplify our notation slightly by agreeing to have the subscript *reg* mean strongly regular rather than regular. The problem then is to understand the set $\Gamma_{\mathrm{reg}}(G)$ of strongly regular conjugacy classes of $G(F)$ in a given stable conjugacy class. (A strongly regular *stable* conjugacy class is by definition the intersection of $G(F)$ with a strongly regular conjugacy class in the group $G(\overline{F})$ of points over an algebraic closure of F.)

Suppose that $\delta \in G(F)$ is strongly regular, with centralizer the maximal torus $T \subset G$ over F, and that $\gamma \in G(F)$ is another element in the stable class of δ. Then γ equals $g^{-1}\delta g$, for some element $g \in G(\overline{F})$. If σ lies in the Galois group $\Gamma_F = \mathrm{Gal}(\overline{F}/F)$, we see that

$$\delta = \sigma(\delta) = \sigma(g\gamma g^{-1}) = \sigma(g)\gamma\sigma(g)^{-1} = t(\sigma)^{-1}\delta t(\sigma),$$

where $t(\sigma)$ is the 1-cocycle $g\sigma(g)^{-1}$ from Γ_F to $T(\overline{F})$. It is easy to check that a second element $\gamma_1 \in G(F)$ in the stable class of δ is $G(F)$-conjugate to γ if and only if the corresponding 1-cocycle $t_1(\sigma)$ has the same image as $t(\sigma)$ in the Galois cohomology group

$$H^1(F,T) = H^1(\Gamma_F,T),$$

which is to say that $t_1(\sigma)t(\sigma)^{-1}$ is of the form $t'\sigma(t')^{-1}$, for some element $t' \in T(\overline{F})$. Conversely, a class in $H^1(F,T)$ comes from an element γ_1 if and only if it is represented by a 1-cocycle of the form $g\sigma(g)^{-1}$. The mapping $\gamma \to \{t(\sigma)\}$ therefore defines a bijection from the set of $G(F)$-conjugacy classes in the stable conjugacy class of δ to the kernel

$$\mathscr{D}(T) = \mathscr{D}(T/F) = \ker(H^1(F,T) \to H^1(F,G)).$$

The codomain $H^1(F,G)$ is only a set with distinguished element 1, since G is generally not abelian. The preimage $\mathscr{D}(T)$ of this element in $H^1(F,T)$ therefore need not be a subgroup. However, $\mathscr{D}(T)$ is contained in the subgroup

$$\mathscr{E}(T) = \mathscr{E}(T/F) = \mathrm{Im}\,(H^1(F,T_{\mathrm{sc}}) \to H^1(F,T))$$

of $H^1(F,T)$, where T_{sc} is the preimage of T in the simply connected cover G_{sc} of the derived group of G. This is because the canonical map $\mathscr{D}(T_{\mathrm{sc}}) \to \mathscr{D}(T)$ is surjective. If $H^1(F,G_{\mathrm{sc}}) = \{1\}$, which is the case whenever F is a nonarchimedean local field [232, §3.2], $\mathscr{D}(T)$ actually equals the subgroup $\mathscr{E}(T)$. This is why Langlands worked with the groups $\mathscr{E}(T)$ in place of $\mathscr{D}(T)$, and why the simply connected group G_{sc} plays an important role in the theory. Langlands introduced these ideas in the initial pages of his foundational article [147], although he had discussed them widely in the years preceding it. We are following some of the discussion in [22, §27].

We have referred to the trace formula regularly in this report, especially in the last three sections. However, to understand the refinements that originate with Langlands' observations above, we need to say something more formal. Continuing with the reductive group G, we now take its field of definition F to be global. The (invariant) trace formula for G is a general identity

$$I_{\mathrm{geom}}(f) = I_{\mathrm{spec}}(f), \qquad f \in C_c^\infty(G(\mathbb{A})), \tag{102}$$

obtained by integrating the geometric and spectral expansions (38) and (39) of the kernel $K(x,y)$ over $x = y$ in $G(F) \setminus G(\mathbb{A})^1$. As we have noted, this cannot be taken literally, since neither integral converges in general. Making sense of it is a long process, but roughly speaking, one truncates the two expansions of $K(x,x)$ in a consistent way so that the integrals converge. One then observes that as functions of the of variable of truncation T, a vector in some translate of a positive cone $\mathfrak{a}_0^+$, the integrals are polynomials in T. One can then set T equal to the polynomial variable at $T_0 \in \mathfrak{a}_0$, a canonical point that depends on a maximal compact subgroup $K_0 \subset G(\mathbb{A})$ and a minimal parabolic subgroup $P_0 \subset G$, both of which are part of the truncation process. The result is a natural identity

$$J_{\text{geom}}(f) = J_{\text{spec}}(f), \qquad f \in C_c^\infty(G(\mathbb{A})), \tag{103}$$

which is independent of the choice of P_0.

However, (103) is only an intermediate step. We recall that a distribution J on $G(\mathbb{A})$ is said to be *invariant* if

$$J(f^y) = J(f), \qquad f \in C_c^\infty(G(\mathbb{A})), y \in G(\mathbb{A}),$$

where

$$f^y(x) = f(yxy^{-1}), \qquad x \in G(\mathbb{A}).$$

The point here is that the linear forms on each side of (103) are noninvariant. One has then to "renormalize" the identity. There is no need to describe this explicitly, although I have been told that it is in the same spirit as a similar (but more complex) operation in quantum field theory that restores the symmetry that was lost in the truncation of divergent intervals. (I may however have misunderstood this. Renormalization seems actually to be a physics analogue of the original truncation process.) In any case, it leads to the identity (102), in which each side is now an invariant distribution. Moreover, the choice of the point T_0 makes each side of (102) independent of K_0 as well as P_0.

We should emphasize that neither (102) nor (103) is just an abstract formula. As in the case of GL(2), each side of (102) represents a rather complex expansion into explicit invariant linear forms, one geometric and one spectral, which can all be decomposed explicitly into their local constituents ([14], [15]). For example, on the geometric side, we have the strongly regular elliptic part

$$I_{\text{ell,reg}}(f) = \sum_{\gamma \in \Gamma_{\text{ell,reg}}(G)} \text{vol}(\gamma)\text{Orb}(\gamma, f), \tag{104}$$

where

$$\text{Orb}(\gamma, f) = \int_{G_\gamma(\mathbb{A}) \backslash G(\mathbb{A})} f(x^{-1}\gamma x)\, dx$$

and

$$\text{vol}(\gamma) = \text{vol}(Z_+ G_\gamma(F) \setminus G_\gamma(\mathbb{A})).$$

Its analogue on the spectral side would be the trace

$$I_2(f) = \sum_{\pi \in \Pi_2(G)} \text{mult}(\pi)\Theta(\pi, f) \tag{105}$$

where

$$\Theta(\pi, f) = \text{tr}(\pi(f)) = \text{tr}\left(\int_{G(\mathbb{A})} f(x)\pi(x)\, dx\right),$$

and $\text{mult}(\pi)$ is the multiplicity with which π occurs discretely in $L^2(Z_+ G(F) \setminus G(\mathbb{A}))$. We are writing Z_+ here for a fixed connected, central subgroup of $G(\mathbb{A})$ that is a complement to the subgroup $G(\mathbb{A})^1$, defined as in §2 in case $F = \mathbb{Q}$.

These terms have been familiar[12] since Selberg introduced his original trace formula for compact quotient. The complementary terms on each side are in some sense just as explicit, but often considerably more complex. We shall not discuss them here. In fact, we will allow ourselves to write

$$I_{\mathrm{ell,reg}}(f) \sim I_2(f) \tag{106}$$

as a heuristic approximation of the invariant trace formula. The right-hand side is what one wants to understand, and left-hand side represents the means by which one hopes to investigate it. It was with this strategy that Langlands created the conjectural theory of endoscopy in the 1970s.

In the larger scheme of things, the invariant trace formula (102) is itself an intermediate step. The final goal was the stable trace formula, which came later [21]. We shall describe it in general terms for further perspective.

First of all, it is useful to keep in mind that there is a simple description of the space of invariant distributions on $G_v = G(F_v)$, for any localization F_v of F. For it is known that this space is the closed linear span (with respect to the weak topology) of *either* the set

$$\mathrm{Orb}(\gamma_v, f_v), \qquad \gamma_v \in \Gamma_{\mathrm{reg}}(G_v),\ f_v \in C_c^\infty(G_v),$$

of strongly regular orbital integrals, *or* the set

$$\Theta(\pi_v, f_v), \qquad \pi_v \in \Pi_{\mathrm{temp}}(G_v),\ f_v \in C_c^\infty(G_v),$$

of irreducible tempered characters. We can also use the first description here to define the notion of a stable distribution. Let $\Delta_{\mathrm{reg}}(G_v)$ be the set of strongly regular stable conjugacy classes in G_v. For any δ_v in this set, we define the *stable* orbital integral as the associated sum

$$\mathrm{SOrb}(\delta_v, f_v) = \sum_{\gamma_v \to \delta_v} \mathrm{Orb}(\gamma_v, f_v), \qquad f \in C_c^\infty(G_v),$$

of orbital integrals over the finite set of conjugacy classes γ_v in δ_v. We then define the subspace of *stable distributions* on G_v to be the closed linear span of the space of stable orbital integrals. The spectral analogue of this description should also be true, but it requires us to know what a stable (tempered) character is. An explicit description of this notion is available in many cases, but not in general. Its general formulation is one of the main goals of the local theory of endoscopy.

[12] In particular, the analogues $J_{\mathrm{ell,reg}}(f)$ and $J_2(f)$ in the original (noninvariant) trace formula were already invariant, as we can recall from the discussion of the special case of GL(2), and are therefore the same as $I_{\mathrm{ell,reg}}(f)$ and $I_2(f)$.

The stable trace formula for a quasisplit[13] group G over F is a refinement

$$S^G_{\text{geom}}(f) = S^G_{\text{spec}}(f), \qquad f \in C^\infty_c(G(\mathbb{A})), \tag{107}$$

of the invariant trace formula (102) in which each side is stable.[14] Its construction, and its role in the broader operation of *stabilization*,[13] is not difficult to describe in general terms. It is in fact quite similar to the stabilization of the Lefschetz trace formula we discussed briefly in Section 8.

Suppose for a moment that F is a local or global field, and that G is any reductive group over F. One of the central notions of endoscopy is the assignment to G of a family of *endoscopic data* $(G', \mathcal{G}', s', \xi')$, where G' is a quasisplit reductive group over F, $\mathcal{G}'$ is a split extension of W_F by $\widehat{G}'$, s' is a semisimple element in $\widehat{G}$, and $\xi' \colon \mathcal{G}' \to {}^LG$ is an L-homomorphism, subject to various conditions.[15] *Equivalence* of endoscopic data is also defined, by a relation that is closely related to conjugation in ${}^LG$ by elements $g \in \widehat{G}$. (See [165, (1.2)].)

This is an admittedly technical part of the theory, but the basic idea is simple enough. Its origins in Langlands' sets $\mathcal{D}(T)$ and $\mathcal{E}(T)$, which we will review presently, are really quite remarkable. Basically one wants to attach smaller quasisplit groups G' to G by taking the dual group $\widehat{G}'$ to be the connected centralizer in $\widehat{G}$ of a semisimple element s', and by constructing the L-group ${}^LG'$, which then determines G' as a quasisplit group, in terms of the centralizer of s' in the larger group ${}^LG$. If the derived group of G is simply connected, one can identify the subgroup $\xi'(\mathcal{G}')$ of ${}^LG$ with ${}^LG'$ [147]. The general case, however, is a little more subtle, and one has to attach some auxiliary data to G' that serve the same purpose. (See [19] for example). As in Section 8, we write G' to represent the full endoscopic datum $(G', \mathcal{G}', s', \xi')$. One says that G' is elliptic if the image $\xi'(\mathcal{G}')$ in ${}^LG$ is contained in no proper parabolic subgroup of ${}^LG$, or equivalently, if $(Z(\widehat{G}')^{\Gamma_F})^0$ is mapped by ξ' onto $(Z(\widehat{G})^{\Gamma_F})^0$. Finally, if F is local, there are only finite many equivalence classes of endoscopic data, while if F is global, there are finitely many classes that are unramified outside a given finite set of places.

At the centre of the theory is the endoscopic transfer $f \to f'$ of functions from G to G', a topic we can revisit after our brief discussions from Sections 6–8. It was defined formally by Langlands and Shelstad [165], following Shelstad's treatment of the case of real groups [221]. If F is local, it is a mapping from functions $f \in C^\infty_c(G(F))$ to smooth functions f' on $\Delta_{\text{reg}}(G')$. The Langlands–Shelstad transfer conjecture was then formulated in [165] as the hypothesis that

$$f'(\delta') = \text{SOrb}(\delta', h'), \qquad \delta' \in \Delta_{\text{reg}}(G'),$$

[13] The stable trace formula is best regarded as a phenomenon for quasisplit groups. On the other hand, *stabilization*, which we will describe presently applies to arbitrary groups.

[14] A stable distribution on $G(\mathbb{A})$ would of course be a continuous linear form that is stable on each of the factors G_v of $G(\mathbb{A})$.

[15] The main conditions are (iv), (a) and (b), on p. 234 of [165]. They identify $\widehat{G}'$ under the restriction of ξ' with the connected centralizer of s' in $\widehat{G}$, and relate $\mathcal{G}'$ under ξ' with the full centralizer s' in ${}^LG$.

for some function $h' \in C_c^\infty(G'(F))$. It had already been established for archimedean F by Shelstad in [221], using the foundations of harmonic analysis on real groups laid out by Harish-Chandra, as we shall discuss later in this section. In the case of nonarchimedian F, the conjecture was reduced to the fundamental lemma by Waldspurger [245], which we have already noted was finally established by Ngô [186], with further contributions [246] from Waldspurger. If F is global, and $f = \prod_v f_v$ lies in $C_c^\infty(G(\mathbb{A}))$, the global mapping is defined by setting $f' = \prod_v f'_v$. One sees that it satisfies the global version of the Langlands–Shelstad conjecture, namely that f' is the image of a function $h' \in C_c^\infty(G'(\mathbb{A}_F))$, by applying the local conjecture to the local functions f_v, and the fundamental lemma at places $v \notin S$ for which f_v is the unramified unit function.

With all of this background, we can now describe, again in quite general terms, the stabilization of the invariant trace formula (102). We are assuming that F is a global field, that G is any reductive group over F, and that f is a function in $C_c^\infty(G(\mathbb{A}_F))$. The stabilization is then represented by decompositions

$$I_{\text{geom}}(f) = \sum_{G'} \iota(G,G')\widehat{S}'_{\text{geom}}(f') \tag{108}$$

and

$$I_{\text{spec}}(f) = \sum_{G'} \iota(G,G')\widehat{S}'_{\text{spec}}(f') \tag{109}$$

of the two sides of (102). These are entirely analogous to the decompositions (73) and (74) of the two sides of the Lefschetz trace formula. The summands are indexed by the equivalence classes of elliptic endoscopic data G' for G, while f' is the Langlands–Shelstad transfer of f to $G'(\mathbb{A}_F)$. The linear forms $S'_{\text{geom}} = S^{G'}_{\text{geom}}$ and $S'_{\text{spec}} = S^{G'}_{\text{spec}}$ are the analogues for the quasisplit groups G' of the linear forms on each side of (107). In particular, they are stable, and therefore have uniquely determined pairings $\widehat{S}'_{\text{geom}}(f')$ and $\widehat{S}'_{\text{spec}}(f')$ with the function f', in the notation of (73) and (74). The coefficients $\iota(G,G')$ attached to G and G' were defined in [150], and given a particularly simple formula in [119]. The expansions (108) and (109) for arbitrary G, and the stable trace formula (107) for quasisplit G, are then proven together.

The basic strategy is quite simple. We emphasize that the linear forms $\widehat{S}'_{\text{geom}}(f')$ and $\widehat{S}'_{\text{spec}}(f')$ in the expansions depend only[16] on G', as a quasisplit group, even though the coefficients $\iota(G,G')$, the function f, the other components $\mathcal{G}'$, s' and ξ' of G' as an endoscopic datum, and the family $\{G'\}$ itself, all depend on G as well. The summands with $G' \neq G$ in (108) and (109) can therefore be assumed inductively to have been defined.

[16] They are also the same as in the summands in the stabilization (72) and (73) of the Lefschetz trace formula.

Suppose first that G is quasisplit. Then $G' = G$ is among the indices of summation in the expansions (108) and (109). We can therefore rewrite them as

$$S_{\mathrm{geom}}(f) = I_{\mathrm{geom}}(f) - \sum_{G' \neq G} \iota(G,G')\widehat{S}'_{\mathrm{geom}}(f')$$

and

$$S_{\mathrm{spec}}(f) = I_{\mathrm{spec}}(f) - \sum_{G' \neq G} \iota(G,G')\widehat{S}'_{\mathrm{spec}}(f').$$

This extends the inductive definition to G, and establishes the formula (107) from its analogues for G' and the formula (102). However, there is still something serious to prove in this case. One must show that the right-hand side of each of these two expansions is a stable linear form in f. Suppose next that G is not quasisplit. Then the sums in (108) and (109) include the term with G' equal to G^*, the quasisplit inner form of G. Assuming that we have already dealt with this case, we may suppose that all of the terms in (108) and (109) are defined. The remaining problem, then, is simply to establish the two identities in this case. Its proof is deep, but turns out to be quite similar to the proof of stability in the quasisplit case.

This discussion is a useful overview, but it is slightly misleading. For once again, the stable trace formula (107) is not to be regarded as just an abstract identity. Like its predecessors (103) and (102), each side of (107) represents a complex expansion into explicit linear forms, one geometric and one spectral, which are now all stable. The stable trace formula (107) is then to be understood as the identity between these two complex expansions. This is how it is proved, and how it is to be used in the applications of endoscopy. The stabilizations (108) and (109), can be regarded heuristically as an identification of the invariant trace formula (102) with a stable trace formula (represented by the summands with $G = G^*$ in (108) and (109)), modulo an unstable error term (represented by the sum over $G \neq G^*$).

These remarks may not be specific enough to be of much help to a general reader. Part of the reason for rehearsing them is for their application to the concrete terms in the heuristic approximation (106) of the invariant trace formula. They are essentially how Langlands constructed a conjectural but explicit stabilization of each side of (106). More precisely, he constructed a stabilization

$$I_{\mathrm{ell,reg}}(f) = \sum_{G'} \iota(G,G')\widehat{S}'_{\mathrm{ell},G\text{-}\mathrm{reg}}(f') \tag{110}$$

of the left-hand side of (106) that adheres to the general principles above, but which was obtained directly from its definition (104) in terms of orbital integrals.[17] Using this for guidance, enhanced by the results established for special cases in [127], Langlands then conjectured a partial stabilization of the right-hand side of (106). Some of his ideas are contained in the expository sections of the volume [150].

[17] The subscript G-reg on the right side of (110) denotes the subset of classes in $\Gamma_{\mathrm{ell,reg}}(G')$ that are images of strongly regular classes in $G(F)$. Its dependence on G is an anomaly that would disappear if we had started on the left with the larger set $\Gamma_{\mathrm{ell}}(G)$ of all (elliptic) semisimple classes in $G(F)$.

To review these things, we can return to the earlier discussion of the sets $\mathscr{D}(T)$ and $\mathscr{E}(T)$ introduced by Langlands to analyze stable conjugacy classes. We are taking G to be a reductive group with quasisplit inner twist G^*, over the local or global field F, with a maximal torus $T \subset G$ over F. The sets $\mathscr{E}(T)$ are to be regarded as geometric objects, for they are clearly founded on the terms on the left-hand side of the (approximate) trace formula (106). It is in their spectral counterparts that class field theory appears, specifically in the Tate–Nakayama duality theory [234].

If F is local, the theory provides a canonical isomorphism

$$H^1(F,T) = H^1(\Gamma,T) \xrightarrow{\sim} \pi_0(\widehat{T}^{\Gamma})^*, \quad \Gamma = \Gamma_F = \mathrm{Gal}(\overline{F}/F),$$

of $H^1(F,T)$ with the group of characters on the component group of the $\mathrm{Gal}(\overline{F}/F)$-invariants in the dual torus $\widehat{T}$. In the case that F is global, it provides a canonical isomorphism

$$H^1(F,T(\overline{\mathbb{A}}_F)/T(\overline{F})) = H^1(\Gamma,T(\mathbb{A}_{\overline{F}})/T(\overline{F})) \xrightarrow{\sim} \pi_0(\widehat{T}^{\Gamma})^*.$$

Using standard techniques, specifically, an application to the short exact sequence

$$1 \to T(\overline{F}) \to T(\mathbb{A}_{\overline{F}}) \to T(\mathbb{A}_{\overline{F}})/T(\overline{F}) \to 1$$

of Γ_F-modules to the isomorphism

$$H^1(F,T(\mathbb{A}_{\overline{F}})) \xrightarrow{\sim} \bigoplus_v H^1(F_v,T)$$

provided by Shapiro's lemma, one then obtains a characterization of the diagonal image of $H^1(F,T)$ in the direct sum over v of the groups $H^1(F_v,T)$. It is given by a canonical isomorphism from the cokernel

$$\mathrm{coker}^1(F,T) = \mathrm{coker}(H^1(F,T) \to \bigoplus_v H^1(F_v,T))$$

onto the image

$$\mathrm{im}\big(\bigoplus_v \pi_0(\widehat{T}^{\Gamma_v})^* \to \pi_0(\widehat{T}^{\Gamma})^*\big)$$

(See [30], [234] and [121, §1–2].)

If these results are combined with their analogues for T_{sc}, they provide similar assertions for the subgroups $\mathscr{E}(T/F)$ of $H^1(F,T)$. In the local case, one has only to replace $\pi_0(\widehat{T}^{\Gamma_v})$ by the group $\mathscr{K}(T/F_v)$ of elements in $\pi_0(\widehat{T}/Z(\widehat{G})^{\Gamma})$ whose image in $H^1(F_v,Z(\widehat{G}))$ is trivial. In the global case, one replaces $\pi_0(\widehat{T}^{\Gamma})$ by the group $\mathscr{K}(T/F)$ of elements in $\pi_0(\widehat{T}/Z(\widehat{G})^{\Gamma})$ whose image in $H^1(F,Z(\widehat{G}))$ is locally trivial, in the sense that their image in $H^1(\Gamma_v,Z(\widehat{G}))$ is trivial for each v. (See [150] and [121].)

Langlands introduced these ideas to be able to construct the stabilization (110) of the strongly regular part of the trace formula. The first step was to write the left-hand side as

$$\begin{aligned} I_{\mathrm{ell,reg}}(f) &= \sum_{\gamma \in \Gamma \mathrm{ell,reg}(G)} \mathrm{vol}(\gamma)\,\mathrm{Orb}(\gamma,f) \\ &= \sum_{\delta \in \Delta_{\mathrm{ell,reg}}(G)} \mathrm{vol}(\delta) \Big(\sum_{\gamma \to \delta} \mathrm{Orb}(\gamma,f) \Big), \end{aligned}$$

where γ is summed in the brackets over the preimage of δ in $\Gamma_{\mathrm{ell,reg}}(G)$, and $\mathrm{vol}(\delta) = \mathrm{vol}(\gamma)$ depends only on δ. We are of course assuming that F is global here. The last sum over γ looks as if it might be stable in f. But stability is a local concept, and there are not enough rational conjugacy classes γ to make this sum stable in each component f_v of f. The problem is the failure of every $G(\mathbb{A}_F)$-conjugacy class in the $G(\mathbb{A}_F)$-stable class of $\delta \in \Delta_{\mathrm{ell,reg}}(G)$ to have a representative γ in $G(F)$. If $T = G_\delta$, the cokernel we denoted by $\mathrm{coker}^1(F,T)$ gives a measure of this failure. Langlands' construction treats the sum $\big(\sum_{\gamma\to\delta}\mathrm{Orb}(\gamma,f)\big)$ as the value at 1 of a function on the finite abelian group $\mathrm{coker}^1(F,T)$. The critical step is to expand this function according to Fourier inversion on $\mathrm{coker}^1(F,T)$. It leads naturally to the definition of endoscopic data $\{G'\}$, and finally the desired stabilization (110).

To simplify the construction, we might as well assume for the present that $G = G_{\mathrm{sc}}$. Then $T = T_{\mathrm{sc}}$, while $\mathscr{E}(T/F) = H^1(F,T)$ and $\mathscr{K}(T/F) = \pi_0(\widehat{T}^\Gamma)$ if F is either local or global. In particular, $\mathscr{K}(T/F) = \widehat{T}^\Gamma$ if T is elliptic in G over F.

With this condition on G, we apply Fourier inversion on $\mathrm{coker}^1(F,T)$. One has to keep track here of the redundancy from the sum $\gamma \to \delta$, given by the set of $G(F)$-conjugacy classes in the $G(\mathbb{A}_F)$-conjugacy class of δ (regarded as a representative in $G(F)$ of the class in $\Delta_{\mathrm{ell,reg}}(G)$). It can be seen (with an appeal to the Hasse principle for $G = G_{\mathrm{sc}}$) that this is bijective with the finite abelian group

$$\ker^1(F,T) = \ker\big(H^1(F,T) \to \bigoplus_v H^1(F_v,T)\big).$$

It then follows that

$$I_{\mathrm{ell,reg}}(f) = \sum_{\delta \in \Delta_{\mathrm{ell,reg}}(G)} \iota(T)\,\mathrm{vol}(\delta) \sum_{\kappa \in \widehat{T}^\Gamma} \mathrm{Orb}^\kappa(\delta,f), \tag{111}$$

where $T = G_\delta$ is the centralizer of (some fixed representative of) δ, $\iota(T)$ equals the product of $(\widehat{T}^\Gamma)^{-1}$ with $|\ker^1(F,T)|$, and

$$\mathrm{Orb}^\kappa(\gamma,f) = \sum_{\{\gamma_{\mathbb{A}} \in \Gamma(G(\mathbb{A}_F)) : \gamma_{\mathbb{A}} \sim \delta\}} \mathrm{Orb}(\gamma_{\mathbb{A}},f)\,\kappa(\gamma_{\mathbb{A}}).$$

The last sum is over the $G(\mathbb{A}_F)$-conjugacy classes $\gamma_{\mathbb{A}}$ in the stable class of δ in $G(\mathbb{A}_F)$. For any such $\gamma_{\mathbb{A}}$, its local component γ_v is $G(F_v)$-conjugate to δ_v for almost all v, from which it follows that $\gamma_{\mathbb{A}}$ maps to an element $t_{\mathbb{A}}$ in the direct sum of the groups $H^1(F_v,T)$. This in turn maps to a point in the cokernel (110), and hence to a character in $(\widehat{T}^\Gamma)^*$. The coefficient $\kappa(\gamma_{\mathbb{A}})$ is the value of this character at κ.

The expression (111) is a step closer to the desired stabilization of $I_{\mathrm{ell,reg}}(f)$. In particular, it contains the origins of the endoscopic data G' in (110). For suppose that T and κ are as in (111). One chooses an admissible[18] embedding $\widehat{T} \subset \widehat{G}$ of its dual group, taking then $s' \in \widehat{G}$ to be the resulting image of $\kappa \in \widehat{T}$, and $\widehat{G'} = \widehat{G}_{s'}$ the connected centralizer of s'. It is known that there is also an L-embedding

$$^LT = \widehat{T} \rtimes W_F \to {}^LG = \widehat{G} \rtimes W_F,$$

of the L-group of T into that of G, which restricts to the given embedding of $\widehat{T}$ into $\widehat{G}$. (This is a little more subtle and entails some choices to which the embedding is sensitive. See [165, (2.6)].) In any case, for a fixed such embedding, the product

$$\mathcal{G}' = {}^LT \cdot \widehat{G'}$$

is an L-subgroup of ^LG, which commutes with s', and provides a split extension

$$1 \to \widehat{G'} \to \mathcal{G}' \to W_F \to 1$$

of W_F by $\widehat{G'}$. In particular, it determines an action of W_F on $\widehat{G'}$ by outer automorphisms, which factors through a finite quotient of Γ_F. We take G' to be a quasisplit group over F for which $\widehat{G'}$, with the given action of Γ_F, is a dual group. Finally, if we let ξ' be the identity L-embedding of $\mathcal{G}'$ into ^LG, the 4-tuple $(G', s', \mathcal{G}', \xi')$ becomes an endoscopic datum for G. We have thus obtained a correspondence

$$(T, \kappa) \to (G', s', \mathcal{G}', \xi'),$$

from the pairs (T, κ) taken from (111), to the endoscopic data derived from them as above.

There is another ingredient to the last correspondence. Given the pair (T, κ), we can choose a maximal torus $T' \subset G'$ over F, together with an isomorphism from T' to T over F that is admissible, in the sense that the associated isomorphism $\widehat{T'} \to \widehat{T}$ of dual groups is the composition of an admissible embedding $\widehat{T'} \subset \widehat{G'}$ (as in the footnote 18) with an inner automorphism of $\widehat{G'}$ that takes $\widehat{T'}$ to $\widehat{T}$. Let $\delta' \in T'(F)$ be the associated preimage of the original point δ. The tori T and T' are the centralizers in G and G' of δ and δ', so we can regard δ and δ' as the primary objects. They become part of a larger correspondence

$$(\delta, \kappa) \to (G', \delta') = ((G', \mathcal{G}', s', \xi'), \delta'). \tag{112}$$

Elements δ' obtained in this way are called *images from G* [165, (1.3)].

[18] This means that it is the mapping assigned to a choice of some pair $(\widehat{B}, \widehat{T})$ in $\widehat{G}$, and some Borel subgroup B of G containing T.

Now suppose that G is arbitrary. The general form of the expansion (111) is derived the same way, and takes an almost identical form

$$I_{\text{reg,ell}}(f) = \sum_{\delta \in \Delta_{\text{ell,reg}}(G)} \iota(T,G)\,\text{vol}(\delta) \sum_{\kappa \in \mathscr{K}(T/K)} \text{Orb}^{\kappa}(\delta, f),$$

where

$$\iota(T,G) = |\ker(\mathscr{E}(T/F) \to \bigoplus_v \mathscr{E}(T,F_v))||\kappa(T/F)|^{-1}$$

and $\text{Orb}^{\kappa}(\delta, f)$ is defined as in (111). The correspondence (112) remains in place, and is easily seen to have an inverse, which in general extends to a bijection

$$\{(G', \delta')\} \xrightarrow{\sim} \{(\delta, \kappa)\}. \tag{113}$$

The domain is the set of equivalence classes of pairs (G', δ'), where G' is an elliptic endoscopic datum for G, δ' is a strongly G-regular, elliptic element in $G'(F)$ that is an image from G, and equivalence is defined by isomorphisms of endoscopic data. The range is the set of equivalence classes of pairs (δ, κ), where δ belongs to $\Delta_{\text{ell,reg}}(G)$, κ lies in $\mathscr{K}(G_\delta/F)$, and equivalence is defined by conjugation by elements in $G(\overline{F})$. (See [150] and [121, Lemma 9.7].) Given (G', δ'), we set

$$f'(\delta') = f_G^{\kappa}(\delta), \tag{114}$$

where to emphasize the bijection, and to keep us mindful of its essential simplicity, we have written $f_G^{\kappa}(\delta)$ in place of $\text{Orb}^{\kappa}(\delta, f)$. In other words

$$f_G^{\kappa}(\delta) = \sum_{\{\gamma_{\mathbb{A}} \in \Gamma(G(\mathbb{A})) : \gamma_{\mathbb{A}} \sim \delta\}} f_G(\gamma_{\mathbb{A}})\kappa(\gamma_{\mathbb{A}}), \tag{115}$$

with

$$f_G(\gamma_{\mathbb{A}}) = \text{Orb}(\gamma_{\mathbb{A}}, f).$$

We can then write

$$I_{\text{ell,reg}}(f) = \sum_{G' \in \mathscr{E}_{\text{ell}}(G)} |\text{Out}_G(G')|^{-1} \sum_{\delta' \in \Delta_{\text{ell},G\text{-reg}}} \text{vol}(\delta')\iota(G_\delta, G) f'(\delta')$$

for the finite group

$$\text{Out}_G(G') = \text{Aut}_G(G')/\text{Int}(G')$$

of outer automorphisms of G' as an endoscopic datum, and with the understanding that $f'(\delta') = 0$ if δ' is not an image from G. Langlands showed that for any pair (G', δ'), the number

$$\iota(G,G') = (\iota(G_\delta, G)\iota(G'_{\delta'}, G')^{-1})|\text{Out}_G(G')|^{-1}$$

is independent of δ' and δ. (Kottwitz later expressed the product in the brackets on the right as a quotient $\tau(G)\tau(G')^{-1}$ of Tamagawa numbers [119, Theorem 8.3.1].)

Set

$$\widehat{S'}_{\mathrm{ell},G\text{-reg}}(f') = \sum_{\delta' \in \Delta_{\mathrm{ell},G\text{-reg}}} \mathrm{vol}(\delta')\iota(G'_{\delta'},G')f'(\delta'). \tag{116}$$

It then follows that

$$I_{\mathrm{ell,reg}}(f) = \sum_{G' \in \mathscr{E}_{\mathrm{ell}}(G)} \iota(G,G')\widehat{S'}_{\mathrm{ell},G\text{-reg}}(f'), \tag{117}$$

since

$$\mathrm{vol}(\delta) = \mathrm{vol}(G_\delta(F) \setminus G_\delta(\mathbb{A}_F)^1) = \mathrm{vol}(G'_{\delta'}(F) \setminus G'_{\delta'}(\mathbb{A}_F)^1) = \mathrm{vol}(\delta').$$

We have sketched how Langlands derived a version (117) of the desired formula (110). However, the term $f'(\delta')$ in (116) is defined in (114) only as a function on the rational classes $\Delta_{\mathrm{ell},G-\mathrm{reg}}(G')$. As we have said earlier, one wants it to be the adelic stable orbital integral of a function in $C_c^\infty(G'(\mathbb{A}_F))$. But to this point, we do not yet have a well defined candidate for its *local* orbital integrals. The sum in (115) is over adelic products $\gamma_{\mathbb{A}} = \prod_v \gamma_v$, in which γ_v is a conjugacy class in $G(F_v)$ that lies in the stable class of the image δ_v of δ in $G(F_v)$. It follows that if $f = \prod_v f_v$, then

$$f'(\delta') = f_G^\kappa(\delta) = \prod_v f_v^\kappa(\delta_v)$$

where

$$f_v^\kappa(\delta_v) = \prod_{\gamma_v \sim \delta_v} f_{v,G}(\gamma_v)\kappa(\gamma_v). \tag{118}$$

But this is not quite the local definition we are looking for. The problem is that we have been treating δ as both a stable class in $\Delta_{\mathrm{ell,reg}}(G)$ and a representative in $G(F)$ of that class. The distinction has not mattered up until now, since $f'(\delta') = f_G^\kappa(\delta)$ depends only on the class of δ. However, the coefficients $\kappa(\gamma_v)$ in the local functions (118) are defined in terms of the "relative position" of γ_v and δ_v, a notion that comes from the original pairing between $H^1(F_v,T)$ and $\pi_0(\widehat{T}^{\Gamma_v})$, and is sensitive to how γ_v and δ_v are situated within their local conjugacy classes.

The solution for Langlands and Shelstad was to replace $\kappa(\gamma_v)$ with a function $\Delta_G(\delta'_v,\gamma_v)$ that they called a *transfer factor*. This is the deepest part of the theory, and it is the content of the papers [165] and [166]. The function is defined as a product of $\kappa(\gamma_v)$ with some subtle factors that depend on δ'_v and δ_v, but not on γ_v. The product $\Delta_G(\delta'_v,\gamma_v)$ then turns out to be independent of the choice of δ_v, and depends therefore only on the local stable class of δ'_v in $G'(F_v)$ and the local conjugacy class of γ_v in $G(F_v)$. Moreover, if δ'_v is the local image of $\delta' \in \Delta_{\mathrm{ell},G\text{-reg}}(G')$ for each v, the product over v of the corresponding local transfer factors is equal to the coefficient $\kappa(\gamma_{\mathbb{A}})$ in (115).

Transfer factors play the role of a kernel in the local transfer of functions. After first introducing them, Langlands and Shelstad defined the transfer to G'_v of a function $f_v \in C_c^\infty(G_v)$ on G_v as an "integral transform"

$$f'_v(\delta'_v) = \sum_{\gamma_v \in \Gamma_{\mathrm{reg}}(G_v)} \Delta_G(\delta'_v, \gamma_v) f_{v,G}(\gamma_v), \quad \delta'_v \in \Delta_{G\text{-reg}}(G'_v), \tag{119}$$

where

$$f_{v,G}(\gamma_v) = |D(\gamma_v)|^{\frac{1}{2}} \mathrm{Orb}(\gamma_v, f_v), \qquad \gamma_v \in \Gamma_{G\text{-reg}}(G_v), \tag{120}$$

is now a *normalized*[19] orbital integral, and $f'_v(\delta'_v) = f_v^{G'}(\delta'_v)$ is the analogue for G'_v of the normalized stable orbital integral

$$f_v^G(\delta_v) = |D(\delta_v)|^{\frac{1}{2}} \mathrm{SOrb}(\delta_v, f_v), \qquad \delta_v \in \Delta_{G\text{-reg}}(G_v), \tag{121}$$

for G_v. The normalizing factor is the absolute value of the Weyl discriminant

$$D(\gamma_v) = D_G(\gamma_v) = \det(1 - \mathrm{Ad}(\gamma_v))_{\mathfrak{g}_v/\mathfrak{t}_v},$$

for the Lie algebras $\mathfrak{g}_v$ and $\mathfrak{t}_v$ of G_v and $T_v = G_{\gamma_v}$. It was only then that they could pose their local transfer conjecture. It became part of the global conjecture (together with the fundamental lemma for the nonarchimedean unit function), which we recall was established later.

Thus, the function $f'(\delta')$ in (116) really is the stable orbital integral at δ' of a function h' in $C_c^\infty(G'(\mathbb{A}))$. It is at this point that one can treat the left-hand side of (116) as the pairing of a stable distribution $S'_{\mathrm{ell},G\text{-reg}}$, defined by the inductive procedure[20] from (117) outlined earlier, with the function in $C_c^\infty(G'(\mathbb{A}))$, rather than just the sum over rational points δ' on the right-hand side of (116).

There are still a couple of technical points that we should at least mention. The Langlands–Shelstad transfer factor depends on a choice of L-embedding of ${}^LG'$ into ${}^LG$. If G_{der} equals G_{sc}, such embeddings exist, and to fix one, it suffices to choose an L-isomorphism from $\mathcal{G}'$ to ${}^LG'$. If not, $\mathcal{G}'$ might not be L-isomorphic to ${}^LG'$. (It is a question whether a certain 2-cocycle with values in $Z(\widehat{G})$ splits.) In this case, minor adjustments have to be made, which entail choosing a central extension

$$1 \to \widetilde{C}' \to \widetilde{G}' \to G' \to 1,$$

and taking f' to be a function on $\widetilde{G}'(\mathbb{A})$ with a certain central character on $\widetilde{C}'(F) \setminus \widetilde{G}'(\mathbb{A})$. (See [165, (4.4)], which involves also taking a central extension $\widetilde{G}$

[19] This is not in conflict with the global notation in (114), thanks to the product formula

$$|D(\gamma)| = \prod_v |D(\gamma_v)_v| = 1, \qquad \gamma \in G(F).$$

Langlands and Shelstad put the quotient $|D_G(\gamma_v)|^{\frac{1}{2}} |D_{G'}(\delta'_v)|^{-\frac{1}{2}}$ into their transfer factor as term $\Delta_{\mathrm{IV}}(\delta'_v, \gamma_v)$ in [165, §3.6]. However, it is instructive to use it to normalize the orbital orbitals, as we will observe later in the section, even as we continue to use the notation of [165] for the transfer factor in (119)

[20] One does not actually need the general inductive definition in the simple case here of the G-regular elliptic terms. One obtains the stable distribution $S^G_{\mathrm{ell},G\text{-reg}}$ (G being quasisplit) directly as a line combination of stable, adelic orbital integrals from the construction of Langlands we have just described.

of G, or [22, p. 202], which is based on the adjustment made in [125] for *twisted* transfer factors.)

Another point is the subscript "G-reg" in the summands on the right-hand side of (117). This is a minor logical violation of our general inductive definitions in the case of the stable linear form $S^G_{\mathrm{ell,reg}}$ on $G(\mathbb{A})$, since the stable distributions S' in (117) are supposed to depend only on G' (and not G). We have already remarked in the footnote 20 that we do not need the general inductive definitions in this concrete case. Still, to be consistent, why don't we just replace the subscripts "G-reg" by "reg", and replace the equality sign in (117) by the "heuristic approximation" symbol $\sim$ we have already used in (106). This in any case is philosophically sound since the complement of $\Delta_{G\text{-reg}}(G')$ in $\Delta_{\mathrm{reg}}(G')$ is sparse.

The stabilization of the strongly regular, elliptic part of the trace formula becomes

$$I_{\mathrm{ell,reg}}(f) \sim \sum_{G' \in \mathscr{E}_{\mathrm{ell}}(G)} \iota(G,G')\widehat{S'}_{\mathrm{ell,reg}}(f'). \tag{122}$$

We could combine this with (106), and the elementary induction arguments that precede (108) and (109). We would then obtain a stabilization

$$I_2(f) \sim \sum_{G' \in \mathscr{E}_{\mathrm{ell}}(G)} \iota(G,G')\widehat{S'}_2(f') \tag{123}$$

of the L^2-discrete part of the trace formula. These arguments all apply to "approximate" identities, which means that (123) is something we expect to be true. Langlands reviewed such arguments, and would then have used (123) to guess at some of the spectral implications of the theory of endoscopy [150]. These include most notably versions of his conjectural classification of representations into local and global L-packets.

We have concluded our discussion of the explicit stabilization (122) of the regular elliptic part (103) of the invariant trace formula. It must seem rather murky to a nonspecialist. It is helpful to think of the bijection (113) as the centre of the process. We can illustrate the transition schematically as follows.

$$\begin{array}{rcl} & I_{\mathrm{ell,reg}}(f) & \qquad (103) \\ \text{Galois cohomology} & \big\downarrow & \\ & \{(G',\delta')\} & \qquad (113) \\ \text{transfer factors} & \big\downarrow & \\ & \sum_{G'} \iota(G,G')\widehat{S'}_{\mathrm{ell,reg}}(f') & \qquad (122) \end{array}$$

Before going on, we recall that there are a number of further topics we promised to take up in this section. We may not be able to give them all the attention they deserve, but there is one critical paper that would in any case be the next step in this discussion. It is Langlands' classification of the representations of real groups

[151], which together with subsequent work of Shelstad, represents an essential link[21] between the earlier work of Harish-Chandra on representation theory and the emerging theory of endoscopy. We shall take this opportunity for a short digression on the work of Harish-Chandra, in which we assume that G is a reductive algebraic group over $\mathbb{R}$.

We have alluded to the construction by Harish-Chandra of the discrete series [86], [88], those representations of a real group that occur discretely[22] in $L^2(G(\mathbb{R}))$, but we have not described it. It was a climax in his long and comprehensive study of the harmonic analysis on a general (semisimple) real group $G(\mathbb{R})$.

Harish-Chandra's harmonic analysis also represents an interplay between the geometric objects and the spectral objects on $G(\mathbb{R})$. These are the orbital integrals and the irreducible characters, whose global versions became the heart of the trace formula. They were both introduced by Harish-Chandra in the early stages of his career. Both are fundamental and deep. It was of course the theory of characters that became a foundation for the discrete series.

Recall that an irreducible unitary representation π of $G(\mathbb{R})$ is infinite-dimensional (unless it is 1-dimensional or attached to a representation of compact factor on $G(\mathbb{R})$). It was not initially clear how it could have a character, since the trace of an infinite-dimensional unitary matrix $\pi(x)$ is not defined. Harish-Chandra's idea was to make systematic use of the general theory of distributions that had just been introduced by Laurent Schwartz [203], [204]. For any irreducible π, Harish-Chandra proved that the operator

$$\pi(f) = \int_{G(\mathbb{R})} f(x)\pi(x)\,\mathrm{d}x$$

attached to a function $f \in C_c^\infty(G(\mathbb{R}))$ was of trace class, and that the linear form

$$f \to \Theta(\pi, f) = \mathrm{tr}(\pi(f))$$

was a distribution (which is to say, continuous for the usual topology on $C_c^\infty(G(\mathbb{R}))$). This is what he called the *character* of π. The proof was not particularly difficult as these things go. Much deeper was a second theorem on characters, his so-called *regularity theorem*. It asserts that any (irreducible) character $\Theta(\pi)$ is a locally integrable function $x \to \Theta(\pi, x)$ on $G(\mathbb{R})$, which is to say that

$$\Theta(\pi, f) = \int_{G(\mathbb{R})} \Theta(\pi, x) f(x)\,\mathrm{d}x, \qquad f \in C_c^\infty(G(\mathbb{R})). \tag{124}$$

[21] As it could also be argued that Langlands' manuscript on Eisenstein series represents a link between Harish-Chandra's investigations into the Plancherel formula and a future trace formula, even though Langlands' monumental volume stands on its own, and in fact also influenced the subsequent course of Harish-Chandra's work.

[22] We have been using the term *relative discrete series* (or *square integrable representations*) to describe the representations that occur discretely modulo the centre $Z(\mathbb{R})$ of $G(\mathbb{R})$. These are slightly more general. They are the representations of Levi subgroups used in the parabolic induction process that yields all the tempered representations, the ones that occur in the full spectral decomposition of $L^2(G(\mathbb{R}))$.

The regularity theorem is really about the differential equations

$$z\Theta = \chi(\Theta, z)\Theta, \qquad z \in \mathscr{Z}_G, \tag{125}$$

satisfied by any invariant eigendistribution Θ of the centre $\mathscr{Z}_G$ of the universal enveloping algebra $\mathscr{U}_G$ of the complex Lie algebra of $G(\mathbb{R})$. This is a property that holds for any character $\Theta = \Theta(\pi)$, by an infinite-dimensional version of Schur's lemma previously established by Harish-Chandra. In this case, the homomorphism

$$\chi(\Theta) : z \to \chi(\Theta, z), \qquad z \in \mathscr{Z}_G,$$

from $\mathscr{Z}_G$ to $\mathbb{C}^*$ is called the *infinitesimal character* of π. Like many results in this area, Harish-Chandra's argument yields not only the existence of the function $\Theta(\pi,x)$, but frequently also an interesting, explicit formula that it satisfies. The first (and easier) half of the proof uses the elliptic regularity theorem for differential equations to prove that the restriction of $\Theta(\pi,x)$ to the open, dense subset $G_{\mathrm{reg}}(\mathbb{R})$ of (strongly) regular elements in $G(\mathbb{R})$ is a (real) analytic function of x. The second half classifies the singularities of the *normalized* character

$$\Phi(\pi,x) = |D(x)|^{\frac{1}{2}}\Theta(\pi,x), \qquad x \in G_{\mathrm{reg}}(\mathbb{R}), \tag{126}$$

by the Weyl discriminant $D(x)$, at the hypersurfaces in the complement of $G_{\mathrm{reg}}(\mathbb{R})$ in $G(\mathbb{R})$. It establishes that any left invariant derivative of $\Phi(\pi,x)$ remains bounded as x approaches a singular hypersurface, and for many π, also gives an explicit formula for the "jump" of the function as it crosses the hypersurface. This yields an interesting boundary value problem satisfied by $\Phi(\pi,x)$ on the closures of the open connected components of $G_{\mathrm{reg}}(\mathbb{R})$. Its solution is what provides the explicit formula for $\Phi(\pi,x)$.

Since characters are invariant distributions, their functions $\Theta(\pi,x)$ are conjugacy invariant in x. This can be combined with the Weyl integration formula

$$\int_{G(\mathbb{R})} h(x)\,\mathrm{d}x = \sum_{\{T\}} |W(G(\mathbb{R}),T(\mathbb{R})|^{-1} \int_{T_{\mathrm{reg}}(\mathbb{R})} \left(|D(t)| \int_{T(\mathbb{R})\backslash G(\mathbb{R})} h(x^{-1}tx)\,\mathrm{d}x \right) \mathrm{d}t$$

for the change of variables used to express the integral of a function $h \in C_c(G(\mathbb{R}))$ as an integral of its averages over conjugacy classes. Here $\{T\}$ is the set of $G(\mathbb{R})$-conjugacy classes of maximal tori in $G(\mathbb{R})$, $W(G(\mathbb{R}),T(\mathbb{R}))$ is the normalizer of $T(\mathbb{R})$ in $G(\mathbb{R})$ modulo its centralizer $T(\mathbb{R})$, $T_{\mathrm{reg}} = T \cap G_{\mathrm{reg}}$, and

$$D(t) = \det((1 - \mathrm{Ad}(t))_{\mathfrak{g}/\mathfrak{t}})$$

is again the Weyl discriminant. It then follows from (124) that

$$\Theta(\pi,f) = \sum_{\{T\}} |W(G(\mathbb{R}),T(\mathbb{R})|^{-1} \int_{T_{\mathrm{reg}}} \Phi(\pi,t) f_G(t)\,\mathrm{d}t, \tag{127}$$

for the normalized character $\Phi(\pi,t)$ from (126) and the normalized orbital integrals $f_G(t)$ from (120). Harish-Chandra used this formula repeatedly in his development of the discrete series.

Here in general terms is what he proved. First of all, a (connected) reductive group G over $\mathbb{R}$ has a discrete series of representations π if and only if it has a maximal torus T over $\mathbb{R}$ that is *anisotropic*, which means that $T(\mathbb{R})$ is compact. We should recall that for *any* maximal torus $T \subset G$ in G over $\mathbb{R}$, we have a chain of three Weyl groups

$$W(G(\mathbb{R}),T(\mathbb{R})) \subset W_{\mathbb{R}}(G,T) \subset W(G,T),$$

in which $W(G,T)$ is the full (complex) Weyl group and $W_{\mathbb{R}}(G,T)$ is the subgroup of elements that stabilize $T(\mathbb{R})$, while $W(G(\mathbb{R}),T(\mathbb{R}))$ is as in (127), the subgroup of elements in $W_{\mathbb{R}}(G,T)$ induced from $G(\mathbb{R})$. In the case here that $T(\mathbb{R})$ is compact, $W(G,T)$ equals $W_{\mathbb{R}}(G,T)$, but $W(G(\mathbb{R}),T(\mathbb{R}))$ is generally a proper subgroup of $W_{\mathbb{R}}(G,T)$. This last circumstance is responsible for some of the complexity of discrete series representations. The second main property he established is that π is completely determined by the restriction $\Theta(\pi,t)$ of its character to the anisotropic torus $T(\mathbb{R})$. Harish-Chandra in fact showed that this restriction $\Theta(\pi,t)$ satisfies an explicit formula that is more complicated than, but nevertheless reminiscent of, the Weyl character formula.

We recall that the Weyl character formula applies to the special case that G is anisotropic, which means that $G(\mathbb{R})$ itself is compact. The discrete series accounts for all of the irreducible representations in this case, and they are all finite-dimensional. According to Weyl's classification, they are parametrized by orbits $\{\chi\}$ of characters χ on $T(\mathbb{R})$ under the Weyl group $W(G,T)$. To state the Weyl character formula for any such representation, we have to choose an order on the roots $\{\alpha\}$ of (G,T). This gives a corresponding set $\{\alpha > 0\}$ of positive roots as well as the associated linear form

$$\rho = \tfrac{1}{2}\sum_{\alpha>0}\alpha \tag{128}$$

on the Lie algebra $\mathfrak{t}(\mathbb{R})$ of $T(\mathbb{R})$, and from each $W(G,\mathbb{R})$-orbit $\{\chi\}$, a unique character χ whose differential $d\chi$ lies in the closure of the associated positive chamber in the dual space $\mathfrak{t}^*(\mathbb{R})$. The Weyl character formula for the representation π_χ attached to the $W(G,T)$-orbit of χ is then

$$\Theta(\pi_\chi,t) = \sum_{w\in W(G(\mathbb{R}),T(\mathbb{R}))}\left(\frac{\varepsilon(w)\chi(w\cdot \exp H)e^{\rho(wH-H)}}{\Delta(\exp H)}\right),$$

for any point

$$t = \exp H$$

in $T_{\mathrm{reg}}(\mathbb{R})$, and for

$$\Delta(\exp H) = \prod_{\alpha>0}\left(1 - e^{-\alpha(H)}\right).$$

It is a simple matter to check that the right-hand side of the formula remains the same if either H or χ is replaced by a Weyl translate wH or $w\chi$, for any $w \in W(G,T)$. The former property is needed for the function $\Theta(\pi_\chi,\cdot)$ on $G(\mathbb{R})$ to be invariant under conjugation, the latter for it to depend only of the $W(G,T)$-orbit of χ.

Harish-Chandra's discrete series had of course to generalize this. The obvious structural difference is that it is the subgroup $W(G(\mathbb{R}),T(\mathbb{R}))$ of

$$W(G,T) = W_{\mathbb{R}}(G,T),$$

acting on the anisotropic torus T, that ought to reflect the $G(\mathbb{R})$-invariance of characters. The similarity is that they would still turn out to be determined by their values on $T(\mathbb{R})$. The natural guess, in retrospect at least, would be that the discrete series are parametrized by suitably defined $W(G(\mathbb{R}),T(\mathbb{R}))$-orbits. This is precisely what Harish-Chandra established, but only after years of concentrated study of the underlying harmonic analysis. All that we need to specify his classification is his formula [86, Theorem 3] and [88, Theorem 18] for their characters on $T_{\mathrm{reg}}(\mathbb{R})$. We shall state the version of it formulated by Langlands on p. 134 of [151], which is closer to the Weyl character formula, and is also compatible with Langlands' ideas on the broader classification of representations.

Given the group G with anisotropic maximal torus T, we can choose characters χ on $T(\mathbb{R})$ and an order on the roots $\{\alpha\}$, as in the special case of anisotropic G above. Langlands represents the order by the corresponding linear form ρ in (128), and therefore considers pairs (χ,ρ) in which χ lies in the closure of the positive chamber in $\mathfrak{t}^*$ attached to ρ. Harish-Chandra's classification of discrete series is then given by a bijection

$$(\chi,\rho) \to \pi_{\chi,\rho}$$

from the $W(G(\mathbb{R}),T(\mathbb{R}))$-orbits of such pairs onto the equivalence classes of discrete series representations such that

$$\Theta(\pi_{\chi,\rho},t) = (-1)^{q_G} \sum_{w\in W(G(\mathbb{R}),T(\mathbb{R}))}\left(\frac{\varepsilon(w)\chi(w\cdot\exp H)e^{\rho(wH-H)}}{\Delta(\exp H)}\right), \qquad (129)$$

for any point $t = \exp H$ in $T_{\mathrm{reg}}(\mathbb{R})$, where $q_G = \frac{1}{2}\dim(G(\mathbb{R})/K_{\mathbb{R}})$ is one-half of the dimension of the symmetric space attached to $G(\mathbb{R})$.

We note in passing that the values of $\Theta(\pi_{\chi,\rho},t_1)$ on a general maximal torus $T_1 \subset G$ can be reduced according to the theory developed by Harish-Chandra to its values (129) on $T(\mathbb{R})$. This is because the singularities of the normalized character $\Phi(\pi_{\chi,\rho},t_1)$, expressed in Harish-Chandra's jump conditions at a singular hypersurface T_{01}, are given in terms of the limits at points $t_{01} \in T_{01}(\mathbb{R})$ of its values $\Phi(\pi_{\chi,\rho},t_0)$ on a maximal torus[23] T_0 that shares the hypersurface T_{01} with T_1, but whose anisotropic part is of one dimension greater than that of T_1. Increasing the

[23] One says that T_1 is a *Cayley transform* of T_0.

anisotropic dimension $d_a(T_1)$ of a maximal torus T_1 makes the corresponding character values simpler. For this reason, the solution of the boundary value problem for $\Phi(\pi_{\chi,\rho},t_1)$ follows by decreasing induction on $d_a(T_1)$, using the differential equations (125), the jump conditions, the basic explicit formula (129) on $T(\mathbb{R})$, and the fact that $\Phi(\pi_{\chi,\rho},t_1)$ is bounded on $T_1(\mathbb{R})$, a consequence in turn of Harish-Chandra's proof that the characters $\Theta(\pi_\chi,\rho)$ of discrete series are tempered distributions. (See the formulas of Harish-Chandra [86] and their simplifications in [95].) We recall from Section 8 that such formulas arose later in the invariant trace formula [15], [17], and were then used by Morel [183] for the geometric boundary terms in the Lefschetz trace formula for the Shimura varieties attached to $\mathrm{GSp}(2n)$.

Langlands arrived at his version (129) of Harish-Chandra's formula in the process of relating the irreducible representations of $G(\mathbb{R})$ to his ideas for parameterizing the local components of automorphic representations. The obvious lesson to be taken from Harish-Chandra's discrete series is that these representations occur naturally in finite sets. Each such set corresponds to an irreducible finite-dimensional representation of the complex group $G(\mathbb{C})$, or equivalently, the representation π_χ of the compact real form of $G(\mathbb{C})$ with highest weight $d\chi$ in the Weyl classification. It consists of representations with the same infinitesimal character, and can be parameterized by the cosets in $W(G(\mathbb{R}),T(\mathbb{R}))\setminus W(G,T)$. These finite sets became known as L-packets, a term we have used regularly throughout the report without actually defining it.

To see its meaning, we note that the first two sections of Langlands' paper [151] were devoted to something quite different. For any group G over $\mathbb{R}$, Langlands considered local L-homomorphisms

$$\phi\colon W_{\mathbb{R}} \to {}^LG$$

from the Weil group to the L-group, with the property[24] that if the image of ϕ is contained in some parabolic subgroup ${}^LP$ of ${}^LG$, the corresponding parabolic subgroup P of G is defined over $\mathbb{R}$. He wrote $\Phi(G)$ for the set of $\widehat{G}$-conjugacy classes of such parameters ϕ, as we have noted earlier, and $\Phi_2(G)$ for the subset of such classes such that the image of ϕ lies in no proper parabolic subgroup ${}^LP$ of ${}^LG$. For the local field $\mathbb{R}$, these are reasonably elementary objects. Langlands calculated them directly in terms of simple data within ${}^LG$. He then observed that for the group G with anisotropic torus T, the set $\Phi_2(G)$ was naturally bijective with the L-packets of discrete series. On the other hand, for any representation

$$r\colon {}^LG \to \mathrm{GL}(n,\mathbb{C}),$$

[24] This condition is often called *relevance*. The reason we have not encountered it before is that we have usually been working with quasisplit groups, where the condition is automatic. We are also using the more streamlined notation of [119, §1] rather than the original formulation in §1–2 of [151]. As a matter of fact, these days one often formulates matters in terms of Vogan's *pure inner forms* [243], in which different inner forms are treated as components of the same object, and where the condition of relevance is only implicit.

one can attach L-functions $L(s, r \circ \phi)$ and ε-factors $\varepsilon(s, r \circ \phi, \psi)$ to parameters $\phi \in \Phi_2(G)$, according to the prescription described in [237, (3.1) and (3.3.1)]. The representations π in the corresponding L-packets Π_ϕ then have the property that their L-functions $L(s, \pi, r)$ and ε-factors $\varepsilon(s, \pi, r, \psi)$ for any r match those of ϕ. This has been proven in cases where the representation theoretic functions have an independent meaning. In cases where they do not, it can be taken simply as their definition.

The Langlands classification was of course for all irreducible representations of a group $G(\mathbb{R})$, not just the discrete series. Langlands first extended his parametrization of L-packets of discrete series by parameters $\phi \in \Phi_2(G)$ to L-packets of square integrable representations (relative discrete series), the case that G has a maximal torus that is anisotropic modulo the centre of G, at the bottom of p. 134 of [151]. This condition holds by definition for the Levi component of any *cuspidal* parabolic subgroup $P = N_P M$ of G. He then observed that a general parameter ϕ could be represented as the image in $\Phi(G)$ of a parameter $\phi_M \in \Phi_2(M)$ under the embedding ${}^L M \subset {}^L G$ attached to some cuspidal parabolic subgroup $P = N_P M$ of G. One can refine this embedding to a two stage embedding

$$\phi_M \to \phi_L \to \phi, \qquad \phi_M \in \Phi_2(M),$$

for parabolic subgroups $P = N_P M \subset Q = N_Q L$ defined as follows. One first writes ϕ_M uniquely as a twist

$$\phi_M = \phi_{M,\mathrm{temp},\lambda},$$

where $\phi_{M,\mathrm{temp}} \in \Phi_2(M)$ has bounded image in $\widehat{M}$, $\lambda \in \mathfrak{a}_M^*$ is a uniquely determined, real-valued linear form on the real vector space $\mathfrak{a}_M$, and $\phi_{M,\mathrm{temp},\lambda}$ is the parameter whose L-packet is the set of representations

$$\pi_{M,\lambda}(m) = \pi_M(m) \mathrm{e}^{\lambda(H_M(m))}, \qquad m \in M(\mathbb{R}),$$

such that π_M lies in the L-packet of $\phi_{M,\mathrm{temp}}$. One then chooses P such that λ lies in the *closure* of the corresponding chamber $(\mathfrak{a}_P^*)^+$, and $Q \supset P$ so that λ lies in the *open* chamber $(\mathfrak{a}_Q^*)^+$, regarded as a convex cone in the closure of $(\mathfrak{a}_Q^*)^+$. The set $\Phi(G)$ can then be identified with the set of $\widehat{G}$-orbits of triplets (ϕ_M, P, Q) of this form.

Langlands' goal in [151] was to define an explicit partition of $\Pi(G)$ into a disjoint union over $\phi \in \Phi(G)$ of finite L-packets Π_ϕ. In particular, the L-packets for G would also be indexed by $\widehat{G}$-orbits of triplets. His answer, which is not hard to describe, is an elegant reformulation of some of Harish-Chandra's fundamental results.

Given a triplet (ϕ_M, P, Q), we can define the L-packet $\Pi_{\phi_M} \subset \Pi_2(M)$ by Langlands' parametrization of Harish-Chandra's relative discrete series. For the next step, we form the parabolic subgroup $R = L \cap P$ of L with Levi component M. For each $\pi_M \in \Pi_{\phi_M}$, we then take $\mathscr{I}_R^L(\pi_M)$, a representation of $L(\mathbb{R})$ parabolically induced from the representation π_M in the relative discrete series of $M(\mathbb{R})$ that is unitary modulo the centre of $L(\mathbb{R})$. Let Π_{ϕ_L, π_M} be its set of irreducible constituents, a finite set of representations of $L(\mathbb{R})$ that are tempered modulo the centre. (The induced representation is generically irreducible, but when is it not, its irreducible constituents are of considerable interest.) Langlands defined the L-packet of ϕ_L in $\Pi(L)$ to be the set

$$\Pi_{\phi_L} = \bigcup_{\pi_M} \Pi_{\phi_L,\pi_M}, \qquad \pi_M \in \Pi_{\phi_M},$$

a union that was known to be disjoint. Finally, for any representation π_L in Π_{ϕ_L}, one can take the induced representation $\mathscr{I}_Q^G(\pi_L)$. This is a nontempered induced representation of $G(\mathbb{R})$, which in general is reducible. However, Langlands proved that it has a unique irreducible quotient $\pi(\pi_L)$. He then defined the L-packet of ϕ to be the set

$$\Pi_\phi = \{\pi = \pi(\pi_L) : \pi_L \in \Pi_{\phi_L}\}$$

of all these *Langlands quotients*, establishing at the same time that these representations were all disjoint.

This is the Langlands classification for real groups. When he introduced it in 1973, the irreducible constituents of the (essentially) tempered induced representations $\mathscr{I}_R^L(\pi_R)$ were not completely understood. However, they were classified soon afterwards by Knapp and Zuckerman [114], [115], thus providing in particular an explicit classification

$$\Pi_{\text{temp}}(G) = \coprod_{\phi \in \Phi_{\text{temp}}(G)} \Pi_\phi$$

of the irreducible *tempered* representations of $G(\mathbb{R})$ in terms of L-packets parameterized by the bounded Langlands parameters $\Phi_{\text{temp}}(G)$. This is the special case of the general classification in which the groups Q in the triplets (ϕ_M, P, Q) are all equal to G.

It was observed later in the 1970s that the Langlands classification would apply in principle also to p-adic groups. In this case, there is still no explicit classification of the relative discrete series $\Pi_2(M)$, or of the irreducible constituents of the induced tempered representations $\mathscr{I}_R^L(\pi_M)$ (although much is known, especially about this second question). However, the general classification, including the properties Langlands established for the quotients that bear his name, remains in force.

Langlands' classification established his conjecture of Local Functoriality for $F = \mathbb{R}$, as stated in Question 4 or 6 of [138], or as *Local Functoriality* stated in Section 5 here. In fact, it gives affirmative answers to all the questions in [138], insofar as they apply to the local field $F = \mathbb{R}$. As we noted in the special case of discrete series above, it assigns local L-functions and ε-factors to representations $\pi \in \Pi(G)$ by setting

$$L(s,\pi,r) = L(s, r\circ\phi), \qquad \pi \in \Pi_\phi,$$

and

$$\varepsilon(s,\pi,r,\psi) = \varepsilon(s, r\circ\phi, \psi), \qquad \pi \in \Pi_\phi,$$

for any $\phi \in \Phi(G)$ and $r\colon {}^LG \to \mathrm{GL}(n,\mathbb{C})$, and again for the functions on the right defined as in [237]. Moreover, the functorial correspondence $\pi' \to \pi$ of local representations is defined explicitly in terms of their L-packets by the transfer of parameters

$$\pi'\colon \Pi_{\phi'} \to \pi \in \Pi_\phi, \qquad \phi = \rho' \circ \phi',\, \phi' \in \Phi(G'),$$

for $\rho'\colon {}^LG' \to {}^LG$ as in the statement of Local Functoriality. This answers the local version of the original questions we posed at the beginning of this section, in the case that $F_v = \mathbb{R}$.

The Langlands classification for real groups suggests a spectral analogy with the theory of local endoscopy we have described. The Langlands parameters $\phi \in \Phi_{\mathrm{temp}}(G)$ ought to be analogues of (strongly) regular stable conjugacy classes $\Delta_{\mathrm{reg}}(G)$ over $\mathbb{R}$. Moreover the packets themselves ought to be analogues of the set of conjugacy classes $\gamma \in \Gamma_{\mathrm{reg}}(G)$ in a stable class δ. However, there is more structure than this on the geometric side. The elements in a "geometric packet" are bijective with the explicit set $\mathscr{D}(T)$, where T is the centralizer of a chosen base point δ in the stable class. We do have their dual analysis in terms of Tate–Nakayama duality, with its ties to endoscopic groups. But there is also the more refined structure given by the Langlands–Shelstad transfer factors, together with the associated transfer conjecture and its ultimate proof. Were there spectral analogues of any of these things?

The question was answered for real groups by Shelstad. We must not forget that she had first to introduce the archimedean transfer factors that became the inspiration for [165], and establish the associated transfer of functions. These remain basic links to the general theory, but they also have foundations in the work of Harish-Chandra. We should discuss them briefly before we describe their spectral consequences. In so doing, we need to step back in history, say to the year 1975. What was available then were the basic ideas of Langlands on stable conjugacy and endoscopic groups, his new preprint on the classification for real groups, and of course, the work of Harish-Chandra.

Shelstad's transfer factors are closely related to some curious factors in a refined normalization of orbital integrals that had been forced on Harish-Chandra. He defined a real group G to be *acceptable* if the linear form (128) on the Lie algebra of any maximal torus T over $\mathbb{R}$ lifts to a quasicharacter ξ_ρ on $T(\mathbb{C})$. This condition is independent of the underlying system of positive roots on T, and holds whenever G_{der} is simply connected. Harish-Chandra often worked with this assumption, with the understanding that adjustments for the general case were easy to add separately. (See for example [90, §8].) Under this condition, he normalized the orbital integrals on $T_{\mathrm{reg}}(\mathbb{R})$ by setting

$$F_f(t) = \varepsilon_{\mathbb{R}}(t)\xi_\rho(t)\Delta(t)\mathrm{Orb}(t,f), \tag{130}$$

for $\Delta(t)$ as in (129) and $\varepsilon_{\mathbb{R}}(t) \in \{\pm 1\}$ the locally constant sign function

$$\mathrm{sign}\Big(\prod_{\alpha \in P_{\mathbb{R}}} (1 - \xi_\alpha(t^{-1}))\Big)$$

on $T_{\text{reg}}(\mathbb{R})$ (with $P_{\mathbb{R}}$ being the set of positive real roots on T) [87, §22], [200, p. 5]. Since

$$|\varepsilon_{\mathbb{R}}(t)\xi_{\rho}(t)\Delta(t)| = |D(t)|^{\frac{1}{2}},$$

this does represent a refinement of our normalization $f_G(t)$ from (120). It was chosen by Harish-Chandra to have the property that if $T(\mathbb{R})$ is compact, and f is a matrix coefficient of a discrete series representation, a function in his Schwartz space $\mathscr{C}(G(\mathbb{R}))$ on $G(\mathbb{R})$ defined in [88], then $F_f(t)$ extends from $T_{\text{reg}}(\mathbb{R})$ to a smooth function on $T(\mathbb{R})$.

Before commenting on Shelstad's transfer factors, we should first include a couple of remarks that further illustrate the dual nature of orbital integrals and irreducible characters. Orbital integrals satisfy differential equations

$$F_{zf}(t) = \gamma(z)F_f(t), \qquad t \in T_{\text{reg}}(\mathbb{R}), z \in \mathscr{Z}_G, \tag{131}$$

where $z \to \gamma(z)$ is the Harish-Chandra homomorphism from $\mathscr{Z}_G$ to the algebra of invariant differential operators on $T(\mathbb{R})$. Also, any left-invariant derivative of $F_f(t)$ remains bounded on $T_{\text{reg}}(\mathbb{R})$ as t approaches a singular hypersurface [87, Theorem 3], and has an explicit formula for the jump as this function crosses the hypersurface [90, Theorem 1]. These are dual to the properties we have described for irreducible characters (which are actually simpler when stated with the normalization (126) replaced by the analogue of Harish-Chandra's refined normalization). In particular, they lead to a boundary value problem for each function $F_f(t)$ on the closure of a connected component of $T_{\text{reg}}(\mathbb{R})$. The only difference with what happens for irreducible characters is that the torus that shares the singular hypersurface with T has anisotropic dimension one *less* than that of T. In other words, it is a Cayley transform of T, rather than other way around. It is consequently a decrease in the anisotropic dimension $d_a(T)$ that makes the associated orbital integrals simpler.

Shelstad's transfer factors for the real group G serve as a bridge to the general Langlands–Shelstad transfer factors in [165]. These were defined in §3 of [165] as products

$$\Delta_G(\delta',\gamma) = \Delta_I(\delta',\gamma)\Delta_{II}(\delta',\gamma)\Delta_{III}(\delta',\gamma)\Delta_{IV}(\delta',\gamma), \tag{132}$$

in which the third term comes with a further decomposition

$$\Delta_{III}(\delta',\gamma) = \Delta_{III_1}(\delta',\gamma)\Delta_{III_2}(\delta',\gamma) = \Delta_1(\delta',\gamma)\Delta_2(\delta',\gamma).$$

The Langlands–Shelstad transfer factors are complex and subtle, but they can be illuminated in their specialization to real groups, and the relations the latter bear to the quotients of Harish-Chandra's normalizing factors for G and G'. In commenting briefly on this, we might as well assume that G_{der} is simply connected.

The term Δ_{IV} in (132) is the quotient of our original normalizing factor $|D(\gamma)|^{\frac{1}{2}} = |D_G(\gamma)|^{\frac{1}{2}}$ by $|D_{G'}(\delta')|^{\frac{1}{2}}$, as we agreed in the footnote 19. The term $\Delta_1 = \Delta_{III_1}$ is essentially the local form of the character $\kappa(\gamma)$ with which we began the original construction. The term $\Delta_2 = \Delta_{III_2}$ deals with the contribution $\xi_\rho(\gamma)\xi_{\rho'}(\delta')^{-1}$ of the function $\xi_\rho(\gamma)$ in (130). It normalizes the choice of the L-

isomorphism from ${}^LG'$ to $\mathscr{G}'$ that makes ξ' an L-embedding of ${}^LG'$ into ${}^LG$. The term Δ_{II} in (132) addresses the contribution of the function $\varepsilon_{\mathbb{R}}(\gamma)\Delta(\gamma)$ in (130), or rather the contribution to $\Delta(\delta',\gamma)$ of its quotient by the factor $|D(\gamma)|^{\frac{1}{2}}$ that was put into Δ_{IV}. Finally the term Δ_I is a sign, which is independent of γ, and compensates for various other noncanonical choices that had to go into the previous terms.

This is a necessarily superficial description of transfer factors. Shelstad actually worked with Harish-Chandra's later normalization ${}'F_f(\gamma)$ [90, §17] of orbital integrals, which makes sense for any G, and gives a more complete motivation for her work, but which is also a little more complicated to describe. From now on, let us just assume without comment that every endoscopic datum G' attached to a given G (over a local or global field F) is such that the group $\mathscr{G}'$ is L-isomorphic to the L-group ${}^LG'$. As we have noted above, this condition always holds if G_{der} is simply connected [147, Proposition 1].

Having introduced the transfer factors $\Delta(\delta',\gamma) = \Delta_G(\delta',\gamma)$ for the real group G with endoscopic datum $(G',\mathscr{G}',s',\xi')$, Shelstad defined the transform

$$f'(\delta') = \sum_{\gamma\in\Gamma_{\mathrm{reg}}(G)} \Delta_G(\delta',\gamma) f_G(\gamma), \quad f \in C_c^\infty(G(\mathbb{R})),\, \delta' \in \Delta_{\mathrm{reg}}(G'),$$

that became the archimedean precursor of the general local transfer mapping (119). She then applied the harmonic analysis of Harish-Chandra systematically to its study, making use of the adjoint relation (127) between characters and orbital integrals, the differential equations (125) satisfied by characters and their analogues (131) for orbital integrals, and the accompanying boundary conditions in each case. She used these techniques first to establish the real form of what became the general Langlands–Shelstad transfer conjecture. She then applied them to the spectral question raised above. Her results, which of course depend also on the Langlands classification for real groups, are as follows

We assume first that G is quasisplit over $\mathbb{R}$. For every tempered Langlands parameter $\phi \in \Phi_{\mathrm{temp}}(G)$, set

$$f^G(\phi) = \sum_{\pi\in\Pi_\phi} f_G(\pi), \qquad f \in C_c^\infty(G(\mathbb{R})), \tag{133}$$

for

$$f_G(\pi) = \Theta(\pi,f) = \mathrm{tr}(\pi(f)).$$

The first (spectral) result of Shelstad is that the linear form $f \to f^G(\phi)$ is a stable distribution. It is called the *stable character* of π, and is clearly the spectral analogue of a stable orbital integral. Given the validity of the transfer conjecture for G and G' (for any G), this proves that the pairing $f'(\phi')$ is well defined for any tempered parameter $\phi' \in \Phi_{\mathrm{temp}}(G')$ for the quasisplit group G'.

The second spectral result applies to any group G over $\mathbb{R}$. It is an expansion

$$f'(\phi') = \sum_{\pi\in\Pi_\phi} \Delta_G(\phi',\pi) f_G(\pi), \tag{134}$$

for complex coefficients $\Delta_G(\phi',\pi)$ supported on the subset of $\pi \in \Pi_\phi$ in $\Pi_{\text{temp}}(G)$, in which $\phi = \xi' \circ \phi'$ is the image of ϕ' in $\Pi_{\text{temp}}(G)$. These coefficients are spectral analogues of Shelstad's (geometric) transfer factors, and can be called *spectral transfer factors*.

Shelstad's third spectral result makes these coefficients explicit, in a sense that depends on a chosen base point π_1 in the packet Π_ϕ. Following Labesse and Langlands [127], [218], and similar definitions we have seen in our earlier sections, she defined $S_\phi = \text{Cent}(\phi(W_F),\widehat{G})$, the centralizer in $\widehat{G}$ of the image of a parameter $\phi \in \Pi_{\text{temp}}(G)$, and $\mathbf{S}_\phi = S_\phi/S_\phi^0 Z(\widehat{G})^\Gamma$, the group of connected components in S_ϕ modulo the Galois invariants in the centre of $\widehat{G}$. To state the result, we use the spectral analogue of the bijection (113), or rather the local spectral analogue of (113) for $F = \mathbb{R}$. Its inverse is a bijection

$$(G',\phi') \xrightarrow{\sim} (\phi,s), \tag{135}$$

from the isomorphism classes of pairs (G',ϕ'), in which G' is an endoscopic datum for G and ϕ' lies in $\Pi_{\text{temp}}(G)$, onto isomorphism classes of pairs (ϕ,s), where ϕ is a parameter in $\Pi_{\text{temp}}(G)$ and s is a semisimple element in S_ϕ. Shelstad's third spectral result asserts that for any (ϕ,s), and for $\pi_1 \in \Pi_\phi$ fixed and $\pi \in \Pi_\phi$ arbitrary, the quotient of $\Delta(\phi',\pi)$ by $\Delta(\phi',\pi_1)$ depends only on the image x of s in $\mathbf{S}_\phi$, and that the resulting mapping

$$x \to \langle x,\pi|\pi_1\rangle = \Delta(\phi',\pi)\Delta(\phi',\pi_1)^{-1} \tag{136}$$

is an *injection* from Π_ϕ to the group of characters on the abelian 2-group $\mathbf{S}_\phi$. This is parallel to what happens for the geometric transfer factors, where (x,π,π_1) would be replaced by (κ,γ,γ_1).

We have completed our brief review of Shelstad's work. Her results first appeared in the papers [219], [217], [220], [221], but they were expanded into a somewhat more expository treatment in the papers [222], [224], [223].

Suppose now that G is a group over any local field F of characteristic 0, which we take to be quasisplit. We have alluded to the conjectural local Langlands correspondence for $\Pi_{\text{temp}}(G)$ in past sections. It seems to have evolved with Langlands' ideas in the early 1970s, based on his experience with GL(2), and then the group $G = \text{SL}(2)$. Its conjectural premises are close to the results of Shelstad for $F = \mathbb{R}$, but there is one significant difference. As we have noted earlier, the Weil group W_F has to be replaced by the local Langlands group L_F. It thus remains equal to W_F if F is archimedean, but is taken to be the product $W_F \times \text{SU}(2)$ if F is nonarchimedean. This is to account for the Steinberg representation of $G(F)$, and more generally, the representations in $\Pi_2(G)$ whose matrix coefficients do not have compact support modulo $Z(G)$. Moreover, despite the fact that the conjectural assertions are otherwise similar to those of Shelstad for $F = \mathbb{R}$, any general proof seems unlikely to be the same. What is missing is Harish-Chandra's explicit classification of the discrete series. Without it, we do not know in general how to attach L-packets to parameters $\phi \in \Phi(G)$ and hence how to construct candidates $f^G(\phi)$ for the basic stable distributions.

Although perhaps already clear to the reader, it does not hurt to emphasize the two-fold nature of the final classification for real groups. There is the Langlands classification, based on Harish-Chandra's discrete series, and then there is Shelstad's endoscopic extension based on this and other work of Harish-Chandra. Given the apparent difficulty of an independent classification of general supercuspidal representations, we would hope to establish the local Langlands correspondence without having an explicit construction of the representations in the packets Π_ϕ.

If $G = \mathrm{GL}(n)$, the local Langlands correspondence was proved by Harris and Taylor [92], Henniart [94] and Scholze [202]. It was established by global means, taken from the theory of Shimura varieties. Since stable conjugacy is the same as conjugacy in this case, there is no endoscopy. The local correspondence becomes a canonical bijection $\phi \to \pi_\phi$ from $\Phi_{\mathrm{temp}}(G)$ to $\Pi_{\mathrm{temp}}(G)$ (or between the larger sets $\Phi(G)$ and $\Pi(G)$). It is characterized by the requirement that the two kinds of local Rankin–Selberg L-functions and ε-factors attached to the representations

$$r\colon\ \mathrm{GL}(n_1,\mathbb{C}) \times \mathrm{GL}(n_2,\mathbb{C}) \to \mathrm{GL}(n_1 n_2,\mathbb{C})$$

coincide for parameters (ϕ_1,ϕ_2) and representations (π_1,π_2).

The local Langlands correspondence was established for a quasisplit symplectic or special orthogonal group over the nonarchimedian local field F in Chapter 6 of [23]. The methods are again global, but here they come from the stabilization of the trace formula. For there is considerable endoscopy to contend with in this case. However, there is also a natural way to construct the basic stable distributions $f^G(\phi)$ attached to parameters $\phi \in \Pi_{\mathrm{temp}}(G)$. They are twisted transfers from corresponding twisted invariant distributions on a general linear group $\mathrm{GL}(N)$, relative to the standard outer automorphism $x \to {}^t x^{-1}$. This is because G is a *twisted endoscopic group* for $\mathrm{GL}(N)$, where $N = 2n+1$ if $G = \mathrm{Sp}(2n)$ and $2n$ if G equals either $\mathrm{SO}(2n+1)$ or $\mathrm{SO}(2n)$. The assertions are similar to those of Shelstad for real groups, but there is no need to state them at this point, since we will soon describe their generalizations that accompany the global classification. We note that similar methods were used by Mok [181] to establish the local Langlands correspondence for quasisplit unitary groups G.

The global endoscopic classification is deeper. There were hints of what form it should take in Langlands' paper [127] with Labesse. However, it goes beyond the global Question 7 of [138], whose local version Question 6 was a foundation for the local Langlands correspondence. This is because the global Weil group was known even for $\mathrm{GL}(2)$ to provide only a sparse set of cuspidal automorphic representations. (The same might be said of the local Weil group W_F in Question 6 for a nonarchimedian field F. But its extension was easily accommodated, either as the Weil–Deligne group [237], or equivalently, as the local Langlands group $L_F = W_F \times \mathrm{SU}(2)$ we have taken here.) For a conjectural global classification, one would need to replace the global Weil group by the hypothetical global Langlands group L_F discussed in the last section.

There is another ingredient that has also to be included in any global classification. It consists of the conjectures introduced in [18], [13], and discussed in the treatment of Shimura varieties in Section 8, for describing those automorphic representations in the discrete spectrum L^2_{disc} that are not locally tempered. These represent the counterexamples to what would be the natural extension of Ramanujan's conjecture. For a quasisplit group G over a global field F, the global parameters are the L-homomorphisms

$$\psi\colon L_F \times \mathrm{SL}(2,\mathbb{C}) \to {}^LG \tag{137}$$

whose restriction to L_F has bounded image in $\widehat{G}$, taken as usual up to $\widehat{G}$-conjugacy. As the essential global objects, these parameters force us to account also for their local analogues

$$\psi_v\colon L_{F_v} \times \mathrm{SL}(2,\mathbb{C}) \to {}^LG_v.$$

For if one is to obtain a global classification of automorphic representations for G, as they occur in the discrete spectrum L^2_{disc}, one would at the same time have to establish a generalization of the local Langlands correspondence for these local parameters.

We assume now that G is a quasisplit symplectic or special orthogonal group, this time over a global field F. This is the group for which the endoscopic classification of automorphic representations was established in [23]. The results, which take up the entire monograph, rest on the stabilization of the trace formula for G, as well as the twisted trace formula for $\mathrm{GL}(N)$, and at some points even the twisted trace formula for $\mathrm{SO}(2N)$. The stabilization of the ordinary (invariant) trace formula completed in [21], is of course a special case of the stabilization of the general twisted trace formula in the monographs [179], [180]. These in turn depend on other things, including the general Langlands–Shelstad transfer conjecture, its twisted analogue by Kottwitz and Shelstad [125], the fundamental lemma and its twisted analogue, and finally, a weighted fundamental lemma and its twisted analogue [247], [46], [47] required for terms in the complement of $I_{\mathrm{ell,reg}}(f)$ in $I_{\mathrm{geom}}(f)$. These results have now all been established, except possibly for the twisted, weighted fundamental lemma, which presumably would follow from the methods of [45] and [46].

In [23], an ad hoc substitute $\mathcal{L}_\psi$ for the hypothetical Langlands group L_F was introduced in §1.4, as well as an ad hoc set of L-homomorphisms

$$\psi\colon \mathcal{L}_\psi \times \mathrm{SL}(2,\mathbb{C}) \to {}^LG,$$

in order to be able to formulate the global classification unconditionally. The global results were then stated in §1.5. The local results were stated in §2.3 and established in Chapter 7. Their proof depends on a special case, the actual local Langlands correspondence, established in Chapter 6 and described briefly above. It also relies on some properties of the intertwining operators between induced representations formulated in §2.3–2.4. These were established also in §7, apart from two references [A25] and [A26] from [23] that have still to be written, but which I expect will be completed soon.

As endoscopic identities for the localizations ψ_v of global parameters, the local results are similar to the local Langlands correspondence for parameters ϕ_v. In particular, they resemble our summary of Shelstad's results for real groups. However, there are also a couple of differences.

One is that spectral transfer factors here can be normalized in terms of the Whittaker models inherited from $\mathrm{GL}(N)$. This is actually a simplification. It allows us to take the analogue of the factor $\Delta(\phi',\pi_1)$ in (136) to be 1. The analogue of (136) becomes an injection

$$x_v \to \langle x_v,\pi_v\rangle = \Delta(\psi_v,\pi_v), \qquad \pi_v \in \Pi_{\psi_v}, \tag{138}$$

from a packet of representations Π_{ψ_v} to the group of characters on the abelian 2-group $\mathbf{S}_{\psi_v} = S_{\psi_v}/S^0_{\psi_v}Z(\widehat{G})^{\Gamma_v}$. Another difference is a complication. This occurs in the construction of the packets Π_{ψ_v}. For a general local parameter $\psi_v \in \Psi(G_v)$, the packet becomes a finite set of *reducible* representations

$$\pi_v = \pi_v^1 \oplus \cdots \oplus \pi_v^k, \qquad \pi_v^i \in \Pi(G_v),$$

of $G(F_v)$. Moreover, while some of the irreducible constituents π_v^i of these finite sums are tempered, others are not. However, they are all unitary. A second complication, minor but nonetheless interesting, concerns the coefficients $\langle x_v,\pi_v\rangle = \Delta(\psi_v',\pi_v)$ that ought to occur in the sum (134). What actually occurs are coefficients $\langle s_{\psi_v}x_v,\pi\rangle$, in which s_{ψ_v} is the image in $\mathbf{S}_{\psi_v}$ of the element $s_\psi = \psi\begin{pmatrix}-1 & 0\\ 0 & -1\end{pmatrix}$. The analogues for ψ_v of (133) and (134) then become

$$f_v^{G_v}(\psi_v) = \sum_{\pi_v\in\Pi_{\psi_v}} \langle s_{\psi_v},\pi_v\rangle f_{G_v}(\pi_v) \tag{139}$$

and

$$f_v'(\psi_v') = \sum_{\pi_v\in\Pi_{\psi_v}} \langle s_{\psi_v}x_v,\pi_v\rangle f_{G_v}(\pi_v). \tag{140}$$

The local parameters $\psi_v \in \Psi(G_v)$ with trivial restrictions to the factor $\mathrm{SL}(2,\mathbb{C})$ are the usual Langlands parameters $\phi_v \in \Psi_{\mathrm{temp}}(G_v)$. The corresponding analogues of (139), (140) and (138) describe the local (endoscopic) Langlands correspondence for the group G_v. In this case, the representations π_v in a packet Π_{ϕ_v} are irreducible and expected to be tempered, while the element s_{ψ_v} in (139) and (140) equals 1. This is clearly close to Shelstad's endoscopic classification for real groups. However, as we have noted, a more complicated global proof is required because of the lack of any p-adic analogue of the Harish-Chandra classification of discrete series. We refer the reader also to the earlier, clearly written volume [1] by Adams, Barbasch and Vogan for *archimedean* parameters ψ_v, in which the conjectures were established for *any* real group, but without our defining property by twisted transfer to $\mathrm{GL}(N)$ that was needed for the global (and p-adic) theory.

Having stated the analogues of the Langlands correspondence for the localizations ψ_v of the global parameters[25] $\psi \in \Psi(G)$, we can describe the global endoscopic classification for G. We write $\Psi_2(G)$ for the subset of global parameters $\psi \in \Psi(G)$ such that $S_\psi^0 = \{1\}$, which is to say that the image of ψ does not lie in any proper parabolic subgroup of ${}^LG$. For any such ψ, we can then form the global packet

$$\Pi_\psi = \{\pi = \bigotimes_v \pi_v : \pi_v \in \Pi_{\psi_v}, \langle \cdot, \pi_v \rangle = 1 \text{ for } v \notin S\}$$

of representations of $G(\mathbb{A})$ that are unramified at almost every place v of F. (The local construction is such that if the function $\langle x_v, \pi_v \rangle$ equals 1, the representation π_v is irreducible and unramified.) For any $\pi \in \Pi_\psi$, the function

$$\langle x, \pi \rangle = \prod_v \langle x_v, \pi_v \rangle$$

is then defined. The main global result is Theorem 1.5.2 of [23]. It asserts that

$$L^2_{\text{disc}}(G(F) \backslash G(\mathbb{A})) \cong \bigoplus_{\psi \in \Psi_2(G)} \Big(\bigoplus_{\pi \in \Pi_\psi(\varepsilon_\psi)} m_\psi \pi \Big). \tag{141}$$

Here m_ψ equals 1 or 2, and

$$\varepsilon_\psi \colon \mathbf{S}_\psi \to \{\pm 1\}$$

is a linear character defined explicitly in terms of global symplectic ε-factors, while $\Pi_\psi(\varepsilon_\psi)$ is the subset of representations in the global packet Π_ψ such that the character $\langle \cdot, \pi \rangle$ on $\mathbf{S}_\psi$ equals ε_ψ.

This completes our very brief summary of the endoscopic classification of representations of the quasisplit symplectic or special orthogonal group G. For more information, a reader could begin with the introduction in [23], and then go to Chapter 1 and perhaps the first few sections of Chapter 2 and Chapter 4. Similar results have been established for quasisplit unitary groups by Mok [181]. We recall also that some of the geometric implications of these matters were discussed in the last two sections.

I should say that the description I have given here is not quite correct as stated. There are technical adjustments required for the case that G equals the group $\mathrm{SO}(2n)$. Suppose for example that G is split over a completion F_v, and that

$$\phi_v^1, \phi_v^2 \colon W_{F_v} \to \mathrm{SO}(2n, \mathbb{C}) = \widehat{G}$$

is a pair of distinct, irreducible, special orthogonal representations of W_{F_v} that are conjugate under the action of $\mathrm{O}(2n, \mathbb{C})$. The local results above imply that $\{\phi_v^1, \phi_v^2\}$ corresponds to a pair $\{\pi_1, \pi_2\}$ of irreducible representations in $\Pi_2(G_v)$. What they

[25] Bear in mind that ψ represents one of our ad hoc parameters on a product $\mathscr{L}_\psi \times \mathrm{SL}(2,\mathbb{C})$. But the group $\mathscr{L}_\psi$ was constructed in [23, 1.4] to contain the local Langlands group L_{F_v}, so its restriction to ψ_v is a homomorphism from $L_{F_v} \times \mathrm{SL}(2,\mathbb{C})$ to ${}^LG_v$.

do not give is the actual bijection (from the two possible choices) between these two sets of order 2 implicit in the local Langlands correspondence. Such ambiguity has to be built into the assertions of [23]. (Their global manifestations are closely related to the multiplicity $m_\psi \in \{1,2\}$ in (141).) I have omitted them deliberately in my summary above, in the hopes of better conveying the essence of what is going on. A reader can easily restore them from the definitions of Chapter 1 of [23]. In any case, this section has gone on long enough!

11 Beyond Endoscopy

The theory of endoscopy we have just discussed has the potential to establish interesting cases of functoriality. They arise from the elliptic endoscopic data $(G', s', \mathscr{G}', \xi')$ attached to a quasisplit group G, and the L-embedding

$$\rho' : {}^LG' \to {}^LG$$

obtained from ξ' and the choice of an L-isomorphism from ${}^LG'$ to the group $\mathscr{G}'$.

For example, suppose that G is a quasisplit classical group, and that G' is a product of two quasisplit classical groups $G_1 \times G_2$ for which the canonical direct product ${}^LG_1 \times {}^LG_2$ is a maximal L-subgroup of ${}^LG$. Functoriality for this case follows from the results of [23] and [181]. The proof depends on the stabilization of the trace formula of G [21], which among other things yields the stable trace formula for G. It also depends on the twisted stabilization of $\mathrm{GL}(N)$, with respect to the standard outer automorphism. We have not discussed twisted endoscopy very much, but the formal definitions are similar to those of ordinary endoscopy. The twisted stabilization of the ordinary (twisted) trace formula was established in complete generality in the two volumes [179], [180], apart from the provisio mentioned earlier on the twisted, weighted fundamental lemma. The classical groups G are themselves twisted endoscopic groups for general linear groups. A consequence of this, functoriality for the natural embedding of ${}^LG$ into $\mathrm{GL}(N,\mathbb{C})$, was also a part of the results in [23] and [181]. We have discussed these matters already, at the end of the last section.

There are certainly other interesting cases of functoriality that come from endoscopy, but most of these are presently out of reach. And at any rate, the examples of functoriality attached in one way or another to endoscopy are pretty sparse compared to the general case.

Beyond Endoscopy is a strategy proposed by Langlands around 2000 for attacking the general Principle of Functoriality. The ideas represent a departure from anything that has gone before. They do involve a comparison of trace formulas, stable trace formulas in fact. However, they entail something else as well, the automorphic L-functions

$$L(s,\pi,r), \qquad \pi \in \Pi_2(G),\ r : {}^LG \to \mathrm{GL}(n,\mathbb{C}),$$

attached to G. Langlands' proposal was to refine the stable trace formula for G by inserting a supplementary factor into the stable multiplicity of π on the spectral side, namely the order of the pole of $L(s,\pi,r)$ at $s=1$. For fixed r, this would vary with π, or rather the global packet of π, and according to what is expected about functoriality, would give information on the "functorial lineage" of π. We refer the reader to Section 1 of [155] for a basic introduction to the ideas, together with a number of critical examples.

To my view, this strategy of Langlands is fundamental, and of the greatest significance. It is also deep and difficult, much more so even than the theory of endoscopy. Despite the fact that Langlands' proposal is now twenty years old, its study is still in the very early stages.

We shall generally assume for the rest of the section that G is the general linear group $\mathrm{GL}(n+1)$ over the field $F=\mathbb{Q}$. The stable trace formula for G is then the same as the invariant trace formula (102). The fundamental problem is to understand how the representations $\pi\in\Pi_2(G)$ in the discrete spectrum are related to functoriality, or more precisely, how they might arise as functorial images of triplets (G',π',ρ'), for L-homomorphisms

$$\rho' : {}^LG' \to {}^LG = \mathrm{GL}(n+1,\mathbb{C}).$$

It is best to discard the nontempered representations π, which one might in any case expect to be able to treat by induction. We therefore restrict our consideration to the subset $\Pi_1(G)=\Pi_{\mathrm{cusp},2}(G)$ of cuspidal automorphic representations in $\Pi_2(G)$. Langlands noted that the functorial preimages (G',π',ρ') of $\pi\in\Pi_1(G)$ should be closely related to the poles at $s=1$ of the L-functions $L(s,\pi,r)$ attached to r. This of course presupposes the meromorphic continuation of $L(s,\pi,r)$ to a half space $\mathrm{Re}\,(s)>1-\varepsilon$, $\varepsilon>0$, something that is not known in general.

Let us assume for a moment that functoriality holds, say for all general linear groups $G=\mathrm{GL}(n+1)$ and $G'=\mathrm{GL}(m+1)$. This implies the meromorphic continuation of the L-functions $L(s,\pi,r)$, and allows us to set

$$m_\pi(r) = -\mathrm{ord}_{s=1}L(s,\pi,r) = \mathrm{res}_{s=1}\left(-\frac{\mathrm{d}}{\mathrm{d}s}\log L(s,\pi,r)\right),$$

the order of the pole of $L(s,\pi,r)$ at $s=1$. For any r, we are then free to define

$$I^r_{\mathrm{cusp}}(f) = \sum_{\pi\in\Pi_1(G)} m_\pi(r)\cdot \mathrm{mult}(\pi)\cdot\Theta(\pi,f), \qquad (142)$$

the contribution of the representations $\pi\in\Pi_1(G)$ to the primary spectral part of (106), but weighted by these integers. For example, if r equals the trivial representation 1_G of ${}^LG$, $L(s,\pi,r)$ is just the completed Riemann zeta function

$$L(s,1) = \pi^{-s/2}\Gamma(s/2)\zeta(s), \qquad \pi = 3.1416...,$$

for each representation π, which has a simple pole at $s=1$. In this case $I^r_{\mathrm{cusp}}(f)$ is itself just the cuspidal part

$$I_{\mathrm{cusp}}(f) = \sum_{\pi \in \Pi_1(G)} \mathrm{mult}(\pi)\Theta(\pi, f) \tag{143}$$

of (106). Langlands suggested the possibility of finding a geometric expansion for $I^r_{\mathrm{cusp}}(f)$ for any r, thereby giving a refinement of the invariant trace formula (102).

The idea was thus to try to find a trace formula whose spectral side is the modified cuspidal expansion (142). To see how this might be possible, we note that $m_\pi(r)$ is the residue at $s=1$ of

$$\begin{aligned}
&-\frac{\mathrm{d}}{\mathrm{d}s}(\log(L^S(s,\pi,r))) \\
&= -\frac{\mathrm{d}}{\mathrm{d}s}(\log(\prod_{p \notin S} \det(1 - r(c(\pi_p))p^{-s})^{-1})) \\
&= \sum_{p \notin S} \frac{\mathrm{d}}{\mathrm{d}s}(\log(\det(1 - r(c(\pi_p)))p^{-s})) \\
&= \sum_{p \notin S} \sum_{k=1}^{\infty} \log(p)\,\mathrm{tr}(r(c(\pi_p))^k)p^{-ks}.
\end{aligned}$$

We can discard the terms with $k \geq 2$ in this last sum, since the function they define would be holomorphic at $s=1$. It follows that

$$m_\pi(r) = \mathrm{res}_{s=1}\left(\sum_{p \notin S} \log(p)\,\mathrm{tr}(r(c(\pi_p)))p^{-s}\right).$$

A familiar application of the Wiener–Ikehara theorem[26] would then give a formula

$$m_\pi(r) = \lim_{N \to \infty}\left(|S_N|^{-1} \sum_{p \notin S_N} \log(p)\,\mathrm{tr}(r(c(\pi_p)))\right),$$

where

$$S_N = \{p \notin S : p \leq N\}.$$

(See [208, p. I-29].)

To exploit this last formula, one can write the test function $f \in C_c^\infty(G(\mathbb{A}))$ as a product $f_S \cdot \mathbf{1}^S$, for $f_S \in C_c^\infty(G(F_S))$ and $\mathbf{1}^S$ the characteristic function of the compact open subgroup $G(\widehat{\mathbb{Z}}^S) = \prod_{p \notin S} G(\mathbb{Z}_p)$ in $G(\mathbb{A}^S)$. One can then enrich this function by adding a factor at any $p \notin S$. We set

[26] We need to assume here that π is locally tempered, and hence that the generalized Ramanujan conjecture holds for G. This implies that the L-series $L(s,\pi,r)$ converges *absolutely* for $\mathrm{Re}\,(s) > 1$, and therefore satisfies conditions (a) and (b) of the theorem stated in [208]. This is really part of functoriality, on which we are basing the present motivational argument. (The assumption should really have been explicit in the discussion of these matters in [24, §2]).

$$f_p^r(x) = f(x) \cdot h_p^r(x_p), \qquad x \in G(\mathbb{A}),$$

where x_p is the component of x in $G(\mathbb{Q}_p)$, and h_p^r is the function in the unramified Hecke algebra $\mathscr{H}(G(\mathbb{Z}_p) \setminus G(\mathbb{Q}_p)/G(\mathbb{Z}_p))$ on $G(\mathbb{Q}_p)$ whose Satake transform equals

$$\widehat{h}_p^r(c_p) = \mathrm{tr}(r(c_p)),$$

for any semisimple conjugacy class c_p in $\widehat{G} = \mathrm{GL}(n+1, \mathbb{C})$, which is to say that

$$\mathrm{tr}(\pi_p(h_p^r)) = \mathrm{tr}(r(c(\pi_p))),$$

for any unramified representation π_p of $G(\mathbb{Q}_p)$. Then

$$\Theta(\pi, f_p^r) = \mathrm{tr}(\pi(f))\,\mathrm{tr}(\pi_p(h_p^r)) = \Theta(\pi, f)\,\mathrm{tr}(r(c(\pi_p))),$$

for any $\pi \in \Pi_1(G)$ such that π_p is unramified. Combining this with the last formula for $m_\pi(r)$ and the definition (142) of $I_{\mathrm{cusp}}^r(f)$, one sees that $I_{\mathrm{cusp}}^r(f)$ would equal

$$\sum_{\pi \in \Pi_1(G)} \Big(\lim_{N \to \infty} |S_N|^{-1} \sum_{p \notin S_N} \log(p)\,\mathrm{tr}(\pi_p(h_p^r)) \Big) \mathrm{mult}(\pi) \Theta(\pi, f)$$
$$= \lim_{N \to \infty} \Big(|S_N|^{-1} \sum_{p \notin S_N} \log(p) \Big(\sum_{\pi \in \Pi_1(G)} \mathrm{mult}(\pi) \Theta(\pi, f_p^n) \Big) \Big).$$

In other words,

$$I_{\mathrm{cusp}}^r(f) = \lim_{N \to \infty} \Big(|S_N|^{-1} \sum_{p \notin S_N} \log(p) I_{\mathrm{cusp}}(f_p^r) \Big). \tag{144}$$

On the other hand, we can formally rewrite the invariant trace formula (102) as

$$I_{\mathrm{geom,temp}}(f) = I_{\mathrm{cusp}}(f), \tag{145}$$

where

$$I_{\mathrm{geom,temp}}(f) = I_{\mathrm{geom}}(f) - (I_{\mathrm{spec}}(f) - I_{\mathrm{cusp}}(f)). \tag{146}$$

The subscript *temp* indicates that the distribution should be locally tempered. This is because the representations in $\Pi_1(G)$ would satisfy the analogue of the Ramanujan conjecture, according to our assumption and the argument of Langlands sketched at the end of [138], as we observed in footnote 26. The Dirichlet series for $L(s, \pi, r)$ would then converge *absolutely* for $\mathrm{Re}\, s > 1$, an implicit condition for the Wiener–Ikehara theorem we applied above. The result would then be an r-trace formula

$$I_{\mathrm{geom,temp}}^r(f) = I_{\mathrm{cusp}}^r(f), \tag{147}$$

for any r, where

$$I_{\mathrm{geom,temp}}^r(f) = \lim_{N \to \infty} \Big(|S_N|^{-1} \sum_{p \notin S_N} \log(p) I_{\mathrm{geom,temp}}(f_p^r) \Big). \tag{148}$$

This would provide a large family of refined trace formulas from which one might try to deduce functoriality.

However, this was all under the assumption of meromorphic continuation for the L-function $L(s,\pi,r)$. The way to establish this, according to Langlands' conjectures, is to apply functoriality to the homomorphisms

$$\rho' = r\colon \widehat{G} = \mathrm{GL}(n+1,\mathbb{C}) \to \mathrm{GL}(N,\mathbb{C}).$$

We cannot very well assume functoriality when it is what we ultimately want to prove. Langlands' idea was to use the spectral expression (142) initially only to motivate the proposed limit (148). For it does tell us that the limit (148) ought to exist. Langlands' hope is that it will eventually be possible to prove independently that the limit does exist, and to express it in terms of a reasonably explicit geometric expansion. One could then work on trying to establish a spectral expansion for the limit akin to (142).

It is clearly an enormous problem. The first major step would be the formidable task of finding a geometric expansion for the difference $I_{\mathrm{geom,temp}}(f)$ in (146). To see what this might entail, we shall consider the formal approximation (106) of the full trace formula (102) given by

$$\begin{aligned} I_{\mathrm{ell,reg}}(f) &= \sum_{\gamma\in\Gamma_{\mathrm{ell,reg}}(G)} \mathrm{vol}(\gamma)\mathrm{Orb}(\gamma,f) \\ \sim \quad I_2(f) &= \sum_{\pi\in\Pi_2(G)} \mathrm{mult}(\pi)\Theta(\pi,f). \end{aligned}$$

Its analogue for the cuspidal trace formula (145) is the approximation

$$I_{\mathrm{ell,reg,temp}}(f) \sim I_{\mathrm{cusp}}(f), \tag{149}$$

where

$$I_{\mathrm{ell,reg,temp}}(f) = I_{\mathrm{ell,reg}}(f) - \sum_{\pi\notin\Pi_1(G)} \mathrm{mult}(\pi)\Theta(\pi,f), \tag{150}$$

the sum being over the complement of $\Pi_1(G)$ in $\Pi_2(G)$. We are retaining the subscript *temp* in the distribution $I_{\mathrm{ell,reg,temp}}(f)$ to emphasize that it is supposed to be an approximation of $I_{\mathrm{geom,temp}}(f)$. The difference between the two distributions is the sum of the supplementary geometric terms in the full trace formula (102) minus the sum of the supplementary spectral terms. We have generally avoided discussing these auxiliary terms (except for GL(2) in Sections 6 and 7), but if G is not equal to GL(2), there are some locally nontempered distributions $I_M(\pi,f)$ among the supplementary spectral terms. The distribution $I_{\mathrm{ell,reg,temp}}(f)$ therefore cannot be locally tempered. It is to be regarded as we have said, simply as a formal approximation of the distribution $I_{\mathrm{geom,temp}}(f)$ we do expect to be tempered.

The point to be made is that the regular elliptic part $I_{\mathrm{ell,reg}}(f)$ of the trace formulas conceals some secrets. It is very familiar, having been known (if not always in adelic form) since Selberg first introduced his trace formula for compact quotient. But despite the fact that the individual orbital integrals in its summands are locally tempered (a fact proved by Harish-Chandra for real groups in 1966 [88], and for p-adic groups somewhat later), their sum is not. The main obstruction is simple enough, the sum over π in (150), but this is its spectral form. We are thus asking for an explicit, and we hope reasonably simple, geometric expansion for the difference (150). As far as I know, this natural question was never considered (except perhaps for $G = \mathrm{GL}(2)$) before 2000. It is the essential part of what we called the first major step above. It will demand much effort, supported no doubt by experience gained from experiments in special cases.

Langlands discussed these and other ideas, with various examples, in Part I of his foundational article [155]. In Part II he examined various terms in the trace formula for GL(2). Part III of [155] is devoted to actual experiments, using computer calculations to estimate some of the quantities from Part II. Part IV contains among other things a few remarks on general groups. A significant part of the article is devoted to a topic that will have to be understood after the initial questions have been answered. It is the supplementary geometric part of the trace formula for GL(2), represented by the noninvariant terms (iv) and (v) on p. 516–517 of [103]. The contributions of these terms will be locally tempered for GL(2), as will the noninvariant contributions of the spectral terms (vi), (vii) and (viii) from [103]. It is a simple enough example for one to be able to ask what influence all of these terms might have on the limit (148) in the case of GL(2). Langlands studied them in some detail [155, §2.4, §4.3 and Appendix C], and found some interesting cancellations among their contributions to the limit.

Langlands' initial article on Beyond Endoscopy was actually the unpublished precursor [154] of [155]. This paper contains some of his first ideas on the new program, with some comparisons to the developing theory of endoscopy. It represents an informal introduction to the main article [155]. The successor to [155] was the report [156], in which Langlands reviewed some of the constructions and calculations from [155]. He then described how they could be formulated in the case of function fields, that is, global fields of positive characteristic.

The next paper was Langlands' 2010 article [74] with Frenkel and Ngô. It contains three critical suggestions for the analysis of $I_{\mathrm{ell,reg,temp}}(f)$. The paper encompasses both number fields and function fields, opening the possibility of extending to number fields techniques that had been exploited by Ngô in his recently completed proof of the fundamental lemma for Lie algebras of positive characteristic. One of these became the first of the three suggestions. It was to parameterize the classes $\gamma \in \Gamma_{\mathrm{ell,reg}}(G)$ that index the summands in $I_{\mathrm{ell,reg}}(f)$ by points in what the authors called the base of the Steinberg–Hitchin fibration. For the case of $G = \mathrm{GL}(n+1)$ here, this amounts to a parameterization of the semisimple classes γ by their characteristic polynomials. It represents a significant change of perspective despite its simplicity.

The base of the Steinberg–Hitchin fibration for $G = \mathrm{GL}(n+1)$ is the product

$$\mathscr{A}(n) = \mathscr{B}(n) \times \mathbb{G}_m$$

of affine n-space $\mathscr{B}(n)$ with the multiplicative group $\mathbb{G}_m = \mathrm{GL}(1)$. The characteristic polynomial

$$\det(\lambda I - \gamma) = \lambda^{n+1} - a_1\lambda^n + \cdots + (-1)^n a_n\lambda + (-1)^{n+1} a_{n+1} = p_a(\lambda)$$

of a class $\gamma \in \Gamma_{\mathrm{ell,reg}}(G)$ has rational coefficients. It gives a bijection $\gamma \to a$ from $\Gamma_{\mathrm{ell,reg}}(G)$ onto the subset $\mathscr{A}_{\mathrm{irred}}(n, \mathbb{Q})$ of elements

$$a = (a_1, \ldots, a_n, a_{n+1})$$

in $\mathscr{A}(n, \mathbb{Q})$ such that $p_a(\lambda)$ is irreducible over $\mathbb{Q}$. The suggestion was thus to rewrite the sum over $\gamma \in \Gamma_{\mathrm{ell,reg}}(G)$ in $I_{\mathrm{reg,ell}}(G)$ as a sum over elements

$$a = (b, a_{n+1}), \qquad b \in \mathbb{Q}^n,\ a_{n+1} \in \mathbb{Q}^*,$$

in $\mathscr{A}_{\mathrm{irred}}(n, \mathbb{Q})$.

The second suggestion from [74] was to try to apply the Poisson summation formula to the sum over $b \in \mathbb{Q}^n$. This is certainly not immediately possible, for a variety of reasons. The problem would be to modify the resulting expression for $I_{\mathrm{ell,reg}}(f)$ in such a way that Poisson summation could be applied to the rearranged sum over b in $\mathbb{Q}^n$ (regarded of course as a lattice in $\mathbb{A}^n$). The third suggestion was to try then to account for the nontempered representations π in (150) directly in terms of the summands attached to the dual variables $\xi \in \mathbb{Q}^n$. The authors conjectured in particular that the contribution of the trivial one-dimensional representation π_1 of $G(\mathbb{A}^n)$ was contained in the dual summand with $\xi = 0$.

With these ideas, the hidden structure in $I_{\mathrm{ell,reg}}(f)$ becomes more compelling. The summands on the right-hand side of (150) are parameterized by, among other things, unipotent conjugacy classes in $\widehat{G}$. The suggestions in [74] are that their contributions to $I_{\mathrm{ell,reg}}(f)$ might have an unexpectedly explicit form. (See [26] for a conjectural description in the case of general linear groups.) The proposed phenomena become all the more intriguing for groups G other than $\mathrm{GL}(n+1)$. The authors discussed their suggestions in general terms in the first three sections of [74]. In the remaining sections, they offered some evidence.

They devoted some time to describing how best to normalize the invariant measures on the various spaces in play, the most important being the additive adelic space $\mathscr{B}(n, \mathbb{A}) \cong \mathbb{A}^n$ that would be the domain of the Fourier transform from Poisson summation. The next step was essentially to construct a function on this space from the local orbital integrals of f, for which the global orbital integrals in $I_{\mathrm{ell,reg}}(f)$ represent the values on the lattice $\mathscr{B}(n, \mathbb{Q}) = \mathbb{Q}^n$. The original function itself was quite unsuitable. However, the authors used an idea of J. Getz in §4 of [74] to truncate it in a certain way so as to make it amenable to Poisson summation. They then showed that in the resulting sum of Fourier transforms over ξ in the dual lattice, the contribution of the 1-dimensional representations, the most highly nontempered representations from the right-hand side of (150), is indeed contained in the summand with $\xi = 0$. The sum over these Fourier transforms with $\xi \neq 0$ therefore removes these representations from the difference (150). The authors in fact showed that it is asymptotically smaller in a natural sense than the sum of the 1-dimensional representations. Estimates of this sort are exactly what are being sought, even if in this case the results are too weak to apply. However, they constitute striking evidence for the general proposals in [74], especially since they apply to any (quasisplit) group G.

The heart of Beyond Endoscopy would be a general comparison of trace formulas. However, this is really something for the future. For in general it could come only after the three suggestions from [74] have been successfully carried out, and indeed, only after the stable generalizations

$$S^r_{\mathrm{geom,temp}}(f) = \lim_{N\to\infty} \big(|S_N|^{-1} \sum_{p\notin S_N} \log(p) S_{\mathrm{geom,temp}}(f^r_p)\big)$$

and

$$S^r_{\mathrm{geom,temp}}(f) = S^r_{\mathrm{cusp}}(f) \tag{151}$$

of the limit (148) and the r-trace formula (147) have been established for any G. The right-hand side of (151) would have to be defined as the general analogue

$$S^r_{\mathrm{cusp}}(f) = \lim_{N\to\infty} \big(|S_N|^{-1} \sum_{p\notin S_N} \log(p) S_{\mathrm{cusp}}(f^r_p)\big)$$

of (144), whose existence would be a consequence of the existence of the limit $S^r_{\mathrm{geom,temp}}(f)$. (The stable analogue $S_{\mathrm{cusp}}(f)$ of $I_{\mathrm{cusp}}(f)$ in this limit would be a sum as in (143), but over global L-packets *all* of whose constituents are cuspidal, or equivalently, all of whose constituents are expected to be locally tempered.) The goal of the comparison would be to provide information about the stable distributions $S^r_{\mathrm{cusp}}(f)$ akin to the stable analogue of the right-hand side of (142). We reiterate that a priori, we would know nothing about the right-hand side of (142). The role of this formula was only to motivate the limit (144) with which the heart of the argument would begin. Once one has versions of the formulas (142) for various r, and comparisons of them with analogues for other groups G', one could finally begin what would presumably be the last stage of the argument, the search for confirmation of functoriality.

Our experience with endoscopy can inform what might be the new comparison. The "Beyond Endoscopic" comparison suggested by Langlands in [155] will have some features in common with endoscopic comparison (better known as stabilization), and some features that are quite different. In general, we could imagine a "Beyond Endoscopic datum" attached to a quasisplit group G over a number field F simply as a pair (G',ρ'), for a reductive group G' over F, and an L-homomorphism

$$\rho' \colon {}^L G' \to {}^L G$$

whose restriction to $\widehat{G}'$ is an embedding. As stated, this is much broader than an endoscopic datum, to the extent that $\widehat{G}'$ is not required to be the connected centralizer in $\widehat{G}$ of a semisimple element $s' \in \widehat{G}$. It is also incomplete, in the sense that (G',ρ') should at least be replaced by a triplet $(G',\mathcal{G}',\xi')$ that satisfies the same conditions as an endoscopic datum $(G', s', \mathscr{G}', \xi')$ (without the point s'), while ρ' could represent the choice of an L-isomorphism from $\mathscr{G}'$ to G'. Langlands proposed a transfer

$$f = \prod_v f_v \to \prod_v f'_v = f', \qquad f \in C_c^\infty(G(\mathbb{A})),$$

of functions from $C_c^\infty(G(\mathbb{A}))$ to functions in $C_c^\infty(G'(\mathbb{A}))$, to be known as stable transfer since it should depend only on the stable orbital integrals of *both* f and f'. This would be quite different from endoscopic transfer.

If we assume the local Langlands correspondence for both G and G' at the places v of F, something that could well be available before this stage of development of Beyond Endoscopy, stable transfer would be easy to define. For any function $f_v \in C_c^\infty(G_v)$, it would be enough to specify the value $f'_v(\phi'_v) = f_v^{G'}(\phi'_v)$ of any tempered local Langlands parameter $\phi'_v \in \Phi(G'_v)$. We would do so by setting

$$f'_v(\phi'_v) = f_v^G(\rho'_v \circ \phi'_v), \tag{152}$$

for the localization ρ'_v of ρ'. The global transfer f' at a function $\prod_v f_v$ in $C_c^\infty(G(\mathbb{A}_F))$ would then simply be defined as the product $\prod_v f'_v$ of local transfers. With this definition, the real problem would then be to determine the orbital integrals of f'_v in terms of those of f_v. In other words, one would like to determine the value $f'(a'_v)$ at an F_v-valued point $a'_v \in \mathscr{A}(G'_v, F_v)$ of the Steinberg–Hitchin base for G'_v in terms of the values $f_v^G(a_v)$ of f_v at F_v-valued points $a_v \in \mathscr{A}(G_v, F_v)$ of the Steinberg–Hitchin base for G_v. According to some version of the Schwartz kernel theorem, we would expect to be able to write

$$f'_v(a'_v) = \int_{\mathscr{A}(G,F_v)} \Delta(a'_v, a_v) f_v^G(a_v)\, \mathrm{d}a_v,$$

for some integral kernel $\Delta(a'_v, a_v)$. The integral is to be understood as the pairing of the "Schwartz function" $f_v^G(\cdot)$ on $\mathscr{A}(G_v, F_v)$ with the "tempered distribution" $\Delta(a'_v, \cdot)$, notions that would have to be suitably interpreted. The qualitative difference between this and the Langlands–Shelstad transfer factor $\Delta(\delta'_v, \gamma_v)$ is manifest.

With a theory of Beyond Endoscopic transfer, one could then finally consider comparing stable trace formulas. There are many possibilities, and it is hard to know in advance which of these might be best. For example, one might try to compare the r-trace formula of G with the stable trace formula of G', or more likely, a linear combination over Beyond Endoscopic data G' of stable trace formulas of G'. One proposed model for such a comparison, founded on the structure of Langlands' automorphic Galois group L_F proposed in Section 9, was described in [24, §2] and [25, §4], where it was called the *primitization* of the r-trace formula. However, what is needed now is more understanding of the many fundamental questions that would have to be answered first, if not in general then at least for natural examples.

This last discussion is intended as background for [158], the article on singularities and transfer Langlands wrote as a continuation of his fundamental paper with Frenkel and Ngô. Our remarks on stable transfer and comparison of trace formulas are taken largely from the early pages of [158]. However, the ostensible purpose of the paper was to explore them in detail for the group $G = \mathrm{SL}(2)$, and the dihedral groups G' of elements of norm 1 in quadratic extensions E of F, together with the degenerate case of $G' = \mathrm{GL}(1)$. Langlands proved the existence of stable transfer $f_v \to f'_v$ in each of these cases, for localizations F_v of F with $q_v \neq 2$. To do so, he relied on general properties of stable characters for SL(2) in [127], and explicit formulas for irreducible characters on $\mathrm{SL}(2, F_v)$, by Harish-Chandra for archimedean v and by Sally and Shalika [197][27] for nonarchimedean v. The proof was given in Section 1 of [158], the first half of the paper, by answering Questions A and B from Section 1.1. We note that there is a recent extension to $\mathrm{SL}(n)$ by Daniel Johnstone [109], using some interesting new methods.

In the second half of the paper, Langlands studied two functions

$$\theta_f(a) = \left(\prod_{v \in S} f_v^G(a_v) \right) \mathbf{1}^S(a^S)$$

and

$$\phi_f(a') = \left(\prod_{v \in S} f'_v(a'_v) \right) \mathbf{1}^S((a')^S),$$

for regular adelic variables $a, a' \in \mathbb{A}_F$ in $\mathscr{A}(G)$, and a large (variable) finite set S of places. The second function is really a composite of the transfers from G to all of the dihedral groups G', with their local Steinberg–Hitchin bases $\mathscr{A}(G'_v)$ embedded in $\mathscr{A}(G_v)$. The regular set in $\mathscr{A}(G_v)$ becomes a disjoint union of (the regular points in) their domains, and $f'_v(a'_v)$ is then the value of the transfer mapping for the group G'_v attached to the domain that contains a'_v. Langlands was interested in the singularities of these functions. For archimedean v, these were obtained from special cases of the general results of Harish-Chandra, mentioned for GL(2) as Harish-Chandra families in Section 7 of this article, and reviewed briefly for general groups in the last section. For nonarchimedean v, the singularities are more complex for general groups. However, their qualitative behavior is well understood in terms of the Shalika germs introduced in [216].

Langlands' interest was in the specialization to $G = \mathrm{SL}(2)$ of the results established for general (simply connected) G in the second half of [74]. These consist of the truncated form of Poisson summation, and its application to the contribution of the trivial automorphic representation of G to the geometric side of the trace formula. Combining various arguments, Langlands was able to establish a more tractable form of Poisson summation for both of the functions θ_f and ϕ_f. The singularities of the local components $f_v^G(a_v)$ and $f'_v(a'_v)$ were of course an important part of this. Once Poisson summation was in place, the singularities could then be used to analyze the asymptotic behavior of the Fourier transforms

---

[27] As Langlands remarks, the proofs for this announcement are not published. They were originally circulated as a long preprint by Sally and Shalika, part of which was later published as the fundamental article [216] on Shalika germs. However, it was the complementary part that contained the p-adic characters, and that was unfortunately never published.

$$\widehat{\theta}_f(a), \widehat{\phi}_f(a), \qquad a \in \mathbb{A}.$$

With the comparison of trace formulas for G and $\{G'\}$ in mind, Langlands then considered the difference

$$S_{\mathrm{reg,ell}}(f) - \sum_{G'} S'_{\mathrm{reg,ell}}(f') \sim \sum_{b \in \mathscr{A}(G,F)} \theta_f(b) - \phi_f(b), \tag{153}$$

a linear form in f that becomes

$$\sum_{b \in \mathscr{A}(G,F)} \left(\widehat{\theta}_f(b) - \widehat{\phi}_f(b)\right)$$

after the application of Poisson summation [158, (5.5) and (5.12)]. In the rest of the paper, he added some remarks on possible next steps, which would include the removal of the contribution $\widehat{\theta}_f(0)$ of the trivial automorphic representation of $\mathrm{SL}(2)$ from (153). His main concern would be to gain information about the proposed limits (148) and their variants.

My discussion of the paper [158] has been superficial, as is no doubt clear from my use of the symbol $\sim$ in (153). In particular, I have said nothing of the truncated Poisson summation formula from [74]. However, it does seem that (153) is a natural identity, with important implications for the general Beyond Endoscopic comparison of stable trace formulas.

A different approach to Poisson summation was introduced by Ali Altuğ in his 2003 thesis, written under the supervision of Langlands and Peter Sarnak. It was published later, with applications, in the three papers [3], [4], [5]. It will be instructive for us to discuss each of these papers.

In contrast to [74], where Poisson summation was established for a general class of test functions f on a general group G, Altuğ's methods apply to a restricted class of functions f on the particular group $G = \mathrm{GL}(2)$. However, they lead to much sharper estimates. Moreover, they would seem in principle to be applicable to more general groups, even if the technical problems in extending them will be formidable. We shall describe the first paper [3], which is devoted to Poisson summation, in some detail.

Following [3], we take G to be the group $\mathrm{GL}(2)$, $F = \mathbb{Q}$, and

$$f = f_\infty \cdot f^\infty = f_\infty \cdot f^{\infty,p} \cdot f_p^k, \quad p \text{ prime}, k \in \mathbb{Z}^{\geq 0},$$

to be a test function in the space

$$C_c^\infty(Z_+ \backslash G(\mathbb{A})), \qquad Z_+ = A_G(\mathbb{R})^0 = \left\{ \begin{pmatrix} u & 0 \\ 0 & u \end{pmatrix} : u > 0 \right\},$$

as follows. The archimedean component f_∞ is any function in $C_c^\infty(Z_+ \backslash G(\mathbb{R}))$, while $f^{\infty,p}$ is the characteristic function of the standard open, maximal compact subgroup $K^{\infty,p}$ of $G(\mathbb{A}^{\infty,p})$, and f_p^k is the product of $p^{-k/2}$ with the characteristic function of the open compact subset

$$\{X \in M_{2\times 2}(\mathbb{Z}_p) : |\det X|_p = p^{-k}\}$$

of $G(\mathbb{Q}_p)$. This is of course a very special case, but it was well suited to illustrating Altuğ's techniques.

For the general regular elliptic part $I_{\text{ell,reg}}(f)$ of the trace formula at f, we are interested in the value of the normalized orbital integral $f_G(\gamma)$ of f at a point $\gamma \in \Gamma_{\text{ell,reg}}(\mathbb{Q})$. It is a product

$$f_G(a) = f_{\infty,G}(a) \cdot |D(\gamma)|^{-\frac{1}{2}} \text{Orb}(\gamma, f^\infty),$$

where

$$|D(\gamma)| = |D(\gamma)|_\infty = (|D(\gamma)|^\infty)^{-1}$$

while

$$a = (b, a_2), \qquad b \in \mathbb{Q},\, a_2 \in \mathbb{Q}^*,$$

is the bijective image of γ in $\mathscr{A}_{\text{irred}}(1,\mathbb{Q})$, and $f_{\infty,G}(a) = f_{\infty,G}(\gamma)$. One sees from the properties of the unramified orbital integral

$$\text{Orb}(\gamma, f^\infty) = \text{Orb}(\gamma, f_p^k)\text{Orb}(\gamma, f^{\infty,p}),$$

and the definition of the function f_p^k, that $f_G(a)$ vanishes unless the irreducible monic, quadratic polynomial $p_a(\lambda)$ has integral coefficients, with constant term a_2 equal to $\pm p^k$. It then follows that $I_{\text{ell,reg}}(f)$ is equal to the expression

$$\sum_{\varepsilon\in\{\pm 1\}} \sum_{b\in\mathbb{Z}^\varepsilon_{\text{irred}}} f^\varepsilon_{\infty,G}(b) \cdot |D(\gamma)|^{-\frac{1}{2}} \text{vol}(\gamma)\text{Orb}(\gamma, f^\infty), \tag{154}$$

where we have written $\mathbb{Z}^\varepsilon_{\text{irred}} = \mathscr{B}^\varepsilon_{\text{irred}}(1,\mathbb{Z})$ for the set of integers $b \in \mathbb{Z}$ such that the pair $a = (b, \varepsilon p^k)$ lies in $\mathscr{A}_{\text{irred}}(1,\mathbb{Z})$, and

$$f^\varepsilon_{\infty,G}(b) = f_{\infty,G}(b, \varepsilon p^k) = f_{\infty,G}(a).$$

We are following the discussion in §3 of the paper [3]. As was noted there, the expansion (154) of $I_{\text{ell,reg}}(f)$ is where any discussion of Poisson summation would begin. The inner sum is of course over a set of integers $b \in \mathbb{Z}$. Is there a natural extension of the values of the summands to a Schwartz function on $\mathbb{R}$ to which Poisson summation could be applied? The answer is clearly negative. For a start, we must not forget that Poisson summation applies to the sum of the values of a suitable function on $\mathbb{R}$ over the lattice $\mathbb{Z}$, not a linear combination.

There are in fact a number of obstacles. The most immediately daunting is perhaps the volume factor $\text{vol}(\gamma)$ in (154). It depends very much on b as an integer, and in particular, on the splitting field E_a over $\mathbb{Q}$ of the quadratic polynomial

$$p_a(\lambda) = p^\varepsilon_b(\lambda), \qquad a = (b, \varepsilon p^k),$$

over $\mathbb{Z}$. The same goes for the factor $\mathrm{Orb}(\gamma, f^{\infty})$. It is a product of q-adic orbital integrals of the global spherical function

$$\left(\prod_{q\neq p} f_q\right) f_p^k$$

at γ. It too depends on the splitting field of $p_b^{\varepsilon}(\lambda)$. The archimedean factor $f_{\infty,G}^{\varepsilon}(b)$ is more amenable. It does extend to the natural function of $b \in \mathbb{R}$ given by the normalized archimedean orbital integrals of f_∞. This should in principle be the function to which one would try to apply Poisson summation. However, it comes with its own problems as a function of $\mathbb{R}$, namely the singularities given by Harish-Chandra's jump conditions. These occur at points $b \in \mathbb{R}$ at which the characteristic polynomials $p_b^{\varepsilon}(\lambda)$ have repeated factors over $\mathbb{R}$, which is to say that the corresponding discriminants vanish. And finally, there is the problem that the sum over b is not over the full lattice $\mathbb{Z}$ in $\mathbb{R}$, but the subset $\mathbb{Z}_{\text{irred}}^{\varepsilon}$ of $\mathbb{Z}$. Altuğ dealt with all of these problems.

For the volume coefficient $\mathrm{vol}(\gamma)$ in (154), the first step was to apply the Dirichlet class number formula for the quadratic extension $E_\gamma = E_a$ of $\mathbb{Q}$. Since the volume is essentially the regulator of $E_\gamma/\mathbb{Q}$, the formula can be written

$$\mathrm{vol}(\gamma) = |D_\gamma|^{\frac{1}{2}} L\Big(1, \Big(\frac{D_\gamma}{\cdot}\Big)\Big),$$

where D_γ is the discriminant of the quadratic extension E_γ, $\left(\frac{D_\gamma}{\cdot}\right)$ is the Kronecker symbol, and $L(\cdot, \left(\frac{D_\gamma}{\cdot}\right))$ is the Dirichlet L-function of $E_\gamma/\mathbb{Q}$ [3, §2.2.2]. At first glance, this seems to replace one problematic coefficient with two new ones!

However, as the discriminant of the field E_γ, D_γ is closely related to the discriminant $D(\gamma)$ of the irreducible quadratic polynomial

$$p_\gamma(\lambda) = p_a(\lambda) = p_b^{\varepsilon}(\lambda).$$

This occurs in its own right as the factor $|D(\gamma)|^{-\frac{1}{2}}$ in (154). Their joint contribution equals

$$|D(\gamma)|^{-\frac{1}{2}} |D_\gamma|^{\frac{1}{2}} = |s_\gamma^2|^{-\frac{1}{2}} = s_\gamma^{-1},$$

for a positive integer s_γ. This last integer is in turn closely related to the product $\mathrm{Orb}(\gamma, f^{\infty})$ of q-adic orbital integrals in (154). Langlands had already computed the product in his initial paper on the subject [155, Lemma 1 and §2.5] as

$$\mathrm{Orb}(\gamma, f^{\infty}) = p^{-\frac{k}{2}} \prod_{f | s_\gamma} f \cdot \Big(\prod_{q|f} \Big(1 - \Big(\frac{D_\gamma}{q}\Big) q^{-1}\Big)\Big). \tag{155}$$

Its product with s_γ^{-1} and the other factor $L(1, \left(\frac{D_\gamma}{\cdot}\right))$ above equals

$$p^{-\frac{k}{2}}\sum_{f|s_\gamma}(f/s_\gamma)\Big(\prod_{q|f}\Big(1-\Big(\frac{D_\gamma}{q}\Big)\Big)\Big)L\Big(1,\Big(\frac{D_\gamma}{\cdot}\Big)\Big),$$

which becomes

$$p^{-\frac{k}{2}}\sum_{f|s_\gamma}(1/f)L\Big(1,\Big(\frac{D(\gamma)/f^2}{\cdot}\Big)\Big)$$

with a change of variable in the sum over f, and the definition

$$L\Big(1,\Big(\frac{D(\gamma)/f^2}{\cdot}\Big)\Big)=\prod_{q|f}\Big(L_q\Big(1,\Big(\frac{D_\gamma}{q}\Big)\Big)^{-1}\Big)\cdot L\Big(1,\Big(\frac{D_\gamma}{\cdot}\Big)\Big)$$

of the Dirichlet L-function for the discriminant $D(\gamma)/f^2$. The expression (154) for $I_{\mathrm{reg,ell}}(f)$ then becomes

$$p^{-\frac{k}{2}}\sum_{\varepsilon}\sum_{b}f^{\varepsilon}_{\infty,G}(b)\Big(\sum_{f|s_\gamma}(1/f)L\Big(1,\Big(\frac{D(\gamma)/f^2}{\cdot}\Big)\Big)\Big). \tag{156}$$

(See [3, (2) and (3)].)

Altuğ was then able to treat the Dirichlet L-values at 1 with what is known as the approximate functional equation. To this end, he attached the more general Dirichlet series

$$L(s,\delta)=\sum_{f^2|\delta}{}' \frac{1}{f^{2s-1}}L\Big(s,\Big(\frac{\delta/f^2}{\cdot}\Big)\Big)$$

to any discriminant δ. Thus, δ is an integer congruent to 0 or 1 modulo 4, while $\big(\frac{\delta/f^2}{\cdot}\big)$ is the Kronecker symbol again, and $\sum'_{f^2|\delta}$ stands for the sum over the integers f such that δ/f^2 is also congruent to 0 or 1 modulo 4 [3, §3.1]. Its value at $\delta=D(\gamma)$ and $s=1$ equals the expression in (156) in brackets. In particular, it is constructed as a product of the Dirichlet L-value $L\Big(1,\big(\frac{D_\gamma}{\cdot}\big)\Big)$ with the values of finitely many q-adic orbital integrals (with $k=0$). This enhanced L-function was introduced by Zagier in 1977 [255], who established a functional equation linking its values at s and $1-s$. He would probably not have been aware of its interpretation in terms of q-adic orbital integrals on $\mathrm{GL}(2,\mathbb{Q}_q)$. However, it is interesting to think that these objects were implicit in his quite different setting not long after they had also become a part of Langlands' study of base change for GL(2), with its implications for the Artin conjecture and ultimately Fermat's Last Theorem.

The approximate functional equation applies to any reasonable Dirichlet series with functional equation [99, §10.6]. Altuğ used it to express the sum $L(1,D(\gamma))$ in the brackets of (156) by an ungainly but more tractable expression. The process takes the value $L(s,\delta)$ of the general L-function above back to a Dirichlet series, but one whose summands over n are weighted so as to converge absolutely, in which the original coefficients are averaged against a well behaved test function F. There is a second term, a contour integral that would give trouble on its own. However, a change of contour, together with an application of the actual functional equation for $L(s,\delta)$ transforms the troublesome integral to a weighted Dirichlet series for $L(1-s,\delta)$, in which the original coefficients are averaged over a second well behaved function H. (See the discussion of [3], and in particular, Proposition 3.4 and Corollary 3.5). With this, the seemingly insurmountable problem of the arithmetic dependence on γ of the original coefficients $\mathrm{vol}(\gamma)$ in (154) can be controlled.

We note in passing that that the product $\zeta(s)L(s,\delta)$ is called the *Dedekind zeta function of the (monogenic) order*

$$R_\delta = \mathbb{Z}[\lambda]/(p_\delta(\lambda))$$

in the quadratic field

$$E_\delta = \mathbb{Q}[\lambda]/(p_\delta(\lambda)),$$

where $p_\delta(\lambda)$ is the irreducible characteristic polynomial attached to δ. (We are assuming here that $\delta = D(\gamma)$ as above.) Its analogue for $\mathrm{GL}(n+1)$ over a number field F is the topic of the paper [254] of Z. Yun. He established its functional equation and class number formula as Theorem 1 of the paper. To do so, he defined the local factors as functions of s that satisfy their own functional equations [254, Theorem 2.5]. Yun then expressed the values at $s=1$ of these local factors in terms of local orbital integrals at γ [254, Corollary 4.6]. This paper is likely to be important in any attempt to generalize the methods of Altuğ. However, it does not provide an analogue for $\mathrm{GL}(n+1)$ of the explicit formula (155) of Langlands for the orbital integrals on $\mathrm{GL}(2)$. Something of the sort appears to be essential for higher rank, but there are some qualitative differences even between the cases of $\mathrm{GL}(2)$ and $\mathrm{GL}(3)$. (See [117].) I have been told that the perverse sheaves attached to unramified orbital integrals by Ngô acquire singularities in generalizing from $\mathrm{GL}(2)$ to $\mathrm{GL}(3)$.

Returning to the regular elliptic expansion (154) for $\mathrm{GL}(2)$, we recall the other two obstacles mentioned above. The first concerned the singularities of the function $f^{\varepsilon}_{\infty,G}(b) = f_{\infty,G}(a)$ at the zero set of the discriminant $D(\gamma) = D(a)$. To deal with it, Altuğ observed that for any Schwartz function Φ in $\mathbb{R}$ and any $\alpha > 0$, the product

$$\Phi(|D(a)|^{-\alpha}) f^{\varepsilon}_{\infty,G}(b), \qquad b \in \mathbb{R},$$

is also a Schwartz function of $\mathbb{R}$ ([3, Proposition 4.1]). Altuğ built this into the Schwartz functions F and H he obtained from the approximate functional equation. He had taken care to include a supplementary parameter in each of the arguments in [3, (4′)] (A^{-1} for F and A for H), which he then simply set equal to $|D(\gamma)|^{\alpha}$, for any number α with $0 < \alpha < 1$. The resulting expression in the brackets of (156) was then sufficient to make the complementary factor $f^{\varepsilon}_{\infty,G}(b)$ in (156) the restriction to $\mathbb{Z}^{\varepsilon}_{\mathrm{irred}}$ of a well defined Schwartz function on $\mathbb{R}$.

The second and last obstruction was that the sum over b in (156) was only over the subset $\mathbb{Z}^{\varepsilon}_{\mathrm{irred}}$ of the lattice $\mathbb{Z}$ of $\mathbb{R}$ (corresponding to *irreducible* characteristic polynomials). Altuğ simply added the complementary set of orbital integrals $f^{\varepsilon}_{\infty,G}(b)$ to the sum in (156). The trace formula actually calls for *weighted* orbital integrals (the term (iv) on page 516 of [103]) to be taken here, but we have been dealing with approximations ever since we identified $I_{\text{reg,ell}}(f)$ with the geometric side of the trace formula. We write $\bar{I}_{\text{reg,ell}}(f)$ for the expression (156) with b summed over $\mathbb{Z}$ rather than the subset $\mathbb{Z}^{\varepsilon}_{\text{irred}}$.

For the record, we write out the expression for (156) obtained by Altuğ for $\bar{I}_{\text{reg,ell}}(f)$ (with the extra summands for $b \in \mathbb{Z}$), even though our discussion is still lacking some of the finer details. It is

$$p^{-\frac{k}{2}} \sum_{\varepsilon} \sum_{b \in \mathbb{Z}} f_G^{\varepsilon}(b) \sum_{f^2 | D^{\varepsilon}(b)} \frac{1}{f} \sum_{\ell=1}^{\infty} \frac{1}{\ell} \left(\frac{D^{\varepsilon}(b)/f^2}{\ell} \right) \cdot \left[F\left(\frac{\ell f^2}{|D^{\varepsilon}(b)|^{\alpha}} \right) + \ell f^2 |D^{\varepsilon}(b)|^{-\frac{1}{2}} H\left(\frac{\ell f^2}{|D^{\varepsilon}(b)|^{1-\alpha}} \right) \right].$$

(This is essentially the first expression given in the proof of Theorem 4.2 from [3].) Since everything converges absolutely, the sum over $b \in \mathbb{Z}$ can be taken inside the sum over f and ℓ. It is still not yet quite possible to apply Poisson summation to the sum over b. For while the last expression in the square brackets extends to a Schwartz function of $b \in \mathbb{R}$, there are also the coefficients $\left(\frac{D^{\varepsilon}(b)/f^2}{\ell} \right)$ that depend arithmetically on $b \in \mathbb{Z}$. Altuğ's solution was to break this sum over b into a double sum over the finite subset

$$C(\ell, f) = \{ b (\text{mod } 4\ell f^2) : D^{\varepsilon}(b) \equiv 0 (\text{mod } f^2), D^{\varepsilon}(b)/f \equiv 0, 1 (\text{mod } 4) \}$$

of congruence classes modulo $4\ell f^2$, and the infinite affine lattice

$$\{ m \in \mathbb{Z} : m \equiv b (\text{mod } 4\ell f^2) \}.$$

Since

$$\left(\frac{D^{\varepsilon}(b+m)/f^2}{\ell} \right) = \left(\frac{D^{\varepsilon}(b)/f^2}{\ell} \right),$$

the coefficients could then be taken outside the sum over m, allowing him then to apply Poisson summation to the sum.

The final result is then

$$\bar{I}_{\text{ell,reg}}(f) = \sum_{\xi \in \mathbb{Z}} \widehat{\bar{I}}_{\text{ell,reg}}(\xi, f), \tag{157}$$

where $\widehat{\bar{I}}_{\text{ell,reg}}(\xi, f)$ equals an expression

$$\sum_{\varepsilon} \sum_{f=1}^{\infty} \frac{1}{f^3} \sum_{\ell=1} \frac{1}{\ell^2} \widehat{f}^{\varepsilon}_{G,\ell,f}(\xi) K_{\ell,f}(\xi, \varepsilon p^k),$$

for a Fourier transform

$$\widehat{f}^{\varepsilon}_{G,\ell,f}(\xi) = \int_{\mathbb{R}} (f^{\varepsilon}_{G}(x)\phi^{F,H}_{\ell,f}(x)\exp\left(\frac{-x\xi}{2\ell f^2 p^{-\frac{k}{2}}}\right) \mathrm{d}x,$$

with the function

$$\phi^{F,H}_{\ell,f}(x) = F(\ell f^2 |D^{\varepsilon}(x)|^{-\alpha}) + (\ell f^2)|D^{\varepsilon}(x)|^{-\frac{1}{2}} H(\ell f^2 |D^{\varepsilon}(x)|^{\alpha-1}),$$

and for a finite exponential, Kloosterman-like sum

$$\mathrm{Kl}_{\ell,f}(\xi,\varepsilon p^k) = \sum_{b\in C(\ell,f)} \left(\frac{D^{\varepsilon}(b)/f^2}{\ell}\right) \exp\left(\frac{b\xi}{4\ell f^2}\right).$$

This is Theorem 4.2 of [3], one of the two main results of the paper.

The second main result of Altuğ's paper [3] is Theorem 6.1. It is a reworking of the formula for the constant term $\widehat{\overline{I}}_{\mathrm{ell,reg}}(0,\xi)$ in the expansion (157), obtained from the functional equations for the L-functions $L(s,\delta)$, some elementary but quite elaborate identities established in §5 of [3] (Lemmas 5.1–5.3, Corollary 5.4), and various changes in the contour integral over x in the original formula. (Altuğ writes the formula for $\widehat{\overline{I}}(0,f)$, and the integral over x that it contains, as the expression $(13)_{\xi=0}$ displayed prior to the statement of Theorem 6.1.)

The formula in the assertion of Altuğ's Theorem 6.1, which we will not quote, seems to make his formula Theorem 4.2 (distilled in (157) above) look simple by comparison! However, it has a compelling logic. It expresses $\widehat{\overline{I}}_{\mathrm{ell,reg}}(0,f)$ as a sum of two simple integrals over $x\in\mathbb{R}$, together with a third complicated integral over x. In his last result [3, Lemma 6.2], Altuğ identifies the two simple integrals as the character $\mathrm{tr}(\mathbf{1}(f))$ of the trivial one-dimensional representation of $G(\mathbb{A})$ at f, and the supplementary term[28] on the spectral side given by (vi) on p. 647 of [103]. The hope is that the contribution of the third integral vanishes in the putative limits (148) attached to (irreducible) representations $r\neq 1$ of $\mathrm{GL}(2,\mathbb{C})$. Altuğ puts Theorem 4.2 and Theorem 6.1 together as a complex formula for $I_{\mathrm{reg,ell}}(f)$, which he states as Theorem 1.1[28] in the introduction to [3].

This completes our discussion of (157), Altuğ's version of Poisson summation for $\mathrm{GL}(2)$. It is surprisingly complex, even as it applies only to a special case within $\mathrm{GL}(2)$. We have emphasised it in our discussion because such detail seems to be a necessary prelude for the kind of estimates that will ultimately be needed. The role of the more elementary version of Poisson summation in [74] and [158], which applies to any group G, is different. It was offered simply as evidence for the basic idea, that of using some form of Poisson summation on the Steinberg–Hitching base to recognize the ultimate contribution of the noncuspidal discrete spectrum to the geometric side of the trace formula.

[28] What I have stated here is not strictly correct. The second simple integral in Theorem 6.1 actually equals *twice* the supplementary term (vi). However, the excess really belongs naturally in the complicated third integral, as Altuğ observes in a restatement of Theorem 1.1 for his subsequent paper. (See Theorem 4.1 in [4].)

In his second paper [4], Altuğ wrote f^p for the function $f = f_\infty \cdot f^{\infty,p} \cdot f_p^k$ from [3] with $k = 1$. He then showed that its value at the third (complicated) integral in the formula for $\widehat{\overline{I}}(0, f^p)$ from Theorem 6.1 of [3] (adjusted[28] as in the statement of Theorem 4.1 of [4]) is bounded in p [4, formula (00) from Corollary 4.3]. This was actually relatively simple. Much deeper were his estimates of the terms $\widehat{\overline{I}}_{\mathrm{ell,reg}}(\xi, f)$ with $\xi \neq 0$ in (157). In Theorem 4.4, he established an estimate

$$|\sum_{\xi \neq 0} \widehat{\overline{I}}(\xi, f^p)| \leq c_\infty p^{\frac{1}{4}},$$

where c_∞ is a constant that depends only on f_∞, but not p. This was a consequence of analytic results on asymptotic properties of Fourier transforms in Appendix A [4, Theorems A.14 and A.15, and Corollary 11.16], as well as arithmetic properties of the character sums $\mathrm{Kl}_{\ell,f}(\xi, \pm p)$ in Appendix B [4, Corollary B.8]. Combined with a straightforward analysis in §3 in [4] of the remaining terms in the full trace formula for $\mathrm{tr}(R_{\mathrm{cusp}}(f^p))$, he arrived at his main result, the estimate

$$\mathrm{tr}(R_{\mathrm{cusp}}(f^p)) = O(p^{\frac{1}{4}}) \tag{158}$$

of [4, Theorem 1.1].

The estimate (158) represents a partial bound towards the Ramanujan conjecture for Hecke eigenvalues of Maass forms. It is the same estimate that had been established in 1980 by Kuznetsov [126] by what is now regarded as a special case of the relative trace formula. As Altuğ remarks, the importance of (158) is in its method of proof by the Arthur–Selberg trace formula, which has more structure, and has been established for general groups. The full Ramanujan conjecture would be a consequence of functoriality, but is still far from known. It would amount to a bound

$$\mathrm{tr}(R_{\mathrm{cusp}}(f^p)) = O(p^{\varepsilon})$$

for every $\varepsilon > 0$. The estimate (158) is thus intermediate between the full Ramanujan conjecture and the elementary bound

$$\mathrm{tr}(R_{\mathrm{cusp}}(f^p)) = O(p^{\frac{1}{2}})$$

represented by the 1-dimensional representation of $\mathrm{GL}(2, \mathbb{A})$. We recall from our discussion of the Langlands–Shahidi method that in establishing functoriality for the irreducible 4- and 5-dimensional representations of $\widehat{G} = \mathrm{GL}(2, \mathbb{C})$, Shahidi and Kim obtained bounds that are sharper than (158). However the full Ramanujan conjecture (for Hecke operators of Maass forms) would require functoriality for all irreducible representations of $\widehat{G}$.

The third paper [5] of Altuğ established an r-trace formula (147) in the special case of $G = \mathrm{GL}(2)$ and $F = \mathbb{Q}$ at hand. For this, he restricted the basic function $f = f_\infty f^\infty$ further, by taking f^∞ to be the characteristic function of the standard maximal compact subgroup K^∞ of $G(\mathbb{A})$, and f_∞ to be a *cuspidal* function on $G(\mathbb{R})$, with tempered characters supported on the discrete series representation parameterized by a fixed integer $k \geq 3$. With these data, Altuğ considered the proposed limit (148) when r is the standard, two-dimensional representation of $\widehat{G} = \mathrm{GL}(2,\mathbb{C})$. Combining the intricate analytic results of [4], he established that the limit exists, and equals 0. In other words, the r-trace formula in this case is

$$I^r_{\mathrm{cusp}}(f) = 0,$$

according to the definition (147). This is what is expected. For the only cuspidal automorphic representations π of $G = \mathrm{GL}(2)$ that are *proper* functorial images (which is to say, not *primitive* in the language of §9), should be of "CM" or "Galois type", attached to two-dimensional representations of the Weil group $W_{\mathbb{Q}}$. This implies that $\pi(f)$ actually vanishes,[29] given the choice of f^∞ and the fact that the class number of $\mathbb{Q}$ is 1. Therefore π should contribute nothing to $I^r_{\mathrm{cusp}}(f)$. It is a general fact that a primitive representation π would also contribute nothing to $I^r_{\mathrm{cusp}}(f)$, for any irreducible nontrivial representation r.

We should include a historical remark at this point before concluding our discussion of Altuğ's work. When Langlands first introduced his ideas in the precursor [154] of his published article [155], Sarnak had reservations about the form of the proposed limit (148) obtained from the order of poles at $s = 1$ of the L-functions $L(s,\pi,r)$. He was concerned that proving and calculating a limit (148), difficult under any circumstances, might be even more intractable with the form of the right-hand side. He suggested in the letter [199] to Langlands replacing the sum over p on the right-hand side of (148) by a sum over $n \in \mathbb{N}$. This amounts essentially to a change of the weighting coefficient $m_\pi(r)$ in (142), the order of the pole of $L(s,\pi,r)$ at $s = 1$, to another coefficient $n_\pi(r)$, the *residue* of $L(s,\pi,r)$ at $s = 1$. It would then open the possibility of applying Poisson summation again, this time to the new sum over n. The Tauberian theorem we quoted from [208] would also hold in this context, applied to the Dirichlet series for $L(s,\pi,r)$ rather than its logarithmic derivative. Langlands himself appears to have been ambivalent about this suggestion. For the new coefficients would no longer be additive in r, complicating the anticipated spectral (primitive) Beyond Endoscopic decomposition of what we are calling the r-trace formula.

The suggestion was first taken up by A. Venkatesh, a student of Sarnak at the time. He actually worked with the Kuznetsov formula for GL(2), a special case of Jacquet's proposed "relative trace formula", rather than the trace formula itself. The technical difficulties simplify in this setting. Using the Poisson summation formula proposed by Sarnak, Venkatesh considered the more complex case of the three-dimensional, symmetric square representation r_2 of GL(2). In his 2002 thesis [241]

[29] This property should remain valid if r is replaced by an $(m+1)$-dimensional symmetric power r_m, for $m > 1$. We would therefore again expect that $I^{r_m}_{\mathrm{cusp}}(f) = 0$, even though this would be much more difficult to prove. On the other hand, if f^∞ is a more general function, or $\mathbb{Q}$ is replaced by a more general field F, we would not expect $I^{r_m}_{\mathrm{cusp}}(f)$ to vanish. For a table of multiplicities

$$-m_\pi(r_m) = [1_{^LG} : r_m \circ \phi],$$

where π corresponds to a 2-dimensional Galois representation ϕ of $\Gamma_{\mathbb{Q}}$, see page 6 of [154].

and a subsequent paper [242], he was able to separate the contribution of the relevant forms of CM-type, which is to say the automorphic representations π of GL(2) attached to two-dimensional representations of the Weil group W_F, by using the residues at $s = 1$ of the L-functions $L(s, \pi, r_2)$. His results were also more general in that they applied to many number fields F in place of $\mathbb{Q}$, and automorphic forms that were ramified, which is to say of level greater than 1. The work of Venkatesh was a breakthrough. It would be very interesting to make a careful comparison of his techniques with those of Altuğ. In this report, we will settle for a few general remarks at the end on the possible future roles on the two kinds of "trace" formulas.

Altuğ took up Sarnak's suggestion in his third paper [5]. The Tauberian theorem for the new weighting coefficient attached to any G, π and r would take the general form

$$n_\pi(r) = \lim_{X \to \infty} |X|^{-1} \sum_{n<X} \operatorname{tr}(T(n, \pi, r))$$

where[30]

$$T(n, \pi, r) = \pi^S(h_n^r), \qquad S = \{\infty\},$$

is the (unramified) Hecke operator for π^S, r and n such that

$$L^\infty(s, \pi, r) = L^S(s, \pi, r) = \sum_{n=0}^{\infty} \operatorname{tr}(T(n, \pi, r))n^{-s}.$$

Altuğ again confined himself to the case that $r = r_1$ is the standard two-dimensional representation of the group ${}^LG = \widehat{G} = \mathrm{GL}(2, \mathbb{C})$. It is likely that his methods could be extended to $r = r_2$, and to the more general fields F and functions f treated by Venkatesh, but he did not attempt to deal with the increased complexity that would arise from the classical trace formula for GL(2) in this paper. In fact, as we noted earlier, he restricted $f = f_\infty f^\infty$ further so as to be supported on the unramified automorphic representations $\pi = \pi_\infty \pi^\infty$ such that π_∞ corresponds to an automorphic form in the space S_k of (holomorphic) cusp forms of weight k (and level 1), for fixed $k \geq 1$. Then

---

30 Motivated by notation from the beginning of the section, we have written

$$h_n^r = \bigotimes_p h_{n,p}^r, \qquad n = \prod_{p \notin S} p^{n_p},$$

here for the function in the unramified Hecke algebra

$$C_c(K^S \setminus G(\mathbb{A}^S)/K^S) = \widetilde{\bigotimes_{p \notin S}} C_c(\mathbf{1}_p \setminus G(\mathbb{Q}_p)/\mathbf{1}_p)$$

whose Satake transform $\widehat{h}_n^r$ at a family of semisimple classes $c = \{c_p : p \notin S\}$ in ${}^LG$ equals the product

$$\widehat{h}_n^r(c^s) = \prod_p \widehat{h}_{n,p}(c_p) = \prod_p \operatorname{tr}((S_{n_p} r)(c_p))$$

of characters of symmetric powers $S_{n_p} r$ of r at the points of c_p. (See [44].)

$$T_k(n) = \bigoplus_{\pi} T(n,\pi,r)$$

is the nth Hecke operator acting on the (finite-dimensional) complex vector space S_k.

Altuğ's main result, Theorem 1.1 of [5], is an estimate

$$\sum_{n<X} \operatorname{tr}(T_k(n)) = O_{k,\varepsilon}(X^{\frac{31}{32}+\varepsilon}),$$

for any $\varepsilon > 0$. In particular, for any such π, the function

$$|X|^{-1} \sum_{n<X} \operatorname{tr}(T(n,\pi,r))$$

converges strongly to 0 as X approaches ∞. In Corollary 1.2 of [5], Altuğ specialized this to an estimate for the Ramanujan Δ-function (of weight 12 and level 1), or rather a property of the corresponding L-function $L(s,\Delta)$. In the remarks following the statement of the corollary [5, p. 3–5], he discussed other interesting consequences, including the original r-trace formula.

The rest of Altuğ's third paper [5] is devoted to the proof of Theorem 1.1. As he observed prior to the statement of the theorem, it puts together all of the work of his previous two papers. In Section 2 he gave a broad outline of the proof of the theorem, and the proof of its Corollary 1.2. In Section 3, he dealt with the supplementary (nonelliptic, noncuspidal) terms in the trace formula for GL(2). Section 4 concerns the elliptic terms, which remain the basic objects. It is the heart of the paper. Section 4.1 is a short review and further analysis of the results of [3], notably the Poisson summation formula (157) for the sums over m introduced prior to its statement. Section 4.2.1 begins with a heuristic discussion of estimates provided by subsequent Theorems 4.9, 4.11 and 4.13, and then proceeds with their proof. Section 4.2.2 contains the final estimates given by the last Theorem 4.15 and its Corollary 4.16. It is in their proof that the critical second application of Poisson summation comes, the one for the sum over n. (The character sum $\mathrm{Kl}_{\ell,f}(\xi,n)$ is observed here to be periodic in n modulo $4\ell f^2$, allowing among other things, a change from the sum over $n > 0$ to other sums over $n \neq 0$.) The final Section 5 contains further real and p-adic analysis. This was used in the proofs of estimates (Proposition 5.2 and Corollary 5.9) postponed from Section 4.

This completes our discussion of the work of Altuğ. As a last word, it might be worthwhile to recapitulate each of his three papers in a sentence or two. The first one [3] establishes Poisson summation, and shows how the nontempered, trivial representation occurs naturally in the Fourier transform term with $\xi = 0$. The second paper [4] is an estimate that asserts that the complementary part of the $\xi = 0$ term as well as the remaining $\xi \neq 0$ terms are exponentially smaller than the trivial (one-dimensional) character. However, the error bound is still exponentially larger than the tempered singular term ((vi) in [103]) that was also removed from the summand with $\xi = 0$. The first two papers apply to a single formula, for a function f^p that depends on p and $k = n_p$. The third paper [5] applies to a weighted average of such formulas, parameterized by positive integers $n = \prod_p p^{n_p}$. The averaging process decreases the error term considerably further, to more than what it would be for any isolated tempered automorphic representation on the spectral side, and in particular, for the singular tempered representations from term (vi) of [103] that were removed from the original summand with $\xi = 0$.

The amount of space we have devoted to Altuğ's work might appear surprising. However, it is likely to serve as a concrete foundation for the future study of Beyond Endoscopy, at least as it follows Langlands' ideas for using the (stable) trace formula. Our discussion has also given us opportunities to illustrate aspects of the basic strategy, as it might apply in practice. To be sure, Altuğ's papers are rather imposing, particularly when one begins with the statement of the main result [3, Theorem 1.1] of his first paper, and its refinement [4, Theorem 4.1] for the second. But despite their complexity, the terms in the stated formulas are elementary. Their derivation and later application depend more than anything on basic analysis, despite the fact that they are leading to new techniques. There is also a suggestive unity to the three papers. Each one solves a specific problem in the simplest of cases, using only the methods of the trace formula. Taken in succession, they match sequential steps laid out by Langlands in his general strategy for Beyond Endoscopy.

An interesting problem would be to relate Altuğ's work with the second half of Langlands' paper [158]. In general, one might eventually want to establish a primitive (stable) trace formula, whose spectral side contains only the cuspidal, tempered representations that are not proper functorial images. One could imagine an inductive definition for any G obtained by subtracting from its stable trace formula the primitive trace formulas attached to all proper beyond endoscopic data.[31]

For $G = \mathrm{GL}(2)$ and G' a proper beyond endoscopic datum, the derived group G'_{der} would then be the trivial group 1_G, so $(G', \mathcal{G}', \xi')$ would therefore correspond to an irreducible two-dimensional representation of the Weil group. As we have noted, every such object ramifies over $\mathbb{Q}$, and therefore contributes nothing. To carry out the proposal, one would therefore want to extend Altuğ's results to more general functions $f = f_\infty f^\infty$, or more general number fields F, or ideally, both. From the resulting sum over $\xi \neq 0$, one could then subtract the expression obtained by Langlands by Poisson summation on the sum of the two-dimensional characters on W_F (informed perhaps by the more transparent formula described on p. 1 of [109]). Langlands' expression does not actually include characters on W_F with finite image, those given by the irreducible characters of dihedral, tetrahedral, octahedral and icosahedral type on the Galois group Γ_F. This might be the real point. Is it too much to hope that we might recognize something of the sum of these characters in the more concrete difference of the original two sums over ξ? We are now talking about the deepest aspect of Beyond Endoscopy, the one that Langlands regarded as the true essence of the problem.

[31] I have yet to think carefully about this. As suggested in the discussion of stable transfer earlier in the section, it would presumably follow the proposed construction of L_F from Section 9. In particular, the beyond endoscopic groups G' should perhaps have G'_{der} simply connected and $\dim G'_{\mathrm{der}} < \dim G_{\mathrm{der}}$.

A part of Langlands' later paper [157] is pertinent to this last point. (The paper is described as a prologue, but the article on Functoriality and Reciprocity to which it refers has not been written.) First of all, Section 4 contains his reflections on the letter of Sarnak [199], and in particular, on the relative merits of working with the residues of either the L-functions $L(s,\pi,\rho)$ themselves or of their logarithmic derivatives. Altuğ's proof of an r-trace formula for GL(2), described in the remarks in §1.1 of [5], is encouraging. It suggests that we might be able to have the best of both worlds.

It is the larger Section 5 of [157] that concerns Langlands' thoughts on complex Galois representations. He describes it as the arithmetic side of Beyond Endoscopy, as opposed perhaps to the analytic side later studied by Altuğ. Langlands discusses in some detail a paper [57] of Dedekind on quaternionic extensions F of $\mathbb{Q}$, the extensions with Galois group the quaternionic group Q_8 of order 8 that the reader will recall (with K/L in place of $F/\mathbb{Q}$) from the diagram at the end of §7. His thoughts on this might be described as "hard arithmetic", as opposed to "hard analysis". They can be taken as a reflection of Langlands' view, expressed in several places, that functoriality for complex Galois representations will be at the heart of Beyond Endoscopy. They are where the subject began, as the attempt by Langlands to extend the Artin reciprocity law, and thereby create a nonabelian class field theory. The Principle of Functoriality was of course his answer. However, the precepts of Beyond Endoscopy do not so far include arithmetic techniques with the power to make further progress on Artin's conjecture.

Another part of the paper [157] is devoted to an entirely different topic. It represents the beginnings of what Langlands called the *geometric theory*, as distinct from what is often referred to as the geometric Langlands program. Very roughly speaking, if the original (arithmetic) Langlands program concerns finite extensions of a global field F, a number field or the field of rational functions on a nonsingular projective curve over a finite field, these geometric programs are over the complex numbers. They concern the finite extensions of the field F of meromorphic functions on a compact nonsingular Riemann surface X. Their starting point is the set $\mathrm{Bun}_X(G)$ of (equivalence classes of) holomorphic principal G-bundles on X, for a complex reductive group G. For any $x \in X$, F_x is used to denote the field of formal complex Laurent series at x, and $\mathcal{O}_x$ is the subring of formal power series at x. There is then a natural bijection

$$\mathrm{Bun}_X(G) \cong G(F) \backslash G(\mathbb{A}_F)/K_F, \tag{159}$$

where

$$G(\mathbb{A}_F) = \widetilde{\prod_{x \in X}} G(F_x)$$

and

$$K_F = \prod_{x \in X} G(\mathcal{O}_x).$$

It is clear from this that the theory should bear at least some formal resemblance to automorphic representation theory, even as its content ought to be primarily geometric.

The original geometric Langlands program was not initiated by Langlands. It is founded on notions from abstract algebraic geometry, accompanied by various sophisticated techniques from category theory. In particular, $\mathrm{Bun}_X(G)$ is treated as an algebraic stack, while the analogue of an automorphic representation takes the form of a perverse sheaf on the stack. Rather than try to comment further on these notions, let me refer the reader to the clearly written Bourbaki lecture [76] by Gaitsgory, who credits ideas of Beilinson, Deligne, Drinfeld and Laumon for the origins of the subject, and who is himself responsible for more recent progress. We note also that "Geometric Langlands" has had a significant influence in string theory. (See the Bourbaki lecture of Frenkel [73], and references there.)

Langlands wanted to understand the geometric theory in more concrete terms. In particular, he wanted to apply the methods of differential geometry and harmonic analysis in place of abstract algebraic geometry. His goal was to attach explicit (unramified) Hecke operators to the moduli space (159), together with corresponding explicit Hecke eigenvalues. This was not done in the prologue [157]. Langlands worked six additional years, at length posting the long paper [160] with his results in Russian, intending it especially for the Russian-American mathematicians in the field. He then worked further to convert it to the paper [162] in English, which he posted finally in 2020. I have read only the short note [161] describing his aims and results in quite general terms, but I shall look forward to reading the long paper.

The background for the paper is interesting. Langlands was motivated by the early paper [28] of Atiyah, written before his much better known work on the index and fixed point theorems. In it, Atiyah presented a concrete classification of the set of complex principal bundles of dimension n over a complex torus, which is to say, the set (159) with $G = \mathrm{GL}(n)$ and X an elliptic curve [28, Theorem 7]. This would be an *exact* fundamental domain for the space (159), rather than an approximate fundamental domain of the kind attached to a Siegel set in the arithmetic theory. Informed by this paper (which I am told is hard going), Langlands turned next to Atiyah's widely read paper [29] with Bott. He then set about studying the space of complex plane bundles over an elliptic curve, the special case of Atiyah's classification with $n = 2$. Langlands has reported that the desired Hecke operators and their eigenvalues appeared, in what seems to have been rather dramatic fashion, only at the very end.

In his note [161], Langlands has proposed a number of ways to attempt to extend these results. The first step might be to establish them for $\mathrm{GL}(n)$-bundles over X, the setting of Atiyah's classification. One could then attempt to replace $\mathrm{GL}(n)$ by an arbitrary complex group G, and X by an arbitrary (compact, nonsingular) Riemann surface.[32] Langlands suggests that these last extensions will be much more difficult, since they would require, among other things, major extensions of Atiyah's classi-

---

[32] The case that X is the Riemann sphere is presumably easy. As Atiyah noted at the beginning of his article, Grothendieck has shown that any vector bundle over a rational curve is a direct sum of line bundles.

fication. One would of course also have to be armed with a firm command of the Langlands paper [162], a serious prerequisite to be sure. Assuming all of this could be established, it would then be interesting to compare the results with the geometric Langlands program based on abstract algebraic geometry. Finally, one could consider "ramification", or rather what would be ramification in the arithmetic theory. That is, one would consider G-bundles with extra structure, just as one takes elliptic curves with level structure in the classical theory of modular forms. It would entail replacing the group K_F by a (normal) subgroup of finite co-dimension, defined by the analogue of congruence conditions ([157, p. 55]). A solution would have particular interest, since the question was initially not widely investigated through abstract algebraic geometry. (See however [8] for remarkable recent progress in this subject.)

Langlands suggests that each of these extensions will be difficult (if not as difficult as the many arithmetic problems in Beyond Endoscopy). However, they would probably have wide appeal. They seem to be revealing a new hidden structure in the objects commonly studied by differential geometers and topologists, which in turn is suggestive of the arithmetic structure in the original Langlands program.

We return to the discussion of Beyond Endoscopy with some final comments on the different possible approaches to functoriality. I have not mentioned the results of Braverman and Kazhdan [39], and the subsequent contributions of Ngô and others [187], [48], [38]. The idea is to try to establish the analytic continuation and functional equation for various automorphic L-functions $L(s,\pi,r)$ directly, and then perhaps use converse theorems to establish functoriality. This has direct roots in the work of Hecke and his original converse theorem, which we discussed in Section 6 in our review of Jacquet–Langlands [103]. It is the opposite of Langlands' fundamental idea of using functoriality to establish the general analytic continuation and functional equation from the basic case of principal L-functions for $\mathrm{GL}(n)$. I have not read these papers, and can make no further comment, even though I have enjoyed lectures on the subject. Braverman and Kazhdan's results also motivated L. Lafforgue. He has conjectured a nonlinear Poisson summation formula, which would be a consequence of functoriality, but which conversely would imply functoriality [128], [129].

We shall finish the report with some speculative remarks on the strategic differences between the (Arthur–Selberg) trace formula and the (Jacquet) relative "trace" formula.[33] Langlands' strategy applied to the former, while the work of Venkatesh for $\mathrm{GL}(2)$ on the latter faced fewer technical difficulties. (See also [194], [195], [196].) We are of course a long way from realizing any of the general goals of Beyond Endoscopy, but it is reasonable to try to plan how best to move forward. I

[33] This is not really a trace formula. In general, it would actually be a formula for periods of automorphic forms, rather than for characters of automorphic representations. Let me take the liberty of calling it (and any one of its variants) the *period formula* in what follows. We should note that the periods here differ from the Grothendieck periods attached to motives discussed in §9. They are defined as integrals of automorphic forms on G against cycles defined by subgroups H of G, which can sometimes represent parings between cohomology classes and homology classes. However, they make sense also for automorphic forms that are not motivic. We can call them *automorphic periods*, as opposed to the *motivic periods* from §9.

believe strongly in the trace formula (perhaps not surprisingly), and I will try to express some of my reasons for this. However, I have limited experience with the period formula, so the reader should keep an open mind. Besides, there is a different point of view that I will express at the very end.

The trace formula is more highly developed. It has a clear general structure, and each of its terms has a natural, well defined source. It is reasonable to think that each of the terms will also have well defined role in Beyond Endoscopy. It is actually the stable trace formula that would be applied to groups other than GL_n. This is more sophisticated, but it is completely general, and has an equally rigid structure.

The period formula, at least in the Kuznetsov formula that was applied by Venkatesh to $\mathrm{GL}(2)$, depends on the fact that a cuspidal automorphic representation of $\mathrm{GL}(2)$ has a Whittaker model. For a quasisplit classical group, it is now known that every tempered, cuspidal L-packet contains a representation with a Whittaker model [23, §8.3]. To exploit this, however, one would want to work with local and global L-packets, which leads us back to the stable trace formula. For exceptional groups, the property is not known, even though it is widely expected. However, a proof would seem to require a theory of endoscopy for general groups. It may be that general endoscopy would have to be a consequence of Beyond Endoscopy but this is speculative. Keep in mind, however, that one really would want a full theory of Beyond Endoscopy for *all* groups, including exceptional groups. For it would be needed just to be able to classify the automorphic representations of, say, general linear groups, in terms of functorial images of primitive representations of smaller groups. This is closely related to the question of an explicit construction of the automorphic Galois group L_F as described in Section 9.

For some more serious speculation (!), consider the following. Beyond Endoscopy is a proposal for attacking the Principle of Functoriality. What about its companion, Langlands' Reciprocity Conjecture? Is it possible that some extension of Beyond Endoscopy might have to be used to establish the two conjectures together? Langlands suggests something of the sort in his interesting and suggestive article [159]. It seems quite plausible to me. If so, the theory of general Shimura varieties would assume a central role beyond what it already holds. As a highly developed and mature theory, ultimately based on the arithmetic theory of reductive groups, it would presumably become a foundation for Reciprocity rather than just a source of interesting examples, much as the theory of Shimura varieties attached to $\mathrm{GL}(2)$ is a foundation of the STW-conjecture for elliptic curves over $\mathbb{Q}$. Of course it would also demand fundamental new techniques, of which Wiles' proof, and the more recent work of Taylor and collaborators, would represent a beginning.

Take for example the role of automorphic representations that are *finite*.[34] These are the automorphic representations that ought to give nonabelian class field theory, surely the ultimate motivation. As we have already observed, they are to be considered the objects at the heart of Beyond Endoscopy, and for which the proof of

[34] By this I mean the analogue for any G over F of classical modular forms of weight 0 or 1, sometimes called modular forms of type A_{00}. In general, they correspond to homomorphisms of the global Weil group W_F to the L-group ${}^L G_F$ that factor through the quotient Γ_F, or equivalently, whose image in $\widehat{G}$ is finite.

Functoriality would presumably be the deepest. But they also represent Artin motives, the most basic of motives. The Reciprocity Conjecture in this case becomes the corresponding case of Functoriality. Since Beyond Endoscopy is supposed to lead to a general proof of Functoriality, it would be surprising if it did not have some major role in the proof of Reciprocity.

It is not hard to think of other analogies between Functoriality and Reciprocity that might be hinting at a common proof. My point is simply this. If the intuition we are taking is valid, we would be well advised to formulate arguments in terms of the stable trace formula. For we would have to come to the problems with a deep understanding of the automorphic properties of general Shimura varieties. As we have seen, these are almost as closely tied to the stable trace formula as are the automorphic representations themselves.

As a final thought, let me consider a different way of viewing the original question. It is a philosophical query, related to the duality between automorphic forms and automorphic representations. Automorphic forms go back to the end of the nineteenth century with Poincaré, followed by the successive generalizations of Hilbert, Siegel and Harish-Chandra. The notion of an automorphic representation came much later, after the adelic language had become common. As we have noted, the term itself seems to have first appeared in Borel's 1976 Bourbaki lecture. But it was Langlands who emphasized the dichotomy between the two notions. As we have seen, this is expressed in the Corvallis papers [35], [145]. The two formulas we are considering reflect this dichotomy. Their parallel origins are clearly viewed in the following simple diagram

$$
\begin{array}{ccc}
\text{automorphic representations} & \longleftrightarrow & \text{automorphic forms} \\
\downarrow & & \downarrow \\
\text{automorphic characters} & \longleftrightarrow & \text{automorphic periods} \\
\downarrow & & \downarrow \\
\text{trace formula} & \longleftrightarrow & \text{period formula}
\end{array}
$$

Do the two formulas (or classes of formulas) play dual roles? Or is the diagram a red herring? Do they give information that is sometimes complementary, or will they ultimately reduce to the same identities, whatever the circumstances? The two formulas are in any event sufficiently different that they could both be applied separately to Beyond Endoscopy, and then compared. It would not matter if the results overlap. In fact, it would be very useful to have a clear understanding of their common properties. This would give us a broader perspective for when we run into difficult problems that demand new ideas, as we most surely will!

References

1. J. Adams, D. Barbasch, and D. A. Vogan, Jr. *The Langlands classification and irreducible characters for real reductive groups*, volume 104 of *Progress in Mathematics*. Birkhäuser Boston, Inc., Boston, MA, 1992.
2. J. Adams and J. F. Johnson. Endoscopic groups and packets of nontempered representations. *Compositio Math.*, 64(3):271–309, 1987.
3. S. A. Altuğ. Beyond endoscopy via the trace formula, I: Poisson summation and isolation of special representations. *Compos. Math.*, 151(10):1791–1820, 2015.
4. S. A. Altuğ. Beyond endoscopy via the trace formula, II: Asymptotic expansions of Fourier transforms and bounds towards the Ramanujan conjecture. *Amer. J. Math.*, 139(4):863–913, 2017.
5. S. A. Altuğ. Beyond endoscopy via the trace formula, III: the standard representation. *J. Inst. Math. Jussieu*, 19(4):1349–1387, 2020.
6. G. W. Anderson. Cyclotomy and an extension of the Taniyama group. *Compositio Math.*, 57(2):153–217, 1986.
7. Y. André. Galois theory, motives and transcendental numbers. In *Renormalization and Galois theories*, volume 15 of *IRMA Lect. Math. Theor. Phys.*, pages 165–177. Eur. Math. Soc., Zürich, 2009.
8. D. Arinkin, D. Gaitsgory, D. Kazhdan, S. Raskin, N. Rozenblyum, and Y. Varshavsky. The stack of local systems with restricted variation and geometric Langlands theory with nilpotent singular support. *arXiv:2010.01906*, 2010.
9. J. Arthur. The Selberg trace formula for groups of F-rank one. *Ann. of Math. (2)*, 100:326–385, 1974.
10. J. Arthur. A trace formula for reductive groups. I. Terms associated to classes in $G(\mathbf{Q})$. *Duke Math. J.*, 45(4):911–952, 1978.
11. J. Arthur. A trace formula for reductive groups. II. Applications of a truncation operator. *Compositio Math.*, 40(1):87–121, 1980.
12. J. Arthur. The trace formula in invariant form. *Ann. of Math. (2)*, 114(1):1–74, 1981.
13. J. Arthur. On some problems suggested by the trace formula. In *Lie group representations, II (College Park, Md., 1982/1983)*, volume 1041 of *Lecture Notes in Math.*, pages 1–49. Springer, Berlin, 1984.
14. J. Arthur. The invariant trace formula. I. Local theory. *J. Amer. Math. Soc.*, 1(2):323–383, 1988.
15. J. Arthur. The invariant trace formula. II. Global theory. *J. Amer. Math. Soc.*, 1(3):501–554, 1988.
16. J. Arthur. The local behaviour of weighted orbital integrals. *Duke Math. J.*, 56(2):223–293, 1988.
17. J. Arthur. The L^2-Lefschetz numbers of Hecke operators. *Invent. Math.*, 97(2):257–290, 1989.
18. J. Arthur. Unipotent automorphic representations: conjectures. *Astérisque*, (171-172):13–71, 1989. Orbites unipotentes et représentations, II.
19. J. Arthur. On local character relations. *Selecta Math. (N.S.)*, 2(4):501–579, 1996.
20. J. Arthur. A note on the automorphic Langlands group. *Canad. Math. Bull.*, 45(4):466–482, 2002. Dedicated to Robert V. Moody.
21. J. Arthur. A stable trace formula. III. Proof of the main theorems. *Ann. of Math. (2)*, 158(3):769–873, 2003.
22. J. Arthur. An introduction to the trace formula. In *Harmonic analysis, the trace formula, and Shimura varieties*, volume 4 of *Clay Math. Proc.*, pages 1–263. Amer. Math. Soc., Providence, RI, 2005.
23. J. Arthur. *The endoscopic classification of representations: Orthogonal and symplectic groups*, volume 61 of *American Mathematical Society Colloquium Publications*. American Mathematical Society, Providence, RI, 2013.

24. J. Arthur. Problems beyond endoscopy. In *Representation theory, number theory, and invariant theory*, volume 323 of *Progr. Math.*, pages 23–45. Birkhäuser/Springer, Cham, 2017.
25. J. Arthur. Functoriality and the trace formula. In *Relative aspects in representation theory, Langlands functoriality and automorphic forms*, volume 2221 of *Lecture Notes in Math.*, pages 319–339. Springer, Cham, 2018.
26. J. Arthur. A stratification related to characteristic polynomials. *Adv. Math.*, 327:425–469, 2018.
27. J. Arthur and L. Clozel. *Simple algebras, base change, and the advanced theory of the trace formula*, volume 120 of *Annals of Mathematics Studies*. Princeton University Press, Princeton, NJ, 1989.
28. M. F. Atiyah. Vector bundles over an elliptic curve. *Proc. London Math. Soc. (3)*, 7:414–452, 1957.
29. M. F. Atiyah and R. Bott. The Yang–Mills equations over Riemann surfaces. *Philos. Trans. Roy. Soc. London Ser. A*, 308(1505):523–615, 1983.
30. M. F. Atiyah and C. T. C. Wall. Cohomology of groups. In *Algebraic Number Theory (Proc. Instructional Conf., Brighton, 1965)*, pages 94–115. Thompson, Washington, D.C., 1967.
31. A. Borel. Introduction to automorphic forms. In *Algebraic Groups and Discontinuous Subgroups (Proc. Sympos. Pure Math., Boulder, Colo., 1965)*, pages 199–210. Amer. Math. Soc., Providence, R.I., 1966.
32. A. Borel. Formes automorphes et séries de Dirichlet (d'après R. P. Langlands). In *Séminaire Bourbaki (1974/1975: Exposés Nos. 453–470), Exp. No. 466*, pages 183–222. Lecture Notes in Math., Vol. 514. 1976.
33. A. Borel. Automorphic L-functions. In *Automorphic forms, representations and L-functions (Proc. Sympos. Pure Math., Oregon State Univ., Corvallis, Ore., 1977), Part 2*, Proc. Sympos. Pure Math., XXXIII, pages 27–61. Amer. Math. Soc., Providence, R.I., 1979.
34. A. Borel and W. Casselman. L^2-cohomology of locally symmetric manifolds of finite volume. *Duke Math. J.*, 50(3):625–647, 1983.
35. A. Borel and H. Jacquet. Automorphic forms and automorphic representations. In *Automorphic forms, representations and L-functions (Proc. Sympos. Pure Math., Oregon State Univ., Corvallis, Ore., 1977), Part 1*, Proc. Sympos. Pure Math., XXXIII, pages 189–207. Amer. Math. Soc., Providence, R.I., 1979. With a supplement "On the notion of an automorphic representation" by R. P. Langlands.
36. A. Borel and N. R. Wallach. *Continuous cohomology, discrete subgroups, and representations of reductive groups*, volume 94 of *Annals of Mathematics Studies*. Princeton University Press, Princeton, N.J.; University of Tokyo Press, Tokyo, 1980.
37. M. V. Borovoĭ. Langlands' conjecture concerning conjugation of connected Shimura varieties. *Selecta Math. Soviet.*, 3(1):3–39, 1983/84. Selected translations.
38. A. Bouthier, B. C. Ngô, and Y. Sakellaridis. On the formal arc space of a reductive monoid. *Amer. J. Math.*, 138(1):81–108, 2016.
39. A. Braverman and D. Kazhdan. γ-functions of representations and lifting. *Geom. Funct. Anal.*, (Special Volume, Part I):237–278, 2000. With an appendix by V. Vologodsky, GAFA 2000 (Tel Aviv, 1999).
40. C. Breuil, B. Conrad, F. Diamond, and R. Taylor. On the modularity of elliptic curves over $\mathbf{Q}$: wild 3-adic exercises. *J. Amer. Math. Soc.*, 14(4):843–939, 2001.
41. F. Brown. Mixed Tate motives over $\mathbb{Z}$. *Ann. of Math. (2)*, 175(2):949–976, 2012.
42. F. Brown. Feynman amplitudes, coaction principle, and cosmic Galois group. *Commun. Number Theory Phys.*, 11(3):453–556, 2017.
43. H. Carayol. Sur les représentations l-adiques associées aux formes modulaires de Hilbert. *Ann. Sci. École Norm. Sup. (4)*, 19(3):409–468, 1986.
44. B. Casselman. Symmetric powers and the Satake transform. *Bull. Iranian Math. Soc.*, 43(4):17–54, 2017.
45. W. Casselman and J. Shalika. The unramified principal series of p-adic groups. II. The Whittaker function. *Compositio Math.*, 41(2):207–231, 1980.
46. P.-H. Chaudouard and G. Laumon. Le lemme fondamental pondéré. I. Constructions géométriques. *Compos. Math.*, 146(6):1416–1506, 2010.

47. P.-H. Chaudouard and G. Laumon. Le lemme fondamental pondéré. II. Énoncés cohomologiques. *Ann. of Math. (2)*, 176(3):1647–1781, 2012.
48. S. Cheng and B. C. Ngô. On a conjecture of Braverman and Kazhdan. *Int. Math. Res. Not. IMRN*, 20:6177–6200, 2018.
49. V. I. Chernousov. The Hasse principle for groups of type E_8. *Dokl. Akad. Nauk SSSR*, 306(5):1059–1063, 1989.
50. L. Clozel. Nombre de points des variétés de Shimura sur un corps fini (d'après R. Kottwitz). *Astérisque*, (216):Exp. No. 766, 4, 121–149, 1993. Séminaire Bourbaki, Vol. 1992/93.
51. L. Clozel, M. Harris, and R. Taylor. Automorphy for some l-adic lifts of automorphic mod l Galois representations. *Publ. Math. Inst. Hautes Études Sci.*, 108:1–181, 2008. With Appendix A, summarizing unpublished work of Russ Mann, and Appendix B by Marie-France Vignéras.
52. L. Clozel and C. S. Rajan. Solvable base change. *arXiv e-prints*, page arXiv:1806.02513, Jun 2018.
53. J. Cogdell. On Artin L-functions. `https://people.math.osu.edu/cogdell.1/artin-www.pdf`. [Online; accessed 1-July-2021].
54. A. Connes and M. Marcolli. *Noncommutative geometry, quantum fields and motives*, volume 55 of *American Mathematical Society Colloquium Publications*. American Mathematical Society, Providence, RI; Hindustan Book Agency, New Delhi, 2008.
55. H. Darmon. A proof of the full Shimura–Taniyama–Weil conjecture is announced. *Notices Amer. Math. Soc.*, 46(11):1397–1401, 1999.
56. H. Darmon. Andrew Wiles's marvelous proof. *Notices Amer. Math. Soc.*, 64(3):209–216, 2017.
57. R. Dedekind. Konstruktion von Quaternionkörpern. *Gesammelte mathematische Werke, Bd*, 2:376–384, 1931.
58. P. Deligne. Formes modulaires et représentations l-adiques. In *Séminaire Bourbaki. Vol. 1968/69: Exposés 347–363*, volume 175 of *Lecture Notes in Math.*, pages Exp. No. 355, 139–172. Springer, Berlin, 1971.
59. P. Deligne. Théorie de Hodge. I. In *Actes du Congrès International des Mathématiciens (Nice, 1970), Tome 1*, pages 425–430. Gauthier-Villars, 1971.
60. P. Deligne. Théorie de Hodge. II. *Inst. Hautes Études Sci. Publ. Math.*, 40:5–57, 1971.
61. P. Deligne. Travaux de Shimura. In *Séminaire Bourbaki, 23ème année (1970/71), Exp. No. 389*, pages 123–165. Lecture Notes in Math., Vol. 244. Springer-Verlag, 1971.
62. P. Deligne. Les constantes des équations fonctionnelles des fonctions L. In *Modular functions of one variable, II (Proc. Internat. Summer School, Univ. Antwerp, Antwerp, 1972)*, pages 501–597. Lecture Notes in Math., Vol. 349, 1973.
63. P. Deligne. La conjecture de Weil. I. *Inst. Hautes Études Sci. Publ. Math.*, 43:273–307, 1974.
64. P. Deligne. Théorie de Hodge. III. *Inst. Hautes Études Sci. Publ. Math.*, 44:5–77, 1974.
65. P. Deligne. Le groupe fondamental de la droite projective moins trois points. In *Galois groups over* $\mathbf{Q}$ *(Berkeley, CA, 1987)*, volume 16 of *Math. Sci. Res. Inst. Publ.*, pages 79–297. Springer, New York, 1989.
66. P. Deligne, J. S. Milne, A. Ogus, and K. Shih. *Hodge cycles, motives, and Shimura varieties*, volume 900 of *Lecture Notes in Mathematics*. Springer-Verlag, Berlin-New York, 1982.
67. P. Deligne and J.-P. Serre. Formes modulaires de poids 1. *Ann. Sci. École Norm. Sup. (4)*, 7:507–530 (1975), 1974.
68. K. Doi and H. Naganuma. On the algebraic curves uniformized by arithmetical automorphic functions. *Ann. of Math. (2)*, 86:449–460, 1967.
69. M. Duflo and J.-P. Labesse. Sur la formule des traces de Selberg. *Ann. Sci. École Norm. Sup. (4)*, 4:193–284, 1971.
70. B. Dwork. On the Artin root number. *Amer. J. Math.*, 78:444–472, 1956.
71. M. Eichler. Quaternäre quadratische Formen und die Riemannsche Vermutung für die Kongruenzzetafunktion. *Arch. Math.*, 5:355–366, 1954.

72. D. Flath. Decomposition of representations into tensor products. In *Automorphic forms, representations and L-functions (Proc. Sympos. Pure Math., Oregon State Univ., Corvallis, Ore., 1977), Part 1*, Proc. Sympos. Pure Math., XXXIII, pages 179–183. Amer. Math. Soc., Providence, R.I., 1979.
73. E. Frenkel. Gauge theory and Langlands duality. *Astérisque*, (332):Exp. No. 1010, ix–x, 369–403, 2010. Séminaire Bourbaki. Volume 2008/2009. Exposés 997–1011.
74. E. Frenkel, R. Langlands, and B. C. Ngô. Formule des traces et fonctorialité: le début d'un programme. *Ann. Sci. Math. Québec*, 34(2):199–243, 2010.
75. J. Fresán and P. Jossen. Exponential motives. *Preprint*, 2018.
76. D. Gaitsgory. Progrès récents dans la théorie de Langlands géométrique. *Astérisque*, (390):Exp. No. 1109, 139–168, 2017. Séminaire Bourbaki. Vol. 2015/2016. Exposés 1104–1119.
77. S. Gelbart and H. Jacquet. A relation between automorphic representations of GL(2) and GL(3). *Ann. Sci. École Norm. Sup. (4)*, 11(4):471–542, 1978.
78. S. Gelbart and F. Shahidi. *Analytic properties of automorphic L-functions*, volume 6 of *Perspectives in Mathematics*. Academic Press, Inc., Boston, MA, 1988.
79. I. M. Gelfand, M. I. Graev, and I. I. Pyatetskii-Shapiro. *Generalized functions. Vol. 6.* AMS Chelsea Publishing, Providence, RI, 2016. Representation theory and automorphic functions, Translated from the 1966 Russian original [MR0220673] by K. A. Hirsch, Reprint of the 1969 English translation [MR0233772].
80. S. G. Gindikin and F. I. Karpelevič. Plancherel measure for symmetric Riemannian spaces of non-positive curvature. *Dokl. Akad. Nauk SSSR*, 145:252–255, 1962.
81. R. Godement and H. Jacquet. *Zeta functions of simple algebras*. Lecture Notes in Mathematics, Vol. 260. Springer-Verlag, Berlin-New York, 1972.
82. M. Goresky, R. Kottwitz, and R. MacPherson. Discrete series characters and the Lefschetz formula for Hecke operators. *Duke Math. J.*, 89(3):477–554, 1997.
83. F. Q. Gouvêa. "A marvelous proof". *Amer. Math. Monthly*, 101(3):203–222, 1994.
84. G. Harder. *Eisensteinkohomologie und die Konstruktion gemischter Motive*, volume 1562 of *Lecture Notes in Mathematics*. Springer-Verlag, Berlin, 1993.
85. G. Harder, R. P. Langlands, and M. Rapoport. Algebraische Zyklen auf Hilbert-Blumenthal-Flächen. *J. Reine Angew. Math.*, 366:53–120, 1986.
86. Harish-Chandra. Discrete series for semisimple Lie groups. I. Construction of invariant eigendistributions. *Acta Math.*, 113:241–318, 1965.
87. Harish-Chandra. Invariant eigendistributions on a semisimple Lie group. *Trans. Amer. Math. Soc.*, 119:457–508, 1965.
88. Harish-Chandra. Discrete series for semisimple Lie group. II. Explicit determination of the characters. *Acta Math.*, 116:1–111, 1966.
89. Harish-Chandra. *Automorphic forms on semisimple Lie groups*. Notes by J. G. M. Mars. Lecture Notes in Mathematics, No. 62. Springer-Verlag, Berlin-New York, 1968.
90. Harish-Chandra. Harmonic analysis on real reductive groups. I. The theory of the constant term. *J. Functional Analysis*, 19:104–204, 1975.
91. M. Harris, N. Shepherd-Barron, and R. Taylor. A family of Calabi-Yau varieties and potential automorphy. *Ann. of Math. (2)*, 171(2):779–813, 2010.
92. M. Harris and R. Taylor. *The geometry and cohomology of some simple Shimura varieties*, volume 151 of *Annals of Mathematics Studies*. Princeton University Press, Princeton, NJ, 2001. With an appendix by Vladimir G. Berkovich.
93. H. Hasse. History of class field theory. In *Algebraic Number Theory (Proc. Instructional Conf., Brighton, 1965)*, pages 266–279. Thompson, Washington, D.C., 1967.
94. G. Henniart. Une preuve simple des conjectures de Langlands pour GL(n) sur un corps p-adique. *Invent. Math.*, 139(2):439–455, 2000.
95. R. A. Herb. Characters of averaged discrete series on semisimple real Lie groups. *Pacific J. Math.*, 80(1):169–177, 1979.
96. T. Honda. Isogeny classes of abelian varieties over finite fields. *J. Math. Soc. Japan*, 20:83–95, 1968.

97. Y. Ihara. Hecke Polynomials as congruence ζ functions in elliptic modular case. *Ann. of Math. (2)*, 85:267–295, 1967.
98. L. Illusie. *Cohomologie l-adique et fonctions L.* Lecture Notes in Mathematics, Vol. 589. Springer-Verlag, Berlin-New York, 1977. Séminaire de Géometrie Algébrique du Bois-Marie 1965–1966 (SGA 5), Edité par Luc Illusie.
99. H. Iwaniec and E. Kowalski. *Analytic number theory*, volume 53 of *American Mathematical Society Colloquium Publications*. American Mathematical Society, Providence, RI, 2004.
100. H. Jacquet. Fonctions de Whittaker associées aux groupes de Chevalley. *Bull. Soc. Math. France*, 95:243–309, 1967.
101. H. Jacquet. *Automorphic forms on* GL(2). *Part II.* Lecture Notes in Mathematics, Vol. 278. Springer-Verlag, Berlin-New York, 1972.
102. H. Jacquet. Principal L-functions of the linear group. In *Automorphic forms, representations and L-functions (Proc. Sympos. Pure Math., Oregon State Univ., Corvallis, Ore., 1977), Part 2*, Proc. Sympos. Pure Math., XXXIII, pages 63–86. Amer. Math. Soc., Providence, R.I., 1979.
103. H. Jacquet and R. P. Langlands. *Automorphic forms on* GL(2). Lecture Notes in Mathematics, Vol. 114. Springer-Verlag, Berlin-New York, 1970.
104. H. Jacquet, I. I. Piatetski-Shapiro, and J. Shalika. Automorphic forms on GL(3). II. *Ann. of Math. (2)*, 109(2):213–258, 1979.
105. H. Jacquet, I. I. Piatetski-Shapiro, and J. Shalika. Relèvement cubique non normal. *C. R. Acad. Sci. Paris Sér. I Math.*, 292(12):567–571, 1981.
106. H. Jacquet, I. I. Piatetskii-Shapiro, and J. A. Shalika. Rankin-Selberg convolutions. *Amer. J. Math.*, 105(2):367–464, 1983.
107. H. Jacquet and J. A. Shalika. On Euler products and the classification of automorphic forms. II. *Amer. J. Math.*, 103(4):777–815, 1981.
108. H. Jacquet and J. A. Shalika. On Euler products and the classification of automorphic representations. I. *Amer. J. Math.*, 103(3):499–558, 1981.
109. D. L. Johnstone. *A Gelfand–Graev Formula and Stable Transfer Factors for SL 2(F)*. ProQuest LLC, Ann Arbor, MI, 2017. Thesis (Ph.D.)–The University of Chicago.
110. A. A. Kirillov. Infinite-dimensional unitary representations of a second-order matrix group with elements in a locally compact field. *Dokl. Akad. Nauk SSSR*, 150:740–743, 1963.
111. M. Kisin. Mod p points on Shimura varieties of abelian type. *J. Amer. Math. Soc.*, 30(3):819–914, 2017.
112. S. L. Kleiman. Algebraic cycles and the Weil conjectures. In *Dix exposés sur la cohomologie des schémas*, volume 3 of *Adv. Stud. Pure Math.*, pages 359–386. North-Holland, Amsterdam, 1968.
113. S. L. Kleiman. Motives. In *Algebraic geometry, Oslo 1970 (Proc. Fifth Nordic Summer-School in Math., Oslo, 1970)*, pages 53–82, 1972.
114. A. W. Knapp and G. J. Zuckerman. Classification of irreducible tempered representations of semisimple groups. *Ann. of Math. (2)*, 116(2):389–455, 1982.
115. A. W. Knapp and G. J. Zuckerman. Classification of irreducible tempered representations of semisimple groups. II. *Ann. of Math. (2)*, 116(3):457–501, 1982.
116. M. Kontsevich and D. Zagier. Periods. In *Mathematics unlimited—2001 and beyond*, pages 771–808. Springer, 2001.
117. R. E. Kottwitz. Unstable orbital integrals on SL(3). *Duke Math. J.*, 48(3):649–664, 1981.
118. R. E. Kottwitz. Shimura varieties and twisted orbital integrals. *Math. Ann.*, 269(3):287–300, 1984.
119. R. E. Kottwitz. Stable trace formula: cuspidal tempered terms. *Duke Math. J.*, 51(3):611–650, 1984.
120. R. E. Kottwitz. Base change for unit elements of Hecke algebras. *Compositio Math.*, 60(2):237–250, 1986.
121. R. E. Kottwitz. Stable trace formula: elliptic singular terms. *Math. Ann.*, 275(3):365–399, 1986.
122. R. E. Kottwitz. Tamagawa numbers. *Ann. of Math. (2)*, 127(3):629–646, 1988.

123. R. E. Kottwitz. Shimura varieties and λ-adic representations. In *Automorphic forms, Shimura varieties, and L-functions, Vol. I (Ann Arbor, MI, 1988)*, volume 10 of *Perspect. Math.*, pages 161–209. Academic Press, Boston, MA, 1990.
124. R. E. Kottwitz. Points on some Shimura varieties over finite fields. *J. Amer. Math. Soc.*, 5(2):373–444, 1992.
125. R. E. Kottwitz and D. Shelstad. Foundations of twisted endoscopy. *Astérisque*, (255):vi+190, 1999.
126. N. V. Kuznecov. The Petersson conjecture for cusp forms of weight zero and the Linnik conjecture. Sums of Kloosterman sums. *Mat. Sb. (N.S.)*, 111(153)(3):334–383, 479, 1980.
127. J.-P. Labesse and R. P. Langlands. L-indistinguishability for $\mathrm{SL}(2)$. *Canadian J. Math.*, 31(4):726–785, 1979.
128. L. Lafforgue. Noyaux du transfert automorphe de Langlands et formules de Poisson non linéaires. *Jpn. J. Math.*, 9(1):1–68, 2014.
129. L. Lafforgue. Du transfert automorphe de Langlands aux formules de Poisson non linéaires. *Ann. Inst. Fourier (Grenoble)*, 66(3):899–1012, 2016.
130. K. F. Lai. Tamagawa number of reductive algebraic groups. *Compositio Math.*, 41(2):153–188, 1980.
131. K. Lakkis. Die Galoisschen Gauss'schen Summen von Hasse. In *Algebraische Zahlentheorie (Ber. Tagung Math. Forschungsinst. Oberwolfach, 1964)*, pages 155–158. Bibliographisches Institut, Mannheim, 1967.
132. R. P. Langlands. Letter to A. Weil. `https://publications.ias.edu/rpl/paper/43`, 1961. [Online; accessed 1-July-2021].
133. R. P. Langlands. Dimension of spaces of automorphic forms. In *Algebraic Groups and Discontinuous Subgroups (Proc. Sympos. Pure Math., Boulder, Colo., 1965)*, pages 253–257. Amer. Math. Soc., Providence, R.I., 1966.
134. R. P. Langlands. Eisenstein series. In *Algebraic Groups and Discontinuous Subgroups (Proc. Sympos. Pure Math., Boulder, Colo., 1965)*, pages 235–252. Amer. Math. Soc., Providence, R.I., 1966.
135. R. P. Langlands. The volume of the fundamental domain for some arithmetical subgroups of Chevalley groups. In *Algebraic Groups and Discontinuous Subgroups (Proc. Sympos. Pure Math., Boulder, Colo., 1965)*, pages 143–148. Amer. Math. Soc., Providence, R.I., 1966.
136. R. P. Langlands. Letter to Deligne. `http://publications.ias.edu/node/57`, 1969. [Online; accessed 1-July-2021].
137. R. P. Langlands. *On the functional equation of the Artin L-functions*. Citeseer, 1970.
138. R. P. Langlands. Problems in the theory of automorphic forms. In *Lectures in modern analysis and applications, III*, pages 18–61. Lecture Notes in Math., Vol. 170. Springer-Verlag, 1970.
139. R. P. Langlands. *Euler products*. Yale University Press, New Haven, Conn.-London, 1971. A James K. Whittemore Lecture in Mathematics given at Yale University, 1967, Yale Mathematical Monographs, 1.
140. R. P. Langlands. Modular forms and ℓ-adic representations. In *Modular functions of one variable, II (Proc. Internat. Summer School, Univ. Antwerp, Antwerp, 1972)*, pages 361–500. Lecture Notes in Math., Vol. 349, 1973.
141. R. P. Langlands. *On the functional equations satisfied by Eisenstein series*. Lecture Notes in Mathematics, Vol. 544. Springer-Verlag, Berlin-New York, 1976.
142. R. P. Langlands. Some contemporary problems with origins in the Jugendtraum. In *Mathematical developments arising from Hilbert problems (Proc. Sympos. Pure Math., Vol. XXVIII, Northern Illinois Univ., De Kalb, Ill., 1974)*, pages 401–418, 1976.
143. R. P. Langlands. Shimura varieties and the Selberg trace formula. *Canadian J. Math.*, 29(6):1292–1299, 1977.
144. R. P. Langlands. Automorphic representations, Shimura varieties, and motives. Ein Märchen. In *Automorphic forms, representations and L-functions (Proc. Sympos. Pure Math., Oregon State Univ., Corvallis, Ore., 1977), Part 2*, Proc. Sympos. Pure Math., XXXIII, pages 205–246. Amer. Math. Soc., Providence, R.I., 1979.

145. R. P. Langlands. On the notion of an automorphic representation. A supplement to the preceding paper. Automorphic forms, representations and L-functions, Proc. Symp. Pure Math. Am. Math. Soc., Corvallis/Oregon 1977, Proc. Symp. Pure Math. 33, 1, 203-207 (1979)., 1979.
146. R. P. Langlands. On the zeta functions of some simple Shimura varieties. *Canadian J. Math.*, 31(6):1121–1216, 1979.
147. R. P. Langlands. Stable conjugacy: definitions and lemmas. *Canadian J. Math.*, 31(4):700–725, 1979.
148. R. P. Langlands. Sur la mauvaise réduction d'une variété de Shimura. In *Journées de Géométrie Algébrique de Rennes. (Rennes, 1978), Vol. III*, volume 65 of *Astérisque*, pages 125–154. Soc. Math. France, Paris, 1979.
149. R. P. Langlands. *Base change for* GL(2), volume 96 of *Annals of Mathematics Studies*. Princeton University Press, Princeton, N.J.; University of Tokyo Press, Tokyo, 1980.
150. R. P. Langlands. *Les débuts d'une formule des traces stable*, volume 13 of *Publications Mathématiques de l'Université Paris VII [Mathematical Publications of the University of Paris VII]*. Université de Paris VII, U.E.R. de Mathématiques, Paris, 1983.
151. R. P. Langlands. On the classification of irreducible representations of real algebraic groups. In *Representation theory and harmonic analysis on semisimple Lie groups*, volume 31 of *Math. Surveys Monogr.*, pages 101–170. Amer. Math. Soc., Providence, RI, 1989.
152. R. P. Langlands. Comments on Shimura varieties and the Selberg trace formula. *Canadian Mathematical Society Selecta 2*, 1995.
153. R. P. Langlands. Representations of abelian algebraic groups. *Pacific J. Math.*, Special Issue:231–250, 1997. Olga Taussky-Todd: in memoriam.
154. R. P. Langlands. Endoscopy and beyond. `http://publications.ias.edu/sites/default/files/Endoscopy-and-beyond-rpl.pdf`, 2000. [Online; accessed 1-July-2021].
155. R. P. Langlands. Beyond endoscopy. In *Contributions to automorphic forms, geometry, and number theory*, pages 611–697. Johns Hopkins Univ. Press, Baltimore, MD, 2004.
156. R. P. Langlands. Un nouveau point de repère dans la théorie des formes automorphes. *Canad. Math. Bull.*, 50(2):243–267, 2007.
157. R. P. Langlands. A prologue to "Functoriality and reciprocity" Part I. *Pacific J. Math.*, 260(2):582–663, 2012.
158. R. P. Langlands. Singularités et transfert. *Ann. Math. Qué.*, 37(2):173–253, 2013.
159. R. P. Langlands. An appreciation. `http://publications.ias.edu/sites/default/files/QingZou.pdf`, 2014. [Online; accessed 1-July-2021].
160. R. P. Langlands. On the analytic form of the geometric theory of automorphic forms, 2018. In Russian.
161. R. P. Langlands. Comments on the previous text, 2019.
162. R. P. Langlands. On the analytic form of the geometric theory of automorphic forms, 2020.
163. R. P. Langlands and D. Ramakrishnan. The description of the theorem. In *The zeta functions of Picard modular surfaces*, pages 255–301. Univ. Montréal, Montreal, QC, 1992.
164. R. P. Langlands and M. Rapoport. Shimuravarietäten und Gerben. *J. Reine Angew. Math.*, 378:113–220, 1987.
165. R. P. Langlands and D. Shelstad. On the definition of transfer factors. *Math. Ann.*, 278(1-4):219–271, 1987.
166. R. P. Langlands and D. Shelstad. Descent for transfer factors. In *The Grothendieck Festschrift, Vol. II*, volume 87 of *Progr. Math.*, pages 485–563. Birkhäuser Boston, Boston, MA, 1990.
167. S. Lefschetz. On the fixed point formula. *Ann. of Math. (2)*, 38(4):819–822, 1937.
168. E. Looijenga. L^2-cohomology of locally symmetric varieties. *Compositio Math.*, 67(1):3–20, 1988.
169. G. W. Mackey. On induced representations of groups. *Amer. J. Math.*, 73:576–592, 1951.
170. B. Mazur. Deforming Galois representations. In *Galois groups over* **Q** *(Berkeley, CA, 1987)*, volume 16 of *Math. Sci. Res. Inst. Publ.*, pages 385–437. Springer, New York, 1989.

171. J. S. Milne. Points on Shimura varieties over finite fields: the conjecture of Langlands and Rapoport.
172. J. S. Milne. Points on Shimura varieties mod p. In *Automorphic forms, representations and L-functions (Proc. Sympos. Pure Math., Oregon State Univ., Corvallis, Ore., 1977), Part 2*, Proc. Sympos. Pure Math., XXXIII, pages 165–184. Amer. Math. Soc., Providence, R.I., 1979.
173. J. S. Milne. The action of an automorphism of $\mathbf{C}$ on a Shimura variety and its special points. In *Arithmetic and geometry, Vol. I*, volume 35 of *Progr. Math.*, pages 239–265. Birkhäuser Boston, Boston, MA, 1983.
174. J. S. Milne. The conjecture of Langlands and Rapoport for Siegel modular varieties. *Bull. Amer. Math. Soc. (N.S.)*, 24(2):335–341, 1991.
175. J. S. Milne. Shimura varieties and motives. In *Motives (Seattle, WA, 1991)*, volume 55 of *Proc. Sympos. Pure Math.*, pages 447–523. Amer. Math. Soc., Providence, RI, 1994.
176. J. S. Milne. Introduction to Shimura varieties. In *Harmonic analysis, the trace formula, and Shimura varieties*, volume 4 of *Clay Math. Proc.*, pages 265–378. Amer. Math. Soc., Providence, RI, 2005.
177. C. Moeglin and J.-L. Waldspurger. Le spectre résiduel de $\mathrm{GL}(n)$. *Ann. Sci. École Norm. Sup. (4)*, 22(4):605–674, 1989.
178. C. Moeglin and J.-L. Waldspurger. *Spectral decomposition and Eisenstein series*, volume 113 of *Cambridge Tracts in Mathematics*. Cambridge University Press, Cambridge, 1995. Une paraphrase de l'Écriture.
179. C. Moeglin and J.-L. Waldspurger. *Stabilisation de la formule des traces tordue. Vol. 1*, volume 316 of *Progress in Mathematics*. Birkhäuser/Springer, Cham, 2016.
180. C. Moeglin and J.-L. Waldspurger. *Stabilisation de la formule des traces tordue. Vol. 2*, volume 317 of *Progress in Mathematics*. Birkhäuser/Springer, Cham, 2016.
181. C. P. Mok. Endoscopic classification of representations of quasi-split unitary groups. *Mem. Amer. Math. Soc.*, 235(1108):vi+248, 2015.
182. S. Morel. Complexes pondérés sur les compactifications de Baily-Borel: le cas des variétés de Siegel. *J. Amer. Math. Soc.*, 21(1):23–61, 2008.
183. S. Morel. Cohomologie d'intersection des variétés modulaires de Siegel, suite. *Compos. Math.*, 147(6):1671–1740, 2011.
184. M. R. Murty. Fermat's last theorem: an outline. *Gazette Sc. Math. Québec*, 16:4–13, 1993.
185. J. Neukirch. The Beĭlinson conjecture for algebraic number fields. In *Beĭlinson's conjectures on special values of L-functions*, volume 4 of *Perspect. Math.*, pages 193–247. Academic Press, Boston, MA, 1988.
186. B. C. Ngô. Le lemme fondamental pour les algèbres de Lie. *Publ. Math. Inst. Hautes Études Sci.*, 111:1–169, 2010.
187. B. C. Ngô. Hankel transform, Langlands functoriality and functional equation of automorphic L-functions. *Jpn. J. Math.*, 15(1):121–167, 2020.
188. H. Ooguri. Interview with Edward Witten. *Notices Amer. Math. Soc.*, 62(5):491–506, 2015.
189. I. I. Piatetski-Shapiro. Multiplicity one theorems. In *Automorphic forms, representations and L-functions (Proc. Sympos. Pure Math., Oregon State Univ., Corvallis, Ore., 1977), Part 1*, Proc. Sympos. Pure Math., XXXIII, pages 209–212. Amer. Math. Soc., Providence, R.I., 1979.
190. M. Rapoport and E. Viehmann. Towards a theory of local Shimura varieties. *Münster J. Math.*, 7(1):273–326, 2014.
191. H. Reimann and T. Zink. Der Dieudonnémodul einer polarisierten abelschen Mannigfaltigkeit vom CM-Typ. *Ann. of Math. (2)*, 128(3):461–482, 1988.
192. K. A. Ribet. On modular representations of $\mathrm{Gal}(\overline{\mathbf{Q}}/\mathbf{Q})$ arising from modular forms. *Invent. Math.*, 100(2):431–476, 1990.
193. H. Saito. Automorphic forms and algebraic extensions of number fields. *Proc. Japan Acad.*, 51(4):229–233, 1975.
194. Y. Sakellaridis. Beyond endoscopy for the relative trace formula I: Local theory. In *Automorphic representations and L-functions*, volume 22 of *Tata Inst. Fundam. Res. Stud. Math.*, pages 521–590. Tata Inst. Fund. Res., Mumbai, 2013.

195. Y. Sakellaridis. Beyond endoscopy for the relative trace formula II: Global theory. *J. Inst. Math. Jussieu*, 18(2):347–447, 2019.
196. Y. Sakellaridis and A. Venkatesh. Periods and harmonic analysis on spherical varieties. *Astérisque*, (396):viii+360, 2017.
197. P. J. Sally, Jr. and J. A. Shalika. Characters of the discrete series of representations of SL(2) over a local field. *Proc. Nat. Acad. Sci. U.S.A.*, 61:1231–1237, 1968.
198. L. Saper and M. Stern. L_2-cohomology of arithmetic varieties. *Ann. of Math. (2)*, 132(1):1–69, 1990.
199. P. Sarnak. Comments on Robert Langland's Lecture: "Endoscopy and Beyond", 2001.
200. G. Schiffmann. Introduction aux travaux d'Harish-Chandra. In *Séminaire Bourbaki, Vol. 10*, pages Exp. No. 323, 153–177. Soc. Math. France, Paris, 1995.
201. W. Schmid. On a conjecture of Langlands. *Ann. of Math. (2)*, 93:1–42, 1971.
202. P. Scholze. The local Langlands correspondence for GL_n over p-adic fields. *Invent. Math.*, 192(3):663–715, 2013.
203. L. Schwartz. *Théorie des distributions. Tome I.* Actualités Sci. Ind., no. 1091 = Publ. Inst. Math. Univ. Strasbourg 9. Hermann & Cie., Paris, 1950.
204. L. Schwartz. *Théorie des distributions. Tome II.* Actualités Sci. Ind., no. 1122 = Publ. Inst. Math. Univ. Strasbourg **10**. Hermann & Cie., Paris, 1951.
205. A. Selberg. Harmonic analysis and discontinuous groups in weakly symmetric Riemannian spaces with applications to Dirichlet series. *J. Indian Math. Soc. (N.S.)*, 20:47–87, 1956.
206. A. Selberg. On discontinuous groups in higher-dimensional symmetric spaces. In *Contributions to function theory (internat. Colloq. Function Theory, Bombay, 1960)*, pages 147–164. Tata Institute of Fundamental Research, Bombay, 1960.
207. A. Selberg. Discontinuous groups and harmonic analysis. In *Proc. Internat. Congr. Mathematicians (Stockholm, 1962)*, pages 177–189. Inst. Mittag-Leffler, Djursholm, 1963.
208. J.-P. Serre. *Abelian l-adic representations and elliptic curves.* McGill University lecture notes written with the collaboration of Willem Kuyk and John Labute. W. A. Benjamin, Inc., New York-Amsterdam, 1968.
209. J.-P. Serre. Motifs. *Astérisque*, (198-200):11, 333–349 (1992), 1991. Journées Arithmétiques, 1989 (Luminy, 1989).
210. J.-P. Serre. Propriétés conjecturales des groupes de Galois motiviques et des représentations l-adiques. In *Motives (Seattle, WA, 1991)*, volume 55 of *Proc. Sympos. Pure Math.*, pages 377–400. Amer. Math. Soc., Providence, RI, 1994.
211. F. Shahidi. On certain L-functions. *Amer. J. Math.*, 103(2):297–355, 1981.
212. F. Shahidi. On the Ramanujan conjecture and finiteness of poles for certain L-functions. *Ann. of Math. (2)*, 127(3):547–584, 1988.
213. F. Shahidi. A proof of Langlands' conjecture on Plancherel measures; complementary series for p-adic groups. *Ann. of Math. (2)*, 132(2):273–330, 1990.
214. F. Shahidi. Automorphic L-functions and functoriality. In *Proceedings of the International Congress of Mathematicians, Vol. II (Beijing, 2002)*, pages 655–666. Higher Ed. Press, Beijing, 2002.
215. F. Shahidi. *Eisenstein series and automorphic L-functions*, volume 58 of *American Mathematical Society Colloquium Publications*. American Mathematical Society, Providence, RI, 2010.
216. J. A. Shalika. A theorem on semi-simple $\mathscr{P}$-adic groups. *Ann. of Math. (2)*, 95:226–242, 1972.
217. D. Shelstad. Characters and inner forms of a quasi-split group over **R**. *Compositio Math.*, 39(1):11–45, 1979.
218. D. Shelstad. Notes on L-indistinguishability (based on a lecture of R. P. Langlands). In *Automorphic forms, representations and L-functions (Proc. Sympos. Pure Math., Oregon State Univ., Corvallis, Ore., 1977), Part 2*, Proc. Sympos. Pure Math., XXXIII, pages 193–203. Amer. Math. Soc., Providence, R.I., 1979.
219. D. Shelstad. Orbital integrals and a family of groups attached to a real reductive group. *Ann. Sci. École Norm. Sup. (4)*, 12(1):1–31, 1979.

220. D. Shelstad. Embeddings of L-groups. *Canadian J. Math.*, 33(3):513–558, 1981.
221. D. Shelstad. L-indistinguishability for real groups. *Math. Ann.*, 259(3):385–430, 1982.
222. D. Shelstad. Tempered endoscopy for real groups. I. Geometric transfer with canonical factors. In *Representation theory of real reductive Lie groups*, volume 472 of *Contemp. Math.*, pages 215–246. Amer. Math. Soc., Providence, RI, 2008.
223. D. Shelstad. Tempered endoscopy for real groups. III. Inversion of transfer and L-packet structure. *Represent. Theory*, 12:369–402, 2008.
224. D. Shelstad. Tempered endoscopy for real groups. II. Spectral transfer factors. In *Automorphic forms and the Langlands program*, volume 9 of *Adv. Lect. Math. (ALM)*, pages 236–276. Int. Press, Somerville, MA, 2010.
225. H. Shimizu. On discontinuous groups operating on the product of the upper half planes. *Ann. of Math. (2)*, 77:33–71, 1963.
226. H. Shimizu. On traces of Hecke operators. *J. Fac. Sci. Univ. Tokyo Sect. I*, 10:1–19 (1963), 1963.
227. H. Shimizu. On zeta functions of quaternion algebras. *Ann. of Math. (2)*, 81:166–193, 1965.
228. G. Shimura. Correspondances modulaires et les fonctions ζ de courbes algébriques. *J. Math. Soc. Japan*, 10:1–28, 1958.
229. G. Shimura. Moduli of abelian varieties and number theory. In *Algebraic Groups and Discontinuous Subgroups (Proc. Sympos. Pure Math., Boulder, Colo., 1965)*, pages 312–332. Amer. Math. Soc., Providence, R.I., 1966.
230. G. Shimura. *Introduction to the arithmetic theory of automorphic functions*. Publications of the Mathematical Society of Japan, No. 11. Iwanami Shoten, Publishers, Tokyo; Princeton University Press, Princeton, N.J., 1971. Kanô Memorial Lectures, No. 1.
231. T. Shintani. On liftings of holomorphic cusp forms. In *Automorphic forms, representations and L-functions (Proc. Sympos. Pure Math., Oregon State Univ., Corvallis, Ore., 1977), Part 2*, Proc. Sympos. Pure Math., XXXIII, pages 97–110. Amer. Math. Soc., Providence, R.I., 1979.
232. T. A. Springer. Galois cohomology of linear algebraic groups. In *Algebraic Groups and Discontinuous Subgroups (Proc. Sympos. Pure Math., Boulder, Colo., 1965)*, pages 149–158. Amer. Math. Soc., Providence, R.I., 1966.
233. T. A. Springer. Reductive groups. In *Automorphic forms, representations and L-functions (Proc. Sympos. Pure Math., Oregon State Univ., Corvallis, Ore., 1977), Part 1*, Proc. Sympos. Pure Math., XXXIII, pages 3–27. Amer. Math. Soc., Providence, R.I., 1979.
234. J. Tate. Duality theorems in Galois cohomology over number fields. In *Proc. Internat. Congr. Mathematicians (Stockholm, 1962)*, pages 288–295. Inst. Mittag-Leffler, Djursholm, 1963.
235. J. Tate. Endomorphisms of abelian varieties over finite fields. *Invent. Math.*, 2:134–144, 1966.
236. J. Tate. Fourier analysis in number fields, and Hecke's zeta-functions. In *Algebraic Number Theory (Proc. Instructional Conf., Brighton, 1965)*, pages 305–347. Thompson, Washington, D.C., 1967.
237. J. Tate. Number theoretic background. In *Automorphic forms, representations and L-functions (Proc. Sympos. Pure Math., Oregon State Univ., Corvallis, Ore., 1977), Part 2*, Proc. Sympos. Pure Math., XXXIII, pages 3–26. Amer. Math. Soc., Providence, R.I., 1979.
238. R. Taylor. Automorphy for some l-adic lifts of automorphic mod l Galois representations. II. *Publ. Math. Inst. Hautes Études Sci.*, 108:183–239, 2008.
239. R. Taylor and A. Wiles. Ring-theoretic properties of certain Hecke algebras. *Ann. of Math. (2)*, 141(3):553–572, 1995.
240. J. Tunnell. Artin's conjecture for representations of octahedral type. *Bull. Amer. Math. Soc. (N.S.)*, 5(2):173–175, 1981.
241. A. Venkatesh. *Limiting forms of the trace formula*. ProQuest LLC, Ann Arbor, MI, 2002. Thesis (Ph.D.) Princeton University.
242. A. Venkatesh. "Beyond endoscopy" and special forms on GL(2). *J. Reine Angew. Math.*, 577:23–80, 2004.

243. D. A. Vogan, Jr. The local Langlands conjecture. In *Representation theory of groups and algebras*, volume 145 of *Contemp. Math.*, pages 305–379. Amer. Math. Soc., Providence, RI, 1993.
244. D. A. Vogan, Jr. and G. J. Zuckerman. Unitary representations with nonzero cohomology. *Compositio Math.*, 53(1):51–90, 1984.
245. J.-L. Waldspurger. Le lemme fondamental implique le transfert. *Compositio Math.*, 105(2):153–236, 1997.
246. J.-L. Waldspurger. Endoscopie et changement de caractéristique. *J. Inst. Math. Jussieu*, 5(3):423–525, 2006.
247. J.-L. Waldspurger. Endoscopie et changement de caractéristique: intégrales orbitales pondérées. *Ann. Inst. Fourier (Grenoble)*, 59(5):1753–1818, 2009.
248. A. Weil. Numbers of solutions of equations in finite fields. *Bull. Amer. Math. Soc.*, 55:497–508, 1949.
249. A. Weil. Sur certains groupes d'opérateurs unitaires. *Acta Math.*, 111:143–211, 1964.
250. A. Weil. Über die Bestimmung Dirichletscher Reihen durch Funktionalgleichungen. *Math. Ann.*, 168:149–156, 1967.
251. A. Weil. *Adeles and algebraic groups*, volume 23 of *Progress in Mathematics*. Birkhäuser, Boston, Mass., 1982. With appendices by M. Demazure and Takashi Ono.
252. H. Weyl. David Hilbert and his mathematical work. *Bull. Amer. Math. Soc.*, 50:612–654, 1944.
253. A. Wiles. Modular elliptic curves and Fermat's last theorem. *Ann. of Math. (2)*, 141(3):443–551, 1995.
254. Z. Yun. Orbital integrals and Dedekind zeta functions. In *The legacy of Srinivasa Ramanujan*, volume 20 of *Ramanujan Math. Soc. Lect. Notes Ser.*, pages 399–420. Ramanujan Math. Soc., Mysore, 2013.
255. D. Zagier. Modular forms whose Fourier coefficients involve zeta-functions of quadratic fields. In *Modular functions of one variable, VI (Proc. Second Internat. Conf., Univ. Bonn, Bonn, 1976)*, pages 105–169. Lecture Notes in Math., Vol. 627, 1977.
256. T. Zink. Isogenieklassen von Punkten von Shimuramannigfaltigkeiten mit Werten in einem endlichen Körper. *Math. Nachr.*, 112:103–124, 1983.
257. S. Zucker. L_2 cohomology of warped products and arithmetic groups. *Invent. Math.*, 70(2):169–218, 1982/83.

List of Publications for Robert P. Langlands

1960

[1] Some holomorphic semi-groups. *Proc. Nat. Acad. Sci. U.S.A.*, 46:361–363.
[2] On Lie semi-groups. *Canad. J. Math.*, 12:686–693.

1961

[3] Dirichlet series associated with quadratic forms. Preprint, Princeton University.

1963

[4] The dimension of spaces of automorphic forms. *Amer. J. Math.*, 85:99–125.

1966

[5] The volume of the fundamental domain for some arithmetical subgroups of Chevalley groups. In *Algebraic Groups and Discontinuous Subgroups (Proc. Sympos. Pure Math., Boulder, Colo., 1965)*, pages 143–148. Amer. Math. Soc., Providence, RI.
[6] Eisenstein series. In *Algebraic Groups and Discontinuous Subgroups (Proc. Sympos. Pure Math., Boulder, Colo., 1965)*, pages 235–252. Amer. Math. Soc., Providence, RI.
[7] Dimension of spaces of automorphic forms. In *Algebraic Groups and Discontinuous Subgroups (Proc. Sympos. Pure Math., Boulder, Colo., 1965)*, pages 253–257. Amer. Math. Soc., Providence, RI.

1970

[8] Problems in the theory of automorphic forms. pages 18–61. Lecture Notes in Math., Vol. 170. Also available in *Matematika*, Moscow, 15(2):57–83, 1971 (in Russian).
[9] On the functional equation of the Artin L-functions. Preprint, Yale University.
[10] (with H. Jacquet). *Automorphic forms on* $\mathrm{GL}(2)$. Lecture Notes in Mathematics, Vol. 114. Springer-Verlag, Berlin-New York. Also available in Russian.

H. Holden, R. Piene (eds.), *The Abel Prize 2018-2022*, The Abel Prize,
https://doi.org/10.1007/978-3-031-33973-8_3

1971

[11] On Artin's *L*-functions. *Rice Univ. Stud.*, 56(2):23–28.
[12] Automorphic forms on GL(2). In *Proc. Int. Congr. Math. (NIce), 1970, Vol. 2*, pages 327–329. Gauthier-Villars, Paris.
[13] *Euler products*, volume 1 of *Yale Mathematical Monographs*. Yale University Press, New Haven, Conn.-London. Also available in *Matematika, Moscow*, 15(1):14–43, 1971 (in Russian).

1973

[14] Modular forms and ℓ-adic representations. In *Modular functions of one variable, II (Proc. Internat. Summer School, Univ. Antwerp, Antwerp, 1972)*, pages 361–500. Lecture Notes in Math., Vol. 349.

1976

[15] Some contemporary problems with origins in the Jugendtraum. In *Mathematical developments arising from Hilbert problems (Proc. Sympos. Pure Math., Vol. XXVIII, Northern Illinois Univ., De Kalb, Ill., 1974)*, pages 401–418. Amer. Math. Soc., Providence, RI.
[16] *On the functional equations satisfied by Eisenstein series*, volume 544 of *Lecture Notes in Mathematics*. Springer-Verlag, Berlin-New York.
[17] Review of $SL_2(\mathbb{R})$ by S. Lang. *Bull. Amer. Soc.*, 82(5):688–691.

1977

[18] Shimura varieties and the Selberg trace formula. *Canad. J. Math.*, 29(6):1292–1299. Also available in *Can. Math. Soc. 1945–1995: Selecta*, Vol. 2, pages 265–295. Society, Ottawa, ON, 1996.

1979

[19] Stable conjugacy: definitions and lemmas. *Canad. J. Math.*, 31(4):700–725.
[20] (with J.-P. Labesse). *L*-indistinguishability for SL(2). *Canad. J. Math.*, 31(4):726–785.
[21] On the notion of an automorphic representation. Supplement in: A. Borel and H. Jacquet. Automorphic forms and authomorphic representations. In *Automorphic forms, representations and L-functions (Proc. Sympos. Pure Math., Oregon State Univ., Corvallis, Ore., 1977), Part 1*, Proc. Sympos. Pure Math., XXXIII, pages 189–207. Amer. Math. Soc., Providence, RI.
[22] Automorphic representations, Shimura varieties, and motives. Ein Märchen. In *Automorphic forms, representations and L-functions (Proc. Sympos. Pure Math., Oregon State Univ., Corvallis, Ore., 1977), Part 2*, Proc. Sympos. Pure Math., XXXIII, pages 205–246. Amer. Math. Soc., Providence, RI.
[23] On the zeta functions of some simple Shimura varieties. *Canad. J. Math.*, 31(6):1121–1216.

[24] Sur la mauvaise réduction d'une variété de Shimura. In *Journées de Géométrie Algébrique de Rennes. (Rennes, 1978), Vol. III*, volume 65 of *Astérisque*, pages 125–154. Soc. Math. France, Paris.

1980

[25] *L*-functions and automorphic representations. In *Proceedings of the International Congress of Mathematicians (Helsinki, 1978)*, pages 165–175, Helsinki. Acad. Sci. Fennica.

[26] *Base change for* GL(2), volume 96 of *Annals of Mathematics Studies*. Princeton University Press, Princeton, N.J.; University of Tokyo Press, Tokyo.

1983

[27] *Les débuts d'une formule des traces stable*, volume 13 of *Publications Mathématiques de l'Université Paris VII*. Université de Paris VII, U.E.R. de Mathématiques, Paris.

[28] Orbital integrals on forms of SL(3). I. *Amer. J. Math.*, 105(2):465–506.

[29] Review of *The theory of Eisenstein Systemts* by M.S. Osborne and G. Warner. *Bull. Amer. Soc.*, 9(3):351–361.

1984

[30] (with D. Shelstad). On principal values on *p*-adic manifolds. In *Lie group representations, II (College Park, Md., 1982/1983)*, volume 1041 of *Lecture Notes in Math.*, pages 250–279. Springer, Berlin.

[31] Speech printed in *Harish-Chandra, 1923–1983*. Pamphlet of the Conference on Harmonic Analysis and the Representation Theory of Reductive Groups, April 23–27, 1984, pages 19–23. Institute for Advanced Study, Princeton. Available as `https://www.ias.edu/sites/default/files/library/Harish-Chandra_1923-1983.pdf`.

1985

[32] Review of *Harish-Chandra, Collected Papers*. Bull. London Math. Soc., 17(2):175–204.

[33] Harish-Chandra. *Biographical Memoirs of Fellows of the Royal Society*, 31:198–225.

[34] Pure mathematics. *The Canadian Encyclopedia*. Edmonton, AB.

1986

[35] (with G. Harder and M. Rapoport). Algebraische Zyklen auf Hilbert–Blumenthal–Flächen. *J. Reine Angew. Math.*, 366:53–120.

1987

[36] The Dirac monopole and induced representations. *Pacific J. Math.*, 126(1): 145–151.

[37] (with M. Rapoport). Shimuravarietäten und Gerben. *J. Reine Angew. Math.*, 378:113–220.

[38] (with D. Shelstad). On the definition of transfer factors. *Math. Ann.*, 278 (1-4):219–271.

1988

[39] Representation theory and arithmetic. In *The mathematical heritage of Hermann Weyl (Durham, NC, 1987)*, volume 48 of *Proc. Sympos. Pure Math.*, pages 25–33. Amer. Math. Soc., Providence, RI.

[40] On unitary representations of the Virasoro algebra. In *Infinite-dimensional Lie algebras and their applications (Montreal, PQ, 1986)*, pages 141–159. World Sci. Publ., Teaneck, NJ.

1989

[41] Eisenstein series, the trace formula, and the modern theory of automorphic forms. In *Number theory, trace formulas and discrete groups (Oslo, 1987)*, pages 125–155. Academic Press, Boston, MA.

[42] On the classification of irreducible representations of real algebraic groups. In *Representation theory and harmonic analysis on semisimple Lie groups*, volume 31 of *Math. Surveys Monogr.*, pages 101–170. Amer. Math. Soc., Providence, RI. Previously available as preprint, Institute for Advanced Study, 1973.

[43] The factorization of a polynomial defined by partitions. *Comm. Math. Phys.*, 124(2):261–284.

[44] (with D. Shelstad). Orbital integrals on forms of SL(3). II. *Canad. J. Math.*, 41(3):480–507.

1990

[45] Harish-Chandra (1923–1983). In *Some eminent Indian mathematicians of the twentieth century, Vol. III*, pages 45–56. Math. Sci. Trust Soc., New Delhi.

[46] Representation theory: its rise and its role in number theory. In *Proceedings of the Gibbs Symposium (New Haven, CT, 1989)*, pages 181–210, Providence, RI. Amer. Math. Soc.

[47] (with D. Shelstad). Descent for transfer factors. In *The Grothendieck Festschrift, Vol. II*, volume 87 of *Progr. Math.*, pages 485–563. Birkhäuser Boston, Boston, MA.

[48] Rank-one residues of Eisenstein series. In *Festschrift in honor of I.I. Piatetski-Shapiro on the occasion of his sixtieth birthday, Part II (Ramat Aviv, 1989)*, volume 3 of *Israel Math. Conf. Proc.*, pages 111–125. Weizmann, Jerusalem.

1992

[49] (co-edited with D. Ramakrishnan). *The zeta functions of Picard modular surfaces*, Univ. Montréal, Montreal, QC.

[50] (with D. Ramakrishnan). The description of the theorem. In *The zeta functions of Picard modular surfaces*, pages 255–301. Univ. Montréal, Montreal, QC.

[51] Remarks on Igusa theory and real orbital integrals. In *The zeta functions of Picard modular surfaces*, pages 335–347. Univ. Montréal, Montreal, QC.

[52] (with C. Pichet, Ph. Pouliot, and Y. Saint-Aubin). On the universality of crossing probabilities in two-dimensional percolation. *J. Statist. Phys.*, 67 (3-4):553–574.

1993

[53] Dualität bei endlichen Modellen der Perkolation. *Math. Nachr.*, 160:7–58.

[54] Harish-Chandra (11 October 1923–16 October 1983). *Current Sci.*, 65(12):922–936.

1994

[55] (with Ph. Pouliot and Y. Saint-Aubin). Conformal invariance in two-dimensional percolation. *Bull. Amer. Math. Soc. (N.S.)*, 30(1):1–61.

[56] (with M.-A. Lafortune). Finite models for percolation. In *Representation theory and analysis on homogeneous spaces (New Brunswick, NJ, 1993)*, volume 177 of *Contemp. Math.*, pages 227–246. Amer. Math. Soc., Providence, RI.

[57] Review of *Elliptic curves* by A.W. Knapp. *Bull. Amer. Soc.*, 30(1):96–100.

1995

[58] (with Y. Saint-Aubin). Algebro-geometric aspects of the Bethe equations. In *Strings and symmetries (Istanbul, 1994)*, volume 447 of *Lecture Notes in Phys.*, pages 40–53. Springer, Berlin.

1996

[59] (with F.R.K. Chung). A combinatorial Laplacian with vertex weights. *J. Combin. Theory Ser. A*, 75(2):316–327.

[60] An essay on the dynamics and statistics of critical field theories. In *Canadian Mathematical Society. 1945–1995, Vol. 3*, pages 173–209. Canadian Math. Soc., Ottawa, ON.

1997

[61] Where stands functoriality today? In *Representation theory and automorphic forms (Edinburgh, 1996)*, volume 61 of *Proc. Sympos. Pure Math.*, pages 457–471. Amer. Math. Soc., Providence, RI.

[62] (with Y. Saint-Aubin). Aspects combinatoires des équations de Bethe. In *Advances in mathematical sciences: CRM's 25 years (Montreal, PQ, 1994)*, volume 11 of *CRM Proc. Lecture Notes*, pages 231–301. Amer. Math. Soc., Providence, RI.

[63] Representations of abelian algebraic groups. *Pacific J. Math.*, 181(3):231–250. Previously appeared as preprint, Yale University, 1968.

1998

[64] André Weil (1906–98). *Nature*, 395:848.

2000

[65] (with M.-A. Lewis and Y. Saint-Aubin). Universality and conformal invariance for the Ising model in domains with boundary. *J. Statist. Phys.*, 98(1-2):131–244.

[66] Harish-Chandra memorial talk. In *The mathematical legacy of Harish-Chandra (Baltimore, MD, 1998)*, volume 68 of *Proc. Sympos. Pure Math.*, pages 47–49. Amer. Math. Soc., Providence, RI.

2001

[67] The trace formula and its applications: an introduction to the work of James Arthur. *Canad. Math. Bull.*, 44(2):160–209.

2002

[68] Euclid's windows and our mirrors. *Notices Amer. Math. Soc.*, 49(5):554–565.

2003

[69] Benim tamıdığım Cahit Arf, *Matematik Dünyası*, 56(Winter Issue):61–63.

2004

[70] Beyond endoscopy. In *Contributions to automorphic forms, geometry, and number theory*, pages 611–697. Johns Hopkins Univ. Press, Baltimore, MD.

2005

[71] The renormalization fixed point as a mathematical object. In *Twenty years of Bialowieza: a mathematical anthology*, volume 8 of *World Sci. Monogr. Ser. Math.*, pages 185–216. World Sci. Publ., Hackensack, NJ.

[72] Descartes ile Fermat. *Matematik Dünyası*, 2:54–61.

2007

[73] Un nouveau point de repère dans la théorie des formes automorphes. *Canad. Math. Bull.*, 50(2):243–267.

[74] Review of *p-adic automorphic forms on Shimura varieties* by H. Hida. *Bull. Amer. Soc.*, 44(2):291–308.

[75] Le programme de Langlands. *Pour la Science*, 361:122–128.

2010

[76] (with E. Frenkel and B.C. Ngô). Formule des traces et fonctorialité: le début d'un programme. *Ann. Sci. Math. Québec*, 34(2):199–243.

2011

[77] Reflexions on receiving the Shaw Prize. In *On certain L-functions*, volume 13 of *Clay Math. Proc.*, pages 297–308. Amer. Math. Soc., Providence, RI.

2012

[78] A prologue to "Functoriality and reciprocity" Part I. *Pacific J. Math.*, 260(2): 582–663.

2013

[79] Singularités et transfert. *Ann. Math. Qué.*, 37(2):173–253.
[80] J.-P. Labesse and J.-L. Waldspurger, *La formule des traces tordue d'après le Friday Morning Seminar*, CRM Monograph Series, vol. 31. Amer. Math. Soc., Providence, RI. With a foreword by R. Langlands.
[81] Is there beauty in mathematical theories? In *The many faces of beauty*, pages 23–78. University of Notre Dame Press, Notre Dame, IN.

2014

[82] Z. Qing, *Mathematical conjecture and issues affecting the world from Kummer to Langlands: Langlands conjecture history* (in Chinese). Harbin Institute of Technology Press. With "An appreciation" by R. Langlands.

2015

[83] Automorphic L-functions. In *Emil Artin and Beyond — Class Field Theory and L-Functions*, Heritage of European Mathematics, pages 163–209. Eur. Math. Soc., Zürich.
[84] *Robert Langlands'tan Türk Okurlarına Mektup*. A letter to Turkish readers that appeared at the beginning of the Turkish translation of Edward Frenkel's *Love and Math*. Paloma Yayınevi, Istanbul.

2018

[85] L. Ji, *Langlands's program and his mathematical world* (in Chinese). With a preface (in English) by R. Langlands. Higher Education Press, Beijing.

Curriculum Vitae for Robert Phelan Langlands

Born:	October 6, 1936 in New Westminister, British Columbia, Canada
Degrees/education:	Bachelor of Science, University of British Columbia, 1957 Master of Science, University of British Columbia, 1958 PhD, Yale University, 1960
Positions:	Instructor, Princeton University, 1960–1961 Lecturer, Princeton University, 1961–1962 Associate Professor, Princeton University, 1962–1967 Professor, Yale University, 1967–1972 Hermann Weyl Professor, Institute for Advanced Study, 1972–2007 Professor Emeritus, Institute for Advanced Study, 2007–

H. Holden, R. Piene (eds.), *The Abel Prize 2018-2022*, The Abel Prize,
https://doi.org/10.1007/978-3-031-33973-8_4

Visiting positions: Miller Research Fellow, University of California, Berkeley, 1964–1965
Middle East Technical University, 1967–1968

Memberships: Royal Society of Canada, 1972
Royal Society of London, 1981
American Academy of Arts and Sciences, 1990
National Academy of Sciences, 1993
American Mathematical Society, Fellow, 2013
Russian Academy of Sciences, Foreign Member, 2011
Bilim Akademisi, Turkey, Honorary Member, 2013
London Mathematical Society, Honorary Member, 2015
Norwegian Academy of Science and Letters, 2018

Awards and prizes: Speaker at the International Congress of Mathematicians, 1970, 1978 (plenary)
Wilbur L. Cross Medal, Yale University, 1975
Jeffery–Williams Prize, Canadian Mathematical Society, 1980
Cole Prize in Number Theory, American Mathematical Society, 1982
Common Wealth Award, 1984
National Academy of Sciences, Award in Mathematics, 1988
Wolf Prize, 1995
Grande Médaille de l'Académie des sciences, 2000
Leroy P. Steele Prize for Seminal Contribution to Research, 2005
Nemmers Prize, 2006
Shaw Prize, 2007
Abel Prize, 2018
Companion of the Order of Canada, 2019

Honorary degrees: University of British Columbia, 1985
McMaster University, 1985
City University of New York, Graduate Center, 1985
University of Waterloo, 1988
University of Paris, VII, 1989
McGill University, 1991
University of Toronto, 1993
Université de Montréal, 1997
Université Laval, 2002
University of Madras, 2005
University of Chicago, 2011

Part II
2019 Karen K. Uhlenbeck

"for her pioneering achievements in geometric partial differential equations, gauge theory and integrable systems, and for the fundamental impact of her work on analysis, geometry and mathematical physics"

THE
ABEL
PRIZE

Citation

The Norwegian Academy of Science and Letters has decided to award the Abel Prize for 2019 to **Karen Keskulla Uhlenbeck**, University of Texas at Austin,

> *"for her pioneering achievements in geometric partial differential equations, gauge theory and integrable systems, and for the fundamental impact of her work on analysis, geometry and mathematical physics."*

Karen Keskulla Uhlenbeck is a founder of modern Geometric Analysis. Her perspective has permeated the field and led to some of the most dramatic advances in mathematics in the last 40 years.

Geometric analysis is a field of mathematics where techniques of analysis and differential equations are weaved with the study of geometrical and topological problems. Specifically, one studies objects such as curves, surfaces, connections and fields which are critical points of functionals representing geometric quantities such as energy and volume. For example, minimal surfaces are critical points of the area and harmonic maps are critical points of the Dirichlet energy. Uhlenbeck's major contributions include foundational results on minimal surfaces and harmonic maps, Yang–Mills theory, and integrable systems.

Minimal surfaces and bubbling analysis

An important tool in global analysis, preceding the work of Uhlenbeck, is the Palais–Smale compactness condition. This condition, inspired by earlier work of Morse, guarantees the existence of minimisers of geometric functionals and is successful in the case of 1-dimensional domains, such as closed geodesics.

Uhlenbeck realised that the condition of Palais–Smale fails in the case of surfaces due to topological reasons. The papers of Uhlenbeck, co-authored with Sacks, on the energy functional for maps of surfaces into a Riemannian manifold, have been extremely influential and describe in detail what happens when the Palais–Smale condition is violated. A minimising sequence of mappings converges outside a finite set of singular points and by using rescaling arguments, they describe the behaviour near the singularities as bubbles or instantons, which are the standard solutions of the minimising map from the 2-sphere to the target manifold.

In higher dimensions, Uhlenbeck in collaboration with Schoen wrote two foundational papers on minimising harmonic maps. They gave a profound understanding of singularities of solutions of non-linear elliptic partial differential equations. The singular set, which in the case of surfaces consists only of isolated points, is in higher dimensions replaced by a set of codimension 3.

The methods used in these revolutionary papers are now in the standard toolbox of every geometer and analyst. They have been applied with great success in many other partial differential equations and geometric contexts. In particular, the bubbling phenomenon appears in many works in partial differential equations, in the study of the Yamabe problem, in Gromov's work on pseudoholomorphic curves, and also in physical applications of instantons, especially in string theory.

Gauge theory and Yang–Mills equations

After hearing a talk by Atiyah in Chicago, Uhlenbeck became interested in gauge theory. She pioneered the study of Yang–Mills equations from a rigorous analytical point of view. Her work formed a base of all subsequent research in the area of gauge theory.

Gauge theory involves an auxiliary vector bundle over a Riemannian manifold. The basic objects of study are connections on this vector bundle. After a choice of a trivialisation (gauge), a connection can be described by a matrix-valued 1-form. Yang–Mills connections are critical points of gauge-invariant functionals. Uhlenbeck addressed and solved the fundamental question of expressing Yang–Mills equations as an elliptic system, using the so-called Coulomb gauge. This was the starting point for both Uhlenbeck's celebrated compactness theorem for connections with curvature bounded in L^p, and for her later results on removable singularities for Yang–Mills equations defined on punctured 4-dimensional balls. The removable singularity theory for Yang–Mills equations in higher dimensions was carried out much later by Gang Tian and Terence Tao. Uhlenbeck's compactness theorem was crucial in Non-Abelian Hodge Theory and, in particular, in the proof of the properness of Hitchin's map and Corlette's important result on the existence of equivariant harmonic mappings.

Another major result of Uhlenbeck is her joint work with Yau on the existence of Hermitian Yang–Mills connections on stable holomorphic vector bundles over complex n-manifolds, generalising an earlier result of Donaldson on complex surfaces. This result of Donaldson–Uhlenbeck–Yau links developments in differential geometry and algebraic geometry, and is a foundational result for applications of heterotic strings to particle physics.

Uhlenbeck's ideas laid the analytic foundations for the application of gauge theory to geometry and topology, to the important work of Taubes on the gluing of self-dual 4manifolds, to the ground-breaking work of Donaldson on gauge theory and 4-dimensional topology, and many other works in this area. The book written by Uhlenbeck and Dan Freed on "Instantons and 4-Manifold Topology" instructed and inspired a generation of differential geometers. She continued to work in this area, and in particular had an important result with Lesley Sibner and Robert Sibner on non self-dual solutions to the Yang–Mills equations.

Integrable systems and harmonic mappings

The study of integrable systems has its roots in 19th century classical mechanics. Using the language of gauge theory, Uhlenbeck and Hitchin realised that harmonic mappings from surfaces to homogeneous spaces come in 1-dimensional parametrised families. Based on this observation, Uhlenbeck described algebraically harmonic mappings from spheres into Grassmannians relating them to an infinite-dimensional integrable system and Virasoro actions. This seminal work led to a series of further foundational papers by Uhlenbeck and Chuu-Lian Terng on the subject and the creation of an active and fruitful school.

The impact of Uhlenbeck's pivotal work goes beyond geometric analysis. A highly influential early article was devoted to the study of regularity theory of a system of non-linear elliptic equations, relevant to the study of the critical map of higher order energy functionals between Riemannian manifolds. This work extends previous results by Nash, De Giorgi and Moser on regularity of solutions of single non-linear equations to solutions of systems.

Karen Uhlenbeck's pioneering results have had fundamental impact on contemporary analysis, geometry and mathematical physics, and her ideas and leadership have transformed the mathematical landscape as a whole.

Mathematical Meanderings

Karen Keskulla Uhlenbeck

Childhood and Education

I was born during World War II as the first of four children of an artist mother and an engineer father. My parents had come of age during the depression, which shaped their lives and my early years. My father worked for the Aluminum Company of America, and my parents moved from New Jersey to Ohio during the war at least in part because the aircraft industry was relocated to the midwest.

Fig. 1: The Keskulla family in 1944 (left) and 1946 (right). (Photo: private)

I was not an exceptional child, although I remember telling an adult acquaintance very seriously that I was not allowed to learn to read before I entered school because of "the Dewey decimal system." I had confused the theories of the educator John Dewey with the numbering of books in the library due to Melvin Dewey. One of my vivid memories is of getting my library card at the age of eight, and starting on a

K. Uhlenbeck
Institute of Advanced Study, School of Mathematics, 1 Einstein Drive, Princeton, NJ 08540, USA,
e-mail: uhlen@math.utexas.edu

H. Holden, R. Piene (eds.), *The Abel Prize 2018-2022*, The Abel Prize,
https://doi.org/10.1007/978-3-031-33973-8_5

career lasting me well into adulthood of reading everything I could get my hands on. I read every book in the house at least three times. I particularly recall the Modern Library editions of Freud and of the plays by Ibsen. I believe Swann's Way defeated me at the time.

Some of my even earlier memories are of playing seriously with blocks; I still remember their coloring and shapes. I was lucky to have a brother two years younger, so our house was equipped with tinker toys, erector sets and Lincoln logs. He did not play much with them, but I spent hours designing elaborate structures. I also absolutely loved jigsaw puzzles: I remember still the shapes of pieces from those puzzles of my early years. It seems in line with these memories that I played elaborate games of double solitaire throughout my childhood. This absorption with books, blocks, jigsaw puzzles, maps and cards was the despair of my mother, who thought I should be doing something constructive.

My childhood was a rich one, full of music, art, plays, gardening, camping, canoeing and hiking. I must have been a very good student, as I was valedictorian of my high school class, but if I had ambition, it was the ambition to learn rather than to be something. My intelligence was more an embarrassment than an asset in high school. However, my father borrowed books from the library which started me reading popular science at home; I recall particularly the popular books by the British astronomer Fred Hoyle. My first mathematical achievement was to follow the argument that there were several different kinds of infinity, which is in George Gamow's book "One, Two, Three – Infinity". Mathematics to me was logic puzzles like the Prisoner's Dilemma, which did not interest me, then or now. I recall signing up for advanced Latin instead of honors mathematics which was taught at the same time. I was fascinated by the physics and cosmology I had discovered at home and started college at the University of Michigan as a physics major.

Fig. 2: Karen in 1952 (left) and 1958 (right). (Photo: private)

My love affair with mathematics started a couple weeks into my first term in honors calculus. The existence of this honors class and many of the other programs I attended as a student was due to the US response to Sputnik (1957). The need for scientists was deemed so great that women and minorities were explicitly welcomed. The TA for the course stole a march on the professor by showing us in the evening help session how to rigorously define a derivative by taking limits. I recall turning to a fellow students and saying with great excitement "Are you allowed to do that?" Some months later we got to the Heine–Borel theorem and I still remember the technique of little boxes, if not the theorem. A fortuitous circumstance put my studies into high gear. I lived in New Jersey and did not go home for vacations, which I dutifully spent according to my upbringing checking out things like the local art museum. That first year I made the acquaintance of a mathematics professor Dan Hughes in front of a large canvas in the small campus museum. Suddenly I was grading linear algebra (without taking it), sitting in graduate classes as a sophomore, as well as babysitting for several professors as a result of this encounter. It helped to acquire a boyfriend who was a graduate student in mathematics.

I like to tell a story about my first graduate class, which was in algebra. As I recall, the high point of the course was a proof of the Wedderburn Theorems. As I had not a very good grasp of the canonical forms for real matrices, I did not understand it very well. Either out of kindness or actual achievement, I got a B for the semester. The experience did not discourage me a bit as I had enjoyed the course. Three years later when I came to take prelim exams in graduate school, what I had heard as a sophomore made perfect sense and I was able to pass the algebra prelim without taking another algebra course. Learning is not linear.

As a complement to the already excellent education I was receiving at the University of Michigan, I enrolled in a junior year abroad program through Wayne State University and spent my junior year in Munich. The formal lectures at the university were an eye opening contrast to the small mathematics classes I had had in Michigan, but I also gained confidence in discovering that the education I was getting at Michigan was as good or better than that of the other students, who came from all over the US. A true child of my parents, I also fell in love with opera, appreciated classical German drama, checked out the art museums in Italy at spring break, and learned to ski.

Back in the US, I took several graduate classes and decided what to do next. I still had no idea of what I wanted to be, but seven years after Sputnik, money and opportunity were readily available for graduate study. My parents had some idea that I would be able to teach with this background and were agreeable. Besides, my (male) friends were all going to graduate school. One of the other women students had applied to Princeton, which did not accept women at that time, but I was not interested in challenging the status quo. Of course she was not admitted. I picked NYU, which I knew had a solid record in turning out women PhDs. Living in New York for a year was a wonderful experience, and I felt very much at home in the department. One of the best experiences was second semester complex variables with Cathleen Morawetz. This is the only full length mathematics course I have sat in throughout my life which was taught by a woman, although I continued to audit mathematics courses through my years as a post doc and faculty member.

At the end of my first year at NYU I married my boyfriend Olke Uhlenbeck (who was a graduate student at Harvard in biophysics) and transferred to Brandeis. This was the best thing that could have happened to me. I had an NSF graduate fellowship and most likely would have been accepted at MIT or Harvard, but I did not even apply. The last thing I wanted to do was to be on the front lines of the second wave of feminism. The very thought was distasteful. When I did meet that difficulty in Berkeley as a post doc, I was at least better prepared mathematically.

Right away I was drawn to the course on the calculus of variations taught by Richard Palais, who became my thesis advisor. I had wandered around liking one subject after another, but the subject of global analysis, where they all fit together, caught me. At the time, global analysis was new and different, the fashionable subject. Many of the familiar theorems from calculus and finite dimensional topology carry over into infinite dimensions and it was believed were on their way to becoming a powerful tool. The Atiyah–Singer index theorem connecting the index of a system of partial differential equations with the topology of the symbol dates from 1963, as does the ground breaking paper of Eells and Sampson on harmonic maps into manifolds of non-positive curvature. Smale's infinite-dimensional version of Sard's theorem gave a whole new meaning to the word generic, and the Palais–Smale condition gave directly a route to Morse theory for multiple integrals in the calculus of variations. There was a weekly (Monday evening as I recall) joint Harvard–MIT–Brandeis global analysis seminar which I faithfully attended along with many of the established mathematicians. Mathematics was exciting!

Dick was a wonderful thesis advisor, who organized and polished his thoughts and funneled many basic papers in PDEs through to me. I had a bent for the technical analysis, and started a career of understanding, writing down and loving inequalities. Dick's approach to proving the existence of a minimum was to write down a continuous non-negative function on a compact connected set, and to show it was never zero and appeal to a theorem. Mine was to find the lower bound. The contrast of approaches served us very well. I still remember going into his office and asking him about the heat equation, and getting a beautiful synopsis which served me well for years. My thesis was hardly earth shattering. I showed that some functionals (like the p-harmonic or the biharmonic functional) on maps between manifolds satisfied the Palais–Smale condition, put some basic metric structures on manifolds between maps, and proved some regularity theorems.

I finished my PhD in 1968 after three years at Brandeis, and the question of what to do next came up. My husband had not yet finished his PhD, and no one (including me) thought I should sit around and do nothing, so Dick looked for a temporary teaching position for me. I was about to take one at Boston College, when MIT came through and, somewhat as an afterthought, offered me an instructorship, which I promptly accepted. I had already met Tony Tromba, Ulrich Koschorke and Mike Shub my last year at Brandeis, and now my peer group grew to include Nancy Kopell, John Franks and the rest of the Moore Instructors at the time. I think I audited a course taught by Victor Guillemin, but am a bit hazy on this. I spent the year writing up my thesis for publication and trying to start up a research program of my own.

In order to write this autobiography, I took a look at my early papers. They grew directly out of my graduate studies, give a good picture of the mathematics I learned from my advisor, and appeared in print over the next few years. I had almost forgotten them. I had been frustrated by the paper of Eells and Sampson, as I saw no reason why the (to me) cumbersome technique of solving the heat equation should be necessary to find the harmonic maps found by Eells and Sampson. In "Harmonic Maps, a Direct Method in the Calculus of Variations" (1970), as Sacks and I did again later, I introduced a perturbation and considered

$$\int_M (|\mathrm{d}s|^2 + \varepsilon|\mathrm{d}s|^p)\,\mathrm{d}\mu$$

for $s\colon M \to N$ and let $\varepsilon \to 0$. For positive ε and $p > \dim M$, this integral does satisfy the Palais–Smale condition. If the image manifold has non-negative curvature, estimates are available, and a limit harmonic map can be found this way. This paper (really a note) received a positive review by Eells, but got no other attention. Since the integrals I had dealt with in my thesis were on Banach manifolds, not Hilbert manifolds, "Morse Theory on Banach Manifolds" (1972) defined a non-degenerate critical point of finite index and showed that a handle-body decomposition of a Banach manifold could be done in the absence of the Morse lemma (which cannot be true except in a Hilbert manifold). The paper "The Morse Index Theorem in Hilbert Space" (1973) grew out of a paper of Smale's (reviewed by my advisor Palais) on the Morse Index theorem, showing that for an integral over a manifold with boundary, the index of a critical map could be computed using the zeros of the second variation restricted to a sweep out. I was always slow to publish and agonized over writing, although I never had any difficulty in thinking up projects or getting papers accepted for publication.

Midcareer

The question of what to do when my husband finished his PhD was more serious. Olke was offered a Miller fellowship at Berkeley, which he accepted enthusiastically. I was not sure what I would do, but again through the influence of my thesis advisor, several weeks after he accepted the position, I was offered a lectureship, which I promptly accepted. The move to Berkeley in 1969 proved indeed exciting. This was the period of the Vietnam war protests, campus disruption, volatile political discussion and issues surrounding the changing position of women. Despite (or because of) all this, my mathematics flourished. I came under the influence of Abe Taub, who had a distinguished career in a number of different areas (relativity, fluid mechanics and shock waves, the large Univac computer) and was now lecturing on quantum mechanics and quantum field theory. I also attended, and as I recall was asked to organize, the analysis seminar. It was my good fortune to listen to Steve

Smale in his magnificent course on celestial mechanics. I loved it and was well prepared. I recall showing the notetakers for the course how to compute the second variation of a constrained variational problem. I later regretted not taking part in the differential geometry centered around S.S. Chern. S.T. Yau, who was at Berkeley at the same time I was, remembers me, but sadly I do not remember him. Of my peer group, which included Blaine Lawson and Alan Weinstein, I had the most contact with Jerry Marsden and to a lesser extent Arthur Fisher.

Fig. 3: Karen in 1969. (Photo: private)

I started two mathematics projects during this time. This first began with a fascination with eigenvalues and eigenfunctions and originated with the course on ODEs I taught at MIT. The eigenfunctions of a sphere seemed magical, and I tried to understand some of the group theory behind their structure. Since I did not have the sense to ask anybody about it, I did not get very far. I knew that Sturm–Liouville theory was one-dimensional, and found the page in Courant–Hilbert about nodal domains in higher dimension. This led me to think about generic properties, and I was able to prove some nice results about generic behavior of eigenvalues and eigenfunctions with a simple application of the Sard–Smale theorem. I really wanted to understand how to codify the fact that the eigenfunctions got bumpier and more complex as the eigenvalue grew; I talked enthusiastically at length about the problem to anybody who would listen. Recently I found out that at that time exactly these types of questions were being addressed by applied mathematicians, but the chasm between applied and pure mathematics was wide at that time.

My second project came out of the course given by Ray Sacks on general relativity that I audited as a lecturer at Berkeley. I confess that I learned basic differential geometry for the first time in this course. But in thinking about how light waves bend about the sun, I invented a way to count their images using Morse theory. To do this project I spent a great deal of time reading the textbooks and basic papers of

Hawking, Ellis and Penrose. I thought of turning my knowledge of PDEs to general relativity. When I discovered that, in addition to the Einstein equations one had to incorporate fluid flow equations, I gave it up as simply too difficult.

When it came time for the serious business of applying for tenure track jobs, things got complicated. I am glad I do not know what went on in faculty meetings when I was hired as a post doc or considered as an assistant professor; I would guess that very little of it had to do with my merits as a mathematician. My husband was interested in staying in Berkeley, and his department was enthusiastic, but when I asked the mathematics department about possibilities for me, the answer was "Well, there is always an epsilon possibility." So we applied elsewhere. My husband received many offers, but at least one mathematics department told me I should apply for a suitable position at a women's college. So we took jobs at the one place that offered me a job – the University of Illinois. In the end, Berkeley did offer me an assistant professorship, and spent the next few years listing me as a faculty member after I turned them down (with great pleasure). I found the next few years of trying to be a married woman professor difficult enough anyway, and it would have been impossible in the political climate at Berkeley.

There was a second difficulty in my job search. Global analysis had been the fad in the sixties, and many young mathematicians had enthusiastically written theses in the subject. At that time the dynamics group and the functional analysis group were linked, and in both cases, tensions arose between the more classical fields of partial and ordinary differential equations and the new approaches. Skepticism arose doubting the claims of global analysis, and this was reflected in how and where the younger mathematicians could get jobs. Remnants of this tension still exist in dynamics, and S.T. Yau, when he burst on the scene, renamed global analysis geometric analysis, to distinguished it from this "soft" predecessor. Despite being trained by a differential topologist, I always had a bent for hard analysis, so I came out of this in the end rather well.

At first my mathematical progress was better than my personal one. In my first years at Champaign-Urbana, I finally solved the regularity problem for p-harmonic maps that I had been working on steadily since graduate school. Harmonic maps between manifolds are critical points of an integral of the norm squared of the derivative of a map. This satisfies the Palais–Smale condition only in dimension 1. However, the critical points of the integral of the derivative raised to the p-th power, do satisfy the Palais–Smale condition for p greater than the dimension of the domain manifold. Unfortunately, these maps are in Sobolev spaces and not very smooth. To be any good for geometry, they should have at least continuous derivatives. This is a problem in analysis that has little to do with the geometry. It was a theorem of De Giorgi–Nash–Moser for one-dimensional images. I was able to show the critical maps were Lipschitz in my thesis. I finally published the proof of smoothness in the case of maps between Euclidean spaces, and others including Martin Fuchs published the full result in the next few years.

I was able to do this only because at some point, maybe at the end of my graduate career, I met Jurgen Moser. He sat down with me, talked over what we now call the Moser iteration scheme, and sent me reprints of his papers (which I carefully read).

Later, when I succeeded in finding a complete set of integrals for the classical Neumann problem in looking at the dimensional reduced equation for minimal spheres, I was embarrassed at how much he praised me in seminars. I still feel grateful to him.

Before I move on mathematically, I should mention a couple of other important people in my career. My father-in-law George Uhlenbeck was a famous physicist who with Goudsmit had discovered spin in the lab (1925). He and his wife Else were courtly Europeans, charming and encouraging without being critical or pushy. From him I heard all about the inner workings of academia, the difficulties of getting students jobs and the opinions physicists had of mathematicians. I also remember still his advice on teaching "At the end of every lecture, tell them what you are going to do in the next lecture, at the beginning of every lecture, tell what you did in the previous lecture, and in the middle give the lecture." As a student, I had found that redundant, but coming from George Uhlenbeck, I put it into practice.

I also met Lesley Sibner at a conference in Trieste. Later I learned she had been studying to be an actress when she took a required calculus course and fell in love with mathematics. Her poise, her charm, her enthusiasm, her encouragement and her mathematical abilities were something special to many people, particularly to a younger woman trying to carve out a place in the mathematics community. We became fast friends mathematically and personally. I visited her and her husband Bob (also a mathematician) in New York often. Her influence can be seen in the Acta paper where I wrote up the regularity theorem for p-harmonic maps. We did later write a paper together on gauge theory. She was the first mathematician I truly wanted to emulate.

The next event in my mathematical life came when Jonathan Sacks came as a post doc to the University of Illinois. He had just received his PhD from the University of California, Berkeley under the direction of Blaine Lawson, and made a point to seek me out. Since I had not talked much to the geometers at Berkeley, we started with little in common, but he brought to my attention the questions about minimal surfaces in the air. He was enthusiastic and convincing and I certainly owe my foray into minimal surfaces to him. I was intrigued by the ordinary differential equation for equivariant spheres, which happens to be the classical Neumann problem. The paper by Neumann is the one and only mathematics paper I have ever seen in Latin! Looking back, I now sense a relationship between the endless solitaire I played as a child and the manipulations which led me to find the integrals for the problem. More importantly Jonathan and I realized how close the $n = 2$ dimension for harmonic maps was to the Palais–Smale condition. If only the dimension were $2 - \varepsilon$ or the power were $2 + \varepsilon$. Not being physicists, we took the second route and considered the integral

$$\int_M (1 + |\mathrm{d}s|^2)^\alpha \, \mathrm{d}\mu$$

for $s\colon M \to N$ and tracked the behavior as $\alpha = (1 + \varepsilon/2) \to 1$ from above. We realized right away that the limit could easily be a map to a point in N, but by looking at the points where the energy concentrated, we obtained a map $s\colon S^2 = \mathbb{R}^2 \cup \infty \to N$. This map is a conformally parameterized minimal sphere. Of course, at first we had no idea that this phenomenon is universal for scale invariant problems. Neither can we recall exactly who used the term "bubbling" but I began to get invitations to give colloquia and was always trying hard to give colloquial explanations.

The next step in my trajectory was four talks given by Michael Atiyah at the University of Chicago on gauge theory. I no longer remember whether this was before or after I moved to Chicago, but they made a deep impression on me. The importance and origin in physics was emphasized. I understood very little of what he said, and struggled through his little book "The Geometry of Yang–Mills Fields" without being much the wiser. At the time, there were two textbooks on the geometry of fibre bundles (now called gauge theory); volumes by Steenrod and Husemöller. They were very hard going (pity the poor physicists). It was not hard to see what was missing, as the linear model for these equations is the cohomology theory given by Hodge theory. The curvature was roughly the exterior derivative of the connection one form and needed to be paired with the equation making the connection co-closed to form an elliptic system. This was basically the way to choose a good representative for the connection. How to do this? The difficulties were well known in physics as the Gribov ambiguity. Examples showed there could be many coordinate systems (gauges) in which the connection one form was co-closed. There is a variational problem, which unfortunately looks like the harmonic map variational problem for which the critical dimension is 2, considerably less than 4 where we needed it. I mulled over this for over a year. I recall the sudden moment of "Eureka." The space of connections in a ball with a small enough norm on curvature is connected, open and closed. Very little geometry, topology or fancy techniques in PDEs. Just some nice estimates. It was the same continuity method used by S.T. Yau at about the same time in his solution of the Calabi conjecture which rocketed him into well-deserved fame.

It is worth mentioning the developing relationship between theoretical physics and geometry. Mathematicians had never given up trying to understand quantum mechanics and general relativity, but I recall very definitely from my graduate school days a mathematician saying to his physicist colleague "Let me tell you what is really going on (in your physics)." I knew that this was not useful interaction. Things changed, partly as a result of the physicists' discovery of group theory, partly as the physics models became more geometric, partly for many other reasons documented by many other people. When I learned about black holes in the 60s, they seemed to be a mathematical fabrication rather than one of the basic concepts in cosmology. With the change, physicists wanted to use if not understand the mathematics, but their physical insights gave new insight into the mathematics. Most of my experience has been with the fruitful interactions of subjects within mathematics with each other, but I never forget that the mathematicians could have written down the Yang–Mills equations. They just didn't. I hung around the fringes of physics for many years, along with a lot of mathematicians, but I never got any insight as to their thought processes. It is a good deal easier for a theoretical physicist to learn techniques from mathematicians than it is for a mathematician to gain physical intuition.

This stage of my career culminated with Simon Donaldson's application of gauge theory to the topology of 4 manifolds. I had made one of the necessary contributions to his theorem by applying the idea of bubbling to the Yang–Mills equations, which are scale invariant in dimension four, and at this point I became very well known for this contribution.

Let me catch up on my personal life. My marriage had not survived being a faculty wife in Champaign-Urbana, and I moved to Chicago to be with my present husband, Bob Williams, also a mathematician. Primarily due to the efforts of Susan Friedlander, I got a tenured offer from the University of Illinois in Chicago, where I was delighted to have other women colleagues: Vera Pless, Louise Hay, Bhama Srinivasan and of course Susan. I was lucky enough to have the office next to Howie Masur, and through him continued the interest in Teichmüller theory and Bill Thurston's reformulation of it that I started with Bill Abikoff in Urbana-Champaign. (Did I mention that Bill Thurston had been the head TA in the calculus course in Berkeley I was in charge of?)

Fig. 4: Karen in 1979. (Photo: private)

The academic year 1979–80 was spent at the Institute for Advanced Study in Princeton. The year organized by S.T. Yau was an exciting one. Up until that time, I had more or less been on the fringes of the mathematics community, but here I met Rick Schoen, Leon Simon, Peter Li, Jean-Pierre Bourguignon and Chuu-Lian Terng. I have a vivid memory of trying to geometrically understand the fibration of hyperbolic 3-manifolds with Chuu-Lian Terng, when Robert Langlands came and slammed the door to my office because we were so loud. At the end of the year, Rick Schoen and I began a project to look at minimal harmonic maps from manifolds of dimension greater than 2. This dimension is above the critical dimension, and we found a new kind of bubbling that occurred along more complicated sets than the points that occurred at the critical dimension. This grew to three papers, although my interest flagged a bit after we understood the basic phenomenon.

At this point I arrived, despite the fact that I never knew where I was going. I began to have graduate students, and accepted an offer from the University of Chicago because I hoped to have more. Everything seemed to happen at once; the offer from Chicago, a MacArthur Fellowship and election to the National Academy. Moreover, S.T. Yau, who had become one of my staunchest supporters, enlisted me on a project to prove the existence of Hermitian Yang–Mills fields on a compact Kaehler manifold. I still think it a very nice theorem about a non-linear system of PDEs, although the write up makes it look much harder than it really is. The algebraic geometers much prefer the proof of Simon Donaldson with a much more algebraic approach. In the wake of my name appearing on these theorems in gauge theory, the younger theoretical physicists now recognized the name Uhlenbeck as belonging to me, not my former father-in-law. I found this rather sad, but his name has reappeared in the Ornstein–Uhlenbeck process so important in financial mathematics. While this is a magnificent illustration of how mathematics developed to describe physical phenomena becomes basic in applications far from physics, I am not sure what George Uhlenbeck would have thought.

This era ended with a visit to the University of California, San Diego in the winter of 1986. Mike Freedman had been there some years, S.T. Yau had moved there, Richard Hamilton and Rick Schoen had joined him, and Leon Simon was willing to come. I took surfing lessons and very much enjoyed meeting the younger post docs and students surrounding the group. We (my husband and I) became enthusiastic. In the end, the project of forming a geometric analysis group around S.T. Yau fell through, and we all went elsewhere. But the idea of moving had come up, and the following year Bob and I accepted jobs at the University of Texas. My graduate students were horrified.

Texas and Beyond

My move to Texas was not greeted with cheers by my professional colleagues, who were skeptical of the (they thought typically Texan) strategy for improving the ranking of the department. Peter O'Donnell had endowed money for special chairs in physics, chemistry, computer science and mathematics, which came with a very good salary and funds which could be spent on professional activities with very few strings attached. I was attracted by the presence of Steven Weinberg in theoretical physics, and Bob by the knot theorist Cameron Gordon in mathematics and the experimental physicist Harry Swinney in physics. Bob also had family in Texas. We moved in 1987, we bought a house in the hill country outside Austin and a pickup truck, and I started attending courses and seminars in the physics department. Four students came with me from Chicago, and all four and the later students I had in Austin have done well professionally. I had too few postdocs, but all but one have done very well.

Two years later John Tate and Dan Freed joined the department. Both had profound influences on me. John was the elder statesman I sought to emulate and I have written about that elsewhere. Dan arrived, full of energy and enthusiasm, and essentially pressuring me into countless projects: first of all writing a book, then building a geometry group modeled on physics groups, founding the Park City Mathematics Institute, establishing Saturday Morning Math and much more. Bob and I camped in Big Bend or took bicycling trips in the winter, camped on the beach at Matagorda Island in the spring and went out West to the Aspen Center for Physics, the Park City Mathematics Institute and later to Montana State University in Bozeman in the summer.

Fig. 5: Big Bend national park. (Photo: private)

Professional life in Texas suited me. The position of women was different here. This was the Texas of Ma Ferguson, Ann Richards, Lady Bird Johnson, Liz Carpenter, Molly Ivens and Barbara Jordan. Cécile DeWitt in the physics department helped as well. The entire staff of the math department became my cheerleaders. During my tenure in Austin, I met wonderful colleagues in many departments through lunches with women faculty in the College of Natural Sciences and Engineering. Later on, at the suggestion of one of Dan Freed's graduate students Orit Davidovitch, we started the Distinguished Women Lecture series, and a good number of the senior women in mathematics in the country visited. I was able to support these efforts and many male and female students and post docs on the Chair funds which came with my position.

A description of this part of my life would be incomplete without a mention of my trips with MacArthur Fellows. I had received the MacArthur "genius" award in 1983, and while I was at first quite intimidated by the other fellows, I had always wanted to be a member of an intellectual elite. I had read everything I could about

Fig. 6: Montana. (Photo: private)

Bloomsbury (with particular attention to Virginia Woolf). So I dove in. This is probably the only time in my life I regretted being a mathematician, since I did not have the confidence to explain mathematical concepts to even physicists. I have a wonderful memory of John Schwarz talking about string theory to a dozen MacArthur Fellows stretched out in deck chairs under the stars in Hawaii. Groups of MacArthur Fellows went on trips to experience and learn about the work of other fellows. We paid for ourselves, but we did have grant money from the MacArthur foundation to spend as we liked. An incomplete list of the places we went includes Hawaii, Appalachia, Santa Fe, the Galapagos, the Brasilian rain forest, a Montana dinosaur dig, and Madagascar. Bob went to a satellite launch. Mathematicians rarely attended, but the three women mathematicians, Nancy Kopell, Ingrid Daubechies and myself with our spouses went to every event we could. I made many friends in many disciplines during this period.

In 1990 I became the second woman to give a plenary lecture at the International Congress of Mathematicians. The first had been Emmy Noether (one of my heroes) in 1932. This still seems the most unreal moment of my career. After all, by comparison the Abel Prize has been around less than twenty years. Under the influence of the MacArthur Fellows and a growing awareness of the difficulties younger women were having, I became involved in activities that were not strictly mathematics, but I continued searching for problems in mathematics. Despite my hanging around physicists for years, I never acquired any intuition. Through reading the papers of Louise Dolan, I discovered that the classical Bäcklund transformations were manifestations of the action of Birkhoff factorization on scattering data, but never understood how either the physicists or the applied mathematicians had found them. This project led me to join forces with Chuu-Lian Terng, a classical differential geometer who was also a student of Palais and had worked under Chern. Together over the next decades we wrote a series of papers on integrable systems, founded

Fig. 7: Karen, Nancy Kopell and Ingrid Daubechies in Amazonia 1996. (Photo: private)

the Women and Math Program at the Institute for Advanced Study and became lifelong friends. I remain grateful to IAS for their support of the program for women before it became fashionable, and I remain close friends with Nancy Hingston and Antonella Grassi who worked with Chuu-Lian and myself in this program for many years. This program benefitted many women, but in allowing me to mix friendship with mentoring and mathematics, I may have benefitted the most.

Fig. 8: Karen in 1992. (Photo: private)

When I was asked to lead a special year at the Institute for Advanced Study in 1997–98, I immediately chose to pursue these interests in integrable systems further. I had been learning the basics of non-linear wave and non-linear Schroedinger equations. I felt there must be some connection between the algebraic approach from integrable systems and all the lovely inequalities involved in the Wick rotated counterparts to Yang–Mills and harmonic maps. The year was a success because of the quality of the post docs who attended. But my project has so far never led anywhere. Many of the post docs arrived with their own agenda, more connected to my previous work than my present interests. A few did switch gears. I subsequently wrote a few papers in dispersive systems, but in the end I found that the subject, despite its attractive collection of inequalities, was not something I could contribute much to. At present, the probabilistic approach started by Bourgain seems to catch at least some of the essence of the behavior of solutions to the equations. I am still talking to one of my collaborators, Andrea Nahmod, who is part of a group studying the Gross–Pitaevsky hierarchy. This is clearly integrable. The problem is to figure out exactly what this means. It is hard to tease the geometry out of the algebra and analysis. The subject of integrable systems has remained fractured and confusing without ever catching the fancy of the mathematics community the way other subjects have done. I am still hoping for bridges between the PDEs, the algebra and the physics. KdV and non-linear Schroedinger appear over and over again. Is this because there are underlying connections, or just that the human mind can grasp only a limited number of patterns?

Fig. 9: Karen, age 5, by Carolyn Keskulla (her mother).

The latter part of my career is not as interesting as the earlier part. I never did come to feel comfortable in the wider professional circles of mathematics. Until I came to write this autobiography, I had regrets about not forming an influential group of mathematicians around me, not carrying my weight in professional organizations, not supporting my students far enough into their careers, not carrying through on the problems I had started, and giving up on the questions I still had. I never learned geometric measure theory, what the Ornstein–Uhlenbeck process was, or why integrable systems appear in conformal field theory. In the last decade I developed health problems which made it difficult for me to have new ideas, although I still appreciated mathematics. Allergies forced my move from Texas. I am grateful to the Institute for Advanced Study for providing a new home for me.

The Abel prize comes at a good time for me to look back over my career. Many thanks to the Abel Committee and the Norwegian government. Some of the monetary award has gone into EDGE and IAS programs for underrepresented minorities. I have changed my medication, and have been able to do mathematics again. Whenever a query comes my way as to what my greatest difficulty has been, I answer "ill health." My two present collaborators, Penny Smith and George Daskalopoulos have enabled me to prove theorems again. I am as excited as I have ever been about the project with George to connect the topological structures of Thurston and his school with ∞ harmonic and least gradient maps. My talents and abilities fit well into the development of mathematics in the last fifty years, and I still get the pleasure from doing mathematics that I got as a child playing with blocks. What more can one ask?

A journey through the mathematical world of Karen Uhlenbeck

Simon Donaldson

Contents

Department of Mathematics, Imperial College London.
Simons Center for Geometry and Physics, Stony Brook University.
e-mail: s.donaldson@imperial.ac.uk

H. Holden, R. Piene (eds.), *The Abel Prize 2018-2022*, The Abel Prize,
https://doi.org/10.1007/978-3-031-33973-8_6

1 Introduction

In this article we discuss some of Karen Uhlenbeck's most prominent mathematical results. Uhlenbeck's publications range across many mathematical areas, including differential geometry and geometric analysis, elliptic and hyperbolic partial differential equations and integrable systems. In this article we only attempt to describe some part of this range. The main omissions are that we say nothing about her work on wave and Schroedinger maps, and very little about integrable systems. The core of the article is contained in Sections 3, 4, 5 and 6, which give an account of some highlights of Uhlenbeck's work on the analysis of harmonic maps and Yang–Mills connections. In each case we begin with the theory for the "critical dimension"—in Sections 3 and 5—before going on to higher dimensions (in Sections 4 and 6). This body of work has been absolutely fundamental in the developments of geometric analysis over the past 40 years and with an impact that extends to many fields, from symplectic geometry and low-dimensional topology to Quantum Field Theory and the mathematics of liquid crystals. At the beginning and end of the article we discuss two other contributions of Uhlenbeck which take somewhat different directions to that in the core; each very influential and highly-cited. The first (in Section 2) is a paper on nonlinear elliptic PDE theory and the other (in Section 7) is on integrable systems aspects of harmonic maps from surfaces to Lie groups.

The author has written another review [15] of some of Karen Uhlenbeck's mathematical work, which focused on variational methods. While there is overlap with the current article we have made the focus here different and sought to avoid duplication. At some points in this article we refer to [15] for further discussion of literature and background.

2 Nonlinear systems and p-harmonic functions

2.1 A regularity theorem

We begin our tour by discussing the 1977 *Acta Mathematica* paper [57] of Uhlenbeck which was one of her first papers with a focus on "hard" PDE theory. To set the scene for this, recall that the Laplace operator Δ on functions on $\mathbf{R}^n$ is the Euler–Lagrange operator associated to the Dirichlet energy, the integral of $|du|^2$. The solution of the boundary value problem for a harmonic function on a domain with prescribed boundary values minimises the Dirichlet energy over the set of all functions with those boundary values. A generalisation is to take any $p > 1$ and the functional defined by the integral of $|du|^p$. The associated Euler–Lagrange equation is the nonlinear *p-Laplace* equation

$$d^*(|du|^{p-2}\, du) = 0. \tag{1}$$

The existence of weak solutions to this equation, lying in the Sobolev space L_1^p and with prescribed boundary values, is relatively straightforward but the question of the regularity of these weak solutions is very subtle. The equation (1) is a degenerate elliptic equation at points where the derivative of u vanishes. We can write the equation as

$$\Delta u + (p-2)\sum \nu_i \nu_j \frac{\partial^2 u}{\partial x_i \partial x_j} = 0,$$

where ν is the unit vector field

$$\nu = \frac{1}{|\mathrm{d}u|}\mathrm{d}u,$$

and ν will usually be discontinuous at zeros of $\mathrm{d}u$. This means that one cannot expect solutions of the p-Laplace equation to be smooth at such zeros. For example the function $u(x) = |x|^\beta$ with $\beta = (p-n)/(p-1)$ is a solution.

Uhlenbeck's *Acta* paper established a central result on this regularity question, as a particular case of a more general theory, showing that the derivative of a p-harmonic function satisfies a $C^{,\alpha}$ Hölder estimate for some α depending on p, n. This particular result had been obtained before by Ural'ceva, appearing in Russian [65]. But the theory developed in Uhlenbeck's paper covers much more than this model case, as we will now explain.

Let M be a smooth manifold, with its complex of differential forms

$$\Omega^0 \xrightarrow{\mathrm{d}} \Omega^1 \xrightarrow{\mathrm{d}} \Omega^2 \ldots$$

If M is compact and Riemannian then in Hodge theory the harmonic representative ω of a k-dimensional de Rham cohomology class is characterised as the minimiser of the L^2 norm over all representatives of that class. It satisfies the equations $\mathrm{d}\omega = 0, \mathrm{d}^*\omega = 0$. In the spirit of the discussion above, it is natural to consider the generalisation of this where one takes a positive function g on $\mathbf{R}$ and minimises

$$\int_M g(|\omega|).$$

For example we could take $g(|\omega|) = |\omega|^p$. For a small variation $\omega + \mathrm{d}\alpha$ in the fixed cohomology class

$$g(|\omega + \mathrm{d}\alpha|) = g(|\omega|) + (\mathrm{d}\alpha, \rho(|\omega|)\omega) + O(\alpha^2)$$

where ρ is the function $\rho(t) = g'(t)/t$. So a minimiser satisfies the Euler–Lagrange equation

$$\mathrm{d}^*(\rho(|\omega|)\omega) = 0, \tag{2}$$

in addition to the closed condition $\mathrm{d}\omega = 0$.

This nonlinear generalisation of Hodge Theory was studied by the Sibners [46] who established that for a large class of functions g there is indeed a unique minimiser, giving a weak solution of the equation (2). When $k = 1$ the closed 1-form ω

can be written locally as the derivative of a function u and, when $g(|\omega|) = |\omega|^p$, we get back to the p-harmonic equation. The equations derived from other functions g arise in the theory of gas dynamics, as explained in [46].

Uhlenbeck's main theorem in [57] asserts that, for a large class of functions g, these weak solutions are Hölder continuous. In fact her result is formulated for more general elliptic complexes, such as the $\overline{\partial}$-complex. The conditions imposed on the function g are, roughly speaking, that it should have the character of $(|\omega|^2 + c)^{p/2}$ for some $c \geq 0$. In the case when $c = 0$ the equation becomes degenerate at the zeros of ω, just as we saw for the p-harmonic equation. But even in the easier case when $c > 0$ the result was new. The force of the result in that case is that it applies to *systems* of PDEs rather than to an equation for a single function. Problems 19 and 20 in Hilbert's 1900 problem list asked about the existence and regularity of solutions to variational problems. In the 1950s, De Giorgi and Nash obtained very general results on the regularity of weak solutions to elliptic variational problems for a single function, but examples show that these results do not extend to systems: we refer to the discussion in [20], Chapter II. Such regularity questions form a theme running through much of Uhlenbeck's work discussed in this article.

In the remainder of this section we sketch some of the main parts of Uhlenbeck's arguments in [57]. To simplify our presentation we will consider only the case of the de Rham complex and the function $g(|\omega|) = |\omega|^p$. (Uhlenbeck's results are stated for domains in $\mathbf{R}^n$ but the proofs should extend to general Riemannian manifolds.) The theorem we are discussing then is:

Theorem 2.1. *Let ω be a k-form on the domain $U \subset \mathbf{R}^n$ with coefficients in L^p which is a weak solution of the equations*

$$d\omega = 0 \quad d^*(|\omega|^{p-2}\omega) = 0. \tag{3}$$

Then ω is Hölder continuous on compact subsets of U.

2.2 A differential inequality

The foundation of Uhlenbeck's proof is an idea which we will meet many other times below: artful use of differential inequalities for *functions* can produce important information about solutions of complicated *systems* of PDEs. To set things up, given an exterior k-form ν with $|\nu| = 1$ define a symmetric matrix (a_{ij}) by the inner products

$$a_{ij} = (dx_i \wedge \nu, dx_j \wedge \nu) \tag{4}$$

This is clearly a positive symmetric matrix $(a_{ij}) \geq 0$ and we also have an upper bound $(a_{ij}) \leq (\delta_{ij})$. Indeed if we define (b_{ij}) by

$$b_{ij} = (I_i \nu, I_j \nu),$$

where I_i is the operation of contraction with $\frac{\partial}{\partial x_i}$ then $(b_{ij}) \geq 0$ and it is a basic fact of exterior algebra that $a_{ij} + b_{ij} = \delta_{ij}$. Now, given a k-form ω on $U \subset \mathbf{R}^n$ we apply this at each point where $\omega \neq 0$, taking $v = \omega/|\omega|$, so we get functions a_{ij}, defined away from these zeros. Let Λ be the linear differential operator, depending on ω,

$$\Lambda(f) = \sum_{ij} \frac{\partial}{\partial x_i}\left(a_{ij}\frac{\partial f}{\partial x_j}\right),$$

and define an operator L by

$$L = \Delta + \frac{p-2}{p-1}\Lambda.$$

(In this article we use the "analysts" sign convention for the Lapacian: $\Delta = \sum \frac{\partial^2}{\partial^2 x_i^2}$.)

Proposition 2.1. *Let ω be an L^p_{loc} solution of the equations* (3) *and define $H = |\omega|^p$ and $\theta = |\omega|^{(p-2)/2}\omega$. Then, with the operator L as defined above,*

$$L(H) \geq c_p |\nabla\theta|^2,$$

where $c_p = 4p/(p-1)(p+2)$.

(In what follows we calculate as though all derivatives are defined in the elementary sense. Of course one has to make precise the meaning of the formula when ω is *a priori* only in L^p, but we will ignore such technicalities here.)

To establish the inequality we begin with the formula

$$\nabla_i|\omega|^p = p|\omega|^{p-2}(\nabla_i\omega, \omega).$$

Replacing p by $(p-2)$ we get

$$\nabla_i\left(|\omega|^{p-2}\omega\right) = |\omega|^{p-2}\nabla_i\omega + (p-2)|\omega|^{p-4}(\nabla_i\omega, \omega)\omega,$$

so

$$(\ \omega, \nabla_i(|\omega|^{p-2}\omega)\ ) = (p-1)|\omega|^{p-2}(\omega, \nabla_i\omega) = \frac{p-1}{p}\nabla_i|\omega|^p.$$

Thus

$$\Delta|\omega|^p = \sum \nabla_i\nabla_i|\omega|^p == \frac{p}{p-1}\sum \nabla_i\left(\omega, \nabla_i(|\omega|^{p-2}\omega)\right). \tag{5}$$

Now recall the basic fact of Hodge theory, that the Laplacian on k-forms on $\mathbf{R}^n$ has two expressions $\Delta = -(\mathrm{d}^*\mathrm{d} + \mathrm{d}\mathrm{d}^*) = \sum \nabla_i\nabla_i$. (This is the same as the statement that $a_{ij} + b_{ij} = \delta_{ij}$ in the preceding discussion.) From (5) we get

$$\Delta|\omega|^p = \frac{p}{p-1}(P+Q) \tag{6}$$

where

$$P = (\omega, \Delta(|\omega|^{p-2}\omega))$$

and

$$Q = (\nabla_i \omega, \nabla_i(|\omega|^{p-2}\omega)).$$

To understand the term P it is convenient to consider a compactly supported test function f and the L^2 inner product $\langle P, f\rangle_{L^2}$. This is

$$\langle f\omega, \Delta|\omega|^{p-2}\omega\rangle_{L^2}.$$

Since $\mathrm{d}^*(\omega|^{p-2}\omega) = 0$ we can write this as

$$\langle \mathrm{d}(f\omega), \mathrm{d}(|\omega|^{p-2}\omega\rangle_{L^2},$$

and since $\mathrm{d}\omega = 0$ this becomes

$$\langle \mathrm{d}f \wedge \omega, \mathrm{d}(|\omega|^{p-2}) \wedge \omega\rangle_{L^2}.$$

Now

$$\mathrm{d}|\omega|^{p-2} = \frac{p-2}{p}|\omega|^{-2}\mathrm{d}|\omega|^p = \frac{p-2}{p}\mathrm{d}H$$

so we can write this as

$$\langle P, f\rangle_{L^2} = \frac{p-2}{p}\langle \mathrm{d}f \wedge \frac{\omega}{|\omega|}, \mathrm{d}H \wedge \frac{\omega}{|\omega|}\rangle_{L^2}.$$

By the definition of the (a_{ij}) this equation is

$$\langle P, f\rangle_{L^2} = \int \sum a_{ij}\nabla_i f \nabla_j H.$$

Since this is true for all f we have

$$P = -\frac{p-2}{p}\sum \nabla_i(a_{ij}\nabla_j H) = \frac{p-2}{p}\Lambda(H).$$

Now (6) becomes $L(H) = Q$. Turning attention to the term Q, we have

$$Q = |\omega|^{p-2}\left(|\nabla\omega|^2 + (p-2)\sum_i (\frac{\omega}{|\omega|}, \nabla_i\omega)^2\right).$$

While, for $\theta = |\omega|^{p/2-1}\omega$,

$$|\nabla\theta|^2 = |\omega|^{p-2}\left(|\nabla\omega|^2 + ((p-2) + \left(\frac{p-2}{2}\right)^2)\sum_i (\frac{\omega}{|\omega|}, \nabla_i\omega)^2\right),$$

and, comparing the two, we see that $Q \geq \frac{4}{p+2}|\nabla\theta|^2$, completing the verification of Proposition 2.1.

The significance of this Proposition 2.1 is that the divergence-form operator L is uniformly elliptic with bounded measurable coefficients (we assume that (a_{ij}) is defined almost everywhere). That is, the eigenvalues of the coefficient matrix $\delta_{ij} + \frac{p-2}{p-1} a_{ij}$ of L are bounded between $1 - |(p-2)/(p-1)|$ and $1 + |(p-2)/(p-1)|$. This opens the way to apply the deep theory from the 1950s of elliptic operators with measurable coefficients, which were the foundation for the results of di Giorgi and Nash mentioned above.

2.3 Outline of proof of Theorem 2.1

One first issue is to show that the form ω in Theorem 2.1 is bounded on compact subsets of the domain U, but we will pass over this to focus on Uhlenbeck's proof of Hölder continuity. (Her proof of boundedness uses related arguments.)

For background, we review some relatively elementary results for the standard Laplace operator. Let $B' \subset B$ be balls in $\mathbf{R}^n$, for example the unit ball and the concentric ball of half the radius. Let h be a positive function on B with $\Delta h \geq 0$. Let M, M' be the suprema of h on the balls B, B' respectively. Then by the definition $M \geq M'$ and the maximum principle implies that $M = M'$ if and only if h is a constant, in which case $\Delta h = 0$. The next proposition gives two quantitative versions of this.

Proposition 2.2. *There are constants C_1, C_2 such that if $\Delta h \geq \rho \geq 0$ on B then*

(1)

$$\int_{B'} \rho \leq C_1 (M - M');$$

(2)

$$\int_{B'} (M' - h) \leq C_2 (M - M').$$

We give a proof of the first item of Proposition 2.2. Let f be the solution of $\Delta f = -\rho$ in B with $f = 0$ on the boundary of B and let $g = h + f$. Then $\Delta g \geq 0$ and so the maximum principle implies that the supremum of g on B' is at most that on ∂B, which is M. For $x \in B'$ we have

$$f(x) = \int_B G(x,y) \rho(y) \, \mathrm{d}y,$$

where G is the Green's function, which is positive in the interior of B. So there is an $\varepsilon > 0$ such that for $x, y \in B'$ we have $G(x,y) \geq \varepsilon$, which implies that

$$f(x) \geq \varepsilon \int_{B'} \rho.$$

So for $x \in B'$

$$h(x) = g(x) - f(x) \leq M - \varepsilon \int_{B'} \rho,$$

so

$$M' \leq M - \varepsilon \int_{B'} \rho,$$

which is the desired inequality with $C_1 = \varepsilon^{-1}$.

It is also easy to deduce the first item from the second, with a slightly different choice of balls. Let χ be a cut-off function supported in B', equal to 1 on some smaller ball $B'' \subset B'$. Then

$$\int_{B''} \rho \leq \int_{B'} \chi \Delta (h - M') = \int_{B'} (\Delta \chi)(h - M') \leq c \int_{B'} (M' - h)$$

where $c = \max|\Delta \chi|$.

Now suppose that ω is a k-form on the ball B satisfying the equation (3) in Theorem 2.1 and set $H = |\omega|^p, \theta = |\omega|^{(p-2)/2}\omega$ as above. So Proposition 2.1 gives $L(H) \geq \rho$ with $\rho = c_p|\nabla\theta|^2$. Let M, M' be the suprema of H on B, B'. Uhlenbeck shows that an inequality of the same nature as the first item in Proposition 2.2 holds in this situation, so that, for a suitable constant C,

$$\int_{B'} |\nabla\theta|^2 \leq C(M - M'). \tag{7}$$

The proof uses results of Moser—part of the theory of operators with bounded coefficients mentioned above—and many substantial additional arguments. In fact Moser's result gives the analogue of the first item in Proposition 2.2 for the operator L and Uhlenbeck obtains the analogue of the first item in the manner indicated above, but additional arguments are required to carry this through because L depends on ω. So we do not have the same control of $|L(\chi)|$ for the cut-off function χ. Of course, in all this the techniques required to treat the operator L are quite different from the elementary techniques which suffice for the Laplace operator.

The conclusion is that, in going from the ball B to the smaller ball B', either the supremum of H decreases substantially or θ is approximately constant on B', in the sense that the L^2 norm of $\nabla\theta$ is small.

The other main component in Uhlenbeck's proof is a "perturbation theorem" for solutions which are close to a constant. We state this over a fixed pair of balls $B'' \subset B'$.

Proposition 2.3. *There are $\varepsilon, \kappa > 0$ such that if ω_0 is a constant form with norm* 1 *and ω is a solution of* (3) *over B' with $M' \leq 2$ such that*

$$\int_{B'} |\omega - \omega_0|^2 \leq \varepsilon, \tag{8}$$

then ω satisfies a Hölder estimate

$$|\omega(x) - \omega(y)| \leq \kappa |x - y|^{1/2}$$

for $x, y \in B''$.

This is of the same flavour as "small energy" results which we will encounter throughout this article. In fact one can go on from this, with a sufficiently small ε, to obtain estimates on all derivatives of ω over B''. The general idea is that the constraint (8) keeps the solution in the regime where the nonlinear equation is well approximated by its linearisation.

We now outline how Uhlenbeck puts these components together to prove Theorem 2.1. Notice that our equations (3) are preserved by translations and dilations of $\mathbf{R}^n$ and also by multiplying the solution ω by a non-zero constant. So the statements above for fixed pairs of balls and—in Proposition 2.3— for ω_0 of unit norm, scale to corresponding result on balls of arbitrary size and for any non-zero constant form ω_0.

Suppose again that ω is a solution over B and normalise so that $M = 1$. Suppose that M' is close to M, so that the L^2 norm of $\nabla\theta$ is small by (7). Let θ_0 be the average of θ over B'. The Poincaré inequality implies that the L^2 norm of $\theta - \theta_0$ is small and it also follows from the hypotheses of M, M' that $|\theta_0|$ will be close to 1. This is not immediately what is needed to apply Proposition 2.3, because that needs control of the L^2 norm of $|\omega - \omega_0|$ for a constant form ω_0. But, with additional arguments, Uhlenbeck achieves this control, for $\omega_0 = |\theta_0|^{2/p-1}\theta_0$.

The conclusion is that there is some fixed small $\lambda > 0$ such that if $M' > (1-\lambda)M$ then ω satisfies the hypotheses of the perturbation theorem (after rescaling) over B' and hence a $\frac{1}{2}$-Hölder estimate over the interior ball B''. This number λ will determine the Hölder exponent achieved in Theorem 2.1. (Uhlenbeck remarks on page 238 of [57] *"it looks like λ will be rather small!"*)

To prove Theorem 2.1 we suppose that the domain U contains the unit ball $B = B_1$. It suffices to estimate $|\omega(x) - \omega(0)|$ for small x, say $|x| \leq \frac{1}{4}$. (Strictly, ω is *a priori* only defined almost everywhere, so some extra words are needed to make sense of pointwise values, but we are ignoring such technicalities here.) Let M_j be the supremum of H on the 2^{-j} ball centred at the origin. If $M_1 \geq (1-\lambda)M_0$ we get a Hölder estimate on ω over the $\frac{1}{4}$-ball $B_{\frac{1}{4}}$ and we are done. If not, we have some definite decrease in the supremum: $M_1 \leq (1-\lambda)M_0$. Now we consider the same alternative for M_1 and M_2. If $M_2 \geq (1-\lambda)M_1$ we have our Hölder estimate over $B_{\frac{1}{8}}$ and in addition we know that for x in $B_{\frac{1}{4}}$

$$|\omega(0) - \omega(x)| \leq 2M_2^{1/p} \leq 2(1-\lambda)^{2/p}M_0^{1/p},$$

where we have used the facts that M_2 is the supremum of $|\omega|^p$ over the $\frac{1}{4}$-ball and $|\omega(0) - \omega(x)| \leq |\omega(0)| + |\omega(x)|$. Continue in the same way: either $M_{j+1} \leq (1-\lambda)M_j$ for all j or there is some k such that $M_{j+1} \leq (1-\lambda)M_j$ for all $j < k$ but $M_{k+1} \geq (1-\lambda)M_k$. In either case we get an estimate on $|\omega(x) - \omega(0)|$ in the manner above and a little bookkeeping shows that this yields the desired Hölder estimate. For example, consider the first situation when $M_j \leq (1-\lambda)^j M_0$ for all j. Then we must have $\omega(0) = 0$ and for x with $2^{-k-1} \leq |x| \leq 2^{-k}$ we have

$|\omega(x)| \leq M_0^{1/p}(1-\lambda)^{k/p}$. This gives

$$|\omega(x)| \leq K|x|^{\alpha},$$

with $\alpha = -p^{-1}\log_2(1-\lambda)$ and $K = M_0^{1/p}2^{\alpha}$.

There is an enormous literature in this area, especially on the p-harmonic equation. (At the time of writing, the *Acta* paper of Uhlenbeck has 316 citations on MathSciNet.) One subtle question is the optimal Hölder exponent. For example, in the case of p-harmonic functions in dimension $n = 2$, Iwaniec and Manfredi show in [26] that the optimal exponent (for $p \neq 2$) is

$$\frac{1}{6}\left(\frac{p}{p-1} + \sqrt{1 + \frac{14}{p-1} + \frac{1}{(p-1)^2}}\right).$$

There are also many papers on the limiting cases $p = 1, \infty$. One recent paper of Daskalopoulos and Uhlenbeck [12] makes connections between ∞-harmonic functions and Thurston's theory of homeomorphisms between hyperbolic surfaces minimising the Lipschitz constant.

3 Harmonic maps of surfaces

3.1 Background

Let (M,g) and (N,h) be Riemannian manifolds. The harmonic mapping equation for a map $f: M \to N$ is the Euler–Lagrange equation associated to the energy functional

$$E(f) = \int_M |\mathrm{d}f|^2,$$

where the norm $|\mathrm{d}f|$ is the standard one defined by g, h. Familiar cases are when M is 1-dimensional, where we get geodesics in N, and when N is 1-dimensional where we get harmonic functions on M. Written explicitly in local coordinates x^i on M and y^{α} on N the equations are

$$\Delta_M y^{\alpha} + \Gamma^{\alpha}_{\beta\gamma} y^{\beta}_{,i} y^{\gamma}_{,j} g^{ij},$$

where $\Gamma^{\alpha}_{\beta\gamma}$ are the Christoffel symbols on N. For analysis, it is often convenient to take N to be isometrically embedded in some large Euclidean space V, which is possible by Nash's embedding theorem. This is never essential but we will use that set-up in this article. Thus f can be thought of as a vector-valued function on M, constrained to lie in $N \subset V$. The harmonic mapping condition is that the projection of $\Delta_M f$ to the tangent bundle of N in V is zero. At each point y of N we have a second fundamental form B_y which is a symmetric bilinear map $TN_y \times TN_y \to \nu_y$, where ν is the normal bundle. The harmonic mapping equation can be written as

$$\Delta_M f + A_f(\mathrm{d}f, \mathrm{d}f) = 0, \tag{9}$$

where A_f is the symmetric bilinear map, at a point $x \in M$,

$$A_f : \mathrm{Hom}(TM_x, TN_{f(x)}) \times \mathrm{Hom}(TM_x, TN_{f(x)}) \to V,$$

obtained from $B_{f(x)}$, the metric on TM_x and the inclusion $\nu \subset V$. It will sometimes be convenient to extend A_f, using orthogonal projection from V to TN, to a bilinear map $\mathrm{Hom}(TM_x, V) \times \mathrm{Hom}(TM_x, V) \to V$, depending on $x \in M$ and $f(x) \in N$.

The dimension of M plays a crucial role in the theory of harmonic maps. In the words of Eells and Lemaire in the Introduction to [17] "*we imagine M made of rubber and N made of marble... the map is harmonic if it constrains M to lie on N in a position of elastic equilibrium*". In that picture we could say that higher-dimensional rubber is weaker and is inclined to tear when searching for an equilibrium position. When M is the 1-dimensional circle there is a geodesic in each homotopy class (a rubber band) minimising energy, but the analogue is not true in higher dimensions. For example it is easy to show that if M is a sphere of dimension 3 or more then the infimum of energy in any homotopy class of maps from M to N is zero. The critical dimension in the theory is $\dim M = 2$. This is bound up with Sobolev inequalities. Regarding N as isometrically embedded in the Euclidean space V, the energy functional is just the square of the usual L^2 norm of the derivative of a map $f : M \to V$, but restricted to maps with image in N. If $\dim M = 1$ the Sobolev inequalities state that maps with derivative in L^2 are continuous, in fact Hölder continuous with exponent $\frac{1}{2}$. A sequence of maps convergent in the Sobolev space L^2_1 converges pointwise and the constraint that the map takes values in N is preserved in the limit. In higher dimensions this is not true: dimension 2 is the borderline where a map with derivative in L^2 is in L^p for all p but not necessarily continuous: evaluation at a point is not well-defined for such a map—it is only defined up to sets of measure zero. The critical nature of dimension two is related to conformal invariance of the energy. In a general dimension $\dim M = n$, if we multiply the metric g by a conformal factor λ we change $|\mathrm{d}f|^2$ by λ^{-1} and the volume element by $\lambda^{n/2}$, so when $n = 2$ these factors cancel.

The main topic of this Section 3 is the paper [39] of Sacks, which opened up the theory of harmonic maps in the critical dimension 2.

3.2 Bubbling

We begin with an illuminating example of maps from a flat 2-torus M to the standard round 2-sphere S^2, both oriented. Consider the homotopy class of maps of degree 1 from M to S^2. There is a simple lower bound on the energy of such maps. We can consider M and S^2 as Riemann surfaces with area forms ω_M, ω_{S^2}. Then we have at each point of M:

$$f^*(\omega_{S^2}) \leq \frac{1}{2}|\mathrm{d}f|^2 \omega_M \tag{10}$$

with equality if and only if $\mathrm{d}f$ is complex linear. This is a simple calculation with 2×2 matrices. If f has degree 1 then the integral of $f^*(\omega_{S^2})$ is the area of S^2. So we get the lower bound

$$E(f) \geq 2\,\mathrm{Area}(S^2).$$

This lower bound is not achieved, because if it were the map would be holomorphic and by elementary Riemann surface theory there is no degree 1 holomorphic map from a torus to the Riemann sphere. On the other hand we can construct maps with energy arbitrarily close to this lower bound. Let D_r be a small disc of radius r in M centred at a point x_0 and identify it isometrically with the standard r-disc in $\mathbf{C}$. Now take a very large disc $D_R \subset \mathbf{C}$ and consider it as a subset of S^2 via the usual description $S^2 = \mathbf{C} \cup \{\infty\}$. So the complement of D_R in S^2 is a small disc centred at the point at infinity. Let $F : S^2 \to S^2$ be a map which is the identity on most of D_R but which collapses the boundary of D_R to the point at infinity. It is clear that when R is large we can do this in such a way that the energy of F is as close as we please to that of the identity map, which is $2\mathrm{Area}(S^2)$. Finally, define a map $f : M \to S^2$ which sends the complement $M \setminus D_r$ of D_r to the point $\infty \in S^2$ and on D_r is the composite $F \circ \underline{\lambda}$ where $\underline{\lambda} : D_r \to D_R$ is multiplication by $\lambda = R/r$.

The energy of this map f is exactly the same as that of F. This follows immediately from the fact that the energy is a conformal invariant for 2-dimensional domains, it only depends on the conformal class of the metric. Thus we get a "minimising sequence" f_i of degree-1 maps from M to S^2 whose energy tends to the infimum $2\mathrm{Area}\, S^2$ by making the construction above with a sequence $R_i \to \infty$. For large i the image of a small disc in M covers most of S^2 and away from x_0 the maps approach the constant harmonic map.

What Sacks and Uhlenbeck established is, roughly speaking, that this is the only way that things can go wrong when trying to apply variational arguments to the energy functional on surfaces. More precisely, they consider a 1-parameter family of deformations of the functional, with parameter $\alpha \geq 1$:

$$E_\alpha(f) = \int_M (1 + |\mathrm{d}f|^2)^\alpha.$$

(Some formulae would be neater if one used the integral of $|\mathrm{d}f|^{2\alpha}$ but this would lead to a degenerate equation and extra difficulties of the kind discussed in Section 2.) When $\alpha = 1$ the functional E_α is equal to $E(f)$, up to a constant. For $\alpha > 1$ the functional controls the $L^{2\alpha}$ norm of the derivative and one has the favourable Sobolev embedding $L^{2\alpha}_1 \to C^0$. This means that there is a complete "Palais–Smale" variational theory, something which was worked out in the earlier paper [56] of Uhlenbeck. So the functional E_α attains its minimum in each homotopy class and more generally there must be sufficient critical points to account for the topology of the mapping space, by minimax and Morse theory arguments. The Sacks and

Uhlenbeck strategy is to seek critical points of E_1 as limits of critical points of the E_α as $\alpha \to 1$. The advantage of this approach, compared with studying minimising or minimax sequences for E directly, is that the critical points of E_α satisfy an elliptic equation and this improves the convergence properties, as we will see. Even if a minimiser for E_1 exists there will always be minimising sequences which only converge in a weak sense, not in C^∞.

We can now state more precisely one of the main results of Sacks and Uhlenbeck for maps between compact manifolds M,N with dim $M = 2$.

Theorem 3.1. *Let $\alpha_i \geq 1$ with $\alpha_i \to 1$ as $i \to \infty$ and let $f_i : M \to N$ be critical points of E_{α_i} with $E_{\alpha_i}(f_i) \leq E_{\max}$ for some fixed $E_{\max}$. Then, after perhaps passing to a subsequence $\{i'\}$, there is a finite set $S = \{q_1, \dots q_d\} \subset M$ and a harmonic map $f : M \to N$ such that $f_{i'}$ converge to f in C^∞ on compact subsets of $M \setminus S$. In addition there are harmonic maps $F_1, \dots, F_d : S^2 \to N$ such that for each $j \in \{1, \dots, d\}$ a suitable sequence of rescalings of $f_{i'}$ near q_j converge on compact subsets of $\mathbf{C} = S^2 \setminus \{\infty\}$ to F_j.*

To explain the last statement; we mean that there are points $p_{i'j}$ converging to q_j and scale factors $\lambda_{i'j}$ tending to ∞ with i' so that if we identify a small disc centred at p_{ij} with a small disc in $\mathbf{C}$ and compose with a scaling map $\underline{\lambda}_{ij}$ of the kind discussed above the resulting maps converge to F_j. One says that the sequence of maps $f_{i'}$ is "bubbling" at the points q_j.

The statement of our Theorem 3.1 here does not capture all that Sacks and Uhlenbeck established. For example, they show that the homotopy classes of f_i and f in $[M,N]$ differ by a class in $[S^2,N]$. But the statement of Theorem 3.1 gives the general idea. A complete discussion involves the notion of a "bubble tree" of maps, which was worked out later; see for example [34].

This theorem of Sacks and Uhlenbeck implies the existence of harmonic maps in many specific situations. For example, if it is known that there is no harmonic map from S^2 to N then for any surface M there is a minimising harmonic map in any homotopy class $[M,N]$. One early and famous application came in the proof by Siu and Yau of the Frankel conjecture [48]. The conjecture was that projective spaces are the only compact complex manifolds admitting Kähler metric with positive biholomorphic sectional curvature. The result of Sacks and Uhlenbeck shows that there is a nonconstant minimising harmonic map from S^2 to such a manifold. Siu and Yau proved, by studying the second variation formula and using the curvature condition, that this map is holomorphic and then the geometry of the resulting family of holomorphic curves shows that the manifold is a projective space. Another important application of the Sacks–Uhlenbeck result, this time in Riemannian geometry, was the "sphere theorem" of Micallef and Moore [30], discussed in [15].

By far the greatest impact of the phenomena uncovered by Sacks and Uhlenbeck came in the special case of the holomorphic maps introduced as a tool in symplectic topology by Gromov in 1987. Here we consider a symplectic manifold (N,ω) with a compatible almost-complex structure J and resulting Riemannian metric $|\xi|^2 = \omega(\xi, J\xi)$. For any oriented Riemannian surface $f : M \to N$ there is an inequality

$$2\int_M f^*(\omega) \leq E(f),$$

with equality if and only if f is holomorphic (i.e. the derivative at each point is complex linear with respect to the Riemann surface structure on M and the almost-complex structure J). This is a generalisation of (10), in the case when N is a surface. This inequality implies that a holomorphic map minimises energy in its homotopy class, so in particular is harmonic. Theorem 3.1, with all $\alpha_i = 1$, describes the convergence behaviour of sequences of these holomorphic maps and becomes the foundation for all of the applications to symplectic topology such as Gromov–Witten invariants, Lagrangian Floer homology, Fukaya categories.... There are many expositions of the theory in this restricted context of holomorphic maps, for example [29], [69].

3.3 Small energy

For simplicity we just discuss the proof of Theorem 3.1 in the case when all α_i are 1: the general case does not involve major extra difficulties. The proof has two main components. The first is a "small energy" estimate.

Theorem 3.2. *Let N be a compact Riemannian manifold and D be the unit disc in $\mathbf{C}$. There are ε, C such that if $f : D \to N$ is harmonic with $E(f) \leq \varepsilon$ then*

$$|\mathrm{d}f(0)|^2 \leq CE(f).$$

More generally, we can choose ε so that if $E(f) \leq \varepsilon$ then $E(f)$ controls all derivatives of f on a fixed interior disc, say the $\frac{1}{2}$-sized disc. The conformal invariance of the energy implies that these estimates apply *with the same small energy threshold* to discs of any size: if f has energy less than ε on the disc D_r we get

$$|\nabla^k f| \leq c_k r^{-k} \sqrt{E(f)}$$

on $D_{r/2}$.

The second component is the removability of point singularities.

Theorem 3.3. *If $f : D \setminus \{0\} \to N$ is a harmonic map with $E(f) < \infty$ then f extends smoothly to D.*

Given these local statements the proof of Theorem 3.1 is relatively straightforward, using arguments of a kind which we will see several other times in this article. We have $f_i : M \to N$ harmonic with energy less than a fixed number H. After passing to a subsequence we can suppose that the energy densities $|\mathrm{d}f_i|^2$ converge as Radon measures: that is, for any continuous function ϕ on M the integrals of $\phi|\mathrm{d}f_i|^2$ over M have a limit as $i \to \infty$. Fix a non-increasing cut-off function σ on $[0, \infty)$, equal to 1 on $[0, 1]$ and supported in $[0, 2]$. For $x \in M$ and $r > 0$, let $\chi_{r,x}$ be the function on M

$$\chi_{r,x}(y) = \sigma(r^{-1}d(x,y)),$$

where $d(\ ,\ )$ is the Riemannian distance. So $\chi_{r,x}$ is a smoothing of the characteristic function of the r-disc $D_{r,x}$ about x. Define $\mu(x,r)$ by

$$\mu(x,r) = \lim_{i\to\infty} \int_M \chi_{r,x}\ |\mathrm{d}f_i|^2.$$

Then $\mu(x,r)$ is an increasing function of r and has a limit $\mu(x)$ as $r \to 0$. By construction

$$\int_{D_{x,r}} |\mathrm{d}f_i|^2 \leq \int \chi_{r,x}|\mathrm{d}f_i|^2 \leq \int_{D_{x,2r}} |\mathrm{d}f_i|^2. \tag{11}$$

Let S be the set of points x in M where $\mu(x) > \varepsilon/2$. The right-hand inequality in (11) implies that there are at most $2E_{\max}/\varepsilon$ points x in S (by taking $2r$ less than half the distance between the points, so that the $2r$ discs with these centres are disjoint). On the other hand, taking ε as in Theorem 3.2, if $\mu(x) < \varepsilon/2$ then for some sufficiently small $r = r(x) > 0$ and all large enough i the left-hand inequality in (11) gives

$$\int_{D_{x,r}} |\mathrm{d}f_i|^2 \leq \varepsilon/2,$$

and the small energy theorem gives estimates on all derivatives of the f_i in $D_{x,r/2}$.

From these arguments we get a finite set $S = \{q_1, \dots q_d\}$ in M such that all derivatives of the f_i are bounded on compact subsets of the complement $M \setminus S$. Taking a subsequence we can assume that the maps converge on the complement to a harmonic map on the punctured manifold with energy at most C, and the removal of singularities theorem implies that this extends to a smooth harmonic map f from M to N, as stated in the first part of Theorem 3.1.

The second part of Theorem 3.1 involves the rescaling construction. Fix a point q_j. For all large i there must be points near q_j where the derivative of f_i is large. Let p_{ij} be a point near q_j where $|\mathrm{d}f_i|$ is maximal. Define λ_{ij} to be these local maximal values. Then after rescaling by these factors with centre p_{ij} we get a sequence of harmonic maps F_{ij} defined on a sequence of large discs in $\mathbf{C}$ which exhaust $\mathbf{C}$ as $i \to \infty$. These maps have bounded derivative and energy so by the same arguments, perhaps passing to a suitable subsequence, they converge to a harmonic map from S^2 to N. That is, we first get a harmonic map from $\mathbf{C}$ to N and then apply the removal of singularities theorem at the point at infinity in S^2. The limiting map is not constant since by construction the derivative of the map F_{ij} at the origin has size 1.

We proceed to discuss the proof of the small energy result Theorem 3.2, leaving that of Theorem 3.3 to the next subsection. To recap, we have a V-valued function f on the disc D which satisfies the PDE $\Delta f = A_f(\mathrm{d}f, \mathrm{d}f)$ and takes values in the compact submanifold N. The discussion below applies to any PDE of this shape, for a smooth map A from $D \times V$ to symmetric bilinear maps $\mathrm{Hom}(T,V) \times \mathrm{Hom}(T,V) \to V$, where T denotes the tangent space of the disc. (In the case at hand we could always extend A in some way to fit into this framework. The fact that, in the case at hand, f maps into the submanifold N is only used in that it gives a bound on $|f|$.)

A basic fact of elliptic PDE theory is that for any $q > 1$ there is a constant K_q such that for all compactly supported functions ϕ on D we have

$$\|\phi\|_{L^q_2} \leq K_q \|\Delta\phi\|_{L^q}. \tag{12}$$

(Here, and throughout this article we write L^q_k for the Sobolev space based on the L^q norm of all derivatives of order $\leq k$. We write the norm as $\| \ \|_{L^q_k}$ or sometimes $\| \ \|_{q,k}$ to improve readability.)

Before going on to the proof of Theorem 3.2 we consider a different situation where we suppose given a solution of the equation with bound on the $L^{2\alpha}$ norm of $\mathrm{d}f$ for some α with $\alpha > 1$. Then we can do a straightforward "bootstrapping" argument. (Here, and in various other parts of this article, we use the convention that c is a constant that can change from line to line.)

By choice of the origin in V we may suppose that the integral of f over the disc is 0. Let χ be a compactly supported function on the disc, equal to 1 on an interior disc D'. Then we have

$$\Delta(\chi f) = \chi \Delta f + 2\mathrm{d}\chi \cdot \mathrm{d}f + \Delta\chi \ f,$$

and

$$\chi A(\mathrm{d}f, \mathrm{d}f) = A(\chi \, \mathrm{d}f, \mathrm{d}f) - A(\mathrm{d}\chi \otimes f, \mathrm{d}f).$$

Thus we have a pointwise bound

$$|\Delta(\chi f)| \leq c\left(|\mathrm{d}(\chi f)| \ |\mathrm{d}f| + |f| + |\mathrm{d}f| + |f| \ |\mathrm{d}f|\right), \tag{13}$$

where c depends only on χ and the given map A. From this we readily obtain, applying the Cauchy–Schwarz inequality,

$$\|\Delta(\chi f)\|_\alpha \leq c\left(\|f\|^2_{2\alpha,1} + \|f\|_{2\alpha,1} \ \|f\|_{2\alpha} + \|f\|_\alpha\right).$$

Since the disc has finite area the $L^{2\alpha}$ norm of f controls the L^α norm and the fact that the integral of f vanishes means that the $L^{2\alpha}$ norm of $\mathrm{d}f$ controls that of f. So we get an inequality

$$\|\Delta(\chi f)\|_\alpha \leq c\left(\|f\|^2_{2\alpha,1} + \|f\|_{2\alpha,1}\right),$$

and hence by the elliptic inequality (12),

$$\|\chi f\|_{\alpha,2} \leq c\left(\|f\|^2_{2\alpha,1} + \|f\|_{2\alpha,1}\right).$$

If $\alpha < 2$ we have a Sobolev embedding $L^\alpha_2 \to L^r_1$ with $r = 2\alpha/(2-\alpha)$. Thus the inequality above gives an L^r bound on $\mathrm{d}(\chi f)$ and so an L^r bound on $\mathrm{d}f$ over the interior disc D'. Since $r > 2\alpha$ this is an improvement on the $L^{2\alpha}$ bound that we started with. If $\alpha > 2$ we have a Sobolev embedding $L^\alpha_2 \to C^{1,\mu}$ for $\mu = 1 - 2/\alpha$.

Starting with our $L^{2\alpha}$ bound on $\mathrm{d}f$, for any $\alpha > 1$ we can iterate such arguments, working on a decreasing sequence of discs, to get interior bounds on all derivatives of f in terms of the $L^{2\alpha}$ norm of $\mathrm{d}f$ over D. In the same fashion we get a *regularity* statement: if we only know at the outset that f is in $L^{2\alpha}_1$ we show that in fact it is smooth in the interior of the disc.

Now we go on to the proof of Theorem 3.2. In the situation above *any* bound on $\|\mathrm{d}f\|_{L^{2\alpha}}$ gives bounds on higher derivatives in the interior. The difference in Theorem 3.2 is that we only get such a bound when $\|\mathrm{d}f\|_{L^2}$ is small. Taking $\alpha \in (1,2)$ and $r = 2\alpha/(2-\alpha)$ we observe that $1/\alpha = 1/r + 1/2$. Thus we can apply Hölder's inequality to (13) to get

$$\|\Delta(\chi f)\|_\alpha \leq c_1 \|\mathrm{d}(\chi f)\|_r \|\mathrm{d}f\|_2 + c_2 \left(\|\mathrm{d}f\|_2 \|f\|_r + \|f\|_\alpha\right).$$

By the elliptic inequality and the Sobolev embedding, $\|\Delta(\chi f)\|_{L^\alpha}$ controls $\|\mathrm{d}(\chi f)\|_{L^r}$ and the assumption that f has integral zero means that $\|\mathrm{d}f\|_{L^2}$ controls both $\|f\|_{L^\alpha}$ and $\|f\|_{L^r}$. So we have

$$\|\mathrm{d}(\chi f)\|_r \leq c_3 \|\mathrm{d}f\|_2 \|\mathrm{d}(\chi f)\|_r + c_4(\|\mathrm{d}f\|_2 + \|\mathrm{d}f\|_2^2).$$

Take $\sqrt{\varepsilon} = 1/(2c_3)$. Then if $\|\mathrm{d}f\|_{L^2} \leq \sqrt{\varepsilon}$ we re-arrange to get

$$\|\mathrm{d}(\chi f)\|_r \leq 2c_4(\|\mathrm{d}f\|_2 + \|\mathrm{d}f\|_2^2).$$

We have $r > 2$ so, replacing D by the smaller disc D', we are in the position considered before and we can go on to estimate all higher derivatives in the interior.

Arguments with the same structure as this will appear often in this article so we give them a name: "*critical quadratic re-arrangement*". The crux is that we get the same exponent in Hölder's inequality $L^2 \times L^r \to L^\alpha$ and in the Sobolev embedding $L^\alpha_1 \to L^r$. This is not a coincidence: it can be traced back to the scaling behaviour of the norms. The small-energy threshold ε produced by this argument is computable. That number might be rather small but it follows from the further development of the theory, as in the proof of Theorem 3.1 above, that in Theorem 3.2 ε can be taken to be any number less than the least energy of a harmonic map from S^2 to N.

3.4 The stress energy tensor and removal of point singularities

Before beginning the proof of Theorem 3.3 we make a digression to review some background which will be used in the proof and also later in this article.

Suppose that we have some functional $\mathscr{F}$ which depends on a Riemannian metric g on a manifold M and other "fields" (in the case at hand the fields are maps from M to the fixed Riemannian manifold N and the functional is the energy). The variation of $\mathscr{F}$ with respect to the fields, holding the metric g fixed, produces Euler–Lagrange equations like the harmonic map equation. But we can also consider variations of the metric g, holding the fields fixed. By general principles the first variation of $\mathscr{F}$ can be written as

$$\delta_g \mathscr{F} = \int_M (T, \delta g).$$

where the tensor T is a section of $s^2 T^* M$ called the stress-energy tensor, which depends on the fields and the metric. Any natural functional arising in differential geometry will be diffeomorphism invariant. It follows that if the fields satisfy the Euler–Lagrange equations generated by $\mathscr{F}$ then if δg is defined by an infinitesimal diffeomorphism—i.e. δg is the Lie derivative $L_v g$ of g along a vector field v on the manifold—then $\delta_g \mathscr{F} = 0$. This is the identity div $T = 0$ or in index notation

$$T^{ij}_{;j} = 0. \tag{14}$$

When the functional $\mathscr{F}$ is conformally invariant the tensor T is trace-free. If v is a conformal Killing field on the manifold (M, g) (that is, $L_v g = \mu g$ for some function μ on M) then the contraction of T by v is a co-closed 1-form. In index notation

$$(T^{ij} v_i)_{;j} = T^{ij}_{;j} v_i + T^{ij} v_{i;j}.$$

The first term on the right-hand side vanishes by (14) and the second vanishes because T^{ij} is symmetric and trace-free and the Lie derivative $L_v g$ is the symmetrisation $v_{i;j} + v_{j;i}$.

In this section we will apply this discussion in the case of the harmonic maps energy with 2-dimensional oriented domain M, which can also be viewed as a Riemann surface. Taking the real part gives an isomorphism between the tensor square of $T^* M$, regarded as a complex line bundle, and the trace-free symmetric tensors. So we have a quadratic differential τ with $T = \mathrm{Re}\,\tau$. The equation (14) goes over to the condition that τ be a holomorphic quadratic differential; this is the *Hopf differential* defined by a harmonic map from a Riemann surface. In a local complex coordinate $z = x + iy$

$$\tau = ((f_x, f_x) - (f_y, f_y) + 2i(f_x, f_y))\, \mathrm{d}z^2.$$

Here we are writing $f_x = \frac{\partial f}{\partial x}$ etc. Similarly, a conformal Killing field v can be viewed as a holomorphic vector field v and the contraction of v with τ is a holomorphic—hence closed and co-closed—1-form. This completes our digression.

The Sacks and Uhlenbeck proof of the removal of singularities theorem goes through a differential inequality for the energy on small discs. Let

$$E(r) = \int_{D_r} |\mathrm{d}f|^2,$$

so clearly $E(r)$ is an increasing function of r and tends to zero as $r \to 0$. We may suppose that $E(\frac{3}{2})$ is less than the small energy value ε of Theorem 3.2. Applying that result to the disc of radius $|z|/2$, say, centred at a point z we get,

$$|\mathrm{d}f(z)|^2 \leq CE(3|z|/2)\ |z|^{-2}, \tag{15}$$

so $|\mathrm{d}f|$ is $o(|z|^{-1})$. The removal of singularities theorem is proved by showing that $|\mathrm{d}f|$ is $O(|z|^{\delta-1})$ for some $\delta > 0$. If we know this then $\mathrm{d}f$ is in $L^{2\alpha}$ for some $\alpha > 1$ and we can apply the regularity theory discussed in the previous section to see that f is smooth across the origin.

Using (15), we see then that it suffices to show that $E(r)$ is $O(r^\kappa)$ for some $\kappa > 0$. The differential inequality to be established is that

$$\kappa E(r) \leq r\frac{\mathrm{d}}{\mathrm{d}r}E(r). \tag{16}$$

If we know this then it follows by a simple comparison argument that

$$E(r) \leq r^\kappa E(1),$$

as required.

It is convenient to exploit the conformal invariance of the problem and to work on the cylinder $(-\infty, 0] \times S^1$ with coordinates (s, θ). So $r = \mathrm{e}^{-s}$ and we now write

$$E(S) = \int_{s\leq S}\int f_s^2 + f_\theta^2 \ \mathrm{d}\theta\, \mathrm{d}s,$$

where subscripts denote partial derivatives and we are writing f_s^2 for (f_s, f_s). We know that f is bounded and that the derivatives f_s, f_θ tend to zero as $s \to -\infty$. We want to show that for some $\kappa > 0$

$$\kappa E \leq \frac{\mathrm{d}E}{\mathrm{d}S}.$$

By translation invariance it suffices to prove this when $S = 0$, that is:

$$\kappa \int_{s\leq 0}\int f_s^2 + f_\theta^2 \ \mathrm{d}\theta\, \mathrm{d}s \leq \int_{s=0} f_2^2 + f_\theta^2 \ \mathrm{d}\theta. \tag{17}$$

By the general theory reviewed above, the contraction of the Hopf differential with the Killing field $\frac{\partial}{\partial s}$ gives the closed 1-form $(f_s^2 - f_\theta^2)\,\mathrm{d}\theta$. It follows that the integral

$$\int_0^{2\pi} f_s^2 - f_\theta^2 \ \mathrm{d}\theta$$

is independent of s and since the integrand tends to zero as $s \to -\infty$ the integral vanishes. In other words, for each fixed s,

$$\int f_s^2 \ \mathrm{d}\theta = \int f_\theta^2 \ \mathrm{d}\theta. \tag{18}$$

For purposes of exposition, let us consider for a moment the case when $A=0$, so f is an ordinary harmonic function: $\Delta f = 0$. Then we have the usual integration-by-parts formula over a finite cylinder:

$$\int_{S_0\leq s\leq 0}\int f_s^2 + f_\theta^2 \,\mathrm{d}\theta\,\mathrm{d}s = \int_{s=0}(f,f_s)\,\mathrm{d}\theta - \int_{s=S_0}(f,f_s)\,\mathrm{d}\theta.$$

Since f_s tends to 0 as $s\to-\infty$ and f is bounded the boundary term at $s=S_0$ tends to zero as $S_0\to-\infty$ and we get

$$\int_{s\leq 0}\int f_s^2 + f_\theta^2 \,\mathrm{d}\theta\,\mathrm{d}s = \int_{s=0}(f,f_s)\,d\theta. \tag{19}$$

The left-hand side of this formula is unchanged if we add a constant to f, so we can suppose that the integral of f over the boundary $\{s=0\}$ vanishes. Now for any function g on the circle of integral zero we have an inequality

$$\int g^2\,\mathrm{d}\theta \leq \int g_\theta^2\,\mathrm{d}\theta. \tag{20}$$

This is clear from the Fourier series. Thus, combining with Cauchy–Schwarz,

$$\left(\int_{s=0}(f,f_s)\,\mathrm{d}\theta\right)^2 \leq \int_{s=0} f_s^2\,\mathrm{d}\theta \int_{s=0} f_\theta^2\,\mathrm{d}\theta. \tag{21}$$

Combining (19) and (21) we have

$$\int_{s\leq 0}\int f_s^2 + f_\theta^2\,\mathrm{d}\theta\,\mathrm{d}s \leq \tfrac{1}{2}\int_{s=0} f_s^2 + f_\theta^2\,\mathrm{d}\theta,$$

which is the required inequality with $\kappa=2$. This gives the growth rate $E(r)=O(r^2)$, which is indeed what will occur for smooth maps.

The idea now is to modify this discussion to take account of the nonlinear term $A(df,df)$ at the cost of changing the constant κ. So suppose again that f is our harmonic map with $\Delta f = A(\mathrm{d}f,\mathrm{d}f)$. Taking the inner product of this equation with f and integrating over the cylinder we get an extra term

$$\int_{s\leq 0}\int (f,A(\mathrm{d}f,\mathrm{d}f))\,\mathrm{d}\theta\,\mathrm{d}s$$

which is bounded in modulus by the integral of $c_1|f|\,|\mathrm{d}f|^2$ for some c_1. If we knew that, over the cylinder, we have $|f|\leq \sigma c_1^{-1}$ for some $\sigma<1$ this would give the desired inequality with $\kappa = 2/(1-\sigma)$. The problem is that at this stage we know that f is bounded on the cylinder but we do not know how to make this bound arbitrarily small. To overcome this, take the average values

$$F(S) = \frac{1}{2\pi}\int_{s=S} f(s,\theta)\,\mathrm{d}\theta,$$

and write $g(S,\theta) = f(S,\theta) - F(S,\theta)$. Then we have

$$\int_{s\leq 0}\int (g,\Delta f)\,\mathrm{d}\theta\,\mathrm{d}s = \int_{s\leq 0}\int (\mathrm{d}g,\mathrm{d}f)\,\mathrm{d}\theta\,\mathrm{d}s + \int_{s=0}(g,f_s)\,\mathrm{d}\theta, \tag{22}$$

since one sees as before that the other boundary term for a finite cylinder tends to 0. Exactly the same argument as before gives

$$\int_{s=0}(g,f_s) \leq (1/2)\int_{s=0} f_s^2 + f_\theta^2\,\mathrm{d}\theta$$

We have $(\mathrm{d}g,\mathrm{d}f) = |\mathrm{d}f|^2 - (f_s,F_s)$ so we get

$$\int_{s\leq 0}\int f_s^2 + f_\theta^2\,\mathrm{d}\theta\,\mathrm{d}s \leq (1/2)\int_{s=0} f_s^2 + f_\theta^2\,\mathrm{d}\theta + I + II,$$

where

$$I = \int_{s\leq 0}\int (g,\Delta f)\,\mathrm{d}\theta\,\mathrm{d}s,$$

and

$$II = \int_{s\leq 0}\int (f_s,F_s)\,\mathrm{d}\theta\,\mathrm{d}s.$$

The integrand in I is bounded by $c_1|g||\mathrm{d}f|^2$. For each fixed s the integral of g is zero and there is an inequality in the same vein as (20)

$$|g|^2 \leq c_2\int g_\theta^2\,\mathrm{d}\theta = c_2\int f_\theta^2\,\mathrm{d}\theta.$$

(This is the Sobolev embedding $L_1^2 \to C^0$ in dimension 1.) So we can suppose that $|g|$ is as small as we please; say $|g| \leq \sigma c_1^{-1}$ for $\sigma < \frac{1}{2}$. Then the term I is bounded by σ times the energy.

Turning to the term II: consider the integral over θ for fixed $s = S$. This is

$$(2\pi)^{-1}\left|\int_{s=S} f_s\,\mathrm{d}\theta\right|^2,$$

which is bounded by

$$\int_{s=S} f_s^2\,\mathrm{d}\theta = (1/2)\int_{s=S} f_s^2 + f_\theta^2\,\mathrm{d}\theta,$$

using (20) again. Putting things together we get

$$(\tfrac{1}{2}-\sigma)\int_{s\leq 0}\int f_s^2 + f_\theta^2\,\mathrm{d}\theta\,\mathrm{d}s \leq (1/2)\int_{s=0} f_s^2 + f_\theta^2\,\mathrm{d}\theta,$$

which is the desired inequality with $\kappa = 1 - 2\sigma$.

In the case of pseudoholomorphic curves the proof of the differential inequality is simpler, using the fact that the energy is twice the integral of $f^*(\omega)$, where ω is the symplectic form. If one can write $\omega = \mathrm{d}\alpha$ for a 1-form α over a neighbourhood of the image of f then Stokes' theorem expresses the energy as a boundary integral. (We will use an argument like this in the proof of Theorem 6.2 below.)

4 Harmonic maps in higher dimensions

4.1 Monotonicity of normalised energy

The main topic of this Section 4 is Uhlenbeck's work with Schoen in the paper [41], on weak solutions to the harmonic map equation, but we begin in this subsection with some background and results for smooth maps.

Consider again the variational theory on an n-dimensional manifold M of a functional $\mathscr{F}(g,\Phi)$ given by the integral of an n-form $F(g,\Phi)$. Suppose that this functional has a non-zero scaling weight w under conformal change of the metric in that

$$F(\lambda g, \Phi) = \lambda^w F(g,\Phi).$$

It follows that the trace of the energy momentum tensor is $wL(g,\Phi)\mathrm{vol}_g$. Let v be a conformal vector field, so $v_{i;j} + v_{j;i} = 2\mu\, g_{ij}$ for some function μ. Then we have

$$(v_i T^{ij})_j = (v_{i;j} T_{ij}) = w\mu F(g,\Phi).$$

So we can write

$$w\mu F(g,\Phi) = \mathrm{d}\eta,$$

where η is the $(n-1)$-form $*(v_i T^{ij})$. Thus if U is a domain in M with compact closure and smooth boundary we have

$$w \int_U \mu\, F(g,\Phi) = \int_{\partial U} \eta. \tag{23}$$

Apply this discussion to the harmonic maps energy functional on a manifold M of dimension $n > 2$, so the field Φ is $f : M \to N$ and $F(g,f) = |\mathrm{d}f|^2 \mathrm{vol}_g$ which has a conformal weight $w = n/2 - 1$. Suppose that $M = \mathbf{R}^n$, so we have a conformal vector field $r\frac{\partial}{\partial r}$ as before and the function μ is the constant 1. The stress-energy tensor is

$$T_{ij} = (\nabla_i f, \nabla_j f) - \tfrac{1}{2}|\mathrm{d}f|^2 \delta_{ij}$$

One sees then that the restriction of the $(n-1)$-form η to the unit sphere is

$$\eta|_{S^{n-1}} = \left(\tfrac{1}{2}|\mathrm{d}f|^2 - |\nabla_r f|^2\right) \mathrm{dvol}_{S^{n-1}},$$

where ∇_r is the radial derivative. So the identity (23) is

$$(n-2)\int_B |\mathrm{d}f|^2 = \int_{\partial B} |\mathrm{d}f|^2 - 2|\nabla_r f|^2, \tag{24}$$

(which agrees with (18) in the case $n=2$). Thus

$$(n-2)\int_B |\mathrm{d}f|^2 \leq \int_{\partial B} |\mathrm{d}f|^2$$

with equality if and only if $\nabla_r f = 0$ on ∂B. If we apply the same argument to the ball B_r of radius r we get

$$(n-2)\int_{B_r} |\mathrm{d}f|^2 \leq r\int_{\partial B} |\mathrm{d}f|^2.$$

Let $\widehat{E}(r)$ be the *normalised energy*

$$\widehat{E}(r) = \frac{1}{r^{n-2}}\int_{B_r} |\mathrm{d}f|^2.$$

The inequality above is equivalent to the *monotonicity condition* $\mathrm{d}\widehat{E}/\mathrm{d}r \geq 0$ and in fact for $r_1 < r_2$

$$\widehat{E}(r_1) = \widehat{E}(r_2) - 2\int_{r_1<r<r_2} r^2|\nabla_r f|^2.$$

A good way to think about the normalised energy is through rescaling. Map the unit ball B to the ball B_r by $x \mapsto rx$ and compose with the restriction of f to B_r to get $\tilde{f}: B \to N$. Then the normalised energy of f on B_r is the energy of $\tilde{f}$ on the unit ball B. The monotonicity condition says that the map f "looks better"—in the sense of having smaller energy—if we look at it on smaller and smaller scales in this manner. Another useful observation is that if f is the composite of a map $\underline{f}$ from $\mathbf{R}^2$ to N with an orthogonal projection from $\mathbf{R}^n$ to $\mathbf{R}^2$ then the normalised energy for f agrees with the ordinary energy for $\underline{f}$, up to a factor.

One important consequence of the monotonicity property is a small energy result. The statement is essentially the same as in the 2-dimensional case of the previous section. A difference is that the equations now depend essentially on the Riemannian metric on M so we formulate the statement a bit differently.

Proposition 4.1. *Let M,N be compact Riemannian manifolds. There are $\varepsilon, r_0, C > 0$ such that if $B_{x,r}$ is a metric r-ball in M with $r \leq r_0$ and the normalised energy $\widehat{E}(B_r)$ of f on B_r is less than ε then on the half-sized ball $B_{x,r/2}$ we have*

$$|\mathrm{d}f|^2 \leq Cr^{-2}E(B_r).$$

As before, once we have the L^∞ bound on $\mathrm{d}f$ we can go on to get estimates on all higher derivatives. This small energy statement can be proved in a manner similar to the proof above of Theorem 3.2 but using Morrey spaces (which we will encounter in subsection 6.3 below) in place of L^p spaces. We will discuss here the proof of

Schoen in [40], using a "worst point" argument, which could also be used in the 2-dimensional case of Theorem 3.2.

For simplicity we suppose that M is locally Euclidean, so the discussion above applies to give a monotonicity formula for balls of sufficiently small size. For a general Riemannian manifold M we do not have exact formulae for the normalised energy but the equations hold with extra error terms which can be made as small as we please, since the geometry is close to Euclidean on small scales, and the same arguments work with minor modifications. By scale invariance and adjustment of constants we can suppose that the ball $B_{x,r}$ is the unit ball B in $\mathbf{R}^n$ and that f has normalised energy at most $E \leq \varepsilon$ on any interior ball.

For a point $x \in B$ let $D(x)$ be the distance to the boundary of B, i.e. $D(x) = 1 - |x|$. The idea is to consider the quantity

$$M = \max_{x \in B} D(x)|\mathrm{d}f(x)|.$$

Since the function D vanishes on the boundary, the maximum is achieved at some interior point x_0. For $\rho \leq 1$, rescale the ball of radius $\rho D(x_0)/2$ with centre x_0 to unit size to get a map $\tilde{f}_\rho$ on the unit ball B with the properties

- $\int_B |\mathrm{d}\tilde{f}_\rho|^2 \leq E$;
- $|\mathrm{d}\tilde{f}_\rho| \leq 4M\rho$ on B;
- $|\mathrm{d}\tilde{f}_\rho(0)| = 2M\rho$;

where the first item uses the small energy property of f on the interior ball. Now we have

$$|\Delta \tilde{f}_\rho| \leq c|\mathrm{d}\tilde{f}_\rho|^2.$$

Elliptic theory gives an inequality

$$|\mathrm{d}\tilde{f}_\rho(0)| \leq c\left(\|\Delta \tilde{f}_\rho\|_{L^\infty} + \|\mathrm{d}\tilde{f}_\rho\|_{L^2}\right),$$

so we get

$$|\mathrm{d}\tilde{f}_\rho(0)| \leq c_1\rho^2 M^2 + c_2\sqrt{E},$$

and

$$\rho M \leq c_3\rho^2 M^2 + c_4\sqrt{\varepsilon}.$$

If we choose ε small enough the equation

$$y = c_3 y^2 + c_4\sqrt{E}$$

will have two solutions: a small solution y_0, approximately $c_2\sqrt{E}$, and a large solution y_1, approximately c_1^{-1}. Fix such an ε. Then the inequality implies that either $\rho M \leq y_0$ or $\rho M \geq y_1$. For very small ρ the first alternative must hold and by continuity it must continue to hold for all $\rho \leq 1$. So we conclude that $M \leq \mathrm{const.}\sqrt{E}$, which establishes the Proposition.

Another result about smooth harmonic maps that can be proved using a similar approach is:

Proposition 4.2. *There is a constant M_0 such that if the harmonic map f satisfies a Hölder bound on the unit ball B:*

$$|f(x)-f(y)| \leq |x-y|^{\alpha}$$

then $|\mathrm{d}f(x)| \leq M_0 D(x)^{-1}$, where $D(x)$ is the distance to the boundary, as above.

As usual, we can go on to get estimates on all derivatives of f in the interior depending only on the Hölder bound. To prove this proposition we define M and x_0 as above. If $M \leq 2$ we can take $M_0 = 2$. If $M \geq 2$ rescale the ball of radius M^{-1} centered at x_0 to unit size to get a harmonic map $\tilde{f}$ on the unit ball B with

(1) $|\tilde{f}(x)-\tilde{f}(y)| \leq M^{-\alpha}|x-y|^{\alpha}$;
(2) $|\mathrm{d}\tilde{f}| \leq 2$ on B;
(3) $|\mathrm{d}\tilde{f}(0)| = 1$.

The harmonic map equation and item (2) give a bound on $|\Delta \tilde{f}|$ over B, which gives a $C^{,\alpha}$ bound on $\mathrm{d}\tilde{f}$ in the half-sized ball. It then follows from item (3) that for some computable number κ we can choose a ray $\{tv\}$ through the origin such that $|\tilde{f}(tv)-\tilde{f}(0)| \geq t/2$ for $t \leq \kappa$. Then item (1) implies that $M \leq 2^{\alpha}\kappa^{1-1/\alpha}$.

The small energy result gives a partial compactness property, extending what we have seen for surfaces in Section 3. Let M and N be compact and let $f_i : M \to N$ be a sequence of harmonic maps with energy bounded by a fixed constant $E_{\max}$. As in Section 2 we can suppose that the energy densities $|\mathrm{d}f_i|^2$ converge as Radon measures. Now we define

$$\mu_i(x,r) = r^{2-n}\int \chi_{x,r}|\mathrm{d}f_i|^2,$$

and $\mu(x,r) = \lim_{i\to\infty}\mu_i(x,r)$. We assume that the $\mu_i(x,r)$ are increasing functions of r. If M is locally Euclidean this follows from the monotonicity property and in general it will be true up to an unimportant error term. Then $\mu(x,r)$ is increasing and has a limit $\mu(x)$ as r tends to 0. As before, we define the set $S \subset M$ to be the set of points where $\mu(x) \geq \varepsilon/2$, for the constant ε in the small energy result, Proposition 4.1. Just as before we get, after passing to a subsequence, a limiting harmonic map $f : M\setminus S \to N$ and the f_i converge to f in C^{∞} on compact subsets of $M\setminus S$. Given a small $\delta > 0$, choose a maximal collection of disjoint $\delta/2$ balls $\{B_{\alpha}\}$ centred at points x_{α} of S. Let A be the number of balls. Then the δ-balls with the same centres cover S and

$$\int_{B_{\alpha}} |\mathrm{d}f_i|^2 \geq (\delta/4)^{n-2}\mu_i(x_{\alpha},\delta/4).$$

Since the B_{α} are disjoint we have

$$E_{\max} \geq \sum_{\alpha}(\delta/4)^{n-2}\mu_i(x_{\alpha},\delta/4),$$

and taking the limit we can replace $\mu_i(x_{\alpha,\delta/4})$ by $\mu(x_{\alpha,\delta/4})$. But $\mu(x_{\alpha},\delta/4) \geq \mu(x_{\alpha}) \geq \varepsilon/2$ so we get a bound on the number A of balls B_{α}

$$A \leq C\delta^{2-n}$$

with $C = 2^{2n-3}\varepsilon^{-1}E_{\max}$. So the set S is covered by at most $C'\delta^{2-n}$ balls of radius δ. This implies that S has Hausdorff dimension at most $(n-2)$ and the $(n-2)$-dimensional Hausdorff measure is bounded by C. (In fact we get a stronger statement, that the $(n-2)$-dimensional "Minkowski content" is finite. This is because the balls in our cover have the same radius δ: in the definition of Hausdorff measure one is allowed to cover by balls of varying radii.)

In sum we have:

Proposition 4.3. *For M,N compact a sequence of harmonic maps from M to N with a fixed energy bound has a subsequence which converges off a set of Hausdorff codimension at least* 2.

4.2 Minimising maps

In the previous subsection we have considered *smooth* harmonic maps. The thrust of the Schoen and Uhlenbeck paper [41] is different because they consider the much more formidable case of a class of weak solutions in L^2_1. These can have singularities, and the great achievement of Schoen and Uhlenbeck was to make these singularities somewhat tractable.

More precisely, for N embedded in the Euclidean space V, let $L^2_1(M,N)$ be the set of maps $f: M \to V$ which are in L^2_1 in the usual sense and which map almost all points of M to N. (Of course, for dim $M > 1$, such a map f is only defined almost everywhere.) The energy functional is defined on these maps and f is called a weak solution if the first variation of the energy vanishes, which is equivalent to f being a weak solution of the equation (9). This notion makes sense because the nonlinear term $A_f(\mathrm{d}f,\mathrm{d}f)$ is in L^1. These weak solutions can be bizarre: there are examples which are not continuous at any point of M [38]. Schoen and Uhlenbeck showed that if one restricts to the class of *energy minimising* maps the situation is much better. For our discussion, we could take the definition of energy minimising to be that there is some $\rho > 0$ such that for all balls $B_\rho \subset M$ if $g \in L^2_1(M,N)$ is equal to f outside B_ρ then $E(g) \geq E(f)$. (Any smooth harmonic map is energy-minimising in this sense.) In fact, in [41] Schoen and Uhlenbeck consider a more general class of equations, adding a perturbation term to the energy, and in [42] they extend the theory to the Dirichlet problem on a manifold M with boundary.

The foundation of the Schoen–Uhlenbeck work is to establish versions of monotonicity and the small energy property for energy-minimising harmonic maps. In the end the statements are essentially the same as for the smooth case but the proofs are different because many of the constructions discussed above do not make sense in this wider class.

For our discussion we assume that M is locally Euclidean, so by scaling we may regard f as being defined on the unit ball B in $\mathbf{R}^n$ and we can assume that any variation supported in B increases energy. The Schoen–Uhlenbeck proof of monotonicity proceeds as follows. Suppose that the restriction of f to the unit sphere $S^{n-1} = \partial B$ is also in L^2_1 and define a map g to be equal to f outside B and by $g(x) = f(\frac{x}{|x|})$ for $|x| \leq 1$. Of course g is not defined at the origin but when $n > 2$ it is an L^2_1 map. Simple calculus gives

$$\int_B |\mathrm{d}g|^2 = \int_{S^{n-1}} |\mathrm{d}_{S^{n-1}} g|^2 \int_0^1 r^{n-3}\, \mathrm{d}r = (n-2)^{-1} \int_{S^{n-1}} |\mathrm{d}_{S^{n-1}} g|^2.$$

Here the notation $\mathrm{d}_{S^{n-1}}$ refers to the derivative of the restriction of the map to the sphere. The energy-minimising property gives

$$(n-2) \int_B |\mathrm{d}f|^2 \leq \int_{S^{n-1}} |\mathrm{d}f|^2 - \int_{S^{n-1}} |\nabla_r f|^2.$$

(Notice that the term involving the radial derivative enters with a different factor compared with (24), but this will not matter.) Let $E(r)$ be the energy of f on the ball B_r. Then $E(r)$ is an increasing function of r and so differentiable at almost all r. Similarly for almost all r the restriction of $|\mathrm{d}f|^2$ to the boundary of the ball B_r is in L^2 and at such values of r

$$E'(r) = \int_{\partial B_r} |\mathrm{d}f|^2.$$

For such r we can apply the preceding discussion for the unit ball, after rescaling, and we obtain the inequality

$$(n-2)E(r) \leq rE'(r).$$

As before, this is the monotonicity statement that the normalised energy $\widehat{E}(r) = r^{2-n}E(r)$ is increasing. Moreover we have, for $r_1 < r_2$,

$$\widehat{E}(r_1) \leq \widehat{E}(r_2) - \int_{r_1 < r < r_2} r^2 |\nabla_r f|^2 \tag{25}$$

The work of Schoen and Uhlenbeck develops an important analogy between the theories of harmonic maps and of minimal submanifolds, and more general volume-minimising sets. The analogue of the argument above in the latter case, for a d-dimensional volume-minimising set $X \subset \mathbf{R}^m$ and a point x in X, is to consider, for small r, the intersection Y_r of X with the sphere of radius r centred at x. Let CY_r be the cone over Y_r with vertex at x and let $\tilde{X}_r$ be the set obtained by removing the intersection $X \cap B_{x,r}$ from X and replacing it with the cone CY_r. Comparing $\tilde{X}_r$ with X, the volume-minimising property shows that $\mathrm{Vol}(X \cap B_{x,r}) \leq \mathrm{Vol}\, CY_r$ and this leads to the monotonicity of the normalised volumes $r^{-d}\mathrm{Vol}(B_{x,r} \cap X)$ with respect to r, for fixed x.

The Schoen–Uhlenbeck proof of the small energy result for energy-minimising maps is more involved and we will outline it in subsection 4.3 below. But before that we discuss the general structural results and overall picture which Schoen and Uhlenbeck obtained. The first consequence is that a minimising map f is smooth outside a closed singular set Σ with $\dim \Sigma < n-2$. For example when $n=2$ this says that Σ is empty, which is immediate from the small energy result. For $n>2$ it is proved by a covering argument similar to the one we described above for Proposition 4.3. A refined result is

Theorem 4.1.

(1) $\dim \Sigma \leq n-3$.
(2) *Suppose that for some $k \geq 2$ and for all ν with $2 \leq \nu \leq k$ there is no smooth minimising harmonic map from the sphere S^ν to N. Then* $\dim \Sigma \leq n-k-2$.

In the second item here, if $k=n-1$ then the statement is that Σ is empty, so the map is smooth.

The proofs of these refined results depend on the important notion of a "tangent map", introduced by Schoen and Uhlenbeck. This is analogous to the notion of a tangent cone in submanifold geometry. Suppose for simplicity that the domain M of the minimising map f is the unit ball in $\mathbf{R}^n$ and for $\lambda > 1$ let f^λ be the composite of f with the scaling map; so $f^\lambda : \lambda B \rightarrow N$. Take any sequence $\lambda_i \rightarrow \infty$, so for any compact set $K \subset \mathbf{R}^n$ the map f^{λ_i} is defined over K for large enough i. Monotonicity implies that the energies of the f^{λ_i} are bounded on compact sets so, possibly passing to a subsequence, there is a weak $L^2_{1,\mathrm{loc}}$ limit f_∞. The inequality (25) implies that f_∞ is *radially homogeneous* in that $\nabla_r f_\infty = 0$ almost everywhere. Such a map is called a *tangent map* to f at 0.

Uniqueness of the tangent map, i.e. that one gets the same limit for any sequence of scalings, is a major question in general—there are examples [68] where it is not unique— but is not directly relevant to the discussion here. The main difficulties are that the convergence obtained is only in the weak topology and that it is not clear that the limit will again be minimising. Much of the work in the paper of Schoen and Uhlenbeck goes into overcoming these difficulties. They show that the convergence can be improved to $L^2_{1,\mathrm{loc}}$ and that, at least in certain restricted situations, the limit is minimising. Glossing over many details, we illustrate the argument for the case when $n=3$. Then the statement (1) of Theorem 4.1 can be improved to the statement that Σ is a discrete set. To see this, let p be a singular point and consider a tangent map at p. By radial homogeneity, this is equivalent to a map $g : S^2 \rightarrow N$. Schoen and Uhlenbeck show that g is minimising, so the singular set is empty by the previous discussion in dimension $n=2$. Thus the tangent map f_∞ has an isolated singularity at the origin and this implies that the singularity p of f is isolated. For another illustration of the argument, consider the case when there are no minimising harmonic maps $S^\nu \rightarrow N$ for $2 \leq \nu \leq n-1$. Suppose there were a singular point p of f. From the tangent map we get a map $g : S^{n-1} \rightarrow N$. Assume for simplicity that this is minimising. The hypothesis implies that g cannot be smooth, so we can go to a singular point of g in S^{n-1} and take a second tangent map there. This gives a map from S^{n-2} to N and the hypothesis implies that this cannot be smooth, so we can find a tangent map at a singular point. Continuing in this way, with these iterated tangent maps, we get a contradiction to the existence of p, so the original map is smooth.

4.3 Small energy

In this subsection we outline the Schoen and Uhlenbeck proof of the small energy result.

First, it was established before that a Hölder continuous weak harmonic map is smooth [24]. (This is similar to Proposition 4.2 in that *a posteriori* the estimate in that proposition holds, but of course the proof is much harder.) Next, it suffices to get a bound for the growth of the normalised energy function. Morrey's Lemma states that if a function g on $\mathbf{R}^n$ with weak derivative in L^1 satisfies an estimate:

$$r^{-n}\int_{B_{x,r}} |\mathrm{d}g| \leq Cr^{-\beta},$$

for all balls $B_{x,r}$ then g is in $C^{,1-\beta}$. For our map f

$$\left(\int_{B_{x,r}} |\mathrm{d}f|\right)^2 \leq \int_{B_{r,x}} |\mathrm{d}f|^2 \,\mathrm{Vol}\, B_{x,r},$$

so if the normalised energy on all $B_{x,r}$ is less than Cr^{α} then f is in $C^{,\alpha/2}$.

The essential statement in Schoen and Uhlenbeck's proof of the small energy result is then:

Theorem 4.2. *There is an $\varepsilon_0 > 0$ and $\theta_0 \in (0,1)$ such that if $f : B \to N$ is an energy minimising map with normalised energy less than ε_0 on all interior balls then*

$$\widehat{E}(\theta_0) \leq \tfrac{1}{2}\widehat{E}(1) = \tfrac{1}{2}E.$$

(The factor $\frac{1}{2}$ here could be replaced by any fixed number in $(0,1)$.) Given this, it follows from an elementary argument (similar to that in subsection 2.3 above) that $\widehat{E}(r) \leq Cr^{\alpha}$ for a suitable α, depending on θ.

The proof by Schoen and Uhlenbeck of Theorem 6 involves the choice of four parameters $\theta_0, \theta, \tau, h$. Here θ_0 will be as in the statement of the Theorem, and we will choose $\theta_0 < \frac{1}{8}$ say. The parameter θ will be chosen in the interval $[\theta_0, 2\theta_0]$. The parameter τ will be much smaller than θ_0. Given θ, τ we write A for the annulus $A = \{x : \theta \leq |x| \leq \theta + \tau\}$. The idea is to construct a comparison map $\tilde{f} : B \to N$ such that $\tilde{f}(x) = f(x)$ for $|x| \geq \theta + \tau$. Then the minimising property of f gives

$$\int_{B_\theta} |\mathrm{d}f|^2 \leq I + II \tag{26}$$

where

$$I = \int_{B_\theta} |\mathrm{d}\tilde{f}|^2 \quad , \quad II = \int_A |\mathrm{d}\tilde{f}|^2.$$

and the task will be to bound these terms I and II.

We write E for the energy of f on the unit ball B, so $E \leq \varepsilon_0$. and we use the convention that c is a constant which changes from line to line. We also fix a tubular neighbourhood $\Omega \subset V$ of N in V and let $\pi : \Omega \to N$ be the standard projection.

The construction of $\tilde{f}$ goes through three other maps f_1, f_2, f_3 and depends on the parameter $h > 0$, which is a small smoothing parameter.

The map f_1

The map $f_1 : B \to V$ is the smooth map obtained as a standard mollification of f by convolution with a function supported in the ball of radius h. (More precisely, f_1 will be defined on a slightly smaller ball than B_1 but this does not matter.) Thus if ϕ is positive function on V with integral 1 and supported in the unit ball:

$$f_1(x) = h^{-n} \int \phi(h^{-1}(x-y)) f(y)\, \mathrm{d}y = \int \phi(z) f(x - hz)\, \mathrm{d}z,$$

where the two formulae are related by the change of variable $y = x - hz$.

The map f_2

The map f_1 does not map into N and the distance between $f_1(x)$ and $f(x)$ need not be small *a priori*. But we will see that we can arrange, by choosing ε_0 small, that f_1 maps into the tubular neighbourhood Ω. Then we define $f_2 = \pi \circ f_1 : B \to N$.

The map f_3

We modify the smoothing construction of f_1 to make a new map $f_3 : B \to V$ equal to f outside $B_{\theta+\tau}$ and to f_2 in B_θ. Let η be a function on B equal to the constant h on B_θ and to 0 outside $B_{\theta+\tau}$ and set

$$f_3(x) = \int \phi(z) f(x - \eta(x) z)\, \mathrm{d}z.$$

We will arrange that f_3 maps into the neighbourhood Ω and we define $\tilde{f} = \pi \circ f_3$.

Bounds on the convolution

The first business is to arrange that f_1 and f_3 map into the tubular neighbourhood Ω of N. This is a crucial insight in the Schoen and Uhlenbeck proof and depends on the following lemma.

Lemma 4.3. *For ϕ as above there is a constant C such that all functions g on the unit ball B with*

$$\int_B \phi(y)g(y) = 0$$

satisfy $\|g\|_{L^2} \leq C\|\mathrm{d}g\|_{L^2}$.

If ϕ were replaced by a constant this becomes the Poincaré Lemma. The proof of the generalisation is straightforward. An immediate consequence is that for any function g on B

$$\|g - g_*\|_{L^2} \leq C\|\mathrm{d}g\|_{L^2}$$

where g_* is the constant function equal to

$$g_* = \int_B \phi(y)g(y)\,\mathrm{d}y.$$

In particular there is some y in B such that

$$|g(y) - g_*| \leq \frac{C}{\sqrt{\mathrm{Vol}B}}\|\mathrm{d}g\|_{L^2}.$$

Now consider $x_0 \in B$ and apply this to $g(y) = f(x_0 - hy)$ so $g_* = f_1(x_0)$. We have

$$\|\mathrm{d}g\|_{L^2}^2 = h^{2-n}\int_{|x-x_0|\leq h} |\mathrm{d}f|^2 \leq \varepsilon_0.$$

We deduce that there is some x with $|x - x_0| \leq h$ such that $|f(x) - f_1(x_0)| \leq c\,\varepsilon_0^{1/2}$. So f_1 maps into the $c\,\varepsilon_0^{1/2}$ neighbourhood of N. This argument also applies to the map f_3, because x_0 is fixed.

The standard convolution formula shows that $\|\mathrm{d}f_1\|_{L^2}^2 \leq \|\mathrm{d}f\|_{L^2}^2 = E$. We also have a pointwise bound

$$|\mathrm{d}f_1| \leq c\varepsilon_0^{1/2}h^{-1}, \tag{27}$$

which follows easily from the bound on the normalised energy.

The term I

By construction

$$I = \int_{B_\theta} |\mathrm{d}f_2|^2.$$

Composition with the projection π can increase the norm of the derivative by at most a small factor so it suffices to bound the integral of $|df_1|^2$ over B_θ. We need a better bound than that given by (27). To achieve this, Schoen and Uhlenbeck consider the harmonic function v on the ball $B_{\frac{1}{2}}$ with the same boundary values as f_1. Then v minimises the Dirichlet energy over all functions with these boundary values so

$$\int_{B_{\frac{1}{2}}} |dv|^2 \leq \int_{B_{\frac{1}{2}}} |df_1|^2 \leq E.$$

Standard theory promotes this to a pointwise bound on the interior ball $B_\theta \subset B_{\frac{1}{4}} \subset B_{\frac{1}{2}}$ so

$$\int_{B_\theta} |dv|^2 \leq cE \text{ Vol } B_\theta = cE\ \theta^n. \tag{28}$$

Now write $w = f_1 - v$, so $\Delta w = \Delta f_1$. Recall that f_1 is the convolution $\phi_h * f$ of f with a function ϕ_h of L^1 norm 1. The Laplace operator Δ on $\mathbf{R}^n$ commutes with convolution so

$$\Delta f_1 = \phi_h * (\Delta f) = \phi_h * (A(df, df)),$$

and hence

$$\|\Delta f_1\|_{L^1} \leq \|\phi_h\|_{L^1} \|A(df, df)\|_{L^1} = \|A(df, df)\|_{L^1}.$$

Clearly $\|A(df, df)\|_{L^1} \leq cE$ so we get

$$\|\Delta w\|_{L^1} \leq cE.$$

Hence

$$\int_{B_{\frac{1}{2}}} |dw|^2 = \int_{B_{\frac{1}{2}}} (w, \Delta w) \leq cE \ \sup_{B_{1/2}} |w|.$$

The bound (27) and maximum principle considerations imply that $|w| \leq c\varepsilon_0^{1/2} h^{-1}$ on $B_{\frac{1}{2}}$. So we conclude that

$$\int_{B_\theta} |dw|^2 \leq \int_{B_{\frac{1}{2}}} |dw|^2 \leq c\ \varepsilon_0 h^{-1} E.$$

Combined with (28) this gives a bound on the L^2 norm of df_1 and hence of $\tilde{f}$ over B_θ:

$$I = \int_{B_\theta} |d\tilde{f}|^2 \leq c(\varepsilon_0 h^{-1} + \theta^n) E. \tag{29}$$

The term II

As before, it suffices to work with f_3. Recall that A is the annulus $\theta \leq |x| \leq \theta + \tau$. Let A_+ be the h neighbourhood of A, so the maps f_3 and $\tilde{f}$ on A are determined by the restriction of f to A_+.

Lemma 4.4. *Suppose that the derivative of η is bounded by $|\mathrm{d}\eta| \leq k$. Then*

$$\int_A |\mathrm{d}f_3|^2 \leq c(1+k)^2 \int_{A_+} |\mathrm{d}f|^2.$$

We have

$$f_3(x) = \int \phi(y) f(x - \eta(x)y)\, \mathrm{d}y.$$

When we differentiate with respect to x we get a term from the derivative of η, which is bounded by k. This gives

$$|\mathrm{d}f_3(x)| \leq (1+k) \int \phi(y)\, |\mathrm{d}f|(x - \eta(x)y)\, \mathrm{d}y.$$

A quick route from here is to use the theory of the maximal function. For points x where $\eta(x) > 0$

$$\begin{aligned} |\mathrm{d}f_3(x)| &\leq (1+k) \int_{|y|\leq 1} |\mathrm{d}f|(x - \eta(x)y)\, \mathrm{d}y \\ &\leq (1+k)\eta(x)^{-n} \int_{B_{x,\eta(x)}} |\mathrm{d}f| \leq (1+k) M(|\mathrm{d}f|), \end{aligned}$$

where $M(|\mathrm{d}f|)$ is the maximal function of $|\mathrm{d}f|$. Then

$$\|\mathrm{d}f_3\|_{L^2} \leq c(1+k)\|M(df)\|_{L^2} \leq c(1+k)\|\mathrm{d}f\|_{L^2},$$

and the lemma follows. (Schoen and Uhlenbeck give a direct calculus proof of this lemma.)

As before, the projection π only changes the energy by small amount, so Lemma 4.4 implies that, if we choose η so that $|\mathrm{d}\eta| \leq 1$, we have

$$II = \int_A |\mathrm{d}\tilde{f}|^2 \leq c \int_{A^+} |\mathrm{d}f|^2.$$

The right-hand side here is bounded by cE but this does not suffice since the constant c could be large. To overcome this Schoen and Uhlenbeck bring in another idea: the choice of θ. The condition that $|\mathrm{d}\eta| \leq 1$ implies that h cannot be more than the annulus thickness τ. Let us now fix $h = \tau/2$ say. So A_+ is an annulus of thickness 5τ. Recall that τ is to be much smaller than θ_0. There are approximately $Q = \theta_0/5\tau$ disjoint annuli of the form A_+ for different values of θ in $[\theta_0, 2\theta_0]$. So we can make a choice of one of these such that

$$\int_{A_+} |\mathrm{d}f|^2 \leq E/Q \leq cE\tau/\theta_0.$$

Making this choice, combining with the bound (29) for the term I and setting $h = \tau/2$ we get

$$I + II \leq c(\varepsilon_0/\tau + \theta_0^n + \tau/\theta_0)E.$$

(By adjusting constants it does not matter if we write θ or θ_0 here since $\theta_0 < \theta < 2\theta_0$.) Then (26) gives a bound on the normalised energy

$$\widehat{E}(\theta) \leq c\left(\varepsilon_0/(\tau\theta_0^{n-2}) + \theta_0^2 + \tau/\theta_0^{n-1}\right) E.$$

By making θ_0 small, then τ, then ε_0, we get a $\theta \in [\theta_0, 2\theta_0]$ with $\widehat{E}(\theta) \leq \frac{1}{2}E$ and by monotonicity $\widehat{E}(\theta_0) \leq \frac{1}{2}E$.

4.4 Some further developments

The influence of the work of Schoen and Uhlenbeck has been immense and extends in many directions (the paper [41] has 303 citations on MathSciNet at the time of writing). Singularities of the kind which came to the fore in their paper arise in various models in Mathematical Physics, for point singularities in $\mathbf{R}^3$. They also appear in complex algebraic geometry as meromorphic maps (related to the work of Uhlenbeck and Yau that we discuss in subsection 6.1 below). In Hardt's survey [21] of developments on singularities of harmonic maps in the decade following the Schoen and Uhlenbeck paper he writes *"the paper [of Schoen and Uhlenbeck] has many ideas and techniques that have proved to have wide influence in geometric analysis."*

Stationary maps form another important subclass of weak harmonic maps. Such a map is called stationary if the first variation of the energy vanishes for variations induced by 1-parameter families of compactly supported diffeomorphisms of the domain. This includes the minimising maps considered by Schoen and Uhlenbeck but forms a larger class. Many of Schoen and Uhlenbeck's results were later extended to stationary maps. An important paper [4] on maps with point singularities in $\mathbf{R}^3$ includes examples of tangent maps which are not stationary or mininimising. If $f : \mathbf{R}^3 \setminus \{0\}$ is the radial extension of a degree 1 holomorphic (hence harmonic) map ϕ from S^2 to S^2 then f is stationary if and only if ϕ is a rotation; otherwise the first variation of the energy under motion of the singularity (with fixed boundary values) does not vanish.

The paper [31] of Naber and Valtorta is one notable more recent development in the study of singularities of harmonic maps. A tangent map from $\mathbf{R}^n$ to N is called k-symmetric if it factors through an orthogonal projection $\mathbf{R}^n \to \mathbf{R}^{n-k}$. Given a minimising weakly harmonic map f let $\Sigma_k \subset M$ be the set of points in x such that no tangent map is $(k+1)$-symmetric. Thus $\Sigma_0 \subset \Sigma_1 \subset \cdots \subset \Sigma$ where Σ is the singular set of f: the Σ_k give a stratification of the singular set. Schoen and Uhlenbeck's work, in the proof of Theorem 4.1 above, shows that Σ_k has Hausdorff dimension at most k. Naber and Valtorta prove the stronger statement that Σ_k is

k-rectifiable (in fact they show this for stationary maps f). They also prove a weak L^3 result for minimising maps f:

$$\text{Vol}\,\{x \in M : |\mathrm{d}f| \geq \varepsilon^{-1}\} \leq C\varepsilon^3.$$

The ideas and techniques in the analysis of harmonic maps which we have discussed in Section 3 and this Section 4 have been important in other branches of differential geometry and PDE theory. For the latter we just mention the large body of work, for example [5], on bubbling phenomena in critical exponent problems. As we have mentioned, there are close analogues between harmonic maps and minimal submanifold theory. The next two sections of this article will describe analogues in gauge theory. Another area is Riemannian geometry. In the case of 4-dimensional manifolds the L^2 norm of the Riemann curvature serves as an energy functional which has analogous properties to the harmonic maps energy in dimension 2. Partial compactness results for solutions of the Einstein equations on 4-manifolds satisfying suitable bounds on the volume, diameter and this L^2 norm were obtained by Anderson [1] and Nakajima [33]. These results follow a similar pattern to the Sacks–Uhlenbeck theory, with "bubbling" at a finite set of points. The results were extended to metrics on 4-manifolds satisfying various other equations such as extremal Kähler metrics, assuming a bound on the Sobolev constant, by Tian and Viaclovsky [55].

The L^2-norm of the Riemann curvature is much less effective in higher dimensions. But for limits of manifolds satisfying Ricci curvature bounds a theory analogous to that of Schoen and Uhlenbeck was developed by Cheeger and Colding [7]. Here the volume ratio of metric balls plays a role analogous to the normalised energy.

5 Gauge Theory

5.1 Background

In the mid-1970s "gauge theory' or "Yang–Mills theory" entered mathematics as a new subject, propelled by interactions with physics, and this subject became the scene for many of Uhlenbeck's most prominent achievements.

We begin by reviewing the basic differential geometry. To simplify notation slightly, we will nearly always consider connections on vector bundles, say complex vector bundles, usually with Hermitian metrics on the fibres. For such a bundle $E \to M$ a connection A can be identified with a covariant derivative, a differential operator

$$\nabla_A : \Omega^0(E) \to \Omega^1(E),$$

and we will often not distinguish between A and ∇_A. In a local trivialisation of E and local coordinates x_i on M the connection is represented by a matrix-valued 1-form $\underline{A} = \sum \underline{A}_i \mathrm{d}x_i$ and the covariant derivative, thought of as acting on vector-valued functions via the trivialisation, has components

$$\nabla_i^{\underline{A}} = \frac{\partial}{\partial x_i} + \underline{A}_i.$$

If the trivialisation is unitary then the $\underline{A}_i$ take values in the skew-adjoint matrices (i.e. in the Lie algebra of the unitary group). The curvature of the connection is a bundle valued 2-form $F_A \in \Omega^2(\mathrm{End}E)$ and if the connection is unitary it lies in $\Omega^2(\mathrm{ad}_E)$, where ad_E is the bundle of skew-adjoint endomorphisms. In a local trivialisation, as above, the curvature is the operator given by the commutator

$$[\nabla_i^{\underline{A}}, \nabla_j^{\underline{A}}] = \frac{\partial \underline{A}_j}{\partial x_i} - \frac{\partial \underline{A}_i}{\partial x_j} + [\underline{A}_i, \underline{A}_j].$$

Written as a matrix-valued 2-form

$$F = \mathrm{d}\underline{A} + \underline{A} \wedge \underline{A}. \tag{30}$$

A change in local trivialisation is given by a map g to the structure group $U(r)$. This acts on the covariant derivative by conjugation and changes $\nabla^{\underline{A}}$ to

$$g\nabla^{\underline{A}}g^{-1} = \frac{\partial}{\partial x_i} + g\underline{A}_i g^{-1} - (\mathrm{d}g)g^{-1}. \tag{31}$$

For any bundle-valued 1-form $a \in \Omega^1(\mathrm{End}(E))$ the operator $\nabla_A + a$ is again a covariant derivative. We regard the space of connections A as an affine space and we just write the new connection as $A + a$. A slight variant of the preceding discussion is to consider an automorphism g of the bundle $E \to M$ covering the identity on M. This acts on covariant derivatives by conjugation and we can write

$$g(A) = A - (\mathrm{d}_A g)g^{-1}. \tag{32}$$

In the same vein the global version of the formula (30) is

$$F(A+a) = F(A) + \mathrm{d}_A a + a \wedge a. \tag{33}$$

In (32) and (33) d_A denotes the coupled exterior derivative defined by the connection (so on bundle-valued 0-forms we could write this also as ∇_A).

The Yang–Mills equations (for the structure group $U(r)$) arise from the functional on the space of unitary connections over a Riemannian or pseudo-Riemannian manifold M

$$\mathscr{E}(A) = \int_M |F(A)|^2.$$

Here $|F|^2$ is computed using the standard norm on skew-adjoint matrices and the quadratic form on 2-forms induced by the Riemannian or pseudo-Riemannian structure. The Yang–Mills equations are the Euler–Lagrange equations associated to this functional which have the form

$$d_A^* F_A = 0 \tag{34}$$

where d_A^* is the formal adjoint of d_A. This follows from (33) since

$$\langle F(A+a), F(A+a)\rangle = \langle F(A), F(A)\rangle + 2\langle d_A a, F(A)\rangle + O(a^2),$$

which is

$$\langle F(A), F(A)\rangle + 2\langle a, d_A^* F(A)\rangle + O(a^2).$$

In the case of a rank 1 bundle, with structure group the circle $U(1)$, the Yang–Mills equations are linear. When M is space-time, with the Lorentzian metric, we get Maxwell's equations for the electromagnetic field, but henceforth in this article we will consider only Riemannian base manifolds.

In the Riemannian case the functional $\mathscr{E}$ is a positive "energy" functional. Just as the harmonic maps functional can be thought of loosely as measuring the deviation of a map from a constant so, at least over a simply connected manifold M, the Yang–Mills functional can be thought of as measuring the deviation from a product connection. The analogy with harmonic maps has been an important guiding theme in the development of Yang–Mills theory, and one emphasised in Uhlenbeck's work, as we will see below. The critical dimension for the base manifold M in the Yang–Mills case is 4, analogous to the critical domain dimension 2 for harmonic map theory. The Yang–Mills functional is conformally invariant in dimension 4. Said in another way, working over a ρ-ball B_ρ in $M = \mathbf{R}^n$, if we rescale the ball to unit size the energy changes by a factor ρ^{4-n}.

Another connection between Yang–Mills theory and harmonic maps comes in the question of "gauge fixing". The local representation $\underline{A}$ of a connection over an open set $\Omega \subset M$ depends on a choice of bundle trivialisation. By changing the trivialisation we can make $\underline{A}$ as "bad" as we like; conversely we would like to choose a good representation for a given connection. A natural way to do this is to seek a "Coulomb gauge" in which $d^*\underline{A} = 0$. This is traditional in electromagnetic theory. Staying in positive signature, we view a magnetic field $\mathbf{B}$ on $\mathbf{R}^3$ as the curvature of a connection on a Hermitian complex line bundle. Transferring to vector-field notation, $\underline{A}$ becomes the magnetic potential $\mathbf{A}$ with curl $\mathbf{A} = \mathbf{B}$ and the Coulomb gauge condition is div $\mathbf{A} = 0$. Then $\Delta\mathbf{A}$ is the current $\mathbf{J} =$ curl $\mathbf{B}$ and $\mathbf{A} = G * \mathbf{J}$ for the Newton potential G. Differentiating, this gives the Biot–Savart formula for the magnetic field generated by a current.

Going back to the general situation, if we start with some arbitrary representation $\underline{A}_0$ for the connection over $\Omega \subset M$ a representation in Coulomb gauge corresponds to a solution of the equation

$$d^*(g\underline{A}_0 g^{-1} - dg g^{-1}) = 0, \tag{35}$$

for a map $g : \Omega \to U(r)$. This is the Euler–Lagrange equation associated to the functional

$$\|g\underline{A}_0 g^{-1} - \mathrm{d}g g^{-1}\|_{L^2}^2.$$

When $\underline{A}_0 = 0$ this is the harmonic maps energy for the map $g : M_0 \to U(r)$ and the equation (35) is the harmonic map equation. For general $\underline{A}_0$ we have a deformation of that equation.

5.2 The 1982 papers in Commun. Math. Phys.

The title of this subsection refers to the two papers of Uhlenbeck [58], [59]. Along with work of Taubes from around the same time, such as [51], these papers initiated the study of the analytic and PDE aspects of Yang–Mills theory, to set alongside the developments of that period of a more differential-geometric and algebro-geometric nature.

The paper [59] bears on the gauge-fixing problem indicated at the end of the previous subsection. While the overall aim is to obtain global results, the main work takes place locally, for connections over a ball. There is a simple way to fix a gauge (the "exponential gauge") for a connection over a ball using parallel transport along rays through the origin. In other words, in polar coordinates the connection form $\underline{A}$ is determined (up to an overall conjugation) by the condition that it contains no $\mathrm{d}r$ component. This is convenient for many purposes but is not well-suited to elliptic analysis. The curvature depends on one derivative of the connection so we would hope, roughly speaking, that we can choose a gauge in which $\underline{A}$ gains one derivative compared with the curvature. But, for example, an L^∞ bound on the curvature gives only an L^∞ bound on the connection form in an exponential gauge, it does not control the derivatives. Similarly, the Yang–Mills equations for $\underline{A}$ are not elliptic in exponential gauge. On the other hand, if the Coulomb gauge condition $\mathrm{d}^*\underline{A} = 0$ is satisfied then, in Sobolev spaces, the leading term $\mathrm{d}\underline{A}$ in the curvature does control roughly speaking one more derivative of $\underline{A}$ and the Yang–Mills equations are elliptic. There is a parallel discussion in Riemannian geometry, with the traditional geodesic coordinates compared with harmonic coordinates. In the latter the Einstein equations for the metric tensor are elliptic.

The central result of [59] is a "small energy" theorem.

Theorem 5.1. *For $p > 1$ there are $\varepsilon, C > 0$ such that if A is a connection over B^n with $\|F\|_{L^{n/2}} \leq \varepsilon$ then there is local trivialisation in which the connection form $\underline{A}$ has the following properties:*

(1) $\mathrm{d}^*\underline{A} = 0$,

(2) *On the boundary, the contraction of $\underline{A}$ by the normal vector vanishes,*

(3)

$$\|\underline{A}\|_{p,1} \leq C\|F\|_{L^p}.$$

Thus the curvature does control one more derivative of the connection form in this L^p sense. Uhlenbeck proves a stronger statement, for connections which are only in L^p_1 (for $p \geq n/2$) but in our discussion we work with smooth connections, which makes things a bit simpler. Also, to simplify notation we write the proof for the case $n = 4$.

Uhlenbeck uses a continuity argument to establish this result. The estimates are another instance of what we are calling "critical quadratic rearrangement", as in the proof of Theorem 3.2 for harmonic maps. For $\rho \in [0,1]$ let A_ρ be the connection over the unit ball obtained by pulling back the restriction of A to the ρ ball by the dilation map. Then the scaling behaviour of the L^2 norm on 2-forms in dimension 4 shows that $\|F(A_\rho)\|_{L^2} \leq \|F(A)\|_{L^2}$, so we have a path of connections joining A to the trivial connection, all with $\|F\|_{L^2} \leq \varepsilon$. The strategy is to construct a corresponding path $\underline{A}_\rho$ satisfying the conditions in the statement. To set this up we consider a variant of the third condition in Theorem 5.1 (for $p = 2$), depending on a small number η to be chosen below. The variant is

$$\|\underline{A}\|_{2,1} < \eta, \qquad (3')$$

which is an open condition.

Suppose that for some ρ we have an $\underline{A}_\rho$ satisfying (1), (2) of Theorem 5.1 and (3′), and just write $\underline{A}_\rho = \underline{A}$. The boundary condition is elliptic for the operator $\mathrm{d}^* \oplus \mathrm{d}$ and if a is a 1-form satisfying the boundary condition in (2) of Theorem 5.1 and $\mathrm{d}^* a = \mathrm{d}a = 0$ then $a = 0$. (For we can write $a = \mathrm{d}f$ and then f satisfies the Laplace equation with Neumann boundary conditions and hence is constant, so $a = 0$.) Then elliptic theory gives estimates

$$\|\underline{A}\|_{p,1} \leq K_p \|\mathrm{d}\underline{A}\|_{L^p}. \tag{36}$$

For $p < 4$ the Sobolev embedding $L^p_1 \to L^q$ with $q = 4p/(4-p)$ combined with (36) tells us that

$$\|\underline{A}\|_{L^q} \leq K'_p \|\mathrm{d}\underline{A}\|_{L^p}.$$

Now the formula $\mathrm{d}\underline{A} = F - \underline{A} \wedge \underline{A}$ leads to

$$\|\underline{A}\|_{p,1} \leq c_1 \|F\|_{L^p} + c_2 \|\underline{A}\|_{p,1} \|\underline{A}\|_{2,1}, \tag{37}$$

where we have used Hölder's inequality, exploiting the fact that $1/p = 1/q + 2$.

Now for the crucial step take $p = 2$. Then (37) gives

$$\|\underline{A}\|_{L^2_1} \leq c_3 \|F\|_{L^2} + c_4 \|\underline{A}\|^2_{L^2_1}.$$

Thus if η is chosen so that $c_4 \eta < \frac{1}{2}$, say, we get $\|\underline{A}\|_{L^2_1} \leq C \|F\|_{L^2}$ with $C = 2c_3$. In other words (1), (2) and (3′) imply (3). Now choose $\varepsilon < \eta/2C$ so that condition (3) implies (3′). Going back to (37) and applying the same rearrangement argument for the quadratic term we get L^p_1 bounds on $\underline{A}$ for all $p < 4$ (provided ε is chosen suitably small). A similar argument works for $p \geq 4$. Note that we are *not* supposing that the curvature is small in L^p for $p > 2$, so $\underline{A}$ could be large in L^p_1, but we have some bound. Similarly, if we want to stay in the smooth category, we can estimate higher derivatives. Then it is straightforward to show that the set of $\rho \in [0,1]$ for which a solution exists is closed. The point is that the open condition (3′) cannot be violated in taking a limit because it is implied by the closed condition (3). (In this case of dimension $n=4$ it is not essential to invoke L^p theory for $p \neq 2$ to prove the main result; the proof can be done in Sobolev spaces L^2_k, as in [16].)

The openness part of the continuity proof uses, as usual, the implicit function theorem. Condition (3′) is open by its nature so we just have to deform the solution to the equations (1), (2). This requires some technical work to set up due to the boundary condition. At a solution $\underline{A}=\underline{A}_\rho$ the linearised equation for a matrix-valued function ψ is

$$d^*(d\psi + [\underline{A}, \psi]) = \sigma,$$

with Neumann boundary condition and where the integral of the given σ is zero. The operator on the left-hand side can be written as $\Delta\psi + \{\underline{A}, d\psi\}$, where $\{\ ,\ \}$ combines the inner product on 1-forms with the matrix bracket. This is treated as a deformation of the ordinary Poisson equation with Neumann boundary conditions. Sobolev estimates similar to those used above show that the linearised equation is soluble and the implicit function theorem can be applied.

The global result in Uhlenbeck's paper [59] concerns the "subcritical" case, with an L^p bound on the curvature for $p > n/2$.

Theorem 5.2. *Let A_i be a sequence of unitary connections on a bundle E over a compact Riemannian n-manifold M satisfying a bound $\|F(A_i)\|_{L^p} \leq C$ for some $p > n/2$. There is a subsequence $\{i'\}$ and bundle automorphisms $g_{i'}$ such that the transformed connections $g_{i'}(A_{i'})$ converge weakly in L^p_1 to an L^p_1 limit A_∞.*

This is a relatively elementary consequence of Theorem 5.1. First, the scaling behaviour of the L^p norm means that there is an r_0 such that if the restriction of A_i to any r_0-ball in M is pulled back to the unit ball $B \subset \mathbf{R}^n$ via geodesic coordinates then the pulled back connection satisfies the small-curvature hypothesis of Theorem 5.1. Cover the manifold M by a finite collection of such small balls. Then after passing to a subsequence and applying a sequence of gauge transformations *over each ball* we can suppose that the connections converge weakly in L^p_1 over the ball. The problem is to convert this local convergence to the global result.

It is convenient to take a different point of view here and consider a bundle-with-connection over a manifold M presented by the data

- an open cover $M = \bigcup U_\alpha$;
- on each overlap $U_\alpha \cap U_\beta$ a transition function $g^{\alpha\beta}$, taking values in the unitary group, such that $g^{\alpha\gamma} = g^{\alpha\beta} g^{\beta\gamma}$ on $U_\alpha \cap U_\beta \cap U_\gamma$;
- on each U_α a connection form $\underline{A}^\alpha$, such that

$$dg_{\alpha\beta} = g^{\alpha\beta}\underline{A}^\alpha - \underline{A}^\beta g^{\alpha\beta} \tag{38}$$

on $U_\alpha \cap U_\beta$.

(Note that (38) is equivalent to (32).)

Now suppose that we have a sequence of such data (for a fixed cover) $g_i^{\alpha\beta}$, $\underline{A}_i^\alpha$ and that the $\underline{A}_i^\alpha$ converge over U_α to $\underline{A}_\infty^\alpha$. For the moment let us suppose that this is C^∞ convergence. Since the unitary group is compact the formula (38) implies that the derivatives $\mathrm{d}g_i^{\alpha\beta}$ are bounded, so there is a subsequence $\{i'\}$ such that the $g_{i'}^{\alpha\beta}$ converge in C^0 on compact subsets of $U_\alpha \cap U_\beta$. Differentiating (38) shows that this convergence is in C^∞ on compact subsets. We can slightly shrink the U_α so, without loss of generality, we can suppose that the $g_i^{\alpha\beta}$ converge in C^∞ on $U_\alpha \cap U_\beta$ to a limit $g_\infty^{\alpha\beta}$. This system of data $(g_\infty^{\alpha\beta}, \underline{A}_\infty^\alpha)$ satisfies all the conditions to define a bundle-with-connection. Let E_i be the bundle defined by the transition functions $g_i^{\alpha\beta}$, for i finite or infinite, and A_i the connection on E_i. What we want to show is that for large enough i there is a bundle isomorphism $h_i : E_\infty \to E_i$ such that the pull-backs $h_i^*(A_i)$ converge to A_∞. This boils down to the problem of choosing $U(r)$-valued functions h_i^α on possibly slightly smaller sets U'_α (which still cover) such that

$$h_i^\alpha g_i^{\alpha\beta} = g_\infty^{\alpha\beta} h_i^\beta \tag{39}$$

on the intersections, and with $h_{\alpha_i} \to 1$ as $i \to \infty$.

By an induction argument it suffices to treat the case when the cover is by two open sets $M = U_\alpha \cup U_\beta$. We choose $h_\beta = 1$ so the condition to solve is

$$h_i = g_\infty g_i^{-1},$$

where we write $h = h_\alpha, g_i = g_i^{\alpha\beta}, g_\infty = g_\infty^{\alpha\beta}$. Now $g_i \to g_\infty$ in C^0 so for i large $g_\infty g_i^{-1}$ takes values in a small neighbourhood of the identity in $U(r)$ and we can write

$$g_\infty g_i^{-1} = \exp(L_i),$$

for a matrix-values function L_i on $U_\alpha \cap U_\beta$. We take a suitable shrunken cover $U'_\alpha \cup U'_\beta$ and a cut-off function χ so that χL_i is equal to L_i on $U'_\alpha \cap U'_\beta$ and χL_i can be extended smoothly by 0 over U'_α. Then we can take $h_i = \exp(\chi L_i)$ and clearly $h_i \to 1$ as $i \to \infty$.

In the setting of Theorem 5.2 we do not have C^∞ convergence of the connection forms $\underline{A}_i^\alpha$ but only weak L_1^p convergence. However this implies weak L_2^p convergence of the $g_i^{\alpha\beta}$ (after taking a subsequence) which implies C^0 convergence, since evaluation at a point is bounded on L_2^p when $p > n/2$. Then the whole argument goes through unchanged. This is the crucial point where the condition $p > n/2$ is required.

We now turn to the other *Commun. Math. Phys.* paper [58]. The main result is the removability of point singularities:

Theorem 5.3. *A Yang–Mills connection over the punctured ball $B^4 \setminus \{0\}$ with curvature in L^2 extends to a smooth Yang–Mills connection over B^4.*

More precisely, if A is a finite-energy Yang–Mills connection on a bundle $E \to B^4 \setminus \{0\}$ then there is a bundle $\tilde{E} \to B^4$ and an isomorphism $\iota : \tilde{E}|_{B^4 \setminus \{0\}} \to E$ such that $\iota^*(A)$ extends smoothly over the origin.

This is the analogue of Sack's and Uhlenbeck's Theorem 3.3 for harmonic maps of the punctured disc and Uhlenbeck's strategy of proof is similar. For $r \leq 1$ let

$$\mathscr{E}(r) = \int_{|x| \leq r} |F|^2.$$

The strategy is to derive differential inequalities relating $\mathscr{E}$ and $\frac{\mathrm{d}\mathscr{E}}{\mathrm{d}r}$.

This paper [58] introduced a number of important techniques and results, on the way to the proof of Theorem 5.3. One was the use of exponential gauges discussed above. Another was a small energy result:

Theorem 5.4. *There are $\varepsilon, C > 0$ such if A is a Yang–Mills connection over the unit ball $B \subset \mathbf{R}^4$ with energy $\mathscr{E}$ less than ε then $|F|^2 \leq C\,\mathscr{E}$ on $B_{\frac{1}{2}}$.*

Again, this is the analogue of what we discussed for harmonic maps. It can be proved using Theorem 5.1 above to find a Coulomb gauge and then applying elliptic estimates, very much like the argument for Theorem 3.2. But the proof in [58] is different. It does not require gauge fixing and introduces another important technique. Recall that the Yang–Mills equations are $\mathrm{d}_A^* F = 0$. The curvature of any connection satisfies the Bianchi identity $\mathrm{d}_A F = 0$, so $\Delta_A F = 0$ where Δ_A is the coupled "Hodge" Laplace operator

$$\Delta_A = -(\mathrm{d}_A\, \mathrm{d}_A^* + \mathrm{d}_A^*\, \mathrm{d}_A).$$

There is another Laplace-type operator $-\nabla_A^* \nabla_A$ acting on the bundle-valued forms. In a local trivialisation this is

$$\sum \left(\nabla_i^A\right)^2.$$

The two are related by a Weitzenbock formula which, over a flat base manifold, is

$$\Delta_A \phi = \nabla_A^* \nabla_A \phi + \{F, \phi\},$$

where the pointwise bilinear operation $\{\ ,\ \}$ combines the bracket on bundle endomorphisms with the map $\Lambda^2 \otimes \Lambda^k \to \Lambda^k$ furnished by the derivative of the action of the orthogonal group on k-forms.

The upshot of this differential geometry is that the curvature of a Yang–Mills connection (over a flat base manifold) satisfies the equation

$$\nabla_A^* \nabla_A F = \{F, F\}. \tag{40}$$

One computes that $|\{F, F\}| \leq 4|F|^2$ and then (40) leads to a differential inequality for $|F|$:

$$\Delta |F| \geq 4|F|^2. \tag{41}$$

(The function $|F|$ may not be smooth at zeros of F but this difficulty can be got around in standard ways. The exact constant 4 in (41) will not be important and depends on conventions for defining the norm $|F|$.)

One can then apply the Nash–Moser iteration technique to derive interior estimates on $f = |F|$ from (41). At the first step, let χ be a fixed cut-off function, equal to 1 on the ball of radius $\frac{3}{4}$ say. Then multiplying by $\chi^2 f$ and integrating by parts:

$$\int_{B^4} (\nabla(\chi^2 f), \nabla f) \leq 4 \int \chi^2 f^3.$$

We have

$$(\nabla(\chi^2 f), \nabla f) = |\nabla(\chi f)|^2 - |\nabla \chi|^2 f^2,$$

so we get

$$\int |\nabla(\chi f)|^2 \leq 4 \int (\chi f)^2 f + \int |\nabla \chi|^2 f^2.$$

Invoking the Sobolev embedding $L^2_1 \to L^4$ in dimension four and Hölder's inequality, we see that if the L^2 norm of f is sufficiently small we can apply quadratic rearrangement to get an L^4 bound on χf, hence an L^4 bound on f in the $\frac{3}{4}$-ball. The relevant manipulation is, again, similar to that in the proof of Theorem 3.2. Repeating the process, with a suitable sequence of concentric balls and cut-off functions and keeping track of the constants, leads to an L^∞ bound on f in the $\frac{1}{2}$-ball. If one only needs to work with Yang–Mills solutions then it is usually possible to avoid the use of the sharp Coulomb gauge-fixing result of [59], using this alternative approach from [58].

Returning to the removal of singularities problem, Uhlenbeck explains in [58], for a connection over $B^n \setminus \{0\}$, the critical nature of the curvature decay condition $|F| \leq C|x|^{-2}$. If we take a non-trivial Yang–Mills connection over S^{n-1} (for example the Levi-Civita connection on the tangent bundle) and pull it back by radial projection we get a Yang–Mills connection over the punctured ball whose curvature is exactly $O(|x|^{-2})$. In dimension $n > 4$ this curvature is in L^2 but when $n = 4$ it is in L^p for any $p < 2$ but not in L^2. For $n = 4$ the small energy result, applied to a ball of radius $|x|/2$ centred at a point x, leads to a bound

$$|F(x)| = o(|x|^{-2}) \tag{42}$$

so we get a little above the critical $O(|x|^{-2})$ threshold but, as in the case of Theorem 3.3 for harmonic maps, more is needed. A differential inequality

$$(1-\delta)\, \mathscr{E} \leq \frac{1}{4r} \frac{\mathrm{d}\mathscr{E}}{\mathrm{d}r}$$

implies that $\mathscr{E} = O(r^{4-4\delta})$, which gives $|F| = O(r^{-2\delta})$ and then F is in L^p for $p < 2/\delta$. Uhlenbeck establishes a more complicated inequality

$$\left(1-\omega\,\mathscr{E}(2r)^{\frac{1}{2}}\right)\mathscr{E}(r)\leq\frac{1}{4r}\frac{\mathrm{d}\mathscr{E}}{\mathrm{d}r}, \tag{43}$$

for a constant ω, from which she is able to deduce that $|F|$ is bounded. Then an exponential gauge produces a bounded connection form, which can be adjusted to satisfy the Coulomb condition, and elliptic regularity shows that the connection extends smoothly over the origin. (See the further discussion in subsection 6.3 below.)

We will not go much further into the details of Uhlenbeck's proof in [58], partly because we will discuss another proof, of Uhlenbeck and Smith, in subsection 6.3 below. One important idea in the proof of [58] is that on any small annulus the connection is close to flat, in the sense that when the annulus is scaled to standard size the curvature is small, by (42) above. This means that the nonlinear equation can be approximated by its linearisation, provided a suitable gauge is used over the annulus. The construction of these gauges takes up much of the work in the paper. In the case of the Abelian gauge group $U(1)$ we have $F=\mathrm{d}\underline{A}$ and the Yang–Mills equation is $\mathrm{d}^*\underline{A}=0$. Then over a domain Ω:

$$\int_{\Omega}|F|^2=\int_{\partial\Omega}\underline{A}\wedge *F.$$

In the general non-Abelian case there is a similar formula with lower order terms, and by estimating these Uhlenbeck obtains the differential inequality (43). The crucial constant 4 in (43) appears as the first eigenvalue of the Laplacian on co-closed 1-forms on S^3.

5.3 Applications

Much of the original motivation for the removal of singularities theorem was to answer a question raised by physicists. A finite-energy Yang–Mills $U(r)$ connection over $\mathbf{R}^4$ extends smoothly to S^4, in particular it has a topological invariant Chern number. In a later paper [60] Uhlenbeck showed that for any finite energy connection the Chern number, defined by integrating the Chern–Weil form, is an integer. This used the full force of Theorem 5.1, in fact extended to L^2_1 connections.

The results of these two papers of Uhlenbeck were foundational in the development of Yang–Mills theory over Riemannian manifolds of dimension at most 4. In one direction they opened the way to the use of variational methods. In the subcritical case, over manifolds of dimension 2 and 3, the analysis is relatively straightforward. In particular any bundle admits a connection which minimises the Yang–Mills functional. This is an easy consequence of Theorem 5.2 which implies that there is a minimising sequence which converges in L^2_1 and in these dimensions the bundle theory works in a straightforward way with L^2_1 connections and L^2_2 gauge transformations because L^2_2 maps are continuous. The existence of minimisers is not the end of the story: one would like to go on to the relate the Yang–Mills connections to the topology of the space of connections modulo gauge equivalence. There has been a lot of work on this in the case of bundles over surfaces, following Atiyah and Bott, including a paper of Daskalopoulus and Uhlenbeck [11]. Rade (who, like Daskalopoulos, was a PhD student of Uhlenbeck) obtained complete results about the Yang–Mills gradient flow in these dimensions [37].

The variational theory in dimension 4 is much harder. One early and clear cut result was obtained by Sedlacek (another PhD student of Uhlenbeck) [43]. Let G be a compact Lie group. A principal G-bundle P over a compact oriented 4-manifold is determined by two characteristic classes $\kappa(P), w(P)$ where $\kappa(P) \in H^4(X, \pi_3(G))$ and $w(P) \in H^2(X, \pi_1(G))$. For example if $G = \mathrm{SU}(m)$ then κ is the second Chern class and w is trivial, since $\mathrm{SU}(m)$ is simply connected, while if $G = \mathrm{SO}(m)$ for $m = 3$ or $m \geq 5$ then κ is a multiple of the first Pontrayagin class and w is the second Stiefel–Whitney class in $H^2(X; \mathbf{Z}/2)$. Now let X have a Riemannian metric, so the Yang–Mills equations are defined. Sedlacek's result is that for any P there is a Yang–Mills connection on a G-bundle $P' \to X$ with $w(P) = w(P')$. It might happen that P' is isomorphic to P but the result allows the possibility that they are different. (This is the Yang–Mills analogue of the fact that for any homotopy class of maps from a surface there is a harmonic map inducing the same homomorphism on fundamental groups, but possibly in a different homotopy class.)

To prove this result, Sedlacek considers a minimising sequence A_i for the Yang–Mills functional on P. A covering argument just like that for Theorem 3.1 shows that after passing to a subsequence i', there is a finite subset $S \subset X$ (possibly empty) such that each point in $X \setminus S$ is the centre of a ball on which the connections $A_{i'}$ satisfy the small energy condition for the Coulomb gauge fixing result, Theorem 5.1. Then we arrive at a situation like that we considered in the proof of Theorem 5.2, with a subsequence i', a cover B_α of $X \setminus S$ and connection 1-forms $\underline{A}^\alpha_{i'}$ in Coulomb gauge and with L^2_1 limits $\underline{A}^\alpha_\infty$ and L^2_2 limits $g^{\alpha\beta}_\infty$ of the transition functions $g^{\alpha\beta}_i$. Sedlacek shows that these limits $\underline{A}^\alpha_\infty$ are weak solutions of the Yang–Mills equations and then, using ellipticity in Coulomb gauge, that they are in fact smooth.

The equation (38), and the cocycle conditions, are preserved in the limit so the smoothness of the $\underline{A}^\alpha_\infty$ implies that of the $g^{\alpha\beta}_\infty$. Thus the limiting data defines a smooth connection A_∞ on a bundle P_∞ over $X \setminus S$. This is a finite-energy Yang–Mills connection so the removal of singularities theorem shows that the bundle and connection extend smoothly over the finite set S to a connection on a bundle P'. The remaining step is to show that the characteristic classes $w(P), w(P')$ are equal. The point here is that P' may not be isomorphic to P over X. Even over $X \setminus S$ the L^2_2 convergence of the transition functions does not give the C^0 convergence, so the last part of the proof of Theorem 5.2 does not extend to this situation (although, by algebraic topology, it does turn out in the end that the bundles are isomorphic over $X \setminus S$).

Consider for example the case when $X = \mathbf{CP}^2$, with its standard metric and orientation, and the gauge group $G = \mathrm{SU}(2)$. If P is the bundle with $c_2(P) = 1$ there is no minimiser of the Yang–Mills functional: a minimising sequence will converge to the flat connection away from a point in $\mathbf{CP}^2$, displaying the same kind of bubbling behaviour that we described for harmonic maps from T^2 to S^2 in Section 3. If $c_2(P) = -1$ on the other hand then minimisers can be constructed (they are instantons, see below). But there is a non-compact moduli space of these minimisers and we could choose a "bad" minimising sequence exhibiting bubbling over a point.

The variational theory was developed much further in a sequence of papers by Taubes such as [52, 53], relating the solutions to the topology of the space of connections modulo equivalence. We refer to the article [15] for a discussions of those developments.

Perhaps the largest impact of Uhlenbeck's papers [58, 59] came in the study of the "instanton" solutions to the Yang–Mills equations in 4 dimensions. Instantons, over an oriented, Riemannian 4-manifold X are connections whose curvature is self-dual or anti-self-dual (the two being interchanged by switching orientation). In the analogy between 4-dimensional Yang–Mills theory and harmonic maps of surfaces, these correspond to holomorphic maps to a Kähler (or just almost-Kähler) manifold. The instantons have consequences in 4-manifold topology, analogous to those of the mapping theory in symplectic topology. The analogue of Theorem 3.1 for sequences of instantons leads to the "Uhlenbeck compactifications" of instanton moduli spaces. These are made up of pairs $([A], D)$ where $[A]$ is the gauge equivalence class of an instanton and D is a formal sum of points q_j of X with multiplicities κ_j. A sequence of instantons converges to $([A], D)$ if the connections converge on the complement of the points q_j and exhibit "bubbling" over the q_j with κ_j units of energy (suitably normalised) concentrating at q_j. We refer to the books [16, 19] for detailed accounts of these developments.

6 The Yang–Mills equations in higher dimensions

6.1 Hermitian Yang–Mills connections on stable bundles

Much of the work involving Yang–Mills theory over manifolds of dimension greater than four focuses on manifolds with some extra structure, as opposed to general Riemannian manifolds. One of the most important such developments came in work of Uhlenbeck and Yau [64], establishing the "Kobayashi–Hitchin conjecture" in complex differential geometry.

Let X be a compact Kähler manifold of complex dimension m with Kähler form ω. The forms on X decompose into bi-type and the metric defines a contraction operator $\Lambda : \Omega^{1,1} \to \Omega^0$. This is just the trace with respect to the metric ω. We recall three identities:

(1) For a function f: $\Delta f = -2i\Lambda\overline{\partial}\partial f$.
(2) For a $(0,1)$ form α: $i\Lambda(\alpha \wedge \overline{\alpha}) = -|\alpha|^2$.
(3) For a $(1,1)$ form θ: $\theta \wedge \omega^{n-1} = (\Lambda\theta)\, \omega^m/m = (m-1)!(\Lambda\theta)$ vol.

Now consider a unitary connection on a complex vector bundle $E \to X$. The curvature decomposes into $F = F^{0,2} + F^{1,1} + F^{2,0}$ and we define $\widehat{F} = i\Lambda F^{1,1}$. So $\widehat{F}$ is a section of the bundle of self-adjoint endomorphisms of E. The connection is called a *Hermitian-Yang–Mills connection* if $F^{0,2}$ and $F^{2,0}$ vanish and

$$\widehat{F} = \mu 1_E, \tag{44}$$

for a constant μ. The constant μ is determined by topology. By the third item above

$$(m-1)! \left(\operatorname{Tr} \widehat{F}\right) \mathrm{vol} = i\operatorname{Tr} F \wedge \omega^{m-1},$$

and by Chern–Weil theory the 2-form $(i/2\pi)\mathrm{Tr}\ F$ represents the first Chern class $c_1(E)$. So if a solution to (44) exists, we have

$$\mu = \frac{2\pi}{(m-1)!\mathrm{Vol}\ (X)} \frac{\deg(E)}{\mathrm{rank} E},$$

where the degree $\deg(E)$ is defined to be the pairing $(c_1(E) \cup \omega^{m-1})[X]$, which is a topological invariant of the bundle $E \to X$ and the Kähler class $[\omega]$. The ratio $\deg\ (E)/\mathrm{rank}\, E$ is called the *slope* of the bundle E. Hermitian-Yang–Mills connections are Yang–Mills connections: in fact they are absolute minimisers of the Yang–Mills functional on the given bundle. If E is the tangent bundle of X and the connection is the Levi-Civita connection then $\widehat{F}$ is the Ricci tensor, so (44) is related to the Einstein equations and solutions are often called Hermitian–Einstein connections in the literature.

The significance of the condition that the curvature F has type $(1,1)$ is that this implies that the connection is compatible with a holomorphic structure on the bundle E. For any connection ∇ we can write

$$\nabla = \partial_\nabla \oplus \overline{\partial}_\nabla : \Omega^0(E) \to \Omega^{1,0}(E) \oplus \Omega^{0,1}(E).$$

The sheaf of local solutions of the equation $\overline{\partial}_\nabla s = 0$ is a sheaf of modules over the structure sheaf of the complex manifold X: the local holomorphic functions. But when $\dim_{\mathbb{C}} X > 1$ the equation $\overline{\partial}_\nabla s = 0$ is overdetermined and for a general connection the only solution will be $s = 0$. The condition $F^{0,2} = 0$ is the integrability condition for this equation, which implies the existence of solutions generating the bundle E, thus defining a holomorphic structure on the bundle. For a unitary connection the component $F^{2,0}$ is $-(F^{0,2})^*$ so the vanishing of one implies the same for the other.

We have then the existence question: given a holomorphic bundle E does it admit a compatible Hermitian-Yang–Mills connection? The Kobayashi–Hitchin conjecture, formulated independently by Kobayashi and Hitchin around 1980, is that such a connection exists if and only if E is a direct sum of *stable* holomorphic bundles of equal slope. The notion of stability here was introduced before by algebraic geometers in the context of moduli problems. A holomorphic bundle of V is defined to be stable if every non-trivial coherent subsheaf $\mathcal{S}$ of rank less than $\mathrm{rank} V$ satisfies the condition

$$\text{slope }(\mathscr{S}) < \text{slope }(V). \tag{45}$$

(By general theory, such a subsheaf is given by a proper subbundle of V outside a singular set of complex codimension 2 or more, so the first Chern class of $\mathscr{S}$ is defined in $H^2(X)$.)

Part of the evidence for this conjecture came from results of Narasimhan and Seshadri from the 1960s which covers the case when X is a complex curve. The fact that the existence of a Hermitian-Yang–Mills connection implies that the bundle is a sum of stable bundles is relatively straightforward and was proved by Kobayashi [27]. In this subsection we discuss Uhlenbeck and Yau's proof of the existence result, for general Kahler manifolds (X,ω). The essential statement can be put in the form:

Theorem 6.1. *If a holomorphic bundle E does not admit a Hermitian-Yang–Mills connection then there is a subsheaf $\mathscr{S}$, as above, with* slope $\mathscr{S} \geq$ slope (V).

There are two ways of setting up differential geometry on holomorphic vector bundles. In one—which is the traditional point of view in complex differential geometry—one has a fixed holomorphic bundle and varies the Hermitian metric. A metric defines a unique compatible connection (often called the "Chern connection"). In the other, we fix the metric $(\, , \,)$ on a C^∞ bundle E and vary the connection. Let ∇_0 be some unitary reference connection and write $\nabla_0 = \partial_0 + \overline{\partial}_0$. If g is any automorphism of E, not necessarily unitary, we define a new covariant derivative by

$$\nabla^g = (g^*)^{-1} \circ \partial_0 \circ g^* + g \circ \overline{\partial}_0 \circ g^{-1}.$$

That is, we conjugate $\overline{\partial}_0$ by g and define the $(1,0)$ part in the unique way to make a unitary connection ∇^g. If g is unitary then $g = (g*)^{-1}$ and we have the ordinary gauge transformation, so ∇^g geometrically equivalent to ∇_0. In general, the $\overline{\partial}$-operators of ∇^g, ∇_0 are equivalent, in the sense that they define isomorphic holomorphic structures on the bundle, but the connections are essentially different. Working modulo the unitary gauge transformations, we can restrict attention to self-adjoint automorphisms, which we write as h. Then

$$\nabla^h = h \circ \overline{\partial}_0 \circ h^{-1} + h^{-1} \circ \partial_0 \circ h.$$

To match up with the first point of view, it is equivalent to fix the holomorphic structure with $\overline{\partial}$-operator $\overline{\partial}_0$ and vary the metric to $(hs, hs) = (s, h^2 s)$. We will use this second point of view, which means that some formulae will look different from those in [64].

The curvature of the connection ∇^h is

$$F(\nabla^h) = F_0 + \overline{\partial}_0(h^{-1}\,\partial_0 h) - \partial_0(\overline{\partial}_0 h\, h^{-1}) - \left(h^{-1}\partial_0 h\, \overline{\partial}_0 h h^{-1} + \overline{\partial}_0 h\, h^2\, \partial_0 h\right), \tag{46}$$

where F_0 is the curvature of ∇_0.

Write $h = \mathrm{e}^u$, so u is a section of the bundle of self-adjoint endomorphisms of E. To gain understanding of the nature of the Hermitian-Yang–Mills equation we can consider the linearisation about $u = 0$. Using the connection between the ∂ and $\overline{\partial}$ operators and the Laplacian one sees that this is

$$\widehat{F}(\mathrm{e}^u) = \widehat{F_0} + i\Lambda(\overline{\partial}_0\partial_0 - \partial_0\overline{\partial}_0)u + O(u^2) = \widehat{F_0} + \nabla_0^*\nabla_0 u + O(u^2). \quad (47)$$

So the linearisation is the coupled Laplacian. The Hermitian-Yang–Mills equation has many similarities with the harmonic equation for a map into the space of hermitian matrices: the nonlinear term is quadratic in the first derivatives of h.

The proof by Uhlenbeck and Yau of Theorem 6.1 uses a continuity method, with the family of equations, for $t \geq 0$:

$$\widehat{F}(\mathrm{e}^u) = -tu. \quad (48)$$

To set things up they prove:

Proposition 6.1. (1) *The equation* (48) *has a solution for large t.*
(2) *The set of $t \in [0,\infty)$ for which a solution to* (48) *exists is open.*
(3) *If there is a smooth family of solutions u_t to* (48) *for t in an interval (t_0,t_1) satisfying a bound $\|u_t\|_{L^\infty} \leq C$ then the solution extends to the closed interval $[t_0,t_1]$.*

Item (1) is proved by Uhlenbeck and Yau with an auxiliary continuity argument. It can also be established using the implicit function theorem, writing $u = \varepsilon\widehat{F_0} + w$ with $\varepsilon = t^{-1}$. When $\varepsilon = 0$ there is a trivial solution $w = 0$ and this can be deformed to a solution for small ε.

The proof of item (2) also uses the implicit function theorem, in a standard way once one knows the invertibility of the linearised operator. This involves slightly complicated calculations, extending the formula (47), which we pass over here.

Uhlenbeck and Yau use an interesting technique to prove item (3), based on an interpolation inequality. For large p:

$$\|v\|^2_{L_1^{2p}} \leq c\|v\|_{L_2^p}\,\|v\|_{L^\infty}. \quad (49)$$

Let $\dot{u}, \dot{h}$ be the t-derivatives of u and $h = \mathrm{e}^u$ on the interval (t_0,t_1). Thus $\dot{u}$ satisfies the linear equation obtained by differentiating (48). Applying the maximum principle to this equation they show that $\|\dot{u}\|_{L^\infty}$ satisfies a fixed bound (this step is related to the invertibility of the linearised operator for item (2)). The formula (46) for the curvature leads to an expression for $\nabla_0^*\nabla_0\dot{h}$ in terms of $h, \widehat{F}(h)$ and the derivatives of h. The hypothesis means that $\widehat{F}(h)$ is bounded and one gets

$$|\nabla_0^*\nabla\dot{h}| \leq c\left(1 + |\nabla_0 h||\nabla_0\dot{h}| + |\nabla_0 h|^2|\dot{h}|\right).$$

The L^∞ bound on $\dot{u}$ gives one on $\dot{h}$ and we get

$$\|\dot{h}\|_{p,2} \leq c\left(1 + \|h\|_{2p,1}\|\dot{h}\|_{2p,1} + \|h\|^2_{2p,1}\right).$$

Let $M(t) = \|h_t\|_{p,2}$, so $|\frac{\mathrm{d}M}{\mathrm{d}t}| \leq \|\dot{h}\|_{p,2}$. The interpolation inequality (49), applied to h and to $\dot{h}$, gives

$$\left|\frac{\mathrm{d}M}{\mathrm{d}t}\right| \leq c\left(1+M+\sqrt{M\left|\frac{dM}{dt}\right|}\right),$$

which implies that

$$\left|\frac{\mathrm{d}M}{\mathrm{d}t}\right| \leq c(1+M).$$

This gives a bound on $\|h_t\|_{L^p_2}$ over the finite interval (t_1,t_2). Then it is straightforward to obtain bounds on all higher derivatives, which implies that the solution extends to the end points.

Uhlenbeck and Yau show that a solution to (48) for $t>0$ satisfies an *a priori* L^∞ bound

$$\|u_t\|_{L^\infty} \leq ct^{-1}, \tag{50}$$

(see item (1) of Proposition 6.2 below). Then Proposition 6.1 implies that solutions exist for all $t>0$. If there is a C such that $\|u_t\|_{L^\infty} \leq C$ for all small $t>0$ then Proposition 6.1 implies that the solution exists for $t=0$; so we have a Hermitian-Yang–Mills connection. The plan of the proof of Theorem 6.1 is to show that if there is no such C—so there is a sequence $t_i \to 0$ such that $\|u_{t_i}\|_{L^\infty} \to \infty$—then there is a subsheaf $\mathscr{S}$ with slope $\mathscr{S} \geq$ slope (V).

Uhlenbeck and Yau establish the following *a priori* estimates for a solution $u=u_t$ of (48), writing (for convenience below) $f = 2^{-\frac{1}{2}}|u|$.

Proposition 6.2. (1) $\|f\|_{L^\infty} \leq ct^{-1}$;
(2) $\|f\|_{L^\infty} \leq c\|f\|_{L^1}$;
(3) $\|\nabla f\|^2_{L^2} \leq c\|f\|_{L^\infty}$;
(4) $\|\nabla_0 u\|^2_{L^2} \leq c(1+\|f\|^2_{L^\infty})$.

For simplicity, we will discuss the proofs in the case of a rank 2 bundle E with trivial determinant. Then we can suppose that the trace of $\widehat{F_0}$ is zero and we restrict to trace-free u. We recall the differential geometric theory for a bundle E decomposed as an orthogonal direct sum $E_I \oplus E_{II}$. A unitary connection on E is defined by connections on E_i and a second fundamental form B, which is a 1-form with values in $\mathrm{Hom}(E_{II}, E_I)$. In matrix notation we can write our connection as

$$\begin{pmatrix} \nabla_I & B \\ -B^* & \nabla_{II} \end{pmatrix}. \tag{51}$$

The curvature is

$$\begin{pmatrix} F_I - BB^* & \mathrm{d}_{I,II}B \\ -\mathrm{d}_{I,II}B^* & F_{II} - B^*B \end{pmatrix} \tag{52}$$

where $\mathrm{d}_{I,II}$ is the coupled exterior derivative defined by ∇_I and ∇_{II}. Over a complex manifold we can write the $\overline{\partial}$-operator on the direct sum as

$$\begin{pmatrix} \overline{\partial}_I & \beta \\ \gamma & \overline{\partial}_{II} \end{pmatrix}$$

where β, γ are bundle-valued $(0,1)$-forms. Then $B = \beta - \gamma^*$. For a connection defining a holomorphic structure on E the component β vanishes if and only if E_{II} is a holomorphic subbundle and similarly for γ and E_I. The (1,1) parts of the quadratic terms in (52) are

$$(BB^*)_{1,1} = \beta\beta^* + \gamma^*\gamma \ , \ (B^*B)_{1,1} = \gamma\gamma^* + \beta^*\beta.$$

The crucial point for us is that the constituents have a definite sign. Using the second of the three formulae stated at the beginning of this subsection we get

$$|\beta|^2 = -\mathrm{Tr}(i\Lambda(\beta\beta^*)) = \mathrm{Tr}(i\Lambda\beta^*\beta) \quad , \quad |\gamma|^2 = \mathrm{Tr}\ (i\Lambda(\gamma\gamma^*))| = -\mathrm{Tr}\ (i\Lambda(\gamma^*\gamma)).$$

When E_I is a holomorphic subbundle, so $\gamma = 0$, this is an aspect of the principle that "curvature decreases in holomorphic subbundles and increases in holomorphic quotients".

To apply this in our situation, work initially over the open set Ω in X where $u \neq 0$. Then u has eigenvalues $-f, f$ and the bundle E is decomposed into a sum of eigenspace line bundles E_I, E_{II}. Thus

$$h = \begin{pmatrix} \mathrm{e}^{-f} & 0 \\ 0 & \mathrm{e}^{f} \end{pmatrix}.$$

We find that

$$\overline{\partial}^h = \begin{pmatrix} \overline{\partial}_I & \mathrm{e}^{-2f}\beta \\ \mathrm{e}^{2f}\gamma & \overline{\partial}_{II} \end{pmatrix},$$

and

$$\widehat{F}(u) = \widehat{F_0} - \begin{pmatrix} P & Q \\ -Q^* & -P \end{pmatrix},$$

where

$$P = -\Delta f + (\mathrm{e}^{4f} - 1)|\gamma|^2 + (1 - \mathrm{e}^{-4f})|\beta|^2.$$

So we have

$$-\Delta f + (\mathrm{e}^{4f} - 1)|\gamma|^2 + (1 - \mathrm{e}^{-4f})|\beta|^2 + tf = p(\widehat{F_0}), \tag{53}$$

where $p(\widehat{F_0})$ is the component of $\widehat{F_0}$ in $\mathrm{End} E_I$.

The maximum principle applied to this equation (53) implies the first item in Proposition 6.2. We have $|p(\widehat{F_0})| \leq C$ for some C depending only on ∇_0 and we get $\max f \leq Ct^{-1}$. Now we can feed this back into (53) to get the the differential inequality

$$-\Delta f + (\mathrm{e}^{4f} - 1)|\gamma|^2 + (1 - \mathrm{e}^{-4f})|\beta|^2 \leq 2C. \tag{54}$$

In particular we have $-\Delta f \leq 2C$ and while we have derived this over the open set where $u \neq 0$ it is straightforward to see that the inequality holds in a weak sense over the whole manifold. If f attains its maximum at a point $p \in X$ then, by considering a comparison function and applying the maximum principle, we see that it is close to the maximum over a ball of fixed size about p, so the L^1 norm of f is comparable to the L^∞ norm, which gives the second item of Proposition 6.2. Next, taking the L^2 inner product of (54) with f we have

$$\int |\nabla f|^2 \leq c \int f \leq c\|f\|_{L^\infty},$$

which is the third item.

Over Ω we have

$$\nabla_0 u = \begin{pmatrix} -\nabla f & f(\beta - \gamma^*) \\ f(\beta^* - \gamma) & \nabla f \end{pmatrix},$$

so

$$|\nabla_0 u|^2 = 2\big(|\nabla f|^2 + f^2(|\beta|^2 + |\gamma|^2)\big).$$

This formula holds over the whole of X, since f vanishes at the points where β, γ are undefined. Similarly for the inequality (54). Write $N = \|f\|_{L^\infty}$. Then it is elementary that for a suitable κ

$$f^2 \leq \kappa(1+N)^2(\mathrm{e}^{4f} - 1) \ , \ f^2 \leq \kappa(1+N)^2(1 - \mathrm{e}^{-4f}),$$

so

$$|\nabla_0 u|^2 \leq 2|\nabla f|^2 + 2\kappa(1+N)^2\big((\mathrm{e}^{4f} - 1)|\gamma|^2 + (1 - \mathrm{e}^{-4f})|\beta|^2\big).$$

Comparing with (54) and integrating over X (so the term Δf in (54) integrates to zero) we get

$$\int |\nabla_0 u|^2 \leq 2 \int |\nabla f|^2 + 2C\kappa(1+N^2)\mathrm{Vol}\, X,$$

which gives item (4) of Proposition 6.2.

Define $v_t = N(t)^{-1} u_t$, where $N(t) = \|f_t\|_{L^\infty} = 2^{-1/2}\|u\|_{L^\infty}$, as above. So the v_t are bounded in L^∞ and their L^1 norm has a strictly positive lower bound by item (2) of Proposition 6.2. By item (4) of Proposition 6.2 the v_t are bounded in L^2_1. Suppose that there is a sequence t_i such that $N(t_i) \to \infty$. Passing to a subsequence, we can suppose that the v_{t_i} have a weak L^2_1 limit v_∞. This is not zero because of the lower bound on the L^1 norms. By item (3) of Proposition 6.2 the derivatives of $|v_i|$ tend to zero in L^2 and it follows that the derivative of $|v_\infty|$ is zero and $|v_\infty|$ is a non-zero constant, ρ say. (Most likely $\rho = \sqrt{2}$, by our normalisation.) It follows then that the L^2_1 bundle endomorphism π defined by $\pi = 2\sqrt{2}\rho^{-1} v_0 - 1$ is a rank 1 orthogonal projection with $\pi^* = \pi$ and $\pi^2 = \pi$.

The image of π is the limit, in a suitable sense, of the small-eigenvalue eigenspaces of the u_{t_i}. To visualise what is going on here, recall that the space $\mathscr{H}$ of positive self-adjoint 2×2 matrices with determinant 1 is a model for hyperbolic 3-space and has a natural compactification to a closed 3-ball, with a 2-sphere at infinity. More intrinsically, if $\mathscr{H}$ consists of Hermitian forms on a 2-dimensional complex vector space V then the sphere at infinity is $\mathbf{P}(V)$. A sequence $H_i \in \mathscr{H}$ tends to a point $[z] \in \mathbf{P}(V)$ if $H_i(z,z) \to 0$. In our situation we are considering sections h_t of a bundle $\mathscr{H}_E$ over X with fibre $\mathscr{H}$ which is compactified by adjoining $\mathbf{P}(E)$. The conclusion above is that if the h_t do not have a finite limit, as a section of $\mathscr{H}_E$—which would give a Hermitian-Yang–Mills connection—they have a limit "at infinity" which is a section of $\mathbf{P}(E)$, i.e. a subbundle of E.

Suppose that $L \subset E$ is a holomorphic subbundle. Orthogonal projection onto L is a smooth self-adjoint section ϖ of End E with

$$\varpi^2 = \varpi \quad \text{and} \quad (1-\varpi)\overline{\partial}_0\varpi = 0.$$

From another point of view, the section ϖ is equivalent to a section of the projectivised bundle $\mathbf{P}(E)$ and the equation $(1-\varpi)\overline{\partial}_0\varpi = 0$ is equivalent to the Cauchy–Riemann equation for this section. More generally, if $\mathscr{S}$ is a rank-1 subsheaf of E we get a meromorphic section of $\mathbf{P}(E)$. The basic example, in local coordinates z_1, z_2 on a complex surface X and a local holomorphic trivialisation of the bundle E, is the sheaf $\mathscr{S}$ which is defined by the image of the bundle map $s : \mathscr{O} \to \mathscr{O} \oplus \mathscr{O}$ with $s(z_1,z_2) = (z_1,z_2)$. Then the meromorphic section is given by the standard rational map $(z_1,z_2) \mapsto [z_1,z_2]$ from $\mathbf{C}^2$ to $\mathbf{P}^1$, undefined at the origin.

For the L^2_1 projection π constructed above, $(1-\pi)\overline{\partial}_0\pi$ is defined in L^2. The proof of Uhlenbeck and Yau is completed by showing three things.

(1) π satisfies the equation $(1-\pi)\overline{\partial}_0\pi = 0$.
(2) This defines a meromorphic section of $\mathbf{P}(E)$ and a coherent subsheaf of E.
(3) The degree of this subsheaf is ≥ 0.

To see the idea of the proof we consider the simple situation when the u_t do not vanish anywhere and the v_t converge in C^∞ to v_0. In that case we have a smooth π_t defined by projection onto the small eigenspace E_I of u_t and π_t is the C^∞ limit of the π_t. Then

$$|(1-\pi_t)\overline{\partial}_0\pi_t| = |\gamma_t|$$

while

$$\int |\gamma_t|^2(\mathrm{e}^{4f_t}-1) \leq 2C\mathrm{Vol}M.$$

Our simplifying assumptions imply that $\min_X f_t \to \infty$ as $t \to 0$, so clearly $(1-\pi_t)\overline{\partial}_0\pi_t$ tends to 0 in L^2. In this simple situation item (2) is trivial so we turn to item (3). By the Chern–Weil formula the degree of E_I is

$$(2\pi)^{-1}\int_M p(\widehat{F_0}) + |\gamma_t|^2 - |\beta_t|^2,$$

while

$$p(\widehat{F_0}) + (\mathrm{e}^{4f_t}-1)|\gamma_t|^2 + (1-\mathrm{e}^{-4f_t})|\beta_t|^2 = tf.$$

So the degree of E_I is

$$(2\pi)^{-1}\int_M tf_t + \mathrm{e}^{4f_t}|\gamma_t|^2 - \mathrm{e}^{-4f_t}|\beta_t|^2 \geq (2\pi)^{-1}\left(\int_M tf_t - \int_M \mathrm{e}^{-4f_t}|\beta_t|^2\right).$$

Under our simplifying assumptions the last term tends to zero as $t \to 0$ and we see that $\deg E_I \geq 0$.

The proofs of items (1) and (3) in the general case requires some more careful analysis but no fundamental difficulties. Item (2) is of a different order. Uhlenbeck and Yau write about their whole proof: *The technical part of the proof is quite straightforward except for one point. We obtain the [subbundles] as "holomorphic" in a very weak sense...obtaining enough regularity to describe them as sheaves is more difficult.*

Uhlenbeck and Yau gave two largely independent treatments of this crucial difficulty. One uses complex analysis techniques. It is equivalent to show that an L^2_1 map into a complex Grassmannian which is a weak solution of the Cauchy–Riemann equations is meromorphic. Thus it fits into the same general realm of the regularity of weak harmonic maps we discussed in Section 4. The other uses gauge theory techniques, which we will postpone to subsection 6.2 below.

The circle of ideas around this correspondence between the existence of solutions to the Hermitian-Yang–Mills equations and the algebro-geometric notion of stability has been extremely fruitful and influential in developments in complex differential geometry over the past four decades and we only mention a few aspects of this. In the case of a complex projective manifold, with integral Kähler class, the author gave alternative proofs in [13, 14] exploiting a variational point of view. Soon after, Simpson gave another proof which combined the variational point of view with the Uhlenbeck–Yau techniques [47]. Simpson considered a more general problem, involving a holomorphic bundle and additional fields, and there have been huge developments in that direction, one pioneer being Uhlenbeck's student Bradlow [3]. Li and Yau extended the correspondence to the case of a general Hermitian base manifold in [28]. In place of the Uhlenbeck–Yau continuity method, most subsequent work has involved the natural nonlinear heat equation—the Yang–Mills flow—associated to the existence question. (As Uhlenbeck and Yau state in [64], the two methods are closely related.) Very complete results have been obtained, by Sibley and Wentworth [45] and other authors, on the limiting behaviour of this flow and connections with the algebro-geometric Harder–Narasimhan filtrations.

In a different direction the questions of existence of Kähler–Einstein metrics (in the Fano case) and more generally of "extremal" metrics turn out to fit into the same conceptual picture as the Hermitian-Yang–Mills theory and this has been the scene for much activity. Meanwhile, on the algebraic geometry side, there have been vast extensions of the notion of stability, starting with the work of Bridgeland, and there is also much activity relating these developments to other equations in complex differential geometry.

6.2 Connections with small normalised energy

We first discuss higher-dimensional generalisations of the small energy results of Theorem 5.4—the gauge theory analogue of part of subsection 4.1. Then we go back to explain their relevance to the Uhlenbeck–Yau proof of Theorem 6.1.

The basic small-energy result for Yang–Mills connections in any dimension is just the same as in dimension 4. For a Yang–Mills connection over the unit ball with sufficiently small energy that energy controls all derivative of the connection, in a suitable gauge, over an interior ball. For the application to the Hermitian-Yang–Mills problem we want to consider a more general situation.

Theorem 6.2. *Let $B \subset \mathbf{C}^m$ be the unit ball and B' an interior ball and let q be an exponent with $m < q < 2m$. There are $\varepsilon, \eta, C > 0$ such that if A is a unitary connection over B with curvature of type $(1,1)$ and such that*

$$\|F(A)\|_{L^2}^2 \leq \varepsilon \quad , \quad \|\widehat{F}\|_{L^q} \leq \eta \tag{55}$$

then there is an L^q bound on the curvature over B'

$$\|F\|_{L^q(B')} \leq C(\|f\|_{L^2(B)} + \|\widehat{F}\|_{L^q(B)}).$$

Given this, we can apply Uhlenbeck's gauge fixing Theorem 5.1, once F is sufficiently small in L^2 and $\widehat{F}$ in L^q, to get a connection form $\underline{A}$ over B' with a bound on $\|\underline{A}\|_{L^q_1(B')}$.

Uhlenbeck's proof of such a result was written in the unpublished manuscript [63]. A description of the proof, and generalisations, can be found in the recent paper [9]. For Yang–Mills connections (including the case of Hermitian-Yang–Mills connections, where $\widehat{F}$ is a constant multiple of the identity) a proof was given by Nakajma [32], following the same lines as Schoen's proof of the corresponding result for harmonic maps (Proposition 4.1 above). It uses a monotonicity formula going back to Price [36] for the normalised Yang–Mills energy in real dimension n

$$\widehat{\mathscr{E}} = r^{4-n} \int_{B_r} |F|^2.$$

The proof of monotonicity, for smooth Yang–Mills connections, goes exactly as in subsection 4.1. (In this section the reader should keep in mind that we are working in complex dimension m so the real dimension is $n = 2m$, to fit in with our previous notations.) We give a proof of Theorem 6.2 on the same lines as Nakajima's proof here.

The first thing is to obtain a monotonicity-type property for the normalised energy of connections with curvature of type $(1,1)$ and with an L^q bound on $\widehat{F}$. For a connection with curvature of type $(1,1)$ we have an identity

$$|F|^2 \operatorname{vol} = \frac{1}{(m-2)!} \operatorname{Tr}(F^2) \wedge \omega^{m-2} + \frac{1}{m} |\widehat{F}|^2 \operatorname{vol}. \tag{56}$$

Write the flat Kähler metric on the unit ball as $\omega = d\lambda$, where the 1-form λ is half the contraction of ω with the radial vector field $r\partial_r$. Then we have

$$\int_B \mathrm{Tr}(F^2)\wedge\omega^{m-2} = \int_{\partial B} \mathrm{Tr}(F^2)\wedge\lambda\wedge\omega^{m-3}.$$

Calculation shows that there is a pointwise bound on ∂B:

$$\frac{2}{(m-3)!}\mathrm{Tr}(F^2)\wedge\lambda\wedge\omega^{m-3} \leq |F|^2 \mathrm{vol}_{\partial B}.$$

So we get

$$2(m-2)\int_B |F|^2 \leq \int_{\partial B} |F|^2 + \frac{2(m-2)}{m}\int_B |\widehat{F}|^2.$$

Applying the same argument to balls of radius $r<1$ we get the inequality for the derivative of normalised energy

$$\frac{d\widehat{\mathscr{E}}}{dr} \geq -\frac{2(m-2)}{m} r^{3-2m}\int_{B_r} |\widehat{F}|^2.$$

This gives

$$\frac{d\widehat{\mathscr{E}}}{dr} \geq -cr^{3-4m/q}\left(\int_{B_r} |\widehat{F}|^q\right)^{2/q},$$

for a suitable constant c. Since $q>m$ we have $3-4m/q>-1$ and we can integrate to get, for all $r\leq 1$,

$$\widehat{\mathscr{E}}(r) \leq \widehat{\mathscr{E}}(1) + c\|\widehat{F}\|^2_{L^q}.$$

It follows from this that, in the setting of Theorem 6.2, we can suppose that the normalised energy on all interior balls is as small as we please, by making ε and η suitably small.

Lemma 6.3. *There are $\theta, c>0$ such that if $\underline{A}$ is a connection form over the unit ball $B\subset \mathbf{C}^m$ with curvature of type $(1,1)$, satisfying $d^*\underline{A}=0$ and the boundary condition of Theorem 5.1, and with $\|\underline{A}\|_{L^m_1}\leq\theta$, then the restriction of $\underline{A}$ to the ball $B_{\frac{3}{4}}$ satisfies*

$$\|\underline{A}\|_{L^q_1(B_{\frac{3}{4}})} \leq c\left(\|F\|_{L^2(B)} + \|\widehat{F}\|_{L^q(B)}\right). \tag{57}$$

The proof is similar to part of the proof of Theorem 5.1. Just as in there we have an estimate (once $\underline{A}$ is small in L^m_1):

$$\|\underline{A}\|_{L^2_1} \leq c\|F\|_{L^2},$$

and the Sobolev embedding maps L^2_1 to L^ν, with $\nu = 2m/(m-1)$, so

$$\|\underline{A}\|_{L^\nu} \leq c\|F\|_{L^2}.$$

For a real 2-form Ω let $\pi(\Omega) = (\Omega^{0,2}, \widehat{\Omega})$. So the conditions on $\underline{A}$ give

$$|\pi d\underline{A}| \leq |\widehat{F}| + |\underline{A} \wedge \underline{A}|.$$

and we also have $\mathrm{d}^*\underline{A} = 0$. The point now is that the operator $D = \mathrm{d}^* \oplus \pi \mathrm{d}$ on 1-forms is (overdetermined) elliptic. In fact it can be identified with

$$\overline{\partial}^* \oplus \overline{\partial} : \Omega^{0,1} \to \Omega^{0,2} \oplus \Omega^0.$$

Let χ be a cut-off function equal to 1 on $B_{\frac{3}{4}}$. We have

$$|D(\chi\underline{A})| \leq |\nabla\chi||\underline{A}| + |(\chi\underline{A}) \wedge \underline{A}| + |\widehat{F}|.$$

Elliptic theory gives

$$\|\chi\underline{A}\|_{\nu_1} \leq c\|D(\chi\underline{A})\|_{L^\nu}.$$

So

$$\|\chi\underline{A}\|_{\nu,1} \leq c\left(\|\chi\underline{A} \wedge \underline{A}\|_{L^\nu} + \|\widehat{F}\|_{L^\nu} + \|\underline{A}\|_{L^\nu}\right).$$

Then we can employ critical quadratic rearrangement, once $\underline{A}$ is sufficiently small in L^m_1 (which implies that it is small in L^{2m}). Assuming that $m \geq 3$ we have $\nu \leq m < q$ and the L^ν norm of $\widehat{F}$ is dominated by a multiple of the L^q norm. Using also the bound we have on the L^ν norm of $\underline{A}$ we get

$$\|\chi\underline{A}\|_{\nu,1} \leq c\left(\|F\|_{L^2} + \|\widehat{F}\|_{L^q}\right).$$

Over the ball where $\chi = 1$ this is an improvement on the L^2_1 bound we had before, since $\nu > 2$. Now we can repeat this process with another cut-off function, supported on the region where $\chi = 1$ and equal to 1 on $B_{\frac{3}{4}}$. After a finite number of such steps we get a bound on the L^q_1 norm of $\underline{A}$ over the $\frac{3}{4}$ ball.

Note that the difference in this proof, compared with that in Theorem 5.1, is that the boundary condition does not combine well with the elliptic operator D, which is why we have to introduce cut-off functions and we only get an interior estimate.

Combining this lemma with Uhlenbeck's Theorem 5.1 we have a small constant θ' such that any connection over B with curvature of type $(1,1)$ and $\|F\|_{L^m} \leq \theta'$ has a connection form satisfying (57).

To prove Theorem 6.2, for x in the unit ball B let $D(x)$ be the distance to the boundary, as before. Given a connection A with curvature F of type $(1,1)$ over $B \subset \mathbf{C}^m$, define $r(x)$ to be the supremum of the $r < D(x)$ such that the L^m norm of F over the r-ball $B_{x,r}$ is less than or equal to θ'. Define

$$M(A) = \max_{x \in B} \frac{D(x)}{r(x)}.$$

If we have a bound $M(A) \leq M_0$ for any fixed M_0 we can apply Lemma 6.3 to get estimates on the L^q norm of F over any interior ball, thus proving Theorem 6.2. We will show that if $\|\widehat{F}\|_{L^q} \leq \eta$ and the normalised energy of A over any interior ball is $\leq \widehat{\varepsilon}$, for sufficiently small $\widehat{\varepsilon}$ and η we have $M(A) \leq 3$.

Suppose, arguing for a contradiction, that $M = M(A) > 3$ and let x_0 be a point where the maximum is attained, so $D(x_0) > 3r(x_0)$. Write $r = r(x_0)$ and let x_1 be any point on the boundary of $B_{x_0,r}$. From the definition, $r(x_1) \geq \frac{2}{3}r$. By the scale invariance of the L^m norm on 2-forms in dimension $2m$ we can apply Lemma 6.3 to the restriction of A to the ball B_{x_0r}. That is, if A' is the rescaled connection over the unit ball, Lemma 6.3 gives an L^q bound on the curvature of A' over the $\frac{3}{4}$-ball. Since $q > m$ this implies an L^m, and by scale invariance this is an L^m bound on the curvature of A over $B_{x_0,3r/4}$. This can be made as small as we please by choosing $\widehat{\varepsilon}, \eta$ small. In just the same way we get an L^m bound on the curvature over $B_{x_1,r/2}$. Now take a finite number of such boundary points like x_1, whose union covers the annulus $B_{x_0,5r/4} \setminus B_{x_0,3r/4}$. Then by making $\widehat{\varepsilon}$ and η small we can arrange that the L^m norm of F over $B_{x_0,5r/4}$ is less than θ', which is a contradiction to the definition of $r = r(x_0)$.

Theorem 6.2 leads to a global consequence for a sequence A_i of unitary connections with curvature of type $(1,1)$ on a fixed bundle E over a compact Kähler manifold X, with a bound on $\|\widehat{F}\|_{L^q}$ (for some $q > m$). After possibly passing to a subsequence i' there is a closed set $S \subset X$ of finite $(2m-4)$-dimensional Hausdorff measure such that $A_{i'}$ converge in $L^q_{1,\mathrm{loc}}$ over the complement $M \setminus S$. To see this we first go back to equation (56). By Chern–Weil theory, the integral of Tr $(F^2) \wedge \omega^{m-2}$ is a topological invariant of the bundle, so an L^2 bound on $\widehat{F}$ is equivalent to one on F. So since $q > 2$ we have an L^2 bound on the curvatures of the A_i. Then the proof is just as for the harmonic maps case discussed in subsection 4.1 above, with gauge transformations constructed as in subsection 5.2. The only additional point is the constraint $\|\widehat{F}\|_{L^q} \leq \eta$ in Theorem 6.2, for the connection over the unit ball. But this will be true for connections obtained by rescaling A_i over sufficiently small balls in X, since $q > m$.

We return now to the Uhlenbeck–Yau proof of Theorem 6.1. Item (1) of Proposition 6.2 shows that on the continuity path (48) the $\widehat{F}$ satisfies a uniform L^∞ bound, so also an L^q bound. So for our sequence $t_i \to 0$ the discussion above applies to the connections $A_i = \nabla^{e^{u_i}}$. Thus without loss of generality there is a weak $L^q_{1,\mathrm{loc}}$ limit convergence outside a codimension-4 set S. The limiting connection is in L^q_1 and has curvature of type $(1,1)$. The usual integrability theorem extends to such connections, so the connection defines a holomorphic bundle E_∞ over $X \setminus S$. We now regard the connections A_i over $V \setminus S$ as a convergent sequence of connections on the bundle E_∞ and the h_i as holomorphic bundle maps from $(E, \overline{\partial}_A)$ to $(E_\infty, \overline{\partial}_{A_i})$. Standard arguments show that, after suitable scalings, these maps converge to a non-trivial holomorphic bundle map from $(E, \overline{\partial}_A)$ to $(E_\infty, \overline{\partial}_{A_\infty})$. The kernel of this map is a coherent subsheaf of E over $X \setminus S$ and one sees that this is the same as the weak L^2_1 subbundle $\mathrm{Im}\pi$ discussed before.

To sum up, Uhlenbeck and Yau use this gauge theory argument to show that the weak subbundle is a coherent sheaf at least outside a set of real codimension 4, and the proof of the regularity theorem for the weak L^2_1 solution π with this extra information is much simpler.

6.3 Removal of codimension 4 singularities

In his paper [54], Tian developed a theory for Yang–Mills connections analogous to that of Schoen and Uhlenbeck for harmonic maps. This included the notion of tangent cones at singular points. An analogue of the Schoen–Uhlenbeck small energy result in the singular case was proved by Tao and Tian in [50]. Let B be the unit ball in $\mathbf{R}^n$ and B' an interior ball. A smooth finite energy Yang–Mills connection A defined on the complement $B \setminus \Sigma$ of a closed set Σ of finite $(n-4)$ dimensional Hausdorff measure is called *admissible*. Just as for harmonic maps, the connection is called *stationary* if for any vector field v with compact support in the interior of B generating a 1-parameter group of diffeomorphisms Φ_t we have

$$\frac{\mathrm{d}}{\mathrm{d}t}\mathscr{E}(\Phi_t^*(A))|_{t=0} = 0.$$

The monotonicity of normalised energy holds for stationary Yang–Mills connections. Tao and Tian's result is

Theorem 6.4. *There is an $\varepsilon > 0$ such that any stationary, admissible connection A over B with energy less than ε extends smoothly over B'.*

When $n = 4$ this is equivalent to Uhlenbeck's Theorem 5.2 on the removal of point singularities. In the 2018 paper [49], Smith and Uhlenbeck give a different proof of Theorem 6.4 which is in many ways simpler than Tao and Tian's. In fact, Smith and Uhlenbeck prove a more general result for solutions of Yang–Mills–Higgs equations, with additional fields. In this subsection 6.3 we will discuss the Smith and Uhlenbeck proof of Theorem 6.4.

A consequence of Theorem 6.4 is that the singular set of a finite energy stationary Yang–Mills connection has dimension strictly less than $(n-4)$. Such singular sets do arise in interesting examples. In particular, an extension by Bando and Siu of the Uhlenbeck–Yau theorem gives the existence of Hermitian-Yang–Mills connections on stable reflexive sheaves over a Kähler manifold [2]. Such sheaves are vector bundles outside a singular set of complex codimension at least three and the Hermitian-Yang–Mills connection has singularities at this set. Recent work of Chen and Sun [8] gives an algebro-geometric description of the tangent cone, in Tian's sense, of the Hermitian-Yang–Mills connection at a singular point.

A key component in Smith and Uhlenbeck's proof is a refinement of the differential inequality (41) for $|F|$, where F is the curvature of a Yang–Mills connection. The derivation of this used what is sometimes called the Kato inequality: $|\nabla_A |F|| \leq |\nabla_A F|$. Similar to the calculations in subsection 2.2 one has

$$|F|\Delta|F| = (\nabla_A^* \nabla_A F, F) + |\nabla_A F|^2 - |\nabla|F|\,|^2,$$

and the Kato inequality gives (41). This can be improved via a "refined Kato inequality", which we now review. (See also the survey [6].)

In general, suppose that V is a Euclidean vector bundle of rank greater than 1 over a Riemannian manifold M with a compatible covariant derivative ∇ and that s is a section of V which does not vanish at a point $p \in M$. Then $|\nabla|s|| = |\nabla s|$ at p if and only if the image of ∇s, regarded as a linear map from TM_p to V_p, lies in the 1-dimensional subspace spanned by $s(p)$. In other words $\nabla s = \theta \otimes s$ for some $\theta \in T^*M_p$. Let $D : \Gamma(V) \to \Gamma(W)$ be a first-order linear differential operator which is the composite of the covariant derivative with a bundle map $\sigma : T^*M \otimes V \to W$, so σ is the symbol of D. Alternatively, for each $\theta \in T^*M$ we have a $\sigma_\theta : V \to W$. Suppose that D is overdetermined elliptic in the sense that σ_θ is injective for all non-zero θ. Then if s is a non-trivial solution of the linear equation $Ds = 0$ we cannot have $\nabla s = \theta \otimes s$, since this would imply that s lies in the kernel of σ_θ. It follows then from general considerations that there is some $k < 1$ such that solutions of the equation $Ds = 0$ satisfy the refined inequality

$$|\nabla|s|| \leq k|\nabla s| \tag{58}$$

for some $k < 1$. In the case at hand, V is the bundle $\Lambda^2 \otimes \mathrm{ad}_E$, the operator D is

$$\mathrm{d}_A \oplus \mathrm{d}_A^* : \Omega^2(\mathrm{ad}) \to \Omega^3(\mathrm{ad}) \oplus \Omega^1(\mathrm{ad})$$

and s is the curvature $F = F_A$. Uhlenbeck and Smith show the constant k is $\sqrt{(n-1)/n}$. It follows then that away from the zeros of F we have an improvement of the differential inequality (41) which it is convenient to write for $f = |F|/4$ as

$$\Delta f \geq \alpha f^{-1} |\nabla f|^2 - f^2, \tag{59}$$

with $\alpha = 1/(n-1)$. We can also write this as

$$\Delta f^{1-\alpha} \geq (\alpha - 1) f^{2-\alpha}.$$

Setting $\overline{f} = f^{1-\alpha}$, this is

$$\Delta \overline{f} \geq (\alpha - 1)\, f\, \overline{f}. \tag{60}$$

As in our previous discussion in subsection 6.3, these inequalities hold in a weak sense over the zeros of F. The value of α is not of fundamental importance, but the proof will be simplified a bit by knowing that $\alpha < \frac{1}{2}$.

To outline the Smith and Uhlenbeck proof we begin with the 4-dimensional case, so we suppose that A is a smooth Yang–Mills connection over the punctured ball $B^4 \setminus \{0\}$ with small energy $\|F\|_{L^2}^2 \leq \varepsilon$. Our goal is to show that F is in L^p for all $p > 0$. As in subsection 6.3 (and below), once we have this it is relatively straightforward to show that the connection extends smoothly over the origin. We divide the argument into four main steps.

Step 1

We claim that the function $\overline{f}$ is a weak solution of the inequality (60) over the 4-ball. In other words, if σ is a smooth positive test function of compact support in B^4 then

$$\int_{B^4} \Delta\sigma\,\overline{f} + (1-\alpha) f\,\overline{f}\sigma \geq 0\cdot \tag{61}$$

If χ_δ is a standard cut-off function, with $\chi_\delta(x) = 1$ for $|x| > 2\delta$ and vanishing for $|x| < \delta$, then multiplying the inequality by χ_δ and integration by parts gives

$$\int_{B^4} \Delta(\chi_\delta\sigma)\overline{f} + (1-\alpha) f\overline{f}\chi_\delta\sigma \geq 0.$$

The kth derivatives of χ_δ are $O(\delta^{-k})$, so we have

$$|\Delta(\chi_\delta\sigma) - \chi_\delta\Delta\sigma| \leq K_\sigma \delta^{-2},$$

for some K_σ depending on σ. Thus we get

$$\int_{B^4} \chi_\delta\left(\Delta\sigma\,\overline{f} + (1-\alpha) f\,\overline{f}\sigma\right) \geq -K_\sigma \delta^{-2} \int_{B_{2\delta}} \overline{f}. \tag{62}$$

We know that f is in L^2 so $\overline{f}$ is in $L^{2/(1-\alpha)}$ and this implies that the integral of $\overline{f}$ over the ball $B_{2\delta}$ is $O(\delta^{2+2\alpha})$. So the right-hand side of (62) tends to 0 with δ, which establishes the claim. (Note that this step also works with $\alpha = 0$.)

Step 2

Consider the linear operator

$$T_f(u) = -\Delta u - (1-\alpha) f u.$$

We consider this as an operator on L^p_2 for $p < 2$. Sobolev embedding gives $L^p_2 \to L^r$ for $r = 2/(2-p)$. This is the same exponent as that given by Hölder's inequality for the multiplication

$$L^2 \times L^r \to L^p,$$

so, since $f \in L^2$, the operator T_f is bounded from L^p_2 to L^p. Furthermore if f is sufficiently small in L^2, which we can suppose, the operator T_f is a small perturbation of $-\Delta$. Linear elliptic theory then shows that we can solve the Dirichlet problem. That is, for $\rho \in L^p$ of compact support in B^4 there is a unique solution of the equation $T_f(u) = \rho$ vanishing on the boundary, and $\|u\|_{L^p_2} \leq C\|\rho\|_{L^p}$.

Step 3

Take a cut-off function ζ of compact support in the unit ball and equal to 1 on the half-sized ball. (This could be $\chi_{\frac{1}{2}}$ but for the later discussion we prefer to use a different symbol.) Set

$$\rho = \Delta(\zeta \overline{f}) - \zeta \Delta \overline{f} = 2\nabla \zeta . \nabla \overline{f} + (\Delta \zeta)\overline{f}. \tag{63}$$

This is supported in an annulus on which the connection A is smooth, so certainly ρ is in L^p. In fact the standard estimates we discussed in subsection 5.2 show that $\|\rho\|_{L^p} \leq C\sqrt{\varepsilon}$. Applying the linear theory, we find a function g such that $T_f(g) = \rho$ with $\|g\|_{L^p_2} \leq C\sqrt{\varepsilon}$. By construction and the inequality (60)

$$T_f(\zeta \overline{f} - g) \leq 0$$

in the weak sense.

Step 4

Write $h = \zeta \overline{f} - g$, so $T_f h \leq 0$. In a case where we had $T_f h = 0$ we would deduce from the uniqueness of the solution to the Dirichlet problem that $\zeta f = g$, so $\overline{f}$ is in L^p_2 on the $\frac{1}{2}$-ball and Sobolev embedding gives $\overline{f} \in L^r$. Here r can be made as large as we please by taking p close to 2. Then f would also be in L^r for all r, which was what we set out to prove.

The final, and critical, step is a result of maximum principle type for the operator T_f, to handle the situation where we have the inequality $T_f h \leq 0$.

Lemma 6.5. *If f is sufficiently small in L^2 then for any function h such that $h \in L^{2q}$ for some $q > 1$, with h smooth on $B^4 \setminus \{0\}$, vanishing on the boundary of B^4 and satisfying $T_f(h) \leq 0$ we have $h \leq 0$.*

This gives the desired conclusion because it implies that $\overline{f} \leq g$ over $B_{\frac{1}{2}}$ and since $\overline{f}$ is positive we get $\overline{f} \in L^r$.

A first step in proving Lemma 6.5 is to reduce to the case when $h \geq 0$, and then show that in fact $h = 0$. This can be done by replacing h by $\max(h, 0)$ and showing that the same differential inequality holds. This is not hard to establish if 0 is a regular value of h, so the zero-set is a submanifold. In any case there are arbitrarily small positive regular values τ: replace h by $\max(h, \tau) - \tau$ and take a limit as $\tau \to 0$.

For later convenience, write $q=4\beta$ and $q/2=2\beta=(1-\alpha)^{-1}=1+\gamma$. So h is $L^{4\beta}$. Let $\chi=\chi_\delta$ be the cut-off function as before. The idea is to control the integral of $|\nabla(\chi h^\beta)|^2$.

There is an identity:

$$|\nabla(\chi h^\beta)|^2+\frac{\beta^2}{\gamma}\chi^2 h^\gamma\,\Delta h=d^*J+V_\delta h^{2\beta}$$

where:

- the 1-form J is a linear combination

$$J=a_1h^{2\beta}\chi\,\mathrm{d}\chi+a_2h^{2\beta-1}\chi^2\,\mathrm{d}h;$$

- the function V_δ is a linear combination

$$V_\delta=a_3|\mathrm{d}\chi|^2+a_4\chi\,\Delta\chi$$

for suitable constants a_i. (This is the crucial point in the proof where the fact that $\alpha>0$ is used. If we took $\alpha=0$ then $\gamma=0$ and the identity breaks down because γ appears in the denominator.)

Integrating over B^4, the term from d^*J vanishes and substituting the differential inequality $-\Delta h\le(1-\alpha)fh$ gives

$$\int|\nabla(\chi h^\beta)|^2\le\frac{(1-\alpha)\beta^2}{\gamma}\int f\,(\chi h^\beta)^2+\int V_\delta\,h^{2\beta}.$$

In dimension 4 we have a Sobolev embedding $L^2_1\to L^4$ so (using the fact that h vanishes on the boundary of B^4) we get

$$\|\chi h^\beta\|^2_{L^4}\le C_1\|f\|_{L^2}\|\chi h^\beta\|^2_{L^4}+C_2\int V_\delta\,h^{2\beta}.$$

If $\|f\|_{L^2}\le(2C_1)^{-1}$ we have

$$\|\chi h^\beta\|^2_{L^4}\le 2C_2\int V_\delta\,h^{2\beta}.$$

The same kind of estimates as in Step 1 show that the right-hand side tends to 0 as $\delta\to0$. This uses the fact that $h\in L^{4\beta}$.

We conclude that $h=0$, as desired. Hence proving Lemma 6.5.

Now we have shown $F\in L^r$ for all r and this completes our discussion of the case $n=4$.

In dimensions bigger than four, Smith and Uhlenbeck follow the same scheme but using *Morrey spaces* in place of Sobolev spaces. For a function ψ on $\mathbf{R}^n$, define the Morrey norm $\| \ \|_{\mathcal{M}^p}$ by

$$\|\psi\|^p_{\mathcal{M}^p} = \sup_{x,r} r^{4-n} \int_{B^n_{x,r}} |\psi|^p.$$

(The notation here is not standard. Smith and Uhlenbeck write $\| \ \|_{X_k}$ for our $\| \ \|_{\mathcal{M}^p}$, with $k = 4/p$.)

Thus a bound on the $\mathcal{M}^2$ Morrey norm of the curvature of a connection is the same as a bound on the normalised energy over all balls. Similar to the normalised energy, we have two properties of these norms:

(1) For functions pulled back from $\mathbf{R}^4$ by orthogonal projection $\mathbf{R}^{n-4} \times \mathbf{R}^4 \to \mathbf{R}^4$ the $\mathcal{M}^p$ norm agrees with the L^p norm on $\mathbf{R}^4$, up to a factor.
(2) For $q = np/4$, Hölder's inequality shows that the L^q norm controls the $\mathcal{M}^p$ norm,

In the vein of (1), in the case when $\Sigma = \mathbf{R}^{n-4} \cap B^n$ and the connection is pulled back from a connection over $B^4 \setminus \{0\}$ by orthogonal projection the n-dimensional proof essentially reduces to that in four dimensions above. In the vein of item (2), the Morrey norm $\mathcal{M}^p$ can be viewed for many purposes as a slightly weakened version of the $L^{np/4}$ norm. There is an elliptic theory so that, for example, for compactly supported functions on $\mathbf{R}^n$,

$$\|\nabla^2 \psi\|_{\mathcal{M}^p} \leq C \|\Delta \psi\|_{\mathcal{M}^p}$$

and also analogues of the Sobolev embeddings so that, for $p < 2$,

$$\|\psi\|_{\mathcal{M}^r} \leq C \|\Delta \psi\|_{\mathcal{M}^p} \tag{64}$$

for $r = 2p/(2-p)$. For $p > 2$

$$\|\psi\|_{C^{,\mu}} \leq C \| \leq C \|\Delta \psi\|_{\mathcal{M}^p}, \tag{65}$$

where the Hölder exponent $\mu = 2 - 4/p$.

The foundation of the Smith and Uhlenbeck argument is the fact, derived from monotonicity, that we can suppose the curvature F is small in $\mathcal{M}^2$-norm. In Steps (1), (4) we need a suitable family of cut-off functions χ_δ, equal to 1 outside the 2δ-neighbourhood of the singular set Σ, vanishing in the δ-neighbourhood and with $|\nabla^k \chi_\delta| \leq c\delta^{-k}$. We need to know that the volumes of these tubular neighbourhoods are $O(\delta^4)$. This follows from the assumption of the dimension of Σ. (In fact it appears to the author that what is required here is the assumption that Σ has finite $(n-4)$-dimensional "Minkowski content", see the remark preceding Proposition 4.3 in subsection 4.2 above.) With these cut-off functions in hand Step 1 works in a similar way so $\overline{f}$ is a weak solution of the inequality (60).

For Step 2, Smith and Uhlenbeck consider the operator, for $p<2$,

$$T_f : \mathcal{M}^p_2 \to \mathcal{M}^p$$

where $\mathcal{M}^p_2$ can be defined as the space of functions ψ on the ball, vanishing on the boundary, with $\Delta\psi \in \mathcal{M}^p$. More precisely, Smith and Uhlenbeck work over cubes rather than balls, which means that the boundary condition can be handled by reflection, but we will ignore this technicality. We have a Sobolev embedding $\mathcal{M}^p_2 \to \mathcal{M}^r$ with $r = 2p/(2-p)$ and multiplication is defined

$$\mathcal{M}^r \times \mathcal{M}^2 \to \mathcal{M}^p.$$

The upshot is that, when f is small in $\mathcal{M}^2$, the operator T_f can be regarded as a small perturbation of the Laplacian and is invertible on these spaces. So for any $\rho \in \mathcal{M}^p$ there is a solution $g \in \mathcal{M}^p_2$ of the equation $-\Delta g + (\alpha-1)fg = \rho$, vanishing on the boundary.

One new feature occurs in Step 3. With our cut-off function ζ we have $T_f(\zeta\overline{f}) \leq \rho$ where

$$\rho = \Delta\zeta\,\overline{f} + 2\nabla\zeta.\nabla\overline{f}.$$

But now the singular set Σ can intersect the annulus on which $\nabla\zeta$ is supported, so it is not obvious that ρ is in $\mathcal{M}^p$. That requires an estimate on the $\mathcal{M}^p$ norm of $\nabla\overline{f}$. Because of this the proof goes by iteration on p. For the first iteration we take $p = \frac{4}{3}$, so $2p/(2-p) = 4$. We begin knowing that $|F| \in \mathcal{M}^2$ (and is small in that norm) and after following through Steps 1-4 we get $|F|^{1-\alpha} \in \mathcal{M}^{\frac{4}{3}}_2 \subset \mathcal{M}^4$ so $|F| \in \mathcal{M}^{4(1-\alpha)}$ (and is small in that norm). Since $\alpha < \frac{1}{2}$ this is an improvement. Repeating the process sufficiently many times gives $|F| \in \mathcal{M}^p$ for all p, or equivalently F is in L^p for all p.

We now return to the problem of estimating ρ for the first iteration, where we want to bound $\|\nabla\overline{f}\|_{\mathcal{M}^{\frac{4}{3}}}$. For this, Smith and Uhlenbeck go back to the inequality (58):

$$-\Delta f + \alpha\frac{|\nabla f|^2}{f} \leq f^2.$$

Let $\tilde{\zeta}$ be another cut off function supported in an interior r-ball $B_{r,x}$ and equal to 1 on the half-sized ball. Multiplying the inequality by ζ and integrating gives

$$\alpha\int_{B_{r,x}} \tilde{\zeta}\frac{|\nabla f|^2}{f} \leq \int_{B_{r,x}} (\Delta\tilde{\zeta})f + \tilde{\zeta}f^2.$$

Knowing that $f \in \mathcal{M}^2$ the right-hand side is easily shown to be $O(r^{n-4})$, so we get an $\mathcal{M}^2$ bound on $f^{-\frac{1}{2}}|\nabla f|$. Now write

$$|\nabla\overline{f}| = |\nabla f^{1-\alpha}| = (1-\alpha)f^{\frac{1}{2}-\alpha}\left(\frac{|\nabla f|}{f^{\frac{1}{2}}}\right).$$

The $\mathscr{M}^2$ bound on f gives an $\mathscr{M}^{4/(1-2\alpha)}$ bound on $f^{\frac{1}{2}-\alpha}$. Then the multiplication:

$$\mathscr{M}^{4/(1-2\alpha)} \times \mathscr{M}^2 \to \mathscr{M}^{4/(3-2\alpha)}$$

gives a bound on $\nabla \overline{f}$ in $\mathscr{M}^{4/(3-2\alpha)}$ and therefore in $\mathscr{M}^{\frac{4}{3}}$.

For the crucial step 4, Smith and Uhlenbeck establish an analogue of Lemma 6.5 in Morrey spaces but we will pass over that and move on to outline their argument for the removal of singularities, given a bound on the L^p norm of the curvature for large p.

Let x_0 be a point in the complement of the singular set Σ. Define the *shadow* of Σ to be the set of x such that for some $t \in (0,1]$ the point $tx + (1-t)x_0$ lies in Σ. Then for a connection defined on the complement of Σ the exponential gauge construction that we discussed in subsection 5.2, using rays emanating from x_0, defines a connection form $\underline{A}$ over the complement of the shadow. The fact that Σ has codimension at least 4 implies that the shadow has codimension at least 3. Uhlenbeck and Smith show that $\underline{A}$ and $\mathrm{d}\underline{A}$ are in L^p for all p, where the latter is interpreted as a distribution. By working over a small ball centred at x_0 and rescaling one can assume that the L^p norms are small. Then the implicit function theorem can be applied to choose a new gauge in which the Coulomb condition is satisfied, and in this gauge elliptic regularity shows that connection is smooth over the small ball. The fact that $\mathrm{d}\underline{A}$ is in L^p, as a distribution, uses crucially the codimension condition. By contrast suppose we had a singular set of codimension 2, so we can have a non-trivial flat connection on the complement. Then the shadow would have codimension 1, the connection form $\underline{A}$ would have a discontinuity across the shadow and the distribution $\mathrm{d}\underline{A}$ would have a singular component supported on the shadow. Thus the same argument would not work in that case, in agreement with the fact that the singularity is not removable.

7 Harmonic maps to Lie groups

7.1 Harmonic maps, flat connections and loop groups

In this section we discuss harmonic maps from surfaces to unitary groups (many of the constructions extend to other compact Lie groups). The main topic is an important paper of Uhlenbeck [61] which, among other things, gives semi-explicit constructions for the general solution in terms of geometric data when the surface is the Riemann sphere. There are many connections with gauge theory, related to the Coulomb gauge condition. The results fit into many large circles of ideas. The fundamental observation of the zero-curvature interpretation of the harmonic map equations is attributed by Uhlenbeck to Pohlmeyer [35], within the integrable systems literature. The equations are, as we explain below in this subsection 7.1 and in subsection 7.2, related to Hitchin's equations on Riemann surfaces and to the Yang–Mills instanton equations in four dimensions. Thence the integrable nature of the harmonic map equations can be related to "twistor" constructions. We will only touch on a small fraction of all these ideas here and there are important parts of Uhlenbeck's paper that we do not discuss. In subsection 7.3 we go back to analysis and results of Hélein on regularity questions, which use somewhat related ideas.

For any simply connected manifold M the harmonic map equations for a map from M to the group $U(r)$ can be formulated as a system of three equations for a pair (A,ψ) where A is a $U(r)$ connection on a bundle E over M (in fact the trivial bundle) and $\psi \in \Omega^1(\mathrm{ad}_E)$. These equations are

$$\mathrm{d}_A^*\psi = 0 \qquad \mathrm{d}_A\psi = 0 \qquad F(A)+\psi\wedge\psi = 0. \tag{66}$$

Given a solution (A,ψ) to (66) the last two equations state that $F(A\pm\psi)=0$, so $A_1 = A+\psi$ and $A_{-1} = A-\psi$ are flat connections and hence gauge equivalent, since M is simply connected. We always have $\mathrm{d}_{A+\psi}^*\psi = \mathrm{d}_A^*\psi$ and so the first equation states that $\mathrm{d}_{A_1}^*\psi = 0$. Choose a trivialisation in which A_1 is the product connections: i.e. the connection 1-form $\underline{A}_1$ is zero. If $\underline{A}_{-1}$ is the connection 1-form of A_{-1} in this trivialisation, the first equation states that $\mathrm{d}^*\underline{A}_{-1} = 0$. The fact that A_{-1} is flat means that $\underline{A}_{-1} = -\mathrm{d}gg^{-1}$ for some $g: M \to U(r)$ and then f satisfies the harmonic map equation $\mathrm{d}^*(\mathrm{d}gg^{-1}) = 0$. Conversely, given such a harmonic map g, let A be the connection on the trivial bundle defined by the connection 1-form $-\frac{1}{2}\mathrm{d}gg^{-1}$ and let $\psi = \frac{1}{2}\mathrm{d}gg^{-1}$. Then $A\pm\psi$ are flat connections and we get a solution of the equations (66).

If we change the sign in the third equation in (66) to $F(A)-a\wedge a=0$ we have a similar discussion, with flat $\mathrm{GL}(n,\mathbf{C})$-connections $A\pm i\psi$ and we get a correspondence with harmonic maps to the dual symmetric space $\mathrm{GL}(r,\mathbf{C})/U(r)$.

In the case when M is a Riemann surface there is an additional symmetry between the first two equations in (66). Recall that ad_E is the real subbbundle of End E consisting of the skew adjoint endomorphisms. Over a Riemann surface M we can write $\psi = \Phi - \Phi^*$ where Φ is a $(1,0)$ form with values in End$E$. Written in terms of Φ, the equations (66) are

$$\overline{\partial}_A\Phi = 0 \ , \ F(A) - (\Phi\wedge\Phi^* + \Phi^*\wedge\Phi) = 0, \tag{67}$$

where $\overline{\partial}_A$ is the coupled $\overline{\partial}$-operator, which is complex linear. Clearly if (A,Φ) is a solution to (67) and λ is a complex number of modulus 1 then $(A,\lambda^{-1}\Phi)$ is also a solution. The equations with the opposite sign of the quadratic term in Φ are *Hitchin's equations* [25]. In the Riemann surface case, one solution of the equations (66) gives a circle of solutions corresponding to $\lambda^{-1}\Phi$ where λ is a complex number with $|\lambda| = 1$; so we have we a flat connection A_λ. More generally, for any $\lambda \in \mathbf{C}^*$ we can define a flat $\mathrm{GL}(r,\mathbf{C})$ connection

$$A_\lambda = A + \lambda^{-1}\Phi - \lambda\Phi^*.$$

Fix a trivialisation of the bundle in which A_1 is the product connection. For each $\lambda \in \mathbf{C}^*$ we have a map $G_\lambda : M \to \mathrm{GL}(n,\mathbf{C})$ such that $A_\lambda = -\mathrm{d}G_\lambda G_\lambda^{-1}$. If we fix a basepoint $p \in M$ then G_λ is determined uniquely by the condition that $G_\lambda(p) = 1$.

Define antiholomorphic involutions of $\mathbf{C}$ and $\mathrm{GL}(r,\mathbf{C})$ by

$$\sigma(\lambda) = \overline{\lambda}^{-1} \quad , \quad \sigma(g) = (g^*)^{-1}.$$

Then the family G_λ has the equivariance property

$$G_{\sigma\lambda} = \sigma(G_\lambda)$$

and in particular G_λ maps into $U(r)$ when $|\lambda| = 1$. The map G_{-1} is the harmonic map g which we associated to the solution of the equations (66).

We can also regard this family as a single map $G : M \times \mathbf{C}^* \to \mathrm{GL}(n,\mathbf{C})$. So any harmonic map $g : M \to U(r)$ defines an "extended map" G.

Uhlenbeck's first result is a characterisation of these extended maps.

Proposition 7.1. *For a simply connected Riemann surface M there is a (1-1) correspondence between harmonic maps $g : M \to U(r)$ and families $G_\lambda : M \to \mathrm{GL}(r,\mathbf{C})$, for $\lambda \in \mathbf{C}^*$ such that*

- *G_λ is holomorphic in λ*
- *$G_{\sigma\lambda} = \sigma(G_\lambda)$,*
- $$\frac{1}{1-\lambda^{-1}} G_\lambda^{-1} \partial G_\lambda$$ *is independent of λ.*

Given such a family G_λ, we define the matrix-valued (1,0) form on M by $\Phi = -(1-\lambda^{-1})^{-1} G_\lambda^{-1} \partial G_\lambda$ and a connection on the trivial bundle with connection 1-form $\Phi^* - \Phi$ to get back to a solution of (67).

Let $\Omega U(r)$ be the based loop group of smooth maps $\gamma : S^1 \to U(r)$ with $\gamma(1) = 1$ and let $\varepsilon : \Omega U(r) \to U(r)$ be evaluation at -1. The restriction of the family G_λ to the unit circle can be regarded as a map $\tilde{G} : M \to \Omega U(n)$ so we have a canonical lift of a harmonic map g over the evaluation map ε:

$$M \stackrel{\tilde{G}}{\to} \Omega U(r) \stackrel{\varepsilon}{\to} U(r). \tag{68}$$

In [44] Segal gives an interpretation of the extended map conditions in Proposition 7.1 in terms of the geometry of the loop group $\Omega U(r)$. This loop space has an infinite-dimensional Kähler structure, preserved by left multiplication of the group. So the complex structure is determined by a complex structure on the tangent space T at the identity. This tangent space T consists of maps ξ from the circle $|\lambda| = 1$ to skew adjoint matrices with $\xi(1) = 0$. Such a map has a Fourier series

$$\xi(\lambda) = \sum_{k=-\infty}^{\infty} a_k \lambda^k$$

(for matrix-valued coefficients a_k) with $\sum a_k = 0$ and $a_k = -a^*_{-k}$. Thus the vector space T is identified with the set of rapidly-decreasing sequences $(a_{-1}, a_{-2}, \dots)$

of complex matrices a_k. The complex structure on T is the obvious one defined by the usual complex structure on these matrix coefficients. (At the group level, this corresponds to the identification of $\Omega U(r)$ with the quotient of the loops in $\mathrm{GL}(r,\mathbf{C})$ by the subgroup of loops which extend holomorphically over the disc, which has a visible complex structure.)

Let V be the subspace of T, of complex dimension r^2, corresponding to sequences $(a_{-1},0,0\dots)$. Extend this by left translation to a subbundle $\underline{V}$ of the tangent space of $\Omega U(r)$, Segal's formulation of Uhlenbeck's correspondence is

Theorem 7.1. *For a simply connected Riemann surface M there is a (1-1) correspondence between harmonic maps from M to $U(r)$ and holomorphic maps $\tilde{G}: M \to \Omega U(r)$ whose derivative at each point maps into $\underline{V}$. The harmonic map corresponding to $\tilde{G}$ is $\varepsilon \circ \tilde{G}$.*

Given $\tilde{G}: M \to \Omega U(r)$, write $G_\lambda : M \to U(r)$ for the corresponding family of maps, with $\lambda \in S^1$. The condition that $\tilde{G}$ is holomorphic is equivalent to saying that, at each point in M, the $(0,1)$ form $G_\lambda^{-1}\overline{\partial} G_\lambda$ extends holomorphically over the unit disc, say

$$G_\lambda^{-1}\overline{\partial} G_\lambda = \sum_{k\geq 0} b_k \lambda^k.$$

Similarly, the condition that the derivative maps into $\underline{V}$ is equivalent to saying that

$$G_\lambda^{-1}\partial G_\lambda = \alpha(1-\lambda^{-1}) + Z(\lambda)$$

for some constant matrix α and $Z(\lambda)$ holomorphic over the disc with $Z(0)=0$. The condition that the G_λ map into $U(r)$ implies that

$$G_\lambda^{-1}\partial G_\lambda = -(G_\lambda^{-1}\overline{\partial} G_\lambda)^* = \sum_{k\geq 0} b_k^* \lambda^{-k}.$$

So Z is zero and $G_\lambda^{-1}\partial G_\lambda = \alpha(\lambda^{-1}-1)$. Thus $(\lambda^{-1}-1)^{-1} G_\lambda^{-1}\partial G_\lambda$ is independent of λ and we can reconstruct (A,Φ) just as before. (This also shows that the family G_λ extends to $\lambda \in \mathbf{C}^*$.)

The preceding discussion is essentially local in the Riemann surface M, and we now focus on the case when M is the Riemann sphere. In that case Uhlenbeck proves an important finiteness result, that the extended maps G_λ have a finite Laurent series. Here it become convenient to drop the normalisation of the extended map G using a base point $p \in M$ (but keeping the other conditions in Proposition 7.1). Then the extended map can be chosen of the form

$$G_\lambda = \sum_{k=0}^{n} T_k \lambda^k, \tag{69}$$

where the T_k are matrix-valued functions on M. Uhlenbeck calls the least possible number n the *uniton number*. (If $n=0$ the map is a constant.) One of the most important ideas in [61] is a construction called "uniton addition"—a form of Bäcklund transformation, taking one solution to another—which we will describe in the next subsection. Uhlenbeck showed that all solutions can be obtained by iterating this construction, and one of her main results is:

Theorem 7.2. *Let $g : S^2 \to U(r)$ be a harmonic map.*

- *The uniton number n of g is strictly less than r.*
- *There is a unique harmonic map $\underline{g} : S^2 \to U(r)$ of uniton number $n-1$ such that g is obtained from $\underline{g}$ by the operation of uniton addition.*

In other words any harmonic map from S^2 to $U(r)$ is constructed in a canonical way by repeating the uniton addition construction at most $(r-1)$ times. The maps with uniton number 1 are the holomorphic maps from S^2 to the Grassmann manifolds $\mathrm{Gr}_k(\mathbf{C}^r)$ of k-dimensional subspaces of $\mathbf{C}^r$ (and, of course, translates by left and right multiplication in $U(r)$). These Grassmann manifolds are isometrically embedded in $U(r)$ by the map which takes a subspace $W \subset \mathbf{C}^r$ to the unitary map $R_W = \pi - \pi^{\perp}$, where π is projection to W and $\pi^{\perp}$ to the orthogonal complement.

This theorem puts the problem of describing the holomorphic maps into the realm of geometry and Uhlenbeck (and subsequent authors) obtained a variety of specific results. These include a derivation of the description by Eells and Wood in [18] of all harmonic maps from S^2 to complex projective space $\mathbf{CP}^n$. These maps have uniton number 2. In Segal's treatment [44] he shows that when $M = S^2$ the holomorphic maps $\tilde{G}$ map into explicit finite-dimensional complex submanifolds—generalised flag manifolds—of the loop group. The problem then comes down to understanding the horizontality condition for these complex curves.

7.2 Uniton addition and instantons on $\mathbf{R}^{2,2}$

Uniton addition can be defined over any Riemann surface M. It changes a harmonic map $g : M \to U(r)$ to a new one of the form $g' = g\rho$, defined using the group structure on $U(r)$, where $\rho : M \to U(r)$ maps into the space of reflections. That is, we have a map $W : M \to \mathrm{Gr}_k(\mathbf{C}^r)$ for some k and $\rho(z) = R_{W(z)}$ in the notation of the previous subsection. The remarkable thing is that finding the maps W for which $g\rho$ is harmonic essentially involves solving only linear PDE. Clearly if we perform the construction again, starting with g' and using the map $-\rho$, we recover g.

We go back to the data (A, Φ) where A is a connection on a bundle E and $\Phi \in \Omega^{1,0}(\mathrm{EndE})$, satisfying equation (67). Using the flat trivialisation for the connection $A_1 = A + \Phi - \Phi^*$ we can regard a map $W : M \to \mathrm{Gr}_k(\mathbf{C}^r)$ as a subbundle $E_I \subset E$.

Lemma 7.3. *If E_I is a holomorphic subbundle of E with respect to the holomorphic structure defined by $\overline{\partial}_A$ which is preserved by Φ in that $\Phi E_I \subset \Omega^{1,0}(E_I)$ then the corresponding map $g\rho$ is harmonic.*

The equation $\overline{\partial}_A \Phi = 0$ is the integrability condition for being able to find such a subbundle locally. The lemma can be proved by direct calculation but we will take a more roundabout route which puts the construction in a wider context. However we postpone that and first discuss an interesting point of view from the work of Valli [66], involving the harmonic maps energy.

The homotopy group $\pi_2(U(r))$ vanishes so there is no obvious topological invariant of a harmonic map from S^2 to $U(r)$. However $\pi_2(\Omega U(r)) = \pi_3(U(r)) = \mathbf{Z}$ so there is an integer degree of the map $\tilde{G}: S^2 \to \Omega U(r)$. As Segal shows in [44], the energy of the harmonic map is equal to twice the degree. In fact this is true locally, in that the energy density of g is equal to twice the pull-back by $\tilde{G}$ of a standard closed 2-form on $\Omega U(r)$ representing the generator of $H^2(\Omega U(r))$. Valli obtained a related formula in [66]. Suppose that g' is obtained from g by uniton addition as above, with a subbundle E_I. Then Valli showed that

$$E(g') = E(g) - 2\deg\,(E_I),$$

where the degree deg E_I is the first Chern class, regarded as an integer. So if deg $(E_I) > 0$ then the energy of g' is strictly smaller. As Valli explained, the existence of a Φ-invariant subbundle E_I of positive degree is a simple consequence of the algebro-geometric classification of holomorphic vector bundles over the Riemann sphere. Applying this repeatedly gives a proof of a slightly weaker form of Uhlenbeck's Theorem 7.2, since we can keep on performing these uniton additions until the energy is 0 and we have a constant map.

There is a intriguing connection here with notions of stability, like those we encountered for the Uhlenbeck–Yau theorem in subsection 6.1. Recall that Hitchin's equations for (A,ϕ) are obtained by changing the sign in (67). A pair (E,ϕ) consisting of a holomorphic bundle E (with $c_1(E) = 0$) over any compact Riemann surface M and a holomorphic $\phi \in \Omega^{1,0}(\mathrm{End}V)$ is called stable if any ϕ-invariant holomorphic subbundle has strictly negative degree. Hitchin showed that if (E,ϕ) is stable there is a corresponding solution of his equations. This is analogous to the Uhlenbeck–Yau Theorem 6.1 and involves solving a similar PDE for a metric on the bundle E. When $M = S^2$, a solution to Hitchin's equations would give a harmonic map to the noncompact symmetric space $\mathrm{GL}(n,\mathbf{C})/U(n)$ and it is easy to see that these are all constant, so there are no stable pairs. If we start with an arbitrary metric and solve a natural gradient flow equation or use a variant of the Uhlenbeck–Yau continuity path the solutions will diverge in the manner we discussed in subsection 6.1 corresponding to a subbundle of E and in fact this will be a ϕ-invariant holomorphic subbundle of strictly positive degree. So attempting to solve the equations with the reversed sign tells us how to build the general solution of (67) by repeated uniton addition.

Before returning to the proof of Lemma 7.3 we make a further digression to discuss an aspect of the equations (67) which were explored in another paper [62] of Uhlenbeck. In subsection 5.3 we mentioned the Yang–Mills instanton equation in four dimensions. These are $F^+(\mathbf{A}) = 0$, where F^+ denotes the self-dual part of the curvature. Consider connections $\mathbf{A}$ over $\mathbf{R}^4$ which are translation-invariant in two directions. In terms of coordinates (u_1,u_2,v_1,v_2) we can write such a connection as

$$\mathbf{A} = A + \Phi_1\, dv_1 + \Phi_2\, dv_2,$$

where A is a connection on a bundle E over $\mathbf{R}^2$ (with coordinates u_1,u_2), lifted to $\mathbf{R}^4$ by projection, and Φ_1,Φ_2 are sections of ad_E over $\mathbf{R}^2$. Then the instanton equation for $\mathbf{A}$ becomes Hitchin's equations for (A,Φ) where $\Phi = (\Phi_1 + i\Phi_2)(du_1 + i\,du_2)$. Now, as explained in [62], change the signature of the metric in four dimensions to get $\mathbf{R}^{2,2}$ with metric $du_1^2 + du_2^2 - dv_1^2 - dv_2^2$. The notion of anti-self-duality makes sense and the instanton equation for translation-invariant solutions become (67). From this point of view, uniton addition is a special case of a more general construction for solutions of the instanton equation over $\mathbf{R}^{2,2}$ (related to the PhD thesis of Uhlenbeck's student Crane [10]). It is easiest to explain this first in a complexified setting, with the coordinates (x_1,x_2,ξ_1,ξ_2) on $\mathbf{C}^4$ and the quadratic form $dx_1\, d\xi_1 + dx_2\, d\xi_2$. The anti-self-duality equations for a 2-form

$$F = F_{x_1x_2}\, dx_1\, dx_2 + F_{\xi_1\xi_2}\, d\xi_1\, d\xi_2 + \sum F_{x_i\xi_j}\, dx_i\, d\xi_j$$

are

$$F_{x_1x_2} = 0 \quad , \quad F_{\xi_1\xi_2} = 0 \quad , \quad F_{x_1\xi_1} + F_{x_2\xi_2} = 0. \tag{70}$$

Let L,R be holomorphic 1-forms on $\mathbf{C}^4$ of the shape

$$L = L_{\xi_1}\, d\xi_1 + L_{\xi_2}\, d\xi_2 \quad , \quad R = R_{x_1}\, dx_1 + R_{x_2}\, dx_2. \tag{71}$$

Define

$$\tilde{L} = -L_{\xi_2}\, dx_1 + L_{\xi_1}\, dx_2 \quad \tilde{R} = R_{x_2}\, d\xi_1 - R_{x_1}\, d\xi_2.$$

Then $L \wedge R$ and $\tilde{L} \wedge \tilde{R}$ have the same self-dual component

$$(L\wedge R)^+ = (\tilde{L}\wedge\tilde{R})^+ = \frac{1}{2}\left(L_{\xi_1}R_{x_2} + L_{\xi_2}R_{x_1}\right)(d\xi_1\, dx_1 + d\xi_2\, dx_2). \tag{72}$$

Write d^+ for the self-dual part of the exterior derivative. The equation $d^+L = 0$ has two components

$$L_{\xi_1,x_1} + L_{\xi_2,x_2} = 0 \quad , \quad L_{\xi_1,\xi_2} - L_{\xi_2,\xi_1},$$

where commas denote partial derivatives. This is the same as the equation $d^+\tilde{L} = 0$ and similarly for $R,\tilde{R}$.

Now consider a holomorphic connection ∇ on a holomorphic bundle V over a domain in $\mathbf{C}^4$ with $V = V_I \oplus V_{II}$. So the connection is given by connections on V_I, V_{II} and second fundamental forms L,R which are holomorphic 1-forms with values in $\mathrm{Hom}(V_I,V_{II}), \mathrm{Hom}(V_{II},V_I)$ respectively. In the notation we used in subsection 6.1

$$\nabla = \begin{pmatrix} \nabla_I & R \\ L & \nabla_{II} \end{pmatrix}, \tag{73}$$

and the curvature is

$$\begin{pmatrix} F_I + R\wedge L & \mathrm{d}_{I,II}R \\ \mathrm{d}_{I,II}L & F_{II} + L\wedge R \end{pmatrix}. \tag{74}$$

The anti-self duality equation for the connection on V has four components

$$F_I^+ + (R\wedge L)^+ = 0\,,\ \mathrm{d}_{I,II}^+ L = 0\,,\ \mathrm{d}_{I,II}^+ R = 0\,,\ F_{II}^+ + (L\wedge R)^+ = 0.$$

Define $\tilde{L}, \tilde{R}$ by the same formulae as above, extended to bundle-valued forms, and use $\tilde{L}, \tilde{R}$ as second fundamental forms defining a new connection $\tilde{\nabla}$ on V, with the same connections on V_I, V_{II}. That is:

$$\tilde{\nabla} = \begin{pmatrix} \nabla_I & \tilde{R} \\ \tilde{L} & \nabla_{II} \end{pmatrix}. \tag{75}$$

The formulae above—extended to bundle valued forms in an obvious way—show that $\tilde{\nabla}$ is an anti-self-dual connection if and only if ∇ is.

Now we look at the real forms of this construction, dealing with unitary connections. We would like an anti-linear involution σ of $\mathbf{C}^4$ such that restricted to the fixed points of σ the forms satisfy $R = -L^*, \tilde{R} = -\tilde{L}^*$. Change notation to write $x_1 = z$, $\xi_1 = \overline{z}$, $x_2 = w, \xi_2 = -\overline{w}$ and let σ be defined by complex conjugation as indicated, so the fixed points of σ are the set where $z + \overline{z}$, $w + \overline{w}$ are real. On this set z, w become complex coordinates and the metric is $\mathrm{d}z\,\mathrm{d}\overline{z} - \mathrm{d}w\,\mathrm{d}\overline{w}$. So the metric has signature $(2,2)$ and matches up with our previous discussion when we take $z = u_1 + iu_2, w = v_1 + iv_2$. In the complex coordinates z, w we have

$$L = L_{\overline{z}}\mathrm{d}\overline{z} + L_{\overline{w}}\mathrm{d}\overline{w}\ \ ,\ \ R = R_z\,\mathrm{d}z + R_w\,\mathrm{d}w,$$

and

$$\tilde{L} = R_{\overline{w}}\mathrm{d}z + R_{\overline{z}}\mathrm{d}w\ \ ,\ \ \tilde{R} = R_w\,\mathrm{d}\overline{z} + R_z\,\mathrm{d}\overline{w}.$$

Thus $R = -L^*$ if and only if $\tilde{R} = -\tilde{L}^*$. The upshot is the following. Suppose that ∇ is a unitary anti-self-dual connection on a bundle V over a domain Ω in $\mathbf{R}^{2,2}$ and identify $\mathbf{R}^{2,2}$ with $\mathbf{C}^2$ as above. Then ∇ defines a holomorphic structure on V. Suppose that V_I is a holomorphic subbundle of $V \to \Omega$. The construction above produces a new unitary anti-self-dual connection $\tilde{\nabla}$ on V, but now the orthogonal complement $V_I^\perp$ is a holomorphic subbundle, with respect to the holomorphic structure defined by $\tilde{\nabla}$.

When restricted to translation-invariant connections this becomes Uhlenbeck's uniton-addition construction. In that case the second fundamental form of a subbundle E_I over $\mathbf{R}^2$ has a component $\beta \in \Omega^{0,1}(\mathrm{Hom}(E_{II}, E_I))$ and Φ has a component $\Phi_{I,II} \in \Omega^{1,0}(\mathrm{Hom}(E_{II}, E_I))$. The construction takes $\Phi^*_{I,II} \in \Omega^{0,1}(\mathrm{Hom}(E_I, E_{II}))$ to build the new second fundamental form and takes $\beta^* \in \Omega^{1,0}(\mathrm{Hom}(E_I, E_{II}))$ to build the new Φ-field.

There is also a "twistor" description of this construction. The Ward correspondence relates instantons on a domain $U \subset \mathbf{R}^{2,2}$ to holomorphic bundles on a three-dimensional twistor space Z. The choice of a compatible complex structure on $\mathbf{R}^{2,2}$ gives a complex surface $D \subset Z$ and the opposite structure another $D' \subset Z$. In complex geometry there is a general construction, sometimes called the Hecke transform. Let $\mathscr{E}$ be a holomorphic bundle over a complex manifold $\mathscr{X}$ and $\mathscr{D}$ be a hypersurface in $\mathscr{X}$. Suppose given a holomorphic subbundle $\mathscr{F}$ of the restriction $\mathscr{E}|_{\mathscr{D}}$. Then we define a new holomorphic bundle $\mathscr{E}' \to \mathscr{X}$ whose local holomorphic sections correspond to sections of $\mathscr{E}$ which lie in $\mathscr{F}$ when restricted to $\mathscr{D}$. The twistor description of the construction is to apply this transform to the bundle over the twistor space Z, using subbundles over D, D'.

7.3 Weak solutions to the harmonic map equation on surfaces

A question left open in our discussion of the regularity theory for harmonic maps in Sections 3 and 4 is whether a weakly harmonic map from a surface—without additional minimising or stationary hypotheses—is smooth. In the case when the target space N has a large isometry group this smoothness was established by Hélein in [22]. The proof was later extended to general target spaces [23] but here we just consider the case of a unitary group, where the proof is particularly simple. The proof depends on the particular structure of the equations, in a similar vein to the preceding discussion in this section.

Proposition 7.2. *Let $g : D \to U(r)$ be a weak solution to the harmonic map equations on the unit disc D in $\mathbf{C}$ with derivative in L^2. Then g is continuous.*

Once the continuity is established smoothness follows from general theory as in [24].

The proof of Proposition 7.2 depends on the following result of H. Wente [67].

Proposition 7.3. *Suppose that $u_1, \dots, u_k, v_1, \dots, v_k$ are functions on the disc $D \subset \mathbf{C}$ with derivatives in L^2. If ϕ satisfies $\Delta\phi = *\sum(\mathrm{d}u_i \wedge \mathrm{d}v_i)$ then ϕ is continuous.*

Of course $*\sum(\mathrm{d}u_i \wedge \mathrm{d}v_i)$ is in L^1, but in dimension 2 it is not the case that any function ϕ with $\Delta\phi$ in L^1 is continuous. That is the point of the result. This is the borderline situation; if $\Delta\phi$ is in L^p for some $p > 1$ then ϕ is continuous.

Let us now see how Wente's result implies Proposition 7.2. The harmonic map equation is $\mathrm{d}^*(\mathrm{d}gg^{-1}) = 0$. So $\mathrm{d}*(\mathrm{d}gg^{-1}) = 0$ and we can write $*\mathrm{d}gg^{-1} = \mathrm{d}\lambda$ for a matrix-valued function λ. Now go back to the equation $\mathrm{d}^*(\mathrm{d}gg^{-1}) = 0$, which is

$$\sum_{i=1}^{2} \frac{\partial}{\partial x_i}\left(\frac{\partial g}{\partial x_i} g^{-1}\right) = 0.$$

Expanding out and multiplying on the right by g we get

$$\Delta g = \sum_{i=1}^{2} \frac{\partial g}{\partial x_i} g^{-1} \frac{\partial g}{\partial x_i} g = \frac{\partial \lambda}{\partial x_2}\frac{\partial g}{\partial x_1} - \frac{\partial \lambda}{\partial x_1}\frac{\partial g}{\partial x_2}.$$

In other words

$$\Delta g = *(\mathrm{d}\lambda \wedge \mathrm{d}g).$$

Thus each matrix entry g_{ab} of g satisfies

$$\Delta g_{ab} = \sum_c \mathrm{d}\lambda_{ac} \wedge \mathrm{d}g_{cb}.$$

Since g takes values in the unitary group the operation of multiplication by g preserves the standard norm on matrices. So the matrix-valued 1-forms $\mathrm{d}\lambda$ and $\mathrm{d}g$ are in L^2 and hence the same for each entry. So we are in the situation considered in Proposition 7.3 and the g_{ab} are continuous.

The proof of Wente's result, Proposition 7.3, has some common features with the Sacks–Uhlenbeck proof of removal of singularities in Section 3. Recall that the Green's function on $\mathbf{R}^2$ is $(2\pi)^{-1}\log r$. It suffices to prove that there is a constant C such that

$$\left|\int_D \log r * (\mathrm{d}u \wedge \mathrm{d}v)\right| \leq C\|\mathrm{d}u\|_{L^2}\|\mathrm{d}v\|_{L^2} \tag{76}$$

for all smooth compactly supported functions u, v on D. Transfer to the cylindrical picture with coordinates (s, θ) where $r = \mathrm{e}^s$. So u, v are now defined on the half-cylinder $(-\infty, 0] \times S^1$, vanishing on $\{0\} \times S^1$, and the left-hand side of (76) is the modulus of

$$I = \int_{(-\infty,0]\times S^1} s \ \mathrm{d}u \wedge \mathrm{d}v.$$

The L^2 norms of $\mathrm{d}u$ and $\mathrm{d}v$ are the same computed in the disc or the cylinder. The fact that u, v are smooth on the disc implies that they are bounded and their derivatives decay exponentially as $s \to -\infty$, measured in the cylinder metric. We have

$$s \ \mathrm{d}u \wedge \mathrm{d}v = \mathrm{d}(s\ u\ \mathrm{d}v) - u\ \mathrm{d}s \wedge \mathrm{d}v$$

and the exponential decay means that it is valid to apply Stokes' Theorem, so that

$$I = -\int_{[-\infty,0]\times S^1} u\ v_\theta\ \mathrm{d}\theta\, \mathrm{d}s.$$

Let $U(s)$ be the average value of u over the circle, for fixed s. Then for each fixed s

$$\int u\ v_\theta\ \mathrm{d}\theta = \int (u - U)\ v_\theta\ \mathrm{d}\theta,$$

and

$$\int (u-U)^2 \mathrm{d}\theta \leq \int u_\theta^2 \,\mathrm{d}\theta.$$

So

$$\left|\int u\ v_\theta\ \mathrm{d}\theta\right|^2 \leq \int u_\theta^2\ \mathrm{d}\theta \int v_\theta^2\ \mathrm{d}\theta.$$

Now, applying the Cauchy–Schwarz inequality to the s integral, we get

$$|I| \leq \|\mathrm{d}u\|_{L^2}\|\mathrm{d}v\|_{L^2},$$

which is (76), with $C = 1$.

References

1. Anderson, M. *Ricci curvature bounds and Einstein metrics on compact manifolds* J. Amer. Math. Soc. 2 (1989) 455–490.
2. Bando, S. and Siu, Y-T. *Stable sheaves and Hermitian–Einstein metrics* Geometry and analysis on complex manifolds World Scientific 1994 39–50.
3. Bradlow, S. *Vortices in holomorphic line bundles over closed Kähler manifolds* Comm. Math. Phys. 135 (1990) 1–17.
4. Brézis, H., Coron, J-M. and Lieb, E. *Harmonic maps with defects* Comm. Math. Phys. 107 (1986) 649–705.
5. Brézis, H. and Nirenberg,L. *Positive solutions of nonlinear elliptic equations involving critical Sobolev exponents* Comm. Pure Appl. Math. 36 (1983) 437–477.
6. Calderbank, D., Gauduchon, P. and Herzlich, M. *On the Kato inequality in Riemannian geometry* Global analysis and harmonic analysis (Marseille-Luminy, 1999), Sémin. Congr., 4, Soc. Math. France, Paris, 2000 95–113.
7. Cheeger, J. and Colding, T. *The structure of spaces with Ricci curvature bounded below, I* Jour. Differential Geometry 45 (1997) 406–480.
8. Chen, X. and Sun, S. *Reflexive sheaves, Hermitian-Yang–Mills connections and tangent cones* Inventiones Math. 225 (2021) 73–129.
9. Chen, X. and Wentworth, R. A. *Compactness for Ω-Yang–Mills connections* arxiv 2106.0913.
10. Crane, L. *Action of the loop group on the self-dual Yang–Mills equation* Comm. Math. Phys. 110 (1987) 391–414.
11. Daskalopoulos, G. and Uhlenbeck, K. *An application of transversality to the topology of the moduli space of stable bundles* Topology 34 (1995) 203–215.
12. Daskalopoulos, G. and Uhlenbeck, K. *Transverse measures and best Lipschitz and least-gradient maps* arxiv 2010.06551.
13. Donaldson, S. *Anti self-dual Yang–Mills connections over complex algebraic surfaces and stable vector bundles* Proc. London Math. Soc. 50 (1985) 1–26.
14. Donaldson, S. *Infinite determinants, stable bundles and curvature* Duke Math. J. 54 (1987) 231–247.
15. Donaldson, S. *Karen Uhlenbeck and the calculus of variations* Notices Amer. Math. Soc. 66 (2019) 303–313.
16. Donaldson, S. and Kronheimer, P. *The geometry of four-manifolds* Oxford U. P. 1990.
17. Eells, J. and Lemaire, L. *A report on harmonic maps* Bull. Lond. Math. Soc. 10 (1978) 1–68.
18. Eells, J. and Wood, J. *Harmonic maps from surfaces to complex projective spaces* Advances in Math 49 (1983) 217–263.
19. Freed, D. and Uhlenbeck, K. *Instantons and four-manifolds* MSRI Publications 1. Springer-Verlag 1984.
20. Giaquinta, M. *Multiple integrals in the calculus of variations and nonlinear elliptic systems* Annals of Math Studies 105, Princeton U.P 1983.
21. Hardt, R. *Singularities of harmonic maps* Bull. Amer. Math. Soc 34 (1997) 15–34.
22. Hélein, F. *Regularity of weakly harmonic maps from a surface to a manifold with symmetries* Manuscripta Math 70 (1991) 203–218.
23. Hélein, F. *Regularité des applications faiblement harmonique entre une surface et une variete Riemannenne* C. R. Acad. Sci Paris Ser I Math 312 (1991) 591–596.

24. Hildebrandt, S., Kaul, H. and Widman, K-O. *An existence theorem for harmonic mappings of Riemannian manifolds* Acta Math. 138 (1977) 1–16.
25. Hitchin, N. *The self-duality equations on a Riemann surface* Proc. Lond. Math. Soc 55 (1987) 59–126.
26. Iwaniec, T. and Manfredi, J. *Regularity of p-harmonic functions on the plane* Revista Mat. Iberoamericana 5 (1989) 1–19.
27. Kobayashi, S. *Curvature and stability of vector bundles* Proc. Japan Acad. Ser. A Math. Sci. 58 (1982) 158–162.
28. Li, J. and Yau, S-T. *Hermitian-Yang–Mills connections on non-Kähler manifolds* Mathematical aspects of string theory (San Diego, 1986) Adv. Ser. Math. Phys. 1 World Scientific 560–573.
29. McDuff, D. and Salamon, D. *J-holomorphic curves and quantum cohomology* American Mathematical Society Colloquium Publications, 52 (2004).
30. Micallef, M. and Moore, J. *Minimal two-spheres and the topology of manifolds with positive curvature on totally isotropic two-planes* Annals of Math. 127 (1988) 199–227.
31. Naber, A. and Valtorta, D. *Rectifiable-Reifenberg and the Regularity of Stationary and Minimizing Harmonic Maps* Annals of Math. 185 (2017) 131–227.
32. Nakajima, H. *Compactness of the moduli space of Yang–Mills connections in higher dimensions* Jour. Math. Soc. Japan 40 1988 383–392.
33. Nakajima, H. *Hausdorff convergence of Einstein 4-manifolds* J. Fac. Sci. Univ. Tokyo Sect. IA Math. 35 (1988), no. 2, 411–424.
34. Parker, T. *Bubble tree convergence for harmonic maps* Jour. Differential Geometry 44 (1996) 595–633.
35. Pohlmeyer, K. *Integrable Hamiltonian systems and interactions through constraints* Comm. Math. Phys, 46 (1976) 207–221.
36. Price, P. *A Monotonicity formula for Yang–Mills fields* Manuscripta Math. 43 (1983) 131–166.
37. Rade, J. *On the Yang–Mills heat equation in two and three dimensions* J. Reine Angew. Math. 431 (1992), 123–163.
38. Rivière, T. *Everywhere discontinuous harmonic maps into spheres* Acta Math. 175 (1995) 197–226.
39. Sacks, J. and Uhlenbeck, K. *The existence of minimal immersions of 2-spheres* Annals of Math. 113 (1981) 1–24.
40. Schoen, R. *Analytic aspects of the harmonic map problem* Seminar on nonlinear partial differential equations, MSRI publications 2 Springer 1984 321–358.
41. Schoen, R. and Uhlenbeck, K. *A regularity theory for harmonic maps* Jour. Differential Geometry 17 (1982) 307–335.
42. Schoen, R. and Uhlenbeck, K. *Boundary regularity and the Dirichlet problem for harmonic maps* Jour. Differential Geometry 18 (1983) 253–268.
43. Sedlacek, S. *A direct method for minimising the Yang–Mills functional over 4-manifolds* Comm. Math. Phys. 86 (1982) 515–517.
44. Segal, G. *Loop groups and harmonic maps* In: Advances in homotopy theory, Lond. Math. Soc. Lecture Notes 139 Cambridge U.P 1989 153–164.
45. Sibley, B. and Wentworth, R. *Analytic cycles, Bott-Chern forms, and singular sets for the Yang–Mills flow on Kähler manifolds* Adv. Math. 279 (2015) 501–531.
46. Sibner, L. and Sibner R. *A non-linear Hodge-de Rham theorem* Acta Math. 125 (1970) 57–73.
47. Simpson, C. *Constructing variations of Hodge structure using Yang–Mills theory and applications to uniformization* Jour. Amer. Math. Soc 1 (1988) 867–918.
48. Siu, Y-T. and Yau S-T. *Compact Kähler manifolds of positive bisectional curvature* Invent. Math. 59 (1980) 189–204.
49. Smith, P. and Uhlenbeck, K. *Removeability of a codimension four singular set for solutions of a Yang–Mills–Higgs equation with small energy* arxiv 1811.03135.
50. Tao, T. and Tian, G. *A singularity removal theorem for Yang–Mills fields in higher dimensions* Jour. Amer. Math. Soc. 17 (2004) 557–593.
51. Taubes, C. *Self-dual Yang–Mills connections on non-self-dual 4-manifolds* J. Differential Geometry 17 (1982) 139–170.

52. Taubes, C. *Min-max theory for the Yang–Mills–Higgs equations* Comm. Math. Phys. 97 (1985) 473–540.
53. Taubes, C. *The stable topology of self-dual moduli spaces* J. Differential Geom. 29 (1989) 163–230.
54. Tian, G. *Gauge theory and calibrated geometry* Annals of Math. 151 (2000) 193–268.
55. Tian, G. and Viaclovsky, J. *Moduli spaces of critical Riemannian metrics in dimension four* Adv. Math. 196 (2005) 346–372.
56. Uhlenbeck, K. *Harmonic maps; a direct method in the calculus of variations* Bull. Am. Math. Soc. 78 (1970) 1082–1087.
57. Uhlenbeck, K. *Regularity for a class of non-linear elliptic systems* Acta Math. 48 1977 217–238.
58. Uhlenbeck, K. *Removable singularities in Yang–Mills fields* Commun. Math. Phys. 83 (1982) 11–29.
59. Uhlenbeck, K. *Connections with L^p bounds on curvature* Commun. Math. Phys. 83 (1982) 31–42.
60. Uhlenbeck, K. *The Chern classes of Sobolev connections* Commun. Math. Phys. 101 (1985) 449–457.
61. Uhlenbeck, K. *Harmonic maps into Lie groups (classical solutions of the chiral model)* Jour. Differential Geometry 30 (1982) 1–50.
62. Uhlenbeck, K. *On the connection between harmonic maps and the self-dual Yang–Mills and sine-Gordon equations* Jour. Geometry and Physics 8 (1992) 283–316.
63. Uhlenbeck, K. *A priori estimates for Yang–Mills fields* Unpublished manuscript.
64. Uhlenbeck, K. and Yau, S-T. *On the existence of Hermitian-Yang–Mills connections in stable vector bundles* Commun. Pure Appl. Math 39 1986 Supplement S257–S293.
65. Ural'ceva, N. *Degenerate quasilinear elliptic systems* Zap. Naucn. Sem. Leningrad Otdel Mat. Inst. Steklov 7 (1968) 184–222 [Russian].
66. Valli, G. *On the energy spectrum of harmonic 2-spheres in unitary groups* Topology 27 (1988) 129–136.
67. Wente, H. *An existence theorem for surfaces of constant mean curvature* Jour. Math. Analysis and Applications 26 (1969) 318–344.
68. White, B. *Nonunique tangent maps at isolated singularities of harmonic maps* Bull. Amer. Math. Soc. 26 (1992) 125–129.
69. Wolfson, J. *Gromov's compactness of pseudo-holomorphic curves and symplectic geometry* Jour. Differential Geometry 28 (1988) 383–405.

List of Publications for Karen K. Uhlenbeck

1970

[1] Morse theory on Banach manifolds. *Bull. Amer. Math. Soc.*, 76:105–106.

[2] Integrals with nondegenerate critical points. *Bull. Amer. Math. Soc.*, 76:125–128.

[3] Harmonic maps; a direct method in the calculus of variations. *Bull. Amer. Math. Soc.*, 76:1082–1087.

[4] Regularity theorems for solutions of elliptic polynomial equations. In *Global Analysis (Proc. Sympos. Pure Math., Vol. XVI, Berkeley, Calif., 1968)*, pages 225–231. Amer. Math. Soc., Providence, RI.

1972

[5] Eigenfunctions of Laplace operators. *Bull. Amer. Math. Soc.*, 78:1073–1076.

[6] Bounded sets and Finsler structures for manifolds of maps. *J. Differential Geometry*, 7:585–595.

[7] Morse theory on Banach manifolds. *J. Functional Analysis*, 10:430–445.

1973

[8] A new proof of a regularity theorem for elliptic systems. *Proc. Amer. Math. Soc.*, 37:315–316.

[9] The Morse index theorem in Hilbert space. *J. Differential Geometry*, 8:555–564.

1974

[10] Lorentz geometry. In *Global analysis and its applications (Lectures, Internat. Sem. Course, Internat. Centre Theoret. Phys., Trieste, 1972), Vol. III*, pages 235–242. Internat. Atomic Energy Agency, Vienna.

1975

[11] A Morse theory for geodesics on a Lorentz manifold. *Topology*, 14:69–90.

H. Holden, R. Piene (eds.), *The Abel Prize 2018-2022*, The Abel Prize,
https://doi.org/10.1007/978-3-031-33973-8_7

1976

[12] Generic properties of eigenfunctions. *Amer. J. Math.*, 98(4):1059–1078.

1977

[13] (with J. Sacks). The existence of minimal immersions of two-spheres. *Bull. Amer. Math. Soc.*, 83(5):1033–1036.
[14] Regularity for a class of non-linear elliptic systems. *Acta Math.*, 138(3-4):219–240, 1977.

1979

[15] Removable singularities in Yang–Mills fields. *Bull. Amer. Math. Soc. (N.S.)*, 1(3):579–581, 1979.

1981

[16] (with J. Sacks). The existence of minimal immersions of 2-spheres. *Ann. of Math. (2)*, 113(1):1–24.
[17] Morse theory by perturbation methods with applications to harmonic maps. *Trans. Amer. Math. Soc.*, 267(2):569–583.

1982

[18] Variational problems for gauge fields. In *Seminar on Differential Geometry*, volume 102 of *Ann. of Math. Stud.*, pages 455–464. Princeton Univ. Press, Princeton, N.J.
[19] Removable singularities in Yang–Mills fields. *Comm. Math. Phys.*, 83(1):11–29.
[20] Connections with L^p bounds on curvature. *Comm. Math. Phys.*, 83(1):31–42.
[21] (with J. Sacks). Minimal immersions of closed Riemann surfaces. *Trans. Amer. Math. Soc.*, 271(2):639–652.
[22] (with R. Schoen). A regularity theory for harmonic maps. *J. Differential Geom.*, 17(2):307–335. Correction, *ibid.* 18(2):329, 1983.
[23] Equivariant harmonic maps into spheres. In *Harmonic maps (New Orleans, La., 1980)*, volume 949 of *Lecture Notes in Math.*, pages 146–158. Springer, Berlin-New York.

1983

[24] (with R. Schoen). Boundary regularity and the Dirichlet problem for harmonic maps. *J. Differential Geom.*, 18(2):253–268.
[25] Conservation laws and their application in global differential geometry. In *Emmy Noether in Bryn Mawr (Bryn Mawr, Pa., 1982)*, pages 103–115. Springer, New York-Berlin.
[26] Closed minimal surfaces in hyperbolic 3-manifolds. In *Seminar on minimal submanifolds*, volume 103 of *Ann. of Math. Stud.*, pages 147–168. Princeton Univ. Press, Princeton, NJ.

[27] Minimal spheres and other conformal variational problems. In *Seminar on minimal submanifolds*, volume 103 of *Ann. of Math. Stud.*, pages 169–176. Princeton Univ. Press, Princeton, NJ.

1984

[28] (with D.S. Freed). *Instantons and four-manifolds*, volume 1 of *Mathematical Sciences Research Institute Publications*. Springer-Verlag, New York. (Translated into Russian). Second edition, 1991.

[29] (with R. Schoen). Regularity of minimizing harmonic maps into the sphere. *Invent. Math.*, 78(1):89–100.

[30] Variational problems for gauge fields. In *Proceedings of the International Congress of Mathematicians, Vol. 1, 2 (Warsaw, 1983)*, pages 585–591. PWN, Warsaw.

[31] Variational problems in nonabelian gauge theories. In *Proceedings of the 1981 Shanghai symposium on differential geometry and differential equations (Shanghai/Hefei, 1981)*, pages 443–471. Sci. Press Beijing, Beijing.

1985

[32] The Chern classes of Sobolev connections. *Comm. Math. Phys.*, 101(4):449–457.

1986

[33] (with S.-T. Yau). On the existence of Hermitian–Yang–Mills connections in stable vector bundles. *Comm. Pure Appl. Math.*, 39(S, suppl.):S257–S293. Also available in *Complex Geometry from Riemann to Kähler–Einstein and Calabi–Yau*, volume 38 of *Advanced Lectures in Mathematics*, pages 453–486. International Press, Somerville, MA, 2018.

1989

[34] (with S.-T. Yau). A note on our previous paper: "On the existence of Hermitian–Yang–Mills connections in stable vector bundles" 33. *Comm. Pure Appl. Math.*, 42(5):703–707. Also available in *Selected works of Shing-Tung Yau. Part 1. 1971–1991. Vol. 5*, pages 231–235, International Press, Boston, MA, 2019.

[35] Harmonic maps into Lie groups: classical solutions of the chiral model. *J. Differential Geom.*, 30(1):1–50.

[36] Commentary on "analysis in the large". In *A century of mathematics in America, Part II*, volume 2 of *Hist. Math.*, pages 357–359. Amer. Math. Soc., Providence, RI.

[37] (with L.M. Sibner and R.J. Sibner). Solutions to Yang–Mills equations that are not self-dual. *Proc. Nat. Acad. Sci. U.S.A.*, 86(22):8610–8613.

1991

[38] Applications of nonlinear analysis in topology. In *Proceedings of the International Congress of Mathematicians, Vol. I, II (Kyoto, 1990)*, pages 261–279. Math. Soc. Japan, Tokyo.

1992

[39] On the connection between harmonic maps and the self-dual Yang–Mills and the sine-Gordon equations. *J. Geom. Phys.*, 8(1-4):283–316.

[40] Instantons and their relatives. In *American Mathematical Society centennial publications, Vol. II (Providence, RI, 1988)*, pages 467–477. Amer. Math. Soc., Providence, RI.

1993

[41] Preface. Global analysis: a subject before its time. In *Global analysis in modern mathematics (Orono, ME, 1991; Waltham, MA, 1992)*, pages vii–xvii. Publish or Perish, Houston, TX.

1994

[42] (with M. Atiyah, A. Borel, G.J. Chaitin, D. Friedan, J. Glimm, J. Gray, M. Hirsch, S. Mac Lane, B. Mandelbrot, D. Ruelle, A. Schwarz, R. Thom, E. Witten, and C. Zeeman). Responses to: A. Jaffe and F. Quinn, "Theoretical mathematics: toward a cultural synthesis of mathematics and theoretical physics" [Bull. Amer. Math. Soc. (N.S.) **29** (1993), no. 1, 1–13]. *Bull. Amer. Math. Soc. (N.S.)*, 30(2):178–207.

1995

[43] (with G.D. Daskalopoulos). An application of transversality to the topology of the moduli space of stable bundles. *Topology*, 34(1):203–215.

[44] (with G. Daskalopoulos and R. Wentworth). Moduli of extensions of holomorphic bundles on Kähler manifolds. *Comm. Anal. Geom.*, 3(3-4):479–522.

[45] (with R. Mazzeo and D. Pollack). Connected sum constructions for constant scalar curvature metrics. *Topol. Methods Nonlinear Anal.*, 6(2):207–233.

1996

[46] (with R. Mazzeo and D. Pollack). Moduli spaces of singular Yamabe metrics. *J. Amer. Math. Soc.*, 9(2):303–344.

[47] Coming to grips with success: a profile of Karen Uhlenbeck. *Math Horiz.*, 3(4):14–17.

1998

[48] (with C.-L. Terng). Introduction. In *Surveys in differential geometry: integral systems*, volume 4 of *Surv. Differ. Geom.*, pages 5–19. Int. Press, Boston, MA.

[49] (with C.-L. Terng). Poisson actions and scattering theory for integrable systems. In *Surveys in differential geometry: integral systems*, volume 4 of *Surv. Differ. Geom.*, pages 315–402. Int. Press, Boston, MA.

2000

[50] (with C.-L. Terng). Bäcklund transformations and loop group actions. *Comm. Pure Appl. Math.*, 53(1):1–75.
[51] (with C.-L. Terng). Geometry of solitons. *Notices Amer. Math. Soc.*, 47(1):17–25.
[52] (with N.-H. Chang and J. Shatah). Schrödinger maps. *Comm. Pure Appl. Math.*, 53(5):590–602.
[53] (with J.A. Viaclovsky). Regularity of weak solutions to critical exponent variational equations. *Math. Res. Lett.*, 7(5-6):651–656.

2003

[54] (with A. Nahmod and A. Stefanov). On Schrödinger maps. *Comm. Pure Appl. Math.*, 56(1):114–151. Erratum, *ibid.* 57(6):833–839, 2004.
[55] (with A. Nahmod and A. Stefanov). On the well-posedness of the wave map problem in high dimensions. *Comm. Anal. Geom.*, 11(1):49–83.

2004

[56] (with C.-L. Terng). $1+1$ wave maps into symmetric spaces. *Comm. Anal. Geom.*, 12(1-2):345–388.

2006

[57] (with C.-L. Terng). Schrödinger flows on Grassmannians. In *Integrable systems, geometry, and topology*, volume 36 of *AMS/IP Stud. Adv. Math.*, pages 235–256. Amer. Math. Soc., Providence, RI.
[58] (with B. Dai and C.-L. Terng). On the space-time monopole equation. In *Surveys in differential geometry. Vol. X*, volume 10 of *Surv. Differ. Geom.*, pages 1–30. Int. Press, Somerville, MA.

2007

[59] (with A. Gonçalves). Moduli space theory for constant mean curvature surfaces immersed in space-forms. *Comm. Anal. Geom.*, 15(2):299–305.

2008

[60] (with M. Vajiac). Virasoro actions and harmonic maps (after Schwarz). *J. Differential Geom.*, 80(2):327–341.

2011

[61] (with C.-L. Terng). The $n \times n$ KdV hierarchy. *J. Fixed Point Theory Appl.*, 10(1):37–61.

2012

[62] (with M. Gagliardo). Geometric aspects of the Kapustin–Witten equations. *J. Fixed Point Theory Appl.*, 11(2):185–198.

2016

[63] (with C.-L. Terng). Tau function and Virasoro action for the $n \times n$ KdV hierarchy. *Comm. Math. Phys.*, 342(1):81–116.
[64] (with C.-L. Terng). Tau functions and Virasoro actions for soliton hierarchies. *Comm. Math. Phys.*, 342(1):117–150.

2018

[65] (with H.L. Bray, W.P. Minicozzi, II, M. Eichmair, L.-H. Huang, S.-T. Yau, R. Kusner, F. Codá Marques, C. Mese, and A. Fraser). The mathematics of Richard Schoen. *Notices Amer. Math. Soc.*, 65(11):1349–1376.

2019

[66] (with P. Smith). Removability of a codimension four singular set for solutions of a Yang–Mills–Higgs equation with small energy. *Surveys in Differential Geometry*, 4:257–291.
[67] (with S.T. Yau). On the existence of Hermitian–Yang–Mills connections in stable vector bundles. In *Selected works of Shing-Tung Yau.* Part 1. 1971–1991. Vol. 5, International Press, pages 1–37.

2022

[68] My personal interaction with AWM. In *Fifty years of women in mathematics—reminiscences, history, and visions for the future of AWM*, Assoc. Women Math. Ser., Vol. 28, Springer, pp. 273–275.

Curriculum Vitae for Karen Keskulla Uhlenbeck

Born: August 24, 1942 in Cleveland, Ohio, USA

Degrees/education: B.S., University of Michigan, 1964
M.A., Brandeis University, 1966
Ph.D., Brandeis University, 1968

Positions: Instructor, Massachusetts Institute of Technology, 1968–1969
Lecturer, University of California, Berkeley, 1969–1971
Associate Professor, University of Illinois Urbana-Champaign, 1971–1976
Associate Professor, University of Chicago, 1977–1983
Professor, University of Chicago, 1983–1988
Professor and Sid W. Richardson Foundation Regents Chair in Mathematics, University of Texas at Austin, 1988–2014

H. Holden, R. Piene (eds.), *The Abel Prize 2018-2022*, The Abel Prize,
https://doi.org/10.1007/978-3-031-33973-8_8

Visiting positions: Visiting Associate Professor, Northwestern University, 1976
Chancellor's Distinguished Visiting Professor, University of California, Berkeley, 1979
Member/Visiting Professor, School of Mathematics, Institute for Advanced Study, 1979–80, 1997–98, 2014–2018
Visiting Member, Mathematical Sciences Research Institute, Berkeley, 1982
Visiting Professor, Harvard University, 1983
Visiting Professor, Max-Planck-Institut für Mathematik, Bonn, 1985
Visiting Professor, University of California, San Diego, 1986
Visitor, Institut des Hautes Études Scientifiques, Bures-sur-Yvette, 1987
Visiting Professor and Sid W. Richardson Foundation Regents Chair in Mathematics, University of Texas at Austin, 1987–88
Visitor, Mathematics Research Centre, Warwick University, 1992
Distinguished Visiting Professor, Institute for Advanced Study, Princeton, 2019–

Memberships: American Academy of Arts and Science, 1985
National Academy of Sciences, 1986
American Philosophical Society, 2007
London Mathematical Society, Honorary Member, 2008
American Mathematical Society, Fellow, 2013
Norwegian Academy of Science and Letters, 2019
Royal Spanish Mathematical Society, Honorary member, 2021

Awards and prizes: Speaker at the International Congress of Mathematicians, 1983, 1990 (plenary)
MacArthur Prize Fellowship, 1983
Common Wealth Award of Distinguished Service, 1995
National Medal of Science, 2000
Leroy P. Steele Prize for Seminal Contribution to Research, 2007
Abel Prize, 2019
Leroy P. Steele Prize for Lifetime Achievement, 2020

Honorary degrees: Knox College, Illinois, 1988
University of Illinois Urbana-Champaign, 2000
University of Ohio, 2001
University of Michigan, 2004
Harvard University, 2007
Brandeis University, 2008
Princeton University, 2012

Part III
2020 Hillel Furstenberg and Grigoriy Margulis

"for pioneering the use of methods from probability and dynamics in group theory, number theory and combinatorics"

THE
ABEL
PRIZE

Citation

The Norwegian Academy of Science and Letters has decided to award the Abel Prize for 2020 to **Hillel Furstenberg**, Hebrew University of Jerusalem, Israel, and **Gregory Margulis**, Yale University, New Haven, Connecticut, USA

> *"for pioneering the use of methods from probability and dynamics in group theory, number theory and combinatorics."*

A central branch of probability theory is the study of random walks, such as the route taken by a tourist exploring an unknown city by flipping a coin to decide between turning left or right at every junction. Hillel Furstenberg and Gregory Margulis invented similar random walk techniques to investigate the structure of linear groups, which are for instance sets of matrices closed under inverse and product. By taking products of randomly chosen matrices, one seeks to describe how the result grows and what this growth says about the structure of the group.

Furstenberg and Margulis introduced visionary and powerful concepts, solved formidable problems and discovered surprising and fruitful connections between group theory, probability theory, number theory, combinatorics and graph theory. Their work created a school of thought which has had a deep impact on many areas of mathematics and applications.

Starting from the study of random products of matrices, in 1963, Hillel Furstenberg introduced and classified a notion of fundamental importance, now called the Furstenberg boundary. Using this, he gave a Poisson type formula expressing harmonic functions on a general group in terms of their boundary values. In his works on random walks at the beginning of the '60s, some in collaboration with Harry Kesten, he also obtained an important criterion for the positivity of the largest Lyapunov exponent.

Motivated by Diophantine approximation, in 1967, Furstenberg introduced the notion of disjointness of ergodic systems, a notion akin to that of being coprime for integers. This natural notion turned out to be extremely deep and have applications to a wide range of areas including signal processing and filtering questions in electrical engineering, the geometry of fractal sets, homogeneous flows and number theory. His " $\times 2 \times 3$ conjecture" is a beautifully simple example which has led to many further developments. He considered the two maps taking squares and cubes on the complex unit circle, and proved that the only closed sets invariant under both these maps are either finite or the whole circle. His conjecture states that the only invariant measures are either finite or rotationally invariant. In spite of efforts by many mathematicians, this measure classification question remains open. Classification of measures invariant by groups has blossomed into a vast field of research influencing quantum arithmetic ergodicity, translation surfaces, Margulis's version of Littlewood's conjecture and the spectacular works of Marina Ratner. Considering invariant measures in a geometric setting, Furstenberg proved in 1972 the unique ergodicity of the horocycle flow for hyperbolic surfaces, a result with many descendants.

Using ergodic theory and his multiple recurrence theorem, in 1977, Furstenberg gave a stunning new proof of Szemerédi's theorem about the existence of large arithmetic progressions in subsets of integers with positive density. In subsequent works with Yitzhak Kaztnelson, Benjamin Weiss and others, he found higher dimensional and far-reaching generalisations of Szemerédi's theorem and other applications of topological dynamics and ergodic theory to Ramsey theory and additive combinatorics. This work has influenced many later developments including the works of Ben Green, Terence Tao and Tamar Ziegler on the Hardy–Littlewood conjecture and arithmetic progressions of prime numbers.

Gregory Margulis revolutionised the study of lattices of semi-simple groups. A lattice in a group is a discrete subgroup such that the quotient has a finite volume. For semi-simple groups, Margulis classified these lattices in his "superrigidity" and "arithmeticity" theorems in the mid-1970s. Armand Borel and Harish-Chandra constructed lattices in semi-simple groups using arithmetic constructions, essentially as the group of integer-valued matrices in a large matrix group. Margulis proved that all lattices in rank 2 or higher arise from this arithmetic construction, as conjectured by Atle Selberg. In 1978, Margulis unveiled the structure of these lattices in his "normal subgroup theorem". Central to his techniques is the amazing and surprising use of probabilistic methods (random walks, Oseledets' theorem, amenability, the Furstenberg boundary) as well as Kazhdan's property (T).

In his 1970 dissertation, Margulis constructed the so-called "Bowen–Margulis measure" of a compact Riemannian manifold of strictly negative variable curvature. Using the mixing property of geodesic flows with respect to this measure, he proved an analogue of the prime number theorem, an asymptotic formula for the number of closed geodesics shorter than a given length. Before this, the only such counting result was via the Selberg trace formula, which works only for locally symmetric spaces. Since then, numerous counting and equidistribution problems have been studied using Margulis' mixing approach.

Another spectacular application of his methods is the proof in 1984 of the decades-old Oppenheim conjecture in number theory: a non-degenerate quadratic form with 3 or more variables either takes a dense set of values on the integers or is a multiple of a form with rational coefficients.

In graph theory, Margulis's creativity resulted in his construction in 1973 of the first known explicit family of expanders, using Kazhdan's property (T). An expander is a graph with high connectivity. This notion, introduced by Mark Pinsker, comes from the study of networks in communications systems. Expander graphs are now a fundamental tool in computer science and error-correcting codes. In 1988 Margulis constructed optimal expanders, now known as Ramanujan graphs, which were discovered independently by Alex Lubotzky, Peter Sarnak and Ralph Phillips. The influence of Furstenberg and Margulis extends way beyond their results and original fields. They are recognised as pioneers by a wide community of mathematicians, from Lie theory, discrete groups and random matrices to computer science and graph theory. They have demonstrated the ubiquity of probabilistic methods and the effectiveness of crossing boundaries between separate mathematical disciplines, such as the traditional dichotomy between pure and applied mathematics.

Autobiography

Hillel Furstenberg

I was born in Berlin in 1935 shortly after the rise of Nazism in Germany. My parents were both born in Germany, their parents having emigrated from Poland and Russia respectively. My father, largely self-taught, served as an advocate's assistant prior to his marriage, and afterwards he was manager of a furniture store. One of my few memories of Germany was the morning after Kristallnacht in November 1938, when the synagogue adjacent to our basement apartment was vandalized, and our own apartment seriously damaged. As Jews without longstanding ties to Germany, we were given expulsion orders. After unsuccessfully applying to various countries for refuge, we were admitted for temporary residence in England, with plans to continue on to the U.S. My father was particularly eager to immigrate to the U.S. where he hoped to join my mother's brother who owned a poultry farm in New Jersey. Concerned that immigration authorities in the U.S. would not allow him entry on account of an ailment he had contracted, he underwent surgery in a London hospital which unfortunately did not succeed. The surgery brought about his death, and he was buried in a London cemetery, leaving my mother to manage with two children, myself and an older sister. We were fortunate that my uncle in New Jersey was able to arrange for our immigration to the U.S. where we arrived in 1940, and our first year in the new country was spent on my uncle's farm.

My earliest school experience was at a small local school near the farm, where kindergarten, first and second grades, were housed in one classroom. As a result my kindergarten year was richer educationally than it is for most children. In addition I had the advantage of tutelage at the hands of my sister, older than myself by three years. So in the later grades, when my classmates were learning addition, I was learning multiplication, and while they were making progress in arithmetic, I was learning algebra. Having a head start in elementary school can build up the self-confidence which is an essential ingredient for further advancement in mathematics.

H. Furstenberg
The Hebrew University of Jerusalem, Edmond J. Safra Campus, Einstein Institute of Mathematics, Givat Ram. Jerusalem, 9190401, Israel, e-mail: Hillel.Fursten@mail.huji.ac.il

H. Holden, R. Piene (eds.), *The Abel Prize 2018-2022*, The Abel Prize,
https://doi.org/10.1007/978-3-031-33973-8_9

Fig. 1: With my mother and sister.

After our relatively short stay in New Jersey we moved to Washington Heights in New York. In Washington Heights I went to schools maintaining a double program — religious Jewish studies together with the standard secular studies. My interest in mathematics became more pronounced as a high school student learning algebra and euclidean geometry. I enjoyed the challenge of the geometry exercises, and was fascinated by — and was somewhat skeptical about — the introduction of imaginary numbers. I filled notebook after notebook with calculations, expecting to discover an inconsistency. Naturally I failed, but the exercise itself was instructive, and failing to reach my desired conclusion was also a useful experience. There was a particular geometry problem which turned out to be of significance for my career. The problem was showing that a triangle with two angle bisectors of equal length is isosceles. Shlomo Sternberg, a lifelong friend, ultimately a member of the Harvard mathematics department, was in my high school class, and the two of us took up the challenge of this problem. With some effort we both came up with solutions, giving two different proofs. The high school we were attending, now called Yeshiva University High School, was housed together with Yeshiva College, and the head of the mathematics department was Professor Jekuthiel Ginsburg, a leading historian of mathematics, who edited a journal called *Scripta Mathematica*. Proud at having solved this problem, Shlomo and I felt this was worthy of the attention of Professor Ginsburg, and we made our way to his office. Professor Ginsburg was duly impressed, and while Sternberg had his own mathematics mentor, Professor Ginsburg took it upon himself to help promote my own mathematical skills. Circumstances at home made it necessary for me to earn money, and so that this should not deter my mathematical pursuit, Professor Ginsburg arranged for me to receive a salary working for his journal, doing translations of papers sent in French and German, and drawing diagrams that were to appear in *Scripta Mathematica*. He also arranged for me to receive a monthly stipend from one of the wealthy supporters of Yeshiva University. The journal, devoted primarily to the history and philosophy of mathematics, also treated "recreational" aspects of mathematics, and my exposure to this helped develop my taste for problem solving.

After high school it was natural for me to continue studies at Yeshiva College, where, in addition to financial assistance, Professor Ginsburg's support enabled me to sidestep some of the usual college requirements, enabling me to focus on mathematics. Although at that time Yeshiva University didn't have a graduate program, it was my additional good fortune that during those years, Professor Ginsburg had decided to bring graduate mathematics to Yeshiva University by inviting mathematicians of note from other institutions to give advanced courses. Among the lecturers were Samuel Eilenberg from Columbia, Jesse Douglas from City College and Abraham Gelbart from Syracuse University. Professor Gelbart's lectures on functional analysis and Hilbert space made a particularly strong impression on me. In addition to the novelty of the subject matter, Professor Gelbart's enthusiasm for Hilbert's profound insight of doing analysis by way of infinite-dimensional spaces was infectious and inspiring. At this stage much of my knowledge came from books as well, and I should mention the two treatises most prominent among these: Whittaker and Watson's "Modern Analysis" and Riesz and Sz.-Nagy's "Leçons d'Analyse Fonctionelle." I also made the acquaintance at this time of an older mathematics student, Seymour Haber, who studied at Syracuse University, but lived in New York, and shared with me many insights. In particular I first learned from him about the theory of almost periodic functions, a chapter from harmonic analysis that would be in the background of much of my later work.

As I indicated earlier, the Yeshiva College curriculum included religious Jewish studies, and these were intended to prepare students for careers as rabbis, serving as religious ministers in Jewish communities. I actually began my studies at Yeshiva College with this goal in mind, and with mathematics taking second place. This had changed by the time of my graduation, and I made the decision to pursue a career in mathematics. The mathematics department at Princeton appealed to me, and with some encouragement from Professor Salomon Bochner from that department, I chose Princeton for my graduate degree.

Fig. 2: Pondering the next step of a calculation.

I had met Professor Bochner when I was interviewed for admission to the department, and I learned that he had also come from an orthodox Jewish background. His father was a recognized Judaica scholar, a good part of whose library had in fact made its way to Bochner's office. Perhaps memories of his father whom he much admired drew his attention to me, and he encouraged me to come to Princeton, where he also served as my advisor. This was most natural, his specialty being harmonic analysis. In fact one of his interests was the theory of almost periodic functions, to which he had made important contributions. As to my doctoral work, which we will describe later, he let me follow my own inclinations, but his insights and intuition have surely impacted on my own leanings and attitudes. While Bochner was encouraging for the most part, he actually chided me once for publishing an early paper entitled "On the inverse operation in groups," which uses the binary relation $(a,b) \to ab^{-1}$ as a model for a novel structure along the lines of a semigroup and also generalizing that of a group. Bochner disdained artificial abstractions, and today I find myself also drawing a line between the natural and the artificial in mathematics.

My interest in harmonic analysis led me to study probability theory, in which the Fourier transform of the distribution function of a random variable plays an important role. I attended the lectures of Professor Willy Feller and became acquainted with the theory of stochastic processes, and in particular, that of Markov processes. It was at this time that Feller was developing his "boundary" theory for Markov chains, and although this wasn't the subject matter of his lectures, the idea of studying the behavior of random walks as time tends to infinity filtered down to me. This was in the background of my later work on Poisson boundaries for groups.

Although Bochner was a devotee of "hands-on" harmonic analysis, a book that came out at this time and had a strong impact on my later work was that of L.H. Loomis, "Abstract Harmonic Analysis." Together with a small group of fellow students, we organized a "baby" seminar — as they were called — with the aim of studying Loomis' book. Here I learned of Gelfand's theory of C^*-algebras which was to serve as a basic tool in my subsequent work. Particularly striking was the algebraization of Wiener's Tauberian theorem and its proof. The representation of an algebra of functions on one space as functions on a compact space was to be of significance in my dissertation, and also in the establishment of an "ergodic correspondence principle," as well as in the construction of the Poisson boundary of a group some years later.

The title of my doctoral dissertation was "Stationary Processes and Prediction Theory." Norbert Wiener had inaugurated the mathematical treatment of prediction for a known stationary process, from a given past to the future, showing the connection to harmonic analysis. In the general (non-deterministic) case, what is sought is a function of "the past" representing the expectation of a future variable. This turns out to be defined "almost everywhere" in terms of the past. The problem I set myself was to find situations where this expectation is well defined for a specific past and without knowledge of the underlying stationary process. A simple example of such a situation is when the time series in question is an almost periodic function of time, where the past has a unique extension to the future as an almost periodic function. A first step in dealing with this problem is to identify the stationary process in question, and here an "ergodic correspondence principle" appears which was to have repercussions in my later work. The dissertation eventually appeared as Volume 44 in the Princeton Annals series. As far as prediction is concerned, the book had little impact. But some of the topics treated had some ramifications, particularly for topological dynamics.

While at Princeton I benefited from the presence of Leon Ehrenpreis, who was visiting the Institute for Advanced Study. Ehrenpreis was a harmonic analyst in the spirit of Laurent Schwartz, and he served as a mathematical "big brother" for me. He was willing to think about any topic and give feedback, and since the direction my work was taking was somewhat offbeat for Princeton at that time, his companionship was encouraging.

After receiving my doctoral degree in 1958, I spent one more year at Princeton with a teaching position, after which I spent two years at MIT as a Moore Instructor. During my stay at Princeton, Harry Kesten visited from Cornell. He shared with me a problem he had heard of from Mark Kac, namely, establishing laws of large numbers for random products of matrices. Our collaboration on this led to a paper in the *Annals of Mathematical Statistics*, which, because of interest to physicists, is probably my most cited paper. At MIT I took advantage of the presence of Norbert Wiener, who was still giving lectures. His course was always entitled "Lectures on the Fourier Integral," but he spoke on whatever struck his fancy at the moment. This was always interesting, and more than appreciating his mathematical prowess, I was impressed by his inquisitive, broad ranging mind, and I saw in him more of a scientist at large than a mathematician.

By this time I was married, having met my wife, Rochelle, during my last year as a graduate student. Rochelle grew up in Chicago, and spent that year in New York. She had studied philosophy, and while we were in the Boston area, she received her M.A. degree in philosophy from Boston University. Eventually her interests turned to literary criticism, which she was to pursue later when we settled in Israel. We moved to Israel seven years after our marriage, fulfilling a dream of my wife, who had spent a pleasurable year in Israel after finishing high school.

For my penultimate position, we moved from Boston to Minneapolis, where I joined the mathematics department of the University of Minnesota. My colleagues in this department put me in touch with developments in topological dynamics; in particular, with the foundational work of Gottschalk and Hedlund, and Robert Ellis. I learned of the notion of "distality" for dynamical systems, and succeeded in establishing a general structure theorem for minimal systems with this property. This turned out to be of significance for my later work on the phenomenon of recurrence in ergodic theory, where a related structure theorem played a crucial role. The distal structure theorem came about by generalizing a particular example of a distal transformation on the 2-torus: $(x,y) \to (x+a, y+f(x))$, easily seen to be distal, to a tower of arbitrary height of "isometric extensions." It turns out that, interpreted correctly, this is the most general minimal distal system. At Minnesota I also returned to the study of noncommuting random products; this time focusing on the qualitative rather than the quantitative behavior. The motivation actually came from the

Fig. 3: With my wife and five children.

dynamical questions I had been studying. The foregoing distal transformation is an example of a "skew product transformation": $T(x,y) = (S(x), R(x)(y))$, where S is a fixed transformation, and $R(x)$ is a transformation on the y-coordinate depending on x. When we iterate this transformation we are led to product transformations of the form $R(S^{n-1}x)\cdots R(S^2x)R(Sx)R(x)$. Assuming the transformations $R(x)$ come from a given group of transformations, we can hope to analyze this product in terms of properties of the group. The complexity of the expression leads one to consider the case where the successive $R^n(x)$ are independent random variables. By this somewhat circuitous route, we came to study random walks on groups, and the associated "Poisson boundaries."

At the University of Minnesota there was a strong probability group which encouraged addressing problems of a probabilistic nature. Monroe Donsker was one of the probabilists in the department, and he suggested to me finding an application of probability theory to algebra. This was in the context of a volume being put out by his acquaintance, Peter Ney, on applications of probability theory to other areas of mathematics. I took up the proposal with the idea of giving an application to group theory. The application was in the form of a "rigidity theorem" regarding lattices in a Lie group. Intuitively one can expect that the boundary occurring for random walks on a lattice would be closely related to that occurring for the ambient group. This led to a result which was a very special case of Margulis' strong rigidity theorem, It also served as a tool in Margulis' "normal subgroup theorem."

After my tenure at the University of Minnesota, we moved to Israel, and I took a position at the Hebrew University in Jerusalem. There were several more institutions in the U.S. which I visited regularly before and after moving to Israel, enabling me to stay abreast of contemporary activity in my areas of interest. One of these was Yale University. At Yale I met Benoit Mandelbrot, who had coined the term "fractal," and had essentially launched a new area in geometry to which I became attracted. Yale was also a center of attraction for mathematicians in the northeast U.S., and a number of ergodic theorists would visit, notably Michael Keane who, with his broad range of interests, was a source of fresh problems and ideas. Another institution

Fig. 4: Visiting Grisha Margulis at Yale University.

which I visited on a regular basis was the University of Maryland. The mathematics department at the University of Maryland was a center for the study of dynamics in general, with topological dynamics as part of its repertory. Joseph Auslander, a student of Gottschalk, was one of the mathematicians responsible for developing this field at Maryland, and among other things, I learned from him to appreciate the algebraic approach to topological dynamics pioneered by Robert Ellis. One aspect of this was availing oneself of the theory of compact semigroups, a subject which would also appear afterwards in linking dynamics to combinatorics.

Fig. 5: With Joe Auslander and his wife, Barbara Meekers at the University of Maryland in 2014.

It was after moving to Israel that I would travel to the U.S. to spend summers at Stanford University. It had become a tradition for ergodic theorists to come to Stanford during the summer to join students and colleagues of Don Ornstein furthering the development of the "new" ergodic theory, that is, ergodic theory beyond ergodic theorems. These sessions were very productive, and also served as a breeding ground for a generation of ergodic theorists.

In fall of 1965 we moved to Israel with our two children. I took a full time position at the Hebrew University in Jerusalem, and a part time position at Bar-Ilan University, where the department of mathematics was still at a formative stage. The department at the Hebrew University was small but possessing a distinguished tradition, with Edmund Landau and Abraham Fraenkel among the faculty in its early years. In 1965 the mathematics faculty included students of the "founding fathers": Aryeh Dvoretzky, who was a student of the analyst M. Fekete, Shimshon Amitsur, a student of the algebraist J. Levitski, Azriel Levy, student of Fraenkel and Robinson in set theory. Two more analysts in the department were Shmuel Agmon and Yitzhak Katznelson, who had studied under Sz. Mandelbrojt in Paris. In mathematical logic there was Michael Rabin whom I had met in Princeton where we overlapped as graduate students, and who later did foundational work in computer science. Rabin was chairman of the department when we came to Israel, and he was instrumental both in my joining the department, and in ensuring that my wife and I were comfortably settled in our new home. At that time, to encourage new immigrant faculty, the Hebrew University would purchase apartments in Jerusalem, requesting a very nominal rent from the new faculty member. Rabin took it upon himself to help us find comfortable living quarters to our liking, even when the cost exceeded the usual allotment prescribed by the university.

At the time of my arrival in Jerusalem, the mathematician whose interests came closest to my own was Yitzhak Katznelson, who specialized in harmonic analysis. Some years later he moved to Stanford, still remaining close to activity in Jerusalem. Katznelson and I worked on many problems together; perhaps the best known of which is the proof of a "density" version of the Hales–Jewett theorem. Two years after my arrival, Benjamin Weiss also came to Israel and joined our department. He and I had much in common; we both were undergraduates at Yeshiva College and did our graduate work at Princeton, and our mathematical interests ran together. In addition to authoring several papers together with me, Benjy — as he was known to everyone — became a mainstay of essentially all my subsequent mathematical work. The area of mathematics in which the three of us, Katznelson, Weiss and myself, were interested in, was ergodic theory, and with Benjy's arrival, this subject gained momentum in our department, which resulted in attracting ergodic theorists from around the world to visit Jerusalem.

This came to the fore in 1975 when the Israeli Institute for Advanced Studies was established at the Hebrew University. This institute had no permanent members, but served as the venue for semester-long, or year-long, programs of intensive international activity in a specialized area. Each year there were to be two programs, one in the sciences and one in the humanities. For the first year of the institute, we proposed organizing a program in ergodic theory. This was accepted together with another program in Jewish mysticism and Talmud, and we could arrange for an impressive line-up of ergodic theorists to spend the year, or part of it, in Jerusalem. This included Donald Ornstein, Daniel Rudolph and Jack Feldman from the U.S., Jean-Paul Thouvenot from France, Mike Keane from Holland, and others. A good part of the activity within our group concerned various notions of equivalence of ergodic systems patterned after Ornstein's isomorphism theory. My own involvement in the program came about due to the serendipitous visit of Konrad Jacobs from Erlangen to Jerusalem. Jacobs had many interests, and while having authored one of the earliest texts in ergodic theory, when he came to Israel he was more interested in combinatorial theory. He was invited to lecture and he chose for his topic the recent resolution by E. Szemerédi of the Erdős–Turan conjecture on arithmetic progressions in sets of integers with positive density. I hadn't been aware of the Erdős–Turan conjecture or Szemerédi's work, but hearing of it at Jacob's talk brought to mind the "ergodic correspondence principle" first made use of in the context of prediction theory. This principle gives a precise rendering of the heuristics whereby the integers are seen as a measure space, density of a set representing its measure and the shift operation, $x \to x+1$, a measure-preserving transformation. The existence of an arithmetic progression $\{a+\mathrm{id}, i=0,1,\ldots,k\}$ in a set of positive density takes on the meaning of a point in a set of positive measure recurring in the set after a power of the measure-preserving transformation is applied $k+1$ times in succession. Made precise, the realisation was that Szemerédi's theorem is equivalent to a "multiple recurrence" phenomenon for measure-preserving transformations, extending the "simple" recurrence phenomenon of Poincaré. This became what is now called the ergodic Szemerédi theorem, and its ergodic-theoretic proof was one of the outcomes of the "ergodic theory year" at the Institute for Advanced Study. To achieve the proof, one makes use of a structure theorem valid for arbitrary ergodic systems, and patterned partially after the structure theorem for distal systems, with a version of "isometric extensions" again serving as the basic constituent of the structure. The proof of this structure theorem was a collaborative effort, involving Benjy Weiss and Izzy Katznelson as well as myself. A number of variations on this multiple recurrence theme were to follow, in which Weiss and Katznelson were again involved.

The academic year '71–'72 was the year of our first sabbatical, and we spent the year at the University of California in Berkeley. At Berkeley I met Joseph Wolf, differential geometer and Lie group theorist, and also Calvin Moore, who, together with Roger Howe, had obtained some of the basic results in ergodic theory for homogeneous flows. During the year I was asked to speak at the weekly colloquium. I knew that an open problem in homogeneous dynamics was to determine whether the horocycle flow on a compact surface of constant negative curvature, proved by Hedlund to be minimal, and by the theorem of Howe–Moore, to be ergodic, was uniquely ergodic. I decided to think about this question and thinking I had a proof, I gave as the title for my talk, "The unique ergodicity of the horocycle flow." The proof I had in mind had a gap, and somewhat under pressure I did find a proof based on harmonic function theory. This result was later subsumed under the far more general results of Marina Ratner on unipotent flows.

In my 1967 paper on "disjointess" of dynamical systems, I had taken note of another phenomenon in the particular case of endomorphisms of the 1-torus, $\mathbb{R}/\mathbb{Z}$; namely if multiplication by p and multiplication by q are two of these, where p and q are multiplicatively independent, then a closed set invariant under both transformations is either finite or everything. This led to a notion of "transversality of actions," reflecting an incommensurability of the two actions, which should be reflected in the transversality of sets respectively invariant under the two actions, this making sense wherever a notion of set dimension is defined. I formulated a conjecture in this regard in a lecture I gave at a symposium honoring Professor Bochner in 1969, the conjecture having just recently been proved independently by M. Wu and P. Schmerkin. In my own unsuccessful attempt to prove the conjecture, I found that ergodic-theoretic ideas could be used in studying issues involving the Hausdorff dimension of fractals. The beginnings of such a theory appears in the paper based on the Bochner symposium lecture, and some of these ideas play a role in Wu's proof of my conjecture. In 2011, I was invited to Kent State University to deliver the CBMS series of lectures, and I chose as my topic "Ergodic Theory and Fractal Geometry." This rounds out the use of ergodic theory in number theory and combinatorics to include geometry.

Fig. 6: Celebrating the Israel Prize in 1993.

Fig. 7: Receiving the Emet Prize in 2004.

I was officially retired from both the Hebrew University and Bar-Ilan University in 2003, having reached the university retirement age of 68. I continued to teach at the Hebrew University, giving advanced courses, mostly related to my work, and this too ended in 2018. Although most of my doctoral students were from the Hebrew University, I didn't regret having taught at Bar-Ilan, where several of the students I had taught also received doctorates. Notable among these was Alex Lubotzky, who began his teaching career at Bar-Ilan, later joining the mathematics department at the Hebrew University, where he became one of the mainstays of the activity in group theory in the department.

Fig. 8: Students and colleagues at my retirement conference in 2003.

Fig. 9: Joined by Jack Feldman and Anatole Katok at my retirement conference in 2003.

Fig. 10: Vitaly Bergelson and Alex Lubotzky, two of my students, at my retirement conference in 2003.

Fig. 11: My wife and I are dining with Ilya Piatetski-Shapiro at Yale University, joined by Steven Miller and Louis Rowen with his daughter (circa 2008).

In a Talmudic passage one rabbi describes how he came to acquire his knowledge: "I have learned something from each of my teachers, more from my colleagues, but most of all from my students." Without making comparisons, I have learned from my teachers, from my colleagues and, of course, from my students. A striking example of learning from one's students involved my student Vitaly Bergelson, who coined the term Ergodic Ramsey Theory, and served over the years as a leading practitioner in the field. At one point he had suggested that it should be possible to go beyond arithmetic progressions, and prove Szemerédi type results for "polynomial" patterns. I firmly discouraged him, claiming that this was a goal for the distant future. Not heeding my warning, Bergelson worked on the problem and indeed did show that it was tractable, thus broadening the scope of ergodic Ramsey theory. Another past student to whom I'm directly indebted is Shmuel (Eli) Glasner. Eli and I joined forces in elaborating the theory of "stationary dynamical systems." This is an extension of measure-preserving dynamics, enabling, among other things, extending ergodic Ramsey theory to non-amenable groups. Other students from whom I've learnt have followed their own inclinations, going into fields disjoint from my own, achieving prominence in these fields. Altogether I find myself privileged in all of the three categories: teachers, colleagues and students.

Fig. 12: Matching wits at a game of chess with my granddaughter.

Autobiography

Grigoriy Margulis

I was born on February 24, 1946, in Moscow, an only child. This was shortly after the end of World War II, and the living conditions, in general, were not so good. But my family was relatively well off, and my parents protected me from all troubles which were around. I lived in Moscow until August of 1991 when I moved to the USA. Since then I have lived in New Haven, Connecticut.

I was interested in mathematics from an early age. My father, who was a mathematician, actually encouraged me to do mathematics. He probably realized that I had some mathematical talent. When I was 7 or 8, I was able to multiply two-digit numbers in my head. My parents also encouraged me to play chess. I combined my interests in mathematics and chess until I became a student in Moscow State University, but after that, I realized that it would be very difficult for me to continue to do so. So I concentrated on mathematics and played chess only occasionally.

Fig. 1: With my parents at the age of 9. Photo: Private.

Department of Mathematics, Yale University, e-mail: margulis@math.yale.edu

H. Holden, R. Piene (eds.), *The Abel Prize 2018-2022*, The Abel Prize,
https://doi.org/10.1007/978-3-031-33973-8_10

As I mentioned, my father was a mathematician. But he was mostly interested in mathematical education. He wrote his candidate dissertation (roughly equivalent to a PhD thesis) under the direction of Alexander Khintchin, who was a famous probabilist. The title of his dissertation was "Infinity and its symbolics in the course of school mathematics" (*"Beskonechnost i ejo simvolika v kurse shkolnoj matematikie"* in Russian). My father was born in 1913 and in 1940 he became a candidate of pedagogical sciences (*"kandidat pedagogicheskih nauk"* in Russian). For many years he worked at pedagogical institutes. He also wrote multiple papers about teaching mathematics in high school (*"v srednei shkole"* in Russian), mostly published in the journal "Mathematics in school" (*"Matematika v shkole"* in Russian). Since 1961 my father served for many years as the deputy chair of the high school section (*"sektsii srednei shkoly"* in Russian) of the Moscow Mathematical Society.

Starting in the 7th grade, when I was approximately 12 years old, I participated in mathematical circles. It was quite informal. I don't remember many details, but it was run by students, sometimes by graduate students, from Moscow University. There was also supervision by more senior mathematicians. We discussed various problems, and we were encouraged to solve them. At that time mathematical circles in Moscow University were connected to the Moscow mathematical olympiads. From the 7th to 10th grade I participated in those olympiads, and every year I was one of the winners. I remember that the prizes in the Moscow mathematical olympiads consisted of collections of mathematical books, which by the way, were extremely cheap during Soviet times. In 1962 I was also a winner of the olympiad in physics and mathematics which took place at the Moscow Institute for Physics and Technology. Also in 1962 I won one of two first prizes in the Republican mathematical olympiad, and because of that I was included in the Soviet team for participation in the International mathematical olympiad. There I won one of twelve silver medals.

During my school years I also regularly met with some senior mathematicians. Among them I should mention Vadim Efremovich. He was a well-known geometer/topologist. His main mathematical interest was equimorphisms. An *equimorphism* is a one-to-one map $f\colon X \to Y$ between two metric spaces X and Y such that both f and f^{-1} are uniformly continuous. In the case of Riemannian manifolds being an equimorphism is a slightly stronger condition than being a pseudo-isometry. Among other things, Efremovich proved that if X and Y are hyperbolic spaces then any equimorphism between X and Y can be extended by continuity to a homeomorphism between the boundaries of X and Y in their standard compactifications. The proof is based on the Morse Lemma, which says that any quasigeodesic in the hyperbolic space is at bounded distance from a geodesic. (Efremovich didn't know about the Morse Lemma, and he reproved it.) Later Dan Mostow, not knowing about Efremovich's result or about the Morse Lemma, proved that the homeomorphism between the boundaries of X and Y is a quasiconformal map, and he used this fact in the proof of his famous strong rigidity theorem. I interacted with Efremovich until the early 1970s. He was a very good teacher and mentor. One of my principles in approaching mathematical and especially geometric problems was *"look at infin-*

ity". I believe that one of the first motivations for this principle came to me during a discussion with Efremovich in 1961–62.

Now I can confess that, during my high school and undergraduate years, I spent a lot of time trying to solve great unsolved problems, including the standard collection of *Fermat's Last Theorem, the Four Color Problem*, etc. Of course, I was unsuccessful. In my case this experience probably had some positive effects, improving my ability to concentrate on difficult problems. But I certainly would not recommend to any young mathematician to do the same.

In 1962 I became an undergraduate student in Moscow State University, which is subdivided into so-called faculties (*"fakultets"* in Russian). I was a student in the division of mathematics of the faculty of mechanics and mathematics (*"mehaniko-matematicheskii fakultet"* or shortly *"mehmat"* in Russian). There was an enormous number of seminars during that time, and starting in the first year I attended quite a few of them, maybe too many.

Fig. 2: Moscow University, Kronrod seminar, during the academic year 1960–61. From left to right, first row: G. Margulis, L. Makar-Limanov, D. Kazhdan, A. Libin, M. Khanchachan and A. Katok; second row: S. Gelfand, V. Fishman, B. Pranov and A.S. Kronrod. Photo: Private.

It was not unusual for undergraduate students in Moscow University to do research. During the academic year 1964–65, when I was a third year undergraduate student, I attended Dynkin's seminar on Martin boundaries. At some point Dynkin conjectured that, under some mild conditions, all positive harmonic functions on nilpotent groups are invariant under translations by the commutator subgroup. I was able to prove this conjecture and wrote my first paper "Positive harmonic functions on nilpotent groups". It was submitted for publication in May of 1965 and published in 1966. The paper seems to have some influence even now.

Starting at the end of my third undergraduate year Sinai became my adviser. His influence on my formation as a mathematician is hard to overestimate. My second paper was about the exponential growth of the fundamental group of a compact 3-dimensional manifold admitting Anosov flows. It was published as an appendix to a survey paper by Anosov and Sinai. The title of my candidate dissertation is "On some aspects of theory of Anosov systems". It was written in 1970 and published only in 2003. The dissertation, among other things, contains the asymptotic formula for the number of closed geodesics on compact manifolds of negative curvature.

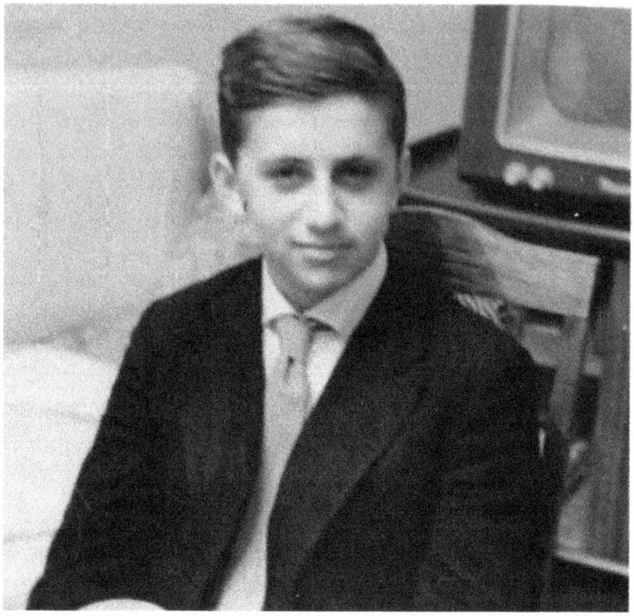

Fig. 3: Me aged 17. Photo: Private.

Dima (now David) Kazhdan was my classmate. At the end of our undergraduate studies in May–June of 1967, he and I proved the existence of unipotent elements in non-uniform lattices in semisimple Lie groups. This was a conjecture by Selberg. More precisely, we proved the following. Let G be a linear semisimple Lie group without compact factors. Then any non-uniform lattice in G contains a non-trivial unipotent element. We also proved the existence of a neighborhood W of the identity element e in the group G such that for any discrete subgroup Γ of G there exists an element g of the group G such that $g\Gamma g^{-1} \cap W = \{e\}$. As a corollary we deduce the existence of a positive lower bound for the covolume of Γ in G which depends only on G and does not depend on a discrete subgroup Γ of G.

The work with Kazhdan was published as a paper in 1968. The paper became quite well known. In particular, Armand Borel gave a talk about it at a Bourbaki seminar. Also after this paper I became very much involved in the theory of discrete subgroups of Lie groups. In particular I started to work on the problem of arithmeticity of non-uniform lattices. Eventually I proved arithmeticity of a higher rank irreducible non-uniform lattice Γ but it took several years. Though I didn't realize it at the time, I followed the general strategy introduced by Selberg and Piatetski-Shapiro. This strategy is based on the study of unipotent elements in the lattice Γ and the subgroups of Γ associated with them. A more detailed account will be given in Appendix 1.

Around 1970 I realized that there are analogs in Riemannian geometry of some statements from the paper with Kazhdan. I told Misha Gromov about it, and he later called these observations the *Margulis Lemma.*

In the late sixties I learned about Mostow's fundamental work on strong rigidity. Thinking about it, I at some point realized that it would be possible to prove arithmeticity of higher rank irreducible lattices if one could prove a statement that is now called *superrigidity*. I believe that superrigidity was a new phenomenon that had not been discovered before my work. The first proof of superrigidity was based on a combination of methods from ergodic theory and algebraic group theory, and one of the ingredients was Oseledec's multiplicative ergodic theorem. A more detailed account will be provided in Appendix 1. Here I want to emphasize that the arithmeticity statement is formulated in arithmetic and algebraic terms but its proof involved ergodic theory. Apparently it was a big surprise for most mathematicians. My work on arithmeticity took about seven years, approximately from 1967–74.

It seems strange now, but when I worked on superrigidity I was not influenced by Furstenberg's work, because I was essentially not familiar with it. It is indeed strange because many ideas and methods introduced by Furstenberg are very similar in style to what I used. I learned about Furstenberg's work only around 1974, and his boundary theory influenced me very much. In particular, one of the basic statements in this theory played a very important role in my proof of the so-called *normal subgroup theorem*.

The normal subgroup theorem says that if G is a connected semisimple Lie group of $\mathbb{R}$-rank at least 2 without compact factors and with finite center, and if Γ is an irreducible lattice in G, then any normal subgroup N of Γ either belongs to the center of Γ or has finite index in Γ. A special case is, for example, the group $G = \mathrm{SL}(3,\mathbb{R})$ of real 3×3 matrices with determinant 1. Take a discrete subgroup Γ of G which has finite covolume. Then any non-trivial normal subgroup of Γ has finite index in Γ.

My proof of the normal subgroup theorem consists of two parts. First consider the case when Γ/N is an amenable group. This case can be treated using representation theory arguments, in particular Kazhdan's property (T). Then consider the case when Γ/N is a non-amenable group. Somehow I realized that one can use algebras of measurable sets. A crucial tool was one of the initial lemmas from Furstenberg's paper on boundary theory. Another crucial tool was a generalization of the density point theorem from measure theory. I cannot explain how I came upon the idea behind the second part of the proof — it was some sort of intuition. Also, the idea of subdividing the proof into the case when Γ/N is amenable and the case when Γ/N is non-amenable was quite new. For all these reasons I consider my proof of the normal subgroup theorem to be the best proof I have done.

The statement of the normal subgroup theorem was known to be true in certain cases, for example for $\mathrm{SL}(n,\mathbb{Z})$, $n \geq 3$. The proofs were based on algebraic methods. Maybe it was natural to assume that the statement was true in general. However the proof in the uniform case was obtained partially by using measure theory, as I alluded to above. So even though the theorem is stated in purely algebraic terms, the proof in the general case is mostly non-algebraic.

I was an undergraduate student at Moscow University from 1962–67 (undergraduate education in the Soviet Union lasted five years). From 1967–70 I was a graduate student, also at Moscow University. My dissertation adviser (scientific director; *"nauchnyi rukovoditel"* in Russian) was Sinai. In 1970 I finished my dissertation and became a candidate of science. After that and formally several months before, I began to work at the Institute for Problems in Information Transmission (*"Institut Problem Peredachi Informatsii"* or *"IPPI"* in Russian).

I didn't get a position at Moscow University or the Steklov Institute, but IPPI was one of the institutes of the Soviet Academy of Sciences. By Soviet standards at that time, IPPI was relatively small, only two hundred researchers, but by Western standards it would probably be considered a huge institution. My immediate boss was Roland Dobrushin, who was a famous probabilist and mathematical physicist. There was also another group there, which was headed by Mark Pinsker. He was famous for his work in information theory. In a sense, I was lucky to be there, because my most well known and widely cited paper on expander graphs, published in 1973, was written under Pinsker's influence. Expander graphs, which incidentally have many applications in computer science, were first defined by Pinsker. Their existence was first proved by Pinsker in the early 1970s. My paper gave the first explicit construction of an infinite family of expander graphs. In that paper the proof was based on the theory of discrete subgroups of Lie groups. In particular I used arguments related to Kazhdan's work on property (T). This was probably unexpected for people working in computer science.

Later on, I constructed an infinite family of graphs using quaternions. This was in 1984 and at that time I was mostly interested in studying the girth of these graphs and, in particular, finding estimates for the girth size. This interest was partially motivated by applications to algebraic coding theory and, in particular, the construction of low density codes. In graph theory, the girth of a graph is the length of the shortest cycle contained in the graph. There was a probabilistic construction due to Erdős and Sachs that gave an upper asymptotic estimate $2\log_p n$ for the girth of a $(p+1)$-regular graph with n vertices (this is simple), while the asymptotic lower estimate was $\log_p n$. Quite surprisingly, my explicit construction gave, in the case where p is a prime, an asymptotic lower estimate $4/3\log_p n$. I believe that up until now there hasn't been any probabilistic construction that goes beyond $\log_p n$.

At the same time as I studied these explicitly constructed graphs I realized, based on some deep work by Deligne, that they were also expander graphs which are in a certain sense better than the previous constructions. Slightly later and completely independently, Lubotzky, Phillips and Sarnak gave basically the same construction, but with some variations. They also used the work of Deligne, and they called the graphs which come from this construction *Ramanujan graphs* because it is related to a work by Ramanujan.

As I mentioned earlier, my immediate boss at IPPI was Dobrushin. At some point he suggested to me to look at probabilistic characteristics of graphs with large connectivity. In 1974 I obtained the following result. Let G be a connected non-oriented finite graph. The *connectivity* $\nu(G)$ of the graph G is defined as the smallest number of edges which should be deleted for the graph G to become non-connected.

Fig. 4: Wedding photo (August 30, 1972). Photo: Private.

Let $f_G(p)$ denote the probability that the graph G becomes non-connected if each edge is deleted with probability p. It is clear that $f_G(p)$ is a differentiable non-decreasing function, $f_G(0) = 0$ and $f_G(1) = 1$. For every $0 < \varepsilon < 1$ let

$$t_1(\varepsilon) = \inf\{p\colon f_p \geq \varepsilon\}, t_2(\varepsilon) = \sup\{p\colon f_G(p) \leq 1 - \varepsilon\}, t(\varepsilon) = t_2(\varepsilon) - t_1(\varepsilon).$$

I proved that for every $\varepsilon > 0$ and $\delta > 0$ one can find n such that if $\nu(G) > n$ then $t(\varepsilon) < \delta$. In other words, if the connectivity $\nu(G)$ is "sufficiently large" then the function $f_G(p)$ jumps "almost immediately" from being "almost 0" to becoming "almost 1".

When I came to IPPI, it was essentially divided into three parts: (A) theoretical/mathematical; (B) engineering, mostly related to communications, computer vision, speech recognition, etc; (C) applications of computers to biology and medicine. Part (A) consisted of two groups headed by Dobrushin and Pinsker. The emphasis in this part was on information theory, coding theory, and what is now called *theoretical computer science*. My work on expanders and probabilistic characteristics of graphs was closely related to these subjects, and in some sense it was done under pressure to do research related to the main direction of part (A). But I had the opportunity of spending most of my time on my own research. I had more time to do research than an average professor in the US does.

Fig. 5: With my wife and son in 1975. Photo: Private.

In 1978 I was awarded a Fields Medal. Of course it was very encouraging. Unfortunately I was not allowed to attend the ICM in Helsinki to receive the award, mostly due to the opposition of an influential part of the top Soviet mathematical establishment at that time. (More details will be provided in Appendix 3.) But apparently because I could not come to the Helsinki Congress, I was allowed in 1979 to come to the West for the first time. It was a three-month visit to Bonn from the beginning of July until the beginning of October, arranged by Hirzebruch. During that visit Jacques Tits came to Bonn and at a small ceremony presented the award.

In September of 1979 during my visit to Bonn I met Gopal Prasad, who told me about Raghunathan's remarkable observation relating the Oppenheim conjecture and the theory of unipotent flows. After that I started to work on the conjecture and eventually proved it. The short version of the proof was published in January of 1987, and more detailed versions were published later. But I remember that I gave some kind of oral presentations already in 1984. Later for many years I continued to work on various generalizations of the Oppenheim conjecture, mostly in collaboration with other mathematicians. A rather detailed account of this work will be given in Appendix 2.

In May or June of 1980 I was at a conference in Poland. There I met Rindler, an Austrian mathematician, who mentioned the problem of Banach and Ruziewicz on the uniqueness of $SO(n)$-invariant means for the algebra of all Lebesgue measurable subsets of the unit sphere in $\mathbb{R}^n$ and a reformulation of this problem due to Rosenblatt, who explained that it would follow from the statement about small almost invariant sets. I almost immediately realized that for $n \geq 5$ this statement can be deduced from property (T) for certain subgroups of $SO(n)$. I used subgroups which can be considered as S-arithmetic lattices in the product of $SO(n)$ and the group of p-adic points of SO_n. Independently and around the same time, Dennis

Sullivan gave a similar proof using slightly different subgroups, but also using work by Rosenblatt.

The Banach–Ruziewich problem was for $n \geq 3$ (for $n = 2$ the corresponding statement is not true). In 1982 I settled an analog of the Banach–Ruziewich problem for $\mathbb{R}^n$ for all $n \geq 3$. In this analog $\mathrm{SO}(n)$-invariant means are replaced by means invariant under the group of isometries of $\mathbb{R}^n$. Instead of property (T), I used a generalization of property (T) which I called *relative property* (T). For $n = 3$ and $n = 4$, the group $\mathrm{SO}(n)$ doesn't contain subgroups having property (T). In 1984 Drinfeld settled the Banach–Ruziewich problem for these remaining two cases, using very deep results of Deligne.

According to a classical theorem of Bieberbach, if Γ is a discrete subgroup of the group of isometries of $\mathbb{R}^n$ such that $\mathbb{R}^n/\Gamma$ is compact then the subgroup Γ contains a subgroup of finite index consisting of parallel translations and therefore is virtually abelian. Let $A(n)$ denote the group of affine transformations of $\mathbb{R}^n$, and let Γ be a discrete subgroup of $A(n)$ acting properly discontinuously on $\mathbb{R}^n$. In 1964 L. Auslander conjectured that if $\mathbb{R}^n/\Gamma$ is compact then Γ is virtually solvable. This statement can be considered as an analog for $A(n)$ of the Bieberbach theorem. In 1977 J. Milnor asked if Γ should be virtually solvable without the assumption that $\mathbb{R}^n/\Gamma$ is compact. This is obviously true for $n = 1$ and is not difficult to prove for $n = 2$. In 1983 I showed that for $n = 3$ the answer to Milnor's question is negative by constructing a free (non-commutative) discrete subgroup of $A(3)$ which acts properly discontinuously on $\mathbb{R}^3$. Actually this example came as a surprise to me and not only to me. For many years I continued and still continue to work on various questions related to properly discontinuous actions of discrete groups of affine transformations, mostly in collaboration with H. Abels and G. Soifer.

Around 1981–82 Jacques Tits suggested to Joachim Heinze, who was in charge of mathematics at Springer Verlag, that I should write a book on the subject of discrete subgroups of Lie groups. Heinze passed this suggestion to me, and eventually I agreed. Approximately at the beginning of 1984 I started to write the book "Discrete Subgroups of Semisimple Lie Groups". At that time I thought that by making relatively small changes to my doctoral dissertation I would be able to finish writing the book in a few months. I was completely wrong. The book was almost three times bigger than the dissertation, and it took more than two years to write it. I wrote the book in Russian and it was translated into English by Valery Mishkun. Many constructive remarks were made by Alan Huckleberry and Gopal Prasad. My book was published in 1991. I believe that it became a standard reference in the subject together with the earlier books "Discrete Subgroups of Lie Groups" by M.S. Raghunathan, published in 1972, and "Ergodic Theory and Semisimple Groups" by Robert Zimmer, published in 1984.

With only one exception, the above-mentioned three-month visit to Bonn in 1979, I was not allowed to travel outside of the so-called "socialist camp" (*"sotsialisticheskii lager"* in Russian) until 1987. But it was much easier to get permission to visit countries inside the "socialist camp". Starting in 1976 I traveled several times to Hungary, twice to Poland, and one time to Bulgaria. These visits were quite fruitful. I met many foreign mathematicians. In 1984 at conferences in Bulgaria and

Hungary I gave oral presentations about my proof of the Oppenheim conjecture, and during my visit to Poland in 1986 I wrote a short version of this proof.

Starting in 1987 it became much easier for me (and not only for me) to get permission to travel abroad. In July of 1987 I came to Oslo, Norway, to participate in the symposium "Number Theory, Trace Formulas and Discrete Groups" in honor of A. Selberg. The title of my talk at the symposium was "Discrete Subgroups and Ergodic Theory". In spite of the quite general title, I essentially talked only about my proof of the Oppenheim conjecture. Most participants of the symposium were number theorists. For them it was a big surprise that a long standing conjecture from number theory could be proved using dynamical methods.

In 1988 I came to Bonn again. This time it was a four-month visit to Max Planck Institute which started approximately at the beginning of March. With the help of G. Harder I arranged for S.G. Dani to visit there for a month. During his visit we started our fruitful collaboration on various generalizations and stronger versions of the Oppenheim conjecture and on related problems in the theory of unipotent flows. The collaboration continued for several years. In the winter of 1988–89 S.G. Dani came to Moscow for a month, and in March of 1989 I came to India, also for a month. These two visits were a part of the Indo-Soviet exchange program between the Department of Science and Technology of the Government of India on the Indian side and the USSR Academy of Sciences on the Soviet side. In the summer of 1991 we worked together for a month at the University of Göttingen. And finally, S.G. Dani came to Yale in 1992 for two weeks.

In Spring of 1989 I came to the USA for the first time. The visit was rather short (a couple of weeks). It started with my participation in a CBMS conference called "Discrete Groups, Expanding Graphs, and Invariant Measures" that was organized by Andy Magid with Alex Lubotzky as the main speaker. It took place in Norman at the University of Oklahoma May 29 through June 2, 1989. My talk at the conference was the first talk I had given at a US university. If I remember correctly, after the conference I visited Stanford, Harvard, and Yale (each university for 2–3 days), and, before returning to Moscow, stayed shortly with a friend in New York.

In the winter of 1989–90 I came to France for three months. It was a combination of a two-month visit to IHES in November-December of 1989 and a one-month visit to College de France in January of 1990, where I gave a series of lectures. It was the first lecture series in my life. During my stay in France there were dramatic political changes in Europe. The Eastern Block was disintegrating, the Berlin wall fell, etc. It was unexpected. For example when I was in Germany in March-June of 1988, I never met anyone who envisioned this sequence of events in the near future.

In April-May of 1990 I came to Israel for a couple of weeks. I presented the Schur lectures at Tel Aviv University and was at the ceremony where de Giorgi and Piatetski-Shapiro were awarded the Wolf Prize. I also met many old friends.

In September of 1990 I came to the USA for a very long visit. First I was at Harvard for a semester from September 1990 until January 1991 and then I stayed at IAS for four months in February–May of 1991. During that visit I received offers from Harvard, Princeton, the University of Chicago, and Yale. I chose Yale. Many people were surprised by this choice. But in retrospect it is clear to me now that from a professional point of view it was the right decision.

After my stay at IAS I came to Germany for two months, and at the end of July I went to Moscow to prepare to move with my family (wife and son) to the USA. We were scheduled to fly to New York on August 24. On August 19 there was a coup in Moscow which eventually failed on August 21. We left Moscow on August 24 as planned, but I remember how frightened I was during the period August 19–24 and especially the period August 19–21.

In the second half of 1991 I became a professor at Yale. This was the start of a new period in my mathematical and non-mathematical life. The style of my work changed. When I was in Moscow I was the single author of most of my papers (certainly more than 75 or 80 percent). One notable exception was the joint paper with Kazhdan. Actually for me it was quite challenging to write up papers, so it took a lot of time. But after I moved to Yale in 1991 almost all of my papers were joint ones. It was a completely different environment. I started to work with many mathematicians, mostly younger than me.

In Moscow I didn't have graduate students (I was the dissertation adviser for G. Soifer but he was not my graduate student). But at Yale I had more than twenty graduate students. I also had several postdocs (often their official position was Gibbs Instructor or Gibbs Assistant Professor). Working with graduate students and postdocs became a very important and rewarding part of my life.

Dima (or officially Dmitry) Kleinbock was my first PhD student at Yale. He graduated in 1996. Dima and I wrote a paper, "Flows on homogeneous spaces and Diophantine approximation on manifolds", which was published in 1998 in the *Annals of Mathematics*. The paper was highly praised by many mathematicians. In that paper we proved long-standing conjectures of Baker and Sprindzuk, which are more often called Sprindzuk's conjectures, from the metric theory of Diophantine approximation. To prove the conjectures, we introduced a new approach which was based on using methods from homogeneous dynamics (more details are provided in Appendix 2).

There is an interesting story related to the just-mentioned paper with Kleinbock. S.G. Dani established the correspondence between Diophantine properties of vectors in $\mathbb{R}^n$ and the asymptotic behavior of translations by diagonal subgroups of certain points in $\mathrm{SL}(n+1,\mathbb{R})/\mathrm{SL}(n+1,\mathbb{Z})$. Dima was working on some applications of this correspondence. Once I was walking in the library of the Yale mathematics department and I noticed the book "Metric Theory of Diophantine Approximation" by V. Sprindzuk. I asked Dima to look at the book. Rather soon we realized that the methods from homogeneous dynamics can be applied to study problems described in Sprindzuk's book. And this eventually led to the proof of Sprindzuk's conjectures.

Approximately from 1993 until 2012 I visited the University of Bielefeld in Germany almost every year, usually for two months in June and July. The visits were first arranged by Herbert Abels and later also by Friedrich Götze. In Bielefeld, besides collaboration with Abels and Götze, I also worked and had mathematical discussions with many other visitors. In particular, a significant part of the joint work with A. Eskin and S. Mozes on the quantitative Oppenheim conjecture was done there. I should also mention the collaboration during my Bielefeld visits with V. Beresnevich, V. Bernik, and D. Kleinbock on various aspects of the metric theory of Diophantine approximation and with G. Soifer on discrete groups of affine transformations.

Fig. 6: With Sinai in 2006. Photo: Private.

After I moved to Yale, my research was mostly focused on applications of homogeneous dynamics to number theory and Diophantine approximation. I already mentioned the joint paper with Kleinbock on Sprindzuk's conjectures. It was followed by several other joint papers, including the following: "Khintchine-type theorems: the convergence case for standard and multiplicative versions" (jointly with V. Bernik and D. Kleinbock), "Metric Diophantine approximation: the Khintchin–Groshev theorem for nondegenerate manifolds" (jointly with V. Beresnevich, V. Bernik, and D. Kleibock). "Non-planarity and metric Diophantine approximation for systems of linear forms" (jointly with V. Beresnevich and D. Kleinbock). Another topic was the distribution of values of indefinite quadratic forms at integral points (it is also often called the quantitative Oppenheim conjecture or the quantitative version of the Oppenheim conjecture). The following joint papers are related to this topic: "Limit distribution of orbits of unipotent flows and values of quadratic forms" (jointly with S.G. Dani), "Upper bounds and asymptotics in a quantitative version of the Oppenheim conjecture" (jointly with A. Eskin and S. Mozes), "Quadratic forms of signature $(2,2)$ and eigenvalue spacing on rectangular 2-torus" (jointly with A. Eskin and S. Mozes), "Quantitative version of the Oppenheim conjecture for inhomogeneous quadratic forms" (jointly with A. Mohammadi), "Distribution of values of quadratic forms at integral points" (jointly with P. Buterus, F. Götze, and T. Hille). (A more detailed account of the work mentioned in this paragraph will be given in Appendix 2.)

Fig. 7: My son Boris (1973–2021) in the summer of 2021. Photo: Private

There are two types of results from homogeneous dynamics which are used in applications to number theory and Diophantine approximation. The first type is, roughly speaking, related to the *behavior at infinity* of certain curves, while the second type involves methods from topological dynamics and ergodic theory. Because of that the results of the first type can be made in a certain sense *effective* rather easily, but for the results of the second type it is very difficult. For many years I was involved in the project to make the results of the second type effective. This led to effective estimates in corresponding number-theoretic problems. In this project I collaborated with M. Einsiedler, E. Lindenstrauss, A. Mohammadi, N. Shah and A. Venkatesh.

I also continued to work on problems related to my earlier work on arithmeticity and superrigidity. First I want to mention the series of three joint papers with D. Fisher "Local rigidity for cocycles", "Almost isometric actions, property (T), and local rigidity", and "Local rigidity of affine actions of higher rank groups and lattices" of combined length more than 150 pages. In another direction, A. Mohammadi and I proved in a 2019 paper that a 3-dimensional compact hyperbolic manifold containing infinitely many closed 2-dimensional geodesic subspaces is arithmetic. (More details are given in Appendix 2.)

I collaborated with some of my students while they were still in graduate school. I already mentioned D. Kleinbock (work on Sprindzuk's conjectures), A. Mohammadi (the work on inhomogeneous quadratic forms), and T. Hille, who is one of the coauthors of the paper by P. Buterus, F. Götze, T. Hille, and G. Margulis. I also collaborated with A. Karlsson on the paper "A multiplicative ergodic theorem and nonpositively curved spaces" and Han Li on the paper "Effective estimates on integer quadratic forms: Masser's conjecture, generators of orthogonal groups, and bounds in reduction theory".

I want to finish the main part of this autobiography with some general remarks. I was asked quite a few times (e.g., in the Abel interview) where to place my mathematics. I received the Wolf Prize mostly for my contributions to algebra. My NSF grants were in the analysis program, and the Abel Prize focused on probability and dynamics. I mostly consider myself to be a geometer. I once talked to Jacques Tits, and he said "I am a geometer and you are a geometer". There are many types of geometers. I am a geometer in the sense that my mathematical thinking is mostly based on (geometric) imagination and intuition.

I was influenced by many mathematicians. I should especially mention (in alphabetical order) Kazhdan, Piatetski-Shapiro, Sinai, and Vinberg during my early career and Furstenberg, Mostow, M.S. Raghunathan, and Tits in later years.

I had a long career in mathematics. My first paper was published in 1966, more than 55 years ago, and I am still active in mathematical research. I always tried to work on difficult problems, sometimes unsuccessfully and sometimes successfully. Success was achieved in some cases by more or less straightforward approaches and in other cases by discovering new connections between various fields of mathematics.

Fig. 8: With my granddaughter Maya at the Abel banquet on May 24, 2022. Photo: Private

Appendix 1. Arithmeticity and superrigidity

A discrete subgroup Λ of a Lie group F is called a *lattice* if the volume of F/Λ with respect to the Haar measure is finite. The lattice Λ is called *uniform* if F/Λ is compact and *non-uniform* otherwise. A lattice Δ in a connected semisimple Lie group H is called *reducible* if there exist connected infinite normal subgroups H_1 and H_2 such that $H_1 \cap H_2 \subset Z(H)$ where $Z(H)$ denotes the center of H, $H_1 \cdot H_2 = H$, and $(H_1 \cap \Delta) \cdot (H_2 \cap \Delta$ has finite index in Δ; otherwise the lattice Δ is called *irreducible*.

Let $\mathbf{H}$ be a linear algebraic group defined over $\mathbb{R}$. The group $\mathbf{H}(\mathbb{R})$ of $\mathbb{R}$-points of $\mathbf{H}$ can be considered as an algebraic subgroup of $\mathrm{SL}(m, \mathbb{R})$ for some m. If $\mathbf{H}$ is defined over $\mathbb{Q}$ (i.e., $\mathbf{H}$ is the set of zeros of polynomials with rational coefficients) then the subgroup $\mathbf{H}(\mathbb{Z})$ consisting of all matrices in $\mathbf{H}(\mathbb{R})$ with integer coefficients is discrete. According to a theorem of A. Borel and Harish-Chandra, if $\mathbf{H}$ is semisimple then $\mathbf{H}(\mathbb{Z})$ is a lattice in $\mathbf{H}(\mathbb{R})$. This lattice is uniform if and only if it does not contain unipotent elements.

Let $\mathbf{G}$ be a connected semi-simple linear algebraic group defined over $\mathbf{R}$, and let G denote the connected component $\mathbf{G}(\mathbb{R})^0$ of the identity in the group $\mathbf{G}(\mathbb{R})$. Let us assume that $\mathbf{G}$ has no non-trivial $\mathbb{R}$-anisotropic factors or, equivalently, that the connected semi-simple Lie group G has no compact factors. Let Ad denote the adjoint representation of the group $\mathbf{G}$. A lattice Λ in the group G is called *arithmetic* if there exist a connected semi-simple $\mathbb{Q}$-group $\mathbf{H}$ and an epimorphism $\varphi \colon \mathbf{H} \to \mathrm{Ad}\,\mathbf{G}$ defined over $\mathbb{R}$ such that the group $(\mathrm{Ker}\,\varphi)(\mathbb{R})$ is compact and the subgroups $\varphi(\mathbf{H}(\mathbb{Z}))$ and $\mathrm{Ad}\,\Lambda$ are commensurable. (Recall that two subgroups are called *commensurable* if their intersection has a finite index in each of them.) This definition is originally due to Piatetski-Shapiro. It is rather complicated. But for irreducible non-uniform lattices the definition can be simplified. For such lattices we can assume that the kernel of φ is finite (after a suitable choice of $\mathbf{H}$ and φ). Also, in the case where the center of $\mathbf{G}$ is trivial, we can assume that φ is an isomorphism or, equivalently, a non-uniform irreducible lattice Δ in G is arithmetic if and only if there exists a $\mathbb{Q}$-form of $\mathbf{G}$ such that the subgroups $\mathbf{G}(\mathbb{Z})$ and Δ are commensurable. The statement about the existence of a $\mathbb{Q}$-form is also true when $\mathbf{G}$ is simply connected in the sense of algebraic group theory.

The $\mathbb{R}$-rank of G is defined as the dimension of a maximal diagonalizable over $\mathbb{R}$ subgroup of G or, equivalently, as the dimension (in the sense of algebraic geometry) of a maximal $\mathbb{R}$-split torus in $\mathbf{G}$. The $\mathbb{R}$-rank of G is denoted by $\mathrm{rank}_{\mathbb{R}}\, G$. If the $\mathbb{R}$-rank of G is greater than 1 then any lattice in G is called a *higher rank lattice*. The following theorem is my main arithmeticity result.

Arithmeticity theorem for higher rank lattices. If $\mathrm{rank}_{\mathbb{R}}\, G > 1$ and Γ is an irreducible lattice in G then the lattice Γ is arithmetic.

This theorem proves the conjecture by Selberg and Piatetski-Shapiro. Selberg stated the conjecture only for non-uniform lattices. For uniform lattices the conjecture is due to Piatetski-Shapiro who, as was mentioned above, gave the definition of an arithmetic uniform lattice. The proofs in the case of non-uniform lattices and in the case of uniform lattices are very different. For non-uniform lattices the proof is essentially algebraic, and for uniform lattices the proof is mostly non-algebraic.

Arithmeticity of non-uniform lattices

Selberg and Piatetski-Shapiro developed a strategy to prove arithmeticity in the case where Γ is non-uniform. This strategy is based on the study of unipotent elements in Γ and subgroups of Γ associated with them. The proof consists of several steps. Step 1 was done in the above-mentioned joint 1968 paper with Kazhdan, where we proved the existence of non-trivial unipotent elements in Γ (for some special cases it had been known before). Step 2 is to prove that if U is a maximal unipotent subgroup of Γ then the Zariski closure of U in $\mathbf{G}$ is a horospherical subgroup of $\mathbf{G}$. Recall that an algebraic subgroup of $\mathbf{G}$ is called *horospherical* if it coincides with the unipotent radical $R_u(\mathbf{P})$ of a parabolic subgroup $\mathbf{P}$ of $\mathbf{G}$.

In Steps 1 and 2 the condition $\text{rank}_{\mathbb{R}} G > 1$ is not used. It is used in the next steps. The following statement is Step 3. There exist proper opposite parabolic $\mathbb{R}$-subgroups $\mathbf{P}_1$ and $\mathbf{P}_2$ of $\mathbf{G}$ such that, for both $i = 1$ and $i = 2$, we have:

(A) the subgroup $\mathbf{U}_i \cap \Gamma$ is Zariski dense in $\mathbf{U}_i$ or, equivalently, the quotient $\mathbf{U}_i(\mathbb{R})/(\mathbf{U}_i \cap \Gamma)$ is compact, where $\mathbf{U}_i$ denotes the unipotent radical of $\mathbf{P}_i$;
(B) the subgroup $\mathbf{U}_i \cap \Gamma$ has an infinite index in $\mathbf{P}_i \cap \Gamma$ or, equivalently, $\dim \mathbf{M}_i > \dim \mathbf{U}_i$, where $\mathbf{M}_i$ denotes the connected component of the identity of the Zariski closure of $\mathbf{P}_i \cap \Gamma$;
(C) $\mathbf{M}_i$ contains all unipotent elements from $\mathbf{P}_i(\mathbb{R})$;
(D) if $\gamma \in \Gamma$ is such that the intersection $\gamma \mathbf{P}_{3-i} \gamma^{-1} \cap \mathbf{P}_i$ is a Levy subgroup in $\mathbf{P}_i$ then (1) $\gamma \mathbf{M}_{3-i} \gamma^{-1} \cap \mathbf{M}_i$ is a Levy subgroup in $\mathbf{M}_i$ and (2) the intersection $\gamma \mathbf{M}_{3-i} \gamma^{-1} \cap \mathbf{M}_i \cap \Gamma$ is a lattice in the subgroup $(\gamma \mathbf{M}_{3-i} \gamma^{-1} \cap \mathbf{M}_i)(\mathbb{R})$.

Let us define two classes Φ and Ψ of algebraic $\mathbb{R}$-subgroups of $\mathbf{G}$. The class Φ consists of subgroups $\gamma \mathbf{M}_i \gamma^{-1}$ where $\gamma \in \Gamma$ and i is either 1 or 2, and the class Ψ consists of subgroups $\gamma_1 \mathbf{M}_1 \gamma_1^{-1} \cap \gamma_2 \mathbf{M}_2 \gamma_2^{-1}$ where $\gamma_1, \gamma_2 \in \Gamma$ are such that the intersection of two parabolic subgroups $\gamma_1 \mathbf{P}_1 \gamma_1^{-1}$ and $\gamma_2 \mathbf{P}_2 \gamma_2^{-1}$ is a Levy subgroup in each of them. Assume that the center of $\mathbf{G}$ is trivial. It is proved in Step 4 that if $\mathbf{D}$ is a subgroup from one of these two classes then there exists a unique $\mathbb{Q}$-structure on $\mathbf{D}$ such that the subgroups $\mathbf{D}(\mathbb{R}) \cap \Gamma$ and $\mathbf{D}(\mathbb{Z})$ are commensurable. In addition, if $\mathbf{H}$ is a subgroup from the class Φ and $\mathbf{F}$ is a subgroup from the class Ψ such that $\mathbf{F}$ is a Levy subgroup in $\mathbf{H}$ then $\mathbf{F}$ is a $\mathbb{Q}$-subgroup in $\mathbf{H}$ and the $\mathbb{Q}$-structure on $\mathbf{F}$ is induced from the $\mathbb{Q}$-structure on $\mathbf{H}$. We also have that this system of $\mathbb{Q}$-structures is invariant under conjugation by elements $\gamma \in \Gamma$. Using the properties (A)–(D), the properties of the system of $\mathbb{Q}$-structures on the subgroups from the classes Φ and Ψ, and an algebraic construction from representation theory, I was able to prove the following theorem.

Rationality theorem. *Assume that the center of* $\mathbf{G}$ *is trivial. Then there exists a* $\mathbb{Q}$*-structure on* $\mathbf{G}$ *such that the following conditions are satisfied: (1)* $\Gamma \subset \mathbf{G}(\mathbb{Q})$*; (2)* $\mathbf{G}$ *is* $\mathbb{Q}$*-simple of* $\mathbb{Q}$*-rank* ≥ 1*; (3) if* $\mathbf{P}$ *is a proper parabolic* $\mathbb{Q}$*-subgroup of* $\mathbf{G}$ *then the subgroups* $\mathbf{P}(\mathbb{Z})$ *and* $\mathbf{P} \cap \Gamma$ *are commensurable.*

Remark. The construction from representation theory mentioned before the formulation of the rationality theorem was used first by Hee Oh in the 1990s and later by Yves Benoist and others in the study of the Zariski dense discrete subgroups Λ of G (not necessarily lattices) such that $\Lambda \cap \mathbf{U}$ is Zariski dense in $\mathbf{U}$ for some non-trivial horospherical subgroup $\mathbf{U}$ of $\mathbf{G}$.

An important ingredient in my proof of the rationality theorem is the statement (ND) below about the non-divergence of orbits of unipotent flows. The importance of this statement in some approaches to the proof of arithmeticity of non-uniform lattices was first realized by Piatetski-Shapiro.

Statement (ND). *Let* $\{u(t)\}$ *be a one-parameter group of unipotent linear transformations of* $\mathbb{R}^n$*, and let* Λ *be a unimodular lattice in* $\mathbb{R}^n$*. Then there exists an* $\varepsilon > 0$ *such that the set*

$$\{t > 0 \colon \|u(t)v\| > \varepsilon \text{ for every non-zero } v \in \Lambda\}$$

is not bounded or, equivalently, the orbit $\{u(t)\Lambda\}$ *does not diverge to infinity in the space* Ω_n *of unimodular lattices in* $\mathbb{R}^n$*.*

The statement (ND) was conjectured (or, more precisely, stated as a theorem without a proof) in 1966 by Piatetski-Shapiro and slightly later by Garland and Raghunathan. I announced the proof of (ND) and of the rationality theorem in 1969. The proof of (ND) was given in a paper published in 1971. Using (ND), I gave a complete proof of the rationality theorem in a very long paper submitted for publication in 1971 and published only in 1975. (A shorter proof was given in my other paper which was published in 1974.) A couple of years later I completed the proof of arithmeticity of (irreducible higher rank) non-uniform lattices using results about subgroups generated by unipotent elements in arithmetic groups. I should mention that M.S. Raghunathan gave a proof of the rationality theorem without using the statement (ND) in a long paper submitted for publication in 1973 and published in 1975.

The statement (ND) looks quite technical. But it and especially its proof became quite influential. M.S. Raghunathan wrote me that analysis of my proof of (ND) was one of the inspirations for stating his conjecture about the closures of orbits of unipotent flows on homogeneous spaces. More details about various generalizations of the statement (ND) are given in Appendix 2.

Arithmeticity of uniform lattices

As I mentioned before, the strategies for proving the non-uniform case and the uniform case are very different. For the non-uniform case there are non-trivial unipotent elements in Γ, and there are some 'building blocks' which allow us to build the structure of an arithmetic subgroup on Γ. For uniform lattices there are no such building blocks. For non-uniform lattices the method of proof is of an algebraic and geometric nature, but for uniform lattices transcendental methods are used. Let me quote A. Borel: "the text of an invited address" was "sent by Margulis to the 1974 ICM in Vancouver (which he was not allowed to attend), where he sketched a proof of arithmeticity in the cocompact case using entirely new ideas. [...] The work of Margulis was based on a new principle soon christened "superrigidity" by Mostow (who presented Margulis's paper orally at the Vancouver Congress)."

Superrigidity theorem. *Assume that G and Γ satisfy the same assumptions as in the formulation of the arithmeticity theorem for higher rank lattices. Let k be a locally compact field of characteristic zero, and let $T\colon \Gamma \to \mathrm{SL}(n,k)$ be a homomorphism. Assume that the Zariski closure $\mathbf{H}$ of $T(\Gamma)$ is connected, absolutely simple, and has trivial center.*

(i) *If k is a finite extension of $\mathbb{Q}_p$ then $T(\Gamma)$ is bounded in the k-topology.*
(ii) *If k is $\mathbb{R}$ or $\mathbb{C}$ and $T(\Gamma)$ is not bounded in the k-topology then T extends to a continuous homomorphism of G to $\mathrm{SL}(n,k)$.*

The statement (i) (resp. (ii)) is called *non-Archimedean superrigidity* (resp. *Archimedean superrigidity.*) We can assume that $\Gamma \subset \mathrm{SL}(n,K)$ where K is a finitely generated extension of $\mathbb{Q}$. Every embedding σ of K into a locally compact field k induces a homomorphism $T_\sigma\colon \Gamma \to \mathrm{SL}(n,k)$. Roughly speaking, the arithmeticity of Γ is proved by applying the superrigidity theorem to such homomorphisms T_σ.

The proof of the superrigidity theorem is based on the study of a certain type of equivariant measurable maps. First we need some definitions. Let $\mathbf{V}$ be an algebraic $\mathbb{R}$-variety, D a subset of $\mathbf{V}(\mathbb{R})$, μ a measure on D, k a locally compact field, and $\mathbf{M}$ an algebraic k-variety. A map $f\colon D \to \mathbf{M}(k)$ is called *rational* if either 1) k is $\mathbb{C}$ or $\mathbb{R}$ and f is a restriction to D of a rational map of the Zariski closure of D to $\mathbf{M}$, or 2) k is a finite extension of $\mathbb{Q}_p$ and f is a map to a point. A measurable (with respect to μ) map $f'\colon D \to \mathbf{M}(k)$ is called *μ-rational* if it coincides almost everywhere (with respect to μ) with a rational map.

A k-rational action of an algebraic k-group $\mathbf{F}$ on an algebraic k-variety $\mathbf{X}$ is called *strongly k-effective* if $\mathbf{F}$ acts effectively on every orbit $\mathbf{F}x, x \in \mathbf{X}(k)$, or, in other words, for every $x \in \mathbf{X}(k)$ one can find $h \in \mathbf{F}$ such that $hx \neq x$. If $\mathbf{F}$ is (absolutely) simple then this condition is equivalent to the condition that no point in $\mathbf{X}(k)$ is fixed by $\mathbf{F}$.

Let $\mathbf{P}$ be a minimal parabolic $\mathbb{R}$-subgroup of $\mathbf{G}$, and let P denote the connected component $\mathbf{P}(\mathbb{R})^0$ of the identity in the group $\mathbf{P}(\mathbb{R})$. Then G/P is a real algebraic variety. Fix a G-quasiinvariant measure μ on G/P. Let the k-group $\mathbf{H}$ act k-rationally on a k-variety $\mathbf{X}$. We say that a μ-measurable map $\varphi : G/P \to \mathbf{X}(k)$ is (Γ,T)-*equivariant* if

$$\varphi(\gamma y) = T(\gamma)\varphi(y)$$

for all $\gamma \in \Gamma$ and almost all (with respect to the measure μ) $y \in G/P$.

The superrigidity theorem rather easily follows from the following two statements: (EM) (existence of equivariant measurable maps) and (RM) (rationality of equivariant measurable maps).

(EM) *If $T(\Gamma)$ is not bounded in k-topology then there exist a strongly k-effective k-rational action of the k-group* $\mathbf{H}$ *on an algebraic k-variety* $\mathbf{M}$ *and a $(\Gamma,T))$-equivariant μ-measurable map $\varphi : G/P \to \mathbf{M}(k)$.*

(RM) *If we are given a k-rational action of the k-group* $\mathbf{H}$ *on a k-variety* $\mathbf{M}$ *then any (Γ,T)-equivariant μ-measurable map $\varphi : G/P \to \mathbf{M}(k)$ is μ-rational.*

The assumption that $\operatorname{rank}_{\mathbb{R}} G > 1$ is used only to prove (RM) but is not used to prove (EM). My first proof of (EM) was for the uniform case and it was mostly based on using Oseledec's multiplicative ergodic theorem and certain integrability estimates for cocycles. My argument could be extended to the non-uniform case. Actually, for that case, I had to use arithmeticity, or at least the rationality theorem to obtain these estimates. One or two years later Furstenberg gave a different proof of (EM) — both in the uniform and non-uniform cases — using his boundary theory. The superrigidity theorem can be generalized, after certain modifications, to the case where the Zariski closure $\mathbf{H}$ of $T(\Gamma)$ is not semisimple. It should be noted that in this case the approach based on the multiplicative ergodic theorem can be applied to prove (EM) but the approach based on the boundary theory cannot be applied to do that.

For groups G of $\mathbf{R}$-rank 1, the lattice Γ is not necessarily arithmetic. There are three infinite series of groups G of $\mathbf{R}$-rank 1 and one exceptional group. They correspond to: (i) real hyperbolic spaces of dimension n; (ii) complex hyperbolic spaces of complex dimension n; (iii) quaternionic hyperbolic spaces of quaternionic dimension n; (iv) the octonian hyperbolic space. In the case (i), around 1965 Makarov and Vinberg gave examples of non-arithmetic lattices in dimensions 3, 4, and 5. Their examples are groups generated by reflections. In 1988 Gromov and Piatetski-Shapiro gave examples of non-arithmetic lattices for all n using the so-called *hybrid construction*. In case (ii), there are examples of non-arithmetic lattices in complex dimensions 2 and 3. In 1980 Mostow gave examples of non-arithmetic lattices for $n = 2$ using groups generated by complex reflections. Using monodromies of hypergeometric functions, Deligne and Mostow constructed in 1986 examples of non-arithmetic lattices for $n = 2$ and $n = 3$. In the cases (iii) and (iv), K. Corlette proved Archimedean superrigidity using harmonic maps. Shortly after that, Gromov and Schoen proved non-Archimedean superrigidity using a similar approach. Thus in cases (iii) and (iv) all lattices are arithmetic.

Let

$$\mathrm{Comm}_G(\Gamma) = \{g \in G \colon g\Gamma g^{-1} \text{ and } \Gamma \text{ are commensurable}\}$$

denote the *commensurator of Γ in G*. If $\mathbf{F}$ is a connected algebraic $\mathbb{Q}$-group then the closure of $\mathbf{F}(\mathbb{Q})$ in $\mathbf{R}$ contains $\mathbf{F}(\mathbb{R})^0$. It easily implies that if the lattice Γ is arithmetic then $\mathrm{Comm}_G(\Gamma)$ is dense in G. The converse is also true. I proved this at the same time as I proved the arithmeticity of Γ for the case $\mathrm{rank}_{\mathbb{R}}\, G > 1$. The arithmeticity again is deduced from some version of superrigidity, and the proof of superrigidity follows the same strategy as in the case of $\mathrm{rank}_{\mathbf{R}}\, G > 1$. Using harmonic maps, A. Karlsson, T. Gelander and I gave in 2006 a relatively short proof of superrigidity in the case when $\mathrm{Comm}_G(\Gamma)$ is dense in G.

At the beginning of 2019, answering a question of A. Reid and C. McMullen, A. Mohammadi and I proved the following theorem:

Theorem. *Let $M = \mathbb{H}^3/\Gamma$ be a closed hyperbolic 3-manifold. If M contains infinitely many totally geodesic surfaces, then M is arithmetic (that is Γ is arithmetic).*

Our proof uses, among other things, the multiplicative ergodic theorem and the martingale convergence theorem. Shortly after we proved the above theorem, Bader, Fisher, Miller and Stover proved that if a finite volume hyperbolic manifold $\mathbb{H}^n/\Gamma$ contains infinitely many maximal totally geodesic subspaces of dimension at least two, then Γ is arithmetic. Their proof and ours both use superrigidity to prove arithmeticity, but their proof of superrigidity and ours are quite different.

At the end of this Appendix I want to make the following remark. My work on superrigidity had a quite big impact on the work of other mathematicians. In particular I should mention the work of R. Zimmer on cocycle superrigidity and his program of non-linear superrigidity. But in the early seventies my goal was to prove arithmeticity in the uniform case, and I considered superrigidity only as a tool.

Appendix 2. Homogeneous dynamics and number theory/diophantine approximation

This Appendix consists of two parts. The first part is mostly about the Oppenheim conjecture and its quantitative versions, and the second is mostly about Baker–Sprindzuk conjectures.

Distribution of values of indefinite quadratic forms at integral points

We say that a real quadratic form is *rational* if it is a multiple of a form with rational coefficients and *irrational* otherwise. Let Q be a real quadratic form in n variables. In 1929 A. Oppenheim conjectured that if Q is irrational and $n \geq 3$ then for every

$\varepsilon > 0$ one can find a non-zero vector $\mathbf{x} \in \mathbb{Z}^n$ such that $|Q(\mathbf{x})| < \varepsilon$. Oppenheim was motivated by Meyer's theorem that if Q is rational and $n \geq 5$ then Q represents zero over $\mathbb{Z}$ non-trivially. Because of that he originally stated the conjecture only for $n \geq 5$. I proved the Oppenheim conjecture in the mid-eighties. The short version of my proof was written at the end of 1986 and was published in 1987, and more detailed versions were published later. But, as I mentioned before, I remember that I gave some kind of oral presentations already in 1984. Before the Oppenheim conjecture was proved it was extensively studied mostly using analytic number theory methods. In particular it was proved in 1946–59 in a series of papers by H. Davenport and his coauthors for diagonal forms where $n \geq 5$ and for general quadratic forms where $n \geq 21$.

In the mid-seventies M.S. Raghunathan made a remarkable observation relating the Oppenheim conjecture and the theory of unipotent flows. He also stated a conjecture about the closures of orbits of unipotent subgroups or, more generally, of subgroups generated by unipotent elements. Let G be a connected Lie group, Γ a lattice in G, and H a closed connected subgroup of G. Raghunathan conjectured that if H is generated by unipotent elements then the closure of any orbit Hx, $x \in G/\Gamma$, is an orbit Fx of a closed subgroup F of G. I proved the Oppenheim conjecture by proving a very special case of the Raghunathan conjecture. This is the statement $(*)$ below, which is actually equivalent to the statement of the Oppenheim conjecture.

$(*)$: *Let* $\Omega_3 \cong \mathrm{SL}(3,\mathbb{R})/\mathrm{SL}(3,\mathbb{Z})$ *denote the space of unimodular lattices in* $\mathbb{R}^3$, *and* $H = \mathrm{SO}(2,1)^0$ *the connected component of the identity of the orthogonal group* $\mathrm{SO}(2,1)$. *If* $x \in \Omega_3$ *and the orbit* Hx *is bounded in the space* Ω_3 *then* Hx *is closed in* Ω_3.

The proof of $(*)$ is based on the technique which involves finding orbits of larger subgroups inside closed sets invariant under unipotent subgroups by studying the minimal invariant sets, and the limits of orbits of sequences of points tending to a minimal invariant set. Further developing this technique, S.G. Dani and I proved in one of our joint papers that, in the notation of $(*)$, every orbit Hx is either closed or dense in Ω_3. In another paper we proved the Raghunathan conjecture in the case when $G = \mathrm{SL}(3,\mathbb{R})$ and $H = \{u(t)\}$ is a one-parameter unipotent subgroup of G such that $u(t) - 1$ has the rank 2 for all $t \neq 0$. Though this is only a very special case, the proof given in the just mentioned paper suggests an approach for proving the Raghunathan conjecture in general.

In the general case Raghunathan's conjecture and its quantitative analogs were proved by M. Ratner in a fundamental series of four papers published in 1990–91. In the first three papers Ratner proves the *measure classification theorem*, which says that if U is a closed connected unipotent subgroup of G then any U-ergodic U-invariant probability measure μ on G/Γ is *algebraic* in the sense that μ is the Haar measure on a closed orbit Fx, $x \in G/\Gamma$, of a closed subgroup $F \supset U$ of G. (The total length of Ratner's proof is more than 150 pages; a much shorter and rather different proof was later given by Tomanov and myself.) Using the measure classification theorem and a quantitative version of the statement (ND) from Appendix 1, Ratner proved the *uniform distribution theorem* for unipotent flows, which says that if $U = \{u(t)\}$ is a one-parameter unipotent subgroup of G and $x \in G/\Gamma$ then there exists an algebraic measure μ_x on G/Γ such that

$$\lim_{T\to\infty} \frac{1}{T} \int_0^T f(u(t)x)\,\mathrm{d}t = \int_{G/\Gamma} f\,\mathrm{d}\mu_x$$

for every bounded continuous function f on G/Γ.

A rather simple argument shows how to prove the Raghunathan conjecture using the uniform distribution theorem.

Let ν be a positive continuous function on the unit sphere in $\mathbb{R}^n$, and let $\Omega = \{v \in \mathbb{R}^n : \|v\| < \nu(v/\|v\|)\}$. Let us denote by $N_{Q,\Omega}(a,b,T)$ the cardinality of the set

$$\{x \in \mathbb{Z}^n : x \in T\Omega \text{ and } a < Q(x) < b\}$$

and by $V_{Q,\Omega}(a,b,T)$ the volume of the set

$$\{x \in \mathbb{R}^n : x \in T\Omega \text{ and } a < Q(x) < b\}.$$

It is easy to verify that if $n \geq 3$ then asymptotically, as $T \to \infty$,

$$V_{Q,\Omega}(a,b,T) \sim \lambda_{Q,\Omega}(b-a)T^{n-2},$$

where

$$\lambda_{Q,\Omega} = \int_{L\cap\Omega} \frac{\mathrm{d}A}{\|\nabla Q\|},$$

L is the light cone $Q = 0$ and $\mathrm{d}A$ is the area element on L.

Let $\mathscr{O}(p,q)$ denote the space of indefinite quadratic forms in $n \geq 3$ variables of signature (p,q) and discriminant ± 1. We note that $p+q = n$ and assume that $p \geq q > 0$. Let (a,b) be an interval. Using the uniform distribution theorem, a quantitative version of the statement (ND) from Appendix 1, and the so-called *linearization technique*, S.G. Dani and I proved in 1992 a refined version of Ratner's uniform distribution theorem and using that obtained the following asymptotically precise lower bounds for $N_{Q,\Omega}(a,b,T)$ when Q is irrational:

$$\liminf_{T\to\infty} \frac{N_{Q,\Omega}(a,b,T)}{V_{Q,\Omega}(a,b,T)} \geq 1.$$

Moreover, this bound is uniform over compact sets of forms: if K is a compact subset of $\mathscr{O}(p,q)$ which consists of irrational forms, then

$$\liminf_{T\to\infty} \inf_{Q\in K} \frac{N_{Q,\Omega}(a,b,T)}{V_{Q,\Omega}(a,b,T)} \geq 1.$$

The situation with the asymptotics and upper bounds for $N_{Q,\Omega}(a,b,T)$ is more delicate. Rather surprisingly here the answer depends on the signature of Q. In a paper published in 1998, A. Eskin, S. Mozes and I proved that if $p \geq 3$, $q \geq 1$ then, as $T \to \infty$,

$$N_{Q,\Omega}(a,b,T) \sim \lambda_{Q,\Omega}(b-a)T^{n-2}$$

for any irrational form $Q \in \mathcal{O}(p,q)$, where $n = p+q$ and $\lambda_{Q,\Omega}$ is the same as in the asymptotic formula for $V_{q,\Omega}(a,b,T)$. This asymptotic formula is not true for an arbitrary irrational form Q of signature $(2,1)$ or $(2,2)$. But in another joint paper by Eskin, Mozes and myself, published in 2005, we showed that in the $(2,2)$ case the set of exceptions is "extremely small". It consists of so-called *extremely well approximable by split rational forms* (EWAS) forms, and the set of (EWAS) forms has Hausdorff dimension 0 in the space $\mathcal{O}(2,2)$. In a paper published in 2011, A. Mohammadi and I extended the results of the joint papers by Eskin, Mozes and myself in the setting of inhomogeneous quadratic forms. It is probably true that in the $(2,1)$ case the set of exceptions is also "extremely small". Regarding upper bounds for an arbitrary irrational quadratic form Q of the signature $(2,1)$ or $(2,2)$, one has to add the additional $\log T$ factor, and this essentially cannot be improved.

My paper on the Oppenheim conjecture and the subsequent joint papers with S.G. Dani on the topological approach to Raghunathan's conjecture use the notion of a minimal set. Because of that the proofs of results obtained in those papers are not "effective". The proof of Ratner's uniform distribution theorem is even "less effective" because it uses various ergodic theorems. The desired effective statements should look like the following. Assume that the form Q satisfies certain diophantine conditions. Then (a) there exists a positive integer $m = m(Q)$ such that for every $\varepsilon < 1/2$ one can find a non-zero vector $\mathbf{x} \in \mathbb{Z}^n$ such that $|Q(\mathbf{x})| < \varepsilon$ and $\|\mathbf{x}\| < \varepsilon^{-m}$; (b) for the domain Ω with the smooth boundary and for some positive $\alpha = \alpha(Q,\Omega)$, the error term

$$|N_{Q,\Omega}(a,b,T) - \lambda_{Q,\Omega}(b-a)T^{n-2}|$$

should be of the order $T^{n-2-\alpha}$. At the moment it is not clear how to achieve (a) and (b) using dynamical/ergodic methods. (The best result in the direction of (a) is due to E. Lindenstrauss and me where, in the estimate for $\|x\|$, ε^{-m} is replaced by the exponent of ε^{-m}.) But for $n \geq 5$ a completely different approach allows us to achieve (a) and (b). This approach is based on Götze's method which F. Götze developed, partially in collaboration with V. Bentkus, in his earlier work and on arguments analogous to some arguments used in my joint papers with Eskin and Mozes. The related paper by P. Buterus, F. Götze, T. Hille, and myself was published in 2022 but a preliminary version was written approximately twenty years before that.

Diophantine approximation on manifolds

We need a lot of notation and terminology in this part of Appendix 2. For $\mathbf{x}, \mathbf{y} \in \mathbb{R}^n$ we let

$$\mathbf{x} \cdot \mathbf{y} = \sum_{i=1}^{n} x_i y_i, \quad \|\mathbf{x}\| = \max_{1 \leq i \leq n} |x_i|,$$

$$\Pi(\mathbf{x}) = \prod_{i=1}^{n} |x_i| \quad \text{and} \quad |\Pi_+(\mathbf{x}) = \prod_{i=1}^{n} |x_i|_+,$$

where $|x|_+$ stands for $\max(|x|, 1)$. A vector $\mathbf{y} \in \mathbb{R}^n$ is called *very well approximable*, to be abbreviated as VWA, if the following two equivalent conditions are satisfied:

(i): for some $\varepsilon > 0$ there are infinitely many $\mathbf{q} \in \mathbb{Z}^n$ such that

$$|\mathbf{q} \cdot \mathbf{y} + p| \cdot \|\mathbf{q}\|^n \leq \|\mathbf{q}^{-n\varepsilon}$$

for some $p \in \mathbb{Z}$;

(ii): for some $\varepsilon > 0$ there are infinitely many $q \in \mathbb{Z}$ such that

$$\|q\mathbf{y} + \mathbf{p}\| \leq |q|^{-\varepsilon}$$

for some $\mathbf{p} \in \mathbb{Z}^n$.

In 1932 K. Mahler conjectured that all points in the curve

$$F = \{(x, x^2, \ldots, x^n) \ : \ x \in \mathbb{R}\}$$

are not VWA. V. Sprindzuk proved Mahler's conjecture in 1964. He also suggested the following terminology. A submanifold $M \subset R^n$ is called *extremal* if almost all points in M are not VWA. The following conjecture was made by Sprindzuk in 1980:

Conjecture A. *Let $f_1, \ldots, f_n$ be real analytic functions in $x \in U, U$ a domain in $\mathbb{R}^d$, which together with 1 are linearly independent over $\mathbb{R}$. Then the manifold $M = \{\mathbf{f}(\mathbf{x}) \colon \mathbf{x} \in U\}$ is extremal.*

Sprindzuk's result can be reformulated as the statement that the curve F is extremal. However, there exists a stronger version of Conjecture A which Sprindzuk did not prove even for the curve F. Namely, a vector $\mathbf{y} \in \mathbb{R}^n$ is called *very well multiplicatively approximable*, to be abbreviated as VWMA, if the following two equivalent conditions are satisfied:

(i): for some $\varepsilon > 0$ there are infinitely many $q \in \mathbb{Z}^n$ such that

$$|\mathbf{q} \cdot \mathbf{y} + p| \cdot \Pi_+(\mathbf{q}) \leq \Pi_+(\mathbf{q})^{-\varepsilon}$$

for some $p \in \mathbb{Z}$;

(ii): for some $\varepsilon > 0$ there are infinitely many $q \in \mathbb{Z}$ such that

$$\Pi(q\mathbf{y} + \mathbf{p}) \cdot |q| \leq |q|^{-\varepsilon}$$

for some $\mathbf{p} \in \mathbb{Z}^n$.

A manifold M is said to be *strongly extremal* if almost all points all $\mathbf{y} \in M$ are not VWMA. In his book published in 1975, A. Baker raised the question if the curve F from the Mahler conjecture is strongly extremal. Later in 1980 Sprindzuk stated

Conjecture B. *Any manifold $M \subset \mathbb{R}^n$ satisfying the assumptions of Conjecture A is strongly extremal.*

In a paper published in 1998, D. Kleinbock and I proved Sprindzuk's conjectures, even in the more general setting of smooth manifolds. Let $\mathbf{f} = (f_1, \ldots, f_n)$ be an n-tuple of C^k functions on an open subset V of $\mathbb{R}^d$. We say that the map $\mathbf{f} : V \to \mathbb{R}^n$ is *l-nondegenerate*, $l \leq k$, at $\mathbf{x} \in V$ if the space $\mathbb{R}^n$ is spanned by partial derivatives of $\mathbf{f}$ at $\mathbf{x}$ of order up to l. The n-tuple $\mathbf{f}$ is *nondegenerate* at $\mathbf{x}$ if it is l-nondegenerate at $\mathbf{x}$ for some l. We say that $\mathbf{f}$: $V \to \mathbb{R}^n$ is *nondegenerate* if it is nondegenerate at almost every point of V. Note that if the functions $f_!, \ldots, f_n$ are analytic and the set V is connected, the nondegeneracy of $\mathbf{f}$ is equivalent to the linear independence of $1, f_1, \ldots, f_n$ over $\mathbb{R}$.

Theorem S. *Let $\mathbf{f}$: $V \to \mathbb{R}^n$ be a nondegenerate C^k map of an open subset V of $\mathbb{R}^d$ into $\mathbb{R}^n$. Then $\mathbf{f}(\mathbf{x})$ is not VWMA (hence not VWA either) for almost every point $\mathbf{x}$ of V.*

Our proof of Theorem S was based on the correspondence, first established by S.G. Dani, between diophantine properties of vectors $\mathbf{y} = (y_1, \ldots, y_n) \in \mathbb{R}^n$ and the behavior of certain orbits in the space of unimodular lattices in $\mathbb{R}^{n+1}$. More precisely, let

$$U_{\mathbf{y}} = \begin{pmatrix} 1 & y_1 & y_2 & \ldots & y_n \\ 0 & 1 & 0 & \ldots & 0 \\ 0 & 0 & 1 & \ldots & 0 \\ \vdots & \vdots & \vdots & \ddots & \vdots \\ 0 & 0 & 0 & \ldots & 1 \end{pmatrix} \in \mathrm{SL}(n+1, \mathbb{R}).$$

Thus $U_{\mathbf{y}}$ is a unipotent matrix with all rows, except the first one, the same as in the identity matrix. We also have to introduce some diagonal matrices. Let

$$g_s = \begin{pmatrix} e^{ns} & 0 & \ldots & 0 \\ 0 & e^{-s} & \ldots & 0 \\ \vdots & \vdots & \ddots & \vdots \\ 0 & 0 & \ldots & e^{-s} \end{pmatrix} \in \mathrm{SL}(n+1, \mathbb{R}); \quad s \geq 0$$

and

$$g_{\mathbf{t}} = \begin{pmatrix} e^{t} & 0 & \ldots & 0 \\ 0 & e^{-t_1} & \ldots & 0 \\ \vdots & \vdots & \ddots & \vdots \\ 0 & 0 & \ldots & e^{-t_n} \end{pmatrix} \in \mathrm{SL}(n+1, \mathbb{R}); \quad \mathbf{t} = (t_1, \ldots, t_n), \quad t_i \geq 0, \quad t = \sum_{i=1}^{n} t_i.$$

As before, let $\Omega_m \cong \mathrm{SL}(m, \mathbb{R}) / \mathrm{SL}(m, \mathbb{Z})$ denote the space of unimodular lattices in $\mathbb{R}^m$. Define a function δ on Ω_m by

$$\delta(\Lambda) = \inf_{\mathbf{v} \in \Lambda \setminus \{0\}} \|\mathbf{v}\|; \quad \Lambda \in \Omega_m.$$

It is not difficult to prove that for any very well multiplicatively approximable vector $\mathbf{y} \in \mathbb{R}^n$ there exists $\gamma > 0$ and infinitely many $\mathbf{t} \in \mathbb{Z}_+^n$ such that

$$\delta(g_{\mathbf{t}} U_y \mathbb{Z}^{n+1}) \leq e^{-\gamma t}.$$

Because of that and in view of Borel–Cantelli, to prove Theorem S it is enough to show that if $\mathbf{f}$ is nondegenerate at $\mathbf{x}_0 \in V$ then there exists a neighborhood B of $\mathbf{x}_0$ in V such that, for every $\gamma > 0$,

$$\sum_{\mathbf{t} \in \mathbb{Z}_+^n} |E_{\mathbf{t},\gamma}| < \infty,$$

where $|A|$ denotes the Lebesgue measure of a set $A \subset \mathbb{R}^d$, and

$$E_{\mathbf{t},\gamma} = \{\mathbf{x} \in B \colon \delta(g_{\mathbf{t}} U_{\mathbf{f}(\mathbf{x})} \mathbb{Z}^{n+1}) \leq e^{-\gamma t}\}.$$

But this easily follows from the following:

Proposition. *Let* $\mathbf{f} \colon V \to \mathbb{R}^n$ *be a* C^k *map of an open subset* V *of* $\mathbb{R}^d$ *into* $\mathbb{R}^n$*, and let* $\mathbf{x}_0 \in V$ *be such that* $\mathbb{R}^n$ *is spanned by partial derivatives of* $\mathbf{f}$ *at* $\mathbf{x}_0$ *of order up to* k*. There exist a ball* $B \subset V$ *centered at* $\mathbf{x}_0$ *and positive constants* D *and* ρ *such that*

$$|\{\mathbf{x} \in B \colon \delta(g_{\mathbf{t}} U_{\mathbf{f}(\mathbf{x})} \mathbb{Z}^{n+1}) \leq \varepsilon\}| \leq D \left(\frac{\varepsilon}{\rho}\right)^{1/dk} |B|.$$

The proof of this proposition is based on a modification of the technique used in proofs of earlier results on the nondivergence of unipotent flows in the space Ω_m. Strengthening the statement (ND) from the Appendix 1, S.G. Dani proved that for any $c > 0$ and any $\Lambda \in \Omega_m$ there exists an $\varepsilon > 0$ such that for any unipotent subgroup $\{u(t)\}$ of $\mathrm{SL}(m, \mathbb{R})$ one has

$$|\{t \in [0, T] \colon \delta(u(t)\Lambda), \varepsilon\}| \leq cT.$$

In the proposition the orbit $\{u(t)\Lambda\}$, parametrized by $t \in \mathbb{R}$, is replaced by the set $g_{\mathbf{t}} U_{\mathbf{f}(\mathbf{x})} \mathbb{Z}^{n+1}$, parametrized by $\mathbf{x} \in V$.

Appendix 3. Fields Medal

In 1978 I was awarded a Fields Medal. For me it was not completely unexpected. Starting in the early Fall of 1977 I heard rumors that there was a possibility that I would receive the award. Also already in 1974, after I proved arithmeticity of higher rank lattices in the uniform case, Piatetski-Shapiro mentioned to me that, in his opinion, I deserved a Fields Medal.

In early March 1978 I received a letter, dated February 28, 1978, from the President of the International Mathematical Union, Dean Montgomery, which informed me that I had "been chosen to receive a Fields Medal at the International Congress of Mathematicians to be held in Helsinki in August 1978". He also wrote: "I hope you will accept the award and that you will be able to be present at Helsinki to receive it". According to the rules in the Soviet Union at the time, to accept the award I needed to get the permission from the Soviet Academy of Sciences. This permission was granted, and on March 28 I wrote to Montgomery that "I am extremely grateful ... for the high honor shown me by the award of a Fields Medal which I accept with great appreciation". I also wrote that "I shall write you again as soon as I am informed of my participation in the Helsinki Congress." On March 30 I wrote a letter to the head of the section of mathematics of the Academy of Sciences N.N. Bogolubov asking for his advice on what to write Montgomery about my participation in the Helsinki Congress. Later Bogolubov wrote a letter in support of the inclusion of me and also of Dobrushin and Sinai in the Soviet delegation to the Congress. (Dobrushin and Sinai were invited as plenary speakers.) On the other hand, the chair of the National committee of Soviet mathematicians I.M. Vinogradov wrote a letter in opposition to the inclusion of Dobrushin, Margulis and Sinai in the Soviet delegation. Both letters were sent to the so-called department of external interactions (*"upravlenie vneshnich snoshenij"* in Russian) of the Academy. The head of the delegation was L.S. Pontriagin, and the decision whom to include in the delegation was made by the National Committee of Soviet Mathematicians. On July 24 a letter was sent to D. Montgomery from the Presidium of the Soviet Academy of Sciences. In the letter, one of the members of the Presidium thanked Montgomery and the International Mathematical Union for awarding a Fields Medal to G.A. Margulis and also wrote that, unfortunately, G.A. Margulis would not be able to come to Helsinki to participate in the International Mathematical Congress. Finally, on July 31 I wrote to Montgomery: "I confirm the acceptance of a Fields Medal. Unfortunately, I shall not be apparently able to attend the Helsinki Congress". (A similar letter I also wrote to the Chair of the Organizing committee of the Helsinki Congress Olli Lehto.)

As I wrote in the main part of this autobiography, I was not allowed to attend the ICM in Helsinki mostly due to the opposition of an influential part of the top Soviet mathematical establishment. The clear evidence of that is presented in the previous paragraph. Actually, the President of the Soviet Academy of Sciences and other higher ups in the Academy were quite unhappy about what happened. They got a scandal that they did not want and did not need.

I could not go to the ICM in Helsinki on my own, because in the Soviet Union traveling abroad was considered to be a privilege and not a right. Let me describe my own experience based on attempts to get authorization to visit mathematical institutions in the West. As an example I take an invitation to IHES. First, the visit to IHES had to be included in the plan of the international scientific cooperation of the Academy of Sciences. After that I had to get so-called characteristics (*"harakteristika"* in Russian). This was some kind of description of my character, and its main purpose was to guarantee my loyalty and to explain that I would not cause any trouble while abroad. It was usually ended with the words "politically literate and morally stable". The "harakteristika" had to be signed at my institute IPPI by the director of IPPI, the secretary of the party committee of IPPI, and the chair of the IPPI branch of the professional union, and it had to be approved by the district committee of the Communist Party. After that the "harakteristika" together with several supporting documents was sent to the Academy of Sciences. There my case was considered by several people and, if the decision was positive, it was sent for final approval to the apparatus of the Central Committee of the Communist Party. Thus it was a multi-step process that could be stopped at every step. Until 1987, this process worked for me only once. This was regarding my visit to Bonn in 1979 mentioned in the main part of this autobiography. But in 1981 I did not get permission to visit France because of "the extremely negative opinion of the National Committee of Soviet Mathematicians on this issue".

When I received a Fields Medal, it was met with frustration (and even outrage) by a significant part of the Soviet mathematical establishment (see, for example, the report on page 205 of the book "Mathematics without borders" by Olli Lehto about remarks made by L.S. Pontriagin during the meeting of the Executive Committee of IMU in May 1978). I do not want to discuss the various reasons which caused this frustration, but just want to mention the following episode. I was told that a mathematician from the Institute of Mathematics in Novosibirsk said something like: "What is going on? A candidate of science gets a Fields Medal".

Appendix 4. Dissertations

There were two scientific degrees in the Soviet Union: *candidate of science* and *doctor of science*. This is still true in Russia. The candidate of science degree roughly corresponds to PhD in USA. But the doctor of science degree is much higher, and I do not think that there is a suitable equivalent for this degree in the West. Doctors of science usually held higher positions than candidates of science did. There was also one important additional difference. To become a graduate student advisor, a candidate of science had to get special permission in each individual case.

To get a scientific degree, you had to submit a dissertation (together with some supporting documents) to a scientific council affiliated with a research or educational institution. In each field, there were such scientific councils, several for doctoral degrees and many for candidate degrees. After the submission of the dissertation the scientific council appointed so-called *official opponents*, two for candidate degrees and three for doctoral degrees. These official opponents had to write reports about the dissertation. The scientific council also sent the dissertation for refereeing to a so-called *leading organization* (*"veduschee predprijatie"* in Russian). A month before the dissertation defense the scientific council sent approximately fifty copies of so-called *"avtoreferat"* to experts in the field, libraries, and other scientific councils. This "avtoreferat" contained a short description of the dissertation, written by you, and also some additional information such as the date of the dissertation defense, the names of the official opponents, the leading organization, etc.

The dissertation defense was a very serious event in the Soviet Union. You had to give a talk in which you described the main results in the dissertation. The official opponents read their reports. Then the report from the leading organization was read.

After that, there was a general discussion open to everyone who came. In the end there was a secret ballot by the members of the scientific council with the possible vote yes, abstain, and no. The dissertation was approved if at least 2/3 of all votes were "yes". If the vote was positive, the dissertation was sent for final approval to the High Attestation Committee (*"Visshaya Attestatsionaya Kommisiya"* or *VAK* in Russian) at the Department of Education.

Now let me go to my personal story. I was a graduate student in *"mehmat"*, which stands for the faculty of mechanics and mathematics at Moscow State University. I submitted my candidate dissertation "On some aspects of the theory of Anosov systems" to the scientific council in mathematics at "mehmat". The official opponents were D.V. Anosov and V.M. Alexeev, and the leading organization was Leningrad State University. Anosov wrote a report but he was out of town at the time of the dissertation defense. Because of that, according to the rules, there was an additional official opponent. It was V.I. Arnold. The result of the vote was: "no" 0, "abstain" 2, all other "yes". This vote was more than enough for the approval. I became a candidate of science in June or July of 1970. The two "abstain" votes were not personal. In that scientific council at the time, it was quite common for there to be some non-yes votes for the dissertation (even for the candidate dissertation) of a Jewish mathematician.

With the doctoral dissertation, it was much more complicated. During the 1970s and several years after that, there was significant discrimination against Jewish mathematicians in the scientific council at "mehmat". I know about two cases (Grigoriy Eskin and Boris Vainberg) when, during the dissertation defense, the number of "yes" votes to approve the doctoral dissertation of a Jewish mathematician was less than the necessary two thirds of all votes. In general, there was some kind of understanding that during that period the doctoral dissertation of a Jewish mathematician cannot be approved by the scientific council at "mehmat". But being awarded a Fields Medal, I thought that there would be no opposition in my case. So I submitted my doctoral dissertation "Discrete subgroups of semisimple algebraic groups" at the end of 1979, shortly after my return from Bonn. The dissertation consisted of three chapters: (1) superrigidity; (2) the normal subgroup theorem; (3) arithmeticity. The process of moving to the dissertation defense procedure started but quite soon it was stopped. They used various excuses, but the main reason was the opposition of a significant number of members of the scientific council. For example, I was told that an influential member of the scientific council said to a group of people: "they overestimated him and we will underestimate him". Here "they overestimated him" meant that they awarded me a Fields Medal and " we will underestimate him" meant that we will not approve my doctoral dissertation. After more than two years I decided to withdraw my dissertation from the scientific council at "mehmat" and submitted it to the scientific council at the Institute of Mathematics in Minsk, where V.P. Platonov was the director of the Institute and the chair of the scientific council. The dissertation defense there was in March 1983. The official opponents were S.P. Novikov, A.A. Kirillov, and V.E. Voskresenskii, and the leading organization was the Leningrad branch of the Institute of Mathematics of the Soviet Academy of Sciences. All 12 votes were "yes". The dissertation was finally approved in VAK after several months, and I became a doctor of science in October 1983. Thus the entire process from the initial submission to the final approval took almost four years.

The work of Hillel Furstenberg and its impact on modern mathematics

Vitaly Bergelson, Eli Glasner and Benjamin Weiss

Contents

Introduction

In the following article we have attempted to give some account of the extraordinary work of Hillel Furstenberg and its impact on modern mathematics. His influence goes far beyond his published papers. He shared his ideas freely and many of them appear in papers written by others. No attempt has been made to be exhaustive, but we have tried to mention his major works. The first two sections are due to Eli Glasner who was one of Hillel's first doctoral students. The third section is due to

Vitaly Bergelson
Department of Mathematics
The Ohio State University
Ohio
USA, e-mail: vitaly@math.ohio-state.edu

Eli Glasner
Department of Mathematics
Tel Aviv University
Tel Aviv
Israel, e-mail: glasner@math.tau.ac.il

Benjamin Weiss
Institute of Mathematics
Hebrew University of Jerusalem
Jerusalem
Israel, e-mail: weiss@math.huji.ac.il

H. Holden, R. Piene (eds.), *The Abel Prize 2018-2022*, The Abel Prize,
https://doi.org/10.1007/978-3-031-33973-8_11

Benjamin Weiss who has been a colleague of Hillel since 1967, and the last section is due to Vitaly Bergelson who was also a doctoral student of Hillel. As the reader will see no attempt has been made to give a unified style to the various parts.

1 Topological dynamics

During his time at Princeton as a graduate student Furstenberg was attracted to analysis and harmonic analysis, with measure theory as a focal point. He was also influenced by a probability course he took with William Feller. His Ph.D. thesis, supervised by Salomon Bochner, dealt with stochastic processes and prediction theory. Also as an instructor at Princeton, a year after his Ph.D. (1958), he collaborated with Harry Kesten in writing a seminal work on random products of matrices [57].

As a graduate student Furstenberg spent a summer at the University of Chicago where Littlewood was visiting and there was an emphasis on analysis. Some mathematicians there were trying to prove a Fatou theorem for harmonic functions for so-called Cartan domains which can be realized as bounded open sets in euclidean space (like the disc), and having a natural topological boundary. The issue was to show that for a bounded harmonic function, as one approaches the boundary, the function converges to a boundary value. They were not successful. Eventually this led Furstenberg to create his pioneering Poisson–Furstenberg boundary theory. Thanks to this theory we now know that this representation of the domain is deceptive when it comes to boundary theory. The euclidean boundary is a factor of the universal one – unlike the classical situation where the universal boundary G/P is the circle bounding the euclidean disc. The group-theoretic approach gives the right boundary, and Fatou's theorem is valid.

In 1961 Furstenberg joined the strong probability group at the University of Minnesota. At the suggestion of one of his colleagues, Monroe Donsker, he contributed to a volume on applications of probability theory to other fields, a chapter on applications to group theory. Specifically he showed how boundary theory could be used for the "rigidity theory" of lattices in a Lie group. While the theorem he proved was subsumed under the general strong rigidity theorem of Margulis [80], it also served as a tool in Margulis' "normal subgroup theorem". These are just a few examples from the beginning of his career, but in fact, one can say that harmonic analysis, probability theory, and probabilistic intuition are behind most of Furstenberg's works.

However, from the very beginning of his research (starting perhaps with his very early famous work [31]) one can see a certain 'twist' which intertwines this intuition with tools from topology and functional analysis.

To see this 'twist' more clearly we remark (see Furstenberg's autobiography)[1] that Loomis' 1953 book 'Introduction to abstract harmonic analysis' [78], which was a new book at that time, had a great influence on him as a graduate student

[1] Appearing in this volume.

(although, perhaps this was not looked upon so favourably by his adviser Salomon Bochner, who preferred a "hands on" approach in his researches in harmonic analysis). The following quote from the preface to Loomis' book is instructive:

> This book is an outcome of a course given at Harvard first by G. W. Mackey and later by the author. The original course was modelled on Weil's book [48] and covered essentially the material of that book with modifications. As Gelfand's theory of Banach algebras and its applicability to harmonic analysis on groups became better known, the methods and content of the course inevitably shifted in this direction, and the present volume concerns itself almost exclusively with this point of view. Thus our development of the subject centers around Chapters IV and V, in which the elementary theory of Banach algebras is worked out, and groups are relegated to the supporting role of being the principal application.

Next let's go to Furstenberg's Ph.D. thesis [32], which was later published as a Princeton Annals of Mathematics Studies [33]. In the introduction to his thesis Furstenberg describes two procedures for defining canonical probability distributions on future values of a given time series, that of *simple predictability* and that of *statistical predictability*. He then says: "Our major problem will then be to show that various classes of sequences will be predictable in one sense or the other". In the next quote from this introduction we can already discern his future approach to many problems in probability and ergodic theory, as well as an early form of his, now famous, correspondence principle.

> Our principal tool is the theory of commutative C^*-algebras which provides us with a convenient means for summing up the algebraic and topological properties of the sequence under consideration.

And another quote from the introduction of [33]:

> Although our problem is formulated for individual time series it is not independent of the theory of stochastic processes. In fact, the class of sequences to which our analysis will be applicable will be such that to each sequence there corresponds a stationary stochastic process for which the given sequence is a "typical" sample sequence.
>
> ...
>
> Our first chapter will therefore be concerned with the question of determining when a sequence can be thought of as occurring from observations on a stationary mechanism, or its abstract counterpart, a stationary process. A sequence of this kind will be said to be "regular" and it determines a "generic" point of the associated process. To make these notions precise we shall have to develop systematically the relevant theory of stationary stochastic processes. We remark that a useful tool in this part of the exposition will be the theory of commutative C^*-algebras; a stationary process will be described in an "invariant" manner as a C^*-algebra with certain special properties.

So, with this very original and innovative approach, Furstenberg, who at that time was not aware of the fundamental monograph of Gottschalk and Hedlund "Topological dynamics" [68], recreated the topic of abstract topological dynamics and started an interplay between this new theory and ergodic theory. In fact, the notions of topological ergodicity, generic points, skew-product transformations, and many other basic notions of topological dynamics can already be found in Furstenberg's thesis.

We recall that a *topological dynamical system* is a triple (X,G,ρ), where X is a compact space, G a topological group, and $\rho : G \to \mathrm{Homeo}(X)$ is a continuous group homomorphism from G into the group of self-homeomorphisms of X equipped with the compact open topology. We usually omit the homomorphism ρ from our notation (even when ρ is not one-to-one) and for $g \in G$ and $x \in X$, we write gx instead of $\rho(g)(x)$. The *orbit* of the point $x \in X$ is the set $Gx = \{gx : g \in G\}$ and the system is *minimal* when $\overline{Gx} = X$ for every $x \in X$. It is *point transitive* if there is some dense orbit, and a point whose orbit is dense is called a *transitive point*.

A *homomorphism* between two systems is a continuous map $\pi : X \to Y$ that intertwines the actions: $g\pi(x) = \pi(gx)$, for every $g \in G$ and $x \in X$. When π is surjective one then says that (Y,G) is a *factor* of (X,G), or that (X,G) is an *extension* of (Y,G). Inverse limits in this category are defined as usual.

When the acting group is $G = \mathbb{Z}$ — the group of integers — the system is called a *cascade* and we usually denote it as (X,T), where T is the transformation $\rho(1)$.

A pair of points (x,x') is said to be *proximal* if there is a net $g_i \in G$ and a point $z \in X$ such that $\lim g_i x = \lim g_i x' = z$. The system (X,G) is said to be *distal* when it has no nontrivial proximal pairs; i.e. when $P = \Delta$. Here $P \subset X \times X$ is the *proximal relation* comprising the proximal pairs, and $\Delta = \{(x,x) : x \in X\}$ is the *diagonal relation*. The system (X,G) is *equicontinuous* when the G-action is equicontinuous; i.e. when for every neighborhood V of Δ in $X \times X$ there is another such neighborhood U and $gU \subset V$ for each $g \in G$. This is the case iff the closure of $\pi(G)$ in $\mathrm{Homeo}(X)$ is a compact group. (Also, when the system is metric; i.e. the phase space X is metrizable, the system is equicontinuous iff there is a compatible G-invariant metric on X.) Thus it follows that equicontinuity implies distality.

If $x_0 \in X$ is a transitive point then the map

$$j_{x_0} : C(X) \to \mathrm{BRUC}(G), \quad (j_{x_0}F)(g) = F(gx_0)$$

from the C^*-algebra $C(X)$ into the algebra $\mathrm{BRUC}(G)$ of bounded right uniformly continuous complex-valued functions on G, is an isometric isomorphism, and the Gelfand theory can be used to translate statements about point transitive dynamical systems to the language of actions of G as a group of automorphisms of the corresponding C^*-algebra. This is of course the meaning of the references to commutative C^*-algebras in the above quotations.

In general a dynamical system may not admit any G-invariant probability measure. However, when the group G is amenable such a measure always exist. A system is called *uniquely ergodic* if it admits a unique G-invariant probability measure.

The paper [34] titled 'Strict ergodicity and transformation of the torus' continues this research direction and proves many, by now classical, results regarding the notions of minimality and strict ergodicity in a class of systems called today "Furstenberg systems".

As an example of an important result of this paper consider the following (special case) of [34, Theorem 2.3].

Theorem 1.1. *Let K denote the circle group $\{\zeta \in \mathbb{C} : |\zeta| = 1\}$, and let $X = K \times K$ be the two-torus. Let $T : X \to X$ be given by*

$$T(\omega, \zeta) = (\omega_0 \omega, g(\omega)\zeta),$$

with $\omega_0 \in K$ not a root of unity and $g : K \to K$ a continuous map. Assume further that g is an essential map and satisfies a Lipschitz condition

$$|g(\omega) - g(\omega')| < M|\omega - \omega'|.$$

Then the dynamical system (X, T) is minimal and uniquely ergodic.

Another outstanding and much cited result of this work is the construction of a minimal distal transformation on the 2-torus which is not uniquely ergodic.

At about the same time, Robert Ellis, who completed his Ph.D. thesis in 1958, under the supervision of Walter Gottschalk at the University of Pennsylvania, had raised the following question.

Question 1.2. *Is it true that every minimal distal system is equicontinuous ?*

In [22] he had shown that this is the case when X is zero-dimensional. In the same work he proved the following key results concerning distal systems: (i) Every distal system (X, G) is *semisimple*; i.e. X is the union of its minimal subsystems. (ii) (X, G) is distal iff the product system $(X \times X, G)$ is semisimple. In the following few years his efforts to answer that question led him to develop a beautiful algebraic theory of dynamical systems whose main tool was the functor that associates to every dynamical system (X, G) its *enveloping semigroup* (also often called the *Ellis semigroup*) $E(X, G)$ (see [22, 23, 27] and [24]). The enveloping semigroup, which by Ellis' result is actually a group when (X, G) is distal, is a crucial tool in the proof of Furstenberg's structure theorem (Theorem 1.4 below).

Definition 1.3. *Let (X, G, π) be a dynamical system. Its enveloping semigroup $E(X, G)$ is the closure of the collection $\{\pi(g) : g \in G\}$ in the compact space X^X.*

The set $E = E(X, T)$ is in fact a compact *right topological semigroup*; i.e. a semigroup where for each $p \in E$ right multiplication by p, $R_p : E \to E$, $q \mapsto qp$ $(q \in E)$, is a continuous map (the multiplication here is of course the composition of functions). It was shown in [22] (although the term 'enveloping semigroup' was formulated in the later work [23]) that a system (X, G) is distal iff $E(X, G)$ is a group.

Furstenberg learned about these new developments from his colleagues at Minnesota and Maryland and in 1963 came out with his astounding paper 'The structure of distal flows' [36]. This pioneering paper created a 'new topological dynamics' and, as we shall see, has since had a tremendous impact on the future of both topological dynamics and ergodic theory.

The main result of [36] is as follows:

An extension $\pi(X,G) \to (Y,G)$ is called an *isometric extension* if there exists a continuous function $d: R_\pi \to \mathbb{R}$ such that for every $y \in Y$ the function d restricted to $\pi^{-1}(y) \times \pi^{-1}(y)$ is a metric and for every pair $(x,x') \in R_\pi$, and $g \in G, d(gx,gx') = d(x,x')$. We say that a (metrizable) minimal system (X,G) is *quasi-isometric* if there is a (countable) ordinal η and a family of systems $\{(X_\theta,G)\}_{\theta \le \eta}$ such that (i) X_0 is the trivial system, (ii) for every $\theta < \eta$ there exists an isometric homomorphism $\phi_\theta : X_{\theta+1} \to X_\theta$, (iii) for a limit ordinal $\lambda \le \eta$ the system X_λ is the inverse limit of the systems $\{X_\theta\}_{\theta<\lambda}$, $(X_\lambda, G) = \underleftarrow{\lim}(X_\theta, G)$, and (iv) $(X_\eta, G) = (X,G)$. We call such a directed set of systems an *I-tower*.

Theorem 1.4 (Furstenberg's distal structure theorem). *A minimal metric system is distal iff it is quasi-isometric. Moreover, when (X,G) is minimal and distal it always admits a canonical I-tower where each extension is maximal.*

The *distal rank* of a system (X,G) is the ordinal η associated to the canonical I-tower. (In [7] Beleznay and Foreman show that, at least for $\mathbb{Z}$-systems, for every countable ordinal η there in fact exists a minimal distal system (X,T) whose distal rank is exactly η.)

Note that the system (X_1,G) is, by definition, equicontinuous and must be nontrivial when (X,G) is nontrivial. Thus, as a corollary we see that every nontrivial metric minimal and distal system admits a nontrivial equicontinuous factor.

In the introduction to his paper Furstenberg presents a simple example (originally studied by Anzai in his paper [2], which was unknown to Furstenberg at that time) of a minimal distal system which is not equicontinuous, thus providing a negative answer to Ellis' problem:

On $X = \mathbb{T}^2$, let $T(x,y) = (x+\alpha, y+x)$ for an irrational α. The system (X,T) is minimal distal but not equicontinuous. The map $\pi(x,y) = x$ defines a factor map onto the maximal equicontinuous factor of (X,T) which is a circle (hence an isometric) extension. [The distality follows easily by considering first pairs of points of the form $((x,y),(x,y'))$ and then pairs of the form $((x,y),(x',y'))$ with $x \ne x'$. For minimality one shows that if M is a *proper* minimal subset of $X \times X$ then $H = \{\beta \in \mathbb{T} : R_\beta M \cap M \ne \varnothing\}$ is the finite subgroup $H = \{0, 1/n, 2/n, \dots, (n-1)/n\}$ for some positive integer n, where $R_\beta(x,y) = (x, y+\beta)$. It follows that the set $\{(x,ny) : (x,y) \in M\}$ is a graph of a continuous function $f : \mathbb{T} \to \mathbb{T}$; then one uses the invariance of M to get $f(x+\alpha) - f(x) = nx$, which is impossible for $n \ge 1$. Finally for non-equicontinuity, one shows that for a suitable sequence $n_i \to \infty$ with $T^{n_i}((0,0)) \to (0,0)$, we have $T^{n_i}((0,0),(1/(2n_i),0)) \to ((0,0),(0,1/2))$.]

In the same year other counterexamples, in the class of minimal nil-systems of rank > 1, were shown in [3] to be minimal distal flows which are not equicontinuous. (Among the Furstenberg systems, studied in [34], there are also counterexamples to Ellis' problem, but this aspect was not addressed there.)

When eventually Robert Ellis learned about these counterexamples he said, half jokingly, that he does not want to believe them, as the search for a positive answer to his question was at the basis of many of his previous achievements.

Another important corollary of the structure theorem is the following:

Theorem 1.5. *Every metric distal G-system (for an arbitrary group G) admits a G-invariant probability measure. If in addition the system is minimal there is a canonically defined such measure.*

As was shown in [34] this measure need not always be ergodic.

Finally, towards the end of the paper Furstenberg provides an alternative proof — á la von Neumann, using the existence of the invariant measure and applying a Hilbert–Schmidt kernel operator — of the fact that, at least for an abelian acting group G, a minimal distal system possesses a non-constant continuous eigenfunction. This idea was later used by Keynes and Robertson [75] in their proof of the following result, motivated by the more famous analogous result in ergodic theory.

Theorem 1.6. *A minimal metric system (X,G) with abelian G is not topologically weakly mixing iff the system (X,G) possesses a non-constant continuous eigenfunction.*

In [49] this method was developed further, building a theory of bundles of Hilbert spaces and Hilbert Schmidt operators, a tool which was then used to give an alternative full proof of the structure theorem.

In 1963-4 Furstenberg was again in Princeton, just in time to communicate his new ideas on topological dynamics to William Veech, another Ph.D. student of Bochner. Bochner suggested to Veech to study the subject of *almost automorphic functions*. After struggling for some time with this subject, using traditional tools of harmonic analysis, Veech indeed applied the approach suggested by Furstenberg and was able to achieve in this way a beautiful and complete analysis of this topic [95] by showing that these functions are exactly those which arise from *almost automorphic* cascades, that is, those minimal cascades which are almost one-to-one extensions of their maximal equicontinuous factor.

In 1970 Veech published his work on the structure of point distal systems [96] (complemented by Ellis in [25]) which generalized Furstenberg's theorem. A minimal system (X,G) is called *point distal* if there is a point $x \in X$ such that the only point $y \in X$ proximal to x is x itself ($P[x] = \{x\}$). Every almost automorphic system is point distal.

The ultimate structure theorem for minimal dynamical systems was then developed by stages in a series of works: [26], [97], [81] and [66].

In his work on Szemerédi's theorem [44], the main tool used in the ergodic proof of this theorem is a structure theorem for ergodic measure-preserving dynamical systems. The latter was, without doubt inspired by the structure theorem for distal systems. Independently of Furstenberg this structure theorem for ergodic measure-preserving systems was obtained by Robert Zimmer [100], [101].

In the topological part of his 1967 work 'Disjointness in ergodic theory, minimal sets, and a problem in Diophantine approximation' [38], Furstenberg extends his new ideas in topological dynamics beyond the class of minimal systems. This seminal work has a wealth of innovative concepts, surprising new results, as well as new examples and counterexamples. Joinings, disjointness, recurrence sets which form 'Furstenberg families' (see e.g. [1]), and many more new concepts, help the author to create a broad picture of topological dynamics and a road map for the study of topologically transitive dynamical systems.

Furstenberg's famous (still open) question, whether Lebesgue's measure is the only continuous probability measure on the unit circle $\mathbb{R}/\mathbb{Z}$ that is invariant under multiplication by both 2 and 3, can be found in this work, following a proof, based on a disjointness argument, that the semigroup generated by the multiplications by 2 and 3 acts almost minimally on the circle, in the sense that every orbit is either finite or dense.

We end this brief review of Furstenberg's contribution to topological dynamics with a short list of some of Furstenberg's more prominent papers, written on this subject after the disjointness paper [38].

In [58] the authors construct the first example of a prime minimal cascade. The work [43] establishes the unique ergodicity of the horocycle flow. In the very influential work [60] Furstenberg and Weiss prove several new theorems in topological dynamics and derive from them results in combinatorial number theory. These include van der Waerden's theorem, Rado's theorem on regular systems of equations, Hindman's finite sums theorem and more. Again the main tool here is the construction of the topological dynamical system generated by a bounded sequence.

In [62] the authors show that a broad class of extensions of measure-preserving systems, in the context of ergodic theory, can be realized by topological models for which the extension is "almost one-to-one". In particular they construct a minimal almost automorphic action of the free group that has no invariant measure, thus answering an old question of Veech.

2 Stationary dynamical systems and the Poisson–Furstenberg boundary

In this most fruitful year 1963, Furstenberg published another path-breaking paper: 'A Poisson formula for semi-simple Lie groups', [35]. In his autobiography Furstenberg describes the "circuitous route" that led him to this work.

> At Minnesota I also returned to the study of noncommuting random products; this time focusing on the qualitative rather than the quantitative behavior. The motivation actually came from the dynamical questions I had been studying. The foregoing distal transformation is an example of a "skew product transformation": $T(x,y) = (S(x), R(x)(y))$, where S is a fixed transformation, and $R(x)$ is a transformation on the y-coordinate depending on x. When we

iterate this transformation we are led to product transformations of the form

$$R(S^{n-1}x)\cdots R(S^2x)R(Sx)R(x).$$

Assuming the transformations $R(x)$ come from a given group of transformations, we can hope to analyse this product in terms of properties of the group. The complexity of the expression leads one to consider the case where the successive $R(S^n x)$ are independent random variables. By this somewhat circuitous route, we came to study random walks on groups, and the associated "Poisson boundaries".

In order to describe the main results of this seminal work we need to introduce some background and basic definitions as follows.

The notion of a distal action was considered by Hilbert (see [102]) in an attempt to give a topological characterization of the concept of a rigid group of motions. (The names 'distal' and 'proximal' were introduced by Gottschalk in his 1956 paper [67] and are also mentioned in [68].) At the other, chaotic side of the landscape of dynamical systems we find notions like weak mixing and proximality. Weak mixing is prevalent in actions of non-compact groups, but a minimal proximal cascade ($\mathbb{Z}$ system) must be a point. More generally, the class of *strongly amenable* groups comprises those groups for which every minimal proximal system is necessarily trivial (see [64]; this class includes all the nilpotent groups and, for countable discrete groups, was recently characterized as the class of groups with only trivial ICC quotients, [30]). A topological group G is *amenable* iff every G-dynamical system admits at least one G-invariant probability measure, iff every minimal strongly proximal G-system is necessarily trivial. See [35] and [63], where the term 'strong proximality' was coined and systematically studied.

Definition 2.1. *A dynamical system* (X,G) *is said to be a* boundary *if it is minimal and* strongly proximal, *i.e. for every probability measure* ν *on* X*, there is a net* $g_i \in G$*, such that* $\lim g_i\nu$ *is a point mass.*

It is not hard to see that every topological group G has a uniquely defined universal boundary $B(G)$; i.e. the system $(B(G),G)$ is a boundary and if (X,G) is a boundary then there is a unique factor map $\pi : (B(G),G) \to (X,G)$.

A dynamical system (Q,G) is called *affine* when the compact set Q is a closed and convex subset of a locally convex linear topological space and G acts on Q by affine homeomorphisms. With every system (X,G) there always is associated the affine system on the space $P(X)$ of probability measures on X. We say that the affine system (Q,G) is *irreducible* if it contains no proper closed, convex and invariant subset. Furstenberg shows that an affine system (Q,G) is irreducible iff $X = \overline{\mathrm{ext}\, Q}$, the closure of the set of extreme points, is a boundary. It then follows that the affine system $P(M(G))$ is the universal irreducible affine flow (see e.g. [64]).

Invariant measures, when they exist, are very useful tools when one studies dynamical properties of a dynamical system. However, it turns out that there are similar objects, called G-stationary measures, that are always available in every dynamical system, and which can for many purposes replace the invariant measures.

Let G be an arbitrary topological group and let (X,G) be a dynamical system. Let μ be a fixed probability measure on G. A measure ν on X such that

$$\mu * \nu = \int g_* \nu \, \mathrm{d}\mu(g) = \nu$$

is called a *μ-stationary measure*. Here $g_*\nu$ is the push-forward of the measure ν under the homeomorphism of X that is given by $g \in G$. The convolutional powers of μ are defined by

$$\mu^n = \mu * \mu * \cdots * \mu \quad (n \text{ times}).$$

To justify the assertion that every system admits a μ-stationary measure, given an arbitrary (metric) G-system (X,G) and a point $x \in X$, we form the sequence

$$\nu_n = \frac{1}{N} \sum_{n=1}^{N} \mu^n * \delta_x$$

and then observe that any limit point ν of this sequence is μ-stationary.

Let now G be a connected semisimple Lie group with finite center. Let $G = KAN$ be an Iwasawa decomposition of G; so that K is a maximal compact subgroup of G, A closed and abelian, N closed and nilpotent, and the map $K \times A \times N \to G$, $(k,a,n) \mapsto kan$ is a surjective analytic diffeomorphism. Let P be the normalizer of N in G, and let M be the centralizer of A in K. It then follows that $P = MAN$ so that $G/P \cong K/M$. The group P and its conjugates are the *minimal parabolic subgroups* of G. Let $B = G/P$. The action of G on the homogeneous space B is strongly proximal, so that (B,G) is a boundary. As we shall see (Theorem 2.2 below) $B = G/P$ is in fact the universal boundary for G. Moreover, all the G-boundaries arise as G/Q where Q ranges over the parabolic subgroups of G.

For the group $G = \mathrm{SL}(n,\mathbb{R})$ it is easy to check that $B = G/P$ is isomorphic to the *flag manifold*:

$$\mathfrak{F}_n = \{(V_1,V_2,\ldots,V_{n-1}) : V_1 \subset V_2 \subset \cdots \subset V_{n-1} \text{ are subspaces of } \mathbb{R}^n \text{ and } \dim V_i = i\}.$$

Let $D = G/K$ be the *symmetric space* of G. The compact group K acts transitively on B and so there is a unique probability measure m on B invariant under K. Furstenberg shows that the only elements of G preserving m are those of K, so that the correspondence $gK \mapsto gm$ defines a one-one map from D onto Gm. This representation yields a compactification of D by imbedding Gm in the compact space of probability measures on B. Let us denote by $\overline{D}$ the closure of D regarded as a set of measures. The significance of B being a boundary, in the usual sense of the word, lies in the fact that the closure of D includes the point measures of B. Thus B itself may be imbedded into $\overline{D}$ as a portion of the boundary. When this is done, the Poisson formula obtained for harmonic functions on D in terms of functions on B may be interpreted as the solution to a boundary value problem, the space B being seen as part of the boundary of D.

Let G be any locally compact, second countable, unimodular topological group with Haar measure λ (e.g. a non-commutative free group, or a non-compact semi-simple Lie group), and suppose that μ is an absolutely continuous probability measure on G. A bounded measurable function $f \in L^\infty(G,\lambda)$ is said to be *μ-harmonic* when it satisfies the functional equation

$$(f*\mu)(g) = f(g) = \int f(gg')\,\mathrm{d}\mu(g').$$

For G a semisimple group with finite center, we say that a bounded, real-valued, measurable function f on G is *harmonic* when it is μ-harmonic for some absolutely continuous, K-invariant probability measure μ on G. It turns out that this definition does not depend on the particular choice of μ. A function $f(p)$ on D is *harmonic* if the function $\tilde{f}(g) = f(gK)$ is a harmonic function on G.

We can now state two of the main results obtained in [35] (see also [82] and [83]).

Theorem 2.2. *The universal boundary $(B(G),G)$ is isomorphic to the homogenous system $(B,G) = (G/P,G)$. Moreover, for a closed subgroup H of G, the homogeneous space G/H is a boundary iff H contains a conjugate of P.*

Theorem 2.3. *Let m denote the unique K invariant probability measure on the maximal boundary $B(G)$ of G. If $f(g)$ is a bounded harmonic function on G, then there exists a bounded function $\widehat{f}$ on $B(G)$ with*

$$f(g) = \int_{B(G)} \widehat{f}(g\omega)\,\mathrm{d}m(\omega) = \int_{B(G)} \widehat{f}(\omega)\frac{\mathrm{d}gm(\omega)}{\mathrm{d}m(\omega)}\,\mathrm{d}m(\omega). \tag{1}$$

Furthermore, if $\widehat{f}(\omega)$ is any bounded measurable function on $B(G)$, (1) defines a harmonic function on G.

The classical Poisson integral formula for a harmonic function h on the unit disc $\mathbb{D}$, with boundary the unit circle $\mathbb{T}$,

$$h(re^{i\theta}) = \frac{1}{2\pi}\int_0^{2\pi} \widehat{h}(\mathrm{e}^{i\phi})\frac{1-r^2}{1-2r\cos(\theta-\phi)+r^2}\,\mathrm{d}\phi,$$

is of course the special case where $G = \mathrm{SL}(2,\mathbb{R})$, $\mathbb{D} \cong D$ is the unit disk, and $m = \mathrm{d}\phi$ is Lebesgue measure on the circle $\mathbb{T} \cong B(G)$.

In the last chapter (Chapter V) of [35], Furstenberg shows how to construct a 'Poisson boundary' Π_μ for an arbitrary absolutely continuous probability measure μ on a semisimple Lie group G. The way Π_μ is defined is by considering the Banach space $\mathcal{H}$ of bounded left uniformly continuous μ-harmonic functions on G and defining, by means of the μ random walk on G, a natural multiplication on $\mathcal{H}$ which makes it a commutative C^*-algebra. Then, the space Π_μ is defined as the Gelfand space of this C^*-algebra. This construction is very general and may be applied for any locally compact group.

In several subsequent works, like [40, 41, 42], Furstenberg develops this general theory of Poisson spaces and obtains a variety of applications to random walks, harmonic functions on general groups and much more. Another characterization, more probabilistic and very useful, of these Poisson boundaries was introduced by Kaimanovich and Vershik in [77].

Again, the paper [35] was very influential. Many important works followed, and were based on ideas and techniques that were developed in it. Notably, let us mention the works of Rosenblatt [88], Kaimanovich–Vershik [77], Nevo–Zimmer [84, 85], Bader–Shalom [4] and Eskin–Mirzakhani [28]. Recently, new surprising connections were found between Furstenberg boundaries and the theory of C^*-algebras. For example, in [74] the authors show that a discrete group G is C^*-simple if and only if the G-action on the Furstenberg boundary is topologically free.

Furstenberg returned to the theory of stationary dynamics in his paper [46] titled 'Stiffness of group actions'. In it he introduced the notion of *stiffness*. If ν is a probability measure on a group G then an action of G on a space X is ν-*stiff* if every ν-stationary measure on X is invariant. Furstenberg showed that for carefully chosen ν on $\mathrm{SL}(d,\mathbb{Z})$, namely probability measures ν so that the corresponding stationary measure on the boundary of $\mathrm{SL}(d,\mathbb{Z})$ is absolutely continuous with respect to Lebesgue, the action of $\mathrm{SL}(d,\mathbb{Z})$ on $\mathbb{T}^d$ is ν-stiff. He conjectured that this should be true for any measure whose support generates $\mathrm{SL}(d,\mathbb{Z})$. This intuition was more than confirmed about ten years later in [19, 20], and further greatly generalized in [18]. In [50] Furstenberg and Glasner embark on a general study of stationary dynamical systems. They develop a general theory of factors, extensions and conditional measures. They then prove a structure theorem and use it to establish a theorem of Szemerédi type for the group $\mathrm{SL}(2,\mathbb{R})$. In a further paper [51] they use the structure theorem to prove a version of multiple recurrence for sets of positive measure in a general stationary dynamical system. In [52] the authors study affinely prime dynamical systems and show that, when the group in question is $G = \mathrm{PSL}(2,\mathbb{R})$, there is a unique – up to equivalence – irreducible affine representation for G which is realized in the regular representation of G on $L^\infty(G)$, where in fact the irreducible closed convex subspaces are in one-one correspondence with bounded harmonic functions on the upper half-plane.

3 Probability, ergodic theory and fractal geometry

I shall begin with a brief description of Hillel's first major work, which was an expanded version of his thesis, published in 1960 in the Princeton Annals of Mathematical Studies under the title "Stationary processes and prediction theory".

In classical prediction theory we are given a stationary stochastic process $\{X_n\}$ on a probability space $(\Omega, \mathcal{B}, P)$, and the best prediction of X_0 given the past $\{X_n \mid n < 0\}$ is taken to be the conditional expectation of X_0 with respect to this "past" σ-algebra. By its very nature this is a function which is defined only up to a set of probability zero. This means that for a specific sequence of past observations there is no well-defined prediction. Hillel's goal was to develop a theory of prediction for individual time series. Of course this can only make sense if the individual time series has some statistical regularity. Now the pointwise ergodic theorem of Birkhoff implies that for a stationary process almost every sample sequence possesses just this kind of regularity. However in order for the process to be "continuously predictable" or more generally "statistically predictable" further restrictions are necessary, and the class of processes for which his theory is developed is restricted, although it does contain m-state Markov processes and finite functions of them.

The focus on individual sample paths led to a host of new ideas which Hillel elaborated on in his later work. Because of his interest in individual points uniquely ergodic systems come to the fore. A topological dynamical system (X, T) consists of a homeomorphism T of a compact space X. It is a classical theorem that there always exists at least one T-invariant probability measure. This follows easily from the weak* compactness of the space of probability measures by taking some limit point of a set $\{\nu_n\}_{n=1}^{\infty}$

$$\nu_n = \frac{1}{n} \sum_{i=1}^{n} T^i \nu,$$

where ν is any probability measure on X.

If there is a unique invariant measure μ then it must be ergodic and furthermore for all continuous functions f and all points $x \in X$ we have

$$\frac{1}{n} \sum_{i=1}^{n} f(T^n x) \to \int f(x) \, d\mu(x).$$

In fact the convergence is even uniform. This means that the pointwise ergodic theorem is valid everywhere for all continuous functions. The basic example of such uniquely ergodic systems are irrational rotations of the circle. Hillel gave in his thesis a criterion, which can be verified in many cases, for when a compact group extension of a uniquely ergodic system is uniquely ergodic. A further paper [34] studied these skew products on tori in great detail and included examples of minimal systems that are distal but not uniquely ergodic.

A very basic contribution of Hillel to modern probability theory was [57] (joint with Harry Kesten) which analyzed the limiting behavior of the products of a sequence of independent, identically distributed random matrices. A law of large numbers was established in this non-commutative setting. This paper laid the foundations for the modern study of products of random matrices which has turned out to have many applications in mathematics, physics and computer science. One of the main results here was the following:

If the expectation of $\log^+ \|X_i\|$ is finite, then

$$\lim_{n\to\infty} \|X_1 X_2 \cdots X_n\|^{1/n}$$

exists with probability one. If the variables $\{X_i\}$ are independent with common distribution μ, the limit is a constant depending only on μ, say $\beta(\mu)$. In a later joint paper with Yuri Kifer [59] he returned to this theme and expressed $\beta(\mu)$ in terms of μ and some auxiliary measures on the $(m-1)$-dimensional projective space. With these auxiliary measures they were also able to study the asymptotic behavior of the vector norms $\|X_1 X_2 \cdots X_n v\|$ for $v \in \mathbb{R}^m$.

Fifty years ago probability theory was on the sidelines in what was considered to be the central mathematical theories. Hillel was one of the pioneers in applying probabilistic methods to classical topics in group theory. In 1971 he published "Random walks and discrete subgroups of Lie groups" [40], in which he gave a new application of probability theory to group theory. A discrete subgroup Γ of a non-compact connected Lie group G is called a **lattice** if the quotient space G/Γ has finite measure. This means that, there is a subset $D \subset G$ with finite left-invariant Haar measure and the translates $D\gamma$, $\gamma \in \Gamma$, cover G. Hillel's main result was that a lattice subgroup of $\mathrm{SL}(d,\mathbf{R})$ with $d \geq 3$ cannot be isomorphic to a subgroup of $\mathrm{SL}(2,\mathbf{R})$ (discrete or not). His proof goes via the study of the maximal boundaries and the Poisson boundaries. He constructs a random walk on a lattice subgroup Γ of $\mathrm{SL}(d,\mathbf{R})$ whose Poisson boundary coincides with the maximal boundary of $\mathrm{SL}(d,\mathbf{R})$. This work is related to the well-known Mostow rigidity and the methods of Hillel influenced the work of G. Margulis especially in his proof of the Normal Subgroup Theorem.

Turning to ergodic theory, one of Hillel's major contributions was his great paper titled "Disjointness in ergodic theory, minimal sets, and a problem in Diophantine approximation" [38]. We can do no better in describing the impact of this work than by quoting from the review by Bill Parry (who was at that time one of the leading figures in ergodic theory and topological dynamics):

> The approach to ergodic theory in this remarkable paper is complementary to the one developed, mainly by the Russian school, associated with numerical and group invariants. In fact, the relationship investigated here between two measure-preserving transformations (processes) and between two continuous maps (flows) is disjointness, an extreme form of non-isomorphism. The concept seems rich enough to warrant quite a few papers, and these papers will no doubt be largely stimulated by the present one. An interesting aspect of the paper, apart from the new results it contains, is the entirely novel demonstration of a number of established theorems.

Needless to say his prediction has been more than fulfilled. In this paper Hillel inaugurated the study of **joinings** of measure-preserving systems. A joining of two systems $(X,\mathcal{B},\mu,T),(Y,\mathcal{C},\nu,S)$ is a measure λ on the product space $X \times Y$ which projects onto μ and ν and is invariant under $T \times S$. At least one such joining always exists, namely the product measure $\mu \times \nu$. If this is the only joining then we say that the systems are disjoint. Hillel used this concept to prove results in filtering noisy signals, and unify the structural theory of various classes of measure-preserving systems. Here are some examples of his results. The class of weakly mixing systems

$\mathcal{W}$ and ergodic pure point spectrum systems, called Kronecker systems, $\mathcal{C}$ are mutually disjoint. The same mutual disjointness holds for the classes of Kolmogorov automorphisms $\mathcal{K}$ and zero entropy systems $\mathcal{Z}$. This notion of joinings was widely used, for example Dan Rudolph gave an ingenious new joining proof in [90] of Bourgain's return time theorem. This was a surprising theorem which showed that the return times of a typical point to a set of positive measure was a good sequence for the Birkhoff theorem to be valid for all measure-preserving systems. Eli Glasner developed much of ergodic theory using this concept in his influential book [65].

In another part of this paper he also introduces disjointness in topological dynamics, with a similar definition. He used this in a remarkable application of these new ideas to Diophantine approximation which I will now describe. Among the equidistribution theorems that H. Weyl proved in his famous 1916 paper [98] is the result that for any increasing sequence of integers a_n the fractional parts of $a_n x$ are equidistributed for a.e. real number x. For $a_n = n$ the only exceptions are rational numbers. However if a_n is a lacunary sequence, i.e. the ratios a_{n+1}/a_n are bounded away from one, then there are uncountably many exceptions. While the semigroup p^n is lacunary, the semi-group generated by co-prime integers p, q, i.e. $\{p^n q^m\}$ is easily seen to be non-lacunary. Hillel showed that for any non-lacunary semi-group Σ of integers and every irrational θ the fractional parts of $\{\sigma\theta : \sigma \in \Sigma\}$ are dense in $[0,1]$. Another way to formulate this is that the only closed invariant sets for the action of Σ on the unit circle $\mathbf{S} = \{z \in \mathbb{C} : |z| = 1\}$ by $z \to z^\sigma$ are finite sets of roots of unity and $\mathbf{S}$ itself. In contrast, for the action on $\mathbf{S}$ which is generated by a single integer p, representing a lacunary semigroup, there is a host of distinct invariant closed sets. These invariant sets can have arbitrary Hausdorff dimension between zero and one. This was the first "rigidity result" that exhibited the dramatic difference between the action of a single transformation and an action by several commuting transformations.

In the setting of ergodic theory Hillel suggested that an analogous result should be true. Namely that the only purely non-atomic measure on $\mathbf{S}$ invariant under such a semigroup Σ is Lebesgue measure. This stubborn problem remains open until today. The best partial results were obtained more than 30 years ago in works by R. Lyons [79], D. Rudolph [89] and A. Johnson [73]. Without entering into the detailed story, the final result they achieved was that if the semigroup leaves a measure μ invariant, and μ has positive entropy under the action of some element of the semigroup then μ equals Lebesgue measure.

In most of the more recent works on rigidity of higher rank actions of algebraic origin there is a similar important role played by the entropy. It is these results in homogeneous dynamics that have had many applications to number theory and Diophantine approximation not to mention quantum unique ergodicity.

Hillel realized that the Diophantine result suggests some sort of transversality between the actions of multiplication by p and multiplication by q. He explored this in a lecture in 1969 at a symposium honoring his mentor S. Bochner [39] where he again developed some new ideas and formulated several conjectures which have stimulated many further developments. The study of the Hausdorff dimension of sets of the type that Hillel was considering in this paper was quite a narrow field

50 years ago. The work of B. Mandelbrot on fractals popularized the field and was one of the stimuli to many attempts to settle these conjectures. Quite recently one of these conjectures was established independently by P. Shmerkin [92] and M. Wu [99] and I shall describe this in some detail as a nice illustration of the depth of Hillel's insights.

Let dim denote the Hausdorff dimension function for subsets of (X,d), a compact metric space. Two closed subsets A,B of X are defined by Hillel to be **transverse** if

$$\dim(A\cap B) = \max\{0, \dim A + \dim B - \dim X\}.$$

Two continuous mappings T,S from X to X are said to be **transverse** if for all closed sets A and B that are T and S invariant, respectively the sets A,B are transverse.

The conjecture in question is:

Conjecture: (1970) *Two positive integers a,b are said to be **multiplicatively independent**, if the ratio of their logarithms is irrational. For an integer m denote by T_m the map of the unit interval to itself defined by $T_m(x) = mx \pmod 1$. The mappings T_a, T_b are transverse for all pairs of multiplicatively independent integers a,b.*

The two proofs of this conjecture by Shmerkin and Wu are quite different. Wu's proof makes use of one of the novel tools developed by Hillel in [39] to obtain some partial results towards his conjecture. Hillel introduced there spaces of measures on trees and Markov processes on these spaces. These ideas were later elaborated and given a geometric form as "CP-processes". Rather than giving a detailed definition of these processes I will quote Hillel's abstract to his 2008 paper titled "Ergodic fractal measures and dimension conservation" [47]:

> A linear map from one Euclidean space to another may map a compact set bijectively to a set of smaller Hausdorff dimension. For 'homogeneous' fractals (to be defined), there is a phenomenon of 'dimension conservation'. In proving this we shall introduce dynamical systems whose states represent compactly supported measures in which progression in time corresponds to progressively increasing magnification. Application of the ergodic theorem will show that, generically, dimension conservation is valid. This 'almost everywhere' result implies a non-probabilistic statement for homogeneous fractals.

Hillel gave a wonderful series of lectures at Kent State University on his ideas in fractal geometry which appeared in [48]. In it he develops a theory of **mini-sets** and **micro-sets** of a closed subset A of $\mathbf{R}^d$. A mini-set of A is just the intersection of A with a small square that is re-scaled to be of unit size, while the micro-sets of A are the limits in the Hausdorff metric of mini-sets of A. There is a new notion of dimension called the star-dimension and ergodic theory is used to show that the star-dimension of a set A is the maximal Hausdorff dimension of a micro-set of A. There are further results connected to preservation of dimension for homogeneous fractals and connections with ergodic theory.

Another fundamental result of Hillel in ergodic theory was developed in his ergodic-theoretic proof of Szemerédi's theorem on arithmetic progressions [44]. At the end of 1975 Konrad Jacobs gave a colloquium talk in Jerusalem on this newly proved theorem. Following this talk Hillel realized that his theorem asserting that weak mixing implies multiple weak mixing proves an ergodic theoretic version of Szemerédi's theorem for weakly mixing systems, while for the disjoint class of Kronecker systems the theorem is quite easy. In order to combine these two cases he developed a basic structure theorem for all ergodic systems. The first part is a measure-theoretic analogue of distality. A **measure distal** system is one which can be represented as a tower of isometric extensions beginning with a Kronecker system. This is modelled on his structure theorem for distal transformations in the setting of topological dynamics. Next he formulated a relativized version of weakly mixing. A measure-preserving system $(X, \mathcal{B}, \mu, T)$ is an **extension** of a measure-preserving system $(Y, \mathcal{C}, \nu, S)$ if there is a measurable mapping $\pi : X \to Y$ mapping μ to ν and satisfying $\pi T = S\pi$. The extension is **relatively weakly mixing** if the relative product of X with itself over Y is ergodic. The exact definition of relative product is somewhat technical and I will omit it. Hillel's[2] structure theorem can then be formulated as follows.

Theorem 3.1. *An arbitrary ergodic system is a relatively weakly mixing extension of a measure distal system.*

In 1978 Hillel delivered the first of the Porter Lectures at Rice University. Based on these lectures he wrote a monograph "Recurrence in Ergodic Theory and Combinatorial Number Theory" [45], which is a masterpiece of exposition and remains, until today, the best introduction to the field. The published work of Hillel resembles the tip of an iceberg in the following sense. Generations of students and colleagues have benefited from his ideas spanning a range of mathematics going far beyond what I have touched on in the above. I will illustrate this with an example coming from the study of topological Markov chains. Bill Parry published the first result on finite equivalence of topological Markov chains having the same entropy [86]. As he says at the end of his introduction: "Furstenberg's Lemma 2 is crucial to the proof of our main theorem." This lemma is simple enough to state:

Lemma. *If A and A' are irreducible non-negative integral matrices with the same maximum eigenvalue λ then $UA = A'U$ for some strictly positive integral matrix U.*

I should repeat that this is but one example drawn from a multitude of others. Finally I will end on a personal note of thanks to Hillel for all that I have learned from him over the years, beginning with his course on probability theory (1963–64) which I attended as a graduate student in Princeton. It was in this course that Hillel first introduced his notion of disjointness, under the name **absolute independence**, and used it to obtain a new result in the theory of filtering of noisy signals.

[2] A similar result was obtained independently at the same time by R. Zimmer [101].

4 Multiple recurrence and applications to combinatorics and number theory

The classical van der Waerden's theorem [94] states that for any finite partition of the integers, $\mathbb{N} = \bigcup_{i=1}^{r} C_i$, one of the sets C_i contains arithmetic progression of arbitrary length. Given a set $E \subset \mathbb{N}$, define its *upper density* $\bar{d}(A)$ by

$$\bar{d}(E) = \limsup_{n \to \infty} \frac{|E \cap \{1, 2, \ldots, N\}|}{N}.$$

Having positive upper density is a natural notion of largeness and it is natural to ask (as P. Erdős and P. Turán did in [29]) whether this notion of largeness is responsible for the validity of van der Waerden's theorem. In 1975 Szemerédi [91] showed that this is indeed so: any set $E \subset \mathbb{N}$ with $\bar{d}(E) > 0$ contains arbitrarily long arithmetic progressions.

In his groundbreaking paper [44], Furstenberg introduced a startling ergodic approach to Szemerédi's theorem. Here is the formulation of Furstenberg's "ergodic Szemerédi theorem" (EST), a far-reaching extension of the classical Poincaré recurrence theorem (which corresponds to the case $k = 1$).

Theorem 4.1 ([44, Theorem 1.4]). *Let $(X, \mathcal{B}, \mu, T)$ be a measure-preserving system and $B \in \mathcal{B}$ with $\mu(B) > 0$. For any $k \geq 1$ there exists an $n \neq 0$ such that*

$$\mu(B \cap T^{-n} B \cap \cdots \cap T^{-kn} B) > 0.$$

It is not hard to see that Szemerédi's theorem can be formulated in the following equivalent (but ostensibly stronger) form.

Theorem 4.2. *If $E \subset \mathbb{N}$ and $\bar{d}(E) > 0$, then for every $n \geq 1$ there is an $n \neq 0$ such that*

$$\bar{d}(E \cap (E - n) \cap \cdots \cap (E - kn)) > 0. \tag{2}$$

While $\bar{d}$ is certainly not a measure on the power set $\mathcal{P}(\mathbb{N})$, it has the translation invariance property: for any $n \in \mathbb{Z}$, $\bar{d}(E - n) = \bar{d}(E)$. It is tempting to interpret formula (2) as a multiple recurrence theorem for the make-believe "combinatorial measure-preserving system" $(\mathbb{N}, \mathcal{P}(\mathbb{N}), \bar{d}, x \mapsto x + 1)$. Furstenberg's correspondence principle, with the help of which he derived Szemerédi's theorem from EST (Theorem 4.1 above), gives a justification for this kind of intuition.

Theorem 4.3 (Furstenberg's correspondence principle). *Let $E \subset \mathbb{Z}$ be such that for some sequence of intervals (I_N) with increasing length,*

$$\bar{d}_{(I_N)}(E) := \limsup_{N \to \infty} \frac{|E \cap I_N|}{|I_N|} > 0.$$

Then there is an invertible measure-preserving system $(X,\mathcal{B},\mu,T)$ *and a set* $A\in\mathcal{B}$ *with* $\mu(A)=\bar{d}_{(I_N)}$, *such that for every* $k\in\mathbb{N}$ *and any* $n_1,n_2,\dots,n_k\in\mathbb{Z}$ *one has*

$$\bar{d}_{(I_N)}(E\cap(E-n)\cap\cdots\cap(E-n_k))\geq\mu(A\cap T^{-n_1}A\cap\cdots\cap T^{-n_k}A).$$

Furstenberg's correspondence principle has a variety of useful variants. For example, if one replaces $\bar{d}_{(I_N)}$ by

$$d^*(E):=\limsup_{N-M\to\infty}\frac{|E\cap\{M,\dots,N-1\}|}{N-M},$$

the measure-preserving system $(X,\mathcal{B},\mu,T)$ which appears in Theorem 4.3 can be guaranteed to be ergodic, a fact which leads to interesting applications (see for example, [10], [6], and [9]).

Furstenberg's correspondence principle was not born with his ergodic-theoretic proof of Szemerédi's theorem. Indeed, a form of it appears already in his thesis [32], where it was used as a tool to reconstruct a stationary process from its past. More concretely, the seminal idea utilised in [32] was to replace the approximate system $(\mathbb{Z},\mathcal{P}(\mathbb{Z}),\bar{d},x\mapsto x+1)$ by a genuine measure-preserving system, namely the orbit closure of the sequence $(\mathbf{1}_E(n))_{n\in\mathbb{Z}}$ in $\{0,1\}^{\mathbb{Z}}$, where $\mathbf{1}_E$ corresponds to the given time series.

The versatility of Furstenberg's correspondence principle and its algebraic nature can be best perceived via Gelfand's representation theorem. We recall that the role of commutative C^*-algebras as a useful tool was stressed in [32]; the possibility of utilizing Gelfand's representation theorem was explicitly mentioned in [44].

By Gelfand's theorem any commutative, unital, countably generated C^*-algebra $\mathcal{A}$ is topologically and algebraically isomorphic to the algebra of continuous complex-valued functions $C(X)$, where X is a compact metric space. In our situation such a C^*-algebra is a subalgebra of $\ell^\infty(\mathbb{Z})$ which is naturally generated by the family $(\mathbf{1}_{E-n})_{n\in\mathbb{Z}}$, where E satisfies $d_{(I_N)}(E):=\lim_{N\to\infty}\frac{|E\cap I_N|}{|I_N|}$ for some sequence of intervals (I_N) with $|I_N|\to\infty$. Let us denote this C^*-algebra by $\mathcal{A}_E$. Let $\mathcal{B}_E$ be the boolean algebra generated by the family $(E-n)_{n\in\mathbb{Z}}$. Refining, if needed, the sequence of intervals (I_N), we get a subsequence (I_{N_i}) with the property that for any $F\in\mathcal{B}_E$, $d_{(I_{N_i})}(F)$ is well defined (meaning that $\lim_{i\to\infty}\frac{|E\cap I_{N_i}|}{|I_{N_i}|}$ exists).

The shift-invariant density $d_{(I_{N_i})}(\cdot)$, defined on $\mathcal{B}_E$, induces a shift-invariant mean on $\mathcal{A}_E$, i.e. a positive functional $L:\mathcal{A}_E\to\mathbb{C}$, such that $L(\mathbf{1})=1$ and for any $F\in\mathcal{B}_E$, $L(\mathbf{1}_F)=d_{(I_{N_i})}(F)$. Now, by Gelfand representation theorem, there is a compact metric space X such that our C^*-algebra $\mathcal{A}_E$ is algebraically and topologically isomorphic to $C(X)$. Let $\tilde{L}:C(X)\to\mathbb{C}$ be the positive linear functional induced by L on $C(X)$. By the Riesz representation theorem, the functional $\tilde{L}$ is given by a Borel probability measure μ on X. Let $r:\mathcal{A}_E\to C(X)$ denote the Gelfand isomorphism. Then, for all $\phi\in\mathcal{A}_E$ we have

$$L(\phi) = \tilde{L}(r(\phi)) = \int_X r(\phi)\,\mathrm{d}\mu.$$

Since r is, in particular, an algebraic isomorphism, and since $\mathbf{1}_E^2 = \mathbf{1}_E$, the image $r(\mathbf{1}_E) \in C(X)$ is an idempotent, and hence of the form $\mathbf{1}_A$ for some clopen subset $A \subset X$. We have

$$\mu(A) = \tilde{L}(\mathbf{1}_A) = \tilde{L}(r(\mathbf{1}_E)) = d_{(I_N)}(E).$$

Now, the L-invariant shift operator $\phi(n) \mapsto \phi(n+1)$, $\phi \in \mathcal{A}_E$, is a C^*-isomorphism of $\mathcal{A}_E$ that induces a C^*-isomorphism of $C(X)$, which, in turn, by a theorem due to Banach ([5, Chapter XI]), is induced by a μ-preserving homeomorphism $T : X \to X$.

So we have

$$\begin{aligned} d_{(I_N)}(E \cap (E-n_1) \cap \cdots \cap (E-n_k)) &\geq d_{(I_{N_i})}(E \cap (E-n_1) \cap \cdots \cap (E-n_k)) \\ &= L(\mathbf{1}_{E\cap(E-n_1)\cap\cdots\cap(E-n_k)}) = \mu(A \cap (A-n_1) \cap \cdots \cap (A-n_k)), \end{aligned}$$

which completes the proof of Theorem 4.3.

We now turn our discussion to a crucial ingredient of the proof of EST, namely, Furstenberg's structure theorem for measure-preserving systems. This structure theorem can be seen as a sophisticated analogue of Furstenberg's distal structure theorem (Theorem 1.4 above) which was established in [36]. To formulate the structure theorem from [44] we need to introduce first some definitions and terminology. Given two measure-preserving systems $\mathbf{X} = (X, \mathcal{B}, \mu, T)$ and $\mathbf{Y} = (Y, \mathcal{D}, \nu, S)$, a measurable and measure-preserving map $\pi : X \to Y$ is called a *homomorphism* if for a.e. $x \in X$, $S\pi(x) = \pi(Tx)$. The system $\mathbf{Y}$ is called a *factor* of $\mathbf{X}$ and $\mathbf{X}$ is an *extension* of $\mathbf{Y}$.

A measure-preserving system is called *Kronecker* if it has the form $(Z, \mathcal{D}, m, R_\alpha)$, where Z is a compact abelian group, $\mathcal{D} =$ Borel sets, $m =$ Haar measure and R_α is defined by $z \mapsto z + \alpha$, where $\alpha \in Z$ generates a dense cyclic subgroup.

A measure-preserving system $(X, \mathcal{B}, \mu, T)$ is called *weakly mixing* if it has no nontrivial eigenfunctions, i.e. functions satisfying $f(Tx) = \lambda f(x)$, where $\lambda \in \mathbb{C}$, $|\lambda| = 1$ and $f \in L^2(X)$, $f \neq 1$. It is not hard to show that if an ergodic system $(X, \mathcal{B}, \mu, T)$ admits a nontrivial eigenfunction then it has a nontrivial Kronecker factor. Moreover, a classical result going back to [76] states that an ergodic system is weakly mixing if and only if its Kronecker factor is trivial.

A natural starting point for proving EST is to verify that it holds for Kronecker systems and for weakly mixing systems. The following theorem in [44] takes care of the weakly mixing case.

Theorem 4.4 ([44, Corollary 2.4]). *If $(X, \mathcal{B}, \mu, T)$ is a weakly mixing system, then for any $k \in \mathbb{N}$ and $A \in \mathcal{B}$, one has*

$$\lim_{N-M\to\infty} \frac{1}{N-M} \sum_{n=M}^{N-1} \left| \mu(A \cap T^{-n}A \cap \cdots \cap T^{-nk}A) - \mu(A)^{k+1} \right| = 0.$$

The proof of EST for Kronecker systems can be done in a variety of ways. The following argument has the advantage of being adaptable for "measure distal systems" which appear in Furstenberg's structure theorem.

Assume that $(X,\mathcal{B},\mu,T)$ is Kronecker, let $A\in\mathcal{B}$ with $\mu(A)>0$, and let $f=\mathbf{1}_A$. It is not hard to see that under our assumptions the orbit closure $\overline{\{T^nf, n\in\mathbb{Z}\}}$ in $L^2(X,\mathcal{B}.\mu)$ is norm compact. Let $\varepsilon>0$ and let $\{T^{n_1}f,\dots,T^{n_k}f\}$ be an $\frac{\varepsilon}{k}$-separated set (i.e. $\|T^{n_i}f-T^{n_j}f\|\geq\frac{\varepsilon}{k}$ for $i\neq j$) which has maximal cardinality. Then, for any n, the set $\{T^{n+n_1}f,\dots,T^{n+n_k}f\}$ is again $\frac{\varepsilon}{k}$-separated and has the same cardinality. This implies that, for some $i\in\{1,\dots,k\}$, $\|T^{n+n_i}f-f\|<\frac{\varepsilon}{k}$ and hence $\|T^nf-f\|<\frac{\varepsilon}{k}$ for a set of n with bounded gaps. This implies

$$\liminf_{N-M\to\infty}\frac{1}{N-M}\sum_{n=M}^{N-1}\mu(A\cap T^{-n}A\cap\dots\cap T^{-nk}A)=$$
$$\liminf_{N-M\to\infty}\frac{1}{N-M}\sum_{n=M}^{N-1}\int f\cdot T^nf\cdots T^{kn}f\,\mathrm{d}\mu>0.$$

The structure theorem established in [44] allows us to, so to speak, interpolate between the Kronecker and the weakly mixing cases.

For a compact metric space M, let $\mathrm{Iso}\,(M)$ denote the compact group of isometries of M. A system $\mathbf{X}=(X,\mathcal{B},\mu,T)$ is *an isometric extension* of a factor $\mathbf{Y}=(Y,\mathcal{D},\nu,S)$ if it can be represented, up to isomorphism, as $X=Y\times M$ for a compact metric space M, with $\mu=\nu\times m_M$, where m_M is a probability Borel measure on M which is invariant under isometries, and $T(y,u)=(Sy,\rho(y)u)$, where $\rho:Y\to\mathrm{Iso}\,(M)$ is measurable. Note that when $\mathbf{Y}$ is a trivial system and $\mathbf{X}$ is an ergodic isometric extension, then $\mathbf{X}$ is Kronecker.

A system $\mathbf{X}$ is *measure distal* if it has a sequence of factors indexed by ordinals $\mathbf{X}_\eta$, $\eta\leq\eta_0$, with $\mathbf{X}=\mathbf{X}_{\eta_0}$, $\mathbf{X}_0=$ trivial system on one point, and such that for $\xi<\eta\leq\eta_0$, $\mathbf{X}_\xi$ is a factor of $\mathbf{X}_\eta$, with $\mathbf{X}_{\eta+1}$ being an isometric extension of $\mathbf{X}_\eta$, and finally $\mathbf{X}_\xi=\lim_{\leftarrow}\{\mathbf{X}_\theta:\theta<\xi\}$ for ξ a limit ordinal (see [44, Definition 8.3]).

Given a factor $\mathbf{Y}$ of a system $\mathbf{X}$ one has a disintegration of the measure μ with respect to ν given by $\mu=\int\mu_y\,d\nu(y)$, so that for almost all $y\in Y$, $\mu Sy=S\mu_y$. Given two extensions $\mathbf{X}_1=(X_1,\mathcal{B}_1,\mu_1,T_1)$ and $\mathbf{X}_2=(X_2,\mathcal{B}_2,\mu_2,T_2)$ of a system $\mathbf{Y}=(Y,\mathcal{D},\nu,S)$, let us denote by $\mu_1\underset{\nu}{\times}\mu_2$ the measure on $X_1\times X_2$ defined by

$$\mu_1\underset{\nu}{\times}\mu_2(A)=\int\mu_{1.y}\times\mu_{2,y}(A)\,\mathrm{d}\nu(y),\quad A\in\mathcal{B}_1\times\mathcal{B}_2,$$

where for $i=1,2$, $\mu_i=\int\mu_{i,y}\,\mathrm{d}\nu(y)$ is the disintegration of the measures μ_i over ν.

The measure space $X_1\underset{Y}{\times}X_2=(X_1\times X_2,\mathcal{B}_1\times\mathcal{B}_2,\mu_1\underset{\nu}{\times}\mu_2)$ is called the *relative product* of $\mathbf{X}_1$ and $\mathbf{X}_2$ with respect to $\mathbf{Y}$. A system $\mathbf{X}$ is a *weakly mixing extension* of a factor $\mathbf{Y}$ if the system $\mathbf{X}_1\underset{\mathbf{Y}}{\times}\mathbf{X}_2$ is ergodic. Here is now the structure theorem that was proved in [44].

Theorem 4.5. *Every ergodic system* $\mathbf{X}$ *has a maximal distal factor* $\mathbf{X}_D$*. Moreover,* $\mathbf{X}$ *is a weakly mixing extension of* $\mathbf{X}_D$*.*

One can briefly describe the proof of Furstenberg's EST as follows. For starters, in view of the ergodic decomposition we can assume that the system $\mathbf{X} = (X, \mathcal{B}, \mu, T)$ is ergodic. Call a system MR (for multiple recurrence) if Theorem 4.1 holds for all sets $A \in \mathcal{B}$ with $\mu(A) > 0$ and all k. By Theorem 4.4 any weakly mixing system is MR. One can also show that, more generally, if $\mathbf{X}$ is a weakly mixing extension of an MR system then $\mathbf{X}$ is MR.

In view of Theorem 4.5 this reduces the proof of EST to distal systems. Now the case of general distal systems can be reduced to distal systems of finite rank (a distal system $\mathbf{X}$ is of rank n if it has a succession of factors $\mathbf{X} = \mathbf{X}_1 \to \cdots \to \mathbf{X}_n \to 1$ point, where each extension is isometric), and for these the property MR can be established via an intricate argument which may be viewed as a "relativization" of the proof for Kronecker systems described above.

Van der Waerden's theorem has a natural multidimensional version which was proved in the 1930s by Tibor Grünwald (who later changed his name to Gallai).

Theorem 4.6 (Multidimensional van der Waerden's theorem [87, page 123]). *For any finite partition of $\mathbb{Z}^d = \cup_{i=1}^N \mathcal{C}_i$, one of $\mathcal{C}_i$, $i = 1, \ldots, r$, contains an affine image of any finite set $F \subset \mathbb{Z}^d$. In other words, there exists an i, $1 \leq i \leq r$, such that for any finite set $F \subset \mathbb{Z}^d$ there exist $u \in \mathbb{Z}^d$ and $a \in \mathbb{N}$ such that $u + F = \{u + ax : x \in F\} \subset \mathcal{C}_i$.*

For a fuller perspective we will formulate a topological version of Gallai's theorem which was proved in [60] and served as a motivation for the multidimensional extension of EST that was established in [53] and will be discussed below.

Theorem 4.7 ([60]). *Let $T_1, T_2, \ldots, T_k$ be commuting homeomorphisms of a compact metric space X to itself. Assume that the dynamical system (X, G), where G is the group generated by $T_1, T_2, \ldots, T_k$, is minimal. Then for any nonempty open set $U \subset X$ there exists an $n \in \mathbb{N}$, such that*

$$U \cap T_1^n U \cap \cdots \cap T_k^n U \neq \varnothing.$$

For a set $S \subset \mathbb{Z}^d$, its *upper Banach density* is defined by the formula

$$d^*(S) = \limsup_{N_i - M_i \to \infty,\ 1 \leq i \leq d} \frac{|S \cap \prod_{i=1}^d \{M_i, M_i + 1, \ldots, N_i - 1\}|}{\prod_{i=1}^d (N_i - M_i)}.$$

The natural question now is whether it is true that any set $S \subset \mathbb{Z}^d$ with $d^*(S) > 0$ contains an affine image of any finite set $F \subset \mathbb{Z}^d$. In [53], Furstenberg and Katznelson answered this question affirmatively by deducing the answer from the following version of Furstenberg's EST.

Theorem 4.8 ([53]). *Let $(X, \mathcal{B}, \mu)$ be a measure space with $\mu(X) < \infty$, let T_1, T_2, $\ldots$, T_k be commuting measure-preserving transformations of X and let $A \in \mathcal{B}$ with $\mu(A) > 0$. Then*

$$\liminf_{N\to\infty}\frac{1}{N}\sum_{n=0}^{N-1}\mu(A\cap T_1^{-n}\cap T_2^{-n}A\cap\cdots\cap T_k^{-n}A)>0.$$

Corollary 4.9 ([53, Theorem B]). *Let $S\subset\mathbb{Z}^k$ be a subset with $d^*(S)>0$, and let $F\subset\mathbb{Z}^k$ be a finite configuration. Then there exists a positive integer n and a vector $u\in\mathbb{Z}^k$ such that $u+F\subset S$.*

The proof of Theorem 4.8, while following the general lines and ideas of the proof of EST in [44], had two novel features. First, notice that while Theorem 4.1 is about the joint behaviour of k commuting transformations of a special form, namely $T,T^2,\ldots,T^k$, in Theorem 4.8 one has to deal with k commuting transformations which are, so to say, in general position. This complicates the underlying structure theory, which has to be adjusted to reflect the more complicated situation when different operators in the group generated by $T_1,\ldots,T_k$ have different dynamical properties. Accordingly, a new kind of extensions was introduced in [53]. These extensions, which the authors call primitive, use the splitting of the group generated by $T_1,\ldots,T_k$ into direct sum of two subgroups, one of which is responsible for "relative weak mixing", and the other represents the phenomenon of "relative compactness". The following two results play a crucial role in the proof of Theorem 4.8.

Theorem 4.10. *If $\mathbf{X}=(X,\mathcal{B},\mu,(T_g)_{g\in\mathbb{Z}^k})$ is an extension of $\mathbf{Y}=(Y,\mathcal{D},\nu,(S_g)_{g\in\mathbb{Z}^k})$, then there is an intermediate factor $\mathbf{Z}$ such that $\mathbf{Z}$ is a primitive extension of $\mathbf{Y}$.*

Theorem 4.11. *If $\mathbf{X}=(X,\mathcal{B},\mu,(T_g)_{g\in\mathbb{Z}^k})$ is a primitive extension of $\mathbf{Y}=(Y,\mathcal{D},\nu,(S_g)_{g\in\mathbb{Z}^k})$ and $\mathbf{Y}$ has the MR property then $\mathbf{X}$ has the MR property.*

The second important novelty in [53] was the utilization (in the proof of Theorem 4.11) of a partition result, namely the multidimensional van der Waerden's theorem (see Theorem 4.6).

It is important to emphasize that partition results such as Theorem 4.6 and a more general result, the Hales–Jewett theorem (to be described bellow), play a fundamental role in the theory of multiple recurrence. On the one hand, partition results motivate the quest for density results. On the other hand, they prove useful in establishing multiple recurrence theorems.

One more instance of mutually enhancing interaction between Ramsey theory and ergodic theory is provided by the theory of IP-recurrence which was developed in [60], [45] and [54]. An important role in this theory is played by the following result due to Neil Hindman.

Theorem 4.12 (). *[72] For any finite partition $\mathbb{N}=\bigcup_{i=1}^N C_i$, one of C_i contains an IP-set, that is a set of the form*

$$\{n_{i_1}+n_{i_2}+\cdots+n_{i_k}:i_1<i_2<\cdots<i_k,\ k\in\mathbb{N}\}.$$

The notion of IP systems and their importance for recurrence issues arose in the course of joint work with Benjamin Weiss [60], during the 1975–1976 special ergodic theory program at the newly established Institute for Advanced Studies at the

Hebrew University. The initials IP refer to idempotence, which is a closely related notion. For example, a proof of Hindman's theorem given in [60] utilizes the idempotents in Ellis's enveloping semigroup. Also, Hindman's theorem is, essentially, a Poincaré recurrence theorem for idempotent ultrafilters (see, for example, [8, Theorem 3.4]).

An $\mathcal{F}$-sequence in an arbitrary space Y is a sequence $(y_\alpha)_{\alpha\in\mathcal{F}}$ indexed by the collection $\mathcal{F}$ of the finite sets of $\mathbb{N}$. If Y is a (multiplicative) semigroup, one says that an $\mathcal{F}$-sequence defines an IP-system if for any $\alpha = \{i_1, i_2, \dots, i_k\} \in \mathcal{F}$, one has $y_\alpha = y_{i_1} y_{i_2} \cdots y_{i_k}$. IP-systems should be viewed as generalized semigroups. Indeed, $y_{\alpha\cup\beta} = y_\alpha y_\beta$ whenever $\alpha \cap \beta = \varnothing$.

Here is an IP version of Theorem 4.7.

Theorem 4.13 (IP van der Waerden [60, Theorem 3.2]). . *Let X be a compact metric space and G a commutative group of its homeomorphisms such that the dynamical system (X,G) is minimal. For any nonempty open set $U \subset X$, any $k \in \mathbb{N}$, any IP-systems $(T_\alpha^{(1)})_{\alpha\in\mathcal{F}}, \dots, (T_\alpha^{(k)})_{\alpha\in\mathcal{F}}$ in G and any $\alpha_0 \in \mathcal{F}$, there exists an $\alpha \in \mathcal{F}$, $\min\alpha > \max\alpha_0$, such that*

$$U \cap T_\alpha^{(1)} U \cap \dots \cap T_\alpha^{(k)} U \neq \varnothing.$$

Clearly Theorem 4.7 is a special case of Theorem 4.13. But IP van der Waerden has many other important corollaries. For example, it follows from Theorem 4.13 that for any finite partition $\mathbb{N} = \bigcup_{i=1}^N C_i$ and for any $\ell \in \mathbb{N}$, one of the C_i has the property that the set of differences of length ℓ arithmetical progressions contained in C_i is rather large, namely it has a nontrivial intersection with any IP set. Sets with this property are called IP* sets. It is not hard to see that any IP* set has bounded gaps. Moreover, one can show with the help of Hindman's theorem that if $S_1, S_2, \dots, S_m$ are IP* sets, then $S_1 \cap S_2 \cap \dots \cap S_m$ also is an IP* set.

The theory of IP-recurrence found an interesting application in [61] where Furstenberg and Weiss established the following result of a diophantine nature (see also [11, Theorem A] and [14, section 0.34]).

Theorem 4.14 ([61, Thorem 15]). *Let $\varepsilon > 0$ and for each $k = 1, \dots, n$, let f_k be a real polynomial of k unknowns, vanishing at zero. Then the system*

$$\begin{gathered}
|f_1(x) - y_1| < \varepsilon \\
|f_2(x, y_1) - y_2| < \varepsilon \\
|f_3(x, y_1, y_2) - y_3| < \varepsilon \\
\dots \\
|f_n(x, y_1, y_2, \dots, y_{n-1}) - y_n| < \varepsilon
\end{gathered}$$

has a non-trivial integer solution. Indeed, the set of $x \in \mathbb{Z}$ for which there is some solution $(x, y_1, \dots, y_n)$ is IP.*

In [54], Furstenberg and Katznelson established the following fundamental IP Szemerédi theorem, which may be viewed as the analogue of Theorem 4.13 for measure-preserving systems.

Theorem 4.15 ([54, Theorem A]). *Let $(X,\mathcal{B},\mu)$ be a probability space and G an abelian group of measure-preserving transformations of X. For any $k \in \mathbb{N}$, any IP-systems $(T_\alpha^{(1)})_{\alpha\in\mathcal{F}},\dots,(T_\alpha^{(k)})_{\alpha\in\mathcal{F}}$ in G and any $A \in \mathcal{B}$ with $\mu(A) > 0$, there exists an $\alpha \in \mathcal{F}$ such that*

$$\mu(A\cap T_\alpha^{(1)}A\cap\cdots\cap T_\alpha^{(k)}A) > 0.$$

The proof of the IP-Szemerédi theorem is achieved via a sophisticated structure theory which can be viewed as an IP variation on the theme of primitive extensions discussed above. Curiously, it is not the IP van der Waerden but the more powerful Hales–Jewett theorem which has to be used when dealing with the IP version of compact extensions.

In the words of the authors of [69], the Hales–Jewett theorem "strips van der Waerden's theorem of its unessential elements and reveals the heart of Ramsey theory. It provides a focal point from which many results can be derived and acts as a cornerstone for much of the more advanced work". To formulate the Hales–Jewett theorem we introduce some definitions.

Let A be a finite alphabet, $A = \{a_1,a_2,\dots,a_k\}$. It is convenient to identify the elements of the nth Cartesian power A^n with the set of words of length n over the alphabet A, which, in turn, can be viewed as an abstract n-dimensional vector space (over A). Let $W_n(A,t)$ be that set of words of length n from the alphabet $A\cup\{t\}$, where t is a letter not belonging to A, which will serve as a variable. If $w \in W_n(A,t)$ is a word in which the variable t actually occurs, then the set $\{w(t)\}_{t\in A} = \{w(a_1),w(a_2),\dots,w(a_k)\}$ is called a combinatorial line.

Theorem 4.16 (Hales–Jewett theorem [71]). *Let $r,k,m \in \mathbb{N}$. There exists a $c = c(r,k,m)$ such that if $n \geq c$, then for any r-coloring of the set $W_n(A)$ of words of length n over the k-letters alphabet $A = \{a_1,a_2,\dots,a_k\}$, there is a monochromatic combinatorial line.*

Taking $A = \{a_1,a_2,\dots,a_\ell\}$ and interpreting $W_n(A)$ as integers in base ℓ, having at most n digits in their base ℓ expansion, we see that, in this situation, the elements of a combinatorial line form an arithmetic progression of length ℓ (with difference of the form $d = \sum_{i=1}^{k-1}\varepsilon_i\ell^i$, where $\varepsilon_i = 0$ or 1). Thus van der Waerden's theorem is a corollary of Theorem 4.16.

One of the signs of the fundamental nature of the Hales–Jewett theorem is that one can easily derive from it its multidimensional version. Let $t_1,t_2,\dots,t_m$ be m variables and let $w(t_1,t_2,\dots,t_m)$ be a word of length n over the alphabet $A\cup\{t_1,t_2,\dots,t_m\}$. If for some n, $w(t_1,t_2,\dots,t_m)$ is a word of length n in which all the variables $t_1,t_2,\dots,t_m$ occur, the result of the substitution

$$\{w(t_1,t_2,\dots,t_m)\}_{(t_1,t_2,\dots,t_m)\in A^m} = \{(w(a_{i_1},a_{i_2},\dots,a_{i_m}) : a_{i_j}\in A,\ j=1,2,\dots,m\}$$

is called a combinatorial m-space.

Observe now that if we replace the original alphabet A by A^m, then a combinatorial line in $W_n(A^m)$ can be interpreted as an m-space in $W_{nm}(A)$. Thus we have the following ostensibly stronger theorem as a corollary of Theorem 4.16.

Theorem 4.17. *Let $r,k,m \in \mathbb{N}$. There exists a $c = c(r,k,m)$ such that if $n \geq c$, then for any r-coloring of the set $W_n(A)$ of words of length n over the k-letter alphabet A, there is a monochromatic m-space.*

It is not hard to see that Theorem 4.17 implies the multidimensional van der Waerden theorem (Theorem 4.6) It also obviously implies the following result of a geometric nature.

Theorem 4.18. *Let F be a finite field. For any $r,m \in \mathbb{N}$, there exists a $c = c(r,m)$ such that if V is a vector space over F having dimension at least c, then for any r-coloring $V = \cup_{r=1}^{r} \mathcal{C}_i$, one of the $\mathcal{C}_i$ contains an m-dimensional affine space.*

One is naturally led to the question of whether a density version of the Hales–Jewett theorem holds (see for instance Conjecture, [69], page 53). The positive answer was obtained in the masterly paper [56]. Here is one of the few equivalent formulations of dHJ, the density version of Hales–Jewett theorem.

Theorem 4.19 ([56, Theorem E]). *There is a function $R(\varepsilon,k)$, defined for all $\varepsilon > 0$ and $k \in \mathbb{N}$, so that if A is a set with k elements, $W_N(A)$ consists of words in A with length N, and if $N \geq R(\varepsilon,k)$, then any subset $S \subset W_N(A)$ with $|S| \geq \varepsilon k^N$ contains a combinatorial line.*

In order to formulate the ergodic counterpart of Theorem 4.19 which was proved in [56], we shall need the following definition.

Definition 4.20. *Let $W(k)$ denote the free semigroup over the k-element alphabet $\{1,2,\ldots,k\}$. Given k sequences $(T_n^{(1)})_{n=1}^{\infty}, (T_n^{(2)})_{n=1}^{\infty}, \ldots, (T_n^{(k)})_{n=1}^{\infty}$ of invertible measure-preserving transformations of a probability space $(X,\mathcal{B},\mu)$, define, for each $w = (w(1),w(2),\ldots,w(k)) \in W(k)$,*

$$T(w) = T_1^{w(1)} T_2^{w(2)} \cdots T_k^{w(k)}.$$

The family $\{T(w), w \in W(k)\}$ is called a $W(k)$-system.

Here is now the ergodic formulation of dHJ.

Theorem 4.21 ([56, Proposition 2.7]). . *Let $\{T(w), w \in W(k)\}$ be a $W(k)$-system of invertible measure-preserving transformations of a probability space $(X,\mathcal{B},\mu)$. For any $A \in \mathcal{B}$ with $\mu(A) > 0$, there exists a combinatorial line $(\ell(t))_{t\in\{1,2,\ldots,k\}}$ in $W(k)$ such that*

$$\mu(T(\ell(1))^{-1}A \cap T(\ell(2))^{-1}A \cap \cdots \cap T(\ell(k))^{-1}A) > 0.$$

The proof of Theorem 4.21, while following the general scheme of other proofs discussed above, is significantly more involved, mainly due to the fact that the transformations forming the $W(k)$ system need not commute.

As a matter of fact, in the case when the $W(k)$ system is comprised of commuting transformations, the situation is reduced to the IP Szemerédi theorem. Yet, despite the absence of commutativity, the proof of Theorem 4.21 has a strong IP-flavor. Also this proof has an important novel feature. Namely, the authors use an infinitary combinatorial result which they obtained in [55] and which is a simultaneous extension of the Hindman and the Hales–Jewett theorems. (This result was also obtained in [21].)

We will briefly discuss now polynomial generalizations of some of the results mentioned above. We start with the polynomial Szemerédi theorem, which was obtained in [12] and can be viewed as a polynomial extension of Furstenberg–Katznelson's theorem.

Theorem 4.22 ([12, Theorem A]). *Let $(X, \mathcal{B}, \mu)$ be a probability space, $T_1, T_2, \ldots, T_t$ be commuting measure-preserving invertible transformations of X, let*

$$p_{1,1}(n), \ldots, p_{1,t}(n), p_{2,1}(n), \ldots, p_{2,t}(n), \ldots, p_{k,1}(n), \ldots, p_{k,t}(n)$$

be polynomials with rational coefficients taking integer values on the integers and satisfying $p_{i,j}(0) = 0$, $i = 1, \ldots, k$, $j = 1, \ldots, t$, and let $A \in \mathcal{B}$ with $\mu(A) > 0$. Then

$$\liminf_{n\to\infty} \frac{1}{N} \sum_{n=0}^{N-1} \mu\left(\prod_{j=1}^{t} T_j^{p_{1,j}(n)} A \cap \prod_{j=1}^{t} T_j^{p_{2,j}(n)} A \cap \cdots \cap \prod_{j=1}^{t} T_j^{p_{k,j}(n)} A\right) > 0.$$

Here is a combinatorial corollary of Theorem 4.22.

Theorem 4.23 ([12, Theorem B']). *Let $P : \mathbb{Z}^r \to \mathbb{Z}^\ell$, $r, \ell \in \mathbb{N}$, be a polynomial mapping satisfying $P(0) = 0$, let $F \subset \mathbb{Z}^r$ be a finite set and let $S \subset \mathbb{Z}^\ell$ be a set of positive upper Banach density. Then for some $n \in \mathbb{N}$ and $u \in \mathbb{Z}^\ell$ one has $u + P(nF) \subset S$.*

Similarly to the fact that in the proof of Theorem 4.8 the multidimensional van der Waerden theorem was used, an instrumental role in the proof of Theorem 4.22 is played by a polynomial version of Theorem 4.6, which we will formulate in a dynamical form, thereby stressing the connection with Furstenberg–Weiss' Theorem 4.13.

Theorem 4.24 ([12, Theorem C]). *Let (X, d) be a compact metric space, $T_1, \ldots, T_t$ commuting homeomorphisms of X and*

$$p_{1,1}(n), \ldots, p_{1,t}(n), p_{2,1}(n), \ldots, p_{2,t}(n), \ldots, p_{k,1}(n), \ldots, p_{k,t}(n)$$

be polynomials with rational coefficients taking integer values on the integers and satisfying $p_{i,j}(0) = 0$, $i = 1, \ldots, k$, $j = 1, \ldots, t$. Then, for any positive ε, there exist $x \in X$ and $n \in \mathbb{N}$ such that

$$d(T_1^{p_{i,1}(n)} T_2^{p_{i,2}(n)} \cdots T_t^{p_{i,t}(n)} x, x) < \varepsilon, \quad \textit{for all } i = 1, \ldots, k$$

simultaneously. Moreover, the integer n can be chosen from any IP-set. If the system $(X, T_1, \ldots, T_t)$ *is minimal, then for any nonempty open set* $U \subset X$*, the set* $S = \{n : g_i^{-1}(n)U \cap U \neq \varnothing\}$ *is an IP*-set, where* $g_i(n) = T_1^{p_{i,1}(n) \ldots p_{i,t}(n)}$, $i = 1, 2, \ldots, k$.

It is worth mentioning that so far Theorem 4.23 does not have a combinatorial (non-ergodic) proof.[3]

Here is a brief summary of some additional results of a polynomial nature which were motivated by Theorem 4.23.

(i) In [16] a general IP-Szemerédi theorem was proved. This result forms a natural IP analogue of Theorem 4.23 for measure-preserving systems.

(ii) In [16] a general form of Theorem 4.23 for countable modules over integral domains was obtained. The main result in [16] shows that the structure theory of measure-preserving group actions extends to polynomial actions which leads to a Furstenberg-style polynomial multiple recurrence theorem. Among the combinatorial corollaries of this result is a polynomial Szemerédi theorem for finite fields.

(iii) In [13] a polynomial Hales–Jewett theorem was proved. This result naturally leads to the conjecture that a density polynomial Hales–Jewett theorem is also true (see [8, pp. 56–57]). A density polynomial Hales–Jewett theorem would form a natural generalization of the "linear" Furstenberg–Katznelson's density Hales–Jewett theorem.

For additional results of a polynomial nature, see [15], [103].

The fundamental work of Furstenberg and his coauthors has served and keeps serving as a powerful impetus for many impressive developments. As an illustration, we will describe in conclusion some of the remarkable results pertaining to patterns in primes.

A long standing conjecture of Erdős states that if a set $A \subset \mathbb{N}$ has the property $\sum_{n \in A} \frac{1}{n} = \infty$, then A contains arbitrarily long arithmetic progressions. Clearly, if true, this conjecture leads to a strong refinement of Szemerédi's theorem on arithmetic progressions. In the special case where A is the set of primes, this conjecture was confirmed by Green and Tao in their spectacular paper [70]. Some of the crucial features of the proof were inspired by Furstenberg's correspondence principle and Furstenberg's structure theorem which were discussed above.

In [93] Tao and Ziegler obtained an "upgrade" of the Green–Tao theorem to polynomial configurations. More precisely, Tao and Ziegler show that for any integer-valued polynomials $P_1, \ldots, P_k \in \mathbb{Z}[m]$, with $P_1(0) = \cdots = P_k(0) = 0$, and any $\varepsilon > 0$, there are infinitely many integers x, m with $1 \leq m \leq x^{\varepsilon}$, such that $x + P_1(m), \ldots, x + P_k(m)$ are simultaneously prime. Furstenberg's correspondence principle and the polynomial Szemerédi theorem (Theorem 4.23) play a crucial role in this paper.

[3] Recently Hillel reminded me that when I had first raised the possibility of such results he told me not to waste my time because "mathematics wasn't ready for such complex considerations".

Hillel Furstenberg's ideas permeated and influenced vast areas of mathematics. Ergodic Ramsey theory, which started with the publication of the groundbreaking paper [44], serves as an excellent example of Furstenberg's impact. I was fortunate to be Hillel's student at the time of inception and early development of this beautiful field. It is both my duty and pleasure to acknowledge Hillel's stimulating influence on my mathematics and to express gratitude to him for being an outstanding role model.

References

1. Akin, Ethan, *Recurrence in topological dynamics. Furstenberg families and Ellis actions.* The University Series in Mathematics. Plenum Press, New York, 1997.
2. Anzai, Hirotada. Ergodic skew product transformations on the torus. *Osaka Math. J.* 3 (1951), 83–99.
3. L. Auslander, L. Green and F. Hahn, *Flows on homogeneous spaces.* Annals of mathematics studies 53, Princeton University Press, Princeton, New Jersey, 1963.
4. Bader, Uri and Shalom, Yehuda, Factor and normal subgroup theorems for lattices in products of groups. *Invent. Math.* 163 (2006), no. 2, 415–454.
5. Banach, Stefan, *Théorie des opérations linéaires, Monografie Matematyczne* (in French). Tom 1. Warszawa, 1932. English translation: *Theory of Linear Operations* (Dover Books on Mathematics, 2009).
6. Beiglböck, Mathias; Bergelson, Vitaly; Fish, Alexander, Sumset phenomenon in countable amenable groups. *Adv. Math.* 223 (2010), no. 2, 416–432.
7. F. Beleznay and M. Foreman, The collection of distal flows is not Borel, *Amer. J. of Math.* 117 (1995), 203–239.
8. Bergelson, Vitaly, Ergodic Ramsey theory – an update. In: *Ergodic theory of* Z^d *actions (Warwick, 1993–1994)*, 1–61, London Math. Soc. Lecture Note Ser., 228, Cambridge Univ. Press, Cambridge, 1996.
9. Bergelson, Vitaly; Ferré Moragues, Andreu, An ergodic correspondence principle, invariant means and applications. *Israel J. Math.* 245 (2021), no. 2, 921–962.
10. Bergelson, Vitaly; Host, Bernard; Kra, Bryna, Multiple recurrence and nilsequences. With an appendix by Imre Ruzsa. *Invent. Math.* 160 (2005), no. 2, 261–303.
11. Bergelson, V; Knutson, I.H.; McCutcheon, R., Simultaneous Diophantine approximation and VIP-systems. *Acta Arith.* 116 (2005), 13–24.
12. Bergelson, V.; Leibman, A., Polynomial extensions of van der Waerden's and Szemeredi's theorems. *J. Amer. Math. Soc.* 9 (1996), no. 3, 725–753.
13. Bergelson, V.; Leibman, A., Set-polynomials and polynomial extension of the Hales–Jewett theorem. *Ann. of Math. (2)* 150 (1999), no. 1, 33–75.
14. Bergelson, V.; Leibman, A., Distribution of values of bounded generalized polynomials. *Acta Math.* 198 (2007), no. 2, 155–230.
15. Bergelson, V.; Leibman, A.; Lesigne, E., Intersective polynomials and the polynomial Szemeredi theorem. *Adv. Math.* 219 (2008), no. 1, 369–388.
16. Bergelson, V.; Leibman, A.; McCutcheon, R., Polynomial Szemeredi theorems for countable modules over integral domains and finite fields. *J. Anal. Math.* 95 (2005), 243–296.
17. Bergelson, V.; McCutcheon, R., An ergodic IP polynomial Szemeredi theorem. *Mem. Amer. Math. Soc.* 146 (2000), no. 695, viii+106 pp.
18. Benoist, Yves; Quint, Jean-François, Mesures stationnaires et fermés invariants des espaces homogénes. (French) [Stationary measures and invariant subsets of homogeneous spaces] *Ann. of Math. (2)* 174 (2011), no. 2, 1111–1162.

19. Jean Bourgain, Alex Furman, Elon Lindenstrauss and Shahar Mozes. Invariant measures and stiffness for non-abelian groups of toral automorphisms, *C. R. Math. Acad. Sci. Paris* 344(12) (2007), 737–742.
20. Jean Bourgain, Alex Furman, Elon Lindenstrauss and Shahar Mozes, Stationary measures and equidistribution for orbits of nonabelian semigroups on the torus, *J. Amer. Math. Soc.* 24(1) (2011), 231–280.
21. Carlson, Timothy J., Some unifying principles in Ramsey theory. *Discrete Math.* 68 (1988), no. 2-3, 117D169.
22. Ellis, Robert, Distal transformation groups. *Pacific J. Math.* 8 (1958), 401–405.
23. Ellis, Robert, A semigroup associated with a transformation group. *Trans. Amer. Math. Soc.* 94 (1960), 272–281.
24. Ellis, Robert. Point transitive transformation groups. *Trans. Amer. Math. Soc.* 101 (1961), 384–395.
25. Ellis, Robert, The Veech structure theorem. *Trans. Amer. Math. Soc.* 186 (1973), 203–218 (1974).
26. Ellis, Robert; Glasner, Shmuel; Shapiro, Leonard, Proximal-isometric ($\mathcal{PJ}$) flows. *Advances in Math.* 17 (1975), no. 3, 213–260.
27. Ellis, Robert; Gottschalk, W. H., Homomorphisms of transformation groups. *Trans. Amer. Math. Soc.* 94 (1960), 258–271.
28. Eskin, Alex, Mirzakhani, Maryam, Invariant and stationary measures for the $\mathrm{SL}(2,\mathbb{R})$ action on moduli space. *Publ. Math. Inst. Hautes Études Sci.* 127 (2018), 95–324.
29. Erdős, Paul; Turán, Paul, On some sequences of integers, *J. London Math. Soc.* 11 (1936), 261–264.
30. Frisch, Joshua; Tamuz, Omer; Ferdowsi, Pooya Vahidi, Strong amenability and the infinite conjugacy class property. *Invent. Math.* 218 (2019), no. 3, 833–851.
31. Furstenberg, Harry, On the infinitude of primes. *Amer. Math. Monthly* 62 (1955), 353.
32. Furstenberg Harry, *Prediction theory*, Princeton University, Ph.D. thesis, 1958.
33. Furstenberg, Harry, *Stationary processes and prediction theory*. Annals of Mathematics Studies, No. 44 Princeton University Press, Princeton, N.J. 1960.
34. Furstenberg, Harry, Strict ergodicity and transformation of the torus. *Amer. J. Math.* 83 (1961), 573–601.
35. Furstenberg, Harry, A Poisson formula for semi-simple Lie groups. *Ann. of Math. (2)* 77 (1963), 335–386.
36. Furstenberg, Harry, The structure of distal flows. *Amer. J. Math.* 85 (1963), 477–515.
37. Furstenberg, Harry, Poisson boundaries and envelopes of discrete groups. *Bull. Amer. Math. Soc.* 73 (1967), 350–356.
38. Furstenberg, Harry, Disjointness in ergodic theory, minimal sets, and a problem in Diophantine approximation. *Math. Systems Theory* 1 (1967), 1–49.
39. Furstenberg, Harry, Intersections of Cantor sets and transversality of semigroups. In: *Problems in analysis, (Sympos. Salomon Bochner, Princeton Univ., Princeton, N.J., 1969)* 41–59, 1970
40. Furstenberg, Harry, Boundaries of Lie groups and discrete subgroups. In: *Actes du Congrés International des Mathématiciens (Nice, 1970)*, Tome 2, pp. 301–306. Gauthier-Villars, Paris, 1971.
41. Furstenberg, Harry, Boundaries of Riemannian symmetric spaces. In: *Symmetric spaces (Short Courses, Washington Univ., St. Louis, Mo., 1969–1970)*, pp. 359–377. Pure and Appl. Math., Vol. 8, Dekker, New York, 1972.
42. Furstenberg, Harry, Boundary theory and stochastic processes on homogeneous spaces. In: *Harmonic analysis on homogeneous spaces (Proc. Sympos. Pure Math., Vol. XXVI, Williams Coll., Williamstown, Mass., 1972)*, 193–229. Amer. Math. Soc., Providence, R.I., 1973.
43. Furstenberg, Harry, The unique ergodicity of the horocycle flow. In: *Recent advances in topological dynamics (Proc. Conf., Yale Univ., New Haven, Conn., 1972; in honor of Gustav Arnold Hedlund)*, pp. 95–115. Lecture Notes in Math., Vol. 318, Springer, Berlin, 1973.

44. Furstenberg, Harry, Ergodic behavior of diagonal measures and a theorem of Szemerédi on arithmetic progressions. *J. Analyse Math.* 31 (1977), 204–256.
45. Furstenberg, Harry, *Recurrence in ergodic theory and combinatorial number theory*. M. B. Porter Lectures. Princeton University Press, Princeton, N.J., 1981. xi+203 pp.
46. Furstenberg, Hillel, Stiffness of group actions. In: *Lie groups and ergodic theory (Mumbai, 1996)*, 105–117, Tata Inst. Fund. Res. Stud. Math., 14, Tata Inst. Fund. Res., Bombay, 1998.
47. Furstenberg, Hillel, Ergodic fractal measures and dimension conservation, *Ergodic Theory Dynam. Systems* **28** (2008), no. 2, 405–422.
48. Furstenberg, Hillel, *Ergodic theory and fractal geometry*, CBMS Regional Conference Series in Mathematics, **120**. American Mathematical Society, Providence, RI, 2014. x+69 pp.
49. Furstenberg, Harry; Glasner, Shmuel, On the existence of isometric extensions. *Amer. J. Math.* 100 (1978), no. 6, 1185–1212.
50. Furstenberg, Hillel; Glasner, Eli, Stationary dynamical systems. In: *Dynamical numbers–interplay between dynamical systems and number theory*, 1–28, Contemp. Math., 532, Amer. Math. Soc., Providence, RI, 2010.
51. Furstenberg, Hillel; Glasner, Eli, Recurrence for stationary group actions. In: *From Fourier analysis and number theory to Radon transforms and geometry*, 283–291, Dev. Math., 28, Springer, New York, 2013.
52. Furstenberg, Hillel; Glasner, Eli; Weiss, Benjamin, Affinely prime dynamical systems. *Chin. Ann. Math. Ser. B* 38 (2017), no. 2, 413–424.
53. Furstenberg, H.; Katznelson, Y., An ergodic Szemeredi theorem for commuting transformations. *J. Analyse Math.* 34 (1978), 275–291 (1979).
54. Furstenberg, H.; Katznelson, Y., An ergodic Szemeredi theorem for IP-systems and combinatorial theory. *J. Analyse Math.* 45 (1985), 117–168.
55. Furstenberg, H.; Katznelson, Y., Idempotents in compact semigroups and Ramsey theory. *Israel J. Math.* 68 (1989), no. 3, 257–270.
56. Furstenberg, H.; Katznelson, Y., A density version of the Hales–Jewett theorem. *J. Anal. Math.* 57 (1991), 64–119.
57. Furstenberg, H.; Kesten, H., Products of random matrices. *Ann. Math. Statist.* 31 (1960), 457–469.
58. Furstenberg, Harry; Keynes, Harvey; Shapiro, Leonard, Prime flows in topological dynamics. *Israel J. Math.* 14 (1973), 26–38.
59. Furstenberg, Hillel and Kifer, Yuri, Random matrix products and measures on projective spaces, *Israel J. of Math.* **46** (1983), 12–32.
60. Furstenberg, H.; Weiss, B., Topological dynamics and combinatorial number theory. *J. Analyse Math.* 34 (1978), 61–85 (1979).
61. Furstenberg, H; Weiss, B., Simultaneous Diophantine approximation and IP-sets, *Acta Arith.* 49 (1988), 413–426.
62. Furstenberg, Hillel; Weiss, Benjamin, On almost 1-1 extensions. *Israel J. Math.* 65 (1989), no. 3, 311–322.
63. Glasner, Shmuel, Topological dynamics and group theory. *Trans. Amer. Math. Soc.* 187 (1974), 327–334.
64. Glasner, Shmuel. *Proximal flows*, Lecture Notes in Math. 517, Springer-Verlag, 1976.
65. Glasner, Eli, *Ergodic theory via joinings*, AMS Mathematical Surveys and Monographs, **101**, 2003.
66. Glasner, Eli, Topological weak mixing and quasi-Bohr systems. Probability in mathematics. *Israel J. Math.* 148 (2005), 277–304.
67. Gottschalk, W. H., Characterizations of almost periodic transformation groups. *Proc. Amer. Math. Soc.* 7 (1956), 709–712.
68. Gottschalk, Walter Helbig; Hedlund, Gustav Arnold, *Topological dynamics*. American Mathematical Society Colloquium Publications, Vol. 36 American Mathematical Society, Providence, R. I., 1955.

69. Graham, Ronald L.; Rothschild, Bruce L.; Spencer, Joel H., *Ramsey theory*. Second edition. Wiley-Interscience Series in Discrete Mathematics and Optimization. A Wiley-Interscience Publication. John Wiley & Sons, Inc., New York, 1990.
70. Green, B.; Tao, T., The primes contain arbitrarily long arithmetic progressions. *Ann. of Math.* 167 (2008), 481–547.
71. Hales, A. W.; Jewett, R. I., Regularity and positional games. *Trans. Amer. Math. Soc.* 106 (1963), 222–229.
72. Hindman, Neil, Finite sums from sequences within cells of a partition of *N*. *J. Combinatorial Theory Ser. A* 17 (1974), 1–11.
73. Johnson, A., Measures on the circle invariant under multiplication by a nonlacunary subsemigroup of the integers, *Israel J. Math.* **77** (1992), 211–240.
74. Kalantar, Mehrdad; Kennedy, Matthew, Boundaries of reduced C^*-algebras of discrete groups. *J. Reine Angew. Math.* 727 (2017), 247–267.
75. Keynes, Harvey B.; Robertson, James B., Eigenvalue theorems in topological transformation groups. *Trans. Amer. Math. Soc.* 139 (1969), 359–369.
76. Koopman B.O.; von Neumann, J., Dynamical systems of continuous spectra, *Proc. Nat. Acad. Sci. U.S.A.* 18 (1932), 255D263.
77. Kaimanovich, V. A.; Vershik, A. M., Random walks on discrete groups: boundary and entropy. *Ann. Probab.* 11 (1983), no. 3, 457–490.
78. Loomis, Lynn H., *An introduction to abstract harmonic analysis.* D. Van Nostrand Company, Inc., Toronto-New York-London, 1953.
79. Lyons, R. On measures simultaneously 2- and 3-invariant, *Israel J. Math.* **61** (1988), 219–224.
80. Margulis, G. A. Discrete groups of motions of manifolds of nonpositive curvature. (Russian) In: *Proceedings of the International Congress of Mathematicians (Vancouver, B.C., 1974)*, Vol. 2, pp. 21–34. Canad. Math. Congress, Montreal, Que., 1975
81. McMahon, D. C., Relativized weak disjointness and relatively invariant measures, *Trans. Amer. Math. Soc.* **236** (1978), 225–237.
82. Moore, Calvin C., Compactifications of symmetric spaces. *Amer. J. Math.* 86 (1964), 201–218.
83. Moore, Calvin C., Amenable subgroups of semisimple groups and proximal flows. *Israel J. Math.* 34 (1979), no. 1-2, 121–138 (1980).
84. Nevo, Amos; Zimmer, Robert J., Homogenous projective factors for actions of semi-simple Lie groups. *Invent. Math.* 138 (1999), no. 2, 229–252.
85. Nevo, Amos; Zimmer, Robert J., A structure theorem for actions of semisimple Lie groups. *Ann. of Math. (2)* 156 (2002), no. 2, 565–594.
86. Parry, W., A finitary classification of topological Markov chains and sofic systems, *Bull. London Math. Soc.* **9** (1977), 86–92.
87. Rado, R., Note on combinatorial analysis. *Proc. London Math. Soc.* (2) 48 (1943), 122–160.
88. Rosenblatt, Joseph, Ergodic and mixing random walks on locally compact groups. *Math. Ann.* 257 (1981), no. 1, 31–42.
89. Rudolph, R., $\times 2$ and $\times 3$ invariant measures and entropy, *Ergodic Theory Dynam. Syst.* 10 (1990), 395–406.
90. Rudolph, R., A joinings proof of Bourgain's return time theorem, *Ergodic Theory Dynam. Syst.* 14, (1994), 197–203.
91. Szemerédi, E., On sets of integers containing no k elements in arithmetic progression, *Acta Arith.* 27 (1975), 199–245.
92. Shmerkin, Pablo, On Furstenberg's intersection conjecture, self-similar measures, and the L^q norms of convolutions, *Ann. of Math. (2)* 189 (2019), no. 2, 319–391.
93. Tao, T.; Ziegler, T., The primes contain arbitrarily long polynomial progressions. *Acta Math.* 201 (2008), no. 2, 213–305.
94. Van der Waerden, Bartel, Beweis einer Baudetschen Vermutung, *Nieuw. Arch. Wisk.* 15 (1927), 212–216.

95. Veech, W. A., Almost automorphic functions on groups. *Amer. J. Math.* 87 (1965), 719–751.
96. Veech, W. A., Point-distal flows. *Amer. J. Math.* 92 (1970), 205–242.
97. Veech, W. A., Topological dynamics, *Bull. Amer. Math. Soc.* 83, (1977), 775–830.
98. Weyl, H., *Über die Gleichverteilung von Zahlen mod Eins*, *Math. Ann.* 77 (1916), 313–352.
99. Wu, Meng, A proof of Furstenberg's conjecture on the intersections of $\times p$ - and $\times q$-invariant sets, *Ann. of Math. (2)* **189** (2019), no. 3, 707–751
100. R. J. Zimmer, Extensions of ergodic group actions, *Illinois J. Math.* 20, (1976), 373–409.
101. R. J. Zimmer, Ergodic actions with generalized discrete spectrum, *Illinois J. Math.* 20, (1976), 555–588.
102. Zippin, Leo, Transformation groups. In: *Lectures in Topology*, pp. 191–221. University of Michigan Press, Ann Arbor, Mich., 1941.
103. Zorin-Kranich, P., A nilpotent IP polynomial multiple recurrence theorem. *J. Anal. Math.* 123 (2014), 183–225.

The work of G. A. Margulis

Alex Eskin, David Fisher and Dmitry Kleinbock

Acknowledgments

A.E. was supported by an Investigator grant from the Simons Foundation. D.F. was supported by NSF grant DMS-1906107 and DMS-2208430. D.K. was supported by NSF grant DMS-1900560 and the Institute for Advanced Study School of Mathematics.

Alex Eskin
Department of Mathematics,
University of Chicago,
5734 University Avenue,
Chicago, IL 60637-1514,
USA,
e-mail: eskin@math.uchicago.edu

David Fisher
Department of Mathematics,
Rice University,
6100 Main St MS 136,
Houston, TX 77005,
USA,
e-mail: davidfisher@rice.edu

Dmitry Kleinbock
Department of Mathematics,
Brandeis University,
Goldsmith 207,
Waltham, MA 02454-9110,
USA,
e-mail: kleinboc@brandeis.edu

H. Holden, R. Piene (eds.), *The Abel Prize 2018-2022*, The Abel Prize,
https://doi.org/10.1007/978-3-031-33973-8_12

Contents

1 General introduction

The work of Margulis is unusual not just for its importance and depth, but for spanning numerous areas of mathematics. In addition, Margulis has a somewhat singular style of solving problems by bringing together ideas from relatively unrelated areas. He frequently revolutionized areas he choose to study. This include the structure of discrete subgroups of Lie groups, where his arithmeticity, superrigidity and normal subgroup theorems brought a wide array of new dynamical ideas to a field that had previously been dominated by algebra and geometry, see Sections 2 and 3. To the area of theoretical computer science he introduced powerful tools from the representation theory of Lie groups, see Section 4. And perhaps most consistently, he developed deep connections between homogeneous dynamics and number theory, see Sections 7–8 and 10–12. While it is clear that Margulis sees himself more as solving problems than as building theories, each of these remarkable insights and connections has laid the groundwork for new theories and even subfields of mathematics.

This is far from a complete survey; the results covered were selected according to the personal tastes of the authors. A much more detailed account of Margulis's work up to 2008 is given in [53]. See also the surveys [85] and [86] on particular topics written by Margulis himself. The recent book [44], besides giving a taste of the enormous impact of Margulis on the field he essentially created, also covers much more of Margulis's output than we do here. The reader should view this text as a "survey of surveys", as we try to present a glimmer of the work and then refer the reader to more detailed surveys as appropriate.

2 Arithmeticity and superrigidity

In this section $\mathbb{G}$ will be a connected semisimple real algebraic group without non-trivial $\mathbb{R}$-anisotropic factors; we will write $G = \mathbb{G}(\mathbb{R})^0$, the connected component of the identity in $\mathbb{G}(\mathbb{R})$; then G is a connected semisimple Lie group without compact factors. One can more generally consider semisimple algebraic groups over a local field k, in which case we will denote $G = \mathbb{G}(k)$. It is also possible to consider products of these groups defined over different fields, as in [84, Introduction]. We will let $\Gamma \subset G$ be a *lattice* (this means that Γ is a discrete subgroup of G, and the quotient G/Γ has finite Haar measure), and assume Γ is *irreducible*, that is, for any proper non-trivial connected normal subgroup H of G the product $H\Gamma$ is dense in G. A lattice Γ is said to be *uniform* if G/Γ is compact, and *non-uniform* otherwise.

We now give the definition of arithmeticity.

Definition 2.1. A group $\Gamma \subset G$ is *arithmetic* if there exists a semisimple algebraic $\mathbb{Q}$-group $\mathbf{G}'$ and an epimorphism $\pi : \mathbf{G}'(\mathbb{R}) \to G$ such that

1. $\ker(\pi)$ is a compact group and
2. $\pi\big(\mathbf{G}'(\mathbb{Z})\big)$ is commensurable to Γ.

This definition is due to Selberg for Γ being irreducible and non-uniform, and to Piatetski-Shapiro in general. In a series of papers in the 1970s Margulis proved:

Theorem 2.2. *Let G be as above with* $\operatorname{rank}_{\mathbb{R}}(G) \geq 2$, *and let Γ be an irreducible lattice in G; then Γ is arithmetic.*

Theorem 2.2 settled a conjecture due to Selberg [99] and Piatetski-Shapiro [92]. Margulis first proved the non-uniform case [74] using a study of unipotent elements, specifically his work on non-divergence of unipotent orbits. (See Section 7 for a detailed discussion of non-divergence results.) In particular, this result depends on the existence of unipotent elements in non-uniform lattices, a fact established earlier in a landmark paper of Kazhdan and Margulis [55]. At the time this approach was known to other mathematicians, including Selberg, Piatetski-Shapiro and Raghunathan, and the latter even completed a proof of Theorem 2.2 for non-uniform lattices at roughly the same time, modulo the result on non-divergence of unipotent orbits (Theorem 7.1 below), which he was unable to establish, see [93].

In a remarkable development a few years later, Margulis also proved Theorem 2.2 by an entirely different approach via what became known as his superrigidity theorem. While the superrigidity theorem and its proof are, in a loose sense, inspired by Mostow's strong rigidity theorem [89], it was not expected at the time. The name "superrigidity" was coined by Mostow. In Margulis' first proof of superrigidity, the lattice was assumed cocompact, an assumption used to apply Oseledec's multiplicative ergodic theorem to a certain cocycle. This limitation was soon overcome in the work of Margulis, Furstenberg and Zimmer. The latter then generalized the superrigidity theorem to apply in a more general context of cocycles over group actions, further extending and developing the reach of this profound result.

We state this result in a couple of different ways to emphasize different forms and to allow us to discuss the proof. Let us start with a definition.

Definition 2.3. Let $\Gamma \subset G$ be a lattice and let $\rho : \Gamma \to H$ be a homomorphism, where H is a topological group. We say that ρ *almost extends to* G if there is a continuous homomorphism $\pi : G \to H$ and another homomorphism $\pi_c : \Gamma \to H$, such that

1. $\overline{\pi_c(\Gamma)} = C \subset H$ is compact,
2. C and $\pi(G)$ commute,
3. $\rho(\gamma) = \pi(\gamma)\pi_c(\gamma)$ for every γ in Γ.

We will call a semisimple Lie group higher rank if its real rank is at least 2. One can then state Margulis' superrigidity theorem as:

Theorem 2.4 (Superrigidity I). *Let G be a higher rank semisimple Lie group with finite center, and let $\Gamma \subset G$ be an irreducible lattice; then any linear representation of Γ over any local field almost extends to G.*

Margulis' proof of this result actually gives much more information on the representation π_c and completely classifies linear representations of Γ modulo finite image representations. Much of the difficulty in the proof of Theorem 2.4 already occurs in the proof of this special case:

Theorem 2.5 (Superrigidity II). *Let G and Γ be as in Theorem 2.4 and let $\mathbf{H}$ be a simple algebraic group with trivial center over a local field k. Assume $\rho : \Gamma \to H = \mathbf{H}(k)$ is a homomorphism with Zariski dense, unbounded image; then ρ extends to G.*

The proof of Theorem 2.4 uses Theorem 2.5. For the proof of Theorem 2.2, Theorem 2.5 suffices. Margulis announced all these results and gave detailed indications of the proofs in [76]. Fuller proofs first appeared in Russian in an appendix to the Russian translation of Raghunathan's book; there the argument did not follow precisely the same outline as in the ICM address and, in particular, covered the case of non-uniform as well as uniform lattices for Theorem 2.5 and Theorem 2.2. This account eventually appeared in English in [78].

In addition in [76], Margulis gave a form of both superrigidity and arithmeticity for lattices with dense commensurator without any assumption on the rank of G. Margulis gave a complete account of his theorems in full generality in his book [84]. Somewhat earlier an account of many of these results of Margulis appeared in a book of Zimmer [105]. We now discuss briefly the results on rigidity of groups with dense commensurator.

Given a discrete group $\Gamma \subset G$ we define the *commensurator* of Γ in G as:

$$\mathrm{Comm}(\Gamma) := \{g \in G : [(g\Gamma g^{-1} \cap \Gamma) : \Gamma] < \infty \text{ and } [(g\Gamma g^{-1} \cap \Gamma) : g\Gamma g^{-1}] < \infty\}.$$

It is easy to verify that if $g \in \mathrm{Comm}(\Gamma)$, then there is a finite cover X of $K\backslash G/\Gamma$, where g acts as an isometry of X. For this reason elements of the commensurator are often viewed as hidden symmetries of $K\backslash G/\Gamma$ or Γ. In the same paper [76], Margulis proved

Theorem 2.6. *Let G be a simple Lie group and $\Gamma \subset G$ a lattice. If $[\Gamma : \mathrm{Comm}(\Gamma)] = \infty$, then Γ is arithmetic.*

The proof of Theorem 2.6 follows the same general outline as the proof of Theorem 2.2. Namely it depends on a superrigidity theorem which is an analogue of Theorem 2.5 but only for representations of Γ that are assumed to extend to $\mathrm{Comm}(\Gamma)$. This is sufficient since the representations of Γ one studies to prove arithmeticity from superrigidity are defined in such a way that it is easy to see they extend to $\mathrm{Comm}(\Gamma)$. In this context, Margulis also proved an analogue of Theorem 2.4 for representations of Γ that extend to $\mathrm{Comm}(\Gamma)$.

We briefly discuss the original philosophy of the proof of superrigidity. It is usually viewed as breaking into two parts. In the first part, one constructs an equivariant measurable map between a homogeneous variety for G and a homogeneous variety for H. By now this can be done by a variety of methods. In Margulis' original proof, he constructed the map using the Oseledec multiplicative ergodic theorem. This method will apply as soon as one knows the lattice is *integrable*. This is a defect of the proof, as it is impossible to verify integrability without knowing some structure of the lattice, though it is straightforward to verify integrability of arithmetic lattices of higher rank. In all known proofs of Theorem 2.4, one actually needs to first prove Theorem 2.2 using Theorem 2.5 exactly to verify that the lattice is integrable. There are cases of Theorem 2.4 where one can only construct the equivariant measurable map using the Oseledec theorem. By now there are several other approaches to constructing equivariant measurable maps that suffice for the proof of Theorems 2.5 and 2.2, including the one due to Margulis [43, 78, 105]. The second part of the proof is where higher rank is used and involves showing that the measurable equivariant map is in fact rational. From there it is relatively easy to verify the map is G-equivariant. Unlike the proof of the first step, where multiple variations of proofs appeared quite quickly, Margulis' original proof of rationality was revisited and refined quite recently by Bader and Furman [7].

We mention here that the most general version of Margulis' superrigidity theorem and also the most general version of Zimmer's cocycle superrigidity theorem, at least in characteristic zero, appear in joint work of the second-named author with Margulis [45]. In particular, one need not assume that G has finite center, which is quite an unnatural assumption in the case where G is a Lie group.

There is however, in [84], a significantly different approach to superrigidity in terms of studying spaces of measurable equivariant sections of certain vector bundles over G/Γ. The next section of this paper is devoted to a sketch of this proof.

Before moving on to that, we mention the second-named author's survey [42] for a discussion of some of the many further results inspired by Margulis' work on arithmeticity and superrigidity.

We will now outline some key ideas both in a proof of superrigidity and in the proof that superrigidity implies arithmeticity.

2.1 *Proof of superrigidity*

We sketch here the proof of Margulis' superrigidity theorem that is contained in [84]. The first step is to view a linear representation $\rho : \Gamma \to \mathrm{SL}_n(k)$ as giving rise to a G-action on a vector bundle E over G/Γ. The vector bundle and the G-action are defined by letting $V = k^n$, taking the $(G \times \Gamma)$-action on $G \times V$ given by $g'(g,v)\gamma = (g'g\gamma^{-1}, \rho(\gamma)v)$ and dividing by the Γ-action to obtain $(G \times V)/\Gamma$. The proof involves the study of spaces of sections of this bundle under the action of various subgroups of G. We state the main ingredients here as two lemmas.

Lemma 2.7. *Let A be a Cartan subgroup of G, and let $a \in A$. Then there exist two A-invariant subbundles $V', V'' \subset V$ such that $V = V' \oplus V''$.*

The subbundles V', V'' are produced using the Oseledec theorem once one knows that the Lyapunov splitting of the a-action on G/Γ is non-trivial. This step can be done in different ways depending on the Zariski closure of $\rho(\Gamma)$ in $\mathrm{SL}_n(k)$. For the proof of Theorem 2.5, Margulis originally used the spectral gap of the G-action on G/Γ to prove this non-triviality when the Zariski closure is simple and non-compact, and $\rho(\Gamma)$ is unbounded. We do not discuss the other cases here.

We now want to convert these invariant subbundles into an a-invariant section of some other vector bundle. This is easily done by taking the projection along V'' to V' in the bundle $\mathrm{End}(V)$. The proof is completed by using the structure of G along with an iterated application of the following lemma:

Lemma 2.8. *Let $A, B \subset G$ be two unbounded closed subgroups that commute. For any vector bundle F over G/Γ and any finite-dimensional space W of A-invariant sections of F, the space*

$$\overline{\mathrm{Span}\{b \cdot w : b \in B, w \in W\}}$$

is also finite-dimensional.

Unboundedness of A and B is used only to verify ergodicity of the A- and B-actions on G/Γ. Margulis in fact proves a more general version of the lemma for actions on vector bundles over ergodic group actions.

The remainder of the proof is fairly simple. First one finds a normal form for any element of G of the form $b_k b_{k-1} \cdots b_1 a$, where a is as in Lemma 2.7 and each b_i lies in a group B_i with the properties that

1. B_1 commutes with a;
2. B_i commutes with B_{i-1};
3. each B_i is unbounded.

Then an iterated application of Lemma 2.8 gives a finite-dimensional space of sections of F that are G-invariant. It is relatively easy to verify that this representation extends the original representation of Γ to one of G.

2.2 Proof of arithmeticity

The proof of arithmeticity from superrigidity follows by studying a sequence of natural representations of Γ obtained by simple algebraic modifications of the defining representation $\rho_0 : \Gamma \subset G$. Here we think of $G = \mathbb{G}(\mathbb{R})$ as the real points of a real algebraic group which we assume for simplicity is simple and center-free. We also assume that Γ is finitely generated, which is elementary in the case when Γ is cocompact and follows in general from the fact that Γ has property (T) of Kazhdan. We now consider the subfield $k \subset \mathbb{R}$ generated by matrix entries of elements of $\rho_0(\Gamma)$. It follows from the work of Vinberg that, up to conjugacy, one can assume that k is the adjoint trace field of G.

The first step is to show that $k \subset \overline{\mathbb{Q}}$. For a contradiction assume the contrary, that is, there exists an element γ of Γ with transcendental trace of the adjoint. Since $\mathrm{Aut}(\mathbb{C})$ is transitive on transcendentals, letting $\rho_i = \rho_0 \circ \sigma_i$ for some sequence of elements $\sigma_i \in \mathrm{Aut}(\mathbb{C})$, we can see that the trace of $\rho_i(\gamma)$ is unbounded. But each ρ_i is forced by superrigidity to either extend to G or have precompact image, and it is easy to check that in either case the adjoint trace of each $\rho_i(\gamma)$ is bounded. A careful reader will note here that ρ_i is considered as a representation into $\mathbb{G}(\mathbb{C})$, and that we use the fact that ρ_0 is $\mathbb{C}$-Zariski dense in $\mathbb{G}(\mathbb{C})$ and therefore so are all the ρ_i.

The second step is to construct the group G' in Definition 2.3. Here we apply restriction of scalars to the group $\mathbb{G}(k)$ to obtain a $\mathbb{Q}$-group

$$\mathbb{G}' = \mathrm{Res}_{k/\mathbb{Q}}(\mathbb{G}),$$

and all that needs to be shown is that $\mathbb{G}'(\mathbb{R})$ is a compact extension of $\mathbb{G}$. This is once again an application of superrigidity, this time to the image of Γ in $\mathbb{G}'(\mathbb{Q}) \subset \mathbb{G}'(\mathbb{R})$. We first note that Γ is Zariski dense because Γ is Zariski dense in G by the Borel density theorem and restriction of scalars preserves Zariski density. If there were another non-compact factor F of $\mathbb{G}'(\mathbb{R})$, then the map of Γ into F would extend to G; this is easily seen to contradict the Zariski density of Γ, since $\Gamma \subset \mathrm{Diag}(G) \times$

$F' \subset G \times F \times F'$, where F' is the collection of other simple factors of $\mathbb{G}''(\mathbb{R})$. Note that here we are in fact using quite strongly that we are considering Lie groups, so that representations with bounded image are contained in compact Lie subgroups which are automatically algebraic subgroups. It is not too difficult to modify this step in the other cases where Margulis' theorem holds.

In the final step, one wants to show that a finite index subgroup of Γ lies not just in $\mathbb{G}'(\mathbb{Q})$ but in $\mathbb{G}'(\mathbb{Z})$. To see this one assumes the contrary; hence there exists a prime p that occurs. One then considers the representation into $\mathbb{G}'(\mathbb{Q}_p)$ and verifies that the hypotheses imply that there is a continuous extension from Γ to G of the representation into some simple factor of $\mathbb{G}'(\mathbb{Q}_p)$. But this is impossible, because G is connected and $\mathbb{Q}_p$ is totally disconnected.

On several occasions Margulis has remarked to the second-named author that he was unsure at the time whether this proof that superrigidity implies arithmeticity was really new. After consulting other experts, it became quite clear that it was. Perhaps in part because no one expected superrigidity to hold at this level of generality, even though there were roughly contemporaneous proofs of very special cases [12]. Margulis has also been quite insistent that his intention was to prove arithmeticity and that superrigidity was a byproduct. By now it is clear that the byproduct, particularly in the form of the cocycle superrigidity theorems originally developed by Zimmer, greatly increased the importance of the proof of arithmeticity. One can see an instance of this in Margulis' own work in Section 5 below.

3 Normal subgroup theorem

While arithmeticity had been conjectured, Margulis' next breakthrough in the theory of lattices in semisimple Lie groups was not as expected.

Theorem 3.1. *Let G be a semisimple center-free Lie group with real rank at least 2, and let $\Gamma \subset G$ be an irreducible lattice. Then any non-trivial normal subgroup of Γ has finite index in Γ.*

The assumption of trivial center is only necessary to avoid the presence of non-trivial central normal subgroups in Γ, i.e. to have a particularly simple statement. The basic idea of the proof of this theorem is quite striking: to prove Γ/N is finite one proves that it is both amenable and has property (T). It is easy to check that any countable group with those two properties is finite. For many Γ, for instance whenever G is simple, property (T) holds for Γ and all its quotients by the work of Kazhdan or Wang. In this case one only needs to prove amenability. The proof of this is also striking: from non-amenability of Γ one produces relatively easily a non-trivial measurable Γ-quotient $G/P \to X$, where the Γ-action on X is via Γ/N. Margulis then proves that all measurable Γ-quotients of G/P are of the form G/Q, a contradiction since the Γ-action on G/Q is faithful. This rigidity of measurable quotients has also had a broad and diverse mathematical impact, playing a role in the study of general ergodic G-actions [91] and uniformly thin subgroups [48], and, slightly more indirectly, in the proof of Zimmer's conjecture [19, 8]. For a somewhat different approach to the existence of projective factors than Margulis' original one that is closer in spirit to its use in the proof of Zimmer's conjecture, see [18]. In Margulis' original proof of the factor theorem, the proof is cast in the language of understanding invariant subalgebras of the algebra of measurable functions on G/P. In [18], the authors recast the proof in terms of studying invariant measures instead. In addition to being related to the work in [19], this also connects the proof of the factor theorem to Margulis' work on rigidity of invariant measures in homogeneous dynamics.

4 Expanders, relative property (T) and lattices

In this section we mention an important innovation of Margulis in theoretical computer science in terms of expander graphs.

We consider here only finite regular graphs, which we denote by $X = X(V,E)$, with a set V of n vertices each of degree k and a set E of $\frac{kn}{2}$ edges. Given a subset A of V, we define the *boundary* of A to be $\partial A = \{y \in V : d(A,y) = 1\}$, where $d(\cdot,\cdot)$ is the standard graph distance obtained by making each edge length one.

Definition 4.1. We say X as above is an (n,k,c)*-expander* if for every $A \subset V$ we have

$$|\partial A| \geq c\left(1 - \frac{|A|}{n}\right)|A|.$$

An infinite sequence X_i of graphs is a called an *expander family* if each X_i is a $(|V_i|,k,c)$-expander for fixed k and c.

We can also define expander graphs in terms of the graph Laplacian Δ. This is the standard operator that averages functions on $L^2(V)$ over nearest neighbors. There is a well-known relationship between expansion as defined above and having a lower bound on the first non-trivial eigenvalue of Δ on $L^2(V)$.

Prior to Margulis' work, the only known constructions of expander families were random and did not give explicit examples. Margulis constructed the first explicit families of expanders by using variants of Kazhdan's property (T) and group theory. The study of expanders and of group-theoretic constructions of expanders remains a broad and vibrant field, see [16, 51].

Definition 4.2. Let G be a locally compact topological group. G has *property (T) of Kazhdan* if there is an $\varepsilon > 0$ and a compact set $K \subset G$ such that for every continuous unitary representation π of G on a Hilbert space $\mathcal{H}$ without invariant vectors and every $v \in \mathcal{H}$ there is a $k \in K$ such that $\|\pi(k)v - v\| > \varepsilon\|v\|$.

It is perhaps easier to understand the contrapositive of this definition. To make this clear say that π *almost has invariant vectors* if for every $\varepsilon > 0$ there exists a vector $v \in \mathcal{H}$ such that $\|\pi(k)v - v\| < \varepsilon\|v\|$ for every k in K. Then the definition says that G

has property (T) if and only if whenever π almost has invariant vectors, it actually has non-trivial invariant vectors.

An early observation of Margulis is the following. Fix a discrete group Γ with property (T), for example $\mathrm{SL}_n(\mathbb{Z})$ for $n \geq 3$, and a finite generating set S of Γ. In addition fix a family of finite index subgroups N_i of Γ, for instance the kernels of reduction mod primes in $\mathrm{SL}_n(\mathbb{Z})$. The the family of Cayley graphs for Γ/N_i with respect to the fixed generating set S is an expander family. This can be verified in either definition of expander family mentioned above.

In fact, Margulis constructed an even more explicit family of expanders using a notion he introduced called relative property (T). The point of this family was not only to have an explicit construction but a very concrete one.

Definition 4.3. Let G be a locally compact topological group and $H \subset G$ a closed subgroup. We say the pair (G,H) has *relative property* (T) if there exist $\varepsilon > 0$ and a compact set $K \subset G$ such that for every continuous unitary representation π of G on a Hilbert space $\mathcal{H}$ without non-trivial H-invariant vectors and every $v \in \mathcal{H}$ there is a $k \in K$ such that $\|\pi(k)v - v\| > \varepsilon\|v\|$.

This is strictly weaker than property (T) unless $G = H$. It says that if a representation π of G on $\mathcal{H}$ almost has invariant vectors, then it has a vector invariant under the subgroup H.

Margulis proved that $\mathrm{SL}_2(\mathbb{Z}) \ltimes \mathbb{Z}^2$ has relative property (T) with respect to the subgroup $\mathbb{Z}^2$ and used this to produce a very explicit graph on the vertex set $(\mathbb{Z}/n\mathbb{Z})^2$ [75]. Explicitly, the vertex (x,y) is joined by an edge to the following 8 vertices

$$(x \pm 2y, y),\ \big(x \pm (2y+1), y\big),\ (x, y \pm 2x),\ \big(x, y \pm (2x+1)\big).$$

These particular graphs were later studied in more detail by Gabber and Galil and are often termed Margulis–Gabber–Galil expanders [49].

Somewhat later Margulis gave a construction of optimal expanders which are now called *Ramanujan graphs* and which were discovered independently by Lubotzky, Phillips and Sarnak [72, 80].

5 Local rigidity of group actions

This section concerns work of the second-named author with Margulis on rigidity of group actions on compact manifolds, and in particular local rigidity of actions of higher rank semisimple groups and their lattices. This general area of research, largely initiated by Zimmer, was inspired by certain kinds of generalizations of Margulis' superrigidity theorem. The easier one to describe is simply to think of extending superrigidity to targets that are diffeomorphism groups of compact manifolds; in other words, to attempt to classify homomorphisms $\rho : \Gamma \to \mathrm{Diff}(M)$, where M is a compact manifold, G is a simple Lie group of higher rank and $\Gamma \subset G$ is a lattice. In general this question is quite difficult and perhaps even intractable, but in particular cases striking results can be obtained. The work of Fisher and Margulis concentrates on the particular question of *local rigidity.*

Definition 5.1. Let D be a topological group, Γ a discrete group and $\rho : \Gamma \to D$ a homomorphism. Then ρ is *locally rigid* if any homomorphism ρ' that is close to ρ in the compact open topology is conjugate to ρ by a small element of D.

The point is that for any ρ, there are trivially nearby ρ' defined by conjugating ρ by small elements of D to obtain ρ'. Local rigidity says that the only nearby representations are those that occur for this trivial reason. Local rigidity in the context of homomorphisms into Lie groups has a long history and was first studied by Selberg, Weil, Calabi, Raghunathan and others in the 1960s. The main impetus for the study of local rigidity of the defining inclusion of a lattice in a Lie group was that it implied the lattice could be conjugated to have matrix entries in a number field; this was observed by Selberg in [99]. This result of Selberg was a major motivation for the conjecture that become Margulis' arithmeticity theorem.

In a series of three papers, the second-named author and Margulis proved some very broad theorems on local rigidity of group actions in the context of the Zimmer program. The first is very general and simple to state [46].

Theorem 5.2. *Let Γ be a discrete group with property (T), and let M be a compact manifold. Then any homomorphism $\rho : \Gamma \to \mathrm{Isom}(M,g)$ is locally rigid when viewed as a homomorphism into* $\mathrm{Diff}(M)$.

The theorem both applies and is non-trivial even when ρ is trivial. There are also many non-trivial isometric actions of groups with property (T) on compact manifolds. Earlier work of Zimmer and Benveniste had provided partial results towards Theorem 5.2.

The more general results obtained by Margulis and the second-named author in [47] require an additional definition. Let $M = H/\Lambda$ be a homogeneous manifold. A diffeomorphism f of M is called *affine* if it can be written as a composition $L_h \circ A$, where A is an automorphism of H preserving Λ and L_h is left translation by an element h in H. An action $\rho : D \to \mathrm{Diff}(H/\Lambda)$ is called affine if each $\rho(d)$ is affine.

Theorem 5.3. *Let H be a Lie group and $\Lambda \subset H$ a cocompact lattice, so that $H/\Lambda =$ M is a compact manifold. Let G be a semisimple Lie group with no compact factors and all simple factors of higher rank, let $\Gamma \subset G$ be a lattice, and let $D = G$ or Γ. Then any affine action $\rho : D \to \mathrm{Diff}(H/\Lambda)$ is locally rigid.*

In [45], the second-named author and Margulis classify the affine actions of D as in Theorem 5.3 on manifolds of the form H/Λ. This is a somewhat involved algebraic computation based on Margulis' superrigidity theorem. Examples constructed by Katok–Lewis and Benveniste show that the assumption that the initial action is affine is in some sense necessary: there are modifications of affine actions using blow up constructions that are not locally rigid [54, 5].

The proof of Theorem 5.3 is quite long and complicated and spans three papers. The first paper [45] proves the correct general version of Zimmer's cocycle superrigidity theorem and then shows that perturbations of constant cocycles have a particularly nice form. The second paper [46] proves both Theorem 5.2 as well as a much more difficult foliated version of it. This paper also makes several generalizations of property (T) that have had a large impact on later research, both by considering Banach spaces that are not Hilbertian and by considering actions that are not isometric or even not globally defined. Finally, the paper [47] combines the ingredients from the first two papers along with several new ideas to prove Theorem 5.3. An additional key ingredient comes from the work of Hirsch, Pugh and Shub on stability of partially hyperbolic actions [52].

The key difference between Theorem 5.3 and prior work is that it makes no assumptions on hyperbolicity of the group action. The proof does involve factoring the action in a certain sense into hyperbolic and isometric parts, but the fact that such a factorization exists in the perturbed action is highly non-trivial and must be established. These papers have had a profound influence on further progress in the Zimmer program, including a quite direct impact of ideas in [46] on the work of Brown, Fisher and Hurtado on Zimmer's Conjecture [8].

6 Dynamical systems on homogeneous spaces: an introduction

In the remaining part of this survey we present an exposition of the contributions of Margulis in the area of homogeneous dynamics, that is, dynamical and ergodic properties of actions on homogeneous spaces of Lie groups. The reader is referred to [67] for a detailed survey of the field. Given a Lie group G and a closed subgroup $\Gamma \subset G$, one can consider the left action of any subgroup $F \subset G$ on G/Γ:

$$x \mapsto gx, \; x \in G/\Gamma, \; g \in F.$$

When F is a one-parameter subgroup, the action thus obtained is called a *homogeneous (one-parameter) flow*. Classical examples are given by geodesic and horocycle flows on surfaces of constant negative curvature, extensively studied in the 1930s–50s using geometric and representation-theoretic methods.

We remark that geodesic flows on surfaces of constant negative curvature are prototypical examples of Anosov flows, and orbits of horocycle flows are stable and unstable leaves relative to the geodesic actions.

Margulis' famous PhD Thesis "On some properties of Anosov flows" (or rather of U-systems, as they were called by Anosov back then), written in 1969 and published in 2004 [87], made a foundational contribution to the theory. However, we will not be covering it here as we are limiting the scope to homogenous dynamics.

The space of lattices in $\mathbb{R}^n$. Here we describe a family of homogeneous spaces particularly important for number-theoretic applications. Let $G = \mathrm{SL}_n(\mathbb{R})$, and let $\mathscr{L}_n$ denote the space of unimodular lattices in $\mathbb{R}^n$. (By definition, a lattice Δ is unimodular iff the volume of $\mathbb{R}^n/\Delta$ is equal to 1.) G acts on $\mathscr{L}_n$ as follows: if $g \in G$ and $\Delta \in \mathscr{L}_n$ is the $\mathbb{Z}$-span of the vectors $v_1, \dots v_n$, then gv is the $\mathbb{Z}$-span of $gv_1, \dots, gv_n$. This action is clearly transitive. The stabilizer of the standard lattice $\mathbb{Z}^n$ is $\Gamma = \mathrm{SL}_n(\mathbb{Z})$. This gives an identification of $\mathscr{L}_n$ with G/Γ.

One can consider a Haar measure on G (both left and right invariant) and the corresponding left-invariant measure on $\mathscr{L}_n$, which, as is well known, happens to be finite; that is, Γ is a lattice in G. An important feature of the quotient topology on G/Γ is that $\mathscr{L}_n$ is not compact (in other words, Γ is non-uniform). More precisely, Mahler's Compactness Criterion says that a subset Q of $\mathscr{L}_n$ is bounded if and only if there exists an $\varepsilon > 0$ such that for any $\Lambda \in Q$ one has $\inf_{\mathbf{x} \in \Lambda \smallsetminus \{0\}} \|\mathbf{x}\| \geq \varepsilon$. In other words, for $\varepsilon > 0$ let $\mathscr{L}_n(\varepsilon) \subset \mathscr{L}_n$ denote the set of lattices whose shortest non-zero vector has length at least ε. Then for any $\varepsilon > 0$ the set $\mathscr{L}_n(\varepsilon)$ is compact.

Unipotent, quasi-unipotent and partially hyperbolic flows. Recall that a square matrix is called *unipotent* (resp. *quasi-unipotent*) if all its eigenvalues are equal to 1 (resp. of absolute value 1). In a general Lie group an element is unipotent/quasi-unipotent if so is its adjoint (acting on the Lie algebra). We will say that a subgroup is unipotent/quasi-unipotent if all its elements are. Examples of unipotent one-parameter subgroups:

$$\left\{ \begin{pmatrix} 1 & t \\ 0 & 1 \end{pmatrix} : t \in \mathbb{R} \right\} \subset \mathrm{SL}_2(\mathbb{R}) \tag{1}$$

(the action of this subgroup on $\mathrm{SL}_2(\mathbb{R})/\Gamma$ induces the horocycle flow on the unit tangent bundle to the quotient of the hyperbolic plane by Γ), and

$$\left\{ \begin{pmatrix} 1 & t & t^2/2 \\ 0 & 1 & t \\ 0 & 0 & 1 \end{pmatrix} : t \in \mathbb{R} \right\} \subset \mathrm{SL}_3(\mathbb{R}).$$

As was mentioned in Section 2, unipotent flows entered the research of Margulis in the beginning of the 1970s, when their non-divergence property became an important ingredient in the proof of arithmeticity of non-uniform lattices. Then in the mid 1970s Raghunathan, Dani and Margulis made a series of conjectures describing rigidity properties of actions of unipotent one-parameter subgroups, as well as those generated by unipotent elements, eventually proved in full generality by Ratner.

We note that quasi-unipotent flows are not far from the unipotent ones in terms of their dynamical properties (see [102] or [67, §4.2]). Let us say that an element (or a subgroup) of G is *partially hyperbolic* it is not quasi-unipotent. The dichotomy "quasi-unipotent vs. partially hyperbolic" can also be characterized in terms of the speed of mixing (at most polynomial vs. exponential) or entropy (zero vs. positive). Thus it is not surprising that the properties of partially hyperbolic flows are drastically different from those of unipotent flows. In the next few sections we describe the contributions of Margulis and his co-authors to both unipotent and partially hyperbolic dynamics on homogeneous spaces, as well as plentiful connections between dynamics and number theory.

7 Quantitative non-divergence

Let $U = \{u_t\}_{t\in\mathbb{R}}$ be a unipotent one-parameter subgroup of G. Consider the left action of U of $\mathrm{SL}_n(\mathbb{R})$ on $\mathscr{L}_n$. When $n = 2$, every unipotent subgroup is conjugate to (1), and it is easy to show, see §7.1, that every orbit spends relatively little time outside $\mathscr{L}_2(\varepsilon)$. In [73, 77] Margulis proved the following result:

Theorem 7.1. *For any one-parameter subgroup $\{u_t\}$ of $\mathrm{SL}_n(\mathbb{R})$, the orbit of every lattice in $\mathscr{L}_n$ under the semi-group $\{u_t : t \geq 0\}$ does not diverge to infinity.*

This was used by Margulis to prove arithmeticity of higher rank lattice subgroups of semisimple Lie groups. Several years later Dani [21] obtained a quantitative strengthening of the initial nondivergence result by showing that such orbits return into a suitably chosen compact set with positive frequency. To be more precise, Dani proved that for any $\Lambda \in \mathscr{L}_n$ there are $0 < \varepsilon, c < 1$ such that for any $T > 0$ one has

$$\left|\{t \in [0,T] : u_t\Lambda \notin \mathscr{L}_n(\varepsilon)\}\right| < cT. \tag{2}$$

(Here and hereafter for a set $E \subset \mathbb{R}$, $|E|$ denotes the Lebesgue measure of E.) These ideas were developed during later work on the Oppenheim conjecture and related topics, see [22, 23, 25, 27, 28, 30, 31]. In this section we only present the result of [61] which gives an explicit dependence of c on ε in (2) and includes the earlier results on quantitative non-divergence as special cases. A more detailed account is given in [58, 11].

7.1 An elementary non-divergence result

Even though it does not capture much of the difficulty of the problem, we start with the $\mathrm{SL}_2(\mathbb{R})$ case as a motivation.

Lemma 7.2. *Suppose $T > 0$, $\Lambda \in \mathscr{L}_2$ and $0 < \rho < 1/\sqrt{2}$ are such that*

$$\forall v \in \Lambda \smallsetminus \{0\} \qquad \sup_{t\in[0,T]} \|u_t v\| \geq \rho. \tag{3}$$

Then for any $\varepsilon < \rho$,

$$|\{t \in [0,T] \ : \ u_t\Lambda \notin \mathscr{L}_2(\varepsilon)\}| \leq 2\left(\frac{\varepsilon}{\rho}\right) T. \tag{4}$$

This can be interpreted as follows. Suppose ρ is the length of the shortest vector in Λ. Then (3) holds. Thus for any $\varepsilon < \rho$ the lemma gives the quantitative statement (4), which says that the trajectory $\{u_t\Lambda\}$, where t ranges from 0 to T, spends little time outside of $\mathscr{L}_2(\varepsilon)$.

Proof. Recall that a vector $v \in \Lambda$ is said to be primitive in Λ if $\mathbb{R}v \cap \Lambda$ is generated by v as a $\mathbb{Z}$-module. Now for $r > 0$ and a primitive $v \in \Lambda$ consider

$$B_v(r) \stackrel{\text{def}}{=} \{t \in B \ : \ \|u_t v\| < r\},$$

where $\|\cdot\|$ is the supremum norm. Let $v = \begin{pmatrix} a \\ b \end{pmatrix} \in P(\Lambda)$ be such that $B_v(\varepsilon) \neq \varnothing$. Then, since $u_t v = \begin{pmatrix} a+bt \\ b \end{pmatrix}$, it follows that $|b| < \varepsilon$, and (3) implies that b is nonzero. Therefore, if we define $f(t) = a + bt$, we have

$$B_v(\varepsilon) = \{t \in [0,T] \ : \ |f(t)| < \varepsilon\} \quad \text{and} \quad B_v(\rho) = \{x \in [0,T] \ : \ |f(t)| < \rho\}.$$

Clearly the ratio of lengths of intervals $B_v(\varepsilon)$ and $B_v(\rho)$ is bounded from above by $2\varepsilon/\rho$ (by looking at the worst case when $B_v(\varepsilon)$ is close to one of the endpoints of B). Since

$$\begin{array}{c}\text{a unimodular lattice in } \mathbb{R}^2 \text{ cannot contain} \\ \text{two linearly independent vectors each of length } < \rho,\end{array} \tag{5}$$

the sets $B_v(\rho)$ are disjoint for different primitive $v \in \Lambda$. Also it is clear that $u_t\Lambda \notin \mathscr{L}_2(\varepsilon)$ whenever $t \in B_v(\rho) \smallsetminus B_v(\varepsilon)$ for some primitive $v \in \Lambda$. Thus we conclude that

$$\begin{aligned}|\{x \in [0,T] \ : \ u_x\Lambda \notin \mathscr{L}_2(\varepsilon)\}| &\leq \sum_v |B_v(\varepsilon)| \\ &\leq 2\frac{\varepsilon}{\rho}\sum_v |B_v(\rho))| \leq 2\frac{\varepsilon}{\rho}T.\end{aligned}$$ □

7.2 *The general case*

In this section, we present a generalization of Lemma 7.2, which is in particular valid for any dimension.

First, we note that the group structure of $U = \{u_t : t \in \mathbb{R}\}$ is not used in the proof of Lemma 7.2. In fact it was already observed in [73] that the feature important for the proof is the polynomial nature of the map $t \mapsto u_t$. More generally, the third-named author and Margulis introduced the following definition:

Definition 7.3. If C and α are positive numbers and B is a subset of $\mathbb{R}^d$, let us say that a function $f : B \mapsto \mathbb{R}$ is *(C,α)-good on B* if for any open ball $J \subset B$ and any $\varepsilon > 0$ one has

$$|\{x \in J : |f(x)| < \varepsilon\}| \leq C\left(\frac{\varepsilon}{\sup_{x\in J}|f(x)|}\right)^{\alpha}|J|. \tag{6}$$

This definition captures the property of unipotent orbits used in the proof of Lemma 7.2.

Lemma 7.4. (a) *f is (C,α)-good on $B \Leftrightarrow$ so is $|f| \Rightarrow$ so is cf $\forall c \in \mathbb{R}$;*
(b) *f_i, $i=1,\dots,k$, are (C,α)-good on $B \Rightarrow$ so is $\sup_i |f_i|$;*
(c) *If f is (C,α)-good on B and $c_1 \le \left|\frac{f(x)}{g(x)}\right| \le c_2$ for all $x \in B$, then g is $\big(C(c_2/c_1)^\alpha,\alpha\big)$-good on B;*

The notion of (C,α)-good functions was introduced in [61] in 1998, but the importance of (6) for measure estimates on the space of lattices was observed earlier. For instance, the next proposition can be traced to [31, Lemma 4.1].

Proposition 7.5. *For any $k \in \mathbb{N}$, any polynomial of degree not greater than k is $\big(k(k+1)^{1/k}, 1/k\big)$-good on $\mathbb{R}$.*

As a corollary, we have:

Corollary 7.1. *For any $n \in \mathbb{N}$ there exist (explicitly computable) $C = C(n)$, $\alpha = \alpha(n)$ such that for any one-parameter unipotent subgroup $\{u_t\}$ of $\mathrm{SL}_n(\mathbb{R})$, any $\Lambda \in \mathscr{L}_n$ and any subgroup Δ of Λ, the function $t \mapsto \|u_t\Delta\|$ is (C,α)-good.*

The following is the main non-divergence result of the third-named author and Margulis. In particular, it is a generalization of Lemma 7.2 to the case of arbitrary n.

Theorem 7.6 ([61]). *Suppose $d,n \in \mathbb{N}$, a lattice $\Lambda \subset \mathbb{R}^n$, a ball $B = B(x_0,r_0) \subset \mathbb{R}^d$, $C,\alpha > 0$, $0 < \rho < 1/n$ and a continuous map $h : \tilde{B} \to \mathrm{SL}_n(\mathbb{R})$ are given, where $\tilde{B} = B(x_0, 3^n r_0)$. Assume that for any primitive subgroup $\Delta \subset \Lambda$,*

(i) *the function $x \mapsto \|h(x)\Delta\|$ is (C,α)-good on $\tilde{B}$, and*
(ii) $\sup_{x\in\tilde{B}} \|h(x)\Delta\| \ge \rho$.

Then for any $\varepsilon < \rho$,

$$|\{x \in B \ :\ h(x)\Lambda \notin \mathscr{L}_n(\varepsilon)\}| \le Cc(n,d)\left(\frac{\varepsilon}{\rho}\right)^\alpha |B|$$

where $c(n,d)$ is an explicit constant.

Here is one of the key ideas used in the proof of Theorem 7.6.

Lattices, subspaces and flags. Let Λ be a lattice in $\mathbf{b}R^n$. We say that a subspace L of $\mathbf{b}R^n$ is *Λ-rational* if $L \cap \Lambda$ is a lattice in L. For any Λ-rational subspace L, we denote by $d_\Lambda(L)$ or simply by $d(L)$ the volume of $L/(L\cap\Lambda)$. Let us note that $d(L)$ is equal to the norm of $e_1 \wedge \cdots \wedge e_\ell$ in the exterior power $\bigwedge^\ell(\mathbf{b}R^n)$, where $\ell = \dim L$ and $(e_1,\dots,e_\ell)$ is a basis over $\mathbb{Z}$ of $L \cap \Lambda$. If $L = \{0\}$ we write $d(L) = 1$.
Recall that a flag is an ascending chain of subspaces $V_1 \subset \cdots \subset V_k$ of $\mathbb{R}^n$. We say that a flag is Λ-rational if all of its subspaces are.
We now present the substitute for (5) needed to work with $n > 2$. The following definition is taken from [59] but it is implicit in [61] and also in [73, 77].

Definition 7.7. Suppose $\Lambda \subset \mathbb{R}^n$ is a lattice, $0 < \varepsilon < \eta$ are constants, and F is a Λ-rational flag. We say that Λ is *marked* by (F, ε, η) if the following hold:

(M1) For any subspace $V \in F$, $d_\Lambda(V) \leq \rho$.
(M2) For any subspace $V \in F$, $d_\Lambda(V) \geq \varepsilon$.
(M3) F is maximal among all the Λ-rational flags satisfying (M1).

The higher-dimensional analogue of (5) is the following:

Proposition 7.8. *Suppose Λ is a lattice, $0 < \varepsilon < \eta < 1$, and suppose there exists a Λ-rational flag F such that Λ is marked by (F, ε, η). Then $\Lambda \in \mathscr{L}_n(\varepsilon)$.*

Proof. Write $F = (V_1, \dots, V_m)$. Then, by (M1) and (M2), for $1 \leq i \leq m$,

$$\varepsilon \leq d_\Lambda(V_i) \leq \eta < 1.$$

Suppose $\Lambda \notin \mathscr{L}_n(\varepsilon)$, then there exists a $v \in \Lambda$ such that $\|v\| < \varepsilon$. Let i be such that $v \in V_i$, $v \notin V_{i+1}$, and let $V = V_i + \mathbb{R}v$. Then V is a Λ-rational subspace. Let $v_1, \dots, v_{k-1}$ be a basis for $V_i \cap \Lambda$, and let v_k be such that $v_1, \dots, v_k$ is a basis for $V \cap \Lambda$. Then,

$$d_\Lambda(V) = \|v_1 \wedge \cdots \wedge v_k\| \leq \|v_1 \wedge \cdots \wedge v_{k-1}\| \|v_k\|.$$

Thus, one has

$$d_\Lambda(V) \leq d_\Lambda(V_i)\|v\| < \varepsilon.$$

In particular, $d_\Lambda(V) < \rho$, and thus by (M3), $V = V_{i+1}$. Then $d_\Lambda(V_{i+1}) < \varepsilon$, contradicting (M1). □

Proof of Theorem 7.6. The proof is a complicated inductive argument based on the idea of the proof of Lemma 7.2 and on Proposition 7.8. We refer the reader to [61], [59] or [58] for details. □

7.3 Applications to Diophantine approximation on manifolds

In this section we list some of the applications of Theorem 7.6 to metric Diophantine approximation. A much more detailed and comprehensive survey is given in [11].

Mahler's conjecture, proved by Sprindžuk in 1964 [101], is the following statement: for any $n \in \mathbb{N}$, any $\varepsilon > 0$ and for almost any real number x, the inequality

$$|p + q_1 x + q_2 x^2 + \cdots + q_n x^n| < \|\mathbf{q}\|^{-n(1+\varepsilon)} \tag{7}$$

has only finitely many solutions $(p, \mathbf{q}) \in \mathbb{Z} \times \mathbb{Z}^n$, where $\mathbf{q} = (q_1, \dots, q_n)$ and $\|\mathbf{q}\| = \max_{1 \leq i \leq n} |q_i|$.

After Sprindžuk's result, several conjectural improvements were proposed. Baker [3] proposed replacing $\|\mathbf{q}\|^n$ in (7) by $\Pi_+(\mathbf{q})^{-(1+\varepsilon)}$, where $\Pi_+(\mathbf{q}) = \prod_{i=1}^n \max(1,|q_i|)$. (This is indeed an improvement of (7) since $\Pi_+(\mathbf{q}) \le \|\mathbf{q}\|^n$.) Sprindžuk proposed replacing the powers of x in (7) by arbitrary analytic functions, which together with 1 are linearly independent over $\mathbb{R}$. In [61], the third-named author and Margulis prove a combination of both of these conjectures. In fact, their result applies to a more general class of functions which need not be real analytic. We thus make the following:

Definition 7.9. Let $\mathbf{f} = (f_1,\dots,f_n) : U \to \mathbb{R}^n$ be a map defined on an open subset U of $\mathbb{R}^d$. Given a point $x_0 \in U$, we say that $\mathbf{f}$ is *ℓ-non-degenerate at* x_0 if $\mathbf{f}$ is ℓ times continuously differentiable on some sufficiently small ball centered at x_0 and the partial derivatives of $\mathbf{f}$ at x_0 of orders up to ℓ span $\mathbb{R}^n$. The map $\mathbf{f}$ is called *non-degenerate at* x_0 if it is ℓ-non-degenerate at x_0 for some $\ell \in \mathbb{N}$; $\mathbf{f}$ is called *non-degenerate almost everywhere* (in U) if it is non-degenerate at almost every $x_0 \in U$ with respect to Lebesgue measure. The non-degeneracy of differentiable submanifolds of $\mathbb{R}^n$ is defined via their parameterisation(s).

Note that a real analytic map $\mathbf{f}$ defined on a connected open set is non-degenerate almost everywhere if and only if $1, f_1, \dots f_n$ are linearly independent over $\mathbb{R}$.

We are now ready to state the main result of [61], which solves the Baker and Sprinžuk conjectures in full generality, and also applies to non-degenerate maps.

Theorem 7.10 ([61, Theorem A]). *Let $\mathbf{f} = (f_1,\dots,f_n)$ be a map defined on an open subset U of $\mathbb{R}^d$ which is non-degenerate almost everywhere. Then for any $\varepsilon > 0$, for almost every $x \in U$, the inequality*

$$|p + q_1 f_1(x) + \cdots + q_n f_n(x)| < \Pi_+(\mathbf{q})^{-1-\varepsilon} \tag{8}$$

has only finitely many solutions $(p,\mathbf{q}) \in \mathbb{Z} \times \mathbb{Z}^n$.

Strategy of proof of Theorem 7.10. Define

$$u_{\mathbf{f}(x)} = \begin{pmatrix} 1 & \mathbf{f}(x) \\ 0 & I_n \end{pmatrix} \in \mathrm{SL}_{n+1}(\mathbb{R}),$$

where I_n is the $n \times n$ identity matrix. Also for $\mathbf{t} = (t_1,\dots,t_n) \in \mathbb{Z}^n_{\ge 0}$, define

$$g_{\mathbf{t}} = \begin{pmatrix} e^t & & & \\ & e^{-t_1} & & \\ & & \ddots & \\ & & & e^{-t_n} \end{pmatrix}, \text{where } t = t_1 + \cdots + t_n.$$

Given a solution $(p,\mathbf{q})$ to (8), define $t_i \in \mathbb{Z}^n_{\ge 0}$ be the smallest integers such that

$$e^{-t_i} \max(1,|q_i|) \le \Pi_+(\mathbf{q})^{-\varepsilon/(n+1)}.$$

Then, an elementary computation using (8) shows that

$$e^t|p+q_1f_1(x)+\cdots+q_nf_n(x)| < e^{n(1+\gamma)}e^{-\gamma t}, \tag{9}$$

where $\gamma = \varepsilon/(n+1+n\varepsilon)$.

For $\Lambda \in \mathscr{L}_{n+1}$, let $\delta(\Lambda)$ denote the length of the shortest non-zero vector in Λ. (Thus, $\Lambda \in \mathscr{L}_{n+1}(\varepsilon)$ if and only if $\delta(\Lambda) \geq \varepsilon$.) Therefore, it follows from (9) that if $(p,\mathbf{q})$ is a solution to (8) and t, $\mathbf{t}$ are as above, then

$$\delta(g_{\mathbf{t}}u_{\mathbf{f}(x)}\mathbb{Z}^{n+1}) < e^{n(1+\gamma)}e^{-\gamma t}. \tag{10}$$

Thus it is enough to prove that for any sufficiently small ball B centered at any point x_0 on which $\mathbf{f}$ is non-degenerate,

$$\sum_{\mathbf{t}} |\{x \in B \ : \ \delta(g_{\mathbf{t}}u_{\mathbf{f}(x)}\mathbb{Z}^{n+1}) < e^{n(1+\gamma)}e^{-\gamma t}\}| < \infty. \tag{11}$$

Indeed, if (11) holds, then the Borel–Cantelli Lemma ensures that for almost all $x \in B$, (10) holds only for finitely many $\mathbf{t}$. Equation (11) is proved in [61] by verifying the conditions of Theorem 7.6.

Khintchine–Groshev type results. The following generalization of Theorem 7.10 is proved in [14] by Bernik, the third-named author and Margulis:

Theorem 7.11. *Let $\mathbf{f} = (f_1,\ldots,f_n)$ be a map defined on an open subset U of $\mathbb{R}^d$ which is non-degenerate almost everywhere. Let $\Psi : \mathbb{Z}^n \to \mathbb{R}_+$ be any function such that*

$$\Psi(q_1,\ldots,q_i,\ldots,q_n) \leq \Psi(q_1,\ldots,q_i',\ldots,q_n) \qquad \text{if } |q_i| > |q_i'| \text{ and } q_iq_i' > 0.$$

Suppose that

$$\sum_{\mathbf{q}\in\mathbb{Z}^n} \Psi(q) < \infty.$$

Then for almost every $x \in U$, the inequality

$$|p+q_1f_1(x)+\cdots+q_nf_n(x)| < \Psi(\mathbf{q})$$

has only finitely many solutions $(p,\mathbf{q}) \in \mathbb{Z}^{n+1}$.

Results of this type for almost all vectors in $\mathbb{R}^n$ (without restricting to submanifolds) are due to Khintchine and Groshev; hence the name "Khintchine–Groshev Theorems" reserved for such statements (see also Section 10.2 below). Note that $\Psi(\mathbf{q}) = \Pi_+(\mathbf{q})^{-1-\varepsilon}$ for $\varepsilon > 0$ satisfies the conditions of Theorem 7.11, and thus Theorem 7.11 is indeed a generalization of Theorem 7.10. We note the following corollaries:

Corollary 7.12. *Let* $\mathbf{f}$ *be as in Theorem 7.11 and suppose* $\psi : \mathbb{R}_+ \to \mathbb{R}_+$ *is any monotonic function satisfying*

$$\sum_{k=1}^{\infty} \psi(k) < \infty.$$

Let $\mathbf{s} = (s_1, \ldots, s_n) \in \mathbb{R}^n_{>0}$ *be such that* $s_1 + \cdots + s_n = 1$*. For* $\mathbf{q} \in \mathbb{Z}^n$*, write* $\|\mathbf{q}\|_{\mathbf{s}} = \max_{1 \le i \le n} |q_i|^{1/s_i}$*. Then, for almost every* $x \in U$*, the equation*

$$|p + q_1 f_1(x) + \cdots + q_n f_n(x)| < \psi(\|\mathbf{q}\|_{\mathbf{s}})$$

has only finitely many solutions $(p, \mathbf{q}) \in \mathbb{Z}^{n+1}$*.*

Note that if $\mathbf{s} = (1/n, \ldots, 1/n)$ then $\|\mathbf{q}\|_{\mathbf{s}} = \|\mathbf{q}\|^n$. In this case Corollary 7.12 was proved previously by Beresnevich in [6] by a different method which does not involve quantitative non-divergence. However, without new ideas, it does not seem to be possible to extend this approach to in order to prove the full version of Corollary 7.12 or Corollary 7.13 below.

Corollary 7.13. *Let* $\mathbf{f}$ *be as in Theorem 7.11 and suppose* $\psi : \mathbb{R}_+ \to \mathbb{R}_+$ *is any monotonic function satisfying*

$$\sum_{k=1}^{\infty} (\log k)^{n-1} \psi(k) < \infty.$$

Then, for almost every $x \in U$*, the equation*

$$|p + q_1 f_1(x) + \cdots + q_n f_n(x)| < \psi\big(\Pi_+(\mathbf{q})\big)$$

has only finitely many solutions $(p, \mathbf{q}) \in \mathbb{Z}^{n+1}$*.*

Strategy of Proof of Theorem 7.11. The idea is to break up into two cases, depending on the size of the gradient $\nabla(\mathbf{q} \cdot \mathbf{f})$. If $\|\nabla(\mathbf{q} \cdot \mathbf{f})\|$ is large, a direct argument is used. If it is small, the problem can be reduced to Theorem 7.6 by choosing an appropriate function h.

We note that the estimates from [14] were used in a follow-up paper [4] of Margulis with Beresnevich, Bernik and the third-named author to establish the divergence case of the Khintchine–Groshev Theorem for non-degenerate manifolds. Some partial results towards a matrix analogue of Theorem 7.10 can be found in the work of Margulis with Beresnevich, Wang and the third-named author [66, 15]; this task was later accomplished by Aka, Breuillard, Rozenzweig and de Saxce in [1]. See [11] for a plethora of more recent applications of quantitative non-divergence estimates to Diophantine approximation.

8 Conjectures of Oppenheim and Raghunathan

In the 1970s unipotent flows made a dramatic appearance with regards to another, seemingly unrelated, problem rooted in number theory. Let

$$Q(x_1,\dots,x_n) = \sum_{1\le i\le j\le n} a_{ij}x_ix_j$$

be an indefinite quadratic form in n variables. It is clear that if Q is a multiple of a form with rational coefficients, then the set of values $Q(\mathbb{Z}^n)$ is a discrete subset of $\mathbb{R}$. Much deeper is the following conjecture:

Conjecture 8.1 (Oppenheim, 1931). *Suppose Q is not proportional to a rational form and $n \ge 5$. Then for every $\varepsilon > 0$ there exists an $x \in \mathbb{Z}^n \smallsetminus \{0\}$ such that $|Q(x)| < \varepsilon$.*

This conjecture was later extended by Davenport to $n \ge 3$. Note that it is easy to construct counterexamples when $n = 2$; see e.g. [40, Proposition 1.3].

In the mid 1970s Raghunathan observed a remarkable connection between the Oppenheim Conjecture and flows on the space of lattices $\mathscr{L}_n = G/\Gamma$, where $G = \mathrm{SL}_n(\mathbb{R})$ and $\Gamma = \mathrm{SL}_n(\mathbb{Z})$. (Implicitly this observation was made several decades earlier by Cassels and Swinnerton-Dyer, see [20].) It can be summarized as follows:

Observation 8.2 (Raghunathan). *Let Q be an indefinite quadratic form, and let $H = \mathrm{SO}(Q)$ denote its orthogonal group. Consider the orbit of the standard lattice $\mathbb{Z}^n \in \mathscr{L}_n$ under H. Then the following are equivalent:*

(a) *The orbit $H\mathbb{Z}^n$ is not relatively compact in $\mathscr{L}_n$.*
(b) *For all $\varepsilon > 0$ there exists $x \in \mathbb{Z}^n \smallsetminus \{0\}$ such that $|Q(x)| < \varepsilon$.*

Proof. Suppose (a) holds, so some sequence $h_k\mathbb{Z}^n$ leaves all compact sets. Then in view of the Mahler compactness criterion there exist $v_k \in h_k\mathbb{Z}^n \smallsetminus \{0\}$ such that $\|v_k\| \to 0$. Then, also by continuity, $Q(v_k) \to 0$. But then $h_k^{-1}v_k \in \mathbb{Z}^n \smallsetminus \{0\}$, and $Q(h_k^{-1}v_k) = Q(v_k) \to 0$. Thus (b) holds.

On the other hand, assuming (b) we get a sequence of nonzero integer vectors x_k such that $Q(x_k) \to 0$ as $k \to \infty$; then, using the transitivity of the H-action on the level sets of Q one find $h_k \in H$ such that $h_kx_k \to 0$ as $k \to \infty$, proving (a). □

Raghunathan also explained why the case $n = 2$ is different: in that case $H = \mathrm{SO}(Q)$ has no unipotent elements. On the other hand, H is generated by its unipotent one-parameter subgroups when $n > 2$. The aforementioned reduction of the Oppenheim conjecture to a problem in homogeneous dynamics motivated Raghunathan to make a far-reaching conjecture on the behavior of unipotent flows on homogeneous spaces:

Theorem 8.3 (Raghunathan's topological conjecture). *Let G be a Lie group, $\Gamma \subset G$ a lattice, and $U \subset G$ a one-parameter unipotent subgroup. Suppose $x \in G/\Gamma$. Then there exists a subgroup F of G (generated by unipotents) such that the closure $\overline{Ux}$ of the orbit Ux is Fx.*

The above theorem in this generality is due to Ratner, but we will first describe earlier developments in which significant special cases were proven by Margulis and Dani–Margulis. In the literature this conjecture was first stated in the paper [22] and in a more general form in [81] (when the subgroup U is not necessarily unipotent but generated by unipotent elements). Margulis's proof of the Oppenheim conjecture, given in [79, 81, 82], uses Raghunathan's observation and proceeds by showing that any relatively compact orbit of $\mathrm{SO}(2,1)$ in $\mathscr{L}_3$ is compact; this implies the Oppenheim Conjecture and can be easily derived from Theorem 8.3. For an account which stays reasonably close to Margulis's original proof see [13, Chapter 6].

Dani and Margulis were able to establish Theorem 8.3 in the special case when $G = \mathrm{SL}_3(\mathbb{R})$ and $U = \{u_t\}$ is a "generic" one-parameter unipotent subgroup of G; that is, such that $u_t - I$ has rank 2 for all $t \neq 0$. The work done in [28], together with the methods developed in [79, 81, 82, 27], suggested an approach for proving the Raghunathan conjecture in general by studying minimal invariant sets and limits of orbits of points tending to a minimal invariant set.

This strategy can be outlined as follows: Let x be a point in G/Γ, and U a connected unipotent subgroup of G. Denote by X the closure of Ux and consider a minimal closed U-invariant subset Y of X. Suppose that Ux is not closed (equivalently, X is not equal to Ux). Then X should contain "many" translations of Y by elements from the normalizer $N(U)$ of U not belonging to U. After that one can try to prove that X contains orbits of bigger and bigger unipotent subgroups until one reaches horospherical subgroups. The basic tool in this strategy is the following fact. Let y be a point in X, and let g_n be a sequence of elements in G such that g_n converges to 1, g_n does not belong to $N(U)$, and $y_n = g_n y$ belongs to X. Then X contains AY where A is a nontrivial connected subset in $N(U)$ containing 1 and "transversal" to U. To prove this one has to observe that the orbits Uy_n and Uy are "almost parallel" in the direction of $N(U)$ most of the time in "the intermediate range".

In fact the set AU as a subset of $N(U)/U$ is the image of a nontrivial rational map from U into $N(U)/U$. Moreover this rational map sends 1 to 1 and also comes from a polynomial map from U into the closure of G/U in the affine space V containing G/U. This affine space V is the space of the rational representation of G such that V contains a vector the stabilizer of which is U (the Chevalley theorem). Some elements of this proof are key to the current program of Lindenstrauss, Mohammadi, Margulis and Shah [71] of giving a fully effective version of Ratner's theorems.

Raghunathan's conjecture was eventually proved in full generality by Ratner, see [95]. It is worth pointing out that Ratner derived Theorem 8.3 from her measure classification theorem, conjectured earlier by Dani. Loosely speaking, it says that all U-invariant ergodic measures are very nice.

Theorem 8.4 (Ratner's measure classification theorem [94]). *Let G be a Lie group, $\Gamma \subset G$ a lattice. Let U be a one-parameter unipotent subgroup of G. Then any ergodic U-invariant measure is algebraic; namely, there exists $x \in G/\Gamma$ and a subgroup F of G such that Fx is closed, and μ is the F-invariant probability measure supported on Fx. (Also the group F is generated by unipotent elements and contains U).*

We note that following the publication of Ratner's papers, Margulis and Tomanov [88] gave a different proof of the measure classification theorem which in particular made use of entropy considerations. This proof turned out to be extremely influential for further developments in the area.

9 Linearization

In this section, we give a partial description of the "linearization" technique introduced in [31] and used for the proof of the lower bounds in the quantitative version of the Oppenheim conjecture. This technique, and in particular Theorem 9.3 below, is used in a multitude of applications of the theory of unipotent flows.

9.1 Non-ergodic measures invariant under a unipotent

The collection $\mathscr{H}$. Let G be a Lie group, Γ a discrete subgroup of G, and $\pi : G \to G/\Gamma$ the natural quotient map. Let $\mathscr{H}$ be the collection of all closed subgroups F of G such that $F \cap \Gamma$ is a lattice in F and the subgroup generated by unipotent one-parameter subgroups of G contained in F acts ergodically on $\pi(F) \cong F/(F \cap \Gamma)$ with respect to the F-invariant probability measure. This collection is countable (see [94, Theorem 1.1] or [31, Proposition 2.1] for different proofs of this result). Note that up to conjugation, $\mathscr{H}$ is the collection of groups which appear in the definition of algebraic measure in the statement of Theorem 8.4.

Let U be a unipotent one-parameter subgroup of G and let $F \in \mathscr{H}$. Define

$$N(F,U) = \{g \in G : U \subset gFg^{-1}\}$$
$$S(F,U) = \bigcup \{N(F',U) : F' \in \mathscr{H}, F' \subset F, \dim F' < \dim F\}.$$

It is clear that if $g \in N(F,U)$ and $F \in \mathscr{H}$, then the orbit $U\pi(g)$ is contained in the closed subset $\pi(gF)$. More precisely, it is possible to prove the following (cf. [90, Lemma 2.4]):

Lemma 9.1. *Let $g \in G$ and $F \in \mathscr{H}$. Then $g \in N(F,U) \smallsetminus S(F,U)$ if and only if the group gFg^{-1} is the smallest closed subgroup of G which contains U and whose orbit through $\pi(g)$ is closed in G/Γ. Moreover in this case the action of U on $g\pi(F)$ is ergodic with respect to a finite gFg^{-1}-invariant measure.*

As a consequence of this lemma, one has

$$\pi\big(N(F,U)\smallsetminus S(F,U)\big) = \pi\big(N(F,U)\big)\smallsetminus \pi\big(S(F,U)\big) \qquad \forall F\in\mathscr{H}.$$

Theorem 8.4 states that given any U-ergodic invariant probability measure on G/Γ, there exists $F\in\mathscr{H}$ and $g\in G$ such that μ is $g^{-1}Fg$-invariant and $\mu\left(\pi(F)g\right) = 1$. Now decomposing any finite invariant measure into its ergodic component and using Lemma 9.1, one obtains the following description for any U-invariant probability measure on G/Γ (see [90, Theorem 2.2]).

Theorem 9.2 (Ratner). *Let U be a unipotent one-parameter subgroup of G and let μ be a finite U-invariant measure on G/Γ. For every $F\in\mathscr{H}$, let μ_F denote the restriction of μ to $\pi\big(N(F,U)\smallsetminus S(F,U)\big)$. Then μ_F is U-invariant, and any U-ergodic component of μ_F is a gFg^{-1}-invariant measure on the closed orbit $g\pi(F)$ for some $g\in N(F,U)\smallsetminus S(F,U)$.*

In particular, for all Borel measurable subsets A of G/Γ,

$$\mu(A) = \sum_{F\in\mathscr{H}^*} \mu_F(A),$$

where $\mathscr{H}^*\subset\mathscr{H}$ is a countable set consisting of one representative from each Γ-conjugacy class of elements in $\mathscr{H}$.

Remark. One often uses Theorem 9.2 in the following form: suppose μ is any U-invariant measure on G/Γ which is not G-invariant. Then there exists an $F\in\mathscr{H}$ such that μ gives positive measure to some compact subset of $N(F,U)\smallsetminus S(F,U)$.

9.2 The theorem of Dani–Margulis on uniform convergence

The "linearization" technique of Dani and Margulis was devised to understand which measures give positive weight to compact subsets subsets of $N(F,U)\smallsetminus S(F,U)$. Using this technique Dani and Margulis proved the following theorem, which is important for many applications.

Theorem 9.3 ([31, Theorem 3]). *Let G be a connected Lie group and let Γ be a lattice in G. Let μ be the G-invariant probability measure on G/Γ. Let $U=\{u_t\}$ be an Ad-unipotent one-parameter subgroup of G and let f be a bounded continuous function on G/Γ. Let $\mathscr{D}$ be a compact subset of G/Γ and let $\varepsilon>0$ be given. Then there exist finitely many proper closed subgroups $F_1=F_1(f,\mathscr{D},\varepsilon),\dots,F_k=F_k(f,\mathscr{D},\varepsilon)$ such that $F_i\cap\Gamma$ is a lattice in F_i for all i, and compact subsets $C_1=C_1(f,\mathscr{D},\varepsilon),\dots,C_k=C_k(f,\mathscr{D},\varepsilon)$ of $N(F_1,U),\dots,N(F_k,U)$ respectively, for which the following holds: For any compact subset $\mathscr{K}$ of $\mathscr{D}\smallsetminus\bigcup_{1\le i\le k}\pi(C_i)$ there exists a $T_0\ge 0$ such that for all $x\in\mathscr{K}$ and $T>T_0$*

$$\left|\frac{1}{T}\int_0^T f(u_t x)\,\mathrm{d}t - \int_{G/\Gamma} f\,\mathrm{d}\mu\right| < \varepsilon. \tag{12}$$

This theorem can be informally stated as follows: Fix f and $\varepsilon > 0$. Then (12) holds uniformly in the base point x, as long as x is restricted to compact sets away from a finite union of "tubes" $N(F,U)$; the latter are associated with orbits which do not become equidistributed in G/Γ, because their closure is strictly smaller.)

Note that only finitely many F_k are needed in Theorem 9.3. This has the remarkable implication that if $F \in \mathscr{H} \smallsetminus \{F_1,\dots,F_k\}$, then (12) holds for $x \in N(F,U)$ even though Ux is not dense in G/Γ (the closure of Ux is Fx). Informally, this means that the non-dense orbits of U are themselves becoming equidistributed as they get longer.

In the rest of this subsection, we present some of the ideas developed for the proof of Theorem 9.3.

Linearization of neighborhoods of singular subsets. Let $F \in \mathscr{H}$. Let $\mathfrak{g}$ denote the Lie algebra of G and let $\mathfrak{g}$ denote its Lie subalgebra associated to F. For $d = \dim \mathfrak{g}$, put $V_F = \wedge^d \mathfrak{g}$, and consider the linear G-action on V_F via the representation $\wedge^d \operatorname{Ad}$, the d-th exterior power of the Adjoint representation of G on $\mathfrak{g}$. Fix $p_F \in \wedge^d \mathfrak{f} \smallsetminus \{0\}$, and let $\eta_F : G \to V_F$ be the map defined by $\eta_F(g) = g \cdot p_F = (\wedge^d \operatorname{Ad} g) \cdot p_F$ for all $g \in G$. Note that

$$\eta_F^{-1}(p_F) = \{g \in N_G(F) : \det(\operatorname{Ad} g|_{\mathfrak{f}}) = 1\}.$$

The idea of Dani and Margulis is to work in the representation space V_F (or more precisely $\bar{V}_F$, which is the quotient of V_F by the involution $v \to -v$) instead of G/Γ. In fact, for most of the argument one works only with the orbit $G \cdot p_F \subset V_F$. The advantage is that F is collapsed to a point (since it stabilizes p_F). The difficulty is that the map $\eta_F : G \to \bar{V}_F$ is not Γ-equivariant, and so becomes multivalued if considered as a map from G/Γ to V_F. Dani and Margulis showed that the orbit $\Gamma \cdot p_F$ is discrete in V_F [31, Theorem 3.4], and that

$$\eta_F^{-1}(A_F) = N(F,U) \tag{13}$$

[31, Prop. 3.2], where A_F be the linear span of $\eta_F\big(N(F,U)\big)$ in V_F.

Let $N_G(F)$ denote the normalizer in G of F. Put $\Gamma_F = N_G(F) \cap \Gamma$. Then for any $\gamma \in \Gamma_F$, we have $\gamma\pi(F) = \pi(F)$, and hence γ preserves the volume of $\pi(F)$. Therefore $|\det(\operatorname{Ad}\gamma|_{\mathfrak{f}})| = 1$, and thus $\gamma \cdot p_F = \pm p_F$. Now define

$$\bar{V}_F = \begin{cases} V_F/\{\mathrm{Id}, -\mathrm{Id}\} & \text{if } \Gamma_F \cdot p_F = \{p_F, -p_F\} \\ V_F & \text{if } \Gamma_F \cdot p_F = p_F. \end{cases}$$

The action of G factors through the quotient map of V_F onto $\bar{V}_F$. Let $\bar{p}_F$ denote the image of p_F in $\bar{V}_F$, and define $\bar{\eta}_F : G \to \bar{V}_F$ as $\bar{\eta}_F(g) = g \cdot \bar{p}_F$ for all $g \in G$. Then $\Gamma_F = \bar{\eta}_F^{-1}(\bar{p}_F) \cap \Gamma$. Let $\bar{A}_F$ denote the image of A_F in $\bar{V}_F$. Note that the inverse image of $\bar{A}_F$ in V_F is A_F.

For every $x \in G/\Gamma$, define the set of representatives of x in $\bar{V}_F$ to be

$$\operatorname{Rep}(x) = \bar{\eta}_F\big(\pi^{-1}(x)\big) = \bar{\eta}_F(x\Gamma) \subset \bar{V}_F.$$

The following lemma allows us to understand the map Rep in a special case:

Lemma 9.4. *If $x = \pi(g)$ and $g \in N(F,U) \smallsetminus S(F,U)$*

$$\mathrm{Rep}(x) \cap \bar{A}_F = \{g \cdot p_F\}.$$

Thus x has a single representative in $\bar{A}_F \subset V_F$.

Proof. Indeed, using (13),

$$\mathrm{Rep}\,(\pi(g)) \cap \bar{A}_F = (g\Gamma \cap N(F,U)) \cdot \bar{p}_F.$$

Now suppose $\gamma \in \Gamma$ is such that $g\gamma \in N(F,U)$. Then g belongs to $N(\gamma F \gamma^{-1}, U)$ as well as to $N(F,U)$. Since $g \notin S(F,U)$, we must have $\gamma F \gamma^{-1} = F$, so $\gamma \in \Gamma_F$. Then $\gamma \bar{p}_F = \bar{p}_F$, so $(g\Gamma \cap N(F,U)) \cdot \bar{p}_F = \{g \cdot \bar{p}_F\}$ as required. □

We extend this observation in the following result (cf. [100, Prop. 6.5]).

Proposition 9.5 ([31, Corollary 3.5]). *Let D be a compact subset of $\bar{A}_F$. Then for any compact set $\mathscr{K} \subset G/\Gamma \smallsetminus \pi(S(F,U))$ there exists a neighborhood Φ of D in $\bar{V}_F$ such that any $x \in \mathscr{K}$ has at most one representative in Φ.*

Using this proposition, one can uniquely represent in Φ the parts of the unipotent trajectories in G/Γ lying in $\mathscr{K}$. Then one also has a "polynomial divergence" estimate similar to the ones used in §7:

Proposition 9.6 ([31, Proposition 4.2]). *Let a compact set $C \subset \bar{A}_F$ and an $\varepsilon > 0$ be given. Then there exists a (larger) compact set $D \subset \bar{A}_F$ with the following property: For any neighborhood Φ of D in $\bar{V}_F$ there exists a neighborhood Ψ of C in $\bar{V}_F$ with $\Psi \subset \Phi$ such that the following holds: For any unipotent one parameter subgroup $\{u_t\}$ of G, an element $w \in \bar{V}_H$ and an interval $I \subset \mathbb{R}$, if $u(t_0)w \notin \Phi$ for some $t_0 \in I$ then,*

$$|\{t \in I : u_t w \in \Psi\}| \le \varepsilon \cdot |\{t \in I : u_t w \in \Phi\}|.$$

As a consequence, Dani and Margulis derive the following result of independent interest:

Theorem 9.7 ([31, Theorem 1]). *Let G be a connected Lie group and let Γ be a discrete subgroup of G. Let U be any closed connected subgroup of G which is generated by the Ad-unipotent elements contained in it. Let $\mathscr{K}$ be a compact subset of $G/\Gamma \smallsetminus \bigcup_{F \in \mathscr{H}} N(F,U)$. Then for any $\varepsilon > 0$ there exists a neighbourhood Ω of $\bigcup_{F \in \mathscr{H}} N(F,U)$ such that for any Ad-unipotent one-parameter subgroup $\{u_t\}$ of G, any $x \in \mathscr{K}$ and any $T \ge 0$,*

$$|\{t \in [0,T] : u_t x \in \Omega\}| < \varepsilon T.$$

Proof of Theorem 9.3. The proof relies on Ratner's measure classification theorem (Theorem 8.4) as well as on a refined version of Theorem 9.7, which carefully handles trajectories of points in some $N(F,U)$. □

10 Partially hyperbolic flows and Diophantine approximation

Let us start this section by describing a particular important example of a partially hyperbolic homogeneous flow. Take $G = \mathrm{SL}_{m+n}(\mathbb{R})$, and consider a subgroup $\{g_t\}$ of G, where

$$g_t = \mathrm{diag}(\underbrace{\mathrm{e}^{t/m}, \ldots, \mathrm{e}^{t/m}}_{m \text{ times}}, \underbrace{\mathrm{e}^{-t/n}, \ldots, \mathrm{e}^{-t/n}}_{n \text{ times}}). \tag{14}$$

A connection between the behavior of g_t-trajectories on $\mathscr{L}_n$ and Diophantine approximation was implicitly observed by Davenport and Schmidt [32] in the late 1960s, and explicitly spelled out by Dani in 1985 [24]. Later this connection, in a more general form, was called "Dani Correspondence" by the third-named author and Margulis [62].

10.1 Exceptional trajectories

Let us state one of Dani's observation from [24]: for an $m \times n$ matrix A define

$$U_A := \begin{pmatrix} I_m & A \\ 0 & I_n \end{pmatrix} \in G; \tag{15}$$

then the trajectory $\{g_t U_A \mathbb{Z}^{m+n} : t \geq 0\}$, with g_t as in (14), is bounded in $\mathscr{L}_n$ if and only if A is *badly approximable*, that is, there exists a $c > 0$ such that

$$\|A\mathbf{q} + \mathbf{p}\|^m \|\mathbf{q}\|^n \geq c \quad \forall \mathbf{p} \in \mathbb{Z}^m,\ \mathbf{q} \in \mathbb{Z}^n \smallsetminus \{0\}.$$

Badly approximable systems of linear forms constitute a classical object of study by number theorists. In particular, it was proved by Schmidt [96, 98] that they form a set of full Hausdorff dimension. This, and the fact that $\{U_A\}$ is the expanding horospherical subgroup of G relative to g_1, enabled Dani to conclude that the set of points in the space of lattices with bounded g_t-trajectories has full Hausdorff dimension. In a follow-up paper [26] Dani used a modification of Schmidt's argument to prove a similar statement for homogeneous spaces of Lie groups of real rank 1. Such conclusions are in sharp contrast with the behavior of unipotent trajectories: indeed, it can be shown using the methods described in the previous section that for a unipotent subgroup U of G the set of points with non-dense U-orbits is contained in a countable union of proper submanifolds of G/Γ. In particular, its Hausdorff dimension is strictly smaller than the dimension of G.

Fast forward to 1990, when Margulis gave a plenary talk at the ICM in Kyoto titled "Dynamical and ergodic properties of subgroup actions on homogeneous spaces with applications to number theory". There, in addition to discussing Raghunathan's conjectures and their generalizations, he stated two conjectures highlighting the chaotic behavior of partially hyperbolic actions. Let us state them here:

Conjecture (A). Let G be a Lie group, Γ a lattice in G, and $\{g_t\}$ a partially hyperbolic subgroup of G. Then for any non-empty open subset Ω of G/Γ, the set

$$\{x \in \Omega : \text{ the } \{g_t\}\text{-orbit of } x \text{ is bounded}\}$$

has Hausdorff dimension equal to the dimension of G.

Conjecture (B). Let G, Γ and $\{g_t\}$ be as in Conjecture (A), and let Y be a finite subset of G/Γ. Then for any non-empty open subset Ω of G/Γ, the set

$$\left\{ x \in \Omega \;\middle|\; \begin{array}{c} \text{the } \{g_t\}\text{-orbit of } x \text{ is bounded, the closure of this orbit} \\ \text{is of Hausdorff dimension less than } \dim(G), \\ \text{and the intersection of this closure with } Y \text{ is empty} \end{array} \right\}$$

has Hausdorff dimension equal to the dimension of G.

Conjecture (A) was proved shortly thereafter by Margulis and the third-named author [60]. In fact its statement had to be adjusted to account for a possibility to have a product of two flows, one being unipotent and another partially hyperbolic. To state a theorem to which the general case of this conjecture can be easily reduced, let us denote by G^+ the expanding horospherical subgroup of G relative to g_1.

Theorem 10.1. *Let G be a connected semisimple Lie group without compact factors, Γ an irreducible lattice in G, and $\{g_t\}$ a partially hyperbolic one-parameter subgroup of G. Then for any $x \in G/\Gamma$ there exist a sequence of neighborhoods V_i of identity in G^+ and a sequence of compact subsets K_i of G/Γ such that* $\operatorname{diam}(V_i) \to 0$ *as $i \to \infty$ and*

$$\dim\big(\{h \in V_s \mid \{g_t hx : t \geq 0\} \subset C_s\}\big) \to \dim G^+ \textit{ as } i \to \infty.$$

The proof of the above theorem comes from equidistribution of g_t-translates of G^+-orbits in G/Γ, which is known to be a consequence of mixing of the flow by the argument essentially going back to the Ph.D. Thesis of Margulis [87]. More specifically, one considers natural "rectangular" partitions of G^+ (called tessellations in [60]) and studies their behavior under the automorphism $h \mapsto g_t h g_{-t}$ of G^+ for some large value of $t > 0$. Equidistribution of translates is used to show than one can cover the set of "bad" points (those getting out of a large compact subset of G/Γ) by a relatively small number of rectangles. Then those rectangles are being thrown out to create a Cantor set consisting of points with trajectories staying within a compact subset of G/Γ.

Note that the case when g_t has a nontrivial unipotent part in its Jordan decomposition poses an additional difficulty, namely one has to deal with polynomial expansion in the neutral foliation; this was done in [60] by making use of the exponential mixing of the flow, which was shown to imply the exponentially fast equidistribution of g_t-translates of unstable leaves. See Section 10.3 for further developments along these lines.

We remark that Conjecture (B) was later proved by the third-named author and Weiss [65] by a completely different method, namely by a modification of the notion of Schmidt games developed in [96, 98].

10.2 Cusp excursions

Dani's correspondence between badly approximable systems of linear forms and bounded trajectories was generalized in [62] to describe arbitrary rates of approximation. For a function $\psi : \mathbb{N} \to \mathbb{R}_+$, let us say that an $m \times n$ matrix A is *ψ-approximable* if there are infinitely many $\mathbf{q} \in \mathbb{Z}^n$ such that

$$\|A\mathbf{q}+\mathbf{p}\|^m \leq \psi(\|\mathbf{q}\|^n) \quad \text{for some } \mathbf{p} \in \mathbb{Z}^m. \tag{16}$$

Clearly A is badly approximable if it is not $c\psi_1$-approximable for some $c > 0$, where $\psi_1(x) := \frac{1}{x}$. To state a theorem generalizing Dani's correspondence to ψ-approximable matrices, one needs a "change of variables" lemma:

Lemma 10.2 ([62, Lemma 8.3]). *Fix $m, nl \in \mathbb{N}$ and $x_0 > 0$, and let $\psi : [x_0, \infty) \mapsto (0, \infty)$ be a non-increasing continuous function. Then for some $t_0 \in \mathbb{R}$ there exists a unique continuous function $\varepsilon : [t_0, \infty) \mapsto (0, \infty)$ such that*

$$\textit{the function} \quad t \mapsto \mathrm{e}^t \varepsilon(t)^n \quad \textit{is strictly increasing and unbounded}, \tag{17}$$

$$\textit{the function} \quad t \mapsto \mathrm{e}^{-t} \varepsilon(t)^m \quad \textit{is nonincreasing}, \tag{18}$$

and

$$\psi(\mathrm{e}^t \varepsilon(t)^n) = \mathrm{e}^{-t} \varepsilon(t)^m \quad \forall t \geq t_0. \tag{19}$$

Conversely, given $t_0 \in \mathbb{R}$ and a continuous function $\varepsilon : [t_0, \infty) \mapsto (0, \infty)$ such that (17) *and* (18) *hold, there exists a unique continuous non-increasing function $\psi : [x_0, \infty) \mapsto (0, \infty)$ satisfying* (19).

In many cases one can explicitly solve (19) to express $\varepsilon(\cdot)$ knowing $\psi(\cdot)$ and vice versa. For example the choice $\psi = c\psi_1(x)$ corresponds to $\varepsilon \equiv$ const, as in the original Dani correspondence. The choice $\psi(x) = \frac{1}{x^\nu}$, where $\nu > 1$, corresponds to *very well approximable* matrices; those appear in the work of Margulis and the third-named author on Diophantine approximation on manifolds, see Section 7.3. Here is a generalized form of the Dani correspondence:

Theorem 10.3 ([62, Theorem 8.5]). *An $m \times n$ matrix A is ψ-approximable if and only if there exist arbitrarily large positive t such that*

$$\delta(g_t U_A \mathbb{Z}^{m+n}) < \varepsilon(t),$$

where $\{g_t\}$ is as in (14), *U_A as in* (15), *and $\varepsilon(\cdot)$ is the function corresponding to ψ as in the previous lemma.*

Loosely speaking, good rational approximations for A correspond to far excursions of the g_t-trajectory of $\Lambda_A = U_A\mathbb{Z}^{m+n}$ into the "cusp neighborhoods" $\Omega_{m+n}(\varepsilon)$. In [62] such cusp excursions were studied in the generality of arbitrary partially hyperbolic actions on finite volume homogeneous spaces. More generally, the following was proved there:

Theorem 10.4. *Let G be a connected semisimple Lie group without compact factors, Γ an irreducible lattice in G, $X = G/\Gamma$, μ the G-invariant probability measure on X, $g \in G$ a partially hyperbolic element. Also let $\{B_n\}$ be a "sufficiently regular" sequence of subsets of X (sufficient regularity here means that the sets can be uniformly approximated by smooth functions from above and below only losing a fraction of their measure). Then*

$$\mu\left(\{x \in X : g^n x \in B_n \textit{ for infinitely many } n \in \mathbb{N}\}\right)$$
$$= \begin{cases} 0 & \textit{if} \quad \sum_{n=1}^{\infty} \mu(B_n) < \infty, \\ 1 & \textit{if} \quad \sum_{n=1}^{\infty} \mu(B_n) = \infty. \end{cases}$$

The proof is based on the exponential rate of mixing for smooth functions on X. Among other things, Theorem 10.4 generalized Sullivan's logarithm law for geodesics in finite volume hyperbolic manifolds [104] and, through the use of Theorem 10.3, provided an alternative proof of the Khintchine–Groshev Theorem: Lebesgue almost every A is (resp., is not) ψ-approximable if and only if the series $\sum_k \psi(k)$ diverges (resp., converges).

10.3 Effective equidistribution

An interesting direction in simultaneous Diophantine approximation is a modification of the standard set-up as in (16) by assigning weights to the individual variables and linear forms. We have seen one example of this in Corollary 7.12. More generally, one can choose $\mathbf{r} = (r_1, \ldots, r_m) \in \mathbb{R}^m_{>0}$ and $\mathbf{s} = (s_1, \ldots, s_n) \in \mathbb{R}^n_{>0}$ such that $r_1 + \cdots + r_m = s_1 + \cdots + s_n = 1$ and, with the notation $\|\mathbf{x}\|_{\mathbf{r}} = \max_{1 \le i \le m} |x_i|^{1/r_i}$ and $\|\mathbf{y}\|_{\mathbf{r}} = \max_{1 \le j \le n} |y_j|^{1/s_j}$, study the solvability (in integers $\mathbf{p}$, $\mathbf{q}$) of the inequality

$$\|A\mathbf{q} + \mathbf{p}\|_{\mathbf{r}} \le \psi(\|\mathbf{q}\|_{\mathbf{s}})$$

instead of (16). This, as shown in [56], can be understood through the action of the weighted diagonal one-parameter subgroup

$$g_t^{\mathbf{r},\mathbf{s}} := \mathrm{diag}(e^{r_1 t}, \ldots, e^{r_m t}, e^{-s_1 t}, \ldots, e^{-t s_n})$$

of $G = \mathrm{SL}_{m+n}(\mathbb{R})$. However, for the use of dynamics in the unweighted case (14) it was important that the group

$$H := \{U_A\}$$

was precisely the expanding horospherical subgroup relative to g_1. This way, for example, one could use mixing to deduce equidistribution of g_t-translates of unstable leaves, in particular deriving the fact that the trajectory $\{g_t\Lambda_A : t \geq 0\}$ is dense in X for almost every A. However, when some weights are different, the group H becomes a proper subgroup of the expanding horospherical subgroup relative to $g_1^{\mathbf{r},\mathbf{s}}$, and studying translates of its orbits poses additional difficulties.

In the process of investigating the problem of improving Dirichlet's theorem in the weighted setting, the third-named author and Weiss proved the equidistribution of $g_t^{\mathbf{r},\mathbf{s}}$-translates of H-orbits in the space of lattices. The proof was based on Rather's Theorem and used the linearization method described in Section 9. Shortly thereafter Margulis came up with an idea that made it possible to establish the effective version of the aforementioned equidistribution result. The following is a special case of the main result of [63]:

Theorem 10.5. *Let G, H and $g_t^{\mathbf{r},\mathbf{s}}$ be as above, let ν be a Haar measure on H, and let μ be the G-invariant probability measure on $X = \mathscr{L}_{m+n}$. Then there exists a $\gamma > 0$ such that for any $f \in C^\infty_{comp}(H)$, $\psi \in C^\infty_{comp}(X)$, for any compact $L \subset X$ and for all $z \in L$ and $t \geq 0$ one has*

$$\left| \int_H f(h)\psi(g_t^{\mathbf{r},\mathbf{s}}hz)\,\mathrm{d}\nu(h) - \int_H f\,\mathrm{d}\nu \int_X \psi\,\mathrm{d}\mu \right| \ll_{f,\psi,L} \mathrm{e}^{-\gamma t}.$$

The implicit constant in the above inequality can be estimated in terms of derivatives of f and ψ and the injectivity radius of L. This was later done explicitly in [64] and used for estimating the Hausdorff dimension of the set of matrices A such that the trajectories $\{g_t^{\mathbf{r},\mathbf{s}}\Lambda_A\}$ miss a given open set, with an application to Diophantine approximation with weights.

The proof of Theorem 10.5 happens to be perhaps even more important than the result itself. The main trick is to write $g_t^{\mathbf{r},\mathbf{s}}$ as a product of two elements of G:

$$g_t^{\mathbf{r},\mathbf{s}} = g_{t'}g'',$$

where $g_{t'}$ is as defined in (14). Then one uses exponentially fast equidistribution of $g_{t'}$-translates, available since H is expanding horospherical with respect to g_1, together with quantitative non-divergence of g''-translates. This idea was later exploited in [68, 10] where multiple effective equidistribution of such translates was established in much bigger generality. See also [69] for an applications to proving a Khintchine-type theorem for improvement of weighted Dirichlet theorem in simultaneous Diophantine approximation.

11 A quantitative version of the Oppenheim Conjecture

We now return to the set-up of the Oppenheim conjecture stated in §8, and describe some ideas involved in the proof of its quantitative version.

Fix an indefinite quadratic form Q. Let ν be a continuous positive function on the sphere $\{v \in \mathbb{R}^n \,:\, \|v\| = 1\}$, and let

$$\Omega := \{v \in \mathbb{R}^n \,:\, \|v\| < \nu(v/\|v\|)\}.$$

We denote by $T\Omega$ the dilate of Ω by $T \in \mathbb{R}_+$. Define the following set:

$$V_{(a,b)}(\mathbb{R}) := \{x \in \mathbb{R}^n \,:\, a < Q(x) < b\}.$$

Also let $V_{(a,b)}(\mathbb{Z}) := \{x \in \mathbb{Z}^n \,:\, a < Q(x) < b\}$. The set $T\Omega \cap \mathbb{Z}^n$ consists of $O(T^n)$ points, and the set of values $Q(T\Omega \cap \mathbb{Z}^n)$ is contained in an interval of the form $[-\mu T^2, \mu T^2]$, where $\mu > 0$ is a constant depending on Q and Ω. Thus one might expect that for any interval (a,b), as $T \to \infty$,

$$|V_{(a,b)}(\mathbb{Z}) \cap T\Omega| \sim c_{Q,\Omega}(b-a)T^{n-2}, \tag{20}$$

where $c_{Q,\Omega}$ is a constant depending on Q and Ω. This may be interpreted as "uniform distribution" of sets $Q(\mathbb{Z}^n \cap T\Omega)$ in the real line. The main result of this section is that (20) holds if Q is not proportional to a rational form, and has signature (p,q) with $p \geq 3$, $q \geq 1$. We also determine the constant $c_{Q,\Omega}$.

If Q is an indefinite quadratic form in n variables, Ω is as above and (a,b) is an interval, it can be shown that there exists a constant $\lambda = \lambda_{Q,\Omega}$ so that as $T \to \infty$,

$$\mathrm{Vol}(V_{(a,b)}(\mathbb{R}) \cap T\Omega) \sim \lambda_{Q,\Omega}(b-a)T^{n-2} \tag{21}$$

One of the main results of the paper [34] by the first-named author, Margulis and Mozes is the following:

Theorem 11.1. *Let Q be an indefinite quadratic form of signature (p,q), with $p \geq 3$ and $q \geq 1$. Suppose that Q is not proportional to a rational form. Then for any interval (a,b), as $T \to \infty$,*

$$|V_{(a,b)}(\mathbb{Z}) \cap T\Omega| \sim \lambda_{Q,\Omega}(b-a)T^{n-2}, \tag{22}$$

where $n = p+q$, and $\lambda_{Q,\Omega}$ is as in (21).

The asymptotically exact lower bound was proved in [31] with the help of the linearization method described in the previous section.

11.1 Passage to the space of lattices

Here we relate the counting problem of Theorem 11.1 to a certain integral involving the orthogonal group of the quadratic form and the space $\mathscr{L}_n$. Roughly this is done as follows. Let f be a bounded function on $\mathbb{R}^n \smallsetminus \{0\}$ vanishing outside a compact subset. For a lattice $\Lambda \in \mathscr{L}_n$ let

$$\tilde{f}(\Lambda) = \sum_{v \in \Lambda \smallsetminus \{0\}} f(v) \tag{23}$$

(the function $\tilde{f}$ is called the *Siegel transform* of f). The proof is based on the identity of the form

$$\int_K \tilde{f}(a_t k \Lambda)\, \mathrm{d}k = \sum_{v \in \Lambda \smallsetminus \{0\}} \int_K f(a_t k v)\, \mathrm{d}k \tag{24}$$

obtained by integrating (23). In (24) $\{a_t\}$ is a certain diagonal subgroup of the orthogonal group of Q, and K is a maximal compact subgroup of the orthogonal group of Q. Then for an appropriate function f, the right-hand side is related to the number of lattice points $v \in [\mathrm{e}^t/2, \mathrm{e}^t]\partial\Omega$ with $a < Q(v) < b$. The asymptotics of the left-hand side is then established using the ergodic theory of unipotent flows and some other techniques. Namely it is shown in [34] that Theorem 11.1 can be reduced to the following theorem:

Theorem 11.2. *Suppose $p \geq 3$, $q \geq 1$. Let $\Lambda \in \mathscr{L}_n$ be a unimodular lattice such that $H\Lambda$ is not closed. Let ν be any continuous function on K. Then*

$$\lim_{t \to +\infty} \int_K \tilde{f}(a_t k \Lambda) \nu(k)\, \mathrm{d}m(k) = \int_K \nu\, \mathrm{d}m \int_{\mathscr{L}_n} \tilde{f}(\Delta)\, \mathrm{d}\mu(\Delta). \tag{25}$$

Note that if we replace $\tilde{f}$ by a bounded continuous function ϕ, then (25) follows easily from Theorem 9.3. (This was the original motivation for Theorem 9.3.) The fact that Theorem 9.3 deals with unipotents and Theorem 11.2 deals with large spheres is not a serious obstacle, since large spheres can be approximated by unipotents. In fact, the integral in (25) can be rewritten as

$$\int_B \left(\frac{1}{T(x)} \int_0^{T(x)} \phi(u_t x)\, \mathrm{d}m(k) \right) \mathrm{d}x,$$

where B is a suitable subset of G and U is a suitable unipotent. Now by Theorem 9.3, the inner integral tends to $\int_{G/\Gamma} \phi$ uniformly as long as x is in a compact set away from an explicitly described set E, where E is a finite union of neighborhoods of sets of the form $\pi(C)$ where C is a compact subset of some $N(F,U)$. By direct calculation one can show that only a small part of B is near E, hence (25) holds.

However, for a non-negative bounded continuous function f on $\mathbb{R}^n$, the function $\tilde{f}$ defined in (23) is unbounded (it is in $L^s(\mathscr{L}_n)$ for $1 \leq s < n$). As was done in [31], it is possible to obtain asymptotically exact lower bounds by considering bounded continuous functions $\phi \leq \tilde{f}$. But to prove the upper bounds in the theorems stated above one needs to examine carefully the situation at the “cusp” of G/Γ, i.e. outside of compact sets.

The functions α_i and α. Let Λ be a lattice in $\mathbb{R}^n$. Recall that the notion of a Λ-rational subspace and the function d_Λ was defined in §7 (following the statement of Theorem 7.6). Let us introduce the following notation:

$$\alpha_i(\Lambda) = \sup\left\{ \frac{1}{d_\Lambda(L)} \,\middle|\, L \text{ is a } \Lambda\text{-rational subspace of dimension } i \right\},\ 0 \le i \le n,$$
$$\alpha(\Lambda) = \max_{0\le i\le n} \alpha_i(\Lambda). \tag{26}$$

By [97, Lemma 2], for any bounded compactly supported function f on $\mathbf{b}R^n$ there exists a positive constant $c = c(f)$ such that $\tilde{f}(\Lambda) < c\alpha(\Lambda)$ for any $\Lambda \in \mathscr{L}_n$. The upper bound in Theorem 11.1 is proved by combining the above observation with the following integrability estimate:

Theorem 11.3 ([34]). *If $p \ge 3$, $q \ge 1$ and $0 < s < 2$, or if $p = 2$, $q \ge 1$ and $0 < s < 1$, then for any $\Lambda \in \mathscr{L}_n$*

$$\sup_{t>0} \int_K \alpha(a_t k\Lambda)^s \,\mathrm{d}m(k) < \infty.$$

The upper bound is uniform as Λ varies over compact subsets of $\mathscr{L}_n$.

This result can be interpreted as follows. For $\Lambda \in \mathscr{L}_n$ and $h \in H$, let $f(h) = \alpha(h\Lambda)$. Since α is left-$\widehat{K}$ invariant, f is a function on the symmetric space $X = K\backslash H$. Theorem 11.3 is the statement that if $p \ge 3$, then the averages of f^s, $0 < s < 2$, over the sets Ka_tK in X remain bounded as $t \to \infty$, and the bound is uniform as one varies the base point Λ over compact sets.

11.2 Margulis functions

We now present some ideas from the proof of Theorem 11.3 and Theorem 11.10. We recall the notation from §8 and §11.1: $G = \mathrm{SL}_n(\mathbb{R})$, $\Gamma = \mathrm{SL}_n(\mathbb{Z})$, $\widehat{K} \cong \mathrm{SO}(n)$ is a maximal compact subgroup of G, $H \cong \mathrm{SO}(p,q) \subset G$, and $K = H \cap \widehat{K} = \mathrm{SO}(p) \times \mathrm{SO}(q)$ is a maximal compact subgroup of H. Let $m(\cdot)$ denote the normalized Haar measure on K. Let $\{a_t : t \in \mathbf{b}R\}$ be a self-adjoint one-parameter subgroup of $\mathrm{SO}(2,1)$, where $\mathrm{SO}(2,1)$ is embedded into $\mathrm{SO}(p,q)$, so that a_t is conjugate to the diagonal matrix with entries $(\mathrm{e}^t, 1, \ldots, 1, \mathrm{e}^{-t})$.

The strategy of the proof is to construct what we now call a *Margulis function.* This idea has been extremely influential, see for example the survey [41].

Let Y be a space on which H acts. (In our case, $Y = \mathscr{L}_n$). For $t > 0$, let A_t be the averaging operator taking a function $\phi : Y \to \mathbb{R}$ to the function $A_t\phi : Y \to \mathbb{R}$ defined by

$$(A_t\phi)(x) = \int_K \phi(a_t kx)\,\mathrm{d}m(k). \tag{27}$$

Definition 11.4. A K-invariant function $f : Y \to [1,\infty]$ is called a *Margulis function* (for the averages A_t) if it satisfies the following properties:

(a) There exists a$\sigma > 1$ such that for all $0 \le t \le 1$ and all $x \in Y$,

$$\sigma^{-1} f(x) \le f(a_t x) \le \sigma f(x). \tag{28}$$

(This holds if $\log f$ is uniformly continuous along the H-orbits.)

(b) For every $c_0 > 0$ there exist $\tau > 0$ and $b_0 > 0$ such that for all $x \in Y$,

$$A_\tau f(x) \leq c_0 f(x) + b_0. \tag{29}$$

(c) f is bounded on compact subsets of Y. For any $\ell > 0$, the set $\overline{\{x : f(x) \leq \ell\}}$ is a compact subset of Y.

We have the following abstract lemma:

Lemma 11.5. *Suppose f is a Margulis function on Y. Then, for all $c < 1$ there exists $t_0 > 0$ (depending on σ and c) and $b > 0$ (depending only on b_0, c_0 and σ) such that for all $t > t_0$ and all $x \in Y$,*

$$(A_t f)(x) \leq c f(x) + b. \tag{30}$$

A more general version of this lemma is proved in [34, §5.3]. The reader may also refer to a simplified proof in [41, §3], specialized to the case $H = \mathrm{SL}_2(\mathbb{R})$.

From the proof of Lemma 11.5, one can deduce the following variant:

Lemma 11.6. *For every $\sigma > 1$ there exists a $c_0 > 0$ such that the following holds. Suppose $f : Y \to [1,\infty)$ is a K-invariant function satisfying* (a) *and* (c) *of Definition 11.4, and let A_t be as in* (27)*. Suppose also that there exists $\tau > 0$ and $b_0 > 0$ such that* (29) *holds. Then f is a Margulis function for the averages A_t.*

For a wider perspective on Margulis functions and many related results, see the survey [41].

Strategy of the proof of Theorem 11.3. Suppose $0 < s < 2$. If the function α^s were a Margulis function on G/Γ, then Theorem 11.3 would follow immediately from (30). Even though this is not true, the idea is to construct a Margulis function f on G/Γ which is within a bounded multiple of α^s.

If $p \geq 3$ and $0 < s < 2$, or if $(p,q) = (2,1)$ or $(2,2)$ and $0 < s < 1$, it is shown in [34, §5.3] that for any $c > 0$ there exist $t > 0$ such that the functions α_i^s satisfy the following system of integral inequalities:

$$A_t \alpha_i^s \leq c_i \alpha_i^s + \mathrm{e}^{2t} \max_{0<j\leq \min(n-i,i)} \sqrt{\alpha_{i+j}^s \alpha_{i-j}^s}, \tag{31}$$

where A_t is the averaging operator $(A_t f)(\Delta) = \int_K f(a_t k \Lambda)$ and $c_i \leq c$. If $(p,q) = (2,1)$ or $(2,2)$ and $s = 1$, then (31) also holds (for suitably modified functions α_i), but some of the constants c_i cannot be made smaller than 1.

In §11.4 we will show that if (31) holds, then for any $\varepsilon > 0$, the function $f = f_{\varepsilon,s} = \sum_{0\leq i\leq n} \varepsilon^{i(n-i)} \alpha_i^s$ is the desired Margulis function, and it follows from (26) that the ratio of α^s and f is uniformly bounded between two positive constants.

We now outline the proof of (31).

11.3 A system of inequalities

By a direct calculation one can prove the following:

Proposition 11.7. *Let* $\{a_t \mid t \in \mathbf{b}R\}$ *be a self-adjoint one-parameter subgroup of* $\mathrm{SO}(2,1)$. *Let* $p,q \in \mathbb{N}$ *and let* $0 < i < p+q = n$. *Let*

$$F(i) = \{x_1 \wedge \cdots \wedge x_i \mid x_1,\ldots,x_i \in \mathbf{b}R^n\} \subset \textstyle\bigwedge^i(\mathbf{b}R^n).$$

Then, if $p \geq 3$, *or if* $p = 2$, $q = 2$ *and* $i \neq 2$, *for any* $0 < s < 2$ *one has*

$$\lim_{t\to\infty} \sup_{v\in F(i),\ \|v\|=1} \int_K \frac{\mathrm{d}m(k)}{\|a_t k v\|^s} = 0. \tag{32}$$

where $K = \mathrm{SO}(p) \times \mathrm{SO}(q)$ *and* $\mathrm{SO}(2,1)$ *is embedded into* $\mathrm{SO}(p,q)$. *If* $p = 2$ *and* $q = 1$, *or if* $p = 2$, $q = 2$ *and* $i = 2$, *then* (32) *holds for any* $0 < s < 1$.

Lemma 11.8. *Let* $\{a_t\}$, p, q *and* n *be as in Proposition 11.7. Denote* $\mathrm{SO}(p)\times\mathrm{SO}(q)$ *by* K. *Suppose* $p \geq 3$, $q \geq 1$ *and* $0 < i < n$, *or* $p = 2$, $q = 2$ *and* $i = 1$ *or* 3. *Then for any* $0 < s < 2$, *and any* $c > 0$ *there exist* $t > 0$ *such that for any* $\Lambda \in \mathscr{L}_n$,

$$\int_K \alpha_i(a_t k\Lambda)^s \,\mathrm{d}m(k) < \frac{c}{2}\alpha_i(\Lambda)^s + \mathrm{e}^{2t} \max_{0<j\leq\min\{n-i,i\}} \left(\sqrt{\alpha_{i+j}(\Lambda)\alpha_{i-j}(\Lambda)}\right)^s. \tag{33}$$

If $p = 2$, $q = 1$ *and* $i = 1,2$, *or if* $p = 2$, $q = 2$ *and* $i = 2$, *then for any* $0 < s < 1$ *and any* $c > 0$ *there exist* $t > 0$ *such that* (33) *holds.*

Proof. Fix $c > 0$. In view of Proposition 11.7 one can find $t > 0$ such that

$$\int_K \frac{\mathrm{d}m(k)}{\|a_t k v\|^s} < \frac{c}{2}\cdot\frac{1}{\|v\|^s} \tag{34}$$

for any $v \in F(i) \smallsetminus \{0\}$. Let $\Lambda \in \mathscr{L}_n$. There exists a Λ-rational subspace L_i of dimension i such that

$$\frac{1}{d_\Lambda(L_i)} = \alpha_i(\Lambda). \tag{35}$$

Inequality (34) therefore implies

$$\int_K \frac{\mathrm{d}m(k)}{d_{a_t k\Lambda}(a_t k L_i)^s} < \frac{c}{2}\cdot\frac{1}{d_\Lambda(L_i)^s}. \tag{36}$$

Observe that

$$\mathrm{e}^{-t} \leq \frac{\|a_t v\|}{\|v\|} \leq \mathrm{e}^t \qquad \forall 0 < j < n \text{ and } \forall v \in F(j) \smallsetminus \{0\}. \tag{37}$$

Let us denote by Ψ_i the set of Λ-rational subspaces L of dimension i with $d_\Lambda(L) < \mathrm{e}^{2t} d_\Lambda(L_i)$. We get from (37) that for a Λ-rational i-dimensional subspace $L \notin \Psi_i$

$$d_{a_t k\Lambda}(a_t kL) > d_{a_t k\Lambda}(a_t kL_i), \qquad k \in K. \tag{38}$$

It follows from (36), (38) and the definition of α_i that

$$\int_K \alpha_i(a_t k\Lambda)^s \, dm(k) < \frac{c}{2}\alpha_i(\Lambda)^s \text{ if } \Psi_i = \{L_i\}. \tag{39}$$

Assume now that $\Psi_i \neq \{L_i\}$, and let $M \in \Psi_i \setminus \{L_i\}$. Then $\dim(M + L_i)$ is equal to $i + j$ where $j > 0$. Now using (35), (37) and the fact that

$$d_\Lambda(L)d_\Lambda(M) \geq d_\Lambda(L \cap M)d_\Lambda(L + M)$$

(see [34, Lemma 5.6]), we get that for any $k \in K$

$$\begin{aligned}\alpha_i(a_t k\Lambda) < \mathrm{e}^t \alpha_i(\Lambda) &= \frac{\mathrm{e}^t}{d_\Lambda(L_i)} < \frac{\mathrm{e}^{2t}}{\sqrt{d_\Lambda(L_i)d_\Lambda(M)}} \\ &\leq \frac{\mathrm{e}^{2t}}{\sqrt{d_\Lambda(L_i \cap M)d_\Lambda(L_i + M)}} \leq \mathrm{e}^{2t}\sqrt{\alpha_{i+j}(\Lambda)\alpha_{i-j}(\Lambda)}.\end{aligned}$$

Hence if $\Psi_i \neq \{L_i\}$

$$\int_K \alpha_i(a_t k\Lambda)^s \, \mathrm{d}m(k) \leq \mathrm{e}^{2t} \max_{0<j\leq \min\{n-i,i\}} \left(\sqrt{\alpha_{i+j}(\Lambda)\alpha_{i-j}(\Lambda)}\right)^s. \tag{40}$$

Combining (39) and (40), we obtain (33). □

11.4 Averages over large spheres

In this subsection we complete the proof of Theorem 11.3.

Proof of Theorem 11.3. It is easy to see that each of the functions α_i^s is K-invariant and has properties (a) and (c) of Definition 11.4. In particular, there exists a $\sigma > 1$ such that for all $1 \leq i \leq n$, equation (28) holds for α_i^s. Let c_0 be such that Lemma 11.6 holds for this σ.

Applying Lemma 11.8, we see that there exists a $\tau > 0$ such that for any $0 < i < n$

$$A_\tau \alpha_i^s < \frac{c_0}{2}\alpha_i^s + \mathrm{e}^{2\tau} \max_{0<j\leq \min\{n-i,i\}} \sqrt{\alpha_{i+j}^s \alpha_{i-j}^s}. \tag{41}$$

Let us define $q(i) = i(n-i)$. Then by a direct computation

$$2q(i) - q(i+j) - q(i-j) = 2j^2.$$

Therefore we get from (41) that for any $0 < i < n$, and any $0 < \varepsilon < 1$,

$$\begin{aligned} &A_\tau(\varepsilon^{q(i)}\alpha_i^s) \\ &< \frac{c_0}{2}\varepsilon^{q(i)}\alpha_i^s + \mathrm{e}^{2\tau} \max_{0<j\le\min\{n-i,i\}} \varepsilon^{q(i)-\frac{q(i+j)+q(i-j)}{2}} \sqrt{\varepsilon^{q(i+j)}\alpha_{i+j}^s\varepsilon^{q(i-j)}\alpha_{i-j}^s} \\ &\le \frac{c_0}{2}\varepsilon^{q(i)}\alpha_i^s + \varepsilon\mathrm{e}^{2\tau} \max_{0<j\le\min\{n-i,i\}} \sqrt{\varepsilon^{q(i+j)}\alpha_{i+j}^s\varepsilon^{q(i-j)}\alpha_{i-j}^s}. \end{aligned} \tag{42}$$

Consider the linear combination

$$f_{\varepsilon,s} = \sum_{0\le i\le n} \varepsilon^{q(i)}\,\alpha_i^s.$$

The function $f_{\varepsilon,s}$ then also has properties (a) and (c) of Definition 11.4. Since $\varepsilon^{q(i)}\alpha_i^s < f_{\varepsilon,s}, \alpha_0 = 1$ and $\alpha_n = 1$, inequalities (42) imply the following inequality:

$$A_\tau f_{\varepsilon,s} < 2 + \frac{c_0}{2} f_{\varepsilon,s} + n\varepsilon\mathrm{e}^{2\tau} f_{\varepsilon,s}.$$

Taking $\varepsilon = \frac{c_0}{2n}\mathrm{e}^{-2\tau}$, we see that there exists a $\tau > 0$ such that

$$A_\tau f_{\varepsilon,s} < c_0 f_{\varepsilon,s} + 2.$$

Then, by Lemma 11.6, $f_{\varepsilon,s}$ is a Margulis function on G/Γ. Since

$$\alpha_i^s \le \varepsilon^{-q(i)} f_{\varepsilon,s},$$

Lemma 11.5 implies that there exists a constant $B > 0$ so that for each i and all $t > 0$,

$$\int_K \alpha_i(a_t k\Lambda)^s \, \mathrm{d}m(k) < B,$$

and that the bound is uniform as Λ varies over compact subsets of G/Γ. From this the theorem follows. □

11.5 Signatures $(2,1)$ *and* $(2,2)$

If the signature of Q is $(2,1)$ or $(2,2)$, then no universal formula like (20) holds. In fact, the following can be shown:

Theorem 11.9. *Let Ω_0 be the unit ball, and let $q = 1$ or 2. Then for every $\varepsilon > 0$ and every interval (a,b) there exists a quadratic form Q of signature $(2,q)$ not proportional to a rational form, and a constant $c > 0$ such that for an infinite sequence $T_j \to \infty$,*

$$|V_{(a,b)}(\mathbb{Z}) \cap T\Omega_0| > cT_j^q(\log T_j)^{1-\varepsilon}.$$

The case $q=1$, $b \leq 0$ of Theorem 11.9 was noticed by Sarnak and worked out in detail in [17]. The quadratic forms constructed are of the form $x_1^2+x_2^2-\alpha x_3^2$, or $x_1^2+x_2^2-\alpha(x_3^2+x_4^2)$, where α is extremely well approximated by squares of rational numbers.

As was observed in [34], the crucial difference between the cases $\max(p,q)>2$ and $\max(p,q)=2$ is that in the latter case Theorem 11.3 does not hold even for $s=1$. The following result is a replacement:

Theorem 11.10 ([34]). *If* $\max(p,q)=2$, *then for any* $\Lambda \in \mathscr{L}_n$

$$\sup_{t>1} \frac{1}{t} \int_K \alpha(a_t k \Lambda)\, \mathrm{d}m(k) < \infty,$$

with a uniform upper bound i as Λ *varies over compact subsets of* $\mathscr{L}_n$.

This produces an upper bound for $|V_{(a,b)}(\mathbb{Z}) \cap T\Omega|$ of the form $c(b-a)T^{n-2}\log T$, where c is a positive constant depending on Q.

In the follow-up paper [35] the first-named author, Margulis and Mozes studied the case $(p,q)=(2,2)$ in detail. Recall that a subspace of $\mathbb{R}^n$ is called *isotropic* if the restriction of the quadratic form to the subspace is identically zero. Observe also that whenever a form of signature $(2,2)$ has a rational isotropic subspace L, then the number of integer points in $L \cap T\Omega$ is of the order of T^2 integral points x for which $Q(x)=0$, hence $|V_{(\varepsilon,\varepsilon)}(\mathbb{Z}) \cap T\Omega| \geq cT^2$ independently of the choice of ε. Thus to obtain an asymptotic formula similar to (22) in the signature $(2,2)$ case, one must exclude the contribution of rational isotropic subspaces. We remark that an irrational quadratic form of signature $(2,2)$ may have at most 4 rational isotropic subspaces (see [35, Lemma 10.3]).

The space of quadratic forms in 4 variables is a linear space of dimension 10. Fix a norm $\|\cdot\|$ on this space.

Definition 11.11. (EWAS) A quadratic form Q is called *extremely well approximable by split forms (EWAS)* if for any $N>0$ there exists a split integral form Q' and $2 \leq k \in \mathbb{R}$ such that

$$\left\| Q - \frac{1}{k} Q' \right\| \leq \frac{1}{k^N}.$$

The main result of [35] is:

Theorem 11.12. *Suppose* Ω *is as above. Let* Q *be an indefinite quadratic form of signature* $(2,2)$ *which is not EWAS. Then for any interval* (a,b), *as* $T \to \infty$,

$$|\tilde{V}_{(a,b)}(\mathbb{Z}) \cap T\Omega| \sim \lambda_{Q,\Omega}(b-a)T^2,$$

where the constant $\lambda_{Q,\Omega}$ *is as in* (21), *and* $\tilde{V}_{(a,b)}(\mathbb{Z})$ *consists of points not contained in isotropic subspaces.*

As was extensively discussed in [35], Theorem 11.12 has implications for eigenvalue spacings on flat 2-dimensional tori.

12 Effective estimates

In this section, we present some work of Margulis and his co-authors regarding effective equidistribution and effective estimates for the number of solutions of certain Diophantine inequalities. This is far from a comprehensive account; we choose to focus on just a few results. A much more detailed and comprehensive survey is given in [33].

12.1 Periodic orbits of semisimple groups

Let G be a real Lie group, let Γ be a lattice in G, and let H be a connected subgroup of G generated by unipotent elements. Using Ratner's theorem together with the linearization technique (§9) and results on nondivergence of unipotent orbits (§7), Mozes and Shah [90] proved that the nonzero weak-$*$ limits of H-invariant ergodic probability measures on G/Γ are again ergodic, hence are Haar measures on closed orbits of closed subgroups $S \supset H$. In [37], Einsiedler, Margulis and Venkatesh quantify the aforementioned approximation procedure with a polynomial rate in the case when H is semisimple. Here is one of the main results of that paper:

Theorem 12.1. *Let G be a connected component of the identity of the group of real points of a connected semisimple algebraic $\mathbb{Q}$-group, let Γ be a congruence lattice in G, and let $H \subset G$ be a semisimple subgroup without any compact factors, generated by unipotent elements and with a finite centralizer in G. Then there exist $\delta > 0$ and $\ell \in \mathbb{N}$ depending only on G, H and $T_0 > 0$ depending only on G, Γ, H with the following property: for any periodic H-orbit Hx in $X = G/\Gamma$ and any $T \geq T_0$ there exists an intermediate subgroup $H \subset S \subset G$ for which Sx is a closed S-orbit with volume $\leq T$ and such that the Haar measure μ_{Hx} on Hx is (T, δ)-close to the Haar measure μ_{Sx} on Hx. The latter means that for any $f \in C^\infty_{comp}(X)$ one has*

$$\left| \int f \, \mathrm{d}\mu_{Hx} - \int f \, \mathrm{d}\mu_{Hx} \right| \ll_{G,\Gamma,H} \|f\|_{W^\ell} T^{-\delta},$$

where $\|f\|_{W^\ell}$ stands for the L^2-Sobolev norm of degree ℓ.

The general strategy of the proof of the above theorem consists of effectively acquiring certain "almost invariance" properties for the measure μ_{Hx}. In qualitative form this strategy was used by Margulis and Dani–Margulis in their proof of various versions of the Oppenheim conjecture [81, 27]. A crucial ingredient in the proof of Theorem 12.1, which is responsible for the polynomial rate of approximation, is a uniform spectral gap for congruence quotients of semisimple algebraic groups. These methods were further advanced in [2] as well as in the recent work of Margulis with Einsiedler, Mohammadi and Venkatesh [36]. For another recent development, namely a polynomially effective version of the linearization results and techniques that we surveyed in §9, see the new preprint by Margulis with Lindenstrauss, Mohammadi and Shah [71].

12.2 Effective solution of the Oppenheim Conjecture

The following was proven by Dani and Margulis in [27]:

Theorem 12.2. *Let Q be an indefinite ternary quadratic form which is not proportional to an integral form. Then the set*

$$\{Q(v) \ : \ v \in \mathbb{Z}^3, \ v \text{ primitive}\}$$

is dense in $\mathbb{R}$.

In the following, we implicitly assume all integral quadratic forms we consider are primitive in the sense that they are not a nontrivial integer multiple of another integral quadratic form.

The main result of the paper [70] is the following quantification of Theorem 12.2.

Theorem 12.3. *Let Q_1 be an indefinite, ternary quadratic form with $\det Q_1 = 1$, and suppose $\varepsilon > 0$. Then for any $T \geq T_0(\varepsilon)\|Q1\|^{K_1}$ at least one of the following holds:*

(i) *There is an integral quadratic form Q_2 with $|\det(Q_2)| < T$ and $\|Q_1 - \lambda Q_2\| \ll \|Q_1\| T^{-1}$, where $\lambda = |\det(Q_2)|^{-1/3}$.*
(ii) *For any $\xi \in [-(\log T)\kappa_2, (\log T)\kappa_2]$ there is a primitive integer vector $v \in \mathbb{Z}^3$ with $0 < \|v\| < T^{K_3}$ satisfying*

$$|Q_1(v) - \xi| \ll (\log T)^{-\kappa_2}$$

(with K_1, κ_2, K_3, and the implicit constants absolute).

Though there are significant differences, the strategy which is used in the paper has many similarities with the strategy which was used by Margulis in [79, 81] and subsequent papers by Dani and Margulis [27, 28, 29, 30]. The main ingredient in these strategies is to prove that an orbit closure contains orbits of additional subgroups. In the original approach, this is achieved using minimal sets for appropriately chosen subactions, while in [70] the beginning point of the orbit of the new subgroup is moving. To make this approach work, the authors need to control how this base point changes so it remains sufficiently generic in an appropriate quantitative sense.

12.3 Power law estimates in dimension at least 5

Note that in the above result the dependence on the parameter T is logarithmic. If the number of variables d is greater than or equal to 5, power estimates are possible.

We now present the main result of [10], which is based in part on earlier work of Götze and Margulis.

To state the result we use the following notation. Denote by Q the symmetric matrix in $\mathrm{GL}_d(\mathbb{R})$ associated with the form $Q(x) := \langle x, Qx\rangle$, where $\langle\cdot,\cdot\rangle$ is the standard Euclidean scalar product on $\mathbb{R}^d$. Let Q_+ denote the unique positive symmetric matrix such that $Q_+^2 = Q^2$ and let $Q_+(x) = \langle x, Q_+x\rangle$ denote the associated positive form with eigenvalues being the eigenvalues of Q in absolute value. Let q, resp. q_0, denote the largest, resp. smallest, of the absolute value of the eigenvalues of Q and assume $q_0 \geq 1$. In the Oppenheim conjecture, we are concerned with the inequality $|Q(m)| < \varepsilon$; we can replace the form Q by Q/ε and consider the inequality $|Q(m)| < 1$. The following effective estimate is proved in [10]:

Theorem 12.4. *For all indefinite and non-degenerate quadratic forms Q of dimension $d \geq 5$ and signature (r,s) there exists for any $\delta > 0$ a non-trivial integral solution $m \in \mathbb{Z}^d \setminus \{0\}$ to the Diophantine inequality $|Q(m)| < 1$ satisfying*

$$\|Q_+^{1/2} m\| \ll_{\delta,d} (q/q_0)^{\frac{d+1}{d-2}} q^{1/2+\max\{\rho d+2,d+1\}/(d-4)+\delta},$$

where the dependency on the signature (r,s) is given by

$$\rho := \rho(r,s) = \begin{cases} \frac{1}{2}\frac{r}{s} & \text{for } r \geq s+3 \\ \frac{1}{2}\frac{s+2}{s-1} & \text{for } r = s+2 \text{ or } r = s+1 \\ \frac{1}{2}\frac{s+1}{s-2} & \text{for } r = s \end{cases}$$

In particular, for indefinite non-degenerate forms in $d \geq 5$ variables of signature (r,s) and eigenvalues in absolute value contained in a compact set $[1,C]$, i.e. $1 \leq q_0 \leq q \leq C$, Theorem 12.4 yields non-trivial solutions $m \in \mathbb{Z}^d$ of $|Q(m)| < \varepsilon$ of size bounded by

$$\|m\| \ll_{C,\delta} \varepsilon^{-\max\{\rho d+2,d+1\}/(d-4)-\delta}.$$

The proof of Theorem 12.4 relies on Götze's analytic approach [50] via Theta series, translating the lattice point counting problem into averages of certain functions on the space of lattices, for which the authors modify the method of integral inequalities introduced by the first-named author with Margulis and Mozes in [34] and described in detail in §11.

References

1. M. Aka, E. Breuillard, L. Rosenzweig and N. de Saxcé. Diophantine approximation on matrices and Lie groups. *Geom. Funct. Anal.* 28(1):1–57, 2018.
2. M. Aka, M. Einsiedler, H. Li and A. Mohammadi. On effective equidistribution for quotients of $\mathrm{SL}(d,\mathbb{R})$. *Israel J. Math.* 236(1):365–391, 2020.
3. A. Baker. *Transcendental number theory*. Cambridge University Press, London-New York, 1975.

4. V. Beresnevich, V. I. Bernik, D. Kleinbock and G. A. Margulis. Metric Diophantine approximation: the Khintchine-Groshev theorem for nondegenerate manifolds. *Mosc. Math. J.* 2(2):203–225, 2002.
5. E. J. Benveniste. *Rigidity and deformations of lattice actions preserving geometric structures.* Thesis (Ph.D.)–The University of Chicago, 1996.
6. V. Beresnevich. A Groshev type theorem for convergence on manifolds, *Acta Math. Hungar.* 94(1–2):99–130, 2002.
7. U. Bader and A. Furman. An extension of Margulis's Superrigidity theorem. In: *Dynamics, geometry, number theory—the impact of Margulis on modern mathematics*, 47–65, Univ. Chicago Press, Chicago, IL, 2022.
8. A. Brown, D. Fisher and S. Hurtado. Zimmer's conjecture: Subexponential growth, measure rigidity, and strong property (T), *Ann. of Math.* 2022, to appear.
9. M. Björklund and A. Gorodnik. Central limit theorems for Diophantine approximants. *Math. Ann.* 374(3-4):1371–1437, 2019.
10. P. Buterus, F. Götze, T. Hille and G. Margulis. Distribution of values of quadratic forms at integral points. *Invent. Math.* 227(3):857–961, 2022.
11. V. Beresnevich and D. Kleinbock. Quantitative non-divergence and Diophantine approximation on manifolds. In: *Dynamics, geometry, number theory—the impact of Margulis on modern mathematics*, 303–341, Univ. Chicago Press, Chicago, IL, 2022.
12. H. Bass, J. Milnor and J.-P. Serre. Solution of the congruence subgroup problem for SL_n and Sp_{2n}. *Inst. Hautes Etúdes Sci. Publ. Math.* 33:59–137, 1967.
13. M. Bekka and M. Mayer. *Ergodic theory and topological dynamics of group actions on homogeneous spaces.* London Math. Soc. Lecture Note Series, Vol. 269, Cambridge University Press, Cambridge, 2000.
14. V. I. Bernik, D. Kleinbock and G. A. Margulis. Khintchine-type theorems on manifolds: the convergence case for standard and multiplicative versions. *Internat. Math. Res. Notices* 2001(9):453–486.
15. V. Beresnevich, D. Kleinbock and G. A. Margulis. Non-planarity and metric Diophantine approximation for systems of linear forms. *J. Théor. Nombres Bordeaux* 27(1):1–31, 2015.
16. E. Breuillard and A. Lubotzky. Expansion in simple groups. In: *Dynamics, geometry, number theory—the impact of Margulis on modern mathematics*, 246–275, Univ. Chicago Press, Chicago, IL, 2022.
17. T. Brennan. *Distribution of Values of Diagonal Quadratic Forms at Integer Points.* Princeton University undergraduate thesis, 1994.
18. A. Brown, F. Rodriguez Hertz and Z. Wang. The normal subgroup theorem through measure rigidity. In: *Dynamics, geometry, number theory—the impact of Margulis on modern mathematics*, 66–91, Univ. Chicago Press, Chicago, IL, 2022.
19. A. Brown, F. Rodriguez Hertz and Z. Wang. Invariant measures and measurable projective factors for actions of higher-rank lattices on manifolds. *Ann. of Math.* 2022, to appear.
20. J. W. S. Cassels and H. P. F. Swinnerton-Dyer. On the product of three homogeneous linear forms and the indefinite ternary quadratic forms, *Philos. Trans. Roy. Soc. London Ser. A.* 248:73–96, 1955.
21. S.G. Dani. On invariant measures, minimal sets and a lemma of Margulis. *Invent. Math.* 51: 239–260, 1979.
22. S. G. Dani. Invariant measures and minimal sets of horoshperical flows. *Invent. Math.* 64:357–385, 1981.
23. S. G. Dani. On orbits of unipotent flows on homogeneous spaces. *Ergod. Theor. Dynam. Syst.* 4:25–34, 1984.
24. S. G. Dani. Divergent trajectories of flows on homogeneous spaces and Diophantine approximation. *J. Reine Angew. Math.* 359:55–89, 1985.
25. S. G. Dani. On orbits of unipotent flows on homogenous spaces II. *Ergod. Theor. Dynam. Syst.* **6** (1986), 167–182.

26. S. G. Dani. Bounded orbits of flows on homogeneous spaces. *Comment. Math. Helv.* **61** (1986), 636–660.
27. S. G. Dani and G. A. Margulis. Values of quadratic forms at primitive integral points. *Invent. Math.* 98:405–424, 1989.
28. S. G. Dani and G. A. Margulis. Orbit closures of generic unipotent flows on homogeneous spaces of $\mathrm{SL}(3,\mathbb{R})$. *Math. Ann.* 286:101–128, 1990.
29. S. G. Dani and G. A. Margulis. Values of quadratic forms at integral points: an elementary approach. *Enseign. Math.* (2) 36(1-2):143–174, 1990.
30. S. G. Dani and G. A. Margulis. Asymptotic behaviour of trajectories of unipotent flows on homogeneous spaces. *Indian. Acad. Sci. J.* 101:1–17, 1991.
31. S. G. Dani and G. A. Margulis. Limit distributions of orbits of unipotent flows and values of quadratic forms. in: *I. M. Gelfand Seminar*, 91–137, Adv. Soviet Math. **16**, Part 1, Amer. Math. Soc., Providence, RI, 1993.
32. H. Davenport and W. M. Schmidt. *Dirichlet's theorem on diophantine approximation*, in: *Symposia Mathematica, Vol. IV (INDAM, Rome, 1968/69)*, 113–132, Academic Press, London, 1970.
33. M. Einsiedler and A. Mohammadi. Effective arguments in unipotent dynamics. In: *Dynamics, geometry, number theory—the impact of Margulis on modern mathematics*, 426–451, Univ. Chicago Press, Chicago, IL, 2022.
34. A. Eskin, G. A. Margulis and S. Mozes. Upper bounds and asymptotics in a quantitative version of the Oppenheim conjecture. *Ann. Math.* (2), 147:93–141, 1998.
35. A. Eskin, G. A. Margulis and S. Mozes. Quadratic forms of signature $(2,2)$ and eigenvalue spacings on rectangular 2-tori. *Ann. Math.* (2) 161(2):679–725, 2005.
36. M. Einsiedler, G. A. Margulis, A. Mohammadi and A. Venkatesh. Effective equidistribution and property (τ). *J. Amer. Math. Soc.* 33(1):223–289, 2020.
37. M. Einsiedler, G. A. Margulis and A. Venkatesh. Effective results for closed orbits of semisimple groups on homogeneous spaces. *Invent. Math.* 177(1):137–212, 2009.
38. A. Eskin, S. Mozes and N. Shah. Unipotent flows and counting lattice points on homogeneous varieties. *Ann. Math.* 143:253–299, 1996.
39. A. Eskin, S. Mozes and N. Shah, Nondivergence of translates of certain algebraic measures, *Geom. Funct. Anal.* 7(1):48–80, 1997.
40. A. Eskin. Unipotent flows and applications. In: *Homogeneous flows, moduli spaces and arithmetic*, 71–129, Clay Math. Proc. 10, Amer. Math. Soc., Providence, RI, 2010.
41. A. Eskin and S. Mozes. Margulis functions and their applications, In: *Dynamics, geometry, number theory—the impact of Margulis on modern mathematics*, 342–361, Univ. Chicago Press, Chicago, IL, 2022.
42. D. Fisher. Superrigidity, arithmeticity, normal subgroups: results, ramifications, and directions. In: *Dynamics, geometry, number theory—the impact of Margulis on modern mathematics*, 9–46, Univ. Chicago Press, Chicago, IL, 2022.
43. Furstenberg, Harry *Rigidity and cocycles for ergodic actions of semisimple Lie groups (after G. A. Margulis and R. Zimmer)*. Bourbaki Seminar, Vol. 1979/80, pp. 273–292, Lecture Notes in Math., 842, Springer, Berlin-New York, 1981.
44. *Dynamics, geometry, number theory—the impact of Margulis on modern mathematics*, D. Fisher, D. Kleinbock and G. Soifer, eds., Univ. Chicago Press, Chicago, IL, 2022.
45. D. Fisher and G. A. Margulis. Local rigidity for cocycles. In: *Surveys in differential geometry, Vol. VIII (Boston, MA, 2002)*, 191–234, Int. Press, Somerville, MA, 2003.
46. D. Fisher and G. A. Margulis. Almost isometric actions, property (T), and local rigidity. *Invent. Math.* 162(1):19–80, 2005.
47. D. Fisher and G. A. Margulis. Local rigidity of affine actions of higher rank groups and lattices. *Ann. of Math.* (2) 170(1):67–122, 2009.
48. M. Fraczyk and T. Gelander. Infinite volume and infinite injectivity radius. *Ann. of Math.*, 2022, to appear.
49. O. Gabber and Z. Galil. Explicit constructions of linear-sized superconcentrators. Special issued dedicated to Michael Machtey. *J. Comput. System Sci.* 22(3):407–420, 1981.

50. F. Götze. Lattice point problems and values of quadratic forms. *Invent. Math.* 157(1):195–226, 2004.
51. S. Hoory, N. Linial and A. Wigderson. Expander graphs and their applications. *Bull. Amer. Math. Soc. (N.S.)* 43(4):439–561, 2006.
52. M. W. Hirsch, C. C. Pugh and M. Shub. *Invariant manifolds.* Lecture Notes in Mathematics, Vol. 583, Springer-Verlag, Berlin-New York, 1977.
53. L. Ji. A summary of the work of Gregory Margulis. *Pure Appl. Math. Q.* 4(1):1–69, 2008.
54. A. Katok and J. Lewis. Global rigidity results for lattice actions on tori and new examples of volume-preserving actions. *Israel J. Math.* 93:253–280, 1996.
55. D. Kazhdan and G. A. Margulis. A proof of Selberg's hypothesis. *Mat. Sb. (N.S.)* 75(117):163–168, 1968.
56. D. Kleinbock. Flows on homogeneous spaces and Diophantine properties of matrices. *Duke Math. J.* 95(1):107–124, 1998.
57. D. Kleinbock. Some applications of homogeneous dynamics to number theory. In: *Smooth ergodic theory and its applications*, Proc. Sympos. Pure Math., Vol. 69, 639–660, Amer. Math. Soc., Providence, RI, 2001.
58. D. Kleinbock, Quantitative Nondivergence and its Diophantine Applications. In: *Homogeneous flows, moduli spaces and arithmetic*, 131–153, Clay Math. Proc.., Vol. 10, Amer. Math. Soc., Providence, RI, 2010.
59. D. Kleinbock, E. Lindenstrauss and B. Weiss. On fractal measures and diophantine approximation. *Selecta Math.* 10:479–523, 2004.
60. D. Kleinbock and G. A. Margulis. Bounded orbits of nonquasiunipotent flows on homogeneous spaces. *Amer. Math. Soc. Translations* 171:141–172, 1996.
61. D. Kleinbock and G. A. Margulis. Flows on homogeneous spaces and Diophantine approximation on manifolds, *Ann. Math.* 148:339–360, 1998.
62. D. Kleinbock and G. A. Margulis. Logarithm laws for flows on homogeneous spaces. *Invent. Math.* 138(3):451–494, 1999.
63. D. Kleinbock and G. A. Margulis. On effective equidistribution of expanding translates of certain orbits in the space of lattices. In: *Number Theory, Analysis and Geometry*, 385–396, Springer, New York, 2012.
64. D. Kleinbock and S. Mirzadeh. Dimension estimates for the set of points with non-dense orbit in homogeneous spaces. *Math. Z.* 295:1355–1383, 2020.
65. D. Kleinbock and B. Weiss. Modified Schmidt games and a conjecture of Margulis. *J. Mod. Dyn.* 7(3):429–460, 2013.
66. D. Kleinbock, G. A. Margulis and J. Wang. Metric Diophantine approximation for systems of linear forms via dynamics. *Int. J. Number Theory* 6(5):1139–1168, 2010.
67. D. Kleinbock, N. Shah and A. Starkov, Dynamics of subgroup actions on homogeneous spaces of Lie groups and applications to number theory, in: *Handbook on Dynamical Systems, Volume 1A*, 813–930, Elsevier Science, North Holland, 2002.
68. D. Kleinbock, R. Shi and B. Weiss. Pointwise equidistribution with an error rate and with respect to unbounded functions. *Math. Ann.* 367(1-2):857–879, 2017.
69. D. Kleinbock, A. Strömberggson and S. Yu. A measure estimate in geometry of numbers and improvements to Dirichlet's theorem. *Proc. London Math. Soc.* 125(4):778–824, 2022
70. E. Lindenstrauss and G. Margulis. Effective estimates on indefinite ternary forms. *Israel J. Math.* 203(1):445–499, 2014.
71. E. Lindenstrauss, G. Margulis, A. Mohammadi and N. Shah. Quantitative behavior of unipotent flows and an effective avoidance principle. *Preprint*, `arXiv:1904.00290`, 2019..
72. A. Lubotzky, R. Phillips and P. Sarnak. Ramanujan graphs. *Combinatorica* 8(3):261–277, 1988.
73. G. A. Margulis. The action of unipotent groups in a lattice space. *Mat. Sb. (N.S.)* 86(128):552–556, 1971.

74. G. A. Margulis. Arithmeticity of nonuniform lattices. (Russian) *Funkcional. Anal. i Priložen.* 7(3):88–89, 1973.
75. G. A. Margulis. Explicit constructions of expanders. (Russian) *Problemy Peredači Informacii* 9(4):71–80, 1973.
76. G. A. Margulis. Discrete groups of motions of manifolds of nonpositive curvature. (Russian) *Proceedings of the International Congress of Mathematicians (Vancouver, B.C., 1974)*, Vol. 2, 21–34. Canad. Math. Congress, Montreal, Que., 1975.
77. G. A. Margulis. On the action of unipotent groups in the space of lattices. In: *Lie Groups and their representations, Proc. of Summer School in Group Representations*, Bolyai Janos Math. Soc., Akademai Kiado, Budapest, 1971, 365–370, Halsted, New York, 1975.
78. G. A. Margulis. Arithmeticity of the irreducible lattices in the semisimple groups of rank greater than 1. *Invent. Math.* 76(1):93–120, 1984.
79. G. A. Margulis. Formes quadratiques indèfinies et flots unipotents sur les spaces homogènes. *C. R. Acad. Sci. Paris Ser. I* 304:247–253, 1987.
80. G. A. Margulis. Explicit group-theoretic constructions of combinatorial schemes and their applications in the construction of expanders and concentrators. (Russian) *Problemy Peredachi Informatsii* 24(1):51–60, 1988; translation in *Problems Inform. Transmission* 24(1):39–46, 1988.
81. G. A. Margulis. Discrete Subgroups and Ergodic Theory. in: *Number theory, trace formulas and discrete subgroups, a symposium in honor of A. Selberg*, 377–398, Academic Press, Boston, MA, 1989.
82. G. A. Margulis. *Indefinite quadratic forms and unipotent flows on homogeneous spaces.* In: *Dynamical systems and ergodic theory*, Vol. 23, 399–409, Banach Center Publ., PWN – Polish Scientific Publ., Warsaw, 1989.
83. G. A. Margulis. Dynamical and ergodic properties of subgroup actions on homogeneous spaces with applications to number theory. in: *Proceedings of the International Congress of Mathematicians (Kyoto, 1990)*, 193–215, Math. Soc. of Japan and Springer, 1991.
84. G. A. Margulis. *Discrete subgroups of semisimple Lie groups.* Ergebnisse der Mathematik und ihrer Grenzgebiete (3) [Results in Mathematics and Related Areas (3)], Vol. 17, Springer-Verlag, Berlin, 1991.
85. G. A. Margulis. Oppenheim conjecture. In: *Fields Medalists' lectures*, 272–327, World-Sci. Publishing, River Edge, NJ, 1997.
86. G. A. Margulis. Diophantine approximation, lattices and flows on homogeneous spaces. In: *A panorama of number theory or the view from Baker's garden*, 280–310, Cambridge Univ. Press, Cambridge, 2002.
87. G. A. Margulis. *On some aspects of the theory of Anosov systems, With a survey by Richard Sharp: Periodic orbits of hyperbolic flows*, Springer Monographs in Mathematics. Springer-Verlag, 2004.
88. G. A. Margulis and G. Tomanov. Invariant measures for actions of unipotent groups over local fields on homogeneous spaces. *Invent. Math.* 116:347–392, 1994.
89. G. D. Mostow. *Strong rigidity of locally symmetric spaces.* Annals of Mathematics Studies, No. 78, Princeton University Press, Princeton, N.J.; University of Tokyo Press, Tokyo, 1973.
90. S. Mozes and N. Shah. On the space of ergodic invariant measures of unipotent flows. *Ergodic Theory Dynam. Systems* 15(1):149–159, 1995.
91. A. Nevo and R. J. Zimmer. A structure theorem for actions of semisimple Lie groups. *Ann. of Math.* (2) 156(2):565–594, 2002.
92. I. I. Piatetkski-Shapiro. Automorphic functions and arithmetic groups. (Russian) 1968 *Proc. Internat. Congr. Math. (Moscow, 1966)*, 232–247 , Izdat. "Mir", Moscow.
93. M. S. Raghunathan. Discrete groups and $\mathbb{Q}$-structures on semi-simple Lie groups. In: *Discrete subgroups of Lie groups and applications to moduli (Internat. Colloq., Bombay, 1973)*, 225–321, Oxford Univ. Press, Bombay, 1975.

94. M. Ratner. On Raghunathan's measure conjecture. *Ann. of Math.* 134:545–607, 1991.
95. M. Ratner. Raghunathan's topological conjecture and distributions of unipotent flows. *Duke Math. J.* 63(1):235–280, 1991.
96. W. M. Schmidt, On badly approximable numbers and certain games. *Trans. Amer. Math. Soc.* **123** (1966), 178–199.
97. W. Schmidt. Asymptotic formulae for point lattices of bounded determinant and subspaces of bounded height. *Duke Math. J.* 35: 327–339, 1968.
98. W. M. Schmidt, Badly approximable systems of linear forms. *J. Number Theory* **1** (1969), 139–154.
99. A. Selberg. On discontinuous groups in higher-dimensional symmetric spaces. In: *Contributions to function theory (Internat. Colloq. Function Theory, Bombay, 1960)*, 147–164, Tata Institute of Fundamental Research, Bombay, 1960.
100. N. A. Shah. Uniformly distributed orbits of certain flows on homogeneous spaces. *Math. Ann.* 289:315–334, 1991.
101. V. G. Sprindžuk. *Mahler's problem in metric number theory. Translated from the Russian by B. Volkmann*. Translations of Mathematical Monographs, Vol. 25, Amer. Math. Soc., Providence, R.I., 1969.
102. A. N. Starkov. The structure of orbits of homogeneous flows and the Raghunathan conjecture. *Russian Math. Surveys* 45(2):227–228, 1990).
103. A. N. Starkov. Ergodic decomposition of flows on homogenous spaces of finite volume. *Math. USSR-Sb.* 68(2):483–502, 1991.
104. D. Sullivan. Disjoint spheres, approximation by imaginary quadratic numbers, and the logarithm law for geodesics. *Acta Math.* 149:215–237, 1982.
105. R. J. Zimmer. *Ergodic theory and semisimple groups*. Birkhaüser, 1984.

List of Publications for Hillel Furstenberg

1953

[1] Note on one type of indeterminate form. *Amer. Math. Monthly*, 60:700–703.

1954

[2] An extension of Menelaus' theorem. *Scripta Math.*, 20:238–239.

1955

[3] On the infinitude of primes. *Amer. Math. Monthly*, 62:353.
[4] The inverse operation in groups. *Proc. Amer. Math. Soc.*, 6:991–997.

1960

[5] (with H. Kesten). Products of random matrices. *Ann. Math. Statist.*, 31:457–469.
[6] *Stationary processes and prediction theory*. Annals of Mathematics Studies, No. 44. Princeton University Press, Princeton, NJ.

1961

[7] Strict ergodicity and transformation of the torus. *Amer. J. Math.*, 83:573–601.

1963

[8] A Poisson formula for semi-simple Lie groups. *Ann. of Math. (2)*, 77:335–386.
[9] The structure of distal flows. *Amer. J. Math.*, 85:477–515.
[10] Noncommuting random products. *Trans. Amer. Math. Soc.*, 108:377–428.

1965

[11] Translation-invariant cones of functions on semi-simple Lie groups. *Bull. Amer. Math. Soc.*, 71:271–326.

H. Holden, R. Piene (eds.), *The Abel Prize 2018-2022*, The Abel Prize,
https://doi.org/10.1007/978-3-031-33973-8_13

1967

[12] Poisson boundaries and envelopes of discrete groups. *Bull. Amer. Math. Soc.*, 73:350–356.
[13] Disjointness in ergodic theory, minimal sets, and a problem in Diophantine approximation. *Math. Systems Theory*, 1:1–49.
[14] Algebraic functions over finite fields. *J. Algebra*, 7:271–277.

1970

[15] Intersections of Cantor sets and transversality of semigroups. In *Problems in analysis (Sympos. Salomon Bochner, Princeton Univ., Princeton, N.J., 1969)*, pages 41–59.

1971

[16] Random walks and discrete subgroups of Lie groups. In *Advances in Probability and Related Topics, Vol. 1*, pages 1–63. Dekker, New York.
[17] (with I. Tzkoni). Spherical functions and integral geometry. *Israel J. Math.*, 10:327–338.
[18] Boundaries of Lie groups and discrete subgroups. In *Actes du Congrès International des Mathématiciens (Nice, 1970), Tome 2*, pages 301–306.

1972

[19] Boundaries of Riemannian symmetric spaces. In *Symmetric spaces (Short Courses, Washington Univ., St. Louis, Mo., 1969–1970)*, pages 359–377. Pure and Appl. Math., Vol. 8.

1973

[20] (with H. Keynes and L. Shapiro). Prime flows in topological dynamics. *Israel J. Math.*, 14:26–38.
[21] Boundary theory and stochastic processes on homogeneous spaces. In *Harmonic analysis on homogeneous spaces (Proc. Sympos. Pure Math., Vol. XXVI, Williams Coll., Williamstown, Mass., 1972)*, pages 193–229.
[22] The unique ergodicity of the horocycle flow. In *Recent advances in topological dynamics (Proc. Conf., Yale Univ., New Haven, Conn., 1972; in honor of Gustav Arnold Hedlund)*, pages 95–115. Lecture Notes in Math., Vol. 318.

1976

[23] A note on Borel's density theorem. *Proc. Amer. Math. Soc.*, 55(1):209–212.

1977

[24] Ergodic behavior of diagonal measures and a theorem of Szemerédi on arithmetic progressions. *J. Analyse Math.*, 31:204–256.

1978

[25] (with B. Weiss). The finite multipliers of infinite ergodic transformations. In *The structure of attractors in dynamical systems (Proc. Conf., North Dakota State Univ., Fargo, N.D., 1977)*, volume 668 of *Lecture Notes in Math.*, pages 127–132. Springer, Berlin.

[26] (with S. Glasner). On the existence of isometric extensions. *Amer. J. Math.*, 100(6):1185–1212.

[27] (with B. Weiss). Topological dynamics and combinatorial number theory. *J. Analyse Math.*, 34:61–85 (1979).

[28] (with Y. Katznelson). An ergodic Szemerédi theorem for commuting transformations. *J. Analyse Math.*, 34:275–291 (1979).

1979

[29] (with A. Bellow). An application of number theory to ergodic theory and the construction of uniquely ergodic models. *Israel J. Math.*, 33(3-4):231–240 (1980).

1980

[30] Random walks on Lie groups. In *Harmonic analysis and representations of semisimple Lie groups*, pages 467–489. Lect. NATO Adv. Study Inst., Liège 1977.

1981

[31] *Recurrence in ergodic theory and combinatorial number theory.* Princeton University Press, Princeton, N.J.

[32] Poincaré recurrence and number theory. *Bull. Amer. Math. Soc. (N.S.)*, 5(3):211–234.

[33] Rigidity and cocycles for ergodic actions of semisimple Lie groups (after G. A. Margulis and R. Zimmer). In *Bourbaki Seminar, Vol. 1979/80*, volume 842 of *Lecture Notes in Math.*, pages 273–292. Springer, Berlin-New York.

1982

[34] (with Y. Katznelson and D. Ornstein). The ergodic theoretical proof of Szemerédi's theorem. *Bull. Amer. Math. Soc. (N.S.)*, 7(3):527–552.

1983

[35] (with Y. Kifer). Random matrix products and measures on projective spaces. *Israel J. Math.*, 46(1-2):12–32.

[36] Poincaré recurrence and number theory. In *The mathematical heritage of Henri Poincaré, Proc. Symp. Pure Math. 39, Part 2, Bloomington/Indiana 1980*, pages 193–216. AMS, Providence, RI.

[37] (with Y. Katznelson and D. Ornstein). The ergodic theoretical proof of Szemerédi's theorem. In *The mathematical heritage of Henri Poincaré, Proc. Symp. Pure Math. 39, Part 2, Bloomington/Indiana 1980*, pages 217–242. AMS, Providence, RI.

1984

[38] IP-systems in ergodic theory. In *Conference in modern analysis and probability (New Haven, Conn., 1982)*, volume 26 of *Contemp. Math.*, pages 131–148. Amer. Math. Soc., Providence, RI.

1985

[39] (with Y. Katznelson). An ergodic Szemerédi theorem for IP-systems and combinatorial theory. *J. Analyse Math.*, 45:117–168.

1986

[40] (with Y. Katznelson). IP_r-sets, Szemerédi's theorem, and Ramsey theory. *Bull. Amer. Math. Soc. (N.S.)*, 14(2):275–278.

1988

[41] (with B. Weiss). Simultaneous Diophantine approximation and IP-sets. *Acta Arith.*, 49(4):413–426.

1989

[42] (with Y. Katznelson). A density version of the Hales-Jewett theorem for $k = 3$. volume 75, pages 227–241. 1989. Graph theory and combinatorics (Cambridge, 1988).

[43] (with B. Weiss). On almost 1-1 extensions. *Israel J. Math.*, 65(3):311–322.

[44] (with Y. Katznelson). Idempotents in compact semigroups and Ramsey theory. *Israel J. Math.*, 68(3):257–270.

[45] (with V. Bergelson, N. Hindman, and Y. Katznelson). An algebraic proof of van der Waerden's theorem. *Enseign. Math. (2)*, 35(3-4):209–215.

1990

[46] Nonconventional ergodic averages. In *The legacy of John von Neumann (Hempstead, NY, 1988)*, volume 50 of *Proc. Sympos. Pure Math.*, pages 43–56. Amer. Math. Soc., Providence, RI.

[47] (with Y. Katznelson and B. Weiss). Ergodic theory and configurations in sets of positive density. In *Mathematics of Ramsey theory*, volume 5 of *Algorithms Combin.*, pages 184–198. Springer, Berlin.

1991

[48] Recurrent ergodic structures and Ramsey theory. In *Proceedings of the International Congress of Mathematicians, Vol. I, II (Kyoto, 1990)*, pages 1057–1069. Math. Soc. Japan, Tokyo.

[49] (with Y. Katznelson). A density version of the Hales–Jewett theorem. *J. Anal. Math.*, 57:64–119.

1993

[50] (with H.B. Keynes, N.G. Markley, and M. Sears). Topological properties of $\mathbf{R}^n$ suspensions and growth properties of $\mathbf{Z}^n$ cocycles. *Proc. London Math. Soc. (3)*, 66(2):431–448.

1994

[51] (with J. Auslander). Product recurrence and distal points. *Trans. Amer. Math. Soc.*, 343(1):221–232.

1995

[52] (with Y. Peres and B. Weiss). Perfect filtering and double disjointness. *Ann. Inst. H. Poincaré Probab. Statist.*, 31(3):453–465.

1996

[53] A polynomial Szemerédi theorem. In *Combinatorics, Paul Erdős is eighty, Vol. 2 (Keszthely, 1993)*, volume 2 of *Bolyai Soc. Math. Stud.*, pages 253–269. János Bolyai Math. Soc., Budapest.
[54] (with B. Weiss). A mean ergodic theorem for $(1/N)\sum_{n=1}^{N} f(T^n x)g(T^{n^2} x)$. In *Convergence in ergodic theory and probability (Columbus, OH, 1993)*, volume 5 of *Ohio State Univ. Math. Res. Inst. Publ.*, pages 193–227. de Gruyter, Berlin.
[55] (with V. Bergelson and R. McCutcheon). IP-sets and polynomial recurrence. *Ergodic Theory Dynam. Systems*, 16(5):963–974.

1998

[56] (with E. Glasner). Robert Ellis and the algebra of dynamical systems. In *Topological dynamics and applications (Minneapolis, MN, 1995)*, volume 215 of *Contemp. Math.*, pages 3–12. Amer. Math. Soc., Providence, RI.
[57] (with E. Glasner). Subset dynamics and van der Waerden's theorem. In *Topological dynamics and applications (Minneapolis, MN, 1995)*, volume 215 of *Contemp. Math.*, pages 197–203. Amer. Math. Soc., Providence, RI.
[58] Stiffness of group actions. In *Lie groups and ergodic theory (Mumbai, 1996)*, volume 14 of *Tata Inst. Fund. Res. Stud. Math.*, pages 105–117. Tata Inst. Fund. Res., Bombay.

2002

[59] From the Erdős–Turán conjecture to ergodic theory—the contribution of combinatorial number theory to dynamics. In *Paul Erdős and his mathematics, I (Budapest, 1999)*, volume 11 of *Bolyai Soc. Math. Stud.*, pages 261–277. János Bolyai Math. Soc., Budapest.

2003

[60] (with B. Weiss). Markov processes and Ramsey theory for trees. volume 12, pages 547–563. 2003.

2004

[61] (with Y. Katznelson). Eigenmeasures, equidistribution, and the multiplicity of β-expansions. In *Fractal geometry and applications: a jubilee of Benoît Mandelbrot. Part 1*, volume 72 of *Proc. Sympos. Pure Math.*, pages 97–116. Amer. Math. Soc., Providence, RI.

2006

[62] (with V. Bergelson and B. Weiss). Piecewise-Bohr sets of integers and combinatorial number theory. In *Topics in discrete mathematics*, volume 26 of *Algorithms Combin.*, pages 13–37. Springer, Berlin.

2008

[63] Ergodic fractal measures and dimension conservation. *Ergodic Theory Dynam. Systems*, 28(2):405–422.

2009

[64] (with V. Bergelson). WM groups and Ramsey theory. *Topology Appl.*, 156(16):2572–2580.

2010

[65] (with E. Glasner). Stationary dynamical systems. In *Dynamical numbers—interplay between dynamical systems and number theory*, volume 532 of *Contemp. Math.*, pages 1–28. Amer. Math. Soc., Providence, RI.

[66] The work of Elon Lindenstrauss. In *Proceedings of the International Congress of Mathematicians. Volume I*, pages 51–54. Hindustan Book Agency, New Delhi.

[67] Ergodic structures and non-conventional ergodic theorems. In *Proceedings of the International Congress of Mathematicians. Volume I*, pages 286–298. Hindustan Book Agency, New Delhi.

2012

[68] (with I. Stewart, D. Mumford, R. Howe, K. Falconer, B.J. West, M.-O. Coppens, N. Cohen, S. Jaffard, M. Berry, and M. Frame). The influence of Benoît B. Mandelbrot on mathematics. *Notices Am. Math. Soc.*, 59(9):1208–1221.

2013

[69] (with E. Glasner). Recurrence for stationary group actions. In *From Fourier analysis and number theory to Radon transforms and geometry*, volume 28 of *Dev. Math.*, pages 283–291. Springer, New York.

2014

[70] *Ergodic theory and fractal geometry*, volume 120 of *CBMS Regional Conference Series in Mathematics*. American Mathematical Society, Providence, RI.

2017

[71] (with E. Glasner and B. Weiss). Affinely prime dynamical systems. *Chin. Ann. Math. Ser. B*, 38(2):413–424.

2019

[72] From invariance to self-similarity: the work of Michael Hochman on fractal dimension and its aftermath. *J. Mod. Dyn.*, 15:437–449.

2022

[73] *IP*-systems and recurrence in ergodic theory: an update. *Israel J. Math.*, 251(2):423–441.

List of Publications for Grigoriy Margulis

1966

[1] Positive harmonic functions on nilpotent groups. *Soviet Math. Dokl.*, 7:241–244. Also available in *Dokl. Akad. Nauk SSSR*, 166:1054–1057 (in Russian).

1968

[2] (with D.A. Každan). A proof of Selberg's hypothesis. *Mat. Sb. (N.S.)*, 75(117): 163–168.

1969

[3] Certain applications of ergodic theory to the investigation of manifolds of negative curvature. *Funct. Anal. Appl.*, 3:335–336. Also available in *Funkts. Anal. Prilozh.*, 3(4):89–90 (in Russian).

[4] On the arithmeticity of discrete groups. *Dokl. Akad. Nauk SSSR*, 187:518–520. Also availble in *Dokl. Akad. Nauk SSSR* 187:518–520 (in Russian).

1970

[5] Discrete subgroups of real semi-simple Lie groups. *Math. USSR, Sb.* 9(1969): 555–568. Also available in *Mat. Sb. (N.S.)*, 80(122):600–615, 1969 (in Russian).

[6] The isometry of closed manifolds of constant negative curvature with the same fundamental group. *Sov. Math., Dokl.*, 11:722–723. Also available in *Dokl. Akad. Nauk SSSR*, 192:736–737 (in Russian).

[7] Certain measures that are connected with U-flows on compact manifolds. *Funct. Anal. Appl.* 4:55–67. Also available in *Funkcional. Anal. i Priložen.*, 4(1):62–76 (in Russian).

1972

[8] The action of unipotent groups in a lattice space. *Math. USSR, Sb.*, 15(1971): 549–554. Also available in *Mat. Sb. (N.S.)*, 86(128):552–556, 1971 (in Russian).

H. Holden, R. Piene (eds.), *The Abel Prize 2018-2022*, The Abel Prize,
https://doi.org/10.1007/978-3-031-33973-8_14

[9] Metric questions in the theory of C-systems. In *Ninth Mathematical Summer School (Kaciveli, 1971) (Russian)*, pages 342–348.

1973

[10] Explicit constructions of expanders. *Problemy Peredači Informacii*, 9(4):71–80.

1974

[11] Arithmeticity of nonuniform lattices. *Funct. Anal. Appl.* 7:245–246. Also available in *Funkcional. Anal. i Priložen.*, 7(3):88–89, 1973 (in Russian).
[12] Arithmeticity and finite-dimensional representations of uniform lattices. *Funct. Anal. Appl.*, 8:258–259. Also available in *Funkcional. Anal. i Priložen.*, 8(3):77–78 (in Russian).
[13] Arithmetic properties of discrete subgroups. *Russ. Math. Surv.* 29(1): 107–156. Also available in *Uspehi Mat. Nauk*, 29(1 (175)):49–98 (in Russian).
[14] Probabilistic characteristics of graphs with large connectivity. *Problemy Peredači Informacii*, 10(2):101–108.

1975

[15] Arithmeticity of nonuniform lattices in weakly noncompact groups. *Funct. Anal. Appl.* 9:31–38. Also available in *Funkcional. Anal. i Priložen.*, 9(1):35–44 (in Russian).
[16] On the action of unipotent groups in the space of lattices. In *Lie groups and their representations (Proc. Summer School, Bolyai, János Math. Soc., Budapest, 1971)*, pages 365–370.
[17] Non-uniform lattices in semisimple algebraic groups. In *Lie groups and their representations (Proc. Summer School on Group Representations of the Bolyai János Math. Soc., Budapest, 1971)*, pages 371–553.

1977

[18] Discrete groups of motions of manifolds of nonpositive curvature. *Am. Math. Soc., Translat., II. Ser.* 109:33–45. Also available in *Proceedings of the International Congress of Mathematicians (Vancouver, B.C., 1974), Vol. 2*, pages 21–34, 1975 (in Russian).
[19] Cobounded subgroups in algebraic groups over local fields. *Funct. Anal. Appl.* 11:119–128. Also available in *Funkcional. Anal. i Priložen.*, 11(2):45–57 (in Russian).
[20] (with G.A. Soĭfer). A criterion for the existence of maximal subgroups of infinite index in a finitely generated linear group. *Sov. Math., Dokl.* 18:847–851. Also available in *Dokl. Akad. Nauk SSSR*, 234(6):1261–1264 (in Russian).

1978

[21] Factor-groups of discrete subgroups. *Sov. Math., Dokl.* 19:1145–1149. Also available in *Dokl. Akad. Nauk SSSR*, 242(3):533–536 (in Russian).

1979

[22] Factor groups of discrete subgroups and measure theory. *Funct. Anal. Appl.* 12:295–305. Also available in *Funktsional. Anal. i Prilozhen.*, 12(4):64–76, 1978 (in Russian).

[23] (with G.A. Soĭfer). Nonfree maximal subgroups of infinite index of the group $\mathrm{SL}_n(\mathbf{Z})$. *Russ. Math. Surv.* 34(4):178–179. Also available in *Uspekhi Mat. Nauk*, 34(4(208)):203–204 (in Russian).

1980

[24] Finiteness of quotient groups of discrete subgroups. *Funct. Anal. Appl.* 13: 178–187. Also available in *Funktsional. Anal. i Prilozhen.*, 13(3):28–39, 1979 (in Russian).

[25] Multiplicative groups of a quaternion algebra over a global field. *Sov. Math., Dokl.*, 21:780–784. Also available in *Dokl. Akad. Nauk SSSR*, 252(3):542–546 (in Russian).

[26] Some remarks on invariant means. *Monatsh. Math.*, 90(3):233–235.

1981

[27] (with G.A. Soĭfer). Maximal subgroups of infinite index in finitely generated linear groups. *J. Algebra*, 69(1):1–23.

[28] On the decomposition of discrete subgroups into amalgams. *Selecta Math. Soviet.*, 1(2):197–213.

1982

[29] Explicit constructions of graphs without short cycles and low density codes. *Combinatorica*, 2(1):71–78.

[30] Finitely-additive invariant measures on Euclidean spaces. *Ergodic Theory Dynam. Systems*, 2(3-4):383–396.

1983

[31] (with A.T. Huckleberry). Invariant analytic hypersurfaces. *Invent. Math.*, 71(1):235–240.

[32] Free completely discontinuous groups of affine transformations. *Sov. Math., Dokl.* 28:435–439. Also available in *Dokl. Akad. Nauk SSSR*, 272(4):785–788 (in Russian).

1984

[33] Arithmeticity of the irreducible lattices in the semisimple groups of rank greater than 1. *Invent. Math.*, 76(1):93–120, 1984.

[34] (with A.S. Omel′ chenko). Deformation of plane curves. *Probl. Inf. Transm.*, 20:257–264. Also available in *Problemy Peredachi Informatsii*, 20(4):31–40 (in Russian).

1986

[35] (with J. Rohlfs). On the proportionality of covolumes of discrete subgroups. *Math. Ann.*, 275(2):197–205.

1987

[36] Complete affine locally flat manifolds with a free fundamental group. *J. Sov. Math.* 36:129–139. Also available in *Zap. Nauchn. Semin. Leningr. Otd. Mat. Inst. Steklova* 134:190–205, 1984 (in Russian).
[37] Formes quadratriques indéfinies et flots unipotents sur les espaces homogènes. *C. R. Acad. Sci. Paris Sér. I Math.*, 304(10):249–253.
[38] (with I.Ya. Gol′ dsheĭd). The condition of simplicity for the spectrum of Lyapunov exponents. *Sov. Math., Dokl.*, 35(2):309–313. Also available in *Dokl. Akad. Nauk SSSR*, 293(2):297–301 (in Russian).

1988

[39] Explicit group-theoretic constructions of combinatorial schemes and their applications in the construction of expanders and concentrators. *Probl. Inf. Transm.*, 24(1):39–46. Also available in *Problemy Peredachi Informatsii*, 24(1):51–60 (in Russian).
[40] Lie groups and ergodic theory. In *Algebra—some current trends (Varna, 1986)*, volume 1352 of *Lecture Notes in Math.*, pages 130–146. Springer, Berlin.
[41] (with F. Grunewald). Transitive and quasitransitive actions of affine groups preserving a generalized Lorentz-structure. *J. Geom. Phys.*, 5(4):493–531.

1989

[42] (with S.G. Dani). Values of quadratic forms at primitive integral points. *C. R. Acad. Sci. Paris Sér. I Math.*, 308(7):199–203.
[43] Discrete subgroups and ergodic theory. In *Number theory, trace formulas and discrete groups (Oslo, 1987)*, pages 377–398. Academic Press, Boston, MA.
[44] (with S.G. Dani). Values of quadratic forms at primitive integral points. *Invent. Math.*, 98(2):405–424.
[45] (with I.Ya. Gol′ dsheĭd). Lyapunov exponents of a product of random matrices. *Russ. Math. Surv.* 44(5):11–71. Also available in *Uspekhi Mat. Nauk*, 44(5(269)):13–60 (in Russian).
[46] Indefinite quadratic forms and unipotent flows on homogeneous spaces. In *Dynamical systems and ergodic theory (Warsaw, 1986)*, volume 23 of *Banach Center Publ.*, pages 399–409. PWN, Warsaw.

1990

[47] (with S.G. Dani). Orbit closures of generic unipotent flows on homogeneous spaces of $\mathrm{SL}(3,\mathbf{R})$. *Math. Ann.*, 286(1-3):101–128.
[48] (with S.G. Dani). Values of quadratic forms at integral points: an elementary approach. *Enseign. Math. (2)*, 36(1-2):143–174.

[49] Orbits of group actions and values of quadratic forms at integral points. In *Festschrift in honor of I. I. Piatetski-Shapiro on the occasion of his sixtieth birthday, Part II (Ramat Aviv, 1989)*, volume 3 of *Israel Math. Conf. Proc.*, pages 127–150. Weizmann, Jerusalem.

1991

[50] *Discrete subgroups of semisimple Lie groups*, volume 17 of *Ergebnisse der Mathematik und ihrer Grenzgebiete (3) [Results in Mathematics and Related Areas (3)]*. Springer-Verlag, Berlin.

[51] Compactness of minimal closed invariant sets of actions of unipotent groups. *Geom. Dedicata*, 37(1):1–7.

[52] (with S.G. Dani). Asymptotic behaviour of trajectories of unipotent flows on homogeneous spaces. *Proc. Indian Acad. Sci. Math. Sci.*, 101(1):1–17.

[53] Lie groups and ergodic theory. In *Algebra and analysis (Kemerovo, 1988)*, volume 148 of *Amer. Math. Soc. Transl. Ser. 2*, pages 73–86. Amer. Math. Soc., Providence, RI.

[54] Dynamical and ergodic properties of subgroup actions on homogeneous spaces with applications to number theory. In *Proceedings of the International Congress of Mathematicians, Vol. I, II (Kyoto, 1990)*, pages 193–215. Math. Soc. Japan, Tokyo.

1992

[55] (with S.G. Dani). On the limit distributions of orbits of unipotent flows and integral solutions of quadratic inequalities. *C. R. Acad. Sci. Paris Sér. I Math.*, 314(10):699–704.

[56] (with G.M. Tomanov). Measure rigidity for algebraic groups over local fields. *C. R. Acad. Sci. Paris Sér. I Math.*, 315(12):1221–1226.

1993

[57] (with S.G. Dani). Limit distributions of orbits of unipotent flows and values of quadratic forms. In *I. M. Gel′fand Seminar*, volume 16 of *Adv. Soviet Math.*, pages 91–137. Amer. Math. Soc., Providence, RI.

1994

[58] (with G.M. Tomanov). Invariant measures for actions of unipotent groups over local fields on homogeneous spaces. *Invent. Math.*, 116(1-3):347–392.

1995

[59] (with G.D. Mostow). The differential of a quasi-conformal mapping of a Carnot–Carathéodory space. *Geom. Funct. Anal.*, 5(2):402–433.

[60] (with H. Abels and G.A. Soĭfer). Semigroups containing proximal linear maps. *Israel J. Math.*, 91(1-3):1–30.

[61] (with A. Eskin and S. Mozes). On a quantitative version of the Oppenheim conjecture. *Electron. Res. Announc. Amer. Math. Soc.*, 1(3):124–130.

1996

[62] (with D.Y. Kleinbock). Bounded orbits of nonquasiunipotent flows on homogeneous spaces. In *Sinaĭ's Moscow Seminar on Dynamical Systems*, volume 171 of *Amer. Math. Soc. Transl. Ser. 2*, pages 141–172. Amer. Math. Soc., Providence, RI.

[63] (with S.P. Novikov, L.A. Bunimovich, A.M. Vershik, B.M. Gurevich, E.I. Dinaburg, V.I. Oseledets, S.A. Pirogov, K.M. Khanin, and N.N. Chentsova). Yakov Grigor′evich Sinaĭ (on the occasion of his sixtieth birthday). *Uspekhi Mat. Nauk*, 51(4(310)):179–191.

[64] (with G.M. Tomanov). Measure rigidity for almost linear groups and its applications. *J. Anal. Math.*, 69:25–54.

1997

[65] (with H. Abels and G.A. Soifer). Properly discontinuous groups of affine transformations with orthogonal linear part. *C. R. Acad. Sci. Paris Sér. I Math.*, 324(3):253–258.

[66] Existence of compact quotients of homogeneous spaces, measurably proper actions, and decay of matrix coefficients. *Bull. Soc. Math. France*, 125(3): 447–456.

[67] Oppenheim conjecture. In *Fields Medallists' lectures*, volume 5 of *World Sci. Ser. 20th Century Math.*, pages 272–327. World Sci. Publ., River Edge, NJ.

1998

[68] (with A. Eskin and S. Mozes). Upper bounds and asymptotics in a quantitative version of the Oppenheim conjecture. *Ann. of Math. (2)*, 147(1):93–141.

[69] (with D.Y. Kleinbock). Flows on homogeneous spaces and Diophantine approximation on manifolds. *Ann. of Math. (2)*, 148(1):339–360.

[70] (with S. Mozes). Aperiodic tilings of the hyperbolic plane by convex polygons. *Israel J. Math.*, 107:319–325.

1999

[71] (with D.Y. Kleinbock). Logarithm laws for flows on homogeneous spaces. *Invent. Math.*, 138(3):451–494. Correction, *ibid.*, 211(2):855–862, 2018.

[72] (with A. Karlsson). A multiplicative ergodic theorem and nonpositively curved spaces. *Comm. Math. Phys.*, 208(1):107–123.

2000

[73] (with È.B. Vinberg). Some linear groups virtually having a free quotient. *J. Lie Theory*, 10(1):171–180.

[74] Problems and conjectures in rigidity theory. In *Mathematics: frontiers and perspectives*, pages 161–174. Amer. Math. Soc., Providence, RI.

[75] (with A. Nevo and E.M. Stein). Analogs of Wiener's ergodic theorems for semisimple Lie groups. II. *Duke Math. J.*, 103(2):233–259.

[76] (with G.D. Mostow). Some remarks on the definition of tangent cones in a Carnot–Carathéodory space. *J. Anal. Math.*, 80:299–317.

[77] (with S.I. Adyan, E.I. Zel′manov, S.P. Novikov, A.S. Rapinchuk, L.D. Faddeev, and V.I. Yanchevskiĭ). Vladimir Petrovich Platonov (on the occasion of his sixtieth birthday). *Russ. Math. Surv.* 55(3):601–610. Also available in *Uspekhi Mat. Nauk*, 55(3(333)):197–204 (in Russian).

[78] (with W.M. Goldman). Flat Lorentz 3-manifolds and cocompact Fuchsian groups. In *Crystallographic groups and their generalizations (Kortrijk, 1999)*, volume 262 of *Contemp. Math.*, pages 135–145. Amer. Math. Soc., Providence, RI.

[79] Free subgroups of the homeomorphism group of the circle. *C. R. Acad. Sci. Paris Sér. I Math.*, 331(9):669–674.

2001

[80] (with N. Qian). Rigidity of weakly hyperbolic actions of higher real rank semisimple Lie groups and their lattices. *Ergodic Theory Dynam. Systems*, 21(1):121–164.

[81] (with V. Bernik and D. Kleinbock). Khintchine-type theorems on manifolds: the convergence case for standard and multiplicative versions. *Internat. Math. Res. Notices*, 9:453–486.

2002

[82] (with H. Abels and G.A. Soifer). On the Zariski closure of the linear part of a properly discontinuous group of affine transformations. *J. Differential Geom.*, 60(2):315–344.

[83] (with V.V. Beresnevich, V.I. Bernik, and D.Y. Kleinbock). Metric Diophantine approximation: the Khintchine–Groshev theorem for nondegenerate manifolds. *Mosc. Math. J.* 2(2):203–225.

[84] Diophantine approximation, lattices and flows on homogeneous spaces. In *A panorama of number theory or the view from Baker's garden (Zürich, 1999)*, pages 280–310. Cambridge Univ. Press, Cambridge.

2003

[85] (with D. Fisher). Local rigidity for cocycles. In *Surveys in differential geometry, Vol. VIII (Boston, MA, 2002)*, volume 8 of *Surv. Differ. Geom.*, pages 191–234. Int. Press, Somerville, MA.

2004

[86] *On some aspects of the theory of Anosov systems*. Springer Monographs in Mathematics. Springer-Verlag, Berlin.

[87] (with A. Eskin). Recurrence properties of random walks on finite volume homogeneous manifolds. In *Random walks and geometry*, pages 431–444. Walter de Gruyter, Berlin.

[88] (with H. Abels). Coarsely geodesic metrics on reductive groups. In *Modern dynamical systems and applications*, pages 163–183. Cambridge Univ. Press, Cambridge.

[89] Random walks on the space of lattices and the finiteness of covolumes of arithmetic subgroups. In *Algebraic groups and arithmetic*, pages 409–425. Tata Inst. Fund. Res., Mumbai.

2005

[90] (with A. Eskin and S. Mozes). Quadratic forms of signature $(2,2)$ and eigenvalue spacings on rectangular 2-tori. *Ann. of Math. (2)*, 161(2):679–725.

[91] (with H. Abels and G.A. Soifer). The Auslander conjecture for groups leaving a form of signature $(n-2,2)$ invariant. *Isr. J. Math.* 148:11–21.

[92] (with D. Fisher). Almost isometric actions, property (T), and local rigidity. *Invent. Math.*, 162(1):19–80.

2008

[93] (with T. Gelander and A. Karlsson). Superrigidity, generalized harmonic maps and uniformly convex spaces. *Geom. Funct. Anal.*, 17(5):1524–1550.

2009

[94] (with M. Einsiedler and A. Venkatesh). Effective equidistribution for closed orbits of semisimple groups on homogeneous spaces. *Invent. Math.*, 177(1): 137–212.

[95] (with D. Fisher). Local rigidity of affine actions of higher rank groups and lattices. *Ann. of Math. (2)*, 170(1):67–122.

[96] (with J.S. Athreya). Logarithm laws for unipotent flows. I. *J. Mod. Dyn.*, 3(3):359–378.

[97] (with W.M. Goldman and F. Labourie). Proper affine actions and geodesic flows of hyperbolic surfaces. *Ann. of Math. (2)*, 170(3):1051–1083.

2010

[98] (with D. Kleinbock and J. Wang). Metric Diophantine approximation for systems of linear forms via dynamics. *Int. J. Number Theory*, 6(5):1139–1168.

[99] (with V.V. Benyash-Krivets, A.B. Zhizhchenko, E.I. Zel'manov, A.A. Mal'tsev, S.P. Novikov, Yu.S. Osipov, G. Prasad, A.S. Rapinchuk, and L.D. Faddeev). Vladimir Petrovich Platonov (on his 70th birthday). *Russ. Math. Surv.*, 65(3):593–596. Also available in *Usp. Mat. Nauk.* 65(3):203–206 (in Russian).

2011

[100] (with A. Mohammadi). Quantitative version of the Oppenheim conjecture for inhomogeneous quadratic forms. *Duke Math. J.*, 158(1):121–160.

[101] (with H. Abels and G.A. Soifer). The linear part of an affine group acting properly discontinuously and leaving a quadratic form invariant. *Geom. Dedicata*, 153:1–46.

[102] Minkowski's theorem for random lattices. *Probl. Inf. Transm.* 47(4):398–402. Also available in *Problemy Peredachi Informatsii*, 47(4):104–108 (in Russian).

2012

[103] (with D.Y. Kleinbock). On effective equidistribution of expanding translates of certain orbits in the space of lattices. In *Number theory, analysis and geometry*, pages 385–396. Springer, New York.

2014

[104] (with A. Mohammadi and H. Oh). Closed geodesics and holonomies for Kleinian manifolds. *Geom. Funct. Anal.*, 24(5):1608–1636.

[105] (with E. Lindenstrauss). Effective estimates on indefinite ternary forms. *Israel J. Math.*, 203(1):445–499.

2015

[106] (with V. Beresnevich and D. Kleinbock). Non-planarity and metric Diophantine approximation for systems of linear forms. *J. Théor. Nombres Bordeaux*, 27(1):1–31.

[107] (with S.I. Adian, V.V. Benyash-Krivets, V.M. Buchstaber, E.I. Zel'manov, V.V. Kozlov, S.P. Novikov, A.N. Parshin, G. Prasad, A.S. Rapinchuk, L.D. Faddeev, and V.I. Chernousov). Vladimir Petrovich Platonov (on his 75th birthday). *Russ. Math. Surv.*, 70(1):197–201. Also available in *Usp. Mat. Nauk.* 70(1):197–200 (in Russian).

[108] (with S.I. Adian, V.V. Benyash-Krivets, V.M. Buchstaber, E.I. Zel'manov, V.V. Kozlov, S.P. Novikov, A.N. Parshin, G. Prasad, A.S. Rapinchuk, L.D. Faddeev, and V.I. Chernousov). Vladimir Petrovich Platonov (to the 75 anniversary). *Chebyshevskiĭ Sb.*, 16(4(56)):6–10.

2016

[109] (with H. Li). Effective estimates on integral quadratic forms: Masser's conjecture, generators of orthogonal groups, and bounds in reduction theory. *Geom. Funct. Anal.*, 26(3):874–908.

2017

[110] (with J.S. Athreya). Logarithm laws for unipotent flows, II. *J. Mod. Dyn.*, 11:1–16.

[111] (with S. Kadyrov, D. Kleinbock, and E. Lindenstrauss). Singular systems of linear forms and non-escape of mass in the space of lattices. *J. Anal. Math.*, 133:253–277.

2018

[112] (with H. Li). New bounds in reduction theory of indefinite ternary integral quadratic forms. *Adv. Math.*, 327:410–424.

[113] (with J.S. Athreya). Values of random polynomials at integer points. *J. Mod. Dyn.*, 12:9–16.

2020

[114] (with M. Einsiedler, A. Mohammadi, and A. Venkatesh). Effective equidistribution and property (τ). *J. Amer. Math. Soc.*, 33(1):223–289.

[115] (with D. Fisher, A. Lubotzky). Foreword. In *Group actions in ergodic theory, geometry, and topology—selected papers*, Univ. Chicago Press, pages ix–xi.

2021

[116] (with G.A. Soifer). Discreteness of deformations of cocompact discrete subgroups. In *Topology, geometry, and dynamics — V. A. Rokhlin-Memorial*, Contemp. Math., 772, Amer. Math. Soc., pages 241–247.

2022

[117] (with T. Pending Gelander, A. Levit). Effective discreteness radius of stabilizers for stationary actions. *Michigan Math. J.*, 72:389–438.

[118] (with P. Pending Buterus, F. Götze, T. Hille). Distribution of values of quadratic forms at integral points. *Invent. Math.*, 227(3):857–961.

[119] (with A. Mohammadi). Arithmeticity of hyperbolic 3-manifolds containing infinitely many totally geodesic surfaces. *Ergodic Theory Dynam. Systems*, 42(3):1188–1219.

Curriculum Vitae for Hillel Furstenberg

Born:	September 29, 1935 in Berlin, Germany
Degrees/education:	Bachelor of Science, Yeshiva University, 1955
	Master of Science, Yeshiva University, 1955
	PhD, Princeton University, 1958
Positions:	C.L.E. Moore Instructor, MIT, 1959–1960
	Assistant Professor, University of Minnesota, 1961–1963
	Professor, University of Minnesota, 1964–1965
	Professor, Hebrew University of Jerusalem, 1965–2003
	Professor, Bar Ilan University, 1966–2003
	Professor Emeritus, Hebrew University of Jerusalem, 2004–
Visiting positions:	Israeli Institute of Advanced Study, 1975–1976

H. Holden, R. Piene (eds.), *The Abel Prize 2018-2022*, The Abel Prize,
https://doi.org/10.1007/978-3-031-33973-8_15

Memberships: Israel Academy of Sciences and Humanities, 1974
National Academy of Sciences, 1989
American Academy of Arts and Sciences, International Honorary Member, 1995
Norwegian Academy of Science and Letters, 2020

Awards and prizes: Rothschild Prize, 1978
Speaker at the International Congress of Mathematicians, 1990, 2010 (plenary)
Israel Prize, 1993
Harvey Prize, 1993
EMET Prize, 2004
Paul Turán Memorial Lecture, 2006
Wolf Prize, 2006/2007
Abel Prize, 2020

Honorary degrees: Yeshiva University, 1990
Ben Gurion University, 2003
University of Rennes, 2008

Curriculum Vitae for Grigoriy Aleksandrovich Margulis

Born: February 24, 1946 in Moscow, USSR (Russia)

Degrees/education: Candidate of Science (PhD), Moscow, 1970
Doctor of Science, Minsk, 1983

Positions: Junior Research Fellow, Institute for Problems in Information Transmission, 1970–1974
Senior Research Fellow, Institute for Problems in Information Transmission, 1974–1986
Leading Research Fellow, Institute for Problems in Information Transmission, 1986–1990
Erastus L. De Forest Professor, Yale University, 1991–2019
Professor Emeritus, Yale University, 2019–

H. Holden, R. Piene (eds.), *The Abel Prize 2018-2022*, The Abel Prize,
https://doi.org/10.1007/978-3-031-33973-8_16

Visiting positions: University of Bonn, 1979
Max Planck Institute for Mathematics, 1988
TIFR, Bombay, 1989
Institut des Hautes Études, 1989, 2003
Collège de France, 1990
Harvard University, 1990–91
Institute for Advanced Study, 1991, 2006
Max Planck Institute, Bonn 1991, 2015, 2018
University of Göttingen, 1991
Columbia University, 1994 (Eilenberg Chair)
Newton Institute, Cambridge, 2000
University of Bielefeld, 1993–2012 (during the summer)

Memberships: American Academy of Arts and Sciences, Honorary Member, 1991
Tata Institute of Fundamental Research, Honorary Member, 1996
National Academy of Sciences, 2001
Connecticut Academy of Science and Engineering, 2004
European Academy of Sciences, 2004
American Mathematical Society, Fellow, 2012
Norwegian Academy of Science and Letters, 2020

Awards and prizes: Prize of Moscow Mathematical Society for young mathematicians, 1968
Fields Medal, 1978
Speaker at the International Congress of Mathematicians, 1974, 1990 (plenary)
Humboldt Prize, 1995
Lobachevsky Prize, 1996
Wolf Prize, 2005
Dobrushin International Prize, 2011
Abel Prize, 2020

Honorary degrees: University of Bielefeld, 1999
École Normale Supériere Paris, 2010
University of York, 2011
University of Lyon, 2013
University of Chicago, 2014

Part IV
2021 László Lovász and Avi Wigderson

"for their foundational contributions to theoretical computer science and discrete mathematics, and their leading role in shaping them into central fields of modern mathematics"

THE ABEL PRIZE

Citation

The Norwegian Academy of Science and Letters has decided to award the Abel Prize for 2021 to **László Lovász** of Eötvös Loránd University in Budapest, Hungary and **Avi Wigderson** of the Institute for Advanced Study, Princeton, USA,

> *"for their foundational contributions to theoretical computer science and discrete mathematics, and their leading role in shaping them into central fields of modern mathematics."*

Theoretical Computer Science (TCS) is the study of the power and limitations of computing. Its roots go back to the foundational works of Kurt Gödel, Alonzo Church, Alan Turing, and John von Neumann, leading to the development of real physical computers. TCS contains two complementary sub-disciplines: algorithm design, which develops efficient methods for a multitude of computational problems; and computational complexity, which shows inherent limitations on the efficiency of algorithms. The notion of polynomial-time algorithms put forward in the 1960s by Alan Cobham, Jack Edmonds, and others, and the famous $P \neq NP$ conjecture of Stephen Cook, Leonid Levin, and Richard Karp had a strong impact on the field and on the work of Lovász and Wigderson.

Apart from its tremendous impact on broader computer science and practice, TCS provides the foundations of cryptography, and is now having a growing influence on several other sciences leading to new insights therein by "employing a computational lens". Discrete structures such as graphs, strings, permutations are central to TCS, and naturally discrete mathematics and TCS have been closely allied fields. While both these fields have benefited immensely from more traditional areas of mathematics, there has been a growing influence in the reverse direction as well. Applications, concepts, and techniques from TCS have motivated new challenges, opened new directions of research, and solved important open problems in pure and applied mathematics.

László Lovász and Avi Wigderson have been leading forces in these developments over the last decades. Their work interlaces in many ways, and, in particular, they have both made fundamental contributions to understanding randomness in computation and in exploring the boundaries of efficient computation.

Along with Arjen Lenstra and Hendrik Lenstra, László Lovász developed the LLL lattice reduction algorithm. Given a high-dimensional integer lattice (grid), this algorithm finds a nice, nearly orthogonal basis for it. In addition to several applications such as an algorithm to factorize rational polynomials, the LLL algorithm is a favorite tool of cryptanalysts, successfully breaking several proposed crypto- systems. Surprisingly, the analysis of the LLL algorithm is also used to design and guarantee the security of newer, lattice-based crypto-systems that seem to withstand attacks even by quantum computers. For some exotic cryptographic primitives, such as homomorphic encryption, the only constructions known are via these lattice-based crypto-systems.

The LLL algorithm is only one among many of Lovász's visionary contributions. He proved the Local Lemma, a unique tool to show the existence of combinatorial objects whose existence is rare, as opposed to the standard probabilistic method used when objects exist in abundance. Along with Martin Grötschel and Lex Schrijver, he showed how to efficiently solve semidefinite programs, leading to a revolution in algorithm design. He contributed to the theory of random walks with applications to Euclidean isoperimetric problems and approximate volume computations of high-dimensional bodies. His paper with Uriel Feige, Shafi Goldwasser, Shmuel Safra, and Mario Szegedy on probabilistically checkable proofs gave an early version of the PCP Theorem, an immensely influential result showing that the correctness of mathematical proofs can be verified probabilistically, with high confidence, by reading only a small number of symbols! In addition, he also solved long-standing problems such as the perfect graph conjecture, the Kneser conjecture, determining the Shannon capacity of the pentagon graph, and in recent years, developed the theory of graph limits (in joint work with Christian Borgs, Jennifer Chayes, Lex Schrijver, Vera Sós, Balázs Szegedy, and Katalin Vesztergombi). This work ties together elements of extremal graph theory, probability theory, and statistical physics.

Avi Wigderson has made broad and profound contributions to all aspects of computational complexity, especially the role of randomness in computation. A randomized algorithm is one that flips coins to compute a solution that is correct with high probability. Over decades, researchers discovered deterministic algorithms for many problems for which only a randomized algorithm was known before. The deterministic algorithm for primality testing, by Agrawal, Kayal and Saxena is a striking example of such a derandomized algorithm. These derandomization results raise the question of whether randomness is ever really essential. In works with László Babai, Lance Fortnow, Noam Nisan and Russell Impagliazzo, Wigderson demonstrated that the answer is likely to be in the negative. Formally, they showed that a computational conjecture, similar in spirit to the $P \neq NP$ conjecture, implies $P = BPP$. This means that every randomized algorithm can be derandomized and turned into a deterministic one with comparable efficiency; moreover the derandomization is generic and universal, without depending on the internal details of the randomized algorithm.

Another way to look at this work is as a trade-off between hardness versus randomness: if there exists a hard enough problem, then randomness can be simulated by efficient deterministic algorithms. Wigderson's subsequent work with Impagliazzo and Valentine Kabanets proves a converse: efficient deterministic algorithms even for specific problems with known randomized algorithms would imply that there must exist such a hard problem.

This work is intimately tied with constructions of pseudorandom (random looking) objects. Wigderson's works have constructed pseudorandom generators that turn a few truly random bits into many pseudorandom bits, extractors that extract nearly perfect random bits from an imperfect source of randomness, and Ramsey graphs and expander graphs that are sparse and still have high connectivity. With Omer Reingold and Salil Vadhan, he introduced the zig-zag graph product, giving an elementary method to build expander graphs, and inspiring the combinatorial proof of the PCP Theorem by Irit Dinur and a memory efficient algorithm for the graph connectivity problem by Reingold. The latter gives a method to navigate through a large maze while remembering the identity of only a constant number of intersection points in the maze!

Wigderson's other contributions include zero-knowledge proofs that provide proofs for claims without revealing any extra information besides the claims' validity, and lower bounds on the efficiency of communication protocols, circuits, and formal proof systems.

Thanks to the leadership of Lovász and Wigderson, discrete mathematics and the relatively young field of theoretical computer science are now established as central areas of modern mathematics.

Autobiography, mostly mathematical

László Lovász

Meeting with mathematics

I was born in Budapest in 1948. This was the year when the Stalinist regime seized total power. From my early childhood, I remember my parents' fear of the secret police, and then in particular the uprising in 1956, the street fight that went on under our windows.

Nevertheless, my younger brother and I had a happy childhood. My father was a surgeon, who provided for us well. He worked very hard in the hospital, and at evenings he wrote scientific papers, earning eventually a degree equivalent to a PhD. I learned a lot of work ethics from him. My mother stayed at home, and we got a tough but loving education from her.

My first eight school years were unremarkable and boring, but in the 8th grade I attended the math club of the school, and met interesting, sometimes challenging math problems, which I loved. When high school applications were due, the teacher of the club visited my parents, and told them about a special class in mathematics in the high school *Fazekas*, which would start for the first time that September. He recommended to enroll me in this class, and also to subscribe to *Középiskolai Matematikai Lapok*, a math monthly for high school students.

These two recommendations changed my life from boring to exciting. I still remember the enormous pleasure of reading the articles and problem sets in the periodical. In one of the first issues I saw there was a paper by Paul Erdős on combinatorial geometry—what a treat!

But it was really the high school that changed my life. I found myself among other youngsters who were as interested in math as I was, and we had endless discussions about mathematics. Many of my classmates became leading mathematicians and math educators. There is no space here to list them, but one of them I

L. Lovász
Alfréd Rényi Institute of Mathematics, Reáltanoda street 13-15, H-1053, Budapest, Hungary,
e-mail: lovasz@renyi.hu

H. Holden, R. Piene (eds.), *The Abel Prize 2018-2022*, The Abel Prize,
https://doi.org/10.1007/978-3-031-33973-8_17

Fig. 1: My parents, younger brother and I in 1953.

have to name: Kati Vesztergombi. We started to date when we were 15, we got married when we were 21, and we have four children and seven grandchildren. Our mathematical interest has been close enough to have a number of joint papers and a couple of joint textbooks, and we have helped each other by listening to (sometimes criticizing) new ideas of each other, proofreading papers and books of each other, discussing issues of teaching, science politics and much more.

Fig. 2: High school classmates (from left to right): Lajos Posa, Miklos Laczkovich, Lászlo Lovász and Istvan Berkes.

Getting back to the high schools year, it was a very special place in many respects. We had 10 classes of math each week. We had excellent teachers—not only in math, but in the sciences and humanities as well. Politically, the school was rather liberal; this did not mean much by today's standards, but in those days it was quite

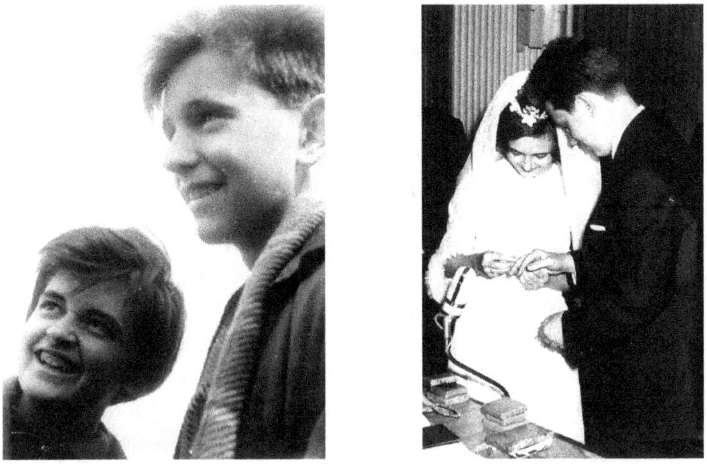

Fig. 3: Left: Kati and I in 1966. Right: Our wedding ceremony in 1969.

unusual to have a teacher who, for example, told us about religion and the life of Jesus.

The special class was the first of its kind, and attracted much attention of the math professors of the Eötvös University, who often came to teach a class or a math club.

Not surprisingly, most memorable were the visits of Paul Erdős. Much has been written about his life and his mathematics, so I restrict myself to a single memory. He stated a lot of unsolved problems about graph theory and other topics in discrete mathematics, some of which were understandable even for high schools students. This is how he got together with my classmate and friend Lajos Pósa, and they wrote several joint papers when Pósa was in elementary school or in the first years of high school. One day Lajos challenged me with one of the results in a joint manuscript of Erdős, Goodman and Pósa, and in a couple of days I could find a proof. Erdős put a footnote in the final version of the paper asserting that I also proved this, independently. The "independence" is disputable (I knew that the theorem was true, which helped a lot), but this is an example of how important Erdős felt it was to promote the careers of his young colleagues.

Moving around in Hungary and in the world

From these high school years on, I feel my life followed a straight line—even though we moved around in the world quite a lot. I attended the Eötvös University for five years, and got a diploma and a doctorate at the same time in 1971. I got a research position at the Eötvös University, and during this, I spent a year in the US, at Vanderbilt University as a postdoc. In Nashville I met Mike Plummer, and he and his wife Sara were helpful in helping us live in that nice, but for us rather alien place. Mike and I started working together on matchings in graphs. Later he visited me in Hungary for a year, and we wrote a monograph on matching theory.

I obtained the Chair of Geometry at the University of Szeged in 1975. My years in Szeged were some of the most productive years of my life. Szeged was a pleasant town, and its university was a quiet place which, nevertheless, put a lot of emphasis on scientific excellence.

In 1982 I moved back to the Eötvös University in Budapest. These last years of the communist regime were turbulent, both politically and economically, and I escaped regularly to the West, as a visiting professor in Bonn, Chicago, Ithaca and Berkeley. In 1987 I accepted a part time position in Princeton, where I spent 1/3–1/2 of my time. In 1993, I got a full time professorship at Yale, and then in 1999, I was offered a job at Microsoft Research in the Theory Group. This was another great time of my life for research, working with outstanding mathematicians, but eventually, in 2006, our hearts drew us back to Budapest. I was teaching at the Eötvös University for the next 8 years.

Fig. 4: Left: Our daughter Anna helping me in 1980 (Márta watching). Right: Laci_2 helping me in 1990. Technology changes but the baby clothes stay!

We have four children: Katalin (born 1972), Márta (1974), Anna (1980) and László Miklós (1990). Following Hungarian tradition, two of them received the first names of their parents. Not the most logical choice (although it is a good source of jokes and puns). To distinguish between them, we call them Kati_2 and Laci_2.

Moving around so much was clearly difficult for our children. On the plus side, they learned perfect English, and became familiar with other cultures. On the minus side, they had to part friends several times, adjust to different school systems, and more. Mathematics was of course always present in the family (frightening away some potential boyfriends, I am afraid), and all our children like it, even Kati_2, who has a PhD in literature, but is teaching folk stitching now, which she says is quite mathematical. Márta is an actuary, Anna has a PhD in economics, and Laci_2 has a PhD in math. I am very proud of them, not only because they have done well in their studies and careers, but also because they are good and honest people, and they

Fig. 5: Left: My daughters $Kati_2$, Anna and Márta singing a Christmas song in 2000. Right: Kati and I with our children ($Kati_2$ with the baby, Márta with two toddlers, Anna behind her and $Laci_2$ on the left), with sons-in-law and some grandchildren in 2008.

maintain strong ties with us and with each other (even though three of them live in the US, and travel restrictions due to Covid have been tough on all of us).

My wife always accuses me that I cannot say no, and perhaps she is right that I should have saved more of my time and energy for research from administration (she accepts teaching). From 2007 to 2010, I served as President of the International Mathematical Union, and I was elected President of the Hungarian Academy of Sciences in 2014, to serve two 3-year terms. This last period was *not* fruitful as mathematical research goes, as to fulfill my duties became more and more difficult politically, with the last half a year overshadowed by the Covid pandemic in addition. Now I am back to research at the Alfréd Rényi Institute of Mathematics, and in spite of all the restrictions imposed by the pandemic, I enjoy doing math research again.

Tarski's problem and graph limits

I was always interested in bridges between various branches of mathematics. Indeed, mathematical ideas have a way of zig-zagging through different branches of mathematics, and I have experienced this with some of my own ideas. In the rest of this autobiography I will sketch three examples of this.

In the early 1960s, graph theory was not highly regarded at all, and during the Summer of 1964, I started to think about how to develop graph theory in a direction more similar to algebra. Can we do operations on a graphs like addition and multiplication? Disjoint union is a good candidate for the sum, and after some thinking I came up with a notion of the product of two graphs. (As it turned out, this operation was introduced before under the name of "strong product".) Several expected properties (commutativity, associativity, distributivity) were easy to verify, but the cancelation law for the product was a hard nut. (This means that if A, B

and C are three graphs and $A \times C \cong B \times C$, then $A \cong B$.) In a month or two I was able to prove this, and in a few more months, I could simplify the proof considerably. The idea was to encode every graph G as an infinite sequence of numbers $\hom(F,G)$, where $\hom(F,G)$ denotes the number of homomorphisms (adjacency-preserving maps) from F to G, and F ranges through all finite graphs. This code of a product of two graphs is just the element-wise product of their codes, and so the cancelation law for graphs reduces to the cancelation law for numbers.

The method extended to other types of products, including products of finite algebraic and relational structures, under mild assumptions about the factor to be canceled. As it turned out, this answered an open problem raised a decade earlier by Alfred Tarski and Bjarni Jonsson. This fact had great benefits for me: Tarski invited me to a couple of Oberwolfach conferences he organized, and (after I got my PhD), Jonsson offered me a postdoc position at Vanderbilt University, which I mentioned above.

Later, in 1979, Paul Erdős, Joel Spencer and I used this encoding of graphs (with a suitable normalization) to study questions in extremal graph theory. Many extremal graph questions concerning a graph G can be phrased as linear inequalities between different entries $\hom(F,G)$ of the code, which lead us to consider the closure of the set of all codes, truncated to graphs F with at most k nodes. We managed to determine the dimension of the closure, but a full characterization of the closure seemed unaccessible at that time.

This characterization came in the early 2000s in a joint paper with Mike Freedman and Lex Schrijver, and it led to the theory of graph limits. I was a researcher at Microsoft Research, and various growing graph sequences, modeling the internet and other very large networks, were a hot topic. With Christian Borgs, Jennifer Chayes, Vera T. Sós, Balázs Szegedy and Kati Vesztergombi we were trying to find the answer to the question: just as there is a Central Limit Theorem for sequences of random variables, is there a similar description of the limiting behavior of a sequence of graphs? The (normalized) code mentioned above offered a definition: A sequence of larger and larger finite graphs (G_n) is *convergent* if the density of any fixed finite graph in G_n has a limit as $n \to \infty$. This notion is trivial unless we talk about dense graphs (where a positive fraction of pairs are connected by edges). Independently (a little earlier) Benjamini and Schramm had developed a limit theory of bounded degree graphs (as it happens, a few doors down the hall at Microsoft Research). The two theories are different, but analogies in their basic goals have been mutually very useful.

This notion of graph limits turned out to be quite lucky, and an elegant, well applicable theory could be built on it. I published a monograph on this subject in 2012, but research has gone on in various directions ever since then. I am still working on the problem of extending the theory to graphs with "intermediate" densities.

Perfect graphs and combinatorial optimization

During my university years my mentor was Tibor Gallai, who told me about the Perfect Graph Conjecture of Claude Berge. A graph is *perfect* if its chromatic number is equal to the size of its largest clique, and the same holds for every induced subgraph of it. The conjecture had a "weak" and a "strong" version; the former had a very simple formulation: *the complement of a perfect graph is also perfect.* In 1972, I could prove the weak version (the strong version was proved in 2006 by Chudnovsky, Robertson, Seymour and Thomas). The proof could be thought of as an extension of the Duality Theorem in linear programming to integer variables (of course, under strong assumptions), and in the following years I was trying to find more general results along these lines.

This was perhaps the first time I got interested in applications of mathematics, in particular in combinatorial optimization. The P-NP theory was well established by then, and it gave an excellent framework to understand the complexity of various problems and of the algorithms solving them. If a problem is NP-hard, then there is no hope to find an efficient (polynomial time) algorithm to solve it exactly; in such cases, finding approximate solutions is a next natural goal.

However, a theory of approximation algorithms was a big open problem at that time (around 1980): few good approximation algorithms were known, and almost no negative results (impossibility of finding even a decent approximate solution) were known. Such a theory came only more than a decade later, and I am happy to have played a minor part in both the positive and negative direction. In the negative direction, I was a co-author of an early paper on the subject, where research lead to the celebrated PCP Theorem (I have watched this development with pleasure, but not taking active part). The positive developments on approximation algorithms are connected with another interesting zig-zag of ideas, so let me say a few sentences on this.

One problem in combinatorial optimization, coming from information theory, was the Shannon capacity problem. This called for determining the maximum number of independent nodes in the product of many copies of a graph. The first unsolved case was the pentagon, an annoyingly simple graph, for which Shannon stated an exact conjecture. In 1978 I could prove the conjecture, using a method of representing the graph in a high-dimensional space. These *orthogonal representations* have branched off in different directions, from semidefinite optimization to quantum physics. Many of the successful approximation algorithms depend on semidefinite optimization.

This connects back to perfect graphs in an unexpected way. Even after the weak perfect graph conjecture was proved, a basic algorithmic question remained. Finding the chromatic number of a general graph is NP-hard, but can we compute the chromatic number of a perfect graph in polynomial time? The breakthrough of 1979 that led to such an algorithm was Khachiyan's polynomial time algorithm for linear programming, called the "Ellipsoid Method". Several teams noticed that this method is flexible enough to be applied beyond explicitly given linear programs, and with Martin Grötschel and Lex Schrijver we could combine it with orthogonal representations to obtain a polynomial time algorithm for computing the chromatic number of a perfect graph.

Fig. 6: Left: Lecturing in Hanoi 2009. Right: Three of us in Oberwolfach (Laci$_2$, Kati and myself).

Algorithmic geometry and cryptography

Martin, Lex and I went on to find many other applications of the Ellipsoid Method in graph theory and combinatorial optimization, eventually writing a monograph on this subject. We developed a general duality principle: in a high-dimensional space, optimizing a linear function over a convex body is equivalent (with respect to polynomial time algorithms) to deciding whether a point belongs to the body.

This work had an unexpected side branch. Trying to extend our results from full-dimensional bodies to bodies lying in unknown lower-dimensional subspaces, we ran into the problem of simultaneous diophantine approximation. The existence of an approximation of a finite number of real numbers by rationals with a common denominator was proved by Dirichlet, but finding these rational numbers efficiently was open. I found a polynomial-time algorithm for this, which used some ideas of Hendrik Lenstra, so I wrote to him about it. He answered that with his brother Arjen, they can use this to factor polynomials with rational coefficients into irreducible factors in polynomial time. This was quite surprising since one would expect that factoring polynomials was more difficult than factoring integers, which is still an unsolved problem. We wrote a joint paper about this. A couple of years later Lagarias and Odlyzko applied this algorithm to break one of the proposed public key crypto-systems, the knapsack code, and since then the algorithm became a standard tool to test crypto-systems for security.

Our work with Lex and Martin provided efficient algorithms for many algorithmic problems for convex bodies, but it left a basic question open: How to compute their volume? Our work on the Ellipsoid Method gave an approximation algorithm for the volume of an n-dimensional convex body, with a multiplicative error of n^n; this looked ridiculously high. I suggested this problem to my PhD student György Elekes, and soon he found an elegant proof that for any polynomial time algorithm that computes an approximation of the volume, there is a convex body in the n-dimensional space for which the result is off by a factor of at least $2^{n/2}$. The lower bound on the error was improved by Bárány and Füredi to $n^{\Theta(n)}$, which is best possible up the constant in the exponent.

So it seemed that computing the volume is hopelessly difficult, until Dyer, Frieze and Kannan came up with a *randomized* algorithm that computed it in polynomial time, with an arbitrarily small relative error in 1989. This did not contradict Elekes's result, which concerned deterministic algorithms. In the years that followed I was involved in improving this algorithm, working with Miki Simonovits, Ravi Kannan and Santosh Vempala. The original algorithm had a complexity of something like n^{26}, way out of the range of practicality. Bringing the exponent down took a lot of work by us and others, where each step involved new ideas from geometry, probability, statistics, and even from the theory of the heat equation. With contributions of several researchers, the bound is now down to $n^{3+o(1)}$ (this follows by combining more recent results of Vempala–Cousins and Chen).

This story illustrates the power of the notion of polynomial time: a polynomial bound of n^{26} is very far from being practical, but it means a structural insight to the problem—and more practical algorithms can be obtained based on its fundamental idea.

Some periods of mathematics are particularly exciting, when it develops so closely together with another science that it is difficult to distinguish which results or even which scientists belong to math or to this area of its application. Think of the development of mathematics and physics in the 18th century: were Newton or Euler physicists or mathematicians? The development of the theory of computing in the second half of the 20th century has a similar feeling. Most of us working on algorithms and their complexity were mathematicians *and* computer scientists (even operations research was closely involved). I feel that I am very lucky to have had the opportunity to live and work through such a great period for mathematics.

Avi Wigderson — a short biography

Avi Wigderson

It is customary for Abel prize winners to write a brief biography in these volumes, and I have done the same below. Writing it, the shortcomings of such an account quickly became apparent. Condensing a person, result, event, experience, or a lesson learned (and there are quite a few of these below) into a sentence or two, leaves so much to be desired and feels like doing an injustice to their meaning for me. But, it had to be done, so here goes.

I was born in 1956 in Haifa, Israel. My parents, Pinchas Wigderson and Shoshana Klagsbrun, were both Holocaust survivors, and most of their families were murdered by the Nazis. They met in Israel, and got married in 1953. My mother worked part-time as a nurse, and my father was an electrical engineer working in the navy shipyards. As is the case with most Holocaust survivors, they rarely talked about this around us, their children. I am the oldest of three brothers. Meir was born a year after me, and Oded, six years later.

We grew up in a tiny apartment in a blue-collar neighborhood called "Ein Hayam" (eye of the sea), a far more pastoral and picturesque name than it is in reality, but for children it was heavenly. Situated a 5 minutes walk from the Mediterranean shore, we could spend plenty of time there, and have a sunset view of the sea every evening from our balcony. I have numerous memories of swimming and snorkeling, something you could do almost all year round, and playing soccer essentially every day. My brother Meir and I also spent hours every day listening to the radio, and every other free minute I spent reading, basically everything I could lay my hands on, from adventure stories to encyclopedias. Our neighborhood was an amazing mix of cultures. Both Arabs and Jews lived there, with almost all Jews being recent immigrants to the new state of Israel. In the eight apartments of our building alone lived families from Egypt, Bulgaria, Morocco, Poland, Greece and Russia! A favorite culinary memory of that mix are the sweets our neighbor from Egypt made, which I consumed in great quantities.

A. Wigderson
School of Mathematics, Institute for Advanced Study, 1 Einstein Drive, Princeton, NJ 08540, USA,
e-mail: avi@math.ias.edu

H. Holden, R. Piene (eds.), *The Abel Prize 2018-2022*, The Abel Prize,
https://doi.org/10.1007/978-3-031-33973-8_18

Fig. 1: My parents (left) and siblings (right). (Private)

My father was definitely a central influence on my love of mathematics (or, what I knew then as puzzles). From a very early age he asked me riddles, and exposed me to a set of puzzle books in Russian he had (mainly on logic and geometry), teaching me enough vocabulary to figure out the riddles myself. Another exposure was the way he described the challenges he faced at work, which had mainly to do with figuring out the reasons for malfunctioning ship engines, and how to fix them, which were always described as giant riddles (these engines were literally huge, and I guess diagnostic tools at the time were very limited). The sense of joy and satisfaction he felt when he figured out a tough one is readily familiar to anyone doing mathematical research when figuring out how to solve a tough problem.

Elementary school was quite easy and boring, but I was a nerd, and enjoyed every type of learning, and even solving boring homework problems. I continued reading a lot of everything, quickly exhausting our neighborhood public library. For high school I was sent to the Re'ali school, one of the best in Israel, that had some excellent teachers and was far more interesting for me. This was my first exposure to kids from a much higher socio-economic class. I made some really good friends there, and started playing chess seriously. It was also the place I was first exposed to math at a higher level. In 11th and 12th grade we had a math teacher, Ya'akov Kaplan, a recent immigrant from Ukraine (who had to learn Hebrew while teaching us), who had a great influence on me. An author of many textbooks, and a teacher in a special math school in Kiev, he taught us math as it is done in college, with axioms and rigorous proofs. He further gave "extra credit" classes that were challenging and fun — by then I knew for sure what I loved best. This class was also when I first met my long-time friend and colleague, Noga Alon, who already then, as now, could solve any problem much faster than anyone else. A special treat was doing a "senior project" under Kaplan's advising; I needed to read some papers, and write a short thesis on "Famous Inequalities", which was my first, very modest, introduction to mathematical research.

Next came the army service, a three-year obligatory part of life for every Israeli. Actually, I was accepted to the coveted "pilot's course", which required a commitment of a seven-year minimum service for anyone graduating from those two years of training. It still amazes me how easy it was for me at the time to sign up for this commitment, and how lucky I was to flunk the course after less than a year, which released me from it. The remaining two years of service were spent at a useful and reasonably interesting office job. Different Israelis have different experiences in the army, which of course depend in part on what they do, and in part on their tolerance of a hierarchical organization with plenty of unreasonable rules. In hindsight, I find that while serving in the army at the ages 18–21 (years which could be spent in much more productive and fun ways) is certainly a waste of time, there are things I (and others) got there which are probably not accessible elsewhere at this age. For instance, I learned that I can run much further than I thought (with full gear on my back), that I can be awake for several nights straight and function afterwards, and that I can tolerate arbitrary orders. I also gained some responsibility, tolerance, maturity and I met many people from different parts of Israeli society. So, given that I had no choice about it, I contributed what I could, gained what I could, and was very happy to be released after three years.

As I was applying to college at the end of my service, I told my parents I would like to study mathematics. Their advice (they never really insisted on anything) was, roughly, that it is not clear what can be done with math once I graduate. How about learning (the relatively new) computer science, where I will anyway be exposed to enough math, but where I will also graduate with a real profession I can use to earn a living. I took their advice, which I still believe was an extremely fortunate decision. This was one of the first years that CS was offered as a major. I learned to program, and for a full year found the wonders of different programming languages (PL1, APL) miraculous (despite having to type the instructions on punch cards). We also programmed on a PDP-8; a microprocessor into which each instruction of a given program was manually fed by setting twelve binary switches and pressing a button, and memory allocation had to be programmed as well. Programming that very low-memory machine was extremely challenging, and I vividly remember one task whose only solution required overwriting part of the program for extra space. In brief, efficient use of computational resources, a topic I spent my life studying, was instilled in me inadvertently by the technology of the time. I learned the usual practical CS courses (again, relevant to that period), which I rather enjoyed. And I took the required calculus and linear algebra courses, which I found amazing and loved every bit, but the CS major requirements crammed into a three-year program meant that I had hardly any time to take more advanced math courses. Most of all, I loved the CS theory courses, automata theory and algorithms. The dominant figure of influence for me during college was Shimon Even (who taught two of these courses); he was an excellent teacher, and had an infectious love of algorithmic problems, which certainly infected me. Besides Shimon, I had an amazing set of classmates, including Oded Goldreich, Hanoch Levy and David Peleg, whose presence in and outside classes was inspiring.

So, while it became completely clear to me what I loved doing best, I still had no clear idea that there is a profession which allows doing this for a living. Luckily, most of the top students in my class applied to graduate school in the US, so I followed suit. I got full scholarship offers from the computer science departments at Princeton and Yale. When I consulted Shimon Even on where should I go, he said that professionally, in theory of CS, they were roughly equal, but that the Princeton campus is much prettier, so I chose Princeton.

During that last year in college I met Edna, a math major, at a fun "puzzle-solving" class. This was Edna's last college course, and for the last meeting of the class she baked a fantastic chocolate cake (I have already mentioned my sweet tooth above), decorated with a solution to the famous "eight queens" puzzle . We fell in love, and after a brief dating period decided to get married. Our first decision was where we should go: Edna had also applied for graduate school in the US before we met, was accepted and planned to go for a PhD at UC Santa Barbara. Edna gave up her plan for mine! When we got to Princeton, she instead applied and got into the MSc program at Rutgers math. We got married in May 1980, just before leaving for the US.

Fig. 2: My wife Edna and I. (Private)

Our way to Princeton was also our honeymoon. We flew to LA (where my uncle lived at the time), bought a used station wagon there, and spent the whole summer of 1980 exploring the US, first the many national parks on the West Coast, and then inland, through many more parks in Utah, Arizona and Colorado, making our way eventually to Princeton. During this period we had our few belongings tied to the car roof, so we could use the back of the car as our bedroom every night. This was a great period to for us to discover lots of things about each other, to learn about the art of car maintenance that we were clueless about and thus were repeatedly taken advantage of, to spend most of the little money we had brought with us, and to see some absolutely amazing sights.

Finally, I was about to learn what research was all about. One way in which one can gauge the maturity of a field is how well-versed in research beginning graduate students are. Today (indeed, in the past quarter-century), it is rare to find in top schools graduate students who did not do some research as undergrads; of course, many start with a handful of published papers under their belts, but most have at least done something equivalent to a senior project in college. At the time, neither I nor my classmates came with any notion of what research was. Who I wanted as an advisor was clear to me — Dick Lipton, who moved to Princeton from Yale just a year before I had arrived, and agreed to take me on as a student. This again I consider extremely lucky, as Dick is the person from whom I learned what research is, and how he does it. His style was "global". He was interested in everything computational! Before I met him he already had major papers on Petri nets, synchronization, program testing, proof verification, graph algorithms, complexity, and much more. And during my three years at Princeton, I saw him change research interests almost as often as people change socks. Dick was generous with his time and ideas. I spent numerous hours with him, and learned about many of these topics — their motivations, state of the art and cool open problems. These 2–3 hour sessions with Dick were so dense that I often went straight to the library to summarize all I had learned lest I forget anything. To learn more, I started chatting with my classmates about their research projects. This led to some joint papers — with Hana Galperin on complexity, with Doug Long on cryptography and with Gopalakrishnan Vijayan on graph algorithms. I got hooked on collaborations, and found that this social style of doing research suited me perfectly. I probably have only a couple of papers which are singly authored — all others are joint.

Another thing I learned from Dick was the value of extensive mathematical knowledge, and how he put this power to use! To date one of the best examples I have of this value is his elegant solution (with Zeke Zalcstein) to the complexity of the word problem in the free group. It is of course an easy problem, with a simple polynomial time algorithm. But can you do it in logarithmic space? Think about it! Again, being in a CS department, there were a bunch of CS courses to take for the qualifying exams, and so the only math courses I took at Princeton were two basic algebra classes (from the legendary Goro Shimura) and a combinatorics course from Doug West. Besides, I browsed many journals, and tried solving as many problems as I could in Laci Lovász' exceptional (in many ways) book of combinatorial exercises. But all in all, I wish I had taken more math courses as a student. While it is possible, and indeed I found necessary, to learn various parts of different areas of mathematics in order to tackle computational complexity problems I was working on, I find that learning "for a purpose" has many shortcomings. The experience of learning complete courses, indeed even sequences of courses, especially when you are a student and have much more time, gives far more intuition and a deeper understanding than what you get from searching for a theorem that would suit your needs. On the other hand, as the theory of computation expands it seems to demand (and influence) ever more diverse parts of mathematics, and so no matter what your school math background, you'll probably need to learn a lot more.

A completely different aspect of research life, which was central to my enjoying it over the years, was revealed to me already early on in graduate school. This was the nature of my research community. During my first year in Princeton I went to my first FOCS conference, and this was a formative experience. I believe that nothing like these FOCS and STOC conferences existed in math in those days, and are still rare now. Semi-refereed by a committee, accepted papers were presented in a way that allowed for fast dissemination of fresh results. Talks in all areas of algorithms and complexity were presented to an audience which was equally varied and for decades would consist of a major portion of this research community. I feel that these venues allowed for superb interaction between subareas, revealing the intricate web of (often surprising) connections between them, propelling the field faster. I attended every talk, and since then, over the decades, attempted to attend as many of these conferences as I could. Even more exciting was the social atmosphere. Everyone was talking to everyone between lectures, and I found there was no distance whatsoever between junior and senior people. I was introduced to Dick Karp, perhaps the most eminent theoretical computer scientist of the era. He took time to ask me about my work (by that time I had only proved one NP-completeness result — these were almost, but not completely, passé by then). This friendly, informal, collaborative atmosphere, clearly created by the leaders of the field at the time, persists to this day (though the field itself, and these conferences in particular, have grown far larger and can't quite accommodate the same intimacy and the possibility of attending all talks). Finally, I found that these venues were great not only for collaborations, but for starting new friendships, which still continue. In short, for me this community became at once an academic and a social home, and I cannot imagine a better one.

Everything I learned about research during graduate school pales in comparison to another learning experience of that period. In December of 1981, a year and a half into our time in Princeton, our son Eyal was born, and with him we started learning how to be parents. This is hardly the proper place to describe the meaning of such an experience. Let me just say that we found it amazing, and also discovered that all other aspects of life could be made compatible with it. Timing was perfect, and Edna returned to her studies after the holiday break. She had classes twice a week at Rutgers, and as she was nursing Eyal, all three of us drove to New Brunswick, and she would nurse him between classes. Despite the usual initial period of sleep-deprived nights, I found that I could think and be productive. At the age of 8 months we sent Eyal to a part time daycare, but most of the time he was with us.

Both Edna and I graduated in the summer of 1983. We started another cross-country trip, now with Eyal, a very patient 18-month-old old baby who tolerated well many hours in the car. Three weeks later we made it to Berkeley, our new home, where I started a postdoc with Dick Karp, and Edna started transitioning from math to programming, which became her profession since then.

Fig. 3: My mentors: Shimon Even (left), Dick Lipton (middle), and Dick Karp (right).

We loved everything about Berkeley: the geographical setting, the culture, the food, the coffee — quite a contrast to Princeton (and we do like Princeton, as is evident from our having spent more time there than anywhere else on Earth)! We stayed for three years in the Bay area, two in Berkeley itself and one at IBM San Jose. The number of people I started collaborating with and the number of research areas I was exposed to during these postdoctoral years was amazing. In particular, working with and observing Dick Karp during this period was a wonderful learning experience for me. Besides research, I was inspired by his meticulous preparation of classes he taught, and more generally with the sense of commitment and dedication he had for every task he undertook.

As I'll start mentioning some research topics and researchers, let me apologize in advance and note that this biography, which is not a technical survey, cannot be complete or precise in content and credits, and therefore many will be omitted. However, for anyone interested, even without background in the theory of computing, let me recommend consulting my book "Mathematics and Computation", which was published by Princeton University Press and is available for free online on my website at `https://www.math.ias.edu/avi/book`. There the reader will find plenty of background, motivation, history of ideas and references to many of the topics I will mention.

Among many topics, extremely significant for me were two very related ones, which computational complexity has transformed into their modern forms mainly in Berkeley in the early 1980s, resting both on computational hardness assumptions. The first was cryptography; I was drawn into it by Oded Goldreich and Silvio Micali, and our works on zero-knowledge proofs and cryptographic protocol design are among my favorites. The second was pseudorandomness. Here, work with Miklós

Ajtai was the start of my life-long interest in the power of randomness in computation. Dramatic computational hardness results which were proved during that period, specifically exponential lower bounds on restricted models like monotone circuits and constant-depth circuits, deeply inspired me to attempt proving some non-trivial but unrestricted hardness results (a first step on the presumably very long journey towards $P \neq NP$). This quest, in which I invested much time and effort during the decades which followed, repeatedly ended in failure. Still, these failed attempts (and what I have learned from them) have been almost as much fun as successful ones in other areas. Great consolation prizes were other restricted lower bounds for different models and the uncovering of the deep connections between hardness and randomness. One more research area on which I spent a lot of time during that postdoc period was parallel computation, or "which problems are inherently sequential?". I worked mainly with Dick Karp and Eli Upfal on algorithms, and with Faith Fich, Friedhelm Meyer auf der Heide and Prabhakar Ragde on lower bounds. Again, the large web of interconnections within algorithms and complexity was unraveling before my eyes. To mention one aspect, sometimes it was possible to remove randomization from probabilistic algorithms without any computational hardness assumptions. Unconditional derandomization results of this type became another passion of mine!

And again, in the middle of this three-year period, in December of 1984, our daughter Einat was born. Now both Edna and I worked full-time at IBM San Jose. At the time, IBM did not give any maternity leave (in the words of their HR person to Edna, "research scientists don't have babies"), and so after a month of the holidays and using sick-leave, Edna went back to work, and Einat joined a full time daycare. Luckily, she was an extremely easy baby, and was perfectly fine with nursing before and after work, as well as during Edna's 48-minute lunch break which she used to drive to Einat's daycare and back, nurse her and bite a sandwich while doing it. This was not as easy for Edna as for Einat. We discovered the wonders and challenges of two kids at home, and their interactions, which again are beyond the scope of this essay.

Without much hesitation, convinced that Israel is *the* place for kids to grow up (just as we did), we applied for jobs in Israel. Both of us got positions at the Hebrew University in Jerusalem, Edna at the computing center, and I in the computer science department. We both grew up in Haifa, and Jerusalem was a completely new experience for us, as it is different from, really, any place on Earth. Its remarkable history makes it one of the most interesting places to visit, and in some ways, often quite difficult to live in. So many empires have conquered it, so many cultures developed there, so many religions consider it their holiest place and so much blood has been shed in these numerous disputes that tensions between various groups are always present, and inevitably every so often flare up. We found a nice apartment in a pretty neighborhood few minutes' walk from campus, so that at least most of the time, our little bubble of life was mostly secluded from the city's influence. We spent a total of about 15 years in Jerusalem, during which we discovered many things about life. We found that indeed social life for kids in Israel was great, but that on the other hand schools were not great at all, which was one trade-off. We discovered that we all love traveling, and have traveled all over the world with the kids at all ages. We also realized that our income in Israel did not quite match our way of life, and that it was good to supplement it abroad — this combined with my professional interests and collaborations in several sabbatical years, all in Princeton, in 1990–1992 at the university, and in 1995 at the Institute for Advanced Study. These many disruptions of life, especially for our kids, having to switch schools, languages, and mostly make new friends, proved both challenging and rewarding in a variety of ways, and with many years of hindsight, I guess all of us view them as positive. Edna's programming skills proved very portable, and during all these trips she found jobs nearby.

In December of 1994 our third child Yuval was born (yes, we seem to have a bias towards December kids...). His brother and sister were 13 and 10 (respectively) by then, so he got the love and attention of the equivalent of two sets of parents. Indeed, Einat, a very determined child, had decided she'd attend Yuval's birth, and prevailed over our doubts and worries of what such an experience may be for a ten-year-old. It ended up a beautiful, unforgettable event, in which she helped the midwife with various chores, and cut the umbilical cord herself! Starting international travel with Yuval even earlier than his siblings, we took him to Berkeley when he was six weeks old. The best sleeping arrangement we found for him in our rental apartment there was our suitcase, where he fit perfectly. This amply prepared him for a year of sleeping in a closet during our 1995 sabbatical, where his crib fit perfectly. I am not sure if these were particularly creative solutions to space problems, but he doesn't seem to have any scars from them. Of the numerous activities and experiences, mundane and exciting, of our family life during these years in Jerusalem I will mention just one more. Our oldest son Eyal discovered a life-long passion. This was climbing. At the age of 14 he joined a group of kids climbing a natural rocky cliff in Jerusalem, got better and better under the guidance of some top climbers and by the age of 18 had already done much rock-climbing in Jordan, Egypt and then Yosemite. Our worries regarding this activity grew with his progress, and little did we know it was just a beginning. After his army service Eyal started climbing ice, a far more dangerous activity which he still persists in, though not as high now as when he was in his 20s, when he scaled peaks in the Himalayas, Pamir and Alaska. For Edna and me this was perhaps the hardest test of a parental principle meeting reality — to what extent should one support a child's passion and talent? This sport looked so benign when he started, but after he was hooked, obviously loving it and excelling at it, it was too late to stop. We learned to silently bite our nails during his expeditions, and celebrate his achievements when he returned safely from them. Einat and Yuval were far kinder to us, and took to much tamer hobbies.

Professionally, these 15 years in Jerusalem meant yet another learning experience for me in several dimensions. This is obvious, as it was my first faculty position. Some aspects of it were to be expected with a short learning curve and mainly requiring time during which one prefers doing other things, but some activities must be done — these include different departmental chores, like committees and responsibilities (e.g. chairing the department — something I did for 2.5 years). Another major new time investment naturally went into teaching. I discovered that I love

Fig. 4: My children. (Private)

teaching, both undergraduate and graduate courses. It helped of course that most students at the Hebrew university were really good, and so, fun to teach. The most challenging novelty for me (and I guess for many new faculty members) was becoming a graduate student advisor. Somehow, there is no standard training for this "job", or even much discussion of it, and indeed, different people view and exercise it differently. This position carries with it responsibilities of a different nature, which I didn't realize as a graduate student. As I have learned, an advisor needs not only to introduce a student to the mysterious activity of research by suggesting research problems and ideas for solving them, but also to help him or her find what topics and methods they like best and are good at (these two are obviously very correlated). And on the emotional side, you need to provide support through the many tough periods that research provides but they are not used to yet — long periods of no progress, set-backs when a proof they have found fails during rigorous write-ups, or is found to be known, tough decisions like when to quit trying after hitting one wall after another for so long, and move to a different problem, and the ultimate worries about how good they are, absolutely and compared to others. All of these evolve in an intimate relationship which lasts several years, and is of course individual, depending on personality, talents, learning curve and so many other aspects. Needless to say, some students don't need or want such close personal contact, and are happy and successful with problem suggestions and other professional guidance (like how to write papers, how to give good lectures, etc.)

Again I was extremely lucky, leaning this art together with my first graduate students at the Hebrew University. As I arrived, quite a few students asked to work with me, and not having a clear idea what it meant besides opportunities to collaborate with more people on the many problems I had been thinking about, I said yes to all of them. I am very grateful to that initial batch of MSc and PhD students, Ron Ben Natan, Aviad Cohen, Yossi Gil, Rafi Heiman, Shlomo Hoory, Mauricio Karchmer, Michal Parnas, Yuri Rabinovich, Ran Raz, and Moti Reif, who in a period of five years taught me almost everything I needed to know about graduate advising.

Needless to say, topping all complex aspects of advising I discussed above, was actually doing research with them. It is a unique collaborative experience in which (unlike collaborations with more senior colleagues) there is a long term commitment of several years to a person and a project. I also learned, from these and many graduate students who I advised in later years at the university, that sometime it is them who supply the topics and questions they want to pursue, with which I was not necessarily very familiar, and so I often learned with them and from them. And so I worked with Mauricio Karchmer and Ran Raz on Boolean circuit complexity, with Eli Ben Sasson on proof complexity, with Yuri Rabinovich on dynamical systems, with Ronen Shaltiel and Noam Nisan (then a visiting student from Berkeley for a semester) on pseudorandomness, with Amir Shpilka on arithmetic complexity, with Dorit Aharonov on quantum computation and more. I have continued to pursue many of these research areas for years or decades, sometimes with the same people! And beyond research, many of these relationships developed into life-long friendships.

I have enjoyed many other aspects of research life in Israel and the Hebrew University. First and foremost, Israel is a phenomenal theory hub — it always was, and of course it is growing and getting stronger all the time. Many wonder at the reason for this. A clear one is that in Israel, CS grew mainly out of math departments, something which ensured a high regard for theory from both mathematicians and practical CS researchers from day one (this is in contrast to the US, where CS grew mostly from electrical engineering departments, and where theory was, and in many places still is, not as highly regarded). At the Hebrew University both the computer science department and the math department are excellent, and having been for many years one and the same school, with people like Michael Rabin, Eli Shamir and Nati Linial having positions in both, created a great collaborative atmosphere which was ready for me to enjoy. Almost all theory courses were cross-listed, and indeed some of my graduate students came from math. The math department had, and still has, a great tradition of top-notch quality teaching standards, and theory students had excellent math backgrounds. Also, Israel is a small country, and it was easy to meet, have seminars, and work with people from all other academic institutions, so some grad courses I taught were attended by students from Tel Aviv and the Weizmann Institute. Finally, there were plenty of international visitors, including many Israelis stopping by for short and long visits. I had fantastic new collaborations and felt extremely productive there. Among my favorite works at the Hebrew university let me mention four. The connection Mauricio Karchmer and I discovered between Boolean circuit depth and communication complexity, which became a major tool in circuit and proof complexity. The pseudo-random generator I developed with Noam Nisan, which became a central tool of conditional derandomization tightening up the relation between hardness and randomness. This culminated, after a series of works, in my paper with Impagliazzo giving a widely believable hardness assumption, which renders efficient deterministic algorithms which are essentially as powerful as probabilistic ones. The general information-theoretic protocol for oblivious computation developed with Michael Ben-Or and Shafi Goldwasser, (providing a parallel to the above-mentioned cryptographic ones we constructed with Goldreich and Micali), and techniques feeding back to crypto (for homomorphic encryption). And finally, with Ben-Or, Goldwasser and Joe Kilian, the introduction of a multi-prover interactive proof model (MIP), which had so many important consequences we could not imagine and were obtained by others, including to the PCP theorem, hardness of approximation, and the recent breakthrough $MIP^* = RE$ on the quantum variant of this model that had surprising applications in pure math.

We were all mostly happy and content in Jerusalem, when a phone call from Phillip Griffiths, then the director of the Institute for Advanced Study, came in early 1999. In a great show of style, after making me the offer to become a faculty member in the school of math, he asked me to pass the phone to Edna, and made her an offer to join the IAS computing group. Perhaps to the mathematical reader of this I should say that it was far from an obvious professional choice to make. There was no computer science at IAS, and indeed to this day I am the only permanent computer scientist. Of course, I had already been there on sabbatical in 1995, and for a couple of semesters after that, and knew of the keen interest of the school to expand into TCS and discrete math. I was also very familiar with the excellent CS department and theory group at Princeton University, where I had many friends and colleagues, but it was clear to me that at the IAS I would need to build from scratch a research environment of the type I liked, to replace the perfect one I really loved in Jerusalem.

But these professional considerations were minor compared to the personal ones. Indeed, it was to be the most complex decision, with the most wide-ranging consequences on our future family life. Of course, going to Princeton meant separating from our extended family, numerous friends, jobs we liked and our home country. But it was much more complicated than that. Eyal was 17, and made clear to us what we knew anyway, namely that he would stay in Israel (which meant being alone for 12th grade and then his army service). Being extremely independent and trustworthy, we did not really worry about his well-being — he could stay in our apartment, use our car and our credit card — conditions which caused envy among his friends and worry among many of ours. Yuval was 4, so we did not worry about him either — it was clear he would come with us. We asked Einat, who was 14 at the time, about what she thought of such an adventure. Having very fond memories from 3 years in previous sabbaticals at Princeton, with friends who went to school with her, she was ready to try moving there for two years (for her 9th and 10th grades), if we agreed we'd let her return then if she didn't like it. So, as it seemed like an exciting new adventure we wanted to embark on, we agreed to give it a try for these two years. Cutting to the end of this drama, Einat discovered after a few weeks that it was a big mistake, as socially it turned out to be far worse than she had imagined. Still, she persevered! Edna and I on the other hand liked it greatly, and decided we would stay. The following two years, Einat's last in high school, we spent mostly in Jerusalem, and then she stayed with Eyal in Jerusalem while we returned with Yuval to Princeton. To this day, Eyal and Einat live in Israel, Yuval and we in the US, with everyone shuttling continuously across the Atlantic to see each other.

The School of Mathematics at the IAS is quite a unique environment. Eight (or fewer) professors, no students at all, and about 70 visitors every year, most of them postdocs in all areas of math, who have no responsibilities at all and are free to concentrate on research. It has a beautiful housing complex for all on campus, with a daycare center for young kids. IAS is placed in the secluded part of Princeton, amidst the Institute woods with their many trails, providing plenty of isolation for reflection if you want that. When you want interaction, there is plenty going on in the School, and if you want more, you are only a ten-minute bike ride away from the University where there are plenty more people to talk to and seminars to attend. Most visitors (or, as we call them, members) indeed find it an ideal place to concentrate on research. So, it seemed like a perfect place to start a postdoc program in TCS and discrete math in the school, which extremely quickly got an excellent reputation and attracted, like all other areas of math, the best graduating students. It has now been running for over 20 years, and has seen over a hundred postdocs pass through it. I was also fortunate to attract fantastic senior people to spend several consecutive years in half- or full-time positions in this program and assist in mentoring the postdocs: Noga Alon, Russell Impagliazzo, Toniann Pitassi, Ran Raz and Sasha Razborov.

I have found that working with postdocs is quite different than with graduate students, for obvious reasons. First, they spend a much shorter time here, a year or two, and so often the relationships with many cannot be as developed as with graduate students who spend 4–5 years with you. Postdocs arrive with significant research experience from graduate school, and often with a clear research agenda. They typically have already some established collaborations which continue while at IAS. So, in many ways, they don't face many of the issues that graduate students do, and the responsibilities of a postdoc advisor are far lighter. While true, one should also remember that the postdoc period can be one of high pressure to many. In some cases, this is the period during which they must both discover and prove independence from their graduate advisor. And for all, this is when they seek their first permanent faculty positions, plan interview trips and lectures, and wait for answers on their applications. In short, there is plenty of personal interaction and responsibility even with postdocs. But most of the interaction revolves around research, and this has been pure pleasure. From my own postdoc years I grew to believe that this period is best spent on expanding horizons, learning new topics and developing new collaborations, and have attempted to create an environment fostering this at IAS. Besides the usual seminars with external speakers, we have internal ones, in which each member teaches all others their specialty in detailed, long lectures. I have learned an enormous amount from these. Needless to say, it is impossible to work with all members, but research interactions with a substantial fraction has taught me much more, and led to many more works I am really happy with. It is impossible to list even a fraction of these, but let me mention a couple of my favorites. With my postdocs Omer Reingold and Salil Vadhan we developed the "zigzag product" of graphs, and with it, a new, combinatorial construction of expander graphs, a pseudo-random object important in numerous areas of math and CS, for which essentially all previous constructions were algebraic in nature. In particular, the construction and proof inspired two spectacular results: the logspace algorithm for graph connectivity of Omer Reingold, and a new proof of the PCP theorem by Irit Dinur. Another is the study of randomness extractors, another important pseudo-random object. Among many facets and related objects, let me mention the introduction of the sum-product theorem in finite fields into the study and explicit construction of extractors and Ramsey graphs, with Russell Impagliazzo, Ronen Shaltiel and my postdocs Boaz Barak and Anup Rao. Curiously (but as often happens), this powerful method was bypassed by stronger ones of a different nature, which provide much better constructions. A very long and broad project is still raging on, with (at the time) postdocs Zeyuan Allen-Zhu, Pavel Hrubes, Visu Makam and Amir Yehudayoff, and Princeton university students Ankit Garg, Rafael Oliveira and continues with them and collaborators Peter Buergisser, Cole Franks and Michael Walter. Very broadly speaking, it has to do with using symmetries of problems to generate efficient algorithms for them, which generalize convex optimization from Euclidean to other geodesically-convex settings. This has been quite unique for me in many ways. While motivated in arithmetic complexity, we discovered that it is closely related to fundamental problems in a a surprising variety of areas of math, physics and optimization. It has demanded that I dive quite deeply into areas I knew nothing about, mainly invariant theory, and some aspects of representation theory and Riemannian geometry. And while all results so far have been algorithmic in nature, I still hope this approach may give a way to prove some unconditional hardness results! Lastly, meaningful for an obvious reason, is recent work with my son Yuval, now a math graduate student, on uncertainty principles (joint work with my two other kids would have been far harder, as Eyal became a neurobiologist and a wine maker, and Einat became a psychologist).

Fig. 5: My wife and I (left) and my children (right). (Private)

With all this in the past already, I feel I have been extremely fortunate to have the blessed life I had, with a loving family, many friends, and an occupation which is my main passion (and also allows me to see the world). I have been in excellent academic institutions, and had great mentors, students, postdocs and collaborators who have taught me so much, and with whom I keep enjoying beating our heads

Fig. 6: With my granddaughters, Tamar (left) and Nuri (right). (Private)

against math problems and discussing anything else. Of all which may be called "achievements" above, there is nothing compared (for both Edna and me) to having watched our children grow and become wonderful, happy people. I hope I'll be able to keep enjoying this kind of life for much longer!

The Mathematics of László Lovász

Martin Grötschel and Jaroslav Nešetřil

Abstract This is an exposition of the contributions of László Lovász to mathematics and computer science written on the occasion of the bestowal of the Abel Prize 2021 to him. Our survey, of course, cannot be exhaustive. We sketch remarkable results that solved well-known open and important problems and that – in addition – had lasting impact on the development of subsequent research and even started whole new theories. Although discrete mathematics is what one can call the Lovász home turf, his interests were, from the beginning of his academic career, much broader. He employed algebra, geometry, topology, analysis, stochastics, statistical physics, optimization, and complexity theory, to name a few, to contribute significantly to the explosive growth of combinatorics; but he also exported combinatorial techniques to many other fields, and thus built enduring bridges between several branches of mathematics and computer science. Topics such as computational convexity or topological combinatorics, for example, would not exist without his fundamental results. We also briefly mention his substantial influence on various developments in applied mathematics such as the optimization of real-world applications and cryptography.

Martin Grötschel
Technische Universität Berlin, Mathematisches Institut, Straße des 17. Juni 135, 10623 Berlin, Germany, e-mail: groetschel@bbaw.de

Jaroslav Nešetřil
Computer Science Institute of Charles University, Faculty of Mathematics of Physics, Charles University, Malostranské nám. 25, 11800 Praha 1, Czech Republic, e-mail: nesetril@iuuk.mff.cuni.cz

H. Holden, R. Piene (eds.), *The Abel Prize 2018-2022*, The Abel Prize,
https://doi.org/10.1007/978-3-031-33973-8_19

Contents

1 Introduction

László Lovász was born in 1948 in Budapest. Laci, as he is called by his friends, attended the Fazekas Mihály Gimnázium in Budapest, a special school for mathematically gifted students and a fertile ground of world-class mathematicians. Katalin Vesztergombi, his wife since 1969, was one of his classmates. Laci's outstanding talent became visible at very young age. He won, for example, several mathematics competitions in Hungary and also won three gold medals in the International Mathematical Olympiad.

Lovász studied mathematics at Eötvös Loránd University (ELTE). He received – with Tibor Gallai as his mentor – his first doctorate (Dr. Rer. Nat.) degree from ELTE in 1971, the Candidate of Sciences (C. Sc.) degree in 1970 and his second doctorate (Dr. Math. Sci.) degree in 1977 from the Hungarian Academy of Sciences. Of great influence for his scientific growth was the outstanding Hungarian combinatorial school (e.g., T. Gallai, A. Hajnal, A. Rényi, M. Simonovits, V. T. Sós, P. Turán, and foremost P. Erdős).

In 1971 Lovász started his professional career as a research associate at ELTE. From 1975 to 1982 he was Docent, later Professor and Chair of Geometry at József Attila University, Szeged; 1983–1993 Chair of Computer Science at ELTE; 1993–1999 Professor of Computer Science at Yale University; and 1999–2006 Senior Researcher, Microsoft Research, Redmond. In 2006 Lovász returned to his hometown Budapest as a Professor and Director of the Mathematical Institute at ELTE from which he retired in 2018. In 2020 he joined the Alfréd Rényi Institute of Mathematics. Lovász served the International Mathematical Union as its President from 2007 to 2010 and the Hungarian Academy of Sciences as its President from 2014 to 2020 during demanding times.

Among the institutions Lovász visited for extended periods of time are Vanderbilt University, University of Waterloo, Universität Bonn, University of Chicago, Cornell University, Mathematical Sciences Research Institute in Berkeley, Princeton University, Princeton Institute for Advanced Study, and ETH Zürich. Five universities bestowed special professorships upon him, he received six honorary degrees and countless high-ranking honors and distinctions, including the Kyoto Prize 2010, see Fig. 2.

Like every scientific discipline, mathematics has become a field with a large number of specializations. The Mathematics Subject Classification (MSC 2020) with its 63 first-level areas and 6,006 specific research areas is a witness of this development. Today, no mathematician has a full understanding of all the mathematical branches. But there are still a few people with broad mathematical knowledge, deep command of their fields of special interest, and the ability to build bridges by transferring results and techniques between fields to expand the mathematical toolboxes and open up new research areas. One of these rare persons is László Lovász. In fact, quite fittingly, two volumes published in his honor at special occasions were entitled *Building Bridges*, see [65] and [12].

Laci's mathematical roots are in combinatorics. But he vastly expanded his reach by employing combinatorial methods in other mathematical fields and bringing, in return, tools from geometry, topology, algebra, analysis, probability theory, information theory, optimization, and even ideas from physics into combinatorics. His deep interest in algorithms led to major advances in modern complexity theory. In his work, Lovász established profound connections between discrete mathematics and computer science. This is reflected in the statement that the Norwegian Academy of

Fig. 1: Lovász and Erdős at dinner in 1977 (Photo: Private)

Fig. 2: In a Tokyo subway station on the way to the Kyoto Prize ceremony: Laci, Kati, and son Laci M. Lovász, András Frank in the back (Photo: Private)

Science and Letters issued in its announcement of the award of the Abel Prize 2021 to him and Avi Wigderson

> for their foundational contributions to theoretical computer science and discrete mathematics, and their leading role in shaping them into central fields of modern mathematics.

At the end of the 1960s and the beginning of the 1970s, graph theory, discrete mathematics, combinatorics, and theoretical computer science were considered peripheral fields of mathematics. This changed completely during Lovász's lifetime. They became central parts of modern mathematics for many reasons. The tremendous development of computer technologies is the most obvious one. Essential factors were also the high quality of the research and the results in these areas and their wide applicability. The solutions of many problems arising in industry, society, other sciences, even in other fields within mathematics critically depend on theories and algorithms invented in discrete mathematics. Many mathematicians and computer scientists contributed to this. László Lovász undoubtedly was and still is one of the key players in this development.

There are other aspects that make László Lovász special. Mathematicians are often divided into "problem solvers" and "theory builders". Graph theory is, in particular, a field to which problem solvers are drawn. Theory builders often see deep and unusual connections, but often leave the difficult exploration of details to others. As we will demonstrate, Lovász is a member of this rare breed of people who possess both talents. Moreover, he brought his talents to bear not only in one field of mathematics, he has also fertilized and inspired significant developments in a wide range of other areas. If asked to formulate the essence of his contributions in few words, we could use the following three:

Depth: Lovász solves many important and widely known problems in a competitive environment. He isolates seemingly special topics and develops them into broad and important calculi.

Elegance: His solutions are often surprisingly (and sometimes seemingly) simple. At the same time, they often are mathematically beautiful and suggest fundamentally new ways to address a problem.

Inspiration: Many of his solutions are the basis of further active research and even the foundations of whole new areas.

László Lovász has published eleven books and more than 300 articles. There is no way to survey his contributions in an article like this. We have chosen to sketch some of the publications and topics that we consider highlights, are not too difficult to explain, had significant impact, moved the frontier of knowledge in the interface of mathematics and computer science substantially, and are of lasting value.

2 Logic and Universal Algebra – Homomorphisms and Tarski's Problem

L. Lovász. Operations with structures. *Acta Mathematica Academiae Scientiarum Hungarica* 18:321–328, 1967.

L. Lovász. On the cancellation law among finite relational structures. *Periodica Mathematica Hungarica* 1:145–156, 1971.

M. Freedman, L. Lovász, L. Schrijver. Reflection positivity, rank connectivity, and homomorphisms of graphs. *Journal of American Mathematical Society* 20(1):37–51, 2007.

Up to the 1960s graph theory was mainly concerned with graphs as objects. Graph parameters were introduced and the structural properties of graphs having these properties were investigated. László Lovász made, as we will outline, very significant contributions to this kind of research, but he left his first fundamental mark, when he was 19 years old, in the more general context of universal algebra.

Intending to step out of the object orientation of graph theory, Lovász got interested in operations with graphs and their algebraic properties. We all know that, for nonzero real numbers a, b, and c, the equation $ac = bc$ implies $a = b$. Suppose we have three graphs A, B, and C, and suppose we have defined a product "$\times$" for which $A \times C = B \times C$ holds, can we infer that $A = B$? Such a question only makes sense if equality "$=$" is replaced by "isomorphic" and the concrete issue to be addressed is: Under what conditions does such a "cancellation law" hold?

Questions of this type were asked by Alfred Tarski, in the context of finite relational structures, to students in Berkeley in the 1960s. Lovász points this out in the following quote, extracted from his article [97], where he states the question and announces his solution:

Our main concern will be in the direct product of finite structures (i.e. $\langle H, R\rangle$ with finite domain H). In [1] the question was discussed under what conditions it is true that any two direct factorizations of a structure have a common refinement. It was mentioned that if the structures A, B have this "refinement-property" then e.g. $A^2 \cong B^2$ implies $A \cong B$. We shall prove a general theorem from which it follows that for finite A, B the last implication always holds. On the other hand, it is easy to see that not all finite structures have the refinement-property (or the unique prime factorization-property).

Fig. 3: Quote from [97]

Reference [1] in the quote above is the paper [25] of Chang, Jónsson, and Tarski of 1964, see also [130].

A finite *graph* G with *vertex set* $V(G)$ and *edge set* $E(G)$ is such a relational structure where $V(G)$ is the ground set and the edges uv define the (binary) relations between vertices u and v. A standard product in graph theory is the *direct* (also named categorial or tensor) *product* $G_1 \times G_2$ of two graphs G_1 and G_2. Its vertex set is $V(G_1) \times V(G_2) = \{(u,v) \mid u \in V(G_1), v \in V(G_2)\}$ and its edge set $E(G_1 \times G_2)$ is defined to be the set of all pairs of vertices $(u_1,u_2),(v_1,v_2) \in V(G_1) \times V(G_2)$ with $u_1v_1 \in E(G_1)$ and $u_2v_2 \in E(G_2)$. The question to be addressed is: Given two graphs G and H and a third graph F, can one conclude that G and H are isomorphic if the direct product $F \times G$ is isomorphic to $F \times H$? This particular question and most of the related problems for finite relational structures were unsolved, despite considerable effort. The earlier solution approaches taken were usually elementary, trying to reduce the problem to known invariants.

Lovász devoted to these problems three of his early papers written in 1967, 1971, 1972. His approach was radically different: He invented a new invariant which solved these problems for the direct product in full generality. His results completely changed this area.

The Lovász argument is easy and can be given here in full. Interestingly, young Lovász formulates his results very generally for finite *relational structures*, i.e., objects of the form $\boldsymbol{A} = (X_A, (R_A; R \in L))$ where R_A is a subset of $X^{\,a(R)}$ ($a(R)$ is the arity of the relational symbol R; L is the fixed set of symbols usually called language). Shortly, we speak about *L-structures*.

A *homomorphism* $f : \boldsymbol{A} \to \boldsymbol{B} = (X_{\boldsymbol{B}}, (R_{\boldsymbol{B}}; \boldsymbol{R} \in L))$ is a mapping $f : X_{\boldsymbol{A}} \to X_{\boldsymbol{B}}$ such that for every $R \in L$, $(x_1, \dots, x_{a(R)}) \in R_{\boldsymbol{A}} \Rightarrow (f(x_1), \dots, fx_{a(R)} \in R_{\boldsymbol{B}}$ holds. The *product* $\boldsymbol{A} \times \boldsymbol{B}$ is defined as $X_{\boldsymbol{A}\times\boldsymbol{B}} = X_{\boldsymbol{A}} \times X_{\boldsymbol{B}}$ where $\boldsymbol{R}_{\boldsymbol{A}\times\boldsymbol{B}}$ is the set of all tuples $((x_1,y_1), \dots, (x_{a(R)}, y_{a(R)}))$ where $(x_1, \dots x_{a(R)}) \in R_{\boldsymbol{A}}$ and $(y_1, \dots, y_{a(R)}) \in R_{\boldsymbol{B}}$.

Note that the projections $\pi_{\boldsymbol{A}} : X_{\boldsymbol{A}\times\boldsymbol{B}} \to X_{\boldsymbol{A}}$ and $\pi_{\boldsymbol{B}} : X_{\boldsymbol{A}\times\boldsymbol{B}} \to X_{\boldsymbol{B}}$ are homomorphisms. Up to an isomorphism, projections uniquely determine the above product. (The whole theory may be restated in categorical terms as worked out in papers by Lovász [102] and Pultr [139].)

Denote by $\hom(\boldsymbol{A},\boldsymbol{B})$ the number of homomorphisms from $\boldsymbol{A}$ to $\boldsymbol{B}$. The key of Lovász's argument is the following statement:

Theorem. *Finite L-structures* $\boldsymbol{A}$ *and* $\boldsymbol{B}$ *are isomorphic if and only if for every other finite structure* $\boldsymbol{C}$ *the following holds:* $\hom(\boldsymbol{C},\boldsymbol{A}) = \hom(\boldsymbol{C},\boldsymbol{B})$.

In other words (and in today's setting), if we take a fixed enumeration $F_1, F_2, \ldots$, $F_n, \ldots$ of all non-isomorphic finite graphs then the vector $L(\boldsymbol{A}) = (\hom(F_i,\boldsymbol{A});\ i = 1,\ldots)$ is the isomorphism invariant, expressed equivalently: $\boldsymbol{A} \cong \boldsymbol{B}$ if and only if $L(\boldsymbol{A}) = L(\boldsymbol{B})$.

Hell and Nešetřil [76] (and others) call this invariant $L(\boldsymbol{A})$ the *Lovász vector.*

This setting is very suitable for the Tarski problem. For example, one immediately obtains that for finite structures $\boldsymbol{A}^k \cong \boldsymbol{B}^{\mathbf{k}}$ holds if and only if $\boldsymbol{A} \cong \boldsymbol{B}$. This follows readily from $\hom(\boldsymbol{C},\boldsymbol{A}^k) = (\hom(\boldsymbol{C},\boldsymbol{A}))^k$.

For brevity we mention another consequence for the special case of graphs. If $\boldsymbol{C}$ is a nonbipartite graph then $\boldsymbol{A} \times \boldsymbol{C} \cong \boldsymbol{B} \times \boldsymbol{C}$ if and only if $\boldsymbol{A} \cong \boldsymbol{B}$. (Note that for bipartite graphs $\boldsymbol{C}$, cancelation need not hold as already for circuits we have $2\boldsymbol{C}_3 \times K_2 \cong \boldsymbol{C}_6 \times K_2$.)

The above theorem is very general and yet the proof is easy. In the nontrivial direction we prove by induction on the cardinality $|X_{\boldsymbol{C}}|$ of the ground set $X_{\boldsymbol{C}}$ that, if $\hom(\boldsymbol{C},\boldsymbol{A}) = \hom(\boldsymbol{C},\boldsymbol{B})$, then also the number of injective homomorphisms coincides, i.e., $\mathrm{inj}(\boldsymbol{C},\boldsymbol{A}) = \mathrm{inj}(\boldsymbol{C},\boldsymbol{B})$.

In the inductive step we have $\hom(\boldsymbol{C},\boldsymbol{A}) = \sum_\theta \mathrm{inj}(\boldsymbol{C}/\theta,\boldsymbol{A})$ where θ is an equivalence on $X_{\boldsymbol{C}}$. Thus by induction assumption we have $0 = \hom(\boldsymbol{C},\boldsymbol{A}) - \hom(\boldsymbol{C},\boldsymbol{B}) = \mathrm{inj}(\boldsymbol{A},\boldsymbol{A}) - \mathrm{inj}(\boldsymbol{A},\boldsymbol{B}) = \mathrm{inj}(\boldsymbol{B},\boldsymbol{B}) - \mathrm{inj}(\boldsymbol{B},\boldsymbol{A})$. But obviously $\mathrm{inj}(\boldsymbol{A},\boldsymbol{A}) > 0$ and $\mathrm{inj}(\boldsymbol{B},\boldsymbol{B}) > 0$, and thus, we have that there are injective homomorphisms from $\boldsymbol{A}$ to $\boldsymbol{B}$ and also from $\boldsymbol{B}$ to $\boldsymbol{A}$. Now as $\boldsymbol{A}$ and $\boldsymbol{B}$ are finite structures we have that $\boldsymbol{A} \cong \boldsymbol{B}$.

Lovász recognized in the Tarski problem a magnificent pearl. His theorem turned out to be very useful. It found many applications and inspired further research. This continues until today, see the articles by Lovász and Schrijver [118], Dvořák [38] and Dawar et al. [34], for example.

The papers [97] and [99] of Lovász belong to the first occurrences of homomorphisms in graph theory. Their successful utilization led to a rich calculus (see, e.g., the books by Hell and Nešetřil [76] and Lovász [112] and the article by Borgs et al. [21]). We outline important parts of this approach.

Lovász already defined in [97] exponential structures $\boldsymbol{A}^{\boldsymbol{B}}$ (and exponential graphs G^H). These played very recently a decisive role in the disproof of the Hedetniemi conjecture which claimed that $\chi(G \times H) = \min(\chi(G), \chi(H))$, see Shitov [147], Wrochna [159], Tardif [156], and Zhu [161].

Another application of homomorphism counting was provided by Lovász in [103], which deals with the following problem: When can one recognize a given finite structure from the collection of all its proper substructures? The special case for undirected graphs is a classical *conjecture of Ulam*, see [157], which may be formulated in our setting as follows:

Do the homomorphism numbers $\hom(F,G)$ *for all graphs F with fewer edges than G determine the graph G?*

This conjecture is known to be true for special classes of graphs (such as trees and maximal planar graphs), and the proofs usually consist of a complicated case analysis. In [103] Lovász gave the first general result: The conjecture is true for graphs that have more edges than their complement (i.e., more than half of all edges).

The proof, although not directly linked to the above theorem proceeds again by clever homomorphism counting. Shortly after, this proof was extended by Müller in [132] (again by homomorphism counting) to graphs with $n \log n$ edges. This is still the best result.

Counting of homomorphisms and the investigation of their structure are cornerstones of further areas of mathematics and theoretical computer science. We just indicate three examples, where they play important roles: Tutte polynomials and their variants, see [46]; constraint satisfaction problems (which can alternatively be viewed as existence theorems for general relational structures), see [52]; and partition functions in statistical physics, see [24, 112, 23].

Let us finally elaborate on partition functions, the last item mentioned above. The concept of graph homomorphisms can be extended to graphs with loops and weights assigned to vertices $\propto_v (G)$ and edges $\beta_{uv}(G)$. For unlabeled graphs F and labeled graphs G, one can define naturally the weight of a mapping $\varphi : V(F) \to V(G)$ and then the total weight of $\hom(F,G)$. Allowing weights on the vertices and edges greatly extends the expressive power of (weighted) homomorphisms. For example, the number $\hom(F,G)$ can express the number of colorings (leading to chromatic and Tutte polynomials), the counting of stable sets (corresponding to the so-called *hard core model* in statistical physics) and also the counting of nowhere zero flows and B-flows (i.e., flows attaining values from a given set B only). All these are parameters of the form $\hom(-,H)$. Freedman, Lovász, and Schrijver [54] provided a structural characterization for all such parameters as follows:

Theorem. *Let f be a (real) graph parameter defined on multigraphs without loops. Then f is equal to* $\hom(-,H)$ *for some weighted graph H on q vertices if and only if $f(K_0) = 1$, the f connection matrix $M(f,k)$ is reflection positive, and its rank satisfies $r(M(f,k)) \leq q^k$ for all $k \geq 0$.*

(Briefly: Above, K_k is the complete graph on k vertices; the connection matrix $M(f,k)$ is defined by values of the parameter f for amalgams of k-multilabeled multigraphs; reflection positivity means that, for all k, such matrices are positive semidefinite.)

This theorem led to many similar results for other classes of graphs and for other types of homomorphism numbers (e.g., in a dual setting with $\hom(F,-)$ instead of $\hom(-,H)$, see [119]). In terms of statistical physics, this theorem can be viewed as a characterization of partition functions of vertex coloring models.

Lovász wrote extensively on this topic and devoted – ten years ago – a monograph [112] to this subject, where the topics indicated here are treated in depth.

3 Coloring Graphs Constructively (on a Way to Expanders)

L. Lovász. On chromatic number of finite set-systems. *Acta Mathematica Academiae Scientiarum Hungaricae*, 19:59–67, 1968.

The *chromatic number* $\chi(G)$ of a graph G is the minimum number of colors which suffice to color all vertices of G such that no two adjacent vertices get the same colour. Alternatively, using the notion of the preceding section, $\chi(G)$ is smallest k for which $\hom(G, K_k) > 0$.

The chromatic number belongs to the most frequently studied combinatorial parameters. The reasons for such attention are that the question of how to color the countries on a map can be easily explained to everyone and that the mathematical modelling of this question can be employed as an appealing introduction to graph theory. The "colorful story of the 4-color conjecture" can be used to shed some light on the rich history of mathematics and the difficulty of finding proofs for problems that appear to be easy. Coloring the vertices of a graph captures the substance and the difficulty of many problems. In a multiple sense, the chromatic number is a difficult concept.

Just consider the easiest question: Are there graphs with large chromatic number? Of course, complete graphs K_n satisfy $\chi\ (K_n) = n$. But are there any other essentially different graphs?

The answer is yes and a classical result, rediscovered several times, states that, for every $k \geq 1$, there are graphs G_k for which $\chi(G_k) = k$ and G_k does not contain K_3 (i.e., the triangle) as a subgraph. This result and its many ramifications, for instance in extremal graph theory, are still in the current focus of coloring research. In fact, any new constructive proof of the existence of such graphs G_k is interesting and attracts great attention. Here is perhaps the simplest proof of this fact: Let us define, for any integer $n \geq 4$, the graph $G = (V, E)$ where V is the set of integer pairs $\{ij\}, 1 \leq i < j \leq n$, and $\{ij, kl\} \in E$ if $i < j = k < l$. Such a graph G is called a *shift graph*. G has no triangles, and it can be shown that $\chi(G) = \lceil \log n \rceil$.

But this is not the end of the story. Graphs may have high chromatic number and very low edge density. P. Erdős showed in [47] that there exist graphs which have arbitrarily large chromatic number and which are locally trees and forests.

Theorem. *For every k, l there exists a graph $G_{k,l}$ such that $\chi(G_{k,l}) \geq k$ and $G_{k,l}$ does not contain circuits of length $\leq l$. (So the shift graph above is a graph of type $G_{k,3}$.)*

Erdős' proof was a landmark. It constitutes one of the key applications of the probabilistic method in graph theory, see, e.g., [6]. The proof shows that the probability of the existence of such graphs $G_{k,l}$ is positive, but does not give any hint how to construct concrete examples of graphs of type $G_{k,l}$. The construction of such graphs has been a longstanding problem with very slow progress (for the historic development and related issues, see, e.g., the Nešetřil article [133]).

The first constructive proof of the theorem above was found by Lovász in one of his early papers [98]. It was one of the highlights of the 1969 conference in Calgary; and through his proof, Lovász again changed the setting of the problem as he constructed the graphs $G_{k,l}$ as special cases of a more general theorem about hypergraphs. His complicated construction was later simplified, the Nešetřil–Rödl construction is perhaps the simplest [135].

But various problems remained.

One of them is the question whether one can provide a construction that uses only graphs. The answer is positive. I. Kříž [90] and more recently N. Alon et al. [3] came up with such constructions, and Ramanujan graphs have to be mentioned here as well.

The existence of graphs $G_{k,l}$ with a large chromatic number and no short circuit is a phenomenon of finite (and of countable) graphs. For graphs with an uncountable number of vertices and uncountable chromatic number, an analogous result does not hold. This was shown by Erdős and Hajnal [48]:

If the chromatic number of a graph is uncountable then it contains every bipartite graph.

A consequence of this result is that such a graph contains every circuit of even length, for example the circuit C_4 of length four.

Graphs $G_{k,l}$ are what can be called difficult examples. They also play an important role in Ramsey theory, extremal combinatorics, topological dynamics, and model theory, to name just a few. In all these areas they are used as examples of complex yet locally simple structures; they are prototypes of local-global phenomena.

It took some time to understand why the construction of graphs $G_{k,l}$ matters, why it is important to know such graphs explicitly. This led to an explosion of theoretical developments combining group theory, number theory, geometry, algebraic graph theory, and, of course, combinatorics. The key notions are now familiar to every student of theoretical computer science: expanders, Ramanujan graphs and sparsification, see Margulis [127], Lubotzky, Phillips, and Sarnak [125], and Spielman and Teng [153].

An expander graph, for instance, is a finite, undirected multigraph (parallel edges are allowed) in which every subset of the vertices that is not "too large" has a "large boundary". There are various formalizations of these notions. Each of them gives rise to a different notion of expanders, e.g., edge expanders, vertex expanders, and spectral expanders. Expander graphs have found applications in the design of algorithms, error correcting codes, pseudorandom generators, sorting networks, robust computer networks and hash functions in cryptography. They also played a role in proofs of important results in computational complexity theory, such as the PCP theorem.

The construction and structure of graphs similar to $G_{k,l}$ continues to be one of the key problems of finite combinatorics and has a character of a saga (see, e.g., Hoory et al. [78] and Nešetřil [133]).

Coloring of graphs and hypergraphs has been a permanent theme of Lovász, and thus, it is mentioned in most sections of our survey. For example, one of the motivations of the next section was the study of 3-chromatic linear hypergraphs, i.e., hypergraphs in which edges meet in at most one vertex, or equivalently, hypergraphs without cycles of length 2.

4 The Lovász Local Lemma

P. Erdős, L. Lovász. Problems and results on 3-chromatic hypergraphs and some related questions. In *Infinite and Finite Sets. Coll. Math. Soc. J. Bolyai*, North Holland: 609–627, 1975.

A *hypergraph* is a collection of sets. The sets are called *edges*, the elements of the edges are *vertices*. The *degree* of a vertex is the number of edges containing it. A hypergraph is called *r-uniform* if every edge has r vertices. The *chromatic number of a hypergraph* is the least number k such that the vertices can be k-colored so that no edge is monochromatic.

Graphs with chromatic number at least 3 are simple to characterize: they must contain an odd circuit. But for hypergraphs, even the characterization of 3-chromatic 3-uniform hypergraphs is difficult (it is an $\mathcal{NP}$-complete problem). Lovász and Woodall had independently shown that every 3-chromatic r-uniform hypergraph contains a vertex of degree at least r. Erdős and Lovász [49] aimed at generalizing this result in various ways. One of the key results of their article is the following:

Theorem. *A $(k+1)$-chromatic r-uniform hypergraph contains an edge which is intersected by at least $k^{r-1}/4$ other edges. Thus, the degree of at least one vertex is larger than $k^{r-1}/(4r)$.*

To prove this theorem, the authors employed probability theory. As pointed out by Erdős, Lovász contributed to the proof a substantial new result of elementary probability. This was later called the *Lovász Local Lemma*.

The motivation for this lemma comes from a well-known observation of elementary probability:

If $X_1, \ldots, X_n$ are random events which are pairwise independent and if the probability of each event X_i is smaller than 1, then the probability that none of the events X_i occurs is positive. The Lovász Local Lemma is a quantitative refinement of this observation for variables which are dependent.

Fig. 4 shows the formulation of the Lovász Local Lemma as stated and proved in the original article [49]. Indeed, it is "just a lemma".

Crystal clear: Not only when the events are independent, but if the dependence graph G has a small degree ($\leq d$) then also none of the events occurs with positive probability. The adjective local in the name of the lemma refers to the situation that each event is dependent only on a small number d of others.

Lemma. *Let G be a (finite) graph with maximum degree d and vertices $v_1, \dots, v_n$. Let us associate an event A_i with v_i $(i = 1, \dots, n)$ and suppose that A_i is independent of the set*

$$\{A_j\colon (v_i, v_j) \in E(G)\}\,.$$

Also suppose

$$P(A_i) \leq \frac{1}{4d}\,. \tag{3}$$

Then

$$P(\bar{A}_1 \dots \bar{A}_n) > 0\,. \tag{4}$$

Proof. We prove more, namely that

$$P(A_1 \mid \bar{A}_2 \dots \bar{A}_n) \leq \frac{1}{2d}\,. \tag{5}$$

This formula makes sense because we may assume by induction

$$P(\bar{A}_2 \dots \bar{A}_n) > 0\,.$$

Then (5) obviously implies (4).

We prove (5) by induction on n. For $n = 1$ it is trivial. Let $v_2, \dots, v_q$ be the points adjacent to v_1, $(q \leq d + 1)$. Then we have

$$P(A_1 \mid \bar{A}_2 \dots \bar{A}_n) = \frac{P(A_1 \bar{A}_2 \dots \bar{A}_q \mid \bar{A}_{q+1} \dots \bar{A}_n)}{P(\bar{A}_2 \dots \bar{A}_q \mid \bar{A}_{q+1} \dots \bar{A}_n)}\,.$$

Here, by (3)

$$P(A_1 \bar{A}_2 \dots \bar{A}_q \mid \bar{A}_{q+1} \dots \bar{A}_n) \leq$$
$$\leq P(A_1 \mid \bar{A}_{q+1} \dots \bar{A}_n) = P(A_1) \leq \frac{1}{4d}\,,$$

and on the other hand

$$P(\bar{A}_2 \dots \bar{A}_q \mid \bar{A}_{q+1} \dots \bar{A}_n) =$$
$$= 1 - P(A_2 + \dots + A_q \mid \bar{A}_{q+1} \dots \bar{A}_n) \geq$$
$$\geq 1 - \sum_{i=2}^{q} P(A_i \mid \bar{A}_{q+1} \dots \bar{A}_n) \geq 1 - (q-1)\frac{1}{2d} \geq \frac{1}{2}$$

by the induction hypothesis. Thus

$$P(A_1 \mid \bar{A}_2 \dots \bar{A}_n) \geq \frac{1}{4d} \Big/ \frac{1}{2} = \frac{1}{2d}\,.$$

This proves the lemma.

Fig. 4: Extracted from [49]

It is hard to overestimate the general importance of this result that just turned up as a "supporting observation" for a proof in the chromatic theory of hypergraphs. It appears again and again in multiple applications, ramifications, and forms. It is not possible to cover here all the applications in Ramsey theory (see Spencer [152]), extremal combinatorics (see Alon and Spencer [6]), number theory, and elsewhere (see, e.g., Ambainis et al. [7], He et al. [75], and Szegedy [154]). It was also discovered, see [145], that the Lovász Local Lemma closely relates to important results of Dobrushin in statistical physics [37]. In fact, the proper setting of the Dobrushin results is in the context of graph limits, see [112], which we discuss in Section 17.

One of the motivations for [49] is the following number-theoretic problem which goes back to Ernst Straus (who was an assistant of Albert Einstein): Is there a function $f(k)$ such that, if S is any set of integers with $|S| = f(k)$, then the integers can be k-colored so that each color meets every translated copy of S (i.e., every set of the form $S + a = \{x + a \mid x \in S\}$)? Lovász and Erdős, already in their paper [49], made use of the Lovász Local Lemma to prove the following more geometric generalization of the question asked by Straus:

For every k, there exists a function $f(k)$, such that $f(k) \leq k \log k$ and for every set S of lattice points in the n-dimensional space E^n with $|S| > f(k)$ there exists a k-coloring of all lattice points such that each translated copy of S contains points of all k colors.

A side remark: There are many variants of coloring problems, and some of them are surprisingly difficult. For example, during a conference in Boulder in 1972 Paul Erdős, Vance Faber, and László Lovász asked whether the vertices of any n-uniform linear hypergraph with n edges can be colored by n colors such that the vertices of any edge get all n colors. This question has many reformulations and turned out to be more difficult than originally thought (even by the authors as Erdős originally offered \$50 for a solution and eventually increased the prize to \$500). About 50 years later the Erdős–Faber–Lovász conjecture was shown to be true for large values of n by D. Y. Kang, T. Kelly, D. Kühn, A. Methuku, and D. Osthus [84].

Nowadays, the Lovász Local Lemma is a “standard trick” which is often taught in basic courses. And it is a very effective trick, as Joel Spencer once remarked: “*Using the Local Lemma one can prove the existence of a needle in a haystack.*”

But the Lovász Local Lemma delivers only existence. The above proof does not yield a method how to find that needle. We only know that certain things exist with positive probability. Only much later a constructive proof was found by Marcus and Tardos [126]. (Remark: A constructive proof for the above Straus’ problem is in Alon et al. [4]; see also J. Beck [14].) Recently, Harvey and Vondrák [72] found another constructive approach to the Lovász Local Lemma.

Investigations of infinite versions (Borel and measurable) of the Lovász Local Lemma started also very recently by A. Bershteyn, G. Kun, O. Pikhurko, and others, see, e.g., [18].)

The Lovász Local Lemma became what one can truly call *a combinatorial principle*. This is László Lovász at its best: Maybe no other Lovász-contribution is so profoundly simple and yet useful and elegant.

5 Coloring Graphs via Topology

L. Lovász. Kneser's conjecture, chromatic number, and homotopy. *Journal of Combinatorial Theory A* 25:319–324, 1978.

Combinatorial questions are often easy to formulate; some also have an elementary solution. But in many cases, the elementary nature of combinatorial problems is just the top of an iceberg, and the hidden complexity must be discovered and tamed before a solution can be found.

A beautiful example of this is the following elementary problem posed in 1955 by Martin Kneser [89], who was working on quadratic forms. In today's language:

Let X be a set with n elements, $n \geq 2k > 0$. Denote by $\binom{X}{k}$ the set of all k-element subsets of X. Then, for every coloring of the sets in $\binom{X}{k}$ by fewer than $n-2k+2$ colors, there are two disjoint sets of the same color.

This problem can be reformulated as a graph theory question as follows. Let $KG(n,k)$ denote the graph (called the *Kneser graph*) whose vertices are all k-element subsets of the set $X = \{1,2,\ldots,n\}$, and in which two vertices are joined by an edge if the corresponding k-element subsets are disjoint. For example, $KG(n,1)$ is the complete graph K_n and $KG(5,2)$ is the famous Petersen graph (the "universal" counterexample to many conjectures in graph theory) shown in Fig. 5.

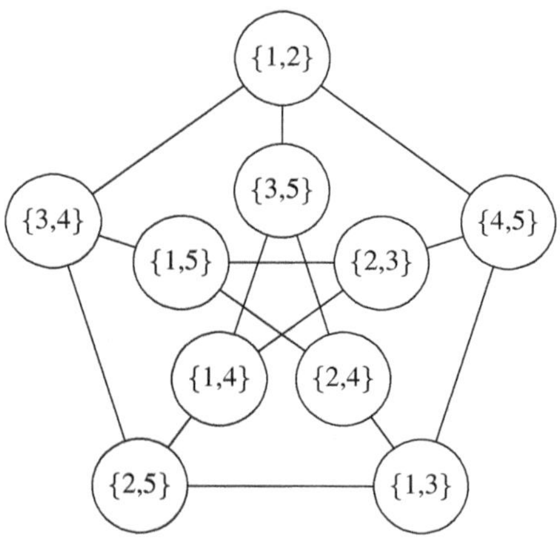

Fig. 5: $KG(5.2)$ = The Petersen graph

Kneser's question reads now: *Does the Kneser graph $KG(n,k)$ have chromatic number $n-2k+2$?*

It is easy to see that $\chi(KG(n,k)) \leq n-2k+2$. However, to find the fitting lower bound for the chromatic number proved to be much harder.

Lovász [98] solved this problem in a surprising way using methods of algebraic topology. The general idea is the following. Lovász associates with any graph G a topological space and establishes a connection between a topological invariant of this space with the chromatic number of G. He then infers properties of the chromatic number of G from properties of the topological invariant of the associated topological space. That this is possible and that topology can yield solutions of difficult graph theory questions was completely unexpected. Lovász's success with this approach was the starting point of a new field: *topological combinatorics*. We briefly sketch the main steps of Lovász's solution of Kneser's problem here.

Lovász proceeds as follows: Given a graph $G=(V,E)$, the neighborhood of a vertex v is composed of all vertices adjacent to v in G. The *neighborhood complex* $N(G)$ of G consist of all the vertices V of the graph G; the simplices of $N(G)$ are sets of vertices with a common neighbor in the graph. Homomorphisms between graphs lead to continuous mappings of neighborhood complexes. From the topological connectivity of $N(KG(n,k))$ it is possible to construct an antipodal continuous mapping between spheres ($N(K_{m+2})$ is an m-dimensional sphere) and one can then apply the Borsuk–Ulam theorem. Thus, Lovász obtained:

Theorem. *If the neighborhood complex $N(G)$ of a graph G is (topologically) k-connected then $\chi(G)\geq k+3$.*

(Topologically k-connected means that there are no holes of dimension $\leq k$. For (simply) connected complexes this is equivalent to the fact that all i-homological groups vanish for $i=0,1,\dots,k$.)

Lovász finally proves a theorem on the connectivity of neighborhood complexes of graphs from which he can infer that the neighborhood complex of a Kneser graph $N(KG(n,k))$ is topologically $(n-2k-1)$-connected. This establishes that the Kneser graph $KG(n,k)$ has chromatic number $n-2k+2$.

This connection (and the whole proof) immediately led to intensive research. Other proofs of this theorem were found (among them "book proofs" of Barany [10] and Green [63]), but all lower bounds for the chromatic number of Kneser graphs use or at least imitate Lovász's topological proof. Matoušek's book [128] surveys in detail various implications and modifications of the proof techniques. For example, it has been shown in [71] that the k-times generalized Mycielski construction has chromatic number $k+2$, and again, topological arguments are the basis of the only known proof of this fact. The paper [71] contains the following interesting construction of graphs G_k.

Put $[k]=\{1,2,\dots,k\}$. The vertices of G_k are all pairs (i,A) where $i\notin A$ and A is a nonempty subset of $[k]$. (i,A) and (j,B) form an edge in G_k if $i\in B, j\in A$ and A and B are disjoint. This "Kneser-like" graph G_k has remarkable properties: Its chromatic number is k, it is critical (i.e., every proper subgraph has a smaller chromatic number) and every strongly k-colorable graph has a homomorphism into it; it is the unique graph with this property. (A *strong coloring* of a graphs is a coloring where the neighborhood of any color class forms a stable set. Such a graph obviously has no triangles.)

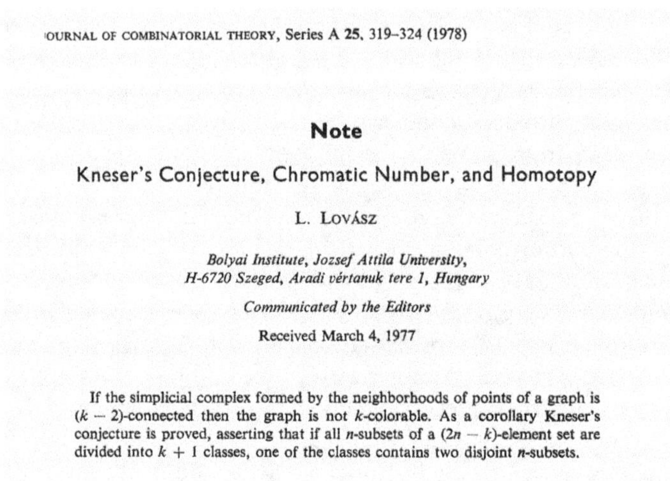
'OURNAL OF COMBINATORIAL THEORY, Series A **25**, 319–324 (1978)

Note

Kneser's Conjecture, Chromatic Number, and Homotopy

L. Lovász

Bolyai Institute, Jozsef Attila University, H-6720 Szeged, Aradi vértanuk tere 1, Hungary

Communicated by the Editors

Received March 4, 1977

If the simplicial complex formed by the neighborhoods of points of a graph is $(k-2)$-connected then the graph is not k-colorable. As a corollary Kneser's conjecture is proved, asserting that if all n-subsets of a $(2n-k)$-element set are divided into $k+1$ classes, one of the classes contains two disjoint n-subsets.

Fig. 6: The beginning of the article [105] starting topological combinatorics

The only known proof of these properties is an adaptation of Lovász's topological proof.

These examples of graphs were instrumental in the recent disproof of the Hedetniemi conjecture (that intended to establish a connection between the direct product of two graphs and their chromatic number, which we mentioned in Section 2; see also [150]) and also in the study of *gap problems* for constraint satisfaction problems. Related questions in this area are called *promised problems*. A typical question here is: How difficult is it to 5-color graphs or hypergraphs under the assumption that we know they are 3-colourable, see [35, 13], and [160].

Lovász's paper opened a whole area whose fruits are still continuing to appear. Matoušek in the preface to [128] rightly wrote that Lovász's proof of the Kneser conjecture is a masterpiece of imagination.

Yet, in typical Lovász style, it was published just as a note (see Fig. 6).

Lovász's solution of the Kneser problem did not exhaust his topological imagination nor the potential of topological methods in combinatorics. He returned to this approach frequently during his career, often in collaboration with Lex Schrijver. We mention just one of the highlights of their cooperation.

Motivated by estimating the maximum multiplicity of the second eigenvalue of Schrödinger operators, Colin de Verdière introduced a new invariant for graphs G, denoted $\mu(G)$, based on spectral properties of matrices associated with G. He proved that $\mu(G) \leq 1$ if and only if G is a disjoint union of paths, that $\mu(G) \leq 2$ if and only if G is outerplanar, and that $\mu(G) \leq 3$ if and only if G is planar.

Robertson, Seymour, and Thomas showed that a graph G is linklessly embeddable if and only if G does not have any of the seven graphs in the Petersen family as a minor. Their combinatorial result implies that $\mu(G) \leq 4$ if G is linklessly embeddable, and they conjectured that $\mu(G) \leq 4$ if and only if G is linklessly embeddable. Lovász and Schrijver, see [117], proved the only if part of this topological characterization. The key ingredient of their proof is a new Borsuk-type theorem on the existence of antipodal links, which is an extension of a polyhedral version of Borsuk's theorem due to Bajmóczy and Bárány. The combination of all these results provides a fascinating characterization of graphs G satisfying $\mu(G) \leq 4$ by means of spectral, combinatorial, and topological properties. Topological methods seem to keep on flourishing in combinatorics and graph theory.

6 Geometric Graphs and Exterior Algebra

L. Lovász. Flats in matroids and geometric graphs. In *Combinatorial Surveys*. Proc. 6 British Comb. Conf. Academic Press, pages 45–86, 1977.

Many of Lovász's proofs deal with graphs (and hypergraphs) and make use of some additional structures. The Shannon Capacity paper, see Section 8, involved a geometric structure which was added (orthogonal representation) so that the problem could be solved. To solve the Kneser problem, discussed in Section 5, Lovász employed results from topology. To recognize that methodology from other mathematical fields can be utilized needs, of course, mathematical maturity, skill, and imagination. We want to highlight that this is a different strategy than merely studying embeddings of graphs (e.g., graphs on surfaces): the special embeddings are being incorporated in proofs as tools in order to solve a (different) problem.

A very special example of this is the Lovász-article [104], which is a remarkable paper for multiple reasons.

The paper was published as an invited lecture in the proceedings of 6th British Combinatorial conference. These proceedings volumes usually contain surveys of recent developments. In contrast, the Lovász paper – full of new ideas – solved an important problem and unleashed research in two different areas: First, it started research in graphs where the vertices form a matroid; Lovász uses here the term *geometric (or pregeometric) graphs*, and this generalization is essential for solving the problem. Secondly, the paper started the application of exterior algebra in combinatorics. Particularly, Lovász defined exterior calculus in matroids and Grassman graded matroids.

Why is Lovász introducing this general machinery? Well, he is explicit about that in the introduction:

> *This paper was intended to deal with the covering problems in graphs. It has turned out, however, that their study becomes much simpler if a more general structure, which we shall call geometric graph, is considered.*

Lovász later on used the term geometric graph in a broader sense, and he recently wrote the book [113] treating the whole area in detail.

What were the "covering problems" of [104]?

The starting point was an old problem due to Tibor Gallai related to τ-critical graphs: The *covering number* of graph $G=(V,E)$, usually denoted by $\tau(G)$, is the minimum cardinality of a set $A \subseteq V$ such that every edge of G meets A. (Such a set A is also called *hitting set*.)

$\tau(G)$ is a "hard" combinatorial parameter (ultimately related to the stability number $\alpha(G)$ and the chromatic number $\chi(G)$).

One approach to gain information about the covering number is to consider graphs that are critical with respect to this parameter. A graph $G=(V,E)$ is *τ-critical* if $\tau(G) > \tau(G-e)$ for every edge $e \in E$. Gallai proved in 1961 that every τ-critical graph G satisfies $|V| \leq 2\tau(G)$. So given τ there are only finitely many τ-critical graphs and this implies a "finite basis theorem".

However, a much stronger statement holds. Let us denote the gap in the above inequality by $\delta(G) := 2\tau(G) - |V(G)|$. Then one can observe that, given a τ-critical graph G, the graph G' obtained from G by subdividing an edge of G by an even number of vertices is also τ-critical, and obviously $\delta(G) = \delta(G')$. Gallai conjectured that this is the only operation that does not destroy τ-criticality and that the number of τ-critical graphs with a given value δ is (essentially) finite. And this was the motivation of Lovász for his paper [104] in which he proved this conjecture.

Theorem. *The number of connected τ-critical graphs G with gap $\delta(G) = 2\tau(G) - |V(G)| = \delta$ and all vertex degrees ≥ 3 is at most $2^{5\delta^2}$.*

The proof of this result is complex. In fact, Lovász develops several new tools. The whole paper makes effective use of geometric graphs (where the vertices form a matroid). This allows Lovász to carry on a subtle refinement of induction procedures. He makes magnificent use of his vast experience with matchings and generalized factors (this was the subject of his doctoral thesis supervised by T. Gallai) which found its way into his early book on matching theory [115] with M. Plummer. The proof also implicitly contains the "skew Bollobás theorem" (in a matroid setting) about an extremal problem for set intersections of pairs of sets and many other inspiring ideas, in particular, the surprising utilization of exterior algebra. This aspect of the paper [104] also generated a whole new theory.

We shall illustrate the use of exterior algebra by the simpler example of the (Prague) dimension of graphs (treated in another Lovász paper [114]).

It is easy to prove that every graph is an (induced) subgraph of the direct product of complete graphs (the product we introduced in Section 2). The smallest number of such a set of complete graphs is called the *dimension* $\dim(G)$ of the graph G.

Thus, $\dim(K_n) = 1$ and $\dim(K_n \times K_n \times \cdots \times K_n) \leq t$ (direct product of t copies of K_n).

It is very nice that we have equality here. The proof in [114] is one of the first applications of exterior algebra in combinatorics, which was initiated in [104].

Theorem. $\dim(K_n^t) = t$ *for every* $t \geq 1$, $n \geq 2$.

K_2^2 is isomorphic to $K_2 + K_2$ and K_2^t is isomorphic to a perfect matching (i.e., disjoint edges) of size 2^{t-1}.

It suffices to prove $\dim(K_2^t) \geq t$. Given a representation $f : K_2^t \to K_N^d$, we put explicitly:

$f(i) = a_i = (a_i^1, \ldots, a_i^d)$ and $f(i') = b_i = (b_i^1, \ldots, b_i^d)$ (we think of matchings having edges $\{i, i'\}$ $i = 1, \ldots, 2^{t-1}$). Clearly all these 2^t vectors are distinct.

The condition that f is an embedding can be then captured by $\prod_{k=1}^d (a_i^k - b_j^k) \neq 0$ if and only if $i = j, \prod_{k=1}^d (a_i^k - a_j^k) = 0$, and $\prod_{k=1}^d (b_i^k - b_j^k) = 0$ for all i, j.

But these expressions can be written even more concisely by means of scalar products of vectors in the exterior algebra, i.e., the same technique which we mentioned above in connection with τ-critical graphs. Towards this end, for a vector $x = (x^1, \ldots, x^d)$, we define 2^d-dimensional vectors

$$x^* = \big(x^*(K) \mid K \subseteq \{1, \ldots, d\}\big),\ x^\# = \big(x^\#(K) \mid K \subseteq \{1, \ldots, d\}\big)$$

$$\text{by} \quad x^*(K) = \prod_{i \varepsilon K} x^i \quad \text{and} \quad x^\#(K) = \prod_{i \notin K} -x^i.$$

The above expressions can then be written as

$$\prod_{k=1}^d (a_i^k - b_j^k) = \sum \Big(\prod_{k \in K} a_i^k \cdot \prod_{k \notin K} -b_j^k \big| K \subseteq \{1, \ldots, d\} \Big) = \sum_K a_i^*(K) \cdot b_j^\#(K) = a_i^* \cdot b_j^\#.$$

Thus $a_i^* \cdot b_j^\# \neq 0$ iff $i = j$. Similarly we have $b_i^* \cdot a_j^\# \neq 0$ iff $i = j$ while $a_i^* \cdot a_j^\# = b_i^* \cdot b_j^\# = 0$ for all $i,\ j$.

It follows then that the set of 2^t vectors a_i^*, b_j^*, $1 \leq i, j \leq 2^t$, is linearly independent in the vector space of dimension 2^d and thus $t \leq d$.

Again, no other (say combinatorial) proof is known.

7 Perfect Graphs and Computational Complexity

L. Lovász. A characterization of perfect graphs. *J. Comb. Theory* 13:95–98, 1972.

L. Lovász. Normal hypergraphs and the perfect graph conjecture. *Discrete Math.* 2:253–267, 1972.

This section addresses a particular class of graphs that is tightly connected with four important parameters. For a graph $G = (V, E)$ with vertex set V and edge set E, a *stable set* (also called independent set) is a set of vertices such that no two vertices are adjacent. The largest size of a stable set of vertices is denoted by $\alpha(G)$ and called the *stability number*. Similarly, the largest size of a *clique* (mutually adjacent vertices) is denoted by $\omega(G)$ and called the *clique number*, the *chromatic number* $\chi(G)$ is the smallest number of stable sets (each stable set is a color class) covering all vertices of G, and the *clique covering number* $\bar{\chi}(G)$ is the smallest number of cliques covering all vertices of G.

If the vertices of a graph are colored so that no two adjacent vertices have the same color then, obviously, the smallest number $\chi(G)$ of colors of such a coloring must be at least as large as the largest number $\omega(G)$ of mutually adjacent vertices, i.e., $\omega(G) \leq \chi(G)$. And similarly, the stability number $\alpha(G)$ cannot be larger than the smallest number $\bar{\chi}(G)$ of cliques covering all vertices of a graph G, i.e., $\alpha(G) \leq \bar{\chi}(G)$.

In the beginning of the 1960s Claude Berge, see [15, 16], called a graph *G perfect* if $\omega(H) = \chi(H)$ holds for all induced subgraphs H of G. In the *complement* $\bar{G}$ of G, two vertices are connected by an edge if and only if they are not connected in G, and thus, $\alpha(G) = \omega(\bar{G})$ and $\chi(G) = \bar{\chi}(\bar{G})$. Berge conjectured:

A graph G is perfect if and only if its complement $\bar{G}$ is perfect.

This conjecture (called the *weak perfect graph conjecture*) started a massive search for classes of perfect graphs. Examples are, for instance, bipartite graphs and their line graphs, interval graphs, parity graphs, and comparability graphs; Schrijver [144] describes many of these graphs in detail in Chapter 66, Hougardy [80] gives a survey of these graphs and provides a list of 120 classes. More importantly, intensive attempts to solve the conjecture began. Fulkerson introduced pluperfect graphs in [56] and, developing in [57] the antiblocking theory for this purpose, he came very close to its solution – as he outlines in [58]. Just a lemma (later called the *replication lemma*) was missing. Lovász [100] solved the conjecture by proving the replication lemma, pointing out, though, that the more difficult step was done first by Fulkerson. In a subsequent paper, Lovász [101] provided a new characterization of perfect graphs as follows:

Theorem. *A graph $G = (V,E)$ is perfect if the following holds: $\omega(H)\alpha(H) \geq |V(H)|$ for all induced subgraphs $H = (V(H),E(H))$ of G.*

This theorem immediately implies the weak perfect graph conjecture since the condition given in it is invariant under taking graph complementation. The perfect graph theorem is also a generalization of the well-known theorems of König on bipartite matching and Dilworth on partially ordered sets. It generated particular interest in the characterization of conditions under which the Duality Theorem of linear programming holds in integer variables and initiated related investigations in polyhedral combinatorics.

Due to its importance and elegance, the Lovász's article [100] was reprinted in the collection *Classic Papers in Combinatorics* [60], edited by I. Gessel and G. C. Rota.

The beginning of the 1970s was a particularly productive time period for László Lovász. He was solving one open problem after the other. These years firmly established his international position as the world foremost researcher in graph theory and combinatorics.

As in many other cases, Lovász was not just looking for a proof of the weak perfect graph conjecture, he looked for a more general mathematical setting for which it is possible to prove farther reaching results that imply the conjecture. In [101] Lovász considered a hypergraph approach. We sketch the construction.

Recall that a hypergraph H is a non-empty finite collection of finite sets called edges; the elements of the edges are the vertices of H. The *chromatic index* of a hypergraph H is the least number of colors with which the edges can be colored so that edges with the same color are disjoint. The number of edges containing a given vertex is called the *degree* of the vertex. The largest degree of a vertex of H is called the *degree of* H.

Clearly, the degree of H is a lower bound on the chromatic index of H. Lovász called a hypergraph H *normal* if the degree and the chromatic index are the same for every partial hypergraph of H. Let us call a set T of vertices a *transversal* (or hitting set) if T meets every edge of H and denote its minimum cardinality by $\tau(H)$. (We just point out that $\tau(H)$ is the hypergraph generalization of $\tau(G)$ for graphs discussed in Section 6.) If we denote by $\nu(H)$ the maximum number of edges of H that are pairwise disjoint, then we obviously have $\nu(H) \leq \tau(H)$. Lovász called a hypergraph H *τ-normal* if this inequality holds with equality for all partial hypergraphs of H. He also introduced procedures to associate with every hypergraph H its edge graph $G(H)$ and with every graph G a hypergraph $H(G)$ and proved the following:

Theorem. *A hypergraph H is normal if and only if its edge graph $G(H)$ is perfect; G is perfect if and only if $H(G)$ is normal; H is τ-normal if and only if $\bar{G}(\bar{H})$ is perfect; $\bar{G}$ is perfect if and only if $H(G)$ is τ-normal.*

Corollary. *A hypergraph is normal if and only if it is τ-normal.*

This hypergraph generalization immediately implies the weak perfect graph conjecture.

A side remark: In Section 4 we mentioned the Erdős–Faber–Lovász conjecture. This appears in this context in the following two equivalent forms: (1) The chromatic index of hypergraphs consisting of n edges such that each edge contains n vertices and any two edges have exactly one vertex in common is n. (2) For graphs G consisting of n cliques of size n so that two of these cliques have one vertex in common, $\omega(G)$ equals $\chi(G)$. As indicated before the conjecture is true for large n, see [84].

Berge [16] also conjectured – later called the *strong perfect graph conjecture* – that a graph is perfect if and only if it does neither contain an odd cycle nor the complement of an odd cycle as an induced subgraph. After a long sequence of contributions of many researchers, this conjecture was finally solved in 2006 by Chudnovsky, Robertson, Seymour, and Thomas [26].

During the early 1970s computational complexity theory took off, see Wigderson's book [158] for an up-to-date survey. The classes of decision problems that can be solved in polynomial time, denoted by $\mathcal{P}$, and those solvable in nondeterministic polynomial time, denoted by $\mathcal{NP}$, were introduced. S. Cook [29] and L. A. Levin [96] independently showed the existence of *$\mathcal{NP}$-complete* problems, which are decision problems in $\mathcal{NP}$ with the property that, if they can be solved with a polynomial time algorithm, then $\mathcal{P} = \mathcal{NP}$. Whether $\mathcal{P}$ is equal to $\mathcal{NP}$ is one of the great open problems in mathematics and computer science.

Optimization problems can be phrased as decision problems by asking whether, for a given value t, there exists a feasible solution with value at least (or at most) t. If the decision problem associated this way to an optimization problem is $\mathcal{NP}$-complete, the optimization problem is called $\mathcal{NP}$-*hard*. For example, if a graph $G = (V,E)$ with rational weights w_v, for every vertex $v \in V$, is given and one wants to find a stable set S in V such that the sum of the weights of the vertices in S is as large as possible, we have a typical combinatorial optimization problem. The associated decision problem asks if there is a stable set whose value is at least t. If this decision problem can be solved in polynomial time, the stable set problem can also be solved in polynomial time by binary search. And vice versa, a polynomial time algorithm for the (weighted) stable set problem would prove that $\mathcal{P} = \mathcal{NP}$.

Karp [86] showed that many graph-theoretical problems, such as computing the value of the four parameters $\alpha(G)$, $\omega(G)$, $\chi(G)$, and $\bar{\chi}(G)$, introduced above, are $\mathcal{NP}$-hard for general graphs G. The immediate question came up: Is this also true for perfect graphs, or can their special structure be exploited to design polynomial time algorithms? This challenge triggered significant developments that we outline later.

Another side remark: Lovász was one of many contributors to one of the most astonishing results in complexity theory, the PCP Theorem. This theorem is the highlight of a long sequence of research on interactive proofs and probabilistically checkable proofs. It states that every decision problem in $\mathcal{NP}$ has probabilistically checkable proofs of constant query complexity using only a logarithmic number of random bits. Nine persons (including Lovász) received the Gödel Prize 2002 "*for the PCP theorem and its applications to hardness of approximation*". A consequence of the PCP Theorem is, for instance, that many well-known optimization problems, including the stable set problem mentioned above and the shortest vector problem for lattices to be introduced subsequently, cannot be approximated efficiently unless $\mathcal{P} = \mathcal{NP}$.

8 The Shannon Capacity of a Graph and Orthogonal Representations

L. Lovász. On the Shannon capacity of graphs. *IEEE Trans. Inform. Theory* 25:1–7, 1979.

L. Lovász. Graphs and geometry. *Amer. Math. Soc.* 2019.

Suppose the vertices of a graph G represent letters of an alphabet and the edges uv of G indicate that the two letters of the alphabet represented by u and v can be confused, e.g., when transmitted over a noisy communication channel. It is obvious that the largest number of one-letter messages that can be sent without danger of confusion is the largest number of vertices mutually not adjacent, i.e., the stability number $\alpha(G)$. Two k-letter words are confusable if their i-th letters, $1 \leq i \leq k$, are confusable or equal.

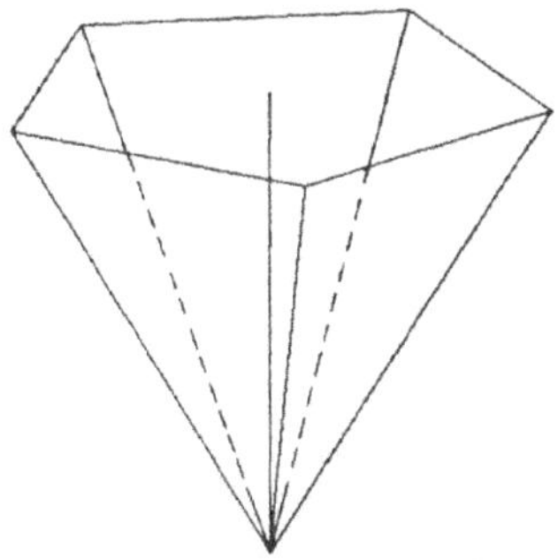

Fig. 7: Orthonormal representation of the 5-cycle in $\mathbb{R}^3$

Let G^k denote the k-th Cartesian product of G. Words with k-letters can be transmitted without danger of confusion if they are unequal and inconfusable in at least one letter. This implies that $\alpha(G^k)$ is the maximum number of inconfusable k-letter words. Forming k-letter words from a stable set of size $\alpha(G)$, one can easily construct $\alpha(G)^k$ inconfusable words. This proves that $\alpha(G)^k \leq \alpha(G^k)$.

Shannon [146] introduced the number

$$\Theta(G) = \sup_k \sqrt[k]{\alpha(G^k)} = \lim_{k\to\infty} \sqrt[k]{\alpha(G^k)},$$

where the second equation follows from $\alpha(G^{k+l}) \geq \alpha(G^k)\alpha(G^l)$. $\Theta(G)$, today called the *Shannon capacity* of G, is a measure of the information that can be transmitted across a noisy communication channel. Shannon proved that $\Theta(G) = \alpha(G)$ for graphs which can be covered by $\alpha(G)$ cliques. Perfect graphs have this property and thus belong to this class. How can one determine $\Theta(G)$ in other cases? Lovász, see [106], invented an ingenious upper bound on the Shannon capacity as follows:

Let $G = (V,E)$ be a graph. An *orthonormal representation* of G is a sequence $(u_i \mid i \in V)$ of $|V|$ vectors $u_i \in \mathbb{R}^N$, where N is some positive integer, such that $\|u_i\| = 1$ for all $i \in V$ and $u_i^T u_j = 0$ for all pairs i, j of nonadjacent vertices. Trivially, every graph has an orthonormal representation (just take all the vectors u_i mutually orthogonal in $\mathbb{R}^V$). Figure 7 shows a less trivial orthonormal representation of the pentagon C_5 in $\mathbb{R}^3$. It is constructed as follows. Consider an umbrella with five ribs of unit length (representing the nodes of C_5) and open it in such a way that nonadjacent ribs are orthogonal. Clearly, this can be achieved in $\mathbb{R}^3$ and gives an orthonormal representation of the pentagon. The central handle (of unit length) is also shown.

Where $(u_i \mid i \in V), u_i \in \mathbb{R}^N$, ranges over all orthonormal representations of G and $c \in \mathbb{R}^N$ over all vectors of unit length, let

$$\vartheta(G,w) := \min_{\{c,(u_i)\}} \max_{i\in V} \frac{w_i}{(c^T u_i)^2}.$$

The quotient has to be interpreted as follows. If $w_i = 0$ then we take $w_i/(c^T u_i)^2 = 0$ even if $c^T u_i = 0$. If $w_i > 0$ but $c^T u_i = 0$ then we take $w_i/(c^T u_i)^2 = +\infty$.

Lovász proved that, if the vertex weights w_i above are all equal to 1 and G is the pentagon graph C_5, i.e., the 5-cycle, then the value of $\vartheta(G,w)$ is $\sqrt{5}$ and equal to the Shannon capacity $\Theta(C_5)$ of C_5.

This looks like a tiny achievement, but at present, this is the only known Shannon capacity of a non-perfect graph. In fact, the complexity of determining the Shannon capacity of a general graph is today still open. Much more important, Lovász provided several different characterizations of the function ϑ (called the *Lovász ϑ-function*) that became, as we show later, important ingredients for proving that the four graph parameters $\alpha(G)$, $\omega(G)$, $\chi(G)$, and $\bar{\chi}(G)$ can be computed in polynomial time for perfect graphs G.

In his recent book [113], Lovász investigated the representation of graphs as geometric objects in great depth. His main message is that such representations are not merely a way to visualize graphs, but important mathematical tools. The range of applications is wide. We mention three examples: rigidity of frameworks and mobility of mechanisms in engineering, learning theory in computer science, the Ising and Fortuin–Kasteleyn model, and conformal invariance in statistical physics. Orthogonal representations of graphs are treated in Chapters 10 to 12. Lovász shows that orthogonal representations are, in addition to the stability and chromatic number, related to several fundamental properties of graphs such as connectivity and tree-width. Among many other aspects, he also discusses a quantum version of the Shannon capacity problem, as well as two further interesting applications of orthogonal representations to the theory of hidden variables and in the construction of strangely entangled states. These are exciting topics in quantum physics that we cannot cover here.

9 The Ellipsoid Method

P. Gács, L. Lovász. Khachiyan's algorithm for linear programming. *Math. Prog. Study* 14:61–68, 1981.

One of the major open complexity problems in the 1970s was the question whether linear programs (LPs) can be solved in polynomial time. The simplex algorithm did (and still does) work well in practice, but for all known variants of this algorithm, there exist sequences of LP-instances for which the running time is exponential. In 1979 Khachiyan indicated in [87] how the ellipsoid method, an algorithm devised for nonlinear nondifferentiable optimization based on work of Shor and Yudin and Nemirovskiĭ, can be modified to check the feasibility of a system of linear inequalities in polynomial time. Employing binary search or a sliding objective function technique, this implies that linear programs are solvable in polynomial time. Linear programs arise almost everywhere in industry, and their fast solution is of economic importance. Thus, Khachiyan's achievement received significant attention in the nonscientific media; it even made it to the front page of the New York Times on November 7, 1979. Most of these statements, though, were exaggerations or misinterpretations.

We sketch the method. Let P be a polyhedron defined by a system of linear inequalities $Ax \leq b$. We assume that P is full-dimensional or empty; and to simplify the exposition, we also assume that P is bounded, i.e., a polytope. The ellipsoid method utilizes the following facts. Given $Ax \leq b$ with rational coefficients, then numbers r and R can be computed in time polynomial in the encoding length of A and b with the following properties. If P is nonempty, the ball B of radius R around the origin contains P, and P contains a ball S of radius r.

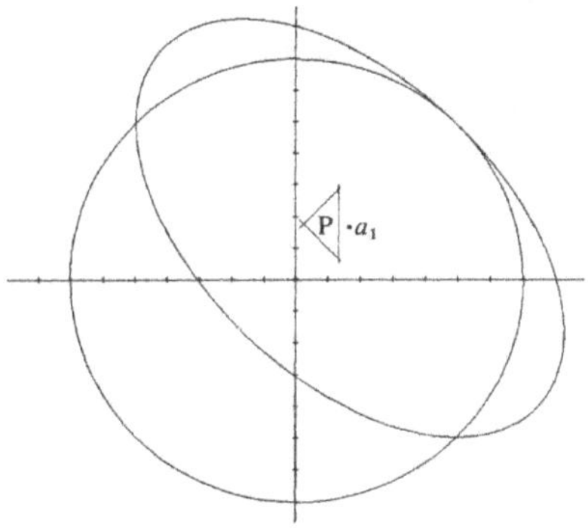

Fig. 8: The first step of the ellipsoid method

The basic ellipsoid method begins with the ball B and center $a_0 = 0$ as initial ellipsoid E_0. In a general step it checks whether the center a_k of the current ellipsoid E_k, $0 \leq k$, is contained in P. If this is the case, a point in P is found and $Ax \leq b$ is feasible. If not, there must be an inequality in the system $Ax \leq b$ that is violated by a_k. Using this inequality, a new ellipsoid E_{k+1} is computed that contains P and has a volume that is – by a constant shrinking rate – smaller than the volume of the previous ellipsoid E_k (cf. Fig. 8). This way a sequence of points a_k and shrinking ellipsoids E_k is created. Using variants of the formulas for determining the Löwner–John-ellipsoid of a convex body, one can prove that the volume shrinking rate satisfies $\mathrm{vol}(E_{k+1})/\mathrm{vol}(E_k) < \mathrm{e}^{-1/(2n)} < 1$ and that the ellipsoid method either discovers a point in P or, after a number N of steps that is polynomial in the encoding length of A and b, the ellipsoid E_N has a volume that is smaller than that of the small ball S. This can only happen if P is empty. All computations carried out can be made with rational numbers of polynomial size in such a way that nonemptiness of P is certified by a finding a feasible solution or the emptiness of P is guaranteed by the mentioned volume argument, see [69] for details.

This method was a total surprise for the linear programming community. A polynomial time termination proof employing shrinking volumes, the combination of geometric and number-theoretic “tricks” (e.g., making a low-dimensional polyhedron full-dimensional, reduction to the bounded case, careful rounding of the real numbers that appear in the update-formulas, and various necessary estimation processes) puzzled the LP-specialists. The brief article by Khachiyan (four pages), written in Russian, needed interpretation. One of the first papers explaining the approach and adding missing details was a preprint by Gács and Lovász [59]. It appeared in the fall of 1979 (and was published in 1981). This paper made Khachiyan's important contribution accessible to a wide audience and had a significant bearing on the boom of follow-up research on the ellipsoid method.

The ellipsoid method, though provably a polynomial time algorithm, performs poorly in practice. Its appearance, however, sparked successful research efforts that led to new LP-algorithms, based on various ideas from nonlinear programming, often also influenced by differential and other types of geometry, that are theoretically and practically fast. They run under the names interior point or barrier methods. New implementations of the simplex algorithm improved its performance significantly as well. The ellipsoid method, on the other hand, turned out to have fundamental theoretical power as an elegant and versatile tool to prove the polynomial time solvability of many geometric and combinatorial optimization problems. The next chapter has details.

10 Oracle-Polynomial Time Algorithms and Convex Bodies

M. Grötschel, L. Lovász, A. Schrijver. The ellipsoid method and its consequences in combinatorial optimization, *Combinatorica* 1:169–197, 1981.

M. Grötschel, L. Lovász, A. Schrijver. *Geometric Algorithms and Combinatorial Optimization*, Springer, 1988.

In a general step of the ellipsoid method, one has to verify that the center of the current ellipsoid is in the polyhedron $P = \{x \in \mathbb{R}_n \mid Ax \leq b\}$. This is usually done by substituting the center into the given inequality system $Ax \leq b$. A reasonable idea is to replace this substitution by an algorithm that checks feasibility and provides a violated inequality in case the center is not in P. Two cases, relevant in real-world applications, where this generalization might be helpful come immediately into mind.

The first is the traditional transformation of combinatorial optimization problems into linear programs. The idea is, for a given combinatorial optimization problem, to define the convex hull of all incidence vectors of feasible solutions and to try to find a linear system describing this polytope, at least partially. The number of facets of such polytopes is often exponentially large in the encoding length of the combinatorial problem. This holds for $\mathcal{NP}$-hard problems and even for some problems solvable in polynomial time. One such instance is the matching problem. This is implied by the result of Rothvoss [142] that the matching problem has "exponential extension complexity". Substituting the ellipsoid center into a linear system of exponential size makes the running time of ellipsoid algorithm exponential. Can one replace the substitution by a polynomial time algorithm?

The second is convex sets, which is even more demanding. Convex sets are intersections of potentially infinitely many halfspaces. Can one optimize over exponentially many linear inequalities in polynomial time?

The roots of this research program were laid by Grötschel, Lovász, and Schrijver in [66] and were fully worked out in [69]. The results were the starting point of what Gritzmann and Klee [64] called an algorithmic theory of convex bodies, or briefly, *computational convexity*. We outline important steps of this approach.

Suppose now that we have some convex set $K \subseteq \mathbb{R}_n$ and we want to obtain information about properties of K. Let us formulate three questions that are typical in this context:

The Strong Optimization Problem (SOPT). *Given a vector* $c \in \mathbb{R}^n$*, find a vector* $y \in K$ *that maximizes* $c^T x$ *on* K*, or assert that* K *is empty.*

The Strong Separation Problem (SSEP). *Given a vector* $y \in \mathbb{R}^n$*, decide whether* $y \in K$*, and if not, find a hyperplane that separates* y *from* K*; more exactly, find a vector* $c \in \mathbb{R}^n$ *such that* $c^T y > \max\{c^T x \mid x \in K\}$.

Fig. 9: A. Schrijver, L. Lovász, M. Grötschel at the International Symposium on Mathematical Programming in Amsterdam, 1991 (Photo: Nationaal Foto-Persbureau B. V.)

The Strong Membership Problem (SMEM). *Given a vector $y \in \mathbb{R}^n$, decide whether $y \in K$.*

It is clear that the strong membership problem can be solved if either the strong optimization or the strong separation problem can be solved. What about the other way around? And what do we have to assume about K, what is the input length of K, and how do we estimate running times? Before addressing these issues, we observe that, if we allow arbitrary convex sets K, the unique solution of an optimization problem over K may have irrational coordinates. To deal with such issues we have to allow margins and to accept approximate solutions. Let us define, for the Euclidean norm and a rational number $\varepsilon > 0$,

$$S(K,\varepsilon) := \{x \in \mathbb{R}^n \mid \|x-y\| \le \varepsilon \text{ for some } y \in K\},\ S(K,-\varepsilon) := \{x \in K \mid S(x,\varepsilon) \subseteq K\}.$$

Points in $S(K,\varepsilon)$ can be viewed as "almost in K", while points in $S(K,-\varepsilon)$ as "deep in K". The exactness requirements of the strong problems above can be softened as follows:

The Weak Optimization Problem (WOPT). *Given a vector $c \in \mathbb{Q}^n$ and a rational number $\varepsilon > 0$, either*

(i) *find a vector $y \in \mathbb{Q}^n$ such that $y \in S(K,\varepsilon)$ and $c^T x \le c^T y + \varepsilon$ for all $x \in S(K,-\varepsilon)$ (i.e., y is almost in K and almost maximizes $c^T x$ over the points deep in K), or*
(ii) *assert that $S(K,-\varepsilon)$ is empty.*

The Weak Separation Problem (WSEP). *Given a vector $y \in \mathbb{Q}^n$ and a rational number $\delta > 0$, either*

(i) *assert that $y \in S(K,\delta)$, or*
(ii) *find a vector $c \in \mathbb{Q}^n$ with $\|c\|_\infty = 1$ such that $c^T x \le c^T y + \delta$ for every $x \in S(K,-\delta)$ (i.e., find an almost separating hyperplane).*

The Weak Membership Problem (WMEM). *Given a vector $y \in \mathbb{Q}^n$ and a rational number $\delta > 0$, either*

(ii) *assert that $y \in S(K,\delta)$, or*
(ii) *assert that $y \notin S(K,-\delta)$.*

We are interested in the algorithmic relations between these problems. To do this we make use of the *oracle algorithm concept*. An *oracle* is a device that solves a certain problem for us. Its typical use is as follows. We feed some input string to the oracle, and the oracle returns another string specifying the solution (which we hope will help to solve our original problem). We make no assumption on the way the oracle finds its solution. An oracle algorithm is an algorithm in the usual sense whose power is enlarged by allowing an oracle to be queried and using the oracle's answer to determine its next computational steps.

If a query to and an answer of the oracle are counted as one step each, we can determine the running time of an oracle algorithm in the usual way. The output of the oracle may, however, be huge so that reading it may take exponential time. Since our aim is to design polynomial time algorithms, we require that for every oracle we have a polynomial q, such that for every query of encoding length at most l, the answer of the oracle has length at most $q(l)$. Under this assumption we say that an oracle algorithm has *oracle-polynomial running time* if its usual running time plus the running time of the interaction with the oracle is bounded by a polynomial in the input length of the original problem. A consequence of this set-up is that, if an oracle can be realized by a polynomial time algorithm on a real computational device, an oracle-polynomial algorithm is in fact a polynomial time algorithm in the usual sense.

For ease of exposition, we restrict ourselves to considering convex bodies K only. A convex set $K \subseteq \mathbb{R}^n$ that is compact and has dimension n is called a *convex body*. To perform computations, we have to assume that the convex body K is given by a mathematical description. Let us briefly call it $\text{Name}(K)$. Then the encoding length of K is defined as the dimension n plus the encoding length of $\text{Name}(K)$. To determine the algorithmic relations between the problems above, we assume that a convex body is given by an oracle for the solution of one of the problems and we investigate whether any of the other problems can be solved employing the oracle. The running times are measured as usual in the size of the input. This is, in the cases described here, the encoding length of K (as defined above) to which we have to add, if they appear in the problem statement, the following: the encoding lengths of the parameters ε and δ, the encoding lengths of the objective function c and the vector y, and moreover the encoding lengths of the additional data (the radii r and R, and the center a_0 of a ball) appearing in the statements of the theorems. The following was proved in [69]:

Theorem. (a) *There exists an-oracle polynomial time algorithm that solves the weak membership problem for every convex body K in $\mathbb{R}^n$ given by a weak optimization or a weak separation oracle.*

(b) *There exists an oracle-polynomial time algorithm that solves the weak separation problem for every convex body K in $\mathbb{R}^n$ given by a weak optimization oracle.*

(c) *There exists an oracle-polynomial time algorithm that solves the weak optimization problem for every convex body K in $\mathbb{R}^n$ given by a weak separation algorithm, provided a radius $R > 0$ of a ball around the origin containing K is given as well.*

(d) *There exists an oracle-polynomial time algorithm that solves the weak optimization problem for every convex body K in $\mathbb{R}^n$ given by a weak membership algorithm, provided the following data are given as well: a vector a_0 and a radius $r > 0$ such that $S(a_0, r) \subseteq K$, and a radius $R > 0$ with $K \subseteq S(0, R)$.*

This theorem establishes the oracle-polynomial time equivalence of WOPT, WSEP, and WMEM under mild additional assumptions. Moreover, the oracle-polynomial time equivalence of the strong versions SOPT, SSEP, and SMEM of these problems can be derived from the results above (assuming, of course, that K is given such that exact answers are possible). One can prove on the other hand that, if we drop one of the additional requirements in the theorem such as the knowledge of radii r or R or the vector a_0, it is impossible to derive oracle-polynomial time algorithms.

A consequence of the last result, see [66] and [69], is the polynomial time solvability of convex function minimization – in the following weak sense:

Theorem. *There exists an oracle-polynomial time algorithm that solves the following problem:*
Input: *A convex body K given by a weak membership oracle, a rational number $\varepsilon > 0$, radii r, $R > 0$, a vector a_0 such that $S(a_0, r) \subseteq K \subseteq S(0, R)$, and a convex function $f : \mathbb{R}^n \to \mathbb{R}$ given by an oracle that, for every $x \in \mathbb{Q}^n$ and $\delta > 0$, returns a rational number t such that $|f(x) - t| < \delta$.*
Output: *A vector $y \in S(K, \varepsilon)$ such that $f(y) < f(x) + \varepsilon$ for all $x \in S(K, -\varepsilon)$.*

This is the first polynomial time solvability result for convex minimization.

11 Polyhedra, Low Dimensionality, and the LLL Algorithm

M. Grötschel, L. Lovász, A. Schrijver. The ellipsoid method and its consequences in combinatorial optimization. *Combinatorica*, 1:169–197, 1981.

A. K. Lenstra, H. W. Lenstra, L. Lovász. Factoring Polynomials with rational coefficients. *Mathematische Annalen* 261(4):515–534, 1982.

The Abel prize citation states (correctly, of course): "*The LLL algorithm is only one among many of Lovász's visionary contributions*". It may be surprising to learn that its invention was triggered by a technical problem arising in the analysis of the ellipsoid method. We explain its origin and usefulness in this context.

Since square roots appear in the update formulas defining the ellipsoid method, computing with irrational numbers is unavoidable. Careful rounding is necessary to reach the desired approximation of an optimal value or solution. In various applications exact solutions can in fact be obtained by appropriate rounding. In integer programming, e.g., the solution vectors are required to have integral entries, and if the objective function is integral, the optimal value v^* is integral as well. If one can tune the ellipsoid method so that it guarantees to find an approximation v of the optimal value v^* such that $|v - v^*| < 1/2$, then one can simply round v to the next integer to find the true optimum value. Such considerations are the key to pass from "weak solutions" to "strong solutions", i.e., derive exact from approximate results. This straightforward rounding unfortunately is often not sufficient.

We sketch the case of optimizing a linear objective function over a polytope $P \subseteq \mathbb{R}^n$. We say that P has *facet-complexity at most* φ if there exists a system of inequalities with rational coefficients that has solution set P and such that the encoding length of each inequality of the system is at most φ. No assumption about the number of inequalities is made. Let us define the encoding length of P to be $n + \varphi$, call such a polyhedron *well-described,* and denote it by $(P;\ n, \varphi)$. One can prove that the encoding length of each vertex of $(P;\ n,\varphi)$ is at most $4n^2\varphi$ and that, if P is full-dimensional, P contains a ball B_P with radius $2^{-7n^3\varphi}$.

To illustrate the annoying "technical problem" that triggered the invention of the LLL algorithm, let us consider a well-described polytope $P \subseteq \mathbb{R}^n$ that is not full-dimensional; for ease of exposition, say P has dimension $n-1$. The ellipsoid method would not work in this case. To get around this problem, one needs to carefully blow P up to a polytope P' that contains P and is full-dimensional such that running the ellipsoid method on P' approximately delivers the desired result for P. This can be done but is technically tedious and requires ugly pre- and post-processing.

Let us instead make a bold step and run the ellipsoid method on P directly. We suppose P is given by a separation oracle. Since P is low-dimensional it is highly unlikely that the ellipsoid method finds a feasible solution in one of its iterations. After a number N of iterations that is polynomial in $n+\varphi$, the N-th ellipsoid E_N contains P and has a volume that is smaller than the volume of B_P, the ball P would contain if P were full-dimensional. This is contradictory. The basic ellipsoid method, assuming a full-dimensional polytope P is given, would conclude now that P is empty. But E_N contains information that one may be able to employ.

Let $H = \{x \in \mathbb{R}^n \mid a^T x = \alpha\}$ be the unique hyperplane containing P. Then $a^T x = \alpha$ is the (up to scaling) unique equation defining H. The last ellipsoid E_N, having such a small volume, must obviously be very "flat" in the direction perpendicular to H. In other words, the symmetry hyperplane F belonging to the shortest axis of E_N must be very close to H. Is it possible to find $a^T x = \alpha$ by rounding the coefficients of the linear equation defining this symmetry hyperplane F? A positive answer would be an elegant way to avoid the blow-up mentioned and the numerical problems associated with it.

The authors of [66] and [69] were at this point in the fall of 1981 and realized that such a rounding can be done – in principle – using the following classical theorem of Dirichlet [36] on the existence of a solution of a simultaneous Diophantine approximation problem.

Theorem. *Given any real numbers $\alpha_1,\ldots,\alpha_n$ and $0 < \varepsilon < 1$, there exist integers $p_1,\ldots,p_n$, and q such that $1 < q < \varepsilon^{-n}$ and $|\alpha_i - p_i/q| < \varepsilon/q$ for $i = 1,\ldots,n$.*

No polynomial algorithm is known to compute such integers. And at the end of their writing session, no progress was achieved. About three months later a letter from L. Lovász arrived (Fig. 10).

Lovász approached the approximation problem via the consideration of (integral) lattices. If $\{b_1,\ldots,b_n\}$ is a basis of $\mathbb{R}^n$, then the set $L = L(b_1,\ldots,b_n)$ that is generated by taking all integral linear combinations of the vectors b_i is called a *lattice* with basis $\{b_1,\ldots,b_n\}$. Integral lattices have been studied in number theory for a very long time (with contributors such as Gauss, Minkowski, Landau, and many others). Clearly, a lattice may have different bases, and it may be interesting to find a "minimal basis" $\{a_1,\ldots,a_n\}$ of L, i.e., a basis such that the product of the norms of the a_i is as small as possible. However, this problem is $\mathcal{NP}$-hard. Lovász introduced the quite technical notion of a *reduced basis*, which we do not explain here, that is a weak form of a minimal basis and proved:

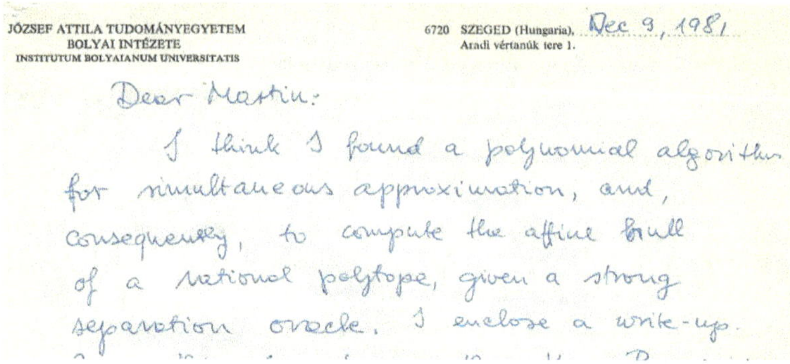

JÓZSEF ATTILA TUDOMÁNYEGYETEM
BOLYAI INTÉZETE
INSTITUTUM BOLYAIANUM UNIVERSITATIS

6720 SZEGED (Hungaria), Dec 9, 1981
Aradi vértanúk tere 1.

Dear Martin:

I think I found a polynomial algorithm for simultaneous approximation, and, consequently, to compute the affine hull of a rational polytope, given a strong separation oracle. I enclose a write-up.

Fig. 10: Beginning of a letter from L. Lovász

Theorem. *There is a polynomial time algorithm that, for any given linearly independent vectors* $\{b_1,\dots,b_n\}$ *in* $\mathbb{Q}^n$*, finds a reduced basis of the lattice* $L(b_1,\dots,b_n)$.

To achieve this, the algorithm, called the LLL algorithm, starts with the Gram–Schmidt orthogonalization and then performs carefully designed exchange operations. Proving polynomiality requires not only controlling the number of steps, but in particular, the estimation of the encoding lengths of all numbers appearing in the course of the algorithm. A consequence of this algorithm is the following weak form of Dirichlet's theorem.

Theorem. *There exists a polynomial time algorithm that, given rational numbers* $\alpha_1,\dots,\alpha_n$ *and* $0<\varepsilon<1$*, computes integers* $p_1,\dots,p_n$*, and* q *such that and* $1\le q\le 2^{n(n+1)/4}\varepsilon^{-n}$ *and* $|\alpha_i q - p_i| < \varepsilon$ *for* $i=1,\dots,n$.

This algorithm, based on computing a reduced basis, made it possible to compute via simultaneous Diophantine approximation the coefficients of the equation $a^T x = \alpha$ defining the hyperplane H containing the well-described polytope $(P;\ n,\varphi)$ as indicated above. By iterating this process, the affine hull of any lower-dimensional polytope can be determined in oracle-polynomial time.

For well-described polyhedra $(P;\ n,\varphi)$, the restriction to the bounded case can also be dropped, and one can show the following:

Theorem. *Any of the following three problems:*

- *strong separation*
- *strong violation*
- *strong optimization*

can be solved in oracle-polynomial time for any well-described polyhedron $(P;\ n,\varphi)$ *given by an oracle for any of the other two problems.*

For a linear program given by a system of rational linear inequalities, the strong separation problem can be trivially solved by substituting a given rational vector y into the inequalities, i.e., linear programs can be solved in polynomial time.

Employing the LLL algorithm and results of András Frank and Éva Tardos [53] one can, in fact, derive a general result about optimization problems for polyhedra and their dual problems in strongly polynomial time. *Strongly polynomial* means that the number of elementary arithmetic operations to solve an optimization problem over a well-described polyhedron and to solve its dual problem does not depend on the encoding length of the objective function. More precisely, the following can be shown:

Theorem. *There exist algorithms that, for any well-described polyhedron* $(P;\, n, \varphi)$ *specified by a strong separation oracle, and for any given vector* $c \in \mathbb{Q}^n$,

(a) *solve the strong optimization problem* $\max\{c^T x \mid x \in P\}$, *and*
(b) *find an optimum vertex solution of* $\max\{c^T x \mid x \in P\}$ *if one exists, and*
(c) *find a basic optimum standard dual solution if one exists.*

The number of calls on the separation oracle, and the number of elementary arithmetic operations executed by the algorithms are bounded by a polynomial in φ. *All arithmetic operations are performed on numbers whose encoding length is bounded by a polynomial in* φ *and the encoding length of the objective function vector* c.

An important application of this theorem is that one can turn many polynomial time combinatorial optimization algorithms into strongly polynomial algorithms.

Summarizing: The search for an elegant proof that avoids tedious numerical estimates was the driving force for the invention of the LLL algorithm.

12 The LLL Algorithm and its Consequences

A. K. Lenstra, H. W. Lenstra, L. Lovász. Factoring polynomials with rational coefficients. *Mathematische Annalen* 261(4):515–534, 1982.

The basis reduction algorithm by L. Lovász to solve a problem that initially looked like a technicality had a significant impact on the book [69] as outlined in Section 11. Its deep impact on other fields was unexpected, even for Lovász himself, as can be inferred from his letter, see Fig. 11.

We consider this as one of the occasional miracles in mathematics where a result that was prompted by the desire to find an elegant solution for a technical detail has consequences that are simply beyond imagination.

Lovász informed not only the coauthors Grötschel and Schrijver of his book [69] about his achievement, but also Hendrik Lenstra. Employing tools from the geometry of numbers, Hendrik had (briefly before) made the substantial discovery that integer programs (IPs) can be solved in polynomial time when the dimension is fixed. Concerning this, he was in discussion with Lovász, who pointed out that some of the steps of Hendrik's IP-algorithm could be improved, see [95].

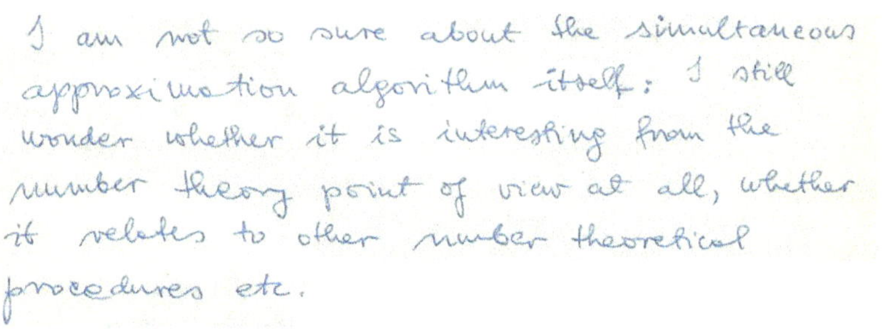
I am not so sure about the simultaneous approximation algorithm itself: I still wonder whether it is interesting from the number theory point of view at all, whether it relates to other number theoretical procedures etc.

Fig. 11: Cutout from a Lovász letter

Hendrik got excited about the news because his brother Arjen was (together with two fellow students) about to implement a method to factor univariate polynomials over algebraic number fields. Zassenhaus had suggested to use the Berlekamp–Hensel approach for this, which, however, could be "*very, very much exponential*" according to Arjen. A few days after Lovász's letter had arrived, Hendrik became convinced that the basis reduction algorithm implies that there is a polynomial time algorithm for factorization in the ring $\mathbb{Q}[X]$ of univariate polynomials over the rational numbers. At that time this looked inconceivable as one did not (and still does not) know a polynomial time algorithm for finding the factors of an integer. After working out the details, Hendrik's observation turned out to be true. The two Lenstra brothers and Lovász combined their contributions and wrote the joint paper [94]. Believing that polynomial time factoring of polynomials over the rational numbers (an unexpected result) is the most important contribution of their work, they agreed to mention only this aspect in the paper title. The full story of this cooperation is nicely described in the article of I. Smeets [151].

It turned out that basis reduction has applications that reach much further than linear programming or polynomial factorization. It is beyond the scope of this article to highlight here the wide range of applications of the basis reduction algorithm, which – in contrast to the ellipsoid method – is usable in practice. We mention two concrete examples.

Odlyzko and te Riele [137] used the basis reduction algorithm to disprove the Mertens conjecture, a conjecture standing in number theory since 1897, which – if true – would have implied the Riemann hypothesis. This disproof was surprising as there was extensive computational evidence that the Mertens conjecture is true.

Lagarias and Odlyzko [92] employed the lattice basis reduction algorithm to launch a polynomial time attack on knapsack-based public-key cryptosystems which made these cryptosystems unsafe.

The LLL algorithm, in fact, created a revolution in cryptography. It is known that the widely used public-key schemes such as the RSA or elliptic-curve cryptosystems can be defeated if Shor's quantum polynomial time factoring algorithm can be implemented on a quantum computer. Many cryptographers are convinced that certain lattice problems cannot be solved efficiently. Based on this, some lattice-based constructions appear to be resistant to attack by both classical and quantum computers.

For surveys, see Regev [141] or Micciancio and Goldwasser [131]. The National Institute of Standards and Technology (NIST) and other institutions are currently preparing cryptography standards for the post-quantum era. The first Quantum-Resistant Cryptographic Algorithms were announced by NIST in July 2022. Lattices play a major role here, and lattice basis reduction algorithms have become standard tools to test the security of cryptosystems.

Instead of attempting to comprehensively document the impact of Lovász's work on basis reduction, we point to the book by Nguyen and Vallée [136] entitled *The LLL Algorithm: Survey and Applications*, which consists of a collection of broad overviews of fields where the LLL algorithm is employed. Chapters, written by specialists in the respective fields, cover, for instance, applications in number theory, Diophantine approximation, integer programming, cryptography, geometry of provable security, inapproximability, and improvements of the LLL algorithm. A reviewer of this book wrote:

> The LLL algorithm embodies the power of lattice reduction on a wide range of problems in pure and applied fields [...] [and] the success of LLL attests to the triumph of theory in computer science.

Finally, the algorithm Lovász designed to find a reduced lattice basis is usually called the LLL algorithm, because it appeared in a paper written by three authors whose last names all start with L. Of course, the Lenstra brothers do not claim that it is their invention, they also attribute it to L. Lovász. But "LLL algorithm" has become the usually employed name of the algorithm.

13 Cutting Planes and the Solution of Practical Applications

> M. Grötschel, L. Lovász, A. Schrijver. *Geometric Algorithms and Combinatorial Optimization*. Springer, Berlin, 1988.

László Lovász has, in addition to inventing a beautiful theory, designed many algorithms, concentrating particularly on polynomial time algorithms. The theory and the algorithms Lovász developed had significant impact on computational practice. Chapter 8 of [69] "Combinatorial Optimization: A Tour d'Horizon" is a highly condensed overview of the applicational potential that arises from combinations of the many insights provided by the ellipsoid method, the LLL algorithm, and further ideas. These have contributed to the astonishing computational success stories that evolved in the last thirty to forty years in combinatorial optimization. We sketch some of these aspects.

In combinatorial optimization, a typical approach is, as indicated before, to attack a problem by transforming it into a linear programming problem with integer variables.

Take the traveling salesman problem, for instance. Given a complete graph $G = (V,E)$ on n vertices and a distance c_e for every edge $e \in E$, we look for a Hamiltonian cycle (briefly: tour) of minimum length. If H is a tour, let $x^H \in \mathbb{R}^E$ be its *incidence vector*, i.e., the e-th component x_e^H of x^H is equal to 1 if $e \in H$, otherwise it is 0. The traveling salesman polytope TSP(G) of G is the convex hull of all incidence vectors of tours in G. TSP(G) is a polytope in $\mathbb{R}^{n(n-1)/2}$. To apply the linear programming approach, we now have to find a linear inequality system, so that the integral solutions of the linear program are exactly the incidence vectors of tours. Such linear programs are called *LP-relaxations*. Let $\delta(W)$ denote the set of edges in E with one endvertex of e in W and the other in $V \setminus W$, and let $x(\delta(W))$ denote the sum over all variables x_e with $e \in \delta(W)$. It is well known that the following linear program:

$$\begin{aligned} 0 \leq x_e \leq 1 &\quad \text{for all } e \in E \\ x(\delta(\{w\})) = 2 &\quad \text{for all } w \in V \\ x(\delta(W)) \geq 2 &\quad \text{for all } W \subseteq V \text{ with } 2 \leq |W| \leq |V| - 2 \end{aligned}$$

is an LP-relaxation of the TSP. An inequality of the third type is called a *subtour elimination constraint*.

Let us call the polytope defined by the linear system above TSPLP(G). All vertices of the traveling salesman polytope TSP(G) are vertices of TSPLP(G). But TSPLP(G) has many nonintegral vertices as well. About 2^n inequalities define TSPLP(G). This renders the straightforward LP-solution approach hopeless. The facet complexity φ of TSPLP(G), however, is small since the entries of every inequality or equation are only 0 or 1 and the right-hand sides are 0, 1, or 2. Thus the facet complexity of TSPLP(G) is linear in the number of variables $|E| = n(n-1)/2$. Due to the oracle-polynomial time equivalence of strong separation and strong optimization, linear programs over TSGLP(G) can be solved in polynomial time – provided, given a vector $y \in \mathbb{Q}^E$, one can find a fast separation algorithm for the subtour elimination constraints.

This can in fact be done, as was observed by Hong [77]. One assigns the value y_e to every edge $e \in E$ as a capacity and computes (this can be done quickly) a minimum nonempty cut $\delta(W^*)$ in this capacitated graph $G = (V, E)$. If $y(\delta(W^*)) < 2$, a violated inequality is found, otherwise y satisfies all subtour elimination constraints. This is an example of a linear program appearing in many practical applications with an exponential number of inequalities that, nevertheless, can be solved in polynomial time. An optimal solution of a linear program over TSPLP(G) is usually nonintegral but provides a very good lower bound on the optimum TSP-value in practice. Finding a provably optimal solution needs additional effort, though.

In 1954 Dantzig, Fulkerson, and Johnson [33] proposed in a seminal paper to solve combinatorial optimization problems such as the traveling salesman problem by starting with some LP-relaxation, checking whether the optimum solution y is the incidence vector of a tour (in this case the problem is solved), and if not searching for inequalities valid for TSP(G) that are violated by y, adding these to the current LP as *cutting planes*, and to continue. This was one of the first proposals to solve linear and integer programs using cutting planes in an iterative process. The cutting plane search in this case was done manually, the LPs were solved by the simplex method. Four years later Gomory [61] invented an automatic cutting plane generation scheme (called *Gomory cuts*) for which he could prove finite termination. This looked like a promising approach to solve integer programs.

However, the computer implementations of this and related approaches in the 1960s and 1970s were not successful in practice. Moreover, theoretical results of Chvátal [27] and others revealed a series of examples for which the number of cutting plane additions cannot be effectively bounded. Hoping for the unimportance of these negative aspects in real-world applications, the idea came up in the 1970s to study combinatorial optimization problems of practical relevance and to look for cutting planes that define facets of the investigated polytopes. These are cuts that cut as deep as possible. The first implementations employing a combination of manual and heuristic searches for facet defining cutting planes at the end of the 1970s indicated practical success. Soon after, the ellipsoid method theory with the principle of polynomial time equivalence of optimization and separation was developed and demonstrated that this approach is a viable idea, and that linear optimization over exponentially large systems of linear inequalities is possible in polynomial time – at least theoretically.

Despite serious attempts, no implementation of the ellipsoid method has shown satisfactory numerical performance in computational practice. By replacing it with new implementations of the dual simplex algorithm, the theoretical polynomial time termination is lost, but astonishing computational results were achieved by many researchers in combinatorial optimization. Of course, lots of additional features (such as presolve techniques, heuristic primal and dual searches, branch and bound, robust numerics, etc.) were implemented as well. The new insights gave a significant push to the theoretical and applied part of combinatorial optimization. Problems with many industrial applications such as linear ordering; set partitioning and packing; knapsack; clustering; various types of matching; connectivity; path, flow and other network problems; max cut; unconstrained Boolean quadratic programming; stable sets; several variations of coloring; and vehicle and passenger routing could be solved for instances of practically relevant sizes. The discovery of new classes of facets and fast separation procedures (exact and heuristic) has been an important ingredient of this solution methodology. To indicate at least one example of practically useful separation algorithms we mention the paper [138] of Padberg and Rao that describes sophisticated and fast separation algorithms for various ramifications of the matching polytope. A large number of separation algorithms are, of course, described in the book [69].

This research activity goes on and brings application relevant instances of many $\mathcal{NP}$-hard combinatorial optimization problems to the realm of practical solvability. For the traveling salesman problem, for example, the "solvability world record" was 42 cities in 1954, it went to 120 in 1977, 2392 in 1987, and in 2017 a TSP with 109,399 cities could be solved to optimality, see the Webpage of Bill Cook [31], his book [30], and the book [8] by Applegate, Bixby, Chvátal, and Cook for comprehensive information. The solution process includes linear programming technology (its theory and implementation) that is able to prove, for example, that a vector in dimension 10^{10} satisfies more than $2^{100,000}$ constraints and is optimal for this system. This is really breathtaking.

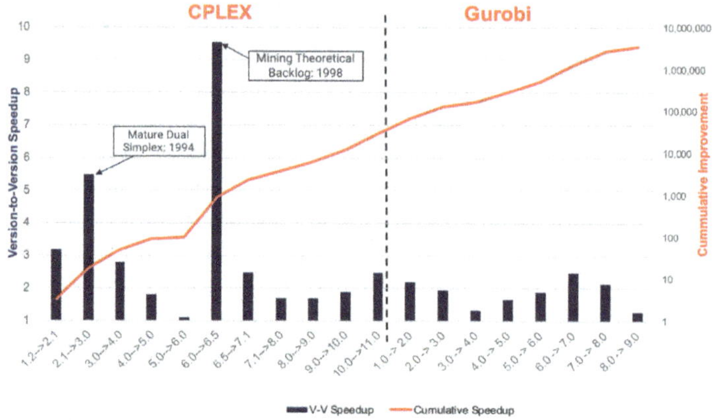

Fig. 12: MIP-code performance 1990–2019 (courtesy Robert E. Bixby)

The success stories indicated above, and the theoretical and practical lessons learned from these began to be harvested and improved by the developers of commercial optimization software in the 1990s. One reason for this is that many mixed-integer optimization problems (MIPs) occurring in industry contain subproblems that are combinatorial optimization problems for which large classes of facet-defining inequalities have been discovered. Efficient separation algorithms for these inequalities were successfully added to the existing MIP-codes. The graphic in Fig. 12, presented with the permission of Bob Bixby, shows the development of the commercial mixed integer programming codes CPLEX and Gurobi in the 30 years from 1990 to 2019. The large bar (pointed at by "Mining Theoretical Backlog") shows an almost tenfold speedup that is obtained from one version of the code to the next in which cutting plane technology (including a fresh implementation of Gomory cuts) was introduced together with various supporting features. The overall message is that the MIP technology in 2019 runs 3.5 million times faster than the codes of 1990. That speedup is due to mathematical and implementation improvements and is independent of the hardware speedup during this period. This is real progress indeed. Cutting plane technology contributed to it significantly.

14 Computing Optimal Stable Sets and Colorings in Perfect Graphs

M. Grötschel, L. Lovász, A. Schrijver. Polynomial algorithms for perfect graphs. *Annals of Discrete Math.* 21:325–256, 1984.

M. Grötschel, L. Lovász, A. Schrijver. Relaxations of vertex packing. *J. Combin. Theory* B 40:330–343, 1986.

The extension of the ellipsoid method to convex bodies outlined in Section 10 was driven by the hope that one could solve the stable set and the coloring problem in perfect graphs in polynomial time with this methodology, see Section 7. The successful attempt is presented in the articles [66, 67], and [68]. We describe the stable set case.

For a graph $G = (V,E)$ and a stable set $S \subseteq V$, one can define the incidence vector x^S in $\mathbb{R}^V$ as follows: the i-th component x_i^S of x^S is equal to 1 if the vertex $i \in V$ is an element of S, and it is 0 otherwise. The stable set polytope of G is the convex hull of all incidence vectors of stable sets S of G, i.e.,

$$\mathrm{STAB}(G) := \mathrm{conv}\{x^S \in \mathbb{R}^V \mid S \subseteq V \text{ stable set}\}.$$

Let $w : V \to \mathbb{Q}$ be any weighting of the vertices of G (we may assume that all weights are positive) and denote the largest weight of a stable set in G by $\alpha(G,w)$. Then $\alpha(G,w)$ is the maximum value of the linear function $w^T x$ for $x \in \mathrm{STAB}(G)$, in other words, $\alpha(G,w)$ can be computed by solving a linear program over $\mathrm{STAB}(G)$. For this observation to be of any use, we have to find inequalities defining $\mathrm{STAB}(G)$. Consider the polytope defined by

$$\begin{aligned}\mathrm{QSTAB}(G) := \{x \in \mathbb{R}^V \mid x_i \geq 0 \quad \forall i \in V,\ x_i + x_j \leq 1 \quad \forall ij \in E,\\ x(Q) \leq 1 \quad \forall Q \subseteq V \text{ clique}\},\end{aligned}$$

where $x(Q)$ denotes the sum of all $x_i, i \in Q$. The corresponding inequality is called a *clique constraint*. Since the intersection of a clique and a stable set contains at most one vertex, all clique constraints are satisfied by all incidence vectors of stable sets. This implies $\mathrm{STAB}(G) \subseteq \mathrm{QSTAB}(G)$ and optimizing over $\mathrm{QSTAB}(G)$ is an LP-relaxation of the stable set problem.

The stable set problem is $\mathcal{NP}$-hard. Therefore, solving linear programs over $\mathrm{STAB}(G)$ is $\mathcal{NP}$-hard as well. For some combinatorial optimization problems, their natural LP-relaxation is solvable in polynomial time. A sobering observation is that, for general graphs, solving linear programs over $\mathrm{QSTAB}(G)$ is also $\mathcal{NP}$-hard. So, in general, nothing is gained algorithmically. For perfect graphs, though, this approach combined with a tighter relaxation delivers the desired result.

Lovász's Shannon capacity article [106] suggests studying a different relaxation of the stable set problem.

Let $(u_i \mid i \in V), u_i \in \mathbb{R}^N$, be any orthonormal representation of G and let $c \in \mathbb{R}^N$ with $\|c\| = 1$. Then for any stable set $S \subseteq V$, the vectors $u_i, i \in S$, are mutually orthogonal and hence,

$$\sum_{i \in S} (c^T u_i)^2 \leq 1.$$

Since $\sum_{i \in V} (c^T u_i)^2 x_i^S = \sum_{i \in S} (c^T u_i)^2$, we see that the inequality

$$\sum_{i \in V} (c^T u_i)^2 x_i \leq 1 \qquad \text{(ORC)}$$

holds for the incidence vector $x^S \in \mathbb{R}^V$ of any stable S set of nodes of G. Thus, (ORC) is a valid inequality for $\text{STAB}(G)$ for any orthonormal representation $(u_i \mid i \in V)$ of G, where $u_i \in \mathbb{R}^N$, and any unit vector $c \in \mathbb{R}^N$. We shall call (ORC) the *orthonormal representation constraints* for $\text{STAB}(G)$.

Utilizing these inequalities, the following set was introduced in [68]. For any graph $G = (V, E)$ let

$$\begin{aligned} \text{TH}(G) := \big\{ x \in \mathbb{R}^V \mid x_i \geq 0 \quad \forall i \in V, \\ \text{and } x \text{ satisfies all orthonormal representation constraints} \big\}. \end{aligned}$$

$\text{TH}(G)$ is the solution set of infinitely many linear inequalities and thus a convex set. Since for every clique Q, its clique constraint appears as an orthonormal representation constraint (given a clique $Q \subseteq V$, let $\{u_i \mid i \in V \setminus Q\} \cup \{c\}$ be mutually orthogonal unit vectors and set $u_j = c$ for $j \in Q$) and every incidence vector of a stable set satisfies all such inequalities, we obtain:

$$\text{STAB}(G) \subseteq \text{TH}(G) \subseteq \text{QSTAB}(G).$$

An important fact is that the Lovász theta function $\vartheta(G, w)$ introduced in Section 8 can also be characterized as follows:

$$\vartheta(G, w) = \max\big\{ w^T x \mid x \in \text{TH}(G) \big\}.$$

$\text{TH}(G)$ is contained in the unit ball, and it is easy to find the center of a ball contained in the interior of $\text{TH}(G)$. Thus, $\text{TH}(G)$ is a convex body satisfying the assumptions required for the oracle-polynomial time equivalence of weak optimization, separation, and membership. The desired result is, of course, the following:

Theorem. *The weak optimization problem for* $\text{TH}(G)$ *is solvable in polynomial time for any graph* $G = (V, E)$.

Lovász, see [106], established several characterizations for his ϑ-function. They can be used in various ways to prove this theorem. One proof, worked out in detail in [66] and [69], is based on the following characterization:

$$\begin{aligned} &\vartheta(G, w) = \max\{\bar{w}^T B \bar{w} \mid B \in \mathcal{K}\}, \\ &\text{where } \mathcal{K} := \{B \in \mathbb{R}^{V \times V} \mid B \in \mathcal{D} \cap \mathcal{M} \text{ and } \text{tr}(B) = 1\}. \end{aligned}$$

Above, $\mathcal{D}$ is the set of positive semidefinite matrices, $\mathcal{M}$ the set of symmetric matrices B that satisfy $b_{ij} = 0$ whenever ij is an edge in G, and $\bar{w}$ denotes the vector whose entries are the square roots of the values $w_i, i \in V$. The main part of the proof consists in showing that the weak membership problem for $\mathcal{K}$ can be solved in polynomial time, and the core of this proof is established by showing whether a symmetric matrix is positive definite.

A by-product of the proof is the first polynomial time algorithm for optimization problems containing positive semidefinite constraints, a major result that led to considerable follow-up research such as the design of polynomial time interior point (and other) algorithms for semidefinite programming.

Another way to establish the above theorem is by utilizing the following fact:

$$\vartheta(G,w) = \min\{\Lambda(A+W) \mid A \in \mathcal{M}^{\perp}\},$$

where Λ denotes the largest eigenvalue, $\mathcal{M}^{\perp}$ the orthogonal complement of $\mathcal{M}$, and W the symmetric $V \times V$-matrix whose entries are the square roots of $w_i w_j$. $\Lambda(A+W)$ is a convex function that ranges over a linear space, and thus, we can obtain $\vartheta(G,w)$ via an unconstrained convex function optimization problem in polynomial time.

A third way to prove the theorem was demonstrated in [116], and this approach turned out to be one of the starting points for a generalization of this technique. Lovász and Schrijver developed in this article a general lift-and-project method that constructs higher-dimensional polyhedra (or, in some cases, convex sets) whose projection approximates the convex hull of 0-1 valued solutions of a system of linear inequalities. An important feature of these approximations is that one can optimize any linear objective function over them in polynomial time. Lift-and-project methods have been extended in many directions and are still an area of intensive research. The recent (not even exhaustive) survey by Fawzi, Gouveia, Parrilo, Saunderson, and Thomas [51] discusses the contributions of almost one hundred articles and illustrates the richness of this topic by presenting examples from many different areas of mathematics and its applications.

We refrain from describing the technically challenging details of this lift-and-project technique and return to stable sets in perfect graphs.

A combination results of Fulkerson [57] and Chvátal [28] yields:

Theorem. $\mathrm{STAB}(G) = \mathrm{QSTAG}(G)$ *if and only if G is perfect.*

And since we already know that $\mathrm{STAB}(G) \subseteq \mathrm{TH}(G) \subseteq \mathrm{QSTAB}(G)$ holds, we obtain:

Corollary. $\mathrm{STAB}(G) = \mathrm{TH}(G) = \mathrm{QSTAG}(G)$ *if and only if G is perfect.*

Since the weak optimization problem for $\mathrm{TH}(G)$ can be solved in polynomial time and since, in case G is perfect, $\mathrm{TH}(G)$ is a well-described polyhedron, the strong optimization problem for $\mathrm{TH}(G)$ can be solved in polynomial time. This yields the desired result:

Theorem. *The stable set problem can be solved in polynomial time for perfect graphs.*

We can now employ the fact that, if a linear program can be solved in polynomial time, the dual linear program can also be solved in polynomial time, see Section 11. By proving that, in this case, an optimum basic solution of the dual program can be transformed in polynomial time into an integral optimum basic solution one can find an optimum solution of the weighted clique covering problem. Since the cliques of a graph G are the stable sets of the complementary graph $\bar{G}$ of G and the colorings of G are the clique covering of $\bar{G}$, we can conclude:

Theorem. *For perfect graphs, the stable set, the clique, the coloring, and the clique covering problem can be solved in polynomial time. This also holds for the weighted versions of these problems.*

15 Submodular Functions

L. Lovász. Submodular functions and convexity. In *Mathematical Programming: The State of the Art* (eds. A. Bachem, M. Grötschel, B. Korte), Springer, pages 235–257, 1983.

Let E be a finite set. A function $f : 2^E \to \mathbb{R}$ is called *submodular* on 2^E (the power set of E) if

$$f(S \cap T) + f(S \cup T) \leq f(S) + f(T) \text{ for all } S, T \subseteq E.$$

Submodular functions play an important role in lattice theory, geometry, graph theory, and particularly, in matroid theory and matroidal optimization problems. The rank function of a matroid, for example, is submodular as well as the capacity function of the cuts in directed and undirected graphs.

Two polyhedra can be associated with a submodular function $f : 2^E \to \mathbb{R}$ in a natural way

$$P_f := \{x \in \mathbb{R}^E \mid x(F) \leq f(F) \text{ for all } F \subseteq E,\ x \geq 0\},$$
$$EP_f := \{x \in \mathbb{R}^E \mid x(F) \leq f(F) \text{ for all } F \subseteq E\}.$$

P_f is called the *polymatroid* associated with the submodular function f, EP_f the *extended polymatroid* associated with f. A deep theorem of Edmonds [43] states that if f and g are two integer-valued submodular functions then all vertices of $P_f \cap P_g$ as well as all vertices of $EP_f \cap EP_g$ are integral. This theorem contains a large number of integrality results in polyhedral combinatorics; in particular, it generalizes the matroid intersection theorem.

To address algorithmic questions concerning the structures introduced above, we assume that a submodular function f is given by an oracle that returns the value $f(S)$ for every query $S \subseteq E$. We also assume that we know an upper bound β on the encoding length of the output of the oracle. With these assumptions we define the encoding length of the submodular function as $|E| + \beta$.

It is well-known that, for any nonnegative linear objective function, the greedy algorithm finds an optimum vertex of EP_f in oracle-polynomial time, and that this vertex is integral provided the submodular function f is integer-valued. Optimizing over polymatroids or the intersections of two polymatroids or the intersections of two extended polymatroids and finding integral optima is more complicated and needs careful analysis. The most important algorithmic problem in this context is:

Submodular Function Minimization. *Given a submodular function $f : 2^E \to \mathbb{Q}$, find a set $S \subseteq E$ minimizing f.*

Lovász has built in [110] a bridge between submodularity and convexity by showing that submodular functions are discrete analogues of convex functions and has thus provided the key to the algorithmic solution of the submodular function minimization problem. The link is established as follows.

Let $f : 2^E \to \mathbb{R}$ be any set function. For every subset $T \subseteq E$, let x^T be its incidence vector and set

$$\widehat{f}(x^T) := f(T).$$

This way $\widehat{f}$ is defined on all 0/1-vectors. Note that every nonzero nonnegative vector $y \in \mathbb{R}^E$ can be expressed uniquely as

$$y = \lambda_1 x^{T_1} + \lambda_2 x^{T_2} + \ldots + \lambda_k x^{T_k},$$
$$\text{such that} \quad \lambda_i > 0, i = 1, \ldots, k \quad \text{and} \quad \varnothing \neq T_1 \subset T_2 \subset \ldots \subset T_k \subseteq E.$$

Then

$$\widehat{f}(y) := \lambda_1 f(T_1) + \lambda_2 f(T_2) + \ldots + \lambda_k f(T_k)$$

is a well-defined extension of the set function f (called the *Lovász extension of f*) to the nonnegative orthant. Lovász proved in [110]:

Theorem. *Let $f : 2^E \to \mathbb{R}$ be any set function and $\widehat{f}$ its extension to nonnegative vectors. Then $\widehat{f}$ is convex if and only if f is submodular.*

Lemma. *Let $f : 2^E \to \mathbb{R}$ be set function with $f(\varnothing) = 0$. Then*

$$\min\{f(S) \mid S \subseteq E\} = \min\{\widehat{f}(x) \mid x \in [0,1]^E\}.$$

Thus, instead of minimizing a set function f over E, it suffices to minimize its Lovász extension $\widehat{f}$ over the unit hypercube. We observe that $\widehat{f}(x)$ can be evaluated in oracle-polynomial time using the oracle defining f and that, if f is submodular, then $\widehat{f}$ is convex. We know already from Section 10 that convex functions can be minimized in oracle-polynomial time. (The assumption $f(\varnothing) = 0$ is irrelevant, if necessary, we can replace f by the function $f - f(\varnothing)$.) This yields:

Theorem. *Let $f : 2^E \to \mathbb{Q}$ be a submodular function. Then a subset S of E minimizing f can be found in oracle polynomial time.*

This theorem implies the polynomial time solvability of many combinatorial optimization problems, including the computation of a minimum capacity cut in a graph. It has various ramifications such as solvability in strongly polynomial time, as outlined in [110] and [69].

The running time of the polynomial time algorithm sketched above makes it, however, infeasible for practical use. New and better polynomial time algorithms, not employing the ellipsoid method, have been devised by Schrijver [143] and Iwata, Fleischer, and Fujishige [81].

16 Volume Computation

L. Lovász. How to compute the volume? *Jber. d. Dt. Math.-Vereinigung*, Jubiläumstagung 1990, B. G. Teubner, Stuttgart, pages 138–151, 1992.

Since the convergence of all versions of the ellipsoid method depends on sequentially shrinking the volume of an ellipsoid containing the given convex body K, it is tempting to ask whether the algorithm can be tuned to provide a reasonable estimate of the volume of K. The key idea in this context is, of course, to come up with an algorithmic version of the Löwner–John theorem, which states that, for a convex body K in $\mathbb{R}^n$, there exists a unique ellipsoid E of minimal volume containing K; moreover, K contains the ellipsoid obtained from E by shrinking it from its center by a factor of n. In formulas, let $E(A,a) := \{x \in \mathbb{R}^n \mid (x-a)^T A^{-1}(x-a) \leq 1\}$ denote the ellipsoid defined by a positive definite matrix A with center $a \in \mathbb{R}^n$, then the Löwner–John theorem states

$$E(n^{-2}A,a) \subseteq K \subseteq E(A,a),$$

if $E(A,a)$ is the Löwner–John ellipsoid E of K. Algorithmically, the following could be achieved in the Grötschel–Lovász–Schrijver book [69].

Theorem. *There exists an oracle-polynomial time algorithm that finds, for any convex body K given by the space dimension n, a weak separation oracle and two real numbers r and R with the property that K is contained in the ball of radius R around the origin and contains a ball of radius r, an ellipsoid E(A,a) such that*

$$E\Big(\frac{1}{n(n+1)^2}A,a\Big) \subseteq K \subseteq E(A,a).$$

With more effort and making additional assumptions such as central symmetry or requiring that a system of defining linear inequalities is explicitly given (in the polytopal case), the factor $1/(n(n+1)^2)$ in front of the matrix A above can be slightly improved, but not fundamentally. If one declares the volume of the interior ellipsoid as an approximation of the volume of K, the relative error turns out to be $2^n n^{3n/2}$, which appears to be outrageously bad.

Surprisingly, the error is not as bad as it looks since subsequently Elekes [45] and others proved that no oracle-polynomial time algorithm can compute, for a convex body K as given above, the volume of K with a much better relative error. We quote a result of Bárány and Füredi [11].

Theorem. *Consider a polynomial time algorithm which assigns to every convex body K given by a membership oracle an upper bound $w(K)$ on its volume* $\mathrm{vol}(K)$. *Then there is a constant $c > 0$ such that in every dimension n there exists a convex body K for which $w(K) > n^{cn}\,\mathrm{vol}(K)$.*

Following up, various authors proved more negative results on the deterministic approximation of the volume, width, diameter and other convexity parameters.

These negative results fueled the investigation of stochastic approaches to estimate the volume of a convex body. Instead of giving a deterministic guarantee, one could try to calculate a number that is close to the true value of the volume with high probability employing a randomized algorithm.

A side remark: Khachiyan [88] and Lawrence [93] proved that, for every dimension n, one can construct systems of rational inequalities defining polytopes P so that the encoding length of the rational number p/q representing the true volume of P requires a number of digits that is exponential in the encoding length of the inequality system. Hence, exact volumes of convex bodies cannot be computed in polynomial time since specifying the exact volume requires exponential space.

A fundamental breakthrough was achieved by Dyer, Frieze, and Kannan [40], who provided a randomized polynomial time approximation scheme for the volume approximation problem where K is given by a membership oracle. The ingredients of their algorithm are a multiphase Monte-Carlo algorithm (using the so-called product estimator) to reduce volume computation to sampling, the utilization of Markov chain techniques for sampling, and the use of the conductance bound on the mixing time, due to Jerrum and Sinclair [82]. The running time of the algorithm is roughly $O(n^{23})$, which is truly prohibitive. The exponent 23 of n was subsequently reduced considerably by adding further techniques and improved estimates to the toolbox of randomized algorithms, including rapid mixing, harmonic functions, connection to the heat kernel, isoperimetric inequalities, discrete forms of the Cheeger inequality, and many more.

Lovász played an important role in the exponent shrinking race. For example, the exponent went down to 16 (Lovász and Simonovits [120]), to 10 (Lovász [111]), to 8 (Dyer and Frieze [39]), to 7 (Lovász and Simonovits [121]), to 5 (Kannan, Lovász, and Simonovits [85]), and to 4 (Lovász and Vempala [124]). A nice survey of the many tricky issues in designing randomized algorithms for volume computation and their analysis is the article by Simonovits [149].

The race for better algorithms has not stopped. On September 3, 2022, the new record was published on arXiv by Jia, Laddha, Lee, and Vempala [83]. The authors show that the volume of a convex body in $\mathbb{R}^n$ defined by a membership oracle can be computed to within relative error ε using $\tilde{O}(n^3\psi^2 + n^3/\varepsilon^2)$ oracle queries, where ψ is the KLS constant. With the current bound of $\psi = \tilde{O}(1)$, this gives an $\tilde{O}(n^3/\varepsilon^2)$ algorithm, improving on the Lovász–Vempala $\tilde{O}(n^4/\varepsilon^2)$ algorithm.

17 Analysis, Algebra, and Graph Limits

L. Lovász, *Large Networks and Graph Limits*. American Mathematical Society, 2012.

We have already indicated that many of the results mentioned in our article seem to be of permanent importance and are used again and again: the Lovász Local Lemma, algorithmic consequences of the ellipsoid method, topological combinatorics, and the LLL algorithm, to name just few. Very recently Lovász's mathematics culminated in a topic that somehow combines this into an all-in-one subject: like a late symphony of a grand composer displaying the experience of the master and an echo of his/her life. We believe that this happened with the subject of graph limits founded and developed by Lovász with co-authors and students in the last 15 years. Here is a brief sketch of this fascinating development.

We have seen in Section 2 that the homomorphism function $\hom(F,G)$ and the Lovász vector $L(G)$ determine every graph G up to an isomorphism. With a proper scaling this leads to the notion of *homomorphism density* $t(F,G)$, which is the probability that a random mapping between sets of vertices of F and G is a homomorphism: $t(F,G) = \frac{\hom(F,G)}{v(G)^{v(F)}}$, where $v(G)$ denotes the number of vertices of graph G.

This definition is close to the sampling density and one motivation for introducing it. One can observe that homomorphism densities do not determine a graph up to an isomorphism but up to a "blowing up of vertices". (This is a procedure by which vertices are replaced by a certain number of twin copies.) It is perhaps more important that one can then define *convergence of a sequence* of finite graphs $G_1, G_2, \ldots, G_n, \ldots$ as the convergence of homomorphism densities $t(F, G_n)$ for every graph F. This convergence concept (and various other notions of convergence) were introduced and investigated in the article [22] of C. Borgs, J. T. Chayes, L. Lovász, V. T. Sós, and K. Vesztergombi.

Hence, a sequence of graphs converges if, for every F, all homomorphism densities (or F-sampling densities) converge. Does this convergence have a real (geometrical) meaning? Are there limit graphs or, perhaps, other limit objects?

Fig. 13: Lovász's Graph Limits book

It appears that these questions have non-trivial yet positive answers and these were the starting point of a very rich and interesting area. In fact, they generated a whole new theory. Here is a sample of some of the results.

Ch. Borgs, J. Chayes, L. Lovász, V.T. Sós and K. Vesztergombi proved the following in [22]:

Theorem. *A sequence of graphs (with unbounded size) is convergent if and only if it converges to a symmetric measurable function* $W : [0,1]^2 \to [0,1]$*. Moreover, up to a measurable bijection, such a function W is uniquely determined.*

Explicitly, this means that for every graph $F = (V,E)$ the homomorphism densities $t(F,G_n)$ are converging to:

$$t(F,W) = \int\limits_{[0,1]^V} \prod_{ij \in E} W(x_i,x_j) \prod_{i \in V} \mathrm{d}x_i.$$

Such functions W are called *graphons*. Graphon is a very intuitive notion and the convergence of a graph sequence to a graphon looks like a movie. It leads to "pixel" pictures like those in the samples shown in Figure 14 (taken from Lovász's book [112]).

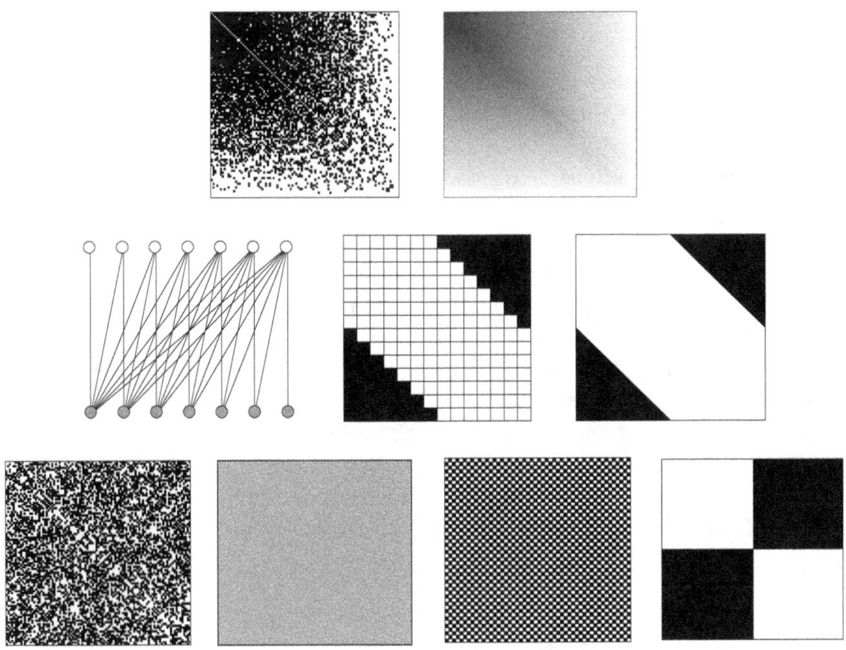

Fig. 14: Samples of graphons

The first row of Fig. 14 shows on the left a randomly grown uniform attachment graph with 100 nodes, and on the right a (continuous) function approximating it. The picture on the right is a grayscale image of the function $U(x,y) = 1 - \max(x,y)$. The second row of Fig. 14 indicates the construction of the graphon for the "halfgraph" (the graph on the left). The bottom part indicates the influence of ordering and the regularity lemma in its simplest form. Note that the sequence of random graphs is converging to a graphon W that is a constant function. It is important that the same is true for "quasirandom graphs".

Graphon is not just an intuitive notion, it has mathematical relevance. This setting extends work of Aldous [2] and Hoover [79] in probability theory on exchangeable random graphs (see, e.g., [9]). Graphons are also not just a generalization. They present a convenient and useful way to study extremal problems for graphs (such as to find maximum number of edges of a graph satisfying given local properties).

These problems then often take the form of linear inequalities. Lovász introduced graph algebras (of "quantum graphs") with nice "pictorial" proofs, see [112]), and independently Alexander Razborov developed "flag algebras" [140], which proved to be a very efficient tool in various extremal problems, see, e.g., [73] and [70].

The graph algebra of Lovász and Razborov was motivated by early examples provided by the Caccetta–Häggkvist conjecture, see [20], the Sidorenko conjecture [148], and the early paper [50] of Erdős, Lovász, and Spencer on topological properties of the graphcopy function.

A typical extremal problem may be expressed as the fact that a certain linear inequality built from homomorphism densities of graphs is nonnegative. This in turn led Lovász to the question whether any such inequality can be deduced from a sum of squares of "quantum graphs". A related question, formulated by Razborov [140], asks whether the validity of any such inequality can be solved by "Cauchy–Schwarz Calculus". Hamed Hatami and Serguei Norin [74] showed that both these questions have a negative answer in general as the related problems are algorithmically undecidable. So, extremal problems may be more difficult than originally thought. This was further supported by the universality results of Cooper, Grzesik, Král, Martins, and L. M. Lovász, see [32] and [70], claiming in particular that every graphon may be extended to a "finitely forcible" graphon.

This approach also provides an understanding of the celebrated Szemerédi regularity lemma. The Szemerédi regularity lemma in this interpretation means an approximation of every graph (and every graphon) by means of a "small" pixel image where almost all entries are constant (but may be different for different pixels).

The key of the approach of [22] is to characterize convergence using the *cut metric* $d_\square(G,H)$ (based on the cut norm introduced by R. Frieze and R. Kannan in [55]). If the homomorphism density is defined by scaled subgraph density, then the cut metric is, somewhat dually, characterized by means of a scaled density of partitions.

The cut metric $d_\square(G,H)$ for finite graphs G,H on the same vertex set V is defined as

$$\max_{S,T\subseteq V} \frac{|e_G(S,T)| - |e_H(S,T)|}{|V \times V|}$$

i.e., as the scaled difference of the-sizes of cuts in G and H; above $e_G(S,T)$ is the number of edges of G between sets S and T. (This definition can be extended to graphs on different vertex sets. This is technical and it takes three full pages in [112]). Interestingly, the cut distance for a graphon W is more easily defined than in the finite case: it is induced by the norm:

$$\|W\| = \sup_{S,T\subseteq[0,1]} \int_{S\times T} W(x,y)\,\mathrm{d}x\,\mathrm{d}y.$$

The cut norm is also very natural and fitting from an algorithmic point of view; and it is bounded by the Grothendieck norm up to a multiplicative constant (as shown by Alon and Naor [5]).

As a culmination of several auxiliary results, one obtains that the convergence is indeed induced by a distance. This is the key fact in many applications and was proved by Lovász and Szegedy in [122]:

Theorem. *If (G_n) is a sequence of graphs of unbounded size, then (G_n) is a converging sequence if and only if (G_n) is a Cauchy sequence with respect to cut distance $d_\square(G_i, G_j)$.*

The following result was proved by Lovász and Szegedy in [123]. Lovász considers it as one of the basic results treated in his book [112].

Theorem. *The space of all graphons W with cut distance is compact.*

This compactness theorem may be viewed as the roof result for the Szemerédi regularity Lemma and its various extensions. It also displays the usefulness of the limit language and of the much more general setting. This area was studied extensively, for instance by Borgs, Chayes, Elek, Lovász, Sós, Szegedy, Vesztergombi, and Tao in [23, 44, 123, 155].

The mathematical richness of this area is best illustrated by Appendix A of [112], which contains the following sections: Möbius functions; the Tutte polynomial; some background in probability and measure theory; moments and the moment problem; ultraproduct and ultralimit; Vapnik–Chervonenkis dimension; nonnegative polynomials; categories. Obviously, it is impossible to present here more than a glimpse of what the book [112] covers.

Note that the above results are interesting for dense graphs. For sparse graphs (for example for graphs with constant degrees) one has to devise a different approach. Limit objects are now called graphings and modelings. For them, results similar to above three theorems are not known. This is treated, e.g., by Benjamini and Schramm [17] and by Nešetřil and Ossona de Mendez [134]; see again [112].

It is amazing that the area of graphs and their limits can be traced back to Lovász's very early algebraic results (mentioned in Section 2). Some forty years later it blossomed in the inspiring climate of the Microsoft Research Theory Group at Redmond in an atmosphere of concentrated research and quality, with persons such as Michael Freedman, Oded Schramm and many other great visitors and with László Lovász as a driving force.

18 Final Remarks

Let us finish the fireworks of beautiful theorems ranging over many parts of mathematics and theoretical computer science by adding a few general remarks.

It happens very rarely that a well-known and long-standing open problem is solved by a novel technique that immediately influences not just that area, but other parts of mathematics as well. Lovász not only accomplished this once. It is unbelievable that Lovász repeatedly offered to the world community exactly such solutions. Some of these proofs are really elegant and were included in the collections of other beautiful "book proofs", see [1] and [129].

In this article we concentrated on Lovász-results which had general influence, led to intensive research by many others, and sometimes spawned the emergence of whole new theories. Work in areas such as combinatorial optimization, applications of the ellipsoid method, algebraic graph theory, graph homomorphisms, topological graph theory, and graph limits is very difficult to imagine without the pioneering accomplishments of László Lovász.

In our Introduction we indicated that Lovász is both a "problem solver" and a "theory builder", and pointed out that the trio *depth, elegance, and inspiration* is a particular signature of his work that makes his achievements unique. We do hope

Fig. 15: Several Lovász-books on a poster (by A. Goodall and J. N.) of the Charles University in Prague (Photo: Private)

that the glimpse into his oeuvre and the scientific influence of his results that we have offered here provides at least a partial proof of our conviction.

To keep this article at a reasonable length we had to omit many topics on which Lovász left his marks. In particular, it was impossible to give adequate attention to the books he has written, see Fig. 15, and the influence they had and still have. To mend this omission, albeit very incompletely, we elucidate the contents and impact of four of his books – extremely briefly, though.

Lovász's third book *Combinatorial Problems and Exercises* [107] became – without any exaggeration – a bible for combinatorialists worldwide. This is a book organized in an unusual way. It has three parts: The first part consists of mostly easily formulated questions and problems, the second part contains hints for the solutions, and the third part thorough proofs with discussions. This of course, makes up the largest part.

Lovász convincingly claims in this book that discrete mathematics, at the time of publication, has grown out of an area with simple questions that are relatively easy to solve without much mathematical knowledge into a structured field with various branches consisting of central concepts and theorems forming a hierarchy and possessing a rich bouquet of proof techniques. Instead of presenting the theories analytically and deductively, Lovász designed his book with the purpose of helping interested readers to learn many of the existing techniques in combinatorics. And as he wrote in the introduction:

> *The most effective (but admittedly very-time consuming) way of learning such techniques is to solve (appropriately chosen) exercises and problems.*

We believe that this book significantly changed the level on which combinatorics (and graph theory in particular) was treated. It caught worldwide attention from the very start (see, e.g., the book review by Bollobás [19]) by combinatorialists, computer scientists, and mathematicians in general. It is remarkable that after more than 40 years of its existence the book, which mirrors the vast experience of the author, is still in print and in use.

A side remark: Combinatorics meetings usually have an open problems session where participants explain questions they are working on and have not solved yet. Lovász, with his wide knowledge of proof techniques, has always been outstanding in being able to solve many of the open problems on the spot.

Matching problems have played a considerable role in the development of graph theory. Well-known and important early results are, e.g., König's Matching Theorems, the Marriage Theorem, and Tutte's f-factor theorem. Matchings, b-matchings, T-joins, etc. have a rich structure theory. The Edmonds–Gallai decomposition is one such example. Various matching problems and their ramifications appear in a large variety of applications of combinatorial optimization (e.g., the Chinese Postman Problem). Many of these are solvable with (highly nontrivial) polynomial time algorithms, for which the pioneering work of J. Edmonds, see [41], laid the basis. Edmonds [42] also achieved a breakthrough in polyhedral combinatorics by providing a linear description of the matching polytope that does not simply follow from total unimodularity. Lovász [108] came up with a new and elegant proof of this result that was later often mimicked for the characterization of other polytopes arising in combinatorial optimization.

The book [115] *Matching Theory*, written by László Lovász and Mike Plummer, provides a broad view of this subject and covers the roughly 40 articles that Lovász has contributed to this field. We just want to highlight Chapters 10 and 11 of this book. Chapter 10 is devoted to the f-factor problem, which asks, for a given graph $G = (V,E)$ and integers $f(v)$ for every vertex $v \in V$, whether there is a spanning subgraph H of G such that the degree of v in H is equal to $f(v)$. In a series of four papers that appeared in 1970–1972, Lovász developed a generalization of the Edmonds–Gallai Structure Theorem to the f-factor problem, providing an elegant answer to the f-factor problem. Chapter 11 introduces further generalizations such as the matroid and polymatroid matching problem, which are interesting (and difficult) combinations of topics in graph and matroid theory. We refer to this chapter of [115] and the article [109] for some of the results that can be shown in this context. Finally, this book contains in the preface a wonderful brief, yet in-depth survey of the historical development of matching theory.

Lovász's book *Large networks and graph limits* [112] aims in a different direction. It is the result of a stay of Lovász at the IAS in Princeton. We have dealt with parts of this book in Section 17. Graph limits became a very active field with contributions ranging from model theory, probability, functional analysis to theoretical computer science, network science and, of course, combinatorics. This theory fits very well with advanced combinatorics; for example, the role of Szemerédi's regularity lemma is highlighted and explained properly in this context. The basic theory of convergent graph sequences is derived in several settings; and multiple applications to parameter and property testing, extremal theory, and other applications are given. The book starts with an informal introduction into large graphs in a network science context, specifying the abundance of real applications, and questions to ask about them. This is followed by a lengthy chapter on the algebra of graph homomorphisms. This chapter can be read independently and is also of independent interest. But one of the main features of this book is to show how this algebra is connected to limit structures and limit distributions. It is amazing how much material was developed in this context in less than a decade. In the very nice preface, Lovász lists the branches of mathematics that come into play in his book and writes:

> *These connections with very different parts of mathematics made it quite difficult to write this book in a readable form* [...] [continuing that he found that] *the most exciting feature of this theory* [...] [is] *its rich connections with other parts of mathematics (classical and non-classical)* [...] [so that he] *decided to explain as many of these connections* [...] [as he] *could fit in the book.*

Summarizing, this book is a real tour de force.

The American Mathematical Society Colloquium Publications were established in 1905. So far 66 books were published in this AMS flagship book series "*offering the finest in scholarly mathematical publishing*". Vol. 60 is the book [112] *Large Networks and Graph Limits* discussed above, Vol. 65 is the book *Graphs and Geometry* [113], so far the last book written by Lovász.

Vol. 60 pictures the emergence and maturation of a new theory while Vol. 65 presents a wide spectrum of geometry related techniques (and tricks) to study graphs. In twenty chapters (and three appendices) Lovász surveys many connections between graph theory and geometry, concentrating on those which lie deeper. These are among others: rubber band representations, coin representations, orthogonal representation, and discrete analytic functions. Interestingly, this book is only about geometry, and thus topology is outside its scope. Nevertheless, the book contains some of the key discoveries of Lovász in a new context.

The Leitmotiv of the whole book [113] is described in the preface:

> *Graphs are usually represented as geometric objects drawn in the plane, consisting of vertices and curves connecting them. The main message of this book is that such a representation is not merely a way to visualize the graph, but an important mathematical tool. It is obvious that this geometry is crucial in engineering if you want to understand rigidity of frameworks and mobility of mechanisms. But even if there is no geometry directly connected to the graph-theoretic problem, a well-chosen geometric embedding has mathematical meaning and applications in proofs and algorithms. This thought emerged in the 1970s, and I found it quite fruitful.*

Lovász has been developing these thoughts for about forty years, observing:

> *Many new results and new applications of the topic have also been emerging, even outside mathematics, like in statistical and quantum physics and computer science (learning theory). At some point I had to decide to round things up and publish this book.*

This finishes his preface. But he returns to these considerations in Chapter 20, "Concluding Thoughts", on page 390 as follows:

> *I am certain that many new results of this nature will be obtained in the future (or are already in the literature, sometimes in a quite different disguise). Whether these will be collected and combined in another monograph, or integrated into science through some other platform provided by the fast changing technology of communication, I cannot predict. But the beauty of nontrivial connections between combinatorics, geometry, algebra and physics will remain here to inspire research.*

When reviewing the book [112] in the Bulletin of the American Mathematical Society, one of us quoted Michel Mendès France, who once told him that envy is the right feeling when reading beautiful mathematics. Yes, this is the feeling one may have when reading Lovász's books such as [112] and [113].

Lovász's exceptional research capabilities and his broad knowledge of mathematics are mirrored in his public presentations and survey articles. He has the ability to explain difficult results in understandable language and, in particular, to display and illustrate connections between seemingly unrelated topics. Examples of that can, e.g., be found in the articles he contributed to the *Handbook of Combinatorics* [62], see also [91]. The titles of some of his survey and motivating articles contain phrases such as *One mathematics* or *Discrete and Continuous: Two sides of the same*. This reflects his philosophy that science is not a collection of independent topics but a tightly connected network to be discovered and understood. He contributed to this conviction also administratively by serving the scientific community in leading positions of the International Mathematical Union and the Hungarian Academy of Sciences.

The unity of mathematics and the role of mathematics in the world have been addressed again and again by László Lovász through many of his activities. Given the outstanding excellence in his own research and the huge experience as a professional in combination with admirable modesty the mathematical community can hardly think of a better representative.

References

1. M. Aigner, G. Ziegler. *Proofs from the Book*. Springer, 1998.
2. D. J. Aldous. Representations of partially exchangeable arrays of random variable. *J. Mutivar. Anal.* 11:581–598, 1981.
3. N. Alon, A. Kostochka, B. Reiniger, D. B. West, X. Zhu. Coloring, sparseness and girth. *Israel J. Math.* 214(1):315–331,2016.
4. N. Alon, I. Kříž, J. Nešetřil. How to color shift hypergraphs. *Studia Scientarum Math Hung.*, 30: 1–11, 1995.
5. N. Alon, A. Naor. Approximating the cut-norm via Grothendieck's inequality. *SIAM J. Comp.* 35:787–803, 2006.
6. N. Alon, J. Spencer. *The Probabilistic Method.* (1st printing 1991) Wiley, 4th edition, 2016.
7. A. Ambainis, J. Kempe, O. Sattath. A quantum Lovász Local Lemma. *In Proceedings of the 42nd Annual ACM Symposium on Theory of Computing (STOC '10).* Cambridge, MA, USA:151–160, 2010.
8. D. Applegate, R. E. Bixby, V. Chvátal, W. J. Cook. *The Traveling Salesman Problem: A Computational Study.* Princeton Series in Applied Mathematics 40, 2007.
9. T. Austin. On exchangeable random variables and the statistic of large graphs and hypergraphs. *Probability Surveys* 5:80–145, 2008.
10. I. Bárány. A short proof of Kneser's conjecture. *J. Comb. Theory*: 84–88, A25 (1978).
11. I. Bárány, Z. Füredi. Computing the volume is difficult. *Proc. of the 18th Annual ACM Symp, Theory Comput.* 442–447, 1986.
12. I. Bárány, G. O. H. Katona, Attila Sali (eds.). *Building Bridges II: Mathematics of László Lovász.* Springer, 2019.
13. L. Barto, M. Kozik. Combinatorial gap theorem and reductions between promise CSPs. In *Proceedings of the 2022 Annual ACM-SIAM Symposium on Discrete Algorithms* (SODA): 1204–1220, 2022.
14. J. Beck. An algorithmic approach to the Lovász Local Lemma I. *Random Str. and Algorithms* 3(4):343–365, 1991.
15. C. Berge. Les problèmes de coloration en théorie des graphes. *Publications de l'Institut de Statistique de l'Université de Paris* 9:123–160, 1960.
16. C. Berge. Some classes of perfect graphs, In *Six Papers on Graph Theory* [related to a series of lectures at the Research and Training School of the Indian Statistical Institute, Calcutta, March–April 1963], Research and Training School, Indian Statistical Institute, Calcutta, pp. 1–21, 1963.
17. I. Benjamini, O. Schramm. Recurrence of distributional limits of finite planar graphs. *Electronic J. Probab.* 6(23):1–13, 2001.
18. A. Bernshteyn. Measurable vertions of the Lovász Local Lemma and measurable graph colorings. *Advances in Mathematics* 353:153–223, 2019.
19. B. Bollobás. Combinatorial problems and exercises by László Lovász. Book Review. *Bull. Amer. Math. Soc.* 4:250, 1981.
20. J. A. Bondy. Counting subgraphs: A new approach to the Caccetta–Häggkvist conjecture. *Discrete Math.* 165–166:71–80, 1997.
21. Ch. Borgs, J. Chayes, L. Lovász, V. T. Sós, K. Vesztergombi. Counting graph homomorphisms. In *Topics in Discrete Mathematics*, edited by M. Klazar, J. Kratochvíl, M. Loebl, J. Matoušek, P. Valtr, R. Thomas, Springer, pp. 315–371, 2006.

22. C. Borgs, J.T. Chayes, L. Lovász, V.T. Sós, K. Vesztergombi. Convergent graph sequences I: Subgraph frequencies, metric properties, and testing. *Advances in Math.* 219:1801–1851, 2008.
23. Ch. Borgs, J. T. Chayes, L. Lovász, V. T. Sos, K. Vesztergombi. Convergent graph sequences II. Multiway cuts and statistical physics. *Annals of Math.* 176:151–219, 2012.
24. Ji-Yi Cai, Xi Chen. *Complexity Dichotomies for Counting Problems.* Cambridge University Press, 2017.
25. C.C. Chang, B. Jónsson, A. Tarski. Refinement properties for relational structures. *Fund. Math.* 55:249–281, 1964.
26. M. Chudnovsky, N. Robertson, P. Seymour, R. Thomas. The strong perfect graph theorem. *Annals of Mathematics*, 164:51–229, 2006.
27. V. Chvátal. Edmonds polytopes and a hierarchy of combinatorial problems. *Discrete Mathematics* 4:305–337, 1973.
28. V. Chvátal. On certain polytopes associated with graphs. *J. Comb. Theory* B 18:138–154, 1975.
29. S. A. Cook. The complexity of theorem-proving procedures. In *Conference Record of Third Annual ACM Symposium on Theory of Computing* (3rd STOC, Shaker Heights, Ohio, 1971), The Association for Computing Machinery, New York, pp. 151–158, 1971.
30. W. J. Cook. *In Pursuit of the Traveling Salesman: Mathematics at the Limits of Computation.* Princeton University Press, 2012.
31. W. J. Cook. Bill Cook's TSP Webpage: `https://www.math.uwaterloo.ca/tsp/`, 2023.
32. J. W. Cooper, D. Král, T. Martins. Finitely forcible graph limits are universal. *Adv. Math.* 340:819–854, 2018.
33. G. Dantzig, R. Fulkerson, S. Johnson. Solution of a large-scale traveling salesman problem. *Journal of the Operations Research Society of America* 2: 393–410, 1954.
34. A. Dawar, T. Jakl, L. Reggio. Lovász-type Theorems and Game Comonads. *36th Annual ACM/IEEE Symposium on Logic in Computer Science*, pages 1–13, 2021.
35. I. Dinur, O. Regev, C. Smyth. The hardness of 3-uniform hypergraphs coloring. *Combinatorica* 25(5):519–535, 2005.
36. G. L. Dirichlet. Verallgemeinerung eines Satzes aus der Lehre von den Kettenbrüchen nebst einigen Anwendungen auf die Theorie der Zahlen. Bericht über die zur Bekanntmachung geeigneten Verhandlungen der Königlich Preussischen Akademie der Wissenschaften zu Berlin, pp. 93–95, 1842. (Reprinted in: L. Kronecker (ed.). *G. L. Dirichlet's Werke*, Vol. I, G. Reimer, Berlin, 1889 (reprinted: Chelsea, New York, 1969), pp. 635–638).
37. R. L. Dobrushin. Estimates of semi-invariants for the Ising model at low temperatures. In *Topics in Statistical and Theoretical Physics*, Amer. Math. Soc Translations 177(2):59–81, 1996.
38. Z. Dvořák. On recognizing graphs by numbers of homomorphisms. *Journal of Graph Theory*: 330–342, 2009.
39. M. Dyer, A. Frieze. Computing the volume of convex bodies: A case where randomness probably helps. In *Probabilistic Combinatorics and Its Applications*, edited by Béla Bollobás, Proceedings of Symposia in Applied Mathematics, Vol. 44, pp. 123–170, 1992.
40. M. Dyer, A. M. Frieze, R. Kannan. A random polynomial time algorithm for approximating the volume of convex bodies. *Journal of the ACM* 38(1):1–17, 1991.
41. J. Edmonds. Paths, trees, and flowers. *Canadian Journal of Mathematics* 17:449–467, 1965.
42. J. Edmonds. Maximum matching and a polyhedron with 0, 1-vertices. *Journal of Research National Bureau of Standards Section B* 69:125–130, 1965.
43. J. Edmonds. Submodular functions, matroids, and certain polyhedral. In *Combinatorial Structures and Their Applications* (Proceedings Calgary International Conference on Combinatorial Structures and Their Applications, Calgary, Alberta, 1969; edited by R. Guy et al.), Gordon and Breach, New York, pp. 69–87, 1970.
44. G. Elek, B. Szegedy. A measure-theory approach to the theory of dense hypergraphs. *Advances in Math.* 231:1731–1772, 2012.

45. G. Elekes. A geometric inequality and the complexity of computing volume. *Discrete and Computational Geometry* 1:289–292, 1986.
46. J. A. Ellis-Monaghan, I. Moffatt. *Handbook of Tutte Polynomial and Related Topics.* CRC Press, 2022.
47. P. Erdős. Graph theory and probability. *Can. J. Math.* 11:34–38, 1959.
48. P. Erdős, A. Hajnal. On chromatic numbers of graphs and set systems. *Acta. Acad. Sci. Hung.* 17:61–99, 1966.
49. P. Erdős, L. Lovász. Problems and results on 3-chromatic hypergraphs and some related questions. In *Infinite and Finite Sets, Coll. Math. Soc. J. Bolyai*, North Holland:609–627, 1975.
50. P. Erdős, L. Lovász, J. Spencer. Strong independence of graphcopy functions. In *Graph Theory and Related Topics*, edited by J. A. Bondy, U. S. R. Murty) Academic Press, pp. 165–172, 1979.
51. H. Fawzi, J. Gouveia, P. A. Parrilo, J. Saunderson, R. R. Thomas. Lifting for simplicity: Concise descriptions of convex sets. *SIAM Review* 64(4): 866–918, 2022.
52. T. Feder, M. Y. Vardi. The computational structure of monotone monadic SNP and constrained satisfaction: A study through datalog and group theory. *SIAM J. Comput.* 28(1):57–104, 1998.
53. A. Frank, É. Tardos. An application of simultaneous diophantine approximation in combinatorial optimization. *Combinatorica* 7:49–65, 1987.
54. M. Freedman, L. Lovász, L. Schrijver. Reflection positivity, rank connectivity, and homomorphisms of graphs. *Journal of American Mathematical Society* 20(1):37–51, 2007.
55. A. Frieze, R. Kannan. Quick approximation to matrices and applications. *Combinatroica* 19:175–220, 1999.
56. D. R. Fulkerson. The perfect graph conjecture and pluperfect graph theorem. In *Proceedings of the Second Chapel Hill Conference on Combinatorial Mathematics and Its Applications*, edited by R. C. Bose, I. M. Chakravarti, T. A. Dowling, D. G. Kelly, K. J. C. Smith, University of North Carolina at Chapel Hill, Chapel Hill, North Carolina, pp. 171–175, 1970.
57. D. R. Fulkerson. Anti-blocking polyhedra. *Journal of Combinatorial Theory*, Series B 12:50–71, 1972.
58. D. R. Fulkerson. On the perfect graph theorem, in: *Mathematical Programming* (Proceedings of an Advanced Seminar, Madison, Wisconsin, 1972; edited by T. C. Hu, S. M. Robinson), Academic Press, New York, pp. 69–76, 1973.
59. P. Gács, L. Lovász. Khachiyan's algorithm for linear programming. *Math. Prog. Study* 14:61–68, 1981.
60. I. Gessel, G.-C. Rota. *Classic Papers in Combinatorics.* Birkhäuser, Boston, MA, 1987.
61. R. E. Gomory. Outline of an algorithm for integer solutions to linear programs. *Bulletin of the American Mathematical Society* 64:275–278, 1958.
62. R. L. Graham, M. Grötschel, L. Lovász (eds). *Handbook of Combinatorics* (2 volumes). Elsevier, 1995.
63. J. E. Green. A new short proof of Kneser's conjecture. *Amer. Math. Monthly* 109:918–920, 2002.
64. P. Gritzmann, V. Klee. Computational convexity. In *Handbook of Discrete and Computational Geometry*, edited by J. E. Goodman et al. CRC Press, Boca Raton, FL., pp. 491–515, 1997.
65. M. Grötschel, G. O. H. Katona (eds.). *Building Bridges: Between Mathematics and Computer Science*. Springer, 2008.
66. M. Grötschel, L. Lovász, A. Schrijver. The ellipsoid method and its consequences in combinatorial optimization. *Combinatorica*, 1:169–197, 1981.
67. M. Grötschel, L. Lovász, A. Schrijver. Polynomial algorithms for perfect graphs. *Annals of Discrete Math.* 21:325–356, 1984.
68. M. Grötschel, L. Lovász, A. Schrijver. Relaxations of vertex packing. *J. Combin. Theory B* 40:330–343, 1986.
69. M. Grötschel, L. Lovász, A. Schrijver. *Geometric Algorithms and Combinatorial Optimization*. Springer, Berlin, 1988.

70. A. Grzesik, D. Král, L. M. Lovász. Elusive extremal graphs. *Proc. London Math. Soc.* 121:1685–1736, 2020.
71. A. Gyarfás, T. Jensen, M. Stiebitz. On graphs with strongly independent color classes. *J. Graph Theory* 46(1):1–14, 2004.
72. N. J. A. Harvey, J. Vondrák. Algorithmic proof of the Lovász Local Lemma via resampling oracles. *SIAM J. Comp* 49(2):394–428, 2020.
73. H. Hatami, J. Hladký, D. Král, S. Norine, A. Razborov. Non-three-colourable common graphs exist. *Combinatorics Probability and Computing* 21(5):734–742, 2012.
74. H. Hatami, S. Norin. Undecidability of linear inequalities in graph homomorphism densities. *J. Am. Math. Soc.* 24(2):547–565, 2011.
75. K. He, Q. Li, X. Sun, J. Zhang. Quantum Lovász Local Lemma: Shearer's bound is tight. In *Proceedings of the 51st Annual ACM Symposium on Theory of Computing (STOC '19), Phoenix, AZ, USA 23–26 June 2019*, 461–472, 2019.
76. P. Hell, J. Nešetřil. *Graphs and Homomorphisms*. Oxford University Press, 2004.
77. S. Hong. A linear Programming Approach for the Traveling Salesman Problem. Ph.D Thesis. The Johns Hopkins University, 1972.
78. S. Hoory, N. Linial, A. Wigderson. Expander graphs and their applications. *Bull. Amer. Math. Soc.* 43:439–561, 2006.
79. D. Hoover. *Relations on Probability Spaces and Arrays of Random Variables.* Institute for Advanced Study, Princeton, 1979.
80. S. Hougardy. Classes of perfect graphs. *Discrete Math.* 306(19–20):2529–2571, 2006.
81. S. Iwata, L. Fleischer, S. Fujishige. A combinatorial strongly polynomial algorithm for minimizing submodular functions. *Journal of the Association for Computing Machinery* 48:761–777, 2001.
82. M. Jerrum, S. Sinclair. Conductance and the rapid mixing property for Markov chains: The approximation of the permanent resolved. *Proc. 20th ACM STOC*, pp. 235–244, 1988.
83. H. Jia, A. Laddha, L. Aditi, T. L. Lee, S. S. Vempala. Reducing isotropy and volume to KLS: An $O(n^3\psi^2)$ volume algorithm. Preprint. arXiv:2008.02146v2, 2022.
84. D. Y. Kang, T. Kelly, D. Kühn, A. Methuku, D. Osthus. A proof of the Erdős–Faber–Lovász conjecture. Preprint. arXiv:2101.04698v3, 2023.
85. R. Kannan, L. Lovász, M. Simonovits. Random walks and an $O^*(n^5)$ volume algorithm for convex bodies. *Random Structures and Algorithms* 11:1–50, 1997.
86. R. M. Karp. Reducibility among combinatorial problems. In *Complexity of Computer Computations* (Proceedings of a symposium on the Complexity of Computer Computations, IBM Thomas J. Watson Research Center, Yorktown Heights, New York, 1972; edited by R. E. Miller, J. W. Thatcher), Plenum Press, New York, pp. 85–103, 1972.
87. L. G. Khachiyan. Polinomialnyĭ algoritm v lineĭnom programmirovanii [Russian], *Doklady Akademii Nauk SSSR* 244:1093–1096, 1979. [English translation: A polynomial algorithm in linear programming, *Soviet Mathematics Doklady* 20:191–194, 1979]
88. L. G. Khachiyan. Complexity of polytope volume computation. In *New Trends in Discrete and Computational Geometry*, edited by J. Pach, Springer, pp. 91–101, 1993.
89. M. Kneser. Aufgabe 360. *Jahresbericht der DMV* 58(2):27, 1955.
90. I. Kříž. A hypergraph – free construction of highly chromatic graphs without short cycles. *Combinatorica* 9(2):227–229, 1989.
91. M. Laczkovich. Random walk in and around mathematics – Interview with László Lovász. *Magyar Tudomány Hung. Sci.* 182:1108–1123, 2021.
92. J. C. Lagarias, A. M. Odlyzko. Solving low-density subset sum problems. *J. Assoc. Comput. Mach.* 32:229–246, 1985.
93. J. Lawrence. Polytope volume computation. *Math. Comput.* 57:259–271, 1991.
94. A. K. Lenstra, H. W. Lenstra, L. Lovász. Factoring polynomials with rational coefficients. *Mathematische Annalen* 261 (4):515–534, 1982.
95. H. W. Lenstra, Jr. Integer programming with a fixed number of variables. *Mathematics of Operations Research* 8:538–548, 1983.
96. L. A. Levin. Universal search problems. *Problemy Peredachi Informatsii* 9(3):115–116, 1973. (English translation Problems of Information Transmission 9(3):265–266, 1973.)

97. L. Lovász. Operations with structures. *Acta Math. Hung.* 18:321–328, 1967.
98. L. Lovász. On chromatic number of finite set-systems. *Acta Math. Hung.* 19:59–67, 1968.
99. L. Lovász. On the cancelation law among finite relational structures. *Acta Math. Hung.* 1:145–156, 1971.
100. L. Lovász. A characterization of perfect graphs. *J. Comb. Theory* 13:95–98, 1972.
101. L. Lovász. Normal hypergraphs and the perfect graph conjecture. *Discrete Math.* 2:253–267, 1972.
102. L. Lovász. Direct product in locally finite categories. *Acta Sci. Math. Szeged* 23:319–322, 1972.
103. L. Lovász. A note on the line reconstruction problem. *J. Comb. Theory* 13:309–310, 1972.
104. L. Lovász. Flats in matroids and geometric graphs. In *Combinatorial Surveys*. Proc. 6 British Comb. Conf. Academic Press, pages 45–86, 1977.
105. L. Lovász. Kneser's Conjecture, chromatic number, and homotopy. *J. Comb. Theory A* 25:319–324, 1978.
106. L. Lovász. On the Shannon capacity of graphs. *IEEE Trans. Inform. Theory* 25:1–7, 1979.
107. L. Lovász. *Combinatorial Problems and Exercises.* North Holland, 1979.
108. L. Lovász. Graph theory and integer programming. In *Discrete Optimization I* (Proceedings Advanced Research Institute on Discrete Optimization and Systems Applications and Discrete Optimization Symposium, Banff, Alta, and Vancouver, B.C., Canada, 1977; edited by P. L. Hammer, E. L. Johnson, B. H. Korte) [Annals of Discrete Mathematics 4], North-Holland, Amsterdam, pp. 141–158, 1979.
109. L. Lovász. Matroid matching and some applications, *J. Comb. Theory* B 28:208–236, 1980.
110. L. Lovász. Submodular functions and convexity. In *Mathematical Programming: The State of the Art*, edited by A. Bachem, M. Grötschel, B. Korte, Springer, pp. 235–257, 1983.
111. L. Lovász. How to compute the volume? *Jber. d. Dt. Math.-Vereinigung*, Jubiläumstagung 1990, B. G. Teubner, Stuttgart, pages 138–151, 1992.
112. L. Lovász. *Large Networks and Graph Limits.* Amer. Math. Soc., 2012.
113. L. Lovász. *Graphs and Geometry*. Amer. Math. Soc., 2019.
114. L. Lovász, J. Nešetřil, A. Pultr. On a product dimension of graphs. *J. Comb. Theory B* 29:47–67, 1980.
115. L. Lovász, M. Plummer. *Matching Theory*. North Holland, 1986.
116. L. Lovász, A. Schrijver. Cones of matrices and set-functions, and 0-1 optimization. *SIAM J. Optim.* 1:166–190, 1991.
117. L. Lovász, A. Schrijver. A Borsuk theorem for antipodal links and a spectral characterization of linklessly embeddable graphs. *Proceedings of the Amer. Math. Soc.* 126:1275–1285, 1998.
118. L. Lovász, L. Schrijver. Semidefinite functions on categories. *Electron. J. Combin.* 16(2), 2009.
119. L. Lovász, L. Schrijver. Dual graph homomorphism functions. *J. Comb. Theory A* 117:216–222, 2010.
120. L. Lovász, M. Simonovits. The mixing rate of Markov chains, an isoperimetric inequality, and computing the volume. *Proc. 31st IEEE Annual Symp. on Found. of Comp. Sci*, pp. 346–354, 1990.
121. L. Lovász, M. Simonovits. Random walks in a convex body and an improved volume algorithm. *Random Struct. and Algorithms* 4:359–412, 1993.
122. L. Lovász, B. Szegedy. Limits for dense graph sequences. *J. Combin. Theory B* 96:933–957, 2006.
123. L. Lovász, B. Szegedy. Szemerédi's Lemma for the analyst. *Geom.func. Anal.* 17:252–270, 2007.
124. L. Lovász, S. S. Vempala. Simulated annealing in convex bodies and an $O^*(n^4)$ volume algorithm. *Journal of Computer and System Sciences* 72:392–417, 2006.
125. A. Lubotzky, R. Phillips, P. Sarnak. Ramanujan graphs. *Combinatorica* 8:261–277, 1988.
126. A. Marcus, G. Tardos. Excluded permutation matrices and the Stanley–Wilf conjecture. *J. Comb. Theory A*, 107:153–223, 2004.
127. G. A. Margulis. Explicit construction of concentratoro. *Probl. Pered. Inform.* 9:71–80, 1973.

128. J. Matoušek. *Using the Borsuk–Ulam Theorem: Lectures on Topological Methods in Combinatorics and Geometry.* Springer, 2003.
129. J. Matoušek. *Thirty-Three Miniatures.* American Mathematical Society, 2010.
130. R. McKenzie. Cardinal multiplication of structures with a reflexive relation. *Fundamenta Matematicae* 70:59–101, 1971.
131. D. Micciancio, S. Goldwasser. *Complexity of Lattice Problems: A Cryptographic Perspective.* Springer International Series in Engineering and Computer Science 671. Springer, Boston, 2002.
132. V. Müller. The edge reconstruction hypothesis is true for graphs with more than $n \log n$ edges. *J. Comb. Theory B* 22:281–283, 1977.
133. J. Nešetřil. A combinatorial classics – Sparse graphs with high chromatic number. In *Erdős Centenial,* Springer, pp. 383–407, 2013.
134. J. Nešetřil, P. Ossona de Mendez. Unified approach to structural limits and limits of graphs with bounded tree depth. *Memoires Amer. Math. Soc.* 263(1272), 2020.
135. J. Nešetřil, V. Rödl. A short proof of the existence of highly chromatic hypergraphs without short cycles. *J. Comb. Th.* B 27(2):225–227, 1979.
136. P. Nguyen, B. Vallée (eds). *The LLL Algorithm: Information Security and Cryptography.* Springer, Berlin, 2010.
137. A. M. Odlyzko, H. J. J. te Riele. Disproof of the Mertens conjecture. *Journal für die Reine und Angewandte Mathematik* 357:138–160, 1985.
138. M. W. Padberg, M. R. Rao, Odd minimum cut-sets and b-matchings. *Mathematics of Operations Research* 7:67–80, 1982.
139. A. Pultr. Isomorphism types of objects in categories determined by numbers of morphisms. *Acta Sci. Math. Szeged* 35:155–160, 1973.
140. A. A. Razborov. Flag algebras. *The Journal of Symbolic Logic* 72(4):1239–1282, 2007.
141. O. Regev. Lattice-based cryptography. In Dwork, C. (eds.), *Advances in Cryptology – CRYPTO 2006*. Lecture Notes in Computer Science, vol 4117, pp. 131–141, Springer, Berlin, 2006.
142. T. Rothvoss. The matching polytope has exponential extension complexity. *Journal of the ACM* 64(6):1–19, 2017.
143. A. Schrijver. A combinatorial algorithm minimizing submodular functions in strongly polynomial time. *Journal of Combinatorial Theory, Series B* 80:346–355, 2000.
144. A. Schrijver. *Combinatorial Optimization. Polyhedra and Efficiency* (3 volumes). Springer, Berlin 2003.
145. A. D. Scott, A. D. Sokal. The repulsive lattice gas, the independent-set polynomial and the Lovász Local Lemma. *J. Stat. Phys.* 118:1151–1261, 2005.
146. C. E. Shannon. The zero error capacity of a noisy channel, in: *1956 Symposium on Information Theory*, IRE Transactions on Information Theory IT-2: 8–19, 1956. [Reprinted in: *Claude Elwood Shannon – Collected Papers*, edited by N. J. A. Sloane, A. D. Wyner, IEEE Press, Piscataway, New Jersey, pp. 221–238, 1993]
147. Y. Shitov. Counterexample to Hedetniemi's conjecture. *Annals of Math.* 190(2):663–667, 2019.
148. A. F. Sidorenko. Inequalities for functionals generated by bipartite graphs. *Diskretnaya Matematika* (3):50–65, 1991. (Russian original, English translation: *Discrete Math. Appl.* 2(5):489–504, 1992.)
149. M. Simonovits. How to compute the volume in high dimension? *Math. Program.* Ser. B 97:337–374, 2003.
150. G. Simonyi, A. Zsbán. On topological relaxations of chromatic conjectures. *European Journal of Combinatoric* 31(8):2110–2119, 2010.
151. I. Smeets. The history of the LLL-algorithm. In *The LLL Algorithm: Information Security and Cryptography*, edited by P. Nguyen, B. Vallée, Springer, Berlin, pp. 1–17, 2010.
152. J. Spencer. Ramsey's theorem – a new lower bound. *J. Comb. Theory A* 18:108–115, 1975.
153. D. A. Spielman, S.-H. Teng. Spectral sparsification of graphs. *SIAM J. on Comp.* 40(4):981–1025, 2011.

154. M. Szegedy. The Lovász Local Lemma – A survey. In *Proceedings of the 8th International Computer Science Symposium in Russia (CSR 2013)* Ekaterinburg, Russia, 25–29 June 2013:1–11, 2013.
155. T. C. Tao. Szemerédi's regularity lemma revisited. *Contrib. Discrete Math.* 1:8–28, 2006.
156. C. Tardif. The chromatic number of the product of 5-chromatic graphs can be 4. Preprint, 2022.
157. S. M. Ulam. *A Collection of Mathematical Problems*, page 29, Wiley Interscience, New York, 1960.
158. A. Wigderson. *Mathematics and Computation.* Princeton, NJ: Princeton University Press, 2019.
159. M. Wrochna. On inverse powers of graphs and topological implications of Hedetniemi's conjecture. *J. Combin. Theory B* 139: 267–295,2019.
160. M. Wrochna. A note on hardness of promise hypergraph coloring. Preprint, arXiv:2205.14719v1, 2022.
161. X. Zhu. Relatively small counterexamples to Hedetniemi conjecture. *J. Comb. Th B* 146:141–150, 2021.

On the works of Avi Wigderson

Boaz Barak, Yael Kalai, Ran Raz, Salil Vadhan and Nisheeth K. Vishnoi

Abstract This is an overview of some of the works of Avi Wigderson, 2021 Abel prize laureate. Wigderson's contributions span many fields of computer science and mathematics. In this survey we focus on four subfields: *cryptography*, *pseudorandomness*, *computational complexity lower bounds*, and the theory of *optimization over symmetric manifolds*. Even within those fields, we are not able to mention all of Wigderson's results, let alone cover them in full detail. However, we attempt to give a broad view of each field, as well as describe how Wigderson's papers have answered central questions, made key definitions, forged unexpected connections, or otherwise made lasting changes to our ways of thinking in that field.

Boaz Barak and Salil Vadhan
School of Engineering and Applied Sciences, Harvard University.

Yael Kalai
Microsoft Research and MIT

Ran Raz
Department of Computer Science, Princeton University.

Nisheeth K. Vishnoi
Department of Computer Science, Yale University.

H. Holden, R. Piene (eds.), *The Abel Prize 2018-2022*, The Abel Prize,
https://doi.org/10.1007/978-3-031-33973-8_20

Contents

1 Introduction

In a career that has spanned more than 40 years, Wigderson has resolved long-standing open problems, made definitions that shaped entire fields, built unexpected bridges between different areas, and introduced ideas and techniques that inspired generations of researchers. A recurring theme in Wigderson's work has been uncovering the deep connections between computer science and mathematics. His papers have both demonstrated unexpected applications of diverse mathematical areas to questions in computer science, and shown how to use theoretical computer science insights to solve problems in pure mathematics. Many of these beautiful connections are surveyed in Wigderson's own book [284].

In writing this chapter, we were faced with a daunting task. Wigderson's body of work is so broad and deep that it is impossible to do it justice in a single chapter, or even in a single book. Thus, we chose to focus on a few subfields and, within those, describe only some of Wigderson's central contributions to these fields.

In Section 2 we discuss Wigderson's contribution to *cryptography*. As we describe there, during the second half of the 20th century, cryptography underwent multiple revolutions. Cryptography transformed from a practical art focused on "secret writing" to a science that protects not only communication but also *computation*, and provides the underpinning for our digital society. Wigderson's works have been crucial to this revolution, vastly extending its reach through constructions of objects such as *zero-knowledge proofs* and *multi-party secure computation*.

In Section 3, we discuss Wigderson's contribution to the field of *pseudorandomness*. One of the great intellectual achievements of computer science and mathematics alike has been the realization that many deterministic processes can still behave in "random-like" or *pseudorandom* manner. Wigderson has led the field in understanding and pursuing the deep implications of pseudorandomness for problems in computational complexity, such as the power of randomized algorithms and circuit lower bounds, and in developing the theory and explicit constructions of "pseudorandom objects" like expander graphs and randomness extractors. Wigderson's work in this field used mathematical tools from combinatorics, number theory, algebra, and information theory to answer computer science questions, and has applied computer science abstractions and intuitions to obtain new results in mathematics, such as explicit constructions of Ramsey graphs.

Section 4 covers Wigderson's contribution to the great challenge of theoretical computer science: proving *lower bounds* on the computational resources needed to achieve computational tasks. Algorithms to solve computational problems have transformed the world and our lives, but for the vast majority of interesting computational tasks, we do not know whether our current algorithms are *optimal* or whether they can be dramatically improved. To demonstrate optimality, one needs to prove such *lower bounds*, and this task has turned out to be exceedingly difficult, with the famous P vs. NP question being but one example. While the task is difficult, there has been some progress in it, specifically in proving lower bounds for restricted (but still very useful and interesting) computational models. Wigderson has been a central contributor to this enterprise.

Section 5 covers a line of work by Wigderson and his co-authors on developing and analyzing continuous optimization algorithms for various problems in computational complexity theory, mathematics, and physics. Continuous optimization is a cornerstone of science and engineering. There is a very successful theory and practice of convex optimization. However, progress in the area of *nonconvex* optimization has been hard and sparse, despite a plethora of nonconvex optimization problems in the area of machine learning. Wigderson and his co-authors, in their attempt to analyze some nonconvex optimization problems important in complexity theory, realized that these are no ordinary nonconvex problems – nonconvexity arises because the objective function is invariant under certain group actions. This insight led them to synthesize tools from invariant theory, representation theory, and optimization to develop a quantitative theory of optimization over Riemannian manifolds that arise from continuous symmetries of noncommutative matrix groups. Moreover, this pursuit revealed connections with and applications to a host of disparate problems in mathematics and physics.

All of us are grateful for having this opportunity to revisit and celebrate Wigderson's work. More than anything, we feel lucky to have had the joy and privilege of knowing Avi as a mentor, colleague, collaborator, and friend.

2 Cryptography

Cryptography has been used for thousands of years, going back to ancient Egypt, Sumeria, and Greece. However, throughout the vast majority of that time, it had two major limitations. First, there was no formal analysis of cryptographic schemes, leading to a "cat and mouse" game in which ciphers are continuously designed and then broken, leading Edgar Allan Poe to say in 1847 that *"Human ingenuity cannot concoct a cipher which human ingenuity cannot resolve."* Second, cryptography was synonymous with "secret writing": the design of schemes that enable two parties that share some secret information (i.e., *secret key*) to communicate by using encryption and decryption.

In the second half of the 20th century, cryptography broke out of these two limitations. First, starting with the work of Shannon [250], cryptography was placed on solid mathematical foundations. Second, with their invention of *public key cryptography*, Diffie and Hellman [78] ushered in a new era where cryptography extended far beyond secret writing. However, neither Shannon nor Diffie and Hellman could imagine how far cryptography would grow. First, in almost all settings, analyzing cryptographic schemes required going beyond the information-theoretic methods of Shannon, and to use *computational complexity* as a basis. Second, in the 1980s, cryptography was extended to protect not only *communication* but also *computation*, with a crowning achievement being "secure multiparty computation" protocols that allow any number of parties to compute arbitrary functions on their secret inputs, controlling precisely what information would be revealed and to whom.

Avi Wigderson played a key role in these developments. He was instrumental in mapping out the computational assumptions required for many cryptography tools and proving the central feasibility result for secure multiparty computation.

In this section, we survey some of Wigderson's contributions to cryptography, focusing on two central themes.

1. Building cryptographic schemes that are secure under *computational assumptions* (which can be viewed as stronger variants of the famous $\mathsf{P} \neq \mathsf{NP}$ conjecture). This line of works is covered in Section 2.1.
2. Building cryptographic schemes that are proven to be secure *unconditionally*, without relying on any computational assumptions, but rather on certain environmental conditions such as a trusted majority of parties. This field is sometimes known as *information-theoretically secure cryptography*, and some of Wigderson's contributions to it are covered in Section 2.2.

Two objects play a central role in both fields: *zero-knowledge proofs*, and *multi-party secure computation*. These are the foundational tools for extending cryptography from only securing *communication* to securing *computation*. Wigderson has made seminal contributions to constructing these objects in both the computational and information-theoretic regimes, enabling many follow-up works that used these tools to achieve a vast range of cryptographic applications. The description below is informal in parts, and many proofs are omitted. However they can be found in Goldreich's excellent textbook [108, 109,]. See also the recent text [272] for more on information-theoretic cryptography.

2.1 Cryptography under computational assumptions

Cryptography is intimately connected to computational complexity. Indeed, achieving most cryptographic goals requires the existence of functions that are *computationally hard to compute*. The necessity of computational hardness for encryption was realized early on by Shannon [250]. However, in the early 1980s, researchers realized that computational hardness is *sufficient* for achieving applications that extend far beyond encryption. Avi Wigderson played a key role in this revolution that vastly expanded the domain of cryptography beyond its classical goals of protecting the confidentiality and authenticity of communications.

2.1.1 Zero knowledge proofs for all languages in NP

Zero-knowledge proofs achieve the seemingly paradoxical notion of convincing a party (known as the "verifier") that a particular statement X is true, without giving any information to the verifier as to *why* that statement is true. For example, a prover that knows the factorization of the number $N = 1{,}013{,}883{,}390{,}263{,}903$ as $N = 32{,}722{,}259 \times 30{,}984{,}517$ can easily prove the statement "N is composite and has a factor with least-significant digit 7" by providing the factorization, but a zero-knowledge proof allows them to prove this fact *without* revealing the prime factors.[1]

In 1982, Goldwasser, Micali and Rackoff [116] defined the notion of zero knowledge proofs, and gave such proofs for particular examples of languages, such as quadratic residuosity, for which no efficient algorithm is known. To do this, [116] needed to extend the notion of proofs to include *interaction*— the prover and verifier exchange messages rather than just a static piece of text— and *randomization*— the verifier's algorithm is randomized, and it is only convinced of the proof validity with high probability.

[1] Simply proving that a number N is composite can be done easily, since the verifier can use an efficient *primality testing* algorithm [2], and so even an empty proof suffices. Also, current classical (i.e., non-quantum) algorithms can be used to efficiently factor numbers with up to a few hundred digits [177, 49].

We now formally define interactive proofs in general and zero-knowledge proofs in particular. We restrict our attention to proof systems for languages in NP, which is the case of most practical relevance in cryptography. Recall that the class NP consists of all languages for which membership can be *efficiently verified*. Formally, we define an NP-relation to be a relation $R \subseteq \{0,1\}^* \times \{0,1\}^*$ such that there is a polynomial-time algorithm to check whether a pair (x,y) is in R and such that there is a polynomial p such that for every pair $(x,y) \in R$, $|y| \le p(|x|)$. For a relation R, we define the *lanaguage* corresponding to R to be $L_R = \{x | (x,y) \in R\}$. The class NP is the set of all languages L such that $L = L_R$ for some NP-relation R. Given a string x, one can *prove* that x is in L by providing the string y such that $(x,y) \in R$. By the definition of a relation NP, the string y will be of length at most polynomial in $|x|$ and the membership of (x,y) in R can be verified efficiently.

An *interactive algorithm* is a (potentially randomized) algorithm that, given a current state s_i and the message received m_i, outputs the updated state s_{i+1} and the message it sends m_{i+1}. An interaction between two interactive algorithms A and B given inputs a,b, respectively, proceeds in the natural way: the initial states of A and B are a,b, respectively. Then we iterate between each algorithm, computing its new state and message sent based on the previous state and message received from the other party. (For concreteness, we assume that A sends the first message, and thus gets the empty message and its initial state as input to compute it.) We say that an interactive algorithm A is *polynomial time* if there are some polynomials p,q such that, letting n be the length of A's input: (1) the total number of rounds A will interact with before halting, as well as the length of each message it sends and its internal state, is at most $p(n)$; (2) A computes every message and updated state using at most $q(n)$ operations.

Definition 2.1 (Interactive proofs for NP languages). Let R be an NP relation and $L = L_R$ the corresponding NP language. An (efficient) *interactive proof system* for R is a pair of interactive randomized polynomial-time algorithms P, V that satisfy the following properties:

Completeness: For every $(x,y) \in R$, if P gets x,y as input and V gets x as input, then at the end of the interaction, V will output 1 with probability 1.

Soundness: For every interactive algorithm P^* (even inefficient one) and every $x \notin L$, if V gets x as input and interacts with P^*, then the probability that V outputs 1 at the end of the interaction is at most $\frac{1}{2}$.[2]

The formal definition of Zero-Knowledge proofs uses the notion of a *simulator*. The idea is that to demonstrate that a verifier V did not learn anything from an interaction with a prover P, we show that V *could have simulated the interaction by itself.*

[2] The probability $1/2$ of error can be reduced to 2^{-k} by the standard trick of repeating the protocol k times sequentially.

Definition 2.2 (Zero-knowledge proofs). Let R be an NP relation and (P,V) be an efficient interactive proof system for R. We say that (P,V) is *zero knowledge* if the following holds. For every polynomial-time interactive algorithm V^* there exists a (non-interactive) randomized polynomial-time algorithm S^* such that: for every $(x,y) \in R$, if we let s^* be the random variable corresponding to V^*'s state, then s^* is *computationally indistinguishable* from the random variable $S^*(x)$.

Let $\{X_\alpha\}_{\alpha\in I}$ and $\{Y_\alpha\}_I$ be two parameterized collections of random variables, with $I \subseteq \{0,1\}^*$, and X_α, Y_α supported on strings of length at most polynomial in $|\alpha|$. We say that $\{X_\alpha\}$ and $\{Y_\alpha\}$ are *computationally indistinguishable*, denoted by $\{X_\alpha\} \approx_c \{Y_\alpha\}$, if there exists some function $\mu : \mathbb{N} \to (0,1]$ such that $\mu(n) = n^{-\omega(1)}$ (i.e., $\lim_{n\to\infty} \frac{\log \mu(n)}{\log n} = -\infty$) and such that for every Boolean circuit C_α of size at most $1/\mu(|\alpha|)$, $|\Pr[C_\alpha(X_\alpha) = 1] - \Pr[C_\alpha(Y_\alpha) = 1]| < \mu(|\alpha|)$. We often omit the subscript α when it is clear from context and so use the notation $X_\alpha \approx_c Y_\alpha$ or simply $X \approx_c Y$. For example, the condition of Definition 2.2 is that there is some function $\mu : \mathbb{N} \to [0,1]$ such that $\mu(n) = n^{-\omega(1)}$ and such that for every Boolean circuit C of size $\leq 1/\mu(|x|)$, $|\Pr[C(s^*) = 1] - \Pr[C(S^*(x)) = 1]| < \mu(|x|)$.

At the time of Goldwasser et al's result, it was not clear that zero-knowledge proofs are not a mere "curiosity" restricted to very specific examples. (Indeed, the [116] paper famously took three years before it was accepted for publication.) In Wigderson's work with Goldreich and Micali [112] they showed that this is decidedly not the case. Rather, they showed that (under standard cryptographic assumptions) *every* language in NP has a zero-knowledge proof.

One way to define NP is that it consists of languages L such that the membership of a string x in L can be proven by an efficient mathematical proof (i.e., a piece of text at most polynomially long in $|x|$, which can be verified in polynomial time). Hence [112]'s result can be thought of as saying that if a statement x has an efficient proof at all, then it also has an efficient proof in which the verifier learns nothing except that the statement is true.[3] The way that [112] proved their theorem was ingenious. They used the celebrated Cook–Levin Theorem, which is typically considered as a *negative* or *impossibility* result, to show a positive result.

We now describe their protocol. The Cook–Levin Theorem says that there are concrete problems that are NP-*complete* in the sense that any other problem in NP reduces to them. One example of an NP-complete language is *three coloring* or $3COL$, which is defined as follows. For a graph $G = (V,E)$, $G \in 3COL$ if and only if there exists $\chi : V \to \{1,2,3\}$ such that for every $\{u,v\} \in E$, $\chi(u) \neq \chi(v)$. A classical result is the following:

---

[3] This way of phrasing it is a bit of a cheat, since the first instance of "proof" corresponds to a standard mathematical proof— a static deterministically-verifiable piece of text— while the second one corresponds to the extended notion of [116], which includes interaction and randomness. A follow-up work [37] extended this by showing that everything that can be proven by a randomized interactive proof also has such a proof which is zero knowledge.

Theorem 2.3 (Cook–Levin–Karp [73, 178, 158]). *3COL is* NP*-complete. That is, for every* $L \in \mathsf{NP}$ *there exists a polynomial-time reduction* $r : \{0,1\}^* \to \{0,1\}^*$ *such that for every* $x \in \{0,1\}^*$,

$$x \in L \Leftrightarrow r(x) \in 3COL\,.$$

Moreover, if R is the NP*-relation corresponding to L, there exists polynomial-time algorithms* r', r'' *such that*

1. *For every* $(x,y) \in R$, $r'(x,y)$ *is a valid 3-coloring for the graph* $r(x)$.
2. *For every* $x \in \{0,1\}^*$ *and* $G = r(X)$, *if* χ *is a valid 3-coloring for G then* $(x, r''(G,\chi)) \in R$.

The "moreover" part of Theorem 2.3 was already implicit in the classical works of [73, 178, 158], and was explicitly discussed by Levin (which is why a triple (r, r', r'') as above is sometimes known as a "Levin reduction"). Wigderson and his coauthors used this insight to show that in order to give a zero-knowledge proof system for all $L \in \mathsf{NP}$, it suffices to give a zero-knowledge proof system for $3COL$. Specifically, for every NP language L, let r'_L, r''_L be the reductions from L to $3COL$ as in Theorem 2.3. Given a zero-knowledge protocol (P_{3COL}, V_{3COL}) for $3COL$ we can obtain a protocol (P_L, V_L) as follows:

- Verifier and prover get x as input, and the prover gets in addition y such that $(x,y) \in R_L$
- Verifier and prover compute $G = r(x)$ and prover computes $\chi = r'(x,y)$.
- Verifier and prover run the protocol (P_{3COL}, V_{3COL}) with inputs (G,χ) and G respectively.

2.1.2 Computationally secure multiparty computation

Obtaining a zero-knowledge proof system for every problem in NP is an intellectually satisfying result on its own merits. But does it have further applications? In another work of Wigderson with Goldreich and Micali [113], they showed that the answer is a resounding *yes*. They introduced a general technique to use zero-knowledge proofs as a way to *compile* protocols that achieve a very weak form of security into ones that achieve a very strong one. Using their technique, [113] proved what is arguably "The Fundamental Theorem of Cryptography"— a protocol for *secure multiparty computation.*

Secure multiparty computation (MPC) is a vast generalization of many tasks in cryptography, including encryption, electronic voting, voting, privacy-preserving data mining, and more. Full proofs and even the precise definition of MPC with all its variants is beyond the scope of this section. Lindell's survey [183] gives an excellent introduction, while the books [109, 74] go into more detail.

The setup is that there are n parties holding private inputs $x_1 \in \mathscr{X}_1, \ldots, x_n \in \mathscr{X}_n$, and they wish to compute a (potentially probabilistic) map

$$F : \mathscr{X}_1 \times \cdots \times X_n \to \mathscr{Y}_1 \times \cdots \times \mathscr{Y}_n$$

such that (roughly speaking) the following properties hold:

Completeness: Every party i will learn the value y_i where $(y_1, \ldots, y_n) = F(y_1, \ldots, y_n)$.

Privacy: A party i will not learn anything else apart from y_i about the private inputs of the other parties. More generally, every adversary that controls some set $A \subseteq [n]$ of parties will not learn more about the private inputs $\{x_i\}_{i \notin A}$ of the other parties beyond what could be derived from the outputs $\{y_i\}_{i \in A}$.

Soundness: An adversary that controls A as above cannot modify the outputs y_i of $i \notin A$ beyond its choice of the inputs $\{x_i\}_{i \in A}$.[4]

Up to considerations of computational efficiency, as well as allowing for interactive communication between parties, we can cast almost any cryptography problem as an instance of MPC. For example, the encryption task can be thought of as computing the function $F(x, \varnothing) = (\varnothing, x)$ where $\varnothing$ is the "empty" input/output. That is, computing F corresponds to ensuring that the second party learns x and that no one learns anything else. Conducting an auction could correspond to computing the function $F(x_1, \ldots, x_n) = (y_1, \ldots, y_n)$ where $y_i = 1_{i = \arg\max\{x_i\}}$. (That is, each party only learns whether or not they were the highest bidder.)

Yao [290] gave a version of MPC that was restricted in two ways. First, Yao's protocol was only for two parties. Second, and more importantly, Yao's protocol assumed a very restricted (and unrealistic) adversary: one that follows precisely the protocol's instructions but tries to extract information from the communication it is involved in. Such adversaries are known in cryptographic parlance as *passive* or *honest-but-curious*. Since in general, we have no reason to expect attackers to obey our protocol's instructions, honest-but-curious is not a realistic model for security.

Wigderson's work [113] solved both issues. First, they gave a general MPC protocol for n parties in the hones-but-curious model. Second, they used zero-knowledge proofs to provide a general transformation or "compiler" from protocols that are only secure against honest-but-curious adversaries into ones that are secure against general (also known as *malicious*) adversaries. Since their work, the general paradigm of using zero-knowledge proofs to "boost" security from passive to active adversaries has found numerous uses in theory and practice.

The details of [113]'s protocol are complex, and we omit them here. However, some of the techniques are illustrated in Section 2.2, which describes a different multiparty secure computation protocol of Wigderson in the *information-theoretic* setting. In both cases, the general idea is that (1) we can describe a general function F as a *Boolean circuit*, which is a composition of simple *gates*, and (2) once we do so, we can achieve a secure computation protocol by performing a gate-by-gate

---

[4] The adversary might also be able to abort the protocol; we ignore this issue of aborts in this section, but it is discussed extensively in the literature.

computation of intermediate values that are "encrypted" in the sense that no party (or strict subset of parties) can recover them on its own.

Arguably, it is [113]'s honest-but-curious to malicious *compiler* which had had the most significant impact. The idea behind this compiler is simple yet ingenious: Every party in the protocol will use zero-knowledge proofs to prove that it has followed the protocol's instructions. For example, suppose that at a given step in the protocol, the party i has private input x_i, and has received messages $m_1,\ldots,m_t$. Suppose that according to the protocol's instructions, the party should compute its next message as

$$m_{t+1} = \Pi_i(x_i, m_1, \ldots, m_t) \quad (1)$$

where Π_i is some known polynomial-time function that is specified by the protocol. A simple way for i to convince the other parties that it computed m_{t+1} correctly is to reveal all the inputs used in the computation, including the private input x_i. But that would, of course, violate i's privacy. However, the statement "there exists x_i that satisfies (1)" is an NP statement. Hence, it can be proved in *zero knowledge*, and so in a way that does not reveal x_i.

While this is the general idea, implementing it involved additional complications, including ensuring consistency (that the same private input x_i is used in all messages), dealing with randomized protocols (that are inherent to cryptography), and more. Using tools such as *commitment schemes* and *coin-tossing protocols*, [113] overcame those obstacles and proved that (under standard cryptographic assumptions), there exists a secure multiparty computation protocol for *every* polynomial time (potentially probabilistic) map $F : \mathscr{X}_1 \times \cdots \times X_n \to \mathscr{Y}_1 \times \cdots \times \mathscr{Y}_n$. This is one of the most fundamental theorems in all of cryptography and shows that if we are willing to allow for (polynomial-time) computation and communication overhead, *every* protocol problem can be solved. Although the road from such a theoretical proof of existence to practical constructions is long and arduous, the results of [113], as well as the techniques they introduced, served as guiding lights for theorists and practitioners alike.

2.2 Information-Theoretic Cryptography

In the previous section we showed how to construct zero-knowledge interactive proofs for all of NP, and how to use them to construct secure multi-party computation protocols. These works [112, 113] rely on cryptographic assumptions (such as the existence of a one-way function) and assume that the malicious parties are computationally bounded and cannot break the underlying cryptographic assumption. In particular, these protocols do not offer *everlasting security*. Namely, even if during the execution of the protocol the parties did not learn any information (beyond the validity of the statement or the output of the computation), if many years later the computers become stronger and manage to break the underlying cryptographic assumption, then at that point information can be leaked.

This motivates the question of whether we can obtain everlasting security. In other words, can we construct a zero-knowledge interactive proof and a secure MPC protocol that do not rely on cryptographic assumptions and provide security against all-powerful adversaries? Wigderson, together with Ostrovsky [214], proved that the answer is no for zero-knowledge interactive proofs. Namely, they showed that one-way functions are necessary for constructing zero-knowledge interactive proofs for all of NP (assuming $\mathsf{NP} \neq \mathsf{BPP}$). Similarly, Wigderson, together with Ben-Or and Goldwasser [39], showed that cryptographic assumptions are necessary for secure MPC (with the security guarantees as presented in Section 2.1).

For Wigderson and his coauthors, these lower bounds were nothing but an invitation to surpass them. In the case of zero-knowledge they managed to do so by changing the model in a clever and interesting way. Specifically, to obtain information-theoretic zero knowledge, Wigderson, together with Ben-Or, Goldwasser and Kilian [38], considered a new proof model: Rather than considering a verifier that is interacting with a single prover, they considered a verifier that is interacting with *two non-communicating provers*. They constructed information-theoretic zero-knowledge 2-prover interactive proofs. We elaborate on this construction and on the immense impact of the 2-prover model in Section 2.2.1. For the case of secure MPC, Wigderson, together with Ben-Or, Goldwasser [39] managed to get an information-theoretic security by restricting the fraction of parties the adversary is allowed to corrupt to be less than $1/2$ in the honest-but-curious setting and less than $1/3$ in the general malicious setting. They constructed an ingenious MPC protocol that is information-theoretic secure assuming the adversary is restricted as above, and assuming that each pair of parties is connected via a secure channel. This result is a true breakthrough and has served as a foundation for numerous subsequent works. We elaborate on it in Section 2.3.1.

2.2.1 Multi-Prover Zero-Knowledge Interactive Proofs

In the multi-prover interactive proof model, there are multiple provers who can co-operate and communicate between them to decide on a common optimal strategy *before* the interaction with the verifier starts. But, once they start to interact with the verifier, they can no longer interact nor can they see the messages exchanged between the verifier and the other provers.

Definition 2.4 (2-prover interactive proof). A 2-prover interactive proof for a language $L \in \mathsf{NP}$ consists of two provers (P_1, P_2) and a probabilistic polynomial time verifier V such that the verifier takes as input an instance x and each prover takes as input both x and a corresponding witness w. The verifier samples two queries (q_1, q_2) and sends q_i to prover P_i. Each prover P_i computes an answer $a_i = P_i(x, w, q_i)$ and sends a_i to the verifier V, who then outputs a verdict bit $b = V(x, q_1, q_2, a_1, a_2)$ indicating accept or reject.

The following two properties are required to hold:

- **Completeness.** For every $x \in L$ and any corresponding witness w s.t. $(x,w) \in R_L$, the verifier $V(x)$, who generates queries q_1 and q_2, accepts the answers $a_1 = P_1(x,w,q_1)$ and $a_2 = P_2(x,w,q_2)$, with probability 1.
- **Soundness.** For every $x \notin L$ and any two (computationally unbounded) cheating provers P_1^* and P_2^*, the probability that the verifier $V(x)$, who generates queries q_1 and q_2, accepts the answers $a_1 = P_1^*(x,q_1)$ and $a_2 = P_2^*(x,q_2)$, is at most $1/2$.

Theorem 2.5. *Every language $L \in \mathsf{NP}$ has a two-prover perfect zero-knowledge interactive proof-system.*

Proof idea. Recall that in Section 2.1.1 we showed how to construct a computational zero-knowledge interactive proof. The computational aspect follows from the use of a commitment scheme, whose hiding property is only computational.

The main new ingredient in the 2-prover zero-knowledge proof is an information-theoretic commitment. Recall that in the zero-knowledge proof presented in Section 2.1.1, in the first step the prover sends a commitment (in the case 3-coloring, this is a commitment to a legal coloring). To achieve zero-knowledge we need this commitment scheme to be hiding and for soundness this commitment must be binding. It is known that any commitment scheme that is statistically binding can only be *computationally hiding*, and this is precisely where the cryptographic hardness assumption comes in.

Wigderson et. al. [38] get around this barrier by constructing a commitment scheme that is both statistically binding and statistically hiding in a model where there are two committers, who are assumed to be non-communicating. In what follows, we present a slightly simplified version of their commitment scheme. We show how to commit to a single bit, and one can commit to arbitrarily many bits by repetition.

2.2.2 Bit commitment scheme in the 2-prover setting

In what follows we show how two provers P_1 and P_2 commit to a bit $b \in \{0,1\}$. First, before the protocol begins, they share a random string $w \leftarrow \{0,1\}^n$ and a single random bit $d \leftarrow \{0,1\}$ which are hidden from the verifier V. n controls the binding failure; taking $n = 1$ will guarantee binding with probability $1/2$ whereas taking a general n will guarantee binding with probability 2^{-n}.

Commitment phase:

1. The verifier V chooses a random string $r \leftarrow \{0,1\}^n$, and sends r to P_1. He sends nothing to P_2.
2. Prover P_1 sends $x = (d \cdot r) \oplus w$ and the prover P_2 sends $z = b \oplus d$.

Opening phase:

1. Prover P_2 sends to the verifier the committed bit b along with w.
2. The verifier V accepts if and only if $x = ((b \oplus z) \cdot r) \oplus w$.

Analysis. In what follows we argue that this commitment scheme is information-theoretic hiding and is also information-theoretical binding (assuming the two provers do not interact).

Hiding: Note that $(x,z) \equiv U_{n+1}$ where U_{n+1} denotes the uniform distribution over $n+1$ bits:

$$(x,z) = ((d \cdot r) \oplus w, b \oplus d) \equiv (U_n, b \oplus d) \equiv U_{n+1}.$$

Binding: We show that any pair of cheating provers can break the binding property with probability at most 2^{-n}. To this end, consider any pair of cheating provers P_1^* and P_2^* that send (x,z) to V, and that later P_2 can open successfully to both 0 using w_0 and to 1 using w_1. This means that

$$(z \cdot r) \oplus w_0 = (1 \oplus z) \cdot r \oplus w_1$$

which in turn implies that

$$w_0 \oplus w_1 = r,$$

and thus P_2^* can predict r, which should happen with probability 2^{-n}.

Information theoretic 2-prover zero-knowledge proof. Equipped with this information-theoretic commitment scheme, the 2-prover zero-knowledge construction is essentially the same as that presented in Section 2.1.1 while replacing the computational commitment scheme with the information-theoretic one presented above.

2.3 The Importance of the Multi-Prover Interactive Proof Model

As mention above, the original motivation of [38] for considering the model of multi-prover interactive proofs was for constructing statistical zero-knowledge proofs, a goal that they achieved with utter success. However, already in their original paper, [38] realized the potential power of such a proof model, and they posed the following open problem: *"It is interesting to consider what is the power of this new model solely with respect to language recognition."* Their intuition for why this model is powerful stems from the fact that *"the verifier can check the provers against each other."* In particular, the example they give is that of suspects that try to cover up a crime. It seems hard to cheat in a consistent manner.

Indeed, Babai et. al. [24] showed that this proof model is extremely powerful. In particular, they showed that the correctness of any (deterministic or non-deterministic) time-T computation can be verified in a 2-prover interactive proof model, where the communication complexity is only polylog(T) and where the running time of the verifier is only polylog(T) plus quasi-linear in the input length.

In particular, a polynomial time verifier can verify the correctness of exponentially long (deterministic or non-deterministic) computations.

The power of this proof model had groundbreaking consequences, leading to the notable PCP theorem [94, 23, 20, 19]. In particular, Fortnow et. al. [95] realized that if in the proof for a time-T computation the messages from the verifier to the provers are of size $O(\log T)$ (which they indeed in the construction of [24]) then each prover can generate a list consisting of its answers to *all* possible verifier messages, and this list will be of size $\mathsf{poly}(T)$. The lists of the two provers can be thought of as a list of size $\mathsf{poly}(T)$ that can be verified by reading only two blocks of size polylog(T).

This simple observation is spectacular. In the context of non-deterministic computations, it means that one can take any proof, and convert it into a new proof which is polynomially longer, but which can be verified by randomly reading only poly-logarithmically many bits of the proof. Indeed this observation created an immense splash in the theory community, leading to the PCP theorem, which says that for any NP language L, with a corresponding NP relation R, it holds that there is an efficient transformation that given any $(x,w) \in R$ generates a *probabilistically checkable proof* (PCP) π of size polynomial in the size of (x,w), such that if $x \notin L$ then after (randomly) reading only *3 bits* of π the verifier will reject the proof with probability $7/8$, and if $(x,w) \in L$ and the proof π was honestly generated then it is accepted by the probabilistic verifier with probability 1.

The 2-prover interactive proof model and the PCP theorem had numerous applications to the theory of computation and beyond. They form the foundation for all known hardness of approximation results, and are at the heart of all known succinct computationally sound proof system (also known as argument systems). Succinct arguments have played an important role in cryptography over the last 15 years. This significance is underscored by the hundreds of papers that have been published, the dozens of systems that have been built, and their deployment by numerous blockchain corporations, including prominent ones like Ethereum.

2.3.1 Information-Theoretic Secure Multi-Party Computation

Recall that in 2.1.2 we elaborated on the work of [113], which showed how a set of n parties can compute any function of their (secret) inputs securely, where the security guarantees *computational*, i.e., security holds against *computationally bounded* adversaries. The focus of this section is on obtaining information-theoretic security (also known as perfect security), where security holds even against *all powerful* adversaries. Indeed, Wigderson, together with Ben-Or and Goldwasser [39], showed that any function can be computed with *perfect security* assuming each pair of parties is connected with a private channel, as long as the malicious adversary controls less than $1/3$ of the parties. If the adversary is restricted to be semi-honest then security is guaranteed as long as the adversary controls a minority of the parties. Moreover, they showed that such a corruption rate is tight, both in the malicious setting and the semi-honest setting.

This result is truly remarkable. Indeed, it is a cornerstone in the field of secure multi-party computation, and has paved the way for a lot of subsequent work, making it a highly influential and a groundbreaking contribution. As is often the case in the literature, we refer to this protocol as the BGW protocol.

High-level overview of the BGW protocol. Suppose a set of n parties wish to securely compute a function f from n inputs to n outputs. Let C be an arithmetic circuit computing f. On a high level, the BGW protocol securely emulates the computation of C in a gate-by-gate manner, starting from the input gates all the way up to the output gates. More specifically, it proceeds with the following steps:

1. **Secret sharing of the inputs.** The protocol begins with the parties sharing their inputs with each other using a secure secret-sharing scheme [249]. If the adversary is assumed to be semi-honest, and thus is assumed to corrupt less than $n/2$ parties, then the parties share each bit of their secret input using the Shamir t-out-of-n secret-sharing scheme [249] with $t = \lceil n/2 \rceil - 1$. Such a scheme ensures that if at most t shares are revealed then no information about the secret is revealed, whereas $t+1$ shares can be efficiently combined to reveal the secret. We elaborate on Shamir's secret-sharing scheme below.
   If the adversary is malicious then one needs to use a *verifiable secret-sharing* (VSS) scheme. The first information-theoretically secure VSS scheme was constructed in the work of [39], and we elaborate on it towards the end of this section.
2. **Gate-by-gate emulation.** The parties then emulate the computation of each gate of the circuit, computing secret shares of the gate's output from the secret shares of the gate's inputs. As we shall see, the Shamir secret-sharing scheme, as well as the VSS scheme of [39], have the property that computing shares corresponding to addition gates can be done *locally*, without any interaction. Thus, the parties only interact in order to emulate the computation of multiplication gates. This step is the most involved part of the protocol, and we elaborate on it below.
3. **Output reconstruction.** Finally, the parties reconstruct the value of each output wire from the shares of the that wire. Namely, if an output wire belongs to party P_i then all the parties send party P_i their shares corresponding to the wire and P_i uses all these shares to reconstruct the output.

We next describe the BGW protocol in more detail. We first focus on the semi-honest setting, and only towards the end of the section we discuss the malicious setting. We start by recalling Shamir's secret-sharing scheme.

Shamir Secret Sharing Scheme. Suppose a party (often referred to as the dealer) wishes to share a secret input among n parties, with the guarantee that any $t+1$ of the parties can use their shares to efficiently reconstruct the secret and yet any t shares do not reveal any information about the secret. In what follows we assume for simplicity, and without loss of generality, that the secret is a single bit (and thus can be embedded in any finite field).

Let $\mathbb{F}$ be a finite field of size greater than n, let $\alpha_1, \ldots, \alpha_n$ arbitrary distinct non-zero elements in $\mathbb{F}$. In order to share a secret $s \in \mathbb{F}$ a random degree t polynomial $p(x) \in \mathbb{F}[x]$ is chosen such that $p(0) = s$. The share of party P_i is set to be $p(\alpha_i)$. By interpolation, given any $t+1$ points it is possible to reconstruct the polynomial p and compute the secret $s = p(0)$. Furthermore, since p is random subject to $p(0) = s$, and thus has t random coefficients, its values at any t or less of the α_i's give no information about the secret s.

Gate-by-Gate Emulation. We next show how to use the structure of Shamir's secret-sharing scheme to do the gate-by-gate emulation. The first observation is that addition gates can be computed locally. That is, given shares $p(\alpha_i)$ and $q(\alpha_i)$ of the two input wires corresponding to an addition gate, it holds that $r(\alpha_i) = p(\alpha_i) + q(\alpha_i)$ is a valid sharing of the output wire. This is due to the fact that the polynomial $r(x) = p(x) + q(x)$ has the same degree as both $p(x)$ and $q(x)$, and $r(0) = p(0) + q(0)$.

Remark 2.6. Note that if the function f is linear, and thus can be computed using a circuit that has only addition gates, then the gate-by-gate emulation step is completely non-interactive.

Regarding multiplication gates, a natural attempt is to compute the product of the shares, namely, party P_i computes $r(\alpha_i) = p(\alpha_i) \cdot q(\alpha_i)$. Indeed, the constant term is $r(0) = p(0) \cdot q(0)$, as desired. However, the degree of $r(x)$ becomes $2t$, as opposed to t. This is a problem since the reconstruction algorithm works as long as the polynomial used for the sharing is of degree at most t. We therefore need to reduce the degree of r down to t. To solve this, Wigderson and his coauthors devised a beautiful and elegant degree reduction protocol. This protocol also ensures that the new degree t polynomial is a *random* degree t polynomial with free coefficient $p(0) \cdot q(0)$. This is crucial for security, since if the polynomial is not random then t shares may reveal undesired information.

Specifically, the degree reduction protocol first randomizes the degree-$2t$ polynomial $r = p \cdot q$ so that it is uniformly distributed, and then it reduces its degree to t while maintaining its uniformity. Specifically, a multiplication gate is computed via the following protocol:

1. **Local multiplication.** Each party locally multiplies its input shares. Namely, party P_i computes $r(\alpha_i) = p(\alpha_i) \cdot q(\alpha_i)$.
2. **Randomizing the polynomial r.** Each party P_i generates a random degree $2t$ polynomial z_i such that $z_i(0) = 0$, and sends to each party P_j the share $z_i(\alpha_j)$. Then, each party P_i adds all the shares it received and the original share it computed to obtain
$$\sum_{j=1}^{n} z_j(\alpha_i) + r(\alpha_i).$$
The resulting shares define a random degree $2t$ polynomial R such that $R(0) = p(0) \cdot q(0)$.
3. **Degree reduction.** The parties run a private protocol where each party P_i converts its share $R(\alpha_i)$ into the share $R_{\mathsf{trunc}}(\alpha_i)$, where R_{trunc} is simply a truncation of the polynomial R to a degree t polynomial. Namely, if $R(x) = \sum_{j=0}^{2t} a_i x^i$ then $R_{\mathsf{trunc}}(x) = \sum_{j=0}^{t} a_i x^i$.

A priori it is not clear how a party P_i can compute $R_{\mathsf{trunc}}(\alpha_i)$ from $R(\alpha_i)$. Indeed, P_i cannot do this on its own, and needs the help of all other parties P_j, who have shares $R(\alpha_j)$. Note that this computation needs to be done in a private manner, which is the problem we are trying to solve in the first place, and thus it seems that we are back to square one! However, the magical observation of [39] is that this truncation function, which converts shares of R to shares of R_{trunc}, is *linear*. As mentioned in Remark 2.6, linear functions we already know how to compute securely since they do not require any multiplication gates! Thus, it remains to argue the linearly of this function, which is argued in the claim below.

Claim. There exists a fixed (public) matrix $A \in \mathbb{F}^{n\times n}$ such that for every degree $2t$ polynomial $R : \mathbb{F} \to \mathbb{F}$ and for every distinct non-zero elements $\alpha_1, \dots, \alpha_n \in \mathbb{F}$,

$$A \cdot (R(\alpha_1), \dots, R(\alpha_n))^T = (R_{\mathsf{trunc}}(\alpha_1), \dots, R_{\mathsf{trunc}}(\alpha_n))^T,$$

where R_{trunc} is defined as above.

Proof. Let $\mathrm{R} = (R_0, R_1, \dots, R_{2t}, 0, \dots, 0) \in \mathbb{F}^n$ denote the vector of coefficients of the polynomial R. Let V_α be the Vondermonde matrix corresponding to $\alpha = (\alpha_1, \dots, \alpha_n)$. Namely, for every $i, j \in [n]$, $V_\alpha(i, j) = \alpha_i^{j-1}$. Note that

$$V_\alpha \cdot \mathrm{R}^T = (R(\alpha_1), \dots, R(\alpha_n))^T$$

It is well known that the Vondermonde matrix V_α is invertible if $\alpha_1, \dots, \alpha_n \in \mathbb{F}$ are all distinct and non-zero. Therefore,

$$\mathrm{R}^T = V_\alpha^{-1} \cdot (R(\alpha_1), \dots, R(\alpha_n))^T. \tag{2}$$

Let P be the linear projection function that takes as input a vector $(a_1, \dots, a_n) \in \mathbb{F}^n$ and outputs $(a_1, \dots, a_{t+1}, 0 \dots, 0) \in \mathbb{F}^n$. Namely, in matrix representation, $P(i, j) = 1$ if $i = j$ and both are in $\{1, \dots, t+1\}$, and $P(i, j) = 0$ otherwise. Thus, denoting by $\mathrm{R}_{\mathsf{trunc}}$ the $t+1$ coefficients of the degree t polynomial R_{trunc} followed by zeros, i.e., $\mathrm{R}_{\mathsf{trunc}} = (R_0, R_1 \dots, R_t, 0, \dots, 0)$, by the definition of P_t and by Equation (2)

$$\mathrm{R}_{\mathsf{trunc}}^T = P \cdot V_\alpha^{-1} \cdot (R(\alpha_1), \dots, R(\alpha_n))^T.$$

This, together with the definition of V_α, implies that

$$(R_{\mathsf{trunc}}(\alpha_1), \dots, R_{\mathsf{trunc}}(\alpha_n))^T = V_\alpha \cdot P \cdot V_\alpha^{-1} \cdot (R(\alpha_1), \dots, R(\alpha_n))^T.$$

We thus conclude the proof of this claim by setting $A = V_\alpha \cdot P \cdot V_\alpha^{-1}$.

This concludes the description of the BGW protocol in the honest-but-curious setting, where the adversary is assumed to follow the protocol. We next show how Wigderson and his co-authors modify this protocol to obtain security against a *malicious* adversary who controls less than 1/3 of the parties and may deviate arbitrarily from the protocol. The main new tool is a *verifiable secret-sharing* scheme.

2.3.2 Verifiable Secret Sharing

A verifiable secret-sharing (VSS) scheme, originally defined by Chor et al. [68], is a secret-sharing scheme that is secure even in the presence of *malicious* adversaries. Recall that a secret-sharing scheme (with threshold t) is made up of two stages: A sharing stage and a reconstruction stage. In the sharing stage, the dealer shares a secret among the n parties so that any $t+1$ parties can efficiently reconstruct the secret from their shares, while any subset of t or fewer shares reveal no information about the secret. In the reconstruction stage, a set of $t+1$ or more parties reconstruct the secret. If we consider Shamir's secret-sharing scheme, much can go wrong if the dealer or some of the parties are malicious. Recall, that to share a secret s, the dealer is supposed to choose a random polynomial q of degree t with $q(0) = s$ and then hand each party P_i its share $q(\alpha_i)$. However, a malicious dealer can choose a polynomial of higher degree, and as a result different subsets of $t+1$ parties may reconstruct different values. Thus, the shared value is not well defined. In addition, in the reconstruction phase a corrupted party can provide an arbitrary malicious value instead of the prescribed value $q(\alpha_i)$, thus effectively changing the value of the reconstructed secret.

A verifiable secret-sharing scheme is aimed at solving precisely these issues. Chor et al. [68] constructed a VSS scheme with *computational* security, i.e., assuming the malicious parties are computationally bounded (and assuming the hardness of some computational problem). Wigerson with his coauthors [39] constructed an information-theoretically secure VSS scheme, which ensures security against *all powerful* adversaries, assuming that at most $t < n/3$ of the parties are corrupted. More specifically, the security guarantee is that the shares received by the honest parties are guaranteed to be $q(\alpha_i)$ for a well-defined degree-t polynomial q, even if the dealer is corrupted. To achieve this guarantee, the secret-sharing stage is followed by a verification stage which is an interactive stage where the parties "correct" their shares if needed. This correction protocol, which we elaborate on below, is extremely beautiful!

Given this security guarantee it is possible to use techniques from the field of error-correcting codes in order to reconstruct q (and thus $q(0) = s$) as long as $n-t$ correct shares are provided and $t < n/3$. This is due to the fact that Shamir's secret-sharing scheme when looked at in this context is exactly a Reed–Solomon code.

VSS **via bivariate polynomials.**

The VSS protocol of [39] consists of three stages.

1. **Secret sharing stage.** Loosely speaking, in this stage the dealer embeds the Shamir secret-sharing polynomial in a *bivariate* polynomial $S(x,y)$. Specifically, to share a secret $s \in \{0,1\}$ the dealer chooses a random bivariate polynomial $S(x,y)$ of degree t in each variable, such that $S(0,0) = s$. Note that $q(z) = S(0,z)$ is a polynomial corresponding to the Shamir secret-sharing scheme and the values $q(\alpha_1),...,q(\alpha_n)$ are the Shamir shares embedded into $S(x,y)$. Similarly, $p(z) = S(z,0)$ is a polynomial corresponding to the Shamir secret-sharing scheme and the values $p(\alpha_1),...,p(\alpha_n)$ are the Shamir shares embedded into $S(x,y)$. The

dealer sends each party P_i *two* univariate polynomials as shares; these polynomials are $f_i(x) = S(x, \alpha_i)$ and $g_i(y) = S(\alpha_i, y)$. The Shamir-share of party P_i is $f_i(0) = S(0, \alpha_i)$, and the polynomials f_i and g_i are given only for the sake of verification.

2. **Verification stage.** At this point the parties engage in an interactive verification protocol. First, each party P_i sends each party P_j the value $s_{i,j} = f_i(\alpha_j)$. Note that if the dealer is honest, then the elements $s_{1,j}, \ldots, s_{n,j}$ sent to party P_j should be the value of the polynomial g_j on $\alpha_1, \ldots, \alpha_n$, respectively. Each party P_j checks that indeed for every $i \in [n]$, $s_{i,j} = g_j(\alpha_i)$. If this is not the case, it broadcasts a request for the dealer to reveal $s_{i,j}$. If P_j has more than t requests then the dealer is clearly malicious, in which case P_j broadcasts a "complaint" thereby asking the dealer to reveal his private shares f_i and g_i. Finally, after the dealer broadcasts all the information requested, each party P_i checks that all the public and private information he received from the dealer are consistent. If P_i finds any inconsistencies he broadcasts a complaint thereby asking the dealer to reveal his private shares. If at this point more than t parties have asked to make their shares public, the dealer is clearly malicious and all the parties pick the default zero polynomial as the dealer's polynomial. Likewise, if the dealer did not answer all the broadcasted requests he is declared malicious. On the other hand, if t or less parties have complained then there are at least $t+1$ honest parties who are satisfied (this follows from the fact that $t < n/3$). The shares of these parties uniquely define a polynomial $S(x,y)$ and this polynomial conforms with all the information that was made public (otherwise one of these honest parties would have complained). In this case the complaining parties take the public information as their share.
3. **Reconstruction stage.** At this point each party sends its updated share $f_i(0)$, and the secret s is reconstructed by running the Reed–Solomon decoding algorithm.

Note that if the dealer is honest then no information about the shares of any honest party is revealed during the verification process. If however the dealer is malicious, we do not need to protect the privacy of his information, and the verification procedure ensures that all the honest parties values lie on some polynomial of degree t.

Gate-by-Gate Emulation in the Malicious Setting

As was done in the honest-but-curious setting, addition gates are computed locally by adding the corresponding shares, whereas computing multiplication gates is significantly more involved. Recall that in the honest-but-curious setting the multiplication step was done by multiplying the shares locally, thus obtaining shares of a degree-$2t$ polynomial. Then the parties rerandomized this polynomial and then truncated it. In the malicious setting, the rerandomization step needs to be made secure against malicious adversaries. In addition, to apply the degree reduction step, we need to argue that the truncation is a linear function, but for this we must make sure that the all the parties use as input to this function their correct point on the product polynomial $h(x) = f(x)g(x)$. To guarantee that this is indeed the case, error correcting codes are used yet again.

3 Pseudorandomness

A major theme in Wigderson's work is to understand the power of randomness in efficient computation, addressing questions such as:

- Can randomized algorithms solve problems much more efficiently than deterministic algorithms, or can every randomized algorithm be converted into a deterministic algorithm with only a small loss in efficiency?
- Can we give explicit, deterministic constructions of combinatorial objects whose existence is proven via the Probabilistic Method?
- Can we convert weak random sources, which may have biases and correlations, into high-quality random bits that can be used for running randomized algorithms or protocols?

In this section, we will survey the answers that Wigderson's work has given to these fundamental questions, and the close connections between the questions that he has helped uncover and exploit. For more details, we recommend the broader surveys of pseudorandomness [110, 276, 131].[5]

3.1 Hardness vs. Randomness

3.1.1 Motivation

In the 1970s and 1980s, randomization was discovered to be an extremely powerful tool in theoretical computer science. By allowing algorithms to "toss coins," we could potentially solve problems much more efficiently than before. In particular, polynomial-time randomized algorithms were found for a number of problems that were only known to have exponential-time deterministic algorithms, such as POLYNOMIAL FACTORIZATION (over finite fields) [42], PRIMALITY TESTING [257, 199, 219], POLYNOMIAL IDENTITY TESTING [76, 245, 292], and APPROXIMATELY COUNTING MATCHINGS in graphs [147]. However, it was unclear whether this apparent exponential savings provided by randomization was real, or just a reflection of our ignorance: could there be polynomial-time deterministic algorithms for these problems that we just hadn't discovered or proven correct yet? For example, already Miller [199] gave a deterministic polynomial-time algorithm for PRIMALITY TESTING based on the Extended Riemann Hypothesis, and three decades later, Agrawal, Kayal, and Saxena [2] gave an unconditional deterministic polynomial-time algorithm. Thus, the following problem remained open:

Open Problem 3.1. *Are there problems that can be solved by randomized algorithms in polynomial time that cannot be solved by deterministic algorithms in polynomial time?*

---

[5] Some of our text is taken verbatim from [276], with permission.

We now formalize this question using complexity classes that capture the power of efficient deterministic and randomized algorithms. As is common in complexity theory, these classes are defined in terms of decision problems, where instances are given by binary strings $x \in \{0,1\}^* \stackrel{\text{def}}{=} \bigcup_{n=0}^{\infty} \{0,1\}^n$ and the set of instances where the answer should be "yes" is specified by a *language* $L \subseteq \{0,1\}^*$, or equivalently a boolean function $f : \{0,1\}^* \to \{0,1\}$. However, the definitions generalize in natural ways to other types of computational problems, such as computing functions or solving search problems.

Recall that we say a deterministic algorithm A *runs in time* $t : \mathbb{N} \to \mathbb{N}$ if A takes at most $t(|x|)$ steps on every input x (where $|x|$ is the length of x in bits), and it *runs in polynomial time* if it runs time $t(n) = O(n^c)$ for a constant c. Polynomial time is a theoretical approximation to feasible computation, with the advantage that it is robust to reasonable changes in the model of computation and representation of the inputs.

Definition 3.2. P is the class of languages L for which there exists a deterministic polynomial-time algorithm A such that for all instances x,

- $x \in L \Rightarrow A(x)$ accepts, and
- $x \notin L \Rightarrow A(x)$ rejects.

Definition 3.3. BPP is the class of languages L for which there exists a probabilistic polynomial-time algorithm A such that for all instances x,

- $x \in L \Rightarrow \Pr[A(x) \text{ accepts}] \geq 2/3$, and
- $x \notin L \Rightarrow \Pr[A(x) \text{ accepts}] \leq 1/3$,

where the probabilities are taken over the random coin tosses of the algorithm A.

The choice of the thresholds $\ell = 1/3$ and $u = 2/3$ is arbitrary, and any two distinct constants $\ell < u$ yields an equivalent definition, since the error probability of a randomized algorithm can be made arbitrarily small by running the algorithm several times and accepting if at least an $(\ell + u)/2$ fraction of the executions accept.

The cumbersome notation BPP stands for "bounded-error probabilistic polynomial-time," due to the unfortunate fact that PP ("probabilistic polynomial-time") refers to the definition where the inputs in L are accepted with probability greater than 1/2 and inputs not in L are accepted with probability at most 1/2. Despite its name, PP is not a reasonable model for randomized algorithms, as it takes exponentially many repetitions to reduce the error probability. BPP is considered the standard complexity class associated with probabilistic polynomial-time algorithms, and thus a driving question of Wigderson's work surveyed in this section is the following formalization of Open Problem 3.1 (negated).

Open Problem 3.4. *Does* $\mathsf{BPP} = \mathsf{P}$*?*

More generally, we are interested in quantifying how much savings randomization provides. One way of doing this is to find the smallest possible upper bound on the deterministic time complexity of languages in BPP. For example, we would like to know which of the following complexity classes contain BPP:

Definition 3.5 (Deterministic Time Classes).

$$\begin{aligned}
\mathsf{DTIME}(t(n)) &= \{L : L \text{ can be decided deterministically in time } O(t(n))\} \\
\mathsf{P} &= \cup_c \mathsf{DTIME}(n^c) \qquad \text{(“polynomial time”)} \\
\tilde{\mathsf{P}} &= \cup_c \mathsf{DTIME}(2^{(\log n)^c}) \quad \text{(“quasipolynomial time”)} \\
\mathsf{SUBEXP} &= \cap_\varepsilon \mathsf{DTIME}(2^{n^\varepsilon}) \qquad \text{(“subexponential time”)} \\
\mathsf{EXP} &= \cup_c \mathsf{DTIME}(2^{n^c}) \qquad \text{(“exponential time”)},
\end{aligned}$$

where the unions and intersections are taken over all $c, \varepsilon \in (0, \infty)$

As a baseline, we can always remove randomization with at most an exponential slowdown:

Proposition 3.6. $\mathsf{BPP} \subseteq \mathsf{EXP}$.

Proof. If L is in BPP, then there is a probabilistic polynomial-time algorithm A for L running in time $t(n)$ for some polynomial t. Let $m(n) \le t(n)$ be an upper bound on the number of random bits used by A on inputs of length n. Thus we can view A as a deterministic algorithm on two inputs — its regular input $x \in \{0,1\}^n$ and its coin tosses $r \in \{0,1\}^{m(n)}$. Writing $A(x;r)$ for A's output on input $x \in \{0,1\}^n$ and coin tosses $r \in \{0,1\}^{m(n)}$, we have

$$\Pr_r[A(x;r) \text{ accepts}] = \frac{1}{2^{m(n)}} \sum_{r \in \{0,1\}^{m(n)}} A(x;r).$$

We can compute the right-hand side of the above expression in deterministic time $2^{m(n)} \cdot t(n)$. □

We see that the enumeration method is *general* in that it applies to all BPP algorithms, but it is *infeasible* (taking exponential time). However, if the algorithm uses only a small number of random bits, it becomes feasible:

Proposition 3.7. *If L has a probabilistic polynomial-time algorithm that runs in time $t(n)$ and uses $m(n)$ random bits, then $L \in \mathsf{DTIME}(t(n) \cdot 2^{m(n)})$. In particular, if $t(n) = \mathsf{poly}(n)$ and $m(n) = O(\log n)$, then $L \in \mathsf{P}$.*

Thus an approach to proving $\mathsf{BPP} = \mathsf{P}$ is to show that the number of random bits used by any BPP algorithm can be reduced to $O(\log n)$. This is the angle of attack pursued in Wigderson's work, as surveyed in the next section. However, to date, Proposition 3.6 remains the best unconditional upper-bound we have on the deterministic time-complexity of BPP.

Open Problem 3.8. *Is* BPP *"closer" to* P *or* EXP*? Is* $\mathsf{BPP} \subseteq \tilde{\mathsf{P}}$*? Is* $\mathsf{BPP} \subseteq \mathsf{SUBEXP}$*?*

3.1.2 Wigderson's Contributions

Derandomization from Circuit Lower Bounds

In the early 1980s, the answer to Open Problem 3.4 seemed very likely to be no, that $\mathsf{BPP} \neq \mathsf{P}$, since there were many examples of problems where randomization provided an exponential speedup over the best deterministic algorithms known at the time. The first evidence that randomization might not be so powerful came from Yao [289], who showed that if there exist "cryptographically secure" pseudorandom generators, as defined by Blum and Micali [45], then $\mathsf{BPP} \subseteq \mathsf{SUBEXP}$. In a series of works, Wigderson and his collaborators obtained much stronger derandomization results, convincing the theoretical computer science community that indeed $\mathsf{BPP} = \mathsf{P}$.

Theorem 3.9 ([211, 25, 141]).

1. *If* EXP *has a function of circuit complexity* $n^{\omega(1)}$, *then* $\mathsf{BPP} \subseteq \mathsf{SUBEXP}$.
2. *If* $\mathsf{E} \stackrel{\text{def}}{=} \mathsf{DTIME}(2^{O(n)})$ *has a function of circuit complexity* $2^{\Omega(n)}$, *then* $\mathsf{BPP} = \mathsf{P}$.

To define the "circuit complexity" referred to in the theorem, we associate a language $L \subseteq \{0,1\}^*$ in EXP or E with its characteristic function $f : \{0,1\}^* \to \{0,1\}$. For each n, we consider the restriction $f_n : \{0,1\}^n \to \{0,1\}$ of f to instances of length n, and ask how many boolean operations (AND, OR, NOT) are needed to compute f_n, i.e. what is the size of the smallest *boolean circuit* computing f_n, as a function of n. (See Section 4.1 for a more formal definition.) An algorithm running in time $t(n)$ can be simulated by boolean circuits of size $\tilde{O}(t(n)) \stackrel{\text{def}}{=} t(n) \cdot \text{polylog}(t(n))$, but the converse is not true, since circuits are a *nonuniform* model of computation, essentially allowing a different program for each input length (rather than a single set of instructions that can solve problems of arbitrary size). Thus Theorem 3.9 can be interpreted as saying "if nonuniformity cannot speed up all (exponential-time) algorithms too much, then randomization never provides too much of a speed up." Or, in more of a "win-win" formulation, "either we can speed up all (exponential-time) algorithms with nonuniformity, or we can efficiently derandomize all probabilistic algorithms."

The two items in Theorem 3.9 are special cases of a more general quantitative result that relates circuit complexity to derandomization. At the "low end," Item 1 says that if there are problems solvable in exponential time that have superpolynomial circuit complexity, then we get a subexponential-time derandomization of BPP. This is the same as Yao's aforementioned result [289] but with a much weaker hypothesis than the existence of cryptographically secure pseudorandom generators. At the "high end," Item 2 says that if instead we have problems with exponential circuit complexity, then in fact we get polynomial-time derandomization. (We also need to make the relatively minor switch from EXP to E. If we used EXP instead, the conclusion would be $\mathsf{BPP} \subseteq \tilde{\mathsf{P}}$.)

These circuit complexity hypotheses are very plausible. The NP-complete problems are promising candidates; they can be solved in exponential time and are conjectured to require superpolynomial and even exponential circuit complexity, though we are very far from proving it. (Such a result would resolve the famous P vs. NP question.) See Chapter 4 for a survey of Wigderson's fundamental contributions to circuit lower bounds.

Sections 3.1.3–3.1.5 give an overview of the proof of Theorem 3.9.

Circuit Lower Bounds from Derandomization

Theorem 3.9 of Wigderson et al. establishes a unidirectional implication between two major projects in theoretical computer science: *if* we can prove circuit lower bounds, *then* we can provably derandomize BPP. Since proving circuit lower bounds is so difficult, it is natural to wonder whether derandomization could be easier. With Impagliazzo and Kabanets [138], Wigderson proved that if we "add nondeterminism" to the complexity classes, then in fact circuit lower bounds are *equivalent* to derandomization.

Theorem 3.10 ([138]). MA *(a randomized analogue of* NP*) has a nontrivial derandomization (namely,* $\mathsf{MA} \neq \mathsf{NEXP}$, *where* NEXP *is an exponential-time analogue of* NP*)* if and only if NEXP *does not have polynomial-sized circuits.*

It follows from Theorem 3.10 that derandomization of ordinary randomized polynomial-time algorithms (without nondeterminism) also implies circuit lower bounds. Specifically, it turns out that if we can derandomize the generalization of BPP to "promise problems" (i.e. partial boolean functions, where we don't define or care about the output on certain inputs), then we can also derandomize MA and hence deduce from Theorem 3.10 that NEXP does not have polynomial-sized circuits. (This implication of Theorem 3.10 also follows from the earlier work of [57].) Building on Theorem 3.10, Kabanets and Impagliazzo [150] proved that even if the specific problem of POLYNOMIAL IDENTITY TESTING (which is in BPP) has a nontrivial derandomization, then NEXP does not have polynomial-sized boolean circuits *or* the PERMANENT does not have polynomial-sized arithmetic circuits, either of which would be breakthroughs in complexity theory. (See Section 4.9 for Wigderson's work on arithmetic circuit lower bounds and Section 5.2.3 for Wigderson's work on variants of POLYNOMIAL IDENTITY TESTING.)

A more positive interpretation of Theorem 3.10 is that we might be able to prove new circuit lower bounds by coming up with new methods for derandomizing algorithms. This possibility was realized in Williams' program of proving circuit lower bounds by designing faster SAT algorithms [286] and his breakthrough result that NEXP does not have polynomial-sized ACC circuits [287], both of which built on the results and techniques of [138].

Optimizing the Hardness vs. Randomness Tradeoff

As mentioned above, there is a full spectrum of "hardness vs. randomness" implications between the "low end" and "high end" derandomizations stated in Theorem 3.9. With Impagliazzo and Shaltiel [140], Wigderson pointed out that in the intermediate regime, the proof of Theorem 3.9 yielded results that were suboptimal in a sense that can be made formal, and initiated a line of work that culminated in optimal hardness vs. randomness tradeoffs achieved by Shaltiel and Umans [247, 274]. More recently, researchers have turned to more finely quantifying how much slowdown is needed to derandomize algorithms. Under suitably strong complexity assumptions, recent works [81, 64] give evidence that every randomized algorithm running in time $T(n)$ can be converted to a deterministic algorithm running in time $n^{1+\varepsilon} \cdot T(n)$ for an arbitrarily small constant $\varepsilon > 0$. Thus, it seems that randomization saves at most an almost-linear factor in runtime!

Derandomization from Uniform Assumptions

Theorem 3.10 and the results discussed after it show that some derandomizations of BPP require novel circuit lower bounds, at least for NEXP. Nevertheless, in a remarkable paper with Impagliazzo [142], Wigderson showed that a nontrivial derandomization of BPP *is* possible under the uniform assumption $\mathsf{EXP} \neq \mathsf{BPP}$. Specifically, for every language L in BPP, they obtain a deterministic subexponential-time algorithm that correctly decides L on all but a $1/\mathsf{poly}(n)$ fraction of inputs of length n, for infinitely many values of n. Moreover, this holds not just for the uniform distribution on instances of length n, but simultaneously for every efficiently samplable distribution on instances.

An intriguing feature of the Impagliazzo–Wigderson uniform derandomization is that it is (necessarily [271]) a "non-black-box" construction. That is, the construction and proof actually make use of the *code* of the programs that compute the assumed hard function $f \in \mathsf{E}$ and that decide the language $L \in \mathsf{BPP}$ that is being derandomized. In contrast, Theorem 3.9, like most results in complexity theory, treats these algorithms as "black boxes," only using the fact that they can be solved by efficient programs to deduce that other programs using them as subroutines are efficient.

The Impagliazzo–Wigderson uniform derandomization of BPP is a "low-end" result; assuming only a superpolynomial lower bound for EXP, we get (only) a subexponential-time derandomization of BPP. It remains an open problem to have a high-end (or nearly high-end) analogue of their result, for example to get a polynomial-time (or quasipolynomial-time) average-case derandomization of BPP under the assumption that $\mathsf{E} \not\subseteq \mathsf{BPTIME}(2^{o(n)})$ (or $\mathsf{EXP} \not\subseteq \mathsf{BPSUBEXP}$). In [271], a uniform analogue of the high-end worst-case-to-average-case hardness for E (Theorem 3.22) was given.

Subsequent works have given uniform average-case derandomizations of randomized algorithms with one-sided error (RP) [149] and constant-round interactive proofs (a.k.a. Arthur–Merlin games, AM) [189, 138, 126, 248]. Recent work has identified "almost-everywhere" uniform hardness assumptions (e.g. computational problems where every uniform probabilistic polynomial-time algorithm fails to solve the problem on all but finitely many inputs) that are *equivalent* to worst-case derandomization of BPP (generalized to "promise problems") [111, 65, 186].

3.1.3 Pseudorandom Generators

The approach to derandomizing algorithms suggested by Yao [289] and pursued by Wigderson is by constructing *pseudorandom generators*. These are defined in terms of computational indistinguishability, which was introduced in Section 2 and will be convenient to reformulate here in a non-asymptotic form:

Definition 3.11 (computational indistinguishability [115]). Random variables X and Y taking values in $\{0,1\}^m$ are (s,ε) *indistinguishable* if for every boolean circuit $T:\{0,1\}^m \to \{0,1\}$ of size at most s, we have

$$|\Pr[T(X)=1]-\Pr[T(Y)=1]| \leq \varepsilon.$$

The left-hand side above is called also the *advantage* of T in distinguishing X and Y.

If we set $s=\infty$ (or even $s=2^m$), then we allow all 2^{2^m} boolean functions T as statistical tests, and (s,ε)-indistinguishability is equivalent to requiring that X and Y have total variation distance at most ε. (See Definition 3.33.) However, by restricting to computationally efficient tests, e.g., with $s=\mathsf{poly}(m)$, then we obtain a significantly relaxed definition, where even random variables X and Y with disjoint supports can be indistinguishable. At the same time, for all efficient purposes (i.e. tasks that can be done by a boolean circuit of size s), X and Y are interchangeable.

A pseudorandom generator is a procedure that stretches a short seed if truly random bits into a long string that is computationally indistinguishable from uniform.

Definition 3.12 (pseudorandom generator [45, 289]). A deterministic function $G:\{0,1\}^d \to \{0,1\}^m$ is an (s,ε) *pseudorandom generator (PRG)* if

1. $d<m$, and
2. $G(U_d)$ and U_m are (s,ε) indistinguishable, where U_k denotes a random variable uniformly distributed over $\{0,1\}^k$.

If a test $T:\{0,1\}^m \to \{0,1\}$ has advantage at most ε in distinguishing $G(U_d)$ from U_m, we say that G *ε-fools T*.

People attempted to construct pseudorandom generators long before this definition was formulated. Their generators were tested against a battery of statistical tests (e.g. the number of 1's and 0's are approximately the same, the longest run is of length $O(\log m)$, etc.), but these fixed set of tests provided no guarantee that the generators would perform well in an arbitrary application. Indeed, most classical constructions (e.g. linear congruential generators, as implemented in the standard C library) are known to fail in some applications.

Intuitively, the above definition guarantees that the pseudorandom bits produced by the generator are as good as truly random bits for *all* efficient purposes (where efficient means computable by a circuit of size at most s). In particular, we can use such a generator to reduce the number of random bits used by any algorithm from m to $d(m)$ provided that the algorithm runs in time at most $t = s/\text{polylog}(s)$, because the behavior of any such algorithm on any input x can be simulated by a boolean circuit of size s. For the resulting algorithm to be efficient, we will also need the generator to be efficiently computable.

Definition 3.13. We say a sequence of generators $\{G_m : \{0,1\}^{d(m)} \to \{0,1\}^m\}$ is *computable in time $t(m)$* if there is a *uniform and deterministic* algorithm M such that for every $m \in \mathbb{N}$ and $x \in \{0,1\}^{d(m)}$, we have $M(m,x) = G_m(x)$ and $M(m,x)$ runs in time at most $t(m)$.

Note that even when we define the pseudorandomness property of the generator with respect to nonuniform boolean circuit, the efficiency requirement refers to uniform algorithms. For readability, we will usually refer to a single generator $G: \{0,1\}^{d(m)} \to \{0,1\}^m$, with it being implicit that we are really discussing a family $\{G_m\}$.

Theorem 3.14. *Suppose that for all m there exists an $(m, 1/8)$ pseudorandom generator $G: \{0,1\}^{d(m)} \to \{0,1\}^m$ computable in time $t(m)$. Then*

$$\mathsf{BPP} \subseteq \bigcup_c \mathsf{DTIME}(2^{d(n^c)} \cdot (n^c + t(n^c))).$$

Proof. Let L be any language in BPP. Then there is a constant c such that L is decided by a bounded-error randomized algorithm in time $t(n) = O(n^{c-1})$ on inputs of length n.

The idea is to replace the random bits used by A with pseudorandom bits generated by G, use the pseudorandomness property to show that the algorithm will still be correct with high probability, and finally enumerate over all possible seeds to obtain a deterministic algorithm.

Claim. For all sufficiently large n and every $x \in \{0,1\}^n$, $A(x; G(U_{d(n^c)}))$ errs with probability smaller than $1/2$.

Proof of claim: Suppose that there exists some $x \in \{0,1\}^n$ on which $A(x; G(U_{d(n^c)}))$ errs with probability at least $1/2$. Then, $T(\cdot) = A(x, \cdot)$ distinguishes $G(U_{d(n^c)})$ from U_{n^c} with advantage at least $1/2 - 1/3 > 1/8$. Since algorithms running in time $t(n)$ can be simulated by boolean circuits of size at most $\tilde{O}(t(n))$, $T(\cdot)$ can be computed by a boolean circuit of size at most n^c, for sufficiently large n. This contradicts the pseudorandomness property of G. □

Now, enumerate over all seeds of length $d(n^c)$ and take a majority vote. There are $2^{d(n^c)}$ of them, and for each we have to run both G and A. □

In the definition of a cryptographic pseudorandom generator used by Yao [289], the requirement was that G is computable in time polynomial in its input length, i.e. $t(m) \leq \mathsf{poly}(d(m))$. This implies that $d(m) \geq t(m)^{\delta} \geq m^{\delta}$ for a constant $\delta > 0$, so the running time of the derandomization in Theorem 3.14 is at least $2^{d(n^c)} \geq 2^{n^{\delta c}}$, and we can at best conclude $\mathsf{BPP} \subseteq \mathsf{SUBEXP}$.

Thus a key to Theorem 3.9 was the realization by Nisan and Wigderson [211] that, for derandomization, we can relax the efficiency requirements of a cryptographic generator in two ways. First, we can afford for the generator to be computable in time exponential in its seed length, since anyway we enumerate over all seeds when derandomizing. Second (and relatedly), we can afford for the generator to run in more time than the algorithms it fools. Indeed, in Theorem 3.14, we only need to fool circuits of size m, but we are happy for a generator computable in time $\mathsf{poly}(m)$. In contrast, a cryptographic generator actually requires fooling circuits of size $m^{\omega(1)}$, ones that are superpolynomially larger than the output length and the running time of the generator. Thus, they proposed the following efficiency requirement:

Definition 3.15 ([211]). A generator $G : \{0,1\}^{d(m)} \to \{0,1\}^m$ is *quick* (a.k.a. *mildly explicit*) if it is computable in time $\mathsf{poly}(m, 2^{d(m)})$.

They demonstrated the benefits of these relaxed requirements with the beautiful pseudorandom generator construction described in the next section (which is a key component of the proof of Theorem 3.9).

3.1.4 The Nisan–Wigderson Generator

The Nisan–Wigderson generator constructs a quick pseudorandom generator from any function in E that is sufficiently *hard on average*:

Definition 3.16. For $s \in \mathbb{N}$ and $\alpha > 0$, a function $f : \{0,1\}^{\ell} \to \{0,1\}$ is (s, α) *average-case hard* if for every boolean circuit A of size at most s, we have

$$\Pr[A(U_\ell) \neq f(U_\ell)] > \alpha.$$

Note that, in contrast to the definition of BPP, here the probabilities are taken over the *input* to the algorithm A, rather than its random coin tosses. When $\alpha = 0$, Definition 3.16 simply says that f has circuit complexity greater than s, but when α is nonzero it is a significantly stronger hardness requirement on f. Note that $\alpha = 1/2$ is impossible, since a constant function (of size $s = 1$) can always compute f correctly on at least half of the inputs.

We now state the Nisan–Wigderson theorem, restricted to the "high-end" regime, where hardness is against circuits of exponential size.

Theorem 3.17 ([211]). *Suppose that there is a constant* $\delta > 0$ *and a function* $f \in \mathsf{E}$ *such that for every input length* $\ell \in \mathbb{N}$, f_ℓ *is* $(2^{\delta\ell}, 1/2 - 1/2^{\delta\ell})$ *average-case hard. Then for every* $m \in \mathbb{N}$, *there is a quick* $(m, 1/m)$ *pseudorandom generator* $G: \{0,1\}^{d(m)} \to \{0,1\}^m$ *with seed length* $d(m) = O(\log m)$. *In particular,* $\mathsf{BPP} = \mathsf{P}$.

Similar to Theorem 3.9, this is a specific instance of a quantitative tradeoff between hardness and derandomization. In particular, if we replace both occurrences of the exponential bound $2^{\delta\ell}$ with a superpolynomial bound $\ell^{\omega(1)}$, we obtain the "low-end" conclusion that $\mathsf{BPP} \subseteq \mathsf{SUBEXP}$. However, the hypothesis in Theorem 3.17 is significantly stronger in that it assumes average-case hardness rather than worst-case hardness, and very strong average-case hardness at that: no small circuit can compute f with probability much better than random guessing. In the next section, we will discuss how Wigderson and collaborators relaxed the average-case hardness assumption to a worst-case one in order to obtain Theorem 3.9.

The starting point for Theorem 3.17 is the realization, implicit in Yao [289], that if f is $(s, 1/2 - \varepsilon)$ average-case hard, then $G(x) = (x, f(x))$ is an $(s - O(1), \varepsilon)$ pseudorandom generator. That is, by applying f once on a uniformly random input, we obtain one pseudorandom bit (beyond the $d = \ell$ truly random bits in the seed). So, to obtain may pseudorandom bits, we can try applying f many times. For this to provide a generator with large stretch (i.e. with output length superlinear in the input length), we cannot evaluate f on independent random inputs, but rather need to generate many correlated inputs, but ensure that the correlations don't destroy the pseudorandomness.

The idea, building on Nisan [207], is to use inputs to f that share very few bits. Specifically, the sets of seed bits used for each input to f will be given by a *design*:

Definition 3.18. $S_1, \cdots, S_m \subseteq [d]$ is an (ℓ, a)-*design* if

1. $\forall i, |S_i| = \ell$
2. $\forall i \neq j, |S_i \cap S_j| \leq a$

It turns out that there exist designs with lots of sets having small intersections over a small universe:

Lemma 3.19. *For every every* $\ell, m \in \mathbb{N}$, *there exists an* (ℓ, a)*-design* $S_1, \cdots, S_m \subseteq [d]$ *with* $d = O\left(\frac{\ell^2}{a}\right)$ *and* $a = \log_2 m$. *Such a design can be constructed deterministically in time* $\mathsf{poly}(m, d)$.

The important points are that intersection sizes are only logarithmic in the number of sets, and the universe size d is linear in ℓ in case we take $m = 2^{\Omega(\ell)}$.

Construction 3.20 (Nisan–Wigderson Generator). *Given a function* $f: \{0,1\}^\ell \to \{0,1\}$ *and an* (ℓ, a)*-design* $S_1, \cdots, S_m \subseteq [d]$, *define the* Nisan–Wigderson generator $G: \{0,1\}^d \to \{0,1\}^m$ *as*

$$G(x) = f(x|_{S_1}) f(x|_{S_2}) \cdots f(x|_{S_m})$$

where if x *is a string in* $\{0,1\}^d$ *and* $S \subseteq [d]$, *then* $x|_S$ *is the string of length* $|S|$ *obtained from* x *by selecting the bits indexed by* S.

This elegant construction is analyzed as follows.

Theorem 3.21. *Let $G : \{0,1\}^d \to \{0,1\}^m$ be the Nisan–Wigderson generator based on a function $f : \{0,1\}^\ell \to \{0,1\}$ and some (ℓ, a) design. If f is $(s, 1/2 - \varepsilon/m)$ average-case hard, then G is a (s', ε) pseudorandom generator, for $s' = s - m \cdot 2^a$.*

Theorem 3.17 follows from Theorem 3.21 by setting $\varepsilon = 1/m$, $a = \log_2 m$, and $s = 2^{\delta\ell}$, and observing that for $\ell = (1/\delta) \cdot \log_2(2m^2) = O(\log m)$, we have

$$s' = s - m \cdot 2^a = 2m^2 - m^2 \geq m,$$

and $\varepsilon/m \leq 1/2^{\delta\ell}$, so we have an $(m, 1/m)$ pseudorandom generator. The seed length is $d = O(\ell^2/\log m) = O(\log m)$.

Proof. Suppose for contradiction that G is not an (s', ε) pseudorandom generator. By the equivalence of pseudorandomness and next-bit unpredictability [289], there is a size s' circuit P such that

$$\Pr[P(f(X|_{S_1})f(X|_{S_2})\cdots f(X|_{S_{i-1}})) = f(X|_{S_i})] > \frac{1}{2} + \frac{\varepsilon}{m}, \tag{3}$$

for some $i \in [m]$ and a uniformly random $X \leftarrow \{0,1\}^d$. From P, we will construct a small circuit A that computes f on a uniformly random input with probability greater than $1/2 + \varepsilon/m$.

Let $Y = X|_{S_i}$. By averaging, we can fix all bits of $X|_{\overline{S_i}} = z$ (where $\overline{S_i}$ is the complement of S) such that the prediction probability remains greater than $1/2 + \varepsilon/m$ over Y. Define $f_j(y) = f(x|_{S_j})$ for $j \in \{1, \cdots, i-1\}$. (That is, $f_j(y)$ forms x by placing y in the positions in S_i and z in the others, and then applies f to $x|_{S_j}$). Then

$$\Pr_Y[P(f_1(Y)\cdots f_{i-1}(Y)) = f(Y)] > \frac{1}{2} + \frac{\varepsilon}{m}.$$

Note that $f_j(y)$ depends on only $|S_i \cap S_j| \leq a$ bits of y. Thus, we can compute each f_j with a look-up table hardwired into our circuit. Indeed, every function on a bits can be computed by a boolean circuit of size at most 2^a. (In fact, size at most $O(2^a/a)$ suffices.)

Then, by considering $A(y) = P(f_1(y)\cdots f_{i-1}(y))$, we deduce that f can be computed with error probability smaller than $1/2 - \varepsilon/m$ by a boolean circuit of size at most $s' + (i-1) \cdot 2^a < s' + m \cdot 2^a = s$. This contradicts the hardness of f. Thus, we conclude G is an (m, ε) pseudorandom generator. □

3.1.5 Pseudorandom Generators from Worst-Case Lower Bounds

As we saw in the previous section, the Nisan–Wigderson construction gives us pseudorandom generators from boolean functions that are very hard on average, where every boolean circuit of size $2^{\delta\ell}$ must err with probability greater than $1/2 - 1/2^{\delta\ell}$

on a random input. In works with Babai, Fortnow, and Nisan [25] and Impagliazzo [141], Wigderson showed how to relax the assumption to worst-case hardness, yielding Theorem 3.9. This was done by showing how to convert worst-case hard functions into average-case hard functions, which again we state only in the high-end regime of parameters:

Theorem 3.22 (worst-case to average-case hardness for E [141]). *Suppose that for a constant $\delta > 0$, there is a function in* E *that has circuit complexity at least $2^{\delta\ell}$ on inputs of length ℓ. Then there is a constant $\delta' > 0$ and a function in* E *that is $(2^{\delta'\ell}, 1/2 - 1/2^{\delta'\ell})$ average-case hard.*

Combining Theorem 3.22 and Theorem 3.17 yields the high-end part of Theorem 3.9.

Beyond the application to pseudorandomness and derandomization, the relationship between worst-case complexity and average-case complexity is a central question in complexity theory. (See the survey [47].) In particular, whether a similar result is true for NP (rather than E) remains a major open problem.

3.2 Expanders, Extractors, and Ramsey Graphs

Another area in which randomness has proved very useful is in the Probabilistic Method [13], whereby mathematical objects with interesting properties are proven to exist by showing that a randomly chosen object has the desired property with high (or at least nonzero) probability. A famous example is Erdős' existence proof for Ramsey graphs — graphs with no large clique or independent set [91].

In such cases, the problem of derandomization becomes one of finding *explicit constructions* of objects with the desired properties. The search for explicit constructions is of pure mathematical interest, as a way of developing and testing our understanding of the mathematical properties at hand. They are also important for many computer science applications, where we need efficient algorithms to describe and work with the objects.

In this section, we survey Wigderson's contributions to explicit constructions, in particular to the constructions of expander graphs, randomness extractors, and Ramsey graphs, as well as identifying and exploiting the connections between these.

3.2.1 Expander Graphs

Expander graphs are graphs that are "sparse" yet very "well-connected." They are ubiquitous in theoretical computer science, with applications including communication and routing networks [216, 215], derandomizing algorithms [4, 235], error-correcting codes [121], lower bounds on circuit complexity [277] and proof complexity [40], integrality gaps for optimization problems [184, 22], data structures [58], fault-tolerant storage [275] and more. A rich mathematical theory has developed around constructing expanders and understanding their properties; we refer to Wigderson's survey with Hoory and Linial [135] as well as [276, 270, 172] for many aspects that we will not be able to cover here.

We will typically interpret the properties of expander graphs in an asymptotic sense. That is, there will be an infinite family of graphs G_i, with a growing number of vertices N_i. By "sparse," we mean that the (maximum or average) degree D_i of G_i should be very slowly growing as a function of N_i. The "well-connectedness" property has a variety of different interpretations, which we will discuss below. Typically, we will drop the subscripts of i and the fact that we are talking about an infinite family of graphs will be implicit in our theorems. We will state many of our definitions for *directed multigraphs* (which we'll call *digraphs* for short), though in the end we will mostly study undirected multigraphs.

The most intuitive definition of expansion is the following.

Definition 3.23. A digraph G is a (K,A) *vertex expander* if for all sets S of at most K vertices, the *(out-)neighborhood* $N(S) \stackrel{\text{def}}{=} \{u | \exists v \in S \text{ s.t. } (u,v) \in E\}$ is of size at least $A \cdot |S|$.

Ideally, we would like graphs with degree $D = O(1)$, and (K,A) vertex expansion with $K = \Omega(N)$ where N is the number of vertices, and A as close to D as possible.

It is often useful to work instead with a linear-algebraic measure of expansion. For simplicity, we restrict attention to regular graphs in presenting the definition.

Definition 3.24. Let G be an N-vertex D-regular digraph with random-walk matrix M (so M_{ij} equals the number of edges from i to j divided by D). Let $\sigma_2(G) \in [0,1]$ denote the second-largest singular value of M. The *spectral expansion* of G is $\gamma(G) = 1 - \sigma_2(G)$.[6]

Ideally, we would like an infinite family of graphs with degree $D = O(1)$ and $\gamma(G) = \Omega(1)$. Alon [9] proved that this linear-algebraic measure of expansion is equivalent to the combinatorial measure of vertex expansion for common parameters of interest.

Theorem 3.25 ([9]). *Let $\mathcal{G}$ be an infinite family of D-regular multigraphs, for a constant $D \in \mathbb{N}$. Then the following two conditions are equivalent:*

- *There is a constant $\delta > 0$ such that every $G \in \mathcal{G}$ is an $(N/2, 1+\delta)$ vertex expander.*
- *There is a constant $\gamma > 0$ such that every $G \in \mathcal{G}$ has spectral expansion at least γ.*

When people informally use the term "expander," they often mean a family of regular graphs of *constant degree D* satisfying one of the two equivalent conditions above. However, we note that the quantitative relationship between vertex expansion and spectral expansion is lossy, so optimizing one of these measures of expansion need not yield optimality with respect to the other.

[6] In some other sources, the term spectral expansion refers to $\sigma_2(G)$ rather than $\gamma(G)$. Here we use $\gamma(G)$, because it has the more natural feature that larger values of γ correspond to the graph being "more expanding".

We can get more intuition for spectral expansion by considering some equivalent formulations of it. Since G is regular, the uniform distribution, written as a row vector $u = (1/N, \dots, 1/N)$, is an eigenvector of the random-walk matrix M of eigenvalue 1, i.e. $uM = u$. By the Perron–Frobenius Theorem, the largest singular value of M equals 1, and thus we have the following variational characterization of spectral expansion.

Lemma 3.26. $1 - \gamma(G) = \sigma_2(G) = \max_{x \perp u} \frac{\|xM\|}{\|x\|} = \max_{\pi} \frac{\|\pi M - u\|}{\|\pi - u\|}$, *where the first maximum is over all nonzero row vectors $x \in \mathbb{R}^N$ that are orthogonal to u, and the second maximum is over all probability distributions $\pi \in [0,1]^N$ (also written as row vectors).*

That is, if we start at any probability distribution π on the vertices of G and take one step of the random walk to end up at probability distribution πM, the ℓ_2 distance to uniform will shrink by at least a factor of $1 - \gamma(G)$. So if $\gamma(G)$ is bounded away from 0, then random walks on G will converge quickly to the uniform distribution.

Another useful characterization of $\gamma(G)$ is as follows.

Lemma 3.27. $\gamma(G) \geq \gamma$ *iff we can write $M = \gamma J + (1-\gamma)E$, where J is the matrix with every entry $1/N$ and $\|E\| \leq 1$, where $\|E\|$ is the spectral norm of E.*

Notice that J is the random-walk matrix for the complete graph on N vertices with self-loops, which is intuitively the most expanding possible graph (albeit not sparse). Thus, Lemma 3.27 says that an expander can be viewed as a sparse approximation of the complete graph.

It can be shown that a random D-regular undirected graph on N vertices is an excellent expander with high probability, for $D = O(1)$ and $N \to \infty$. For example, it achieves spectral expansion $\gamma(G) = 1 - 2\sqrt{D-1}/D + o(1)$ [99], which is optimal up to the $o(1)$ [205], and achieves $(\alpha A, D-1-\varepsilon)$ vertex expansion for any $\varepsilon > 0$ and $\alpha = \alpha(D, \varepsilon)$ [216, 34]. For some applications of expanders, however, we cannot afford to choose the graph at random, because it may be too costly in memory, communication, or randomness. Indeed, some applications even require exponentially large expander graphs, in which case a random graph would be completely infeasible to manipulate. Thus, we seek *explicit constructions* of expanders.

Definition 3.28. Let $\mathcal{G} = \{G_i\}$ be an infinite family of digraphs where G_i has N_i vertices and is D_i-regular. We say that $\mathcal{G}$ is *(fully) explicit* if given N_i, $u \in [N_i]$, and $j \in [D_i]$, the j'th neighbor of u in G_i can be computed deterministically in time $\mathsf{poly}(\log N_i)$.

That is, we require a very efficient local description of the graph, where computing neighbors can be done in time polynomial in the bitlength of vertices, rather than in time polynomial in the number of vertices.

Thus, starting with Margulis [195], there is a long and beautiful line of work on explicit constructions of constant-degree expanders, with one highlight being the optimal spectral expanders of Lubotzky, Phillips, and Sarnak [191] and Margulis [196], known as *Ramanujan graphs*. Many of these constructions were based on deep results from algebra and number theory, and it was of interest to have more combinatorial approaches to constructing expanders.

With Reingold and Vadhan [238], Wigderson gave a combinatorial construction of expanders based on a new graph operation, called the *zig-zag product*. Although these expanders did not match the spectral expansion of Ramanujan graphs, the additional flexibility offered by the construction found numerous applications, which we will survey below.

Specifically, their approach to constructing expanders is to start with a constant-sized expander of appropriate parameters and repeatedly apply graph operations to build larger and larger graphs while preserving the degree and spectral expansion.

Two standard operations on an N-vertex D-regular graph G with random-walk matrix M are the following:

Squaring: G^2 is the graph on N vertices whose random-walk matrix is M^2. That is, edges in G^2 are walks of length 2 in G. If G has spectral expansion at least $\gamma = 1-\sigma$, then G^2 has spectral expansion at least $1-\sigma^2 = 2\gamma - \gamma^2$

Tensoring: $G \otimes G$ is the graph on N^2 vertices whose random-walk matrix is $M \otimes M$ (the Kronecker product). That is, random walks in $G \otimes G$ correspond to two independent random walks in G. If G has spectral expansion at least γ, then $G \otimes G$ also has spectral expansion at least γ.

Squaring has the benefit of improving expansion and tensoring has the benefit of creating larger graphs, but both have the downside of increasing the degree D to D^2. Thus, we need an operation that decreases the degree, without hurting the expansion too much. This is what the zig-zag product achieves.

The Zig-Zag Product

Let G be a D_1-regular digraph on N_1 vertices and H be a regular digraph on D_1 vertices. The *zig-zag product* of G and H, denoted $G \textcircled{z} H$, is defined as follows. The nodes of $G \textcircled{z} H$ are the pairs (u,i) where $u \in V(G)$ and $i \in V(H)$. We think of this each vertex u of G with a copy of $V(H)$, which we refer to as a *cloud*, and associate each vertex of H with one of the edges incident to u. The edges in $G \textcircled{z} H$ then correspond to taking an H-step within a cloud, using a G-step to move between clouds, and an H-step in the resulting cloud. See Figure 1 for an illustration.

Definition 3.29 (Zig-zag Product). Let G be an D_1-regular digraph on N_1 vertices, and H a D_2-regular digraph on D_1 vertices. Then $G \textcircled{z} H$ is the following D_2^2-regular graph on $N_1 D_1$ vertices. The vertices are pairs $(u,i) \in [N_1] \times [D_1]$, and for $a,b \in [D_2]$, the (a,b)'th neighbor of a vertex (u,i) is the vertex (v,j) computed as follows:

1. Let i' be the a'th neighbor of i in H. (That is, take an H-step to move from (u,i) to (u,i').)
2. Let v be the i''th neighbor of u in G, so $e = (u,v)$ is the i''th edge leaving u. Let j' be such that e is the j''th edge entering v in G. (That is, take a G-step to move from (u,i') to (v,j').)

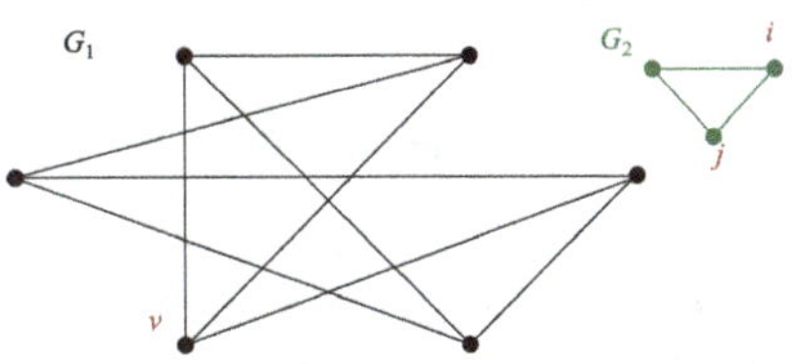

(a) Initial D_1-regular graph G_1 and graph G_2 on D_1 vertices.

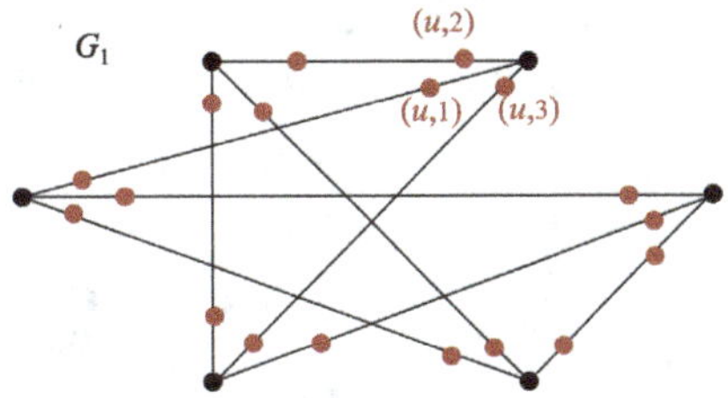

(b) First we construct a cloud with the same shape as G_2 for each vertex $u \in G_1$.

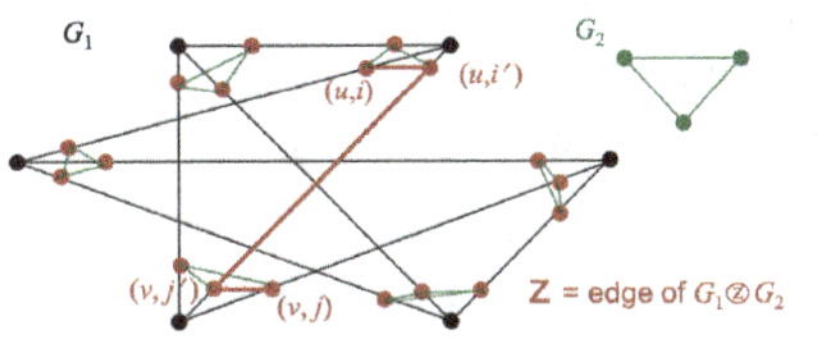

(c) Three steps connecting (u,i) and (v,j) following Definition 3.29

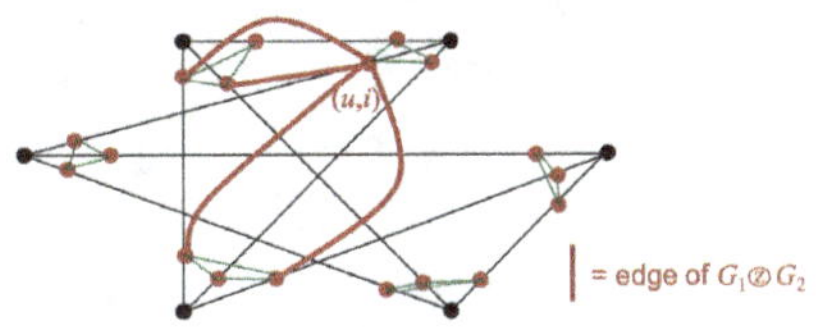

(d) We construct an edge between (u,i) and (v,j) and similarly for other edges.

Fig. 1: Illustration of the edge construction in the Zig-Zag product $G_1 ⓩ G_2$.

3. Let j be the b'th neighbor of j' in H. (That is, take an H-step to move from (v, j') to (v, j).)

Note that the graph $G ⓩ H$ depends on how the edges leaving and entering each vertex of G are numbered. Thus it is best thought of as an operation on labelled graphs. Nevertheless, the following lower bound on its expansion holds regardless of the labelling:

Theorem 3.30 ([238]). *If G has spectral expansion at least γ_1 and H has spectral expansion at least γ_2, then $G ⓩ H$ has spectral expansion at least $\gamma_1 \gamma_2^2$*

G should be thought of as a big graph and H as a small graph, where D_1 is a large constant and D_2 is a small constant. Observe that when $D_1 > D_2^2$ the degree is reduced by the zig-zag product.

Before giving intuition for Theorem 3.30, let's see how it can be used to construct an infinite family of constant-degree expanders.

Construction 3.31 (Zig-Zag Based Expanders). *Let H be a fixed D-regular graph on D^4 vertices with spectral expanion at least 7/8.*[7] *Define*

$$\begin{aligned} G_1 &= H^2 \\ G_t &= G_{t-1}^2 ⓩ H \qquad \text{for } t > 1 \end{aligned}$$

[7] Since the number of vertices is polynomially related to the degree, such graphs are much easier to construct than constant-degree expanders, and there are a number of simple constructions. Alternatively, since we think of D as a constant, H can be found by exhaustive search.

A straightforward induction, using Theorem 3.30 and the properties of squaring, shows that this is an infinite family of expanders:

Proposition 3.32. *For all t, G_t is a D^2-regular graph on D^{4t} vertices with spectral expansion at least* $1/2$.

Although simple to describe, Construction 3.31 does not quite meet our definition of explicitness (Definition 3.28), since the natural recursive way to compute neighbors in G_t (by doing two neighbor computations in G_{t-1}) appears to take time exponential in t, which is polynomial in $N_t = D^{4t}$, rather than polylogarithmic. This can be remedied by tensoring in addition to squaring, so that the number of vertices grows much more quickly than the depth of the recursion.

There are two different intuitions underlying the expansion of the zig-zag product:

1. Given an initial distribution (U,I) on the vertices of $G_1 ⓩ G_2$ that is far from uniform, there are two extreme cases. (Here we use capital letters to denote random variables corresponding to the lower-case values in Definition 3.29.) Either

   a. All the (conditional) distributions $I|_{U=u}$ within the clouds are far from uniform, *or*

   b. All the (conditional) distributions $I|_{U=u}$ within the clouds of size D_1 are uniform (in which case the marginal distribution U on the clouds must be far from uniform).

   In Case 3.2.1, the first H-step $(U,I) \mapsto (U,I')$ already brings us closer to the uniform distribution, and the other two steps cannot hurt (as they are steps on regular graphs). In Case 3.2.1, the first H-step has no effect, but the G-step $(U,I') \mapsto (V,J')$ has the effect of making the marginal distribution on clouds closer to uniform, i.e. V is closer to uniform than U. But note that the joint distribution (V,J') isn't actually any closer to the uniform distribution on the vertices of $G_1 ⓩ G_2$ because the G-step is a permutation. Still, if the marginal distribution V on clouds is closer to uniform, then the conditional distributions within the clouds $J'|_{V=v}$ must have become further from uniform, and thus the second H-step $(V,J') \mapsto (V,J)$ brings us closer to uniform.
   This intuition can be turned into a formal proof, and with a careful analysis (which can be found in [238]) yields slightly better expansion bounds than stated in Theorem 3.30.
2. A second intuition, which follows [240, 236], leads to a very short of Theorem 3.30. Here we think of the expander H as behaving similarly to the complete graph on D_1 vertices, via Lemma 3.27. In the case that H equals the complete graph, then it is easy to see that $G ⓩ H = G \otimes H$, whose spectral expansion is equal to $\gamma(G)$ (since the complete graph has spectral expansion 1). For general H, we use Lemma 3.27 to decompose the random-walk matrix for H into a convex combination of the random-walk matrix for the complete graph and an error matrix of spectral norm at most 1, with the coefficient on the complete graph being $\gamma(H)$. Doing this for both steps on H in the zig-zag product leads to a spectral expansion lower bound of $\gamma(H)^2 \cdot \gamma(G)$.

As mentioned earlier, Construction 3.31 does not achieve an optimal relationship between spectral expansion and degree (which is $\gamma(G) = 1 - \Theta(1/\sqrt{D})$, achieved by random graphs [99] or explicit Ramanujan graphs [191, 196]). However, in subsequent work with Capalbo, Reingold, and Vadhan [62], Wigderson used a variant of the zig-zag product to construct near-optimal directed or bipartite *vertex expanders*, namely constant-degree graphs where sets of size up to $K = \Omega(N)$ expand by a factor of $A = (1-\varepsilon) \cdot D$. (Viewed as bipartite graphs, the expansion is from the left side of the graph to the right side of the graph, corresponding the use of out-neighborhoods in Definition 3.23.) This variant of the zig-zag product comes from viewing expanders as forms of randomness extractors (as discussed in the next section), and builds on the first intuition for the zig-zag product given above. This was the first explicit construction of constant-degree graphs with expansion factor $A > D/2$, which has a qualitative implication that is important in a number of applications: they are also *unique-neighbor expanders*, where every left-set S of size at most K has at least one neighbor (in fact, at least $(1-2\varepsilon)D$ neighbors) on the right that is incident to exactly one vertex in S.

A different variant of the zig-zag product was introduced by Ben-Aroya and Ta-Shma [36] and used to give a combinatorial construction of "almost-Ramanujan" expanders (namely with $\gamma(G) = 1 - 1/D^{1/2-o(1)}$, where the $o(1) \to 0$ as $D \to \infty$). This same variant was then used by Ta-Shma [265] in his breakthrough construction of linear error-correcting codes (aka small-biased sets) that nearly meet the Gilbert–Varshamov bound.

With Alon and Lubotzky [12], Wigderson gave an intriguing algebraic interpretation of the zig-zag product: Under certain conditions, if G and H are Cayley graphs, then $G \textcircled{z} H$ is a Cayley graph for the *semi-direct product* of the underlying groups. Using this connection, they answered a question of Lubotzky and Weiss [192] and proved that expansion of Cayley graphs is *not* a group property: a group can have two constant-sized sets of generators, such that the Cayley graph defined by one is expanding and the other is not. With Meshulam [198] and Rozenman and Shalev [239], Wigderson further used this group-theoretic zig-zag to obtain iterative constructions of expanding Cayley graphs.

Perhaps the most striking application of the zig-zag product is Reingold's algorithm for UNDIRECTED S-T CONNECTIVITY [234], which we will see in Section 3.3, which in turn inspired Dinur's celebrated combinatorial proof of the PCP Theorem [79]. (See Section 2.3 for discussion of the PCP Theorem.)

Wigderson has also formulated and initiated the study of many other variants of expansion, such as expanding hypergraphs [100], monotone expanders [85], and notions of expansion for collections of linear maps [180].

3.2.2 Randomness Extractors

Randomness extractors are functions that extract almost-uniform bits from sources of biased and correlated bits. The original motivation for extractors was to simulate randomized algorithms with weak random sources as might arise in nature. This motivation is still compelling, but extractors have taken on a much wider significance in the years since they were introduced. They have found numerous applications in theoretical computer science beyond this initial motivating one, in areas from cryptography to distributed algorithms to hardness of approximation. (See the surveys [210, 276, 246].) In this section, we will survey Wigderson's numerous contributions to the theory of extractors, their constructions, and their applications. Many of these contributions involve developing and exploiting the close connection between randomness extractors and expander graphs.

We begin with some probability definitions that are needed to introduce randomness extractors.

Definition 3.33. For random variables X and Y taking values in $\mathcal{U}$, their *statistical difference* (also known as *total variation distance*) is

$$\Delta(X,Y) = \max_{T \subseteq \mathcal{U}} |\Pr[X \in T] - \Pr[Y \in T]|.$$

We say that X and Y are *ε-close* if $\Delta(X,Y) \leq \varepsilon$.

Recall that random variables being ε-close is equivalent to them being (∞,ε)-indistinguishable (Definition 3.11).

Definition 3.34 (entropy measures). Let X be a discrete random variable. Then

- the *Shannon entropy* of X is:

$$\mathrm{H}_{Sh}(X) = \mathop{\mathrm{E}}_{x \stackrel{R}{\leftarrow} X}\left[\log \frac{1}{\Pr[X=x]}\right].$$

- the *Rényi entropy* of X is:

$$\mathrm{H}_2(X) = \log\left(\frac{1}{\mathrm{E}_{x \stackrel{R}{\leftarrow} X}[\Pr[X=x]]}\right) \quad \text{and}$$

- the *min-entropy* of X is:

$$\mathrm{H}_\infty(X) = \min_x \left\{\log \frac{1}{\Pr[X=x]}\right\},$$

where all logs are base 2.

Fact 3.35. 1. *For every random variable X,*

$$\mathrm{H}_\infty(X) \leq \mathrm{H}_2(X) \leq \mathrm{H}_{Sh}(X),$$

with equality iff X is uniform on its support.

2. *For every random variable X, $\mathrm{H}_2(X) \leq 2\mathrm{H}_\infty(X)$, and for every $\varepsilon > 0$, there is a random variable X', such that $\mathrm{H}_2(X') \leq \mathrm{H}_\infty(X) + \log(1/\varepsilon)$.*

To illustrate the differences between the three notions, consider a source X such that $X = 0^n$ with probability 0.99 and $X = U_n$ with probability 0.01. Then $\mathrm{H}_{Sh}(X) \geq 0.01n$ (contribution from the uniform distribution), $\mathrm{H}_2(X) \leq \log(1/.99^2) < 1$ and $\mathrm{H}_\infty(X) \leq \log(1/.99) < 1$ (contribution from 0^n). Note that even though X has Shannon entropy linear in n, we cannot expect to extract bits that are close to uniform or carry out any useful randomized computations with one sample from X, because it gives us nothing useful 99% of the time. Thus, we should use the stronger measures of entropy given by H_2 or H_∞. These entropy measures were introduced into the randomness extraction literature by Cohen and Wigderson [70] and Chor and Goldreich [67], respectively.

We will consider the task of extracting randomness from sources where all we know is a lower bound on the min-entropy (which is equivalent to a lower bound on Rényi entropy by Fact 3.35):

Definition 3.36. A random variable X is a *k-source* if $\mathrm{H}_\infty(X) \geq k$, i.e., if $\Pr[X = x] \leq 2^{-k}$ for all x.

A typical setting of parameters is $k = \delta n$ for some fixed δ, e.g., $\delta = 1/10$. We call δ the *min-entropy rate*. Some different ranges that are commonly studied (and are useful for different applications): $k = \mathrm{polylog}(n)$, $k = n^\gamma$ for a constant $\gamma \in (0,1)$, $k = \delta n$ for a constant $\delta \in (0,1)$, and $k = n - O(1)$. The middle two ($k = n^\gamma$ and $k = \delta n$) are the most natural for simulating randomized algorithms with weak random sources.

An ideal goal for a randomness extractor is to take one sample from an unknown k-source as input and output almost-uniformly distributed bits. Unfortunately, this is impossible to achieve:

Proposition 3.37. *For any* $\mathrm{Ext} : \{0,1\}^n \to \{0,1\}$ *there exists an* $(n-1)$*-source* X *so that* $\mathrm{Ext}(X)$ *is constant.*

Proof. There exists a $b \in \{0,1\}$ so that $|\mathrm{Ext}^{-1}(b)| \geq 2^n/2 = 2^{n-1}$. Then let X be the uniform distribution on $\mathrm{Ext}^{-1}(b)$. □

Thus, instead researchers turned to the problem of simulating randomized algorithms with a weak random source. That is, suppose we have a language $L \in \mathsf{BPP}$. The BPP algorithm for L assumes a source of truly uniform and independent bits. Can we decide membership in L in polynomial time if we are instead given one sample from a k-source X with large enough min-entropy k? Of course, the answer is yes if $\mathsf{BPP} = \mathsf{P}$, but here we want unconditional results, not assuming circuit lower bounds like Theorem 3.9. With Cohen [70], Wigderson gave the first positive answer to this question for sources of constant entropy rate, namely $\delta = k/n > 3/4$. This was then improved to any constant entropy rate $\delta > 0$ by Zuckerman [293], and then these approaches were abstracted by Nisan and Zuckerman [213] into the following elegant definition of a randomness extractor:

Definition 3.38 (seeded extractors [213]). A function $\mathrm{Ext} : \{0,1\}^n \times \{0,1\}^d \to \{0,1\}^m$ is a (k,ε)*-extractor* if for every k-source X on $\{0,1\}^n$, $\mathrm{Ext}(X, U_d)$ is ε-close to U_m.

That is, an extractor extracts almost-uniform bits given one sample from a k-source and a *seed* consisting of d truly random bits. The point is that if d is small enough, such as $d = O(\log n)$, we can eliminate the seed entirely by trying all 2^d possibilities rather than choosing it at random, similarly to Proposition 3.7.[8]

Indeed, using the Probabilistic Method, it can be shown that seed length $d = O(\log n)$ is possible:

Theorem 3.39 ([255, 294]). *For every $n \in \mathbb{N}$, $k \in [0,n]$ and $\varepsilon > 0$, there exists a (k,ε)-extractor* $\mathrm{Ext}: \{0,1\}^n \times \{0,1\}^d \to \{0,1\}^m$ *with $m = k + d - 2\log(1/\varepsilon) - O(1)$ and $d = \log(n-k) + 2\log(1/\varepsilon) + O(1)$. Indeed, a randomly chosen function* Ext *with these parameters is a (k,ε)-extractor with high probability.*

Both the lower bound on the output length m and upper bound on the seed length d can be shown to be optimal up to additive constants for almost all settings of parameters [220]. A small constant ε, say $\varepsilon = 1/8$, can be shown to be sufficient for simulating randomized algorithms with a weak random source. In this case, the seed length is $d = \log(n-k) + O(1)$ and we extract all but $O(1)$ of the $k+d$ bits of entropy that is fed is into the extractor as input.

However, like with expanders, for applications of extractors, we typically need explicit constructions, ones where Ext is computable in polynomial time. There was a long line of work giving increasingly improved constructions of extractors, and a milestone was achieved by Wigderson, together with Lu, Reingold, and Vadhan [190], who gave explicit extractors that are optimal up to constant factors.

Theorem 3.40 ([190]). *For all constants $\varepsilon, \alpha > 0$, and all $n, k \in \mathbb{N}$, there is an explicit (k,ε)-extractor* $\mathrm{Ext} : \{0,1\}^n \times \{0,1\}^d \to \{0,1\}^m$ *with $d = O(\log n)$ and $m = (1-\alpha) \cdot k$.*

In fact, the error parameter ε in Theorem 3.40 can be made subconstant, even almost polynomially small. Constructions with no constraint on ε were later given by Guruswami, Umans, and Vadhan [122] and by Dvir and Wigderson [86], the latter being based on Dvir's resolution of the Kakeya problem in finite fields [82]. For taking the entropy loss rate α parameter to be subconstant, the first construction was given earlier than Theorem 3.40 by Wigderson and Zuckerman [285], but had seed length $d = \mathrm{polylog}(n)$ rather than $d = O(\log n)$. Subsequent to Theorem 3.40, Dvir, Kopparty, Saraf, and Sudan [83] achieved $d = O(\log n)$ with $\alpha = 1/\mathrm{polylog}(n)$.

[8] The similarity of this approach to derandomization via pseudorandom generators is not a coincidence. Trevisan [269] showed that Wigderson et al.'s conditional construction of pseudorandom generators from circuit lower bounds (Theorem 3.9) can also be interpreted as an unconditional construction of randomness extractors! Indeed, the same holds for any construction of pseudorandom generators from a "black-box" hard function f, and thus Wigderson's two lines of work on pseudorandom generators and extractors were unified.

Extractors vs. Expanders

The works of Wigderson with Cohen [70] and with Friedman [100] showed that explicit constructions of certain kinds of imbalanced bipartite expanders suffice for simulating randomized algorithms with weak sources of randomness. Building on this connection, the Nisan–Zuckerman definition of seeded extractors [213] can be interpreted graph-theoretically as follows. Given any function $\mathrm{Ext} : \{0,1\}^n \times \{0,1\}^d \longrightarrow \{0,1\}^m$, we can view Ext as a bipartite graph G with $N = 2^n$ vertices on the left, $M = 2^m$ vertices on the right, and left-degree $D = 2^d$, where the y'th neighbor of $x \in \{0,1\}^n$ is $\mathrm{Ext}(x,y)$.

Suppose Ext is a (k,ε)-extractor. Then given any set $S \subseteq \{0,1\}^n$ of size $K = 2^k$, the uniform distribution on S, which we'll denote U_S, is a k-source. The extractor property tells us that $\mathrm{Ext}(U_S, U_{[D]})$ is ε-close to uniform on $[M]$. That is, a random neighbor of a random element of S is ε-close to uniform on the right-hand vertices of G. In particular, $|N(S)| \geq (1-\varepsilon)M$. This property is just like vertex expansion, except that it ensures a large neighborhood for sets of size exactly K (rather than all sets of size at most K). Indeed, this variant of vertex expansion was introduced in graph-theoretic form in [217, 242, 255], and is equivalent to the following relaxation of extractors.

Definition 3.41 (dispersers). A function $\mathrm{Disp} : \{0,1\}^n \times \{0,1\}^d \to \{0,1\}^m$ is a (k,ε)-*disperser* if for every k-source X on $\{0,1\}^n$, $\mathrm{Disp}(X, U_d)$ has a support of size at least $(1-\varepsilon)\cdot 2^m$.

Despite this connection, the parameters most commonly studied for extractors/dispersers and expanders are quite different. Extractors and dispersers typically have polylogarithmic degree (e.g. $D = \mathrm{polylog}(N)$, corresponding to seed length $d = O(\log n)$), are very imbalanced (e.g. $M = N^{\delta}$ for a constant $\delta \in (0,1)$), and often do not actually 'expand' (i.e. $|N(S)| < |S|$, since we are generally satisfied with retaining entropy, not necessarily increasing it). Nevertheless, in the "high min-entropy regime" $k = (1-o(1))n$, extractors and expanders become more closely related, and indeed Goldreich and Wigderson [114] showed that by taking a power of a constant-degree spectral expander, we obtain the following "high min-entropy extractors":

Theorem 3.42 ([114]). *For every $n, k \in \mathbb{N}$ and $\varepsilon > 0$, there is an explicit (k,ε)-extractor* $\mathrm{Ext} : \{0,1\}^n \times \{0,1\}^d \longrightarrow \{0,1\}^n$ *with $d = O(n - k + \log(1/\varepsilon))$.*

Note that the seed length of this extractor is linear rather than logarithmic, but importantly it is linear in $n-k$ rather than just n. So when $k = n - o(\log n)$, the seed length is shorter than that of Theorem 3.40. The origin of Wigderson's zig-zag product described in Section 3.2.1 was in the context of extractors, to compose extractors such as given in Theorem 3.40 and in Theorem 3.42 to obtain a "best of both" seed length of $O(\log(n-k))$ [237].

Wigderson's constant-degree expanders with expansion $(1-\varepsilon)D$ [62] came from considering a common generalization of expanders and extractors. In applying an extractor, any distribution X that has large enough (min-)entropy gets transformed into one that is close to uniform. In contrast, a random step on expander transforms any distribution X that does *not* have too much entropy into one with higher entropy. Formally, a spectral expander can be interpreted as one that increases Rényi entropy (noting that the expression $\mathrm{E}_{x \xleftarrow{R} X}[\Pr[X=x]]$ that appears in the definition of Rényi entropy equals the squared ℓ_2 norm of the probability mass function of X). To bridge the two, we can ask for a function $\mathrm{Con}:\{0,1\}^n \times \{0,1\}^d \to \{0,1\}^m$ such that for every random variable X of min-entropy $k \le k_{max}$, it holds that $\mathrm{Con}(X,U_d)$ is ε-close to having min-entropy at least $k+a$. Such a function is a necessarily a $(K_{max},(1-\varepsilon)A)$ vertex expander (where $K_{max}=2^{k_{max}}$ and $A=2^a$), and in fact if $a=d$, the converse holds as well [266]. A general abstraction of *randomness conductors* that encompasses all of these notions was given in [62], and a zig-zag product for conductors was developed and used to obtain constant-degree bipartite expanders with expansion $(1-\varepsilon)\cdot D$.

The ℓ_2-to-ℓ_1 switch from requiring that Rényi entropy increases to only requiring that the output distribution is ε-close in total variation distance to having higher entropy is crucial for enabling these results. Indeed, it is impossible to derive expansion greater than $D/2$ from spectral expansion alone [152]. Already in Wigderson's earlier work with Zuckerman [285], randomness extractors were used to construct balanced bipartite vertex expanders of non-constant degree that are impossible to derive from spectral expansion.

3.2.3 Multi-source Extractors and Ramsey Graphs

In the previous section, we argued that seeded extractors (Definition 3.38) suffice for simulating randomized algorithms with a single sample from a weak random source because we can enumerate over all possible seeds in polynomial time. However, this trick does not work for a number of other applications of randomness, such as in cryptography, distributed computing, and Monte Carlo simulation, where it is not clear how to combine the results from enumeration. Thus, it is natural to ask whether we can extract almost-uniform bits given *only* access to weak sources of randomness, i.e. with no uniformly random seed.

For example, we could consider extracting randomness from a small number of independent k-sources, a problem first studied by Chor and Goldreich [67]. That is we want a function $\mathrm{Ext}:(\{0,1\}^n)^c \to \{0,1\}^m$ such that for all independent random variables $X_1,X_2,\ldots,X_c$ where each X_i is a k-source, $\mathrm{Ext}(X_1,X_2,\ldots,X_c)$ is ε-close to U_m. Or we could weaken the requirement to that of a disperser, where we only require that the output has support size at least $(1-\varepsilon)\cdot 2^m$.

In addition to their motivation for obtaining high-quality randomness, extractors for $c=2$ independent sources are of interest because of connections to communication complexity and to Ramsey theory. In particular, a disperser for 2 independent k-sources of length n with output length $m=1$ is *equivalent* to a *bipartite Ram-*

sey graph — a bipartite graph with N vertices on each side that contains no $K \times K$ bipartite clique or $K \times K$ bipartite independent set (for $N = 2^n$ and $K = 2^k$): connect left vertex x and right vertex y iff $\mathrm{Disp}(x,y) = 1$. Giving explicit constructions of Ramsey graphs that approach $K = O(\log N)$ bound given by the Probabilistic Method [91] is a long-standing open problem posed by Erdős [92].

Chor and Goldreich [67] gave extractors for 2 independent sources of min-entropy rate δ (i.e. k-sources on $\{0,1\}^n$ with $k = \delta n$) when $\delta > 1/2$, and there was no improvement in this bound for nearly 2 decades. Substantial progress began again in Wigderson's work with Barak and Impagliazzo [30], who used new results in arithmetic combinatorics to construct extractors for a constant number of independent sources of min-entropy rate δ for an arbitrarily small constant $\delta > 0$. Specifically, they used the Sum–Product Theorem over finite fields of Bourgain, Katz, and Tao [50]; this theorem says that for p prime and every subset $A \subseteq \mathbb{F}_p$ whose size is not too close to p, either the set $A + A$ of pairwise sums or the set $A \cdot A$ of pairwise products is of size significantly larger than $|A|$. Using this theorem and other results in additive number theory, Barak, Impagliazzo, and Wigderson show that if A, B, C are random variables distributed in $\mathbb{F}_p$ with min-entropy rate $\delta < .9$, then $A \cdot B + C$ is ε-close to having min-entropy rate $(1+\alpha) \cdot \delta$ for a universal constant $\alpha > 0$. Recursively applying this result reduces the task of extracting from $\mathsf{poly}(1/\delta)$ sources of min-entropy δn to extracting from 2 sources of min-entropy rate larger than $1/2$, which allows for applying the Chor–Goldreich extractor [67].

In subsequent works, Wigderson obtained even better multi-source extractors and dispersers. With Barak, Kindler, Shaltiel, and Sudakov [31], Wigderson constructed explicit extractors for 3 sources of min-entropy $k = \delta n$ [31]. With Barak, Rao, and Shaltiel [32], Wigderson constructed dispersers for 2 sources of min-entropy $k = n^{o(1)}$ [32], or equivalently bipartite Ramsey graphs that avoid $K \times K$ cliques and independent sets of size $K = 2^{(\log N)^{o(1)}}$. This latter result was a major improvement over the previous best explicit construction of Ramsey graphs by Frankl and Wilson [97], which had $K = 2^{\sqrt{n}}$ and only applied to the nonbipartite case. A long line of subsequent work has continued to improve the parameters of 2-source extractors and dispersers, and very recently Li [179] has achieved 2-source extractors for min-entropy $k = O(\log n)$, which is optimal up to a constant factor, and thus bipartite Ramsey graphs for $K = \mathrm{polylog}(N)$, which is optimal up to the constant in the exponent.

3.3 Unconditional derandomization

Theorem 3.9 of Wigderson and collaborators gives strong evidence that randomness does not provide a substantial gain in the efficiency of algorithms, but it assumes circuit lower bounds that we are very far from proving. Thus, together with Ajtai [5], Wigderson asked whether there are large classes of algorithms that we can *unconditionally derandomize*, namely without making any unproven complexity assumptions.[9] They showed that this is indeed possible, giving an unconditional subexponential-time derandomization of probabilistic constant-depth circuits. After that, unconditional derandomization became a huge area of research, which is still flourishing. We refer the reader to the survey by Hatami and Hoza [131] for recent developments in the area.

3.3.1 Undirected S-T Connectivity

One subclass of BPP that has proved amenable to unconditional derandomization is BPL, where we restrict the algorithms to use a logarithmic amount of space. (When we measure the space complexity of an algorithm, we only count the read-write working memory, and do not count the space needed for the read-only input and write-only output.)

Definition 3.43. A language L is in BPL if there exists a randomized algorithm A that always halts, uses space at most $O(\log n)$ on inputs of length n, and satisfies the following for all inputs x:

- $x \in L \Rightarrow \Pr[A(x) \text{ accepts}] \geq 2/3$.
- $x \notin L \Rightarrow \Pr[A(x) \text{ accepts}] \leq 1/3$.

The standard model of a randomized space-bounded machine is one that has access to a coin-tossing box (rather than an infinite tape of random bits), and thus must explicitly store in its workspace any random bits it needs to remember. The requirement that A always halts ensures that its running time is at most $2^{O(\log n)} = \mathsf{poly}(n)$, because otherwise there would be a loop in its configuration space. Thus $\mathsf{BPL} \subseteq \mathsf{BPP}$.

Similarly to the time case (Definition 3.5), we can ask what is the smallest deterministic space bound needed to simulate BPL:

Definition 3.44 (Deterministic Space Classes).

$$\begin{aligned}\mathsf{DSPACE}(s(n)) &= \{L : L \text{ can be decided deterministically in space } O(s(n))\}\\ \mathsf{L} &= \mathsf{DSPACE}(\log n)\\ \mathsf{L}^c &= \mathsf{DSPACE}(\log^c n)\end{aligned}$$

Classic results in complexity theory [48, 148] tell us that $\mathsf{BPL} \subseteq \mathsf{L}^2$; however, this is not really a result about randomized algorithms, since it applies even for the unbounded-error version of BPL (where inputs in L are accepted with probability greater than 1/2 and inputs not in L with probability at most $1/2$). Thus the interesting question is whether we can show $\mathsf{BPL} = \mathsf{L}$ (randomization provides only a constant-factor savings in memory), or at least $\mathsf{BPL} \subseteq \mathsf{L}^c$ for a constant $c < 2$.

[9] The work of Ajtai and Wigderson [5] actually preceded Theorem 3.9, but was instead motivated by Yao's proof [289] that $\mathsf{BPP} \subseteq \mathsf{SUBEXP}$ under the assumption that cryptographic pseudorandom generators exist.

The potential power of randomization for logspace algorithms was first demonstrated in the late 1970s for the following basic problem:

Computational Problem 3.45. UNDIRECTED S-T CONNECTIVITY: *Given an undirected graph G and two vertices s and t, is there a path from s to t in G?*

Basic algorithms like breadth-first or depth-first search solve UNDIRECTED S-T CONNECTIVITY in linear time, but also take linear space. With randomization we can solve the problem in only logarithmic space:

Theorem 3.46 ([6]). UNDIRECTED S-T CONNECTIVITY *is in* BPL.

Proof (sketch). The algorithm simply does a polynomial-length random walk starting at s:

Algorithm 3.47 (UNDIRECTED S-T CONNECTIVITY **via Random Walks).**
Input: (G,s,t), where $G=(V,E)$ has n vertices.

1. Let $v=s$.
2. Repeat $\mathsf{poly}(n)$ times:
   a. If $v=t$, halt and accept.
   b. Else randomly select $v \stackrel{R}{\leftarrow} \{w : (v,w) \in E\}$.
3. Reject (if we haven't visited t yet).

Notice that this algorithm only requires space $O(\log n)$, in order to maintain the current vertex v as well as a counter for the number of steps taken. Clearly, it never accepts when there isn't a path from s to t. It can be shown that in any connected undirected graph, a random walk of length $\mathsf{poly}(n)$ from one vertex will hit any other vertex with high probability. Applying this to the connected component containing s, it follows that the algorithm accepts with high probability when s and t are connected. □

Using Nisan's pseudorandom generator for space-bounded computation [208], Wigderson, together with Nisan and Szemerédi [209], proved that UNDIRECTED S-T CONNECTIVITY is in $\mathsf{L}^{3/2}$. Inspired by that result, Saks and Zhou [241] then proved that $\mathsf{BPL} \subseteq \mathsf{L}^{3/2}$, which remains essentially the best derandomization of BPL to date.[10] Then Wigderson, together with Armoni, Ta-Shma, and Zhou [18], proved that UNDIRECTED S-T CONNECTIVITY is in $\mathsf{L}^{4/3}$. In 2005, Reingold [234] finally resolved the space complexity of UNDIRECTED S-T CONNECTIVITY:

Theorem 3.48 ([234]). UNDIRECTED S-T CONNECTIVITY*is in* L.

---

[10] Recently, Hoza [136] gave a slight improvement, showing that $\mathsf{BPL} \subseteq \mathsf{DSPACE}(\log^{3/2} n/\sqrt{\log\log n})$.

Reingold's Theorem is based on the following two ideas:

- UNDIRECTED S-T CONNECTIVITY can be solved in logspace on constant-degree expander graphs. More precisely, it is easy on constant-degree graphs where every connected component is promised to be an expander (i.e. has spectral expansion bounded away from 0): we can try all paths of length $O(\log N)$ from s in logarithmic space; this works because expanders have logarithmic diameter.
- The same operations that Reingold, Vadhan, and Wigderson [238] used to construct an infinite expander family (described Section 3.2.1) can also be used to turn *any* graph into an expander (in logarithmic space). There, we started with a constant-sized expander and used various operations to build larger and larger expanders. The goal was to increase the size of the graph (which was accomplished by zig-zag and/or tensoring), while preserving the degree and the expansion (which was accomplished by zig-zag and squaring). Here, we want to *improve* the expansion (which is accomplished by squaring), while preserving the degree (as is handled by zig-zag) and ensuring the graph remains of polynomial size (so tensoring is counterproductive and not used).

3.3.2 General Space-Bounded Computation

Like in the time-bounded case, one of the main approaches to derandomizing BPL is to construct pseudorandom generators $G: \{0,1\}^d \to \{0,1\}^n$ such that no randomized $(\log n)$-space algorithm can distinguish $G(U_d)$ from U_n. In order to get derandomizations that are correct on every input x, we require pseudorandom generators that fool *nonuniform* space-bounded algorithms. Since randomized space-bounded algorithms get their random bits as a stream of coin tosses, we only need to fool space-bounded distinguishers that read each of their input bits once, in order. Thus, instead of boolean circuits, we want pseudorandom generators for the following class of distinguishers:

Definition 3.49. An *ordered branching program B* of *width w* and *length n* is given by a *start state* $s_0 \in [w]$, *m transition functions* $B_1, \ldots, B_n : [w] \times \{0,1\} \to [w]$, and a set $A \subseteq [w]$ of *accept states*. On an input $x \in \{0,1\}^n$, B computes by updating its state via the rule $s_i = B_i(s_{i-1}, x_i)$ for $i = 1, \ldots, n$ and accepting iff $s_n \in A$.

The width w of a branching program corresponds to a space bound of $\log w$ bits. Similarly to Theorem 3.14, a family of generators $G_n : \{0,1\}^{d(n)} \to \{0,1\}^n$ that is computable in space $O(d(n))$ and such that $G_n(U_{d(n)})$ cannot be distinguished from U_n by ordered branching of width $w = n$ implies that $\mathsf{BPL} \subseteq \bigcup_c \mathsf{DSPACE}(c \log n + d(n^c))$. (Enumerating all seeds of length $d(m)$ only requires an additive space increase of $d(m)$.) In particular, a pseudorandom generator with seed length $d(n) = O(\log^c n)$ immediately implies $\mathsf{BPL} \subseteq \mathsf{L}^c$.

Unfortunately, the best known pseudorandom generator for general space-bounded computation is Nisan's generator [208], whose seed length of $O(\log^2 n)$ does not improve on the bound $\mathsf{BPL} \subseteq \mathsf{L}^2$. Nevertheless, Saks and Zhou [241] used Nisan's generator as part of a more sophisticated algorithm to obtain their result that $\mathsf{BPL} \subseteq \mathsf{L}^{3/2}$.

Together with Impagliazzo and Nisan [139], Wigderson gave an appealing alternative to Nisan's generator that has been the subject of much subsequent research and improved analyses for restricted models of ordered branching programs:

Definition 3.50. Given a sequence of regular digraphs $\mathscr{H} = (H_1, \ldots, H_\ell)$ where $\deg(H_i) = d_i$ and $|V(H_i)| = 2\prod_{j=1}^{i-1} d_j$, the **INW generator constructed with** $\mathscr{H}$, denoted $\mathrm{INW}_{\mathscr{H}}$ or INW_ℓ when the family is clear, is the function defined recursively where for $x \in \{0,1\}$ we have $\mathrm{INW}_0(x) = x$ and for $x \in V(H_i)$ and $y \in [d_i]$, we have

$$\mathrm{INW}_i(x,y) = (\mathrm{INW}_{i-1}(x), \mathrm{INW}_{i-1}(H_i[x,y])), \tag{4}$$

where $H_i[x,y]$ denotes the y'th neighbor of vertex x in the graph H_i. INW_i thus generates an output of length 2^i using a seed of length $\lceil \log(2\prod_{i=1}^{\ell} d_i) \rceil$.

That is, INW_i correlates the seeds of INW_{i-1} used to generate the first 2^{i-1} bits and the second 2^{i-1} bits as neighbors in the graph H_i. Impagliazzo, Nisan, and Wigderson [139] proved that an instantiation of this generator fools logspace algorithms with a seed length of $O(\log^2 m)$. They did this by analyzing the construction when the graphs H_i are good spectral expanders:

Theorem 3.51 ([139]). *If every graph H_i has spectral expansion at least $1-\sigma$, then INW_ℓ ε-fools ordered branching programs of with w and length $n = 2^\ell$ with error at most $\varepsilon = \sigma \cdot nw$.*

To achieve spectral expansion $1-\sigma$, we can use explicit expanders H_i with degree $d_i = \mathsf{poly}(1/\sigma)$, and hence get seed length

$$\left\lceil \log\left(2\prod_{i=1}^{\ell} d_i\right)\right\rceil = O\left(\log n \cdot \log\left(\frac{1}{\sigma}\right)\right).$$

To achieve error ε by Theorem 3.51, we should set $\sigma = \varepsilon/nw$, and thus we get seed length $O(\log n \cdot \log(nw/\varepsilon))$, exactly matching Nisan [206] and giving seed length $O(\log^2 n)$ when $w = n$ and $\varepsilon = 1/8$ as needed for derandomizing BPL.

To get intuition for Theorem 3.51, notice that if took the graph H_i to be complete graphs with self-loops, then in Expression (4) for INW_i we would be using independent seeds for the left half and right half, so the error (distinguishing advantage) of INW_i should be at most twice the error of INW_{i-1} (since we are using it twice). Furthermore, since an expander with spectral expansion at least $1-\sigma$ approximates the complete graph to within spectral norm at most σ, we incur an additional error of at most σw in the i'th level of recursion, where we pay a factor of w by summing the error over the w possible states of the branching program at the halfway point. Thus the error ε_i for INW_i can be bounded by the recurrence $\varepsilon_i \le 2\varepsilon_{i-1} + \sigma w$, which solves to $\varepsilon_\ell \le (2^\ell - 1) \cdot \sigma w < \sigma \cdot nw$.

Impagliazzo, Nisan, and Wigderson [139] actually proved that the INW generator fools a wider class of algorithms than ordered branching programs, called *network algorithms*. Subsequent developments, however, have focused on obtaining improved analyses for more restricted classes of ordered branching programs, namely *regular* and *permutation* branching programs. An ordered branching program B is a *permutation program* if for every i and bit $x_i \in \{0,1\}$, the transition function $B_i(\cdot, x_i) : [w] \to [w]$ is a permutation on the state set. That is, the transitions are reversible (for any fixed input x). A *regular* branching program is more general and just requires that for every state $s_i \in [w]$, there are exactly two pairs $(s_{i-1}, x_i) \in [w] \times \{0,1\}$ such that $B_i(s_{i-1}, x_i) = s_i$. A more intuitive formulation of regularity comes from thinking of each transition function B_i of the branching program as a bipartite graph with w vertices on each side, where left-vertex s_{i-1} is connected to right-vertices $B_i(s_{i-1}, 0)$ and $B_i(s_{i-1}, 1)$; in this viewpoint, a branching program is regular iff all of its associated bipartite graphs are regular. (They are always 2-leftregular; the additional requirement here is that they are also 2-rightregular.) One motivation for studying pseudorandomness for regular branching programs is that a general ordered branching program of width w and length n can be simulated by an ordered regular branching program of width wn [236, 46, 176].

The UNDIRECTED S-T CONNECTIVITY problem can be reduced to estimating the acceptance probability of an ordered permutation branching program, and it was shown by [240] that an instantiation of the INW generator with seed length $O(\log n)$ can be used to derandomize Algorithm 3.47 on the corresponding graphs and thus give a simpler proof of Reingold's Theorem (Theorem 3.48). Next, it was shown in [55] showed that the INW generator fools ordered regular branching programs with seed length $O(\log n \cdot \log\log n + \log n \cdot \log(w/\varepsilon))$. Note that this seed length is nearly linear rather than quadratic in $\log n$. In [166, 75, 260] it was shown that the INW generator fools ordered permutation branching programs with seed length $O(\mathsf{poly}(w) \cdot \log n \cdot \log(1/\varepsilon))$, which is $O(\log n)$ for constant w and ε. Finally, in [137], it was shown that the INW generator fools ordered permutation branching programs that have a single accept state with seed length $O(\log n \cdot \log\log n + \log n \cdot \log(1/\varepsilon))$, with no dependence on the width w. In [218, 46, 63], the INW generator, with these improved analyses, was also used to construct relaxations of pseudorandom generators (hitting-set generators and weighted pseudorandom generators) for ordered regular and/or permutation branching programs that have an even better dependence on the error parameter ε.

The key to these improved analyses is to show that the error of the INW generator accumulates more slowly for these models of branching programs than given by Theorem 3.51, for example achieving $\varepsilon = O(\sigma \cdot \log n)$ yields the result of [137]. The error analysis of [137] builds on [240] in viewing the composition of the INW generator with an ordered branching program as the result of an iterated graph operation. Note that if B is an ordered permutation branching program of length n and $G : \{0,1\}^d \to \{0,1\}^n$ is any generator, then composing B and G can be viewed as defining a 2^d-regular bipartite multigraph $B \circ G$ with w vertices on each side, where we connect left-vertex $s \in [w]$ to the final state reached when we run B on each of the outputs of G from start state s. The recursive operation (4) defining the INW generator amounts to taking a "product" of the two bipartite graphs $B_L \circ \mathrm{INW}_{i-1}$ and $B_R \circ \mathrm{INW}_{i-1}$, where B_L and B_R are the first and second halves of a program B of length 2^i. If the graph H_i is the complete graph with self-loops, then this is a standard graph product operation, where the edges are obtained by first following an edge in $B_L \circ \mathrm{INW}_{i-1}$ and then following an independent edge in $B_R \circ \mathrm{INW}_{i-1}$. (If the left half and right half are identical, this is simply graph squaring.) When H_i is a sparse expander, then this is a "derandomized product" operation that has a similar spirit to the zig-zag product of Wigderson and collaborators [238]. Analyzing this repeated derandomized product using notions of approximation from spectral graph theory [3] yields an improved analysis of the INW generator.

3.3.3 Constant-depth Circuits and Iterated Restrictions

The first computational model that was studied for unconditional derandomization, in the seminal paper of Ajtai and Wigderson [5], was constant-depth polynomial-size boolean circuits with unbounded fan-in AND and OR gates, also known as AC^0. They gave an unconditional construction of a pseudorandom generator with seed length $O(n^\varepsilon)$ fooling AC^0, for any constant $\varepsilon > 0$. This was improved by Nisan [207] to seed length polylog(n), using a construction that inspired the Nisan–Wigderson generator described in Section 3.1.4. (Indeed, Nisan's generator is the special case of the Nisan–Wigderson generator where the hard function f is the parity function.) Ajtai and Wigderson [5] pointed out a compelling algorithmic application of pseudorandom generators for AC^0, namely to derandomize the Karp–Luby BPP algorithm for approximately counting the number of satisfying assignments to a DNF formula (i.e. a depth 2 AC^0 circuit that is an OR of ANDs of literals) [159]. There are now nearly polynomial-time deterministic algorithms for this problem, all of which use pseudorandom generators along with other algorithmic techniques [211, 194, 193].

Although Nisan [207] dramatically improved upon the seed length of the Ajtai–Wigderson generator, the approach taken by Ajtai and Wigderson—*iterated pseudorandom restrictions*—has undergone a revival over the past decade. The idea of iterated pseudorandom restrictions is to not try to generate all n pseudorandom bits at once, but to use a short seed to select and assign values to a smaller fraction of the bits. If we use a seed of length d_0 to assign a p fraction of the bits, then by iterating, we can use a seed of length $O(d_0 \cdot (\log n)/p)$ to assign all the bits. The benefit of this approach is that when analyzing the pseudorandomness of the pn bits generated in each iteration, we can think of the remaining $(1-p)n$ bits as being chosen uniformly at random. Thus, fooling a test $T : \{0,1\}^n \to \{0,1\}$ reduces to fooling a random restriction ρ of T where we select $(1-p)n$ coordinates to restrict pseudorandomly but assign their values uniformly at random. For many computational models (in particular constant-depth circuits), random restrictions cause substantial simplification, making the restricted function $T|_\rho$ easier to fool.

Over the past decade, iterated pseudorandom restrictions and variants have been used to obtain improved pseudorandom generators for a variety of computational models. One example is the model of *combinatorial rectangles*, which test membership in a set of the form $R_1 \times R_2 \times \cdots \times R_n \subseteq [m]^n$, which can be viewed as a special case of both ordered branching programs and AC^0 formulas. For this model, Wigderson and collaborators gave the first pseudorandom generator whose seed length is logarithmic in m and n for a subconstant error parameter ε [17]. The iterated restriction approach of Ajtai and Wigderson was used in [118] to achieve a seed length that is nearly logarithmic in all the parameters, i.e. $\tilde{O}(\log(mn/\varepsilon))$. Since then, variants of the iterated restrictions approach have been used to obtain improved generators for constant-depth circuits, *arbitrary-order* read-once branching programs, De Morgan formulas, and various restricted versions of these models. The number of works is too large to list here, so we refer the reader to the excellent survey of Hatami and Hoza [131].

4 Computational Complexity Lower Bounds

Proving lower bounds for the resources needed to perform computational tasks, in different computational models, is among the most challenging and most important topics in theoretical computer science. Let us start by quoting the starting paragraph of Wigderson's recently-published monumental book, *Mathematics and Computation: A Theory Revolutionizing Technology and Science* [284]:

> Here is just one tip of the iceberg we'll explore in this book: How much time does it take to find the prime factors of a 1,000-digit integer? The facts are that (1) we can't even roughly estimate the answer: it could be less than a second or more than a million years, and (2) practically all electronic commerce and Internet security systems in existence today rest on the belief that it takes more than a million years!

This paragraph says it all. While computers have revolutionized our world, the resources required to perform computational tasks are poorly understood. Developing a mathematical theory of computation is crucial in our information age, where computers are involved in essentially every part of our life.

Computational complexity, the study of the amount of resources needed to perform computational tasks, is essential for understanding the power of computation and for developing a theory of computation. It is also essential in designing efficient communication protocols, secure cryptographic protocols and in understanding human and machine learning.

We present here some of Wigderson's works on computational complexity theory, focusing on computational complexity lower bounds. We will see that often these works introduced powerful techniques that had substantial impact and many followup works.

4.1 Boolean Circuit Complexity

Boolean circuits are the standard computational model for computing Boolean functions $f : \{0,1\}^n \to \{0,1\}$. Given a Boolean function $f : \{0,1\}^n \to \{0,1\}$, we ask how many Boolean operations are needed to compute f. As the set of allowed Boolean operations, we consider here the set of Boolean logical gates $\{\wedge, \vee, \neg\}$ (also known as De Morgan basis).

Given n input variables $x_1, \dots, x_n \in \{0,1\}$, a Boolean circuit is a directed acyclic graph as follows: All nodes are of in-degree 0 or 2. A node of in-degree 0 (that is, a *leaf*) is labelled with either an input variable x_i or its negation $\neg x_i$. A node of in-degree 2 is labelled with either $\wedge$ or $\vee$ (in the first case the node is an AND gate and in the second case an OR gate). A node of out-degree 0 is called an *output* node. The circuit is called a *formula* if the underlying graph is a (directed) tree.

Each node in the circuit (and in particular each output node) computes a Boolean function from $\{0,1\}^n$ to $\{0,1\}$ as follows. A leaf just computes the value of the input variable or negation of input variable that labels it. For every non-leaf node v, if v is an AND gate it computes the AND of the functions computed by its two children, and if v is an OR gate it computes the OR of the functions computed by its two children. If the circuit has only one output node, the function computed by the circuit is the function computed by the output node.

A Boolean circuit is *monotone* if it doesn't use negation gates. Each node in a monotone Boolean circuit (and in particular each output node) computes a monotone Boolean function from $\{0,1\}^n$ to $\{0,1\}$.

The *size* of a circuit is defined to be the number of nodes in it and the *depth* of a circuit is defined to be the length of the longest directed path from a leaf to an output node in the circuit. For a circuit C, we denote its size by $\mathsf{S}(C)$ and its depth by $\mathsf{D}(C)$. For a Boolean function f, we denote by $\mathsf{S}(f)$ the size of the smallest Boolean circuit for f, usually referred to as the circuit size of f, and by $\mathsf{D}(f)$ the smallest depth of a Boolean circuit for f, usually referred to as the circuit depth of f. For a monotone Boolean function f, we refer to the size of the smallest monotone Boolean circuit for f, as the monotone circuit size of f, and to the smallest depth of a monotone Boolean circuit for f, as the monotone circuit depth of f.

We note that often the unbounded-fanin case is also considered, where the in-degree of a node is not limited to be 0 or 2. For example, this is convenient when studying constant-depth circuits. In these cases, the size of the circuit is usually defined as the number of edges in it, rather than the number of nodes.

Proving lower bounds for the size and depth of Boolean circuits has been a major challenge for many years. In particular, the biggest challenge is to prove super-polynomial lower bounds for the size of Boolean circuits and formulas, for some explicit function. Such bounds would imply lower bounds for essentially all other models of computation. For example, super-polynomial (in n) lower bounds on the size of (a family of) circuits that compute a family of functions $\{f_n : \{0,1\}^n \to \{0,1\}\}_{n \in \mathbb{N}}$ would imply that that family of functions is not in the complexity class P (polynomial time). If in addition the family of functions is in NP (non-deterministic polynomial time), such a result would imply that $\mathsf{P} \neq \mathsf{NP}$.

However, progress on this type of questions has been very limited. The best known lower bounds for the size of Boolean circuits, for an explicit function, are only linear in n [171, 145], and the best known lower bounds for the depth of Boolean circuits, for an explicit function, are only logarithmic in n.

4.2 Communication Complexity

Communication complexity, first introduced by Yao [288], is a central model in complexity theory that studies the amount of communication needed to solve a problem, when the input to the problem is distributed between two (or more) parties.

In the two-player deterministic model, each of two players gets an input, where the two inputs x, y are chosen from some set of possibilities (known to both players). The players' goal is to solve a communication task that depends on both inputs, such as computing a function $f(x,y)$, where $f : \{0,1\}^n \times \{0,1\}^n \to \{0,1\}$ is known to both players and x, y are inputs of length n bits.

The players communicate in rounds, where in each round one of the players sends a message to the other player. At the end of the protocol, in the example given above, both players need to know the value of $f(x,y)$.

The communication complexity of a protocol is the maximal number of bits communicated by the players in the protocol, where the maximum is taken over all possibilities for the inputs. The communication complexity of a communication task is the minimal communication complexity of a protocol that solves that task. For a communication protocol P, we denote its communication complexity by $\mathsf{CC}(P)$. For a communication task G, we denote by $\mathsf{CC}(G)$ the smallest communication complexity of a (deterministic) protocol that solves G. The probabilistic case, where the players are allowed to use a public random string and are allowed to err with some fixed small probability smaller than $\frac{1}{2}$ is often studied as well. For a communication task G, we denote by $\mathsf{CC}_\varepsilon(G)$ the smallest communication complexity of a (probabilistic) protocol that solves G correctly with probability at least $1-\varepsilon$ on every input.

As an example, we give the problem of Set-Intersection, or Set-Disjointness, a central problem in communication complexity. In this problem, each of two players gets a vector in $\{0,1\}^n$ and their goal is to determine whether there exists a coordinate $i \in [n]$ where they both have 1. This simple problem inspired a lot of progress in communication complexity. It has been known for a long time that the probabilistic communication complexity of Set-Intersection is $\Omega(n)$ [155, 231, 29, 54, 53]. The lower bound is trivially tight, up to the multiplicative constant.

4.3 Karchmer–Wigderson Games

Karchmer and Wigderson gave a striking connection between the depth of Boolean circuits and communication complexity. They showed that for every Boolean function $f : \{0,1\}^n \to \{0,1\}$, there is a simple and intuitive communication complexity game G_f such that the smallest depth of a Boolean circuit for f is exactly equal to the deterministic communication complexity of G_f. Moreover, if f is monotone, there is also a communication complexity game M_f such that the smallest depth of a monotone Boolean circuit for f (that is, a Boolean circuit for f that doesn't use negations) is exactly equal to the deterministic communication complexity of M_f. In particular, this reduces the problem of proving lower bounds for the depth of Boolean circuits, a problem that seems hard to understand or analyze, to a problem in communication complexity that seems much more intuitive and easier to work with [157].

Definition 4.1. [157] **(KW Games, G_f):** For every function $f : \{0,1\}^n \to \{0,1\}$, define the communication game G_f as follows: Player 1 gets $x \in \{0,1\}^n$ such that $f(x) = 1$. Player 2 gets $y \in \{0,1\}^n$ such that $f(y) = 0$. The goal of the two players is to find a coordinate $i \in [n]$ such that $x_i \neq y_i$ (note that there is at least one such i since $f(x) \neq f(y)$).

Definition 4.2. [157] **(KW Games, M_f):** For every monotone function $f : \{0,1\}^n \to \{0,1\}$, define the communication game M_f as follows: Player 1 gets $x \in \{0,1\}^n$ such that $f(x) = 1$. Player 2 gets $y \in \{0,1\}^n$ such that $f(y) = 0$. The goal of the two players is to find a coordinate $i \in [n]$ such that $x_i = 1$ and $y_i = 0$ (note that there is at least one such i since $f(x) > f(y)$, and hence since f is monotone, $x \not\leq y$).

Recall that we denote deterministic communication complexity by CC and circuit depth by D. In particular, for a function $f : \{0,1\}^n \to \{0,1\}$, we denote by $\mathsf{D}(f)$ the smallest depth of a Boolean circuit for f. We denote by $\mathsf{CC}(G_f)$ the deterministic communication complexity of the game G_f, and if f is monotone, we denote by $\mathsf{CC}(M_f)$ the deterministic communication complexity of the game M_f.

Theorem 4.3 ([157]). *For every $f : \{0,1\}^n \to \{0,1\}$, $\mathsf{CC}(G_f) = \mathsf{D}(f)$.*

Proof. Let $z_1, \ldots, z_n \in \{0,1\}$ be the n input variables for f and recall that we denote by x, y the inputs for the game G_f.

Proving $\mathsf{CC}(G_f) \leq \mathsf{D}(f)$: Let C be any Boolean circuit for f. We will construct a communication protocol for the game G_f, with communication complexity $\mathsf{D}(C)$. The construction is by induction on $\mathsf{D}(C)$.

Base case: $\mathsf{D}(C) = 0$. In this case, $f(z_1, \ldots, z_n)$ is simply the function z_i or $\neg z_i$, for some i. Therefore, there is no need for communication, since i is a coordinate in which x and y always differ. That is, the two players can give the answer i, for any input pair (x, y). This is a protocol for G_f, with communication complexity 0.

Induction step: Consider the top gate of C. Assume first that the top gate is an AND gate and hence $C = C_1 \wedge C_2$, where C_1, C_2 are the two sub-circuits representing the two children of the top gate of C. Thus, $\mathsf{D}(C_1), \mathsf{D}(C_2) \leq \mathsf{D}(C) - 1$. Denote by f_1 and f_2 the functions computed by C_1 and C_2 respectively. Thus $f = f_1 \wedge f_2$. By the inductive hypothesis, $\mathsf{CC}(G_{f_1}), \mathsf{CC}(G_{f_2}) \leq \mathsf{D}(C) - 1$. We know that $f(x) = 1$ and $f(y) = 0$. Therefore, we know that $f_1(x), f_2(x)$ are both equal to 1 and at least one of $f_1(y)$ or $f_2(y)$ is equal to 0. Let us present the protocol for G_f. In the first step of the protocol, Player 2 sends a value in $\{1,2\}$, indicating which of the functions f_1 or f_2 is equal to 0 on y (or an arbitrary value in $\{1,2\}$ if both are equal to 0). Assume that Player 2 sends 1. In this case, we have $f_1(x) = 1$ and $f_1(y) = 0$. Hence, to solve the game G_f, the players can apply a protocol for G_{f_1}. By the inductive hypothesis, there is such a protocol with communication complexity $\mathsf{CC}(G_{f_1}) \leq \mathsf{D}(C) - 1$. In the same way, if Player 2 sends 2 the players can use the protocol for G_{f_2}. The players used only one additional bit of communication. Hence, we can conclude that

$$\mathsf{CC}(G_f) \leq 1 + \max\{\mathsf{CC}(G_{f_1}), \mathsf{CC}(G_{f_2})\} \leq 1 + (\mathsf{D}(C) - 1) = \mathsf{D}(C).$$

We assumed that $C = C_1 \wedge C_2$. The other case, $C = C_1 \vee C_2$, is proved in the same way, except that Player 1 is the one who sends the first bit, indicating whether $f_1(x) = 1$ or $f_2(x) = 1$.

Since the construction is valid for every circuit C for f, and in particular for the one with smallest depth, we can conclude that $\mathsf{CC}(G_f) \leq \mathsf{D}(f)$.

Proving $\mathsf{CC}(G_f) \geq \mathsf{D}(f)$: For this proof, we define a more general communication game. For any two disjoint sets: $A, B \subseteq \{0,1\}^n$, denote by $G_{A,B}$ the following game: Player 1 gets $x \in A$. Player 2 gets $y \in B$. The goal of the two players is to find a coordinate i such that $x_i \neq y_i$. Note that G_f is the same as $G_{f^{-1}(1), f^{-1}(0)}$.

We will prove the following claim: If $\mathsf{CC}(G_{A,B}) = d$ then there is a function $g : \{0,1\}^n \to \{0,1\}$ such that: $g(x) = 1$, for every $x \in A$; $g(y) = 0$, for every $y \in B$; and $\mathsf{D}(g) \leq d$. That is, the function g separates A from B, and $\mathsf{D}(g) \leq d$. Note that for the game $G_f = G_{f^{-1}(1), f^{-1}(0)}$, the function g must be the function f itself. Hence, we obtain that $\mathsf{D}(f) \leq \mathsf{CC}(G_f)$, as required. The proof of the claim is by induction on $d = \mathsf{CC}(G_{A,B})$.

Base case: $d = 0$. That is, the two players know the answer without any communication. Hence, there is a coordinate i such that, for every $x \in A$ and every $y \in B$, we have $x_i \neq y_i$. Thus, either the function $g(z) = z_i$ or the function $g(z) = \neg z_i$ satisfies the requirements of the claim (depending on whether for every $x \in A$ we have $x_i = 1$, or, for every $x \in A$ we have $x_i = 0$).

Induction step: We have a protocol of communication complexity d for the game $G_{A,B}$. Assume first that Player 1 sends the first bit in the protocol. That bit partitions the set A into two disjoint sets $A = A_0 \cup A_1$ (where A_0 is the set of all inputs x where Player 1 sends 0 and A_1 is the set of all inputs x where Player 1 sends 1). If the first bit sent by Player 1 is 0, the rest of the protocol is a protocol for the game $G_{A_0,B}$. If the first bit sent by Player 1 is 1, the rest of the protocol is a protocol for the game $G_{A_1,B}$. Hence, for both games, $G_{A_0,B}$ and $G_{A_1,B}$, we have protocols with communication complexity at most $d - 1$. By the inductive hypothesis, we have two functions g_0 and g_1 that satisfy: $g_0(x) = 1$, for every $x \in A_0$; $g_1(x) = 1$, for every $x \in A_1$; $g_0(y) = g_1(y) = 0$, for every $y \in B$; and $\mathsf{D}(g_0), \mathsf{D}(g_1) \leq d-1$. We define $g = g_0 \vee g_1$. Thus: For every $x \in A$, we have $g(x) = g_0(x) \vee g_1(x) = 1$; For every $y \in B$, we have $g(y) = g_0(y) \vee g_1(y) = 0$; and $\mathsf{D}(g) \leq 1 + \max\{\mathsf{D}(g_0), \mathsf{D}(g_1)\} \leq d$. That is, g satisfies the requirements.

If Player 2 sends the first bit, B is partitioned into two disjoint sets, $B = B_0 \cup B_1$, and as before, the rest of the protocol is a protocol for the games G_{A,B_0} and G_{A,B_1} (depending on the bit that was sent). By the inductive hypothesis, we have two functions, g_0, g_1, corresponding to the two games G_{A,B_0} and G_{A,B_1}, such that: $g_0(x) = g_1(x) = 1$, for every $x \in A$; $g_0(y) = 0$, for every $y \in B_0$; $g_1(y) = 0$, for every $y \in B_1$. We define $g = g_0 \wedge g_1$. Thus: For every $x \in A$, we have $g(x) = g_0(x) \wedge g_1(x) = 1$; For every $y \in B$, we have $g(y) = g_0(y) \wedge g_1(y) = 0$; and $\mathsf{D}(g) \leq 1 + \max\{\mathsf{D}(g_0), \mathsf{D}(g_1)\} \leq d$. □

For a monotone Boolean function $f : \{0,1\}^n \to \{0,1\}$, denote by $\mathsf{MD}(f)$ the smallest depth of a monotone Boolean circuit for f.

Theorem 4.4 ([157]). *For every monotone* $f : \{0,1\}^n \to \{0,1\}$, $\mathsf{CC}(M_f) = \mathsf{MD}(f)$.

Proof. Similar to the proof of Theorem 4.3. □

Example (k-Clique): Take the Boolean function f to be the $(n/2)$-Clique function in simple graphs with n vertices. That is, the input for f is a simple graph with n vertices and the output is 1 if and only if the graph contains a clique of size at least $n/2$. The games G_f and M_f are defined as follows: In both games, Player 1 gets a graph x (with n vertices) that contains a clique of size at least $n/2$ and Player 2 gets a graph y (with n vertices) that doesn't contain a clique of size at least $n/2$. The goal of the two players in the game M_f is to find an edge in the graph x that is not an edge in the graph y. The goal of the two players in the game G_f is to find an edge in the graph x that is not an edge in the graph y or an edge in the graph y that is not an edge in the graph x.
Theorem 4.3 shows that the communication complexity of the game G_f is exactly equal to the circuit depth of the $(n/2)$-Clique function. In particular, one can try to prove a lower bound for the circuit depth of the $(n/2)$-Clique function, by proving a lower bound for the communication complexity of the game G_f. Note that no lower bound better than $\Omega(\log n)$ has ever been proved for the circuit depth of an explicit Boolean function and such a bound would be a major breakthrough.
Theorem 4.4 shows that the communication complexity of the game M_f is exactly equal to the monotone circuit depth of the $(n/2)$-Clique function. Moreover, it turned out that one can use this connection to prove a lower bound for the monotone circuit depth of the $(n/2)$-Clique function, by proving a lower bound for the communication complexity of the game M_f [227].

We will now present an alternative equivalent way to define the game M_f, in terms of the minterms and maxterms of the monotone Boolean function f. Every monotone Boolean function can be characterized by the set of its minterms and the set of its maxterms.

Definition 4.5. (Minterm, Maxterm): Let $f : \{0,1\}^n \to \{0,1\}$ be a monotone Boolean function. A *minterm* of f is an input $x \in \{0,1\}^n$ such that $f(x) = 1$ and for every input $x' < x$, we have $f(x') = 0$. A *maxterm* of f is an input $y \in \{0,1\}^n$ such that $f(y) = 0$ and for every input $y' > y$, we have $f(y') = 1$.

Definition 4.6. [157] **(KW Games, M_f):** For every monotone function $f : \{0,1\}^n \to \{0,1\}$, define the communication game M_f as follows: Player 1 gets $x \in \{0,1\}^n$ such that x is a minterm of f. Player 2 gets $y \in \{0,1\}^n$ such that y is a maxterm of f. The goal of the two players is to find a coordinate $i \in [n]$ such that $x_i = 1$ and $y_i = 0$ (note that there is at least one such i since $f(x) > f(y)$, and hence since f is monotone, $x \not\leq y$).

We have defined the game M_f in two different ways, once in Definition 4.2 and once in Definition 4.6. While the two definitions do not give the exact same game, the two games are equivalent, so we denote both of them by M_f. To see the equivalence, let M_f' be the game from Definition 4.2 and let M_f'' be the game from Definition 4.6. First, note that M_f'' is a restriction of the game M_f' to a subset of inputs, so any protocol for M_f' is also a protocol for M_f''. On the other hand, the players can use a protocol for M_f'' to solve M_f' as follows: Given an input x such that $f(x) = 1$, Player 1 can find a minterm x' of f such that $x' \leq x$. In the same way, given an input y such that $f(y) = 0$, Player 2 can find a maxterm y' of f such that $y' \geq y$. The players can now apply the protocol for M_f'' on inputs x', y' to find a coordinate i such that $x_i' = 1$ and $y_i' = 0$. Since $x_i \geq x_i'$ and $y_i \leq y_i'$, we also have $x_i = 1$ and $y_i = 0$.

Example (s-t-Connectivity): Take the Boolean function f to be the s-t-Connectivity function in simple graphs with n vertices. That is, the input for f is a simple graph with n vertices, two of which are labeled as s and t, and the output is 1 if and only if the graph contains a path connecting s and t. Obviously, f is a monotone function, since adding edges cannot disconnect an existing path from s to t.
A minterm of f is a graph that contains a path from s to t, and no additional edges. That is, a minterm is just a path from s to t (that does not intersect itself). A maxterm of f is a graph G such that the set of vertices of G can be partitioned into two disjoint sets S and T, with $s \in S$ and $t \in T$ and such that G contains all edges inside S and inside T, but no edge between S and T. We think of a maxterm as a partition of the set of vertices into two sets (S and T), or as a two-coloring of the vertices by the colors 0 and 1 (where S is colored 0 and T is colored 1).
The game M_f is defined as follows: Given n vertices, two of which are labeled by s and t, Player 1 gets a path from s to t and Player 2 gets a coloring of the n vertices by the colors $\{0,1\}$ such that s is colored 0 and t is colored 1. The goal of the two players is to find an edge (u,v) on the path such that u is colored 0 and v is colored 1 (or vice versa).
Theorem 4.4 shows that the communication complexity of the game M_f is exactly equal to the monotone circuit depth of the s-t-Connectivity function. Moreover, it turned out that one can use this connection to prove a lower bound for the monotone

circuit depth of the s-t-Connectivity function, by proving a lower bound for the communication complexity of the game M_f [157].

Unlike the case of general Boolean circuits, where progress in proving lower bounds for explicit Boolean functions has been very limited, there has been a long and very successful line of works that establish strong lower bounds for the monotone circuit size and for the monotone circuit depth of many explicit functions, starting from Razborov's celebrated super-polynomial lower bounds for the size of monotone Boolean circuits [229, 230].

Since their introduction, KW games have had a huge impact on the study of monotone circuit depth and beyond, and have been further studied in numerous works. Already in their original paper, Karchmer and Wigderson used KW games to prove a tight lower bound of $\Omega(\log^2 n)$ for the monotone circuit depth of the ***s*-*t*-**Connectivity function in graphs with n vertices [157]. In particular, this result gave the first super-polynomial separation between monotone circuit size and monotone formula size and separated the monotone versions of the complexity classes NC_1 and NC_2. We present this result in Section 4.4.

Raz and Wigderson used KW games to prove tight lower bounds of $\Omega(n)$ for the monotone circuit depth of the clique and matching functions in graphs with n vertices [227]. We present this result in Section 4.5.

Karchmer, Raz and Wigderson used KW games to outline an approach for proving super-logarithmic lower bounds for the depth of general Boolean circuits [156]. We present this result in Section 4.6.

Raz and McKenzie used KW games to separate the monotone versions of the complexity classes NC and P, as well as NC_i and NC_{i+1} for every i [224]. That paper also introduced a general technique for proving lower bounds for communication complexity, a technique that was later on named by Göös, Pitassi and Watson, the *lifting method*. Göös, Pitassi and Watson initiated the study of the lifting method as a general technique for proving separation results in communication complexity [117], followed by a long line of recent works.

For a long time, KW games have been used mainly to study circuit depth, rather than circuit size. Nevertheless, a recent paper by Garg, Göös, Kamath and Sokolov shows how to use (an extension of) KW games to prove lower bounds for monotone circuit size [101], using Razborov's DAG-like communication protocols [232].

4.4 Lower Bounds for the Monotone Depth of ST-Connectivity

We will now present Karchmer and Wigderson's proof that any monotone circuit for the s-t-Connectivity function in graphs with n vertices is of depth $\Omega(\log^2 n)$ [157]. We deviate from Karchmer and Wigderson's original presentation in various places.

Recall that the game M_f for the s-t-Connectivity function is defined as follows: Given n vertices, two of which are labeled by s and t, Player 1 gets a path from s to t and Player 2 gets a coloring of the n vertices by the colors $\{0,1\}$ such that s is colored 0 and t is colored 1. The goal of the two players is to find an edge (u,v) on the path such that u is colored 0 and v is colored 1 (or vice versa). By Theorem 4.4, the communication complexity of this game is exactly equal to the monotone circuit depth of the s-t-Connectivity function.

It is helpful to first see an upper bound for the communication complexity of the game. A simple protocol for this game is as follows: In the first round, Player 1 sends the name (number) of the middle vertex in the path and Player 2 replies with its color. If the color of the middle vertex is 0 then the players continue with the second half of the path, and if the color is 1 then the players continue with the first half of the path. The players continue to perform a binary search, until they are left with a path of length 1. This path will be an edge (u,v) such that u is colored 0 and v is colored 1. In each round of the protocol, the players communicate $O(\log n)$ bits (the number of the vertex and its color). Since in each step the path is shortened by a factor of 2, the number of rounds will be $O(\log n)$. Altogether, the communication complexity of the protocol is $O(\log^2 n)$. Hence, by Theorem 4.4, the monotone circuit depth of s-t-Connectivity is $O(\log^2 n)$. Next, we present the lower bound.

Theorem 4.7 ([157]). *The monotone circuit depth of s-t-Connectivity is $\Omega(\log^2 n)$.*

Proof. For the proof of the lower bound, we will modify the communication game M_f for the s-t-Connectivity function, and present a variant of the game that we refer to as $STCON(\ell,n)$. In this game, there are two parameters, n and $\ell \le n^{0.1}$. We assume without loss of generality that ℓ is a power of 2. We have $\ell \cdot n$ vertices arranged in ℓ layers with n vertices in each layer, and two additional vertices s and t. We assume that the layers are numbered $(1,\ldots,\ell)$ and the vertices in each layer are numbered $(1,\ldots,n)$. Player 1 gets a path of length $\ell+1$ from s to t that passes through each of the ℓ layers exactly once, in their order. That is, the path starts at s, goes to a vertex in the first layer then a vertex in the second layer and so on, and finally goes from the last layer to the vertex t. Such a path can be presented as $x \in [n]^\ell$, specifying the number of the vertex that the path reaches in each layer. Player 2 gets a coloring of the $\ell \cdot n+2$ vertices by the colors $\{0,1\}$ such that s is colored 0 and t is colored 1. Such a coloring can be presented as $y \in \{0,1\}^{\ell \cdot n}$, specifying the color of each vertex in each layer. The goal of the two players is to find an edge (u,v) on the path such that u is colored 0 and v is colored 1 (or vice versa).

We will show a lower bound of $\Omega(\log \ell \cdot \log n)$ for the communication complexity of $STCON(\ell,n)$. Since $STCON(\ell,n)$ is a restriction to a subset of inputs of the game M_f (for the s-t-Connectivity function with $\ell \cdot n+2$ vertices), such a bound implies a lower bound of $\Omega(\log^2 n)$ for the monotone circuit depth of the s-t-Connectivity function in graphs with n vertices. Next, we give the proof for

$$\mathrm{CC}(STCON(\ell,n)) = \Omega(\log \ell \cdot \log n).$$

Let $X = [n]^\ell$ and $Y = \{0,1\}^{\ell \cdot n}$. For a subset $A \subseteq X$, we define its density as $\alpha = \frac{|A|}{|X|}$ and for a subset $B \subseteq Y$, we define its density as $\beta = \frac{|B|}{|Y|}$. Recall that in the game $STCON(\ell,n)$, the input for Player 1 is viewed as $x \in X$ and the input for Player 2 is viewed as $y \in Y$.

We will consider restrictions of the game $STCON(\ell,n)$ to subsets of inputs $A \subseteq X$ and $B \subseteq Y$ and define the game $STCON(\ell,n,A,B)$ to be the same as $STCON(\ell,n)$, except that the input for Player 1 is $x \in A$ and the input for Player 2 is $y \in B$. We denote by $\mathsf{C}(\ell,n,\alpha,\beta)$ the minimal communication complexity of a game $STCON(\ell,n,A,B)$ with a set $A \subseteq X$ of density α and a set $B \subseteq Y$ of density β.

Fixing n to be a (sufficiently large) integer, and fixing $t \stackrel{\text{def}}{=} \frac{1}{2n^{0.1}}$, we will show that for every $\ell \leq n^{0.1}$, every $\alpha \geq t$ and $\beta \geq 0$,

$$\mathsf{C}(\ell,n,\alpha,\beta) \geq c \cdot \log \ell \cdot \log n + \log(\alpha) + \log(\beta),$$

where $c > 0$ is a (sufficiently small universal) constant and the logarithm is base 2. Hence, $\mathsf{CC}(STCON(\ell,n)) = \mathsf{C}(\ell,n,1,1) = \Omega(\log \ell \cdot \log n)$.

The proof for $\mathsf{C}(\ell,n,\alpha,\beta) \geq c \cdot \log \ell \cdot \log n + \log(\alpha) + \log(\beta)$ is by induction over ℓ, α, β, in this order (and note that since n is fixed there is a finite number of possibilities for ℓ, α, β, so the induction is sound). We will consider two cases: $\alpha \geq 2t$ and $2t > \alpha \geq t$.

Case I: $\alpha \geq 2t$: Let $A \subseteq X$ be a subset of density α and $B \subseteq Y$ be a subset of density β. Consider any protocol P for the game $STCON(\ell,n,A,B)$ and let d be the communication complexity of the protocol. Since $\alpha \geq 2t$, none of the edges of the path x is fixed and hence $d > 0$. We will prove that

$$d \geq c \cdot \log \ell \cdot \log n + \log(\alpha) + \log(\beta).$$

Assume first that Player 1 sends the first bit in the protocol P. That bit partitions the set A into two disjoint sets $A = A_0 \cup A_1$ (where A_0 is the set of all inputs x where Player 1 sends 0 and A_1 is the set of all inputs x where Player 1 sends 1). If the first bit sent by Player 1 is 0, the rest of the protocol is a protocol for the game $STCON(\ell,n,A_0,B)$. If the first bit sent by Player 1 is 1, the rest of the protocol is a protocol for the game $STCON(\ell,n,A_1,B)$. Hence, for both games, $STCON(\ell,n,A_0,B)$ and $STCON(\ell,n,A_1,B)$, we have protocols with communication complexity at most $d-1$. Let α_0 be the density of A_0 and α_1 be the density of A_1. Note that $\alpha_0 + \alpha_1 = \alpha$ and hence at least one of α_0, α_1 is larger than or equal to $\alpha/2$. Hence $\mathsf{C}(\ell,n,\alpha/2,\beta) \leq d-1$. By the inductive hypothesis, $d-1 \geq c \cdot \log \ell \cdot \log n + \log(\alpha/2) + \log(\beta)$, that is

$$d \geq c \cdot \log \ell \cdot \log n + \log(\alpha) + \log(\beta).$$

The case where Player 2 sends the first bit in the protocol P is similar.

Case II: $2t > \alpha \geq t$: Let $A \subseteq X$ be a subset of density α and $B \subseteq Y$ be a subset of density β. Consider any protocol P for the game $STCON(\ell,n,A,B)$ and let d be the communication complexity of the protocol. We will prove that

$$d \geq c \cdot \log \ell \cdot \log n + \log(\alpha) + \log(\beta).$$

Note that we can assume without loss of generality that $\log(\beta) \geq -\log^2 n$, as otherwise the right hand side of the inequality is smaller than 0 (if $c < 1$).

Every path $x \in A$ can be written as $x = (x_L, x_R)$, where $x_L \in [n]^{\ell/2}$ is the left-hand half of the path x (the first $\ell/2$ coordinates of x) and $x_R \in [n]^{\ell/2}$ is the right-hand half of the path x (the last $\ell/2$ coordinates of x). We say that $x_L \in [n]^{\ell/2}$ is significant if there exist at least $\frac{\alpha}{4} \cdot n^{\ell/2}$ extensions $x_R \in [n]^{\ell/2}$ such that $(x_L, x_R) \in A$. Let $A_L \subseteq [n]^{\ell/2}$ be the set of significant paths x_L. We say that $x_R \in [n]^{\ell/2}$ is significant if there exist at least $\frac{\alpha}{4} \cdot n^{\ell/2}$ extensions $x_L \in [n]^{\ell/2}$ such that $(x_L, x_R) \in A$. Let $A_R \subseteq [n]^{\ell/2}$ be the set of significant paths x_R. Let α_L be the density of A_L in $[n]^{\ell/2}$, that is, $\alpha_L = \frac{|A_L|}{n^{\ell/2}}$. Let α_R be the density of A_R in $[n]^{\ell/2}$, that is, $\alpha_R = \frac{|A_R|}{n^{\ell/2}}$. Since for every $(x_L, x_R) \in A$, either x_L is not significant or x_R is not significant or both are significant, we have

$$\alpha \cdot n^{\ell} = |A| \leq n^{\ell/2} \cdot \tfrac{\alpha}{4} \cdot n^{\ell/2} + \tfrac{\alpha}{4} \cdot n^{\ell/2} \cdot n^{\ell/2} + \alpha_L \cdot n^{\ell/2} \cdot \alpha_R \cdot n^{\ell/2},$$

that is

$$\tfrac{\alpha}{2} \leq \alpha_L \cdot \alpha_R.$$

Thus, $\alpha_L \geq \sqrt{\frac{\alpha}{2}}$ or $\alpha_R \geq \sqrt{\frac{\alpha}{2}}$. Without loss of generality

$$\alpha_L \geq \sqrt{\tfrac{\alpha}{2}}.$$

Every coloring $y \in B$ can be written as $y = (y_L, y_R)$, where $y_L \in \{0,1\}^{(\ell/2)\cdot n}$ is the left-hand half of the coloring y (the first $(\ell/2) \cdot n$ coordinates of y) and $y_R \in \{0,1\}^{(\ell/2)\cdot n}$ is the right-hand half of the coloring y (the last $(\ell/2) \cdot n$ coordinates of y). We say that $y_L \in \{0,1\}^{(\ell/2)\cdot n}$ is significant if there exist at least $\frac{\beta}{2} \cdot 2^{\ell \cdot n/2}$ extensions $y_R \in \{0,1\}^{(\ell/2)\cdot n}$ such that $(y_L, y_R) \in B$. Let $B_L \subseteq \{0,1\}^{(\ell/2)\cdot n}$ be the set of significant colorings y_L. Let β_L be the density of B_L in $\{0,1\}^{(\ell/2)\cdot n}$, that is, $\beta_L = \frac{|B_L|}{2^{\ell \cdot n/2}}$. Since for every $(y_L, y_R) \in B$, either y_L is not significant or significant, we have

$$\beta \cdot 2^{\ell \cdot n} = |B| \leq 2^{\ell \cdot n/2} \cdot \tfrac{\beta}{2} \cdot 2^{\ell \cdot n/2} + \beta_L \cdot 2^{\ell \cdot n/2} \cdot 2^{\ell \cdot n/2},$$

that is,

$$\beta_L \geq \tfrac{\beta}{2}.$$

Let T be a subset of vertices (to be determined later) in the last $\ell/2$ layers, that is, layers $\frac{\ell}{2}+1, \ldots, \ell$. Given T, we define the set $A'_L \subseteq A_L$ to be the set of all $x_L \in A_L$ such that there exists an extension $x_R \in [n]^{\ell/2}$ such that $(x_L, x_R) \in A$ and all vertices of the path x_R are in T. Given T, we define the set $B'_L \subseteq B_L$ to be the set of all $y_L \in B_L$ such that there exists an extension $y_R \in \{0,1\}^{(\ell/2)\cdot n}$ such that $(y_L, y_R) \in B$ and all vertices of T are colored 1 by the coloring y_R.

We claim that the protocol P can be used to solve the communication game $STCON(\ell/2, n, A'_L, B'_L)$ and hence $\mathrm{CC}(STCON(\ell/2, n, A'_L, B'_L)) \leq d$. This can be done as follows. Given T and an input $x_L \in A'_L$, Player 1 finds an extension $x_R \in [n]^{\ell/2}$

such that $(x_L, x_R) \in A$ and all vertices of the path x_R are in T. Given T and an input $y_L \in B'_L$, Player 2 finds an extension $y_R \in \{0,1\}^{(\ell/2)\cdot n}$ such that $(y_L, y_R) \in B$ and all vertices of T are colored 1 by the coloring y_R. The players run the protocol P on inputs $x = (x_L, x_R)$, $y = (y_L, y_R)$. Since all vertices of the path x_R are in T, they are all colored 1 by the coloring y_R. Hence, the edge (u,v) returned by the communication protocol P must satisfy $u = s$ or u is in the first $\ell/2$ layers, that is, layers $1, \ldots, \frac{\ell}{2}$. Thus, it is a valid answer for the game $STCON(\ell/2, n)$ on inputs x_L, y_L.

It remains to show that there exists a subset of vertices T, as above, such that A'_L, B'_L are large, say $|A'_L| \geq |A_L|/2$ and $|B'_L| \geq |B_L|/2$. Assume first that there exists such a set T. Then, since $\alpha_L \geq \frac{\sqrt{\alpha}}{2}$ and $\beta_L \geq \frac{\beta}{2}$, we get

$$\mathsf{C}\left(\tfrac{\ell}{2}, n, \tfrac{\sqrt{\alpha}}{4}, \tfrac{\beta}{4}\right) \leq \mathsf{CC}(STCON(\ell/2, n, A'_L, B'_L)) \leq d.$$

Hence, by the inductive hypothesis,

$$\begin{aligned} d &\geq c \cdot \log\left(\tfrac{\ell}{2}\right) \cdot \log n + \log\left(\tfrac{\sqrt{\alpha}}{4}\right) + \log\left(\tfrac{\beta}{4}\right) \\ &= c \cdot \log \ell \cdot \log n - c \cdot \log n + \tfrac{1}{2} \cdot \log(\alpha) + \log(\beta) - 4 \\ &= c \cdot \log \ell \cdot \log n + \log(\alpha) + \log(\beta) - c \cdot \log n - \tfrac{1}{2} \cdot \log(\alpha) - 4 \\ &\geq c \cdot \log \ell \cdot \log n + \log(\alpha) + \log(\beta), \end{aligned}$$

where the last inequality is because $-c \cdot \log n - \frac{1}{2} \cdot \log(\alpha) - 4 \geq 0$, which is true for a sufficiently small constant c, since $\alpha < 2t = \frac{1}{n^{0.1}}$ (by the premise of Case II) and since n is sufficiently large.

Thus, it remains to argue that there exists a subset of vertices T, as above, such that $|A'_L| \geq |A_L|/2$ and $|B'_L| \geq |B_L|/2$. Recall that T is a subset of vertices in the last $\ell/2$ layers, that is, layers $\frac{\ell}{2} + 1, \ldots, \ell$. We will choose T randomly as follows. Take $n^{0.2}$ random paths $x_R \in [n]^{\ell/2}$ (viewed as paths on the last $\ell/2$ layers) and let T be the union of the sets of vertices on all these paths. Equivalently, the restriction of T to each layer (from the last $\ell/2$ layers), is generated by taking $n^{0.2}$ random vertices (with repetitions). Note also that T can be extended to a set $T' \supset T$ of size, say, $2n^{0.2} \cdot (\ell/2)$, such that the distribution of T' is exponentially close to the distribution of a random set of size $2n^{0.2} \cdot (\ell/2)$ of vertices in the last $\ell/2$ layers.

Recall that the set $A'_L \subseteq A_L$ is the set of all $x_L \in A_L$ such that there exists an extension $x_R \in [n]^{\ell/2}$ such that $(x_L, x_R) \in A$ and all vertices of the path x_R are in T. Recall that for every $x_L \in A_L$ there exist at least $\frac{\alpha}{4} \cdot n^{\ell/2}$ extensions $x_R \in [n]^{\ell/2}$ such that $(x_L, x_R) \in A$. Therefore, since $\frac{\alpha}{4} \geq \frac{1}{8n^{0.1}}$, for each $x_L \in A_L$ with probability exponentially close to 1, one of these extensions was chosen among the $n^{0.2}$ random paths that were chosen to generate T. Thus, with probability very close to 1 almost every $x_L \in A_L$ is also in A'_L and in particular $|A'_L| \geq |A_L|/2$.

Recall that the set $B'_L \subseteq B_L$ is the set of all $y_L \in B_L$ such that there exists an extension $y_R \in \{0,1\}^{(\ell/2)\cdot n}$ such that $(y_L, y_R) \in B$ and all vertices of T are colored 1 by the coloring y_R. Recall that for every $y_L \in B_L$ there exist at least $\frac{\beta}{2} \cdot 2^{\ell \cdot n/2}$ extensions $y_R \in \{0,1\}^{(\ell/2)\cdot n}$ such that $(y_L, y_R) \in B$, and recall that we assumed (without loss of generality) that $\beta \geq 2^{-\log^2 n}$. For each extension y_R, we consider the set T_{y_R} of all vertices (in the last $\ell/2$ layers) that y_R colors 1. For every $y_L \in B_L$, we consider the family of sets $F_{y_L} = \{T_{y_R}\}_{y_R:(y_L,y_R)\in B}$. Thus, for every $y_L \in B_L$, we have that $|F_{y_L}| \geq 2^{-\log^2 n-1} \cdot 2^{\ell \cdot n/2}$. By the Kruskal–Katona theorem [167, 160, 188] (or alternatively by information-theoretic arguments), such a large family of sets is guaranteed to contain, with probability close to 1, a set T_{y_R} that contains the random set T' (where the probability is over the choice of T'). Note that if T_{y_R} contains T', the coloring y_R colors all vertices in T' (and hence all vertices in T) by 1. Thus, with probability close to 1 almost every $y_L \in B_L$ is also in B'_L and in particular $|B'_L| \geq |B_L|/2$. □

4.5 Lower Bounds for the Monotone Depth of Clique and Matching

Next, we present Raz and Wigderson's lower bound of $\Omega(n)$ for the monotone circuit depth of the clique and matching functions in graphs with n vertices [227]. The proof establishes a lower bound of $\Omega(n)$ for the communication complexity of the corresponding KW games, by a direct reduction to known lower bounds in communication complexity, namely the lower bound of $\Omega(n)$ for the probabilistic communication complexity of Set-Disjointness [155, 231, 29, 54, 53]. This, in turn, further demonstrates the power of KW games, as well as the power of reductions from Set-Disjointness as a major tool for proving lower bounds in communication complexity and other computational models.

Recall that in the problem of Set-Intersection, or Set-Disjointness, each of two players gets a vector in $\{0,1\}^n$ and their goal is to determine whether there exists a coordinate $i \in [n]$ where they both have 1. Recall that for a communication task G, we denote by $\mathsf{CC}_\varepsilon(G)$ the smallest communication complexity of a probabilistic protocol that solves G correctly with probability at least $1-\varepsilon$ on every input.

Theorem 4.8 ([155]). *For any constant* $\varepsilon > 0$, $\mathsf{CC}_\varepsilon(Disjointness) \geq \Omega(n)$.

We will consider the following communication game, denoted M_1:

Definition 4.9. (Communication game M_1): Let $n = 3k$ and let V be a set of n vertices. Player 1 gets a k-matching x on (a subset of) the set of vertices V, that is, k edges (with vertices in V) that don't touch each other. Player 2 gets a set y of $k-1$ vertices in V. The goal of the two players is to find an edge in x that does not touch any of the vertices in y. (By the pigeonhole principle there must be at least one such edge).

We will prove that the deterministic communication complexity of M_1 is $\Omega(n)$,

$$\mathsf{CC}(M_1) \geq \Omega(n).$$

This bound implies lower bounds for the monotone circuit depth of several functions. We give a few examples:

Theorem 4.10 ([227]). *Let $n = 3k$. Let $Match$ be the (monotone) Boolean function that gets as an input a graph with n vertices and outputs 1 if and only if the graph contains a k-matching (and outputs 0 otherwise). The monotone circuit depth of $Match$ is $\Omega(n)$.*

Proof. Consider an input (x,y) for the game M_1. The k-matching x is a minterm of the function $Match$. The set y of $k-1$ vertices can be viewed as a maxterm of the function $Match$, by considering a graph that contains all possible edges with at least one vertex in y. Any protocol P for the monotone KW game of the function $Match$ can be applied on (x,y) to get an edge in x that doesn't touch y. That is, any protocol P for the monotone KW game of the function $Match$ can be applied also as a protocol for M_1. Since $\mathsf{CC}(M_1) \geq \Omega(n)$, the communication complexity of P is $\Omega(n)$. Hence, by Theorem 4.4, the monotone circuit depth of $Match$ is $\Omega(n)$. □

Theorem 4.11 ([227]). *Let PM be the (monotone) Boolean function that gets as an input a graph with n vertices and outputs 1 if and only if the graph contains a perfect matching (and outputs 0 otherwise). The monotone circuit depth of PM is $\Omega(n)$.*

Proof. Follows by a standard reduction from $Match$ to PM: Given an input graph Z for the function $Match$, where the number of vertices in Z is $n = 3k$, construct a graph Z' by adding k vertices to Z and connecting them to all other vertices. Then, there exists a perfect matching in Z' if and only if there exists a matching of size k in Z. □

Theorem 4.12 ([227]). *Let $n = 3k$. Let Clique be the (monotone) Boolean function that gets as an input a graph with n vertices and outputs 1 if and only if the graph contains a clique of size $2k+1$ (and outputs 0 otherwise). The monotone circuit depth of Clique is $\Omega(n)$.*

Proof. Consider an input (x,y) for the game M_1. Given the set y of $k-1$ vertices, consider the graph y' that contains all edges that do not touch y (that is, a clique in the complement of y). Since y' is a clique of size $2k+1$, the function *Clique* outputs 1 on y'. Given the k-matching x, let the graph x' be the complement of x, that is, the graph that contains all edges except the matching x. The function *Clique* outputs 0 on x'. Any protocol P for the monotone KW game of the function *Clique* can be applied on (y',x') to get an edge in y' that is not an edge in x', that is an edge of x that doesn't touch y. Thus, any protocol P for the monotone KW game of the function *Clique* can be applied also as a protocol for M_1. Since $\mathsf{CC}(M_1) \geq \Omega(n)$, the communication complexity of P is $\Omega(n)$. Hence, by Theorem 4.4, the monotone circuit depth of *Clique* is $\Omega(n)$. □

Using similar arguments, one can establish lower bounds for the monotone depth of several other functions, such as, matching and perfect matching in bipartite graphs and clique functions with different sizes of cliques.

It remains to prove the lower bound for the deterministic communication complexity of M_1.

Theorem 4.13 ([227]). $\mathsf{CC}(M_1) \geq \Omega(n)$.

Proof. Let $n = 3k$ and let V be a set of n vertices. Consider the following communication game, denoted M_2: Player 1 gets a k-matching x on (a subset of) the set of vertices V. Player 2 gets a set y of k vertices in V. The goal of the two players is to output 1 if there is an edge in x that does not touch any of the vertices in y, and output 0 otherwise, that is, if every edge in x touches a vertex in y.

We will first prove that for any constant $\varepsilon > 0$,

$$\mathsf{CC}(M_1) \geq \Omega(\mathsf{CC}_\varepsilon(M_2)). \tag{5}$$

Assume that we have a deterministic communication protocol P_1 for the communication game M_1. We will use P_1 to construct a probabilistic communication protocol P_2 for the communication game M_2, with the same communication complexity as P_1 (up to an additive constant).

First note that, using a common random string, we can assume that the protocol P_1 is a zero-error probabilistic protocol such that for every input (x,y) for the game M_1, the protocol P_1 outputs each correct answer with the exact same probability (that is, if for the input (x,y) there are several correct answers the protocol outputs each of them with the same probability). This can be assumed, since, using the common random string, the players can randomly permute the vertices in V before applying the protocol P_1.

Let (x,y) be an input for the game M_2. Player 2 gets the set y of k vertices, and will randomly choose a vertex $v \in y$ and remove it. Now, Player 2 is left with a set y' of $k-1$ vertices. The two players can now apply the protocol P_1 (for M_1) on the input (x,y') and obtain as an output an edge $e \in x$ that doesn't touch any of the vertices in y'. The players now check if the removed vertex v is on the edge e. If the vertex v is not on the edge e, the protocol P_2 (for M_2) will outputs 1 (as e is an edge that doesn't touch any vertex in y). If the vertex v is on the edge e, the protocol P_2 will output 0 (that is, P_2 assumes that there is no edge in x that doesn't touch y, as such an edge was not found by P_1).

Note that if P_2 outputs 1 there can be no error (as e does not touch v or any other vertex in y, since the protocol P_1 is always correct). On the other hand, if P_2 outputs 0, an error is possible, as there might be a different edge e' in x that doesn't touch any of the vertices in y, and yet the protocol P_1 outputs the edge e that does touch v. However, since the edges do not touch each other, there is at most one edge $e \in x$ that touches v. Since we assume that the protocol P_1 outputs each possible correct answer with the exact same probability, and since an error occurs only if e was the output of P_1 (and not any of the possible edges e'), the probability for an error is at most $1/2$ (for any input (x,y)). (The probability of error may be smaller if there are several edges e' that do not touch any vertex in y).

To further reduce the probability of error to any constant ε, one can repeat the protocol P_2 a constant number of times. This concludes the proof for Equation (5).

□

Next, we consider the following communication complexity game, denoted $3Dist$ (3-Distinctness): Let $n = 3k$. Player 1 and Player 2 get inputs $x, y \in \{a,b,c\}^k$, respectively. That is, each player gets a string of $k = n/3$ letters from $\{a,b,c\}$. The goal is to decide whether there is a coordinate i such that $x_i = y_i$.

We will prove that for any constant $\varepsilon > 0$,

$$\mathsf{CC}_\varepsilon(M_2) \geq \mathsf{CC}_\varepsilon(3Dist). \tag{6}$$

Assume that we have a probabilistic communication protocol P_2 for the communication game M_2. We will use P_2 to construct a probabilistic communication protocol P_3 for the communication game $3Dist$, with the same communication complexity and the same error as P_2.

Let (x,y) be an input for the game $3Dist$. Thus $x, y \in \{a,b,c\}^k$. For each coordinate in $\{1,\ldots,k\}$, we construct a triangle (with different vertices for each triangle) and label its 3 vertices by a,b,c. We label each edge of each triangle by the letter that labels the vertex that it does not touch (that is, the vertex opposite to it).

The players convert their inputs to inputs for the game M_2 in the following way: Player 1 interprets her k coordinates as the corresponding k edges in the k triangles (one edge for each coordinate). That is, each x_i is interpreted as the corresponding edge in the ith triangle. Denote the set of these edges by x'. Player 2 interprets her k coordinates as the corresponding k vertices in the k triangles. That is, each y_i is interpreted as the corresponding vertex in the ith triangle. Denote the set of these vertices by y'. Obviously, there is an edge in x' that doesn't touch y' if and only if there is a coordinate i such that $x_i = y_i$. Thus, the players can use the protocol P_2 on input (x',y') and declare the answer. This gives a probabilistic communication protocol P_3 for $3Dist$ with the same communication complexity and the same error as P_2. This concludes the proof for Equation (6).

Finally, we will prove that for any constant $\varepsilon > 0$,

$$\mathsf{CC}_\varepsilon(3Dist) \geq \Omega(n). \tag{7}$$

This will follow by a reduction from the Set-Disjointness problem and by the lower bound for the probabilistic communication complexity of Set-Disjointness (Theorem 4.8).

Assume that we have a probabilistic communication protocol P_3 for $3Dist$. We will show how to use this protocol to solve the Set-Disjointness problem. Given an input pair (x,y) for the Set Disjointness problem, such that $x, y \in \{0,1\}^k$, the two players will generate inputs x', y' for $3Dist$ as follows. To generate x', Player 1 starts from x and translates 0 to b and 1 to a. That is, for every i, if $x_i = 0$ then $x'_i = b$ and if $x_i = 1$ then $x'_i = a$. To generate y', Player 2 starts from y and translates 0 to c and 1 to a. That is, for every i, if $y_i = 0$ then $y'_i = c$ and if $y_i = 1$ then $y'_i = a$. Obviously, $x_i = y_i = 1$ if and only if $x'_i = y'_i$. Hence, the two players can apply the protocol P_3 on (x',y') and declare the answer. This gives a probabilistic communication protocol for Set-Disjointness, with the same communication complexity and the same error as P_3. By Theorem 4.8, the communication complexity of the protocol is $\Omega(n)$. This concludes the proof for Equation (7).

By Equation (5), Equation (6) and Equation (7), we get $\mathsf{CC}(M_1) \geq \Omega(n)$. □

4.6 The KRW Conjecture

Karchmer, Raz and Wigderson suggested an approach for proving super-logarithmic lower bounds for general Boolean circuit depth [156]. We will briefly outline this approach here.

Let n be an integer and assume for simplicity that $\log\log n$ is also an integer (where the logarithm is base 2). Let $k = \log n$. Let $f : \{0,1\}^k \to \{0,1\}$ be a random Boolean function. Since it's not hard to prove (by a standard counting argument) that a random Boolean function has large circuit depth (with high probability), we can assume that, say, $\mathsf{D}(f) \geq \frac{k}{2}$ (where D denotes circuit depth).

For two Boolean functions, $h : \{0,1\}^r \to \{0,1\}$ and $g : \{0,1\}^m \to \{0,1\}$, define their composition $h \circ g : \{0,1\}^{rm} \to \{0,1\}$ by

$$h \circ g(\mathrm{x}_1, \ldots, \mathrm{x}_\mathrm{r}) = h(g(\mathrm{x}_1), \ldots, g(\mathrm{x}_\mathrm{r})),$$

where $\mathrm{x}_1, \ldots, \mathrm{x}_\mathrm{r} \in \{0,1\}^m$. Define $f^{(d)}$ to be the composition of f with itself d times.

KRW conjectured that for a random function $f : \{0,1\}^k \to \{0,1\}$ and any function $g : \{0,1\}^m \to \{0,1\}$,

$$\mathsf{D}(f \circ g) \geq \varepsilon \cdot \mathsf{D}(f) + \mathsf{D}(g)$$

(with high probability over the choice of f), for some constant $\varepsilon > 0$. There are also various variants of this conjecture.

Assuming that the conjecture holds, we get

$$\mathsf{D}(f^{(d)}) \geq \varepsilon \cdot \mathsf{D}(f) \geq \tfrac{\varepsilon}{2} \cdot d \cdot k,$$

(with high probability over the choice of f). Taking $d = k/\log k$, we get a function $f^{(d)} : \{0,1\}^n \to \{0,1\}$ of super-logarithmic depth. The function $f^{(d)}$ is not explicit, as f is a random function, but since f depends on only $k = \log n$ input variables, its truth table of size n can be given as n additional input variables, so that $f^{(d)}$ is explicitly given.

To prove the conjecture, KRW suggested to use Karchmer–Wigderson games. Given $f : \{0,1\}^k \to \{0,1\}$ and $g : \{0,1\}^m \to \{0,1\}$, the KW game corresponding to $f \circ g$ is as follows:
Player 1 gets $\mathrm{x}_1, \ldots, \mathrm{x}_\mathrm{k} \in \{0,1\}^m$ such that

$$f(g(\mathrm{x}_1), \ldots, g(\mathrm{x}_\mathrm{k})) = 1.$$

Player 2 gets $y_1,\dots,y_k \in \{0,1\}^m$ such that

$$f(g(y_1),\dots,g(y_k)) = 0.$$

The goal of the two players is to find (i,j) such that $x_{ij} \neq y_{ij}$.

To see the intuition behind the conjecture, assume that the inputs $(x_1,\dots,x_k)$ for Player 1 and $(y_1,\dots,y_k)$ for Player 2 satisfy that for every $i \in \{1,\dots,k\}$, if $g(x_i) = g(y_i)$ then $x_i = y_i$. Then, an answer (i,j) for the KW game corresponding to $f \circ g$ gives an answer i for the KW game corresponding to f (with input $((g(x_1),\dots,g(x_k)),(g(y_1),\dots,g(y_k))))$ and an answer j for an instance of the KW game corresponding to g (namely, the KW game corresponding to the ith coordinate, that is, the game played with input (x_i,y_i)).

Finally, we note that while the conjecture is still wide open and seems hard to prove, some steps towards proving the conjecture have been taken in several papers, including the works by Edmonds, Impagliazzo, Rudich and Sgall [89], Håstad and Wigderson [129], Gavinsky, Meir, Weinstein and Wigderson [106], Dinur and Meir [80], Meir [197] (to name a few).

4.7 Communication Complexity of Set-Disjointness

The Set-Disjointness problem that was already mentioned before is a central problem in communication complexity, with numerous applications. We have already seen how strong lower bounds for the monotone depth of Boolean functions (Theorem 4.10, Theorem 4.11 and Theorem 4.12) follow from known lower bounds for the probabilistic communication complexity of Set-Disjointness (Theorem 4.8). The Set-Disjointness problem can be described as follows (equivalently to our previous description). Each of two players gets a subset of $[n]$ and their goal is to determine whether the two subsets intersect.

Since the Set-Disjointness problem is so central, and because of the many applications, it is very interesting to also study variants of this problem. Håstad and Wigderson [130] studied the perhaps most natural variant of the problem, the Set-Disjointness problem with sets of a fixed size k. That is, given n and $k \leq n/2$, Player 1 gets a subset $x \subset [n]$ of size k, Player 2 gets a subset $y \subset [n]$ of size k, and their goal is to determine whether the two subsets x,y intersect. We denote this communication game by D_k^n. In the deterministic case, it is not hard to prove that for every $k \leq n/2$, $\mathsf{CC}(D_k^n) = \Theta(\log\binom{n}{k})$ [130].

Håstad and Wigderson proved that in the probabilistic case, the communication complexity of D_k^n is in fact $O(k)$. (This bound is tight when $k < cn$ for any constant $c < 1/2$).

Theorem 4.14 ([130]). *For any n and $k \leq n/2$ and constant $\varepsilon > 0$, $\mathsf{CC}_\varepsilon(D_k^n) = O(k)$.*

Proof. **(Sketch)** Player 1 gets a subset $x \subset [n]$ of size k and Player 2 gets a subset $y \subset [n]$ of size k. The players run a communication protocol that, assuming that x, y are disjoint, has communication complexity $O(k)$ and at the end of the protocol both players know (with high probability) two disjoint subsets $S, T \subset [n]$ such that $x \subseteq S$ and $y \subseteq T$. The sets S, T can be viewed as a proof for the disjointness of x, y. If after ck bits of communication (when c is a sufficiently large constant), the protocol fails, it follows that (with high probability) x, y are not disjoint.

The protocol works in $O(\log k)$ steps. First define,

$$N_0 = [n], \;\; S_0 = \varnothing, \;\; T_0 = \varnothing, \;\; x_0 = x, \;\; y_0 = y.$$

After each Step i, the players will have subsets $N_i, S_i, T_i, x_i, y_i \subseteq [n]$, where $N_i \cup S_i \cup T_i$ is a partition of $[n]$ and

$$x \cap T_i = \varnothing, \;\; y \cap S_i = \varnothing, \;\; x_i = x \cap N_i, \;\; y_i = y \cap N_i.$$

Moreover,

$$S_{i-1} \subseteq S_i, \;\; T_{i-1} \subseteq T_i.$$

Intuitively, after each Step i, the players have already restricted the possible intersection of x and y to the set N_i and it remains to check if x_i and y_i intersect.

Each Step i is done as follows. Assume without loss of generality that $|x_{i-1}| \leq |y_{i-1}|$. (Otherwise we switch the rolls of the players in Step i). Let $k_{i-1} = |x_{i-1}|$. The players interpret the public random string as a sequence of random subsets $Z_1, Z_2, \ldots \subseteq N_{i-1}$. Player 1 examines the first $2^{ck_{i-1}}$ sets in this sequence (where c is a sufficiently large constant) and sends the index j of the first set Z_j such that $x_{i-1} \subseteq Z_j$ (if such a set exists). Since c is sufficiently large, such a set Z_j exists with high probability, as the probability that a random set Z_j satisfies $x_{i-1} \subseteq Z_j$ is $2^{-k_{i-1}}$. Since the random string is public, both players now know Z_j and update

$$N_i = Z_j, \;\; S_i = S_{i-1}, \;\; T_i = T_{i-1} \cup N_{i-1} \setminus Z_j, \;\; x_i = x_{i-1}, \;\; y_i = y_{i-1} \cap Z_j.$$

Note that all the required properties from the sets N_i, S_i, T_i, x_i, y_i are satisfied. That is, $N_i \cup S_i \cup T_i$ is a partition of $[n]$ and

$$x \cap T_i = \varnothing, \;\; y \cap S_i = \varnothing, \;\; x_i = x \cap N_i, \;\; y_i = y \cap N_i, \;\; S_{i-1} \subseteq S_i, \;\; T_{i-1} \subseteq T_i.$$

Note also that if x, y are disjoint then $|y_i|$ is equal to $|y_{i-1}|/2$ in expectation and as long as $|y_{i-1}|$ is sufficiently large (say, larger than a sufficiently large constant), $|y_i| \leq 0.6 \cdot |y_{i-1}|$ with high probability and hence $|y_i| + |x_i| \leq 0.8 \cdot (|y_{i-1}| + |x_{i-1}|)$ with high probability (where high probability here means 1 minus probability exponentially small in $|y_{i-1}| + |x_{i-1}|$).

If x, y are disjoint then, after repeating this protocol for $O(\log k)$ steps, we get that with high probability the final x_i, y_i are both empty. This is because the sum of their sizes keeps decreasing by a constant factor until it is smaller than a (sufficiently large) constant and then it keeps decreasing by at least 1, with a constant probability in each step. When x_i, y_i are both empty, the protocol stops, and we have two disjoint subsets $S,T \subset [n]$ such that $x \subseteq S$ and $y \subseteq T$ (where S,T are the final S_i,T_i). The communication complexity is $O(k)$, since in each step the communication complexity is $O(|y_{i-1}|+|x_{i-1}|)$ and thus converges to $O(k)$, as $O(|y_{i-1}|+|x_{i-1}|)$ keeps decreasing by a constant factor until it is constant. □

The protocol in the proof of Theorem 4.14 uses an exponential amount of randomness. Nevertheless, the amount of randomness can always be reduced to $O(\log n)$ by a general theorem of Newman [204].

4.8 Quantum versus Classical Communication Complexity

Buhrman, Cleve and Wigderson were the first to study communication complexity advantages of quantum communication protocols over classical ones [56]. A quantum communication protocol is a protocol where the players can send quantum states, rather than just classical bits, and the communication complexity of the protocol is defined to be the total number of qubits sent by the protocol, that is, the sum of the lengths (in qubits) of all the quantum states that are sent by the protocol [291].

Buhrman, Cleve and Wigderson proved a general theorem that shows that any quantum algorithm with small query complexity implies quantum communication protocols for related problems, with small communication complexity. Given a (total or partial) function $f : \{0,1\}^n \to \{0,1\}$, one can define the following two communication complexity problems: Given two inputs, $x,y \in \{0,1\}^n$, where Player 1 gets x and Player 2 gets y, the goal of the two players is to compute $f(x \wedge y)$, or $f(x \oplus y)$ (where $x \wedge y$ and $x \oplus y$ denote a coordinate by coordinate application of $\wedge$ and $\oplus$). Buhrman, Cleve and Wigderson proved that if there is a quantum algorithm for computing the function $f(z)$, with k quantum queries to the input $z = (z_1,\ldots,z_n)$, then there are quantum communication complexity protocols for computing $f(x \wedge y)$ and $f(x \oplus y)$, with communication complexity $O(k \cdot \log n)$ [56].

This general theorem gives a method for translating quantum query complexity upper bounds into quantum communication complexity upper bounds, as well as translating quantum communication complexity lower bounds into quantum query complexity lower bounds. Using this general theorem, Buhrman, Cleve and Wigderson obtained interesting consequences in both directions. They used known lower bounds for quantum communication complexity to obtain new lower bounds for quantum query complexity. They also used the general theorem to establish an exponential separation between zero-error quantum communication complexity, and classical deterministic communication complexity, that is, they gave a communication task that can be solved by a zero-error quantum communication complexity protocol with small communication complexity, and such that any classical deterministic communication complexity protocol for that task, requires exponentially larger communication complexity.

Perhaps the most striking consequence of Buhrman, Cleve and Wigderson's general theorem is that it proves that the Set-Disjointness problem can be solved by a quantum protocol with communication complexity $O(\sqrt{n} \cdot \log n)$ [56].

Theorem 4.15 ([56]). *The quantum communication complexity of Set-Disjointness is* $O(\sqrt{n} \cdot \log n)$.

The proof of Theorem 4.15 follows from the general theorem by using Grover's algorithm for computing the OR of n input variables, using only $O(\sqrt{n})$ quantum queries to the input [119].

Theorem 4.15 stands in contrast to Theorem 4.8 that states that the classical probabilistic communication complexity of Set-Disjointness is $\Omega(n)$. Theorem 4.15 established a quadratic gap between quantum and classical probabilistic communication complexity and was followed by a long line of works that further studied the relative power of quantum and classical communication protocols. We note that this quadratic separation remained essentially the largest known gap between quantum and classical probabilistic communication complexity of total functions, for almost two decades. A line of recent works improved that gap to an almost cubic gap [1, 16, 267, 26, 251]. Proving a super-polynomial gap between quantum and classical probabilistic communication complexity of total functions remains a fascinating and long-standing open problem in communication complexity. For partial functions (promise problems), exponential gaps between quantum and classical probabilistic communication complexity were established by a long line of works [221, 28, 105, 164, 104, 107]. Finally, we note that Theorem 4.15 was proved to be essentially tight by Razborov [233].

4.9 Partial Derivatives in Arithmetic Circuit Complexity

Arithmetic circuits are the standard computational model for arithmetic computations, such as computing the determinant or the permanent of a matrix or the product of two matrices. Given a field $\mathbb{F}$ and an n-variate polynomial $P(x_1,\ldots,x_n)$ over $\mathbb{F}$, we ask how many $+, \times$ operations over $\mathbb{F}$ are needed to compute P.

An arithmetic circuit over $\mathbb{F}$, with input variables $x_1,\ldots,x_n \in \mathbb{F}$, is a directed acyclic graph as follows: Every node of in-degree 0 (that is, a *leaf*) is labelled with either an input variable or a field element or a product of an input variable and a field element. Every node of in-degree larger than 0 is labelled with either $+$ or $\times$ (in the first case the node is a *sum* gate and in the second case a *product* gate). A node of out-degree 0 is called an *output* node. The circuit is called a *formula* if the underlying graph is a (directed) tree.

Each node in the circuit (and in particular each output node) computes a polynomial in the ring of polynomials $\mathbb{F}[x_1,\ldots,x_n]$ as follows. A leaf just computes the value of the input variable, or field element, or product of input variable and field element, that labels it. For every non-leaf node v, if v is a sum gate it computes the sum of the polynomials computed by its children, and if v is a product gate it computes the product of the polynomials computed by its children. If the circuit has only one output node, the polynomial computed by the circuit is the polynomial computed by the output node.

The *size* of a circuit is defined to be the number of wires (edges) in it and the *depth* of a circuit is defined to be the length of the longest directed path from a leaf to an output node in the circuit.

Proving lower bounds for the size of arithmetic circuits has been a major challenge for many years. Super-linear lower bounds for the size of general arithmetic circuits were proven in the seminal works of Strassen [261] and Baur and Strassen [35]. Their method, however, only gives lower bounds of up to $\Omega(n \log d)$, where n is the number of input variables and d is the degree of the computed polynomial. In particular, if the degree $d = d(n)$ is polynomial in n this gives lower bounds of at most $\Omega(n \log n)$. Lower bounds for various restricted classes of arithmetic circuits have also been studied in many works.

In 1997, Nisan and Wigderson suggested a general approach for obtaining lower bounds for restricted classes of arithmetic circuits [212]. The approache is based on measuring the dimension of the vector space spanned by all partial derivatives of the polynomials computed at the nodes of the circuit. (Partial derivatives were previously used to obtain lower bounds for arithmetic circuits in the works of Smolensky [256] and Nisan [206]).

For an n-variate polynomial $f(x_1,\dots,x_n)$, let $D(f)$ denote the set of all partial derivatives, of all orders, of f (including f itself as the partial derivative of order 0), and let $\mathsf{Dim}(f)$ denote the dimension of the vector space spanned by $D(f)$. The main idea is to bound the growth of $\mathsf{Dim}(f)$ from the leaves to the outputs of the circuit and hence show that for an output of the circuit, $\mathsf{Dim}(f)$ is bounded. Thus, the circuit cannot compute a polynomial $P(x_1,\dots,x_n)$ with a larger $\mathsf{Dim}(P)$. The following simple formulas are easily proved and are useful for bounding the growth of $\mathsf{Dim}(f)$,

$$\mathsf{Dim}(h+g) \le \mathsf{Dim}(h) + \mathsf{Dim}(g),$$
$$\mathsf{Dim}(h \times g) \le \mathsf{Dim}(h) \cdot \mathsf{Dim}(g).$$

Nisan and Wigderson used this approach to prove several lower bounds, including exponential lower bounds for the size of depth-3 homogeneous circuits (where an homogeneous circuit is a circuit where all nodes in the circuit compute homogeneous polynomials), and exponential lower bounds for the size of constant-depth set-multilinear circuits (where a set-multilinear circuit is a circuit where the set of variables $\{x_1,\dots,x_n\}$ is partitioned into d subsets $X_1,\dots,X_d$ such that, for every node v in the circuit, each monomial in the polynomial computed by the node v contains at most one variable from each subset X_i) [212].

The partial-derivatives method of Nisan and Wigderson has been very influential on later works. Many subsequent works used this approach as a starting point and further built on these ideas to obtain lower bounds for additional classes of arithmetic circuits. In particular, these ideas have been very important in the study of multilinear circuits (for example, [225, 223, 222, 228, 226, 84, 66, 11]), constant-depth homogeneous circuits (for example, [169, 162, 170]) and bounded-depth arithmetic circuits (for example, [252, 161, 120, 96, 182, 14]).

4.10 Resolution Made Simple

Resolution is a proof system (technically, refutation system) for refuting unsatisfiable CNF formulas, that is, unsatisfiable Boolean formulas in conjunctive normal forms.

Given Boolean variables $x_1,\ldots,x_n \in \{0,1\}$, a *literal* is either a variable, x_i, or a negation of a variable, $\neg x_i$. A clause in these variables is an OR of literals, that is, $\bigvee_{i=1}^{k} z_i$, for some k, where each z_i is a literal. The *Resolution rule* says that if C and D are two clauses and x_i is a variable then any assignment that satisfies both clauses, $C \vee x_i$ and $D \vee \neg x_i$, also satisfies the clause $C \vee D$. Thus, from $C \vee x_i$ and $D \vee \neg x_i$, one can deduce $C \vee D$.

A Resolution refutation for a set of clauses F (equivalently, for a CNF formula F) proves that the clauses in F are not simultaneously satisfiable. For a set of clauses F, a Resolution refutation is a sequence of clauses $C_1, C_2, \ldots, C_s$ such that: (1) Each clause C_j is either a clause in F or obtained by the Resolution rule from two previous clauses in the sequence; and (2) The last clause, C_s, is the empty clause (and is hence unsatisfiable). The *size*, or *length*, of a Resolution refutation is the number of clauses in it.

It is well known that Resolution is a sound and complete propositional proof system, that is, a CNF formula F is unsatisfiable if and only if there exists a Resolution refutation for F. We think of a refutation for an unsatisfiable formula F also as a proof for the tautology $\neg F$. Hence, Resolution refutations are also called Resolution proofs.

Resolution is one of the most widely studied propositional proof systems. Lower bounds for the size of Resolution proofs for many propositional tautologies have been proved, starting from Haken's celebrated exponential lower bounds for the propositional pigeonhole principle [127].

Ben-Sasson and Wigderson suggested a general approach for proving lower bounds for the size of Resolution proofs, an approach that generalized, unified and simplified essentially all previously known lower bounds for Resolution, was used to obtain many additional lower bounds, and ultimately gave a deeper understanding of Resolution as a proof system [40].

The approach focuses on the *width* of a resolution proof. The width of a resolution proof is defined to be the number of literals in the largest clause of the proof. Ben-Sasson and Wigderson argued that Resolution is best studied when the focus is on the width. Their key theorem relates the smallest length of a Resolution proof to the smallest width of a Resolution proof. Informally, the theorem states that if a set of clauses F has a short Resolution refutation then it also has a Resolution refutation with small width. The proof is based on a proof by Clegg, Edmonds and Imagliazzo, who gave similar relations (between size of a proof and degree of a proof) for algebraic proof systems [69].

Theorem 4.16 ([40]). *Let F be a an unsatisfiable CNF formula. Let w_0 be the size of the largest clause in F. Let w be the minimal width of a Resolution refutation for F. Let s be the minimal size of a Resolution refutation for F. Then,*

$$w \leq w_0 + O\left(\sqrt{n \log s}\right).$$

In particular, Theorem 4.16 shows that one can obtain lower bounds for the size of Resolution proofs by proving lower bounds for the width of Resolution proofs (which, in many cases, is easier to analyze).

The size of the clauses of a Resolution proof was implicit in previous works and played a major role in previous lower bounds. Previous lower bounds for the size of Resolution proofs were usually proved in two steps as follows. In the first step, the entire proof was hit by a random restriction of the variables (that is, some of the variables were randomly set to 0, some were randomly set to 1 and some were left untouched), in order to hit and eliminate all large clauses of the proof (assuming for a contradiction that the proof is short). The second step proved that large clauses must exist in any Resolution refutation for the restriction of the unsatisfiable formula under the random restriction from the first step (and hence the proof must be long). The approach of Ben-Sasson and Wigderson simplified essentially all previous proofs, as the random restriction was no longer needed and one could focus on proving lower bounds on the width of Resolution refutations for the original unsatisfiable formula, rather than for a random restriction of it.

5 Complexity, Optimization, and Symmetries

This section presents an overview of work by Wigderson and his co-authors on optimization methods to come up with efficient algorithms for various algorithmic problems in computational complexity theory, mathematics, and physics [185, 102, 103, 8, 7, 61, 59, 60]. A common theme in all these works is the realization that the relevant algorithmic tasks can be formulated as optimization problems over algebraic groups that also have an analytic structure. A representative optimization problem is to find a minimum norm vector in the orbit of a given $\mathrm{GL}_n(\mathbb{C})$ action on a vector space. This viewpoint led Wigderson and his co-authors to deploy tools from invariant theory, representation theory, and optimization to develop a quantitative theory of optimization over Riemannian manifolds that arise from continuous symmetries of noncommutative groups.

The starting point is the work [185] that analyzes the convergence of a matrix scaling algorithm to compute an approximation to the permanent (Section 5.1). This corresponds to the commutative setting where the symmetries corresponded to diagonal subgroups (tori) of a matrix group. The role of symmetries in the analysis of the algorithm, however, was not quite explicit.

In [123], Gurvits extended the results of [185] to the noncommutative setting of "operators". In particular, he studied Edmonds' singularity problem [88] and, motivated by [185], he presented a (deterministic) "operator scaling" algorithm for it. However, he fell short of presenting convergence bounds for this algorithm. Section 5.2 presents the work [102] that gives convergence bounds for Gurvits' operator scaling algorithm. This paper makes the first contact of scaling algorithms to invariant theory. It also demonstrates the applicability of computational problems over group orbits and scaling techniques far beyond complexity theory: to mathematics and physics. Section 5.2.3 presents a result from [102] that gives a deterministic polynomial time algorithm for the noncommutative version of Edmonds' singularity problem. Section 5.2.4 gives an outline of a result from [103] that shows how operator scaling can be used to efficiently compute Brascamp–Lieb constants important in mathematics.

Section 5.3 visits the paper [7] which starts with the realization that the problem of finding a minimum-norm vector over an orbit is a geodesically convex optimization problem over a Riemannian manifold. Subsequently, [7] extend the theory of second-order methods in convex optimization to the setting of geodesically convex optimization and give an algorithm whose running time depends logarithmically on the error in the approximation. The focus here is on introducing geodesic convexity and showing how the capacity of an operator can be captured by a geodesically convex optimization problem.

Finally, Section 5.4, presents results from [60]. Here, the general norm minimization problem is introduced and various variants of it studied by [60] are presented. These problems unify and generalize prior works in this line. Of particular importance is the connection to noncommutative duality in invariant theory which extends linear programming duality and allows one to give conditions on when an optimization problem is feasible. This gives rise to other connections such as moment maps (analog of Euclidean gradients) and a precise notion of geodesic convexity. This paper culminates with the definition and convergence bounds for first-order and second-order algorithms for various optimization problems over noncommutative matrix groups. The convergence bounds are based on novel parameters related to the group action via a synthesis of algebra and analysis. This paper also gives a host of new analytic algorithms for various problems important in invariant theory and complexity theory.

5.1 Permanent and matrix scaling

Let $A \in \mathbb{R}^{n\times n}$ be a square matrix with entries $A_{i,j}$ for $1 \le i,j \le n$. The permanent of A is defined as:

$$\mathrm{Per}(A) := \sum_{\sigma\in S_n} \prod_{i=1}^{n} A_{i,\sigma(i)},$$

where S_n is the set of all permutations over n symbols, i.e., the set of bijections $\sigma : \{1,2,\ldots,n\} \to \{1,2,\ldots,n\}$. The permanent makes its appearance in various branches of science and mathematics and algorithms to compute it are sought after. For instance, permanents of $0,1$-valued matrices are intimately connected to perfect matchings in bipartite graphs. Consider a bipartite graph $G = (L,R,E)$ where L,R is the bipartition of the vertex set of G and E is the set of edges of G. Assume $|L| = |R| = n$ and define an $n \times n$ matrix A (adjacency matrix of G) whose (i,j)th entry is 1 is an edge between the ith vertex of L and the jth vertex of R. It follows from the definition that the permanent of A is equal to the number of perfect matchings in G.

The computational complexity of the permanent has been extensively studied in theoretical computer science. Valiant [278] proved that it is unlikely that there is an efficient algorithm that computes the permanent of a nonnegative matrix – even when the matrix has only $0,1$ entries (the problem is #P−complete). This result, under standard assumptions in complexity theory, rules out an efficient algorithm to compute the permanent of a nonnegative matrix and raises the question of finding approximations to it. Checking if $\mathrm{Per}(A)$ of a nonnegative matrix is zero or not, however, is in P since it reduces to checking if the associated bipartite graph has a perfect matching or not.

5.1.1 Doubly stochastic matrices and their permanents

A special class of nonnegative matrices is doubly-stochastic matrices whose row sums and column sums are all equal to one.

Definition 5.1. (Doubly stochastic matrix) An $n \times n$ matrix A is said to be doubly stochastic if it is nonnegative and its rows and columns sum up to one: For each i, $\sum_{j=1}^{n} A_{i,j} = 1$ and for each j, $\sum_{i=1}^{n} A_{i,j} = 1$.

If a nonnegative matrix A is an adjacency matrix of a graph G each of whose vertices has degree $d \geq 1$, then the matrix $\frac{1}{d}A$ is doubly stochastic. The set of all doubly stochastic matrices is convex and, in fact, a polytope – the Birkhoff polytope [44]. The well-known Birkhoff-von Neumann theorem states that the Birkhoff polytope is a convex hull of $n \times n$ permutation matrices.

The matrix with all entries $\frac{1}{n}$ is doubly stochastic. Its permanent is $\frac{n!}{n^n}$. van der Waerden conjectured that the permanent of any $n \times n$ doubly-stochastic matrix must be at least $\frac{n!}{n^n}$. Interestingly, this lower bound does not depend on the entries of A as long as it is doubly stochastic. Egorychev [90] and Falikman [93] proved the van der Waerden conjecture.

Theorem 5.2. (Permanent of doubly-stochastic matrices [90, 93]) *For any $n \times n$ doubly-stochastic matrix A,* $\mathrm{Per}(A) \geq \frac{n!}{n^n}$.

On the other hand, just for a row-stochastic matrix A, it trivially holds that

$$\mathrm{Per}(A) \leq \prod_{i=1}^{n} \sum_{j=1}^{n} A_{i,j} = 1.$$

Since $\frac{n!}{n^n} \geq e^{-n}$, for a doubly-stochastic matrix, the permanent is between e^{-n} and 1. Hence, if A is doubly stochastic, then we can output 1 and this is an e^n approximation to its permanent.

5.1.2 Matrix scaling

The starting point of the work Linial, Samorodnitsky, and Wigderson [185] (see also the journal version of this paper [185]) is the observation that the permanent (of any matrix) has certain *symmetries*: For positive vectors $x, y \in \mathbb{R}_{>0}$, if we define $B := XAY$ where X and Y are diagonal matrices corresponding to vectors x and y respectively, then we can write down the permanent of B exactly:

$$\mathrm{Per}(B) = \left(\prod_{i=1}^{n} x_i\right) \mathrm{Per}(A) \left(\prod_{j=1}^{n} y_j\right). \tag{8}$$

This operation of left and right multiplying A with diagonal matrices is referred to as (matrix) *scaling*. Thus, in the case A is not doubly stochastic (something that can be efficiently checked), one can try to find a scaling (x, y) of A such that B is doubly stochastic. If so, one can output

$$\frac{1}{\left(\prod_{i=1}^{n} x_i\right)\left(\prod_{j=1}^{n} y_j\right)}$$

as an approximation for $\mathrm{Per}(A)$. From the discussion in the previous section, such an algorithm would be an e^n approximation to the permanent.

An approach to finding such a scaling is to do the following iteratively: Find a vector x that ensures that all the rows of the scaled A sum up to one, and then pick a y that ensures the same for the columns. This matrix scaling algorithm was suggested by Sinkhorn [254]. Franklin and Lorenz [98] analyzed the convergence rate of Sinkhorn's scaling algorithm. They showed that, when a doubly-stochastic scaling of A exists, Sinkhorn's algorithm outputs a matrix B that is ε away (in ℓ_∞-distance) from being doubly stochastic and, to do so, it takes a polynomial number of iterations in the number of bits needed to represent the input matrix A and $\frac{1}{\varepsilon}$. Kalantari and Khachiyan [153] gave a convex-optimization-based algorithm to check if A can be scaled to a doubly-stochastic matrix and, if it can be, then to find an ε approximation to it. The running time of their algorithm is polynomial in the number of bits needed to represent the input matrix A and $\log \frac{1}{\varepsilon}$; thus, giving a deterministic polynomial time algorithm that approximates the permanent of a nonnegative A to within a multiplicative factor of e^n.

The question that [185] studied is if the number of iterations can be made independent of the number of bits needed to represent A. Such an algorithm, whose number of iterations does not depend on the entries of A, is referred to as a *strongly polynomial* time algorithm. At its core, this turns out to be related to the following mathematical question: If $\mathrm{Per}(A) > 0$, then how small can it get as a function of the entries of A? Theorem 5.2 [90, 93] implies that, if A is doubly stochastic, then this cannot get below e^{-n}.

Preprocessing step. The idea in [185] is to augment Sinkhorn's scaling algorithm with a *preprocessing* step that, in the beginning, scales the columns of A to ensure that the permanent of the new matrix is lower bounded by n^{-n}. They do so by first efficiently finding a permutation $\sigma \in S_n$ that maximizes $\prod_{i=1}^n A_{i,\sigma(i)}$. They then show that there is a positive diagonal matrix Y such that $B = AY$ and, for all $1 \le i, j \le n$, $B_{i,\sigma(i)} \ge B_{i,j}$. This ensures that if we normalize the rows of B such that each of them sums up to one, the permanent of the resulting matrix is at least $\frac{1}{n^n}$.

Potential function and measuring progress. To analyze the progress in Sinkhorn's scaling algorithm, [185] consider the permanent itself as the potential function. If A_t is the matrix at the beginning of the tth iteration of Sinkhorn's scaling algorithm, they show that, as long as A_t is *far* from being doubly stochastic,

$$\mathrm{Per}(A_{t+1}) \gtrsim \left(1 + \frac{1}{n}\right) \mathrm{Per}(A_t). \tag{9}$$

They use the following potential function that measures the distance of a matrix B from being doubly stochastic:

$$\mathrm{ds}(B) := \|R(B) - I\|_F^2 + \|C(B) - I\|_F^2. \tag{10}$$

Here $R(B), C(B)$ are diagonal matrices whose (i,i)th entries are the sum of the ith row and ith column respectively.

To gain some intuition why (9) is true, first note that if we have positive numbers $c_1, \ldots, c_n$ that sum up to 1 and are more than δ distance from the all one vector ($\|1 - c\|_2^2 \approx \delta$) then $\prod_{i=1}^n c_i \lesssim 1 - \frac{\delta}{2}$. Hence, if we have a matrix B that is row stochastic and we scale its columns to 1, i.e., consider BC^{-1}, where C is the diagonal matrix corresponding to the column sums of B, then

$$\mathrm{Per}(BC^{-1}) = \frac{\mathrm{Per}(B)}{\prod_{i=1}^n c_i} \gtrsim \mathrm{Per}(B) \cdot (1 + \delta).$$

Thus, as long as $\delta \ge \frac{1}{n}$, the permanent increases by a multiplicative factor of $1 + \frac{1}{n}$.

Termination condition. If after t iterations, $\mathrm{ds}(A_t) \ge \frac{1}{n}$, then

$$\mathrm{Per}(A_{t+1}) \gtrsim \left(1 + \frac{1}{n}\right)^t \mathrm{Per}(A_1).$$

Since the permanent of a row-stochastic matrix is upper bounded by 1, and $\mathrm{Per}(A_1) \ge \frac{1}{n^n}$ due to the preprocessing step, the above cannot continue for more than about n^2 iterations. Thus, after roughly n^2 iterations, $\mathrm{ds}(A_t) < \frac{1}{n}$, and A_t is close to a doubly stochastic matrix. Finally, [185] prove an approximate version of Theorem 5.2 and lower bound the permanent of approximately doubly-stochastic matrices. Roughly speaking, they show that if B is row stochastic and $\mathrm{ds}(B) < \frac{1}{n}$, then

$\mathrm{Per}(B) > \frac{1}{e^{n(1+o(1))}}$. Thus, we can output the matrix produced after about n^2 iterations. This completes the sketch of the proof of the following theorem.

Theorem 5.3. (Approximating permanent via matrix scaling [185]) *There is an algorithm that, given an $n \times n$ nonnegative matrix A, computes a number Z such that* $\mathrm{Per}(A) \leq Z \leq e^{n(1+o(1))} \cdot \mathrm{Per}(A)$ *using $\tilde{O}(n^5)$ elementary operations.*

Subsequent to the work of [185], Jerrum, Sinclair, and Vigoda [146], building upon a long line of work, showed that the Markov Chain Monte Carlo framework can be deployed to obtain a randomized algorithm to estimate the permanent of any nonnegative matrix to within a factor of $1+\varepsilon$ in time that is polynomial in the bit-lengths of A and $\frac{1}{\varepsilon}$. As for deterministic algorithms, in a follow-up work, Gurvits and Samorodnitsky [125] show how scalings can be viewed as solutions to certain convex programs – leading to convex programming relaxations for the permanent and better deterministic approximations; see Section 5.2.5 and [262] for a discussion. This line of work on deterministic approximation algorithms has recently been generalized to a class of general counting and optimization problems; see [263, 15].

5.2 Noncommutative singularity testing and operator scaling

Edmonds [88] considered the following generalization of checking whether the permanent of a nonnegative matrix is zero or not: Given an m-tuple of $n \times n$ complex matrices $A_1, \ldots, A_m$, is there a singular matrix in their linear space (over $\mathbb{C}$) or not? This *singularity* problem is equivalent to deciding if the polynomial

$$p_{A_1,\ldots,A_m}(x_1,\ldots,x_m) := \det(x_1A_1 + \cdots + x_mA_m)$$

is identically zero or not. $p_{A_1,\ldots,A_m}$ is a homogeneous polynomial of degree n and can be efficiently evaluated at any given point. To see how deciding if a bipartite graph has a perfect matching is a special case of Edmonds' singularity problem, we let A_i be the matrix which has a 1 only at the entry corresponding to the ith edge in the associated graph and 0 elsewhere; see [187].

There is a simple and efficient randomized algorithm to test this: Pick independent and random values for each of the variables $x_1, \ldots, x_m$ from the set $\{1, 2, \ldots, 2n\}$ and output the value of $p_{A_1,\ldots,A_m}$ for this input. It can be shown that if $p_{A_1,\ldots,A_m}$ is not identically zero then, with probability at least $\frac{1}{2}$, this algorithm outputs a nonzero value. By repeating an appropriate number of times, this probability can be amplified to any number less than 1. This problem is an instance of the *Polynomial Identity Testing* (PIT) problem where one is given a polynomial and the goal is to check if it is identically zero or not. The randomized algorithm mentioned above works for PIT as well. While for some special cases of PIT deterministic algorithms are known (e.g., the deterministic primality testing algorithm of Agrawal, Kayal, and Saxena [2]), the problem of coming up with an efficient deterministic algorithm for PIT remains open. We mention that Edmonds' singularity problem is almost the same as the fully general PIT problem due to a result of Valiant [278] that establishes the "universality" of the determinant. [151] proved that derandomizing PIT implies arithmetic circuit lower bounds for the complexity class NEXP; tying the goal of derandomizing PIT to one of the central goals of theoretical computer science: that of proving circuit lower bounds.

Gurvits [123] considered a version of Edmonds' singularity problem and reformulated it in terms of *completely positive operators* that take positive definite matrices to positive definite matrices. Subsequently, he generalized the matrix scaling algorithm of Linial, Samorodnitsky, and Wigderson [185] to *operator scaling* for this problem. He introduced a potential function – *capacity* – that can track the progress of the operator scaling algorithm and used it to give deterministic polynomial time algorithms for Edmonds' singularity problem for various special cases (Section 5.2.1). However, he could not prove a bound on the number of iterations of his operator scaling in general. The main result of the paper by Garg, Gurvits, Oliviera, and Wigderson [102] is a bound on the number of iterations of Gurvits' operator scaling algorithm. The key ingredient in their analysis is a lower bound on the capacity of a completely positive operator (Section 5.2.2). This implies that Gurvits' operator scaling algorithm can also approximate the capacity of a completely positive operator to any accuracy in polynomial time. Moreover, [102] show that this algorithm implies a deterministic polynomial time algorithm for testing a *noncommutative* version of Edmonds' problem (Section 5.2.3). Here, prior to the work of [102], the best algorithms (whether randomized or deterministic) required an exponential time algorithm [143]. In a companion paper Garg, Gurvits, Oliviera, and Wigderson [103] show the application of this operator scaling machinery to the various computational problems involving the Brascamp–Lieb inequalities (Section 5.2.4).

5.2.1 Completely positive operator and its capacity

Let $M_n(\mathbb{C})$ denote the set of $n \times n$ matrices with complex entries. Let $\mathrm{GL}_n(\mathbb{C})$ denote the degree n general linear group of $n \times n$ invertible matrices over $\mathbb{C}$. Let $\mathrm{SL}_n(\mathbb{C})$ denote the degree n special linear group of $n \times n$ matrices over $\mathbb{C}$ with determinant 1. Both of the above are groups with respect to ordinary matrix multiplication. Let $H_n(\mathbb{C})$ denote the set $n \times n$ Hermitian matrices. Let S_+^n denote the set of $n \times n$ complex positive semi-definite (PSD) matrices and let S_{++}^n denote the set of $n \times n$ complex positive definite (PD) matrices. For two matrices X, Y their tensor product is denoted by $X \otimes Y$.

Definition 5.4. (Completely positive operator) For positive integers $n_1 \geq n_2$, an operator $T : M_{n_1}(\mathbb{C}) \to M_{n_2}(\mathbb{C})$ is said to be completely positive if there are $n_2 \times n_1$ complex matrices $A_1, \ldots, A_m$ such that, for $X \in S_{++}^{n_1}$, $T(X) = \sum_{i=1}^m A_i X A_i^\dagger$. The dual of T is denoted by T^* and is such that $T^*(Y) = \sum_{i=1}^m A_i^\dagger Y A_i$ for $Y \in S_{++}^{n_2}$.

If $n_1 = n_2 = n$, we say that T is a *square* operator.

Definition 5.5. (Doubly-stochastic completely positive operator) A completely positive operator $T : M_{n_1}(\mathbb{C}) \to M_{n_2}(\mathbb{C})$ is said to be doubly stochastic if $T\left(\frac{n_2}{n_1} I_{n_1}\right) = I_{n_2}$ and $T^*(I_{n_2}) = I_{n_1}$.

[123] introduced the following notion of capacity for completely positive operators.

Definition 5.6. (Capacity of a completely positive operator [123]) For a completely positive operator $T : M_{n_1}(\mathbb{C}) \to M_{n_2}(\mathbb{C})$, its capacity is defined as

$$\mathrm{Cap}(T) := \inf\left\{ \frac{\det\left(\frac{n_2}{n_1} T(X)\right)}{\det(X)^{\frac{n_2}{n_1}}} : X \succ 0 \right\}.$$

We focus on the square case and return to the rectangular (nonsquare) case in Section 5.2.4. In the square case ($n_1 = n_2 = n$),

$$\mathrm{Cap}(T) := \inf\{\det(T(X)) : X \succ 0,\ \det(X) = 1\}.$$

A square operator is said to be *rank decreasing* if there is an $X \succeq 0$ such that $\mathrm{rank}(T(X)) < \mathrm{rank}(X)$. Operators that are not rank decreasing are referred to as *rank nondecreasing*. The analog of this property in the matrix case (for a nonnegative matrix A) is as follows: For every nonnegative vector x, the number of coordinates of the vector Ax that are positive is at least the number of coordinates of x that are positive. This is just Hall's condition and implies that the permanent of A is positive. [123] proved that, for a completely positive operator, $\mathrm{Cap}(T) > 0$ if and only if T is rank nondecreasing. [102] give other conditions that are equivalent for a completely positive operator to be rank nondecreasing. One such condition that is relevant to proving a lower bound on the capacity is that there exist $d \times d$ matrices $F_1, \ldots, F_m$ for some d such that the polynomial

$$\det(F_1 \otimes A_1 + \cdots + F_m \otimes A_m) \neq 0. \tag{11}$$

Similar to the notion of distance to a matrix to being doubly stochastic in Definition 5.1, consider the following distance of a completely positive operator from being doubly stochastic:

$$\mathrm{ds}_O(T) := \mathrm{Tr}((T(I) - I)^2) + \mathrm{Tr}((T^*(I) - I)^2). \tag{12}$$

An analog of Equation (8) that captures the symmetries of the operator setting is as follows: Let T be a completely positive square operator defined by $A_1, \ldots, A_m$ and $B, C \in \mathrm{GL}_n(\mathbb{C})$. Then, if we define $T_{B,C}$ to be the operator defined by $BA_1C, \ldots, BA_mC$, then

$$\mathrm{Cap}(T_{B,C}) = |\det(B)|^2 \cdot \mathrm{Cap}(T) \cdot |\det(C)|^2. \tag{13}$$

If $B, C \in \mathrm{SL}_n(\mathbb{C})$, then $\mathrm{Cap}(T_{B,C}) = \mathrm{Cap}(T)$. This is true because the capacity is defined in terms of determinants and, hence, the symmetries of the determinant arise. It is worth noting that the polynomials in (the l.h.s. of) Equation (11) are invariant when B,C have determinant 1. In fact, these polynomials linearly span the space of all such invariant polynomials.

Moreover, suppose T is a completely positive operator specified by $A_1,\ldots,A_m$ and either $\sum_{i=1}^m A_iA_i^\dagger = I$ (row-stochastic) or $\sum_{i=1}^m A_i^\dagger A_i = I$ (column-stochastic) then it follows from the AM-GM inequality that

$$\mathrm{Cap}(T) \leq \det(T(I)) \leq \left(\frac{\mathrm{Tr}(T(I))}{n}\right)^n = 1. \tag{14}$$

5.2.2 Operator scaling

We present a sketch of the operator scaling algorithm and its analysis. Suppose T is a completely positive operator specified by m $n \times n$ matrices $A_1,\ldots,A_m$, where each entry of each matrix is an integer bounded in absolute value by M. Our goal is to decide if $\mathrm{Cap}(T) > 0$ or not. Or equivalently, to decide if T is rank nondecreasing.

An operator scaling of T is given by positive matrices B,C such that the operator $T_{B,C}$ defined by $B^{\frac{1}{2}}A_1C^{\frac{1}{2}},\ldots,B^{\frac{1}{2}}A_mC^{\frac{1}{2}}$ is doubly stochastic. The left normalization (or scaling) of T, denoted by T_L, is defined as

$$T_L(X) := T(I)^{-\frac{1}{2}}T(X)T(I)^{-\frac{1}{2}}$$

and the right normalization (or scaling) of T is defined as

$$T_R(X) := T(T^*(I)^{-\frac{1}{2}}XT^*(I)^{-\frac{1}{2}}).$$

It follows that $T_L(I) = I$ and $T_R^*(I) = I$.

Gurvits' operator scaling algorithm [123] follows the same outline as the matrix scaling algorithm analyzed in [185]. It first checks if both $T(I)$ and $T^*(I)$ are nonsingular. If not, then T is rank decreasing and the algorithm stops. Else, it keeps performing left and right normalizations on T until the distance to double stochasticity is below $\frac{1}{n}$. If the operator T is rank decreasing, then one can argue that the left and right normalizations cannot make it rank nondecreasing. Thus, the algorithm will always output rank decreasing in this case.

[102] prove that if T is rank nondecreasing, then after a small-enough number of iterations t, the operator T_t is such that $\mathrm{ds}_O(T_t) < \frac{1}{n}$. This is analogous to the matrix case: They show that every iteration such where $\mathrm{ds}_O(T_t) > \frac{1}{n}$,

$$\mathrm{Cap}(T_{t+1}) \gtrsim \left(1 + \frac{1}{n}\right)\mathrm{Cap}(T_t).$$

Since there is an upper bound of 1 on the capacity of a row-stochastic operator, it remains to lower bound $\mathrm{Cap}(T_1)$ when $\mathrm{Cap}(T) > 0$. We note that for the matrix case, we used permanent as a measure, but could have also used an appropriate notion of capacity as defined in Section 5.2.5.

The main technical contribution of [102] is a lower bound on the capacity of a right-normalized completely positive operator. Let T_A be a completely positive operator specified by integer-valued matrices $A_1,\dots,A_m$ each of whose entries is bounded in absolute value by M. Let T be the right normalization of T_A. Then, it follows that

$$\mathrm{Cap}(T)=\frac{\mathrm{Cap}(T_A)}{\det(T^*(I))}.$$

Thus, to lower bound $\mathrm{Cap}(T)$, it is sufficient to lower bound $\mathrm{Cap}(T_A)$ and upper bound $\det(T_A^*(I))$. The latter follows from an upper bound on

$$\mathrm{Tr}(T_A^*(I))=\sum_{i=1}^{m}\mathrm{Tr}\left(A_i^\dagger A_i\right)\le M^2mn^2.$$

Thus, by the AM-GM inequality

$$\det(T^*(I))\le\left(\frac{\mathrm{Tr}(T_A^*(I))}{n}\right)^n\le(Mmn)^n. \tag{15}$$

The original proof of a lower bound on the capacity of a nondecreasing completely positive operator T_A relied on degree bounds in invariant theory; we return to it in the next section. Here we mention their proof based on Alon's Combinatorial Nullstellensatz [10]; see also [279]. Alon's result states that if $p(z_1,\dots,z_\ell)$ is a nonzero polynomial (over $\mathbb{C}$) with the degree of z_i is d_i, then there are nonnegative integers $(a_1,\dots,a_\ell)$ such that $\sum_{i=1}^{\ell}a_i\le d$ and $a_i\le d_i$ such that $p(a_1,\dots,a_\ell)\neq 0$.

From Equation (11), we know that if T is rank nondecreasing, then there exist $d\times d$ matrices $F_1,\dots,F_m$ for some d such that the polynomial

$$\det(F_1\otimes A_1+\cdots+F_m\otimes A_m)\neq 0.$$

Thus, the (ordinary) polynomial $\det(X_1\otimes A_1+\cdots+X_m\otimes A_m)$ (in the variables corresponding to entries of matrices $X_1,\dots,X_m$) is nonzero. Thus, Alon's result implies that there exist integer-valued matrices $D_1,\dots,D_m$ such that

$$\det(D_1\otimes A_1+\cdots+D_m\otimes A_m)\neq 0$$

and, importantly, the sum of the square of all the entries of all the matrices is bounded by n^2d.

Let $X\succ 0$ and define $C_i:=T_A(X)^{-\frac{1}{2}}A_iX^{\frac{1}{2}}$. Thus, $\sum_{i=1}^{m}C_iC_i^\dagger=I$ and, hence $\mathrm{Tr}\left(\sum_{i=1}^{m}C_iC_i^\dagger\right)=n$. Now, let $Y:=D_1\otimes C_1+\cdots+D_m\otimes C_m$. Then, on the one hand, by the AM-GM inequality,

$$\det(YY^\dagger)\le\left(\frac{\mathrm{Tr}(YY^\dagger)}{nd}\right)^{nd}\le\left(\frac{n^3d}{nd}\right)^{nd}=n^{2dn},$$

where one uses the bound on the sum of the square of entries of D_is. On the other hand,

$$\det(YY^\dagger) = |\det(Y)|^2 \geq |\det(D_1 \otimes A_1 + \cdots + D_m \otimes A_m)|^2 \det(X)^d \cdot \det(T_A(X))^{-d}.$$

Since all entries of $D_1 \otimes A_1 + \cdots + D_m \otimes A_m$ are integers and its determinant is nonzero, $|\det(D_1 \otimes A_1 + \cdots + D_m \otimes A_m)| \geq 1$, implying

$$\det(T_A(X)) \geq (\det(YY^\dagger))^{-\frac{1}{d}} \geq n^{-\frac{2dn}{d}} = \frac{1}{n^{2n}}. \tag{16}$$

Thus, combining Equations (15) and (16), we obtain the following theorem.

Theorem 5.7. (Lower bound on the capacity of a rank nondecreasing operator [102]) *Let T be the right-normalized version of a rank nondecreasing and completely positive operator given by $A_1, \ldots, A_m$, where each A_i is an $n \times n$ integer matrix with each entry bounded in absolute value by M. Then,* $\mathrm{Cap}(T) \geq \frac{1}{(Mmn^3)^n}$.

As discussed above, this implies the following theorem to check if a completely positive operator is rank nondecreasing or, equivalently, if its capacity is positive.

Theorem 5.8. (Checking if a completely positive operator is rank nondecreasing [102]) *There is an algorithm that, given a completely positive operator T given by $A_1, \ldots, A_m$, where each A_i is an $n \times n$ integer matrix with each entry bounded in absolute value by M, decides if T is rank nondecreasing or not in time polynomial in n, m, and* $\log M$.

While we did bound the number of iterations needed by Gurvits' operator scaling algorithm for the above theorem, we omitted a discussion on ensuring that the bit complexity of the numbers that arise in the execution of the algorithm remain polynomially bounded in the input bit length; see [102] for details.

[102] also show how an adaptation of Gurvits' operator scaling algorithm can be used to obtain an approximation of the operator capacity. We omit the algorithm and the proof.

Theorem 5.9. (Approximating the capacity of an operator [102]) *There is an algorithm that, given a completely positive operator T on dimension n, and described by b bits, outputs a $1+\varepsilon$ multiplicative approximation to* $\mathrm{Cap}(T)$ *in time polynomial in $n, b, \frac{1}{\varepsilon}$.*

In a subsequent work, Bürgisser, Garg, Oliveira, Walter, and Wigderson [61] present a generalization of operator scaling to *tensor scaling*; we omit the details. We note that, unlike the matrix and operator scaling case, to test scalability, it is not sufficient to take ε which is polynomially small. Currently, there is no known polynomial time algorithm for testing the scalability of tensors.

In another follow-up work, Bürgisser, Franks, Garg, Oliveira, Walter, and Wigderson [59] study the *nonuniform* version of scaling where one is given prescribed *marginals* and an input matrix/operator/tensor, and the goal is to decide if we can scale the input to have the prescribed marginals. For instance, instead of scaling a nonnegative matrix so that the row sums and column sums are all one, one may ask to find a scaling to a specified row sum vector r and a column sum vector c. In the matrix scaling case, the theory of nonuniform scaling is not much different from the theory of uniform scaling. However in the operator and tensor scaling settings, the nonuniformity presents additional challenges; see [59].

5.2.3 Noncommutative singularity and identity testing

Let $A_1,\ldots,A_m \in M_n(\mathbb{C})$ and consider $x_1,\ldots,x_m$ to be noncommutative variables. The algorithmic problem, which is a noncommutative version of Edmonds' singularity problem, is to check if $L := \sum_{i=1}^m x_i A_i$ is invertible (nonsingular) over the skew-field (also known as division ring or field of fractions) of $x_1,\ldots,x_n$. This notion of nonsingularity is nontrivial to define and there are several equivalent ways to do it. Perhaps the simplest is if there is a way of "plugging in" a matrix for each x_i to get an invertible matrix, i.e., do there exist $d \times d$ matrices $B_1,\ldots,B_m$ (for some d) s.t. $\sum_{i=1}^m B_i \otimes A_i$ is invertible.

The connection between the noncommutative singularity problem and the capacity of a completely positive operator is as follows: Consider the completely positive operator $L(X) := \sum_{i=1}^m A_i X A_i^\dagger$ defined by the matrices $A_1,\ldots,A_m$ input to the noncommutative singularity problem. Then, $\sum_{i=1}^m x_i A_i$ is singular over the skew-field if and only if there is an $X \succ 0$ such that $\mathrm{rank}(L(X)) < \mathrm{rank}(X)$, i.e., the completely positive operator L is rank decreasing. Thus, from Theorem 5.8, it immediately follows that the problem of checking noncommutative singularity is in P.

Theorem 5.10. (Noncommutative singularity testing [102]) *There is a deterministic algorithm that, given m $n \times n$ matrices $A_1,\ldots,A_m$ whose entries need at most b bits to represent, decides in time* $\mathrm{poly}(n,m,b)$ *if the matrix $L = \sum_{i=1}^m x_i A_i$ is invertible over the free skew field.*

Polynomial identity testing, in the commutative setting, captures the polynomial and rational function identity test for formulas [278]. The same is not true in the noncommutative setting. However, Cohn [72] proved that there is an efficient algorithm that converts every arithmetic formula $\phi(x)$ in noncommuting variables of size s to a symbolic matrix L_ϕ of size $\mathrm{poly}(s)$, such that the rational expression computed by ϕ is identically zero if and only if L_ϕ is singular. Theorem 5.10 implies that there is a deterministic algorithm that, for any noncommutative formula over $\mathbb{Q}$ of size s and bit complexity b, determines in $\mathrm{poly}(s,b)$ steps if it is identically zero. Thus, the noncommutative rational identity testing problem is in P; see also the works of [144, 128] for different proofs of this result. Note that Theorem 5.10 requires access to the matrices $A_1,\ldots,A_m$. The problem of proving an analogous result when we have only black-box access to $\sum_{i=1}^m x_i A_i$ remains open. We note that, in the noncommutative setting, inversions are nontrivial to handle, while in the commutative setting we can push them out and eliminate them. Indeed, an efficient deterministic algorithm to check if a noncommutative formula *without* inversions is identically zero is known; see Raz and Shpilka [225].

5.2.4 Brascamp–Lieb constants

Let n, m, and $(n_j)_{j\in[m]}$ be positive integers and $p := (p_j)_{j\in[m]}$ be nonnegative real numbers. Let $B := (B_j)_{j\in[m]}$ be an m-tuple of linear transformations where B_j is a surjective linear transformation from $\mathbb{R}^n$ to $\mathbb{R}^{n_j}$. The corresponding Brascamp–Lieb *datum* is denoted by (B,p). The Brascamp–Lieb inequality states that for each Brascamp–Lieb datum (B,p) there exists a constant $C(B,p)$ (not necessarily finite) such that for any selection of real-valued, nonnegative, Lebesgue measurable functions f_j where $f_j : \mathbb{R}^{n_j} \to \mathbb{R}$,

$$\int_{x\in\mathbb{R}^n} \left(\prod_{j\in[m]} f_j(B_j x)^{p_j} \right) \mathrm{d}x \leq C(B,p) \prod_{j\in[m]} \left(\int_{x\in\mathbb{R}^{n_j}} f_j(x)\, \mathrm{d}x \right)^{p_j}. \tag{17}$$

The smallest constant that satisfies (17) for any choice of $f := (f_j)_{j\in[m]}$ satisfying the properties mentioned above is called the Brascamp–Lieb *constant* and we denote it by $\mathrm{BL}(B,p)$. Brascamp–Lieb inequalities generalize many inequalities used in analysis and all of mathematics, such as the Hölder inequality and the Loomis–Whitney inequality; see the paper by Brascamp and Lieb [52].

A Brascamp–Lieb datum (B,p) is called *feasible* if $\mathrm{BL}(B,p)$ is finite, otherwise, it is called *infeasible*. Bennett, Carbery, Christ, and Tao [41] proved that the constant $\mathrm{BL}(B,p)$ is nonzero whenever p belongs to the set $P_B \subseteq \mathbb{R}^m$ defined as follows:

$$P_B := \left\{ p \in \mathbb{R}^m_{\geq 0} : \textstyle\sum_{j=1}^m p_j \dim(B_j U) \geq \dim(U), \text{ for every lin. subspace } U \subseteq \mathbb{R}^n \right\}.$$

Note that the above definition has infinitely many linear constraints on p as V varies over different subspaces of $\mathbb{R}^n$. However, there are only finitely many different linear restrictions as $\dim(B_j V)$ can only take integer values from $[n_j]$. Consequently, P_B is a convex set and, in particular, a polytope. Examples of Brascamp–Lieb polytopes include matroid basis polytopes and linear matroid intersection polytopes; see [103].

A Brascamp–Lieb inequality is nontrivial only when (B,p) is a feasible Brascamp–Lieb datum. Therefore, it is of interest to characterize feasible Brascamp–Lieb data and compute the corresponding Brascamp–Lieb constant. Towards this, Lieb [181] showed that one needs to consider only Gaussian functions as inputs for (17). This result suggests the following characterization of the Brascamp–Lieb constant as an optimization problem.

Theorem 5.11. (Gaussian maximizers [181]) *Let (B,p) be a Brascamp–Lieb datum with $B_j \in \mathbb{R}^{n_j\times n}$ for each $j \in [m]$. Then,*

$$\frac{1}{\mathrm{BL}(B,p)^2} = \inf \left\{ \frac{\det\left(\sum_{j=1}^m p_j B_j^\top Y_j B_j\right)}{\prod_{j=1}^m \det(Y_j)^{p_j}} : Y_j \in \mathbb{R}^{n_j\times n_j}, Y_j \succ 0, j = 1,2,\ldots,m \right\}. \tag{18}$$

One of the computational questions concerning the Brascamp–Lieb inequality is: Given a Brascamp–Lieb datum (B,p), can we compute $\mathrm{BL}(B,p)$ in time that is polynomial in the number of bits required to represent the datum? Since computing $\mathrm{BL}(B,p)$ exactly may not be possible due to the fact that this number may not be rational even if the datum (B,p) is, one seeks an arbitrarily good approximation. Formally, given the entries of B and p in binary, and an $\varepsilon > 0$, compute a number Z such that

$$\mathrm{BL}(B,p) \leq Z \leq (1+\varepsilon)\,\mathrm{BL}(B,p)$$

in time that is polynomial in the combined bit lengths of B and p and $\log \frac{1}{\varepsilon}$.

There are a few obstacles to this problem: (a) Checking if a given Brascamp–Lieb datum is feasible is not known to be in P. (b) The formulation of the Brascamp–Lieb constant by Lieb [181] as in (18) is neither concave nor logconcave in the usual sense. Thus, techniques developed in the context of linear and convex optimization do not seem to be directly applicable.

Garg, Gurvits, Oliviera, and Wigderson [103] gave an algorithm to compute the Brascamp–Lieb constant in polynomial time when the vector p is rational and given in unary. More precisely, the running time of their algorithm to compute $\mathrm{BL}(B,p)$ up to multiplicative error $1+\varepsilon$ has a polynomial dependency to ε^{-1} and the *magnitude* of the denominators in the components of p rather than the number of bits required to represent them. They also presented algorithms with similar running times for checking if a Brascamp–Lieb datum is feasible, or if a given point is approximately in the Brascamp–Lieb polytope. The key idea in [103] is to use Lieb's characterization (Theorem 5.11) to reduce the problem of computing $\mathrm{BL}(B,p)$ to the problem of computing the capacity of a completely positive operator. We note that the special case when the matrices are of rank 1; i.e., $B_j \in \mathbb{R}^{1\times n}$ for every $j=1,2,\ldots,m$ was studied in [264]. By interpreting Brascamp–Lieb constants in the rank-1 regime as solutions to certain entropy-maximization problems, [253, 264] showed that they can be computed, up to a multiplicative precision $\varepsilon > 0$, in time polynomial in m and $\log \frac{1}{\varepsilon}$.

The reduction. Let $p_j = \frac{c_j}{c}$ for integers $(c_j)_{j\in[m]}$ and c. [103] construct a completely positive operator $T_{B,p}$ such that $\mathrm{Cap}(T_{B,p}) = \frac{1}{\mathrm{BL}(B,p)^2}$. Let $m' := \sum_{j=1}^m c_j$ and consider a mapping $\sigma : [m'] \to [m]$ which maps all those i to j that satisfy

$$\sum_{k<j} c_k < i \leq \sum_{k\leq j} c_k.$$

Let M_{ij} be an $n_{\sigma(i)} \times n$ matrix that is zero if $\sigma(i) \neq j$ and $B_{\gamma(i)}$ if $\gamma(i) = j$. Now, for $\ell \in [m']$ define A_ℓ to be the block matrix whose rows are $M_{i\ell}$ for $i \in [m']$. $T_{B,p}$ is now a rectangular completely positive operator from $M_{nc}(\mathbb{C}) \to M_n(\mathbb{C})$ that maps a positive definite X to $\sum_{i\in[m']} A_i^\dagger X A_i$.

Recall the capacity of a nonsquare completely positive operator (Definition 5.6):

$$\mathrm{Cap}(T_{B,p}) := \inf\left\{ \left(\frac{\det(T_{B,p}(X))}{c}\right) : X \succ 0,\ \det(X)^{\frac{1}{c}} = 1 \right\}.$$

Given the block form of each A_i, it follows that

$$T_{B,p}(X) = \sum_{i=1}^{m'} B_{\sigma(i)}^{\dagger} X_i B_{\sigma(i)},$$

where X_i is an appropriate submatrix of X. Thus, it follows from the basic properties of the determinant that we can write

$$\mathrm{Cap}(T_{B,p}) = \inf\left\{ \det\left(\frac{\sum_{i=1}^{m'} B_{\sigma(i)}^{\dagger} X_i B_{\sigma(i)}}{c} \right) : X_i \succ 0,\ \prod_{i=1}^{m'} \det(X_i) = 1 \right\}.$$

Replace $\sum_{i:\sigma(i)=j} X_i$ by $c_j Y_j$ to obtain

$$\mathrm{Cap}(T_{B,p}) = \inf\left\{ \det\left(\frac{\sum_{j=1}^{m} c_j B_j^{\dagger} Y_j B_j}{c} \right) : Y_j \succ 0,\ \prod_{j=1}^{m} \det(Y_j)^{c_j} = 1 \right\} = \frac{1}{\mathrm{BL}(B,p)^2}$$

via Theorem 5.11. To ensure we can use the algorithm developed for capacity, we also need to also prove that $T_{B,p}$ is rank nondecreasing. Towards this, first, we need to extend the notion of rank nondecreasing to nonsquare operators and then show that it satisfies this property; see [103] for the details.

Note that this construction does not lead to an optimization problem whose dimension is polynomial in the input bit length as the size of the constructed operator in the operator scaling problem depends exponentially on the bit lengths of the entries of p. From the geodesic convexity of capacity (discussed in Section 5.3), it follows that the Brascamp–Lieb constant is also a solution to a geodesically convex optimization problem. A succinct geodesically convex formulation was provided in [259].

5.2.5 Polynomial capacity

A basic version of the capacity of polynomials was considered in a paper by Gurvits and Samorodnitsky [125] and then generalized to operators (Definition 5.6) by [124]. Subsequently, Gurvits defined a notion of capacity for hyperbolic polynomials in [124] and used it to prove a generalization of van der Waerden conjecture by Bapat [27] for mixed discriminants. In this section, we present this notion of polynomial capacity just for the setting of the permanent. For an $n \times n$ nonnegative matrix A, consider the polynomial

$$f_A(x_1, \ldots, x_n) := \prod_{i=1}^{n} \sum_{j=1}^{n} A_{i,j} x_j.$$

[124] considered the following notion of capacity:

$$\mathrm{Cap}(f_A) := \inf\left\{ f_A(x_1,\ldots,x_n) : x_i > 0, \prod_{i=1}^{n} x_i = 1 \right\}. \tag{19}$$

It is easily checked that if A is stochastic then $0 \leq \mathrm{Cap}(f_A) \leq 1$, and $\mathrm{Cap}(f_A) = 1$ if and only if A is doubly stochastic. The main result of [124] when specialized for the above polynomial implied that

$$\mathrm{Per}(A) \geq \left(\frac{n!}{n^n}\right) \mathrm{Cap}(f_A),$$

giving an alternate proof of the van der Waerden conjecture (Theorem 5.2). Thus, $\mathrm{Cap}(f_A)$ is an e^{-n} approximation to $\mathrm{Per}(A)$. One can also replace the permanent potential function with the capacity in the proof of [185] presented in Section 5.1. Moreover, after introducing new variables $y_i = \log x_i$ and replacing the objective with $\log f_A(x_1,\ldots,x_n)$, one obtains a convex program that can be solved efficiently; see [124, 253, 262, 264]. This gives an alternate proof of Theorem 5.3. As discussed in previous sections, [124] arrived at this notion of capacity while trying to extend the work of Linial, Samorodnitsky, and Wigderson [185] to Edmonds' singularity problem. The proof technique in [124] relied on the location of the roots of the polynomial under consideration (f_A in the case of permanent). This viewpoint itself has had far-reaching consequences in theoretical computer science and mathematics; see [280].

5.3 Capacity and geodesic convex optimization

In the most general setting, an optimization problem takes the form

$$\inf_{x \in K} f(x),$$

for some set K and some function $f : K \to \mathbb{R}$.[11] When $K \subseteq \mathbb{R}^d$, we can talk about the convexity of K and f. K is said to be convex if any "straight line" joining two points in K is entirely contained in K, and f is said to be convex if, on any such straight line, the average value of f at the endpoints is at least the value of f at the mid-point of the line. When f is "smooth" enough, there are equivalent definitions of convexity in terms of the standard differential structure in $\mathbb{R}^d$: the gradient or the Hessian of f. Thus, convexity can also be viewed as a property arising from the interaction of the function and how we differentiate in $\mathbb{R}^n$; e.g., the Hessian of f at every point in K should be positive semi-definite. When both K and f are convex, the optimization problem is called a convex optimization problem. The fact that the convexity of f implies that any local minimum of f in K is also a global minimum, along with the fact that computing gradients and Hessians is typically

[11] Part of this section draws from [281].

easy in Euclidean spaces, makes it well-suited for developing first-order algorithms such as gradient descent and second-order algorithms such as interior point methods. Analyzing the convergence of these methods boils down to understanding how well-behaved derivatives of the function are, and there is a well-developed theory of algorithms for convex optimization; see [51, 203, 282].

Several optimization problems, however, are nonconvex. An important example is that of the capacity of a completely positive operator (Definition 5.6)

$$\mathrm{Cap}(T) := \inf\{\det(T(X)) : X \succ 0,\ \det(X) = 1\} = \inf\left\{\frac{\det(T(X))}{\det(X)} : X \succ 0\right\}, \quad (20)$$

which is nonconvex as the objective function is nonconvex. However, [102] observed a curious property of the capacity: Consider the following Lagrangian of this optimization problem:

$$f(X,\lambda) := \log\det(T(X)) + \lambda \cdot \log\det X,$$

where λ is the multiplier for the constraint. Then, any X for which $\nabla_X f(X,\lambda) = 0$ is an optimal solution to Equation (20). Bürgisser, Garg, Oliveira, Walter, and Wigderson [61] mention that the capacity optimization problem, while nonconvex, is *geodesically convex*. While the domain of positive definite matrices is convex in the ordinary sense, the key to showing that capacity optimization is geodesically convex is to view this space as a manifold and redefine what it means to be a straight line by introducing a *metric*.

This redefinition of a straight line entails the introduction of a different differential structure. Roughly speaking, a manifold is a topological space that locally looks like Euclidean space. "Differentiable manifolds" are a special class of manifolds that come with a differential structure that allows one to do calculus over them. Straight lines on differential manifolds are called "geodesics", and a set that has the property that a geodesic joining any two points in it is entirely contained in the set is called geodesically convex (with respect to the given differential structure). A function that has this property that its average value at the end points of a geodesic is at least the value of f at the mid-point of the geodesic is called geodesically convex (with respect to the given differential structure). And, when K and f are both geodesically convex, the optimization problem is called a geodesically convex optimization problem. Geodesically convex functions also have key properties similar to convex functions such as the fact that a local minimum is also a global minimum.

Allen-Zhu, Garg, Li, Oliveira, and Wigderson [7] develop first-order and second-order methods for a class of geodesically convex optimization problems that include capacity. In this section, we first introduce the basics of geodesic convexity (Section 5.3.1), show that the capacity optimization problem in Equation (20) is geodesically convex (Section 5.3.2), and give a high-level view of the algorithms in [7] (Section 5.3.3). We do not develop a theory of geodesic convexity here but give the minimal details to ensure that we can argue that the capacity function in (20) is geodesically convex; see [273, 281] for a thorough treatment on geodesic convexity.

5.3.1 The Riemannian geometry of positive definite matrices and geodesic convexity

For simplicity, here we consider the case of real symmetric matrices and symmetric positive definite matrices. Let S^n denote the space of all $n \times n$ real symmetric matrices and let S^n_{++} denote the space of all $n \times n$ symmetric positive definite matrices. S^n_{++} is a smooth manifold and the tangent space at $P \in S^n_{++}$ is $\mathbb{R}^{\frac{n(n+1)}{2}}$ which is homeomorphic to S^n for each $P \in S^n_{++}$. We consider the metric induced by the Hessian of the function: $-\log\det(P)$ for a positive definite matrix P. This function is convex and the metric is

$$g_P(U,W) := \mathrm{Tr}[P^{-1}UP^{-1}W]$$

for $P \in S^n_{++}$ and $U,W \in S^n$. g_P is clearly symmetric, bilinear, and positive definite. It is also nondegenerate as $\mathrm{Tr}[P^{-1}UP^{-1}W] = 0$ for every W implies

$$\mathrm{Tr}[P^{-1}UP^{-1}U] = \mathrm{Tr}[P^{-\frac{1}{2}}UP^{-\frac{1}{2}}P^{-\frac{1}{2}}UP^{-\frac{1}{2}}] = 0$$

or equivalently $P^{-\frac{1}{2}}UP^{-\frac{1}{2}} = 0$. Since P is a nonsingular matrix, $P^{-\frac{1}{2}}UP^{-\frac{1}{2}} = 0$ is equivalent to $U = 0$. Next, we observe that S^n_{++} with g is a Riemannian manifold. This follows from the observation that g_P varies smoothly with P.

Since the metric tensor allows us to measure distances on a Riemannian manifold, there is an alternative, and sometimes useful, way of defining geodesics on it: as length-minimizing curves. Before we can define a geodesic in this manner, we need to define the length of a curve on a Riemannian manifold. This gives rise to a notion of distance between two points as the minimum length of a curve that joins these points. Using the metric tensor we can measure the instantaneous length of a given curve. Integrating along the vector field induced by its derivative, we can measure the length of the curve. And, we can then define the shortest curve – geodesic – that connects two points.

It is well-known that the geodesic with respect to the Hessian of the log-determinant metric that joins P to Q on S^n_{++} can be parameterized as follows (see [43]):

$$\rho(t) := P^{\frac{1}{2}}(P^{-\frac{1}{2}}QP^{-\frac{1}{2}})^t P^{\frac{1}{2}}. \tag{21}$$

Thus, $\rho(0) = P$ and $\rho(1) = Q$.

In general, let (M,g) be a Riemannian manifold. A set $K \subseteq M$ is said to be geodesically convex with respect to g if, for any $p,q \in K$, any geodesic ρ_{pq} that joins p to q lies entirely in K. It follows from Equation (21) that S^n_{++} is a geodesically convex set with respect to the metric defined above.

Definition 5.12 (Geodesically convex function). Let (M,g) be a Riemannian manifold and $K \subseteq M$ be a geodesically convex set with respect to g. A function $f : K \to \mathbb{R}$ is said to be a geodesically convex function with respect to g if for any $p,q \in K$, and for any geodesic $\rho pq : [0,1] \to K$ that joins p to q,

$$\forall t \in [0,1] \;\; f(\gamma_{pq}(t)) \leq (1-t)f(p) + tf(q).$$

$\log\det(X)$ is geodesically both convex and concave on S^n_{++} with respect to the metric $g_X(U,V) := \mathrm{Tr}[X^{-1}UX^{-1}V]$. To see this, let $X, Y \in S^n_{++}$ and $t \in [0,1]$. Then, the geodesic joining X to Y is

$$\rho(t) = X^{\frac{1}{2}}(X^{-\frac{1}{2}}YX^{-\frac{1}{2}})^t X^{\frac{1}{2}}.$$

Thus,

$$\log\det(\rho(t)) = \log\det(X^{\frac{1}{2}}(X^{-\frac{1}{2}}YX^{-\frac{1}{2}})^t X^{\frac{1}{2}}) = (1-t)\log\det(X) + t\log\det(Y).$$

Therefore, $\log\det(X)$ is a geodesically linear function over the positive definite cone with respect to the metric g.

5.3.2 Geodesic convexity of capacity

We now show that the capacity of a completely positive operator T is a geodesically convex optimization problem. First, we show that $T(X)$ is "geodesically convex". In other words, for any geodesic, $\rho : [0,1] \to S^n_{++}$,

$$\forall t \in [0,1], \;\; T(\rho(t)) \preceq (1-t)T(\rho(0)) + tT(\rho(1)). \tag{22}$$

Write $T(X) := \sum_{i=1}^m A_i X A_i^\top$ for some $n \times n$ matrices A_i. Consider the geodesic $\rho(t) := P^{\frac{1}{2}}\exp(tQ)P^{\frac{1}{2}}$ for $P \in S^n_{++}$ and $Q \in S^n$. The second derivative of T along ρ is

$$\frac{\mathrm{d}^2 T(\rho(t))}{\mathrm{d}t^2} = \sum_{i=1}^m A_i P^{\frac{1}{2}} Q \exp(tQ) Q P^{\frac{1}{2}} A_i^\top = T(P^{\frac{1}{2}} Q \exp(tQ) Q P^{\frac{1}{2}}).$$

Since $P^{\frac{1}{2}} Q \exp(tQ) Q P^{\frac{1}{2}}$ is positive definite for any $t \in [0,1]$, $T(P^{\frac{1}{2}} Q \exp(tQ) Q P^{\frac{1}{2}})$ is also positive definite as T is a strictly positive operator. Consequently, $\frac{\mathrm{d}^2}{\mathrm{d}t^2} T(\rho(t))$ is positive definite, and (22) holds.

Now, we argue that $\log\det(T(X))$ is also geodesically convex. We need to show that the Hessian of $\log\det(T(X))$ is positive semi-definite along any geodesic. Let us consider the geodesic $\rho(t) := P^{\frac{1}{2}}\exp(tQ)P^{\frac{1}{2}}$ for $P \in S^n_{++}$ and $Q \in S$, and let $h(t) := \log\det(T(\rho(t)))$. The second derivative of $\log\det(T(X))$ along ρ is:

$$\frac{\mathrm{d}^2 h(t)}{\mathrm{d}t^2} = \mathrm{Tr}\left[-T(\rho(t))^{-1} \frac{\mathrm{d}}{\mathrm{d}t} T(\rho(t)) T(\rho(t))^{-1} \frac{\mathrm{d}}{\mathrm{d}t} T(\rho(t)) + T(\rho(t))^{-1} \frac{\mathrm{d}^2}{\mathrm{d}t^2} T(\rho(t)) \right].$$

Thus, we need to verify that $\left.\frac{\mathrm{d}^2 h(t)}{\mathrm{d}t^2}\right|_{t=0} \geq 0$. In other words, we need to show that

$$\mathrm{Tr}\left[T(P)^{-1} \left(T(P^{\frac{1}{2}} Q^2 P^{\frac{1}{2}}) - T(P^{\frac{1}{2}} Q P^{\frac{1}{2}}) T(P)^{-1} T(P^{\frac{1}{2}} Q P^{\frac{1}{2}}) \right) \right] \geq 0.$$

In particular, if we show that

$$T(P^{\frac{1}{2}}Q^2P^{\frac{1}{2}}) \succeq T(P^{\frac{1}{2}}QP^{\frac{1}{2}})T(P)^{-1}T(P^{\frac{1}{2}}QP^{\frac{1}{2}}),$$

then we are done. Let us define another strictly positive linear operator

$$T'(X) := T(P)^{-\frac{1}{2}}T(P^{\frac{1}{2}}XP^{\frac{1}{2}})T(P)^{-\frac{1}{2}}.$$

If $T'(X^2) \succeq T'(X)^2$, then by picking $X = Q$ we arrive at the conclusion. This inequality is an instance of Kadison's inequality, see [43] for more details. Therefore, $\log\det(X)$ is a geodesically convex function. We can now conclude that $\log\det T(X) - \log\det X$ is geodesically convex as $\log\det X$ is geodesically linear.

Theorem 5.13. (Geodesic convexity of capacity [168, 258]) *Let $T(X)$ be a completely positive linear operator. Then, $\frac{\det(T(X))}{\det(X)}$ is geodesically convex on S^n_{++} with respect to the metric $g_X(U,W) := \mathrm{Tr}[X^{-1}UX^{-1}W]$.*

5.3.3 Computing the capacity via geodesically convex optimization

As discussed in Section 5.2.5, for polynomial capacity, one can make an appropriate change of variables and make the polynomial capacity optimization problem convex with respect to the Euclidean metric. This allows for the deployment of standard convex optimization techniques to obtain algorithms that run in time polynomial in $n, \log M, \log\frac{1}{\varepsilon}$); see [154, 71, 8, 264]. The main result of Allen-Zhu, Garg, Li, Oliviera, and Wigderson [7] is an algorithm which ε-approximates capacity and runs in time polynomial in $n, m, \log M$ and $\log\frac{1}{\varepsilon}$, where M denotes the largest magnitude of an entry of A_i. Thus, it improves upon the result of [102] presented in Section 5.2 which runs in time polynomial in $n, m, \log M, \frac{1}{\varepsilon}$. The algorithm of [7] finds an $X_\varepsilon \succ 0$ such that

$$\log\det(T(X_\varepsilon)) - \log\det(X_\varepsilon) \leq \log\mathrm{Cap}(T) + \varepsilon.$$

Their algorithm is a geodesic generalization of the "box-constrained" Newton's method introduced in [71, 8]. In each iteration, their algorithm expands the objective into its second-order Taylor expansion and then solves it via Euclidean convex optimization; see [51, 203, 282] for Newton's method in Euclidean space. Their algorithm is a general second-order method and applies to any geodesically convex problem (over the space of positive definite matrices) that satisfies a particular "robustness" property. This robustness property asserts that the function behaves like a quadratic function in every "small" neighborhood with respect to the metric, it is weaker than self-concordance, and it was introduced in the Euclidean space in [71, 8].

Roughly speaking, their algorithm starts with an $X_0 = I$ and computes X_{t+1} from X_t by solving a constrained Euclidean convex quadratic minimization problem as follows: For a symmetric matrix H, let $f^t(H) := F(X_t^{\frac{1}{2}} e^H X_t^{\frac{1}{2}})$. Let q^t be the second-order Taylor approximation of f^t around $H = 0$. Since F is geodesically convex, q^t is convex in the ordinary sense. Thus, one can optimize $q^t(H)$ under the constraint $\|H\|_2 \leq \frac{1}{2}$ (this is the *box* constraint). If H_t is the optimizer to this constrained optimization problem, $X_{t+1} := X_t^{\frac{1}{2}} e^{H_t} X_t^{\frac{1}{2}}$. [7] show that after about $R \log \frac{1}{\varepsilon}$ iterations, this algorithms produces an ε-approximate minimizer to F. Here R is a bound on the *distance* of each iterate to the optimal solution.

For the operator scaling problem, the function $F(X) := \log\det\left(\sum_{i=1}^m A_i X A_i^\top\right) - \log\det(X)$ which is geodesically convex over the Riemannian manifold of positive definite matrices. They show how to modify this function slightly and provide a bound for R (or rather an alternative to it).

As an application, [7] present a polynomial time algorithm for an equivalence problem for the left-right group action underlying the operator scaling problem. This yields a deterministic polynomial-time algorithm for (commutative) PIT problems; we omit the details, see [7].

5.4 *The null-cone problem, invariant theory, and noncommutative optimization*

We present a summary of the paper by Bürgisser, Franks, Garg, Oliveira, Walter, and Wigderson [59, 60] that generalizes and unifies many prior works and initiates a systematic development of a theory of noncommutative optimization under symmetries. We start by presenting some basics in Section 5.4.1. In Section 5.4.2, we introduce the general definition of capacity and that of the null cone. In Section 5.4.3, we introduce the notion of a moment map that leads to connections with geodesic convexity and noncommutative duality. Finally, in Section 5.4.4, we mention the computational problems and the algorithmic results from [60].

5.4.1 Groups, orbits, and invariants

We consider a vector space $V \cong \mathbb{C}^m$ for some m. Given a group G, the *action* of G on V is a function $\phi : G \times V \to V$ for which we write $\phi(g, v)$ as just $g \cdot v$. A group action must further satisfy the properties that $g \cdot (h \cdot v) = (gh) \cdot v$ and $e \cdot v = v$, where e is the identity element in G. An *orbit* of $v \in V$ under a given action of G is the set

$$\mathscr{O}_v := \{w \in V : w = g \cdot v \text{ for some } g \in G\}.$$

The closure of an orbit $\mathscr{O}_v$ is denoted by $\overline{\mathscr{O}_v}$.

A group representation π is a map from an element $g \in G$ to an invertible linear transformation $\pi(g)$ of the vector space V (or $\mathrm{GL}(V)$). Enforcing π to be a group homomorphism (i.e., for any $g_1, g_2 \in G$ we have $\pi(g_1 g_2) = \pi(g_1)\pi(g_2)$) implies the action $g \cdot v := \pi(g)v$ is a group action.

Invariant polynomials are polynomial functions on V that are invariant by the action of G. The ring of invariant of polynomials is denoted by $\mathbb{C}[V]^G$ and is finitely generated due to a theorem of Hilbert [132, 133]. It is known that for two vectors $v_1, v_2 \in V$, their orbit-closures intersect if and only if $p(v_1) = p(v_2)$ for all $p \in \mathbb{C}[V]^G$; see [201].

As an example, operator scaling can be viewed as a special case of the left-right action of $G = \mathrm{SL}_n(\mathbb{C}) \times \mathrm{SL}_n(\mathbb{C})$ on $V = (\mathbb{C}^{n \times n})^m$:

$$\pi(C,D) \cdot (A_1, \ldots, A_m) := (CA_1D^\dagger, \ldots, CA_mD^\dagger).$$

Here, the invariants for the left-right action are generated by polynomials of the form $\det(\sum_{i=1}^m E_i \otimes A_i)$, where E_i are complex $d \times d$ matrices for some d. Derksen and Makam [77] prove that $d \leq n^5$ suffices. This implies that, to check if the orbit-closures for two $(A_1, \ldots, A_m)$ and $(B_1, \ldots, B_m)$ under the left-right action of $\mathrm{SL}_n(\mathbb{C}) \times \mathrm{SL}_n(\mathbb{C})$, it suffices to check if $\det(\sum_{i=1}^m Y_i \otimes A_i) = \det(\sum_{i=1}^m Y_i \otimes B_i)$ for all $d \times d$ matrices Y_is on disjoint sets of variables for $d \leq n^5$. This is an instance of the ordinary PIT problem and a deterministic algorithm for this problem is provided by the algorithm in [7] discussed in Section 5.3.3.

5.4.2 Capacity and the null cone

[60] generalize operator scaling and the algorithmic results for it to the case when π is *any* representation of $G = \mathrm{GL}_n(\mathbb{C})$. To do so, one needs to assume that V is equipped with an inner product $\langle \cdot, \cdot \rangle$ which defines a norm $\|v\| := \sqrt{\langle v, v \rangle}$. For a representation π, [60] define the capacity of an element $v \in V$ as

$$\mathrm{Cap}(v) := \inf_{g \in G} \|\pi(g)v\|. \tag{23}$$

In the commutative (torus) case, this is precisely the notion of polynomial capacity introduced by Gurvits [124] (Section 5.2.5). In the left-right action case, the notion of operator capacity (Definition 5.6) and the one in Equation (23) can also be seen to coincide.

A natural question is: For what v is $\mathrm{Cap}(v) = 0$? This brings us to the notion of the *null cone* of V, which is defined as follows:

$$\mathcal{N} := \{v \in V : \mathrm{Cap}(v) = 0\}.$$

Thus, the null cone is the set of all vectors $v \in V$ whose orbit closure contains 0.

5.4.3 Geodesic convexity, moment map, and noncommutative duality

For a representation π of $\mathrm{GL}_n(\mathbb{C})$, a vector $v \in V$, and an $H \in \mathscr{H}(n)$, consider $\log \|\pi(\mathrm{e}^{tH})v\|$ as a function of t. Here, e^{tH} is the matrix exponential and is a geodesic in $\mathrm{GL}_n(\mathbb{C})$ starting at the identity element in the direction H. This function can be proved to be convex in t, making a connection to geodesic convexity; see [60] for details. The derivative of this function at time $t = 0$ gives rise to the *moment map* $\mu(v)$ for $v \in V$ as follows: For an $H \in \mathscr{H}(n)$,

$$\langle \mu(v), H \rangle := \frac{\partial \log \|\pi(\mathrm{e}^{tH})v\|}{\partial t}(0).$$

Thus, a moment map can be viewed as a noncommutative version of the gradient in a suitably defined Riemannian manifold that arises from the symmetries of noncommutative groups [60]. Hence, as $\|\pi(g)v\|$ tends to $\mathrm{Cap}(v)$ with g, $\mu(v)$ tends to zero.

For $H \in H(n)$, let $\mathrm{spec}(H) := (\lambda_1, \ldots, \lambda_n)$, where $\lambda_1 \geq \cdots \geq \lambda_n$ are the eigenvalues of H. The *moment polytope* of v, denoted by $\Delta(v)$, is the closure of the set of eigenvalues of $\mu(w)$ as w varies in the orbit of v:

$$\Delta(v) := \overline{\{\mathrm{spec}(\mu(w) : w \in \mathscr{O}_v\}}.$$

It is a nontrivial result that $\Delta(v)$ is a convex polytope [165, 21, 202].

It was proved by Kempf and Ness [163] that v is not in the null-cone $\mathscr{N}$ if and only if $\mu(w) = 0$ for some w in the orbit closure of v, or $0 \in \Delta(v)$. This is an important result and can be viewed as a noncommutative analog of Farkas' Lemma in the commutative world. Thus, we can draw an analogy to convex optimization: If we view the moment map as the gradient of the action of π at the identity element, then the $\|w\|$ is minimized when the gradient is zero. v is in the null cone if and only if $\mathrm{Cap}(v) = 0$.

One of the key structural results in [60] is a quantitative version of the Kempf–Ness theorem.

Theorem 5.14. (Noncommutative duality [60]) *For a unit vector v in V,*

$$1 - \frac{\|\mu(v)\|}{\gamma(\pi)} \leq \mathrm{Cap}^2(v) \leq 1 - \frac{\|\mu(v)\|^2}{4N(\pi)}.$$

Here, the *weight norm* $N(\pi)$ is defined to be the maximum Euclidean norm of a weight that occurs in π. A weight vector $\lambda \in \mathbb{Z}^n$ occurs in π if one of its irreducible subspaces is of type λ. And, the *weight margin* $\gamma(\pi)$ is the minimum Euclidean distance between the origin and the convex hull of any subset of the weights of π that does not contain the origin. The weights arise in the study of irreducible representations of π and we direct the reader to [60] for a discussion on them.

For matrix scaling (the left-right action by special torus group), it can be shown that $\gamma(\pi) \geq \frac{1}{\text{poly}(n)}$. For operator scaling too (with the left-right action by $\text{SL}_n(\mathbb{C}) \times \text{SL}_n(\mathbb{C})$), it can be shown that $\gamma(\pi) \geq \frac{1}{\text{poly}(n)}$.

5.4.4 Noncommutative optimization under symmetries

[60] study a variety of general and related problems related to orbits of group actions.

1. **Null cone membership problem:** Given (π, v), check if $v \in \mathcal{N}$.
2. **Moment polytope membership problem:** Given (π, v, p), check if $p \in \Delta(v)$.
3. **Norm-minimization problem:** Given (π, v, ε) such that $\text{Cap}(v) > 0$, output a $g \in G$ such that $\log \|\pi(g) \cdot v\| - \log \text{Cap}(v) \leq \varepsilon$.
4. **Scaling problem:** Given (π, v, p, ε) such that $p \in \Delta(v)$, output an element $g \in G$ such that $\|\text{spec}(\mu(\pi(g)v)) - p\| \leq \varepsilon$.

[60] discuss how these problems capture a diverse set of problems in different areas of computer science, mathematics, and physics. We already discussed the application to approximating the permanent (Section 5.1), noncommutative singularity testing (Section 5.2.3), and computing Brascamp–Lieb constants (Section 5.2.4). Other applications include the Horn problem: Do there exist three Hermitian matrices A, B, C with prescribed eigenvalues such that $A + B = C$?, the quantum marginal problem: Given density matrices describing local quantum states, is there a global pure state consistent with the local states? Moreover, these problems also connect to geometric complexity theory (GCT) [200] that formulates a variant of VP vs. VNP question as checking if the (padded) permanent lies in the orbit-closure of the determinant (of an appropriate size), under the action of the general linear group on polynomials induced by its natural linear action on the variables.

[60] also show how, sometimes, these abovementioned problems may reduce to each other and discuss multiple ways in which the input may be specified. For instance, in the operator scaling problem π is fixed (and not part of the input) while, in general, one could be given an oracle to $\pi(g)v$ for a $g \in G$ and an input vector v. p and ε are assumed to be given in binary and they present algorithms that run in time both a polynomial in $\frac{1}{\varepsilon}$ and in $\log \frac{1}{\varepsilon}$. [60] note that techniques from [253, 264] can be used to design polynomial time algorithms for commutative null cone and moment polytope membership in the oracle setting.

Prior works for these problems, including the ones discussed in Section 5.1 and 5.2, the underlying groups need to be products of at least two copies of rather specific linear groups ($\text{SL}(n)$s or tori), to support the algorithms and analysis. More importantly, these actions were *linear* in each of the copies. In [60], arbitrary group actions of GL_n that can be described by a representation, are handled. They develop two general methods, a first-order and a second-order method, which require information about the gradient and the Hessian of the function to be optimized. Their algorithms rely on the connection of the moment map to geodesic convexity and the running time bounds depend on the quantitative parameters – weight norm and weight margin – arising in their quantitative version of noncommutative duality (Theorem 5.14). The main technical work goes into showing how these parameters control convergence to the optimum in each of these methods.

The first-order method of [60] is a natural analog of gradient descent. For the problem of computing $\mathrm{Cap}(v)$, it starts with an element $g_0 = I$ (the identity element in G) and repeats for τ iterations and a suitable "step-size" $\eta > 0$ the following:

$$g_{t+1} := \mathrm{e}^{-\eta\mu(\pi(g_t)v)} g_t.$$

They show that there is a choice of η such that this method, when $\mathrm{Cap}(v) > 0$, finds a g such that $\|\mu(\pi(g)v)\| \le \varepsilon$ for $\tau = \left(\frac{N(\pi)^2}{\varepsilon^2} |\log \mathrm{Cap}(v)|\right)$. This approximately solves the scaling problem for $p = 0$. The generalization to $p \neq 0$ is also presented.

Their second-order method, at a high-level, repeatedly optimizes quadratic Taylor expansions of the objective in a small neighborhood (similar to Newton's method in convex optimization). It is an extension of their method for computing operator capacity mentioned in Section 5.3.3. The number of iterations it takes for the above-mentioned scaling problem is $\tilde{O}\left(\frac{N(\pi)\sqrt{n}}{\gamma(\pi)} \left(|\log \mathrm{Cap}(v)| + \log \frac{n}{\varepsilon}\right)\right)$.

The work of [60] has also led to a host of new challenges in noncommutative optimization. An important one is to design analogs of the "cutting plane" or the "interior point methods" in the noncommutative setting. Such algorithms would likely yield true polynomial time algorithms for Problems 1–4 mentioned above; see [134] for some progress towards the latter goal. Finally, there are several other works where the lens of symmetry has been helpful in the design of nonconvex optimization and sampling algorithms, see [33, 243, 283, 87, 244, 175, 174, 173] and the references therein.

References

1. S. Aaronson, S. Ben-David, and R. Kothari. Separations in query complexity using cheat sheets. In D. Wichs and Y. Mansour, editors, *Proceedings of the 48th Annual ACM SIGACT Symposium on Theory of Computing, STOC 2016, Cambridge, MA, USA, June 18-21, 2016*, pages 863–876. ACM, 2016.
2. M. Agrawal, N. Kayal, and N. Saxena. PRIMES is in P. *Annals of Mathematics. Second Series*, 160(2):781–793, 2004.
3. A. Ahmadinejad, J. Kelner, J. Murtagh, J. Peebles, A. Sidford, and S. Vadhan. High-precision estimation of random walks in small space. In *2020 IEEE 61st Annual Symposium on Foundations of Computer Science*, pages 1295–1306. IEEE Computer Soc., Los Alamitos, CA, [2020] ©2020.
4. M. Ajtai, J. Komlós, and E. Szemerédi. Sorting in $c \log n$ parallel steps. *Combinatorica*, 3(1):1–19, 1983.
5. M. Ajtai and A. Wigderson. Deterministic simulation of probabilistic constant depth circuits. In F. P. Preparata and S. Micali, editors, *Randomness and Computation*, volume 5 of *Advances in Computing Research*, pages 199–223. JAI Press Inc., 1989.

6. R. Aleliunas, R. M. Karp, R. J. Lipton, L. Lovász, and C. Rackoff. Random walks, universal traversal sequences, and the complexity of maze problems. In *20th Annual Symposium on Foundations of Computer Science (San Juan, Puerto Rico, 1979)*, pages 218–223. IEEE, New York, 1979.
7. Z. Allen-Zhu, A. Garg, Y. Li, R. M. de Oliveira, and A. Wigderson. Operator scaling via geodesically convex optimization, invariant theory and polynomial identity testing. In I. Diakonikolas, D. Kempe, and M. Henzinger, editors, *Proceedings of the 50th Annual ACM SIGACT Symposium on Theory of Computing, STOC 2018, Los Angeles, CA, USA, June 25-29, 2018*, pages 172–181. ACM, 2018.
8. Z. Allen Zhu, Y. Li, R. Oliveira, and A. Wigderson. Much faster algorithms for matrix scaling. In *FOCS'17: Proceedings of the 58th Annual IEEE Symposium on Foundations of Computer Science*, 2017.
9. N. Alon. Eigenvalues and expanders. *Combinatorica*, 6(2):83–96, 1986. Theory of computing (Singer Island, Fla., 1984).
10. N. Alon. Combinatorial nullstellensatz. *Combinatorics, Probability and Computing*, 8(1-2):7–29, 1999.
11. N. Alon, M. Kumar, and B. L. Volk. Unbalancing sets and an almost quadratic lower bound for syntactically multilinear arithmetic circuits. *Comb.*, 40(2):149–178, 2020.
12. N. Alon, A. Lubotzky, and A. Wigderson. Semi-direct product in groups and zig-zag product in graphs: connections and applications (extended abstract). In *42nd IEEE Symposium on Foundations of Computer Science (Las Vegas, NV, 2001)*, pages 630–637. IEEE Computer Soc., Los Alamitos, CA, 2001.
13. N. Alon and J. H. Spencer. *The probabilistic method.* Wiley Series in Discrete Mathematics and Optimization. John Wiley & Sons, Inc., Hoboken, NJ, fourth edition, 2016.
14. P. Amireddy, A. Garg, N. Kayal, C. Saha, and B. Thankey. Low-depth arithmetic circuit lower bounds via shifted partials. *Electron. Colloquium Comput. Complex.*, TR22-151, 2022.
15. N. Anari and S. O. Gharan. A generalization of permanent inequalities and applications in counting and optimization. In H. Hatami, P. McKenzie, and V. King, editors, *Proceedings of the 49th Annual ACM SIGACT Symposium on Theory of Computing, STOC 2017, Montreal, QC, Canada, June 19-23, 2017*, pages 384–396. ACM, 2017.
16. A. Anshu, A. Belovs, S. Ben-David, M. Göös, R. Jain, R. Kothari, T. Lee, and M. Santha. Separations in communication complexity using cheat sheets and information complexity. In I. Dinur, editor, *IEEE 57th Annual Symposium on Foundations of Computer Science, FOCS 2016, 9-11 October 2016, Hyatt Regency, New Brunswick, New Jersey, USA*, pages 555–564. IEEE Computer Society, 2016.
17. R. Armoni, M. Saks, A. Wigderson, and S. Zhou. Discrepancy sets and pseudorandom generators for combinatorial rectangles. In *37th Annual Symposium on Foundations of Computer Science (Burlington, VT, 1996)*, pages 412–421. IEEE Comput. Soc. Press, Los Alamitos, CA, 1996.
18. R. Armoni, A. Ta-Shma, A. Wigderson, and S. Zhou. An $O(\log(n)^{4/3})$ space algorithm for (s,t) connectivity in undirected graphs. *Journal of the ACM*, 47(2):294–311, 2000.
19. S. Arora, C. Lund, R. Motwani, M. Sudan, and M. Szegedy. Proof verification and hardness of approximation problems. In *33rd Annual Symposium on Foundations of Computer Science, Pittsburgh, Pennsylvania, USA, 24-27 October 1992*, pages 14–23. IEEE Computer Society, 1992.
20. S. Arora and S. Safra. Probabilistic checking of proofs; A new characterization of NP. In *33rd Annual Symposium on Foundations of Computer Science, Pittsburgh, Pennsylvania, USA, 24-27 October 1992*, pages 2–13. IEEE Computer Society, 1992.
21. M. F. Atiyah. Convexity and commuting Hamiltonians. *Bulletin of the London Mathematical Society*, 14(1):1–15, 1982.
22. Y. Aumann and Y. Rabani. An $O(\log k)$ approximate min-cut max-flow theorem and approximation algorithm. *SIAM Journal on Computing*, 27(1):291–301, 1998.
23. L. Babai, L. Fortnow, L. A. Levin, and M. Szegedy. Checking computations in polylogarithmic time. In C. Koutsougeras and J. S. Vitter, editors, *Proceedings of the 23rd Annual ACM Symposium on Theory of Computing, May 5-8, 1991, New Orleans, Louisiana, USA*, pages 21–31. ACM, 1991.

24. L. Babai, L. Fortnow, and C. Lund. Non-deterministic exponential time has two-prover interactive protocols. In *31st Annual Symposium on Foundations of Computer Science, St. Louis, Missouri, USA, October 22-24, 1990, Volume I*, pages 16–25. IEEE Computer Society, 1990.
25. L. Babai, L. Fortnow, N. Nisan, and A. Wigderson. BPP has subexponential time simulations unless EXPTIME has publishable proofs. *Computational Complexity*, 3(4):307–318, 1993.
26. N. Bansal and M. Sinha. k-forrelation optimally separates quantum and classical query complexity. In S. Khuller and V. V. Williams, editors, *STOC '21: 53rd Annual ACM SIGACT Symposium on Theory of Computing, Virtual Event, Italy, June 21-25, 2021*, pages 1303–1316. ACM, 2021.
27. R. Bapat. Mixed discriminants of positive semidefinite matrices. *Linear Algebra and its Applications*, 126:107–124, 1989.
28. Z. Bar-Yossef, T. S. Jayram, and I. Kerenidis. Exponential separation of quantum and classical one-way communication complexity. *SIAM J. Comput.*, 38(1):366–384, 2008.
29. Z. Bar-Yossef, T. S. Jayram, R. Kumar, and D. Sivakumar. An information statistics approach to data stream and communication complexity. *J. Comput. Syst. Sci.*, 68(4):702–732, 2004.
30. B. Barak, R. Impagliazzo, and A. Wigderson. Extracting randomness using few independent sources. *SIAM Journal on Computing*, 36(4):1095–1118 (electronic), 2006.
31. B. Barak, G. Kindler, R. Shaltiel, B. Sudakov, and A. Wigderson. Simulating independence: new constructions of condensers, Ramsey graphs, dispersers, and extractors. *Journal of the ACM*, 57(4):Art. 20, 52, 2010.
32. B. Barak, A. Rao, R. Shaltiel, and A. Wigderson. 2-source dispersers for $n^{o(1)}$ entropy, and Ramsey graphs beating the Frankl-Wilson construction. *Ann. of Math. (2)*, 176(3):1483–1543, 2012.
33. A. Barvinok and G. Blekherman. Convex geometry of orbits. *Combinatorial and Computational Geometry, Math. Sci. Res. Inst. Publ*, pages 51–77, 2005.
34. L. A. Bassalygo. Asymptotically optimal switching circuits. *Problems of Information Transmission*, 17(3):206–211, 1981.
35. W. Baur and V. Strassen. The complexity of partial derivatives. *Theor. Comput. Sci.*, 22:317–330, 1983.
36. A. Ben-Aroya and A. Ta-Shma. A combinatorial construction of almost-Ramanujan graphs using the zig-zag product. In *40th Annual ACM Symposium on Theory of Computing (Victoria, British Columbia)*, pages 325–334. ACM, 2008.
37. M. Ben-Or, O. Goldreich, S. Goldwasser, J. Håstad, J. Kilian, S. Micali, and P. Rogaway. Everything provable is provable in zero-knowledge. In S. Goldwasser, editor, *Advances in Cryptology - CRYPTO '88, 8th Annual International Cryptology Conference, Santa Barbara, California, USA, August 21-25, 1988, Proceedings*, volume 403 of *Lecture Notes in Computer Science*, pages 37–56. Springer, 1988.
38. M. Ben-Or, S. Goldwasser, J. Kilian, and A. Wigderson. Multi-prover interactive proofs: How to remove intractability assumptions. In *Proceedings of the 20th Annual ACM Symposium on Theory of Computing, May 2-4, 1988, Chicago, Illinois, USA*, pages 113–131, 1988.
39. M. Ben-Or, S. Goldwasser, and A. Wigderson. Completeness theorems for non-cryptographic fault-tolerant distributed computation (extended abstract). In J. Simon, editor, *Proceedings of the 20th Annual ACM Symposium on Theory of Computing, May 2-4, 1988, Chicago, Illinois, USA*, pages 1–10. ACM, 1988.
40. E. Ben-Sasson and A. Wigderson. Short proofs are narrow - resolution made simple. *J. ACM*, 48(2):149–169, 2001.
41. J. Bennett, A. Carbery, M. Christ, and T. Tao. The Brascamp-Lieb inequalities: Finiteness, structure and extremals. *Geometric and Functional Analysis*, 17(5):1343–1415, 2008.
42. E. R. Berlekamp. Factoring polynomials over large finite fields. *Mathematics of Computation*, 24:713–735, 1970.
43. R. Bhatia. *Positive definite matrices*. Princeton University Press, 2009.
44. G. D. Birkhoff. Tres observaciones sobre el algebra lineal. *Universidad Nacional de Tucuman Revista, Serie A*, 5:147–151, 1946.

45. M. Blum and S. Micali. How to generate cryptographically strong sequences of pseudorandom bits. *SIAM Journal on Computing*, 13(4):850–864, 1984.
46. A. Bogdanov, W. M. Hoza, G. Prakriya, and E. Pyne. Hitting sets for regular branching programs. In *37th Computational Complexity Conference*, volume 234 of *LIPIcs. Leibniz Int. Proc. Inform.*, pages Art. No. 3, 22. Schloss Dagstuhl. Leibniz-Zent. Inform., Wadern, 2022.
47. A. Bogdanov and L. Trevisan. Average-case complexity. *Foundations and Trends® in Theoretical Computer Science*, 2(1):1–106, 2006.
48. A. Borodin, S. Cook, and N. Pippenger. Parallel computation for well-endowed rings and space-bounded probabilistic machines. *Information and Control*, 58(1-3):113–136, 1983.
49. F. Boudot, P. Gaudry, A. Guillevic, N. Heninger, E. Thomé, and P. Zimmermann. The state of the art in integer factoring and breaking public-key cryptography. *IEEE Secur. Priv.*, 20(2):80–86, 2022.
50. J. Bourgain, N. Katz, and T. Tao. A sum-product estimate in finite fields, and applications. *Geometric and Functional Analysis*, 14(1):27–57, 2004.
51. S. Boyd and L. Vandenberghe. *Convex optimization*. Cambridge university press, 2004.
52. H. J. Brascamp and E. H. Lieb. Best constants in Young's inequality, its converse, and its generalization to more than three functions. *Advances in Mathematics*, 20(2):151–173, 1976.
53. M. Braverman, A. Garg, D. Pankratov, and O. Weinstein. From information to exact communication. In D. Boneh, T. Roughgarden, and J. Feigenbaum, editors, *Symposium on Theory of Computing Conference, STOC'13, Palo Alto, CA, USA, June 1-4, 2013*, pages 151–160. ACM, 2013.
54. M. Braverman and A. Moitra. An information complexity approach to extended formulations. In D. Boneh, T. Roughgarden, and J. Feigenbaum, editors, *Symposium on Theory of Computing Conference, STOC'13, Palo Alto, CA, USA, June 1-4, 2013*, pages 161–170. ACM, 2013.
55. M. Braverman, A. Rao, R. Raz, and A. Yehudayoff. Pseudorandom generators for regular branching programs. In *51th Annual IEEE Symposium on Foundations of Computer Science, FOCS 2010, October 23-26, 2010, Las Vegas, Nevada, USA*, pages 40–47. IEEE Computer Society, 2010.
56. H. Buhrman, R. Cleve, and A. Wigderson. Quantum vs. classical communication and computation. In J. S. Vitter, editor, *Proceedings of the Thirtieth Annual ACM Symposium on the Theory of Computing, Dallas, Texas, USA, May 23-26, 1998*, pages 63–68. ACM, 1998.
57. H. Buhrman, L. Fortnow, and T. Thierauf. Nonrelativizing separations. In *Thirteenth Annual IEEE Conference on Computational Complexity (Buffalo, NY, 1998)*, pages 8–12. IEEE Computer Soc., Los Alamitos, CA, 1998.
58. H. Buhrman, P. B. Miltersen, J. Radhakrishnan, and S. Venkatesh. Are bitvectors optimal? *SIAM Journal on Computing*, 31(6):1723–1744 (electronic), 2002.
59. P. Bürgisser, C. Franks, A. Garg, R. M. de Oliveira, M. Walter, and A. Wigderson. Efficient algorithms for tensor scaling, quantum marginals, and moment polytopes. In M. Thorup, editor, *59th IEEE Annual Symposium on Foundations of Computer Science, FOCS 2018, Paris, France, October 7-9, 2018*, pages 883–897. IEEE Computer Society, 2018.
60. P. Bürgisser, C. Franks, A. Garg, R. M. de Oliveira, M. Walter, and A. Wigderson. Towards a theory of non-commutative optimization: Geodesic 1st and 2nd order methods for moment maps and polytopes. In D. Zuckerman, editor, *60th IEEE Annual Symposium on Foundations of Computer Science, FOCS 2019, Baltimore, Maryland, USA, November 9-12, 2019*, pages 845–861. IEEE Computer Society, 2019.
61. P. Bürgisser, A. Garg, R. M. de Oliveira, M. Walter, and A. Wigderson. Alternating minimization, scaling algorithms, and the null-cone problem from invariant theory. In A. R. Karlin, editor, *9th Innovations in Theoretical Computer Science Conference, ITCS 2018, January 11-14, 2018, Cambridge, MA, USA*, volume 94 of *LIPIcs*, pages 24:1–24:20. Schloss Dagstuhl - Leibniz-Zentrum für Informatik, 2018.
62. M. Capalbo, O. Reingold, S. Vadhan, and A. Wigderson. Randomness conductors and constant-degree lossless expanders. In *34th Annual ACM Symposium on Theory of Computing (STOC '02)*, pages 659–668, Montréal, CA, May 2002. ACM. Joint session with *CCC '02*.

63. L. Chen, X. Lyu, A. Tal, and H. Wu. New prgs for unbounded-width/adaptive-order read-once branching programs. In *Proceedings of the 50th EATCS International Colloquium on Automata, Languages and Programming (ICALP '23)*, 2023. To appear.
64. L. Chen and R. Tell. Simple and fast derandomization from very hard functions: eliminating randomness at almost no cost. In S. Khuller and V. V. Williams, editors, *STOC '21: 53rd Annual ACM SIGACT Symposium on Theory of Computing, Virtual Event, Italy, June 21-25, 2021*, pages 283–291. ACM, 2021.
65. L. Chen and R. Tell. Hardness vs randomness, revised: uniform, non-black-box, and instance-wise. In *2021 IEEE 62nd Annual Symposium on Foundations of Computer Science—FOCS 2021*, pages 125–136. IEEE Computer Soc., Los Alamitos, CA, [2022] ©2022.
66. S. Chillara, N. Limaye, and S. Srinivasan. Small-depth multilinear formula lower bounds for iterated matrix multiplication with applications. *SIAM J. Comput.*, 48(1):70–92, 2019.
67. B. Chor and O. Goldreich. Unbiased bits from sources of weak randomness and probabilistic communication complexity. *SIAM Journal on Computing*, 17(2):230–261, Apr. 1988.
68. B. Chor, S. Goldwasser, S. Micali, and B. Awerbuch. Verifiable secret sharing and achieving simultaneity in the presence of faults. In *26th Annual Symposium on Foundations of Computer Science (sfcs 1985)*, pages 383–395, 1985.
69. M. Clegg, J. Edmonds, and R. Impagliazzo. Using the groebner basis algorithm to find proofs of unsatisfiability. In G. L. Miller, editor, *Proceedings of the Twenty-Eighth Annual ACM Symposium on the Theory of Computing, Philadelphia, Pennsylvania, USA, May 22-24, 1996*, pages 174–183. ACM, 1996.
70. A. Cohen and A. Wigderson. Dispersers, deterministic amplification, and weak random sources (extended abstract). In *30th Annual Symposium on Foundations of Computer Science (Research Triangle Park, North Carolina)*, pages 14–19. IEEE, 1989.
71. M. B. Cohen, A. Madry, D. Tsipras, and A. Vladu. Matrix scaling and balancing via box constrained Newton's method and interior point methods. In *FOCS'17: Proceedings of the 58th Annual IEEE Symposium on Foundations of Computer Science*, 2017.
72. P. M. Cohn. The embedding of firs in skew fields. *Proceedings of the London Mathematical Society*, s3-23(2):193–213, 1971.
73. S. A. Cook. The complexity of theorem-proving procedures. In M. A. Harrison, R. B. Banerji, and J. D. Ullman, editors, *Proceedings of the 3rd Annual ACM Symposium on Theory of Computing, May 3-5, 1971, Shaker Heights, Ohio, USA*, pages 151–158. ACM, 1971.
74. R. Cramer, I. Damgård, and J. B. Nielsen. *Secure Multiparty Computation and Secret Sharing*. Cambridge University Press, 2015.
75. A. De. Pseudorandomness for permutation and regular branching programs. In *Proceedings of the 26th Annual IEEE Conference on Computational Complexity, CCC 2011, San Jose, California, June 8-10, 2011*, pages 221–231. IEEE Computer Society, 2011.
76. R. A. DeMillo and R. J. Lipton. A probabilistic remark on algebraic program testing. *Information Processing Letters*, 7(4):193–195, 1978.
77. H. Derksen and V. Makam. Polynomial degree bounds for matrix semi-invariants. *Advances in Mathematics*, 310:44–63, 2017.
78. W. Diffie and M. E. Hellman. New directions in cryptography. *IEEE Trans. Inf. Theory*, 22(6):644–654, 1976.
79. I. Dinur. The PCP theorem by gap amplification. *Journal of the ACM*, 54(3):article 12, 44 pages (electronic), 2007.
80. I. Dinur and O. Meir. Toward the KRW composition conjecture: Cubic formula lower bounds via communication complexity. *Comput. Complex.*, 27(3):375–462, 2018.
81. D. Doron, D. Moshkovitz, J. Oh, and D. Zuckerman. Nearly optimal pseudorandomness from hardness. *J. ACM*, 69(6):43:1–43:55, 2022.
82. Z. Dvir. On the size of Kakeya sets in finite fields. *Journal of the American Mathematical Society*, 22(4):1093–1097, 2009.
83. Z. Dvir, S. Kopparty, S. Saraf, and M. Sudan. Extensions to the method of multiplicities, with applications to Kakeya sets and mergers. In *2009 50th Annual IEEE Symposium on Foundations of Computer Science (FOCS 2009)*, pages 181–190. IEEE Computer Soc., Los Alamitos, CA, 2009.

84. Z. Dvir, G. Malod, S. Perifel, and A. Yehudayoff. Separating multilinear branching programs and formulas. In H. J. Karloff and T. Pitassi, editors, *Proceedings of the 44th Symposium on Theory of Computing Conference, STOC 2012, New York, NY, USA, May 19 - 22, 2012*, pages 615–624. ACM, 2012.
85. Z. Dvir and A. Wigderson. Monotone expanders: constructions and applications. *Theory of Computing. An Open Access Journal*, 6:291–308, 2010.
86. Z. Dvir and A. Wigderson. Kakeya sets, new mergers, and old extractors. *SIAM Journal on Computing*, 40(3):778–792, 2011.
87. A. Edelman, T. A. Arias, and S. T. Smith. The geometry of algorithms with orthogonality constraints. *SIAM J. Matrix Anal. Appl.*, 20(2):303–353, Apr. 1999.
88. J. Edmonds. Systems of distinct representatives and linear algebra. *Journal of Research of the National Bureau of Standards*, 71:241–245, 1967.
89. J. Edmonds, R. Impagliazzo, S. Rudich, and J. Sgall. Communication complexity towards lower bounds on circuit depth. *Comput. Complex.*, 10(3):210–246, 2001.
90. G. P. Egorychev. The solution of van der Waerden's problem for permanents. *Advances in Mathematics*, 42(3):299–305, 1981.
91. P. Erdős. Some remarks on the theory of graphs. *Bulletin of the American Mathematical Society*, 53:292–294, 1947.
92. P. Erdős. Problems and results in chromatic graph theory. In *Proof Techniques in Graph Theory (Proc. Second Ann Arbor Graph Theory Conf., Ann Arbor, Mich., 1968)*, pages 27–35. Academic Press, New York, 1969.
93. D. I. Falikman. Proof of the van der Waerden conjecture regarding the permanent of a doubly stochastic matrix. *Mathematical Notes*, 29(6):475–479, 1981.
94. U. Feige, S. Goldwasser, L. Lovász, S. Safra, and M. Szegedy. Approximating clique is almost NP-complete (preliminary version). In *32nd Annual Symposium on Foundations of Computer Science, San Juan, Puerto Rico, 1-4 October 1991*, pages 2–12. IEEE Computer Society, 1991.
95. L. Fortnow, J. Rompel, and M. Sipser. On the power of multi-power interactive protocols. In *Proceedings: Third Annual Structure in Complexity Theory Conference, Georgetown University, Washington, D. C., USA, June 14-17, 1988*, pages 156–161. IEEE Computer Society, 1988.
96. H. Fournier, N. Limaye, G. Malod, and S. Srinivasan. Lower bounds for depth-4 formulas computing iterated matrix multiplication. *SIAM J. Comput.*, 44(5):1173–1201, 2015.
97. P. Frankl and R. M. Wilson. Intersection theorems with geometric consequences. *Combinatorica*, 1(4):357–368, 1981.
98. J. Franklin and J. Lorenz. On the scaling of multidimensional matrices. *Linear Algebra and its Applications*, 114-115:717–735, 1989. Special Issue Dedicated to Alan J. Hoffman.
99. J. Friedman. A proof of Alon's second eigenvalue conjecture and related problems. *Memoirs of the American Mathematical Society*, 195(910):viii+100, 2008.
100. J. Friedman and A. Wigderson. On the second eigenvalue of hypergraphs. *Combinatorica*, 15(1):43–65, 1995.
101. A. Garg, M. Göös, P. Kamath, and D. Sokolov. Monotone circuit lower bounds from resolution. *Theory Comput.*, 16:1–30, 2020.
102. A. Garg, L. Gurvits, R. M. de Oliveira, and A. Wigderson. A deterministic polynomial time algorithm for non-commutative rational identity testing. In I. Dinur, editor, *IEEE 57th Annual Symposium on Foundations of Computer Science, FOCS 2016, 9-11 October 2016, Hyatt Regency, New Brunswick, New Jersey, USA*, pages 109–117. IEEE Computer Society, 2016.
103. A. Garg, L. Gurvits, R. M. de Oliveira, and A. Wigderson. Algorithmic and optimization aspects of Brascamp-Lieb inequalities, via operator scaling. In H. Hatami, P. McKenzie, and V. King, editors, *Proceedings of the 49th Annual ACM SIGACT Symposium on Theory of Computing, STOC 2017, Montreal, QC, Canada, June 19-23, 2017*, pages 397–409. ACM, 2017.
104. D. Gavinsky. Entangled simultaneity versus classical interactivity in communication complexity. *IEEE Trans. Inf. Theory*, 66(7):4641–4651, 2020.

105. D. Gavinsky, J. Kempe, I. Kerenidis, R. Raz, and R. de Wolf. Exponential separation for one-way quantum communication complexity, with applications to cryptography. *SIAM J. Comput.*, 38(5):1695–1708, 2008.
106. D. Gavinsky, O. Meir, O. Weinstein, and A. Wigderson. Toward better formula lower bounds: The composition of a function and a universal relation. *SIAM J. Comput.*, 46(1):114–131, 2017.
107. U. Girish, R. Raz, and A. Tal. Quantum versus randomized communication complexity, with efficient players. *CoRR*, abs/1911.02218, 2019.
108. O. Goldreich. *Foundations of Cryptography: Volume I Basic Tools*. Cambridge University Press, 2001.
109. O. Goldreich. *The Foundations of Cryptography: Volume II Basic Applications*. Cambridge University Press, 2004.
110. O. Goldreich. *A primer on pseudorandom generators*, volume 55 of *University Lecture Series*. American Mathematical Society, Providence, RI, 2010.
111. O. Goldreich. In a world of $\mathbf{P} = \mathbf{BPP}$. In *Studies in complexity and cryptography*, volume 6650 of *Lecture Notes in Comput. Sci.*, pages 191–232. Springer, Heidelberg, 2011.
112. O. Goldreich, S. Micali, and A. Wigderson. How to prove all NP-statements in zero-knowledge, and a methodology of cryptographic protocol design. In A. M. Odlyzko, editor, *Advances in Cryptology - CRYPTO '86, Santa Barbara, California, USA, 1986, Proceedings*, volume 263 of *Lecture Notes in Computer Science*, pages 171–185. Springer, 1986.
113. O. Goldreich, S. Micali, and A. Wigderson. How to play any mental game or A completeness theorem for protocols with honest majority. In *Proceedings of the 19th Annual ACM Symposium on Theory of Computing, 1987, New York, New York, USA*, pages 218–229. ACM, 1987.
114. O. Goldreich and A. Wigderson. Tiny families of functions with random properties: A quality-size trade-off for hashing. *Random Structures & Algorithms*, 11(4):315–343, 1997.
115. S. Goldwasser and S. Micali. Probabilistic encryption. *Journal of Computer and System Sciences*, 28(2):270–299, Apr. 1984.
116. S. Goldwasser, S. Micali, and C. Rackoff. The knowledge complexity of interactive proof-systems (extended abstract). In R. Sedgewick, editor, *Proceedings of the 17th Annual ACM Symposium on Theory of Computing, May 6-8, 1985, Providence, Rhode Island, USA*, pages 291–304. ACM, 1985. Preliminary versions circulated since 1982.
117. M. Göös, T. Pitassi, and T. Watson. Deterministic communication vs. partition number. *SIAM J. Comput.*, 47(6):2435–2450, 2018.
118. P. Gopalan, R. Meka, O. Reingold, L. Trevisan, and S. P. Vadhan. Better pseudorandom generators from milder pseudorandom restrictions. In *53rd Annual IEEE Symposium on Foundations of Computer Science, FOCS 2012, New Brunswick, NJ, USA, October 20-23, 2012*, pages 120–129. IEEE Computer Society, 2012.
119. L. K. Grover. A fast quantum mechanical algorithm for database search. In G. L. Miller, editor, *Proceedings of the Twenty-Eighth Annual ACM Symposium on the Theory of Computing, Philadelphia, Pennsylvania, USA, May 22-24, 1996*, pages 212–219. ACM, 1996.
120. A. Gupta, P. Kamath, N. Kayal, and R. Saptharishi. Approaching the chasm at depth four. *J. ACM*, 61(6):33:1–33:16, 2014.
121. V. Guruswami. Iterative decoding of low-density parity-check codes. *Bulletin of the EATCS*, 90:53–88, October 2006.
122. V. Guruswami, C. Umans, and S. Vadhan. Unbalanced expanders and randomness extractors from Parvaresh–Vardy codes. *Journal of the ACM*, 56(4):1–34, 2009.
123. L. Gurvits. Classical complexity and quantum entanglement. *J. Comput. Syst. Sci.*, 69(3):448–484, 2004.
124. L. Gurvits. Hyperbolic polynomials approach to Van der Waerden/Schrijver-Valiant like conjectures: sharper bounds, simpler proofs and algorithmic applications. In *Proceedings of the thirty-eighth annual ACM symposium on Theory of computing*, pages 417–426. ACM, 2006.

125. L. Gurvits and A. Samorodnitsky. A deterministic algorithm for approximating the mixed discriminant and mixed volume, and a combinatorial corollary. *Discrete & Computational Geometry*, 27:531–550, 2002.
126. D. Gutfreund, R. Shaltiel, and A. Ta-Shma. Uniform hardness versus randomness tradeoffs for Arthur-Merlin games. *Computational Complexity*, 12(3-4):85–130, 2003.
127. A. Haken. The intractability of resolution. *Theor. Comput. Sci.*, 39:297–308, 1985.
128. M. Hamada and H. Hirai. Computing the NC-rank via discrete convex optimization on CAT(0) spaces. *SIAM Journal on Applied Algebra and Geometry*, 5(3):455–478, 2021.
129. J. Håstad and A. Wigderson. Composition of the universal relation. In J. Cai, editor, *Advances In Computational Complexity Theory, Proceedings of a DIMACS Workshop, New Jersey, USA, December 3-7, 1990*, volume 13 of *DIMACS Series in Discrete Mathematics and Theoretical Computer Science*, pages 119–134. DIMACS/AMS, 1990.
130. J. Håstad and A. Wigderson. The randomized communication complexity of set disjointness. *Theory Comput.*, 3(1):211–219, 2007.
131. P. Hatami and W. Hoza. Theory of unconditional pseudorandom generators. *Electron. Colloquium Comput. Complex.*, TR23-019, 2023.
132. D. Hilbert. Über die theorie der algebraischen formen. *Math. Annalen*, 36:473–534, 1890.
133. D. Hilbert. Ueber die vollen invariantensysteme. *Mathematische Annalen*, 42:313–373, 1893.
134. H. Hirai, H. Nieuwboer, and M. Walter. Interior-point methods on manifolds: theory and applications. *CoRR*, abs/2303.04771, 2023.
135. S. Hoory, N. Linial, and A. Wigderson. Expander graphs and their applications. *Bulletin of the AMS*, 43(4):439–561, 2006.
136. W. M. Hoza. Better pseudodistributions and derandomization for space-bounded computation. In M. Wootters and L. Sanità, editors, *Approximation, Randomization, and Combinatorial Optimization. Algorithms and Techniques, APPROX/RANDOM 2021, August 16-18, 2021, University of Washington, Seattle, Washington, USA (Virtual Conference)*, volume 207 of *LIPIcs*, pages 28:1–28:23. Schloss Dagstuhl - Leibniz-Zentrum für Informatik, 2021.
137. W. M. Hoza, E. Pyne, and S. Vadhan. Pseudorandom Generators for Unbounded-Width Permutation Branching Programs. In J. R. Lee, editor, *12th Innovations in Theoretical Computer Science Conference (ITCS 2021)*, volume 185 of *Leibniz International Proceedings in Informatics (LIPIcs)*, pages 7:1–7:20, Dagstuhl, Germany, 2021. Schloss Dagstuhl–Leibniz-Zentrum für Informatik.
138. R. Impagliazzo, V. Kabanets, and A. Wigderson. In search of an easy witness: exponential time vs. probabilistic polynomial time. *Journal of Computer and System Sciences*, 65(4):672–694, 2002.
139. R. Impagliazzo, N. Nisan, and A. Wigderson. Pseudorandomness for network algorithms. In *Proceedings of the Twenty-Sixth Annual ACM Symposium on the Theory of Computing*, pages 356–364, Montréal, Québec, Canada, 23–25 May 1994.
140. R. Impagliazzo, R. Shaltiel, and A. Wigderson. Reducing the seed length in the nisan-wigderson generator. *Combinatorica*, 26(6):647–681, 2006.
141. R. Impagliazzo and A. Wigderson. $P = BPP$ if E requires exponential circuits: Derandomizing the XOR lemma. In *Proceedings of the Twenty-Ninth Annual ACM Symposium on Theory of Computing*, pages 220–229, El Paso, Texas, 4–6 May 1997.
142. R. Impagliazzo and A. Wigderson. Randomness vs time: derandomization under a uniform assumption. *Journal of Computer and System Sciences*, 63(4):672–688, 2001. Special issue on FOCS 98 (Palo Alto, CA).
143. G. Ivanyos, Y. Qiao, and K. V. Subrahmanyam. Non-commutative edmonds' problem and matrix semi-invariants. *Comput. Complex.*, 26(3):717–763, 2017.
144. G. Ivanyos, Y. Qiao, and K. V. Subrahmanyam. Constructive non-commutative rank computation is in deterministic polynomial time. *Comput. Complex.*, 27(4):561–593, 2018.
145. K. Iwama and H. Morizumi. An explicit lower bound of 5n - o(n) for boolean circuits. In *MFCS*, volume 2420 of *Lecture Notes in Computer Science*, pages 353–364. Springer, 2002.
146. K. Jerrum, A. Sinclair, and E. Vigoda. A polynomial-time approximation algorithm for the permanent of a matrix with nonnegative entries. *J. ACM*, 51(4):671–697, July 2004.

147. M. Jerrum and A. Sinclair. Approximating the permanent. *SIAM Journal on Computing*, 18(6):1149–1178, 1989.
148. H. Jung. Relationships between probabilistic and deterministic tape complexity. In *Mathematical foundations of computer science, 1981 (Štrbské Pleso, 1981),*, Lecture Notes in Comput. Sci., 118,, pages 339–346,. ,, 1981.
149. V. Kabanets. Easiness assumptions and hardness tests: trading time for zero error. *Journal of Computer and System Sciences*, 63(2):236–252, 2001.
150. V. Kabanets and R. Impagliazzo. Derandomizing polynomial identity tests means proving circuit lower bounds. *Computational Complexity*, 13(1-2):1–46, 2004.
151. V. Kabanets and R. Impagliazzo. Derandomizing polynomial identity tests means proving circuit lower bounds. *Comput. Complex.*, 13(1-2):1–46, 2004.
152. N. Kahale. Eigenvalues and expansion of regular graphs. *Journal of the ACM*, 42(5):1091–1106, 1995.
153. B. Kalantari and L. Khachiyan. On the complexity of nonnegative-matrix scaling. *Linear Algebra and its Applications*, 240:87–103, 1996.
154. B. Kalantari, I. Lari, F. Ricca, and B. Simeone. On the complexity of general matrix scaling and entropy minimization via the RAS algorithm. *Math. Program.*, 112(2):371–401, 2008.
155. B. Kalyanasundaram and G. Schnitger. The probabilistic communication complexity of set intersection. *SIAM J. Discret. Math.*, 5(4):545–557, 1992.
156. M. Karchmer, R. Raz, and A. Wigderson. Super-logarithmic depth lower bounds via the direct sum in communication complexity. *Comput. Complex.*, 5(3/4):191–204, 1995.
157. M. Karchmer and A. Wigderson. Monotone circuits for connectivity require super-logarithmic depth. *SIAM J. Discret. Math.*, 3(2):255–265, 1990.
158. R. M. Karp. Reducibility among combinatorial problems. In R. E. Miller and J. W. Thatcher, editors, *Proceedings of a symposium on the Complexity of Computer Computations, held March 20-22, 1972, at the IBM Thomas J. Watson Research Center, Yorktown Heights, New York, USA*, The IBM Research Symposia Series, pages 85–103. Plenum Press, New York, 1972.
159. R. M. Karp, M. Luby, and N. Madras. Monte Carlo approximation algorithms for enumeration problems. *Journal of Algorithms*, 10(3):429–448, 1989.
160. G. Katona. A theorem of finite sets. *Classic Papers in Combinatorics*, pages 381–401, 1987.
161. N. Kayal. An exponential lower bound for the sum of powers of bounded degree polynomials. *Electron. Colloquium Comput. Complex.*, TR12-081, 2012.
162. N. Kayal, N. Limaye, C. Saha, and S. Srinivasan. An exponential lower bound for homogeneous depth four arithmetic formulas. *SIAM J. Comput.*, 46(1):307–335, 2017.
163. G. Kempf and L. Ness. The length of vectors in representation spaces. In K. Lønsted, editor, *Algebraic Geometry*, pages 233–243, Berlin, Heidelberg, 1979. Springer Berlin Heidelberg.
164. B. Klartag and O. Regev. Quantum one-way communication can be exponentially stronger than classical communication. In L. Fortnow and S. P. Vadhan, editors, *Proceedings of the 43rd ACM Symposium on Theory of Computing, STOC 2011, San Jose, CA, USA, 6-8 June 2011*, pages 31–40. ACM, 2011.
165. B. Kostant. On convexity, the Weyl group and the Iwasawa decomposition. *Annales scientifiques de l'École Normale Supérieure*, Ser. 4, 6(4):413–455, 1973.
166. M. Koucký, P. Nimbhorkar, and P. Pudlák. Pseudorandom generators for group products: extended abstract. In L. Fortnow and S. P. Vadhan, editors, *STOC*, pages 263–272. ACM, 2011.
167. J. B. Kruskal. The number of simplices in a complex. *Mathematical optimization techniques*, 10:251–278, 1963.
168. F. Kubo and T. Ando. Means of Positive Linear Operators. *Mathematische Annalen*, 246:205–224, 1979.
169. M. Kumar and S. Saraf. Superpolynomial lower bounds for general homogeneous depth 4 arithmetic circuits. *CoRR*, abs/1312.5978, 2013.
170. M. Kumar and S. Saraf. On the power of homogeneous depth 4 arithmetic circuits. *SIAM J. Comput.*, 46(1):336–387, 2017.

171. O. Lachish and R. Raz. Explicit lower bound of *4.5n - o(n)* for boolena circuits. In *STOC*, pages 399–408. ACM, 2001.
172. L.-C. Lau. Cs 860: Eigenvalues and polynomials. `https://cs.uwaterloo.ca/~lapchi/cs860/notes/eigenpoly.pdf`, 2022.
173. J. Leake, C. S. McSwiggen, and N. K. Vishnoi. Sampling matrices from Harish-Chandra-Itzykson-Zuber densities with applications to quantum inference and differential privacy. In S. Khuller and V. V. Williams, editors, *STOC '21: 53rd Annual ACM SIGACT Symposium on Theory of Computing, Virtual Event, Italy, June 21-25, 2021*, pages 1384–1397. ACM, 2021.
174. J. Leake and N. K. Vishnoi. On the computability of continuous maximum entropy distributions: Adjoint orbits of Lie groups. In *arXiv 2011.01851*, 2020.
175. J. Leake and N. K. Vishnoi. On the computability of continuous maximum entropy distributions with applications. *SIAM J. Comput.*, 51(5):1451–1505, 2022.
176. C. H. Lee, E. Pyne, and S. Vadhan. On the power of regular and permutation branching programs, 2023. Manuscript.
177. A. K. Lenstra, W. Hendrik Jr, et al. *The development of the number field sieve*, volume 1554. Springer Science & Business Media, 1993.
178. L. A. Levin. Universal sequential search problems. *Problemy peredachi informatsii*, 9(3):115–116, 1973. English translation in [268].
179. X. Li. Two source extractors for asymptotically optimal entropy, and (many) more. arXiv:2303.06802 [cs.CC], 2023.
180. Y. Li, Y. Qiao, A. Wigderson, Y. Wigderson, and C. Zhang. On linear-algebraic notions of expansion. arXiv:2212.13154 [match.CO], 2023.
181. E. H. Lieb. Gaussian kernels have only Gaussian maximizers. *Inventiones Mathematicae*, 102(1):179–208, 1990.
182. N. Limaye, S. Srinivasan, and S. Tavenas. Superpolynomial lower bounds against low-depth algebraic circuits. In *62nd IEEE Annual Symposium on Foundations of Computer Science, FOCS 2021, Denver, CO, USA, February 7-10, 2022*, pages 804–814. IEEE, 2021.
183. Y. Lindell. Secure multiparty computation. *Commun. ACM*, 64(1):86–96, 2021.
184. N. Linial, E. London, and Y. Rabinovich. The geometry of graphs and some of its algorithmic applications. *Combinatorica*, 15(2):215–245, 1995.
185. N. Linial, A. Samorodnitsky, and A. Wigderson. A deterministic strongly polynomial algorithm for matrix scaling and approximate permanents. In *Proceedings of the thirtieth annual ACM symposium on Theory of computing*, pages 644–652. ACM, 1998.
186. Y. Liu and R. Pass. Leakage-resilient hardness v.s. randomness. *Electron. Colloquium Comput. Complex.*, TR22-113, 2022.
187. L. Lovász. Singular spaces of matrices and their application in combinatorics. *Boletim da Sociedade Brasileira de Matemática - Bulletin/Brazilian Mathematical Society*, 20(1):87–99, 1989.
188. L. Lovász. *Combinatorial problems and exercises (2. ed.)*. North-Holland, 1993.
189. C.-J. Lu. Derandomizing Arthur-Merlin games under uniform assumptions. *Computational Complexity*, 10(3):247–259, 2001.
190. C.-J. Lu, O. Reingold, S. Vadhan, and A. Wigderson. Extractors: optimal up to constant factors. In *Proceedings of the 35th ACM Symposium on Theory of Computing (STOC '03)*, pages 602–611. ACM, 2003.
191. A. Lubotzky, R. Phillips, and P. Sarnak. Ramanujan graphs. *Combinatorica*, 8(3):261–277, 1988.
192. A. Lubotzky and B. Weiss. Groups and expanders. In *Expanding graphs (Princeton, NJ, 1992)*, volume 10 of *DIMACS Ser. Discrete Math. Theoret. Comput. Sci.*, pages 95–109. Amer. Math. Soc., Providence, RI, 1993.
193. M. Luby and B. Veličković. On deterministic approximation of DNF. *Algorithmica*, 16(4-5):415–433, 1996.
194. M. Luby, B. Veličković, and A. Wigderson. Deterministic approximate counting of depth-2 circuits. In *ISTCS*, pages 18–24, 1993.
195. G. A. Margulis. Explicit constructions of expanders. *Problemy Peredači Informacii*, 9(4):71–80, 1973.

196. G. A. Margulis. Explicit group-theoretic constructions of combinatorial schemes and their applications in the construction of expanders and concentrators. *Problemy Peredači Informacii*, 24(1):51–60, 1988.
197. O. Meir. Toward better depth lower bounds: A KRW-like theorem for strong composition. *Electron. Colloquium Comput. Complex.*, TR23-078, 2023.
198. R. Meshulam and A. Wigderson. Expanders in group algebras. *Combinatorica*, 24(4):659–680, 2004.
199. G. L. Miller. Riemann's hypothesis and tests for primality. *Journal of Computer and System Sciences*, 13(3):300–317, Dec. 1976.
200. K. D. Mulmuley and M. Sohoni. Geometric complexity theory I: An approach to the P vs. NP and related problems. *SIAM Journal on Computing*, 31(2):496–526, 2001.
201. D. Mumford, J. Fogarty, and F. Kirwan. *Geometric Invariant Theory*. Ergebnisse der Mathematik und Ihrer Grenzgebiete, 3 Folge/A Series of Modern Surveys in Mathematics Series. Springer Berlin Heidelberg, 1994.
202. L. Ness and D. Mumford. A stratification of the null cone via the moment map. *American journal of mathematics*, 106(6):1281–1329, 1984.
203. Y. Nesterov. *Introductory lectures on convex optimization*, volume 87. Springer Science & Business Media, 2004.
204. I. Newman. Private vs. common random bits in communication complexity. *Inf. Process. Lett.*, 39(2):67–71, 1991.
205. A. Nilli. On the second eigenvalue of a graph. *Discrete Mathematics*, 91(2):207–210, 1991.
206. N. Nisan. Lower bounds for non-commutative computation (extended abstract). In C. Koutsougeras and J. S. Vitter, editors, *Proceedings of the 23rd Annual ACM Symposium on Theory of Computing, May 5-8, 1991, New Orleans, Louisiana, USA*, pages 410–418. ACM, 1991.
207. N. Nisan. Pseudorandom bits for constant depth circuits. *Combinatorica*, 11(1):63–70, 1991.
208. N. Nisan. Pseudorandom generators for space-bounded computation. *Combinatorica*, 12(4):449–461, 1992.
209. N. Nisan, E. Szemerédi, and A. Wigderson. Undirected connectivity in o(log ^1.5 n) space. In *33rd Annual Symposium on Foundations of Computer Science, Pittsburgh, Pennsylvania, USA, 24-27 October 1992*, pages 24–29. IEEE Computer Society, 1992.
210. N. Nisan and A. Ta-Shma. Extracting randomness: A survey and new constructions. *Journal of Computer and System Sciences*, 58(1):148–173, February 1999.
211. N. Nisan and A. Wigderson. Hardness vs randomness. *Journal of Computer and System Sciences*, 49(2):149–167, Oct. 1994.
212. N. Nisan and A. Wigderson. Lower bounds on arithmetic circuits via partial derivatives. *Comput. Complex.*, 6(3):217–234, 1997.
213. N. Nisan and D. Zuckerman. Randomness is linear in space. *Journal of Computer and System Sciences*, 52(1):43–52, Feb. 1996.
214. R. Ostrovsky and A. Wigderson. One-way fuctions are essential for non-trivial zero-knowledge. In *Second Israel Symposium on Theory of Computing Systems, ISTCS 1993, Natanya, Israel, June 7-9, 1993, Proceedings*, pages 3–17, 1993.
215. D. Peleg and E. Upfal. Constructing disjoint paths on expander graphs. *Combinatorica*, 9(3):289–313, 1989.
216. M. Pinsker. On the complexity of a concentrator. In *7th Annual Teletraffic Conference*, pages 318/1–318/4, Stockholm, 1973.
217. N. Pippenger. On networks of noisy gates. In *FOCS*, pages 30–38. IEEE, 1985.
218. E. Pyne and S. Vadhan. Pseudodistributions that beat all pseudorandom generators (extended abstract). In V. Kabanets, editor, *Proceedings of the 36th Computational Complexity Conference (CCC '21)*, volume 200 of *LIPIcs*, pages 33:1–33:15. Schloss Dagstuhl - Leibniz-Zentrum für Informatik, 2021.
219. M. O. Rabin. Probabilistic algorithm for testing primality. *Journal of Number Theory*, 12(1):128–138, 1980.
220. J. Radhakrishnan and A. Ta-Shma. Bounds for dispersers, extractors, and depth-two superconcentrators. *SIAM Journal on Discrete Mathematics*, 13(1):2–24 (electronic), 2000.

221. R. Raz. Exponential separation of quantum and classical communication complexity. In J. S. Vitter, L. L. Larmore, and F. T. Leighton, editors, *Proceedings of the Thirty-First Annual ACM Symposium on Theory of Computing, May 1-4, 1999, Atlanta, Georgia, USA*, pages 358–367. ACM, 1999.
222. R. Raz. Separation of multilinear circuit and formula size. *Theory Comput.*, 2(6):121–135, 2006.
223. R. Raz. Multi-linear formulas for permanent and determinant are of super-polynomial size. *J. ACM*, 56(2):8:1–8:17, 2009.
224. R. Raz and P. McKenzie. Separation of the monotone NC hierarchy. *Comb.*, 19(3):403–435, 1999.
225. R. Raz and A. Shpilka. Deterministic polynomial identity testing in non-commutative models. *Comput. Complex.*, 14(1):1–19, 2005.
226. R. Raz, A. Shpilka, and A. Yehudayoff. A lower bound for the size of syntactically multilinear arithmetic circuits. *SIAM J. Comput.*, 38(4):1624–1647, 2008.
227. R. Raz and A. Wigderson. Monotone circuits for matching require linear depth. *J. ACM*, 39(3):736–744, 1992.
228. R. Raz and A. Yehudayoff. Lower bounds and separations for constant depth multilinear circuits. *Comput. Complex.*, 18(2):171–207, 2009.
229. A. Razborov. Lower bounds on the monotone complexity of some boolean function. In *Soviet Math. Dokl.*, volume 31, pages 354–357, 1985.
230. A. A. Razborov. Lower bounds on monotone complexity of the logical permanent. *Mathematical Notes of the Academy of Sciences of the USSR*, 37:485–493, 1985.
231. A. A. Razborov. On the distributional complexity of disjointness. *Theor. Comput. Sci.*, 106(2):385–390, 1992.
232. A. A. Razborov. Unprovability of lower bounds on circuit size in certain fragments of bounded arithmetic. *Izvestiya: mathematics*, 59(1):205, 1995.
233. A. A. Razborov. Quantum communication complexity of symmetric predicates. *Izvestiya: Mathematics*, 67(1):145, feb 2003.
234. O. Reingold. On black-box separations in cryptography. Tutorial at the Third Theory of Cryptography Conference (TCC '06), March 2006. Slides available from http://research.microsoft.com/en-us/people/omreing/.
235. O. Reingold. Undirected connectivity in log-space. *Journal of the ACM*, 55(4):Art. 17, 24, 2008.
236. O. Reingold, L. Trevisan, and S. Vadhan. Pseudorandom walks in regular digraphs and the RL vs. L problem. In *Proceedings of the 38th Annual ACM Symposium on Theory of Computing (STOC '06)*, pages 457–466, 21–23 May 2006.
237. O. Reingold, S. Vadhan, and A. Wigderson. Entropy waves, the zig-zag graph product, and new constant-degree expanders and extractors. In *Proceedings of the 41st Annual Symposium on Foundations of Computer Science (FOCS '00)*, pages 3–13, Redondo Beach, CA, 17–19 Oct. 2000. IEEE.
238. O. Reingold, S. Vadhan, and A. Wigderson. Entropy waves, the zig-zag graph product, and new constant-degree expanders. *Annals of Mathematics*, 155(1), January 2001.
239. E. Rozenman, A. Shalev, and A. Wigderson. Iterative construction of Cayley expander graphs. *Theory of Computing. An Open Access Journal*, 2:91–120, 2006.
240. E. Rozenman and S. Vadhan. Derandomized squaring of graphs. In *Proceedings of the 8th International Workshop on Randomization and Computation (RANDOM '05)*, number 3624 in Lecture Notes in Computer Science, pages 436–447, Berkeley, CA, August 2005. Springer.
241. M. Saks and S. Zhou. $\mathrm{BP_HSPACE}(S) \subseteq \mathrm{DSPACE}(S^{3/2})$. *Journal of Computer and System Sciences*, 58(2):376–403, 1999.
242. M. Sántha. On using deterministic functions to reduce randomness in probabilistic algorithms. *Information and Computation*, 74(3):241–249, 1987.
243. R. Sanyal, F. Sottile, and B. Sturmfels. Orbitopes. *Mathematika*, 57(2):275–314, 2011.
244. J. Saunderson, P. A. Parrilo, and A. S. Willsky. Semidefinite descriptions of the convex hull of rotation matrices. *SIAM Journal on Optimization*, 25(3):1314–1343, 2015.

245. J. T. Schwartz. Fast probabilistic algorithms for verification of polynomial identities. *Journal of the ACM*, 27(4):701–717, 1980.
246. R. Shaltiel. Recent developments in extractors. In G. Paun, G. Rozenberg, and A. Salomaa, editors, *Current Trends in Theoretical Computer Science*, volume 1: Algorithms and Complexity, pages 189–228. World Scientific, 2004.
247. R. Shaltiel and C. Umans. Simple extractors for all min-entropies and a new pseudo-random generator. *Journal of the ACM*, 52(2):172–216, 2005.
248. R. Shaltiel and C. Umans. Low-end uniform hardness versus randomness tradeoffs for AM. *SIAM Journal on Computing*, 39(3):1006–1037, 2009.
249. A. Shamir. How to share a secret. *Communications of the Association for Computing Machinery*, 22(11):612–613, 1979.
250. C. E. Shannon. Communication theory of secrecy systems. *The Bell system technical journal*, 28(4):656–715, 1949.
251. A. A. Sherstov, A. A. Storozhenko, and P. Wu. An optimal separation of randomized and quantum query complexity. In S. Khuller and V. V. Williams, editors, *STOC '21: 53rd Annual ACM SIGACT Symposium on Theory of Computing, Virtual Event, Italy, June 21-25, 2021*, pages 1289–1302. ACM, 2021.
252. A. Shpilka and A. Wigderson. Depth-3 arithmetic circuits over fields of characteristic zero. *Comput. Complex.*, 10(1):1–27, 2001.
253. M. Singh and N. K. Vishnoi. Entropy, optimization and counting. In *Proceedings of the 46th Annual ACM Symposium on Theory of Computing*, pages 50–59. ACM, 2014.
254. R. Sinkhorn. A relationship between arbitrary positive matrices and doubly stochastic matrices. *The annals of mathematical statistics*, 35(2):876–879, 1964.
255. M. Sipser. Expanders, randomness, or time versus space. *Journal of Computer and System Sciences*, 36(3):379–383, 1988. Structure in Complexity Theory Conference (Berkeley, CA, 1986).
256. R. Smolensky. On interpolation by analytic functions with special properties and some weak lower bounds on the size of circuits with symmetric gates. In *31st Annual Symposium on Foundations of Computer Science, St. Louis, Missouri, USA, October 22-24, 1990, Volume II*, pages 628–631. IEEE Computer Society, 1990.
257. R. Solovay and V. Strassen. A fast Monte-Carlo test for primality. *SIAM Journal on Computing*, 6(1):84–85, 1977.
258. S. Sra and R. Hosseini. Conic geometric optimization on the manifold of positive definite matrices. *SIAM Journal on Optimization*, 25(1):713–739, 2015.
259. S. Sra, N. K. Vishnoi, and O. Yildiz. On geodesically convex formulations for the Brascamp-Lieb constant. In E. Blais, K. Jansen, J. D. P. Rolim, and D. Steurer, editors, *Approximation, Randomization, and Combinatorial Optimization. Algorithms and Techniques, APPROX/RANDOM 2018, August 20-22, 2018 - Princeton, NJ, USA*, volume 116 of *LIPIcs*, pages 25:1–25:15. Schloss Dagstuhl - Leibniz-Zentrum für Informatik, 2018.
260. T. Steinke. Pseudorandomness for permutation branching programs without the group theory. Technical Report TR12-083, Electronic Colloquium on Computational Complexity (ECCC), July 2012.
261. V. Strassen. Die berechnungskomplexität von elementarsymmetrischen funktionen und von interpolationskoeffizienten. *Numerische Mathematik*, 20:238–251, 1973.
262. D. Straszak and N. K. Vishnoi. On convex programming relaxations for the permanent. *CoRR*, abs/1701.01419, 2017.
263. D. Straszak and N. K. Vishnoi. Real stable polynomials and matroids: Optimization and counting. In H. Hatami, P. McKenzie, and V. King, editors, *Proceedings of the 49th Annual ACM SIGACT Symposium on Theory of Computing, STOC 2017, Montreal, QC, Canada, June 19-23, 2017*, pages 370–383. ACM, 2017.
264. D. Straszak and N. K. Vishnoi. Maximum entropy distributions: Bit complexity and stability. In A. Beygelzimer and D. Hsu, editors, *Conference on Learning Theory, COLT 2019, 25-28 June 2019, Phoenix, AZ, USA*, volume 99 of *Proceedings of Machine Learning Research*, pages 2861–2891. PMLR, 2019.

265. A. Ta-Shma. Explicit, almost optimal, epsilon-balanced codes. In H. Hatami, P. McKenzie, and V. King, editors, *Proceedings of the 49th Annual ACM SIGACT Symposium on Theory of Computing, STOC 2017, Montreal, QC, Canada, June 19-23, 2017*, pages 238–251. ACM, 2017.
266. A. Ta-Shma, C. Umans, and D. Zuckerman. Lossless condensers, unbalanced expanders, and extractors. *Combinatorica*, 27(2):213–240, 2007.
267. A. Tal. Towards optimal separations between quantum and randomized query complexities. In S. Irani, editor, *61st IEEE Annual Symposium on Foundations of Computer Science, FOCS 2020, Durham, NC, USA, November 16-19, 2020*, pages 228–239. IEEE, 2020.
268. B. A. Trakhtenbrot. A survey of russian approaches to perebor (brute-force searches) algorithms. *Annals of the History of Computing*, 6:384–400, 1984.
269. L. Trevisan. Extractors and pseudorandom generators. *Journal of the ACM*, 48(4):860–879 (electronic), 2001.
270. L. Trevisan. Lecture notes on graph partitioning, expanders and spectral methods. `https://lucatrevisan.github.io/books/expanders-2016.pdf`, 2017.
271. L. Trevisan and S. Vadhan. Pseudorandomness and average-case complexity via uniform reductions. *Computational Complexity*, 16(4):331–364, December 2007.
272. H. Tyagi and S. Watanabe. *Information-Theoretic Cryptography*. Cambridge University Press, 2023.
273. C. Udriste. *Convex functions and optimization methods on Riemannian manifolds*, volume 297. Springer Science & Business Media, 1994.
274. C. Umans. Pseudo-random generators for all hardnesses. *Journal of Computer and System Sciences*, 67(2):419–440, 2003.
275. E. Upfal and A. Wigderson. How to share memory in a distributed system. *Journal of the ACM*, 34(1):116–127, 1987.
276. S. P. Vadhan. *Pseudorandomness*, volume 7 (1–3) of *Foundations and Trends in Theoretical Computer Science*. now publishers, December 2012. 336 pages.
277. L. G. Valiant. Graph-theoretic properties in computational complexity. *Journal of Computer and System Sciences*, 13(3):278–285, 1976.
278. L. G. Valiant. The complexity of computing the permanent. *Theoretical Computer Science*, 8(2):189–201, 1979.
279. N. K. Vishnoi. An algebraic proof of Alon's combinatorial nullstellensatz. *Congressus Numerantium*, 152:89–91, 2001.
280. N. K. Vishnoi. Zeros of polynomials and their applications to theory: A primer. In *FOCS 2013 Workshop on Zeros of Polynomials and their Applications to Theory*, pages 1–18, 2013.
281. N. K. Vishnoi. Geodesic convex optimization: Differentiation on manifolds, geodesics, and convexity. *CoRR*, abs/1806.06373, 2018.
282. N. K. Vishnoi. *Algorithms for Convex Optimization*. Cambridge University Press, 2021.
283. W. C. Waterhouse. Do symmetric problems have symmetric solutions. *American Mathematical Monthly*, 90(6):378–387, 1983.
284. A. Wigderson. *Mathematics and Computation: A Theory Revolutionizing Technology and Science*. Princeton University Press, 2019.
285. A. Wigderson and D. Zuckerman. Expanders that beat the eigenvalue bound: explicit construction and applications. *Combinatorica*, 19(1):125–138, 1999.
286. R. Williams. Improving exhaustive search implies superpolynomial lower bounds. In *STOC'10—Proceedings of the 2010 ACM International Symposium on Theory of Computing*, pages 231–240. ACM, New York, 2010.
287. R. Williams. Non-uniform ACC circuit lower bounds. In *Proceedings of the 26th Annual IEEE Conference on Computational Complexity, CCC 2011, San Jose, California, June 8-10, 2011*, pages 115–125. IEEE Computer Society, 2011.
288. A. C. Yao. Some complexity questions related to distributive computing (preliminary report). In M. J. Fischer, R. A. DeMillo, N. A. Lynch, W. A. Burkhard, and A. V. Aho, editors, *Proceedings of the 11h Annual ACM Symposium on Theory of Computing, April 30 - May 2, 1979, Atlanta, Georgia, USA*, pages 209–213. ACM, 1979.

289. A. C. Yao. Theory and applications of trapdoor functions (extended abstract). In *23rd Annual Symposium on Foundations of Computer Science*, pages 80–91, Chicago, Illinois, 3–5 Nov. 1982. IEEE.
290. A. C. Yao. How to generate and exchange secrets (extended abstract). In *27th Annual Symposium on Foundations of Computer Science, Toronto, Canada, 27-29 October 1986*, pages 162–167. IEEE Computer Society, 1986. The Garbled circuit protocol was presented in the oral presentation of this paper.
291. A. C. Yao. Quantum circuit complexity. In *34th Annual Symposium on Foundations of Computer Science, Palo Alto, California, USA, 3-5 November 1993*, pages 352–361. IEEE Computer Society, 1993.
292. R. Zippel. Probabilistic algorithms for sparse polynomials. In E. W. Ng, editor, *EUROSAM*, volume 72 of *Lecture Notes in Computer Science*, pages 216–226. Springer, 1979.
293. D. Zuckerman. Simulating BPP using a general weak random source. *Algorithmica*, 16(4/5):367–391, Oct./Nov. 1996.
294. D. Zuckerman. Randomness-optimal oblivious sampling. *Random Structures & Algorithms*, 11(4):345–367, 1997.

List of Publications for László Lovász

1965

[1] On graphs not containing independent circuits. *Mat. Lapok*, 16:289–299 (in Hungarian).

1966

[2] On decomposition of graphs. *Studia Sci. Math. Hungar.*, 1:237–238.

1967

[3] Operations with structures. *Acta Math. Acad. Sci. Hungar.*, 18:321–328.
[4] Über die starke Multiplikation von geordneten Graphen. *Acta Math. Acad. Sci. Hungar.*, 18:235–241 (in German).
[5] On connected sets of points. *Ann. Univ. Sci. Budapest. Eötvös Sect. Math.*, 10:203–204.

1968

[6] On chromatic number of finite set-systems. *Acta Math. Acad. Sci. Hungar.*, 19:59–67.
[7] On covering of graphs. In *Theory of Graphs (Proc. Colloq., Tihany, 1966)*, pages 231–236. Academic Press, New York.
[8] Graphs and set systems. In *Beiträge zur Graphentheorie (Kolloquium, Manebach, 1967)*, pages 99–106. Teubner, Leipzig.

1969

[9] Connections between number theoretic properties of polynomials and their substitutional values. *Mat. Lapok*, 20:129–132 (in Hungarian).

1970

[10] Subgraphs with prescribed valencies. *J. Combinatorial Theory*, 8:391–416, 1970.

H. Holden, R. Piene (eds.), *The Abel Prize 2018-2022*, The Abel Prize,
https://doi.org/10.1007/978-3-031-33973-8_21

[11] The factorization of graphs. In *Combinatorial Structures and their Applications (Proc. Calgary Internat. Conf., Calgary, Alta., 1969)*, pages 243–246. Gordon and Breach, New York.

[12] A generalization of König's theorem. *Acta Math. Acad. Sci. Hungar.*, 21:443–446.

[13] A remark on Menger's theorem. *Acta Math. Acad. Sci. Hungar.*, 21:365–368.

[14] Generalized factors of graphs. In *Combinatorial theory and its applications, II (Proc. Colloq., Balatonfüred, 1969)*, pages 773–781. Bolyai Janos Math. Soc., Budapest.

1971

[15] Unique factorization in certain classes of structures. In *Mini-Conf. Univers.*, pages 24–25. Bolyai Janos Math. Soc., Szeged.

[16] (with K. Győry). Representation of integers by norm forms. II. *Publ. Math. Debrecen*, 17:173–181.

[17] On the cancellation law among finite relational structures. *Period. Math. Hungar.*, 1(2):145–156.

[18] On finite Dirichlet series. *Acta Math. Acad. Sci. Hungar.*, 22:227–231.

1972

[19] A brief survey of matroid theory. *Mat. Lapok*, 22:249–267 (in Hungarian).

[20] On the number of halving lines. *Ann. Univ. Sci. Budapest. Eötvös Sect. Math.*, 14:107–108.

[21] Normal hypergraphs and the perfect graph conjecture. *Discrete Math.*, 2(3): 253–267.

[22] A characterization of perfect graphs. *J. Combinatorial Theory Ser. B*, 13:95–98.

[23] A note on the line reconstruction problem. *J. Combinatorial Theory Ser. B*, 13:309–310.

[24] On the structure of factorizable graphs. I, II. *Acta Math. Acad. Sci. Hungar.*, 23:179–195; ibid. 23 (1972), 465–478.

[25] The factorization of graphs. II. *Acta Math. Acad. Sci. Hungar.*, 23:223–246.

[26] Direct product in locally finite categories. *Acta Sci. Math. (Szeged)*, 33: 319–322.

[27] A note on factor-critical graphs. *Studia Sci. Math. Hungar.*, 7:279–280.

1973

[28] On the sieve formula. *Mat. Lapok*, 23:53–69 (in Hungarian).

[29] (with A. Recski). On the sum of matroids. *Acta Math. Acad. Sci. Hungar.*, 24:329–333.

[30] Connectivity in digraphs. *J. Combinatorial Theory Ser. B*, 15:174–177.

[31] (with P. Major). A note to a paper of Dudley. *Studia Sci. Math. Hungar.*, 8: 151–152.

[32] Independent sets in critical chromatic graphs. *Studia Sci. Math. Hungar.*, 8:165–168.

[33] (with L. Babai). Permutation groups and almost regular graphs. *Studia Sci. Math. Hungar.*, 8:141–150.

[34] Coverings and coloring of hypergraphs. In *Proceedings of the Fourth Southeastern Conference on Combinatorics, Graph Theory, and Computing (Florida Atlantic Univ., Boca Raton, Fla., 1973)*, pages 3–12. Utilitas Mathematica Pub., Winnipeg, MB.

[35] Factors of graphs. In *Proceedings of the Fourth Southeastern Conference on Combinatorics, Graph Theory, and Computing (Florida Atlantic Univ., Boca Raton, Fla., 1973)*, pages 13–22. Utilitas Mathematica Pub., Winnipeg, MB.

[36] (with P. Erdős, A. Simmons and E.G. Straus). Dissection graphs of planar point sets. In *A survey of combinatorial theory (Proc. Internat. Sympos., Colorado State Univ., Fort Collins, Colo., 1971)*, pages 139–149. North-Holland, Amsterdam.

[37] Antifactors of graphs. *Period. Math. Hungar.*, 4:121–123.

[38] (with J. Pelikán). On the eigenvalues of trees. *Period. Math. Hungar.*, 3:175–182.

[39] (with A. Hajnal and L. F. Tóth). Research problems. *Period. Math. Hungar.*, 4:81–82.

1974

[40] (with M.D. Plummer). On a family of planar bicritical graphs. In *Combinatorics (Proc. British Combinatorial Conf., Univ. Coll. Wales, Aberystwyth, 1973)*, volume 13 of *London Math. Soc. Lecture Note Ser.*, pages 103–107. Cambridge University Press, Cambridge.

[41] Valencies of graphs with 1-factors. *Period. Math. Hungar.*, 5:149–151.

[42] (with L. Babai and W. Imrich). Finite homeomorphism groups of the 2-sphere. In *Topics in topology (Proc. Colloq., Keszthely, 1972)*, pages 61–75. Colloq. Math. Soc. János Bolyai, Vol. 8.

[43] (with D. Greenwell). Applications of product colouring. *Acta Math. Acad. Sci. Hungar.*, 25:335–340.

[44] Minimax theorems for hypergraphs. In *Hypergraph Seminar (Proc. First Working Sem., Ohio State Univ., Columbus, Ohio, 1972; dedicated to Arnold Ross)*, volume 411 of *Lecture Notes in Math.*, pages 111–126. Springer, Berlin.

[45] (with V. Chvátal). Every directed graph has a semi-kernel. In *Hypergraph Seminar (Proc. First Working Sem., Ohio State Univ., Columbus, Ohio, 1972; dedicated to Arnold Ross)*, volume 411 of *Lecture Notes in Math.*, page 175. Springer, Berlin.

1975

[46] (with M.D. Plummer). On bicritical graphs. In *Infinite and finite sets (Colloq., Keszthely, 1973; dedicated to P. Erdős on his 60th birthday), Vol. II*, pages 1051–1079. Colloq. Math. Soc. János Bolyai, Vol. 10.

[47] (with R.R. Appleson) A characterization of cancellable k-ary structures. *Period. Math. Hungar.*, 6:17–19.

[48] (with P. Erdős). Problems and results on 3-chromatic hypergraphs and some related questions. In *Infinite and finite sets (Colloq., Keszthely, 1973; dedicated to P. Erdős on his 60th birthday), Vol. II*, pages 609–627. Colloq. Math. Soc. János Bolyai, Vol. 10.

[49] On the ratio of optimal integral and fractional covers. *Discrete Math.*, 13(4): 383–390.

[50] Three short proofs in graph theory. *J. Combinatorial Theory Ser. B*, 19(3):269–271.

[51] Spectra of graphs with transitive groups. *Period. Math. Hungar.*, 6(2):191–195.

[52] 2-matchings and 2-covers of hypergraphs. *Acta Math. Acad. Sci. Hungar.*, 26(3-4):433–444.

[53] (with M.D. Plummer). On a family of planar bicritical graphs. *Proc. London Math. Soc. (3)*, 30:160–176.

[54] (with S.A. Burr and P. Erdős). On graphs of Ramsey type. *Proceedings of the Sixth Southeast. Conf. Comb., Graph Theor., Comput.*, volume 14 of *Congressus Numerantium*, page 643. Utilitas Mathematica Pub., Winnipeg, MB.

1976

[55] (with D.E. Daykin). The number of values of a Boolean function. *J. London Math. Soc. (2)*, 12(2):225–230.

[56] (with M.L. Marx). A forbidden substructure characterization of Gauss codes. *Bull. Amer. Math. Soc.*, 82(1):121–122.

[57] Covers, packings, and some heuristic algorithms. In *Proceedings of the Fifth British Combinatorial Conference (Univ. Aberdeen, Aberdeen, 1975)*, volume 15 of *Congressus Numerantium*, pages 417–429. Utilitas Mathematica Pub., Winnipeg, MB.

[58] (with M. Simonovits). On the number of complete subgraphs of a graph. In *Proceedings of the Fifth British Combinatorial Conference (Univ. Aberdeen, Aberdeen, 1975)*, volume 15 of *Congressus Numerantium*, pages 431–441. Utilitas Mathematica Pub., Winnipeg, MB.

[59] (with S.A. Burr and P. Erdős). On graphs of Ramsey type. *Ars Combin.*, 1(1):167–190.

[60] On two minimax theorems in graph. *J. Combinatorial Theory Ser. B*, 21(2):96–103.

[61] On some connectivity properties of Eulerian graphs. *Acta Math. Acad. Sci. Hungar.*, 28(1-2):129–138.

[62] (with M.L. Marx). A forbidden substructure characterization of Gauss codes. *Acta Sci. Math. (Szeged)*, 38(1-2):115–119.

[63] Chromatic number of hypergraphs and linear algebra. *Studia Sci. Math. Hungar.*, 11(1-2):113–114.

1977

[64] On minimax theorems of combinatorics. *Mat. Lapok*, 26(3-4):209–264 (in Hungarian).

[65] Certain duality principles in integer programming. In *Studies in integer programming (Proc. Workshop, Bonn, 1975)*, volume 1 of *Ann. of Discrete Math.*, pages 363–374. North-Holland, Amsterdam.

[66] Flats in matroids and geometric graphs. In *Combinatorial surveys (Proc. Sixth British Combinatorial Conf., Royal Holloway Coll., Egham, 1977)*, pages 45–86. Academic Press, London.

[67] (with P. Gács). Some remarks on generalized spectra. *Z. Math. Logik Grundlagen Math.*, 23(6):547–554.

[68] (with M.D. Plummer). On minimal elementary bipartite graphs. *J. Combinatorial Theory Ser. B*, 23(1):127–138.

[69] A homology theory for spanning trees of a graph. *Acta Math. Acad. Sci. Hungar.*, 30(3-4):241–251.

[70] (with K. L. Vesztergombi and J. Pelikan). Kombinatorik. Deutsche Übersetzung und wissenschaftliche Redaktion: G. Eisenreich. *Mathematische Schulerbücherei*. No. 90. Leipzig: BSB B. G. Teubner Verlagsgesellschaft. 132 S. m. 62 Abb.; M 6.00 (in German).

1978

[71] (with R.L. Graham). Distance matrix polynomials of trees. *Adv. in Math.*, 29(1):60–88.

[72] (with R.L. Graham). Distance matrix polynomials of trees. In *Theory and applications of graphs (Proc. Internat. Conf., Western Mich. Univ., Kalamazoo, Mich., 1976)*, volume 642 of *Lecture Notes in Math.*, pages 186–190. Springer, Berlin.

[73] (with V. Neumann-Lara and M. Plummer). Mengerian theorems for paths of bounded length. *Period. Math. Hungar.*, 9(4):269–276.

[74] Kneser's conjecture, chromatic number, and homotopy. *J. Combin. Theory Ser. A*, 25(3):319–324.

[75] Some finite basis theorems on graph theory. In *Combinatorics (Proc. Fifth Hungarian Colloq., Keszthely, 1976), Vol. II*, volume 18 of *Colloq. Math. Soc. János Bolyai*, pages 717–729. North-Holland, Amsterdam-New York.

[76] (with K. Vesztergombi). Restricted permutations and Stirling numbers. In *Combinatorics (Proc. Fifth Hungarian Colloq., Keszthely, 1976), Vol. II*, volume 18 of *Colloq. Math. Soc. János Bolyai*, pages 731–738. North-Holland, Amsterdam-New York.

[77] (with R.L. Graham). Distance matrix polynomials of trees. In *Problèmes combinatoires et théorie des graphes (Colloq. Internat. CNRS, Univ. Orsay, Orsay, 1976)*, volume 260 of *Colloq. Internat. CNRS*, pages 189–190. CNRS, Paris.

[78] (with A. Hajnal). An algorithm to prevent the propagation of certain diseases at minimum cost. *Interface between Comput. Sci. and Oper. Res., Proc. Symp.*, 99, 105–108.

1979

[79] Graph theory and integer programming. *Mat. Lapok*, 27(1-2):69–86 (in Hungarian).

[80] On the Shannon capacity of a graph. *IEEE Trans. Inform. Theory*, 25(1):1–7. Also available in *Kibern. Sb.*, Nov. Ser. 19, 5–22 (in Russian).

[81] *Combinatorial problems and exercises*. North-Holland Publishing Co., Amsterdam-New York. Second edition, North-Holland Publishing Co., Amsterdam, second edition, 1993 and AMS Chelsea Publishing, Providence, RI, 2007.

[82] Topological and algebraic methods in graph theory. In *Graph theory and related topics (Proc. Conf., Univ. Waterloo, Waterloo, Ont., 1977)*, pages 1–14. Academic Press, New York-London.

[83] (with P. Erdős and J. Spencer). Strong independence of graphcopy functions. In *Graph theory and related topics (Proc. Conf., Univ. Waterloo, Waterloo, Ont., 1977)*, pages 165–172. Academic Press, New York-London.

[84] Graph theory and integer programming. In *Discrete optimization (Proc. Adv. Res. Inst. Discrete Optimization and Systems Appl., Banff, Alta., 1977), I*, volume 4, pages 141–158. North-Holland, Amsterdam.

[85] On determinants, matchings, and random algorithms. In *Fundamentals of computation theory (Proc. Conf. Algebraic, Arith. and Categorical Methods in Comput. Theory, Berlin/Wendisch-Rietz, 1979)*, volume 2 of *Math. Res.*, pages 565–574. Akademie-Verlag, Berlin.

[86] (with R. Aleliunas, R.M. Karp, R.J. Lipton and C. Rackoff). Random walks, universal traversal sequences, and the complexity of maze problems. In *20th Annual Symposium on Foundations of Computer Science (San Juan, Puerto Rico, 1979)*, pages 218–223. IEEE, New York,

1980

[87] A new linear programming algorithm—better or worse than the simplex method? *Math. Intelligencer*, 2(3):141–146. Also available in *Pokroky Mat. Fyz. Astronom.*, 26(4):193–202, 1981 (in Czech).

[88] Matroid matching and some applications. *J. Combin. Theory Ser. B*, 28(2):208–236.

[89] Matroids and Sperner's lemma. *European J. Combin.*, 1(1):65–66.

[90] Selecting independent lines from a family of lines in a space. *Acta Sci. Math. (Szeged)*, 42(1-2):121–131.

[91] (with J. Nešetřil and A. Pultr). On a product dimension of graphs. *J. Combin. Theory Ser. B*, 29(1):47–67.

[92] Efficient algorithms: an approach by formal logic. In *Studies on mathematical programming (Papers, Third Conf. Math. Programming, Mátrafüred, 1975)*, volume 1 of *Math. Methods Oper. Res.*, pages 119–126. Akad. Kiadó, Budapest.

1981

[93] (with P. Gács). Khachiyan's algorithm for linear programming. *Math. Programming Stud.*, 14:61–68. Also available in *Yingyong Shuxue yu Jisuan Shuxue*, 3:1–4, 1980 (in Chinese).

[94] (with J.A. Bondy). Cycles through specified vertices of a graph. *Combinatorica*, 1(2):117–140.

[95] (with M. Grötschel and A. Schrijver). The ellipsoid method and its consequences in combinatorial optimization. *Combinatorica*, 1(2):169–197. Corrigendum, in *Combinatorica*, 4(4):291–295, 1984. Also available in *IIASA Collab. Proc. Ser.* CP-81-S1, pages 511–546.

[96] The matroid matching problem. In *Algebraic methods in graph theory, Vol. I, II (Szeged, 1978)*, volume 25 of *Colloq. Math. Soc. János Bolyai*, pages 495–517. North-Holland, Amsterdam-New York.

[97] (with B. Korte). Mathematical structures underlying greedy algorithms. In *Fundamentals of computation theory (Szeged, 1981)*, volume 117 of *Lecture Notes in Comput. Sci.*, pages 205–209. Springer, Berlin-New York.

[98] (with A. Schrijver). Some combinatorial applications of the new linear programming algorithm. In *Combinatorics and graph theory (Calcutta, 1980)*, volume 885 of *Lecture Notes in Math.*, pages 33–41. Springer, Berlin-New York.

[99] (with A. Sárközy and M. Simonovits). On additive arithmetic functions satisfying a linear recursion. *Ann. Univ. Sci. Budapest. Eötvös Sect. Math.*, 24:205–215.

1982

[100] (with Y. Yemini). On generic rigidity in the plane. *SIAM J. Algebraic Discrete Methods*, 3(1):91–98.

[101] (with A.K. Lenstra and H.W. Lenstra, Jr.). Factoring polynomials with rational coefficients. *Math. Ann.*, 261(4):515–534. Also available in *Math. Cent.*, IW 195/82, 30 pp.

[102] (with A. Recski). Selected topics of matroid theory and its applications. *Rend. Circ. Mat. Palermo (2)*, (Suppl. 2):171–185.

[103] Bounding the independence number of a graph. In *Bonn Workshop on Combinatorial Optimization (Bonn, 1980)*, volume 16 of *Ann. Discrete Math.*, pages 213–223. North-Holland, Amsterdam-New York.

[104] (with I. Bárány). Borsuk's theorem and the number of facets of centrally symmetric polytopes. *Acta Math. Acad. Sci. Hungar.*, 40(3-4):323–329.

[105] Tibor Gallai is seventy years old. *Combinatorica*, 2(3):203–205.

[106] (with J. Edmonds and W.R. Pulleyblank). Brick decompositions and the matching rank of graphs. *Combinatorica*, 2(3):247–274.

1983

[107] Self-dual polytopes and the chromatic number of distance graphs on the sphere. *Acta Sci. Math. (Szeged)*, 45(1-4):317–323.

[108] Ear-decompositions of matching-covered graphs. *Combinatorica*, 3(1):105–117.

[109] Submodular functions and convexity. In *Mathematical programming: the state of the art (Bonn, 1982)*, pages 235–257. Springer, Berlin.

[110] (with A. Schrijver). Remarks on a theorem of Rédei. *Studia Sci. Math. Hungar.*, 16(3-4):449–454.

[111] (with B. Korte). Structural properties of greedoids. *Combinatorica*, 3(3-4):359–374.

[112] Perfect graphs. In *Selected topics in graph theory, 2*, pages 55–87. Academic Press, London.

[113] (with M. Simonovits). On the number of complete subgraphs of a graph. II. In *Studies in pure mathematics*, pages 459–495. Birkhäuser, Basel.

1984

[114] (with B. Korte). Greedoids and linear objective functions. *SIAM J. Algebraic Discrete Methods*, 5(2):229–238.

[115] (with M. Grötschel and A. Schrijver). Geometric methods in combinatorial optimization. In *Progress in combinatorial optimization (Waterloo, Ont., 1982)*, pages 167–183. Academic Press, Toronto, ON.

[116] (with B. Korte). Greedoids—a structural framework for the greedy algorithm. In *Progress in combinatorial optimization (Waterloo, Ont., 1982)*, pages 221–243. Academic Press, Toronto, ON.

[117] (with W. Cook and A. Schrijver). A polynomial-time test for total dual integrality in fixed dimension. In *Mathematical programming at Oberwolfach, II (Oberwolfach, 1983)*, volume 22 of *Mathematical Programming Studies*, pages 64–69. Springer, Berlin.

[118] (with B. Korte). Shelling structures, convexity and a happy end. In *Graph theory and combinatorics (Cambridge, 1983)*, pages 219–232. Academic Press, London.

[119] Normal hypergraphs and the weak perfect graph conjecture. In *Topics on perfect graphs*, volume 88 of *North-Holland Math. Stud.*, pages 29–42. North-Holland, Amsterdam.

[120] (with M. Grötschel and A. Schrijver). Polynomial algorithms for perfect graphs. In *Topics on perfect graphs*, volume 88 of *North-Holland Math. Stud.*, pages 325–356. North-Holland, Amsterdam.

[121] Algorithmic aspects of combinatorics, geometry and number theory. In *Proceedings of the International Congress of Mathematicians, Vol. 1, 2 (Warsaw, 1983)*, pages 1591–1599. PWN, Warsaw.

[122] (with A. Hajnal and V. T. Sós, editors). Finite and infinite sets. (Sixth Hungarian Combinatorial Colloquium held in Eger, Hungary, July 6-11, 1981). Vols. I, II. *Colloquia Mathematica Societatis János Bolyai*, 37. János Bolyai Mathematical Society, North-Holland. 902 pp.

[123] (with R. Kannan, A. K. Lenstra). Polynomial factorization and nonrandomness of bits of algebraic and some transcendental numbers. In *Proceedings of the Sixteenth Annual ACM Symposium on Theory of Computing*, pages 191–200. ACM, New York.

1985

[124] (with B. Korte). Polymatroid greedoids. *J. Combin. Theory Ser. B*, 38(1): 41–72.

[125] (with B. Korte). A note on selectors and greedoids. *European J. Combin.*, 6(1):59–67.

[126] (with J.A. Bondy). Lengths of cycles in Halin graphs. *J. Graph Theory*, 9(3):397–410.

[127] Vertex packing algorithms. In *Automata, languages and programming (Nafplion, 1985)*, volume 194 of *Lecture Notes in Comput. Sci.*, pages 1–14. Springer, Berlin.

[128] (with B. Korte). Basis graphs of greedoids and two-connectivity. In *Mathematical programming, I*, volume 24 of *Mathematical Programming Studies* , pages 158–165. Springer, Berlin.

[129] (with A. Björner and B. Korte). Homotopy properties of greedoids. *Adv. in Appl. Math.*, 6(4):447–494.

[130] (with B. Korte). Relations between subclasses of greedoids. *Z. Oper. Res. Ser. A-B*, 29(7):A249–A267.

[131] (with B. Korte). Posets, matroids, and greedoids. In *Matroid theory (Szeged, 1982)*, volume 40 of *Colloq. Math. Soc. János Bolyai*, pages 239–265. North-Holland, Amsterdam.

[132] (with U. Faigle, R. Schrader and G. Turán). Search problems in ordered sets. In *Operations research proceedings 1984 (St. Gallen, 1984)*, pages 411–415. Springer, Berlin.

[133] Some algorithmic problems on lattices. In *Theory of algorithms (Pécs, 1984)*, volume 44 of *Colloq. Math. Soc. János Bolyai*, pages 323–337. North-Holland, Amsterdam.

[134] Mathematical notion of complexity. *IFAC Proceedings Series*, 1105–1110.

[135] Computing ears and branchings in parallel. *26th Annual IEEE Symposium on Foundations of Computer Science*, pages 464–467. IEEE Computer Society Press.

1986

[136] (with B. Korte). Noninterval greedoids and the transposition property. *Discrete Math.*, 59(3):297–314.

[137] (with M. Grötschel and A. Schrijver). Relaxations of vertex packing. *J. Combin. Theory Ser. B*, 40(3):330–343.

[138] (with J. Spencer and K. Vesztergombi). Discrepancy of set-systems and matrices. *European J. Combin.*, 7(2):151–160.

[139] (with N. Alon and P. Frankl). The chromatic number of Kneser hypergraphs. *Trans. Amer. Math. Soc.*, 298(1):359–370.

[140] (with K. Cameron and J. Edmonds). A note on perfect graphs. *Period. Math. Hungar.*, 17(3):173–175.

[141] (with M.D. Plummer). *Matching theory*, volume 121 of *North-Holland Mathematics Studies*. North-Holland Publishing Co., Amsterdam; North-Holland Publishing Co., Amsterdam, Annals of Discrete Mathematics, 29. Corrected reprint, AMS Chelsea Publishing, Providence, RI, 2009.

[142] (with U. Faigle, R. Schrader and Gy. Turán). Searching in trees, series-parallel and interval orders. *SIAM J. Comput.*, 15(4):1075–1084.

[143] *An algorithmic theory of numbers, graphs and convexity*, volume 50 of *CBMS-NSF Regional Conference Series in Applied Mathematics*. Society for Industrial and Applied Mathematics (SIAM), Philadelphia, PA.

[144] (with B. Korte). Homomorphisms and Ramsey properties of antimatroids. *Discrete Appl. Math.*, 15:283–290.

[145] Algorithmic aspects of some notions in classical mathematics. In *Mathematics and computer science (Amsterdam, 1983)*, volume 1 of *CWI Monogr.*, pages 51–63. North-Holland, Amsterdam.

[146] (with R. Kannan). Covering minima and lattice point free convex bodies. In *Foundations of software technology and theoretical computer science (New Delhi, 1986)*, volume 241 of *Lecture Notes in Comput. Sci.*, pages 193–213. Springer, Berlin.

[147] Connectivity algorithms using rubber bands. In *Foundations of software technology and theoretical computer science (New Delhi, 1986)*, volume 241 of *Lecture Notes in Comput. Sci.*, pages 394–411. Springer, Berlin.

[148] (with E. Szemerédi, editors). Theory of algorithms. (Colloquium on the Theory of Algorithms held in Pécs, Hungary, July 16-21, 1984). *Colloquia Mathematica Societatis János Bolyai*, North-Holland (Elsevier Science Publishers), 430 pp.

1987

[149] Matching structure and the matching lattice. *J. Combin. Theory Ser. B*, 43(2):187–222.

[150] (with A. Dress). On some combinatorial properties of algebraic matroids. *Combinatorica*, 7(1):39–48.

[151] (with A. Björner). Pseudomodular lattices and continuous matroids. *Acta Sci. Math. (Szeged)*, 51(3-4):295–308.

1988

[152] (with R. Kannan and A.K. Lenstra). Polynomial factorization and nonrandomness of bits of algebraic and some transcendental numbers. *Math. Comp.*, 50(181):235–250.

[153] (with M. Grötschel and A. Schrijver). *Geometric algorithms and combinatorial optimization*, volume 2 of *Algorithms and Combinatorics: Study and Research Texts*. Springer-Verlag, Berlin.

[154] (with N. Linial and A. Wigderson). Rubber bands, convex embeddings and graph connectivity. *Combinatorica*, 8(1):91–102.

[155] (with R. Kannan). Covering minima and lattice-point-free convex bodies. *Ann. of Math. (2)*, 128(3):577–602.

[156] Geometry of numbers: an algorithmic view. In *ICIAM '87: Proceedings of the First International Conference on Industrial and Applied Mathematics (Paris, 1987)*, pages 144–152. SIAM, Philadelphia, PA.

[157] (with C.A.J. Hurkens, A. Schrijver and É. Tardos). How to tidy up your set-system? In *Combinatorics (Eger, 1987)*, volume 52 of *Colloq. Math. Soc. János Bolyai*, pages 309–314. North-Holland, Amsterdam.

[158] (with P. Erdős and K. Vesztergombi). The chromatic number of the graph of large distances. In *Combinatorics (Eger, 1987)*, volume 52 of *Colloq. Math. Soc. János Bolyai*, pages 547–551. North-Holland, Amsterdam.

[159] (with A. Hajnal and V. T. Sós, editors). Combinatorics. (Sixth Hungarian Combinatorial Colloquium held in Eger, Hungary, July 5-10, 1987). *Colloquia Mathematica Societatis János Bolyai*, 52. János Bolyai Mathematical Society, North-Holland. 596 pp.

[160] (with M. Saks). Lattices, Möbius functions and communication complexity. In *Proceedings of the 29th Annual Symposium on Foundations of Computer Science*, pages 81–90. IEEE Computer Society Press, Washington DC.

1989

[161] (with B. Korte). The intersection of matroids and antimatroids. In *Proceedings of the Oberwolfach Meeting "Kombinatorik" (1986)*, volume 73, pages 143–157.

[162] (with M. Saks and A. Schrijver). Orthogonal representations and connectivity of graphs. *Linear Algebra Appl.*, 114/115:439–454. Correction, in *Linear Algebra Appl.*, 313(1-3):101–105, 2000.

[163] (with R. Anderson, P. Shor, J. Spencer, É. Tardos and S. Winograd). Disks, balls, and walls: analysis of a combinatorial game. *Amer. Math. Monthly*, 96(6):481–493.

[164] (with K. Vesztergombi). Extremal problems for discrepancy. In *Irregularities of partitions (Fertőd, 1986)*, volume 8 of *Algorithms Combin. Study Res. Texts*, pages 107–113. Springer, Berlin.

[165] (with M. Saks and W.T. Trotter). An on-line graph coloring algorithm with sublinear performance ratio. *Discrete Math.*, 75(1-3):319–325.

[166] (with P. Erdős and K. Vesztergombi). On the graph of large distances. *Discrete Comput. Geom.*, 4(6):541–549.

[167] (with B. Korte). Polyhedral results for antimatroids. In *Combinatorial Mathematics: Proceedings of the Third International Conference (New York, 1985)*, volume 555 of *Ann. New York Acad. Sci.*, pages 283–295. New York Acad. Sci., New York.

[168] (with O. Goecke and B. Korte). Examples and algorithmic properties of greedoids. In *Combinatorial optimization (Como, 1986)*, volume 1403 of *Lecture Notes in Math.*, pages 113–161. Springer, Berlin.

[169] (with M.D. Plummer). Some recent results on graph matching. In *Graph theory and its applications: East and West (Jinan, 1986)*, volume 576 of *Ann. New York Acad. Sci.*, pages 389–398. New York Acad. Sci., New York.

[170] Geometry of numbers and integer programming. In *Mathematical programming (Tokyo, 1988)*, volume 6 of *Math. Appl. (Japanese Ser.)*, pages 177–201. SCIPRESS, Tokyo.

[171] Singular spaces of matrices and their application in combinatorics. *Bol. Soc. Brasil. Mat. (N.S.)*, 20(1):87–99.

[172] Faster algorithms for hard problems. *Information Processing 1989 (ed. G. X. Ritter)*, pages 135–141. Elsevier.

1990

[173] (with R. Kannan and H.E. Scarf). The shapes of polyhedra. *Math. Oper. Res.*, 15(2):364–380.

[174] A new approach to algorithmic mathematics. *Pokroky Mat. Fyz. Astronom.*, 35(1):12–22 (in Czech).

[175] (with I. Csiszár, J. Körner, K. Marton and G. Simonyi). Entropy splitting for antiblocking corners and perfect graphs. *Combinatorica*, 10(1):27–40.

[176] (with I. Bárány and Z. Füredi). On the number of halving planes. *Combinatorica*, 10(2):175–183.

[177] Communication complexity: a survey. In *Paths, flows, and VLSI-layout (Bonn, 1988)*, volume 9 of *Algorithms Combin.*, pages 235–265. Springer, Berlin.

[178] (with A. Schrijver). Matrix cones, projection representations, and stable set polyhedra. In *Polyhedral combinatorics (Morristown, NJ, 1989)*, volume 1 of *DIMACS Ser. Discrete Math. Theoret. Comput. Sci.*, pages 1–17. Amer. Math. Soc., Providence, RI.

[179] (with M. Simonovits). The mixing rate of Markov chains, an isoperimetric inequality, and computing the volume. In *31st Annual Symposium on Foundations of Computer Science, Vol. I, II (St. Louis, MO, 1990)*, pages 346–354. IEEE Comput. Soc. Press, Los Alamitos, CA.

[180] How to compute the volume? *Jahresbericht der DMV, Jubiläumstag.*, 100 Jahre DMV, 138–151.

[181] (with B. Korte, H. J. Prömel and A. Schrijver, editors). Paths, flows, and VLSI-layout. Proceedings of a meeting held from June 20 to July 1, 1988, at the University of Bonn, Germany. *Algorithms and Combinatorics*, 9. Springer-Verlag. xxii, 383 pp.

1991

[182] The work of A.A. Razborov. In *Proc. Int. Congr. Math. Vol. 1 (Kyoto/Japan 1990)*, pages 37–40. Math. Soc. Japan, Tokyo.

[183] Geometric algorithms and algorithmic geometry. In *Proc. Int. Congr. Math. Vol. 1 (Kyoto/Japan 1990)*, pages 139–154. Math. Soc. Japan, Tokyo.

[184] (with A. Schrijver). Cones of matrices and set-functions and 0-1 optimization. *SIAM J. Optim.*, 1(2):166–190.

[185] (with L. Babai and A.J. Goodman). Graphs with given automorphism group and few edge orbits. *European J. Combin.*, 12(3):185–203, 1991.

[186] (with A. Björner and P.W. Shor). Chip-firing games on graphs. *European J. Combin.*, 12(4):283–291.

[187] The work of A.A. Razborov. In *Proceedings of the International Congress of Mathematicians, Vol. I, II (Kyoto, 1990)*, pages 37–40. Math. Soc. Japan, Tokyo.

[188] Geometric algorithms and algorithmic geometry. In *Proceedings of the International Congress of Mathematicians, Vol. I, II (Kyoto, 1990)*, pages 139–154. Math. Soc. Japan, Tokyo.

[189] (with B. Korte and R. Schrader). *Greedoids*, volume 4 of *Algorithms and Combinatorics*. Springer-Verlag, Berlin.

[190] (with U. Feige, S. Goldwasser, S. Safra and M. Szegedy). Approximating clique is almost NP-complete. In *Proceedings of the 32nd Annual Symposium on Foundations of Computer Science*, pages 2–12. IEEE Computer Society Press, Washington DC.

[191] (with M. Naor, I. Newman and A. Wigderson). Search problems in the decision tree model. In *Proceedings of the 32nd Annual Symposium on Foundations of Computer Science*, pages 576–585. IEEE Computer Society Press, Washington DC.

1992

[192] (with J. Csima). A matching algorithm for regular bipartite graphs. *Discrete Appl. Math.*, 35(3):197–203.

[193] (with H.E. Scarf). The generalized basis reduction algorithm. *Math. Oper. Res.*, 17(3):751–764.

[194] (with I. Bárány and R. Howe). On integer points in polyhedra: a lower bound. *Combinatorica*, 12(2):135–142.

[195] (with A. Björner). Chip-firing games on directed graphs. *J. Algebraic Combin.*, 1(4):305–328.

(with M. Simonovits). On the randomized complexity of volume and diameter. In *33rd Annual Symposium on Foundations of Computer Science*, pages 482–491. IEEE Computer Society, Washington DC.

[196] (with G. Halász, D. Miklós and T. Szönyi, editors). Sets, graphs and numbers. A birthday salute to Vera T. Sós and András Hajnal. *Colloquia Mathematica Societatis János Bolyai*. 60. North-Holland Publishing Company, 752 pp.

[197] (with A. Björner, and A.C.C. Yao). Linear decision trees: Volume estimates and topological bounds. In *Proceedings of the 33rd Annual ACM Symposium on Theory of Computing*, Part F129722, pages 170–177. ACM, New York.

[198] (with U. Felge). Two-prover one-round proof systems: Their power and their problems. In *Proceedings of the 33rd Annual ACM Symposium on Theory of Computing*, Part F129722, pages 733–744. ACM, New York.

1993

[199] (with J. Csima). Dating to marriage. *Discrete Appl. Math.*, 41(3):269–270.

[200] (with N. Karmarkar, R. Karp, R. Lipton and M. Luby). A Monte Carlo algorithm for estimating the permanent. *SIAM J. Comput.*, 22(2):284–293.

[201] (with Á. Seress). The cocycle lattice of binary matroids. *European J. Combin.*, 14(3):241–250.

[202] (with M. Simonovits). Random walks in a convex body and an improved volume algorithm. *Random Structures Algorithms*, 4(4):359–412.

[203] (with P. Winkler). A note on the last new vertex visited by a random walk. *J. Graph Theory*, 17(5):593–596.

[204] (with M. Saks). Communication complexity and combinatorial lattice theory. *J. Comput. Syst. Sci.*, 47(2):322–349.

[205] Paul Erdős is 80. In *Combinatorics, Paul Erdős is eighty, Vol. 1*, Bolyai Soc. Math. Stud., pages 9–11. János Bolyai Math. Soc., Budapest.

[206] (with M. Grötschel and A. Schrijver). *Geometric algorithms and combinatorial optimization*, volume 2 of *Algorithms and Combinatorics*. Springer-Verlag, Berlin, second edition.

[207] Features of computer language: communication of computers and its complexity. *Acta neurochirurgica. Supplementum*, 56:91–95.

1994

[208] (with A. Björner). Linear decision trees, subspace arrangements and Möbius functions. *J. Amer. Math. Soc.*, 7(3):677–706, 1994.

[209] (with A. Björner, S.T. Vrećica and R.T. Živaljević). Chessboard complexes and matching complexes. *J. London Math. Soc. (2)*, 49(1):25–39.

[210] Stable sets and polynomials. *Discrete Math.*, 124(1-3):137–153.

1995

[211] (with M. Naor, I. Newman and Avi Wigderson). Search problems in the decision tree model. *SIAM J. Discrete Math.*, 8(1):119–132.

[212] (with R. Kannan and M. Simonovits). Isoperimetric problems for convex bodies and a localization lemma. *Discrete Comput. Geom.*, 13(3-4):541–559.

[213] Randomized algorithms in combinatorial optimization. In *Combinatorial optimization (New Brunswick, NJ, 1992–1993)*, volume 20 of *DIMACS Ser. Discrete Math. Theoret. Comput. Sci.*, pages 153–179. Amer. Math. Soc., Providence, RI.

[214] (with H. van der Holst and A. Schrijver). On the invariance of Colin de Verdière's graph parameter under clique sums. *Linear Algebra Appl.*, 226/228: 509–517.

[215] (with Á. Seress). The cocycle lattice of binary matroids. II. *Linear Algebra Appl.*, 226/228:553–565.

[216] (with P. Winkler). Exact mixing in an unknown Markov chain. *Electron. J. Comb.*, 2:Research Paper 15, 12 pp. Comments, in *Electron. J. Comb.* 2:Research Paper 15, Comment 1, 251–262.

[217] (with P. Winkler). Mixing of random walks and other diffusions on a graph. In *Surveys in combinatorics, 1995 (Stirling)*, volume 218 of *London Math. Soc. Lecture Note Ser.*, pages 119–154. Cambridge Univ. Press, Cambridge.

[218] (with M. Grötschel). Combinatorial optimization. In *Handbook of combinatorics, Vol. 1, 2*, pages 1541–1597. Elsevier Sci. B. V., Amsterdam.

[219] (with D.B. Shmoys and É. Tardos). Combinatorics in computer science. In *Handbook of combinatorics, Vol. 1, 2*, pages 2003–2038. Elsevier Sci. B. V., Amsterdam.

[220] (with L. Pyber, D.J.A. Welsh and G.M. Ziegler). Combinatorics in pure mathematics. In *Handbook of combinatorics, Vol. 1, 2*, pages 2039–2082. Elsevier Sci. B. V., Amsterdam.

[221] (with P. Winkler). Efficient stopping rules for Markov chains . In *Proceedings of the 27th annual ACM symposium on the theory of computing (STOC)*, pages 76–82. ACM, New York.

[222] (W. Cook and P. Seymour, editors). Combinatorial optimization. Papers from the DIMACS special year. Papers from workshops held at DIMACS at Rutgers University, New Brunswick, NJ, USA, Sept. 1992–Aug. 1993. *DIMACS. Series in Discrete Mathematics and Theoretical Computer Science*. 20. American Mathematical Society (AMS). xi, 441 pp.

[223] (R.L. Graham and M. Grötschel, editors). Handbook of combinatorics. Vol. 1-2. Amsterdam: Elsevier (North-Holland); Cambridge, MA: MIT Press. cii, 2198 pp.

[224] (with C. D. Godsil). Tools from linear algebra. In *Handbook of combinatorics, Vol. 1, 2*, pages 1705–1748. Elsevier Sci. B. V., Amsterdam.

[225] (with J. Pach and M. Szegedy). On Conway's thrackle conjecture. In *Proceedings of the 11th Annual Symposium on Computational Geometry*, pages 147–151. ACM, New York.

1996

[226] Random walks on graphs: a survey. In *Combinatorics, Paul Erdős is eighty, Vol. 2 (Keszthely, 1993)*, volume 2 of *Bolyai Soc. Math. Stud.*, pages 353–397. János Bolyai Math. Soc., Budapest.

[227] (with U. Feige, S. Goldwasser, S. Safra and M. Szegedy). Interactive proofs and the hardness of approximating cliques. *J. ACM*, 43(2):268–292.

[228] (with A. Kotlov). The rank and size of graphs. *J. Graph Theory*, 23(2):185–189.

[229] Information and complexity (how to measure them?). In *The emergence of complexity in mathematics, physics, chemistry and biology (Vatican City, 1992)*, volume 89 of *Pontif. Acad. Sci. Scr. Varia*, pages 65–80. Pontif. Acad. Sci., Vatican City.

1997

[230] The membership problem in jump systems. *J. Combin. Theory Ser. B*, 70(1): 45–66.

[231] (with J. Pach and M. Szegedy). On Conway's thrackle conjecture. *Discrete Comput. Geom.*, 18(4):369–376.

[232] (with D. Aldous and P. Winkler). Mixing times for uniformly ergodic Markov chains. *Stochastic Process. Appl.*, 71(2):165–185.

[233] (with R. Kannan and M. Simonovits). Random walks and an $O^*(n^5)$ volume algorithm for convex bodies. *Random Structures Algorithms*, 11(1):1–50.

[234] (with A. Kotlov and S. Vempala). The Colin de Verdière number and sphere representations of a graph. *Combinatorica*, 17(4):483–521.

1998

[235] (with A. Schrijver). A Borsuk theorem for antipodal links and a spectral characterization of linklessly embeddable graphs. *Proc. Amer. Math. Soc.*, 126(5):1275–1285.

[236] (with P. Winkler). Reversal of Markov chains and the forget time. *Combin. Probab. Comput.*, 7(2):189–204.

[237] (with P. Winkler). Mixing times. In *Microsurveys in discrete probability (Princeton, NJ, 1997)*, volume 41 of *DIMACS Ser. Discrete Math. Theoret. Comput. Sci.*, pages 85–133. Amer. Math. Soc., Providence, RI.

[238] One mathematics. *Mitt. Dtsch. Math.-Ver.*, (2):33–39.

[239] (with A. Beveridge). Random walks and the regeneration time. *J. Graph Theory*, 29(2):57–62.

[240] (with A. Brieden, R. Kannan, P. Gritzmann, V. Klee and M. Simonovits). Approximation of diameters: Randomization doesn't help. In *Proceedings of the 39th Annual Symposium on Foundations of Computer Science*, pages 244–251. IEEE Computer Society Press, Washington DC.

1999

[241] (with H. van der Holst and A. Schrijver). The Colin de Verdière graph parameter. In *Graph theory and combinatorial biology (Balatonlelle, 1996)*, volume 7 of *Bolyai Soc. Math. Stud.*, pages 29–85. János Bolyai Math. Soc., Budapest.

[242] (with A. Schrijver). On the null space of a Colin de Verdière matrix. *Ann. Inst. Fourier*, 49(3):1017–1025.

[243] Hit-and-run mixes fast. *Math. Program.*, 86(3, Ser. A):443–461.

[244] (with F. Chen and I. Pak). Lifting Markov chains to speed up mixing. In *Annual ACM Symposium on Theory of Computing (Atlanta, GA, 1999)*, pages 275–281. ACM, New York.

[245] (with R. Kannan). Faster mixing via average conductance. In *Annual ACM Symposium on Theory of Computing (Atlanta, GA, 1999)*, pages 282–287. ACM, New York.

2000

[246] (with L. Lipták) Facets with fixed defect of the stable set polytope. *Math. Program.*, 88(1, Ser. A):33–44.

[247] Integer sequences and semidefinite programming. *Publ. Math.*, 56(3-4):475–479.

[248] Discrete and continuous: two sides of the same? In *GAFA 2000 (Tel Aviv, 1999)*, pages 359–382. Birkhäuser, Basel.

[249] (with J. Kahn, J.H. Kim and V.H. Vu). The cover time, the blanket time, and the Matthews bound. In *41st Annual Symposium on Foundations of Computer Science (Redondo Beach, CA, 2000)*, pages 467–475. IEEE Comput. Soc. Press, Los Alamitos, CA.

[250] (with A. Gyárfás, G. Katona, A. Recski and L. Székely, editors). Graph theory and combinatorial biology. Based on talks and discussions of the international colloquium on combinatorics and graph theory, Balatonlelle, Hungary, July 1996. *Bolyai Society Mathematical Studies. 7*. János Bolyai Mathematical Society, Budapest. 413 pp.

2001

[251] (with L. Lipták). Critical facets of the stable set polytope. *Combinatorica*, 21(1):61–88.

[252] Energy of convex sets, shortest paths, and resistance. *J. Combin. Theory Ser. A*, 94(2):363–382.

[253] (with N. Alon). Unextendible product bases. *J. Combin. Theory Ser. A*, 95(1):169–179.

[254] Steinitz representations of polyhedra and the Colin de Verdière number. *J. Combin. Theory Ser. B*, 82(2):223–236.

[255] (with A. Brieden, P. Gritzmann, R. Kannan, V. Klee and M. Simonovits). Deterministic and randomized polynomial-time approximation of radii. *Mathematika*, 48(1-2):63–105.

2002

[256] (with K. Vesztergombi). Geometric representations of graphs. In *Paul Erdős and his mathematics, II (Budapest, 1999)*, volume 11 of *Bolyai Soc. Math. Stud.*, pages 471–498. János Bolyai Math. Soc., Budapest.

[257] (with U. Feige and P. Tetali). Approximating min-sum set cover. In *Approximation algorithms for combinatorial optimization*, volume 2462 of *Lecture Notes in Comput. Sci.*, pages 94–107. Springer, Berlin.

[258] (with G. Halász, M. Simonovits and V. T. Sós, editors). Paul Erdős and his mathematics I. Based on the conference, Budapest, Hungary, July 4–11, 1999. *Bolyai Society Mathematical Studies. 11*. Springer, Berlin. 728 pp.

[259] (with G. Halász, M. Simonovits and V. T. Sós, editors). Paul Erdős and his mathematics II. Based on the conference, Budapest, Hungary, July 4–11, 1999. *Bolyai Society Mathematical Studies. 11*. Springer, Berlin. 695 pp.

[260] (with S. Arora and B. Bollobás). Proving integrality gaps without knowing the linear program. In *Proceedings of the 43rd Annual Symposium on Foundations of Computer Science*, pages 313–322. IEEE Computer Society Press, Washington DC.

[261] (with I. Benjamini). Global information from local observation. In *Proceedings of the 43rd Annual Symposium on Foundations of Computer Science*, pages 701–710. IEEE Computer Society Press, Washington DC.

2003

[262] (with J. Pelikán and K. Vesztergombi). *Discrete mathematics. Elementary and beyond*. Undergraduate Texts in Mathematics. Springer-Verlag, New York.

[263] Semidefinite programs and combinatorial optimization. In *Recent advances in algorithms and combinatorics*, volume 11 of *CMS Books Math./Ouvrages Math. SMC*, pages 137–194. Springer, New York.
[264] (with I. Benjamini). Harmonic and analytic functions on graphs. *J. Geom.*, 76(1-2):3–15.
[265] (with N.J.A. Harvey, R.E. Ladner and T. Tamir). Semi-matchings for bipartite graphs and load balancing. In *Algorithms and data structures*, volume 2748 of *Lecture Notes in Comput. Sci.*, pages 294–306. Springer, Berlin.
[266] (with S. Vempala). Logconcave functions: Geometry and efficient sampling algorithms. In *Proceedings of the 44th Annual Symposium on Foundations of Computer Science*, pages 640–649. IEEE Computer Society Press, Washington DC.

2004

[267] (with K. Vesztergombi, U. Wagner and E. Welzl). Convex quadrilaterals and k-sets. In *Towards a theory of geometric graphs*, volume 342 of *Contemp. Math.*, pages 139–148. Amer. Math. Soc., Providence, RI.
[268] (with U. Feige and P. Tetali). Approximating min sum set cover. *Algorithmica*, 40(4):219–234.
[269] (with S. Vempala). Hit-and-run from a corner. In *Proceedings of the 36th Annual ACM Symposium on Theory of Computing*, pages 310–314. ACM, New York.
[270] (with J. Chen, R.D. Kleinberg, R. Rajaraman, R. Sundaram and A. Vetta). (Almost) tight bounds and existence theorems for confluent flows. In *Proceedings of the 36th Annual ACM Symposium on Theory of Computing*, pages 529–538. ACM, New York.
[271] Discrete analytic functions: an exposition. In *Surveys in differential geometry. Vol. IX*, volume 9 of *Surv. Differ. Geom.*, pages 241–273. Int. Press, Somerville, MA.
[272] (with H. J. Prömel, editors). Combinatorics. Oberwolfach Rep. 1, No. 1, Report 1, 5–109.

2005

[273] (with M. Bordewich, M. Freedman and D. Welsh). Approximate counting and quantum computation. *Combin. Probab. Comput.*, 14(5-6):737–754.
[274] (with K. Jain and P.A. Chou). Building scalable and robust peer-to-peer overlay networks for broadcasting using network coding. In *Proceedings of the 24th Annual ACM Symposium on Principles of Distributed Computing*, pages 51–59. ACM, New York.

2006

[275] Graph minor theory. *Bull. Amer. Math. Soc. (N.S.)*, 43(1):75–86.
[276] (with S. Vempala). Hit-and-run from a corner. *SIAM J. Comput.*, 35(4):985–1005.

[277] (with S. Vempala). Simulated annealing in convex bodies and an $O^*(n^4)$ volume algorithm. *J. Comput. System Sci.*, 72(2):392–417.

[278] (with N.J.A. Harvey, R.E. Ladner and T. Tamir). Semi-matchings for bipartite graphs and load balancing. *J. Algorithms*, 59(1):53–78.

[279] (with M. Saks). A localization inequality for set functions. *J. Combin. Theory Ser. A*, 113(4):726–735.

[280] The rank of connection matrices and the dimension of graph algebras. *European J. Combin.*, 27(6):962–970.

[281] (with R. Kannan and R. Montenegro). Blocking conductance and mixing in random walks. *Combin. Probab. Comput.*, 15(4):541–570.

[282] (with I. Benjamini, G. Kozma, D. Romik and G. Tardos). Waiting for a bat to fly by (in polynomial time). *Combin. Probab. Comput.*, 15(5):673–683.

[283] (with C. Borgs, J. Chayes, V.T. Sós and K. Vesztergombi). Counting graph homomorphisms. In *Topics in discrete mathematics*, volume 26 of *Algorithms Combin.*, pages 315–371. Springer, Berlin.

[284] (with B. Szegedy). Limits of dense graph sequences. *J. Combin. Theory Ser. B*, 96(6):933–957.

[285] (with C. Borgs, J. Chayes, V.T. Sós, B. Szegedy and K. Vesztergombi). Graph limits and parameter testing. In *STOC'06: Proceedings of the 38th Annual ACM Symposium on Theory of Computing*, pages 261–270. ACM, New York.

[286] (with S. Arora, B. Bollobás and I. Tourlakis). Proving integrality gaps without knowing the linear program. *Theory Comput.*, 2:19–51.

[287] (with S. Arora, I. Newman, Y. Rabani, Y. Rabinovich and S. Vempala). Local versus global properties of metric spaces (extended abstract). In *Proceedings of the Seventeenth Annual ACM-SIAM Symposium on Discrete Algorithms*, pages 41–50. ACM, New York.

[288] (with E. Győri and G.O.H. Katona, editors) More sets, graphs and numbers. A salute to Vera Sós and András Hajnal. *Bolyai Society Mathematical Studies 15*. Springer, Berlin. 405 pp.

[289] (with S. Vempala). Fast algorithms for logconcave functions: Sampling, rounding, integration and optimization. In *Proceedings of the 47th Annual Symposium on Foundations of Computer Science*, pages 57–66. IEEE Computer Society Press, Washington DC.

2007

[290] (with M. Freedman and A. Schrijver). Reflection positivity, rank connectivity, and homomorphism of graphs. *J. Amer. Math. Soc.*, 20(1):37–51.

[291] (with A. Blokhuis, L. Storme and T. Szőnyi). On multiple blocking sets in Galois planes. *Adv. Geom.*, 7(1):39–53.

[292] (with B. Szegedy). Szemerédi's lemma for the analyst. *Geom. Funct. Anal.*, 17(1):252–270.

[293] (with S. Vempala). The geometry of logconcave functions and sampling algorithms. *Random Structures Algorithms*, 30(3):307–358.

[294] Connection matrices. In *Combinatorics, complexity, and chance*, volume 34 of *Oxford Lecture Ser. Math. Appl.*, pages 179–190. Oxford Univ. Press, Oxford.

[295] (with J. Chen, R.D. Kleinberg, R. Rajaraman, R. Sundaram and A. Vetta). (Almost) tight bounds and existence theorems for single-commodity confluent flows. *J. ACM*, 54(4):Art. 16, 32.

[296] (with K. Jain and P.A. Chou) Building scalable and robust peer-to-peer overlay networks for broadcasting using network coding. *Distrib. Comput. 19*, No. 4, 301–311.

2008

[297] (with A. Gács and T. Szőnyi). Directions in AG$(2,p^2)$. *Innov. Incidence Geom.*, 6/7:189–201.

[298] (with V.T. Sós). Generalized quasirandom graphs. *J. Combin. Theory Ser. B*, 98(1):146–163.

[299] (with A. Schrijver). Graph parameters and semigroup functions. *European J. Combin.*, 29(4):987–1002.

[300] (with C. Borgs, J.T. Chayes, V.T. Sós and K. Vesztergombi). Convergent sequences of dense graphs. I. Subgraph frequencies, metric properties and testing. *Adv. Math.*, 219(6):1801–1851.

[301] In memoriam János Surányi. *Mat. Lapok (N.S.)*, 14(2):1–2.

[302] (with E. Győri and G.O.H. Katona, editors). Horizons of combinatorics. Survey papers related to the conference, Balatonalmádi, Hungary, July 17–21, 2006. *Bolyai Society Mathematical Studies 17*, 280 pp.

[303] (with E. Győri and G.O.H. Katona, editors). Combinatorics. Abstracts from the workshop held January 6–12, 2008. *Oberwolfach Rep. 5*, No. 1, 5–78.

[304] (with S. Lovasz, P. Nyirady and I. Romics). A novel quantitative method for measuring obstruction in the upper urinary tract: The 'obstruction coefficient'. *International Journal of Urology*, 15(6):499–504.

2009

[305] (with B. Szegedy). Contractors and connectors of graph algebras. *J. Graph Theory*, 60(1):11–30.

[306] (with A. Schrijver). Semidefinite functions on categories. *Electron. J. Comb.*, 16(2):Research Paper 14, 16 pp.

[307] Very large graphs. In *Current developments in mathematics, 2008*, pages 67–128. Int. Press, Somerville, MA.

[308] The "little geometer" and the difficulty of computing the volume. *Ann. Univ. Sci. Budapest. Eötvös Sect. Math.*, 52:31–36.

[309] (with W. J. Cook and J. Vygen, editors). Research trends in combinatorial optimization. Papers based on the presentations at the workshop Bonn, Germany, 2008. Dedicated to Bernard Korte on the occasion of the 70th birthday. Springer, Berlin. 562 pp.

2010

[310] (with A. Schrijver). Dual graph homomorphism functions. *J. Combin. Theory Ser. A*, 117(2):216–222.

[311] (with C. Borgs and J. Chayes). Moments of two-variable functions and the uniqueness of graph limits. *Geom. Funct. Anal.*, 19(6):1597–1619.

[312] (with A. Beveridge). Exit frequency matrices for finite Markov chains. *Combin. Probab. Comput.*, 19(4):541–560.

[313] (with B. Szegedy). Testing properties of graphs and functions. *Israel J. Math.*, 178:113–156.

[314] (with I. Faragó). László Simon is 70 years old. *Ann. Univ. Sci. Budapest. Eötvös Sect. Math.*, 53:3–6.

[315] (with B. Szegedy). Regularity partitions and the topology of graphons. In *An irregular mind*, volume 21 of *Bolyai Soc. Math. Stud.*, pages 415–446. János Bolyai Math. Soc., Budapest.

[316] (with I. Smeets, A. Lenstra, H. Lenstra and P. Van Emde Boas). The history of the LLL-algorithm. In *The LLL algorithm. Survey and applications*, pages 1–17. Springer, Dordrecht.

[317] (with G. Palla and T. Vicsek). Multifractal network generator. *Proceedings of the National Academy of Sciences of the United States of America*, 107(17):7640–7645.

2011

[318] (with B. Szegedy). Finitely forcible graphons. *J. Combin. Theory Ser. B*, 101(5):269–301.

[319] Subgraph densities in signed graphons and the local Simonovits-Sidorenko conjecture. *Electron. J. Comb.*, 18(1):Paper 127, 1–21.

[320] (with R.J. Kang, T. Müller and E.R. Scheinerman). Dot product representations of planar graphs. *Electron. J. Comb.*, 18(1):Paper 216, 14 pp.

[321] (with J. Nešetřil, P. Ossona de Mendez and A. Schrijver). Preface [Homomorphisms and limits]. *European J. Combin.*, 32(7):951–953.

[322] (with C. Borgs, J. Chayes, V. Sós and K. Vesztergombi). Limits of randomly grown graph sequences. *European J. Combin.*, 32(7):985–999.

[323] (with J./ Nešerřil, P. Ossona de Mendez and A. Schrijver, editors). Special issue: Homomorphisms and limits. *Eur. J. Comb. 32*, No. 7, 951–1175.

[324] Graph theory over 45 years. In *An invitation to mathematics. From competitions to research*, pages 85–95. Springer,. Berlin.

2012

[325] (with I. Deák). Computational results of an $O^*(n^4)$ volume algorithm. *European J. Oper. Res.*, 216(1):152–161.

[326] (with J. Draisma, D.C. Gijswijt, G. Regts and A. Schrijver). Characterizing partition functions of the vertex model. *J. Algebra*, 350:197–206.

[327] (with S. Arora, I. Newman, Y. Rabani, Y. Rabinovich and S. Vempala). Local versus global properties of metric spaces. *SIAM J. Comput.*, 41(1):250–271.

[328] (with B. Szegedy). Random graphons and a weak Positivstellensatz for graphs. *J. Graph Theory*, 70(2):214–225.

[329] (with C. Borgs, J.T. Chayes, V.T. Sós and K. Vesztergombi). Convergent sequences of dense graphs II. Multiway cuts and statistical physics. *Ann. of Math. (2)*, 176(1):151–219.

[330] *Large networks and graph limits*, volume 60 of *American Mathematical Society Colloquium Publications*. American Mathematical Society, Providence, RI.

[331] (with I.Z. Rusza, and V.T. Sós, editors). Erdős centennial. On the occasion of Paul Erdős 100th anniversary of his birth. *Bolyai Society Mathematical Studies 25*. Springer, Berlin. 730 pp.

2013

[332] (with C. Borgs, J. Chayes and J. Kahn). Left and right convergence of graphs with bounded degree. *Random Structures Algorithms*, 42(1):1–28.

[333] (with K. Vesztergombi). Non-deterministic graph property testing. *Combin. Probab. Comput.*, 22(5):749–762.

[334] Trends in mathematics: how they could change education. *ICCM Not.*, 1(2): 79–84.

[335] (with B. Szegedy, editors). Working session: Limits of structures. Abstracts from the working session held March 31 – April 15, 2013. Oberwolfach Rep. 10, No. 2, 963–1024.

2014

[336] (with H. Hatami and B. Szegedy). Limits of locally-globally convergent graph sequences. *Geom. Funct. Anal.*, 24(1):269–296.

2015

[337] (with B. Szegedy). The automorphism group of a graphon. *J. Algebra*, 421:136–166.

2016

[338] (with J. Nešetřil and A. Schrijver). Preface [Special issue: Recent advances in graphs and analysis]. *European J. Combin.*, 52(part B):245–247.

[339] (with O. Antolín Camarena, E. Csóka, T. Hubai and G. Lippner). Positive graphs. *European J. Combin.*, 52(part B):290–301.

[340] (with K. Alladi, S. Krantz, V.T. Sós, R.L. Graham, J,. Spencer, J.-P. Kahane and M.B. Nathanson, editors). Reflections on Paul Erdős on his birth centenary. *Notices Am. Math. Soc.*, 62(2):121–143.

[341] (with T. Hubai and D. Kunszenti-Kovács). Positive graphs. In *Discrete mathematical days. Extended abstracts of the 10th "Jornadas de matemática discreta y algorítmica" (JMDA)*, volume 54 of *Electronic Notes in Discrete Mathematics*, pages 355–360. Elsevier, Amsterdam.

2017

[342] (with A. Schrijver). Nullspace embeddings for outerplanar graphs. In *A journey through discrete mathematics*, pages 571–591. Springer, Cham.

2019

[343] *Graphs and geometry*, volume 65 of *American Mathematical Society Colloquium Publications*. American Mathematical Society, Providence, RI.

[344] (with D. Kunszenti-Kovács and B. Szegedy). Measures on the square as sparse graph limits. *J. Combin. Theory Ser. B*, 138:1–40.

2020

[345] Compact graphings. *Acta Math. Hungar.*, 161(1):185–196.

[346] Hyperfinite graphings and combinatorial optimization. *Acta Math. Hungar.*, 161(2):516–539.

2021

[347] Flows on measurable spaces. *Geom. Funct. Anal.*, 31(2):402–437.

[348] Discrete quantitative nodal theorem. *Electron. J. Comb.*, 28(3):Research Paper 3.58, 6 pp.

[349] (with G. Ódor, D. Czifra, J. Komjáthy and M. Karsai). Switchover phenomenon induced by epidemic seeding on geometric networks. *Proc. Nat. Acad. Sci.*, 118(41):e2112607118.

2022

[350] (with D. Kunszenti-Kovács and B. Szegedy). Multigraph limits, unbounded kernels, and Banach space labeled graphs. *J. Funct. Anal.* 282(2):Paper No. 109284.

List of Publications for Avi Wigderson

1982

[1] A new approximate graph coloring algorithm. *Proceedings of the Annual ACM Symposium on Theory of Computing*, 1982, 325–329.

[2] The Complexity of the Hamiltonian Circuit Problem for Maximal Planar Graphs. *Technical Report 298, Department of EECS*, Princeton University, February 1982.

[3] (with G. Vijayan). Planarity of Edge Ordered Graphs. *Technical Report 307, Department of EECS*, Princeton University, December 1982.

1983

[4] Improving the performance guarantee for approximate graph coloring. *J. Assoc. Comput. Mach.*, 30(4):729–735.

[5] (with D. Dolev). On the security of multi-party protocols in distributed systems. *Advances in cryptology*, Proc. Workshop Santa Barbara/Calif., 167–175.

[6] (with D. Dolev, C. Dwork and N. Pippenger). Superconcentrators, generalizers and generalized connectors with limited depth. *Conference Proceedings of the Annual ACM Symposium on Theory of Computing*, 42–51.

[7] (with H. Galperin). Succinct representations of graphs. *Inform. and Control*, 56(3):183–198.

[8] (with D.L. Long). How discreet is the discrete log? *Conference Proceedings of the Annual ACM Symposium on Theory of Computing*, 413–420.

[9] (with U. Vishkin). Dynamic parallel memories. *Inform. and Control*, 56(3): 174–182.

1985

[10] (with A. Aggarwal, M. Klawe, D. Lichtenstein and N. Linial). Multi-layer grid embeddings. *Annual Symposium on Foundations of Computer Science (Proceedings)*, 186–196.

H. Holden, R. Piene (eds.), *The Abel Prize 2018-2022*, The Abel Prize, https://doi.org/10.1007/978-3-031-33973-8_22

[11] (with M. Ajtai). Deterministic simulation of probabilistic constant depth circuits. *Advances in Computing Research - Randomness and Computation*, Vol. 5, 199–222. Also available in *Annual Symposium on Foundations of Computer Science (Proceedings)*, 11–19.

[12] (with R.M. Karp). A fast parallel algorithm for the maximal independent set problem. *J. Assoc. Comput. Mach.*, 32(4):762–773. Also available in *Proceedings of the Annual ACM Symposium on Theory of Computing*, 266–272.

[13] (with R.M. Karp and E. Upfal). Are search and decision problems computationally equivalent? *Conference Proceedings of the Annual ACM Symposium on Theory of Computing,* 464–475.

[14] (with R.M. Karp and E. Upfal). Complexity of parallel computation on matroids. *Annual Symposium on Foundations of Computer Science (Proceedings)*, 541–550.

[15] (with F.E. Fich, F. Meyer auf der Heide and P. Ragde). One, two, three ... infinity: lower bounds for parallel computation. *Conference Proceedings of the Annual ACM Symposium on Theory of Computing,* 48–58.

[16] (with F. Meyer auf der Heide). Complexity of parallel sorting. *Annual Symposium on Foundations of Computer Science (Proceedings)*, 532–540. Also available in *SIAM Journal on Computing*, 16(1): 100–107.

[17] (with G. Vijayan). Rectilinear graphs and their embeddings. *SIAM J. Comput.*, 14(2):355–372.

[18] (with U. Vishkin). Trade-offs between depth and width in parallel computation. *SIAM J. Comput.*, 14(2):303–314. Also available in *Annual Symposium on Foundations of Computer Science (Proceedings)*, 146–153.

1986

[19] (with A. Borodin, F. Fich, F. Meyer auf der Heide and E. Upfal). A time-space tradeoff for element distinctness. In *SIAM J. Comput.*, 16(1):97–99 Also available in *STACS 86 (Orsay, 1986)*, volume 210 of *Lecture Notes in Comput. Sci.*, pages 353–358. Springer, Berlin.

[20] (with A. Borodin, F.E. Fich, F. Meyer auf der Heide and E. Upfal). A tradeoff between search and update time for the implicit dictionary problem. In *Automata, languages and programming (Rennes, 1986)*, volume 226 of *Lecture Notes in Comput. Sci.*, pages 50–59. Springer, Berlin. Also available in *Proc. of the 13th ICALP Conference*, July 1986.

[21] (with O. Goldreich and S. Micali). Proofs that release minimum knowledge. In *Mathematical foundations of computer science, 1986 (Bratislava, 1986)*, volume 233 of *Lecture Notes in Comput. Sci.*, pages 639–650. Springer, Berlin.

[22] (with R.M. Karp and M. Saks). On a search problem related to branch-and-bound procedures. *Annual Symposium on Foundations of Computer Science (Proceedings)*, 19–28.

[23] (with R.M. Karp and E. Upfal). Constructing a perfect matching is in Random NC. *Combinatorica*, 6(1):35–48. Also available in *Conference Proceedings of the Annual ACM Symposium on Theory of Computing*, 232.

[24] (with N. Linial and L. Lovasz). Physical interpretation of graph connectivity, and its algorithmic applications. *Annual Symposium on Foundations of Computer Science (Proceedings)*, 39–48.
[25] (with N. Megiddo). On Play by Means of Computing Machines. *Conference on Theoretical Aspects of Reasoning about Knowledge*, 259–274.
[26] (with O. Goldreich and S. Micali). Proofs that yield nothing but their validity and a methodology of cryptographic protocol design. *Journal of the ACM (JACM)*, 38(3):690–728 Also available in *Annual Symposium on Foundations of Computer Science (Proceedings)*, 174–187.
[27] (with M. Perry). Search in a known pattern. *Journal of Political Economy*, Vol. 94, No. 1, 225–230.
[28] (with M. Saks). Probabilistic boolean decision trees and the complexity of evaluating game trees. *Annual Symposium on Foundations of Computer Science (Proceedings)*, 29–38.

1987

[29] (with F.E. Fich and F. Meyer auf der Heide). Lower bounds for parallel random-access machines with unbounded shared memory. In *Advances in computing research, Vol. 4*, pages 1–15. JAI, Greenwich, CT.
[30] (with F. Meyer auf der Heide). The complexity of parallel sorting. *SIAM J. Comput.*, 16(1):100–107.
[31] (with O. Goldreich and S. Micali). Completeness theorem for protocols with honest majority. *Conference Proceedings of the Annual ACM Symposium on Theory of Computing*, 218–229.
[32] (with O. Goldreich and S. Micali). How to play any mental game. *Proc. of the 19th STOC*, 218–229.
[33] (with O. Goldreich and S. Micali). How to prove all NP statements in zero-knowledge and a methodology of cryptographic protocol design (extended abstract). In *Advances in cryptology—CRYPTO '86 (Santa Barbara, Calif., 1986)*, volume 263 of *Lecture Notes in Comput. Sci.*, pages 171–185. Springer, Berlin.
[34] (with E. Upfal). How to share memory in a distributed system. *J. Assoc. Comput. Mach.*, 34(1):116–127. Also available in *Proceedings - Annual IEEE Symposium on Foundations of Computer Science*, FOCS, 1984-October, 171–180.

1988

[35] (with M. Ben-Or, S. Goldwasser and J. Kilian). Completeness theorems for non-cryptographic fault-tolerant distributed computation. *Proceedings of the Annual ACM Symposium on Theory of Computing*, 113–131. Also available in *Proceedings of the Annual ACM Symposium on Theory of Computing*, 1–10.
[36] (with A. Borodin, F.E. Fich, F. Meyer auf der Heide and E. Upfal). A tradeoff between search and update time for the implicit dictionary problem. *Theor. Comput. Sci.*, 58(1-3):57–68.

[37] (with F.E. Fich and P. Ragde). Relations between concurrent-write models of parallel computation. *SIAM J. Comput.*, 17(3):606–627.

[38] (with F.E. Fich and P. Ragde). Simulations among concurrent-write PRAMs. *Algorithmica*, 3(1):43–51. Also available in *Conference on the Principles of Distributed Computation*, August 1984.

[39] (with S. Goldwasser, J. Kilian and M. Ben-Or). Multi- Prover Interactive Proofs: How to Remove Intractability. *Assumptions Proc. of the 20th STOC*, 113–131.

[40] (with B. Just and F. Meyer auf der Heide). On computations with integer division: Extended abstract. *Lecture Notes in Computer Science (including subseries Lecture Notes in Artificial Intelligence and Lecture Notes in Bioinformatics)*, 29–37.

[41] (with R.M. Karp and E. Upfal). The complexity of parallel search. *J. Comput. Syst. Sci.*, 36(2):225–253.

[42] (with N. Linial and L. Lovász). Rubber bands, convex embeddings and graph connectivity. *Combinatorica*, 8(1):91–102.

[43] (with D.L. Long). The discrete logarithm hides $O(\log n)$ bits. *SIAM J. Comput.*, 17(2):363–372.

[44] (with P. Ragde, W. Steiger and E. Szemerédi). The parallel complexity of element distinctness is $\Omega\sqrt{\log n}$. *SIAM J. Discrete Math.*, 1(3):399–410.

1989

[45] (with A. Cohen). Dispersers, deterministic amplification, and weak random sources. *Annual Symposium on Foundations of Computer Science (Proceedings)*, 14–19.

[46] (with B. Just, F. Meyer auf der Heide). On computations with integer division. In *STACS 88, Theoretical aspects of computer science, Proc. 5th Annu. Symp., Bordeaux/France 1988*, volume 294 of *Lecture Notes in Comput. Sci.*, pages 29–37. Springer, Berlin. Also available in *Informatique Théorique et Applications*, Tome 23 (1989) no. 1, pp. 101–111.

[47] (with R. Raz). Probabilistic communication complexity of Boolean relations. *Annual Symposium on Foundations of Computer Science (Proceedings)*, 562–567.

1990

[48] (with N. Alon and M. Karchmer). Linear circuits over GF(2). *SIAM J. Comput.*, 19(6):1064–1067.

[49] (with M. Ben-Or, S. Goldwasser and J. Kilian). Efficient identification schemes using two prover interactive proofs. In *Advances in cryptology—CRYPTO '89 (Santa Barbara, CA, 1989)*, volume 435 of *Lecture Notes in Comput. Sci.*, pages 498–506. Springer, New York.

[50] (with F.E. Fich). Toward understanding exclusive read. *SIAM J. Comput.*, 19(4):718–727. Also available in *Proceedings of the 1st Annual ACM Symposium on Parallel Algorithms and Architectures, SPAA 1989*, 76–82.

[51] (with J. Gil and F. Meyer auf der Heide). Not all keys can be hashed in constant time. *Proc. of the 22nd STOC*, 244–253.

[52] (with R. Heiman and I. Newman). On read-once threshold formulae and their randomized decision tree complexity. In *Fifth Annual Structure in Complexity Theory Conference (Barcelona, 1990)*, pages 78–87. IEEE Comput. Soc. Press, Los Alamitos, CA.

[53] (with M. Karchmer). Monotone circuits for connectivity require super-logarithmic depth. *SIAM J. Discrete Math.*, 3(2):255–265. Also available in *Proceedings of the Annual ACM Symposium on Theory of Computing*, 539–550.

[54] (with I. Newman and P. Ragde). Perfect hashing, graph entropy, and circuit complexity (preliminary version). In *Fifth Annual Structure in Complexity Theory Conference (Barcelona, 1990)*, pages 91–99. IEEE Comput. Soc. Press, Los Alamitos, CA.

1991

[55] Information-theoretic reasons for computational difficulty. In *Proceedings of the International Congress of Mathematicians, Vol. I, II (Kyoto, 1990)*, pages 1537–1548. Math. Soc. Japan, Tokyo.

[56] (with A. Aggarwal, M. Klawe, D. Lichtenstein and N. Linial). A lower bound on the area of permutation layouts. *Algorithmica*, 6(2):241–255.

[57] (with P. Gemmel, R. Lipton, R. Rubinfeld and M. Sudan). Selftesting / correcting for Polynomials and for Approximate Functions. *Proc. of the 23rd STOC*, 32–42.

[58] (with R. Heiman). Randomized vs. deterministic decision tree complexity for read-once Boolean functions. *Comput. Complexity*, 1(4):311–329. Also available in *Proc. of the 6th Structures in Complexity Theory Conference*, 172–179.

[59] (with O. Goldreich and S. Micali). Proofs that yield nothing but their validity, or All languages in NP have zero-knowledge proof systems. *J. Assoc. Comput. Mach.*, 38(3):691–729.

[60] (with P. Ragde). Linear-size constant-depth polylog-threshold circuits. *Inform. Process. Lett.*, 39(3):143–146.

[61] (with Y. Rabinovich). An Analysis of a Simple Genetic Algorithm. *Proc. of the 4th International Conference on Genetic Algorithms*, 215–221.

1992

[62] The complexity of graph connectivity. In *Mathematical foundations of computer science 1992 (Prague, 1992)*, volume 629 of *Lecture Notes in Comput. Sci.*, pages 112–132. Springer, Berlin.

[63] (with J. Gil and W. Steiger). Geometric medians. *Discrete Math.*, 108 (1-3):37–51.

[64] (with N. Nisan and E. Szemeredi). Undirected connectivity in $O(\log^{1.5} n)$ space. In *33rd annual symposium on Foundations of computer science (Pittsburgh, PA, 1992)*, pages 24–29. IEEE, Washington, DC.

[65] (with R. Raz). Monotone circuits for matching require linear depth. *J. Assoc. Comput. Mach.*, 39(3):736–744. Also available in *Proc. of the 22nd STOC*, 287–292.

[66] (with Y. Rabionvich and A. Sinclair). Quadratic dynamical systems. In *33rd annual symposium on Foundations of computer science (Pittsburgh, PA, 1992)*, pages 304–313. IEEE, Washington, DC. Also available in *Proc. of the 33rd FOCS conference*, 24–27.

1993

[67] The fusion method for lower bounds in circuit complexity. In *Combinatorics, Paul Erdős is eighty, Vol. 1*, Bolyai Soc. Math. Stud., pages 453–468. János Bolyai Math. Soc., Budapest.

[68] (with L. Babai, L. Fortnow and N. Nisan). BPP has subexponential time simulations unless EXPTIME has publishable proofs. *Comput. Complexity*, 3(4):307–318. Also available in *Proc. of the 6th Structures in Complexity Theory Conference*, 213–219.

[69] (with M. Karchmer). On span programs. In *Proceedings of the Eighth Annual Structure in Complexity Theory Conference (San Diego, CA, 1993)*, pages 102–111. IEEE Comput. Soc. Press, Los Alamitos, CA.

[70] (with J. Håstad). Composition of the universal relation. In *Advances in computational complexity theory (New Brunswick, NJ, 1990)*, volume 13 of *DIMACS Ser. Discrete Math. Theoret. Comput. Sci.*, pages 119–134. Amer. Math. Soc., Providence, RI.

[71] (with R. Heiman and I. Newman). On read-once threshold formulae and their randomized decision tree complexity. *Theor. Comput. Sci.*, 107(1):63–76.

[72] (with S. Hoory). Universal traversal sequences for expander graphs. *Inform. Process. Lett.*, 46(2):67–69.

[73] (with M. Karchmer). Characterizing non-deterministic circuit size. In *Proceedings of the 25th annual ACM symposium on theory of computing*. STOC '93. San Diego, CA, USA, May 16–18, 1993. New York, NY: Association for Computing Machinery (ACM). 532–540.

[74] (with M. Karchmer, N. Linial, I. Newman and M. Saks). Combinatorial characterization of read-once formulae. *Discrete Math.*, 114(1-3):275–282.

[75] (with M. Luby and B. Velickovic). Deterministic Approximate Counting of Depth-2 Circuits. *Proc. of the 2nd ISTCS (Israeli Symposium on Theoretical Computer Science)*, 18–24.

[76] (with N. Nisan). Rounds in communication complexity revisited. *SIAM J. Comput.*, 22(1):211–219. Also available in *Proceedings of the Annual ACM Symposium on Theory of Computing*, Part F130073, 419–429.

[77] (with R. Ostrovski). Nontrivial Zero-Knowledge implies One–Way Functions. *Proc. of the 2nd ISTCS*, 3–17.

[78] (with A. Orlitsky). Secrecy enhancement via public discussion. *Proceedings of the 1993 IEEE International Symposium on Information Theory*, 155.

[79] (with A. Razborov). $n^{\Omega(\log n)}$ lower bounds on the size of depth-3 threshold circuits with AND gates at the bottom. *Inform. Process. Lett.*, 45(6):303–307.

[80] (with A. Razborov and E. Szemerédi). Constructing small sets that are uniform in arithmetic progressions. *Combin. Probab. Comput.*, 2(4):513–518.

1994

[81] Amazing power of pairwise independence. *Conference Proceedings of the Annual ACM Symposium on Theory of Computing*, 645–647.

[82] Wonders of the digital envelope - a crash course in modern cryptography. *IFIP Transactions A: Computer Science and Technology*, (A-51), 235–238.

[83] $NL/poly \subseteq \oplus L/poly$. *Proceedings of the IEEE Annual Structure in Complexity Theory Conference*, 59–62.

[84] (with S. Ben-David, A. Borodin, R. Karp and G. Tardos). On the power of randomization in on-line algorithms. *Algorithmica*, 11(1):2–14. Also available in *Proc. of the 22nd STOC*, 379–388.

[85] (with M. Karchmer, I. Newman and M. Saks). Non-deterministic communication complexity with few witnesses. *J. Comput. System Sci.*, 49(2):247–257. *Proceedings of the Seventh Annual Structure in Complexity Theory Conference*, 275–281.

[86] (witb R. Impagliazzo and N, Nisan). Pseudorandomness for network algorithms. In *Proceedings of the 26th annual ACM symposium on theory of computing*, STOC '94, Montreal, Canada, May 23–25, 1994. New York, NY: Association for Computing Machinery (ACM). 356–364.

[87] (with R. Impagliazzo and R. Raz). Direct product theorem. *Proceedings of the IEEE Annual Structure in Complexity Theory Conference*, 88–96.

[88] (with N. Nisan). Hardness vs. randomness. *J. Comput. System Sci.*, 49(2):149–167. Also available in *Annual Symposium on Foundations of Computer Science (Proceedings)*, 2–11.

[89] (with N. Nisan). On rank vs. communication complexity. In *35th Annual Symposium on Foundations of Computer Science (Santa Fe, NM, 1994)*, pages 831–836. IEEE Comput. Soc. Press, Los Alamitos, CA.

1995

[90] Computational pseudo-randomness. *Proceedings ISTCS 1995 - 3rd Israel Symposium on the Theory of Computing and Systems*, 218–219.

[91] (with N. Alon, U. Feige and D. Zuckerman). Derandomized graph products. *Comput. Complexity*, 5(1):60–75.

[92] (with I. Damgård, O. Goldreich and T. Okamoto). Honest verifier vs dishonest verifier in public coin zero-knowledge proofs. *Lect. Notes Comput. Sci. 963*, 325–338.

[93] (with J. Friedman). On the second eigenvalue of hypergraphs. *Combinatorica*, 15(1):43–65.

[94] (with N. Nisan). On rank vs. communication complexity. *Combinatorica*, 15(4):557–565.

[95] (with M. Karchmer and R. Raz). Super-logarithmic depth lower bounds via the direct sum in communication complexity. *Comput. Complexity*,

5(3-4):191–204. Also available in *Proc. of the 6th Structures in Complexity Theory Conference*, 299–304.

[96] (with L. Lovász, M. Naor and I. Newman). Search problems in the decision tree model. *SIAM J. Discrete Math.*, 8(1):119–132. Also available in *Annual Symposium on Foundations of Computer Science (Proceedings)*, 576–585.

[97] (with I. Newman). Lower bounds on formula size of Boolean functions using hypergraph entropy. *SIAM J. Discrete Math.*, 8(4):536–542.

[98] (with N. Nisan). On the complexity of bilinear forms, dedicated to the memory of Jacques Morgenstern. In *Proceedings of the 27th annual ACM symposium on the theory of computing*, pages 723–732. ACM, New York.

1996

[99] (with H. Alt, L. Guibas, K. Mehlhorn and R. Karp). A method for obtaining randomized algorithms with small tail probabilities. *Algorithmica*, 16(4-5):543–547.

[100] (with R. Armoni, M. Saks and S. Zhou). Discrepancy sets and pseudorandom generators for combinatorial rectangles. In *37th Annual Symposium on Foundations of Computer Science (Burlington, VT, 1996)*, pages 412–421. IEEE Comput. Soc. Press, Los Alamitos, CA.

[101] (with L. Babai, A. Gál, J. Kollár, L. Rónyai and T. Szabó). Extremal bipartite graphs and superpolynomial lower bounds for monotone span programs. In *Proceedings of the Twenty-eighth Annual ACM Symposium on the Theory of Computing (Philadelphia, PA, 1996)*, pages 603–611. ACM, New York.

[102] (with A. Gál). Boolean complexity classes vs. their arithmetic analogs. In *Proceedings of the Seventh International Conference on Random Structures and Algorithms (Atlanta, GA, 1995)*, volume 9, pages 99–111.

[103] (with J. Gil and F. Meyer auf der Heide). The tree model for hashing: lower and upper bounds. *SIAM J. Comput.*, 25(5):936–955.

[104] (with N. Nisan). Lower bounds on arithmetic circuits via partial derivatives. *Comput. Complexity*, 6(3):217–234. Also available in *36th Annual Symposium on Foundations of Computer Science (Milwaukee, WI, 1995)*, pages 16–25. IEEE Comput. Soc. Press, Los Alamitos, CA.

[105] (with R. Sharan). A new NC algorithm for perfect matching in bipartite cubic graphs. In *Israel Symposium on Theory of Computing and Systems (Jerusalem, 1996)*, pages 202–207. IEEE Comput. Soc. Press, Los Alamitos, CA. Also available in *Proc. of ISTCS 96*, 56–65.

1997

[106] (with O. Goldreich). Tiny families of functions with random properties: a quality-size trade-off for hashing. In *Proceedings of the Workshop on Randomized Algorithms and Computation (Berkeley, CA, 1995)*, volume 11, pages 315–343. Also available in *Proceedings of the 26th annual ACM symposium on theory of computing*, STOC '94, Montreal, Canada, May 23–25, 1994. New York, NY: Association for Computing Machinery (ACM). 574–584.

1998

[107] (with H. Buhrman and R. Cleve). Quantum vs. classical communication and computation. In *STOC '98. Proceedings of the 30th annual ACM symposium on theory of computing (Dallas, TX, 1998)*, pages 63–68. ACM, New York.

[108] (with A. Condon, L. Hellerstein and S. Pottle). On the power of finite automata with both nondeterministic and probabilistic states. *SIAM J. Comput.*, 27(3):739–762. Also available in *Proceedings of the 26th annual ACM symposium on theory of computing*, STOC '94, Montreal, Canada, May 23–25, 1994. New York, NY: Association for Computing Machinery (ACM). 676–685.

[109] (with N. Linial and A. Samorodnitsky). A deteministic strongly polynomial algorithm for matrix scaling and approximate permanents. In *STOC '98. Proceedings of the 30th annual ACM symposium on theory of computing (Dallas, TX, 1998)*, pages 644–652. ACM, New York.

[110] (with P.B. Miltersen, N. Nisan and Shmuel Safra). On data structures and asymmetric communication complexity. *J. Comput. Syst. Sci.*, 57(1):37–49. Also available in *Proceedings of the 27th annual ACM symposium on the theory of computing*. STOC '95. Las Vegas, NV, USA, May 29 – June 1, 1995. New York, NY: ACM. 103–111.

1999

[111] De-randomizing BPP: the state of the art. *Proceedings of the Annual IEEE Conference on Computational Complexity*, 76–77.

[112] Probabilistic and deterministic approximations of the permanent. *Lecture Notes in Computer Science (including subseries Lecture Notes in Artificial Intelligence and Lecture Notes in Bioinformatics)*, 1671.

[113] (with R. Armoni, A. Ta-Shma and S. Zhou). SL $\subseteq$ L$^{4/3}$. In *STOC '97 (El Paso, TX)*, pages 230–239. ACM, New York.

[114] (with L. Babai and A. Gál). Superpolynomial lower bounds for monotone span programs. *Combinatorica*, 19(3):301–319.

[115] (with Z. Bar-Yossef and O. Goldreich) Deterministic amplification of space-bounded probabilistic algorithms. *Proceedings of the Annual IEEE Conference on Computational Complexity*, 188–198.

[116] (with E. Ben-Sasson). Short proofs are narrow—resolution made simple. In *Annual ACM Symposium on Theory of Computing (Atlanta, GA, 1999)*, pages 517–526. ACM, New York. Also available in *Proceedings of the Annual IEEE Conference on Computational Complexity*, 2.

[117] (with O. Goldreich). Improved derandomization of BPP using a hitting set generator. In *Randomization, approximation, and combinatorial optimization (Berkeley, CA, 1999)*, volume 1671 of *Lecture Notes in Comput. Sci.*, pages 131–137. Springer, Berlin.

[118] (with R. Impagliazzo). P = BPP if E requires exponential circuits: derandomizing the XOR lemma. In *STOC '97 (El Paso, TX)*, pages 220–229. ACM, New York.

[119] (with R. Impagliazzo and R. Shaltiel). Near-optimal conversion of hardness into pseudo-randomness. In *40th Annual Symposium on Foundations of Computer Science (New York, 1999)*, pages 181–190. IEEE Computer Soc., Los Alamitos, CA.

[120] (with I. Parnafes and R. Raz). Direct product results and the GCD problem, in old and new communication models. In *STOC '97 (El Paso, TX)*, pages 363–372. ACM, New York.

[121] (with Y. Rabinovich). Techniques for bounding the convergence rate of genetic algorithms. *Random Structures Algorithms*, 14(2):111–138.

[122] (with A. Razborov and A. Yao). Read-once branching programs, rectangular proofs of the pigeonhole principle and the transversal calculus. In *STOC '97 (El Paso, TX)*, pages 739–748. ACM, New York.

[123] (with A. Shpilka). Depth-3 arithmetic formulae over fields of characteristic zero. In *Fourteenth Annual IEEE Conference on Computational Complexity (Atlanta, GA, 1999)*, pages 87–96. IEEE Computer Soc., Los Alamitos, CA.

[124] (with D. Zuckerman). Expanders that beat the eigenvalue bound: explicit construction and applications. *Combinatorica*, 19(1):125–138. Also available in *Proceedings of the 25th annual ACM symposium on theory of computing*. STOC '93. San Diego, CA, USA, May 16–18, 1993. New York, NY: Association for Computing Machinery (ACM). 245–251.

2000

[125] (with M. Alekhnovich, E. Ben-Sasson and A.A. Razborov). Pseudorandom generators in propositional proof complexity. In *41st Annual Symposium on Foundations of Computer Science (Redondo Beach, CA, 2000)*, pages 43–53. IEEE Comput. Soc. Press, Los Alamitos, CA.

[126] (with M. Alekhnovich, E. Ben-Sasson, A.A. Razborov). Space complexity in propositional calculus. In *Proceedings of the Thirty-Second Annual ACM Symposium on Theory of Computing*, pages 358–367. ACM, New York.

[127] (with R. Armoni, A. Ta-Shma and S. Zhou). An $O(\log(n)^{4/3})$ space algorithm for (s,t) connectivity in undirected graphs. *J. ACM*, 47(2):294–311.

[128] (with O. Goldreich). On Pseudorandomness with respect to Deterministic Observers. *Proceedings of the satellite workshops of the 27th ICALP, Carleton Scientific (Proc in Informatics 8)*, 77–84.

[129] (with R. Impagliazzo and R. Shaltiel). Extractors and pseudo-random generators with optimal seed length. In *Proceedings of the Thirty-Second Annual ACM Symposium on Theory of Computing*, pages 1–10. ACM, New York.

[130] (with N. Linial, A. Samorodnitsky). A deterministic strongly polynomial algorithm for matrix scaling and approximate permanents. *Combinatorica*, 20(4):545–568.

[131] (with O. Reingold and R. Shaltiel). Extracting randomness via repeated condensing. In *41st Annual Symposium on Foundations of Computer Science (Redondo Beach, CA, 2000)*, pages 22–31. IEEE Comput. Soc. Press, Los Alamitos, CA.

[132] (with O. Reingold and S. Vadhan). Entropy waves, the zig-zag graph product, and new constant-degree expanders and extractors (extended abstract). In *41st Annual Symposium on Foundations of Computer Science (Redondo Beach, CA, 2000)*, pages 3–13. IEEE Comput. Soc. Press, Los Alamitos, CA.

2001

[133] (with N. Alon and A. Lubotzky). Semi-direct product in groups and zig-zag product in graphs: connections and applications (extended abstract). In *42nd IEEE Symposium on Foundations of Computer Science (Las Vegas, NV, 2001)*, pages 630–637. IEEE Computer Soc., Los Alamitos, CA.

[134] (with E. Ben-Sasson). Short proofs are narrow—resolution made simple. *J. ACM*, 48(2):149–169.

[135] (with O. Goldreich and S. Vadhan). On interactive proofs with a laconic prover. In *Automata, languages and programming*, volume 2076 of *Lecture Notes in Comput. Sci.*, pages 334–345. Springer, Berlin.

[136] (with R. Impagliazzo). Randomness vs time: derandomization under a uniform assumption. *J. Comput. Syst. Sci.*, 63(4):672–688. Also available in *Annual Symposium on Foundations of Computer Science - Proceedings*, 734–743.

[137] (with A. Shpilka). Depth-3 arithmetic circuits over fields of characteristic zero. *Comput. Complexity*, 10(1):1–27.

2002

[138] (with M. Alekhnovich, E. Ben-Sasson and A.A. Razborov). Space complexity in propositional calculus. *SIAM J. Comput.*, 31(4):1184–1211.

[139] (with M. Capalbo, O. Reingold and S. Vadhan). Randomness conductors and constant-degree lossless expanders. In *Proceedings of the Thirty-Fourth Annual ACM Symposium on Theory of Computing*, pages 659–668. ACM, New York. Also available in *Proceedings of the Annual IEEE Conference on Computational Complexity*, 15.

[140] (with E. Friedgut and J. Kahn). Computing graph properties of randomized subcube partitions. In *Randomization and approximation techniques in computer science*, volume 2483 of *Lecture Notes in Comput. Sci.*, 105–113. Springer, Berlin.

[141] (with O. Goldreich). Derandomization that is rarely wrong from short advice that is typically good. In *Randomization and approximation techniques in computer science*, volume 2483 of *Lecture Notes in Comput. Sci.*, pages 209–223. Springer, Berlin.

[142] (with O. Goldreich and S. Vadhan). On interactive proofs with a laconic prover. *Comput. Complexity*, 11(1-2):1–53. Also available in *Proc. of the 28th ICALP*, 334–345.

[143] (with R. Impagliazzo and V. Kabanets). In search of an easy witness: exponential time vs. probabilistic polynomial time. *J. Comput. Syst. Sci.*, 65(4): 672–694. Also available in *Conference on Computational Complexity*, 2–12, 2001.

[144] (with R. Meshulam). Expanders from symmetric codes. In *Proceedings of the Thirty-Fourth Annual ACM Symposium on Theory of Computing*, pages 669–677. ACM, New York. Also available in *Proceedings of the Annual IEEE Conference on Computational Complexity*, 16.

[145] (with A. Razborov and A. Yao). Read-once branching programs, rectangular proofs of the pigeonhole principle and the transversal calculus. *Combinatorica*, 22(4):555–574.

[146] (with O. Reingold and S. Vadhan). Entropy waves, the zig-zag graph product, and new constant-degree expanders. *Ann. of Math. (2)*, 155(1):157–187.

2003

[147] On the work of Madhu Sudan. *Notices Amer. Math. Soc.*, 50(1):45–50. Also available in *Mitt. Dtsch. Math.-Ver.*, (1):64–69, 2003.

[148] Zigzag products, expander constructions, connections, and applications. *Lecture Notes in Computer Science (including subseries Lecture Notes in Artificial Intelligence and Lecture Notes in Bioinformatics)*, 2914:443.

[149] (with A. Ambainis, L.J. Schulman, A. Ta-Shma and U. Vazirani). The quantum communication complexity of sampling. *SIAM J. Comput.*, 32(6):1570–1585. Also available in *Annual Symposium on Foundations of Computer Science - Proceedings*, 342–351.

[150] (with B. Barak and R. Shaltiel). Computational analogues of entropy. In *Approximation, randomization, and combinatorial optimization*, volume 2764 of *Lecture Notes in Comput. Sci.*, pages 200–215. Springer, Berlin.

[151] (with E. Ben-Sasson, M. Sudan and S. Vadhan). Randomness-efficient low degree tests and short PCPs via epsilon-biased sets. In *Proceedings of the Thirty-Fifth Annual ACM Symposium on Theory of Computing*, pages 612–621. ACM, New York.

[152] (with J. Håstad). Simple analysis of graph tests for linearity and PCP. *Random Structures Algorithms*, 22(2):139–160. Also available in *Proceedings of the Annual IEEE Conference on Computational Complexity*, 244–254.

[153] (with C.-J. Lu, O. Reingold and S. Vadhan). Extractors: optimal up to constant factors. In *Proceedings of the Thirty-Fifth Annual ACM Symposium on Theory of Computing*, pages 602–611. ACM, New York.

2004

[154] Average case complexity. In *Computational complexity theory*, volume 10 of *IAS/Park City Math. Ser.*, pages 89–99. Amer. Math. Soc., Providence, RI.

[155] Depth through breadth, or why should we attend talks in other areas? In *Proceedings of the 36th Annual ACM Symposium on Theory of Computing*, page 579. ACM, New York.

[156] (with M. Alekhnovich, E. Ben-Sasson and A.A. Razborov). Pseudorandom generators in propositional proof complexity. *SIAM J. Comput.*, 34(1):67–88.

[157] (with E. Ben-Sasson and R. Impagliazzo). Near optimal separation of tree-like and general resolution. *Combinatorica*, 24(4):585–603.

[158] (with R. Meshulam). Expanders in group algebras. *Combinatorica*, 24(4):659–680.

[159] (with E. Rozenman, A. Shalev). A new family of Cayley expanders (?). In *Proceedings of the 36th Annual ACM Symposium on Theory of Computing*, pages 445–454. ACM, New York.

[160] (with A. Shpilka). Derandomizing homomorphism testing in general groups. In *Proceedings of the 36th Annual ACM Symposium on Theory of Computing*, pages 427–435. ACM, New York.

[161] (wtih S. Rudich, editors). *Computational complexity theory*, volume 10 of *IAS/Park City Mathematics Series*. American Mathematical Society, Providence, RI; Institute for Advanced Study (IAS), Princeton, NJ, 2004.

2005

[162] (with B. Barak, G. Kindler, R. Shaltiel and B. Sudakov). Simulating independence: new constructions of condensers, Ramsey graphs, dispersers, and extractors. In *STOC'05: Proceedings of the 37th Annual ACM Symposium on Theory of Computing*, pages 1–10. ACM, New York.

[163] (with P. Beame, T. Pitassi and N. Segerlind). A direct sum theorem for corruption and the multiparty nof communication complexity of set disjointness. *Proceedings of the Annual IEEE Conference on Computational Complexity*, 52–66.

[164] (with M. Luby). Pairwise independence and derandomization. *Found. Trends Theor. Comput. Sci.*, 1(4):237–301.

[165] (with D. Xiao). A randomness-efficient sampler for matrix-valued functions and applications. *Proceedings - Annual IEEE Symposium on Foundations of Computer Science, FOCS*, 397–406.

2006

[166] Applications of the sum-product theorem in finite fields. *Proceedings of the Annual IEEE Conference on Computational Complexity*, 111.

[167] The power and weakness of randomness in computation. *Lect. Notes Comput. Sci.*, 3887:28–29.

[168] (with B. Barak and R. Impagliazzo). Extracting randomness using few independent sources. *SIAM J. Comput.*, 36(4):1095–1118. Also available in *Proceedings - Annual IEEE Symposium on Foundations of Computer Science*, FOCS, 384–393.

[169] (with B. Barak, A. Rao and R. Shaltiel). 2-source dispersers for sub-polynomial entropy and Ramsey graphs beating the Frankl–Wilson construction. In *STOC'06: Proceedings of the 38th Annual ACM Symposium on Theory of Computing*, pages 671–680. ACM, New York.

[170] (with P. Beame, T. Pitassi and N. Segerlind). A strong direct product theorem for corruption and the multiparty communication complexity of disjointness. *Comput. Complexity*, 15(4):391–432.

[171] (with M. Cheraghchi and A. Shokrollahi). Computational hardness and explicit constructions of error correcting codes. *44th Annual Allerton Conference on Communication, Control, and Computing 2006*, 1173–1179.

[172] (with I. Dinur and M. Sudan). Robust local testability of tensor products of LDPC codes. In *Approximation, randomization and combinatorial optimization*, volume 4110 of *Lecture Notes in Comput. Sci.*, pages 304–315. Springer, Berlin.

[173] (with S. Hoory and N. Linial). Expander graphs and their applications. *Bull. Amer. Math. Soc. (N.S.)*, 43(4):439–561.

[174] (with R. Impagliazzo and R. Shaltiel). Reducing the seed length in the Nisan-Wigderson generator. *Combinatorica*, 26(6):647–681.

[175] (with O. Reingold and R. Shaltiel). Extracting randomness via repeated condensing. *SIAM J. Comput.*, 35(5):1185–1209.

[176] (with E. Rozenman and A. Shalev). Iterative construction of Cayley expander graphs. *Theory Comput.*, 2:91–120.

[177] (with A. Shpilka). Derandomizing homomorphism testing in general groups. *SIAM J. Comput.*, 36(4):1215–1230.

2007

[178] P, NP and mathematics—a computational complexity perspective. In *International Congress of Mathematicians. Vol. I*, pages 665–712. Eur. Math. Soc., Zürich.

[179] (with J. Håstad). The randomized communication complexity of set disjointness. *Theory Comput.*, 3:211–219.

2008

[180] Brief history of the foundations of cryptography [2]. *Theory Comput.*, 4(7): 137–168. Also available in *Notices of the American Mathematical Society*, 6–7.

[181] (with S. Aaronson). Algebrization: a new barrier in complexity theory. In *STOC'08*, pages 731–740. ACM, New York. Also available in *Proceedings of the 40th annual ACM symposium on theory of computing*, STOC 2008. Victoria, Canada, May 17–20, 2008. New York, NY: Association for Computing Machinery (ACM). 731–740.

[182] (with V. Guruswami and J.R. Lee). Euclidean sections of ℓ_1^N with sublinear randomness and error-correction over the reals. In *Approximation, randomization and combinatorial optimization*, volume 5171 of *Lecture Notes in Comput. Sci.*, pages 444–454. Springer, Berlin.

[183] (with R. Impagliazzo, R. Jaiswal and V. Kabanets). Uniform direct product theorems: simplified, optimized, and derandomized. In *STOC'08*, pages 579–588. ACM, New York.

[184] (with G. Kalai). Neighborly embedded manifolds. *Discrete Comput. Geom.*, 40(3):319–324.

[185] (with G. Kindler, A. Rao and R. O'Donnell). Spherical cubes and rounding in high dimensions. *Proceedings - Annual IEEE Symposium on Foundations of Computer Science, FOCS*, 189–198.

[186] (with E. Viola). Norms, XOR lemmas, and lower bounds for polynomials and protocols. *Theory Comput.*, 4(7):137–168. Also available in *Proceedings of the Annual IEEE Conference on Computational Complexity*, 141–154.

[187] (with D. Xiao). Derandomizing the Ahlswede-Winter matrix-valued Chernoff bound using pessimistic estimators, and applications. *Theory Comput.*, 4(3):53–76. Also available in *Electronic Colloquium on Computational Complexity, Report TR06-105, ISSN 1433–8092, 13th Year, 105th Report.*

2009

[188] Randomness extractors—applications and constructions. In *Foundations of software technology and theoretical computer science—FSTTCS 2009*, volume 4 of *LIPIcs. Leibniz Int. Proc. Inform.*, pages 471–473. Schloss Dagstuhl. Leibniz-Zent. Inform., Wadern.

[189] The work of Leslie Valiant. In *STOC'09—Proceedings of the 2009 ACM International Symposium on Theory of Computing*, pages 1–2. ACM, New York.

[190] (with S. Arora and D. Steurer). Towards a study of low-complexity graphs. In *Automata, languages and programming. Part I*, volume 5555 of *Lecture Notes in Comput. Sci.*, pages 119–131. Springer, Berlin.

[191] (with A. Chattopadhyay). Linear systems over composite moduli. In *2009 50th Annual IEEE Symposium on Foundations of Computer Science—FOCS 2009*, pages 43–52. IEEE Computer Soc., Los Alamitos, CA.

[192] (with Z. Dvir and A. Gabizon). Extractors and rank extractors for polynomial sources. *Comput. Complexity*, 18(1):1–58. Also available in *Proceedings - Annual IEEE Symposium on Foundations of Computer Science, FOCS*, 52–62.

[193] (with V. Guruswami and J.R. Lee). Expander codes over reals, Euclidean sections, and compressed sensing. *2009 47th Annual Allerton Conference on Communication, Control, and Computing, Allerton 2009*, 1231–1234.

[194] (with R. Impagliazzo, R. Jaiswal and V. Kabanets). Uniform direct product theorems: simplified, optimized, and derandomized. *SIAM J. Comput.*, 39(4):1637–1665.

[195] (with R. Impagliazzo and V. Kabanets). New direct-product testers and 2-query PCPs. In *STOC'09—Proceedings of the 2009 ACM International Symposium on Theory of Computing*, pages 131–140. ACM, New York.

[196] (with E. Viola). One-way multiparty communication lower bound for pointer jumping with applications. *Combinatorica*, 29(6):719–743. *Proceedings - Annual IEEE Symposium on Foundations of Computer Science, FOCS*, 427–437.

2010

[197] Expanders. In *Princeton Companion to Mathematics.*, pages 196–199. Princeton University Press, Princeton.

[198] (with O. Goldreich). Computational complexity. In *Princeton Companion to Mathematics.*, pages 575–604. Princeton University Press, Princeton.

[199] (with B. Applebaum and B. Barak). Public-key cryptography from different assumptions. In *STOC'10—Proceedings of the 2010 ACM International Symposium on Theory of Computing*, pages 171–180. ACM, New York.

[200] (with B. Barak, G. Kindler, R. Shaltiel and B. Sudakov). Simulating independence: new constructions of condensers, Ramsey graphs, dispersers, and extractors. *J. ACM*, 57(4):Art. 20, 52 pp.

[201] (with X. Chen and N. Kayal). Partial derivatives in arithmetic complexity and beyond. *Found. Trends Theor. Comput. Sci.*, 6(1-2):1–138.

[202] (with Z. Dvir). Monotone expanders: constructions and applications. *Theory Comput.*, 6(12):291–308.

[203] (with P. Hrubeš and A. Yehudayoff). Non-commutative circuits and the sum-of-squares problem [extended abstract]. In *STOC'10—Proceedings of the 2010 ACM International Symposium on Theory of Computing*, pages 667–676. ACM, New York.

[204] (with P. Hrubeš and A. Yehudayoff). Relationless completeness and separations. In *25th Annual IEEE Conference on Computational Complexity—CCC 2010*, pages 280–290. IEEE Computer Soc., Los Alamitos, CA.

[205] (with T. Kaufman). Symmetric LDPC codes and local testing. *Lect. Notes Comput. Sci.* 6390:312–319. Also available in *Proc. of ICS 2010*, 406–421.

2011

[206] The Gödel phenomenon in mathematics: a modern view. In *Kurt Gödel and the foundations of mathematics*, pages 475–508. Cambridge Univ. Press, Cambridge.

[207] (with B. Barak, Z. Dvir and A. Yehudayoff). Rank bounds for design matrices with applications to combinatorial geometry and locally correctable codes [extended abstract]. In *STOC'11—Proceedings of the 43rd ACM Symposium on Theory of Computing*, pages 519–528. ACM, New York.

[208] (with Z. Dvir). Kakeya sets, new mergers, and old extractors. *SIAM J. Comput.*, 40(3):778–792. Also available in *Proceedings — Annual IEEE Symposium on Foundations of Computer Science, FOCS*, 625–633.

[209] (with O. Goldreich). On the circuit complexity of perfect hashing. In *Studies in complexity and cryptography*, volume 6650 of *Lecture Notes in Comput. Sci.*, pages 26–29. Springer, Heidelberg.

[210] O. Goldreich. *Studies in complexity and cryptography*, volume 6650 of *Lecture Notes in Computer Science*. Miscellanea on the interplay between randomness and computation, In collaboration with Lidor Avigad, Mihir Bellare, Zvika Brakerski, Shafi Goldwasser, Shai Halevi, T. Kaufman, L. Levin, N. Nisan, Dana Ron, M. Sudan, Luca Trevisan, S. Vadhan, Avi Wigderson and David Zuckerman. Edited by Goldreich. Springer, Heidelberg.

[211] (with O. Goldreich and N. Nisan). On Yao's XOR-lemma. In *Studies in complexity and cryptography*, volume 6650 of *Lecture Notes in Comput. Sci.*, pages 273–301. Springer, Heidelberg.

[212] (with O. Goldreich and S. Vadhan). Simplified derandomization of BPP using a hitting set generator. In *Studies in complexity and cryptography*, volume 6650 of *Lecture Notes in Comput. Sci.*, pages 59–67. Springer, Heidelberg.

[213] (with P. Hrubeš and A. Yehudayoff). Non-commutative circuits and the sum-of-squares problem. *J. Amer. Math. Soc.*, 24(3):871–898.

2012

[214] (with B. Barak, A. Rao and R. Shaltiel). 2-source dispersers for $n^{o(1)}$ entropy, and Ramsey graphs beating the Frankl-Wilson construction. *Ann. of Math. (2)*, 176(3):1483–1543.

[215] (with Z. Dvir, A. Rao and A. Yehudayoff). Restriction access. In *Proceedings of the 3rd Innovations in Theoretical Computer Science Conference*, pages 19–33. ACM, New York.

[216] (with R. Impagliazzo and V. Kabanets). New direct-product testers and 2-query PCPs. *SIAM J. Comput.*, 41(6):1722–1768.

[217] (with G. Kindler, A. Rao and R. O'Donnell). Spherical cubes: Optimal foams from computational hardness amplification. In *Communications of the ACM*, 55(10):90–97.

[218] (with A. Yehudayoff). Population recovery and partial identification. In *2012 IEEE 53rd Annual Symposium on Foundations of Computer Science—FOCS 2012*, pages 390–399. IEEE Computer Soc., Los Alamitos, CA.

2013

[219] (with B. Barak, Z. Dvir and A. Yehudayoff). Fractional Sylvester-Gallai theorems. *Proc. Natl. Acad. Sci. USA*, 110(48):19213–19219.

[220] (with P. Hrubeš and A. Yehudayoff). An asymptotic bound on the composition number of integer sums of squares formulas. *Canad. Math. Bull.*, 56(1):70–79.

[221] (with G.N. Rothblum and S. Vadhan). Interactive proofs of proximity: delegating computation in sublinear time. In *STOC'13—Proceedings of the 2013 ACM Symposium on Theory of Computing*, pages 793–802. ACM, New York.

2014

[222] Quantum computing since Democritus. *Notices Amer. Math. Soc.*, 61(10): 1218–1220.

[223] Randomness—a computational complexity perspective. In *XVIIth International Congress on Mathematical Physics*, pages 254–263. World Sci. Publ., Hackensack, NJ. Also available in *Computer Science — theory and applications*, volume 5010 of *Lecture Notes in Comput. Sci.*, pages 1–2. Springer, Berlin.

[224] (with A. Ai, Z. Dvir and S. Saraf). Sylvester-Gallai type theorems for approximate collinearity. *Forum Math. Sigma*, 2:Paper No. e3, 23 pp.

[225] (with Z. Dvir and S. Saraf). Breaking the quadratic barrier for 3-LCC's over the reals. In *Proceedings of the 46th annual ACM symposium on theory of computing*, pages 784–793. ACM, New York.

[226] (with Z. Dvir and S. Saraf). Improved rank bounds for design matrices and a new proof of Kelly's theorem. *Forum Math. Sigma*, 2:Paper No. e4, 24 pp.

[227] (with D. Gavinsky, O. Meir and O. Weinstein). Toward better formula lower bounds: an information complexity approach to the KRW composition conjecture. In *STOC'14—Proceedings of the 2014 ACM Symposium on Theory of Computing*, pages 213–222. ACM, New York.

[228] (with O. Goldreich). On derandomizing algorithms that err extremely rarely. In *STOC'14—Proceedings of the 2014 ACM Symposium on Theory of Computing*, pages 109–118. ACM, New York.

[229] (with P. Hrubeš). Non-commutative arithmetic circuits with division. In *ITCS'14—Proceedings of the 2014 Conference on Innovations in Theoretical Computer Science*, pages 49–65. ACM, New York.

2015

[230] (with E. Abbe and A. Shpilka). Reed-Muller codes for random erasures and errors. *IEEE Trans. Inform. Theory*, 61(10):5229–5252. Also available in *STOC'15—Proceedings of the 2015 ACM Symposium on Theory of Computing*, pages 297–306. ACM, New York.

[231] (with O. Goldreich and E. Viola). On randomness extraction in AC0. In *30th Conference on Computational Complexity*, volume 33 of *LIPIcs. Leibniz Int. Proc. Inform.*, pages 601–668. Schloss Dagstuhl. Leibniz-Zent. Inform., Wadern.

[232] (with P. Hrubeš). Non-commutative arithmetic circuits with division. *Theory Comput.*, 11(15):357–393.

[233] (with T. Ma). Sum-of-squares lower bounds for sparse PCA. In *Advances in Neural Information Processing Systems*, 2015-January, 1612–1620.

[234] (with R. Meka and A. Potechin). Sum-of-squares lower bounds for planted clique. In *STOC'15—Proceedings of the 2015 ACM Symposium on Theory of Computing*, pages 87–96. ACM, New York.

[235] (with S. Moran, A. Shpilka and A. Yehudayoff). Compressing and teaching for low VC-dimension. In *2015 IEEE 56th Annual Symposium on Foundations of Computer Science—FOCS 2015*, pages 40–51. IEEE Computer Soc., Los Alamitos, CA.

2016

[236] (with M.A. Forbes, A. Shpilka and I. Tzameret). Proof complexity lower bounds from algebraic circuit complexity. In *31st Conference on Computational Complexity*, volume 50 of *LIPIcs. Leibniz Int. Proc. Inform.*, pages Art. No. 32, 17. Schloss Dagstuhl. Leibniz-Zent. Inform., Wadern.

[237] (with A. Garg, L. Gurvits and R. Oliveira). A deterministic polynomial time algorithm for non-commutative rational identity testing. In *57th Annual IEEE*

Symposium on Foundations of Computer Science—FOCS 2016, pages 109–117. IEEE Computer Soc., Los Alamitos, CA.

[238] (with R. Gelles, B. Haeupler, G. Kol and N. Ron-Zewi). Towards optimal deterministic coding for interactive communication. In *Proceedings of the Twenty-Seventh Annual ACM-SIAM Symposium on Discrete Algorithms*, pages 1922–1936. ACM, New York.

[239] (with P. Gopalan, N. Nisan, R.A. Servedio and K. Talwar). Smooth Boolean functions are easy: efficient algorithms for low-sensitivity functions. In *ITCS'16—Proceedings of the 2016 ACM Conference on Innovations in Theoretical Computer Science*, pages 59–70. ACM, New York.

[240] (with P. Gopalan and R.A. Servedio). Degree and sensitivity: tails of two distributions. In *31st Conference on Computational Complexity*, volume 50 of *LIPIcs. Leibniz Int. Proc. Inform.*, pages Art. No. 13, 23. Schloss Dagstuhl. Leibniz-Zent. Inform., Wadern.

[241] (with T. Kaufman). Symmetric LDPC codes and local testing. *Combinatorica*, 36(1):91–120.

[242] (with A. Yehudayoff). Population recovery and partial identification. *Mach. Learn.*, 102(1):29–56.

2017

[243] Low-depth arithmetic circuits In *Communications of the ACM*, 60(6), 91–92.

[244] (with Z. Allen-Zhu, Y. Li and R. Oliveira). Much faster algorithms for matrix scaling. In *58th Annual IEEE Symposium on Foundations of Computer Science—FOCS 2017*, pages 890–901. IEEE Computer Soc., Los Alamitos, CA.

[245] (with Z. Dvir and S. Saraf). Superquadratic lower bound for 3-query locally correctable codes over the reals. *Theory Comput.*, 13(11), 36 pp.

[246] (with A. Garg, L. Gurvits and R. Oliveira). Algorithmic and optimization aspects of Brascamp-Lieb inequalities, via operator scaling. In *STOC'17—Proceedings of the 49th Annual ACM SIGACT Symposium on Theory of Computing*, pages 397–409. ACM, New York.

[247] (with D. Gavinsky, O. Meir and O. Weinstein). Toward better formula lower bounds: the composition of a function and a universal relation. *SIAM J. Comput.*, 46(1):114–131.

[248] (with S. Moran, A. Shpilka and A. Yehudayoff). Teaching and compressing for low VC-dimension. In *A journey through discrete mathematics*, pages 633–656. Springer, Cham.

2018

[249] (with Z. Allen-Zhu, A. Garg, Y. Li and R. Oliveira). Operator scaling via geodesically convex optimization, invariant theory and polynomial identity testing. In *STOC'18—Proceedings of the 50th Annual ACM SIGACT Symposium on Theory of Computing*, pages 172–181. ACM, New York.

[250] (with P. Bürgisser, C. Franks, A. Garg, R. Oliveira and M. Walter). Efficient algorithms for tensor scaling, quantum marginals, and moment polytopes. In

59th Annual IEEE Symposium on Foundations of Computer Science—FOCS 2018, pages 883–897. IEEE Computer Soc., Los Alamitos, CA.

[251] (with P. Bürgisser, A. Garg, R. Oliveira and M. Walter). Alternating minimization, scaling algorithms, and the null-cone problem from invariant theory. In *9th Innovations in Theoretical Computer Science*, volume 94 of *LIPIcs. Leibniz Int. Proc. Inform.*, Art. No. 24, 20 pp. Schloss Dagstuhl. Leibniz-Zent. Inform., Wadern.

[252] (with K. Efremenko, A. Garg and R. Oliveira). Barriers for rank methods in arithmetic complexity. In *9th Innovations in Theoretical Computer Science*, volume 94 of *LIPIcs. Leibniz Int. Proc. Inform.*, pages Art. No. 1, 19. Schloss Dagstuhl. Leibniz-Zent. Inform., Wadern.

[253] (with A. Garg, L. Gurvits and R. Oliveira). Algorithmic and optimization aspects of Brascamp-Lieb inequalities, via operator scaling. *Geom. Funct. Anal.*, 28(1):100–145.

[254] (with R. Gelles, B. Haeupler, G. Kol and N. Ron-Zewi). Explicit capacity approaching coding for interactive communication. *IEEE Trans. Inform. Theory*, 64(10):6546–6560.

[255] (with E. Viola). Local expanders. *Comput. Complexity*, 27(2):225–244.

[256] On the nature of the Theory of Computation (ToC). *Electron. Colloquium Comput. Complex*, Report No. 72. Also available as Chapter 20 in 257.

2019

[257] *Mathematics and Computation. A Theory Revolutionizing Technology and Science.* Princeton University Press, Princeton, NJ.

[258] (with P. Bürgisser, C. Franks, A. Garg, and M. Walter). Towards a theory of Non-Commutative Optimization: Geodesic 1st and 2nd Order Methods for Moment Maps and Polytopes. *Proceedings - Annual IEEE Symposium on Foundations of Computer Science, FOCS*, 2019-November, 845–861.

[259] (with Z. Dvir, S. Gopi and Y. Gu). Spanoids—an abstraction of spanning structures, and a barrier for LCCs. In *10th Innovations in Theoretical Computer Science*, volume 124 of *LIPIcs. Leibniz Int. Proc. Inform.*, pages Art. No. 32, 20. Schloss Dagstuhl. Leibniz-Zent. Inform., Wadern.

[260] (with A. Garg, V. Makam and R. Oliveira). More barriers for rank methods, via a 'numeric to symbolic' transfer. *Proceedings - Annual IEEE Symposium on Foundations of Computer Science, FOCS*, 2019-November, 824–844.

[261] (with O. Meir). Prediction from partial information and hindsight, with application to circuit lower bounds. *Comput. Complexity*, 28(2):145–183.

[262] (with O. E. Raz). Subspace arrangements, graph rigidity and derandomization through submodular optimization. *Bolyai Soc. Math. Stud.*, 28:377–415.

2020

[263] (with Z. Dvir, S. Gopi and Y. Gu). Spanoids—an abstraction of spanning structures, and a barrier for LCCs. *SIAM J. Comput.*, 49(3):465–496.

[264] (with A. Garg, L. Gurvits and R. Oliveira). Operator scaling: theory and applications. *Found. Comput. Math.*, 20(2):223–290.

[265] (with A. Garg, C. Ikenmeyer, V. Makam and M. Walter). Search problems in algebraic complexity, GCT, and hardness of generators for invariant rings. *Leibniz International Proceedings in Informatics*, 169, 12.

[266] (with O. Goldreich). On the size of depth-three boolean circuits for computing multilinear functions. *Lecture Notes in Computer Science (including subseries Lecture Notes in Artificial Intelligence and Lecture Notes in Bioinformatics)*, 41–86.

[267] (with O. Goldreich). Non-adaptive vs Adaptive Queries in the Dense Graph Testing Model. *Electron. Colloquium Comput. Complex*, Report No. 160.

[268] (with O. Goldreich). Constructing Large Families of Pairwise Far Permutations: Good Permutation Codes Based on the Shuffle-Exchange Network. *Electron. Colloquium Comput. Complex*, Report No. 192.

2021

[269] (with N. Fleming, M. Göös, R. Impagliazzo, and L.Y. Tan). On the power and limitations of branch and cut. *Leibniz International Proceedings in Informatics, LIPIcs*, 200, 6.

[270] (with O. Goldreich). Robustly self-ordered graphs: Constructions and applications to property testing. *Leibniz International Proceedings in Informatics, LIPIcs* 200, 12.

[271] (with P. Bürgisser, M. L. Dogan, V. Makam, and M. Walter). Polynomial time algorithms in invariant theory for torus actions. *Leibniz International Proceedings in Informatics, LIPIcs*, 200, 32.

[272] (with V. Makam). Singular tuples of matrices is not a null cone (and the symmetries of algebraic varieties). *Journal fur die Reine und Angewandte Mathematik*. Also available in *Proceedings - Annual IEEE Symposium on Foundations of Computer Science, FOCS*, 2020-November, 881–888.

[273] (with Y. Wigderson). The uncertainty principle: variations on a theme. *Bull. Amer. Math. Soc. (N.S.)*, 58(2):225–261.

[274] (with M.A. Forbes, A. Shpilka and I. Tzameret). Proof complexity lower bounds from algebraic circuit complexity. *Theory Comput.*, 17, Paper No. 10, 88 pp.

2022

[275] Non-commutative optimization — where algebra, analysis and computational complexity meet. *ISSAC '22 — Proceedings of the 2022 International Symposium on Symbolic and Algebraic Computation*, ACM, pages 13–19.

[276] (with O. Goldreich). Non-adaptive vs adaptive queries in the dense graph testing model. In *2021 IEEE 62nd Annual Symposium on Foundations of Computer Science — FOCS 2021*, IEEE Computer Soc., Los Alamitos, pages 269–275.

Curriculum Vitae for László Lovász

Born: March 9, 1948 in Budapest, Hungary

Degrees/education: C.Math.Sci., Hungarian Academy of Sciences, 1970
Dr.Rher.Nat., Eötvös Loránd University, 1971
Dr.Math.Sci., Hungarian Academy of Sciences, 1977

Positions: Research Associate, Eötvös Loránd University, 1971–1975
Docent, József Attila University, 1975–1978
Professor, Chair of Geometry, József Attila University, 1978–1982
Professor, Chair of Computer Science, Eötvös Loránd University, 1983–1993
Professor, Yale University, 1993–1999
Senior Researcher, Microsoft Research, 1999–2006
Director, Mathematical Institute, Eötvös Loránd University, 2006–2011

H. Holden, R. Piene (eds.), *The Abel Prize 2018-2022*, The Abel Prize,
https://doi.org/10.1007/978-3-031-33973-8_23

Professor, Eötvös Loránd University, 2006–2018
Professor Emeritus, Eötvös Loránd University, 2018–
Research Professor, Alfréd Rényi Institute of Mathematics, 2020–

Visiting positions: Vanderbilt University, 1972–1973
University of Waterloo, 1978–1979
University of Bonn, 1984–1985
University of Chicago, 1985
Cornell University, 1985
MSRI, Berkeley, 1986
Princeton University, 1987, 1989, 1990–1991 and 1992–1993
Institute for Advanced Study, 2011–2012
ETH, 2016

Memberships: Hungarian Academy of Sciences, 1979
European Academy of Sciences, Arts and Humanities, 1981
Academia Europaea, 1991
Rheinland-Westphälische Akademie der Wissenschaften, 1993
Deutsche Akademie der Naturforscher Leopoldina, 2002
Russian Academy of Sciences, 2006
Royal Netherlands Academy of Arts and Sciences, 2006
Royal Swedish Academy of Sciences, 2007
London Mathematical Society, Honorary member, 2009
National Academy of Sciences, 2012
American Mathematical Society, Fellow, 2012
Norwegian Academy of Science and Letters, 2021

Awards and prizes: Géza Prize, 1970
Pólya Prize, 1979
Best Information Theory Paper Award, IEEE, 1981
Fulkerson Prize, 1982
Speaker at the International Congress of Mathematicians, 1983, 1990 (plenary)
State Prize, Hungary, 1985
Szele Medal, 1992
Brouwer Medal, 1993
National Order of Merit of Hungary, 1998
Bolzano Medal, 1998
Knuth Prize, 1999
Wolf Prize, 1999
Corvin Chain, Hungary, 2001
Gödel Prize, 2001
MAA Hedrick Lecturer, 2002
John von Neumann Medal, 2005
John von Neumann Theory Prize, 2006

Bolyai Prize, 2007
Széchenyi Grand Prize, 2008
Kyoto Prize in Basic Sciences, 2011
Fulkerson Prize, 2012
Hypatia Prize, 2019
Order of Saint Stephen of Hungary, 2020
Abel Prize, 2021

Honorary degrees: University of Waterloo, 1992
University of Szeged, 1999
Budapest University of Techonology, 2002
University of Calgary, 2006
Tel Aviv University, 2018
Charles University, 2020

Curriculum Vitae for Avi Wigderson

Born: September 9, 1956 in Haifa, Israel

Degrees/education: BSc, The Technion – Israel Institute of Techonology, 1980
MSE, Princeton University, 1981
MA, Princeton University, 1982
PhD, Princeton University, 1983

Positions: Visiting Assistant Professor, Berkeley, 1983–1984
Visiting Scientist, IBM Research, San Jose, 1984–1985
Fellow, MSRI, 1985–1986
Senior Lecturer, Hebrew University, 1986–1987
Associate Professor, Hebrew University, 1987–1992
Professor, Hebrew University, 1991–2003
Chairman, Computer Science Institute, Hebrew University, 1993–1995

H. Holden, R. Piene (eds.), *The Abel Prize 2018-2022*, The Abel Prize,
https://doi.org/10.1007/978-3-031-33973-8_24

Professor, Institute for Advanced Study, 1999–

Visiting positions: Princeton University, 1990–1992
Institute for Advanced Study, 1995–1996

Memberships: American Academy of Arts and Sciences, 2011
National Academy of Sciences, 2013
Association for Computing Machinery, Fellow, 2018
Norwegian Academy of Science and Letters, 2021

Awards and prizes: President's List of Excellence, The Technion – Israel Institute of Technology, 1977–1980
IBM Graduate Fellowship, Princeton University, 1982–1983
Alon Fellowship, 1986–1989
Bergman Fellowship, 1989
Speaker at the International Congress of Mathematicians, 1990, 1994 (plenary), 2006 (plenary), 2022 (virtual)
Nevanlinna Prize, 1994
Yoram Ben-Porat Presidential Prize for Outstanding Researcher, 1994
Gibbs Lecture, 2008
Conant Prize, 2008
Gödel Prize, 2009
Fields Institute Distinguished Lecture Series, 2010
Rothschild Visiting Fellow, Isaac Newton Institute, 2011
Knuth Prize, 2019
Abel Prize, 2021

Part V
2022 Dennis P. Sullivan

"for his groundbreaking contributions to topology in its broadest sense, and in particular its algebraic, geometric and dynamical aspects"

THE
ABEL
PRIZE

Citation

The Norwegian Academy of Science and Letters has decided to award the Abel Prize for 2022 to **Dennis Parnell Sullivan**, of Stony Brook University, USA, and the Graduate School and University Center of the City University of New York, USA

> *"for his groundbreaking contributions to topology in its broadest sense, and in particular its algebraic, geometric and dynamical aspects."*

Topology was born in the late 19th century, as a new, qualitative approach to geometry. In topology a circle and a square are the same, but the surface of the earth and that of a donut are different. Developing a precise language and quantitative tools for measuring the properties of objects that do not change when they are deformed has been invaluable throughout mathematics and beyond, with significant applications in fields ranging from physics to economics to data science.

Dennis Sullivan has repeatedly changed the landscape of topology by introducing new concepts, proving landmark theorems, answering old conjectures and formulating new problems that have driven the field forwards. He has moved from area to area, seemingly effortlessly, using algebraic, analytic and geometric ideas like a true virtuoso.

His early work was on the classification of manifolds — spaces which cannot be distinguished from Euclidean flat space in the small, but which globally are different (for example, the surface of a sphere is, in the small, roughly a plane). Building on the work of William Browder and Sergei Novikov, he developed an algebraic topological perspective on this problem and invented some brilliant techniques to solve the problems that arise. This included the ideas of "localisation of a space at a prime" and "completion of a space at a prime". These are ideas exported from pure algebra that provide a new language for expressing geometric phenomena, which have become tools for resolving multitudes of other problems. Nowadays it is commonplace to work at one prime at a time, using different methods for different primes.

Another of Sullivan's breakthroughs was the study of what is left when all the primes are ignored — known as *rational homotopy theory*. He and Daniel Quillen gave two different complete algebraic descriptions of what is left from a space in this setting. Sullivan's model is based on differential forms — a concept of multivariable calculus, enabling direct connection to geometry and analysis. This made a major part of algebraic topology suitable for calculation, and has proven revolutionary. The use of differential forms made it especially relevant to algebraic geometry in combination with Hodge theory, as is shown in Sullivan's work with Pierre Deligne, Phillip Griffiths and John Morgan.

To understand smooth manifolds, the completions were necessary, and one of the high points of his work here was his proof of the Adams conjecture, independently of Quillen. Sullivan has also drawn attention to the idea of a *homotopy fixed set*, formulating a central conjecture in homotopy and introducing a widely used tool. The original "Sullivan conjecture" was solved many years later by Haynes Miller.

Sullivan went on to tackle a host of topological, dynamical and analytic problems, always with the idea of a geometric structure on a space playing a central role.

He showed that the topological structure of a manifold of dimension five or more can always be promoted to a *Lipschitz structure*, allowing analytic methods to be brought to bear. His argument uses arithmetic groups to replace Kirby's torus with a hyperbolic manifold immersed in Euclidean space. With Simon Donaldson, he proved such structures need not exist in dimension four.

In dynamics, Sullivan introduced a dictionary between Kleinian groups and iterated rational maps, pivoting on the theory of measurable complex structures. He proved that rational maps have no wandering domains, solving a 60-year-old conjecture by Fatou, and brilliantly drawing a parallel with Ahlfors' finiteness theorem. He went on to use similar methods to provide a conceptual proof of Feigenbaum's universality for cascades of period doublings, recasting these results as the uniqueness of a smooth structure on a strange attractor. Sullivan's dictionary, his rigidity theorem for Kleinian groups, and his a priori bounds for renormalisation are now fundamental principles in conformal dynamics.

In a subsequent return to the development of algebraic structures of manifolds, with Moira Chas, he astonished the field by finding a new invariant of manifolds. With its links to topological field theory, "string topology" has grown quickly into a field of its own.

Dennis Sullivan's insistent probing for fundamental understanding, and his capacity to see analogues between diverse areas of mathematics and build bridges between them, has forever changed the field.

Encounters with Geometry — an Autobiography of Concepts

Dennis Sullivan

When we first learned long division at age eight I had fun working out examples where the problem was put at the top of the page and the long division algorithm computation ran diagonally down the page to finish exactly in the tiny vicinity of the lower right hand corner. I wanted to make the biggest problem that fit on one page. This was a curious metaphor of one's serious practical limitations today in simulating 3D fluid motion on computers.

The family photo below taken in 1950 shows me (age 9) with my brother Michael Sullivan (1942–1957) and my single parent mother Rita Sullivan. My mother always said if asked about schooling that she had "graduated from the school of hard knocks", meaning at most high school. However she made a serious success in advertising, becoming a top account executive in the boom TV/advertising industry of the 50s and 60s. My younger brother died of a brain tumor after a 30 month struggle, 4 days after a fun 15th birthday party. This seems to have had the effect that I later needed to believe in something like mathematics and to never give up.

When I asked Sister Benedict in the tenth grade at Marian High School after her geometry demonstration, "why do those two ninety degree rotations around two perpendicular axes in space not commute?" She answered "That's the way God made it." I wanted to know more about this structure.

When the first year physics Professor at Rice University made some perturbation calculations regarding the stability or not of rotations about symmetry axes of solid bodies, we listened studiously. These calculations showed stability about the smallest and largest angular momentum axes and showed instability about the intermediate angular momentum axis. But when he took up a book with a rubber band wrapped tightly around the intermediate axis and tossed it many times into the air showing perfectly the results of the math calculations, one was deeply impressed by the connection possible between abstract computations and physical reality.

Dennis Sullivan
Math CUNY Grad Center & Prof Math, SUNY, Stony Brook
e-mail: sullivan0212@gmail.com

H. Holden, R. Piene (eds.), *The Abel Prize 2018-2022*, The Abel Prize,
https://doi.org/10.1007/978-3-031-33973-8_25

Fig. 1: My brother Michael (1942–1957) with a broken arm, my mother Rita (1922–2014), and me. The photo is from 1950. (Credit: Private)

When as a sophomore chemical engineering student at Rice taking the required second year calculus course (real and complex, with definitions and proofs), the Professor one day drew a kidney shaped domain in the plane and explained, without proof, the domain can be moved and distorted by a z-differentiable $f(z)$ to the round domain, thus at each point rotating and possibly distorting around that point with a scaling and all of this with an essentially unique $f(z)$, I radically changed my view of mathematics to one of greater admiration and deeper respect going beyond that earned by the interesting calculations and formulae experienced up to that point. This Riemann mapping statement was unexpected, general and deep. It was a different kind of understanding.

On another day the Professor explained, without proof, that any harmonic function on any such domain, like the real part or imaginary part of that $f(z)$, could be calculated at any point in the domain by averaging the values at the boundary of that harmonic function with the probability density of a random path, starting at the point, landing at each boundary point of the domain. I thought "One can actually define in mathematics the notion of a random path and then discuss hitting probabilities? Wow."

In the third year at Rice University, the physical chemistry Professor connected thermodynamics with the statistics of zillions of particles interacting incredibly leaving only two or three numerical quantities to characterize the final state. I was

Fig. 2: Coming from the library at IHES, late 70s. (Credit: IHES)

torn between physical chemistry and mathematics. Mathematics, especially algebra, was difficult for me to do or to understand. Whereas physical chemistry was intuitive, though somewhat ill defined.

In the final fourth year at Rice when the Professor gave a homework problem to the advanced topology math class filled with math majors and PhD students at Rice, "Show that any continuous function from the long line to the real numbers is eventually constant," no one including me could do it before the Friday due date. It was a strange problem and it was an intriguing problem. I continued working on it for two weeks, an incredibly long time in the demanding atmosphere at Rice University in those days. I finally understood enough about the long line to see why a continuous function was eventually constant. This was a first hint to me personally about the nature of studying mathematics. It takes time to understand, it's not really speed and cleverness in contests that count but rather enough diligence over time motivated by being interested that really counts.

Fig. 3: Einstein chair seminar at CUNY. (Credit: Bora Ferlengez)

This lesson was enhanced during the three hour final exam of that same year long topology course which had ended with two weeks of formal algebraic topology. We all assumed this would not really appear much on the exam. We were wrong. Algebraic topology was about half of the final exam centered around one question. The rest of the exam was easy and people started leaving during the first hour and I was finished also with what I understood, besides having memorized the definitions of the last two weeks of the course. After one hour I was alone in the room. Should I leave too, or why not give it a try given that experience with the "long line problem?" I had two hours. The question was "Show a bounded region in Euclidean space divided into generalized triangles has no homology in the top dimension." There were memorized paraphrases of the definitions "first look for cycles and then do something more to get the homology." This could be studied in the special case of two-dimensional space, i.e. the flat plane. Then, by recalling the memorized definition, a cycle is a linear combination of filled in oriented ordinary triangles so that all the boundary edges with their coefficients and orientations taken into account cancel giving a sum of zero. OK, what does this definition of cycle mean geometrically? After some time, one example of cancellation occurred, glue two triangles together along two respective edges and those two edges cancel in the total sum of the six oriented edges of the two triangles. Finally, sitting there much longer pondering this, a picture emerged that is still fresh in memory. One was trying to fit the triangles together so that no exposed edges are left. As one tried to fit the triangles together mentally, a moving image appeared spontaneously, this of triangles rising up out of the plane first to form a kind of bowl and then continuing to close up into a spherical shape. This was the picture of a cycle in dimension two. So, if the triangles were forced to stay in a bounded region of the flat plane there would always be exposed edges. Conclusion, there were no cycles and there was a fortiori no top homology of a bounded region in flat space of any dimension. And one had a picture of a cycle in any dimension, like a closed object that is the higher dimensional analogue of a closed surface.

Finally, there is one more memorable moment with a more significant interpretation in the middle of the third semester of math grad school at Princeton University. This, just before the oral qualifying exam allowing one to begin work on a thesis. The exam was on the set of notes by John Milnor on the theory of closed manifolds (like those abstract cycles in the previous story in any dimension but without any singularities) and now only considered up to cobounding a manifold of one higher dimension (like being connected by a homology of one higher dimension but again with no singularities). This celebrated work was due to René Thom (1954) and was beautifully exposed by Milnor using carefully presented definitions and proofs about transversality and homotopy groups with several theorems, about a dozen substantial proof steps, all in a few dozen pages. While walking to the exam after extensive studying — any specific question could be answered, from the undergraduate perspective it was a done deal — still maybe it was a good idea to have one more look. When that last look was made one of these mental images appeared at one step of the main proof. This was the transversality picture: an image of a long possibly knotted stretched out circular tubular slinky in three-space being compressed

Fig. 4: My home library. (Credit: Clara Sullivan)

down onto a two disk in some external two-sphere. This image was combined with the logical fact related to homotopy that the pictured partially defined mapping of space to the two-sphere extended to the complement outside the tube of the slinky by mapping the complement to the complement of the two-disk in the sphere — and this was actually possible in an essentially unique way, up to deformation, because the complement of that two disk in the sphere was contractible to a point. This slinky transversality picture plus that logical point about the extension was the central key point that was obvious once noticed but after that notice the entire set of notes just fell into simple evident steps ... one could forget everything but that key pair: the picture and the logical point, and easily reproduce logically the entire set of notes describing René Thom's theory. I thought to myself, "This is what it means to understand a piece of mathematics. I want to experience this feeling again."

Dennis Sullivan’s Work on Dynamics

Edson de Faria and Sebastian van Strien

Contents

E. de Faria
Instituto de Matematica e Estatistica, Universidade de Sao Paulo, e-mail: `edson@ime.usp.br`

S. van Strien
Department of Mathematics, Imperial College London, e-mail: `s.van-strien@imperial.ac.uk`

H. Holden, R. Piene (eds.), *The Abel Prize 2018-2022*, The Abel Prize,
https://doi.org/10.1007/978-3-031-33973-8_26

Before going into Dennis Sullivan's work on dynamics, we would like to share some of our reminiscences on the remarkable way in which he influenced a huge number of mathematicians, including the two of us. Both while at IHES and at CUNY, Dennis had an office which came with an anteroom. Our impression is that he would spend most of his time in this anteroom, talking about mathematics with whoever he had invited or whoever was around. Quite often while listening to somebody, he would end up giving a new spin or a new interpretation to what they had been saying. Similarly, he would explain what he was working on, trying out new ideas, and also often explaining results of others. Spending time with him was always an incredible experience.

In this spirit, Dennis explained much of his work on renormalisation to Welington de Melo. In turn, Welington would then try to explain what he had heard and learned to SvS. When he could not convince SvS of some argument, Welington would go back to Dennis and this process would repeat again, sometimes many times. This is how the final chapter of the book *One-Dimensional Dynamics* of SvS with Welington de Melo came into being, see [30]. This chapter contains a full exposition of Dennis' remarkable renormalisation theory, arguably the only place in which it was published with full details.

At the Graduate Center of CUNY, Dennis coordinated the *Einstein Chair Seminar*, more informally known as the Sullivan seminar, bringing in speakers from all over the world. The seminar ran once a week with talks in the afternoon, but the invited speakers usually came in the morning, and intense mathematical discussions ensued through lunch, and oftentimes even after the speaker's talk, following a short break for tea. During the talks, Dennis often asked questions not necessarily to know the answer himself, but because he knew that somebody else in the audience would find the answer helpful. In this way, Dennis took on the role of introducing two people to each other. His presence in the audience would usually make a talk much more accessible and interesting. His questions would often clarify connections that would have remained implicit otherwise.

When Dennis invented or learned about a new mathematical idea, he would push this idea to the limit. For him it was very important to understand what this idea would give, and equally important to find out what the limitations of this idea might be. Moreover, whenever possible, he liked to associate names to arguments such as the *dollar argument*, *smallest interval argument* or the *non-coiling argument* in order to synthesise a complex proof into its core ideas.

Often he mentioned that to understand a proof properly, you should treat it like a three-dimensional object. You should not only look at it from one side, but from all sides. So in this sense, in his view, a proof was about mathematical understanding rather than about 'killing' a theorem.

Indeed, Dennis' interest in a result might not necessarily be in the power of the statement per se, but in the tools that are used in the proof of this result. Once he understands the tools and ideas, then he probably can recover the statement of the results by himself.

So let us turn to the field of dynamical systems. Dennis Sullivan always had a keen interest in the field of dynamical systems, and already in the 1970s published several high impact papers in this area, many of them remarkably short. Let us first highlight a few of his papers on smooth dynamical systems, and then his groundbreaking papers on Kleinian groups and holomorphic dynamics.

1 Smooth dynamics

Although his early papers on dynamics treat a diverse range of problems, they all have an overarching theme: what smoothness (or other) structures are compatible with a particular dynamical setting.

1.1 From topology to dynamics

In the mid-seventies, Dennis started to become more and more interested in dynamics, transitioning from the pure study of structures on manifolds to the study of dynamical objects such as flows and, more generally, foliations on manifolds. One of his first striking results in this direction is the following.

Theorem 1.1 (Counterexample to the periodic orbit conjecture, see [100], [98]). *There exists a flow on a compact five-dimensional manifold all of whose orbits are periodic, and yet the lengths of such orbits are not bounded.*

This theorem[1] was announced in [100], and a more detailed argument given in [98]. Dennis' topological-geometric construction in the latter paper yields a flow on a smooth 5-dimensional manifold M which is Lipschitz, but he states that he sees no reason why the example could not be made smooth. In an addendum at the end of the paper, Dennis briefly explains an idea due to Kuiper that results in an example in which the flow is C^∞. He also explains a beautiful construction due to Thurston which yields a real-analytic flow with the desired properties on the manifold $M = \mathbb{R}^5/(\Gamma \times \mathbb{Z} \times \mathbb{Z})$, where Γ is the group

$$\Gamma = \left\{ \begin{pmatrix} 1 & x & y \\ 0 & 1 & z \\ 0 & 0 & 1 \end{pmatrix} : x, y, z \in \mathbb{Z} \right\}.$$

In other words, Γ is the so-called *discrete Heisenberg group.*

[1] This theorem is also discussed in McMullen's beautiful talk on Dennis' work at MSRI in the spring of 2022, see `https://vimeo.com/702914316?embedded=true&source=vimeo_logo&owner=106107493`

1.2 Rigidity in smooth dynamics

The following paper, joint with Shub, dates back to 1985.

Theorem 1.2 (Expanding maps of the circle, see [94]). *Two C^r, $r \geq 2$, expanding maps of the circle which are absolutely continuously conjugate are C^r conjugate.*

The proof of this result is short and starts by invoking well-known results to reduce the problem to the setting where both maps preserve the Lebesgue measure. Using the assumption that the maps are C^2 and expanding makes it possible to consider an iterate of the maps taking a small interval to big scale with bounded distortion. This then implies that the assumption that the conjugacy h is absolutely continuous gives that h is, in fact, Lipschitz. Going on from there, one obtains that h is C^r.

In some broader sense the main idea of this paper was the starting point for quite a lot of later research. Indeed, the pullback argument that Dennis introduced in the field of holomorphic dynamics is somewhat similar in spirit. Quite a few other papers followed on from this work. For example,

- if the multipliers of corresponding periodic points of two topologically conjugate unimodal interval are equal, then the conjugacy is smooth, see [75, 64] (there are corresponding results in higher dimensions);
- if a conjugacy between two interval maps is smooth at some point, then it is smooth everywhere, see [2];
- there are quite a few very interesting related results for group actions of circle maps, see for example [31];
- when studying the smoothness of a conjugacy between two maps defined on Cantor sets, Sullivan introduced the notion of a scaling function on a Cantor set, see [107] and also [89].

Another paper, joint with Norton, on rigidity in dynamics is concerned with Denjoy examples of C^1 torus diffeomorphisms $T^2 \to T^2$ which are isotopic to the identity. These are maps f which are topologically semi-conjugate to a minimal translation on the torus, i.e., $hf = Rh$, where h is a continuous map of T^2 onto itself, homotopic to the identity, such that the set of $x \in T^2$ for which the cardinality of $h^{-1}(x)$ is greater than 1 is nonempty and countable. Then the interior of any $h^{-1}(x)$, if nonempty, is a wandering domain for f.

Theorem 1.3 (Smoothness and wandering domains for torus maps, see [86]). *Let f be a torus map of Denjoy type, and let $\Gamma \neq T^2$ be its minimal set. Then (i) if f preserves a measurable, essentially bounded conformal structure on Γ, then the maps f^n, restricted to the prime end boundaries of the wandering domains, cannot be uniformly quasiconformal; and (ii) if f preserves a C^2 conformal structure on Γ, then f cannot be C^3.*

One consequence of this theorem is that one cannot have a C^3 Denjoy diffeomorphism so that the iterates of some disc are all disjoint. Of course this theorem is a partial higher-dimensional analogue of Denjoy's famous theorem showing that a C^2 diffeomorphism of the circle without periodic orbits cannot have wandering intervals, and therefore must be topologically conjugate to an irrational rotation. Again quite a few papers followed on from this work, for example:

- it is not possible to have C^1 toral diffeomorphism with wandering round discs, see [82];
- very recently it was shown that there exist smooth and even real analytic diffeomorphisms of Denjoy type on the torus with a wandering topological disc, see [82, 115]; see also [4].
- wandering topological discs were also established for smooth two-dimensional diffeomorphisms, see [56], and even for polynomial maps in higher dimensions [3, 12].

Another theorem Dennis proved, jointly with Gambaudo and Tresser, is the following (informally stated).

Theorem 1.4 (Smoothness and linking number of periodic orbits for diffeomorphism of the disc, see [41]). *Let f be a C^1 diffeomorphism of the disc with periodic orbits p_n, $n \geq 0$, so that for each $n \geq 0$ the periodic orbit p_{n+1} 'cycles as a satellite' around p_n. Then the average* linking number *between p_{n+1} and p_n must converge as $n \to \infty$.*

One of the inspirations for this paper was a question by Smale, who asked whether it was possible to construct a smooth diffeomorphism on the disc with infinitely many hyperbolic periodic saddles, but without periodic sinks or sources (or neutral points). In [10, 40] such examples were constructed in the C^1 respectively C^2 category. It was subsequently shown that one can construct smooth and even real analytic diffeomorphisms with these properties, see [42] and also [22, 67], namely with a Feigenbaum–Coullet–Tresser Cantor attractor. The constructions in those papers build on the renormalisation theory developed for interval maps. The braid type of the periodic orbits is quite different from those in [10, 40], and as far as we know it is not yet known whether there are smooth diffeomorphisms which are topologically conjugate to the ones constructed in those papers (in which it is conjectured that one cannot construct C^3 diffeomorphisms topologically conjugate to their examples).

What the theorems discussed in this section have in common is that they are about invariant structures, and that the full theory in this direction has not yet been completed. For this reason having the additional conformal structure was quite appealing to Dennis. Another reason to start working on holomorphic dynamics in the 1980s was of course that he saw a compelling analogue with the theory on Kleinian groups that he had been working on previously – see Section 2 below.

1.3 Further results

Another research direction we would like to explicitly mention is Dennis' work on currents, see [99, 92]. In his beautiful survey talk at the Abel Prize lectures at the University of Oslo on Dennis' work, Étienne Ghys singles out this work (this talk is available on YouTube).[2]

2 Dynamics and ergodic theory of Kleinian groups

Dennis's interest in the geometric, dynamical and ergodic properties of discrete groups of hyperbolic isometries dates back to the mid to late 1970s. His work in this area was motivated in part by Mostow's rigidity theorem from two decades earlier, and in part by the work of Ahlfors on Kleinian groups, especially his famous finiteness theorem from 1965. Dennis was also greatly influenced by Thurston's work on geometric structures over 3-manifolds. It was Lipman Bers who first told Dennis about the so-called *Ahlfors conjecture*, according to which the limit set of a Kleinian group acting in hyperbolic 3-space either has zero Lebesgue measure in the sphere at infinity, or else it is equal to the entire sphere. This is now a theorem, thanks to the work of several mathematicians – see for instance [72] and references therein.

Let us present a brief account of the contributions of Dennis to this beautiful subject. Before we do that, we need to recall a few preliminary notions. For general background on the geometry of discrete groups and hyperbolic geometry, especially in dimension 3, we recommend [8], [77] and [72]. For a systematic exposition of the work of Dennis (and Patterson) on the ergodic theory of discrete groups, see [85].

The Moebius group

Consider the one-point compactification $\widehat{\mathbb{R}^n} = \mathbb{R}^n \cup \{\infty\}$ of Euclidean n-space. The Moebius group in dimension n is the group $MG(\mathbb{R}^n)$ consisting of all transformations $T : \widehat{\mathbb{R}^n} \to \widehat{\mathbb{R}^n}$ which arise as all possible compositions of (linear) conformal transformations of the form $x \mapsto Ax + b$, where A is a scalar multiple of an orthogonal matrix and $b \in \mathbb{R}^n$, with the inversion $J : \widehat{\mathbb{R}^n} \to \widehat{\mathbb{R}^n}$ given by $J(x) = x/|x|^2$ for $x \neq 0$, $J(0) = \infty$ and $J(\infty) = 0$. The elements of $MG(\mathbb{R}^n)$ are called *Moebius transformations*.

[2] See `https://www.youtube.com/watch?v=reC5-XUeH_4`.

Hyperbolic space

Let us denote by $\mathbb{H}^n$ hyperbolic n-space, which we view as the open unit ball $B^n = \{x : |x| < 1\} \subset \mathbb{R}^n$ endowed with the *hyperbolic metric* (also called *Poincaré metric*) given by

$$ds = \frac{2|dx|}{1-|x|^2} .$$

The *ideal boundary* or *sphere at infinity* of hyperbolic n-space is by definition the sphere $S^{n-1} = \partial B^n$, endowed with the standard conformal structure inherited from $\mathbb{R}^n$. It is customary to denote the sphere at infinity by S_∞. The group $\mathrm{Isom}^+(\mathbb{H}^n)$ of orientation-preserving isometries of this metric consists precisely of all *Moebius transformations* that preserve the unit ball, i.e., those $T \in MG(\mathbb{R}^n)$ such that $T(B^n) = B^n$. Every $T \in \mathrm{Isom}^+(\mathbb{H}^n)$ acts on the sphere at infinity as a *conformal automorphism*. The elements of $\mathrm{Isom}^+(\mathbb{H}^n)$ are classified according to their action on S_∞ as follows. If $T \in \mathrm{Isom}^+(\mathbb{H}^n)$ has exactly one fixed point in S_∞, then T is said to be a *parabolic* transformation. If it has exactly two fixed points in S_∞, then it is called a *loxodromic* transformation. All other elements of $\mathrm{Isom}^+(\mathbb{H}^n)$ are said to be *elliptic*.

Kleinian groups

A (generalized) *Kleinian group* is a discrete subgroup $\Gamma \subset \mathrm{Isom}^+(\mathbb{H}^n)$. Discreteness means in particular that the orbit $\Gamma(x) = \{\gamma x : \gamma \in \Gamma\}$ of any point $x \in B^n$ can only accumulate on the sphere at infinity. The set $\Lambda(\Gamma) \subseteq S_\infty$ of all such accumulation points is the *limit set* of Γ (see figure 1). Its complement $\Omega(\Gamma) = S_\infty \setminus \Lambda(\Gamma)$ is the *domain of discontinuity* or *ordinary set* of Γ. Clearly, both $\Lambda(\Gamma)$ and $\Omega(\Gamma)$ are completely invariant under the action of Γ. When $n = 3$, the ordinary set $\Omega(\Gamma)$ is precisely the *domain of normality* of Γ, that is to say, the set of all points $z \in S_\infty \equiv \widehat{\mathbb{C}}$ having a neighborhood $V_z \subset \widehat{\mathbb{C}}$ such that $\{\gamma|_{V_z} : \gamma \in \Gamma\}$ is a normal family in the sense of Montel (thus, $\Omega(\Gamma)$ is the analogue of the Fatou set for a rational map, and the limit set $\Lambda(\Gamma)$ is the analogue of the Julia set – see §4). A Kleinian group is said to be *non-elementary* if its limit set consists of more than two points.

There are various ways under which the Γ-orbit of a point $x \in B^n$ can accumulate on a point of $\Lambda(\Gamma)$ in the sphere at infinity. The two most important are a *conical* approach and a *horospherical* approach. Let us be more precise.

Definition 2.1. Let $\Gamma \subset \mathrm{Isom}^+(\mathbb{H}^n)$ be a Kleinian group and let $\xi \in \Lambda(\Gamma)$.

(i) We say that ξ is a *conical limit point* of Γ if for each $x \in B^n$ there exists a sequence $\{\gamma_n\} \subset \Gamma$ such that the ratio

$$\frac{|\xi - \gamma_n(x)|}{1 - |\gamma_n(x)|}$$

remains bounded as $n \to \infty$.

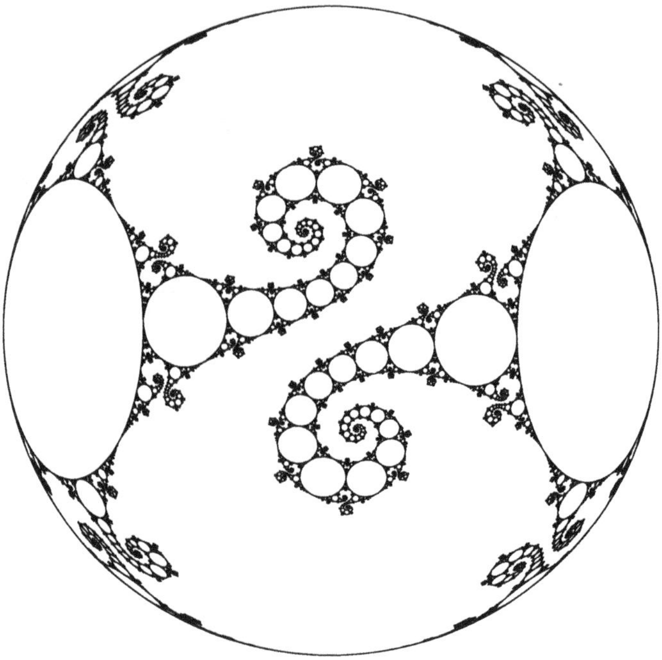

Fig. 1: The limit set of a Kleinian group sitting in the sphere at infinity is oftentimes a fractal object. [Credit: This picture was generated using C. McMullen's program "lim", available at `https://people.math.harvard.edu/~ctm/programs/`.]

(ii) We say that ξ is a *horospherical limit point* of Γ if for each $x \in B^n$ there exists a sequence $\{\gamma_n\} \subset \Gamma$ such that the ratio

$$\frac{|\xi - \gamma_n(x)|^2}{1 - |\gamma_n(x)|}$$

goes to zero as $n \to \infty$.

It is an exercise to show that if ξ is the fixed point of a loxodromic element of Γ, then ξ is a conical limit point of Γ. The set of all conical limit points of Γ is called the *conical limit set*, and it is denoted $\Lambda_c(\Gamma)$. The set of all horospherical limit points of Γ is called the *horospherical limit set*, and it is denoted $\Lambda_h(\Gamma)$. Clearly, these are both Γ-invariant.

An important class of Kleinian groups is the class consisting of so-called *convex co-compact groups*. Given a (non-elementary) Kleinian group Γ, let $\Lambda = \Lambda(\Gamma)$ be its limit set, and consider the convex hull $C(\Lambda)$ of Λ inside hyperbolic space. Then $C(\Lambda)$ is invariant under Γ, and we say that Γ is convex co-compact if the quotient $C(\Lambda)/\Gamma$ is compact. It is not difficult to see that if Γ is convex co-compact, then every element of Λ is a conical limit point – in other words, $\Lambda_c(\Gamma) = \Lambda(\Gamma)$ in this case.

Hyperbolic manifolds

The quotient space $M_\Gamma = \mathbb{H}^n/\Gamma$ of hyperbolic n-space by a Kleinian group Γ is what one calls an *orbifold* (after Thurston). Such a quotient is always a manifold when $n = 2, 3$, but it may fail to be one when $n > 3$. However, if Γ acts *freely and properly discontinuously* on $\mathbb{H}^n$, then M_Γ is indeed a manifold. Such manifolds are called *hyperbolic*. The natural quotient projection $\mathbb{H}^n \to M_\Gamma$ is a proper covering map, and therefore the hyperbolic metric of $\mathbb{H}^n$ descends to M_Γ. Thus, $\mathbb{H}^n$ is the universal covering space of M_Γ, and the fundamental group $\pi_1(M_\Gamma)$ is (isomorphic to) Γ. It is not difficult to see that if two Kleinian groups Γ_1, Γ_2 are *conjugate* subgroups of $\mathrm{Isom}^+(\mathbb{H}^n)$, i.e., if there exists a $\gamma \in \mathrm{Isom}^+(\mathbb{H}^n)$ such that $\Gamma_1 = \gamma^{-1}\Gamma_2\gamma$, then the corresponding orbifolds M_{Γ_i}, $i = 1, 2$, are *isometric*, and conversely.

Quasi-conformal homeomorphisms

A *quasiconformal homeomorphism* $h : S_\infty \to S_\infty$ is a homeomorphism which is differentiable Lebesgue almost-everywhere, and whose derivative at each point of differentiability maps round spheres onto ellipsoids whose ratios between the largest axis and the smallest axis yield a measurable function on the sphere that is essentially bounded. The essential norm of this function is called the quasi-conformal distortion of h, denoted K_h. It turns out that if $K_h = 1$ then h is in fact *conformal*. Every quasi-conformal homeomorphism determines a measurable field of ellipsoids, also known as a *measurable conformal structure* on S_∞. In dimension two, such measurable conformal structures can be integrated to recover h up to post-composition by a conformal map – a famous result known as the measurable Riemann mapping theorem – but no such theorem exists in higher dimensions.

2.1 Sullivan's rigidity theorem

It is a truly remarkable theorem due to G. Mostow [84] that complete finite-volume hyperbolic n-manifolds are determined up to isometry by their fundamental groups when $n \geq 3$. This is the famous *Mostow rigidity theorem*, which can be formally stated as follows.

Theorem 2.2 (Mostow Rigidity). *Let M and N be two complete, finite-volume hyperbolic n-manifolds with $n \geq 3$, and let $\theta : \pi_1(M) \to \pi_1(N)$ be an isomorphism between their fundamental groups. Then there exists an isometry $f : M \to N$ between both manifolds such that the induced isomorphism $f_* : \pi_1(M) \to \pi_1(N)$ agrees with θ.*

In fact, this theorem was proved by Mostow for *closed* manifolds, i.e., compact manifolds without boundary. It was then extended to finite volume manifolds by Marden [71] in dimension $n = 3$, and by Prasad [90] in all dimensions $n \geq 3$.

The way Mostow proved his theorem was by first showing that the manifolds M,N are *pseudo-isometric* in the following sense. A continuous, surjective map $\phi : M \to N$ between two hyperbolic manifolds is a *pseudo-isometry* if (a) it induces an isomorphism between the fundamental groups of both manifolds and moreover (b) there exist constants $K > 1$ and $\delta > 0$ such that

$$\frac{1}{K} \leq \frac{d_N(\phi(x),\phi(y))}{d_M(x,y)} \leq K$$

for each pair of points $x,y \in M$ such that $d_M(x,y) \geq \delta$. Here d_M, d_N denote the hyperbolic distances in M and N, respectively. Condition (b) says that a pseudo-isometry distorts hyperbolic distances between points by a bounded amount, provided these points are sufficiently far apart.

If one lifts a given pseudo-isometry between M and N to their universal covering space, one gets a pseudo-isometry of hyperbolic n-space. Mostow proved in [84] that every pseudo-isometry of $\mathbb{H}^n$ extends continuously to the sphere at infinity as a *quasiconformal homeomorphism* $h : S_\infty \to S_\infty$. The key step in the proof is to show by means of an ergodic argument that if Γ is a finite-volume Kleinian group then there is only one measurable conformal structure on S_∞ which is Γ-invariant. This implies that the quasiconformal homeomorphism h is actually conformal, which in turn means that the two associated Kleinian groups $\Gamma_M \simeq \pi_1(M), \Gamma_N \simeq \pi_1(N)$ are conjugate subgroups of $\mathrm{Isom}^+(\mathbb{H}^n)$, and therefore the hyperbolic manifolds M and N must be isometric.

Here is another way of stating Mostow's theorem. Following [102], we say that a hyperbolic manifold M is *Mostow-rigid* if any pseudo-isometry between M and another hyperbolic manifold N is homotopic to an isometry. Recast in this language, Mostow's theorem states that every complete, finite-volume hyperbolic M is Mostow-rigid.

Dennis proved in [102], in his own words, a *maximal extension* of Mostow's theorem. In order to state his theorem, let us introduce some notation. Given a hyperbolic manifold M, a point $p \in M$ and $r > 0$, let $V_M(p,r)$ denote the hyperbolic volume of the set $\{x \in M : d_M(p,x) < r\}$. For example, for the hyperbolic ball of radius r in hyperbolic n-space, we have

$$V_{\mathbb{H}^n}(0,r) = \omega_n \int_0^{\tanh(r/2)} \frac{2^n |x|^{n-1}\, \mathrm{d}|x|}{(1-|x|^2)^n} = \omega_n \int_0^r \sinh^{n-1} t \, \mathrm{d}t \sim \mathrm{const} \cdot \mathrm{e}^{(n-1)r} .$$

Here, ω_n denotes the euclidean area of $S_\infty = S^{n-1}$.

Lemma 2.3. *Let $M = \mathbb{H}^n/\Gamma$ be a hyperbolic n-manifold. Then the following assertions are equivalent.*

(i) *The ratio $V_M(p,r)/V_{\mathbb{H}^n}(0,r)$ goes to zero as $r \to \infty$.*
(ii) *The Kleinian group Γ acts conservatively on S_∞.*
(iii) *The horospherical limit set of Γ has full Lebesgue measure on S_∞.*

A fourth assertion equivalent to the above three is that the fundamental domain of Γ in $\mathbb{H}^n$ has zero Lebesgue measure on S_∞. We now have all the necessary elements to state the Sullivan rigidity theorem.

Theorem 2.4 (Sullivan Rigidity). *Let M be a complete hyperbolic n-manifold, and suppose that M satisfies any of the equivalent conditions of Lemma 2.3. Then M is Mostow rigid.*

Dennis deduces this theorem from the following result, also due to him.

Theorem 2.5. *Let $\Gamma \subset \mathrm{Isom}^+(\mathbb{H}^n)$ be a Kleinian group, and consider its action on the sphere at infinity. Suppose ν is a measurable conformal structure i.e., a measurable field of ellipsoids) which is Lebesgue almost everywhere invariant under Γ. Then ν agrees a.e. with the standard conformal structure of S_∞ on the limit set Λ_Γ.*

In the case of hyperbolic 3-manifolds, i.e., when $n = 3$, a major consequence of this theorem is obtained by combining it with the Ahlfors finiteness theorem. Recall that a Riemann surface X is said to be of *finite type* if X is obtained from a compact Riemann surface by removing from it a finite set of points.

Theorem 2.6 (Ahlfors Finiteness Theorem). *Let $\Gamma \subset \mathrm{PSL}(2,\mathbb{C})$ be a finitely generated Kleinian group. Then $\Omega(\Gamma)/\Gamma$ is a finite union of Riemann surfaces of finite type.*

In particular, the *Teichmüller space* $\mathrm{Teich}(\Omega(\Gamma)/\Gamma)$ is finite-dimensional. Hence we have the following result.

Corollary 2.7. *Let $\Gamma \subset \mathrm{PSL}(2,\mathbb{C})$ be a finitely generated Kleinian group. Then the space of quasi-conformal deformations of Γ is parametrized by* $\mathrm{Teich}(\Omega(\Gamma)/\Gamma)$, *and is therefore finite-dimensional.*

In particular, if Γ has a dense orbit on the sphere S^2, then the space of quasi-conformal deformations of Γ reduces to a point, i.e., Γ is quasi-conformally rigid.

2.2 Conformal densities and Patterson–Sullivan measures

Let Γ be a non-elementary Kleinian group acting on $\mathbb{H}^n \cup S_\infty$, and as before let $\Lambda(\Gamma) \subseteq S_\infty$ be its limit set. Also as before, let $\Lambda_c(\Gamma) \subseteq \Lambda(\Gamma)$ be its conical limit set. Generalizing work of Patterson for Fuchsian groups [88], Dennis was able to construct, in [101], an invariant measure for the geodesic flow on the unit tangent bundle of the hyperbolic manifold $\mathbb{H}^n/\Gamma$. This measure comes from a *conformal density* μ on the sphere at infinity, and the geodesic flow is either ergodic or dissipative, depending on whether μ assigns positive or zero measure to $\Lambda_c(\Gamma)$, respectively. We proceed to a brief description of the construction. Details can be found either in the original paper by Dennis, or in the book by Nicholls [85].

Conformal densities

Let us start by clarifying what is meant by conformal density. Let M be a smooth manifold, let $\mathscr{R}$ be a non-empty collection of Riemannian metrics on M, and let $\alpha > 0$. Following [101, p. 421], we define a *conformal density of dimension α*, or *α-conformal density*, on M (relative to $\mathscr{R}$) to be a function that assigns to each element $g \in \mathscr{R}$ a positive, finite Borel measure μ_g on M in such a way that, whenever g_1 and g_2 are in the same conformal class (i.e., whenever $g_1 = \varphi g_2$ for some positive function φ), then μ_{g_1} and μ_{g_2} are in the same measure class, and the Radon–Nikodym derivative $\mathrm{d}\mu_{g_1}/\mathrm{d}\mu_{g_2}$ satisfies

$$\frac{\mathrm{d}\mu_{g_1}}{\mathrm{d}\mu_{g_2}} = \left(\frac{g_1}{g_2}\right)^{\alpha} .$$

We are not interested in conformal densities in such vast generality, but rather in the following specific context. We take $M = S_\infty$, and let $\mathscr{R} = \{g_x : x \in B^n\}$, where g_0 is the standard (euclidian) Riemannian metric on the sphere S_∞, and for each $x \in B^n$ the Riemannian metric g_x is obtained by transporting g_0 via any hyperbolic isometry mapping 0 to x. These metrics are all conformally equivalent. Thus, in the present context, we can think of an α-conformal density on the sphere at infinity as an assignment $x \mapsto \mu_x$ from points on hyperbolic space to positive measures on S_∞, all in the same measure class. Shortening the notation to $\mu_x = \mu_{g_x}$, we deduce after a simple calculation that

$$\frac{d\mu_{x_1}}{d\mu_{x_2}}(\xi) = \left(\frac{P(x_1,\xi)}{P(x_2,\xi)}\right)^{\alpha}, \tag{1}$$

for each pair of points $x_1, x_2 \in B^n$ and all $\xi \in S^{n-1} \equiv S_\infty$, where

$$P(x,\xi) = \frac{1-|x|^2}{|x-\xi|^2}$$

is the well-known *Poisson kernel.*

Given a non-elementary Kleinian group Γ, we are interested in conformal densities of the type just described *that are entirely supported in the limit set of Γ and that are Γ-invariant.* More precisely, we want to know whether there exists an α-conformal density $\mu = \{\mu_x : x \in B^n\}$ such that

(1) For each x, the measure μ_x has support in the limit set $\Lambda(\Gamma)$.
(2) For each pair of points $x_1, x_2 \in B^n$, the measures μ_{x_1}, μ_{x_2} are mutually absolutely continuous, and the Radon–Nikodym derivative $d\mu_{x_1}/d\mu_{x_2}$ satisfies (1).
(3) For all $x \in B^n$ and each $\gamma \in \Gamma$, we have $\gamma_* \mu_x = \mu_{\gamma x}$.

Given a Γ-invariant conformal density of dimension α in the sense just described, each of its associated measures μ_x is an *α-conformal measure* in the sense that

$$\mu_x(\gamma(E)) = \int_E |\gamma_x'(\xi)|^{\alpha}\, d\mu_x(\xi)$$

for each Borel set $E \subset S_\infty$ and each $\gamma \in \Gamma$ (cf. the discussion on conformal measures for rational maps in §3.2). The question as to whether such *Patterson–Sullivan measures* exist is examined below.

Patterson–Sullivan measures: construction

The *Poincaré series* of the (non-elementary) Kleinian group Γ is defined as

$$g_s(x,y) = \sum_{\gamma \in \Gamma} e^{-sd(x,\gamma y)}, \tag{2}$$

where $x, y \in \mathbb{H}^n$, d is the hyperbolic metric on $\mathbb{H}^n$, and $s > 0$ is a real parameter. Whether the series (2) converges or not for a given value of s is independent of which points x, y one chooses. In order to state this more precisely, define the *critical exponent* of Γ to be the number

$$\delta(\Gamma) = \inf\{s > 0 : g_s(0,0) < \infty\},$$

which turns out to be strictly positive when Γ is non-elementary[3]. Then it is a fact that the series (2) converges for all $s > \delta(\Gamma)$, and diverges for all $0 < s < \delta(\Gamma)$.

[3] This was first proved by Beardon [7].

It is also not difficult to prove that $\delta(\Gamma) \leq n-1$. What happens when $s = \delta(\Gamma)$? Obviously, only one of two things:

(i) if $g_{\delta(\Gamma)}(0,0) < \infty$, we say that Γ is a group of *convergence type*;
(ii) if $g_{\delta(\Gamma)}(0,0) = \infty$, we say that Γ is a group of *divergence type*.

For the construction to follow, let us fix a point $y \in B^n$ once and for all (for example, we could take $y = 0$). For each $x \in B^n$ and each $s > \delta(\Gamma)$, one constructs a positive Borel measure $\mu_{x,s}$ on the closure of B^n as follows. The rough idea is to place a Dirac mass at each point of the Γ-orbit of y, with weights that depend on the hyperbolic distance between each such point and x in a suitable way. Let us be more precise.

When the group Γ is of *divergence* type, one simply defines[4]

$$\mu_{x,s} = \frac{1}{g_s(y,y)} \sum_{\gamma \in \Gamma} \mathrm{e}^{-sd(x,\gamma y)} \delta_{\gamma x}.$$

With this definition, one can consider the weak limits of such measures when $s \searrow \delta(\Gamma)$. Since $g_s(y,y) \to \infty$ as $s \to \delta(\Gamma)$, the point masses are swept off to the sphere at infinity, and any weak limit will be a measure supported on the sphere (actually on the limit set). The existence of limits is guaranteed by a classical result in real analysis (namely, Helly's theorem).

However, when the group is of *convergence* type, the above will not work, because $g_s(y,y)$ remains bounded as $s \to \delta(\Gamma)$, and whatever limiting measure we get will still have an atom at each point in the Γ-orbit of y (recall that the goal is to obtain measures supported on the limit set of Γ). To circumvent this problem, Dennis borrows an idea due to Patterson [88] (in the Fuchsian case, $n = 2$) and introduces a *mollifier*, called, not surprisingly, the *Patterson auxiliary function*. This is a continuous non-decreasing function $h : \mathbb{R}^+ \to \mathbb{R}^+$ having the following properties:

(1) For each $\varepsilon > 0$ there exists an $r_0 > 0$ such that $h(tr) \leq t^{\varepsilon} h(r)$ for all $r > r_0$ and all $t > 1$.
(2) The series

$$\sum_{\gamma \in \Gamma} \mathrm{e}^{-sd(x,\gamma y)} h(\mathrm{e}^{d(x,\gamma y)})$$

converges for $s > \delta(\Gamma)$ and diverges for $s \leq \delta(\Gamma)$.

Using this function, one defines the *modified Poincaré series*

$$g_s^*(x,y) = \sum_{\gamma \in \Gamma} \mathrm{e}^{-sd(x,\gamma y)} h(\mathrm{e}^{d(x,\gamma y)})$$

This now diverges when $s \searrow \delta(\Gamma)$. Thus, for each $x \in B^n$ and each $s > \delta(\Gamma)$ we may now consider the positive Borel measure $\mu_{x,s}$ defined by

$$\mu_{x,s} = \frac{1}{g_s^*(y,y)} \sum_{\gamma \in \Gamma} \mathrm{e}^{-sd(x,\gamma y)} h(\mathrm{e}^{d(x,\gamma y)}) \delta_{\gamma x}.$$

---

[4] We denote by δ_z the Dirac probability measure concentrated at $z \in B^n$.

As before, the resulting weak limits as $s \searrow \delta(\Gamma)$ are positive Borel measures on the sphere at infinity, and their supports are contained in the limit set $\Lambda(\Gamma)$.

In either case, we have for each $x \in B^n$ a non-empty closed subset $M_x(\Gamma) \subset \mathscr{M}^+(S_\infty)$ of the space of all positive Borel measures on the sphere at infinity (endowed with the topology of weak convergence of measures). It is possible to prove that the $M_x(\Gamma)$'s are all homeomorphic (see for instance [85, Th. 3.4.1]).

We are now ready to summarize some of the main results obtained by Dennis in [101]. That paper is very rich, and we can hardly do any justice to it in such a short exposition.

The first theorem generalizes results obtained by Patterson and Bowen in the Fuchsian case.

Theorem 2.8. (Patterson–Sullivan Measures and Hausdorff Dimension), see [101]) *Let $\Gamma \subset \mathrm{Isom}^+(\mathbb{H}^n)$ be a non-elementary Kleinian group, and let $\delta = \delta(\Gamma)$ be its critical exponent.*

(i) *There exists a δ-conformal density $\mu = \{\mu_x : x \in B^n\}$ on the sphere at infinity which is Γ-invariant and satisfies $\mu_x \in M_x(\Gamma)$ for each x.*
(ii) *We have $\dim_H(\Lambda_c(\Gamma)) \leq \delta$, i.e., the conical limit set of Γ has Hausdorff dimension less than or equal to its critical exponent.*
(iii) *If Γ is convex co-compact, then the δ-conformal density in (i) is unique up to a scalar multiple, and for each euclidian ball $B(\xi, r)$ centered at a point $\xi \in \Lambda_c(\Gamma)$, and each $x \in B^n$, we have $\mu_x(B(\xi,r) \cap \Lambda_c(\Gamma)) \asymp r^\delta$. In particular, $\dim_H(\Lambda_c(\Gamma)) = \delta$, i.e., the Hausdorff dimension of the conical limit set is* equal *to δ in this case.*

The last item in the above statement is a very elegant result which generalizes an equally elegant result for Fuchsian groups due to Bowen [9].

Another striking result obtained by Dennis in [101] states that the *total-mass function* of a Γ-invariant conformal density in dimension $\delta(\Gamma)$ is an eigenfunction of the *hyperbolic Laplacian*. Let us state this result a bit more precisely, explaining the meaning of these terms. The hyperbolic Laplacian in $\mathbb{H}^n \equiv B^n$ (written in generalized polar coordinates) is the second-order differential operator

$$\Delta_h = \frac{(1-r^2)^2}{4}\left[\Delta + \frac{2(n-2)r}{1-r^2}\frac{\partial}{\partial r}\right],$$

where Δ is the standard (euclidian) Laplacian – see for instance [1, p. 56]. If $\mu = \{\mu_x : x \in B^n\}$ is a Γ-invariant conformal density in dimension $\delta(\Gamma)$, its *total-mass function* is the function $\varphi : \mathbb{H}^n \to \mathbb{R}$ given by

$$\varphi(x) = \int_{\partial B^n} \mathrm{d}\mu_x(\xi) = \int_{S_\infty} \mathrm{d}\mu_x(\xi).$$

Theorem 2.9. *The total-mass function φ is an eigenfunction of the hyperbolic Laplacian, i.e., it satisfies*

$$\Delta_h \varphi = \lambda(\Gamma)\varphi,$$

with eigenvalue $\lambda(\Gamma) = \delta(\Gamma)(1 + \delta(\Gamma) - n)$.

This statement is perhaps made more plausible if one takes into account that, for each α, we have

$$\Delta_h (P(x,\xi)^\alpha) = \alpha(\alpha - n + 1)P(x,\xi)^\alpha,$$

a fact that can easily be checked by direct calculation.

Finally, as Dennis explains in [101], a Γ-invariant conformal density μ gives rise to a measure m on the unit tangent bundle of the quotient hyperbolic manifold $\mathbb{H}^n/\Gamma$ which is *invariant under the geodesic flow*. In addition, the normalized probability measures $\varphi(x)^{-1}\mu_x$ can be used to generate a Markovian stochastic process on $\mathbb{H}^n/\Gamma$ akin to Brownian motion. The beautiful synthesis obtained by Dennis in [101] relates the recurrent properties of this Markovian process with the ergodic properties of the geodesic flow on the quotient manifold, and can be informally stated as follows.

Theorem 2.10 (Ergodic Measures for the Geodesic Flow, see [101]). *Let* $\Gamma \subset \mathrm{Isom}^+(\mathbb{H}^n)$ *be a non-elementary Kleinian group, and let* $\delta = \delta(\Gamma)$ *be its critical exponent. Also, let* μ *be a* Γ*-invariant* δ*-conformal density. Consider the following assertions:*

(1) *The conical limit set* $\Lambda_c(\Gamma)$ *has positive* μ*-measure.*
(2) *The action of* Γ *on* $S_\infty \times S_\infty$ *minus the diagonal is ergodic with respect to* $\mu \times \mu$.
(3) *The geodesic flow on the unit tangent bundle to* $\mathbb{H}^n/\Gamma$ *is ergodic with respect to* m_μ.
(4) *The group* Γ *is of divergence type.*
(5) *The Markov process on* $\mathbb{H}^n/\Gamma$ *is recurrent.*

Then (1), (2) and (3) are equivalent, and they imply (4). If in addition $2\delta > n-1$, *then (4) implies (5), and (5) implies all the others (i.e., all five assertions are equivalent in this case).*

2.3 Further results

Dennis has written a number of other very interesting papers on the geometry and dynamics of Kleinian groups. For example, in [105] he extended some of the above results from the convex co-compact case to the case of *geometrically finite groups* – one of the main consequences being the fact that the Hausdorff dimension of the limit set of a geometrically finite group is equal to the critical exponent of the group. In [103], he investigated the excursion of geodesics on hyperbolic surfaces, relating their behaviour near cusps with classical results in Diophantine approximations.

3 Holomorphic dynamics

3.1 The reemergence of holomorphic dynamics in Paris in the 1980s

Right from the start the city of Paris has played a key role in the study of holomorphic dynamical systems. In the 1920s Julia and Fatou developed many results on the theory of iterations of rational maps $f\colon \bar{\mathbb{C}} \to \bar{\mathbb{C}}$. They introduced what is now called the *Fatou set*, the set of points in $\bar{\mathbb{C}}$ which have a neighbourhood N on which the iterates $f^n|N$, $n \in \mathbb{N}$, form a normal family, i.e., are equicontinuous. Similarly, the Julia set $J(f)$ is defined as the complement of $F(f)$. At the time the main tool Julia and Fatou had at their disposal was the Montel theorem, which states that a family of maps $f^n|N$, $n \in \mathbb{N}$, defined on an open set $N \subset \bar{\mathbb{C}}$ is normal if it has the property that there are three points in $\bar{\mathbb{C}}$ which are omitted in $\cup_n f^n(N)$. Among the many results they showed is that the Julia set is the closure of the set of repelling periodic points. They also developed a theory on the local dynamics near periodic points.

In the early 1980s there was a huge revival of this theory in Paris, the main drivers being Dennis Sullivan, Adrien Douady, Hamal Hubbard and Michael Herman. One of the main reasons for this resurgence was that it became increasingly clear that there were *new powerful tools* available, namely the *Measurable Riemann Mapping Theorem* (MRMT) and the notion of *quasiconformal maps*. These would make it possible to complete and go much beyond the theory initiated by Julia and Fatou in the 1920s.

There are quite a few equivalent definitions of the notion of a *quasiconformal map*, and all reflect that such maps are generalisations of conformal maps. One of these definitions is that $h\colon U \to V$ is a quasiconformal map if it is an orientation preserving homeomorphism between two domains U,V on $\mathbb{C}$ so that the Beltrami equation

$$\frac{\partial h}{\partial \bar{z}} = \mu(z)\frac{\partial h}{\partial z}$$

makes sense and so that μ is Lebesgue measurable essentially bounded, i.e., satisfies $||\mu||_\infty < 1$. This means that at each point $z \in U$ at which h is differentiable, the derivative $Dh(z)$ maps circles to ellipses with uniformly bounded eccentricity.

The MRMT implies that each such μ is associated to a quasiconformal map h_μ and, crucially, that h_μ depends analytically on μ.

Dennis was amongst the first to realise the power of the MRMT in the field of holomorphic dynamics, partly because he had previously used it very successfully in the study of Kleinian groups. Parallel to Douady and Hubbard's seminal Orsay notes [32, 33, 34, 35], which are a tour de force through the entire subject of holomorphic dynamics, Dennis proved the following remarkable theorem:

Theorem 3.1 (No-wandering-domains Theorem, see [106]). *Let f be a rational map on the Riemann sphere. Then each component of the Fatou set is eventually periodic.*

More precisely, for each component U of $F(f)$ there exists an $m \geq 0$ such that $V = f^m(U)$ is periodic, i.e., there exists a $p \geq 1$ such that $f^p(V) = V$. Moreover, each periodic component V of $F(f)$ can be classified. Indeed, it contains one of the following:

1. a periodic point of eigenvalue $\lambda = 0$ (called *superattractive*),
2. a periodic point of eigenvalue $0 < |\lambda| < 1$ (called *attractive*), or
3. ∂V contains a periodic point whose multiplier is a root of unity (*rational indifferent*),
4. f^p is analytically conjugate to an irrational rotation on V and either
   a. V is a simply-connected *Siegel disc* or
   b. a doubly-connected *Herman ring*.

To prove the first part of the statement one needs to show that if f is a rational map, then no component of its Fatou set is a wandering domain, i.e., a domain so that all its iterates are pairwise disjoint. In a nutshell Dennis' proof of this theorem goes as follows: suppose by contradiction that f has a wandering domain W. Then this makes it possible to construct an infinite-dimensional space of deformations of f, contradicting that the space of rational maps of a given degree is finite-dimensional.

That the space of rational maps is finite-dimensional is crucial: shortly after the preprint version of [106] appeared, Baker constructed entire functions $f\colon \mathbb{C} \to \mathbb{C}$ which do have wandering domains. Sullivan's no wandering theorem has also been extended to the setting of entire maps (and similar spaces) with a finite number of singular values, see for example [37] and [43]. A very elegant proof of the no wandering domains theorem due to McMullen – which circumvents the use of the MRMT and uses an infinitesimal deformations argument more in line with Ahlfors' original proof of his finiteness theorem – can be found in [78, p. 90].

Unfortunately, as there is no corresponding MRMT in the real one-dimensional case, the analogous theory in the real one-dimensional case requires a careful combinatorial analysis together with an understanding of the non-linearity of the map. For circle diffeomorphisms this goes back to Denjoy in the 1930s and from this paper Dennis learned the *smallest interval argument*: Assume that W is a maximal wandering interval, i.e., that W is not contained in a larger wandering interval. Then for each $n \geq 3$ consider the smallest, say $f^i(W)$, amongst the collection of disjoint intervals $W, \dots, f^n(W)$. Then $f^i(W)$ has neighbours on each side which are larger (or empty space in the case of an interval map). So $f^i(W)$ is well-inside the convex hull W_i' of the two neighbours. Using the way W_i' is chosen, the interval W_i' can be pulled back to an interval $W_0' \supset W$ so that the pullbacks $W_0', \dots, W_i'$ are essentially disjoint. This disjointness and the fact that f is a C^2 diffeomorphism implies that W is δ-well-inside W_0' where δ does not depend on n. Using the maximality of W this gives a contradiction. That $f^i(W)$ is well-inside the convex hull W_i' is often called *Koebe space* and is a property that is often used both in real and holomorphic dynamics.

The terminology *Koebe space* comes from the Koebe Lemma in complex analysis, which states that for each $\delta > 0$ there exists a $K > 0$ such that when $U_0 \subset U$ are topological discs and $\phi : U \to \mathbb{C}$ is a univalent map and the modulus of $U \setminus U_0$ is at least δ then

$$\frac{|D\phi(z)|}{|D\phi(z')|} \leq K \text{ for all } z, z' \in U_0.$$

The power of this lemma is that K does not depend on ϕ. It turns out that in the real case there is an analogous result: for each $\delta > 0$ there exists a $K > 0$ such that when $I_0 \subset I$ are intervals and $g\colon I \to \mathbb{R}$ is a diffeomorphism such that $Sg \geq 0$ then

$$\frac{|Dg(z)|}{|Dg(z')|} \leq K \text{ for all } z, z' \in I_0.$$

In applications, g is usually the inverse of a diffeomorphic branch of an interval map f^n where f is assumed to have negative Schwarzian $Sf < 0$. The reason why this is useful is that the Schwarzian property has the property that $Sf < 0$ implies $Sf^n < 0$ and that $Sf^n < 0$ implies $Sf^{-n} > 0$ (on diffeomorphic branches). Furthermore, if f is a real polynomial with only real critical points then $Sf < 0$. This observation that the negative Schwarzian could be used to bound the number of periodic attractors of an interval map was first made by Singer, see [95], but also appeared at around the same time in for example Herman's work [48]. A version of the Koebe Lemma in this setting was proved for the first time in [114]. That the Schwarzian derivative is related the distortion of cross-ratio was already known by E. Cartan in the 1930s, see the discussion in [30, Sections IV.1 and IV.2]. As the use of the above distortion estimate is so widespread, one often refers to the *Koebe Principle* and the assumption on the domains $U_0 \subset U$ (resp. $I_0 \subset I$) as *Koebe space*.

For interval maps and critical circle maps, the presence of critical points implies that one cannot control the non-linearity of the map and its iterates. Instead, it turns out that it is enough (i) to consider the *cross-ratio distortion* of a triple of adjacent intervals under iterates, (ii) assume that the map has some local symmetry around the critical points (e.g. the maps are non-flat at the critical points) and (iii) a more elaborate combinatorial analysis of orbits of wandering intervals. This was done by Guckenheimer, Yoccoz, Lyubich, Block, de Melo, van Strien, Martens in various generalities. For a history and a full analogue of Theorem 3.1, see [76, 30]. Probably the most elegant way of proving absence of wandering intervals in this setting can be found in [113]. Interestingly, it was Dennis who emphasised and insisted that the right smoothness class for (i) is $C^{1+Zygmund}$, whereas the earlier results required that the map was C^3 and even assumed that the map has negative Schwarzian. See [108], [109] or [30] for the definition of the classes $C^{1+Zygmund}$ and $C^{1+zygmund}$. The analogue of Sullivan's no-wandering Theorem 3.1 is:

Theorem 3.2 (See [76, 30]). *Assume that f is an interval map which is $C^{1+Zygmund}$ and has non-flat critical points. Then f has no wandering interval. Moreover, if f is a $C^{2+zygmund}$ map with non-flat critical points, then there exist $\kappa > 1$ and $n_0 \in \mathbb{N}$ such that*

$$|Df^n(p)| > \kappa$$

for every periodic point p of f of period $n \geq n_0$.

Interestingly, it is not clear to what extent local symmetry around a critical point is crucial. Indeed, consider a map of the form

$$f(x) = \begin{cases} x^{\alpha} + c & \text{for } x > 0 \\ x^{\beta} + c & \text{for } x < 0 \end{cases}$$

with $\alpha \neq \beta, \alpha, \beta > 1$ and c real. It is not known whether such a map can have wandering intervals. For $\alpha = \beta$ the answer is no, due to the previous theorem. For the case that $\alpha \neq \beta$ very little is known, except for the case that $\alpha = 1 < \beta$ and f has Feigenbaum–Coullet–Tresser dynamics, see [61] (and the proof of absence of wandering intervals in that case follows a rather curious approach).

The real bounds that go into the proof of the absence of wandering intervals for real maps, certainly inspired Dennis' proofs of complex bounds which are crucial in his renormalisation theory.

As mentioned, a crucial ingredient in the proof of real bounds is Schwarzian derivative, or more generally the notion of *cross-ratio*. A special cross-ratio inequality was used by Yoccoz to show that smooth circle homeomorphisms with a unique non-flat critical point cannot have wandering intervals, see [118]. More general cross-ratio inequalities were then used in [29] for the interval case and subsequently in [112] for circle endomorphisms, see [30, Sections IV.1 and IV.2] for a discussion of the connection between cross-ratio and Schwarzian derivative. In particular, the cross-ratio distortion arguments (i) suggest the relevance of the Poincaré metric on $(\mathbb{C} \setminus \mathbb{R}) \cup J$, which is the complex analogue of the cross-ratio on a real interval J. Indeed, let $D_r(J)$ be the set of points consisting of the set of points with distance to J of at most r with respect to the Poincaré metric on $(\mathbb{C} \setminus \mathbb{R}) \cup J$. This set is often called a Poincaré disc, and is bounded by two arcs of the circles through a, b. Using the Schwartz inclusion lemma, it then follows that if f is (for example) a real polynomial so that $f\colon J' \to J$ is a diffeomorphism and so that all critical values of f lie in $\mathbb{R} \setminus J$, then the component of $f^{-1}(D_r(J))$ intersecting J' is contained in $D_r(J')$. This turned out to be a key ingredient to the proof of his theorem on complex bounds for renormalisable maps, see Theorem 3.16.

Naturally, Dennis did ask himself whether there are analogues of his no wandering domain theorem in the higher-dimensional case in the smooth category. A partial answer to this question is given by Theorem 1.3 for toral diffeomorphisms of Denjoy type.

3.2 Conformal measures for rational maps

Soon after developing his no wandering theorem, Dennis introduced the notion of *conformal δ-measure* for a rational map f. This is a Borel probability measure m so that

$$m(fA) = \int_A |Df|^{\delta} \, \mathrm{d}m,$$

for every Borel measurable set $A \subset \bar{\mathbb{C}}$ and where it is assumed that $\delta \geq 0$, see [104]. Dennis then showed that one can also construct conformal measures on the Julia set analogous to what he had done before in the setting of Kleinian groups, extending earlier work by Patterson:

Theorem 3.3 (Existence of conformal measures for rational maps, see [104]). *For every rational map there exists a conformal measure. In the hyperbolic case, the exponent δ is positive and is equal to the Hausdorff dimension of the Julia set of the rational map.*

The paper [104] is also shows that for Dennis the theory of dynamical systems is unified: topological, smooth and ergodic aspects are all connected. Moreover, in his view the theory of real and complex one-dimensional systems together with the theory of Kleinian groups all should be viewed as highly interwoven.

3.3 The λ-Lemma

One of Dennis' most used and cited papers on holomorphic dynamics is one in which he, and his coauthors Mañé and Sad, proved that most maps are *stable*. The main technical tool in that paper is the celebrated:

Theorem 3.4 (λ-Lemma, see [70]). *Let A be a subset of $\overline{\mathbb{C}}$, $\mathbb{D}$ the open unit disc and $i_\lambda : A \to \overline{\mathbb{C}}$ a family of maps so that*

1. *for each $z \in A$, $\mathbb{D} \ni \lambda \mapsto i_\lambda(z)$ is analytic;*
2. *$A \ni z \mapsto i_\lambda(z)$ is injective for each $\lambda \in \mathbb{D}$;*
3. *$i_0 = id$.*

Then every $i_\lambda : A \to \overline{\mathbb{C}}$ has a quasiconformal extension to a continuous map $i_\lambda : \overline{A} \to \overline{\mathbb{C}}$, which for fixed λ is a topological embedding, and so that $\mathbb{D} \ni \lambda \mapsto i_\lambda(z)$ is analytic for each fixed $z \in \overline{A}$.

The proof of the λ-Lemma is surprisingly simple, and is based on the Schwarz lemma which states that any analytic map $\xi : \mathbb{D} \to \bar{\mathbb{C}} \setminus \{0, 1, \infty\}$ is contracting w.r.t. the Poincaré metric on these sets. Now consider the cross-ratio distortion of any distinct four points in A. Using that the cross-ratio distortion omits the values $0, 1, \infty$ and this version of the Schwarz lemma, one obtains the above λ-Lemma.

In addition to the MRMT and the theory of quasiconformal homeomorphisms, the λ-Lemma has become one of the most widely used tools in the field of holomorphic dynamical systems.[5]

Later, jointly with Thurston, Dennis improved this λ-Lemma to show that one can extend $i_\lambda : A \to \overline{\mathbb{C}}$ to a qc map $i_\lambda : \overline{\mathbb{C}} \to \overline{\mathbb{C}}$, provided we restrict λ to a suitable disc $\mathbb{D}_0 \Subset \mathbb{D}$ (here the choice of $\mathbb{D}_0$ is universal), see [111].

[5] Independently, Lyubich proved an analogous result, see [69].

3.4 Density of stable maps

The initial motivation for the λ-Lemma, and the main purpose of the paper [70], was to prove that stable maps are dense (within the space of rational maps). As a first step towards proving this, the class of J-stable maps is considered. Here f is called *J-stable* if for each g near f there exists a homeomorphism h of $J(f)$ to $J(g)$ so that $h \circ f = g \circ h$ on $J(f)$ and so that $J(g)$ depends continuously on g in the sense of Hausdorff distance between closed sets. f is called *structurally stable* if the conjugacy holds on $\bar{\mathbb{C}}$.

Consider a family $f_w(z)$ or rational maps depending on $w \in W$, where W is a connected complex submanifold of $\mathbb{C}^{2d+1}$, and so that $(w,z) \to f_w(z)$ is analytic in w, z. Let $H(f) \subset W$ be the set of $w \in W$ for which there exists a neighbourhood V with $w \in V \subset W$ so that each periodic point p_w of f_w depends analytically on $w \in V$ and so that their multiplier satisfies either $\lambda(w) \neq 1$ for all $w \in V$ or $\lambda(w) \equiv constant$ for all $w \in V$.

Theorem 3.5 (J-stability, see [70]). *$H(f)$ is open and dense in W. Moreover, f_w is J-stable if and only if $w \in W$ and the conjugating homeomorphism h_w can be taken to be analytic in w and quasiconformal in z.*

Analogous to the set $H(f)$, the authors introduce the set $C(f) \subset W$ of points w for which there is a neighbourhood V with $w \in V \subset W$ so that each critical point $c_i(w)$ of f_w depends analytically on $w \in V$ and so that any critical relation $f_w^n(c_i(w)) = f_w^m(c_j(w)))$ holds either for all w in V or for none.

Theorem 3.6 (Structurally stable maps are dense, see [70]). *$C(f)$ is an open and dense subset of $H(f)$. Moreover, if $w \in C(f)$ then f is structurally stable.*

In the late 1960s Smale suggested that a similar result should hold for general smooth dynamical systems, but this turned out to be false (due to examples by Newhouse and others).

3.5 Towards the Fatou conjecture: absence of line fields

A map f is said to be *Axiom A* (or *hyperbolic*) if there exist $\rho > 1$ and $C > 0$ such that $|(f^k)'(z)| > C\rho^k$ for all $k \geq 0$ and $z \in J(f)$. It is not hard to see that f is Axiom A if and only if the periodic components of $F(f)$ are superattractive or attractive and if the orbit of every critical point of f is eventually contained in one of these components.

Fatou already stated the following:

Conjecture 3.7 (Fatou Conjecture). *Each rational map can be approximated by an Axiom A rational map of the same degree.*

No doubt Dennis tried to prove this conjecture but to this day nobody has succeeded in doing so. One of the main appealing properties of Axiom A maps is that they are stable, provided they satisfy some mild additional conditions, and that their dynamics is very well-understood. For example, for such a map Lebesque almost every initial point converges under iterates to a periodic attractor.

Amongst many other results in [81], Dennis, together with McMullen, shows that the above conjecture can be reduced to proving the absence of *measurable invariant line fields* supported on Julia sets. Here a measurable line field on a forward invariant subset $K \subset J(f)$ of positive Lebesgue measure is a measurable function $z \mapsto \mu(z)$ on K. Here one can think of $\mu(z)$ as a line through z, and invariance means that $\mu(f(z)) = Df_z \mu(z)$.

Theorem 3.8 (A conditional proof of the Fatou conjecture, see [81]). *Assume that any rational map which supports a measurable invariant line field on its Julia set is a Lattès map. Then the above Fatou Conjecture holds.*

3.6 Monotonicity of entropy and the pullback argument

Another problem which was extensively studied in the early 1980s was whether the topological entropy of the family $f_a \colon [0,1] \to [0,1]$, $a \in [0,4]$, defined by $f_a(x) = ax(1-x)$ is a monotone function in a. This problem was solved by several people independently and using different methods. For a history of this problem, see [62]. Dennis' approach was particularly important because it became a key ingredient in the proof of density of hyperbolicity within interval maps, see Theorem 3.17 below.

Monotonicity follows immediately from the following:

Theorem 3.9 (Monotonicity of entropy). *Consider the family $f_a \colon [0,1] \to [0,1]$, $a \in [0,4]$, defined by $f_a(x) = ax(1-x)$. Then no periodic orbit disappears as a increases.*

If $f_a, f_{a'}$ are two such maps which are topologically conjugate and whose critical points are eventually periodic, then $a = a'$.

This theorem is non-trivial. Indeed, it is not known whether within the family $x \mapsto x^d + c$ with $d > 1$ fixed but not necessarily an integer, bifurcations are monotone in c. Partial results, and monotonicity for other non-trivial families of intervals maps, are given in [62]. One also has monotonicity within families of real polynomials of higher degree, namely each set of parameter for which the topological entropy is constant is connected. For the case of real cubic critical polynomials, see [83] and for the general case of real polynomials with all critical points real, see [11] and also [57].

The rigidity statement in the second part of Theorem 3.9 follows immediately from Thurston's famous theorem, see [36]. The approach proposed by Dennis uses the *pullback argument*. This argument is formalised in the following theorem and also applies to the setting of polynomial-like maps discussed below. Let $P(f)$ be the closure of the forward iterates of critical points of f.

Theorem 3.10 (Pullback argument). *Let h_0 be a quasiconformal homeomorphism such that*

1. $h_0(P(f_a)) = P(f_{a'})$,
2. *there exists a qc map h_1 for which $f_{a'} \circ h_1 = h_0 \circ f_a$ such that $h_1 = h_0$ on $P(f_a)$ and which is homotopic to h_0 rel. $P(f_a)$ and*
3. *h_0 is a conformal conjugacy between f_a and $f_{a'}$ near ∞ (and near their periodic attractor if they exist).*

Then there exists a qc homeomorphism h such that $f_{a'} \circ h = h \circ f_a$. If h is conformal near all periodic attractors (if they exist) and if f_a does not carry an invariant line field on $J(f_a)$ then h is conformal.

Proof. By the Homotopy Lifting Theorem, there exists a sequence of homeomorphisms h_n such that $f_{a'} \circ h_{n+1} = h_n \circ f_a$ which are homotopic to h_0 rel. $P(f_a)$. Since f_a and $f_{a'}$ are conformal, h_{n+1} will have the same qc dilatation as h_n. Moreover, h_{n+1} agrees with h_n on a set F_n such that $F_{n+1} \supset F_n$ so that $\cup F_n$ is dense in $\bar{\mathbb{C}}$. Since the space of qc maps is compact, h_n converges to a qc map h. If the Julia set of f_a has zero Lebesgue measure (or, even more generally, does not carry invariant line fields) then h is conformal.

One can deduce Theorem 3.9 quite easily from the pullback argument, using the *open-closed argument*. Indeed, assume by contradiction that the conclusion of Theorem 3.9 is wrong. Then there exists two topologically conjugate post-critically finite real quadratic maps f_a and $f_{a'}$ with $a \neq a'$. Choose $[a,a']$ 'maximal' i.e., such that there exists no real $a'' \notin [a,a']$ for which $f_{a''}$ is a real quadratic map which again is topologically conjugate to f_a. The pullback argument implies that f_a and $f_{a'}$ are in fact quasiconformally conjugate. Let h be a qc-conjugacy such that $f_{a'} \circ h = h \circ f_a$ and let μ be its Beltrami coefficient. Then the MRMT gives a (normalised) family of qc maps h_t whose Beltrami coefficient is $t\mu$ for $|t| \leq 1+\varepsilon$, provided $\varepsilon > 0$ is small. A simple calculation then shows that each of the maps $g_t := h_t \circ f_a \circ h_t^{-1}$ is again conformal. This is because h_t sends the ellipse field determined by $t\mu$ to a field of circles, and the invariance of the Beltrami coefficient implies that this ellipse field is preserved by Df. Moreover, g_t depends analytically on t. That g_t is in fact quadratic follows from the degree of the map (and a suitable normalisation of h_t). It follows that there exists an $\varepsilon > 0$ such that for $s \in (a-\varepsilon, a'+\varepsilon)$ each f_s is quasiconformally conjugate to f_a. But this contradicts the maximality of the choice of $[a,a']$.

This proof is very interesting because it makes it possible to reduce density of hyperbolicity with real quadratic maps to quasisymmetric rigidity in the unicritical setting, see Section 3.11. In fact, even if in the context of maps with several critical points, quasisymmetric rigidity can be used to derive density of hyperbolicity, see [60, 58, 91].

Another reason that makes Dennis' proof of Theorem 3.9 so interesting is that it also applies to the following setting, introduced by Douady and Hubbard [34].

Definition 3.11. Assume that U and V are simply connected domains in $\mathbb{C}$. Then a holomorphic map $F: U \to V$ is called *quadratic-like* if the closure of U is contained in V and if there exists a unique critical point c of F such that F restricted to $U \setminus \{c\}$ is a covering map of degree two onto $V \setminus \{F(c)\}$. The subset

$$K(F) = \{z \in U : F^n(z) \in U \text{ for all } n \geq 0\}$$

is called the *filled Julia set* of F.

Any quadratic map is quadratic-like (just take U to be some very large disc). Moreover, by the so-called Straightening Theorem, see [34], any quadratic-like map is quasiconformally conjugate to a quadratic map.

One step in the renormalisation theory developed by Dennis (and also in subsequent developments) is to show that certain iterates of a given map have quadratic-like restrictions $f^n: U_n \to V_n$ with the additional property that the modulus $\text{mod}(V_n \setminus U_n)$ is bounded from below uniformly in n. Such bounds are called *a-priori bounds* or *complex bounds*.

3.7 Renormalisation theory for interval maps

Consider the family $f_a(x) = ax(1-x)$. A simple computer simulation shows that this family of maps undergoes a period doubling bifurcation from period 2^n to period 2^{n+1} at some parameter a_n. That these parameters a_n are in fact unique (and increasing) can be deduced from a result similar to Theorem 3.9.

One of the reasons why iterations of interval maps attracted so much attention from the late 1970s was the observation by Feigenbaum and independently by Coullet and Tresser of metric *universality* within a wide class of such families. Namely, it turns out that the parameters a_n converge to some limit value a_∞ at a particular rate $\delta > 1$:

$$\frac{a_{n-1} - a_{n-2}}{a_n - a_{n-1}} \to \delta = 4.669201...$$

Remarkably, if one takes some other family such as $f_a(x) = a\sin(\pi x)$ then the corresponding rate is the same! Moreover, the map f_{a_∞} has an invariant Cantor set and the scaling structure of this Cantor set also displays metric universality.

Feigenbaum, Coullet and Tresser already suggested a mechanism which would be responsible for this universality. The key idea is renormalisation. Indeed, there exists a (real) parameter interval $[u_1, v_1] \ni a_\infty$ such that for each $a \in [u_1, v_1]$ there exists an interval $J_a^1 \ni 1/2$ such that $f_a^2(J_a^1) \subset J_a^1$ and such that $f_a(J_a^1)$ and J_a^1 have disjoint interiors. Such maps are called *2-renormalisable*. Note that $1/2$ is the critical point of f_a. It turns out that there exists an interval $[u_2, v_2] \subset [u_1, v_1]$ such that for each $a \in [u_2, v_2]$ the map $f_a^2|J_a^1$ (rescaled) is again 2-renormalisable. In other words, there exists an interval J_a^2 with $1/2 \in J_a^2 \subset J_a^1$ such that $f_a^4(J_a^2) \subset J_a^2$ and such that $J_a^2, \ldots, f_a^3(J_a^2)$ have disjoint interiors. Moreover, $J_a^2, f_a^2(J_a^2) \subset J_a^1$ and

$f(J_a^2), f_a^3(J_a^2) \subset f(J_a^1)$. So these maps are twice 2-renormalisable. Continuing like this, for each k there exists an interval $[u_k, v_k]$ such that for each $a \in [u_k, v_k]$ the map f_a is 2^k-renormalisable. For $a_\infty \in \cap[u_k, v_k]$ the set

$$\Lambda_{a_\infty} := \bigcap_{k \geq 0} \left(J_a^j \cup \cdots \cup f_a^{2^k-1}(J_a^k)\right)$$

is a Cantor set.

To formalise this one can define, near the limit map f_{a_∞}, the renormalisation operator

$$R(f) = f^2|J \text{ rescaled,}$$

where J is the maximal interval of renormalisation of period two, i.e., the maximal interval such that $f^2(J) \subset J$ and such that $f(J)$ and J have disjoint interiors. This operator is well-defined for all maps which are at least once 2-renormalisable.

More generally one has the following

Definition 3.12. An interval map f is *renormalisable* if there exist $p > 0$ and an interval J around a critical point of f such that $J, \ldots, f^{p-1}(J)$ have disjoint interiors and such that $f^p(J) \subset J$. The operator $R(f) = f^p|J$ rescaled is then called the renormalisation operator. See Figure 2.

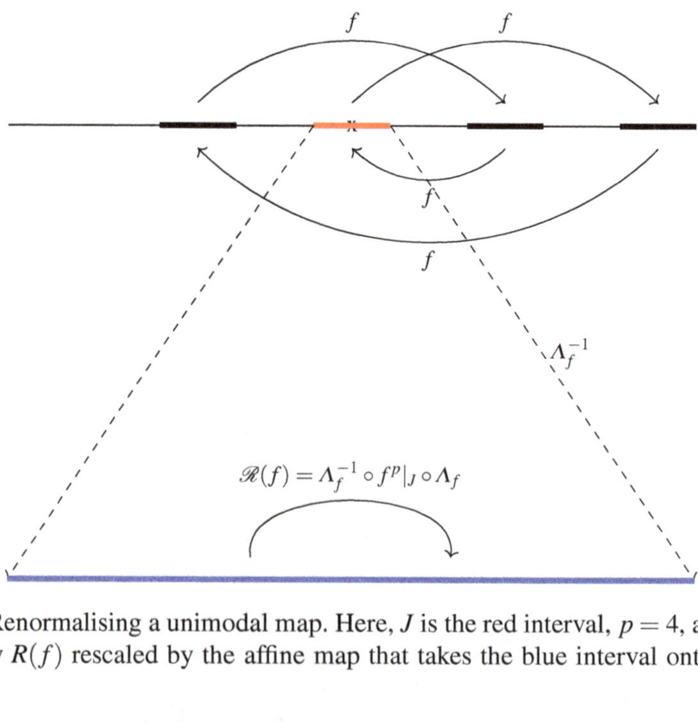

Fig. 2: Renormalising a unimodal map. Here, J is the red interval, $p = 4$, and $\mathscr{R}(f)$ is simply $R(f)$ rescaled by the affine map that takes the blue interval onto the red interval.

If f is unimodal, then $f^p|J$ is again a unimodal map and it makes sense to require that p is minimal and J is maximal with the above properties. If $R(f)$ is again renormalisable, then we say that f is twice renormalisable. Similarly, f is called *infinitely renormalisable* if this process can be repeated infinitely often. If the corresponding integers $p_1, p_2, \ldots$ are all bounded by some number $P < \infty$ then f is called infinitely renormalisable of bounded type.

It turns out that the renormalisation conjectures of Feigenbaum and Coullet & Tresser of metric universality follow from (i) the existence of a fixed point ψ of the operator R, (ii) that the spectrum of the operator $DR\psi$ lies off the unit circle and (iii) that $DR\psi$ has a unique expanding eigenvalue. The universal parameter scaling constant δ is equal to this expanding eigenvalue. The universal dynamical scaling structure of Λ_{a_∞} follows from the largest contracting eigenvalue of DR.

The universality from the Feigenbaum–Coullet–Tresser conjectures then follows from the fact that for any family which crosses the stable manifold of R transversally, the parameter scalings δ and the dynamical scaling can be obtained from the spectrum of $DR(\psi)$.

The existence of the fixed point ψ and an analysis of the spectrum of the linear map DR were established by Lanford before Dennis started working on the renormalisation conjectures. This was done in part using careful rigorous computer estimates. However, there were three limitations to Lanford's results.

Firstly, Lanford's proof did not establish which maps are contained in the stable manifold of R. In other words, it remained unclear whether any 2^∞-infinitely renormalisable unimodal maps (with a quadratic critical point) would be in the stable manifold of the period doubling operator. Secondly, Lanford's proof did not establish a conceptual proof of why this result was true. Finally, Lanford's proof also did not cover the more general situation of maps which are infinitely renormalisable of bounded type, but only of constant type $p_1 = p_2 = \cdots$.

The huge result which Dennis managed to obtain is the following:

Theorem 3.13 (Renormalisation for unimodal interval maps, see [109] and also [30]). *There exists a Cantor set $\mathbb{K}_p$ of infinitely renormalisable maps of bounded type $\leq p$, of real analytic unimodal maps, which form an invariant subset for the corresponding renormalisation operator. The renormalisation operator acts on $\mathbb{K}_p$ as a full shift on finitely many symbols. Each real analytic unimodal map with a quadratic critical point which is infinitely renormalisable map and of bounded type is in the stable manifold of the corresponding renormalisation operator.*

After this, other proofs of the renormalisation conjectures appeared. McMullen [79, 80] and Avila and Lyubich [5] gave easier proofs that the stable manifold of the renormalisation operator contains all the relevant infinitely renormalisable maps using 'towers' respectively based on a Schwarz Lemma. These proofs additionally give that under renormalisation infinitely renormalisable maps of bounded type converge with an exponential rate to the above Cantor set of infinitely renormalisable maps of bounded type. Moreover, Lyubich [66] gave a conceptual proof showing that the renormalisation operator has a unique expanding eigenvalue.

The above proofs require that the maps are real analytic. In the C^r context, these theorems also go through, see [27]. An alternative approach to extend the renormalisation theory for real analytic maps to smooth maps is via asymptotically holomorphic maps, see [17].

In the multimodal setting, the renormalisation picture is not complete yet. In this setting, the conjecture could be

1. Topologically conjugate mappings converge exponentially quickly under renormalisation.
2. The stable manifolds of renormalisation are smooth.
3. The transverse directions to the stable manifolds are exponentially expanded by renormalisation.

Part 2 of this conjecture has been proved in [20]. For the case of bounded combinatorics, see [97].

3.8 Real and complex bounds

The first step towards proving Theorem 3.13 is to show that the space of renormalisable maps is compact. To do this, Dennis established a priori bounds, first in the real and then in the complex setting:

Theorem 3.14 (Real Bounds). *Let f be a real analytic map which is infinitely renormalisable and of bounded type. Then the C^2 norm of the maps $R^n f$ is uniformly bounded.*

The first part of this theorem shows that $R^n f$ is a composition of a quadratic map and a map g whose non-linearity is bounded from above. The main ingredient he used for this is the (real) Koebe Principle and the *smallest interval argument* discussed above. Indeed, let J be the first renormalisation interval such that $f^p(J) \subset J$ with $p \geq 1$ minimal. Then the intervals $J, \dots, f^{p-1}(J)$ are disjoint and one among them, let us say $f^i(J)$, is the smallest. This means that, unless $f^i(J)$ is one of the two extreme intervals in this collection, the interval $f^i(J)$ is contained in an interval $T = [f^l(J), f^r(J)]$ which has the property that both components of $T \setminus f^i(J)$ are not small compared to $f^i(J)$. This simple idea can be used to obtain *Koebe space*, namely that the map $f^{p-1}\colon f(J) \to f^p(J) \subset J$ extends to a diffeomorphism with range $T' \supset J$ so that J is well-inside J. Using the real Koebe Principle the map $f^{p-1}\colon f(J) \to f^p(J)$ then has bounded non-linearity.

Extending this argument, and using the pullback argument from above, one then obtains one of the key steps:

Theorem 3.15 (Quasisymmetric Rigidity in the Renormalisable Case). *Let f, g be two infinitely renormalisable real-analytic maps of bounded type and with quadratic critical points. Then if f, g are topologically conjugate they are in fact quasisymmetrically conjugate.*

To obtain Theorem 3.13 Dennis needed to extend the real maps $f^p \colon J \to J$ to quadratic-like maps and to obtain compactness with the space of such maps:

Theorem 3.16 (Complex Bounds). *Let f be an infinitely renormalisable real-analytic unimodal map of bounded type $p_i \leq p$ for all $i \geq 0$ and with quadratic critical points. Then, for every n sufficiently large, $R^n f$ extends to a quadratic-like map $F \colon U \to V$ such that the modulus of $V \setminus U$ is bounded by some number $\rho > 0$ which does not depend on f but only the upper bound p.*

Moreover, any limit of $R^n f$ has a complex analytic extension which is in the so-called Epstein class.

Real bounds were proved in a much more general context, see (in increasing generality) [74, 93, 113]. Complex bounds for real unicritical maps were proved in [63, 68] and for multicritical maps (in increasing generality) in [96, 21]. Complex bounds do not hold for general (non-real) quadratic maps: there are examples of infinitely renormalisable quadratic maps for which no modulus bounds as in the above theorem hold. On the other hand, for non-renormalisable polynomials maps (with only hyperbolic periodic points) one does have complex bounds, see [60] and [18]. An important ingredient in the latter developments is the quasi-additive lemma by Kahn and Lyubich [54]. This lemma was also used to treat some maps which are infinitely renormalisable, see [52, 53]. It is not known how to extend complex bounds to the case of general rational maps, as in general it is not clear how to construct an initial puzzle partition.

3.9 Riemann surface laminations and the non-coiling lemma

To complete the proof of Theorem 3.13, Dennis introduced a new tool, namely his *non-coiling principle* and his *almost geodesic principle*. To explain this, consider a qc conjugacy H between F_0 and F_1. Its Beltrami coefficient $\mu_H = \bar{\partial} H / \partial H$ is invariant under F_0. It follows that the family of qc maps H_t associated to $\mu_t = t\mu_H$ (coming from the MRMT) defines a family of quadratic like maps $F_t = H_t \circ F_0 \circ H_t^{-1}$ connecting F_0 to F_1. This is called a *Beltrami path* between F_0 and F_1. Dennis' *almost geodesic principle* shows that the Beltrami path corresponding to an *almost extremal vector* does not coil: if the tangent Beltrami vector is almost extremal then the Beltrami path remains almost a geodesic for a long (but a priori fixed) time.

To show that the renormalisation operator is (weakly) contracting he then argued as follows. From the complex bounds discussed in the previous subsection, he obtained that there exists a compact space $\mathcal{K}$ such that, if we take an arc connecting two conjugate maps which are infinitely renormalisable and of bounded type, then after n renormalisations, this arc is mapped in $\mathcal{K}$. Here n depends on the choice of the chosen maps, but $\mathcal{K}$ does not. Now take an almost geodesic path between two maps F_0, F_1. Extend this path to a geodesic path between two maps $\tilde{F}_0, \tilde{F}_1$ which are extremely far apart. Now apply renormalisation. For n sufficiently large, the renormalisations $R^n(\tilde{F}_0), R^n(\tilde{F}_1)$ are in $\mathscr{K}$ and so not far apart. But then, by the *almost geodesic principle*, the renormalisations $R^n(F_0), R^n(F_1)$ are extremely close.

To make all this work, Dennis had to consider germs of quadratic-like maps. For this reason he considered inverse limits of the quadratic-like maps $F_i\colon U_i \to V_i$ which led to the study of *Riemann surface laminations*. This is not the place to go into a full description of the beautiful theory of such objects, but let us at least say a word or two. Roughly speaking, a Riemann surface lamination (RSL) is a space akin to a foliated space in which the chart domains are homeomorphic to $D \times T$, where $D \subset \mathbb{C}$ is a disk and the transversal $T \subset \mathbb{R}$ is typically a Cantor set (or an interval), and the chart transitions are holomorphic along the horizontal leaves. In the present context, the main example is the following. Let $F : U \to V$ be a quadratic-like map, let $K(F) \subseteq \mathbb{C}$ be its filled-in Julia set, which we assume to contain the critical point of F (so that it is connected), set $W = V \setminus K(F)$, and consider the inverse system of holomorphic covering maps:

$$\cdots \to F^{-n-1}W \to F^{-n}W \to \cdots \to F^{-1}W \to W.$$

The inverse limit space $\mathscr{L}(F)$ of this system is a fibration over W, the fiber above each $x \in W$ being a Cantor set (the binary Cantor set at the end of the tree giving the full backward orbit of x). From this it follows that $\mathscr{L}(F)$ is an RSL in a natural way.

The inverse limit map $F_\infty : \mathscr{L}(F) \to \mathscr{L}(F)$ is invertible and acts properly discontinuously on $\mathscr{L}(F)$, and the quotient $X_F = \mathscr{L}(F)/F_\infty$ is a *compact* RSL. In [109], Dennis defined a deformation space or *Teichmüller space* of X_F (and more general RSLs) in such a way that every Beltrami path between two quadratic-like maps F_0 and F_1 as above can be lifted to a Beltrami path between the corresponding laminations X_{F_0} and X_{F_1}, and all deformations are encoded in this fashion. What makes this possible is the fact that the Julia set of an infinitely renormalisable quadratic-like map with complex bounds does not carry any non-trivial quasi-conformal deformations. Dennis then proved the non-coiling principle and the almost-geodesic principle at the level of laminations, transporting the resulting contraction of the Teichmüller distance downstairs, at the level of maps.

We will not go further into Dennis' proof of this tour de force (a full description can be found in [30, Chapter VI], in addition to the papers [109] and [110]). After all, as mentioned, this last step was improved in subsequent proofs which show that the invariant Cantor set of R attracts other maps with an *exponential rate*, see McMullen [79, 80] and Avila and Lyubich [5]. Still, the reader who wants to learn more about RSLs should consult the elegant survey written by Ghys in [15].

3.10 Renormalisation theory for circle maps

While working on the renormalisation problem for unimodal maps, Dennis was well aware that similar experimental discoveries to those made by Feigenbaum and Coullet–Tresser had been made by physicists concerning circle homeomorphisms having a single non-flat critical point (of power-law type). The topological classification of such maps had been accomplished by Yoccoz in [118].

Dennis also knew that, in the circle context, Lanford had formulated a renormalisation conjecture akin to the one for unimodal maps, using the language of commuting pairs. Dennis then suggested to EdF, as a thesis problem, to adapt his holomorphic ideas to the case of such critical circle maps. The key step was to find an analogue of quadratic-like maps in the context of critical circle maps. This was accomplished in [23] (see also [24]) with the notion of *holomorphic commuting pair* (inspired in part by a computer picture drawn by H. Epstein), alongside a proof of complex bounds (as well as a pull-back argument) for such objects, assuming the rotation number of the underlying critical circle map to be of bounded type, and also that the circle maps belonged to a special class of maps known as *Epstein class*. The necessary real bounds had already been established by Herman (unpublished manuscript, but see [49]) and Swiatek [112]. The bounded type assumption was removed by Yampolsky [116], still assuming the Epstein property. The latter was finally removed by EdF and Welington de Melo in [26].

For unicritical circle maps, the fact that, under suitable full-measure conditions on the rotation number, exponential convergence of renormalisations leads to $C^{1+\alpha}$ rigidity was established in [25] (counterexamples to this ansatz for rotation numbers in a special zero-measure class were constructed in the same paper). The analogous conditional statement obtained replacing $C^{1+\alpha}$ by C^1 holds under no restriction on the rotation number (other than being irrational), as shown by [55]. The exponential convergence of renormalisations for *real-analytic* unicritical circle maps with bounded type rotation number was proved in [26]. Using the concept of parabolic renormalisation, Yampolsky [117] was able to remove the bounded type hypothesis, and in fact proved that the renormalisation operator attractor is globally hyperbolic in the analytic context. In the larger space of C^3 unicritical circle maps, exponential convergence towards the attractor was first proved by Guarino in his thesis under de Melo – see [46] – assuming rotation numbers of bounded type only. The bounded type hypothesis was later removed in [47], at the cost of assuming the maps to be C^4.

In recent years, considerable work has been done to extend these rigidity, universality and renormalisation convergence results to multicritical circle maps – see for instance [39] and references therein. An important step towards this goal is to first establish the *quasisymmetric rigidity* of such maps – this was accomplished in [38] for multicritical circle maps with critical points having arbitrary (real) power-law criticalities. For this and much more about multicritical circle maps, see [28].

3.11 The Fatou conjecture in the real setting

As mentioned, Fatou conjectured that each rational map can be approximated by an Axiom A rational map of the same degree. This question is still wide open, even in the quadratic case. However, in the setting of real maps the corresponding result has been answered completely:

Theorem 3.17 (The Real Fatou Conjecture). *Each real polynomial can be approximated by a real Axiom A polynomial of the same degree.*

In the setting of real quadratic polynomials, the real Fatou conjecture was proved independently by Lyubich [65] and Graczyk & Swiątek [44, 45]. The setting of real polynomials of higher degree $d > 2$ was solved (using entirely different tools) by Kozloski, Shen and van Strien, see [59] and [58]. In the non-real non-renormalisable case see also [60] and [18]. An important ingredient in the latter developments is the quasi-additive lemma by Kahn and Lyubich [54]. This lemma was also used to treat some maps which are infinitely renormalisable, see [52, 53].

Although the proofs of these results are not due to Dennis, he played an important role in them. Indeed, his work suggested that to prove density of hyperbolicity that it would be enough to prove the following

Theorem 3.18 (Quasisymmetric Rigidity in one-dimensions). *If f, g are topologically conjugate real polynomials with only real critical points, and all their critical points are quadratic, then these maps are quasisymmetrically conjugate.*

Dennis' renormalisation theory for infinitely renormalisable unimodal (unicritical) maps of bounded type, relied on this result (in this setting). In the quadratic case, the above theorem is due to [65] and Graczyk & Swiątek [44, 45]. Their proof relied on the property that, in this setting, the moduli of certain annuli tends to infinity. This growth of moduli is a deep and subtle result, but this does not hold for unimodal maps with a degenerate critical point nor for multimodal maps with non-degenerate critical points. So for the general case a different approach is needed. The approach by Kozlosvki, Shen and van Strien in [59] uses the enhanced nest, which is a particular choice of a sequence of puzzle pieces that turn out to have Koebe space. An introductory survey on this technique can be found in [18]. The most general quasisymmetric result is contained in joint work of Clark and van Strien, see [19].

3.12 Sullivan's quasisymmetry rigidity programme

Even though there is no analogue of the MRMT in the real setting, one of Dennis' insights was that quasisymmetric rigidity should still be a very powerful tool in addressing questions about the topological structure of conjugacy classes of interval maps. For example, whether such a conjugacy class is a connected manifold. This insight turned out to be justified. Indeed, let $\mathscr{A}^{\underline{\nu}}$ be the space of real analytic maps

Kleinian Groups	**Iterated Analytic Maps**
Kleinian group Γ	Holomorphic map f
Γ finitely generated	f rational map
Γ is Fuchsian	f is a Blaschke product
Domain of discontinuity $\Omega(\Gamma)$	Fatou set $\mathscr{F}(f)$
Limit set $\Lambda(\Gamma)$ $\Lambda(\Gamma) \neq \emptyset$	Julia set $J(f)$ $J(f) \neq \emptyset$
$\Omega(\Gamma)$ has either $0, 1, 2$ or infinitely many components	$\mathscr{F}(f)$ has either $0, 1, 2$ or infinitely many components
Either $\Lambda(\Gamma) = \widehat{\mathbb{C}}$ or $\Lambda(\Gamma)$ has empty interior	Either $J(f) = \widehat{\mathbb{C}}$ or $J(f)$ has empty interior
Ahlfors finiteness theorem	Sullivan's no-wandering-domains theorem
Bers area theorem	Shishikura's bound on the number of non-repelling periodic cycles
Mostow's rigidity theorem	Thurston's uniqueness theorem on post-critically finite rational maps
Patterson–Sullivan measures on $\Lambda(\Gamma)$	Sullivan's conformal measures on $J(f)$
The quotient manifold $\mathbb{H}/\Gamma$	Lyubich–Minsky lamination
Geometrically finite groups with no cusps are dense	Are hyperbolic rational maps dense?

Table 1: Some entries in Sullivan's dictionary

with precisely ν real critical points $c_1 < \cdots < c_\nu$ of order $\ell_1, \ldots, \ell_\nu$. The following theorem was shown by Clark and van Strien [20].

Theorem 3.19. *Let $f \in \mathscr{A}^{\underline{\nu}}$. Then the space $\mathscr{T}_f$ of real analytic maps in $\mathscr{A}^{\underline{\nu}}$ which are topologically conjugate to f forms an analytic manifold. This manifold is connected and simply connected.*

This theorem extends results of Avila–Lyubich–de Melo [6] for the quasi-quadratic unimodal case and of Clark [16] for the more general unimodal case. Their methods fail in the case where there are several critical points. For this reason, the notion of *pruned Julia set* is introduced in [20]. This set is a version of the Julia set (pruned) but depends on where one 'prunes'. A pruned Julia set can be defined for each real analytic map f. The real analytic map f, together with its pruned Julia set, define a real analytic *external map* of the circle *with discontinuities*. Using this external map, one can construct a *a pruned polynomial-like* complex extension of the real analytic map. Finally, from all this one is able to show that topological conjugacy classes are connected (something which was not even known in the general unimodal setting). Even more, this space is contractible and forms an analytic manifold.

4 Sullivan's dictionary

Dennis' wide-range view of Mathematics allows him to draw fruitful analogies between different theories, leading to several conjectures on either side. A case in point is what is now known as the *Sullivan dictionary* between the theory of Kleinian groups (in dimension $n = 3$) on one side and the theory of iterated holomorphic maps on the other side. A sample of entries in this dictionary is shown in Table 1. Note that the last entry in the table has a question mark: that is none other than the famous Fatou Conjecture, widely regarded as the main classical open problem about the dynamics of rational maps.

Not all meaningful analogies, however, deserve to be in the dictionary. For instance, a famous conjecture by Ahlfors in the 1960s stated that the limit set of a finitely generated Kleinian group is either the entire sphere or else has zero Lebesgue measure. As we mentioned in the beginning of Section 2, this is now a theorem, thanks to the combined efforts of several mathematicians. The final piece of the puzzle was laid down by Canary [14], building primarily on previous works by Thurston and Bonahon – see for instance [72] for a description of the whole story, and references therein. The corresponding statement for iterated holomorphic maps – to wit, that the Julia set of a rational map is either the entire sphere or else has zero Lebesgue measure – was thought for a long time to be true, until X. Buff and A. Chéritat [13] found an example of a quadratic polynomial whose Julia set has positive measure.

As Dennis himself has explained to us, he never put the Ahlfors conjecture in the dictionary because – after working on the problem for about 13 months in the late seventies and exhausting all available ergodic arguments that would have solved it – he came to the conclusion that no one would be able to prove it using what was currently known about finitely generated Kleinian groups. After proving his finiteness theorem, what Ahlfors really wanted to know was under which conditions the limit set of a Kleinian group could support non-trivial quasiconformal deformations. He asked the question about the Lebesgue measure of the limit set in the finitely generated case because, if the measure indeed turned out to be zero in that case, there would be no such deformations. Thus, Dennis realized that the real question was not the measure zero question, but rather to describe, if any, the quasiconformal deformations on the limit set. He was able to prove a very general result that states that, given a Kleinian group Γ and any Γ-invariant subset $E \subset \Lambda_\Gamma$ of its limit set, there are quasiconformal deformations of Γ supported in E if and only if there are *positive measure wandering sets* inside E, and when this happens, the space of such nontrivial deformations is infinite-dimensional. In particular, since for finitely generated groups the space of deformations is a-priori known to be finite-dimensional, there are no quasiconformal deformations supported on Λ_Γ when Γ is finitely generated. This holds even if the limit set happens to be the whole sphere. This absence of invariant line fields supported in the limit set is stated in Dennis' Theorem 2.5. After more than 40 years, the corresponding statement for rational maps[6] remains an open problem. And it is a fundamental problem: indeed it is possible to prove that if the statement for rational maps is true, then so is the Fatou conjecture. Thus, the question of absence of invariant line fields certainly deserves its place as an entry in the Sullivan dictionary (albeit being conspicuously absent from Table 1).

Over the years, the Sullivan dictionary has continued to inspire new results. A recent example is provided by the work of Hee Oh. Working on the Kleinian side of the dictionary, in collaboration with Margulis and Mohammadi [73], she examined closed geodesics and holonomies for hyperbolic 3-manifolds. She was then asked by Dennis himself about an analogue of her results for rational maps. This resulted in her paper [87] in collaboration with Winter, in which they establish estimates on the number of primitive periodic orbits of a hyperbolic rational map.

5 Final words

Dennis Sullivan's major contributions to the field of Dynamical Systems, some of which we attempted to describe here, constitute but one facet of his extraordinary work as a mathematician. There are several other facets. Thus, for his fundamental work in Topology, especially regarding the study of geometric and/or algebraic structures on manifolds, see the article by Shmuel Weinberger in the present volume. In more recent years, Dennis has essentially founded, in collaboration with

[6] To wit, that a rational map is either a Lattès example or else carries no invariant line fields in its Julia set.

M. Chas, the sub-field of Topology now known as String Topology. We have heard it said elsewhere that *Dennis Sullivan is a mathematician who has re-invented himself several times*, and that seems to us a very accurate statement.

We have not included anything about Dennis' recent work on fluid dynamics. Nor have we mentioned any work on dynamics that Dennis co-wrote with some of his students and/or post-docs, such as the work with Jiang and Morita [51], his work with Hu [50] or his work with Pinto [89], nor with many other collaborators from the dynamical systems community.

In closing, it is important to add that Dennis has always been extremely generous when sharing his ideas with other researchers, as well as in guiding young mathematicians. According to the Math Genealogy Project, he has had so far 40 students, and a total of 155 descendants. But many more mathematicians, young and old, although not formally his students, have been directly influenced by him. Through his insightful lectures, and his inquisitive quest not merely for results, but for *understanding* Mathematics, Dennis has inspired and will continue to inspire us all. Among his students, those who have written a thesis in dynamics under his supervision include Andre de Carvalho, Adam Epstein, Jun Hu, Yunping Jiang, Curt McMullen, Waldemar Paluba, Guiai Peng, Meiyu Su, as well as one of us (EdF).

Acknowledgements We would like to thank Curt McMullen, Leon Staresinic, Edson Vargas for their useful comments, and especially Dennis Sullivan for explaining to us the origins of his dictionary.

References

1. Ahlfors, L. V. (1981). *Möbius transformations in several dimensions*. Ordway Professorship Lectures in Mathematics. University of Minnesota, School of Mathematics, Minneapolis, Minn.
2. Alves, J. F., Pinheiro, V., and Pinto, A. A. (2014). Explosion of smoothness for conjugacies between multimodal maps. *J. Lond. Math. Soc. (2)*, 89(1):255–274.
3. Astorg, M., Buff, X., Dujardin, R., Peters, H., and Raissy, J. (2016). A two-dimensional polynomial mapping with a wandering Fatou component. *Ann. of Math. (2)*, 184(1):263–313.
4. Avila, A., Fayad, B., Le Calvez, P., Xu, D., and Zhang, Z. (2020). On mixing diffeomorphisms of the disc. *Invent. Math.*, 220(3):673–714.
5. Avila, A. and Lyubich, M. (2011). The full renormalization horseshoe for unimodal maps of higher degree: exponential contraction along hybrid classes. *Publ. Math. Inst. Hautes Études Sci.*, (114):171–223.
6. Avila, A., Lyubich, M., and de Melo, W. (2003). Regular or stochastic dynamics in real analytic families of unimodal maps. *Invent. Math.*, 154(3):451–550.
7. Beardon, A. F. (1968). The exponent of convergence of Poincaré series. *Proc. London Math. Soc. (3)*, 18:461–483.
8. Beardon, A. F. (1983). *The geometry of discrete groups*, volume 91 of *Graduate Texts in Mathematics*. Springer-Verlag, New York.
9. Bowen, R. (1979). Hausdorff dimension of quasicircles. *Inst. Hautes Études Sci. Publ. Math.*, (50):11–25.

10. Bowen, R. and Franks, J. (1976). The periodic points of maps of the disk and the interval. *Topology*, 15(4):337–342.
11. Bruin, H. and van Strien, S. (2015). Monotonicity of entropy for real multimodal maps. *J. Amer. Math. Soc.*, 28(1):1–61.
12. Buff, X. (2018). Wandering Fatou component for polynomials. *Ann. Fac. Sci. Toulouse Math. (6)*, 27(2):445–475.
13. Buff, X. and Chéritat, A. (2012). Quadratic Julia sets with positive area. *Ann. of Math. (2)*, 176(2):673–746.
14. Canary, R. D. (1993). Ends of hyperbolic 3-manifolds. *J. Amer. Math. Soc.*, 6(1):1–35.
15. Cerveau, D., Ghys, E., Sibony, N., and Yoccoz, J.-C. (2003). *Complex dynamics and geometry*, volume 10 of *SMF/AMS Texts and Monographs*. American Mathematical Society, Providence, RI; Société Mathématique de France, Paris. With the collaboration of Marguerite Flexor, Papers from the Meeting "State of the Art of the Research of the Société Mathématique de France" held at the École Normale Supérieure de Lyon, Lyon, January 1997, Translated from the French by Leslie Kay.
16. Clark, T. (2014). Regular or stochastic dynamics in families of higher-degree unimodal maps. *Ergodic Theory Dynam. Systems*, 34(5):1538–1566.
17. Clark, T., de Faria, E., and Van Strien, S. (2022a). Asymptotically holomorphic methods for infinitely renormalizable c^r unimodal maps. *Ergodic Theory and Dynamical Systems*, page 1–49.
18. Clark, T., Drach, K., Kozlovski, O., and van Strien, S. (2022b). The dynamics of complex box mappings. *Arnold Math. J.*, 8(2):319–410.
19. Clark, T. and van Strien, S. (2018). Quasisymmetric rigidity in one-dimensional dynamics.
20. Clark, T. and van Strien, S. (2023). Conjugacy classes of real analytic one-dimensional maps are analytic connected manifolds.
21. Clark, T., van Strien, S., and Trejo, S. (2017). Complex bounds for real maps. *Comm. Math. Phys.*, 355(3):1001–1119.
22. De Carvalho, A., Lyubich, M., and Martens, M. (2005). Renormalization in the Hénon family. I. Universality but non-rigidity. *J. Stat. Phys.*, 121(5-6):611–669.
23. de Faria, E. (1992). *Proof of universality for critical circle mappings*. ProQuest LLC, Ann Arbor, MI. Thesis (Ph.D.)–City University of New York.
24. de Faria, E. (1999). Asymptotic rigidity of scaling ratios for critical circle mappings. *Ergodic Theory Dynam. Systems*, 19(4):995–1035.
25. de Faria, E. and de Melo, W. (1999). Rigidity of critical circle mappings. I. *J. Eur. Math. Soc. (JEMS)*, 1(4):339–392.
26. de Faria, E. and de Melo, W. (2000). Rigidity of critical circle mappings. II. *J. Amer. Math. Soc.*, 13(2):343–370.
27. de Faria, E., de Melo, W., and Pinto, A. (2006). Global hyperbolicity of renormalization for C^r unimodal mappings. *Ann. of Math. (2)*, 164(3):731–824.
28. de Faria, E. and Guarino, P. (2022). *Dynamics of circle mappings*. 33 ° Colóquio Brasileiro de Matemática. Instituto Nacional de Matemática Pura e Aplicada (IMPA), Rio de Janeiro.
29. de Melo, W. and van Strien, S. (1989). A structure theorem in one-dimensional dynamics. *Ann. of Math. (2)*, 129(3):519–546.
30. de Melo, W. and van Strien, S. (1993). *One-dimensional dynamics*, volume 25 of *Ergebnisse der Mathematik und ihrer Grenzgebiete (3) [Results in Mathematics and Related Areas (3)]*. Springer-Verlag, Berlin.
31. Deroin, B., Kleptsyn, V., and Navas, A. (2007). Sur la dynamique unidimensionnelle en régularité intermédiaire. *Acta Math.*, 199(2):199–262.
32. Douady, A. and Hubbard, J. H. (1982). Itération des polynômes quadratiques complexes. *C. R. Acad. Sci. Paris Sér. I Math.*, 294(3):123–126.
33. Douady, A. and Hubbard, J. H. (1984). *Étude dynamique des polynômes complexes. Partie I*, volume 84 of *Publications Mathématiques d'Orsay [Mathematical Publications of Orsay]*. Université de Paris-Sud, Département de Mathématiques, Orsay.

34. Douady, A. and Hubbard, J. H. (1985a). *Étude dynamique des polynômes complexes. Partie II*, volume 85 of *Publications Mathématiques d'Orsay [Mathematical Publications of Orsay]*. Université de Paris-Sud, Département de Mathématiques, Orsay. With the collaboration of P. Lavaurs, Tan Lei and P. Sentenac.
35. Douady, A. and Hubbard, J. H. (1985b). On the dynamics of polynomial-like mappings. *Ann. Sci. École Norm. Sup. (4)*, 18(2):287–343.
36. Douady, A. and Hubbard, J. H. (1993). A proof of Thurston's topological characterization of rational functions. *Acta Math.*, 171(2):263–297.
37. Erëmenko, A. E. and Lyubich, M. Y. (1992). Dynamical properties of some classes of entire functions. *Ann. Inst. Fourier (Grenoble)*, 42(4):989–1020.
38. Estevez, G. and de Faria, E. (2018). Real bounds and quasisymmetric rigidity of multicritical circle maps. *Trans. Amer. Math. Soc.*, 370(8):5583–5616.
39. Estevez, G., Smania, D., and Yampolsky, M. (2022). Renormalization of analytic multicritical circle maps with bounded type rotation numbers. *Bull. Braz. Math. Soc. (N.S.)*, 53(3):1053–1071.
40. Franks, J. and Young, L. S. (1981). A C^2 Kupka-Smale diffeomorphism of the disk with no sources or sinks. In *Dynamical systems and turbulence, Warwick 1980 (Coventry, 1979/1980)*, volume 898 of *Lecture Notes in Math.*, pages 90–98. Springer, Berlin-New York.
41. Gambaudo, J.-M., Sullivan, D., and Tresser, C. (1994). Infinite cascades of braids and smooth dynamical systems. *Topology*, 33(1):85–94.
42. Gambaudo, J.-M., van Strien, S., and Tresser, C. (1989). Hénon-like maps with strange attractors: there exist C^∞ Kupka-Smale diffeomorphisms on S^2 with neither sinks nor sources. *Nonlinearity*, 2(2):287–304.
43. Goldberg, L. R. and Keen, L. (1986). A finiteness theorem for a dynamical class of entire functions. *Ergodic Theory Dynam. Systems*, 6(2):183–192.
44. Graczyk, J. and Świątek, G. (1997). Generic hyperbolicity in the logistic family. *Ann. of Math. (2)*, 146(1):1–52.
45. Graczyk, J. and Świątek, G. (1998). *The real Fatou conjecture*, volume 144 of *Annals of Mathematics Studies*. Princeton University Press, Princeton, NJ.
46. Guarino, P. and de Melo, W. (2017). Rigidity of smooth critical circle maps. *J. Eur. Math. Soc. (JEMS)*, 19(6):1729–1783.
47. Guarino, P., Martens, M., and de Melo, W. (2018). Rigidity of critical circle maps. *Duke Math. J.*, 167(11):2125–2188.
48. Herman, M.-R. (1979). Sur la conjugaison différentiable des difféomorphismes du cercle à des rotations. *Inst. Hautes Études Sci. Publ. Math.*, (49):5–233.
49. Herman, M. R. (2018). *Notes inachevées—sélectionnées par Jean-Christophe Yoccoz*, volume 16 of *Documents Mathématiques (Paris) [Mathematical Documents (Paris)]*. Société Mathématique de France, Paris.
50. Hu, J. and Sullivan, D. P. (1997). Topological conjugacy of circle diffeomorphisms. *Ergodic Theory Dynam. Systems*, 17(1):173–186.
51. Jiang, Y. P., Morita, T., and Sullivan, D. (1992). Expanding direction of the period doubling operator. *Comm. Math. Phys.*, 144(3):509–520.
52. Kahn, J. and Lyubich, M. (2008). A priori bounds for some infinitely renormalizable quadratics. II. Decorations. *Ann. Sci. Éc. Norm. Supér. (4)*, 41(1):57–84.
53. Kahn, J. and Lyubich, M. (2009a). A priori bounds for some infinitely renormalizable quadratics. III. Molecules. In *Complex dynamics*, pages 229–254. A K Peters, Wellesley, MA.
54. Kahn, J. and Lyubich, M. (2009b). The quasi-additivity law in conformal geometry. *Ann. of Math. (2)*, 169(2):561–593.
55. Khanin, K. and Teplinsky, A. (2007). Robust rigidity for circle diffeomorphisms with singularities. *Invent. Math.*, 169(1):193–218.
56. Kiriki, S., Nakano, Y., and Soma, T. (2017). Non-trivial wandering domains for heterodimensional cycles. *Nonlinearity*, 30(8):3255–3270.
57. Kozlovski, O. (2019). On the structure of isentropes of real polynomials. *J. Lond. Math. Soc. (2)*, 100(1):159–182.

58. Kozlovski, O., Shen, W., and van Strien, S. (2007a). Density of hyperbolicity in dimension one. *Ann. of Math. (2)*, 166(1):145–182.
59. Kozlovski, O., Shen, W., and van Strien, S. (2007b). Rigidity for real polynomials. *Ann. of Math. (2)*, 165(3):749–841.
60. Kozlovski, O. and van Strien, S. (2009). Local connectivity and quasi-conformal rigidity of non-renormalizable polynomials. *Proc. Lond. Math. Soc. (3)*, 99(2):275–296.
61. Kozlovski, O. and van Strien, S. (2020). Asymmetric unimodal maps with non-universal period-doubling scaling laws. *Comm. Math. Phys.*, 379(1):103–143.
62. Levin, G., Shen, W., and van Strien, S. (2020). Positive transversality via transfer operators and holomorphic motions with applications to monotonicity for interval maps. *Nonlinearity*, 33(8):3970–4012.
63. Levin, G. and van Strien, S. (1998). Local connectivity of the Julia set of real polynomials. *Ann. of Math. (2)*, 147(3):471–541.
64. Li, S. and Shen, W. (2006). Smooth conjugacy between S-unimodal maps. *Nonlinearity*, 19(7):1629–1634.
65. Lyubich, M. (1997). Dynamics of quadratic polynomials. I, II. *Acta Math.*, 178(2):185–247, 247–297.
66. Lyubich, M. (1999). Feigenbaum-Coullet-Tresser universality and Milnor's hairiness conjecture. *Ann. of Math. (2)*, 149(2):319–420.
67. Lyubich, M. and Martens, M. (2011). Renormalization in the Hénon family, II: the heteroclinic web. *Invent. Math.*, 186(1):115–189.
68. Lyubich, M. and Yampolsky, M. (1997). Dynamics of quadratic polynomials: complex bounds for real maps. *Ann. Inst. Fourier (Grenoble)*, 47(4):1219–1255.
69. Lyubich, M. Y. (1983). Some typical properties of the dynamics of rational mappings. *Uspekhi Mat. Nauk*, 38(5(233)):197–198.
70. Mañé, R., Sad, P., and Sullivan, D. (1983). On the dynamics of rational maps. *Ann. Sci. École Norm. Sup. (4)*, 16(2):193–217.
71. Marden, A. (1974). The geometry of finitely generated kleinian groups. *Ann. of Math. (2)*, 99:383–462.
72. Marden, A. (2007). *Outer circles*. Cambridge University Press, Cambridge. An introduction to hyperbolic 3-manifolds.
73. Margulis, G., Mohammadi, A., and Oh, H. (2014). Closed geodesics and holonomies for Kleinian manifolds. *Geom. Funct. Anal.*, 24(5):1608–1636.
74. Martens, M. (1994). Distortion results and invariant Cantor sets of unimodal maps. *Ergodic Theory Dynam. Systems*, 14(2):331–349.
75. Martens, M. and de Melo, W. (1999). The multipliers of periodic points in one-dimensional dynamics. *Nonlinearity*, 12(2):217–227.
76. Martens, M., de Melo, W., and van Strien, S. (1992). Julia-Fatou-Sullivan theory for real one-dimensional dynamics. *Acta Math.*, 168(3-4):273–318.
77. Maskit, B. (1988). *Kleinian groups*, volume 287 of *Grundlehren der mathematischen Wissenschaften [Fundamental Principles of Mathematical Sciences]*. Springer-Verlag, Berlin.
78. McMullen, C. (2020). *Riemann surfaces, dynamics and geometry*. Harvard University.
79. McMullen, C. T. (1994). *Complex dynamics and renormalization*, volume 135 of *Annals of Mathematics Studies*. Princeton University Press, Princeton, NJ.
80. McMullen, C. T. (1996). *Renormalization and 3-manifolds which fiber over the circle*, volume 142 of *Annals of Mathematics Studies*. Princeton University Press, Princeton, NJ.
81. McMullen, C. T. and Sullivan, D. P. (1998). Quasiconformal homeomorphisms and dynamics. III. The Teichmüller space of a holomorphic dynamical system. *Adv. Math.*, 135(2):351–395.
82. Merenkov, S. (2019). No round wandering domains for C^1-diffeomorphisms of tori. *Ergodic Theory Dynam. Systems*, 39(11):3127–3135.
83. Milnor, J. and Tresser, C. (2000). On entropy and monotonicity for real cubic maps. *Comm. Math. Phys.*, 209(1):123–178. With an appendix by Adrien Douady and Pierrette Sentenac.
84. Mostow, G. D. (1968). Quasi-conformal mappings in n-space and the rigidity of hyperbolic space forms. *Inst. Hautes Études Sci. Publ. Math.*, (34):53–104.

85. Nicholls, P. J. (1989). *The ergodic theory of discrete groups*, volume 143 of *London Mathematical Society Lecture Note Series*. Cambridge University Press, Cambridge.
86. Norton, A. and Sullivan, D. (1996). Wandering domains and invariant conformal structures for mappings of the 2-torus. *Ann. Acad. Sci. Fenn. Math.*, 21(1):51–68.
87. Oh, H. and Winter, D. (2017). Prime number theorems and holonomies for hyperbolic rational maps. *Invent. Math.*, 208(2):401–440.
88. Patterson, S. J. (1976). The limit set of a Fuchsian group. *Acta Math.*, 136(3-4):241–273.
89. Pinto, A. A. and Sullivan, D. (2006). The circle and the solenoid. *Discrete Contin. Dyn. Syst.*, 16(2):463–504.
90. Prasad, G. (1973). Strong rigidity of **Q**-rank 1 lattices. *Invent. Math.*, 21:255–286.
91. Rempe-Gillen, L. and van Strien, S. (2015). Density of hyperbolicity for classes of real transcendental entire functions and circle maps. *Duke Math. J.*, 164(6):1079–1137.
92. Ruelle, D. and Sullivan, D. (1975). Currents, flows and diffeomorphisms. *Topology*, 14(4):319–327.
93. Shen, W. (2003). Bounds for one-dimensional maps without inflection critical points. *J. Math. Sci. Univ. Tokyo*, 10(1):41–88.
94. Shub, M. and Sullivan, D. (1985). Expanding endomorphisms of the circle revisited. *Ergodic Theory Dynam. Systems*, 5(2):285–289.
95. Singer, D. (1978). Stable orbits and bifurcation of maps of the interval. *SIAM J. Appl. Math.*, 35(2):260–267.
96. Smania, D. (2001). Complex bounds for multimodal maps: bounded combinatorics. *Nonlinearity*, 14(5):1311–1330.
97. Smania, D. (2020). Solenoidal attractors with bounded combinatorics are shy. *Ann. of Math. (2)*, 191(1):1–79.
98. Sullivan, D. (1976a). A counterexample to the periodic orbit conjecture. *Inst. Hautes Études Sci. Publ. Math.*, (46):5–14.
99. Sullivan, D. (1976b). Cycles for the dynamical study of foliated manifolds and complex manifolds. *Invent. Math.*, 36:225–255.
100. Sullivan, D. (1976c). A new flow. *Bull. Amer. Math. Soc.*, 82(2):331–332.
101. Sullivan, D. (1979). The density at infinity of a discrete group of hyperbolic motions. *Inst. Hautes Études Sci. Publ. Math.*, (50):171–202.
102. Sullivan, D. (1981). On the ergodic theory at infinity of an arbitrary discrete group of hyperbolic motions. In *Riemann surfaces and related topics: Proceedings of the 1978 Stony Brook Conference (State Univ. New York, Stony Brook, N.Y., 1978)*, volume 97 of *Ann. of Math. Stud.*, pages 465–496. Princeton Univ. Press, Princeton, N.J.
103. Sullivan, D. (1982). Disjoint spheres, approximation by imaginary quadratic numbers, and the logarithm law for geodesics. *Acta Math.*, 149(3-4):215–237.
104. Sullivan, D. (1983). Conformal dynamical systems. In *Geometric dynamics (Rio de Janeiro, 1981)*, volume 1007 of *Lecture Notes in Math.*, pages 725–752. Springer, Berlin.
105. Sullivan, D. (1984). Entropy, Hausdorff measures old and new, and limit sets of geometrically finite Kleinian groups. *Acta Math.*, 153(3-4):259–277.
106. Sullivan, D. (1985). Quasiconformal homeomorphisms and dynamics. I. Solution of the Fatou-Julia problem on wandering domains. *Ann. of Math. (2)*, 122(3):401–418.
107. Sullivan, D. (1988). Differentiable structures on fractal-like sets, determined by intrinsic scaling functions on dual Cantor sets. In *The mathematical heritage of Hermann Weyl (Durham, NC, 1987)*, volume 48 of *Proc. Sympos. Pure Math.*, pages 15–23. Amer. Math. Soc., Providence, RI.
108. Sullivan, D. (1991). Renormalization, Zygmund smoothness and the Epstein class. In *Chaos, order, and patterns (Lake Como, 1990)*, volume 280 of *NATO Adv. Sci. Inst. Ser. B: Phys.*, pages 25–34. Plenum, New York.
109. Sullivan, D. (1992). Bounds, quadratic differentials, and renormalization conjectures. In *American Mathematical Society centennial publications, Vol. II (Providence, RI, 1988)*, pages 417–466. Amer. Math. Soc., Providence, RI.

110. Sullivan, D. (1993). Linking the universalities of Milnor-Thurston, Feigenbaum and Ahlfors-Bers. In *Topological methods in modern mathematics (Stony Brook, NY, 1991)*, pages 543–564. Publish or Perish, Houston, TX.
111. Sullivan, D. P. and Thurston, W. P. (1986). Extending holomorphic motions. *Acta Math.*, 157(3-4):243–257.
112. Świątek, G. (1988). Rational rotation numbers for maps of the circle. *Comm. Math. Phys.*, 119(1):109–128.
113. van Strien, S. and Vargas, E. (2004). Real bounds, ergodicity and negative Schwarzian for multimodal maps. *J. Amer. Math. Soc.*, 17(4):749–782.
114. van Strien, S. J. (1981). On the bifurcations creating horseshoes. In *Dynamical systems and turbulence, Warwick 1980 (Coventry, 1979/1980)*, volume 898 of *Lecture Notes in Math.*, pages 316–351. Springer, Berlin-New York.
115. Wang, J. and Yang, H. (2021). A question of Norton-Sullivan in the analytic case. *Int. Math. Res. Not. IMRN*, (22):17201–17219.
116. Yampolsky, M. (1999). Complex bounds for renormalization of critical circle maps. *Ergodic Theory Dynam. Systems*, 19(1):227–257.
117. Yampolsky, M. (2002). Hyperbolicity of renormalization of critical circle maps. *Publ. Math. Inst. Hautes Études Sci.*, (96):1–41 (2003).
118. Yoccoz, J.-C. (1984). Il n'y a pas de contre-exemple de Denjoy analytique. *C. R. Acad. Sci. Paris Sér. I Math.*, 298(7):141–144.

Sullivan's Juvenilia: Surgery and Algebraic Topology

Shmuel Weinberger

Contents

Dennis Sullivan's early work, despite being strongly motivated by geometric problems, had a strong algebraic nature, certainly much more so than his later work on foliations, Kleinian groups, and dynamics where geometry and analysis play central roles. The aim of this essay is to try to explain to the non-expert some of this early work and indicate some of its impact to the current moment.

Having made my aim to try, I shall succeed. After several attempts, I realized I cannot genuinely explain his work in any detail, nor its implications (the former, because of its difficulty, especially for the non-topologist, and the latter because of its magnitude). Instead, I've decided to shoot for explaining some of the context, the audacity and beauty of the ideas, and then hint about some of the mathematical areas that these ideas opened up.

S. Weinberger
Department of Mathematics,
University of Chicago,
5734 S. University Avenue Chicago, IL 60637-1514,
e-mail: shmuel@math.uchicago.edu

Partially supported by an NSF grant.

H. Holden, R. Piene (eds.), *The Abel Prize 2018-2022*, The Abel Prize,
https://doi.org/10.1007/978-3-031-33973-8_27

I apologize to all (i.e., to frustrated readers and to Dennis) for omissions, obscurities, and inaccuracies.

1 Surgery and its classifying spaces

Dennis Sullivan's thesis and early work was on the subject of surgery theory, which we *define* to be the problem of classifying manifolds in dimension greater than 4. (That everything is different in dimension 4 was discovered in 1981–82 through the amazing work of Freedman and Donaldson, but that is another story.)

The possibility of a classification of manifolds, and that higher dimensions would be more effaceable than low, was demonstrated by Smale [97], with the proof of the high-dimensional Poincaré conjecture.[1] He showed that any smooth manifold of dimension at least 5 that is homotopy equivalent to the sphere is homeomorphic to it (and, indeed the homeomorphism could be taken to be simplicial in a suitable triangulation). On the other hand, Milnor first showed by example that there is a smooth manifold homeomorphic to S^7 that is not diffeomorphic to it, and then with Kervaire classified such manifolds, i.e., determined the differentiable structures on the sphere (in dimension at least 5) — or at least showed that the number is finite, and gave an almost complete reduction to homotopy theory. We will return to the work of Kervaire and Milnor later, but see [95] for a very useful and much more thorough discussion.

Browder and Novikov independently took the next critical step: extending these techniques beyond the sphere, to all simply connected manifolds. Here is one of their theorems:

Theorem. *There are infinitely many closed smooth manifolds homotopy equivalent to a given closed compact smooth simply connected manifold M^n ($n > 4$) iff for some $0 < 4i < n$, $H^{4i}(M;\mathbb{Q}) \neq 0$.*

Indeed, there is a map from the set of such manifolds to $\oplus H^{4i}(M;\mathbb{Q})$, defined using Pontrjagin classes[2], which is finite to one, and whose image is "pretty dense." (In a suitable parametrization, the image, while not a subgroup, contains a lattice, and is contained in one.) In the case of $M = S^n$, this contains the finiteness of the set of the differential structures that is the "main theorem" of [59].

Sullivan's early work, viewed narrowly, is about making this more computable. The following "Sullivan–Wall surgery exact sequence" is a reorganization of this methodology.

---

[1] The key theorem here is the h-cobordism theorem, that gives a condition for a compact simply connected manifold with boundary to be a product $M \times [0,1]$. The nonsimply connected version of this (due to Barden–Mazur–Stallings) is likewise fundamental, and is one of the first deep connections between manifold topology and algebraic-K-theory.

[2] Pontrjagin classes are cohomology classes in $H^{4i}(M;\mathbb{Q})$ that are measures of the nontriviality of the tangent bundle of M (see [76]).

$$\cdots \longrightarrow L_{n+1}(\pi) \longrightarrow S(M) \longrightarrow [M : G/\mathrm{Cat}] \longrightarrow L_n(\pi).$$

Now for some definitions. Cat is a category of manifolds, either smooth or PL at first, but later extended to Top, the category of topological manifolds.

S^{Cat} is the main object of study. It is the $\{(W, f)\,|\,W$ is a Cat manifold, and $f : W \to M$ is a simple homotopy equivalence$\}/$Cat isomorphism and homotopy. In other words, (W, f) and (W', f') represent the same element of S^{Cat} if there is a homeomorphism $F : W \to W'$ so that the composite Ff' is homotopic to f.

G/Cat is a space that Sullivan introduced, and as usual in algebraic topology, $[M : G/\mathrm{Cat}]$ denotes the homotopy classes of maps from M to G/Cat. $L_n(\pi)$ are abelian groups that only depend on the dimension n of M (mod 4) and its fundamental group π, and on which loops in π_1, M is orientable (we suppress this last piece of data from the notation). For the trivial group, the calculation (due to Kervaire–Milnor, although the definition came later, in the work of Browder and Novikov, as we've seen) is

$$L_n(e) = \mathbb{Z}, 0, \mathbb{Z}_2, 0 \quad \text{for } n \cong 0, 1, 2, 3 \bmod 4, \text{ respectively.}$$

The L-groups were defined in [110], and the map $[M : G/\mathrm{Cat}] \to L_n(\pi)$ is called the surgery obstruction map. Understanding L-groups and this map have become the core of surgery theory. We will return to this at the end of this section.

Exactness is a bit unusual in this setting, as, aside from L, all the objects involved are sets and not groups.

These sets all have distinguished elements, for S the identity element, and for $[M : G/\mathrm{Cat}]$ the constant map. $L_{n+1}(\pi) \to S(M)$ is actually an action of the group on the set, and exactness at that point in the sequence means that the isotropy of an element α is the image of $[M \times [0,1] : G/\mathrm{Cat}]$, where the boundary conditions are that the restriction of the function to the boundary be the image of α in $[M : G/\mathrm{Cat}]$.

This is a bit complicated, but it gets better when Cat = PL (and even better when Cat = Top [61]). The upshot will be that one can put abelian group structures on everything in sight, and that they become a sequence of groups and homomorphisms in the conventional sense, but this realization came a bit later from work of Quinn, Ranicki, and Siebenmann. What Sullivan showed was that the space G/PL was completely understandable. More precisely, $[M : G/\mathrm{PL}]$ is an abelian group, and that one can understand this abelian group via its localizations.

More precisely, with $\mathbb{Z}_{(2)}$ denoting rational numbers with odd denominators, $\mathbb{Z}[1/2]$ denoting the ones with denominators a power of 2, and KO the Grothendieck group of real vector bundles (we will return to K-theory in Section 2), we have

$$[M : G/\mathrm{PL}] \otimes \mathbb{Z}[1/2] \cong KO^0(M) \otimes \mathbb{Z}[1/2]$$
$$[M : G/\mathrm{PL}] \otimes \mathbb{Z}_{(2)} \cong [M : T] \oplus \bigoplus_{i>0} H^{4i+2}(M; \mathbb{Z}/2\mathbb{Z}) \oplus H^{4i+4}(M; \mathbb{Z}_{(2)})$$

and $[M : T]$ fits into an exact sequence

$$0 \longrightarrow H^4(M; \mathbb{Z}_{(2)}) \longrightarrow [M : T] \longrightarrow H^2(M; \mathbb{Z}/2\mathbb{Z}) \longrightarrow H^5(M; \mathbb{Z}_{(2)}) \longrightarrow \cdots$$

where the map $H^2(M;\mathbb{Z}/2\mathbb{Z}) \longrightarrow H^5(M;\mathbb{Z}_{(2)})$ is given by the Bockstein of cup-squaring (which is a homomorphism mod 2), i.e., βSq^2.

In the topological category, the answer, according to [61] is even simpler,

$$[M : G/\mathrm{Top}] \otimes \mathbb{Z}[1/2] \cong KO^0(M) \otimes \mathbb{Z}[1/2]$$
$$[M : G/\mathrm{Top}] \otimes \mathbb{Z}_{(2)} \cong \bigoplus_{i>0} H^{4i+2}(M;\mathbb{Z}/2\mathbb{Z}) \oplus H^{4i+4}(M;\mathbb{Z}_{(2)}).$$

An immediate consequence of these calculations is that, for example $S^{\mathrm{PL}}(S^2 \times S^6) = \mathbb{Z}/2\mathbb{Z} \times \mathbb{Z}/2\mathbb{Z}$, or indeed a formula for $S^{\mathrm{PL}}(M)$ for any simply connected manifold M (of dimension at least 5). But, there's much more included in this work, as I'd like to now explain.

1. Sullivan's actual result describes the structure of F/PL in terms of localizations. These tensorings of abelian groups by subrings of the rational numbers are actually localizations in the sense of algebra. Already, Serre had introduced and brilliantly used the idea of doing homotopical calculations when inverting some or all primes [94]. Sullivan here gave an example of an important space, whose description requires such acrobatics. At 2, it is (aside from a low-dimensional blemish, repaired in the topological category) a product of Eilenberg–MacLane spaces, and away from 2 (i.e., at odd primes) it is BO, the infinite-dimensional real Grassmannian.

2. The method for proving these results is also important. (It is worth reading his autobiographical essay in the reprint [101] of the original MIT notes, for the lines of thought leading in this direction. Many beautiful ideas entered into the attempt to understand this space — especially revolving around singular spaces – which were distilled out in the final version where things locked into place, yet recur in other later works.)

In more modern terms, we would say that he was exploiting the fact that F/PL is a module over MSO, the spectrum that represents smooth bordism, i.e., the homology theory whose cycles are oriented manifolds. The structure of unoriented bordism was analyzed by René Thom in the work that earned him the Fields medal; the oriented case was much more intricate and was analyzed by Averbukh, Milnor, Novikov and Wall. In any case, in both cases the spectrum is quite simple at the prime 2 and more complicated at odd primes (although there is no odd torsion in the bordism groups). This difference is reflected in the difference between the behavior of G/PL at 2 and at odd primes.

Sullivan himself phrased this in terms of considering the interaction between the geometric interpretation of $[M : G/\mathrm{PL}]$ and bordism, understanding how obstructions to completing surgery are affected by crossing with a smooth manifold.

3. He discovered that PL (block) bundles have $KO[1/2]$ orientations. (This means that oriented PL manifolds have some kind of Poincaré duality for real K-theory, ignoring issues at the prime 2.) These are new refined characteristic classes for the PL setting. The work of Kirby–Siebenmann shows that they are topological invariants, as well (the rational version of this is Novikov's Fields medal winning theorem on the topological invariance of rational Pontrjagin classes [80]). Sullivan already

knew that in the case of smooth manifolds, these are the symbol of the signature operator on the manifold (for any Riemannian metric) (see [54]).

4. He also gave an a priori geometric way of thinking about maps into G/PL, using what he called the "characteristic variety." A map of M into G/PL is exactly equivalent to a series of elements of $\mathbb{Z}/2$, $\mathbb{Z}/n$ or $\mathbb{Z}$ associated to each element of this characteristic variety.

This is a residue of the geometric thinking that I mentioned before. Two aspects of this, though, are of lasting significance. The first is the use of $\mathbb{Z}/n\mathbb{Z}$ manifolds and their cobordism for probing torsion issues. More generally, this suggests a philosophy of varying problems from manifold settings to different kinds of singular ones to be able to identify more subtle issues – which either crystalize there, or become more less subtle.

The other is that it suggests a way of adding the elements of $[M : F/\mathrm{PL}]$ — by adding the invariants assigned to the characteristic variety, associated to each of the elements. These numbers are signatures (and Kervaire–Arf invariants), and the Hirzebruch signature formula, relating signatures[3] to the usual characteristic classes of bundles, is highly nonlinear [49], so this is a brand new "addition." It turns out that with this addition, the surgery map becomes a group homomorphism.[4]

1.1 Further developments

I'd like to discuss (very selectively) some of the ramifications of these ideas. Needless to say, having a directly computable version of the homotopical part of surgery theory has been foundational for many further directions. For example, the classification of free PL (or topological) group actions of cyclic groups on the sphere in [110] and completed in [65] would have been impossible without this work.[5]

---

[3] If W is a closed oriented manifold of dimension $4k$, a multiple of 4, then $\mathrm{sign}(M)$ is by definition the signature of the symmetric bilinear form on $H^{2k}(W;\mathbb{Q})$ given by cup product.

[4] I do not know the history of this fact — whether Sullivan knew this, or whether it is due to John Morgan (who told some friends of mine); it was also essentially in the final paper by Siebenmann in [61], which is based on ideas of Casson and Quinn, that accomplishes more, namely that the whole sequence can be turned into a sequence of groups and homomorphisms, and is, of course, written in the topological setting.

[5] Unpublished work of Rothenberg managed to deal with the case of free circle actions, and Browder and Livesay gave an approach to $\mathbb{Z}/2\mathbb{Z}$ actions (explained beautifully in [31]) that did not rely on Sullivan's work. Their results, in retrospect, use the PL Poincaré conjecture directly in their work, while for Sullivan that enters in the calculation of the homotopy groups of G/PL (which is a part of the work involved in unraveling its structure).

Localization

Firstly, localization of spaces. The basic idea of taking different spaces at different primes and combining them[6] to produce "monsters" arose in the work of Hilton and Roitberg (together with Mislin, they wrote a very useful tract on localization [53]); the technique is generally called Zabrodsky mixing. Adams [3] described their work memorably as, "It is somewhat reminiscent of the descriptions of monsters in medieval bestiaries; you put the head of a lion on the body of a horse. A more subtle technique is to take parts from the same animal and put them back together a different way round; this is as if you take your lion, cut his head off and then stick the same head back on, but facing to the rear." Their application was to give the first examples of finite-dimensional H-spaces (spaces that admit a continuous multiplication (with identity), that are not homotopy equivalent to Lie groups. For Sullivan, on the other hand, the mixing was not done to create monsters — it was an inevitability necessary to tame them.

This had significant ramification in the theory of group actions. (Here are some typical references [7].) Mixing ideas can be used to establish that the essential problem of understanding group actions for a finite group G, are often concentrated at the primes dividing the order of G. For example, to show that some simply connected mod 2 homology sphere S^{2k+1} has a free involution, one should mix $\mathbb{RP}^{2k+1}$ with the localization of that manifold away from 2. (One can't do this in even dimensions because $\mathbb{RP}^{2k}$ is not a rational homology sphere; but then one can construct orientation preserving actions with 2 fixed points.) That at least does the homotopy theory. Getting this space to be a manifold whose 2-fold cover is the original one is a problem for nonsimply connected surgery (see [17] for this specific result, and [30] for general group free group actions on homology spheres). We will return to the theory of group actions in the next section in the context of *completions*.

Bordism and classifying spaces

The method of analysis of G/PL has also been very influential. Sullivan used the cobordism invariance of surgery obstructions, and how obstructions behave under products, giving homomorphisms

$$\Omega_n(G/\mathrm{PL}) \longrightarrow L_n(e),$$

which factor through, away from 2, a map

$$\Omega_n(G/\mathrm{PL}) \otimes_{\Omega} \mathbb{Z}[1/2] \to L_n(e) \otimes \mathbb{Z}[1/2],$$

where Ω denotes bordism (in the next section, it will denote loop space; please don't get confused) and $\Omega_n(X)$ can be viewed as a $\mathbb{Z}/4$ graded module over smooth bor-

[6] Assuming this is possible. To combine two local spaces (which are local with respect to complementary sets of primes) into an integral one needs them to be rationally compatible.

dism, where an oriented manifold W of dimension $4k$ is identified with the integer $\text{sign}(W)$. The first gives rise to characteristic classes in G/PL at the prime 2 (see [102, 92, 78]). The second gives the map to KO theory, using an analogue of the connection between bordism and K-theory discovered by Conner and Floyd [28].

Given the ubiquity of bordism invariants in topology, many spaces can be analyzed by this method. Wall [111] gave a formula for surgery obstruction by directly adapting Sullivan's argument. Other classifying spaces arising in group actions or classifying spaces for nonlocally flat neighborhoods are also analyzed by this method (see [26, 25]).

Intersection homology[7] [18, 41] is another venue where one has a class of spaces that have a signature (the whole point of IH is its self-duality for, at least, spaces with even codimensional strata). His method leads to half of the beautiful theorem of P. Siegel that Witt spaces (i.e., the polyhedra for which IH is a self dual sheaf) form a cycle theory for $KO[1/2]$. The class defined by the identity map $X \to X$ is Sullivan's $KO[1/2]$ class, when X is a PL manifold. (See [39] for variants.)

KO[1/2] orientation

The $KO[1/2]$ orientation of PL bundles, actually definable for topological microbundles (thanks to Kirby–Siebenamann; formally the point is that Top/PL only has homotopy at the prime 2[8]), is, as mentioned above, the symbol of the signature operator for smooth manifolds.[9] This has equivariant generalizations (see [67, 68, 91]) which play a role in the next subsection. It is also the first hint of a deep systematic connection between surgery theory and the index theory of elliptic operators, a major direction in subsequent decades (see e.g. [49, 50, 51, 52, 113]) especially in light of analytic approaches to the Novikov conjecture.

We will return to this in the final subsection.

Hauptvermutung and triangulation (and beyond)

Sullivan himself applied his theory to the Hauptvermutung. The Hauptvermutung is the question about the uniqueness of the triangulation of any polyhedron.

This question goes back to Poincaré, who seemed to be interested in this for showing the topological invariance of homology. Of course, homotopy invariance is not so hard to show for homology, so the problem perhaps lost some of its urgency during the decades that it was open. Its solution in various setting is of immense significance.

---

[7] Sullivan played an important bit role in the development of intersection homology: he conjectured that theories invented by Cheeger and by Goresky–MacPherson were isomorphic, proved in [42].

[8] Indeed it has only one nonzero homotopy group $\pi_3(\text{Top}/\text{PL}) = \mathbb{Z}/2\mathbb{Z}$.

[9] See [107] for how to define signature operators on PL manifolds, and [106] for Lipschitz Riemannian manifolds, and their connection to Pontrjagin classes.

Milnor [74], using work of Mazur and earlier work of J. H. C. Whitehead, had disproved this for general polyhedra: if one takes homotopy equivalent non-diffeomorphic 3-dimensional lens spaces,[10] say $L_5(1,1)$ and $L_5(2,3)$ then $L_5(1,1) \times D^4 \cup c(L_5(1,1) \times S^3)$ and $L_5(2,3) \times D^4 \cup c(L_5(2,3) \times S^3)$ (cX here denotes the cone on X) are homeomorphic, but are not combinatorially isomorphic, i.e., have no combinatorially isomorphic subdivisions. All homeomorphisms require some "infinite construction."

After this, the question of the Hauptvermutung for manifolds became a central problem. It is tied up with the question of the existence of PL triangulations for manifolds, as well — and of course, the topological invariance (and definability) of combinatorial invariants.

In 1966, Novikov proved that the Pontrjagin classes $p_i(M) \in H^{4i}(M;\mathbb{Q})$ were topological invariants. Let me remind you about why this is so remarkable: These invariants are actually invariants of the tangent bundle of a smooth manifold, and by definition they are Chern classes of the complexified tangent bundle $p_i = c_{2i}(\tau \otimes \mathbb{C})$. Complexification is a very "vector space" linear notion, so it would seem that no part of the definition makes any sense in a topological setting — and so why would these be topological invariants? (Moreover, they are not smooth invariants in $H^{4i}(M;\mathbb{Z})$, although that is where this definition places them.)

His proof actually works to show that the L-classes (the combinations of the Pontrjagin classes entering Hirzebruch's signature theorem) are well defined. Those have denominators, and that is the source of the rationality that arises. He shows that when one approximates a homeomorphism by a smooth map, the signatures of transverse inverse images of certain submanifolds are the same as that of the submanifold. (For a diffeomorphism this is a tautology!)

This work led him to conjecture that certain submanifolds associated to group cohomology of the fundamental group (so for example for manifolds with free abelian fundamental group, these would be the submanifolds whose transverse inverse images of subtori under any smooth map to a torus inducing an isomorphism on π_1) have signatures that are homotopy invariants of the manifold. This is called the "Novikov conjecture" and it, and its cousins the Borel conjecture, the Farrell–Jones conjecture, and the Baum–Connes conjecture have been central in topology, differential geometry, and index theory for fifty years.

Noting that the invariants associated to a homotopy equivalence of simply connected manifolds in $[M : F/\mathrm{PL}]$ are signatures (and Kervaire invariants) of transverse inverse images of $\mathbb{Z}/n$ submanifolds of M, Sullivan readily deduced that for simply connected manifolds of dimension at least 5 with no 2-torsion in $H^4(M;\mathbb{Z})$ any homeomorphism could be homotoped to a PL homeomorphism: a Hauptvermutung.

---

[10] Lens spaces are quotients of the sphere by free linear actions of cyclic groups: in the example here, one uses representations of $\mathbb{Z}/5\mathbb{Z}$, which are all automatically complex — and the eigenvalues are $\exp(2\pi i/5)$ with multiplicity 2 in the first example and $\exp(4\pi i/5)$ and $\exp(6\pi i/5)$ in the second. de Rham, using the theory of Reidemeister torsion, gave the diffeomorphic (= PL, in this case) classification of these manifolds in what is, in retrospect, one of the earliest applications of algebraic K-theory to topology. The book [27] is an excellent introduction to this work.

Around the same time, Lashof and Rothenberg had proven the same for 4-connected manifolds. However, all of this work was completely superseded by the work of Kirby and Siebenmann that completely elucidated the matter:

1. They showed that there was a good triangulation theory for topological manifolds of dimension > 4, so that a triangulation of M exists iff the topological tangent "bundle" of $M \to \mathrm{BTop}$ lifts to BPL.
2. $\mathrm{BPL} \to \mathrm{BTop}$ is an isomorphism of homotopy groups except in dimension 4; consequently the obstruction to the existence of a PL triangulation lies in $H^4(M;\mathbb{Z}/2\mathbb{Z})$ and to the uniqueness (Hauptvermutung) in $H^3(M;\mathbb{Z}/2\mathbb{Z})$.
3. G/Top is the same as G/PL away from 2; at 2 it differs in being even nicer: $[M : G/\mathrm{Top}] \cong \bigoplus_{i\geq 0} H^{4i+2}(M;\mathbb{Z}/2\mathbb{Z}) \oplus H^{4i+4}(M;\mathbb{Z}_{(2)})$.
4. Topological manifolds have handlebody structures (the basic structure Smale discovered using Morse theory for smooth manifolds, that he used for the h-cobordism theorem), and have transversality.

At the end of their magnificent work, topological manifolds were geometrically as understandable as smooth ones, and algebraically much better.

Around this time, Sullivan made an important observation about the proof of Novikov's theorem. Homeomorphism was used in a very weak way. What was relevant about a map $h\colon M \to N$ to guarantee that h^* pulls back rational Pontrjagin classes was that h is a hereditary homotopy equivalence, i.e., for every open subset U of N, the map from $h^{-1}(U) \to U$ is a proper homotopy equivalence. This condition on a map is also referred to as cell-like (abbreviated CE), it boils down to the condition that the inverse image of every point of N is nullhomotopic in every neighborhood of itself. This property for a map is basic to "Bing topology" that was responsible for many beautiful examples in topology (like spaces which have no manifold points, which when crossed with $\mathbb{R}$ are manifolds) and was key to the proof of the four-dimensional Poincaré conjecture by Freedman [39].

In response to this observation, Siebenmann [96] proved the following using the crucial torus trick of Kirby [60] that ultimately underlay all of their work above:

Theorem. *If $f\colon M \to N$ is a CE map between manifolds of dimension $n > 4$, then f is a uniform limit of homeomorphisms.*

Subsequently, Chapman and Ferry [24] gave the metric condition which tells when a map is close to a homeomorphism. The following consequence,[11] due to Ferry [37], conjectured by Siebenmann, is the following:

Theorem. *For every compact closed manifold M (with a metric), there is an $\varepsilon > 0$ such that, if $f\colon M \to N$ is a map to another connected manifold of (at most)* $\dim M$ *such that* $\mathrm{diam}(f^{-1}(n)) < \varepsilon$ *for every point n of N, then f is homotopic to a homeomorphism.*

[11] Ferry proved it in dim > 4, but it turns out to be true in all dimensions because of the work of [39] and [81].

The significance of these developments cannot be overestimated. Ultimately, they paved the way for the re-entrance of geometry into high-dimensional topology.

In a completely different direction, Edwards proved the following astounding extension of Siebenmann's theorem:

Theorem. *Suppose that X is finite-dimensional, and* $\dim X > 4$. *Suppose that a map from a closed manifold* $M \to X$ *is CE, then the map is a uniform limit of homeomorphisms iff X has the disjoint disk property (DDP).*

DDP means this: If φ, ψ are maps of $D^2 \to X$ then φ, ψ can be approximated arbitrarily closely by disjoint embeddings. General position gives this for manifolds of dimension > 4. According to a theorem of Quinn [86, 87] a connected finite-dimensional ANR homology manifold[12] of dim > 4 is a cell-like image of a manifold if, for example, it has an open subset which is. As a result, if P is any integral homology sphere, the suspension of the suspension of P is a manifold (and therefore the sphere). The link of a singular point in this polyhedron is the suspension of P, which is not a manifold (it has 2 singular points with nonsimply connected deleted neighborhoods), so this gives a non-PL triangulation of the sphere S^n, when $n > 4$. (This phenomenon was discovered by Cannon and Edwards before Edwards proved the above theorem.)

3-dimensional homology spheres play a special role in the above theory. The $\mathbb{Z}/2\mathbb{Z}$ Kirby-Siebenmann obstruction is due to the fact that there is a $\mathbb{Z}/2\mathbb{Z}$ obstruction (Rochlin's theorem) to them bounding signature $= 0$ manifolds. The possibility that twice every such homology sphere bounded an acyclic manifold raised the possibility that all manifolds of dim > 4 have (non PL) triangulations, but Manolescu [71] showed that this is not the case using Seiberg–Witten Floer homology, and that there are nontriangulable manifolds in all dimensions > 3.[13]

Remark. The Hauptvermutung for polyhedra in the sense of deciding when polyhedra are homeomorphic involves, besides the work on manifolds already mentioned, "controlled topology" — that is redoing all of geometric topology with attention paid to the sizes of the objects being constructed. (The theorems of Ferry and Chapman–Ferry mentioned above are great examples; the theorem of Quinn on making homology manifolds CE images is a landmark application.)

Besides the Kirby–Siebenmann obstruction, there are obstructions related to the algebraic K-theory groups $K_i(\mathbb{Z}\pi)$ of the fundamental groups of links of strata for $i \leq 1$ as well as a contribution from L-theory to understand the manifold issues (see [6]). (The Milnor example with point singularities relates to K_1 — via the work of de Rham; if one were to consider the 4th suspensions of these lens spaces, although the manifold points are the same as in the Milnor example, these two polyhedra are not homeomorphic.)

---

[12] A space X is a d-dimensional homology manifold, if for all x, the relative homology groups for X, $H^i(X, X - x)$ are the same as those for $\mathbb{R}^d$.

[13] Dimension 4 was a consequence of the work of Freedman and either Taubes, or Casson (see [4]). It now can be deduced from the 3-dimensional Poincaré conjecture.

This "Hauptvermutung" is closely related to another problem to which Sullivan contributed in unpublished work, the nonlinear similarity problem.

We mentioned the work of de Rham earlier on when lens spaces are diffeomorphic. The work of Kirby–Siebenmann (or Chapman [23]) shows that de Rham's obstructions remain valid in the topological category. In other words, two representations of $\mathbb{Z}/n\mathbb{Z}$ that are free (as actions on the unit sphere) are conjugate (as group actions on the sphere) iff they are conjugate as linear representations.

The nonlinear similarity question is whether this is true for all representations of finite (or compact) groups? Rothenberg [90] showed that one can modify de Rham's argument to show that a PL conjugacy always gives rise to a linear conjugacy, so this is a Hauptvermutung type question. Sullivan and Schultz [93] independently used the Sullivan–Kirby–Siebenmann analysis of $G/\mathrm{Top}[1/2]$ to prove that nonlinear similarity implies linear similarity for odd p-groups.

In the early 80s, the general picture was worked out in a way that, I think, is a tribute to Sullivan's instinct. Cappell and Shaneson [19] gave counterexamples for $\mathbb{Z}/4k$, $k > 1$, but Hsiang and Pardon [56] and Madsen and Rothenberg [67, 68] showed that for odd order groups nonlinear and linear similarity are the same. Aside from torsion phenomena (later analyzed by Hambleton and Pederson [48, 47]) the story is this: the equivariant signature operator is a topological invariant (see [91, 89]) — it is an orientation in KO^G, generally, when G is odd order, and the equivariant signature operator is a strong enough invariant to distinguish the representations (based on the same number theory as arose in the work of de Rham, but for reasons closer to the work of Atiyah–Bott [9]). When V and W are indistinguishable in this way, a multiple of V and of W are topologically conjugate.

See [112] for a general theory of surgery on stratified spaces that gives a unification of the considerations that arose in this example, embedding theory, the comments about Witt spaces, and more, together with the perspective of the next subsection.

Surgery revisited

I mentioned earlier that with "characteristic variety addition" the surgery obstruction map $[M;F/\mathrm{Top}] \to L_n(\pi)$ is a homomorphism.

There is another formal issue that can bother one. The L-groups are covariantly functorial in π (and therefore M), but $[M;F/\mathrm{Top}]$ is contravariantly functorial in M.

A way around this is suggested by Sullivan's work, as well. If M is an oriented PL (topological) manifold, then M is orientable for the cohomology theory $[\ ;F/\mathrm{Top}]$, at 2 using the orientation of M, and away from 2 using the Sullivan $KO[1/2]$ orientation. If one perturbs the orientation by lower order terms properly [78], one gets an orientation for this theory. This means we can view (in the oriented case) $[M : F/\mathrm{Top}]$ and $H_m(M;F/\mathrm{Top})$.

The next point is that G/Top looks very much like its fourth loopspace[14] $\Omega^4(G/\mathrm{Top})$ — it is the identity component of that disconnected space. The following result is a correction by [79] of a result of Siebenmann[15] (which had ignored the role of components):

Theorem. *If M is a connected manifold of dimension at least 5, there is an injection $S(M) \to S(M \times D^4 \operatorname{rel} \partial)$ whose image is a subgroup; the cokernel injects into $L_0(e) = \mathbb{Z}$.*

$S(M \times D^4 \operatorname{rel} \partial)$ is made of manifolds with boundary with maps of pairs into $M \times D^4$ which we assume to be a homeomorphism on the boundary $\partial(M \times D^4)$. This is an abelian group, where addition is given by "stacking" (like the definition of higher homotopy groups). The map to $L_0(e) = \mathbb{Z}$ is given by taking the transverse inverse image of D^4, filling in the boundary S^3 with a disk and taking the signature (divided by 8).

Now all the terms of the surgery exact sequence are groups, and all the maps turn into homomorphisms. Even more is true, due to Quinn and Ranicki [88]. It is functorial! This grows out of the definition of spectra [84] $L_n(\pi)$, which are their own 4-fold loopspaces by crossing with $\mathbb{C}P^2$ (a simply connected with signature 1). And that there is a map $F/\mathrm{Top} \to L_0(e)$ which is a homotopy equivalence to the 0-component.

In the setting of homology manifolds, the results are best [15].[16] If we modify $S(M)$ to be the homology manifolds (simple) homotopy equivalent to M, up to s-cobordism, then the periodicity map is actually an isomorphism, and the surgery exact sequence is a 4-periodic functorial sequence of abelian groups and homomorphisms.

Included in this is that one can now define a sequence of abelian groups $S_k(X)$ for any X, that have a geometric interpretation when X is a manifold (and $k = \dim X$). Of critical importance is the case of $X = K(\pi, 1)$, for then we have factorization of the surgery obstruction map $[M : F/\mathrm{Top}] \to L_n(\pi)$ through (Poincaré duality and functoriality) a map $H_n(K(\pi,1); L(e)) \to L_n(\pi)$. This map is called the assembly map.[17]

This immediately explains that essentially the only possible characteristic classes that are homotopy invariant, in the nonsimply connected case, are those identified by Novikov; indeed the Novikov conjecture can be interpreted in terms of the injectivity of the assembly map — explaining more will take us far afield, but there are several books about this topic.

---

[14] This is a form of Bott periodicity; in the real case it is an 8 fold periodicity, but already after the 4th loopspace the periodicity is an isomorphism away from 2.

[15] The periodicity map is given a geometric interpretation in [16].

[16] This paper has a glitch. It is correct as written in dimension = 6, but in general applies to Poincaré complexes that have normal invariants, in particular it describes the homology manifolds that are homotopy equivalent to manifolds.

[17] The reason for this is that it has an interpretation in terms of producing global objects by gluing together local pieces (and this was the way it was first introduced) [85]. In other settings, it is a forgetful map (in "controlled topology") or an index map (in the setting of elliptic operators, see e.g. [58] or [29]).

2 The Adams Conjecture

Arguably the largest impact of the ideas of Quillen and Sullivan within algebraic topology was taking seriously the idea of treating spaces algebraically. Localizations and completions were a first step — those being basic operations on commutative rings. In both of their work, étale cohomology and considerations of Galois symmetry were a second key idea. Nowadays, the whole subject is fractured by the primes – and indeed the chromatic view of the stable category has infinitely many primes (for each ordinary prime). Categorical methods, homotopy limits, and in particular, homotopy fixed points, now play a central role. Algebraic topology from the 1980s is largely a continuation of this deep thrust achieved in the late 60s and 70s.

I highly recommend the books [38] and [11] for much more about the étale ideas in the proof and the localization/completion theory that grows directly from this work.

2.1 Background

The work I will explain in this section is deeply connected to smooth manifolds, and to homotopy theory, so to explain this, I need to give some background going back to classical work of Pontrjagin (for which the best reference is the lovely text [75]).

Suppose that one has a map $f\colon S^N \to S^m$. One can approximate it by a smooth map (which we will still call f). According to Sard's theorem, almost every point p in S^m is a regular value, which means that the differential Df at every inverse image of p is surjective. The implicit function theorem then gives that $f^{-1}(p)$ is a smooth submanifold of S^N (of dimension $N-m$).

There's a bit more data, called a framing, which is an ordered m-tuple of normal vector fields to this submanifold: at each point of the submanifold, consider $Df^{-1}(v_1,\dots,v_m)$ for a positively oriented basis for the tangent space TS^m at p.

On the reverse, given a framed submanifold, using the tubular neighborhood, it is not hard give a map $S^N \to S^m$ producing this submanifold as the transverse inverse image of, say, the south pole.

Smooth homotopies give rise to (and are induced by) framed cobordisms. The cobordisms between submanifolds of S^N take place in $S^N \times [0,1]$ and are codimension $N-m$ framed submanifolds of the cylinder that restrict to the given framed manifolds on the boundary.

This pair of constructions[18] gives an isomorphism between the homotopy group $\pi_N(S^m)$ and the framed cobordism classes of framed submanifolds of S^N. The Whitney embedding theorem that says that (abstractly) d-dimensional submanifolds of S^{2d+1} are exactly the same thing as d-dimensional submanifolds of any dimensional

[18] One of René Thom's achievements was replacing the target sphere by a more complicated space, so that the relevant manifolds that arise in a natural variant of this construction are arbitrary smooth manifolds.

sphere should at least make the Freudenthal suspension theorem that $S^{n+d} \to S^n$ is independent of n, once $n > d+1$. These are the stable homotopy groups of spheres π_d^s.

In this context it is natural to ask which elements are represented by the standard sphere S^d (sitting equatorially in S^{n+d})? The ambiguity is in the framing. There's a framing that comes from it being the equator of the equator of the equator n times, and any other framing differs from that one by a map $S^d \to O(n)$. This map is called the J-homomorphism

$$J\colon \pi_d O(n) \longrightarrow \pi_d^s,$$

where we might as well let n go to infinity. A truly amazing theorem of Bott, at the foundation of K-theory and the index theorem,[19] says that $\pi_d(O)$ is 8-periodic in d, with the groups being given by $\mathbb{Z}/2$, $\mathbb{Z}/2$, 0, $\mathbb{Z}$, 0, 0, 0, $\mathbb{Z}$, starting with $d = 0$.

This gives a wonderful source of potentially nontrivial elements of homotopy groups of spheres. Adams worked out (aside from a factor of 2) what $\mathrm{Im}(J)$ is; the remaining ambiguity being settled by the solution of Adams conjecture, which I will return to. Kervaire and Milnor [59] related $\mathrm{Cok}(J)$ to the differentiable structures on S^d, as I now explain.

Let Θ_d denote the group of oriented homotopy spheres, $d > 4$. (According to Smale's theorem, these manifolds are all homeomorphic to S^d.) It is a group under connected sum. There is a map $\Theta_d \to \pi_d^s/\mathrm{Im}(J)$ because (this is not obvious, and relies on Bott periodicity) every homotopy sphere has a framing. (One mods out by $\mathrm{Im}(J)$ because of the existence of many possible framings.)

Kervaire–Milnor constructed the map $\pi_d^s \to L_d(e)$ whose kernel (modulo $\mathrm{Im}(J)$) is the image of Θ_d. What is left is to understand the homotopy spheres which represent 0 in stable homotopy. These are interpreted in terms of the Arf invariant problem (see [98]) and the question of what are the possible signatures of manifolds that are parallelizable outside of a point.

Of course Θ_d is just another name for $\mathrm{S}^{\mathrm{Diff}}(S^d)$, so the rewriting of surgery theory is then

$$\pi_{d+1}(G/O) \longrightarrow L_{d+1}(e) \longrightarrow \Theta_d \longrightarrow \pi_d(F/O) \longrightarrow L_d(e),$$

which we can relate to the discussion by Kervaire and Milnor when one realizes that there is an exact sequence

$$\cdots \longrightarrow \pi_d(O) \longrightarrow \pi_d^s \longrightarrow \Theta_d \longrightarrow \pi_d(G/O) \longrightarrow \pi_{d-1}(O) \longrightarrow \pi_{d-1}^s \longrightarrow \cdots$$

whose source comes from another interpretation of the stable homotopy groups of spheres.

To explain this, we must review a bit more — in this case, about the classifying spaces of compact Lie groups, or, equivalently the spaces that classify principal K-bundles for a Lie group K. The basic story is that associated to K there is a space BK, so that principal K-bundles over X correspond to homotopy classes of maps $[X : BK]$. (There is a "universal K-bundle" over BK, and given a map, one pulls

---

[19] And related to the periodicity of G/PL, G/Top and surgery theory generally discussed in the last section.

back the universal bundle using the map to get a principal K-bundle over X.) If X is a suspension $= \Sigma Y$, then, since any principal bundle over a contractible space is trivial, one can see that the principal bundles over X are $[Y : K]$ — at each point of the central Y, one compares the two trivializations of the K-bundles over the two cones.

Algebraic topologically, this means that ΩBK is homotopy equivalent to K, and that the homotopy groups of BK are those of K, with a shift by 1.

Now, if H is a subgroup of K, then every H bundle is (or, more correctly, induces) a K-bundle, which gives a map $BH \to BK$. The homotopy fiber of this map is K/H.

The exact sequence above is associated to a map $BO \to BF$, where BO is the classifying space of the limit of the $O(n)$'s, i.e., the classifying space of linear sphere bundles (= vector bundles), and BG is the classifying space of something more homotopy-theoretic — a homotopy-theoretic analogue of a vector bundle, called a spherical fibration. F/O is just defined as the fiber of the map $BO \to BG$, so it is the classifying space of sphere bundles that are trivialized as spherical fibrations, and indeed this is the way it arises in surgery theory.

A d-dimensional spherical fibration over X is a map $p\colon E \to X$ whose homotopy fiber is a sphere S^{d-1}. Two of these are equivalent if there is a homotopy equivalence between the total spaces that makes the obvious diagram commute up to homotopy.

Spherical fibrations can be pulled back, and glued together from spherical fibrations on subspaces that agree on their overlap, and they can be classified by a classifying space BG_d. There's an obvious suspension construction (the union of two copies of the mapping cylinder of p along p); BG is the limit. By thinking about spherical fibrations over spheres, it is not hard to see that their homotopy groups are the stable homotopy groups of spheres.

If one is trying to understand the relation between geometry and homotopy type these classifying spaces are of critical importance, because Atiyah [8] showed that any homotopy equivalence between manifolds gives an equivalence between the stable normal bundles[20] of the manifolds as spherical fibrations. If one wants to understand smooth topology one therefore must understand the map $BO \to BG$, or what is equivalent, "When is a stable vector bundle trivial as a spherical fibration?'

The Adams conjecture answers this question.

2.2 The Statement of the Adams conjecture

Let us introduce a bit of notation. $KO(X)$ is the Grothendieck group of real vector bundles on a space X; $K(X)$ is the Grothendieck group of complex vector bundles. As mentioned above $KO(X) \cong [X : \mathbb{Z} \times BO]$. More generally $KO^{-n}(X) \cong [X \times (D^n \operatorname{rel} S^{n-1}) : BO]$. Using Bott periodicity (e.g. [14]), namely that there is an

[20] The stable normal bundle of a manifold is its normal bundle in a high-dimensional Euclidean space. By Whitney's embedding theorem, this is independent of all choices if the dimension of the ambient Euclidean space is sufficiently high.

isomorphism $\Omega^8 BO \cong \mathbb{Z} \times BO$, i.e., in the above terminology $KO^{n+8} \cong KO^n$. (Similar statements are true for K, except that BO is replaced by BU, and the 8 by a 2.)

K (and KO) are rings with addition being the Whitney sum of vector bundles and multiplication induced by the tensor product. These actually form cohomology theories — one pulls back vector bundles, and, for example, Mayer–Vietoris is geometrically almost obvious.

Characteristic classes (see [76]) relate K-theory to cohomology. For example rationally, K-theory (and KO) are $\bigoplus H^{2i}(X;\mathbb{Q})$ ($\bigoplus H^{4i}(X;\mathbb{Q})$) as $\mathbb{Z}_2$ ($\mathbb{Z}_8$) graded rings. However, integrally K-theory is quite different. More importantly for our story than the ring structure is the algebra of operations. And, for simplicity, we will restrict our attention to K-theory.

In algebraic topology, every time an algebraic invariant is refined, it becomes more powerful. Cohomology is more powerful than homology, although they both rest on the same foundations, because of its algebra structure. Mod 2 cohomology (or mod p cohomology) is richer than rational cohomology because of the presence of Steenrod operations [100], i.e., its structure as module over the Steenrod algebra. (On the other hand, in the next section, we will revel in the simplifications possible rationally because there are no nontrivial operations.) K-theory has a rich algebra of operations that come from natural constructions on vector spaces, such as those coming from an exterior algebra.

The Adams conjecture can be motivated by a combination of three ideas: localization, the splitting principle, and optimism.

The first ingredient is localization. To show that two spherical fibrations are equivalent, it suffices to show that they are isomorphic when localized at every prime separately. Why? To be equivalent at p means that there is a fiberwise map $f\colon \xi \to \xi'$ whose degree is prime to p on each fiber. One can add maps, which will add degrees, provided that ξ is a fiberwise suspension — which can be arranged by stabilizing once. (Remember, we are dealing with stable spherical fibrations here, so we can add trivial bundles at will.)

So one works locally.

The next question is: what is a source of local equivalences? One obvious source comes from complex line bundles. If ξ is a complex line bundle — thought of as a principal S^1-bundle — we can consider $\xi/\mathbb{Z}_k$ the quotient of ξ by multiplication by k^{th} roots of unity. $\xi \to \xi/\mathbb{Z}_k$ is an isomorphism for any prime that does not divide k.

The line bundle associated to $\xi/\mathbb{Z}_k$ is $\xi^{\otimes k}$, the kth tensor power of ξ. Note that for a sum of line bundles, we would get an isomorphism at p between $\xi \oplus \eta$ and $\xi^{\otimes k} \oplus \eta^{\otimes k}$. We would like an operation that on a sum of line bundles is the sum of the kth tensor powers of the constituent line bundles. This is possible, and this is what the Adams operation Ψ^k does.

The precise definition is where the splitting principle comes in.

Although it is not true that every complex vector bundle is a sum of line bundles, and even when it is, such a decomposition need not be unique, one can frequently pretend that it's true, when one has a formula that is symmetric in the line bundles.

For a cohomological example, the nth Chern class of a complex vector bundle is the nth symmetric function of the 1st Chern classes of the n formal line bundles that it would be a sum of, if it were a sum of line bundles. (These formal line bundles, or better their 1st Chern classes, are sometimes called the Chern roots of the bundle.) This Einsatz can be made rigorous, at least for verifying formulae, in a number of ways — one way is to pass to a frame bundle, where the pullback is a sum of line bundles, and check that this pullback map is injective in cohomology. Once one knows this formula for c_n one can take any symmetric function of the line bundles, rewrite this as a function of the symmetric polynomials, and therefore obtain a combination of Chern classes. (The Chern roots of a bundle are then the formal roots of a polynomial whose coefficients are the Chern classes of that bundle.)

The role of the Chern classes in K-theory is taken by the λ^n, the nth exterior powers, of the vector bundle. Once one uses this, then Ψ^k will be the kth Newton polynomial that expresses the sum of the kth powers of the roots of a polynomial in terms of its coefficients (or equivalently, in terms of symmetric powers). Note negative coefficients cause no trouble because we are working (stably) in a Grothendieck group. The Adams operations have a number of useful properties which we list:

- The Ψ^k are algebra homomorphisms.
- $\Psi^k \Psi^l = \Psi^{kl}$.
- $\Psi^k(\xi) = \xi^{\otimes k}$ for ξ a line bundle.

To make these operations stable, one needs to invert k. (Because on the Bott element Ψ^k is multiplication by k.)

The Adams conjecture is the statement that for any bundle $\Psi^k(\xi) - \xi$ is trivial as a spherical fibration, localized at p, and that furthermore, as one varies k to be prime to p, these give all of the isomorphisms among spherical fibrations.[21] For example on $\pi_{4n}(BU) = \mathbb{Z}$ (by Bott periodicity), Ψ^k is multiplication by k^{2n}, so one should have that the kernel of the map $\pi_{4n-1}(U) \to \pi^s_{4n-1}$, localized at p, should be the ideal generated by $((k^{2n} - 1)\,;\, p$ does not divide $k)$. Classical number theory shows that at least the odd part of $\mathrm{Im}(J)$ would then be the denominator of B_n/n where B_n is the nth Bernoulli number.

Adams [1] was able to verify his conjecture for the case of spheres (although for real K-theory, his work missed a factor of 2 that later work did catch), which was strong enough to determine $\mathrm{Im}(J)$ in π_i^s.

In any case, the Adams conjecture is true, and was affirmed by Quillen [83] and Sullivan [103], both using ideas of étale cohomology, and later by Becker and Gottlieb [10], by a very clever argument that essentially reduces vector bundles with structure group $U(n)$ to those whose structure group is $T_n \rtimes \Sigma_n$, the normalizer of its maximal torus, where it follows from Adams' original proof. This then enables one to give a great deal of information about G/O, and indirectly BPL and BTop.

[21] Here one must restrict to real vector bundles; complex vector bundles have "complex conjugates" which are equivalent as real vector bundles. (Stably, including real vector bundles in the complex ones (only) loses 2-torsion.) Adams defined an analogue of Stiefel–Whitney classes to prove that the part not coming from the $(\Psi^k - 1)$s is homotopy invariant.

2.3 Comments on Sullivan's proof

The title of Sullivan's paper "The genetics of homotopy theory and the Adams conjecture" indicates one of the key themes in his approach: the idea of fracturing algebraic topology via completions — that are much less tame than the localizations that already arise implicitly in the work of Serre, and explicitly (at least) in his own work on the structure of F/PL. Sullivan's proof of the Adams conjecture is based on interpreting the Adams operations via a surprising Galois symmetry in the (profinitely completed) homotopy types of BU and BO.

This fracturing of homotopy theory is an analogue of the use of adèles in arithmetic (and indeed Sullivan remarks that the impulse to do this is strongly suggested by the Hasse–Minkowski theorem about quadratic forms). While Quillen's proof used characteristic p ideas and related Adams operations to the Frobenius automorphism of the algebraic closed field of characteristic p, Sullivan instead used Galois automorphisms in the universal $\mathrm{Gal}(\overline{\mathbb{Q}})$ (i.e., the Galois group of the algebraic numbers) that induce the kth power on p^{th} roots of unity, but is the identity on k^{th} roots of unity.

There is a powerful impulse in this work to geometrize abstract algebraic constructions, and then extract stronger geometric statements from this. This tendency has only become more prominent in the intervening decades — with the connections between abstract algebraic geometry and homotopy theory becoming only more profound. No longer can one joke, as Adams did in the Weyl lectures on infinite loop spaces, "The apparatus of definitions, theorems and proofs needed to carry out this programme in detail demands a capital investment of intellectual work which may seem daunting to those not directly concerned; many readers may be able to remember feeling the same way about spectral sequences, sheaf theory or whatever is now their favorite tool; let us be glad we don't work in algebraic geometry" [2].

One beautiful aspect of Sullivan's geometrization of the Adams conjecture is that he was able to prove it *unstably*. Completing at p, and letting k be any integer not divisible by p, he showed that it was possible to define Ψ^k on the unstable $BU(n)$. (As it's an algebraic variety, i.e., a Grassmannian, there is a natural discontinuous action of the Galois group; the magic of étale homotopy theory gives an action on the completed homotopy type.) These are actually homotopy equivalences at p. An induction on n, and some geometry about p-complete spherical fibrations show that the underlying spherical fibration of a vector bundle doesn't change under the operation Ψ^k, i.e., that $\Psi^k - I = 0$ in the p-complete theory.

2.4 Some of its aftermath

Nowadays much of algebraic topology is done in the complete setting. This much almost goes without saying.

Sullivan's paper itself contained other beautiful nuggets: The Galois symmetry on $\mathbb{C}\mathrm{P}^\infty$, completed at p, produced topological group structures on S_p^{2n-1} whenever n

divides $p-1$, i.e., there is a theory of principal bundles for these completed spheres. An unstable Adams operation on the cohomology of $BSU(2)$ gives a self map of this classifying space whose "degree" (i.e., induced map on H^4) is an arbitrary odd square.[22]

Very natural for Sullivan, given his focus on Galois symmetry, was to consider $\mathrm{Gal}(\mathbb{C};\mathbb{R}) = \mathbb{Z}/2$ acting on the étale homotopy of a variety. The thoroughgoing use of categories here led him to speculate that the real points of a variety V would, completed at 2, be determinable from the equivariant homotopy type of the 2-completion of V and its Galois action, and indeed the formula should be

$$V(\mathbb{R})^\wedge_2 \cong \mathrm{Map}_{\mathbb{Z}/2}(S^\infty : V(\mathbb{C})^\wedge_2)^\wedge_2.$$

Here $S^\infty = E\mathbb{Z}/2$ is the universal space with free $\mathbb{Z}/2$ action — it is the infinite-dimensional sphere with the free involution on it. He then conjectured a similar formula for all p-groups acting on all finite complexes. This "Sullivan conjecture" was a major topic in 1980s algebraic topology (proved by H. Miller for spaces with trivial action, and then in general by Carlsson, Lannes, and Miller [20, 63, 73]).

This point of view of comparing fixed point sets to homotopy fixed point sets for quite different classes of groups is now commonplace throughout mathematics in problems ranging from number theory, arithmetic geometry and K-theory to geometric topology, index theory and differential geometry. It is beyond the scope of this article to adumbrate the many places where this idea arises and try to estimate its import. (Moreover it has surely been discovered and rediscovered many times.)[23]

Within homotopy theory there were a number of immediate applications. One extremely elegant one was the following:

Theorem ([72]). *If X is a noncontractible finite simply connected complex, then for infinitely many n, $\pi_n(X)$ has p-torsion.*

This had been conjectured by Serre, who for $p = 2$, showed that for infinitely many n, the localization at 2 was nontrivial.

Another important direction that grew out of the Sullivan conjecture was a deep analysis of the maps between classifying spaces of compact Lie groups, of which we will mention one influential paper.

Theorem ([34]). *If π is a p-group and G is a connected Lie group, then*

$\mathrm{Hom}(\pi : G)/\sim$ *(i.e., homomorphisms from π into G up to conjugacy)* $\to [B\pi : BG]$

is a bijection.

[22] Bernstein showed that the degree was a square and that it must be odd (if nonzero) was shown by G. Cooke.

[23] And can certainly be traced implicitly to sources earlier than Sullivan, as well. However, my point is that the Sullivan conjecture and its centrality in topology made this point explicit at least to the community of topologists.

We note that Sullivan's paper already gave one new result on maps between classifying spaces where the maps do not come from homomorphisms of Lie groups, namely the result on "degrees" of self maps of $BSU(2)$.

Finally, we should mention that Sullivan's theory has immediate implications for surgery theory. In the smooth category, it tells us that, completed at odd primes, F/O is a product of two spaces, BSO and $\mathrm{Cok}(J)$, where $\mathrm{Cok}(J)$ is the part of BF not in the image of the J-homomorphism. Indeed it is a factor of BF (the other factor being a space homotopy equivalent to the fiber of $\Psi^k - I$, for k a generator of the multiplicative group $\mathbb{Z}/p^2$, which is where the J-homomorphism lands). BPL has a similar description: it is the product of $BSO \times B\mathrm{Cok}(J)$. The first isomorphism enables (in principle) a study of surgery on smooth manifolds — although one is stymied by the fact that the surgery obstruction map is not a homomorphism, and the second enables (with more hard work) an analysis of PL bordism (see e.g. [66]).

Epilogue: A return to geometry

A consequence of étale homotopy theory is the following result of Deligne and Sullivan:

Theorem. *Every hyperbolic n-manifold has a finite sheeted cover that is parallelizable (i.e., whose tangent bundle is trivial) in the complement of a point.*

Such manifolds can be immersed into $\mathbb{R}^n$ by Smale–Hirsch immersion theory. Sullivan [105] (see also [108] for a more detailed treatment) made use of this to give a hyperbolic manifold variation on Kirby's torus trick that was critical to the development of the Kirby–Siebenmann theory we discussed before, enabling the following beautiful result:

Theorem. *Every topological manifold of dimension larger than 4 has a unique Lipschitz structure.*

A Lipschitz structure is like a smooth manifold, defined using charts in Euclidean space, except that one now requires that the overlap maps are bi-Lipschitz. (In dimension < 4, one can even find smooth structures.) In dimension 4, Donaldson and Sullivan showed that some 4-manifolds do not have Lipschitz structures [33].

3 Rational homotopy theory

In his paper "Infinitesimal computations in topology," Sullivan [104] created a link between the mathematics related to calculus and that of algebraic topology. Of course, de Rham's theorem already does this, and one can even see earlier precedents in the Cauchy integral theorem and even Gauss' definition of linking number between space curves. But, Sullivan's work shows that these ideas can be developed to the point where the algebra of differential forms gives a faithful description of rational homotopy theory (which is homotopy theory where one localizes by inverting all primes).

One should note that a little earlier, Quillen [82] gave a different, dual, algebraization of rational homotopy theory by a rather indirect string of equivalences of categories. These are of course equivalent to each other (see e.g. [43] for a description of the connection). In some sense, Quillen's approach is "cellular," while Sullivan's is based on Postnikov systems (see below). Our focus will be on sketching Sullivan's theory and a few of its applications.

There are now a number of excellent references for rational homotopy theory and some of its applications; we note [35, 43].

3.1 Sullivan's model

Already Serre's thesis [94] made it clear that homotopy theory over the rationals is fairly comprehensible. The test cases: spaces chosen for the simplest homology or for the simplest homotopy, i.e., the sphere S^n and the Eilenberg–MacLane space $K(\mathbb{Z},n)$, have comprehensible homotopy and cohomology (respectively).

$$\pi_i(S^n)\otimes\mathbb{Q}=0 \text{ unless } i=n \text{ or } n \text{ is even and } i=2n-1, \text{ in which case it is } \mathbb{Q},$$

$$H^i(K(\mathbb{Z},n);\mathbb{Q})=0 \text{ unless } i=n \text{ or } n \text{ is even and } i \text{ is a multiple of } n, \text{ in which case it is } \mathbb{Q}.$$

In the latter case; the cohomology algebra is a polynomial algebra on one n-dimensional generator. Recalling that cohomology is a graded algebra so that the square of any odd-dimensional class must vanish, we can express both cases as saying the rational cohomology of $K(\mathbb{Z},n)$ is a free graded algebra on one n-dimensional generator. (And if V is a finite-dimensional graded vector space, then with the self-evident notation, one quickly sees using the Kunneth formula, that $H^*(K(V))\cong\Lambda(V^*)$, where Λ denotes the free graded commutative algebra (generated by V^*).[24])

There is more to homotopy theory than is visible in the cohomology algebra. For example the famous Hopf map $f:S^3\to S^2$ is trivial in cohomology, but is nontrivial. This can be seen from many points of view, but the point of view most relevant to us is the following. Let ω be a volume form on S^2. Since $H^2(S^3)=0$, $f^*\omega$ is 0 as a cohomology class, i.e., there is a 1-form ν so that $d\nu=f^*\omega$. Consider the

[24] There is a dual statement in homotopy that is the analogous starting point for Quillen's version of rational homotopy. Homotopy groups form a graded Lie algebra under the "Whitehead product," and the homotopy groups of spheres are free graded Lie algebras. The analogue of the cohomology of Eilenberg–MacLane space is the calculation of the homotopy of wedges of spheres, Hilton's theorem (see e.g. Milnor's paper in [2]).

3-form $\nu \wedge f^*\omega$. As $H^1 = 0$, it is well defined as a cohomology class. The integral $\int(\nu \wedge f^*\omega)$ is an analytic expression for the well-known Hopf invariant.

This suggests that there's more to be mined from the cochain level. The first major result of the theory is the following:

Theorem. *The homotopy theory of simply connected commutative differential graded algebras over $\mathbb{Q}$, with finitely generated cohomology groups, is equivalent to the rational homotopy category of simply connected finite complexes.*

Let's unpack this a little. We are looking at graded algebras with differential, so that they are naturally cochain complexes. The standard example would be the de Rham complex.

The simplicial cochain complex is not quite an example — the usual formula for cup product does not give the desired commutativity relations. As is well known (and it is the source of the famous Steenrod algebra!) there is no way to define directly on the ordinary cochain complex (over $\mathbb{Z}$) a genuinely graded commutative algebra structure.

However, it is desirable (indeed necessary) to work over $\mathbb{Q}$ rather than $\mathbb{R}$. Sullivan's patch for this problem is to work with simplicial complexes, and then use polynomial differential forms with $\mathbb{Q}$ coefficients (i.e., a cochain complex which is defined simplicially in this way). It is not hard to prove the relevant Poincaré lemma and de Rham theorem for these, to see that one does get a model for rational cohomology that now has genuine graded commutativity.

So there is a commutative differential graded algebra (CDGA) associated to a space. And maps between spaces given well-defined chain homotopy classes of maps between these (that preserve the algebra structure and the differentials). This algebraic shadow of the space actually capture rationally — according to Sullivan — the whole rational homotopy theory of the space.

Sullivan observed that in this setting, there is a "best" model for a space – a model of each homotopy class of CDGA – that is well defined up to (non-canonical) isomorphism. As is traditional, we will call this $\mathscr{M}$, and it has the following two properties: (i) $\mathscr{M}$ is free as a CGA (i.e., like the cohomology of a product of Eilenberg–MacLane spaces, it is a tensor of polynomial and exterior algebras), and (ii) above dimension 0, the image of d is *decomposable*, i.e., lies in $\mathscr{M}^+ \otimes \mathscr{M}^+$ where $\mathscr{M}^+$ denotes the positive-dimensional part of $\mathscr{M}$.

Theorem. *Every homotopy class of CDGAs has a unique, up to isomorphism, minimal model as above.*

The homotopy groups of X are dual to the indecomposable part of $\mathscr{M}$.

Let's do some examples. These are good motivations as well as illustrations of the theory. And they indicate that this is a very computable theory.

Example 3.1. S^n, n odd. In dimension n we must have $\mathbb{Q}\langle x_n \rangle$, but we don't need anything more to get the cohomology right: Obviously there must be a map from $\mathbb{Q}\langle x_n \rangle$ to any CDGA for S^n, and it will be a chain equivalence (based on cohomology). (We haven't recorded that $\mathrm{d}x = 0$, since it's lying in a 0 group.) So that's it.

Example 3.2. S^n, n even. It starts the same $\mathbb{Q}[x_n]$. However we need something to kill the cohomology class x_n^2, so we add a generator $[y_{2n-1}]$ with $dy = x^2$. At this point, one realizes that the reasoning from Example 3.1 applies, and discovers that this must be the minimal model for S^n (n even). And one has "recovered" the result of Serre about the rational homotopy groups of spheres.[25]

Example 3.3. We will begin the calculation for $S^3 \vee S^3$. We start with $\mathbb{Q}\langle a_3, b_3\rangle$. Of course, for parity reasons, $a_3^2 = b_3^2 = 0$, but a_3b_3 won't be 0, which necessitates an element c_5 with $dc_5 = a_3b_3$. We are not done yet. If we had no more generators for the indecomposable, we'd have two 8-dimensional classes a_3c_5 and b_3c_5 (as you can readily see these are cocycles). This necessitates two more 7-dimensional generators, say e_7 and f_7 with $de_7 = a_3c_5$ and $df_7 = b_3c_5$ and so on ... One clearly can continue this process formally (at the moment, this merely means mechanically with no additional thought). These elements that we've produced give that $\pi_5(S^3 \vee S^3) \otimes \mathbb{Q} = \mathbb{Q}$ and $\pi_7(S^3 \vee S^3) \otimes \mathbb{Q} = \mathbb{Q}^2$. It would not be hard to express the homomorphisms from these homotopy groups to $\mathbb{R}$ via integrals of suitable differential forms entirely analogously to the way we expressed the Hopf invariant.

(Incidentally, the names of these generators, in terms of Whitehead products of the two generators of $\pi_3(S^3 \vee S^3)$ are $[a,b]$, $[a,[a,b]]$ and $[b,[a,b]]$.)

Example 3.4. Suppose we consider building a CW complex with 3 skeleton $S^3 \vee S^3$ but attaching one more cell. If it's a 6-cell, then the attaching map determines what cup product of the duals to homology classes are. (And, rationally, the only issue is whether you've used a nontrivial multiple of $[a,b]$ or not.) More interesting is attaching an 8-cell. The resulting complex always has the same cohomology algebra (all products have no choice but to be 0) but they can have different rational homotopy types. The minimal model of such complexes will look like (in its early 8 stage), for example, $\mathbb{Q}\langle a_3, b_3, c_5, e_7, f_7, g_8 \mid dc_5 = a_3b_3, de_7 = a_3c_5, df_5 = b_3c_5 \dots\rangle$ versus $\mathbb{Q}\langle a_3, b_3, c_5, e_7 \mid dc_5 = a_3b_3$ and $de_7 = a_3c_5\rangle$.

There are many more examples that one can give. Examples 3.1-3.3 were somehow simpler than Example 3.4. In these examples, the cohomology algebra itself determined the minimal model. In Example 3.4, there are different possible spaces (minimal models) with the same cohomology algebra. There is always a simplest DGA with a given cohomology: namely the cohomology algebra itself (with differential d set $= 0$). In particular, any rational DGA comes from a space (that can be algorithmically constructed from the algebra — in short, the minimal model describes a Postnikov system that describes such a space),[26] and there's a simplest one — namely the one which is homotopy equivalent to that DGA. Thus for the cohomology algebra of $S^3 \vee S^3 \vee S^8$ the first minimal algebra we wrote down is the one which is simplest.

[25] Of course, this is an almost completely circular argument. The proof of this equivalence of categories, etc., depends critically on the Serre spectral sequence and the calculation of the cohomology of Eilenberg–MacLane spaces, the same ingredients Serre had used.

[26] For other fields, this is not true. For example, the algebra $\mathbb{F}_2[x_3]/(x_3^3 = 0)$ does not arise for any space (using the Adem relation $Sq^3 = Sq^1Sq^2$).

These simplest homotopy types are called formal.

Definition. A space is called *formal* if its minimal model is homotopy equivalent to its cohomology algebra (viewed as CDGA with 0 differential).

Many spaces arising in mathematics are formal.

Example. If G is a Lie group, or more generally a symmetric space G/H, then one can find harmonic representatives for each cohomology class and the wedge of two harmonic forms is harmonic. Therefore these spaces are formal.[27] (On the other hand, not all homogenous spaces are formal.)

Theorem ([32]). *All Kähler manifolds are formal.*

This, too, is a consequence of Hodge theory. Similarly holomorphic maps between Kähler manifolds are "formal consequences" of their induced maps on cohomology.

The marriage of Hodge theory and Sullivan's rational homotopy theory is indeed a happy one. Morgan [77] (and Hain [46]) have shown that there is a natural mixed Hodge structure on homotopy groups of smooth varieties. Morgan used this to show that there are finitely presented groups that are not homotopy groups of smooth varieties (although, Kapovich and Kollàr [57] have shown that every finitely presented group is the fundamental group of singular affine 3-fold).

Remarkably both the rational and the p-adic parts of Sullivan's theory thus have beautiful and deep connections to algebraic geometry.

In general, cohomology determines much less than homotopy type. A little thought shows that for the automorphisms of a simply connected X, the eigenvalues in homotopy are the same as those in cohomology (although there might be differences that are nilpotent). Another structural result that follows from Sullivan's work (although done independently by a different method by Wilkerson [114]) is the following:

Theorem. *For any finite simply connected complex X, the group of self homotopy equivalences has a quotient by a finite normal subgroup which is an arithmetic group.*

(See [62] for an example where one needs to take this finite quotient, and one cannot merely assert commensurability.)

Note that this theorem is about X, not about the rationalization of X, X_0. The automorphisms of that space is the Lie group over $\mathbb{Q}$ that $\mathrm{Aut}(X)$ is almost a lattice in.

---

[27] We are eliding the prima facie difference between rationally formal versus formal over $\mathbb{R}$.

3.2 *A few words on the proof*

Sullivan's theorem is extremely natural, and the proof is also conceptually very simple. Finite simply connected complexes have finitely generated homotopy groups, and rationalization just tensors these with $\mathbb{Q}$. To set up the equivalence of categories, one inducts on the number of nontrivial homotopy groups.

We've already understood what happens with one homotopy group: the DGA is exactly the free algebra on the dual of that group.

For any space X, one can consider $X_k = \lim(Z \mid Z \supseteq X$ a k-equivalence$)$. This limit is a space which includes X, and for which the inclusion is a k-equivalence. All the homotopy above dimension k vanishes. There are obvious maps $X_{k+1} \to X_k$ whose fibers are Eilenberg–MacLane spaces. Most importantly, these maps are classified by maps to Eilenberg–MacLane spaces.

Let's do an example, Example 3.2 above.

$S^n_n = K(\mathbb{Z},n)$. The fiber of (the rationalization of) $S^n_{2n-1} \to S^n_n$ is $K(\mathbb{Q},2n-1)$. There is a fibration (the so-called "path fibration" of Serre) $K(\mathbb{Q},2n-1) \to * \to K(\mathbb{Q},2n)$, and the fibration $S^n_{2n-1} \to S^n_n$ that we are interested in is the pullback of the path fibration under a suitable map $S^n_n \to K(\mathbb{Q},2n)$.

What is this map? Maps to Eilenberg–MacLane spaces are cohomology classes, so we are looking for a cohomology class in $H^{2n}(S^n_n;\mathbb{Q}) = H^{2n}(K(\mathbb{Q},n);\mathbb{Q})$, and we take the cup square of the n-dimensional class.

Note the obvious connection between the minimal model and this description of the homotopy theory. These spaces, and the cohomology classes that describe how the homotopy groups of X inductively relate to X's homotopy type — elements that are called the k-invariants of X — are exactly the elements of the classical description of a homotopy type, called the Postnikov system of X (see e.g. [99]). The k-invariants are definable in both the setting of spaces and DGAs and set up the equivalence of categories by an inductive argument.

This proof also leads directly to statements relating the homotopy theory of X to its rational homotopy theory. By going to rationalizations, one has finite fibers $[Y,X] \to [Y_{(0)},X_{(0)}]$, although the image is harder to characterize: one needs to see how compatible the rational map is with the lattices inside $\pi_i \otimes \mathbb{Q}$. In any case, these ideas explain where the Sullivan–Wilkerson theorem mentioned in the previous section comes from.

3.3 *A few more applications*

In this subsection, we will mention a few of the applications of rational homotopy theory. The first two we will discuss are consequences of the fact that it makes calculations feasible. Consequently structure that might never have been imagined can come to light. The last application is a direction that is very much work in progress.

3.3.1 Free loop spaces

The application of rational homotopy theory to the free loop space $\Lambda X = \mathrm{Map}(S^1 : X)$, the space of maps from a circle to X, due to Vigue-Poirier and Sullivan [109] is one of the earliest. Besides the beauty of this paper, it showed how directly one can compute minimal models for natural constructions — this can be modified to describe other function spaces, and spaces of sections. It remains important, as ΛX continually recurs in different contexts. Cohomology and equivariant cohomology of loop spaces are related to cyclic homology, which are receptacles of invariants from algebraic K-theory. The motivation for the paper [109] is the question:

Problem. *Does every closed Riemannian manifold have infinitely many closed geodesics?*

We still don't know. A related question was answered by Serre:

Theorem ([94]). *If M is a closed Riemannian manifold and p and q are points, then there are infinitely many geodesics that connect p to q.*

If $\pi_1 M$ is infinite, then this is obvious: the homotopy classes of arcs connecting p to q are in a 1-1 correspondence with this group, and each such class has length minimizing representative. If $\pi_1 M$ is finite, then we might as well work in the universal cover, and assume it's trivial. One can see that if there were only finitely many geodesics, then ΩM would have the homotopy type of a finite complex, but by Hopf's theorem [55] the rational cohomology of ΩM would then be that of a product of odd spheres — which would make the cohomology of M itself infinite-dimensional (by a spectral sequence argument).

For closed geodesics the situation is much more complicated. If $\pi_1 M$ is infinite, then we don't automatically win: the components of ΛM are in a 1-1 correspondence with the *conjugacy classes* of elements of the group — and it's still unknown whether this must be infinite for all infinite finitely presented groups.

But even for the simply connected case, the relation between closed geodesics (i.e., the critical points of the Energy function on ΛM) and the (co)homology of ΛM is not so straightforward. One can rotate closed geodesics (so one might be led to consider ΛM equivariantly), and much more seriously, go through the same geodesic k times — which means that each critical point gives rise to infinitely many others that are geometrically the "same." An important result does connect the Morse theory of ΛM to its homology:

Theorem (Gromoll–Meyer [44]). *If for some field the Betti numbers of ΛM are unbounded, then M has infinitely many closed (geometrically distinct) geodesics for any Riemannian metric on M.*

So one wonders when the conclusion of the Gromoll–Meyer theorem holds; [109] answers this by first describing the model for ΛM from the minimal model of M. (The homotopy of ΛM is simple enough to describe: $\pi_i(\Lambda M) = \pi_i(M) \times \pi_{i+1}(M)$.)

Theorem. *Let* $\mathscr{M} = \langle x_1, x_2, \ldots, x_d \rangle$ *be a minimal model for M. Then a model for* ΛM *is given by* $\mathscr{A} = \langle x_1, x_2, \ldots, y_1, y_2, \ldots, y_{d'} \rangle$, *where the degree of* y_i *equals the degree of* x_{i-1} *and* $'$ *is defined so that* $d'x_i = dx_i$ *and* $d'y_i = -ix_i$, *where i is the unique degree* -1 *derivation from* $\mathscr{M} \to \mathscr{A}$ *with* $d'i + id' = 0$.

In practice the impact on cohomology is not that transparent. For example for ΛS^{2k} the cohomology algebra is nilpotent (in strong contrast to the situation of ΩS^{2k}, which has a polynomial generator in dimension $4k-2$).

This they then apply to show:

Theorem. *The Betti numbers of the free loop space of a simply connected manifold M,* ΛM, *with* $\mathbb{Q}$ *coefficients are bounded iff* $H^*(M;\mathbb{Q})$ *is generated by 1 element. Consequently if* $H^*(M;\mathbb{Q})$ *requires two or more generators, then for any Riemannian metric on M, M has infinitely many geometrically distinct closed geodesics.*

Let me close this subsection by mentioning some work that is definitely not "juvenilia." In [40], Bill Goldman, motivated by work of Wolpert in Teichmüller theory, was led to define a Lie algebra structure on the free abelian group generated by free homotopy classes of oriented curves on an oriented surface. Chas and Sullivan showed that this can be extended to the homology of the free loop space (thinking of a cycle there as a cycle of simple closed curves in M and considering the parametrized family of intersections).

Theorem ([25]). $H^*(\Lambda M)$ *has a product of degree* $-n$ *(where* $n = \dim M$*).*

This gives an intricate structure on the homology of ΛM (a Batalin–Vilkovisky algebra) and a Lie algebra structure on the equivariant homology of ΛM (with respect to the circle action that rotates closed curves). Later this was given a homotopy-theoretic description by Cohen and Jones. [36] give an explicit model of the loop product (rationally) in terms of the minimal model.

Remarkably,[28] this structure arises in symplectic geometry. The book [64] contains a number of articles that explain the connection, Viterbo's theorem (see Abouzaid's paper in that volume), between homology of free loop spaces and the symplectic cohomology of the cotangent bundle of a closed manifold T^*M — which is an isomorphism of BV algebras. We will not explore this direction, leaving the reference to [64] as a good starting point.

3.3.2 The Elliptic/Hyperbolic dichotomy

The amazing computability of the minimal model theory allows for the deep exploration of the homotopy category that is not (yet) possible at a finite prime. The following is one such remarkable example:

Theorem. *If X is a finite simply connected complex of dimension k, then either* $\pi_i(X) \otimes \mathbb{Q} = 0$ *for* $i \geq 2k$, *or the average size of* $\pi_i(X) \otimes \mathbb{Q}$ *grows exponentially. Moreover, it is nontrivial along a sequence of i's of index at most k.*

---

[28] This is probably not that remarkable to someone with a broad enough perspective.

The latter statement means that the rank of $\pi_i \times \cdots \times \pi_{i+k-1}$ is always nontrivial (unless X is rationally contractible). If X is in the first class, it is called **elliptic**; in the second, it is called **hyperbolic**. Elliptic spaces have very special properties, so that the generic situation is the hyperbolic one:

Theorem (see [35]). *Elliptic spaces satisfy Poincaré duality.*
Let $\chi_\pi(X) = \Sigma(-1)^i \dim \pi_i(X) \otimes \mathbb{Q}$, then both $\chi_\pi(X) \leq 0$ and $\chi(X) \geq 0$ and exactly one of these is an equality. Moreover, the sum of the ranks of the odd homotopy groups is $\leq 2\dim(X) - 1$ and the sum of the even ranks is $\leq \dim(X)$. The Poincaré series of X, $\sum b_i(X)t^i$ is term by term $\leq (1+t)\dim(X)$, so the total homology $\leq 2\dim(X)$.

Elliptic spaces are rare, but they do occur. Of course spheres and their products are elliptic. All simply connected homogeneous spaces G/H are elliptic. A conjecture attributed to Bott asserts that all manifolds of nonnegative sectional curvature are.

An easy argument shows that if a torus T^k acts locally freely (i.e., with finite isotropy groups) on such an X, then $k \leq -\chi_\pi(X)$. In particular, this recovers the theorem of Alday–Halperin [5] that for $X = G/K$ a quotient of simply connected compact Lie groups, T^k can only act locally freely if $k \leq \operatorname{rank}(G) - \operatorname{rank}(K)$.

Remark. There is a nice application of rational homotopy theory to noncompact manifolds with nonnegative curvature. According to the Soul Theorem of Cheeger and Gromoll, every such manifold is diffeomorphic to a vector bundle over a compact nonpositively curved manifold. In [11], it is shown that the converse does not hold: there are bundles over nonnegatively curved manifolds that do not have nonnegatively curved metrics. (Interestingly, the examples are not simply connected.)

3.3.3 Quantitative Homotopy Theory

Soon after Sullivan's paper appeared, Gromov wrote a short paper [45] pointing out that, because of rational homotopy theory, one can bound "homotopy theory" using Lipschitz constants. The following is one of the results in the paper:

Theorem. *If Y is a finite simply connected complex and X is a finite complex, then the number of homotopy classes of maps represented by continuous functions with Lipschitz constant at most L grows like $O(L^k)$ for some k.*

(The proof in [45] is complete when X is a sphere; see [70] for a general discussion.) This is in strong contrast with the nonsimply connected situation where there do not seem to be any nontrivial estimates: there the number is $O(\exp(L^{\dim(X)}))$.

Surgery theory, immersion theory, cobordism theory, etc., tend to reduce geometric topological problems to ones of algebraic topology. In order to understand what the solutions actually looks like, one wants to know "how large the Lipschitz constant of a homotopy" has to be when one has homotopic Lipschitz maps.

Using minimal models, one can give examples where this grows faster than linearly [22], but it is always polynomial (in the simply connected case) [69]. The papers [21] and [22] use minimal models to guide the construction of fairly efficient homotopies for some important Ys.

These results are not quite strong enough to prove the even-dimensional case of the following theorem of [13]:

Theorem. *There is a constant $C(m,n)$ such that any two L-Lipschitz maps $f,g\colon S^m \to S^n$ that are homotopic are $C(m,n)L$ homotopic.*

Fascinatingly, they show that, except for a very special class of spaces (the "scalable spaces," which for manifolds are the ones which are formal and have cohomology algebra injecting into the cohomology of a torus of the same dimension [12]), the minimal model theory doesn't give the optimal answer. The following is surely only the first step in a new deep more geometric refinement of rational homotopy theory.

Theorem ([12]). *If M^m is a formal simply connected m-manifold, then there are self maps with Lipschitz constant L and degree $O(L^m)$ if and only if M is scalable. In the nonscalable case, the best degree is smaller by a power of* $\log(L)$.

(For nonformal simply connected manifolds, the maximum degree, if it grows at all, grows like a lower power of L.)

3.3.4 Finite primes and $\mathbb{Z}$

We close by mentioning that, motivated by rational homotopy theory, models of homotopy theory at finite primes and integrally have been constructed. Because of the existence of operations, DGAs are not enough; all of this work depends on the notion of E_∞-algebras and explaining this would take us far afield. We will settle for referring to Mandell [69] for a model of p-complete simply connected finite complexes, and Yuan [115] for an integral theory.

References

1. J. F. Adams. On the groups $J(X)$. I. *Topology*, 2:181–195, 1963.
2. J. F. Adams. *Algebraic topology—a student's guide*. London Mathematical Society Lecture Note Series, No. 4. Cambridge University Press, London-New York, 1972.
3. J. F. Adams. *Infinite loop spaces*. Annals of Mathematics Studies, No. 90. Princeton University Press, Princeton, N.J.; University of Tokyo Press, Tokyo, 1978.
4. S. Akbulut and J. D. McCarthy. *Casson's invariant for oriented homology 3-spheres*, volume 36 of *Mathematical Notes*. Princeton University Press, Princeton, NJ, 1990. An exposition.
5. C. Allday and S. Halperin. Lie group actions on spaces of finite rank. *Quart. J. Math. Oxford Ser. (2)*, 29(113):63–76, 1978.

6. D. R. Anderson and W. C. Hsiang. The functors K_{-i} and pseudo-isotopies of polyhedra. *Ann. of Math. (2)*, 105(2):201–223, 1977.
7. A. H. Assadi and P. Vogel. Actions of finite groups on compact manifolds. *Topology*, 26(2):239–263, 1987.
8. M. F. Atiyah. Thom complexes. *Proc. London Math. Soc. (3)*, 11:291–310, 1961.
9. M. F. Atiyah and R. Bott. A Lefschetz fixed point formula for elliptic complexes. II. Applications. *Ann. of Math. (2)*, 88:451–491, 1968.
10. J. C. Becker and D. H. Gottlieb. The transfer map and fiber bundles. *Topology*, 14:1–12, 1975.
11. I. Belegradek and V. Kapovitch. Obstructions to nonnegative curvature and rational homotopy theory. *J. Amer. Math. Soc.*, 16(2):259–284, 2003.
12. A. Berdnikov, L. Guth, and F. Manin. Degrees of maps and multiscale geometry. 2020.
13. A. Berdnikov and F. Manin. Scalable spaces. *Invent. Math.*, 229(3):1055–1100, 2022.
14. R. Bott. The stable homotopy of the classical groups. *Ann. of Math. (2)*, 70:313–337, 1959.
15. J. Bryant, S. Ferry, W. Mio, and S. Weinberger. Topology of homology manifolds. *Ann. of Math. (2)*, 143(3):435–467, 1996.
16. S. Cappell and S. Weinberger. A geometric interpretation of Siebenmann's periodicity phenomenon. In *Geometry and topology (Athens, Ga., 1985)*, volume 105 of *Lecture Notes in Pure and Appl. Math.*, pages 47–52. Dekker, New York, 1987.
17. S. Cappell and S. Weinberger. A simple construction of Atiyah-Singer classes and piecewise linear transformation groups. *J. Differential Geom.*, 33(3):731–742, 1991.
18. S. E. Cappell and J. L. Shaneson. Piecewise linear embeddings and their singularities. *Ann. of Math. (2)*, 103(1):163–228, 1976.
19. S. E. Cappell and J. L. Shaneson. Nonlinear similarity. *Ann. of Math. (2)*, 113(2):315–355, 1981.
20. G. Carlsson. Equivariant stable homotopy and Sullivan's conjecture. *Invent. Math.*, 103(3):497–525, 1991.
21. G. R. Chambers, D. Dotterrer, F. Manin, and S. Weinberger. Quantitative null-cobordism. *J. Amer. Math. Soc.*, 31(4):1165–1203, 2018. With an appendix by Manin and Weinberger.
22. G. R. Chambers, F. Manin, and S. Weinberger. Quantitative nullhomotopy and rational homotopy type. *Geom. Funct. Anal.*, 28(3):563–588, 2018.
23. T. A. Chapman. Topological invariance of Whitehead torsion. *Amer. J. Math.*, 96:488–497, 1974.
24. T. A. Chapman and S. Ferry. Approximating homotopy equivalences by homeomorphisms. *Amer. J. Math.*, 101(3):583–607, 1979.
25. M. Chas and D. Sullivan. String topology. *arXiv:math/9911159*, 1999.
26. J. Cheeger. On the Hodge theory of Riemannian pseudomanifolds. In *Geometry of the Laplace operator (Proc. Sympos. Pure Math., Univ. Hawaii, Honolulu, Hawaii, 1979)*, Proc. Sympos. Pure Math., XXXVI, pages 91–146. Amer. Math. Soc., Providence, R.I., 1980.
27. M. M. Cohen. *A course in simple-homotopy theory*. Graduate Texts in Mathematics, Vol. 10. Springer-Verlag, New York-Berlin, 1973.
28. P. E. Conner and E. E. Floyd. *The relation of cobordism to K-theories*. Lecture Notes in Mathematics, No. 28. Springer-Verlag, Berlin-New York, 1966.
29. A. Connes. *Noncommutative geometry*. Academic Press, Inc., San Diego, CA, 1994.
30. J. F. Davis and S. Weinberger. Group actions on homology spheres. *Invent. Math.*, 86(2):209–231, 1986.
31. S.L. de Medrano. *Involutions on Manifolds*. Ergebnisse der Mathematik und ihrer Grenzgebiete. 2. Folge. Springer Berlin Heidelberg, 2012.
32. P. Deligne, P. Griffiths, J. Morgan, and D. Sullivan. Real homotopy theory of Kähler manifolds. *Invent. Math.*, 29(3):245–274, 1975.
33. S. K. Donaldson and D. P. Sullivan. Quasiconformal 4-manifolds. *Acta Math.*, 163(3-4):181–252, 1989.
34. W. Dwyer and A. Zabrodsky. Maps between classifying spaces. In *Algebraic topology, Barcelona, 1986*, volume 1298 of *Lecture Notes in Math.*, pages 106–119. Springer, Berlin, 1987.

35. Y. Félix, S. Halperin, and J.-C. Thomas. *Rational homotopy theory*, volume 205 of *Graduate Texts in Mathematics*. Springer-Verlag, New York, 2001.
36. Y. Félix, J.-C. Thomas, and M. Vigué-Poirrier. Rational string topology. *J. Eur. Math. Soc. (JEMS)*, 9(1):123–156, 2007.
37. S. Ferry. Homotoping ε-maps to homeomorphisms. *Amer. J. Math.*, 101(3):567–582, 1979.
38. E. M. Friedlander. *Étale homotopy of simplicial schemes*. Annals of Mathematics Studies, No. 104. Princeton University Press, Princeton, N.J.; University of Tokyo Press, Tokyo, 1982.
39. G. Friedman. K-Witt bordism in characteristic 2. *Arch. Math. (Basel)*, 100(4):381–387, 2013.
40. W. M. Goldman. Invariant functions on Lie groups and Hamiltonian flows of surface group representations. *Invent. Math.*, 85(2):263–302, 1986.
41. M. Goresky and R. MacPherson. Intersection homology theory. *Topology*, 19(2):135–162, 1980.
42. M. Goresky and R. MacPherson. Intersection homology. II. *Invent. Math.*, 72(1):77–129, 1983.
43. P. Griffiths and J. Morgan. *Rational homotopy theory and differential forms*, volume 16 of *Progress in Mathematics*. Springer, New York, second edition, 2013.
44. D. Gromoll and W. Meyer. Periodic geodesics on compact riemannian manifolds. *J. Differential Geometry*, 3:493–510, 1969.
45. M. Gromov. Homotopical effects of dilatation. *J. Differential Geometry*, 13(3):303–310, 1978.
46. R. M. Hain. Iterated integrals and mixed Hodge structures on homotopy groups. In *Hodge theory (Sant Cugat, 1985)*, volume 1246 of *Lecture Notes in Math.*, pages 75–83. Springer, Berlin, 1987.
47. I. Hambleton and E. K. Pedersen. Topological equivalence of linear representations for cyclic groups. II. *Forum Math.*, 17(6):959–1010, 2005.
48. I. Hambleton and E. K. Pedersen. Topological equivalence of linear representations of cyclic groups. I. *Ann. of Math. (2)*, 161(1):61–104, 2005.
49. N. Higson. The Baum-Connes conjecture. In *Proceedings of the International Congress of Mathematicians, Vol. II (Berlin, 1998)*, number Extra Vol. II, pages 637–646, 1998.
50. N. Higson and J. Roe. Mapping surgery to analysis. I. Analytic signatures. *K-Theory*, 33(4):277–299, 2005.
51. N. Higson and J. Roe. Mapping surgery to analysis. II. Geometric signatures. *K-Theory*, 33(4):301–324, 2005.
52. N. Higson and J. Roe. Mapping surgery to analysis. III. Exact sequences. *K-Theory*, 33(4):325–346, 2005.
53. P. Hilton, G. Mislin, and J. Roitberg. *Localization of nilpotent groups and spaces*. North-Holland Mathematics Studies, No. 15. North-Holland Publishing Co., Amsterdam-Oxford; American Elsevier Publishing Co., Inc., New York, 1975.
54. N. Hitchin. The Atiyah–Singer index theorem. In H. Holden and R. Piene, editors, *The Abel Prize, 2003–2007. The First Five Years*, pages 115–150. Springer-Verlag, Berlin, 2010.
55. H. Hopf. Über die Topologie der Gruppen-Mannigfaltigkeiten und ihre Verallgemeinerungen. *Ann. of Math. (2)*, 42:22–52, 1941.
56. W. C. Hsiang and W. Pardon. When are topologically equivalent orthogonal transformations linearly equivalent? *Invent. Math.*, 68(2):275–316, 1982.
57. M. Kapovich and J. Kollár. Fundamental groups of links of isolated singularities. *J. Amer. Math. Soc.*, 27(4):929–952, 2014.
58. G. G. Kasparov. Equivariant KK-theory and the Novikov conjecture. *Invent. Math.*, 91(1):147–201, 1988.
59. M. A. Kervaire and J. W. Milnor. Groups of homotopy spheres. I. *Ann. of Math. (2)*, 77:504–537, 1963.
60. R. C. Kirby. Stable homeomorphisms and the annulus conjecture. *Ann. of Math. (2)*, 89:575–582, 1969.

61. R. C. Kirby and L. C. Siebenmann. *Foundational essays on topological manifolds, smoothings, and triangulations*. Annals of Mathematics Studies, No. 88. Princeton University Press, Princeton, N.J.; University of Tokyo Press, Tokyo, 1977. With notes by John Milnor and Michael Atiyah.
62. M. Krannich and O. Randal-Williams. Mapping class groups of simply connected high-dimensional manifolds need not be arithmetic. *C. R. Math. Acad. Sci. Paris*, 358(4):469–473, 2020.
63. J. Lannes. Sur les espaces fonctionnels dont la source est le classifiant d'un p-groupe abélien élémentaire. *Inst. Hautes Études Sci. Publ. Math.*, (75):135–244, 1992. With an appendix by Michel Zisman.
64. J. Latschev. Appendix to the chapter by J.-L. Loday. In *Free loop spaces in geometry and topology*, volume 24 of *IRMA Lect. Math. Theor. Phys.*, pages 157–163. Eur. Math. Soc., Zürich, 2015.
65. T. Macko and C. Wegner. On the classification of fake lens spaces. *Forum Math.*, 23(5):1053–1091, 2011.
66. I. Madsen and R. J. Milgram. *The classifying spaces for surgery and cobordism of manifolds*. Annals of Mathematics Studies, No. 92. Princeton University Press, Princeton, N.J.; University of Tokyo Press, Tokyo, 1979.
67. I. Madsen and M. Rothenberg. On the classification of G-spheres. I. Equivariant transversality. *Acta Math.*, 160(1-2):65–104, 1988.
68. I. Madsen and M. Rothenberg. On the classification of G-spheres. II. PL automorphism groups. *Math. Scand.*, 64(2):161–218, 1989.
69. M. A. Mandell. Cochains and homotopy type. *Publ. Math. Inst. Hautes Études Sci.*, (103):213–246, 2006.
70. F. Manin and S. Weinberger. Integral and rational mapping classes. *Duke Math. J.*, 169(10):1943–1969, 2020.
71. C. Manolescu. Homology cobordism and triangulations. In *Proceedings of the International Congress of Mathematicians—Rio de Janeiro 2018. Vol. II. Invited lectures*, pages 1175–1191. World Sci. Publ., Hackensack, NJ, 2018.
72. C. A. McGibbon and J. A. Neisendorfer. On the homotopy groups of a finite-dimensional space. *Comment. Math. Helv.*, 59(2):253–257, 1984.
73. H. Miller. The Sullivan conjecture on maps from classifying spaces. *Ann. of Math. (2)*, 120(1):39–87, 1984.
74. J. Milnor. Two complexes which are homeomorphic but combinatorially distinct. *Ann. of Math. (2)*, 74:575–590, 1961.
75. J. W. Milnor. *Topology from the differentiable viewpoint*. Princeton Landmarks in Mathematics. Princeton University Press, Princeton, NJ, 1997. Based on notes by David W. Weaver, Revised reprint of the 1965 original.
76. J. W. Milnor and J. D. Stasheff. *Characteristic classes*. Annals of Mathematics Studies, No. 76. Princeton University Press, Princeton, N. J.; University of Tokyo Press, Tokyo, 1974.
77. J. W. Morgan. The algebraic topology of smooth algebraic varieties. *Inst. Hautes Études Sci. Publ. Math.*, (48):137–204, 1978.
78. J. W. Morgan and D. P. Sullivan. The transversality characteristic class and linking cycles in surgery theory. *Ann. of Math. (2)*, 99:463–544, 1974.
79. A. J. Nicas. Induction theorems for groups of homotopy manifold structures. *Mem. Amer. Math. Soc.*, 39(267):vi+108, 1982.
80. S. P. Novikov. On manifolds with free abelian fundamental group and their application. *Izv. Akad. Nauk SSSR Ser. Mat.*, 30:207–246, 1966.
81. G. Perelman. The entropy formula for the ricci flow and its geometric applications. *arXiv preprint math/0211159*, 2002.
82. D. Quillen. Rational homotopy theory. *Ann. of Math. (2)*, 90:205–295, 1969.
83. D. Quillen. The Adams conjecture. *Topology*, 10:67–80, 1971.
84. F. Quinn. A geometric formulation of surgery. In *Topology of Manifolds (Proc. Inst., Univ. of Georgia, Athens, Ga., 1969)*, pages 500–511. Markham, Chicago, Ill., 1970.

85. F. Quinn. Nilpotent classifying spaces, and actions of finite groups. *Houston J. Math.*, 4(2):239–248, 1978.
86. F. Quinn. Resolutions of homology manifolds, and the topological characterization of manifolds. *Invent. Math.*, 72(2):267–284, 1983.
87. F. Quinn. Corrigendum to: "Resolutions of homology manifolds, and the topological characterization of manifolds" [Invent. Math. **72** (1983), no. 2, 267–284; MR0700771 (85b:57023)]. *Invent. Math.*, 85(3):653, 1986.
88. A. A. Ranicki. *Algebraic L-theory and topological manifolds*, volume 102 of *Cambridge Tracts in Mathematics*. Cambridge University Press, Cambridge, 1992.
89. J. Rosenberg and S. Weinberger. An equivariant Novikov conjecture. *K-Theory*, 4(1):29–53, 1990. With an appendix by J. P. May.
90. M. Rothenberg. Torsion invariants and finite transformation groups. In *Algebraic and geometric topology (Proc. Sympos. Pure Math., Stanford Univ., Stanford, Calif., 1976), Part 1*, Proc. Sympos. Pure Math., XXXII, pages 267–311. Amer. Math. Soc., Providence, R.I., 1978.
91. M. Rothenberg and S. Weinberger. Group actions and equivariant Lipschitz analysis. *Bull. Amer. Math. Soc. (N.S.)*, 17(1):109–112, 1987.
92. C. P. Rourke and D. P. Sullivan. On the Kervaire obstruction. *Ann. of Math. (2)*, 94:397–413, 1971.
93. R. Schultz. On the topological classification of linear representations. *Topology*, 16(3):263–269, 1977.
94. J.-P. Serre. Homologie singulière des espaces fibrés. Applications. *Ann. of Math. (2)*, 54:425–505, 1951.
95. L. Siebenmann. John W. Milnor's work on the classification of differentiable manifolds. In H. Holden and R. Piene, editors, *The Abel Prize 2008–2012*, pages 393–433. Springer, Heidelberg.
96. L. C. Siebenmann. Approximating cellular maps by homeomorphisms. *Topology*, 11:271–294, 1972.
97. S. Smale. Generalized Poincaré's conjecture in dimensions greater than four. *Ann. of Math. (2)*, 74:391–406, 1961.
98. V. P. Snaith. A history of the Arf-Kervaire invariant problem. *Notices Amer. Math. Soc.*, 60(8):1040–1047, 2013.
99. E. H. Spanier. *Algebraic topology*. Springer-Verlag, New York, [1994]. Corrected reprint of the 1966 original.
100. N. E. Steenrod. *Cohomology operations*. Annals of Mathematics Studies, No. 50. Princeton University Press, Princeton, N.J., 1962. Lectures by N. E. Steenrod written and revised by D. B. A. Epstein.
101. D. Sullivan. Geometric topology: localization, periodicity and Galois symmetry. MIT notes.
102. D. Sullivan. Notes from Princeton Geometric Topology seminar.
103. D. Sullivan. Genetics of homotopy theory and the Adams conjecture. *Ann. of Math. (2)*, 100:1–79, 1974.
104. D. Sullivan. Infinitesimal computations in topology. *Inst. Hautes Études Sci. Publ. Math.*, (47):269–331 (1978), 1977.
105. D. Sullivan. Hyperbolic geometry and homeomorphisms. In *Geometric topology (Proc. Georgia Topology Conf., Athens, Ga., 1977)*, pages 543–555. Academic Press, New York-London, 1979.
106. D. Sullivan and N. Teleman. An analytic proof of Novikov's theorem on rational Pontrjagin classes. *Inst. Hautes Études Sci. Publ. Math.*, (58):79–81 (1984), 1983.
107. N. Teleman. Combinatorial Hodge theory and signature operator. *Invent. Math.*, 61(3):227–249, 1980.
108. P. Tukia and J. Väisälä. Quasiconformal extension from dimension n to $n+1$. *Ann. of Math. (2)*, 115(2):331–348, 1982.
109. M. Vigué-Poirrier and D. Sullivan. The homology theory of the closed geodesic problem. *J. Differential Geometry*, 11(4):633–644, 1976.

110. C. T. C. Wall. Surgery of non-simply-connected manifolds. *Ann. of Math. (2)*, 84:217–276, 1966.
111. C. T. C. Wall. Formulae for surgery obstructions. *Topology*, 15(3):189–210, 1976.
112. S. Weinberger. *The topological classification of stratified spaces*. Chicago Lectures in Mathematics. University of Chicago Press, Chicago, IL, 1994.
113. S. Weinberger, Z. Xie, and G. Yu. Additivity of higher rho invariants and nonrigidity of topological manifolds. *Comm. Pure Appl. Math.*, 74(1):3–113, 2021.
114. C. W. Wilkerson. Applications of minimal simplicial groups. *Topology*, 15(2):111–130, 1976.
115. A. Yuan. Integral models for spaces via the higher Frobenius. *J. Amer. Math. Soc.*, 36(1):107–175, 2023.

List of Publications for Dennis P. Sullivan

1966

[1] Triangulating Homotopy Equivalences. Ph.D. Thesis, Princeton University.

1967

[2] Triangulating Homotopy Equivalences. Preprint, University of Warwick.
[3] Smoothing Homotopy Equivalences. Preprint, University of Warwick.
[4] On the Hauptvermutung for manifolds. *Bull. Amer. Math. Soc.*, 73:598–600.

1970

[5] (with M. Cohen). On the regular neighborhood of a two-sided submanifold. *Topology*, 9:141–147.

1971

[6] *Geometric topology. Part I.* MIT, Cambridge, Mass. Localization, periodicity, and Galois symmetry, Revised version. Also available in Russian (*Matematika.* Izdat. Mir, Moscow, 1975). Also available as *Geometric topology: localization, periodicity and Galois symmetry*, volume 8 of *K-Monographs in Mathematics*. Springer, Dordrecht, 2005.
[7] Combinatorial invariants of analytic spaces. In *Proceedings of Liverpool Singularities—Symposium, I (1969/70)*, Lecture Notes in Mathematics, Vol. 192, pages 165–168. Springer, Berlin.
[8] Geometric periodicity and the invariants of manifolds. In *Manifolds–Amsterdam 1970 (Proc. Nuffic Summer School)*, Lecture Notes in Mathematics, Vol. 197, pages 44–75. Springer, Berlin.
[9] Galois symmetry in manifold theory at the primes. In *Actes du Congrès International des Mathématiciens (Nice, 1970), Tome 2*, pages 169–175.
[10] Singularities in spaces. In *Proceedings of Liverpool Singularities Symposium, II (1969/1970)*, pages 196–206. Lecture Notes in Math., Vol. 209.
[11] (with C.P. Rourke). On the Kervaire obstruction. *Ann. of Math. (2)*, 94:397–413.

H. Holden, R. Piene (eds.), *The Abel Prize 2018-2022*, The Abel Prize,
https://doi.org/10.1007/978-3-031-33973-8_28

1974

[12] (with J.W. Morgan). The transversality characteristic class and linking cycles in surgery theory. *Ann. of Math. (2)*, 99:463–544.

[13] Genetics of homotopy theory and the Adams conjecture. *Ann. of Math. (2)*, 100:1–79.

[14] (with M. Shub). A remark on the Lefschetz fixed point formula for differentiable maps. *Topology*, 13:189–191.

1975

[15] Inside and outside manifolds. In *Proceedings of the International Congress of Mathematicians (Vancouver, B. C., 1974), Vol. 1*, pages 201–207.

[16] Differential forms and the topology of manifolds. In *Manifolds—Tokyo 1973 (Proc. Internat. Conf., Tokyo, 1973)*, pages 37–49.

[17] (with M. Shub). Homology theory and dynamical systems. *Topology*, 14(2): 109–132.

[18] On the intersection ring of compact three manifolds. *Topology*, 14(3):275–277.

[19] (with J. Palis, C. Pugh and M. Shub). Genericity theorems in topological dynamics. In *Dynamical systems—Warwick 1974 (Proc. Sympos. Appl. Topology and Dynamical Systems, Univ. Warwick, Coventry, 1973/1974; presented to E. C. Zeeman on his fiftieth birthday)*, pages 241–250. Lecture Notes in Math., Vol. 468.

[20] (with B. Parry). A topological invariant of flows on 1-dimensional spaces. *Topology*, 14(4):297–299.

[21] (with D. Ruelle). Currents, flows and diffeomorphisms. *Topology*, 14(4):319–327.

[22] (with P. Deligne, P. Griffiths and J. Morgan). Real homotopy theory of Kähler manifolds. *Invent. Math.*, 29(3):245–274. Also available in *Uspehi Mat. Nauk*, 32(3(195)):119–152, 247, 1977 (in Russian).

[23] (with B. Kostant). The Euler characteristic of an affine space form is zero. *Bull. Amer. Math. Soc.*, 81(5):937–938.

[24] La classe d'Euler réelle d'un fibré vectoriel à groupe structural $SL_n(Z)$ est nulle. *C. R. Acad. Sci. Paris Sér. A-B*, 281(1):Aii, A17–A18.

[25] (with M. Vigué-Poirrier). Sur l'existence d'une infinité de géodésiques périodiques sur une variété riemannienne compacte. *C. R. Acad. Sci. Paris Sér. A-B*, 281(9):Aii, A289–A291.

[26] (with P. Deligne). Fibrés vectoriels complexes à groupe structural discret. *C. R. Acad. Sci. Paris Sér. A-B*, 281(24):Ai, A1081–A1083.

1976

[27] (with A. Ranicki). A semi-local combinatorial formula for the signature of a 4*k*-manifold. *J. Differential Geometry*, 11(1):23–29.

[28] (with R.F. Williams). On the homology of attractors. *Topology*, 15(3):259–262.

[29] A generalization of Milnor's inequality concerning affine foliations and affine manifolds. *Comment. Math. Helv.*, 51(2):183–189.

[30] A counterexample to the periodic orbit conjecture. *Inst. Hautes Études Sci. Publ. Math.*, 46:5–14.

[31] Cycles for the dynamical study of foliated manifolds and complex manifolds. *Invent. Math.*, 36:225–255.

[32] (with M. Vigué-Poirrier). The homology theory of the closed geodesic problem. *J. Differential Geometry*, 11(4):633–644.

[33] Cartan–de Rham homotopy theory. In *Colloque "Analyse et Topologie" en l'Honneur de Henri Cartan (Orsay, 1974)*, pages 227–254. Astérisque, No. 32–33.

[34] A new flow. *Bull. Amer. Math. Soc.*, 82(2):331–332.

1977

[35] (with R. Edwards and K. Millett). Foliations with all leaves compact. *Topology*, 16(1):13–32.

[36] Infinitesimal computations in topology. *Inst. Hautes Études Sci. Publ. Math.*, 47:269–331.

1978

[37] A foliation of geodesics is characterized by having no "tangent homologies". *J. Pure Appl. Algebra*, 13(1):101–104.

1979

[38] (with L. Siebenmann). On complexes that are Lipschitz manifolds. In *Geometric topology (Proc. Georgia Topology Conf., Athens, Ga., 1977)*, pages 503–525. Academic Press, New York-London.

[39] Hyperbolic geometry and homeomorphisms. In *Geometric topology (Proc. Georgia Topology Conf., Athens, Ga., 1977)*, pages 543–555. Academic Press, New York-London.

[40] A homological characterization of foliations consisting of minimal surfaces. *Comment. Math. Helv.*, 54(2):218–223.

[41] The density at infinity of a discrete group of hyperbolic motions. *Inst. Hautes Études Sci. Publ. Math.*, 50:171–202.

[42] Rufus Bowen (1947–1978). *Inst. Hautes Études Sci. Publ. Math.*, 50:7–9.

1981

[43] Travaux de Thurston sur les groupes quasi-fuchsiens et les variétés hyperboliques de dimension 3 fibrées sur S^1. In *Bourbaki Seminar, Vol. 1979/80*, volume 842 of *Lecture Notes in Math.*, pages 196–214. Springer, Berlin-New York.

[44] For $n > 3$ there is only one finitely additive rotationally invariant measure on the n-sphere defined on all Lebesgue measurable subsets. *Bull. Amer. Math. Soc. (N.S.)*, 4(1):121–123.

[45] A finiteness theorem for cusps. *Acta Math.*, 147(3-4):289–299.
[46] Growth of positive harmonic functions and Kleinian group limit sets of zero planar measure and Hausdorff dimension two. In *Geometry Symposium, Utrecht 1980 (Utrecht, 1980)*, volume 894 of *Lecture Notes in Math.*, pages 127–144. Springer, Berlin-New York.
[47] (with A. Phillips). Geometry of leaves. *Topology*, 20(2):209–218.
[48] (with J.D. Maitland Wright). On lifting automorphisms of monotone σ-complete C^*-algebras. *Quart. J. Math. Oxford Ser. (2)*, 32(127):371–381.
[49] On the ergodic theory at infinity of an arbitrary discrete group of hyperbolic motions. In *Riemann surfaces and related topics: Proceedings of the 1978 Stony Brook Conference (State Univ. New York, Stony Brook, N.Y., 1978)*, volume 97 of *Ann. of Math. Stud.*, pages 465–496. Princeton Univ. Press, Princeton, N.J.

1982

[50] Disjoint spheres, approximation by imaginary quadratic numbers, and the logarithm law for geodesics. *Acta Math.*, 149(3-4):215–237.
[51] Discrete conformal groups and measurable dynamics. *Bull. Amer. Math. Soc. (N.S.)*, 6(1):57–73.
[52] Seminar on conformal and hyperbolic geometry. Preprint, IHES, 92 pp.
[53] Itération des fonctions analytiques complexes. *C. R. Acad. Sci. Paris Sér. I Math.*, 294(9):301–303.

1983

[54] (with R. Mañé and P. Sad). On the dynamics of rational maps. *Ann. Sci. École Norm. Sup. (4)*, 16(2):193–217.
[55] Conformal dynamical systems. In *Geometric dynamics (Rio de Janeiro, 1981)*, volume 1007 of *Lecture Notes in Math.*, pages 725–752. Springer, Berlin.
[56] (with W. Thurston). Manifolds with canonical coordinate charts: some examples. *Enseign. Math. (2)*, 29(1-2):15–25.
[57] (with P. Deligne). Division algebras and the Hausdorff–Banach–Tarski paradox. *Enseign. Math. (2)*, 29(1-2):145–150.
[58] The Dirichlet problem at infinity for a negatively curved manifold. *J. Differential Geom.*, 18(4):723–732.
[59] (with N. Teleman). An analytic proof of Novikov's theorem on rational Pontrjagin classes. *Inst. Hautes Études Sci. Publ. Math.*, 58:79–81, 1983.
[60] (with J.H. Curry and L. Garnett). On the iteration of a rational function: computer experiments with Newton's method. *Comm. Math. Phys.*, 91(2):267–277.

1984

[61] Entropy, Hausdorff measures old and new, and limit sets of geometrically finite Kleinian groups. *Acta Math.*, 153(3-4):259–277.

[62] (with T. Lyons). Function theory, random paths and covering spaces. *J. Differential Geom.*, 19(2):299–323.
[63] (with J. Aaronson). Rational ergodicity of geodesic flows. *Ergodic Theory Dynam. Systems*, 4(2):165–178.
[64] (with H.P. McKean). Brownian motion and harmonic functions on the class surface of the thrice punctured sphere. *Adv. in Math.*, 51(3):203–211.
[65] (with H. Hermes and A. Lundell). Nilpotent bases for distributions and control systems. *J. Differential Equations*, 55(3):385–400.

1985

[66] Quasiconformal homeomorphisms and dynamics. I. Solution of the Fatou–Julia problem on wandering domains. *Ann. of Math. (2)*, 122(3):401–418.
[67] Quasiconformal homeomorphisms and dynamics. II. Structural stability implies hyperbolicity for Kleinian groups. *Acta Math.*, 155(3-4):243–260.
[68] (with M. Shub). Expanding endomorphisms of the circle revisited. *Ergodic Theory Dynam. Systems*, 5(2):285–289.
[69] (with É. Ghys and L.R. Goldberg). On the measurable dynamics of $z \mapsto e^z$. *Ergodic Theory Dynam. Systems*, 5(3):329–335.
[70] On the dynamical structure near an isolated completely unstable ellipitic fixed point. In *Atas do 14 Coloquio Brasileiro de Matematica*, volume II of *Publ. do IMPA*, pages 553–559.

1986

[71] Related aspects of positivity: λ-potential theory on manifolds, lowest eigenstates, Hausdorff geometry, renormalized Markoff processes.... In *Aspects of mathematics and its applications*, volume 34 of *North-Holland Math. Library*, pages 747–779. North-Holland, Amsterdam.
[72] (with B. Weiss and J.D. Maitland Wright). Generic dynamics and monotone complete C^*-algebras. *Trans. Amer. Math. Soc.*, 295(2):795–809.
[73] (with T. Pignataro). Ground state and lowest eigenvalue of the Laplacian for noncompact hyperbolic surfaces. *Comm. Math. Phys.*, 104(4):529–535.
[74] (with W.P. Thurston). Extending holomorphic motions. *Acta Math.*, 157 (3-4):243–257.
[75] The spinor representation of minimal surfaces in space. Preprint Texas 1986 (a letter to H. Rosenberg), 5 pp. Available as `https://www.math.stonybrook.edu/~dennis/publications/PDF/DS-pub-0084.pdf`.
[76] On negative curvature, variable, pinched and constant. Preprint Texas 1986. Available as `https://www.math.stonybrook.edu/~dennis/publications/PDF/DS-pub-0085.pdf`.

1987

[77] Quasiconformal homeomorphisms in dynamics, topology, and geometry. In *Proceedings of the International Congress of Mathematicians, Vol. 1, 2 (Berkeley, Calif., 1986)*, pages 1216–1228. Amer. Math. Soc., Providence, RI.

[78] (with B. Rodin). The convergence of circle packings to the Riemann mapping. *J. Differential Geom.*, 26(2):349–360.

[79] Related aspects of positivity in Riemannian geometry. *J. Differential Geom.*, 25(3):327–351.

[80] (with J. Dodziuk, T. Pignataro and B. Randol). Estimating small eigenvalues of Riemann surfaces. In *The legacy of Sonya Kovalevskaya (Cambridge, Mass., and Amherst, Mass., 1985)*, volume 64 of *Contemp. Math.*, pages 93–121. Amer. Math. Soc., Providence, RI.

1988

[81] Differentiable structures on fractal-like sets, determined by intrinsic scaling functions on dual Cantor sets. In *The mathematical heritage of Hermann Weyl (Durham, NC, 1987)*, volume 48 of *Proc. Sympos. Pure Math.*, pages 15–23. Amer. Math. Soc., Providence, RI.

[82] (with R. Hardt). Variation of the Green function on Riemann surfaces and Whitney's holomorphic stratification conjecture. *Inst. Hautes Études Sci. Publ. Math.*, 68:115–137.

[83] Differentiable structures on fractal like sets, determined by intrinsic scaling functions on dual Cantor sets. In *Nonlinear evolution and chaotic phenomena (Noto, 1987)*, volume 176 of *NATO Adv. Sci. Inst. Ser. B: Phys.*, pages 101–110. Plenum, New York.

1989

[84] (with S.K. Donaldson). Quasiconformal 4-manifolds. *Acta Math.*, 163 (3-4):181–252.

[85] Bounded structure of infinitely renormalizable mappings. In *Universality in Chaos*, (P. Cvitanoviĉ, editor), Routledge, New York, 9 pp.

1990

[86] (with Y. Katznelson and S. Nag). On conformal welding homeomorphisms associated to Jordan curves. *Ann. Acad. Sci. Fenn. Ser. A I Math.*, 15(2):293–306.

1991

[87] Renormalization, Zygmund smoothness and the Epstein class. In *Chaos, order, and patterns (Lake Como, 1990)*, volume 280 of *NATO Adv. Sci. Inst. Ser. B: Phys.*, pages 25–34. Plenum, New York.

1992

[88] Bounds, quadratic differentials, and renormalization conjectures. In *American Mathematical Society centennial publications, Vol. II (Providence, RI, 1988)*, pages 417–466. Amer. Math. Soc., Providence, RI.

[89] (with Y.P. Jiang and T. Morita). Expanding direction of the period doubling operator. *Comm. Math. Phys.*, 144(3):509–520.

[90] (with F.P. Gardiner). Symmetric structures on a closed curve. *Amer. J. Math.*, 114(4):683–736.

1993

[91] (with A. Connes and N. Teleman). Formules locales pour les classes de Pontrjagin topologiques. *C. R. Acad. Sci. Paris Sér. I Math.*, 317(5):521–526.

[92] Linking the universalities of Milnor–Thurston, Feigenbaum and Ahlfors–Bers. In *Topological methods in modern mathematics (Stony Brook, NY, 1991)*, pages 543–564. Publish or Perish, Houston, TX.

1994

[93] (with J.M. Gambaudo and C. Tresser). Infinite cascades of braids and smooth dynamical systems. *Topology*, 33(1):85–94.

[94] (with A. Connes and N. Teleman). Quasiconformal mappings, operators on Hilbert space, and local formulae for characteristic classes. *Topology*, 33(4):663–681.

[95] (with F.P. Gardiner). Lacunary series as quadratic differentials in conformal dynamics. In *The mathematical legacy of Wilhelm Magnus: groups, geometry and special functions (Brooklyn, NY, 1992)*, volume 169 of *Contemp. Math.*, pages 307–330. Amer. Math. Soc., Providence, RI.

1995

[96] Exterior *d*, the local degree, and smoothability. In *Prospects in topology (Princeton, NJ, 1994)*, volume 138 of *Ann. of Math. Stud.*, pages 328–338. Princeton Univ. Press, Princeton, NJ.

[97] (with S. Nag). Teichmüller theory and the universal period mapping via quantum calculus and the $H^{1/2}$ space on the circle. *Osaka J. Math.*, 32(1):1–34.

1996

[98] (with A. Norton). Wandering domains and invariant conformal structures for mappings of the 2-torus. *Ann. Acad. Sci. Fenn. Math.*, 21(1):51–68.

[99] (with I. Biswas and S. Nag). Determinant bundles, Quillen metrics and Mumford isomorphisms over the universal commensurability Teichmüller space. *Acta Math.*, 176(2):145–169.

[100] Triangulating and smoothing homotopy equivalences and homeomorphisms. Geometric Topology Seminar Notes. In *The Hauptvermutung book*, volume 1 of *K-Monogr. Math.*, pages 69–103. Kluwer Acad. Publ., Dordrecht.

[101] (with A.A. Ranicki, A.J. Casson, M.A. Armstrong, C.P. Rourke and G.E. Cooke). *The Hauptvermutung book*, volume 1 of *K-Monographs in Mathematics*. Kluwer Academic Publishers, Dordrecht.

[102] (with G. Cui and Y. Jiang). Dynamics of geometrically finite rational maps. Preprint CUNY, 30 pp. Available as `https://www.math.stonybrook.edu/~dennis/publications/PDF/DS-pub-0106.pdf`.

1997

[103] (with J. Hu). Topological conjugacy of circle diffeomorphisms. *Ergodic Theory Dynam. Systems*, 17(1):173–186.

1998

[104] (with C.T. McMullen). Quasiconformal homeomorphisms and dynamics. III. The Teichmüller space of a holomorphic dynamical system. *Adv. Math.*, 135(2):351–395.

[105] (with R. Body, M. Mimura and H. Shiga). *p*-universal spaces and rational homotopy types. *Comment. Math. Helv.*, 73(3):427–442.

1999

[106] On the foundation of geometry, analysis, and the differentiable structure for manifolds. In *Topics in low-dimensional topology (University Park, PA, 1996)*, pages 89–92. World Sci. Publ., River Edge, NJ.

[107] (with M. Chas). String topology. Preprint, arXiv:math/9911159.

2001

[108] Reminiscences of Michel Herman's first great theorem. In *Michael R. Herman*, No. 8 of *Gazette des Mathematiciens*, pages 91–94. Societe Mathematique de France, Paris.

2002

[109] (with J. Heinonen). On the locally branched Euclidean metric gauge. *Duke Math. J.*, 114(1):15–41.

2003

[110] (with G. Cui and Y. Jiang). On geometrically finite branched coverings. I. Locally combinatorial attracting. In *Complex dynamics and related topics: lectures from the Morningside Center of Mathematics*, volume 5 of *New Stud. Adv. Math.*, pages 1–14. Int. Press, Somerville, MA.

[111] (with G. Cui and Y. Jiang). On geometrically finite branched coverings. II. Realization of rational maps. In *Complex dynamics and related topics: lectures from the Morningside Center of Mathematics*, volume 5 of *New Stud. Adv. Math.*, pages 15–29. Int. Press, Somerville, MA.

2004

[112] René Thom's work on geometric homology and bordism. *Bull. Amer. Math. Soc. (N.S.)*, 41(3):341–350.

[113] (with M. Chas). Closed string operators in topology leading to Lie bialgebras and higher string algebra. In *The legacy of Niels Henrik Abel*, pages 771–784. Springer, Berlin.

[114] Open and closed string field theory interpreted in classical algebraic topology. In *Topology, geometry and quantum field theory*, volume 308 of *London Math. Soc. Lecture Note Ser.*, pages 344–357. Cambridge Univ. Press, Cambridge.

[115] (with A. Pinto). Dynamical systems applied to asymptotic geometry. IMS at Stony Brook University, Preprint No. 6, 29 pp.

2005

[116] Sigma models and string topology. In *Graphs and patterns in mathematics and theoretical physics*, volume 73 of *Proc. Sympos. Pure Math.*, pages 1–11. Amer. Math. Soc., Providence, RI.

[117] A stratified rational homology manifold version of the Atiyah–Bott fixed point theorem. *J. Differential Geom.*, 70(2):325–328.

2006

[118] (with A.A. Pinto). The circle and the solenoid. *Discrete Contin. Dyn. Syst.*, 16(2):463–504.

2007

[119] String topology background and present state. In *Current developments in mathematics, 2005*, pages 41–88. Int. Press, Somerville, MA.

2008

[120] (with J. Simons). Axiomatic characterization of ordinary differential cohomology. *J. Topol.*, 1(1):45–56.

[121] (with R.L. Cohen and J.R. Klein). The homotopy invariance of the string topology loop product and string bracket. *J. Topol.*, 1(2):391–408.

[122] (with J. Simons). Structured bundles define differential K-theory. Géométrie différentielle, physique mathématique, mathématiques et société. I. Number 321, pages 1–3.

2009

[123] Homotopy theory of the master equation package applied to algebra and geometry: a sketch of two interlocking programs. In *Algebraic topology—old and new*, volume 85 of *Banach Center Publ.*, pages 297–305. Polish Acad. Sci. Inst. Math., Warsaw.

[124] (with M. Chas). String topology in dimensions two and three. In *Algebraic topology*, volume 4 of *Abel Symp.*, pages 33–37. Springer, Berlin.

2010

[125] (with J. Simons). Structured vector bundles define differential K-theory. In *Quanta of maths*, volume 11 of *Clay Math. Proc.*, pages 579–599. Amer. Math. Soc., Providence, RI.
[126] (with J. Simons). The Atiyah–Singer index theorem and Chern–Weil forms. *Pure Appl. Math. Q.*, 6(2, Special Issue: In honor of Michael Atiyah and Isadore Singer):643–645.
[127] (with J. Simons). The Mayer–Vietoris Property in Differential Cohomology. Preprint, arXiv:1010.5269v2.

2011

[128] Algebra, topology and algebraic topology of 3D ideal fluids. In *Low-dimensional and symplectic topology*, volume 82 of *Proc. Sympos. Pure Math.*, pages 1–7. Amer. Math. Soc., Providence, RI.
[129] A finite time blowup result for quadratic ODE's. In *Dynamics, games and science. I*, volume 1 of *Springer Proc. Math.*, pages 761–762. Springer, Heidelberg.

2012

[130] (with J. Simons). Differential characters for K-theory. In *Metric and differential geometry*, volume 297 of *Progr. Math.*, pages 353–361. Birkhäuser/Springer, Basel.

2014

[131] 3D incompressible fluids: combinatorial models, eigenspace models, and a conjecture about well-posedness of the 3D zero viscosity limit. *J. Differential Geom.*, 97(1):141–148.
[132] (with R. Lawrence). A formula for topology/deformations and its significance. *Fund. Math.*, 225(1):229–242.
[133] Solenoidal manifolds. *J. Singul.*, 9:203–205.
[134] (with N. Ranade). The cumulant bijection and differential forms. Preprint, arXiv:1407.0422v2.

2015

[135] (with S. Basu, J. McGibbon and M. Sullivan). Transverse string topology and the cord algebra. *J. Symplectic Geom.*, 13(1):1–16.

2016

[136] A discourse on the measurable Riemann mapping theorem & incompressible fluid motion. In *The legacy of Bernhard Riemann after one hundred and fifty years. Vol. II*, volume 35 of *Adv. Lect. Math. (ALM)*, pages 691–702. Int. Press, Somerville, MA.

2017

[137] Simplicity is the point. In *Simplicity: ideals of practice in mathematics and the arts*, Math. Cult. Arts, pages 269–274. Springer, Cham.

[138] (with C. LeBrun, G. Besson, M. Gromov, J. Simons, J. Cheeger, J.-P. Bourguignon, J. Lafontaine, J. Kazdan, M.-L. Michelsohn, P. Pansu, D. Ebin and K. Grove). Marcel Berger remembered. *Notices Amer. Math. Soc.*, 64(11): 1285–1295.

2018

[139] (with J. Simons). Characters for complex bundles and their connections. Preprint, arXiv:1803.07129v1.

2019

[140] (with A. Gorokhovsky and Z. Xie). Generalized Euler classes, differential forms and commutative DGAs. *J. Topol. Anal.*, 11(1):109–118.

2020

[141] Lattice hydrodynamics. In *Some aspects of the theory of dynamical systems: a tribute to Jean-Christophe Yoccoz*, Vol. 1, pages 215–222. Astérisque 415, Société Mathématique de France, Paris.

[142] A decade of Thurston stories. In *What's next?—the mathematical legacy of William P. Thurston*, volume 205 of *Ann. of Math. Stud.*, pages 415–421. Princeton Univ. Press, Princeton, NJ.

[143] (with A. Kwon). Geometry on all prime three-manifolds. Preprint, arXiv: 1906.10820v2.

[144] (with A. Kwon, P. Rao and D. An). Numerical comparison of momentum model and vorticity model. Preprint CUNY, 18 pp.

2021

[145] (with N. Ranade and C. Sadanand). Structure on the top homology and related algorithms. *Topology Appl.*, 289:Paper No. 107397, 11.

[146] (with R. Lawrence and N. Ranade). Quantitative towers in finite difference calculus approximating the continuum. *Q. J. Math.*, 72(1-2):515–545.

[147] Rokhlin's theorem, a problem and a conjecture. In *Topology, geometry, and dynamics—V. A. Rokhlin-Memorial*, volume 772 of *Contemp. Math.*, pages 325–329. Amer. Math. Soc., Providence, RI.

2022

[148] (with A. Kwon). Complément to the Thurston 3D-geometrization picture. *Enseign. Math. (2)*, 68(3-4):379–387.

Curriculum Vitae for Dennis Parnell Sullivan

Born:	February 12, 1941 in Port Huron, Michigan, USA
Degrees/education:	Bachelor of Arts, Rice University, 1963 PhD, Princeton University, 1966
Positions:	NATO Fellowship, University of Warwick, 1966–1967 Miller Research Fellow, University of California, Berkeley, 1967–1969 Sloan Fellow, MIT, 1969–1973 Associate Professor, Paris-Sud University, 1973–1974 Professor, IHÉS, 1974–1997 Albert Einstein Chair in Science, City University of New York, 1981– Professor, Stony Brook University, 1996–

H. Holden, R. Piene (eds.), *The Abel Prize 2018-2022*, The Abel Prize,
https://doi.org/10.1007/978-3-031-33973-8_29

Visiting positions: Visiting Scholar, Institute for Advanced Study, 1967–1968, 1968–1970 and 1975
Stanislaw Ulam Visiting Professor of Mathematics, University of Colorado, 1981–82
Visiting Silver Chair, University of Texas at Austin, 1986

Memberships: National Academy of Sciences, 1983
New York Academy of Sciences, 1983
Brazilian National Academy of Sciences, 1984
London Mathematical Society, Honorary Member, 2012
Irish Royal Society, 2011
Norwegian Academy of Science and Letters, 2022

Awards and prizes: Speaker at the International Congress of Mathematicians, 1970, 1974 (plenary), 1986
Oswald Veblen Prize in Geometry, 1971
Prix Élie Cartan, 1981
King Faisal International Prize for Science, 1994
New York City Mayor's Award for Excellence in Science and Technology, 1997
Ordem Scientifico Nacional, Brazilian Academy of Sciences, 1998
National Medal of Science, 2004
Leroy P. Steele Prize for Lifetime Achievement, 2006
Wolf Prize, 2010
Balzan Prize, 2014
Abel Prize, 2022

Honorary degrees: University of Warwick, 1983
École Normale Supérieure de Lyon, 2001

Part VI
Abel Activities 2018–2022

H. Holden, R. Piene (eds.), *The Abel Prize 2018-2022*, The Abel Prize,
https://doi.org/10.1007/978-3-031-33973-8

Photos

The Abel Prize. (Photo: Abel Prize)

Robert Langlands discussing mathematics with young students at NTNU. Chair of the Abel Board Kristian Ranestad in the back. (Photo: NTNU)

The award ceremony in 2019 in the Aula of the University of Oslo. (Photo: Abel Prize)

The Abel Lectures at the University of Oslo in 2019. (Photo: Ola Gamst Sæther/Abel Prize)

Karen Uhlenbeck with young pupils in 2019 in Bergen. (Photo: Jens Helleland Ådnanes/University of Bergen)

The award ceremony for Hillel Furstenberg in 2021 in the Norwegian Embassy in Tel Aviv. The Ambassador Kåre R. Aas to the right. (Photo: Abel Prize)

The finalists of the Niels Henrik Abel competitions for high-school students in 2022 outside the main building of the NTNU — Norwegian University of Science and Technology, Trondheim. (Photo: Glen Musk)

The award ceremony of Avi Wigderson and Gregory Margulis at the Norwegian Embassay in Washington DC in 2022. The Deputy Chief of Mission Torleiv Opland in the middle. (Photo: Abel Prize)

The Abel Laureates (left to right) Hillel Furstenberg, László Lovász, Dennis Sullivan, Endre Szemerédi, and Gregory Margulis in front of the Abel Monument in the garden of the Royal Palace in 2022. (Photo: Natalia Demina)

Dennis Sullivan discussing with young students. DNVA Secretary General Gunn Elisabeth Birkelund to the left. (Photo: Ola Gamst Sæther/Abel Prize)

Hillel Furstenberg in front of the portraits of the Abel Laureates in the Norwegian Academy of Sciences and Letters in 2022. (Photo: Eirik Furu Baardesen/DNVA)

The Abel Committee

2018

John Rognes (University of Oslo, Norway), chair
Sun-Yung Alice Chang (Princeton University, USA)
Irene Fonseca (Carnegie Mellon University, USA)
Ben J. Green (University of Oxford, UK)
Marie-France Vignéras (Institut de Mathématiques de Jussieu, Paris, France)

2019

Hans Munthe-Kaas (University of Bergen, Norway), chair
Sun-Yung Alice Chang (Princeton University, USA)
Irene Fonseca (Carnegie Mellon University, USA)
Gil Kalai (Hebrew University of Jerusalem, Israel)
François Labourie (Université de Nice, France)

2020

Hans Munthe-Kaas (University of Bergen, Norway), chair
Gil Kalai (Hebrew University of Jerusalem, Israel)
François Labourie (Université de Nice, France)
Sylvia Serfaty (New York University, USA)
Claire Voisin (Collège de France, France)

2021

Hans Munthe-Kaas (University of Bergen, Norway), chair
Alexander Lubotzky (Hebrew University, Israel)
Subhash Khot (New York University, USA)
Sylvia Serfaty (New York University, USA)
Claire Voisin (Collège de France, France)
[Vaughan F.R. Jones † (Vanderbilt University, USA)]

2022

Hans Munthe-Kaas (University of Bergen, Norway), chair
Alexander Lubotzky (Hebrew University, Israel)
Subhash Khot (New York University, USA)
Raman Parimala (Emory University, USA)
Ulrike Tillmann (University of Oxford, UK)

The Niels Henrik Abel Board

2018

Kristian Ranestad (chair)
Anne Borg
Hans Munthe-Kaas
Einar Rønquist
Anne Carine Tanum
Øystein Hov (observer)

2019

John Grue (chair)
Einar Rønquist
Hilde Christiane Bjørnland
Aslak Bakke Buan
Sissel Rogne
Øystein Hov (observer)

2020

John Grue (chair)
Einar Rønquist
Hilde Christiane Bjørnland
Aslak Bakke Buan
Sissel Rogne
Øystein Hov (observer)

2021

John Grue (chair)
Einar Rønquist
Hilde Christiane Bjørnland
Aslak Bakke Buan
Sissel Rogne
Gunn Elisabeth Birkelund (observer)

2022

John Grue (chair)
Einar Rønquist
Hilde Christiane Bjørnland
Aslak Bakke Buan
Sissel Rogne
Gunn Elisabeth Birkelund (observer)

The Abel Lectures

2018

Robert P. Langlands (Institute for Advanced Study, Princeton): *On the geometric theory*
James Arthur (University of Toronto): *The Langlands program: arithmetic, geometry and analysis*
Edward Frenkel (University of California at Berkeley): *Langlands program and unification*

2019

Karen K. Uhlenbeck (University of Texas at Austin): *Some thoughts on the calculus of variations*
Chuu-Lian Terng (University of California at Irvine): *Solitons in geometry*
Robert Bryant (Duke University): *Limits, bubbles, and singularities: An introduction to the fundamental ideas of Karen Uhlenbeck*
Matt Parker (UK): *[Popular lecture] An attempt to visualise minimal surfaces and maximum dimensions*

2020 [video lectures given in 2021]

Hillel Furstenberg (Hebrew University, Jerusalem): *Boundary of groups*
Gregory Margulis (Yale University, New Haven, Connecticut): *Arithmeticity of discrete subgroups and related topics*

2021 [video lectures]

László Lovász (Eötvös Loránd University, Budapest): *Continuous limits of finite structures*
Avi Wigderson (Institute for Advanced Study, Princeton): *The value of errors in proofs*

2022

Dennis P. Sullivan (Stony Brook University, New York): *Gathering chestnuts of math related to fluid motion*
Michael J. Hopkins (Harvard University): *The great wild manifold rodeo: Dennis Sullivan in algebraic topology*
Etienne Ghys (ENS Lyon): *Dynamics à la Dennis Sullivan*
Jim Simons (The Simons Foundation, New York): *A discussion with Nils A. Baas and Nicolai Tangen*

The Abel Laureate Presenters

In March each year when the President of the Norwegian Academy of Science and Letters announces the Abel Laureate and the Chair of the Abel Committee states the reasons for the selection, a scientist presents the work of the Laureate. Below we list the presenters for the period 2018–22:

2018 (Robert P. Langlands) Alexander Bellos, London

2019 (Karen K. Uhlenbeck) Jim Al-Khalili, Surrey

2020 (Hillel Furstenberg and Gregory Margulis) Alexander Bellos, London

2021 (László Lovász and Avi Wigderson) Alexander Bellos, London

2022 (Dennis P. Sullivan) Alexander Bellos, London

The Interviews with the Abel Laureates

Transcripts of parts of the interviews that Bjørn I. Dundas, University of Bergen, and Christian Skau, Norwegian University of Science and Technology, made with each laureate in connection with the Prize ceremonies, can be found in the following publications:

2018 Robert P. Langlands

EMS Newsletter, issue 109 (Sep. 2018) 19–27,
AMS Notices, **66** (2019) 494–503.

2019 Karen K. Uhlenbeck

EMS Newsletter, issue 113 (Sep. 2019) 21–29,
AMS Notices, **67** (2020) 393–403.

2020 Hillel Furstenberg and Gregory Margulis

EMS Newsletter, issue 118 (Dec. 2020) 45–56,
AMS Notices, **68** (2021) 992–997 (Margulis),
AMS Notices, **68** (2021) 1189–1196 (Furstenberg).

2021 László Lovász and Avi Wigderson

EMS Magazine, issue 122 (Dec. 2021) 16–31,
AMS Notices, **69** (2022) 828–843.

2022 Dennis P. Sullivan

EMS Magazine, issue 125 (2022) 20–30,
AMS Notices, **70** (2023) 623–642.

The Abel Banquet 2003–2022

In the evening on the day of the Abel Prize Award Ceremony, the Norwegian Government hosts a banquet at the Akershus Castle in Oslo, normally attended by the King of Norway, Harald V. Below is the list of people who gave a speech at the banquet:

2003 John M. Ball (IMU President & member of Abel Committee)

2004 Sir John Kingman (EMS President)

2005 James Arthur (AMS President)

2006 Jacob Palis (former IMU President & former member of Abel Committee)

2007 László Lovász (IMU President & former member of Abel Committee)

2008 Ari Laptev (EMS President)

2009 Ingrid Daubechies (former member of Abel Committee)

2010 Sir Michael Atiyah (Abel Laureate 2004)

2011 Marta Sanz-Solé (EMS President)

2012 Ivar Ekeland (foreign member of the Norwegian Academy of Science and Letters)

2013 Hendrik W. Lenstra (former member of the Abel Committee)

2014 Jean-Pierre Bourguignon (ERC President)

2015 Shigefumi Mori (IMU President)

2016 Pearl Dykstra (member of European Commission's High Level Group of Scientific Advisors)

2017 Robbert Dijkgraaf (Director of the Institute for Advanced Study)

2018 Caroline Series (LMS President)

2019 Carlos E. Kenig (IMU President)

2020 no banquet

2021 no banquet

2022 Kristin Clemet (former Minister of Education and Research)

Addenda, Errata, and Updates[1]

2003 Jean-Pierre Serre

Publications:

2017

[305] Appeared in *Lie Groups, Geometry, and Representation Theory*. Progr. Math., vol. 326, Birkhäuser/Springer, Cham, 2018, pp. 527–540.

2019

[307] Souvenirs sur Jean-Marc Fontaine. *Gaz. Math.* No. 162, 12–13.
[308] Distribution asymptotique des Valeurs Propres des Endomorphismes de Frobenius [d'après Abel, Chebyshev, Robinson,...]. *Astérisque No. 414, Séminaire Bourbaki. Vol. 2017/2018. Exposés 1136–1150*, Exp. No. 1146, 379–426.

2020

[309] Rational points on curves over finite fields. With contributions by E. Howe, J. Oesterlé, and C. Ritzenthaler. Edited by A. Bassa, E. Lorenzo García, C. Ritzenthaler and R. Schoof. *Documents Mathématiques* 18, x+187 pp.
[310] La vie et l'oeuvre de John Tate. *C. R. Math. Acad. Sci. Paris*. 358, no. 11-12, 1129–133.
[311] The life and work of John Tate. *Resonance* 25, no. 2, 169–175.
[312] La vie et l'oeuvre de Jean-Marc Fontaine. *C. R. Math. Acad. Sci. Paris* 358, no. 9-10, 1045–1046.

2021

[313] (with E. Bayer-Fluckiger). Lines on cubic surfaces, Witt invariants and Stiefel–Whitney classes. *Indag. Math. (N.S.)* 32, no. 5, 920–938.

2022

[314] Groupes de Coxeter finis: involutions et cubes. *Enseign. Math.* 68, no. 1-2, 99–133.

Addendum CV:
Speaker at the International Congress of Mathematicians, 1954, 1962 (plenary)

---

[1] H. Holden, R. Piene (eds.): *The Abel Prize 2003–2007. The First Five Years* (Springer, Berlin, 2010), *The Abel Prize 2008–2012* (Springer, Berlin, 2014), and *The Abel Prize 2013–2017* (Springer, Berlin, 2019).

2004 Sir Michael Atiyah and Isadore M. Singer

Sir Michael Atiyah passed away on 11 January 2019.
Isadore M. Singer passed away on 11 February 2021.

Publications by M. Atiyah:

2016

[275] Appeared in *Internat. J. Modern Phys. A* 33(24) 1830022, 16 pp.

2019

[286] (with Yin Yue Sha) The sharp radius of the neutron, proton, electron, critical photon and the atomic nucleus. *Engineering Technology Open Access* 3, no 1 ETOAJ.MS.ID.555608.
[287] (with Joseph Kouneiher) Todd function as weak analytic function. *Int. J. Geom. Methods Mod. Phys.* 16, no. 6, 1950091, 11 pp.
[288] A problem in Euclidean geometry. In: *Representation Theory, Automorphic Forms & Complex Geometry*, Int. Press, Somerville, MA, pp. 1–2.
[289] Professor S.-T. Yau at 70. *ICCM Not.* 7, no. 1, 2.

2021

[290] (with M. Marcolli) Anyon networks from geometric models of matter. *Q. J. Math.*, 72(1-2), 717–733.

See also Bull. Amer. Math. Soc. 58, no 4 (2021) for a special "Atiyah-issue".

Addendum CV:
M. Atiyah:
Speaker at the International Congress of Mathematicians, 1962, 1966 (plenary), 1970, 1978, 2018 (plenary)

I. Singer:
Speaker at the International Congress of Mathematicians, 1974 (plenary)

2005 Peter D. Lax

Publications:

2018

[238] was published in 2017.

Addendum CV:
Speaker at the International Congress of Mathematicians, 1966, 1970, 1983 (plenary)

2006 Lennart Carleson

Addendum CV:
Speaker at the International Congress of Mathematicians, 1962, 1966 (plenary), 1990

2007 S. R. Srinivasa Varadhan

Publications:

2018

[156] Appeared as: Identification of the polaron measure I: Fixed coupling regime and the central limit theorem for large times. *Comm. Pure Appl. Math.* 73 (2020), no. 2, 350–383. Corrigendum and addendum: *loc. sit.* 75 (2022), no. 7, 1642–1653.
[157] Appeared as: Identification of the polaron measure in strong coupling and the Pekar variational formula. *Ann. Probab.* 48 (2020), no. 5, 2119–2144.
[158] The role of topology in large deviations. *Expo. Math.* 36, no. 3-4, 362–368.

2020

[159] (with C. Mukherjee) The Polaron measure. *Applied probability and stochastic processes*, Infosys Sci. Found. Ser. Math. Sci., Springer, Singapore, pp. 415–419.
[160] (with D. Blasius, T. Digernes, R. Fioresi, R. Gangolli, M. Rapoport) Recollections of V. S. Varadarajan. *Notices Amer. Math. Soc.* 67, no. 9, 1365–1373.

2021

[161] The Polaron problem. In: *A Tribute to the Legend of Professor C. R. Rao—the Centenary Volume*, Springer, Singapore, pp. 25–30.

2022

[162] (with C. Mukherjee) The Polaron problem. In: *The Physics and Mathematics of Elliott Lieb–The 90th Anniversary.* Vol. II, EMS Press, Berlin, pp. 73–77.
[163] *Harmonic Analysis.* Courant Lecture Notes in Mathematics, vol. 31. American Mathematical Society, Providence, RI, vii+101 pp.

Addendum CV:
Speaker at the International Congress of Mathematicians, 1978, 1994 (plenary), 2010 (plenary)

2008 John G. Thompson and Jacques Tits

Jacques Tits passed away on 5 December 2021.

Addendum CV:
J.G. Thompson:
Speaker at the International Congress of Mathematicians, 1962, 1966 (plenary)
J. Tits:
Speaker at the International Congress of Mathematicians, 1962 (plenary), 1970, 1974 (plenary)

2009 Mikhail Gromov

Publications:

2016

[137a] (with N.V. Shabaldina) Construction of a cascade parallel composition of times automata using BALM-II. (Russian) *Model. Anal. Inf. Sist.* 23, no. 6, 715–728.

2017

[141a] (with A.S. Tvardovskiĭ, K. Èl'-Faki, N.V. Evtushenko) Design of tests with guaranteed completeness for nondeterministic timed automata. (Russian) *Model. Anal. Inf. Sist. 24*, no. 4, 496–507.

2019

[146] Mean curvature in the light of scalar curvature. *Ann. Inst. Fourier (Grenoble)* 69, no. 7, 3169–3194.

2020

[147] In memory of Gennadi Henkin. *J. Geom. Anal.* 30, no. 3, 2292.
[148] (with M. Braverman, V.M. Buchstaber, et al.) Mikhail Aleksandrovich Shubin (obituary). *Russian Math. Surveys* 75, no. 6, 1143–1152.
[149] Morse spectra, homology measures, spaces of cycles and parametric packing problems. In: *What's Next?—the Mathematical Legacy of William P. Thurston*, Ann. of Math. Stud., vol. 205, Princeton Univ. Press, pp. 141–205.

[150] (with J.-P. Bourguignon, E. Calabi, J. Eells, O. Garcia-Prada) Where does geometry go? A research and education perspective. In: *Eugenio Calabi—Collected Works*, Springer, Berlin, pp. 803–818.

2022

[151] Pinching constants for hyperbolic manifolds. In: *Collected works of William P. Thurston with Commentary. Vol. I—Foliations, Surfaces and Differential Geometry*, Amer. Math. Soc., Providence, pp. 655–666.

[152] (with Herbert Blaine Lawson Jr. and William Thurston) Hyperbolic 4-manifolds and conformally flat 3-manifolds. *Collected works of William P. Thurston with commentary. Vol. I. Foliations, surfaces and differential geometry*, pp. 667–685, Amer. Math. Soc., Providence, RI.

Addendum CV:
Speaker at the International Congress of Mathematicians, 1970, 1978, 1983, 1986 (plenary)

2010 John T. Tate

John T. Tate passed away on 16 October 2019.

Addendum CV:
Speaker at the International Congress of Mathematicians, 1962, 1970 (plenary)

2011 John W. Milnor

Publications:

2018

[156] was published in *Bulletin Amer. Math. Soc.* (2020) 57, no. 2, 171–267.

[157] (with A. Bonifant, X. Buff) Antipode preserving cubic maps: the fjord theorem. *Proc. Lond. Math. Soc.* (3) 116, no. 3, 670–728.

2021

[158] (with A. Bonifant, S. Sutherland) The W. Thurston algorithm applied to real polynomial maps. *Conform. Geom. Dyn.* 25, 179–199.

2022

[159] (with W. Thurston) Characteristic numbers of 3-manifolds. In: *Collected Works of William P. Thurston with Commentary. Vol. I—Foliations, Surfaces and Differential Geometry*, Amer. Math. Soc., Providence, pp. 619–624.

[160] (with W. Thurston) On iterated maps of the interval. In: *Collected Works of William P. Thurston with Commentary. Vol. III—Dynamics, Computer Science and General Interest*, Amer. Math. Soc., Providence, pp. 7–105.
[161] (with W. Thurston) Chapter 7 in: *The Geometry and Topology of Three-Manifolds. Vol. IV*, American Mathematical Society, Providence.

See also Bull. Amer. Math. Soc. 52(4) 2015.

Addendum CV:
Speaker at the International Congress of Mathematicians, 1958, 1962 (plenary), 2014 (plenary)

2012 Endre Szemerédi

Publications:

2018

[192] (with P. Hajnal) Two geometrical applications of the semi-random method. In: *New Trends in Intuitive Geometry*, Bolyai Soc. Math. Stud., 27, János Bolyai Math. Soc., Budapest, pp. 189–199.
[193] Additive combinatorics and graph theory. In: *European Congress of Mathematics*, Eur. Math. Soc., Zürich, pp. 685–716.

2019

[194] (with C. Reiher, V. Rödl, A. Ruciński, M. Schacht) Minimum vertex degree condition for tight Hamiltonian cycles in 3-uniform hypergraphs. *Proc. Lond. Math. Soc.* (3) 119 (2019), no. 2, 409–439.
[195] (with M. Simonovits) Embedding graphs into larger graphs: results, methods, and problems. In: *Building Bridges II—Mathematics of László Lovász*, Bolyai Soc. Math. Stud., 28, Springer, Berlin, pp. 445–592.

2021

[196] (with I. Ruzsa, G. Shakan, J. Solymosi) On distinct consecutive differences. In: *Combinatorial and Additive Number Theory IV*, Springer Proc. Math. Stat., pp. 425–434.

Addendum CV:
Speaker at the International Congress of Mathematicians, 1974
Order of Saint Stephen of Hungary, 2020

2013 Pierre Deligne

Publications:

2020

[122] (with N. Amend, G. Röhrle) On the $K(\pi, 1)$-problem for restrictions of complex reflection arrangements. *Compos. Math.* 156, no. 3, 526–532.

2021

[123] Le critère d'Abel pour la résolubilité par radicaux d'une équation irréductible de degré premier. *C. R. Math. Acad. Sci. Paris* 359, 919–921.

Further unpublished documents can be found on Deligne's website of Institute for Advanced Study, Princeton.

Addendum CV:
Speaker at the International Congress of Mathematicians, 1970, 1974 (plenary)

2014 Yakov Sinai

Publications:

2010

[264] Reprinted in 2019.

2018

[292] (with I. Vinogradov) Eigenfunctions of Laplacians in some two-dimensional domains. In: *Dynamical Systems, Ergodic Theory, and Probability: in Memory of Kolya Chernov*, Contemp. Math., 698, Amer. Math. Soc., Providence, pp. 195–199.

2020

[293] (with S. Gusein-Zade, Y. Ilyashenko, K. Khanin, S. Shlosman, M. Tsfasman) Roland Lvovich Dobrushin (July 20, 1929–November 12, 1995). *Mosc. Math. J.* 20, no. 4, 641–644.
[294] (with M.E.H. Bahri) Statistical mechanics of freely fluctuating two-dimensional elastic crystals. *J. Stat. Phys.* 180, no. 1-6, 739–748.
[295] (with C. Boldrighini, S. Frigio, P. Maponi, A. Pellegrinotti) An antisymmetric solution of the 3D incompressible Navier–Stokes equations with "tornado-like" behavior. *J. Exp. Theor. Phys.* 131, 356–360.

2021

[296] (with K. Khanin, M. Lyubich, E.D. Siggia) Mitchell Feigenbaum. *Notices Amer. Math. Soc.* 68, no. 5, 757–767.

Addendum CV:
Speaker at the International Congress of Mathematicians, 1962, 1970, 1978, 1990 (plenary)

2015 John Nash Jr. and Louis Nirenberg

John Nash Jr. passed away on 23 May 2015.
Louis Nirenberg passed away on 26 January 2020.

See *Bull. Amer. Math. Soc.* 54, no 2 (2017) for a special "Nash-issue".

Publications by L. Nirenberg:

2018

[166] *Lectures on Differential Equations and Differential Geometry*. CTM. Classical Topics in Mathematics, 7. Higher Education Press, Beijing, ix+174 pp.

The Abel Lectures by T. Rivière and F. Morgan were published as
Rivière: Exploring the unknown: The work of Louis Nirenberg on partial differential equations, *Notices, AMS*, 63, no 2 (2016), pp. 120–125, and *EMS Surv. Math. Sci.*, 9, no 1 (2022), pp. 1–29.
Morgan: Soap bubbles and mathematics, *Eur. Math. Soc. Newsl.*, 97 (2015), pp. 32–36.

See also N. Dencker: Nirenberg's contributions to linear partial differential equations: Pseudo-differential operators and solvability, *Bull. AMS*, 60, no 2 (2023), pp. 159–166.

Addendum CV:
L. Nirenberg:
Speaker at the International Congress of Mathematicians, 1962 (plenary), 1974

2016 Sir Andrew J. Wiles

Addendum CV:
Speaker at the International Congress of Mathematicians, 1994 (plenary)
Common Wealth Award of Distinguished Service, 1996

2017 Yves Meyer

Publications:

2018

[207] Histoire des idées et culture mathématiques: transmettre. *Matapli* No. 117, 59–64.

2019

[208] Three problems on trigonometric sums. *Acta Math. Sin. (Engl. Ser.)* 35, no. 6, 721–727.

2020

[209] Trigonometric series with a given spectrum. *Tunis. J. Math.* 2, no. 4, 881–906.

2021

[210] Iraneo Peral and the rebirth of mathematics in Spain. *Gac. R. Soc. Mat. Esp.* 24, no. 3, 475–479.
[211] Restriction algebras of Fourier–Stieltjes transforms of Radon measures. *J. Geom. Anal.* 31, no. 9, 9131–9142.
[212] A letter by Eli Stein. *J. Geom. Anal.* 31, no. 7, 7297–7303.
[213] From Salomon Bochner to Dan Shechtman. *Transactions of the Royal Norwegian Society of Sciences and Letters*, no 1, 22 pp.
[214] Crystalline measures and mean-periodic functions. *Transactions of the Royal Norwegian Society of Sciences and Letters*, no 2, 26 pp.

2023

[215] (with A. Fan) Trigonometric multiplicative chaos and applications to random distributions. *Sci. China Math.* 66, no. 1, 3–36.
[216] Crystalline measures in two dimensions. *Publ. Mat.* 67, no. 1, 469–480.

Addendum CV:
Speaker at the International Congress of Mathematicians, 1970, 1983, 1990

The Abel Lecture at the International Congress of Mathematicians:

2022 (virtual) Avi Wigderson: "Symmetries, Computation and Math (or, can $P \neq NP$ be proved via gradient descent?)"

The Abel Lecture at the European Congress of Mathematics:

2020 (Portoroz, Slovenia) Lázsló Lovász: "Graph limits and Markov spaces"

In *The Abel Prize 2013–2017* on p. 737 the artist's name should read Erika Klagge (not Erika Kappel).

Printed in the USA
CPSIA information can be obtained
at www.ICGtesting.com
CBHW071259080724
11078CB00017B/103

9 783031 339721